SCIENCE AND CIVILISATION IN CHINA

BY

JOSEPH NEEDHAM, F.R.S.

FELLOW AND PRESIDENT OF CAIUS COLLEGE
SIR WILLIAM DUNN READER IN BIOCHEMISTRY IN THE
UNIVERSITY OF CAMBRIDGE
FOREIGN MEMBER OF ACADEMIA SINICA

With the collaboration of

WANG LING, PH.D.

TRINITY COLLEGE, CAMBRIDGE
ASSOCIATE RESEARCH FELLOW OF ACADEMIA SINICA

VOLUME 4

PHYSICS AND PHYSICAL TECHNOLOGY

PART II: MECHANICAL ENGINEERING

CAMBRIDGE
AT THE UNIVERSITY PRESS
1965

PUBLISHED BY
THE SYNDICS OF THE CAMBRIDGE UNIVERSITY PRESS

Bentley House, 200 Euston Road, London, N.W. 1
American Branch: 32 East 57th Street, New York 22, N.Y.
West African Office: P.O. Box 33, Ibadan, Nigeria

©

CAMBRIDGE UNIVERSITY PRESS

1965

Printed in Great Britain at the University Printing House, Cambridge
(Brooke Crutchley, University Printer)

THE PICTURE OF THE TAOIST GENII PRINTED ON THE COVER

of this book is part of a painted temple scroll, recent but traditional, given to Mr Brian Harland in Szechuan province (1946). Concerning these four divinities, of respectable rank in the Taoist bureaucracy, the following particulars have been handed down. The title of the first of the four signifies 'Heavenly Prince', that of the other three 'Mysterious Commander'.

At the top, on the left, is Liu *Thien Chün*, Comptroller-General of Crops and Weather. Before his deification (so it was said) he was a rain-making magician and weather forecaster named Liu Chün, born in the Chin dynasty about +340. Among his attributes may be seen the sun and moon, and a measuring-rod or carpenter's square. The two great luminaries imply the making of the calendar, so important for a primarily agricultural society, the efforts, ever renewed, to reconcile celestial periodicities. The carpenter's square is no ordinary tool, but the gnomon for measuring the lengths of the sun's solstitial shadows. The Comptroller-General also carries a bell because in ancient and medieval times there was thought to be a close connection between calendrical calculations and the arithmetical acoustics of bells and pitch-pipes.

At the top, on the right, is Wên *Yuan Shuai*, Intendant of the Spiritual Officials of the Sacred Mountain, Thai Shan. He was taken to be an incarnation of one of the Hour-Presidents (*Chia Shen*), i.e. tutelary deities of the twelve cyclical characters (see p. 440). During his earthly pilgrimage his name was Huan Tzu-Yü and he was a scholar and astronomer in the Later Han (b. +142). He is seen holding an armillary ring.

Below, on the left, is Kou *Yuan Shuai*, Assistant Secretary of State in the Ministry of Thunder. He is therefore a late emanation of a very ancient god, Lei Kung. Before he became deified he was Hsin Hsing, a poor woodcutter, but no doubt an incarnation of the spirit of the constellation Kou-Chhen (the Angular Arranger), part of the group of stars which we know as Ursa Minor. He is equipped with hammer and chisel.

Below, on the right, is Pi *Yuan Shuai*, Commander of the Lightning, with his flashing sword, a deity with distinct alchemical and cosmological interests. According to tradition, in his early life he was a countryman whose name was Thien Hua. Together with the colleague on his right, he controlled the Spirits of the Five Directions.

Such is the legendary folklore of common men canonised by popular acclamation. An interesting scroll, of no great artistic merit, destined to decorate a temple wall, to be looked upon by humble people, it symbolises something which this book has to say. Chinese art and literature have been so profuse, Chinese mythological imagery so fertile, that the West has often missed other aspects, perhaps more important, of Chinese civilisation. Here the graduated scale of Liu Chün, at first sight unexpected in this setting, reminds us of the ever-present theme of quantitative measurement in Chinese culture; there were rain-gauges already in the Sung (+12th century) and sliding calipers in the Han (+1st). The armillary ring of Huan Tzu-Yü bears witness that Naburiannu and Hipparchus, al-Naqqāsh and Tycho, had worthy counterparts in China. The tools of Hsin Hsing symbolise that great empirical tradition which informed the work of Chinese artisans and technicians all through the ages.

SCIENCE AND CIVILISATION
IN CHINA

AND seeing that the Arts and Crafts, with other like Feats, whose inventours be contained in this book, are in this Realm of England occupied and daily put in exercise to the profit of many, and ease of all men, it were in mine opinion both a point of detestable unkindnesse, and a part of extream inhumanity, to defraud them of their praise and perpetual memory, that were Authors of so great Benefits to the universal World.

POLYDORE VERGIL
De Rerum Inventoribus (1512)
English edition of Thomas Langley, 1659

中國科學技術史

李約瑟 著

莫朝鼎

REWI ALLEY

New Zealander by birth and citizenship
Chinese by adoption and grace

pioneer of the industrial co-operatives
teacher of engineering
lover of Chinese youth
writer and poet

and

SOLOMON ABRAMOVITCH TRONE

Engineer, sometime adviser on industrialisation in Russia,
China, India and Israel

humanist and man of vision
'Electrifikatsiya! Elektrifikatsiya!'

to these two friends
this volume is
dedicated

*The Syndics of the Cambridge University Press
desire to acknowledge with gratitude certain financial aid
towards the production of this book, afforded by the
Bollingen Foundation*

CONTENTS

LIST OF ILLUSTRATIONS

LIST OF TABLES

LIST OF ABBREVIATIONS

The following abbreviations are used in the text and footnotes. For abbreviations used for journals and similar publications in the bibliographies, see pp. 604 ff.

B & M   Brunet & Mieli, *Histoire des Sciences* (*Antiquité*).

BCFA   Britain–China Friendship Association.

C   Thang Hsin-Yü, *Chi Kung Tzhu Tien* (Dictionary of Mechanical Engineering). Sci. & Technol. Press, Shanghai, 1955.

CCTS   Têng Yü-Han (Johann Schreck) & Wang Chêng, *Chhi Chhi Thu Shuo* (Diagrams and Explanations of Wonderful Machines), +1627.

CPCRA   Chinese People's Association for Cultural Relations with Foreign Countries.

CSHK   Yen Kho-Chün (ed.), *Chhüan Shang-ku San-Tai Chhin Han San-Kuo Liu Chhao Wên* (complete collection of prose literature (including fragments) from remote antiquity through the Chhin and Han Dynasties, the Three Kingdoms, and the Six Dynasties), 1836.

CSS   Fêng Yün-Phêng & Fêng Yün-Yuan, *Chin Shih So* (Collection of Carvings, Reliefs and Inscriptions), 1821.

D   Anon., *Hsin Ting Ying Han Tzhu Tien* (Abridged English–Chinese Dictionary). Commercial Press, Shanghai, 1911.

G   Giles, H. A., *Chinese Biographical Dictionary*.

HCCC   Yen Chieh (ed.), *Huang Chhing Ching Chieh* (monographs by Chhing scholars on classical subjects).

HF   Hsiung San-Pa (Sabatino de Ursis) & Hsü Kuang-Chhi, *Thai Hsi Shui Fa* (Hydraulic Machinery of the West), +1612.

K   Karlgren, *Grammata Serica* (dictionary giving the ancient forms and phonetic values of Chinese characters).

KCCY   Chhen Yuan-Lung, *Ko Chih Ching Yuan* (Mirror of Scientific and Technological Origins), an encyclopaedia of +1735.

KCKW   Wang Jen-Chün, *Ko Chih Ku Wei* (Scientific Traces in Olden Times), 1896.

KCT   Lou Shou, *Kêng Chih Thu* (Pictures of Tilling and Weaving), +1145. See Franke (11); Pelliot (24).

KYCC   Chhüthan Hsi-Ta, *Khai-Yuan Chan Ching* (The Khai-Yuan reign-period Treatise on Astrology and Astronomy), +729.

LSCC   Lü Pu-Wei, *Lü Shih Chhun Chhiu* (Master Lü's Spring and Autumn Annals), a compendium of natural philosophy, −239.

MCPT   Shen Kua, *Mêng Chhi Pi Than* (Dream Pool Essays), +1089.

NCCS   Hsü Kuang-Chhi, *Nung Chêng Chhüan Shu* (Complete Treatise on Agriculture), +1639.

NCNA   New China News Agency.

NS   Wang Chên, *Nung Shu* (Treatise on Agriculture), +1313.

R   Read, Bernard E. *et al.*, Indexes, translations and précis of certain chapters of the *Pên Tshao Kang Mu* of Li Shih-Chen. If the reference is to a plant see Read (1); if to a mammal, see Read (2); if to a bird see Read (3); if to a reptile see Read (4 or 5); if to a mollusc see Read (5); if to a fish see Read (6); if to an insect see Read (7).

SCTS   *Chhin-Ting Shu Ching Thu Shuo* (imperial illustrated edition of the *Historical Classic*) 1905.

SF   Thao Tsung-I (ed.), *Shuo Fu* (Florilegium of (Unofficial) Literature), *c.* +1368.

SSTK   O-Erh-Thai *et al.* (ed.), *Shou Shih Thung Khao* (Complete Investigation of the Works and Days), an imperially commissioned treatise on agriculture and rural crafts, +1742.

STTH   Wang Chhi, *San Tshai Thu Hui* (Universal Encyclopaedia), +1609.

T   Tunhuang Archaeological Research Institute numbers of the Chhien-fo-tung cave-temples. If an identification is given according to the system of Hsieh Chih-Liu in his *Tunhuang I Shu Hsü Lu* (Shanghai, 1955) the Institute number and the Pelliot number are also given, but if a single number is given it is the Institute number. A valuable concordance table of the three systems is given in Hsieh's book, and a still more complete one in Chhen Tsu-Lung (1).

TCKM   Chu Hsi *et al.* (ed.), *Thung Chien Kang Mu* (Short View of the *Comprehensive Mirror (of History) for Aid in Government*), general history of China, +1189, with later continuations.

TCTC   Ssuma Kuang, *Tzu Chih Thung Chien* (Comprehensive Mirror (of History) for Aid in Government), +1084.

TH   Wieger, L. (1), *Textes Historiques*.

TKKW   Sung Ying-Hsing, *Thien Kung Khai Wu* (The Exploitation of the Works of Nature), +1637.

TPYL   Li Fang (ed.), *Thai-Phing Yü Lan* (the Thai-Phing reign-period (Sung) Imperial Encyclopaedia), +983.

TSCC   Chhen Mêng-Lei *et al.* (ed.), *Thu Shu Chi Chhêng*; the Imperial Encyclopaedia of +1726. Index by Giles, L. (2).

TT   Wieger, L. (6), 'Taoïsme', vol. 1, Bibliographie Générale (catalogue of the works contained in the Taoist Patrology, *Tao Tsang*).

TTC   *Tao Tê Ching* (Canon of the Tao and its Virtue).

TW   Takakusu, J. & Watanabe, K., (*Tables du Taishō Issaikyō nouvelle édition (Japonaise) du Canon bouddhique chinoise*). Index-catalogue of the Tripiṭaka.

WCTS   Wang Chêng, *Chu Chhi Thu Shuo* (Diagrams and Explanations of a number of Machines), +1627.

WCTY/CC   Tsêng Kung-Liang (ed.), *Wu Ching Tsung Tao (Chhien Chi)*, military encyclopaedia, first section, +1044.

WHTK   Ma Tuan-Lin, *Wên Hsien Thung Khao* (Comprehensive Study of (the History of) Civilisation), +1319.

WPC   Mao Yuan-I, *Wu Pei Chih* (Treatise on Armament Technology), +1628.

YHSF   Ma Kuo-Han (ed.), *Yü Han Shan Fang Chi I Shu* (Jade-Box Mountain Studio Collection of (reconstituted and sometimes fragmentary) Lost Books), 1853.

ACKNOWLEDGEMENTS

LIST OF THOSE WHO HAVE KINDLY READ THROUGH SECTIONS IN DRAFT

The following list, which applies only to this volume, brings up to date those printed in Vol. 1, pp. 15 ff., Vol. 2, p. xxiii, Vol. 3, pp. xxxix ff. and Vol. 4, pt. 1, p. xxi.

Mr Robert Brittain (Williamstown, Mass.)	Water-raising Machinery.
Miss Alison Burford (Cambridge)	Animal Traction.
Prof. Aubrey Burstall (Newcastle)	All subsections.
Dr J. Coales (Cambridge)	South-pointing Carriage.
Mr J. H. Combridge (London)	All subsections.
Dr A. G. Drachmann (Copenhagen)	Basic Machines, Clockwork, Windmill.
Prof. V. Elisséeff (Paris)	All subsections.
Mr C. H. Gibbs-Smith (London)	Aeronautics.
Dr Falconer Henry (Cambridge)	All subsections.
Mr R. P. N. Jones (Cambridge)	South-pointing Carriage.
Dr Lo Jung-Pang (Seattle)	Vehicles, Animal Traction.
Sir Bennett Melvill Jones, F.R.S. (Cambridge)	Aeronautics.
Dr L. A. Moritz (Achimoto)	Milling.
Dr Dorothy M. Needham, F.R.S. (Cambridge)	All subsections.
Prof. Luciano Petech (Rome)	All subsections.
Dr Ladislao Reti (São Paulo)	Palaeotechnic Machinery.
Mr Raphael Salaman (Harpenden)	Tools and Materials.
Sir George Sansom (Berkeley, Calif.)	Wheelbarrow and Sailing-Carriage.
Ing. Th. Schiøler (Copenhagen)	Water-raising Machinery.
The late Dr Dorothea Singer (Par)	All subsections.
Mr John Stanitz (Cleveland, Ohio)	All subsections.
Dr E. G. Sterland (Bristol)	All subsections.
Mr Rex Wailes (Beaconsfield)	Windmill.
Dr Michael Wood (Cambridge)	All subsections.
Mr Hans Wulff (Kensington, N.S.W.)	All subsections.

AUTHOR'S NOTE

PURSUING our exploration of the almost limitless caverns of Chinese scientific history, so much of which has never yet come to the knowledge and recognition of the rest of the world, we now approach the glittering veins of physics and physical technology; a subject which forms a single whole, constituting Volume Four, though delivered to the reader in three separate volumes. First come the physical sciences themselves (Vol. 4, pt. 1) and then their diverse applications in all the many branches of mechanical engineering (Vol. 4, pt. 2), civil and hydraulic engineering, and nautical technology (Vol. 4, pt. 3).

With the opening chapter we find ourselves at a focal point in the present study, for mechanics and dynamics were the first of all the conquests of modern science. Mechanics was the starting-point because the direct physical experience of man in his immediate environment is predominantly mechanical, and the application of mathematics to mechanical magnitudes was relatively simple. But ancient and medieval China belonged to a world in which the mathematisation of hypotheses had not yet brought modern science to birth, and what the scientific minds of pre-Renaissance China neglected might prove almost as revealing as that which aroused their interest and investigation. Three branches of physics were well developed among them, optics (Section 26g), acoustics (26h), and magnetism (26i); mechanics was weakly studied and formulated, dynamics almost absent. We have attempted to offer some explanation for this pattern but without any great conviction, and better understanding of the imbalance must await further research. The contrast with Europe, at least, where there was a different sort of one-sidedness, is striking enough, for in Byzantine and late medieval times mechanics and dynamics were relatively advanced while magnetic phenomena were almost unknown.

In optics the Chinese of the Middle Ages kept empirically more or less abreast of the Arabs, though greatly hampered in theory by the lack of that Greek deductive geometry of which the latter were the inheritors. On the other hand they never entertained that peculiar Hellenistic aberration according to which vision involved rays radiating from, not into, the eye. In acoustics the Chinese proceeded along their own lines because of the particular and characteristic features of their ancient music, and here they produced a body of doctrine deeply interesting but not readily comparable with those of other civilisations. Inventors of the bell, and of a great variety of percussion instruments not known in the West, they were especially concerned with timbre both in theory and practice; developing their unique theories of melodic composition within the framework of a twelve-note gamut rather than an eight-note scale. At the end of the +16th century Chinese mathematical acoustics succeeded in solving the problem of equal temperament just a few decades before its solution was reached in the West (Section 26h, 10). Lastly, Chinese investigation of magnetic phenomena and their practical application constituted a veritable epic. Men were arguing in China about the cause of the declination of the magnetic needle, and using it at sea, before Westerners even knew of its directive property.

Readers pressed for time will doubtless welcome once more a few suggestions. In the chapters which we now present it is possible to perceive certain outstanding traditions of Chinese physical thought and practice. Just as Chinese mathematics was indelibly algebraic rather than geometrical, so Chinese physics was wedded to a prototypic wave-theory and perennially averse to atoms, always envisaging an almost Stoic continuum; this may be seen in Section 26b and followed through in relation to tension and fracture (c, 3) and to sound vibrations (h, 9). Another constant Chinese tendency was to think in pneumatic terms, faithfully developing the implications of the ancient concept of *chhi* (= *pneuma*, *prāṇa*). Naturally this shows itself most prominently in the field of acoustics (Section 26h, 3, 7, etc.), but it was also connected with some brilliant successes in the field of technology such as the inventions of the double-acting piston-bellows and the rotary winnowing-fan (Section 27b, 8), together with the water-powered metallurgical blowing-machine (27h, 3, 4, direct ancestor of the steam-engine itself). It was also responsible for some extraordinary insights and predictions in aeronautical prehistory (27m, 4). Traditions equally strong and diametrically opposite to those of Europe also make their appearance in the purely technical field. Thus the Chinese had a deep predilection for mounting wheels and machinery of all kinds horizontally instead of vertically whenever possible; as may be followed in Section 27 (h, k, l, m).

Beyond this point, guidance to the reader is not very practicable since so many different preoccupations are involved. If he is interested in the history of land transport he will turn to the discussion of vehicles and harness (Section 27e, f), if he delights, like Leviathan, in the deep waters, a whole chapter (29) will speak to him of Chinese ships and their builders. The navigator will turn from the compass itself (26i, 5) to its fuller context in the haven-finding art (29f); the civil engineer, attracted by a survey of those grand water-works which outdid the 'pyramides of Aegypt', will find it in Section 28f. The folklorist and the ethnographer will appreciate that 'dark side' of history where we surmise that the compass-needle, most ancient of all those pointer-readings that make up modern science, began as a 'chess-man' thrown on to a diviner's board (26i, 8). The sociologist too will already find much of interest, for besides discussing the place of artisans and engineers in feudal-bureaucratic society (27a, 1, 2, 3), we have ventured to raise certain problems of labour-saving invention, man-power, slave status and the like, especially with regard to animal harness (f, 2), massive stone buildings (28d, 1), oared propulsion (29g, 2), and water-powered milling and textile machinery (27h).

Many are the ways in which these volumes link up with those which have gone before. We shall leave the reader's perspicacity to trace how the *philosophia perennis* of China manifested itself in the discoveries and inventions here reported. We may point out, however, that mathematics, metrology and astronomy find numerous echoes: in the origins of the metric system (Section 26c, 6), the development of lenses (g, 5), and the estimation of pitch-pipe volumes (h, 8)—or the rise of astronomical clocks (Section 27 *j*), the varying conceptions of perspective (28 *d*, 5) and the planning of hydraulic works (*f*, 9). Similarly, much in the present volumes points forward to chapters still to come. All uses of metal in medieval Chinese engineering imply what we have yet to say on metallurgical achievements; in the meantime reference may be made to the separate monograph *The Development of Iron and Steel Technology in China*, published as a Newcomen Lecture[a] in 1958. In all mentions of mining and the salt industry it is understood that these subjects will be fully dealt with at a later stage. All water-raising techniques remind us of their basic agricultural purpose, the raising of crops.

As for the discoveries and inventions which have left permanent mark on human affairs, it would be impossible even to summarise here the Chinese contributions. Perhaps the newest and most surprising revelation (so unexpected even to ourselves that we have to withdraw a relevant statement in Vol. 1) is that of the six hidden centuries of mechanical clockwork which preceded the clocks of + 14th-century Europe. Section 27*j* is a fresh though condensed treatment of this subject, incorporating still further new and strange material not available when the separate monograph *Heavenly Clockwork* was written in 1957 with our friend Professor Derek J. de Solla Price, now of Yale University.[b] It still seems startling that the key invention of the escapement should have been made in a pre-industrial agrarian civilisation among a people proverbially supposed by bustling nineteenth-century Westerners to take no account of time. But there are many other equally important Chinese gifts to the world, the development of the magnetic compass (Section 26*i*, 4, 6), the invention of the first cybernetic machine (27*e*, 5), both forms of efficient equine harness (27*f*, 1), the canal lock-gate (28*f*, 9, v) and the iron-chain suspension bridge (28*e*, 4). The first true crank (Section 27*b*, 4), the stern-post rudder (Section 29*h*), the man-lifting kite (Section 27 *m*)—we cannot enumerate them all.

In these circumstances it seems hardly believable that writers on technology have run up and down to find reasons why China contributed nothing to the sciences, pure or applied. At the beginning of a recent popular *florilegium* of passages on the history of technology one comes across a citation from the + 8th-century Taoist book *Kuan Yin Tzu*,[1] given as an example of 'oriental rejection of this world and of worldly activity'. It had been culled from an interesting essay on religion and the idea of progress, well known in the thirties and still stimulating, the author of which, led astray by the old rendering of Fr Wieger, had written: 'It is obvious that such beliefs can afford no basis for social activity and no incentive to material progress.' He was, of course, concerned to contrast the Christian acceptance of the material world with 'oriental' otherworldliness, in which the Taoists were supposed to participate. Yet in almost every one of the inventions and discoveries we here describe the Taoists and Mohists were intimately involved (cf. e.g. Sections 26*c*, *g*, *h*, *i*, 27*a*, *c*, *h*, *j*, 28*e*, etc.). As it happened, we had ourselves studied the same *Kuan Yin Tzu* passage and

---

[a] Needham (32); cf. (31).   [b] Needham, Wang & Price (1); cf. Needham (38).

[1] 關尹子

given parts of it in translation at an earlier stage;[a] from this it can be seen that Wieger's version[b] was no more than a grievously distorted paraphrase. Far from being an obscurantist document, denying the existence of laws of Nature (a concept totally unheard-of by the original writer)[c] and confusing reality with dream, the text is a poem in praise of the immanent Tao, the Order of Nature from which space and time proceed, the eternal pattern according to which matter disperses and reassembles in forms ever new; full of Taoist relativism, mystical but in no way anti-scientific or anti-technological, on the contrary prophesying of the quasi-magical quasi-rational command over Nature which he who truly knows and understands the Tao will achieve. Thus upon close examination, an argument purporting to demonstrate the philosophical impotence of 'oriental thought', turns out to be nothing but a figment of occidental imagination.

Another method is to admit that China did something but to find a satisfying reason for saying nothing about it. Thus a recent compendious history of science published in Paris maintains that the sciences of ancient and medieval China and India were so closely bound to their peculiar cultures that they cannot be understood without them. The sciences of the ancient Greek world, however, were truly sciences as such, free of all subordination to their cultural matrix and fit subjects with which to begin a story of human endeavour in all its abstract purity. It would be much more honest to say that while the social background of Hellenistic science and technology can be taken for granted because it is quite familiar to us from our schooldays onwards, we do not yet know much about the social background of Chinese and Indian science, and that we ought to make efforts to get acquainted with it. In fact, of course, no ancient or medieval science and technology can be separated from its ethnic stamp,[d] and though that of the post-Renaissance period is truly universal, it is no better understandable historically without a knowledge of the milieu in which it came to birth.

Finally, many will be desirous of looking into questions of intercultural contacts, transmissions and influences. Here we may only mention examples still puzzling of inventions which occur almost simultaneously at both ends of the Old World, e.g. rotary milling (Section 27 d, 2) and the water-mill (h, 2). Parallels between China and ancient Alexandria often arise (for instance in Section 27 b) and the powerful influence of Chinese technology on pre-Renaissance Europe appears again and again (26 c, h, i; 27 b, d, e, f, g, j, m; 28 e, f; 29 j). In the realm of scientific thought, as usual, influences were less marked, but one may well wonder whether the implicit wave-conceptions of China did not exert some effect in Renaissance Europe.

In a brilliant *ponencia* at the Ninth International Congress of the History of Sciences at Barcelona in 1959 Professor Willy Hartner raised the difficult question of how far anyone can ever anticipate anyone else. What does it mean to be a precursor or a predecessor? For those who are interested in intercultural transmissions this is a vital point. In European history the problem has assumed acute form since the

[a] Vol. 2, pp. 449 and 444.
[c] Cf. Section 18 in Vol. 2 above.
[b] Originally (4), p. 548.
[d] Cf. Vol. 3, p. 448.

school of Duhem acclaimed Nicholas d'Oresme and other medieval scholars as the precursors of Copernicus, Bruno, Francis Bacon, Galileo, Fermat and Hegel. Here the difficulty is that every mind is necessarily the denizen of the organic intellectual medium of its own time, and propositions which may look very much alike cannot have had quite the same meaning when considered by minds at very different periods. Discoveries and inventions are no doubt organically connected with the milieu in which they arose. Similarities may be purely fortuitous. Yet to affirm the true originality of Galileo and his contemporaries is not necessarily to deny the existence of precursors, so long as that term is not taken to mean absolute priority or anticipation; and in the same way there were many Chinese precursors or predecessors who adumbrated scientific principles later acknowledged—one thinks immediately of Huttonian geology (Vol. 3, p. 604), the comet tail law (p. 432) or the declination of the magnetic needle (Section 26i). So much for science more or less pure; in applied science we need hesitate less. For example, the gaining of power from the flow and descent of water by a wheel can only have been first successfully executed once. Within a limited lapse of time thereafter the invention may have occurred once or twice independently elsewhere, but such a thing is not invented over and over again. All subsequent successes must therefore derive from one or other of these events. In all these cases, whether of science pure or science applied, it remains the task of the historian to elucidate if possible how much genetic connection there was between the precursor and the great figures which followed him. Did they know certain actual written texts? Did they work by hearsay? Did they first conceive their ideas alone and find them unexpectedly confirmed? As Hartner says, the variations range from the certain to the impossible.[a] Often hearsay seems to have been followed by a new and different solution (cf. Section 27j, l). In our work here presented to the reader he will find that we are very often quite unable to establish a genetic connection (for example, between the suspension of Ting Huan and that of Jerome Cardan, in Section 27d, 4; or between the rotary ballista of Ma Chün and that of Leonardo, in Sections 27a, 2 and 30i, 4), but in general we tend to assume that when the spread of intervening centuries is large and the solution closely similar, the burden of proof must lie on those who desire to maintain independence of thought or invention. On the other hand the genetic connection can sometimes be established with a high degree of probability (for example, in the matters of equal temperament, Section 26h, 10, sailing-carriages, Section 27e, 3, and the kite, the parachute and the helicopter, Section 27m). Elsewhere one is left with strong suspicions, as with regard to the water-wheel escapement clock (Section 27j, 6).

Although every attempt has been made to take into account the most recent research in the fields here covered, we regret that it has generally not been possible to mention work appearing after May 1962.

[a] Many a surprise is still in store for us. After the discovery by Al-Ṭaṭāwī in 1924 that Ibn al-Nafīs (+1210 to +1288) had clearly described the pulmonary circulation (cf. Meyerhof (1, 2); Haddad & Khairallah, 1), it was long considered extremely unlikely that any hint of this could have reached the Renaissance discoverer of the same phenomenon, Miguel Servetus (cf. Temkin, 2). But now O'Malley (1) has found a Latin translation of some of the writings of Ibn al-Nafīs published in +1547.

We have not printed a contents-table of the entire project since the beginning of Vol. 1, and it has now been felt desirable to revise it in prospectus form.[a] So much work has now been done in preparation for the later volumes that it is possible to give their outline subheadings with much greater precision than could be done seven years ago. More important, perhaps, is the division into volumes. Here we have sought to retain unaltered the original numbering of the successive Sections, as for the needs of cross-referencing we must. Vol. 4, as originally planned, included physics, all branches of engineering, military and textile technology and the arts of paper and printing. As will be seen, we now entitle Vol. 4 *Physics and Physical Technology*, Vol. 5 *Chemistry and Chemical Technology* and Vol. 6 *Biology and Biological Technology*. This is a logical division, and Vol. 4 concludes very reasonably with Nautics (29), for in ancient and medieval times the techniques of shipping were almost entirely physical. Similarly Vol. 5 starts with Martial Technology (30), for in this field and in those times the opposite was the case; the chemical factor was essential. We found not only that we must embody iron and steel metallurgy therein (hence the slight but significant change of title), but also that without the epic of gunpowder, the fundamental discovery of the first known explosive and its development through five pre-occidental centuries, the history of Chinese military technique could not be written. With Textiles (31) and the other arts (32) the same argument was found to apply, for so many of the processes (retting, fulling, dyeing, ink-making) allied them to chemistry rather than to physics. Of course we could not always consistently adhere to this principle; for instance, no discussion of lenses was possible without some knowledge of glass technology, and this had therefore to be introduced at an early stage in the present volume (26*g*, 5, ii). For the rest, it is altogether natural that Mining (36), Salt-winning (37) and Ceramic Technology (35) should find their place in Vol. 5. The only asymmetry is that while in Vols. 4 and 6 the fundamental sciences are dealt with at the beginning of the first part, in Vol. 5 the basic science, chemistry, with its precursor, alchemy, is discussed in the second part. This probably matters the less because in ready response to the critics who found Vol. 3 too heavy and bulky for comfortable meditative evening reading, the University Press has decided to produce the present volume in three physically separate parts, each being as usual independent and complete in itself. One more point remains. In Vol. 1, pp. 18 ff. we gave details of the plan of the work (conventions, bibliographies, indexes, etc.) to which we have since closely adhered, and we promised that in the last volume a list would be given of the editions of the Chinese books used. It now seems undesirable to wait so long, and thus for the convenience of readers with knowledge of the Chinese language we propose to append to the last part of this Volume an interim list of these editions down to the point then reached.

China to Europeans has been like the moon, always showing the same face—a myriad peasant-farmers, a scattering of artists and recluses, an urban minority of scholars, mandarins and shopkeepers. Thus do civilisations acquire 'stereotypes' of one another. Now, raised upon the wings of the space-ship of linguistic resource and

---

[a] An extract of this contents-table relevant to the present volume will be found on pp. 432–4 below.

riding the rocket of technical understanding (to use an Arabic trope), we intend to see what is on the other side of the disc, and to meet the artisans and engineers, the shipwrights and the metallurgists of China's three-thousand-year-old culture.

In our note at the beginning of Vol. 3 we took occasion to say something of the principles of translation of old scientific texts and of the technical terms contained in them.[a] Since this is the first volume largely devoted to the applied sciences we are moved to insert a few reflections here on the present position of the history of technology, a discipline which has suffered even more perhaps than the history of science itself from that dreadful dichotomy between those who know and those who write, the doers and the recorders. If men of scientific training, with all their handicaps, have contributed far more than professional historians to the history of science and medicine (as is demonstrably true), technologists as a whole have been even less well equipped with the tools and skills of historical scholarship, the languages, the criticism of sources, and the use of documentary evidence Yet nothing can be more futile than the work of a historian who does not really understand the crafts and techniques with which he is dealing, and for any literary scholar it is hard to acquire that familiarity with things and materials, that sense of possibilities and probabilities, that understanding of Nature's ways, in fact, which comes (in greater or lesser measure) to everyone who has worked with his hands whether at the laboratory bench or in the factory workshop. I always remember once studying some medieval Chinese texts on 'light-penetration mirrors' (*thou kuang chien*[1]), that is to say, bronze mirrors which have the property of reflecting from their polished surfaces the designs executed in relief on their backs. A non-scientific friend was really persuaded that the Sung artisans had found out some way of rendering metal transparent to light-rays, but I knew that there must be some other explanation and it was duly found (cf. Section 26g, 3). The great humanists of the past were very well aware of their limitations in these matters, and sought always, so far as possible, to gain acquaintance with what my friend and teacher, Gustav Haloun, used half-wistfully half-ironically to call the *realia*. In a passage we have already quoted (Vol. 1, p. 7), another outstanding sinologist, Friedrich Hirth, urged that the Western translator of Chinese texts must not only translate, he must identify, he must not only know the language but he must also be a collector of the objects talked about in that language. The conviction was sound, but if porcelain or cloisonné could (at any rate in those days) be collected and contemplated with relative ease, how much more difficult is it to acquire an understanding of machinery, of tanning or of pyrotechnics, if one has never handled a lathe, fitted a gear-wheel or set up a distillation.

What is true of living humanists in the West is also true of some of the Chinese scholars of long ago whose writings are often our only means of access to the techniques of past ages. The artisans and technicians knew very well what they were doing, but they were liable to be illiterate, or at least inarticulate (cf. the long and illuminating text which we have translated in Section 27a, 2). The bureaucratic scholars, on the other hand, were highly articulate but too often despised the rude mechanicals whose

---

[1] 透光鑑   [a] Cf. Needham (34).

activities, for one reason or another, they wrote about from time to time. Thus even the authors whose words are now so precious were often more concerned with their literary style than with the details of the machines and processes which they mentioned. This superior attitude was also not unknown among the artists, back-room experts (like the mathematicians) of the officials' *yamens*, so that often they were more interested in making a charming picture than in showing the precise details of machinery when they were asked to limn it, and now sometimes it is only by comparing one drawing with another that we can reach certainty about the technical content. At the same time there were many great scholar-officials throughout Chinese history from Chang Hêng in the Han to Shen Kua in the Sung and Tai Chen in the Chhing who combined a perfect expertise in classical literature with complete mastery of the sciences of their day and the applications of these in artisanal practice.

For all these reasons our knowledge of the development of technology is still in a lamentably backward state, vital though it is for economic history, that broad meadow of flourishing speculation. In a recent letter, Professor Lynn White, who has done as much as anyone else in the field, wrote memorable words with which we fully agree: 'The whole history of technology is so rudimentary that all one can do is to work very hard, and be happy when one's errors are corrected.'[a] On every hand pitfalls abound. On a single page of a recent most authoritative and admirable collective treatise, one of our best historians of technology can first suppose Heron's toy windmill to be an Arabic interpolation, though the *Pneumatica* never passed to us through that language, and a moment later assert that Chinese travellers in +400 saw wind-driven prayer-wheels in Central Asia, a story based on a mistranslation now just 125 years old. The same authoritative treatise says that Celtic wagons of the −1st century had hubs equipped with roller-bearings, and we ourselves at first accepted this opinion. We learnt in time, however, that examination of the actual remains preserved at Copenhagen makes this highly improbable, and that reference to the original paper in Danish clinches the matter—the pieces of wood which came out from the hub-spaces when disinterred were flat strips and not rollers at all. We have often been saved from other such mistakes only by the skin of our teeth, so to say, and it is not in a spirit of criticism, but rather to demonstrate the difficulties of the work, that we draw attention to them.

Certain safeguards one can always try to obtain. There is no substitute for actually seeing for oneself in the great museums of the world, and the great archaeological sites; there is no substitute for personal intercourse with the practising technicians themselves. To be sure the scholarly standard of any particular work must necessarily depend upon the ground which is covered. Only the specialist using intensive methods —a Rosen elucidating the tangled roots of ophthalmic lenses or a Drachmann exploring Roman oil-presses—can afford the time to go into a matter *au fond* and bring truth

[a] Truly all the conclusions in the present volume must be regarded as provisional. Faithfully though we seek to apply the comparative method, our final assessment is often only a bridge built upon wide-spaced and insecure piers looming out of the mists. A single new and crucial fact can sometimes change the whole complexion of what seemed a fairly stable pattern. Our successors will doubtless see more clearly—but how it all happened Allah knoweth best.

wholly out of the well. We have tried to do this only in very few fields, such as that of medieval Chinese clockwork, because our aim is essentially extensive and pioneering. There is no escape, much must be taken on trust. If we are deficient in our knowledge of the objects of occidental archaeology, it is because we have laboured to study *in situ* those of the Chinese culture-area, our primary responsibility. If we had been able to visit the museum in Copenhagen where the Dejbjerg wagons are kept we might have been more wary of accepting current statements about them, but—ὁ βίος βραχύς, ἡ δὲ τέχνη μακρή, the craft is long but life is short. At the other end of the scale a deep debt of gratitude is owing to the President and Council of Academia Sinica for generous facilities which enabled me in 1958, together with Dr Lu Gwei-Djen, to visit or revisit many of the great museums and archaeological sites in China.

But not with archaeologists only must one converse. One must follow the example of little Dr Harvey (of Caius College). In the seventeenth century John Aubrey tells us of a conversation he had with a sow-gelder, a countryman of little learning but much practical experience and wisdom. He told him that he had met Dr William Harvey, who had conversed with him above two or three hours, and 'if he had been', the man remarked, 'as stiff as some of our starched and formall doctors, he had known no more than they'. A Kansu carter threw light upon the harness not only of our own time but indirectly of the Han and Thang, Szechuanese iron-workers were well able to help our understanding of how Chhiwu Huai-Wên in +545 made co-fusion steel, and a Peking kite-maker could reveal with his simple materials those secrets of the cambered wing and the airscrew which lie at the heart of modern aeronautical science. Nor may the technicians of one's own civilisation be neglected, for a traditional Surrey wheelwright can explain how wheels were 'dished' by the artisans of the State of Chhi two thousand years and more ago. A friend in the zinc industry disclosed to us that the familiar hotel cutlery found today all over the world is made essentially of the medieval Chinese alloy *paktong*,[1] a nautical scholar from Greenwich demonstrated the significance of the Chinese lead in fore-and-aft sailing, and it took a professional hydraulic engineer to appreciate at their true value the Han measurements of the silt-content of river-waters. As Confucius put it, *San jen hsing, pi yu wo shih*,[2] 'Where there are three men walking together, one or other of them will certainly be able to teach me something'.[a]

The demonstrable continuity and universality of science and technology prompts a final observation. Some time ago a not wholly unfriendly critic of our previous volumes wrote, in effect: this book is fundamentally unsound for the following reasons. The authors believe (1) that human social evolution has brought about a gradual increase in man's knowledge of Nature and control of the external world, (2) that this science is an ultimate value and with its applications forms today a unity into which the comparable contributions of different civilisations (not isolated from each other as incompatible and mutually incomprehensible organisms) all have flowed and flow as rivers to the sea, (3) that along with this progressive process human society is

---

[a] *Lun Yü*, VII, xxi.

[1] 白銅   [2] 三人行必有我師

moving towards forms of ever greater unity, complexity and organisation. We recognised these invalidating theses as indeed our own, and if we had a door like that of Wittenberg long ago we would not hesitate to nail them to it. No critic has subjected our beliefs to a more acute analysis, yet it reminded us of nothing so much as that letter which Matteo Ricci wrote home in +1595 to describe the various absurd ideas which the Chinese entertained about cosmological questions:[a] (One), he said, they do not believe in solid crystalline celestial spheres, (Item) they say that the heavens are empty, (Item) they have five elements instead of the four so universally recognised as consonant with truth and reason, etc. But we have made our point.

A decade of fruitful collaboration came to an end when early in 1957 Dr Wang Ling[1] (Wang Ching-Ning[2]) departed from Cambridge to the Australian National University at Canberra, where he is now Professorial Fellow in Chinese Studies. Neither of us will ever forget the early years of the project, when our organisation was finding its feet, and a thousand problems had to be solved (with equipment much less adequate than now) as we went along. Dr Wang's partnership was exercised throughout the present volume. The essential continuity of day-to-day collaboration with Chinese scholars was however happily preserved at his departure by the arrival of a still older friend, Dr Lu Gwei-Djen,[3] late in 1956. Among other posts, Dr Lu had been Research Associate at the Henry Lester Medical Institute, Shanghai, Professor of Nutritional Science at Ginling College, Nanking, and later in charge of the Field Cooperation Offices Service in the Department of Natural Sciences at UNESCO headquarters in Paris. With a basis of wide experience in nutritional biochemistry and clinical research, she is now engaged in pioneer work for the biological and medical part of our plan (Vol. 6). Probably no single subject in our programme presents more difficulties than that of the history of the Chinese medical sciences. The volume of the literature, the systematisation of the concepts (so different from those of the West), the use of ordinary and philosophical words in special senses so as to constitute a subtle and precise technical terminology, and the strangeness of certain important branches of therapy—all demand great efforts if the result is to give, as has not yet been given, a true picture of Chinese medicine. It is very fortunate that time permits our excavations to commence from the bedrock upwards. At the same time Dr Lu has participated in the revision of the present volume for press, and in certain fields such as the history of the wheelwrights' art and the history of efficient equine harness has joined another friend, Mr Raphael Salaman, and myself, in active original research outside the field of her own speciality.

A year later (early in 1958) we were joined by Dr Ho Ping-Yü,[4] then Reader in Physics at the University of Malaya, Singapore. Primarily an astro-physicist by training, and translator of the astronomical chapters of the *Chin Shu*, he was happily willing to broaden his experience in the history of science by devoting himself to the study of alchemy and early chemistry, helping thus to lay the foundations for the relevant volume (Vol. 5). Such work had been initiated some years earlier by yet another friend,

[a] Cf. Vol. 3, p. 438.

[1] 王鈴　　[2] 王靜寧　　[3] 魯桂珍　　[4] 何丙郁

Dr Tshao Thien-Chhin,[1] when a Research Fellow of Caius College, before his return to the Biochemical Institute of Academia Sinica at Shanghai. Dr Tshao had been one of my wartime companions, and while in Cambridge made a most valuable study of the alchemical books in the *Tao Tsang*.[a] Dr Ho Ping-Yü was able to extend this work with great success in many directions. Although Dr Ho is now Professor of Chinese at the University of Malaya, Kuala Lumpur, it is my earnest hope that he will be able to rejoin us in Cambridge for the final preparation of the volume on chemistry and chemical technology.

It is good to record that already a number of important subsections of both these volumes (5 and 6) have been written. The publication of some of these in draft form facilitates criticism and aid by specialists in the different fields.

Lastly, an occidental collaborator appears with us on the title-page of the first part of this volume, Mr Kenneth Robinson, one who combines most unusually sinological and musical knowledge. Professionally he is an educationalist, and with a Malayan background in teachers' training, now as Education Officer in Sarawak frequents the villages and long-houses of the Dayaks and other peoples, whose remarkable orchestras seem to him to evoke the music of the Chou and Han. We were fortunate indeed that he was willing to undertake the drafting of the Section on the recondite but fascinating subject of physical acoustics, indispensable because it was one of the major interests of the scientific minds of the Chinese middle ages. He is thus the only participator in this enterprise so far who has contributed direct authorship as well as research activity. Another European colleague, Mr John Combridge, of the Engineering Department of the General Post Office, has greatly added to our understanding of medieval Chinese clockwork, especially by experiments with working models.

Once again it is a pleasure to offer public gratitude to those who have helped us in many different ways. First, our advisers in linguistic and cultural fields unfamiliar to us, notably Prof. D. M. Dunlop for Arabic, Dr Shackleton Bailey for Sanskrit, Dr Charles Sheldon for Japanese, and Mr G. Ledyard for Korean. Secondly, those who have given us special assistance and counsel, Dr H. J. J. Winter in medieval optics, Dr Laurence Picken in acoustics, Mr E. G. Sterland in mechanical engineering, the late Dr Herbert Chatley in hydraulic engineering and Cdr. George Naish in nautics. Thirdly, all those whose names will be found in the adjoining list of readers and kind critics of Sections in draft or proof form (p. xli). But only Dr Dorothy Needham, F.R.S., has weighed every word in these volumes and our debt to her is incalculable.

Once again we renew our warmest thanks to Mr Derek Bryan, O.B.E., and Mrs Margaret Anderson for their indispensable and meticulous help with press work, and to Mr Charles Curwen and Mr Ian McMaster for acting as our agents-general with regard to the ever-increasing flood of current Chinese literature on the history and archaeology of science and technics. Miss Muriel Moyle has continued to provide her very detailed indexes, the excellence of which has been saluted by many reviewers.

[a] Cf. Vol. 1, p. 12.

[1] 曹天欽

As the enterprise continues, the volume of typing and secretarial work seems to grow beyond expectation, and we have had many occasions to recognise that a good copyist is like the spouse in Holy Writ, precious beyond rubies. Thus we most gratefully acknowledge the help of the late Mrs Betty May, Miss Margaret Webb, Miss Jennie Plant, Miss June Lewis, Mr Frank Brand, Mrs W. M. Mitchell, Miss Frances Boughton, Mrs Gillian Rickaysen and Mrs Anne Scott McKenzie.

The part played by publisher and printer in a work such as this, considered in terms either of finance or technical skill, is no less vital than the research, the organisation and the writing itself. Few authors could have more appreciation of their colleagues executive and executant than we for the Syndics and the Staff of the Cambridge University Press. Among the latter formerly was our friend Frank Kendon, for many years Assistant Secretary, whose death occurred after the appearance of Volume 3. Known in many circles as a poet and literary scholar of high achievement, he was capable of divining the poetry implicit in some of the books which passed through the Press, and the form which his understanding took was the bestowal of infinite pains to achieve the external dress best adapted to the content. I shall always remember how when *Science and Civilisation in China* was crystallising in this way, he 'lived with' trial volumes made up in different styles and colours for some weeks before arriving at a decision most agreeable to the author and his collaborators—and what was perhaps more important, equally so to thousands of readers all over the world.

To the Master and Fellows of the Hall of the Annunciation, commonly called Gonville and Caius College, a family of immediate colleagues, I can offer only inadequate words. I do not know where else conditions so perfect for carrying out an enterprise such as this could be found, a peaceful workshop in the topographical centre of the University and all its libraries, between the President's apple-tree and the Porta Honoris. The daily appreciation and encouragement of every one in the Society helps us to surmount all the difficulties of the task. Nor can I omit meed of thanks to the Head of the Department of Biochemistry and its Staff for the indulgent understanding which they show to a colleague seconded, as it were, to another universe.

The financing of the research work for our project has always been difficult and still presents serious problems. We are nevertheless deeply indebted to the Wellcome Trust, whose exceptionally generous support has relieved us of all anxiety concerning the biological and medical volume. We cannot forbear from offering our deepest gratitude for this to its Scientific Consultant, formerly long its Chairman, Sir Henry Dale, O.M., F.R.S. An ample benefaction by the Bollingen Foundation, elsewhere achnowledged, has assured the adequate illustration of the successive volumes. To Dato Lee Kong-Chian of Singapore we are beholden for a splendid contribution towards the expenses of research for the chemical volume, and Dr Ho's work towards this was made possible by sabbatical leave from the University of Malaya. Here we wish to pay a tribute to the memory of a great physician and servant of his country, Wu Lien-Tê, of Emmanuel College, already Major in the Chinese Army Medical

Corps before the fall of the Chhing dynasty, founder long ago of the Manchurian Plague Prevention Service and pioneer organiser of public health work in China. During the last year of his life Dr Wu exerted himself to help in securing funds for our work, and his kindness in this will always be warmly remembered. Some kind well-wishers of our enterprise have now grouped themselves together in a committee of 'Friends of the Project' with a view to securing further necessary financial support, and to our old friend Dr Victor Purcell, C.M.G., who most kindly accepted the honorary secretaryship of this committee, our best appreciation is offered. At various periods during the studies which see the light in these volumes we have also received financial help from the Universities' China Committee and from the Managers of the Ocean Steamship Company acting as Trustees of funds bequeathed by members of the Holt family; for this we record most grateful thanks.

27. MECHANICAL ENGINEERING

(a) INTRODUCTION

AT this point we enter upon the discussion of the wider applications of physical principles in the control of forces and the use of sources of power. It may at first sight be found surprising that this is not preceded by an account of mining and metallurgy in old China. But although metals played so dominant a part in post-Renaissance engineering, this was by no means their role in the Middle Ages, either in East or West. The classification of Mumford[a] will be remembered—our present 'neotechnic' phase of electricity, nuclear energy, alloys and plastics followed upon a 'palaeotechnic' one in which coal and iron were the keynotes, but before that there had been an enormously longer period, the 'eotechnic' phase. In its Chinese manifestation, this was the age of wood, bamboo, and water, and it lasted until the spread of Renaissance technology over the Asian continent. Of course such a periodisation does not mean that metals were not of great importance in old China; one might instance the social significance of bronze for weapons in the Chou, the refined use of bronze for gear-wheels and crossbow-triggers in the Han,[b] an age when cast-iron plough-shares were in general use, and the making of steel implements both then and later. In some respects, such as the mastery of iron-casting,[c] and the first use and knowledge of zinc,[d] the Chinese were much ahead of Europeans. But most engineering constructions of any large size continued to be mainly of wood and stone.

It will be convenient to defer an account of the chief Chinese books on engineering until our consideration of the various fundamental types of machines.[e] The literature is distinctly small, however, perhaps mainly because the constructions of artisans, however ingenious, were too often regarded as unworthy the attention of the Confucian literati. Nevertheless, considerable numbers of illustrations have survived from the $+11$th century onwards (the springtime of printing), and there existed what may be called specific iconographic traditions of engineering drawing, though the draughtsmen or artists evidently did not always clearly understand what it was they were illustrating, and may have thought it beneath their dignity to enquire too closely. On the other hand, a great number of texts, many in the dynastic histories, have also survived.

The difficulty with the pictures (when the veil of scholarly dilettantism is not impenetrable) is that we know where we are technologically, but we cannot always easily get back to the original date of the illustration. Extensive research in Sung editions by persons of engineering as well as sinological competence will be necessary

---

[a] (1), p. 109. Two of the words were first introduced by Patrick Geddes. There are, of course, other classifications, such as Leroi-Gourhan's 'rustic', 'semi-industrial', etc. (1), vol. 1, p. 41. The appraisal of some of these by Mumford (4) in a revaluation of his own work is well worth reading.

[b] See p. 86 below, and Sect. 30e.　　　　　　　　　[c] See Sect. 30d and Needham (32).

[d] See Sect. 33 and Sect. 36.　　　　　　　　　　　[e] See pp. 165 ff., 168 below.

to improve this situation, and even so the limits of what has been preserved may be reached fairly soon. That late pictures may, however, perpetuate correct technological traditions, is strikingly shown by the case of a +17th-century sericultural book, certain illustrations in which portray detail by detail a description in an +11th-century text.[a] Conversely, the difficulty with literary sources alone is that we may have a firm date, which indeed may be quite early, but we are not always sure where we are technologically, either because the description of the mechanism is insufficient,[b] or because there is ground for fearing that the meanings of the technical terms have suffered changes from time to time.[c] The only cure for this is presumably the discovery and analysis of further texts.

Chinese technical literature has sometimes been reproached for a certain vagueness or ambiguity. What weight there is in this derives from the fact that Confucian scholars sometimes had to write about things which did not really interest them, while the technicians themselves, who could really have explained matters, did not write at all. But the opinion of Laufer, set down more than forty years ago, is worth recording.

In studying the same subject [he said] in Chinese and European literature, I find more and more that Chinese literature is not as bad as it is made out to be, even by sinologists, and that European literature is not as good as we like to boast it to be, in comparison with that of the Chinese and other nations. In investigating Chinese things, it is safe to consult Chinese accounts in each and every case. In spite of all their drawbacks (which I readily admit), these are simple, plain, matter-of-fact, and to the point, while the corresponding European notes prove in many cases superficial, misleading, or entirely erroneous.[d]

[a] The machine here referred to is the *sao chhê*[1] (silk-reeling apparatus) for winding off the silk from the cocoons. Its classical description is in the *Tshan Shu*[2] (Book of Sericulture), written by Chhin Kuan[3] about +1090, but this would not be easily understandable without the further explanations and illustrations of Hsü Kuang-Chhi's[4] *Nung Chêng Chhüan Shu*[5] (Complete Treatise on Agriculture) of +1639, especially ch. 33. The essential parts of this were translated into English a little more than two hundred years later (Anon., 39) in a publication which gave much better pictures than those in the +17th-century work. These were taken, in all probability, from the *Tshan Sang Ho Pien*,[6] written by Sha Shih-An[7] *et al.* in 1843 (cf. Liu Ho & Roux (1), p. 23). Similar illustrations, but not quite of such good quality, are found in the *Tshan Sang Chi Yao*,[8] written by Shen Ping-Chhêng[9] in 1869. Both these technical treatises on sericulture show not only the machine itself but all its components before assembly, including driving-belt, eccentric lug, and that forerunner of the 'flyer', the ramping-arm. Although belonging to the +19th century they could have served perfectly as illustrations for the +11th-century book, so faithful to continuous tradition was Chinese eotechnic practice. We shall deal with the whole subject of textile machinery in detail in Vol. 5, pt. 1 (Sect. 31). Meanwhile we shall speak again of the silk-reeling machine in the discussion of the history of the driving-belt (p. 107) and other matters (pp. 116, 301, 382, 404, 601) below. It is highly probable, in view of the antiquity of the silk industry in China, that the reeling-machine was already very old when Chhin Kuan described it in +1090, but occidental historians, e.g. Forti (1), p. 111, followed by Gille (11), have not hesitated to place its invention in Italy, generally favouring Bologna, towards the end of the +13th century. Where textile machinery is concerned, particular caution is needed before ascribing any developments to pre-Renaissance Europe.

[b] A tantalising case is that of Ma Chün's +3rd-century improvements to the loom, cf. below, again in Sect. 31.

[c] Cf. the case of the steering-oar and rudder, below, Sect. 29. There are forgotten glossaries of mechanical terms by Wylie (12) and Giquel (1) but these give mainly the words current in the earlier days of the impact of Western technology on China, and know nothing of ancient and medieval Chinese terms.   [d] (3), p. 26.

[1] 繰車   [2] 蠶書   [3] 秦觀   [4] 徐光啓   [5] 農正全書
[6] 蠶桑合編   [7] 沙式菴   [8] 蠶桑輯要   [9] 沈秉成

Studies of the history of engineering in the Chinese culture-area have been very few, so that those of Chang Yin-Lin (*2, 8*), Liu Hsien-Chou (*1, 4, 5*) and Chatley (*2, 36*) deserve particular gratitude. That of Horwitz (1) already mentioned,[a] which traced some of the illustrations in the *Thu Shu Chi Chhêng* encyclopaedia of +1726 to previous European sources, gave a quite one-sided impression which its author re-deemed in subsequent papers (6, 7). To these and many others by this notable Austrian historian of technology we shall refer in their place. For tools rather than machines there is the unique and valuable book of Hommel (1) on contemporary traditional Chinese practice. The only companion volume to Hommel (so far as we know) is the recent interesting treatise of Than Tan-Chhiung (*1*) on traditional Chinese technology, abundantly illustrated with scale drawings.[b] These secondary sources are all of real value. The book of Julien & Champion, though old, remains a classic, but it belongs rather to the realm of metallurgy and chemical technology.[c] Laufer (3) on the pottery of the Han dynasty contains unexpectedly copious discussions of certain matters in the history of Chinese engineering. We shall mention other aids, both in Chinese and Western languages, from time to time, but it is worth pointing out that the number of translations of original texts in this field was exceedingly small before we began our work.

The study of tools and the simpler machines borders of course upon the realms of anthropology, raising questions which can only be answered by comparative studies of the technologies of all peoples in the eotechnic stage. Leroi-Gourhan, whose book on comparative technology (1) is the most interesting of the kind which we have found,[d] points out how illogical the conventional categories are, indeed how unconsciously Europocentric, when traditional Chinese medicine, for instance, has been treated as part of ethnography, while the parallel material of our own Middle Ages has been accepted as part of the true history of medical science.[e] European music, however primitive, is music—all other music is anthropology. In fact, the technical arts of a people such as the Chinese blend not only into those of other, neighbouring, peoples at much more archaic stages of development, but also, like our own, going back in time, connect with the tools and customary operations of prehistoric man.[f] Hence arise all problems of invention and diffusion, such as are considered in the classical books of Mason (2) and Sayce (1). To these processes of invention and the adoption or neglect of inventions, especially in relation to the social milieu, we shall return in

[a] Vol. 1, p. 152.

[b] This feature renders it complementary to Hommel, and aligns it with the exhaustive studies of Chinese shipbuilding which we owe to Worcester (1, 2, 3) and which we shall draw upon in Sect. 29.

[c] See Sects. 30, 33, 34 and 36. Although this work was well based on such authoritative Chinese sources as the *Thien Kung Khai Wu*, it is sometimes a little difficult to use since paraphrases and quasi-translations are not always clearly distinguished from the statements of the authors themselves.

[d] Though unfortunately lacking all bibliographical reference material.

[e] (1), vol. 1, p. 20.

[f] On this a few references only must suffice: Furon (1); Montandon (1); Kroeber (1); Burkitt (1, 2, 3); McCurdy (1); Childe (1). The best, if not the only, examination of prehistoric tools (scrapers, chisels, hatchets, gouges, drills, etc.) from the practical engineering and physiological point of view is that of Frémont (6, 20). Leroi-Gourhan (1, 2) has attempted something similar for primitive weapons; on this see Sect. 30 below.

Section 49 on limiting factors. Meanwhile the interesting work of Schuhl (1) on the social philosophy of engineering may be strongly recommended as companion reading for what will follow in the present Section.

In the field of biography a good deal of spade-work has been done in China. Arising out of the activities of the Society for Research in the History of Chinese Architecture, a valuable series of collections of short lives of technologists of all kinds, not only builders, has been published. These are due to the work of Chu Chhi-Chhien and his collaborators.[a] More recently there has been a popularisation of such knowledge in the small book of Yen Yü (1) which deals with fifteen engineers and technicians, and eleven scientific men, from antiquity onwards. A similar popular work by Chhien Wei-Chhang (1) on the history of science and engineering in China devotes five chapters to the latter.[b] Although these productions are of a quite elementary character they could profitably be read by specialists in the traditional sinological fields.

Since the greatest works on the history of science, such as those of Sarton (1) and Thorndike (1), are relatively weak on the side of engineering and technology, we have had to rely for the comparative background of Chinese mechanics on other more specialised books. When the first draft of this Section was made we did not have the advantage of the collective *History of Technology* edited by the doyen of British history of science, the late Charles Singer, with a band of collaborators,[c] and written by many individual authors. In framework it is of course confined (albeit so compendious) to the history of technology in Europe and the ancient civilisations of the Near East, but the implications of this were recognised and explained by Singer (10) in a special article at the end of the second volume. The rich content of the whole work has fortunately been available to us in the later stages. A parallel series of volumes, all due to the pen of one writer, Forbes (10–15),[d] has also been at our disposal. These sometimes touch upon East Asian developments and transmissions and sometimes omit them, in neither event always felicitously, however. The less recent collective work edited by Uccelli (1) is on a large scale and excellently illustrated, but very deficient in bibliography and source references. Similar comments apply to the important work of Feldhaus (2), for which some of his other books (7, 8, 9) were preparations or abridgments. At the other extreme of size are the small books of Ducassé (1) and Lilley (3), which can serve only as preliminary guides, though offering interesting points of view. Another contemporary book of repute is that of Forbes (2); it contains much of interest but lacks reliability on all non-European origins, which indeed it seems

---

[a] They draw upon a wide variety of literary sources including of course the biographies of technologists in the official dynastic histories, and their abridgments in *TSCC*, *Khao kung tien*, ch. 5. The first six instalments of the register were prepared by Chu Chhi-Chhien & Liang Chhi-Hsiung (1–6), who were joined by Liu Ju-Lin in the seventh. The later ones bear the names of Chu Chhi-Chhien & Liu Tun-Chen (1, 2).

[b] It is interesting to note that in 1953 these books were reaching a circulation of nearly 100,000 copies. Still greater audiences are attained by the most popular books such as the little pictorial introduction of Ho Chi-Mei & Jui Kuang-Thing (1), which runs over the whole history of technology in the simplest terms, giving due place and part throughout to Chinese inventions and discoveries, but drawing largely also upon the European technological classics of all periods. Such productions must be of the greatest value in educating the rural masses of China to play their part in the modern world.

[c] Singer, Holmyard, Hall & Williams (1).   [d] Crit. Leemans (1).

deliberately to minimise. We are however much indebted to Forbes for his useful bibliography (1) of the literature on the history of technology in West European languages. The best one-volume history of engineering in the Western world probably remains that of Usher (1),[a] but Beckmann's old book on inventions (1) is by no means superseded. Perhaps the most outstanding and intellectually satisfying contribution known to us is the general and systematic history of machinery by T. Beck (1), which covers the whole period from the beginning of the Alexandrians to the end of the Renaissance, providing a large number of very clear diagrams. This work, unfortunately scarce, is quite invaluable, and nothing which has appeared during the past fifty years has in any way replaced it.

For engineering in European antiquity the chief aids are well enough known, including a small but brilliant work by a great scholar, Diels (1); a voluminous treatise in four volumes by Blümner (1); and, intermediate in size, the lively but not too reliable book of Neuburger (1).[b] In addition there are the comments of all the editors of writers such as Heron[c] or Vitruvius,[d] and much valuable secondary literature.[e] On ancient Egyptian industries one turns to the useful book of Lucas (1). Among older works, that of A. Espinas (1) is sociological and speculative, while that of Vierendeel (1), attempts to cover modern as well as ancient times, but not with success. The new book of F. Klemm (1) is here much better; it differs from all others in that it is a florilegium of relevant quotations dating from occidental antiquity to the post-Renaissance period.

For the history of engineering and technology in medieval Europe, important both for comparisons with Chinese eotechnic developments, and for the acceptance of inventions from the East by the West, four outstanding memoirs come to mind, those of Bloch (3), des Noëttes (3), Lynn White (1)[f] and Stephenson (1). The last-named, like that of Horwitz (3), is based on illustrations in a famous MS. of +1023 (cf. Amelli, 1), the *De Rerum Naturis* of Hrabanus Maurus (+776 to +856), founder of the famous school at Fulda in Germany.[g] Gille (4, 11) has essayed general reviews of the technological developments in Europe between +1100 and +1400, and has devoted a good paper (12) to the +15th-century military engineers of Germany,[h] supplementing thus Berthelot's classical descriptions (4, 5, 6). On Leonardo da Vinci and his time the studies of Uccelli (2), Gille (3) and Hart (1, 2, 3, 4) have proved helpful. Concerning the great Renaissance books on machinery, the enquirer may find help in the works

[a] A new edition, revised and enlarged, appeared in 1954. Now we have also Burstall (1).

[b] The English publishers of this qualified for the censure of the republic of letters, since they issued the translation with a drastically abbreviated index and the total omission of the bibliographies appended to each chapter, the general bibliography, and the table of sources of the illustrations.

[c] E.g. Woodcroft (1).   [d] E.g. Morgan (1); Granger (1).

[e] Such as the papers of T. Beck (2, 4, 5); Gille (5), and especially Drachmann (2, 7, 9).

[f] We much regret that Professor Lynn White's *Mediaeval Technology and Social Change* (7), with all its valuable new insights and information, appeared too late to help us in our surveys. But the justice which he does to Asian contributions (as also in 8, 9) is highly gratifying. For the same reason Professor Liu Hsien-Chou's *History of Chinese Engineering Inventions* (7), which contains some interesting material not previously published, could not be taken into account by us.

[g] Cf. Sarton (1), vol. 1, p. 555. Other great medieval Western figures also have their modern compères—Theobald (1), Dodwell (1) and Hawthorne & Smith (1) for +11th-century Theophilus; Lassus & Darcel (1) and Hahnloser (1) for +13th-century Villard de Honnecourt.

[h] See further, pp. 91, 110, 113, etc., below. On Guido da Vigevano see Hall (3).

of Taenzler (1), Davison (1), Gille (6) and Parsons (2), as well as in the fundamental and systematic account of T. Beck (1). Useful albums of late medieval and Renaissance technological illustrations are now coming forward (e.g. that of Schmithals & Klemm, 1). Another of Gille's reviews (13) takes up the engineering problems specific to the +17th century. As for the history of wood-work in building and other branches of technology nothing surpasses the treatise of Moles (1).[a]

Works on the history of engineering in modern times are available in even greater abundance,[b] though the level of scholarship tends to be lower. It is interesting sometimes to follow through in them principles which have been found to originate in eotechnic China. But for tracing the late history of engineering and technology, obsolete textbooks are often more useful than up-to-date ones—containing as they do informative material which was afterwards omitted; hence the value of John Harris's *Lexicon Technicum* of +1708, or Andrew Ure's early +19th-century *Dictionary of Arts, Manufactures and Mines*. We have already long been referring in this book to the encyclopaedia of the history of technology compiled by Feldhaus (1); a unique work, if perhaps more valuable for the excellence and abundance of its illustrations than for the accuracy of its text. The treasury of Feldhaus has been extended, but not at all replaced, by the encyclopaedia of Uccelli (3). It is also very useful to have at hand the chronological register of inventions and discoveries put together by Darmstädter (1).

In quite a different category to any of the foregoing, but not to be overlooked for our purpose, are the books by Western engineers who have themselves worked and travelled in China. They form a very considerable literature, but here we need only mention such examples as that of Parsons (1) with its interesting illustrations, and the more thoughtful description of Esterer (1). Others will be drawn upon as need arises. Related to these, and not at all irrelevant to the present subject, are the treatises on the industries of Japan,[c] Korea, Indo-China,[d] and other neighbouring cultures, the majority of which seem to be by Western students.

This completes the account of the comparative aids to the investigation of the story of Chinese engineering and technology which it seemed desirable to include by way of introduction.

When we trace it back far enough, the history of technology can be seen to originate in the famous +16th- and +17th-century literary quarrel between the supporters of the 'Ancients' and the 'Moderns'. Hardly anyone in the Middle Ages would have noticed that technology had a history, but at the Renaissance it gradually dawned on historians that the ancient Romans did not write on paper, knew nothing of printed books, and used no collar harness, spectacles, explosive weapons or magnetic compasses. The disquiet caused by this realisation was partly the occasion of the controversy,[e] which took its place as an important aspect of the inevitable clash between

---

[a] But see also Latham (1).

[b] E.g. Fleming & Brocklehurst (1); Burlingame (1, 2, 3); Leonard (1); Giedion (1); Burstall (1).

[c] E.g. those of Ninagawa (1); Rein (1); Brinckmann (1), etc.

[d] These are sometimes combined with elementary technical instruction, as in the book of Barbotin (1).

[e] A full account was written by Rigault (1). Bury (1), more briefly, sets the affair in its wider context of the history of the conception of progress.

PLATE CXXIV

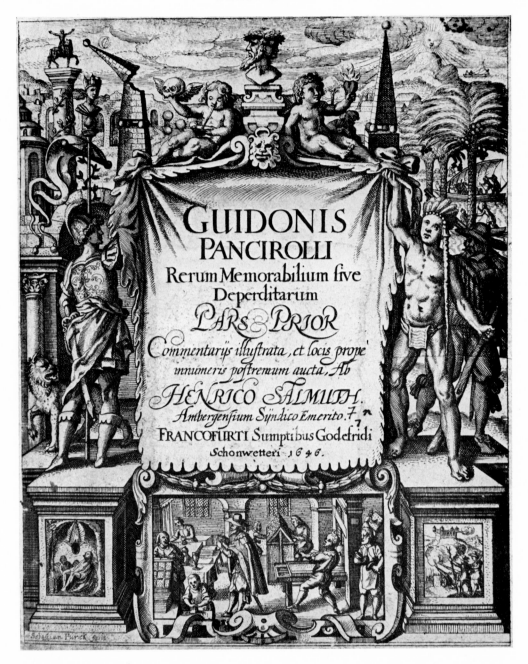

Fig. 351. Title-page of the +1646 edition of Guido Panciroli's book on the history of inventions. The left-hand side of the allegorical picture represents the achievements of the ancient West, but they are placed under the sign of the moon. Beside the toppling obelisk of Roman law, a cherub holds the winged skull of classical learning and blows the bubbles of empty Greek or scholastic philosophy. On the right, in contrast, the sun shines brightly on the achievements of the modern world, with its discoveries of the exotic, strange peoples and strange craft. The cherub at the top, surveyed by Janus' youthful face, holds up the flaming heart of new inspiration, while at the bottom panels display the non-European arts of gunpowder and typography.

PLATE CXXV

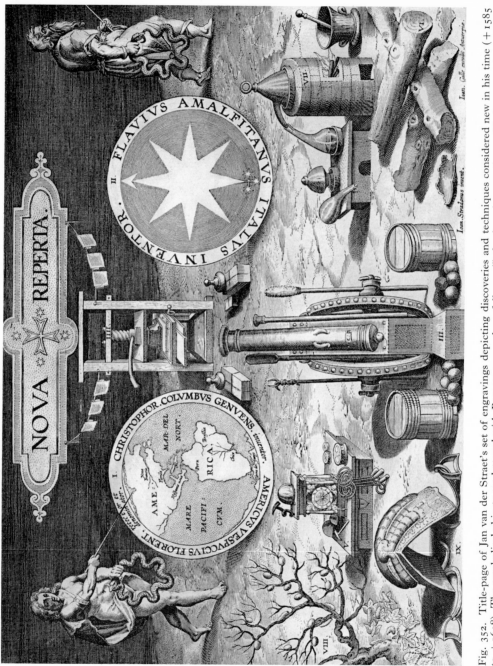

Fig. 352. Title-page of Jan van der Straet's set of engravings depicting discoveries and techniques considered new in his time (+1585 to +1638). The symbolical objects are lettered with Roman numerals as follows: (i) the American continent, (ii) the magnetic compass, (iii) gunpowder weapons, (iv) the printing-press, (v) the mechanical clock, (vi) guaiac wood, (vii) distillation, (viii) silk, the silkworms on the mulberry tree, and (ix) the stirrup. Five of these were directly derived from Chinese culture, and two more indirectly so. Time, growing ever older, and carrying the Worm Ouroboros, passes across the stage from left to right.

the humanistic polymaths and the experimental philosophers. Among the protagonists of the Moderns, Jerome Cardan in +1550 signalised the compass, printing and gunpowder as three inventions to which 'the whole of antiquity has nothing equal to show'.[a] The same argument was urged by Jean Bodin sixteen years later, in his rejection of the theory of the 'Golden Age' and the belief in the degeneration of man.[b] Francis Bacon gave it its most eloquent expression in +1620.[c] Few writers at that time, and few historians later,[d] recognised clearly the non-European origin of the three inventions, or drew from this fact its full implications. But the result was a new branch of learning and a literature still today interesting to read, in which the origins of these new and disturbing discoveries were sought for.

Here we have space to mention only three as representative out of many books, first Polydore Vergil's[e] *De Rerum Inventoribus*, first issued in +1499 but reproduced in a multitude of editions, then Guido Panciroli's[f] bulky *Rerum Memorabilium* of +1599 with its English translation of +1715, and lastly Thomas Powell's[g] slight but amusing *Humane History of most Manual Arts* (+1661). Beckmann was, as it were, the last of this tradition, and at the same time the first of the modern historians of technology. To visualise the new aspects of human knowledge which they valued, we may glance at one of Panciroli's title-pages (Fig. 351). Here the achievements of Greece and Rome are represented on the left, with a little vignette below of the light shining in darkness, but on the right the new and opulent Far East and Far West come forward in the person of an American Indian with a pigtail, behind whom we see palm-trees and lateen sails. Below, significantly, pride of place is given to the printing-press and to a cannonade. But perhaps the most striking of all the symbolical pictures of this period is the title-page which Johannes Stradanus[h] prefixed to his series of engravings entitled *Nova Reperta* (New Discoveries), first issued about +1585, and completed by +1638. We reproduce it here in Fig. 352. It lists the great new discoveries and inventions in the following order: (1) America, (2) the magnetic compass, (3) gunpowder weapons, (4) the printing-press, (5) the mechanical clock, (6) guaiacum,[i] (7) distillation, (8) silk, and (9) the stirrup. We know now that no less than six of these (2, 3, 4, 5, 8, 9) were directly derived, at the very least by stimulus diffusion,[j] from

---

[a] *De Subtilitate*, bk. 3.

[b] *Methodus ad facilem Historiarum Cognitionem*, ch. 7, pp. 359ff.

[c] *Novum Organum*, bk. 1, aphorism 129; Ellis & Spedding ed. p. 300. The passage has already been quoted in full; Vol. 1, p. 19.

[d] Even Bury (1) passed over in silence the fact that the cardinal inventions cited by Cardan, Bodin, Bacon and so many others, were all of Chinese origin.

[e] C. +1470 to +1555; in England from +1501 onwards. A friend of Linacre, More, Colet and Latimer, he wrote a history of England for Henry VII. A bibliography of his book on inventions will be found in John Ferguson (2), vol. 1, pt. 2, vol. 2, pt. 2.

[f] Italian jurist, +1523 to +1599. Bibliography in John Ferguson (2), vol. 1, pt. 2, vol. 2, pt. 1.

[g] Welsh writer, c. +1572 to +1635 (?). Bibliography in John Ferguson (2), vol. 1, pt. 3.

[h] Jan van der Straet, a Fleming from Bruges (+1523 to +1605), who lived most of his life in Florence. His technological illustrations have been republished and edited by Dibner (2).

[i] The resinous waxy wood of an American tree, *Guaiacum officinale*, of great repute in the +16th century as *lignum vitae* or Holy Wood for the treatment of syphilis (cf. da Orta (1), ed. Markham, pp. 380ff.). Any results were probably due to its irritant diuretic and purgative effects (Sollmann (1), p. 692). But it proved to be useful for making gear-wheels and bearings (cf. p. 499 below).

[j] See Vol. 1, pp. 244ff., and on the specific topics, Sects. 26*i*, 30, 32, 27*j* (pp. 440ff. below), and 31.

China. Moreover, as the home of the stern-post rudder[a] as well as the compass, that culture had also had a hand in the geographical discoveries, (1) and their corollary (6). Distillation (7) had in fact been common to Greek and Chinese antiquity,[b] while the last spread quickly, in all probability, among the nomadic peoples of Asia.[c]

Because of the obscurity of the records, the undeveloped state of oriental studies, and the comparatively primitive methods of historical research available in the +16th and +17th centuries, the true Eastern home of most of the 'modern inventions' was for a long time not identified, though surmised by some. In fact, during the Middle Ages, Europe had accepted a great number of Indian, Arabic, Persian, and especially Chinese, techniques, generally without knowing the places of their origin. It was perhaps the very mystery of such innovations that provided the basic stimulus for the birth of the history of technology.[d] All this took place essentially before the time of the Galilean break-through to modern science and the burst of discoveries and theories associated with the Royal Society and the scientific academies, yet the words which Joseph Glanvill in +1661 applied to them held good for the still unrecognised Asian contributions of earlier ages with equal appropriateness:[e] 'That discouraging Maxime, *Nil dictum quod non dictum prius*, hath little room in my estimation, nor can I tye up my belief to the letter of Solomon; I do not think, that all Science is Tautology: these last Ages have shewn us what Antiquity never saw; no, not in a dream.'

[a] See Sect. 29h below.

[b] See Sect. 33 below.

[c] Stradanus' own historical attributions are interesting. He allotted the compass to Italy and the printing-press and gunpowder to Germany in conventional style. Silk had come from 'a city', Serinda (cf. Vol. 1, p. 185, and Sect. 31), and he thought it so important that he added six further engraved plates on it under the title *Vermis Sericus*. Stradanus then strangely denied the water-mill to Western antiquity, and also (with justice this time) the windmill, though he did not know its Persian home. Another technique which he denied to Western antiquity was the polishing of armour by power-driven wheels. In this of course he was quite right, as we may see in the monographs of Schroeder (1) and Woodbury (3) on the history of grinding machinery. Apart from milling techniques, which we shall discuss in detail below (pp. 183ff.), we shall meet with grinding machines proper on a number of occasions, e.g. in connection with the generally recognised earliest occurrence (+9th century) of the crank in Europe (p. 112), or the oldest illustration of a driving-belt (+15th century) in that culture (p. 102). It is perhaps characteristic of the Europocentric bias of historians of technology that not a word is said in these monographs of the outstanding example of ancient and medieval achievement in mechanical grinding—the jade-cutting steel discs of Chinese antiquity (cf. Vol. 3, pp. 667ff.). An engraving by Israel van Meckenem, c. +1485, is accepted as the first clear evidence of a vertical grinding-wheel with treadle and crank, but this probably holds good only for Europe, for as we shall see, all the components were either as old, or older, in China. The first powered grinding-wheels are depicted in Leonardo, but we do not know whether they were actually built and used in his time; in any case they become common in paintings from +1568 onwards. Good drawings are in Zonca (+1621), pp. 33, 36, 39. The history of grinding machinery is of particular interest because it is so closely connected with the development of means for holding the work against the wheel and of guiding their relative motions mechanically. When modern grinding machines intended for obtaining precision shapes, rather than for massive abrasion, sharpening or polishing, developed in the 19th century they could take advantage of all the work that had been done on the lathe—with its strong bed, stocks, chucks, swivel tables and lead screws.

[d] Naturally other streams joined to form this literature, e.g. the Greek and Roman traditions of legendary inventors and culture-heroes. On this cf. the monograph of Kleingunther (1). See also Vol. 1, pp. 51ff., and below, pp. 42ff.

[e] *Scepsis Scientifica*, ed. Owen, ch. 22.

(1) THE NAME AND CONCEPT OF ENGINEER

A few words may not be out of place here regarding the origins of the terms used for engineers in Western languages and in Chinese. To our minds, the word 'engine' has come to have so vivid and precise a meaning that it is hard at first to remember that it derives from that quality of cleverness or ingenuity which is (or was thought to be) inborn in certain people—'ingenium', indwelling genius, innerly generated. Since derivatives of these roots were already in common Roman use for expressing qualities of wit, craft and skill, it is not surprising that 'ingeniarius', as a term in the more restricted sense, is found in Europe with increasing frequency from the +12th century onwards.[a] Not till the +18th was it freed from its primary military connotation. The course of events in China was not quite parallel with this.

From the earliest times the word *kung*[1] implied work of an artisanal character, technical as opposed to agricultural. This is perpetuated in the modern term for engineering, *kung chhêng*,[2] the second of the two characters having originally meant measurement, dimension, quantity, rule, examination, reckoning, etc.[b] Other old words such as *chi*[3] (originally the loom, the machine *par excellence*) and *tien*[4] (originally lightning) came in the course of time to do duty for mechanical and electrical devices respectively.[c] But none of their combinations, as applied to persons, is even medieval. The really old term for artisan-engineer is *chiang*,[5,6] in which an axe or the technical-work character is set in what may be a box (K 741), but more probably is a carpenter's square (*chü*[7,8]). One oracle-bone form of this (K 95d) actually shows a man holding a carpenter's square. The word *kung* itself also derives from a drawing of this instrument. It is safe to conclude that in Chinese culture, primarily eotechnic as it was, engineering work *par excellence* was wood-work. The *Chou Li*[d] calls master-craftsmen Kuo Kung.[9]

K 95d

This is not the whole of the story, however. Technical ability which was particularly skilful and admirable was called *chhiao*[10] (K 1041).[e] The right-hand radical here is interesting because it is related to a number of other forms in which the general meaning is that of 'breathing-out'. Alone, as *khao*,[11] it means, according to the *Shuo Wên*, to sob; but much more familiar is the terminal expiratory exclamatory particle used so much in Chou and early Han poetry, *hsi*.[12] Some relatives, such as *hao*,[13,14] to call out,

---

[a] An elaborate treatment has been given by Schimank (1). Cf. Straub (1), pp. 131, 133; Lynn White (7), p. 160. 'Ailnoth ingeniator', London, +1157; Salzman (1), p. 11.

[b] Cf. the Fang Chhêng[15] chapter of the *Chiu Chang Suan Shu*, which deals with certain forms of calculation; cf. Vol. 3, p. 26 above.

[c] The most classical expression for a machine is a 'strange contrivance' (*chhi chhi*[16]). Another old terms is *chi chieh*,[17] or sometimes *chhi chieh*.[18] These latter phrases have a significant social background, on which see Vol. 2, p. 125 above.

[d] Ch. 11, p. 12b.

[e] For example, the *Hsi Ching Tsa Chi* refers to Ting Huan (see below, p. 233) as a *chhiao kung*,[19] and the *San Kuo Chih* calls Ma Chün (see below, p. 39) a *ming chhiao*.[20]

[1] 工	[2] 工程	[3] 機	[4] 電	[5] 匠	[6] 匚
[7] 巨	[8] 矩	[9] 國工	[10] 巧	[11] 丂	[12] 兮
[13] 号	[14] 號	[15] 方程	[16] 奇器	[17] 機械	[18] 器械
[19] 巧工	[20] 名巧				

are still in common use. The semantic significance of the term for engineering genius in Chinese, therefore, would be identical with that of Latin, but expressed in the opposite way, not emphasising that genius which was born *in*, but that which was manifested and breathed *out*.

Sometimes artisans and engineers were simply called 'makers' or 'doers'. The former verb, *tsao*,[1] is used for constructors of armillary spheres or seismographs, such as Chang Hêng (cf. Vol. 3, p. 627 above); while the title Chiang Tso,[2] Commander of the Builders (lit. Makers or Doers), occurs as early as the Chhin, for officials in charge of artisans and workshops.[a] *Tsao* has a strongly creative undertone, implying invention or innovation. *Tso* is more executive and organisational, as in the title common later, Chiang Tso Ta Chiang,[3] Arch-Craftsman and Commander of the Builders.

Thus besides the semantic contrasts noted above, the associations of the Chinese terms for engineers and artisans seem always to have been more civilian and less military than those of the 'ingeniarii', 'angigneors' and 'engynours' of the West.[b]

(2) ARTISANS AND ENGINEERS IN FEUDAL-BUREAUCRATIC SOCIETY

At the beginning of the Section on astronomy it was necessary to say a good deal about the 'official' character of that science in China.[c] The astronomer (also a State astrologer) was lodged in part of the imperial palace, and the Bureau to which he belonged was an integral part of the civil service. To some extent, and on a lower plane, artisans and engineers also participated in this bureaucratic character, partly because in nearly all dynasties there were elaborate Imperial Workshops and Arsenals, and partly because, during certain periods at least, those trades which possessed the most advanced techniques were 'nationalised', as in the Salt and Iron Authorities under the Former Han.[d] We shall also see later (p. 32) that there was a tendency for technicians to gather around the figure of one or another prominent official who encouraged them, as his personal followers.

In all the preceding discussions of this book, where philosophers, princes, astronomers or mathematicians were mainly concerned, we have been dealing exclusively with the educated part of the Chinese population.[e] But now we reach a turning-point in the survey for we must necessarily step into the obscurer expanses of the trades and husbandries, in fact the application of scientific principles (whether or not always fully formulated) in practical technical work of many different kinds. We can therefore no longer leave out of account the mass of the workers and the conditions under which they laboured. They were the human material without which the planners of irrigation works or bridges or vehicle workshops, or even the designers of astronomical apparatus,

---

[a] *Shih Wu Chi Yuan*, ch. 26, pp. 31*b*ff.
[b] We refrain from enlarging here on the significance of this in relation to the sociological contrast between feudal-bureaucratic and city-state mercantile civilisation.
[c] Vol. 3, pp. 186ff.
[d] From −119 onwards. Cf. Sects. 30*d*, 37 and 48 below.
[e] Educated, that is to say, in the conventional sense, i.e. in literature and classical learning.

[1] 造   [2] 將作   [3] 將作大匠

could have done nothing, and not seldom it was from them that ingenious inventors and capable engineers rose up to leave particular names in history. A brief discussion of the role of the technical workers in feudal-bureaucratic society seems therefore indispensable here. This belongs, it is true, to the 'social background' which Sect. 48 will study, but we find it irresistibly obtruding itself here as part of the foreground.

It is generally allowed that the most important document for the study of ancient Chinese technology is the *Khao Kung Chi*[1] (Artificers' Record) chapter of the *Chou Li* (Record of the Institutions (lit. Rites) of the Chou Dynasty). Though the book in general is a Han compilation, it is supposed to report traditions which came down from the Warring States period (e.g. −4th century), but whether this is true also of the 'Artificers' Record' has been harder to say. All the ancient commentators agree that the original text of that part of the *Chou Li* which described the activities of the Minister of Public Works (Ssu Khung[2]), whose operations were supposed to belong to the winter season,[a] was lost at the beginning of the Han. The substitute for it which now forms the *Khao Kung Chi* was collected by Prince Hsien of Ho-Chien[3] (Liu Tê[4]) (*d.* −130),[b] and must have necessitated an entourage of technicians rather more practical than those who surrounded the Prince of Huai-Nan. Though some have thought that the present text was written by Liu Hsin late in the −1st century, internal evidence is against such a dating. Kuo Mo-Jo (*1*) noticed that the *Khao Kung Chi* mentions products or crafts of all the major States except Chhi, and that some of the weights and measures to which it refers were characteristic of Chhi. Yang Lien-Shêng (*7*) found three places where expressions in the dialect of Chhi occur. The suggestion thus arises that the document stems from an official compilation of the State of Chhi in the Warring States period. This would agree very well with the known eminence of that country in all kinds of techniques and sciences.[c] It would also raise an interesting comparison with the *Kuan Tzu* book, a text which is now thought to have been put together by the Chi-Hsia Academicians of Chhi.[d]

In any case, there is no doubt that although the text of the *Khao Kung Chi* took its present form in the early Han, it embodies a great deal of earlier date. Many commentaries were written on it as the centuries passed; no doubt the most important are those of Chêng Chung[5] about +75, Chêng Hsüan[6] about +180, and Chia Kung-Yen[7] during the +8th century. All are collected in the *Chhin Ting Chou Kuan I Su*,[8] the imperial edition of the *Chou Li* issued in +1748. Among the directors of this project was Fang Pao,[9] who devoted a special study to the 'Artificers' Record', the *Khao Kung Hsi I*.[10] The standard emendation of the text itself is due to Sun I-Jang (*3*). In +1746

---

[a] The word *khung* means 'empty' and here originally signified empty time, i.e. 'leisure' of the people, a surcease from their agricultural employments, during which they were to be mobilised for other purposes.

[b] Cf. Vol. 1, p. 111 above.

[c] Cf. Vol. 1, pp. 91 ff.; Vol. 2, pp. 232 ff., and many other places.

[d] Cf. Vol. 1, p. 95. The great parallel was of course the compilation of the *Lü Shih Chhun Chhiu* in the −3rd century by the scholars and technicians of Chhin (cf. Vol. 3, p. 196).

---

[1] 考工記 [2] 司空 [3] 河間獻王 [4] 劉德 [5] 鄭眾
[6] 鄭玄 [7] 賈公彥 [8] 欽定周官義疏 [9] 方苞 [10] 考工析疑

Tai Chen[1] produced a brilliant critical archaeological analysis[a] of the technology in the *Chou Li*, the *Khao Kung Chi Thu*.[2] His example was followed by Chhêng Yao-Thien[3] in his *Khao Kung Chhuang Wu Hsiao Chi*[4] (Brief Notes on the Specifications), and in his *Thung I Lu*.[5] Other writers made studies of particular sections in the *Khao Kung Chi*, e.g. that of Juan Yuan (2) on the procedures of the wheelwrights and cartwrights.[b]

The opening paragraphs of the *Khao Kung Chi* are so interesting that they are worth giving in full.[c]

The State has six classes of workers, and the hundred artisans form one of them.

There are those who sit to deliberate upon the Tao (of Society) and there are others who take action to carry it on. Some examine the curvature, the form, and the quality (of natural objects) in order to prepare the five raw materials,[d] and to distribute them for making instruments (useful for) the people. Others transport things rare and strange from the four corners (of the world) to make objects of value. Others again devote their strength to augment the products of the earth, or to (weave tissues from) silk and hemp.

Now it is the princes and lords who sit to deliberate upon the Tao, while carrying it into execution is the function of ministers and officials. Examining the raw materials and making the useful instruments is the work of the hundred artisans. Transportation is the affair of merchants and travellers, tilling the soil belongs to the farmers, and weaving is the office of women workers.

In the land of Yüeh, there are no special makers of hoes (*po*[6]) but every man knows how to make one. In the land of Yen, there are no special makers of hide armour (*han*[7]), but every man knows how to make it. In the land of Chhin, there are no special makers of pikestaffs (*lu*[8]), but every man can make them. Among the nomads (Hu) the bow and the chariot are necessities, and all men there are skilled at making them.

Tools and machines were invented by men of wit (*chih chê*[9]), and their traditions maintained by men of skill (*chhiao chê*[10]); those who continue them generation by generation are called artisans (*kung*[11]). So all that is done by the hundred artisans was originally the work of sages. Metal melted (*shuo*[12]) to make swords, clay hardened to make vessels, chariots for going on land and boats for crossing water—all these arts were the work of the sages.

Now heaven has its seasons and earth has its *chhi* (local influences), particular stuffs have their virtues (*mei*[13]) and particular workers have their skills; if these four things are brought together, something good comes out of it. But with good material and good workmen it still may happen that the product is not good; in this case the season has not been suitable, or the *chhi* of the earth has not been obtained.

Take, for example, the sweet-fruited orange (*chü*[14]), when it is transplanted to the north of the Huai River, it turns into the bitter-fruited orange (*chih*[15]).[e] And the crested mynah (bird)[f]

---

[a] On this see the special study of Kondō Mitsuo (1).
[b] *Khao Kung Chi Chhê Chih Thu Chieh*.[16]
[c] Tr. Biot (1), vol. 2, pp. 457ff., eng. auct., mod. The leading ideas on State, society and industry in the *Chou Li* have been reviewed by Shigezawa Toshio (1).
[d] Presumably metal, jade, leather, wood and earth.
[e] This remark will be referred to again below (Sect. 42) in connection with the history of grafting.
[f] *Aethiopsar cristatellus*, R296.

[1] 戴震	[2] 考工記圖	[3] 程瑤田	[4] 考工創物小記	[5] 通藝錄
[6] 鎛	[7] 函	[8] 廬	[9] 知者	[10] 巧者
[11] 工	[12] 爍	[13] 美	[14] 橘	[15] 枳
[16] 考工記車制圖解				

(*chhü yü*[1]) never comes north of the Chi River. And badgers (*ho*[2]) die if they cross the Wên. This is natural because of the *chhi* of the earth.

People prize the knives of Chêng, the axes of Sung, the pen-knives of Lu, and the double-edged swords of Wu and Yüeh. No other places can make these things so well. This is natural because of the *chhi* of the earth. So also the best horn comes from Yen, the best hardwood from Ching, the best arrow-wood from Fên-Hu, and the best gold and tin from Wu and Yüeh. These are the natural virtues of these materials.

Heaven has its seasons of production and destroying; trees and plants a time to live and a time to die. Even stone crumbles (*lê*[3]), and water freezes or flows. These are according to the natural seasons of heaven.

Generally speaking, wood-working comprises seven operations, metal-working six, treatment of skins and furs five, painting five and polishing five, modelling in clay two. Woodwork includes the making of wheels, chariot-bodies, bows, pikestaffs, house-building, cart-making, and cabinet-making with valuable woods. Metal-work includes forging (*chu*[4]),[a] smelting (*yeh*[5]), bell-founding, making measures, containers, agricultural implements and swords. Work with skins includes drying, making hide armour, drums, leather and furs. Painting includes embroidery in one or more colours, the dyeing of feathers, basketry, and silk cleaning. Polishing includes the working of jade, the cutting and testing of arrows, sculpture and the making of stone-chimes. Modelling in clay includes the art of the potter and that of the tile-moulder.

This last was the art most esteemed by the dynasty of Shun. The Hsia gave first place to the art of house-construction, and that of the Yin (Shang) preferred the art of making vessels. But that of the Chou set highest the work of chariot body builders.

[The writer then embarks upon a discussion of chariots.]

By way of remembering where we are, let it be noted in passing that this document was probably written, in consciously archaic form, some time in the −2nd century, but referring to the early part of the −3rd and containing whatever reminiscences could be collected from still earlier periods. The first half of the −3rd century was the time of activity of Euclid, and the beginnings of the Museum at Alexandria; Megasthenes was carrying out his diplomatic mission in India, and Berossos was transmitting as much as possible of Babylonian astronomy to the Greeks. In engineering, the Pharos of Sostratus was more or less contemporary with the Great Wall of Chhin Shih Huang Ti.[b]

---

[a] This word is not the usual one for forging. It means simply to beat, as in the ramming of earth, so that perhaps it should be understood to refer here to the making of moulds of clay or sand for molten metal. *Yeh* implies liquid metal, even when referring to smelting rather than casting. On these questions see Sect. 30*d* below. There we shall find that some iron tools of the Warring States period seem to have been made neither by forging wrought iron nor by pouring cast iron, but by swage-moulding semi-liquid low-carbon steel. Might not the word *chu* here be a reference to this?

[b] For information on Indian craftsmen comparable with the material in the *Chou Li*, and drawn mainly from the Jātaka tales of early Buddhist times (−6th to −3rd centuries), see Foley (2). A particularly close parallel to the *Khao Kung Chi* is found in Ceylon, where the *Janavaṃsa* gives elaborate details of the duties and procedures of the artisans in the royal workshops. The text as we have it, ascribed to Siṃha of Kessellena, is in Sinhalese and dates no earlier than the +15th century, but it claims a Pali original. A good description of it is given by Coomaraswamy (5), pp. 21, 54ff. The institution of caste, strange in a Buddhist country, but persisting there still, is prominent in it. Coomaraswamy also has much to say of another parallel text, the *Sāriputra*, named after its author, and dating

---

[1] 鶪鶋   [2] 貉   [3] 泐   [4] 築   [5] 冶

The passage shows something of the characteristic Chinese love of arbitrary systematisation, but is clearly based largely on fact, and correlative thinking[a] appears only in the eleventh paragraph, where the favourite techniques of the several dynasties are mentioned.[b] The passage opens with a notable statement of the distinction between manual work (*shu*[1]) and mental (*hsüeh*[2]), and goes on (in the fourth paragraph) to suggest that there was a time when there were no specialised manual workers in Chhi. The four conditions of industrial production—season (*shih*[3]), local factors (*chhi*[4]), virtues of materials (*mei*[5]) and skill (*chhiao*[6])—appear in the sixth paragraph, and of all of these except skill the text goes on to give examples. In doing so, it makes some interesting statements on the ecology of plants and animals, explaining the naturalness of local factors. These have certainly always been important in the siting of industries, the presence of ores or coal or forests, the nature of the water,[c] and so on. The penultimate paragraph, which seeks to classify the various techniques, follows closely, in the technical terms which it uses, the names of the various kinds of artificers mentioned in the text of the *Khao Kung Chi* itself. These seem to be very incomplete (for example, there is mention of embroiderers but nothing about different types of weavers, and a special category of feather-dyers, but nothing about those who dye cloth);[d] yet this is perhaps natural in view of what is known about the history of the text. Table 54 reproduces for reference all the classes of artisans mentioned in the *Khao Kung Chi*.[e]

If the tables of officials, with their ranks and classes of assistants, which the *Chou Li* gives for all the other ministries, had been preserved for the Ministry of Works (the 'Winter Office', Tung Kuan[7]), we should perhaps know more than we do about the organisation of the Imperial Workshops, at any rate for the Han time itself. Unfortunately, they were contained in the part which was lost, and those who filled the gap did not supply them. The question arising is one of great importance, namely to what extent the most advanced techniques and industries in ancient times were under centralised bureaucratic control. The Imperial Workshops certainly produced all the ceremonial objects, commodities of daily life, vehicles and machines, required for the courts of the emperor and the princes; and there could be no sharp distinction between such work and the manufacture of arms and equipment for the imperial forces. When the salt and iron industries were 'nationalised', all the artisans concerned in them

---

from the +12th century, (5), pp. 150ff., 161. This is concerned mainly with the making of Buddhist images. The Sinhalese glosses are on a Sanskrit original, believed to derive from the architectural and technological *Silpaśāstra*, a corpus which Coomaraswamy places in the +5th century.

   [a] Cf. Vol. 2, pp. 261ff. above.

   [b] One suspects that there is some connection here with the theory of the five elements and their relation to the successive dynasties (cf. above, Vol. 2, p. 263).

   [c] The Ming book, *Piao I Lu*, has an interesting list (ch. 2, p. 7a) of the different kinds of natural waters—petrifying, chalybeate, sulphurous, copper-depositing, etc.—with tacit reference to siting of industries. Cf. Vitruvius, VIII, iii.

   [d] The dyers (*jan jen*[8]) appear, indeed, among other artisans mentioned elsewhere in the *Chou Li*, ch. 1, p. 9a; ch. 2, p. 35b (Biot (1), vol. 1, pp. 19, 166). They belonged to the inner apartments of the palace together with the textile workers.

   [e] The names may of course be the official titles of the comptrollers or foremen rather than the names of the trades themselves. A special study of them has been made by Yoshida Mitsukuni (3).

[1] 術   [2] 學   [3] 時   [4] 氣   [5] 美   [6] 巧   [7] 冬官
[8] 染人

Fig. 353. A late Chhing representation of artisans at work in the Imperial Workshops (*Shang Fang*). From *SCTS*, ch. 17B, Yüeh Ming (Medhurst (1), p. 172; Legge (1), p. 116), whence the caption: 'Preparations for all eventualities will avert misfortune.'

Table 54. *Trades and industries described in the* Khao Kung Chi *chapter of the* Chou Li

(An asterisk indicates that the section is missing)

			ch.	p.	Biot (1), vol. 2 ch.	p.
(A) WORKERS IN STONE AND JADE			12			
Jade workers[a]	*yü jen*	玉人		1*a*	42	519
Stone carvers	*tiao jen*	雕人		5*a**		530
Stone-chime makers	*chhing shih*	磬氏		5*a*		530
(B) CERAMICS WORKERS						
Potters	*thao jen*	陶人		7*a*		536
Moulders (tiles)	*fang jen*	旗人		7*b*		538
(C) WOOD WORKERS						
Arrow makers	*chieh jen*	榔人		4*b*		530
	shih jen	矢人		5*b*		532
Bow makers[b]	*kung jen*	弓人		24*a*	44	580
Cabinet makers in valuable woods[c]	*tzu jen*	梓人		8*a*	43	540
Weapon handle makers	*lu jen*	廬人		13*a*		548
Surveyors, builders and carpenters	*chiang jen*	匠人		15*a*		553
Agricultural implement handle makers, *see* Cartwrights						
(D) CANAL AND IRRIGATION DITCH BUILDERS (and hydraulic engineers in general)						
Hydraulic workers[d]	*chiang jen*	匠人		18*a*		565
(E) METAL WORKERS (*kung chin chih kung* 攻金之工)						
'Lower alloy' founders[e]	*chu shih*	築氏	11	20*b*, 21*a*	41	490
'Higher alloy' founders[f]	*yeh shih*	冶氏		20*b*, 21*a*		490, 492
Bell-founders	*fu shih*	鳧氏		20*b*, 23*a*		490, 498
Measure makers	*li shih*	㮚氏		20*b*, 25*b*		491, 503
Plough makers	*tuan[g] shih*	段氏		20*b**		491
Sword-smiths	*thao shih*	桃氏		20*b*, 22*b*		491, 496
(F) VEHICLE MAKERS[h]			11	5	40	463
Wheelwrights	*lun jen*	輪人	11	7*a*		466
Master wheelwrights	*kuo kung*	國工		12*b*		475
Body makers	*yü jen*	輿人		14*b*		479
Shaft and axle makers	*chou jen*	輈人		16*a*		482
	chu jen	軸人				
Cartwrights[i]	*chhê jen*	車人	12	21*b*	44	573

Table 54 (cont.)

			ch.	p.	Biot (1), vol. 2	
					ch.	p.
(G) ARMOURERS (of hide, not metal)						
Cuirass makers	han shih	函氏	11	26b	41	506
(H) TANNERS						
Tanners	wei jen	韋人		30b*		514
Skinners	pao jen	鮑人		28a		509
Furriers	chhiu jen	裘人		30b		514
(I) DRUM-MAKERS	yün jen	韗人		29b		511
(J) TEXTILE, DYEING, AND EMBROIDERY						
WORKERS (hua i chih shih 畫繢之事)				30b	42	514
Feather-dyers	chung shih	鍾氏		31b		516
Basket-makers	khuang jen	筐人		32a*		517
Silk-cleanersj	mang shih	㡛氏		33a		517

a Much information about the forms of the various ceremonial pieces, but hardly a word about the techniques of working.

b A long and elaborate section, which makes no mention of the crossbow, nu 弩, though this is spoken of elsewhere, pp. 239, 241, 246.

c Mainly musical instruments and cups. They had a foreman or manager, tzu shih 梓師.

d This section contains valuable information on irrigation canals, cf. Vol. 4, pt. 3.

e 'Lower alloy' bronze was a 3 Cu:2 Sn mixture, said here to be used for writing-knives.

f 'Higher alloy' bronze was a 3 Cu:1 Sn mixture, said here to be used for arrow-heads, lance-heads, etc.   g For tuan 煅.

h Their measurements and dimensions are related to standard weapon lengths.

i Also make handles for agricultural implements.

j Those who remove the gum from the natural silk.

must also have come under immediate government control.[a] At other times, and in other industries during this time, there can be no doubt that the largest part was played by handicraft production independently undertaken by and for the common people. But it may be safe to assume that when any large or unusually complex piece of machinery was constructed (e.g. the early water-mills) this was done either in the imperial workshops or under the close supervision of important provincial officials.[b]

a The most important texts on these Han government factories have been brought together by Chü Chhing-Yuan (1).

b In the Early Han period there were Supervisors of Industry (Kung Kuan[1]) in ten commanderies 'in charge of the production and taxation of goods'. This may have meant only a general control over the handicraft production of numerous families or small workshops but it certainly involved the collection of a market sales tax. The degree of control exercised by the Supervisors of Weaving (Fu Kuan[2]), especially important in Shantung, is also uncertain. But there can be no doubt that the Superintendents of the forty-six Iron Bureaux (Thieh Kuan[3]) and the thirty-five Salt Bureaux (Yen Kuan[4]) were directly responsible for the technical organisation of the production itself, since these were nationalised industries. The Iron Authority had particular importance for agriculture because its factories forged and cast the necessary implements. To these officials there were added, in the Later Han period, Supervisors of Orange Horticulture (Chü Kuan[5]) and of the Timber Industry (Mu Kuan[6]) in Szechuan. On the whole subject, see the study by Lao Kan (4).

[1] 工官   [2] 服官   [3] 鐵官   [4] 鹽官   [5] 橘官   [6] 木官

A thorough monograph which would follow through the history of workshops, imperial workshops and government factories in China is one of the most urgent sinological needs (cf. Fig. 353).

The imperial workshops went by many names, of which Shang Fang[1] was one of the most usual. We have already met with the phrase as a title, in the life of the magician Luan Ta (Vol. 4, pt. 1, p. 315 above), translating it 'Magician and Pharmacist-Royal to the Prince of Chiao-Tung',[a] but the word *fang* applies of course widely to all kinds of techniques. That was in the −2nd century, but by the end of the −1st the expression had come to mean the Imperial Workshops, as is distinctly stated in the biographies of Wang Chi[2] (d. −48)[b] and Tshai Yung (d. +192).[c] About this time phrases such as Chu Chin Chhi Wu[3] (Comptroller of the Forbidden Instruments) occur, indicating that devices and machines (e.g. crossbow-triggers, hodometers, the south-pointing carriage, etc.) were made and conserved in the workshops in the category of 'secret' or 'restricted'.[d] Occasionally the names of the artisans have come down to us. Swann (1) has reproduced a black lacquer lid bearing the following inscription:

In the 3rd year of the Chien-Phing reign-period (−4) this vessel, capacity three pints, and lid, were made at the Western Factory of Szechuan (in the same style of workmanship) as His Majesty's personal cart; lacquered base of hempen cloth, incised designs, painted ornamentation, and handles of gilt. Lacquer was made by the artisan Yu[4] and applied by artisan I;[5] bronze work and gilding by artisan Ku;[6] painting by artisan Fêng;[7] incising by artisan Jung;[8] cleaning by artisan Pao;[9] and finishing by artisan Tsung.[10] Supervision by official Chia[11] in charge of artisans and labour, by chief county magistrate Kuei,[12] his first assistant Hsien,[13] his secretary Kuang,[14] and by official Kuei[15] of the provincial administration.[e]

Significantly, perhaps, five administrators were needed for seven technicians, but it is pleasant, at any rate, to have the names of some of the latter.[f]

All dynasties seem to have had imperial workshops;[g] in the +4th century we seem to hear of them particularly under foreign rulers such as the Hunnish Later Chao,

[a] *Chhien Han Shu*, ch. 25A, p. 23a. A gloss by Yen Shih-Ku gives authority for this interpretation.
[b] Biography in *Chhien Han Shu*, ch. 72, p. 8b.
[c] Biography in *Hou Han Shu*, ch. 90 B, p. 17b; cf. *Shu Hsü Chih Nan*, ch. 6, p. 1a.
[d] It was in the Shang Fang Workshops that Liu Hsiang was ordered by the emperor about −50 to test the methods of Liu An, the Prince of Huai-Nan, for making alchemical gold. A full account of this interesting event will be given in Sect. 33. See *Chhien Han Shu*, ch. 36, pp. 6bff. where the metallurgical and minting functions of the Shang Fang are also referred to.
[e] Tr. Swann (1).
[f] In this connection it is interesting to find the same bureaucratic tendency at the other extreme of Chinese history. When the Fuchow Dockyard was set up in 1866 it was placed in charge of two French engineers (see p. 390 below) and a few Chinese technicians, but no less than a hundred 'subordinate mandarins' were attached as administrators (Chhen Chhi-Thien (3), p. 30). See the account by one of the engineers, Giquel (2).
[g] On the textile workshops of the Han palaces see Yonezawa (1), who has also made a longer study (2) of the Shang Fang during the Wei, Chin and Nan Pei Chhao periods. Many Han mirrors bear inscriptions showing that they originated from the Shang Fang (see *Chin Shih So*, Chin sect., ch. 6, pp. 4bff.). A typical inscription concerning the immortals depicted on the back of one of these (p. 10b) has been translated by Kaltenmark (2), p. 11.

[1] 尙方   [2] 王吉   [3] 主禁器物   [4] 有   [5] 宜   [6] 古
[7] 豐   [8] 戎   [9] 寶   [10] 宗   [11] 嘉   [12] 鬼
[13] 縣   [14] 廣   [15] 癸

who were served by such notable men as Hsieh Fei and Wei Mêng-Pien (cf. p. 257 below). Perhaps these States were less subject to Confucian orthodoxy than purely Chinese governments. When we come to the Thang, there is a great deal of information which has been collected by des Rotours.[a] The workshops were under a central Bureau (Shao Fu Chien[1]) with eight departments: three Workshops (centre, left and right) (Shang Shu[2]), a Weaving and Dyeing Section (Chih Jan Shu[3]), three Foundry Sections including a Mint (Chang Yeh Shu,[4] Chu Yeh Chien,[5] and Chu Chhien Chien[6]), and finally a Bureau for Barter with Foreign Peoples (Hu Shih Chien[7]), in fact a Sales Department. From the description in the *Hsin Thang Shu*,[b] it is clear that every kind of technique known in the +8th century from the purely metallurgical and engineering to the purely artistic was carried on in these institutions. There were periodical examinations as to technical skill, and every object was stamped with the name of the artisan who had made it. At one time from six to seven thousand artisans are spoken of. Unfortunately the Thang history describes the products of the work rather than the techniques employed. All this was separate from other equally important Bureaux: the Chiang Tso Chien[8] (Office of Works), which was mainly concerned with building construction and ceramics, the Chün Chhi Chien[9] (Arsenals Administration), which made crossbows, catapults, and all army equipment, and the Tu Shui Chien[10] (General Water Conservancy), which looked after irrigation works, canals, bridges and the like.

This is perhaps not the place to follow further this interesting subject, which needs more attention than it has yet had. A wealth of information is undoubtedly available, e.g. in Sung books such as the *Shih Wu Chi Yuan*, which tells us[c] about the imperial factories in Wu Tai and Sung times, when another name, that of Tso Fang,[11] became more general; and gives details about the State workshops of the goldsmiths and silver-smiths. Material for the Mongol period existed in the *Yuan Ching Shih Ta Tien*[12] and though this was lost, the relevant parts were copied from the *Yung-Lo Ta Tien* by Wên Thing-Shih[13] and so preserved. Thus we have the *Yuan Tai Hua Su Chi*[14] (Record of the Government Atelier of Painting and Sculpture), and the *Ta Yuan Chan Chi Kung Wu Chi*[15] (Record of the Government Weaving Mills).[d]

The story of the activities of the Ministry of Works during the Ming and Chhing would constitute a book in itself. For those who can read between the lines of massed inventories and requisitions, a mine of information is available in a book compiled by Ho Shih-Chin[16] in +1615; the *Kung Pu Chhang Khu Hsü Chih*[17] (What should be known (to Officials) about the Factories, Workshops and Storehouses of the Ministry of Works). This is of considerable size, and we shall draw upon it later more than once.[e]

---

[a] (1), vol. 1, pp. 458 ff.   [b] Ch. 48 (tr. des Rotours).
[c] Ch. 28, pp. 10*a*, 12*a*; ch. 34, pp. 3*b* ff.
[d] Both these were printed by Wang Kuo-Wei; see Hummel (2), vol. 2, p. 855.
[e] In Sect. 30, for example, with regard to gunpowder compositions.

[1] 少府監	[2] 尚署	[3] 織染署	[4] 掌冶署	[5] 諸冶監
[6] 鑄錢監	[7] 互市監	[8] 將作監	[9] 軍器監	[10] 都水監
[11] 作坊	[12] 元經世大典	[13] 文廷式	[14] 元代畫塑記	
[15] 大元氈罽工物記		[16] 何士晉	[17] 工部廠庫須知	

It may be interesting only to add a passage from the Jesuit Gabriel de Magalhaens (An Wên-Ssu [1]), written about the middle of the +17th century, before the Ming system had been reorganised. He wrote:[a]

The sixth and last superiour Tribunal is called Cum Pu (Kung Pu [2]), or, the Tribunal of the Publick Works. This Tribunal takes care to build and repair the Kings' Palaces, their Sepulchres and Temples, wherein they honour their Predecessors, or where they adore their Deities, the Sun, Moon, Heaven and Earth, etc., as also the Palaces of all the Tribunals throughout the Empire, and those of the great Lords. They are also the Surveyors and Overseers of all the Towers, Bridges, Damms, Rivers, Lakes, and of all things requisite to render Rivers navigable, as High-ways, Wagons, Barks, Boats, and the like. To this Palace belong four more inferiour Courts. The first is called Vin Xen Su (Ying Shan Ssu [3]), which examines and draws the Designs of all the works that are to be done. The second Yu Hem Su (Yü Hêng Ssu [4]), which has the ordering of all the Work-houses and Shops in all the Cities of the Kingdom for the making of warlike Arms and Weapons. The third Tum Xui Su (Tu Shui Ssu [5]), takes care to make the Rivers and Lakes Navigable, to level the High-ways, to build and repair Bridges, and for the making of Wagons and Boats, and other things necessary for the convenience of Commerce. The fourth Ce Tien Su (Thun (Chun) Thien Ssu [6]), are the Overseers of the King's Houses and Lands which he lets out to hire, and of which he has both the Rent and the Fruits of the Harvest.[b]

In general, the provisional conclusion may be justified that a considerable proportion of the most advanced technologists in all ages in China were either directly employed by, or under the close supervision of, administrative authorities forming part of the central bureaucratic government.

But of course not all. The majority of artisans and craftsmen, indeed, must always have been connected with small-scale family workshop production and commerce. Particular localities derived fame from skills which tended to concentrate there; one thinks of the lacquer-makers of Fuchow, the potters of Ching-tê-chen, or the well-drillers of Tzu-liu-ching in Szechuan. But Chinese technical skill tended to wander far and wide. In an earlier volume, we took note of metallurgists and well-drillers in −2nd-century Parthia and Ferghana,[c] and of textile technologists, paper-makers, gold-smiths and painters in +8th-century Samarqand and Kūfah.[d] Over and over again we shall see the respect which environing peoples had for the artisans of China. They did not hesitate to ask for them when circumstances permitted; thus in +1126, when the Sung capital at Khaifêng was being besieged by the Chin Tartars, they demanded from the city all sorts of craftsmen, including goldsmiths and silversmiths, blacksmiths, weavers and tailors; even Taoist priests.[e] When the Taoist Chhiu Chhang-Chhun made his famous journey from Shantung to Samarqand at the request of Chinghiz

---

[a] (1), p. 213. The fullest Jesuit study of the Ministry of Works is, however, that of Cibot (4).
[b] The names of the directorates were easily identified from *TSCC, Khao kung tien*, ch. 3, hui khao 3, pp. 14*b*ff.
[c] Vol. 1, p. 234 (Sect. 7*l*).   [d] Vol. 1, p. 236 (Sect. 7*l*).
[e] *Sung Shih Chi Shih Pên Mo*,[7] ch. 56. For this reference we are indebted to Dr Wu Shih-Chhang. Cf. *Sung Shih*, ch. 23, p. 19*a*.

[1] 安文思   [2] 工部   [3] 營繕司   [4] 虞衡司   [5] 都水司
[6] 屯田司   [7] 宋史紀事本末

Khan in +1221, he met Chinese workmen everywhere. At Chinkhai Balāsāghūn[a] in Outer Mongolia they came in a body to meet him, with banners and bouquets of flowers. When he reached Samarqand he found numbers more of them. On his return two years later, he heard of a great many[b] settled in a northern region, the valley of the Upper Yenisei.[c] As late as +1675, a Russian diplomatic mission officially requested that Chinese bridge-builders should be sent to Russia.[d] What the technical traditions were which all these men represented the body of this Section will show.

Beyond this point one cannot further analyse their social context without becoming involved in difficult questions of status. To follow this path to its conclusion would trespass upon the proper subjects to which Section 48 (on the social and economic background of ancient and medieval Chinese science and technology) will seek to do justice. But here a few words at least must be said, even if only to sketch the background for the question—where did the inventors and the engineers come from? So far, all the technical workers we have mentioned were 'free' plebeians.[e] A wheelwright or a lacquer-worker was a *shu jen*[1] or yeoman, 'of decent birth'; or a *liang jen*,[2] literally a 'goodman'. He belonged to the commoners (*hsiao min*[3]), and for the ancient philosophers would certainly have been a *hsiao jen*[4] (menial man, banausic man) as opposed to a *chün tzu*[5] (magnanimous quasi-aristocratic scholarly official man).[f] As he had a surname he was one of the *pai hsing*[6] (the 'old hundred families'), and belonged to the *pien min*[7] (registered people).[g] Only in very few cases are slaves or semi-servile people mentioned as producers of wealth in the descriptions of the operations of the famous rich industrialists of the Chhin and Han periods.[h] Indeed, a classical passage in the *Yen Thieh Lun* (Discourses on Salt and Iron) of −80 specifies free workers.[i]

[a] Near modern Uliassutai.
[b] These were weavers making fine silks, gauzes, brocades and damasks.
[c] *Chhang-Chhun Chen Jen Hsi Yu Chi*,[8] ch. 1, pp. 13a, 21b, ch. 2, p. 9a; tr. Waley (10), pp. 73, 93, 124.
[d] This was the embassy of Nicolaie Milescu (the Envoy, Spătarul, hence Russ. Spathary), on whom see Panikkar (1), p. 235; Cahen (1), p. 2; and especially Baddeley (2), vol. 2, pp. 351, 385.
[e] The word is thus qualified because ancient and medieval Chinese society cannot be judged by the same standards or in the same terms as that of the West. At a recent symposium on the comparative study of the institution of slavery it was concluded that in China 'the spectrum lacked both ends'. No one was 'free' in the fullest Greek sense and no one was as badly off as those who were unfree in the ancient occidental civilisations. As Pulleyblank has written: 'No conception of positive freedom could exist in an organic society where every man's position was defined by his social relationships, which carried with them not demarcated rights and limited corresponding duties, but moral obligations to superiors (ruler, father, elder brother, husband) by nature absolute and unlimited', (6), p. 205. The Aristotelian definition of property as of two kinds, things and men, was not the sort of definition the Chinese went in for. They recognised in slaves inalienable human qualities, and considered them bound by the five fundamental social relationships just as much as other people. In fact, as Pulleyblank says (p. 217), slaves were not so much a legitimate form of private property in China as the most inferior stratum at the bottom of the social scale. These points are important for the following discussion. For a current study of the concepts of positive and negative freedom cf. Berlin (1).
[f] Cf. Vol. 1, p. 93; Vol. 2, p. 6.
[g] Cf. Pulleyblank (6), p. 216.
[h] Cf. Wilbur (1), p. 218.
[i] On the context of this cf. Vol. 1, p. 105; Vol. 2, pp. 251 ff.

[1] 庶人　　[2] 瓦人　　[3] 小民　　[4] 小人　　[5] 君子　　[6] 百姓
[7] 編民　　[8] 長春眞人西遊記

Formerly the overbearing and powerful great families, obtaining control of the profits of the mountains and lakes, mined iron ore and smelted it with great bellows, and evaporated brine for salt. A single family would assemble a multitude, sometimes as many as a thousand men or more, for the most part wandering unattached plebeians (*fang liu jen min*[1]) who had travelled far from their own villages, abandoning the tombs (of their ancestors). Thus attaching themselves to the great families, they came together in the midst of mountain fastnesses or desolate marshes, bringing about thereby the fruition of businesses based on selfish intrigue (for profit) and intended to aggrandise the power of particular firms and factions.[a]

Needless to say, the speaker is the Lord Grand Secretary, defending the principle of the nationalisation of salt and iron.

But whatever the extent of government-organised production from time to time, the State relied upon an inexhaustible supply of obligatory unpaid labour in the form of the *corvée* (*yao*[2,3], *i*[4] or *yü*[4]; *kung yü*[5]). In Han times every male commoner between the ages of 20 (or 23) and 56 was liable for one month of labour service a year, unless belonging to some specially exempted group.[b] Technical workers certainly performed these obligations in the Imperial Workshops, whether at the capital or in the provinces, or in the factories of such enterprises as the Salt and Iron Authorities. These organisations were never primarily staffed by slaves, and as time went on there naturally grew up the practice of paying dues in lieu of personal service so that a large body of artisans 'permanently on the job' (*chhang shang*[6]) resulted. Chinese historians are now devoting much study to the regulations governing the various technical workers, so far mainly in the later periods. Thus Chü Chhing-Yuan (2) shows how in the Yuan time government artisans (*hsi kuan chiang hu*[7]) were distinguished from the military artisans (*chün chiang*[8]), though both received pay and rations, differing thus from the private civilian artisans (*min chiang*[9]), the services of whom, however, could be requisitioned from time to time. Artisans were always spared by the Mongol conquerors, who assembled them in government factories, e.g. for making mangonel artillery in +1279, and for weaving silk and woollen textiles somewhat later. As in the Sung, artisans could not be conscripted for any service other than their trade. In the Ming, a rota of technical *corvée* labour in government factories according to regular registers (*chiang chi*[10]) again became prominent.[c] Much interesting information is now coming to light from Yuan, Ming and Chhing texts[d] and stone inscriptions[e] about the organisation of the technical trades.

Of the part played by the artisans in political history nearly all remains yet to be written. As one illustrative case, however, we may take the rebellion in Szechuan[f] led

---

[a] Ch. 6, p. 17a, tr. auct. adjuv. Wilbur (1), p. 218; Gale (1), p. 35.

[b] See further Wilbur (1), p. 223.   [c] See Chhen Shih-Chhi (1).

[d] E.g. the paper of Chêng Thien-Thing (1) on the conversations of the scholar Hsü I-Khuei[11] with the silk-weavers of Hangchow towards the end of the Yuan dynasty, about +1361. Cf. the material collected by Tsêng Chao-Yü (1) and her collaborators on the clock-making industry of Nanking and Suchow in the Chhing. For Sung handicraft industry see Wang Fang-Chung (1).

[e] E.g. the papers of Liu Yung-Chhêng (1, 2, 3) on regulations for paper-mills, silk-weaving workshops and handicraft guilds in Chiangsu province during the Chhing period.

[f] For further information see the papers of Chang Yin-Lin (7), Chiang I-Jen (1), and Eichhorn (4).

| [1] 放流人民 | [2] 繇 | [3] 繇 | [4] 役 | [5] 公役 | [6] 常尙 |
| [7] 係官匠戶 | [8] 軍匠 | [9] 民匠 | [10] 匠籍 | [11] 徐一夔 | |

by Wang Hsiao-Po[1] and Li Shun[2] between +993 and +995. It has been variously regarded as a separatist movement designed to perpetuate the independence which the province had enjoyed during the Wu Tai period, and as an uprising of consciously socialist or communist inspiration, but it certainly had much to do with the treatment of the Szechuanese silk industry by the Sung government. Silk brocade (*chin*[3]) manufacture had had an outstanding importance in the province since Chhin times,[a] and beside the great salt-producing centre of Tzu-liu-ching and the timber and orange-growing industries had been one of the economic mainstays of the independence of the region.[b] Among the first actions of the Sung dynasty, however, was the establishment of competitive silk mills and brocade workshops (Chin Kuan[4]) in the east, in the neighbourhood of the new capital at Khaifêng,[c] and at the same time a strict prohibition of the trade of private silk merchants in Szechuan. It seems extremely probable that these steps struck a crushing blow at the industry there, and there can be little doubt that the impoverished silk-workers formed one of the principal sources of man-power for the revolutionary army of Wang and Li. Although the rebellion was subdued, the silk brocade manufacture centred at Chhêngtu gradually recovered in the course of time,[d] and the part which it had played in one of the numerous revolts of peasants and workers in Chinese history[e] was forgotten.

So far we have been thinking only of the artisans who were regular commoners. But this was by no means the lowest social level in ancient and medieval China, for below it there came a number of groups which might almost be termed 'depressed classes' and which certainly contained artisans, indeed sometimes men of skill and parts. The general term for these was *chien min*,[5] 'base' or 'ignoble' people, in contradistinction to the *liang jen*. Thus the question arises whether there was a slave artisanate and how important it was.[f] The majority of sinologists are agreed in believing that the institution of slavery in ancient and medieval China was primarily domestic in character and originally essentially penal.[g] How far then were relatively unfree persons employed in the technical crafts? It can safely be said that a certain proportion of the artisanate was always of this kind, but it was probably generally quite small (less than 10%) in relation to the whole mass of workers in the crafts and

[a] This may be seen from the *Hua Yang Kuo Chih*, ch. 3, which (in +347) treats the silk brocade industry as already old. A special monograph on it was written in the Yuan period by Fei Chu,[6] the *Shu Chin Phu*,[7] still extant.

[b] Taxation lists of all times bear witness to the importance of this Szechuanese manufacture. One of the surviving fragments attributed to Chuko Liang mentions the brocade industry as the chief source of finance for the armies of Shu in the San Kuo period (*TPYL*, ch. 815, p. 8b).

[c] One of these, called the Ling Chin Yuan,[8] was founded by the first Sung emperor in +967, and twice personally visited by the second. Another is recorded by Fei Chu as having had in +1083 a staff of 164 preparers, 54 weavers, 11 dyers, and 110 silk-winders, with a total of 154 machines.

[d] I had the pleasure of studying it myself in Chhêngtu in 1943.

[e] Cf. Anon. (41) and the recent summary by H. Franke (3).

[f] The only specific modern study of this subject by a Western sinologist known to us is the interesting review of Wilbur (3). Biot (20) was a pioneer, as so often.

[g] Cf. Wilbur (1); Erkes (20, 21); H. Franke (13).

[1] 王小波   [2] 李順   [3] 錦   [4] 錦官   [5] 賤民   [6] 費著
[7] 蜀錦譜   [8] 綾錦院

techniques.[a] Moreover, the status of some of the types of relatively unfree persons is rather difficult to define.

There is no doubt about slaves proper, *nu*[1] if male, *pi*[2] if female. The Han followed the Chou in defining the former as condemned to penal servitude (*tsui li*[3]) and the latter to grain-pounding (*chhung kao*[4]).[b] Though war captives (*lu*[5], *fu*[6]) sometimes suffered this fate,[c] the main source of recruitment in ancient times was certainly from criminals (or people regarded as such), and (according to the law of *shou nu*[7] for which the −4th-century State of Chhin was famous) their whole families.[d] A great many of the slaves were owned directly by the imperial State (*kuan nu pi*[8]),[e] but the line is hard to draw because State slaves were often given to high officials and nobles in reward for services or as presents. Their status as property they shared with no other groups, but in fixation of domicile, extra-group marriage restrictions, and restraint from change of occupation, their disabilities applied also to other kinds of *chien min*. In a description of the good government of the State of Chin relating to −563, the *Tso Chuan* says that artisans and menials (*tsao li*[9]) do not think of changing their occupations.[f] Although this is earlier than the systematisation of the *nu li*[10] status, which seems rather a phenomenon of the Warring States period, the general conception of permanence of technical trade was a long-lasting one. There is no doubt either about the position of convicts (*thu jen*[11]),[g] men and women 'enslaved' for a term of years, or for life, with no descendants.[h] As in the ancient West they were sent to the mines, and we hear from a memorial presented by Kung Yü[12] in −44 that more than a hundred thousand people (including officials and *corvée* workers but mostly convicts) were mining copper and iron ore in the mountains.[i] We might well expect technicians among the slaves, the descendants of convicts, since special training and long experience would be more natural among people spending a lifetime in servitude, and indeed it seems likely that

---

a Here we touch upon the question, which is not our immediate focus of interest, as to how far Chinese society was ever essentially 'based on slavery'. This is at present the subject of great debate among Chinese scholars, most of whom are inclined to believe it true for the Shang and early Chou periods. As the problem is primarily quantitative, it is particularly difficult to solve for such ancient times. The reader may be referred to the essays reprinted in Anon. (*14*) and to the book of Wang Chung-Lao (*1*).

b *Chhien Han Shu*, ch. 23, p. 9*b*; *Chou Li*, ch. 9, p. 30*a* (Wilbur (1), p. 258; Biot (1), vol. 2, p. 363; Pulleyblank (6), p. 199).

c Cf. a well-known statement in *Mo Tzu*, ch. 28, p. 16*a*.

d At various times in Chinese history all kinds of enslavement were known, e.g. sale of women and children, self-sale, slavery of debtors, rescue of exposed infants, kidnapping, etc., but these were of minor importance. Etymologically the word *nu* undoubtedly derives from the conception of dependent women and children (cf. *nu*[13,14]).

e There were also *kuan hu*[15] and *tsa hu*,[16] two higher ranks yet still inferior to the freedman or commoner.

f Duke Hsiang, 9th year (Couvreur (1), vol. 2, pp. 238 ff.).

g This word alone is ambiguous as it can also mean *corvée* conscripts or even disciples of philosophers; semantically it implies a lightly-clad or barefoot follower.

h A related word, *hsi*,[17] meant an exile banished to guard the frontiers, i.e. a military convict. A number of well-known scholars suffered this fate, for varying terms of years.

i *Chhien Han Shu*, ch. 72, p. 15*a*.

1 奴	2 婢	3 罪隸	4 舂藁	5 虜	6 俘
7 收奴	8 官奴婢	9 皂隸	10 奴隸	11 徒人	12 貢禹
13 帑	14 孥	15 官戶	16 雜戶	17 徙	

the life of a government slave tradesman was often considerably more comfortable, and certainly more secure, than that of a 'free' yeoman artisan.[a]

We can now begin to ask the technological questions which interest us. What skills had the *nu pi*? A remarkable document dated −59, purporting to be a purchase contract between one Wang Pao[1] and a bearded *nu* named Pien-Liao,[2] has come down to us from the Han.[b] Apart from the products of the slave's labour in garden and orchard, he was supposed to be able to plait straw sandals, hew out cart shafts, make various pieces of furniture and wooden clogs, whittle bamboo writing-tablets, twist rope and weave mats—but besides all this he was to make knives and bows for sale in the neighbouring market. There is a distinctly Aristophanic quality about the recital of the obligations of the slave, but the technical skills demanded are quite clear. Another story records how a young relative of an empress, Tou Kuang-Kuo,[3] accidentally enslaved, was employed about −185 with a gang of a hundred others as a charcoal-burner.[c] And when the great agriculturalist Chao Kuo[4] was made Commissioner of Army Supplies in −87 he 'set skilled and clever male slaves (*kung chhiao nu*[5]) with assistants to manufacture agricultural implements' of improved types in the Iron Bureaux Factories.[d] The general conclusion (applicable also to later centuries) is that among the *nu* slaves there were always some skilled craftsmen, but never as many as among the ranks of the *corvéable* commoners.[e]

A particularly tantalising term is *thung*,[6,7] which has a connotation of youth and may perhaps best be translated 'serving-lads'.[f] It has such a strongly industrial undertone that one is almost tempted to view it as some form of bonded apprenticeship.[g] Though the *Shuo Wên* and other authorities[h] make very little difference between *thung* and *nu*, the former may have been bondsmen for a long term of quasi-educational years. An interesting passage[i] speaks of the *thung* about −100 in terms equivalent to 'hands' at a hiring fair, saying that in a single provincial capital there would be a turnover of 'a thousand fingers of serving-lads a year (*thung shou chih*

---

a It is a striking fact that while revolts hardly ever occurred among slaves, they were quite frequent among convicts, and then generally among those working in the Iron Bureaux Factories, where access to arms was relatively easy (cf. Wilbur (1), p. 224). Conversely the thirty-six ranches of the Han government distributed over the northern and western frontiers were mainly staffed by government slaves, yet no record of any revolt has come down to us, and none of them rode off to join the Huns (cf. Wilbur (1), pp. 227, 233, 405).

b It is preserved in *Chhu Hsüeh Chi*, ch. 19, pp. 18 *b* ff., and has been translated in full by Wilbur (1), pp. 383 ff. Wang Pao was chiefly a literary courtier, but his biography (in *Chhien Han Shu*, ch. 64B, pp. 8 *b* ff.) gives some information on steel-making. Cf. Hawkes (1), pp. 141 ff.

c *Chhien Han Shu*, ch. 97A, p. 7 *b*, tr. Wilbur (1), pp. 275 ff.

d *Chhien Han Shu*, ch. 24A, p. 16 *b*, tr. Wilbur (1), pp. 343 ff.

e In the first half of the −2nd century an extraordinary man named Tiao Hsien[8] made a great success of employing 'bold and spirited male slaves' (*hao nu*[9]) in fishing and salt-refining, and as itinerant and resident merchants. They associated with cavalry generals and governors of commanderies. It was said 'Who would want noble rank when he could be a slave of Tiao?' See *Chhien Han Shu*, ch. 91, p. 9 *b* (tr. Wilbur (1), p. 281; Swann (1), p. 455).

f Cf. the use of the term 'boys' in the Old China Hand dialect of English.

g Swann (1), p. 453, by her translation tacitly concurs in this.

h Cf. Wilbur (1), p. 67, and Pulleyblank (6), p. 205, for quotations and discussion.

i *Chhien Han Shu*, ch. 91, p. 7 *b*, tr. Wilbur (1), p. 336.

¹ 王襃　　²便了　　³竇廣國　　⁴趙過　　⁵工巧奴　　⁶僮
⁷童　　⁸刁閒　　⁹豪奴

chhien[1])'.[a] Then in many important texts the *thung* are connected with industrial operations, notably in those interesting descriptions, already mentioned,[b] of the industrial entrepreneurs or 'capitalists' of the Chhin and early Han (−3rd and −2nd centuries). Thus after the defeat of Chao by Chhin in −228, a Mr Cho,[2] who came of a family long enriched by iron-smelting, was reduced to poverty and deported to Shu (Szechuan), whither he went accompanied only by his wife, pushing a barrow with all their possessions. Yet he had not long been there when he found iron in the mountains and soon became wealthy again to the extent of possessing eight hundred serving-lads (*thung*).[c] Cho Wang-Sun,[3] probably a descendant and also an iron-master, living in the same town nearly a century later, had nearly as many.[d] There has been some hesitation in visualising these young men in the workshops rather than in the rich man's home, but in the Later Han the author of the *Wu Yüeh Chhun Chhiu*[4] had no doubt about their function when he described[e] the three hundred young men and girls (*thung nü thung nan*[5]) working at the bellows and forges of the famous, if semi-legendary, smith Kan Chiang.[6]

Young people of 'serving-lad' status were also closely connected with those units of production, amounting sometimes almost to factories, which grew up within the dwelling-house compounds of the nobles and high officials. For example:[f]

(Chang) An-Shih[7] was honoured among the nobility, and had the income from ten thousand households, yet he dressed himself in coarse black cloth, and his wife personally wound silk (or spun hemp), carried out the twisting (and wove). His seven hundred household serving-lads (*chia thung*[8]) were all skilled in manufactures, so that goods were produced within (his home) and the smallest things saved up; thus he was able to accumulate commodities (for sale, barter or presentation). He was richer than the Commander-in-Chief (Ho) Kuang. The emperor (Hsüan) was afraid of the general and deeply honoured him, but he was really much more fond of (Chang) An-Shih.

This was between −74 and −62. The mention of the powerful Ho Kuang[9] is interesting because in his own household also there was a textile factory, and it is associated in history with improvements to the draw-loom.[g] The *Hsi Ching Tsa Chi* says:[h]

---

[a] Expressions involving 'fingers' and 'hands' used in this abstract sense became a natural part of Chinese labour phraseology, as we shall see, e.g. in Sect. 29 below. Here a thousand 'fingers' is thought to mean a hundred 'head'.

[b] Vol. 2, p. 56.

[c] *Chhien Han Shu*, ch. 91, p. 8*b*; *Shih Chi*, ch. 129, p. 17*a*. A full translation of this text will be given in Sect. 48; in the meantime see Swann (1), p. 452; Wilbur (1), p. 259.

[d] He was the father-in-law of the famous Han writer Ssuma Hsiang-Ju (*d*. −117). See *Chhien Han Shu*, ch. 57A, p. 1*b*, and Wilbur (1), p. 300.

[e] In ch. 4. We give the full translation of the passage in Sect. 30*d* below.

[f] *Chhien Han Su*, ch. 59, p. 11*b*, tr. Wilbur (1), p. 365, mod. Chang An-Shih was the son of Chang Thang[10] whom we shall meet again as a road-builder in Sect. 28.

[g] Archaeological evidence has long established a date around −100 as the first use of the draw-loom for weaving patterned silks, but new excavations of royal tombs of the State of Chhu in the −4th century have now set this beginning considerably further back. It did not appear in Europe until the +4th century at earliest (see Sect. 31).

[h] Ch. 1, p. 4*a*, tr. auct. Cf. Wilbur (1), p. 364.

---

¹ 童手指千	² 卓氏	³ 卓王孫	⁴ 吳越春秋	⁵ 童女童男
⁶ 干將	⁷ 張安世	⁸ 家僮	⁹ 霍光	¹⁰ 張湯

Ho Kuang's (second) wife presented to Shunyü Yen[1] (a woman physician)[a] twenty-four rolls of grape (design) silk brocade and twenty-five rolls of thin silk with a 'scattered flowers' (inwoven pattern). This latter had come from the (textile workshop in) the home of Chhen Pao-Kuang[2] at Chü-lu. (Chhen) Pao-Kuang's wife handed down the technique. Ho Hsien[3] (the wife of Ho Kuang) invited her to come to the Ho family house and asked her to set it up and operate it. The machine had 120 nieh[4] (probably mails on leashes)[b] and in 60 days one roll (of patterned draw-loom silk) worth ten thousand cash was produced.

Thus we gain a glimpse of the substantial domestic production of such things as fine textiles during the Han period, and similar forms of organisation lasted at least until the end of the Thang. The large number of 'concubines' as well as *thung* serving-lads and handy maids in the palaces of emperors, nobles and high officials in ancient and medieval China takes on a somewhat less romantic character when one realises that probably the majority of them were essentially technicians of one sort or another.[c]

Gradually the term *thung* becomes associated with two other terms, *pu chhü*[5] and *kho jen*[6] or *kho nü*,[7] 'bearers' and 'guests', the former rather in the north, the latter in the south.[d] These were non-stigmatised but semi-servile persons, 'retainers' who were transferable from one master to another but could not be bought and sold. Such 'dependants' or 'clients' included at all times a certain number of artisans and secretaries, and were defined as *liang jen*,[8] freedmen with surnames but still attached to a lord.

At first their duties were often largely martial, as 'household men-at-arms', later they had more to do with agriculture, and later still, after the Thang, again became more military.[e] Chinese industrial production always retained, however, a strongly familial character.[f]

What has here been said about slavery, semi-servile groups, free labour and the *corvée* system will urgently present itself later on for general consideration in the context of the history of Chinese technical development. Since this history has not yet been unfolded such a discussion would be out of place at this point. Yet it seems impossible to go further without noting that whatever characteristics ancient and

---

[a] She was a *ju i*,[9] i.e. an obstetrician and pediatrician. In −71 she was persuaded by Ho Hsien to poison the empress *née* Hsü, and the precious textiles which interest us were part of the enormous reward which she received from that wealthy family. Ho Kuang was practically 'king-maker' after −86.

[b] Exactly what this invention was will be discussed in Sect. 31. It probably permitted a design of 120 'moves', and there may have been six or more warp threads passing through each of the rings or 'mails' on the leashes. We do not know any more about Chhen Pao-Kuang and his wife.

[c] Eichhorn (4) is surely right when he says 'Die Textilwerkstätten sind höchstwahrscheinlich hervorgegangen aus den Schneider- und Nähstuben, die an den Höfen der Kaiser und Machthaber bestanden und für die geschickte Näherinnen und Schneiderinnen aus dem ganzen Lande angeworben oder besser gesagt requiriert wurden.... Viele der Palastdamen wurden weniger wegen ihrer weiblichen Reize als vielmehr um bestimmter Kunstfertigkeiten willen in den Palast geholt.' As an example take Hsüeh Ling-Yün[10] (the 'Needle Immortal', Chen Hsien[11]), one of the consorts of Wei Wên Ti (Eichhorn, 5).

[d] Cf. Pulleyblank (6), pp. 215 ff.

[e] See the two studies of Yang Chung-I (1, 2).

[f] Cf. the well-known book of Lang (1).

| [1] 淳于衍 | [2] 陳寶光 | [3] 霍顯 | [4] 鑷 | [5] 部曲 | [6] 客人 |
| [7] 客女 | [8] 瓪人 | [9] 乳醫 | [10] 薛靈芸 | [11] 針仙 | |

medieval Chinese labour conditions may have possessed, they proved no bar to a long series of 'labour-saving' inventions altogether prior to those arising in Europe and Islam. Whether one thinks of the efficient trace harness for horses (from the −4th century onwards), or of the appearance of the still better collar harness in the +5th, or of the simple wheelbarrow in the +3rd (though not in Europe till a thousand years later), one constantly finds that in spite of the seemingly inexhaustible masses of man-power in China, lugging and hauling were avoided whenever possible. How striking it is that in all Chinese history there is no parallel for the slave-manned oared war-galley of the Mediterranean—land-locked though most of the Chinese waters were, sail was the characteristic motive power throughout the ages, and the arrival of great junks at Zanzibar or Kamchatka was only an extension of the techniques of the Yangtze and the Tungthing Lake. When the water-mill appeared early in the +1st century for blowing metallurgical bellows, the records distinctly say that it was considered important as being both more humane and cheaper than man-power or animal-power. And it gives food for thought to find around +1280, three or four centuries before similar developments in Europe, water-power widely applied to textile machinery. Evidently shortage of labour is not in every culture the sole stimulus for labour-saving invention. But of course the problems involved are very complex, and we must ultimately leave to sociological and economic historians the task of interpreting what our investigations will reveal.

However, the contents of the present volume alone suffice to show that in Chinese culture inventions were never rejected because of the fear of technological unemploy-ment. In all our studies we have so far come upon no case of this. Yet in Europe it seems to have been a perennial fear, and industrialisation had to go forward in the teeth of it. Many examples come to mind—Vespasian about +75 declining to make use of a new device for transporting heavy columns to the Capitoline,[a] the city fathers of Nuremberg in the late +16th century suppressing inventions and their transference from trade to trade,[b] Elizabeth of England refusing her patronage to William Lee the inventor of the first frame-knitting machine,[c] or the Council of Venice forbidding multiple spinning machinery 'out of consideration for the livelihood of poor women and their children'.[d] When these contrasts are further explored they may make nonsense of a certain estimate of the Western Middle Ages which Lynn White (1, 8) has poignantly put, and which I myself at times have found seductive. Humanitarian technology, it is said, was not rooted in economic necessity, and found inventive expression only in the Occident.

The labour-saving power-machines of the later [Western] middle ages [he writes], were produced by the implicit theological assumption of the infinite worth of even the most degraded human personality, by an instinctive repugnance towards subjecting any man to a monotonous drudgery which seems less than human in that it requires the exercise neither of intelligence nor of choice. It has often been remarked that the Latin middle ages first discovered the dignity and spiritual value of labour—that to labour is to pray. But the middle

---

[a] Suetonius, *De Vita Caesarum*, ch. 18.   [b] Cf. Klemm (1), pp. 153 ff.
[c] Cf. Norbury (1). Lee died in +1610.
[d] Cardan, *De Subtilitate, c.* +1560, relayed in Thorndike (1), vol. 7, p. 613.

ages went further; they gradually and very slowly began to explore the practical implications of an essentially Christian paradox: that just as the Heavenly Jerusalem contains no temple, so the goal of labour is to end labour.

These are noble words, but can we be so sure that Christendom had a monopoly of humanitarian urges to invention? If China was so often in fact first in the field with labour-saving devices, had Confucian benevolence and Buddhist compassion nothing to do with it? Perhaps it will be better to credit men at all times and everywhere with Mencius' 'human-heartedness', and to look rather at the social and economic circumstances to understand what particular civilisations could or could not in the end accomplish.

These preliminaries having been completed, we can now turn our attention to a very brief survey of the social groups from which inventors and engineers originated.[a] If in the following paragraphs we venture to sketch some provisional social categories and patterns into which the mechanical genius of ancient and medieval China seems to fall, we do it with all reservations. The elucidation of the place of inventors, engineers and men of scientific originality in the society of their times is a special study in itself, and we have not as yet been able systematically to pursue it, partly because it is in a way secondary, the first task being to establish and verify their identity and what they actually did. We are very far indeed from the ideal of a statistical treatment arranged by sciences or by centuries, and it is unlikely that even Section 48 will be able to offer much towards such an analysis. So many are the difficulties; for example, sinologists have not yet systematised the interpretation and translation of official titles, indications of rank, power and duties in the governmental bureaucratic hierarchy, and even when agreement exists, a searching knowledge of the political history of the decades in question may be needed before one can say just how much a particular title or position implied. Where no biographies were entered in the official dynastic histories much *ad hoc* research in contemporary literature would be required to establish the life-stories of some of the engineers whose names and deeds are clear enough, and furthermore it might prove rather unrewarding precisely in the most interesting cases of men who were artisan commoners and not members of the scholar-gentry at all. But with all these limitations in mind it may be worth while to glance at a few of the main life-patterns of technologists,[b] illustrated by examples chosen at random from the Sections before us (in Vols. 4 and 5 of our series), and by one especially relevant +3rd-century essay.

[a] It is important to emphasise here the great fluidity of ancient Chinese society. The great mass of commoners was constantly gaining and losing people to the ranks of the gentry and nobles above, and to those of the servile ranks below. There are innumerable examples of the meteoric rise and fall of families. The usual term for convicts was but six years, and amnesties were frequent. Slaves could be, and were, emancipated in several different ways. Life may have been rather uncertain and surprising, but it held brilliant rewards as well as frightful dangers, and there were no rigid barriers of class or caste.

[b] In the following paragraphs we shall give, for simplicity, only the *floruits* of the persons mentioned. Their full biographical dates, when known, will be found at the places in the text where their work is described in detail; and of course in the Biographical Glossary in Vol. 7.

Let us divide our life-histories into five groups, first high officials, the scholars who had successful and fruitful careers, secondly commoners, third members of the semi-servile groups, fourth those who were enslaved, and fifth, the rather significant group of minor officials, i.e. scholars who were not able to make their way upwards in the ranks of the bureaucracy.[a] It will be seen that the numbers of examples which we can find in the different groups will vary considerably.

First, then, the high officials. Here we may take Chang Hêng[1] (*fl.* +120) and Kuo Shou-Ching[2] (*fl.* +1280) as outstanding representatives of the type. Both we already know well,[b] but a couple of sentences may recall their achievements. Chang Hêng was the inventor of the first seismograph in any civilisation, and the first to apply motive power to the rotation of astronomical instruments; besides this he was a brilliant mathematician and designer of armillary spheres. Kuo Shou-Ching, more than a millennium later, was an equally good mathematician and astronomer but also a most distinguished civil engineer who constructed the Thung Hui Canal and planned most of the Yuan Grand Canal. Both of these men occupied the post of Astronomer-Royal, but in addition Chang Hêng became President of the Imperial Chancellery, while Kuo Shou-Ching was Intendant of Waterways, and Academician. Two entirely comparable figures are Su Sung and Shen Kua, whom we shall mention in a moment. All these men had the good fortune to find appreciation for their scientific and technical talents in their age—others were not so lucky.

Sometimes ability of this kind accompanied military gifts and offices. The Chhin general Mêng Thien[3] (*fl.* −221) comes to mind.[c] He accomplished the fusion and extension of previously existing walls to form the Great Wall, and built a road from Ninghsia to Shansi. Tu Yü[4] (*fl.* +270), one of the generals who reduced the San Kuo State of Wu for the Chin, is prominently associated with the spread of hydraulic trip-hammers and the establishment of multiple geared water-mills for cereal grinding, as well as with the throwing of pontoon bridges across great streams such as the Yellow River.[d] Chhiang Shen[5] (*fl.* +1232), a heroic commander of Jurchen Chin forces, invented a kind of slur-bow and some form of counterweighted trebuchet anticipating the arrival of Arab experts in this powerful artillery.[e] During the long history of the development of gunpowder weapons many military officers appear as inventors,[f] but it is not always easy to determine their rank in terms comparable with the civil administrators.

Sometimes, again, provincial officials are credited with important technical developments. Thus the introduction of the water-powered metallurgical blowing-engine[g] is

---

a This group of 'misfits' must have existed in all pre-Renaissance civilisations, but for different reasons. A classical European case would be that of Roger Bacon, for whom life and scientific work was possible only within the Franciscan order, yet utterly frustrated therein. Even now we can hardly congratulate ourselves on having made conditions of life and work favourable for all who have something to give to humanity.

b Cf. Sects. 19*h*, 20*f*, *g*, *h*, and 24.
d Cf. Sects. 27*h* and 28*e* below.
f See Sect. 30 below.

c Cf. Sect. 28*b*, *c*.
e Cf. Sect. 30*h* below.
g Cf. Sect. 27*h* below.

¹ 張衡   ² 郭守敬   ³ 蒙恬   ⁴ 杜預   ⁵ 强伸

attributed to Tu Shih,[1] who was Prefect of Nanyang in +31, and its further spread, so vital for the technical mastery of cast iron production at this early period, was due to another governor, Han Chi,[2] who was Prefect of Lo-ling in +238. Occasionally a eunuch is prominent in technical advance. The most obvious case is that of Tshai Lun,[3] who began as a confidential secretary to the emperor, was made Director of the Imperial Workshops in +97 and announced the invention of paper[a] in +105. Later he received a marquisate.

On the contributions to science of Chinese princes and the remoter relatives of the imperial houses an interesting monograph could be written. They were particularly favoured with leisure since, though generally well educated, they were in most dynasties ineligible for the civil service, and yet disposed of considerable wealth. Although no doubt the great majority of them throughout the ages did little or nothing for posterity, there were a certain memorable few who devoted their time and riches to scientific pursuits. Here we shall mention only one or two, because for various reasons they tended to interest themselves rather in astronomical, biological or medical than in technical or engineering matters. Liu An,[4] prince of Huai-Nan[5] (fl. −130), with his entourage of naturalists, alchemists and astronomers, is already very familiar to us,[b] but another Han noble, Liu Chhung,[6] prince of Chhen[7] (fl. +173), is more directly relevant, since he was the inventor of grid sights for crossbows, and a famous shot with them as well.[c] In the Thang we meet with Li Kao,[8] prince of Tshao[9] (fl. +784), interested in acoustics and physics,[d] but prominent here because of his successful use about this time of treadmill-operated paddle-boat warships.[e] If similar vessels had not been constructed in China in the +5th century, a matter on which there is some uncertainty, then Li Kao's fleet was the first practical achievement of an idea which (probably quite unknown to him) had been proposed, but not executed, in the West in early Byzantine times. We may add Chu Tsai-Yü[10] (Chêng Shih Tzu,[11] of a princely house in the Ming), already familiar[f] for his work in mathematical acoustics (fl. +1590), and Thopa Yen-Ming,[12] prince of An-Fêng[13] in the Northern Wei (fl. +515) who will recur in a moment in another context.

Curiously, it seems quite exceptional to find an important engineer who attained high office in the Ministry of Works, age-old though this department was in the Chinese bureaucratic pattern. Perhaps this was because the real work was always done by illiterate or semi-literate artisans and master-craftsmen, who could never rise across that sharp gap which separated them from the 'white-collar' literati in the offices of the Ministry above. Perhaps they sometimes felt that they could get on with the job much better if the administrators upstairs were poets or courtiers not too

---

[a] See the account in Sect. 32.
[b] Cf. Sects. 10, 20 and many other references as indexed.
[c] See Sect. 30e below.   [d] Cf. Sects. 26c, 26h (7), etc.
[e] See Sect. 27i below.   [f] In Sect. 26h (10) above.

[1] 杜詩   [2] 韓暨   [3] 蔡倫   [4] 劉安   [5] 淮南王   [6] 劉寵
[7] 陳王   [8] 李皋   [9] 曹王   [10] 朱載堉   [11] 鄭世子
[12] 拓跋延明   [13] 安豐王

uncomfortably familiar with the tools and materials of the trade and mystery. And there may have been other reasons too:

> 'I have taken plank and rope and nail, without the King his leave,
> After the custom of Portesmouth, but I will not suffer a thief.
> Nay, never lift up thy hand at me! There's no clean hands in the trade—
> Steal in measure' quo' Brygandyne, 'There's measure in all things made!'[a]

However that may be, there were exceptions to the rigour of the gap, as is shown by the case of Yüwên Khai[1] (fl. +600), chief engineer of the Sui dynasty for thirty years. He carried out irrigation and conservation works, superintending the construction of the Thung Chi Chhü, part of the Grand Canal,[b] built a large sailing-carriage,[c] and with Kêng Hsün (to whom we shall return in a page or two) devised the standard steelyard clepsydra used throughout the Thang and Sung.[d] He built new capital cities at Chhang-an in +583 and at Loyang in +606, and made a wooden model of the cosmological Ming Thang temple.[e] For many years Minister of Works (Kung Pu Shang Shu[2]), his earlier post had been Director of the Architectural and Engineering Department of the Imperial Palaces (Chiang Tso Ta Chiang[3]). Yüwên Khai must have been a real expert in all the mechanical and constructional arts of his time.

During the Ming dynasty the way upwards for artisans to enter the administrative grades of the Ministry of Works seems to have been more open, and Friese (1) has sketched the careers of a number of men of this kind. Several wood-workers and joiners, notably Khuai Hsiang[4] (fl. +1390 to +1460), Tshai Hsin[5] (fl. +1420) and Hsü Kao[6] (fl. +1522 to +1566), who all showed merit as builders and architects, succeeded in this way, and the last-named rose to be President of the Ministry. Another example is Lu Hsiang[7] (fl. +1380 to +1460), who began as a stone-mason.

If our survey were further extended it would probably contain many more names of men of high official rank. This weightage would be partly due to the enormous social importance of hydraulic engineering works in Chinese society, a fact which rendered skill in this field always highly honourable among scholars and administrators otherwise tending to purely literary accomplishments. But it would also be due to a tendency readily discernible for technicians to cluster in the entourage of a distinguished civil official, who acted as their patron. It is more than probable that the first water-mills and metallurgical blowing-engines were the work of technicians in the service of Tu Shih and Han Chi whose names have not come down to us. Here we find the examples of Su Sung and Shen Kua instructive. Su Sung[8] (fl. +1090), an official of the most distinguished quality, who served as Ambassador and President of the Ministry of Personnel, was responsible, as we shall see,[f] for the construction of the great astronomical clock-tower at Khaifêng and for the greatest horological

---

[a] From Rudyard Kipling's 'King Henry VII and the Shipwrights', in *Rewards and Fairies*.
[b] See Sect. 28*f* below.   [c] See pp. 253 and 279 below.
[d] Cf. p. 480 below, Vol. 3, p. 327, and Needham, Wang & Price (1), pp. 89 ff.
[e] Cf. Vol. 2, p. 287 and elsewhere.
[f] Cf. Sect. 27*j*, especially pp. 445 ff. below.

[1] 宇文愷   [2] 工部尚書   [3] 將作大匠   [4] 蒯祥   [5] 蔡信
[6] 徐杲   [7] 陸祥   [8] 蘇頌

treatise of the Chinese middle ages, but he surrounded himself, in order to accomplish this, with a remarkable band of engineers and astronomers, whose names he preserved and transmitted.[a] Shen Kua[1] (*fl.* +1080), equally gifted and equally successful, was Ambassador and Assistant Minister of Imperial Hospitality, but we think of him chiefly as the author of the most interesting and many-sided scientific book of the Sung period.[b] It is in this work that we find the best authentic statement of the beginnings of printing in China,[c] a statement which introduces us to that great inventive genius Pi Shêng,[2] 'a man in hempen cloth' (*pu i*,[3] that is to say, a commoner, one not dressed in silk) who first devised, about +1045, the art of printing with movable type.[d] 'When Pi Shêng died', says Shen Kua, 'his fount of type passed into the possession of my followers, among whom it has been kept as a precious possession until now.'[e] Thus we have a striking glimpse of the entourage of technicians which an enlightened official could gather around him. Finally, well-known officials would be likely to figure largely in any survey because there were always other reasons for the insertion of their biographies in the dynastic histories. Indeed the facts of most interest to us are often inserted by the historians, almost as an afterthought, at the end of the biographies.

With Pi Shêng we come now to the commoners, the *liang jen*. Men whose names alone we know can probably be safely placed in this group. Ting Huan[4] (*fl.* +180), renowned for his pioneer use of the Cardan suspension and for his construction of rotary fans and ingenious lamps,[f] is termed simply a 'clever artisan' (*chhiao kung*[5]). His contemporary Pi Lan[6] (*fl.* +186), a master-founder who installed water-raising machinery (square-pallet chain-pumps and norias) which brought a piped water-supply system to the palaces and city of Loyang,[g] has no appellation, and was simply 'ordered' to do these and many other things by the chief palace eunuch Chang Jang.[7] Yü Hao[8] (*fl.* +970), that brilliant designer and builder of pagodas,[h] was but a Master-Carpenter (Tu Liao Chiang[9]) and his celebrated *Mu Ching*[10] (Timberwork Manual) was assuredly dictated to a scribe.[i] Of Li Chhun,[11] the great constructor of segmental arch bridges such as were not known elsewhere in the world until seven centuries after his time (*fl.* +610), we know almost nothing, and may presume him a commoner.[j] Now and then an assistant hydraulic engineer or conservancy foreman (*shui kung*[12]), makes a personal appearance, such as Kao Chhao[13] (*fl.* +1047), who was right in his methods for stopping a breached dyke when everyone else was wrong.[k] Sometimes

---

[a] Cf. pp. 448, 464 below, and for fuller details Needham, Wang & Price (1), pp. 16ff.
[b] See on this Vol. 1, pp. 135ff.
[c] *MCPT*, ch. 18, pp. 7aff.; cf. Hu Tao-Ching (1), vol. 2, pp. 597ff.
[d] Pi Shêng used porcelain, but by the +13th century tin was used and by the +14th bronze.
[e] *Shêng ssu, chhi yin wei yü chhün tsung so tê, chih chin pao tsang.*[14]
[f] Cf. p. 233 below.   [g] See Sect. 27g, on p. 345 below.
[h] See Sect. 28d below.
[i] Reference has already been made to this, Vol. 3, p. 153.
[j] His work will be fully described in Sect. 28e (8) below.
[k] The story will be found in Sect. 28f (8) below.

[1] 沈括	[2] 畢昇	[3] 布衣	[4] 丁緩	[5] 巧工	[6] 畢嵐
[7] 張讓	[8] 喻皓	[9] 都料匠	[10] 木經	[11] 李春	[12] 水工
[13] 高超	[14] 昇死其印爲予羣從所得至今寶藏				

we have only the surname of a valued man, e.g. Lacquer-Artisan Wang from Suchow (*Phing-chiang chhi chiang* Wang[1]), who about +1345 devised dismountable boats and collapsible armillary spheres for the imperial court.[a] He rose by +1360 to be Intendant of one of the Imperial Workshops (Kuan Chiang Thi Chü[2]). Sometimes we do not even have the surname—an omission which makes one wonder whether such men were members of one or other of the servile or semi-servile groups in which surnames were not customary. Two cases come to mind: first the old craftsman (*lao kung*[3]) who made astronomical apparatus, yet (as he said to Yang Hsiung[4] about −10) could not transmit his skill and knowledge to his son;[b] and secondly that 'Artisan from Haichow' (*Haichow chiang jen*[5]) who presented to the empress in +692 what was in all probability an anaphoric clock,[c] but who also made Cardan suspensions and other ingenious devices.[d]

In the category of regular commoners we should probably also place minor military officers, and certainly Taoists and Buddhist monks. Among the former we could mention Chhiwu Huai-Wên,[6] a Taoist swordsmith who served in the army of Kao Huan[7] the 'king-maker', founder of the Northern Chhi dynasty, advised him on five-element theory, and took charge of his arsenals (*c.* +543 to +550). Chhiwu Huai-Wên, if not the inventor, was one of the earliest protagonists of the co-fusion method of steel-making, a process ancestral to the Siemens–Martin open-hearth furnace, and he was also a celebrated practitioner of the pattern-welding of swords.[e] Another minor commander who might be instanced for the importance of his technical innovations is Thang Tao,[8] the gallant defender of the city of Tê-an in Hupei during the years +1127 to +1132, when it was repeatedly attacked by the armies of the Jurchen Chin. Together with the civil magistrate Chhen Kuei,[9] apparently an equally inventive mind, he used successfully for the first time a new device called the 'fire-lance' (*huo chhiang*[10]), i.e. a reversed gunpowder rocket held in the hands and employed as a defensive shock weapon. Though the barrels were not of metal, and the composition spattering rather than propulsive in function, this was undoubtedly the real origin of all barrel guns and cannon.[f]

In view of the close association between Taoism and technical arts in ancient China,[g] one would expect to find more Taoist inventors in the middle ages than have so far made their appearance.[h] Nevertheless it is not at all difficult to name some. Of Than

---

a *Shan Chü Hsin Hua*, p. 39*b*; tr. H. Franke (2), no. 107; cf. p. 71 below.
b Cf. Vol. 3, p. 358 above.　　　　　　　　c Cf. p. 469 in Sect. 27*j* below.
d One particular trade, as Franke (15) has shown, was an exception to the prevailing curtain of anonymity, namely the making of ink. The pious care of a group of scholars has preserved for posterity the names and biographies of no less than 325 master-craftsmen in this industry from Han times onwards. Truly it was an honourable trade. But the charisma of the logos, the written word, failed to extend to the paper-makers.
e The co-fusion process will be fully discussed in Sect. 30*d*. Meanwhile see Needham (32).
f The origin, development and legacy of the fire-lance will also be fully discussed in Sect. 30.
g See on this Vol. 2, pp. 53ff., 71ff., 86ff., 121ff.
h By this time, Taoism had of course much degenerated into an ecclesiasticism, yet it was still particularly strong in the fields of alchemy and pharmaceutical naturalism, neither of which are directly relevant here.

[1] 平江漆匠王	[2] 管匠提舉	[3] 老工	[4] 揚雄	[5] 海州匠人
[6] 綦毋懷文	[7] 高歡	[8] 湯璹	[9] 陳規	[10] 火鎗

Chhiao's[1] experiments with lenses about +940 we have already spoken,[a] but as a companion we may take Li Lan[2] (*fl.* +450), who was responsible for a long line of development of 'stopwatch'-clepsydras using jade vessels and mercury, as well as larger water-weighing clepsydras which depended on the pointer-reading property of the steelyard.[b] Among the Taoists we must not forget a woman, Chia Ku Shan Shou[3] (the Valley-loving Mountain Immortal),[c] who engineered mountain roads in Fukien *c.* +1315. But on the whole the Buddhists were more illustrious as technicians in these times. One cannot omit mention of I-Hsing,[4] the greatest astronomer, mathematician, and instrument-designer of his age.[d] Though debarred from all official rank, he was a member of the College of All Sages and the most trusted scientific man at court for nearly twenty years. Of greater engineering interest perhaps are several monks, notably Tao-Hsün[5] (*fl.* +1260) and Fa-Chhao[6] (*fl.* +1050), who built many of the wonderful megalithic stone bridges across the rivers and estuaries in Fukien province.[e] Since the length of the beams is frequently only just below that at which the stone used breaks under its own weight, it would seem that strength-of-materials tests were conducted at this time. To conclude, Huai-Ping[7] (*fl.* +1065) gives another example of a monastic technician—we know him already[f] as the expert in the salvage of sunken vessels.

And now we come to the exceptional cases, men who came down in history as brilliant technologists yet whose social standing in their own time was very low indeed. The only one of clearly semi-servile rank in our registers is Hsintu Fang[8] (*fl.* +525). In his youth he entered the household of a prince of the Northern Wei,[g] Thopa Yen-Ming,[9] as a 'dependant' or 'retainer' (*ju pin kuan*[10]). From +516 onwards this prince (An-Fêng Wang[11]) had been collecting many pieces of scientific apparatus—armillary spheres, celestial globes, trick hydrostatic vessels, seismographs, clepsydras, wind-gauges, etc. He had also inherited a very large library. As a client or pensioner of known scientific skill, Hsintu Fang's position with relation to Thopa Yen-Ming must have been something like that of Thomas Hariot (*d.* +1621) to the ninth Earl of Northumberland, Henry Percy, his patron.[h] It seems that the prince intended to write certain scientific books with the aid of Hsintu, but owing to political and military events felt obliged to flee to the Liang emperor in the south in +528, so that Hsintu Fang had to write the books himself.[i] After this he remained in seclusion, probably

---

[a] Cf. Vol. 4, pt. 1, pp. 116 ff. above. [b] Cf. Vol. 3, pp. 326, also Needham, Wang & Price (1), p. 89.
[c] Further information on her work will be given in Sect. 28*b*.
[d] See Vol. 3, pp. 202 ff. and elsewhere as indexed, also especially pp. 471 ff. below.
[e] A full account will be given in Sect. 28*e* below. [f] Vol. 4, pt. 1, p. 40 above.
[g] Correct accordingly Vol. 3, pp. 358, 633. [h] See the interesting account of Shirley (1).
[i] These are listed in *Wei Shu*, ch. 20, p. 7*b* as (*a*) *Ku Chin Yo Shih*[12] (Musical Instruments, New and Old), (*b*) *Chiu Chang Shih-erh Thu*[13] (Twelve Diagrams for the 'Nine Chapters (of Mathematical Art)'), and (*c*) *Chi Chhi Chun*[14] (Specifications for Scientific Instruments). The first is probably another name for the two books which have come down to us in fragmentary form as *Yo Shu*[15] and *Yo Shu Chu Thu Fa*[16] (cf. Vol. 4, pt. 1, p. 189 above). The third was probably his most important work and its total loss is much to be regretted.

[1] 譚峭	[2] 李蘭	[3] 夾谷山壽	[4] 一行	[5] 道詢
[6] 法超	[7] 懷丙	[8] 信都芳	[9] 拓跋延明	
[10] 入賓館	[11] 安豐王	[12] 古今樂事	[13] 九章十二圖	[14] 集器準
[15] 樂書	[16] 樂書註圖法			

in poverty, till he was called to the court of another potentate, Mujung Pao-Lo,[1] governor of Tungshan, whose younger brother, Mujung Shao-Tsung,[2] recommended him to Kao Huan (the 'king-maker' just mentioned).[a] This great lord he served as Estate Agent, a post which may have exercised his talents in surveying and architecture.[b] Here he was again a 'retainer' (*kuan kho*[3]). From his biographies[c] we learn that Hsintu Fang was of a very humble and abstracted disposition. Mujung Shao-Tsung gave him mules and horses but he would not use them or ride on them, and sent serving-maids (*pi shih*[4]) to test him at night, but he had them ejected from his dwelling. His low social rank did not, it seems, prevent his being consulted by fairly high officials, such as Tsu Thing[5] the poet, Granary Intendant under the Vice-President of Chancellery (presumably in the Eastern Wei, but perhaps at Kao Huan's court before the dynasty of the Northern Chhi was proclaimed). In this case Hsintu's advice was sought concerning a failure of the *hou chhi* technique,[d] for which, it will be remembered, he devised strange rotary fans. Apart from all these services he wrote a number of other books,[e] and at his death about $+550$ was engaged upon a calendrical system (the Ling Hsien Li[6]) which he had undertaken on his own initiative, probably as a contribution to the dynasty about to be established by his patron.[f] Thus this ingenious man of lowly origins found shelter in a troublous time, if not official recognition or high status, as a family retainer in the houses of two unusual patricians.

Examples of technologists who were positively slaves are also proving quite rare. But there is Kêng Hsün[7] (*fl.* $+593$).[g] He began as a client (*kho*[8]) of a governor of Ling-Nan, but when this patron died, Kêng, instead of going home, joined some tribal people in the south and eventually led them in an uprising. When this was defeated and Kêng Hsün captured, the general Wang Shih-Chi,[9] realising his technical ability, saved him from death and admitted him among his slaves (*chia nu*[10]). Here his position was yet not so low that he could not receive instruction from an old friend, Kao Chih-Pao,[11] who had become Astronomer-Royal, and it was as a result of this that Kêng Hsün built an armillary sphere or celestial globe rotated continuously by water-power. It is interesting to find that the emperor rewarded him for this achievement by making

[a] This is the 'Kao Tsu' of the biographies; correct Vol. 3, p. 633 accordingly.

[b] Here he may well have been personally acquainted with Chhiwu Huai-Wên.

[c] The main places are *Wei Shu*, ch. 20, p. 7*b*, ch. 91, pp. 13*b*ff., *Pei Shih*, ch. 89, pp. 13*a*ff., *Pei Chhi Shu*, ch. 49, p. 3*b*, and *Sui Shu*, ch. 16, pp. 9*b*ff. A special study of the life and times of Hsintu Fang, including a conflation of these sources, would be a valuable contribution.

[d] Cf. Vol. 4, pt. 1, pp. 187ff. above, and Bodde (17).

[e] Notably a *Tun Chia Ching*[12] (Manual of Divination by the Denary Cyclical Characters) and a *Ssu Shu Chou Pei Tsung*[13] (The Four Traditional Schools of (Mathematical and Astronomical) Art associated with the (Classic of the) Gnomon and the Circular Paths of Heaven). He also wrote on the hydrostatic 'advisory' vessels (cf. Vol. 4, pt. 1, p. 34 above), but this was probably included in the lost *Chi Chhi Chun*.

[f] His contemporaries regarded him as one of the five leading astronomers and mathematicians of the northern and southern dynasties ($+4$th to $+6$th centuries) together with Chao Fei, Ho Chhêng-Thien, Tsu Chhung-Chih, and Li Yeh-Hsing.

[g] His biography (*Sui Shu*, ch. 78, pp. 7*b*ff. and *Pei Shih*, ch. 89, pp. 31*a*ff.) has been fully translated by Needham, Wang & Price (1), p. 83.

[1] 慕容保樂	[2] 慕容紹宗	[3] 館客	[4] 婢侍	[5] 祖珽
[6] 靈憲曆	[7] 耿詢	[8] 客	[9] 王世積	[10] 家奴
[11] 高智寶	[12] 遁甲經	[13] 四術周髀宗		

him a government slave (*kuan nu*[1]) and attaching him to the Bureau of Astronomy
and Calendar. The rest of his story concerns us less, but the following emperor freed
him (*mien chhi nu*[2] or *fang wei liang min*[3]) and later he occupied posts as Acting Superintendent of the Right Imperial Workshops (Yu Shang-Fang Shu Chien Shih[4]) and
Acting Executive Assistant in the Bureau of Astronomy (Shou Thai Shih Chhêng[5]).
We shall appreciate better in due course his role in the history of horological
engineering,[a] but it is clear at any rate that a long period of slavery was no bar to
official, if not very exalted, position.[b]

We now reach the last of our groups of technicians, and one of the most numerous,
namely that of the minor officials—men who were sufficiently well-educated (even if
of lowly origin) to enter the ranks of the bureaucracy, but whose particular talents or
personalities frustrated all hopes of a brilliant career. These were the kind of men who
could have become famous in science or engineering in a post-Renaissance world.
Take Li Chieh,[6] for instance (*fl.* +1100), the man who, building on the earlier works
of Yü Hao and others, produced the greatest definitive treatise of any age on the
millennial tradition of Chinese architecture;[c] he was only an Assistant in the Directorate of Buildings and Construction (Chiang Tso Chien Chhêng[7]).[d] He did in the end
attain the Directorship itself but only for a year or so since the death of his father
necessitated his retirement to the country. Then, although Li Chieh had been out-
standingly successful as a practical architect as well as a writer, he was posted as
magistrate to a provincial town, Kuochow in Honan, and the emperor's message of
recall to the capital arrived only after his death in +1110.

In this case we are fortunate to have a full biography written by a contemporary,
Chhêng Chü,[8] who had probably himself served under Li Chieh. But in a hundred
others there is no such record. An elaborate specification for hodometers (distance-
measuring vehicles) prepared by Lu Tao-Lung[9] in +1027 has come down to us,[e] but
no information whatever about the life of this engineer. Details about Li Chieh's contemporary Wu Tê-Jen[10] would be even more interesting since his specification for the
more complicated south-pointing carriage[f] in +1107 has stimulated many efforts at

---

[a] See p. 482 below.

[b] In other cases there were distinguished scientific men and technicians who came closely in contact
with the institution of slavery. For example, the astronomer Yü Chi-Tshai,[11] who wrote the *Ling Thai
Pi Yuan*[12] (Secret Garden of the Observatory), one of our oldest extant complete treatises on the subject,
spent a fortune in redeeming relatives and friends who had become enslaved in the wars between the
Liang, Chhen, Northern Chou and Sui dynasties. Yü Chi-Tshai was an Astronomer-Royal of the
Northern Chou, but found favour with Sui Wên Ti after +580, who appointed him one of the Vice-
Presidents of Chancellery and later Grand Imperial Counsellor.

[c] The *Ying Tsao Fa Shih*[13] (Treatise on Architectural Methods).

[d] This organisation, different from the Ministry of Works (Kung Pu[14]) and much smaller, was concerned mainly with imperial palaces and temples. Li Chieh's biography has been carefully studied by
Yetts (8). He started as a very subordinate official in the Bureau of Imperial Sacrifices, but spent most
of his career in the Directorate of Buildings and Construction, first as an Archivist, then Assistant,
and—only after the appearance of his book—Assistant Director.

[e] See on this, p. 283 below.   [f] This is fully translated below, p. 292.

[1] 官奴   [2] 免其奴   [3] 放爲良民   [4] 右尙方署監事
[5] 守太史丞   [6] 李誡   [7] 將作監丞   [8] 程俱   [9] 盧道隆
[10] 吳德仁   [11] 庾季才   [12] 靈臺秘苑   [13] 營造法式   [14] 工部

its reconstruction in modern times. Chang Ssu-Hsün,[1] the Szechuanese who in +976 built the magnificent escapement-wheel clock operated by mercury instead of water and embodying what was probably the first of all power-transmitting chain-drives known to history,[a] was only a Student in the Bureau of Astronomy at the time, and there is no evidence that he ever rose above the position of Assistant in charge of Armillary Spheres and Clocks (Ssu Thien Hun I Chhêng[2]). Sometimes Chinese engineers found that they could make a better career in the service of foreign dynasties, where the pressure of conventional literary culture was less and ingenuity could find spontaneous if unsophisticated admiration and support. Of this phenomenon many examples could be given. Hsieh Fei[3] and Wei Mêng-Pien[4] both served the king of the Hunnish Later Chao dynasty, Shih Hu,[5] around +340. Wei indeed was his Director of Workshops, and together with Hsieh produced south-pointing carriages, floats with complicated mechanical puppets, wagon-mills (camp-mills), revolving seats, fountains, etc.,[b] for all of which artistry Shih Hu had a particular liking. Another 'nomadic' dynasty, the Yen, that of Mujung Chhao[6] (in this case Hsien-pi or proto-Mongol), had the services of the most famous military engineer of his time (c. +410), Chang Kang.[7] He was a great expert on crossbows and perhaps the first inventor of the multiple-spring arcuballistae which were so characteristic of Chinese pre-gunpowder artillery down to the end of the Sung;[c] famous also for his knowledge of fortifications and the modes of attack upon them. In the end, however, he returned to Chinese allegiance and joined the founder of the Liu Sung dynasty. His parallel eight centuries later was Chang Jung,[8] Chief Military Artificer of Chagatai, the son of Chingiz Khan, who built in +1220 a famous floating bridge with a hundred pontoons over the Amu Darya. Also a general of Mongol trebuchet artillery, this engineer constructed a road through the Sung-shu-thou pass east of Kuldja in Sinkiang which had forty-eight trestle bridges able to take two carts abreast.[d] These examples do not mean that the 'nomadic' peoples did not produce remarkable artisans and engineers themselves, for instance the odious Chhihkan A-Li[9] who served the king of the Hunnish State of Hsia (c. +412), Holien Pho-Pho,[10] as Chief Engineer (Chiang Tso Ta Chiang[11]), and whom we shall meet again in connection with brick-making[e] and armoury.[f]

The last four names to be mentioned here are particularly good examples of ingenious men whose posts and occupations seem to have been quite unsuited to their talents. Yen Su[12] (fl. +1030) was a Leonardo-like figure—scholar, painter, technologist and engineer under the Sung emperor Jen Tsung. He devised a type of clepsydra[g] with an overflow tank which remained standard for long afterwards,

---

[a] Cf. the discussion on pp. 111, 457, 471 below.
[b] We shall describe all these different contributions in their proper places below; pp. 159, 256, 287, 552.
[c] See Sect. 30h.
[d] On his life and work see Sect. 28e below.
[e] See Sect. 28c below.　　　　　　　　　　[f] See Sect. 30d (7) below.
[g] Cf. Vol. 3, pp. 324 ff.

[1] 張思訓　　　[2] 司天渾儀丞　　　[3] 解飛　　　[4] 魏猛變　　　[5] 石虎
[6] 慕容超　　　[7] 張綱　　　[8] 張榮　　　[9] 叱干阿利　　　[10] 赫連勃勃
[11] 將作大匠　　　[12] 燕肅

invented special locks and keys,[a] and left specifications for hydrostatic vessels,[b] hodometers[c] and south-pointing carriages.[d] His writings included treatises on time-keeping and on the tides.[e] But most of his life was spent in provincial administrative posts and though he did become an Academician-in-Waiting of the Lung-Thu Pavilion, he never rose above the position of Chief Executive Officer of the Ministry of Rites and had no connection with the Ministry of Works or the other technical directorates. Even more striking are the cases of the two greatest practical men in the pre-European history of the mechanical clock. Liang Ling-Tsan[1] (*fl.* +725), with I-Hsing the inventor of the ancestor of all escapements,[f] was only Chief Secretary of the Left Imperial Guard (Tso Wei Chhang Shih,[2] a minor War Office post), and although his ingenuity was well recognised at court we have no evidence that he ever occupied any more important or suitable position. Han Kung-Lien[3] (*fl.* +1090), Su Sung's principal collaborator in applied mathematics during the construction of the greatest astronomical clock-tower in Chinese history,[g] was an Acting Secretary in the Ministry of Personnel (Li Pu Shou Tang Kuan[4]) when Su Sung found him, and so far as we know that was where he permanently remained.

To end this history the best thing we can do is to present a translation of a +3rd-century essay by the philosopher and poet Fu Hsüan[5] on his friend the engineer Ma Chün[6] (*fl.* +260), perhaps one of the most interesting documents on the social history of ancient and medieval Chinese technology which we have found.[h]

Mr Ma Chün, whose *tzu* was Tê-Hêng,[7] came from Fu-fêng[8] (in the Wei River valley between Wukung and Paochi) and was a man of wide renown for his technical skill (*ming chhiao*[9]). In his youth he travelled into Honan but (at that time) he himself did not yet realise his own talent. Even then his powers of exposition fell far behind his mechanical ingenuity, and I doubt if he could express half of what he knew. Although he had a literary degree (*po shih*[10]) he remained poor. He therefore thought of improving the silk loom (*ling chi*[11]), and thus at last without need of explanations the world recognised his outstanding skill. The old looms had fifty heddles (*tsung*[12]) and fifty treadles (*nieh*[13]). Some even had sixty of each. Mr Ma, fearing to lose the merit of his mourning period by wasting time, changed the design in such a way that it had only twelve treadles. Thus strange new patterns in many wonderful combinations were made, by the inspired conception of the inventor, all arising easily and naturally; indeed it would be impossible to describe the endless permutations of the Yin and Yang (the warp and weft) as the shuttle (*lun phien*[14]) travelled back and forth.[i] But alas, how could he hope to explain to people the (principles of the) improvements which he had made?

[a] On this subject, see pp. 236ff. below.   [b] See Vol. 4, pt. 1, pp. 34ff. above.
[c] This subject is discussed in Sect. 27e on pp. 281ff. below.
[d] Cf. in the same Sect., pp. 286ff. below.
[e] We have dealt with this already in Vol. 3, p. 491.
[f] See further in Sect. 27j on pp. 471ff. below.   [g] Also in Sect. 27j on pp. 448ff. below.
[h] The completest versions, here conflated, are found in *CSHK* (Chin section), ch. 50, pp. 10aff., *San Kuo Chih* (*Wei Shu*), ch. 29, pp. 9aff. and *TSCC, Khao kung tien*, ch. 5, pp. 4bff. Résumés also occur in the *Fu Tzu*[15] book (*Han Wei Tshung Shu*) as an appendix, and in *TPYL*, ch. 752, pp. 7aff.
[i] In Sect. 31 we shall try to analyse the meaning of these inventions, which certainly concerned the draw-loom for weaving patterns in the silk.

[1] 梁令瓚	[2] 左衞長史	[3] 韓公廉	[4] 吏部守當官	[5] 傅玄
[6] 馬鈞	[7] 德衡	[8] 扶風	[9] 名巧	[10] 博士
[11] 綾機	[12] 綜	[13] 躡	[14] 輪扁	[15] 傅子

Mr Ma, being a Policy Review Adviser (Chi Shih Chung[1]) one day fell into a dispute at court with the Permanent Counsellor Kaothang Lung[2] and the Cavalry General Chhin Lang[3] about the south-pointing carriage.[a] They maintained that there had never been any such thing and that the records of it were nonsense. Mr Ma said: 'Of old, there was. You have not thought the matter out. It is really not far from the truth.' But they laughed and said: 'Your name, Chün, means a weight, and your *tzu*, Tê-Hêng, means the balance of virtue. But (if you go on talking like this) your name Chün will be interpreted as a potter's wheel for moulding (empty) vessels, and Hêng will be taken to mean a decision on weight without weighing—in this way there is nothing you could not mould at will!' To this Mr Ma replied: 'Empty arguments with words cannot (in any way) compare with a test which will show practical results.'[b] All this was reported by Kaothang and Chhin to the emperor Ming Ti,[c] whereupon Ma Chün received an order to construct such a vehicle. And he duly made a south-pointing carriage.[d] This was the first of his extraordinary accomplishments. But again it was almost impossible to describe (the principle of it) in words. However, thenceforth the world bowed to his technical skill.

Now in the capital there was some land which could be made into gardens but there was no water with which to irrigate them. Mr Ma (therefore) constructed square-pallet chain-pumps (*fan chhê*[4]).[e] When the serving-lads (*thung erh*[5]) were ordered to work them (by pedalling) the water rushed in and out, turning over and over automatically, and irrigating the gardens. The ingenuity (of the device) was a hundred times that of the ordinary methods. This was the second of his extraordinary accomplishments.

Fu Hsüan next goes on to describe the third and fourth accomplishments, i.e. the mechanical puppet theatre and the rotary ballista. The description of the former we reserve for Sect. 27*c* (p. 158) below. That of the latter will be given in Sect. 30*h* below. He then continues:

(At this time) there was at court a notable scholar, Master Phei (Phei Tzu[6]). When he heard (of Ma's inventions) he laughed, and mocked him with difficult questions. Mr Ma stammered and could not give (satisfactory) replies. As Master Phei could not get the essential ideas (*nan tê chhi yao*[7]) from Ma's explanations, he continued to discredit him. I myself said to Master Phei on one occasion: 'Your great merit is of course eloquence, but where you fall short is in technical skill. Now this is Mr Ma's strong point, but he is not a good talker. For you to attack his inability to express himself is really not fair. On the other hand when you argue with him about those technicalities in which he excels, there must be points which we cannot expect to understand. His special talent is a very rare one in the world. If you insist on raising difficulties about these matters which are so hard to expound, you will go far astray from the truth. For Mr Ma's gifts are all of the mind and not of the tongue. He will never be able to reply (to all you ask of him).'

---

[a] Kaothang Lung was himself an astronomer and calendar expert; we have met with him already in connection with catoptrics (Vol. 4, pt. 1, p. 88 above). Of Chhin Lang we know little more than his interest here.

[b] This reply is strikingly reminiscent of that colloquy between the swordsmith and the sophists which we cited in Vol. 2, p. 72.

[c] R. +227 to +239.

[d] See on this the whole discussion on pp. 286 ff. below.

[e] We shall see the significance of this in Sect. 27*g* (pp. 339 ff. below).

[1] 給事中   [2] 高堂隆   [3] 秦朗   [4] 翻車   [5] 童兒   [6] 裴子
[7] 難得其要

Later on I gave an account of my talk with Master Phei to the Marquis of An-Hsiang.[1] (Unfortunately) the Marquis agreed with Master Phei. So I said: 'The sages selected talent in accordance with ability, and did not carelessly entrust people with the management of affairs. Some obtained distinction because of their spirit, others because of their eloquence. The former had no need of words, for their quality was revealed by their sincerity and virtuous deeds, while the latter had to argue about right and wrong in order to demonstrate their greatness. Look at Tzu-Kung[a] for example. There were also those (who obtained distinction) because of their political ability, like Jan Yu[b] or Chi-Lu,[c] and those who showed literary brilliance, like Tzu-Yu[d] and Tzu-Hsia.[e] Thus even the all-understanding sages used trials and tests in the selection of important personnel. Jan Yu was tested with politics and Tzu-Yu and Tzu-Hsia with learning. Why not apply such tests to lesser persons? What is the point of saying that the principles of things cannot be exhausted in words, and that there can be no end to discussions about the universe, when the truth can so easily be verified by experiment? (*erh shih chih i chih yeh*[2]). Mr Ma is proposing to construct ingenious equipment (*ching chhi*[3]) for the country and the army. All you need to do is to give him ten measures of wood and a couple of workmen, and you will soon know who is right. Why should it be so difficult to (get permission) for (an official) test when the experiment is readily made? (*nan shih i yen*[4]). To discredit extraordinary ability in other people with light words is like (someone) who would impose his own wisdom on the affairs of the universe instead of handling the endless difficulties in accordance with the Tao. This is the path to destruction. What Mr Ma has accomplished has been the result of many modifications, so that what he said in the beginning is not necessarily still true. Must you refuse to make use of Mr Ma because not everything that he has said has proved to be right? How do you expect less well-known technical men to come forward? Among the mass of the people jealousy among competitors and mischief among colleagues is unavoidable. Hence (wise) rulers pay no attention to such things and base their judgment on tests. To throw away (true) standards and refuse to employ (Mr Ma) is like that false assessment which took a piece of perfect jade for common stone from Ching(-Shan). That was why (Pien) Ho[5] was seen hugging an unpolished gem and weeping (at the crossroads).'[f]

After this the Marquis of An-Hsiang saw my point, and talked over the matter with the Marquis of Wu-An.[6] Nevertheless, the (official) tests were never ordered. (Alas, when the government) neglects to arrange such simple trials for a man like Mr Ma whose skill was well known, what hope is there for lesser gems to be brought to light? (I hope) that the rulers in days to come will use this case as a mirror. The ingenuity of Mr Ma could not be surpassed even by Kungshu Phan[7] and Mo Ti[8] in ancient times, nor by Chang Hêng[9] in the Han. As Kungshu Phan and Mo Ti held distinguished positions[g] their talents benefited the

[a] I.e. Tuanmu Tzhu,[10] like all the other men mentioned in this place, one of the disciples of Confucius. Originally a merchant, he was famed for his eloquence and charm.

[b] I.e. Jan Chhiu,[11] who rose to high office in Lu State.

[c] I.e. Chung Yu,[12] soldierly and daring.   [d] I.e. Yen Yen,[13] an educationalist.

[e] I.e. Pu Shang,[14] transmitter of the social and philosophical doctrines of the school.

[f] This is a reference to the story of the −7th-century lapidary who presented a piece of jade to the first two Chou emperors. On false advice, the stone was rejected, and he was each time punished by having a foot cut off, but on the third occasion the value of the jade was recognised and Pien Ho vindicated. See *Shih Chi*, ch. 83, p. 10*b*, and many references in ancient books such as *Han Fei Tzu*, ch. 13, *Lun Hêng*, chs. 8, 15, 29, 30, etc.

[g] This was perhaps somewhat overstating the case. Kungshu Phan was the most famous artisan and engineer of antiquity, but his date (between −470 and −380) was so early that we have little detail

[1] 安鄉侯   [2] 而試之易知也   [3] 精器   [4] 難試易驗   [5] 卞和
[6] 武安侯   [7] 公輸般   [8] 墨翟   [9] 張衡   [10] 端木賜
[11] 冉求   [12] 仲由   [13] 言偃   [14] 卜商

world. But although Chang Hêng was a President of the Imperial Chancellery, and Ma Chün got as far as Policy Review Adviser of a Department (Chi Shih Shêng Chung [1]), neither of them was ever an official of the Ministry of Works, and their ingenuity did not benefit the world. When (authorities) employ personnel with no regard to special talent, and having heard of genius neglect even to test it—is this not hateful and disastrous? [a]

After this remarkably scientific and experimental cry from the +3rd century, further comment would be superfluous, and we leave the last word to Fu Hsüan.

(3) TRADITIONS OF THE ARTISANATE

Only one thing more remains before we can get to the bench, the foundry and the field. Any picture of the eotechnic artisanate would be incomplete without some allusion to its own traditions. About these a good deal has already been said. In Section 3, arising out of our notes on bibliography in general, there was a discussion of the legendary inventors (some, no doubt, originally technic deities) and of the books about them. [b] In Section 13*g*, arising out of a study of the *I Ching* (Book of Changes), an account was given of the primary inventors recounted in its Great Appendix, a text of about the −2nd century. [c] These sages, introducers of the plough, the cart, the boat, the gate, and so on, were doubtless those to whom Confucius was referring in his famous but truncated aphorism: 'The great inventors were seven in number (*Tzu yüeh: tso chê chhi jen i* [2]).' [d] Here we need only add to the above the specifically Taoist traditions enshrined in the *Lieh Hsien Chuan* [3] (Lives of the Famous Hsien). [e] A book with this title certainly existed in the Later Han (+1st century) and was known to Pao Phu Tzu. Traditionally it was attributed to Liu Hsiang [4] (−77 to −6) who was supposed to have based it on a set of pictures, the *Lieh Hsien Thu*, [5] due to one Juan Tshang [6] of the Chhin. Although the text as we now have it was fixed only as late as +1019, when the Taoist Patrology was first printed, internal evidence shows that some portions of it go back to dates such as −35 and +167. Quotations in encyclopaedias further show that a different text, written by Chiang Lu [7] of the Liang (+6th century)

concerning him and what he actually did. There is certainly no evidence that he ever occupied a high official position, in his own State of Lu or elsewhere. Mo Ti is of course the celebrated philosopher of universal love about whom much was said in Vol. 2 (Section 11). His school was certainly connected with a knightly technology of military arts (cf. Sect. 30 below) especially concerning poliorcetics (cf. Sects. 19 and 26 on mathematics and optics), and he became assimilated into alchemical tradition (cf. Sect. 33); but it is doubtful whether he was really ever a minister in the State of Sung.

[a] Tr. auct. with Lu Gwei-Djen. Fu Hsüan's essay ends with a commentary (perhaps by himself, as it is printed as part of the text) which tells us that Master Phei was Phei Hsiu. [8] This is strange, since Phei Hsiu was one of the greatest geographers and cartographers in Chinese history, the father of the rectangular grid (see Vol. 3, pp. 538ff.). The sceptical philosopher Phei Wei [9] is a man whom the cap might fit better (if the commentary is astray), but his dates would make him too young for the role. The commentary also tells us that the Marquis of Wu-An was Tshao Shuang [10] and the Marquis of An-Hsiang was his younger brother Tshao Hsi. [11] In the ruling house of the Wei State, to which they both belonged, scepticism outraged by empirical facts seems to have been quite a tradition (cf. Vol. 3, p. 659).

[b] See Vol. 1, pp. 51ff.   [c] See Vol. 2, p. 327.
[d] *Lun Yü*, XIV, xl; cf. Waley (5), p. 190.   [e] Tr. Kaltenmark (2).

[1] 給事省中   [2] 子曰作者七人矣   [3] 列仙傳   [4] 劉向   [5] 列仙圖
[6] 阮倉   [7] 江祿   [8] 裴秀   [9] 裴頠   [10] 曹爽
[11] 曹羲

魯公輸子先師

Fig. 354. Paper block-print icon of the patron saint of all artisans and engineers, Kungshu Phan. In black on yellow paper, with decorative bands of colour in pink, green, mauve and red. Such icons were commonly pasted on the walls of workshops, with incense-sticks burning before them. Attendants in the foreground bear the tools of the trade, those behind hold technical treatises. In accordance with the bureaucratic character of traditional Chinese culture, Kungshu Phan, like all other tutelary deities, is enthroned as magistrate or governor. Above him the inscription says: 'Master Kungshu of Lu, our Teacher from of Old.'

also circulated for some time. Now many of the immortals in this book (as we should expect from the nature of ancient Taoism)[a] have close connections with the mechanical trades. In the biographies of Chhih Sung Tzu [1] and Ning Fêng Tzu [2] we find traditions of the mastery of fire involved in metallurgy and ceramics,[b] cast iron has a patron in Thao An-Kung,[3] and mirror-polishing with mercury in Mr Fu-Chü.[4] The Taoist patron of makers of mechanical toys was certainly Ko Yu,[5] who animated one and rode away on it, while Lu Phi Kung[6] was the great magician of bridges, ladders and galleries. Even the reel or small windlass of the fishing-rod had its spirit in Tou Tzu-Ming[7] (Lingyang Tzu-Ming[8]). In later sections dealing with chemical industry and pharmacy, dyes and cosmetics, we shall encounter many more of such Taoist patron saints.[c]

Just now we read Fu Hsüan's reference to Kungshu Phan,[9] the greatest of all the tutelary deities of artisans (Fig. 354). In spite of the fact that much of what was handed down about him is clearly legend, there is no reason to doubt his real existence in the State of Lu (hence his other name, Lu Pan[10]) in the − 5th century,[d] and we shall meet him from time to time in connection with kites and other devices.[e] He lives in proverbs, for instance 'brandishing one's adze at the door of Lu Pan (*Lu Pan mên chhien lung fu tzu*[11])', which is as much as to say, in our less elegant idiom, 'teaching one's grandmother to suck eggs'. And as so often in Chinese culture, where everything tends to be one of a pair, like parallel hanging scrolls, Kungshu Phan has a companion Wang Erh,[12] the semi-legendary inventor of curved chisels and graving-tools,[f] who may well have been a master of wood-carving contemporary with him.

Here is the place to mention a curious little work called the *Lu Pan Ching*[13] (Lu Pan's Manual), which circulated widely in the recent past among China's craftsmen.[g] As it seems not to have been studied by any sinologist, some description of it may be appropriate. Its author or compiler, Ssuchêng Wu-Jung,[14] and its editors, Chang Yen[15] and Chou Yen,[16] are quite dateless,[h] but much of the content is so archaic that one gains the impression of dealing with material some of which might well go back at least to the Sung. Anything so traditional will always be hard to date.

---

a Cf. Sect. 10*c*, *d*, *e*, *g*, and *i* above.
b As also, indeed, in those of Hsiao Fu,[17] Shih Mên,[18] and Chieh Tzu Thui.[19] Most of the names of these immortals are artificial, unlikely to have been borne by any living person.
c See the account of him by Liu Ju-Lin in Li Nien (7).
d See the account of him by Liu Ju-Lin in Li Nien (7).
e For example on pp. 96, 189, 238, 313, 573ff below, and one may recall Vol. 2, p. 53 above.
f See Chang Hêng's *Hsi Ching Fu*, in *Wên Hsüan*, ch. 2, p. 5*b* (tr. von Zach (6), p. 5, cf. Hughes (9), p. 117); also *Huai Nan Tzu*, ch. 8, p. 2*a* (tr. Morgan (1), p. 82).
g Its full title is *Kung Shih Tiao Cho Chêng Shih Lu Pan Mu Ching Chiang Chia Ching*[20] (The Timber-work Manual and Artisans' Mirror of Lu Pan, Patron of all Carvers, Joiners and Wood-workers).
h The compiler is described as Superintendent of Imperial Artisans in the Ministry of Works (Peking Division), (Pei-ching Thi-Tu Kung Pu Yü Chiang Ssu[21]).

---

[1] 赤松子	[2] 寧封子	[3] 陶安公	[4] 負局先生	[5] 葛由
[6] 鹿皮公	[7] 竇子明	[8] 陵陽子明	[9] 公輸般	[10] 魯班
[11] 魯班門前弄斧子		[12] 王爾	[13] 魯班經	[14] 司正午榮
[15] 章嚴	[16] 周言	[17] 嘯父	[18] 師門	[19] 介子推
[20] 工師雕斲正式魯班木經匠家鏡			[21] 北京提督工部御匠司	

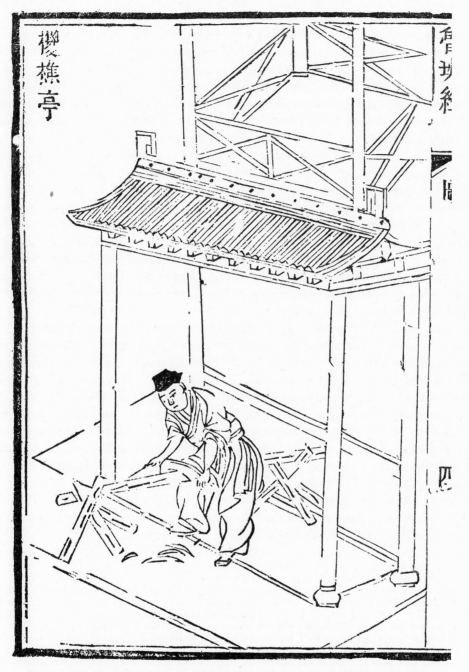

Fig. 355. An illustration from the *Lu Pan Ching*. Using a drawknife (cf. Hommel (1), p. 356; Mercer (1), pp. 97 ff.) in the timber-work of a watch-tower under construction.

The book opens with a series of illustrations (e.g. Fig. 355) showing operations of constructional joinery, sawyers at work, and various kinds of houses, bridges and pavilions, partly built or completed. A comparison might be made here with the well-known Tunhuang fresco of a pavilion under construction (Fig. 356).[a] Among the pictures is one of a kind of tower meant for an astronomical observatory (*ssu thien thai*[1]).[b] Then there follows, after a legendary biography of Kungshu Phan, a mass of detail about the cutting of timber in forests, the erection of pillars and the characteristic king- and queen-post frameworks,[c] the making of granaries, bell-towers, summer-houses, furniture, the wheelbarrow, the square-pallet chain-pump, the piston-bellows, the abacus and many other things. Precise specifications and dimensions are all interspersed with lore about lucky and unlucky days,[d] samples of charms and appropriate sacrifices. As the book proceeds the magical element preponderates more and more over the technical, and thus at the end we find a 'physiognomy' of buildings,[e] directions for exorcistic and luck-bringing incantations,[f] and descriptions of permanent protective cantrips.[g] The whole work, therefore, which deserves serious study, constitutes a unique piece of traditional technology and folklore.

One might find a close parallel to the *Lu Pan Ching* in the *Māyāmataya* of Ceylon,[h] a Sinhalese manual of craftsmanship for artisans and builders with an equally prominent admixture of divination and protective magic. But like the Chinese work it contains shrewd practical advice and throws much light upon the techniques of the past. Another parallel which might be cited as showing how long-enduring certain texts could be when they 'descended' to the common people and became accepted as part of traditional lore, is the *Aristotle's Masterpiece* of Western Europe. This book, which can still be purchased today, has probably formed the main source of instruction on sexual and embryological matters for working-class people during the past several centuries.[i] Yet it originated in an epitome of the opinions of Albertus Magnus (+ 1206 to + 1280) on generation, which bore the title *De Secretis Mulierum* and was still circulating in the + 16th and + 17th centuries. Another source of *Aristotle's Masterpiece* was the book of Fortunius Licetus (*De Monstrorum Causis Natura et Differentiis*, + 1616), illustrations from which are still being published in this 'submerged' form. All in all, one may guess that the *Lu Pan Ching*, though not recognisable in any of the bibliographies of the dynastic histories, has a pretty long past.

[a] Cave no. 445, early Thang in date.
[b] Cf. Vol. 3, pp. 297 ff. for a discussion of these towers.
[c] Full explanations will be found in Sect. 28 d below.
[d] As Dr Lu Gwei-Djen has pointed out to us, the doctrine of lucky and unlucky days (cf. Vol. 2, p. 357), superstition though it was, had a useful social function in a culture which had not borrowed from Israel the institution of a weekly day of rest for the toilers.
[e] Entitled Hsiang Chai Pi Chüeh.[2] On physiognomy in general see Vol. 2, pp. 363 ff.
[f] Entitled Ling Chhü Chieh Fa Tung Ming Chen Yen Pi Shu.[3]
[g] Entitled Pi Chüeh Hsien Chi.[4] Here we find the famous stone slabs so nostalgically familiar to those who have lived in the Chinese countryside, with the inscription *Thai Shan shih kan tang*.[5]
[h] Fully described by Coomaraswamy (5), pp. 120 ff. Its present text claims to have been translated from a Sanskrit work (not yet identified) in 1837. But again the material is so archaic that it must go back at least several centuries, indeed it is part of an age-old traditional lore.
[i] Cf. Needham (2), pp. 73 ff. and the bibliographical study of Ferckel (1).

[1] 司天臺   [2] 相宅秘訣   [3] 靈驅解法洞明眞言秘書   [4] 秘訣仙機   [5] 泰山石敢當

PLATE CXXVI

Fig. 356. Building a pavilion or watch-tower; from cave no. 445 at Chhien-fo-tung near Tunhuang. A cave-temple fresco of the +8th century.

PLATE CXXVII

Fig. 357. Japanese swordsmiths in ritual dress at work in their forge (Anon. 36).

The most outstanding characteristic of all Lu Pan's children down to modern times was perhaps the fact that they worked by knack, by rule of thumb, and by inherited slowly evolved tradition. This we saw already in considerable detail when discussing the artisanate of the ancient Chinese world.[a] The Taoist philosophers, particularly interested in techniques not transmissible by words, were always giving examples of this incommunicable yet learnable skill—swordsmiths for instance, and bellstand-carvers, arrow-makers, buckle-makers and wheelwrights. Two immortal stories remain of them: Ting the butcher (Ting phao jen[1]), who cleaved his bullock carcases according to the structure and pattern of the Tao,[b] and Pien the wheelwright (Pien lun chiang[2]), who told Duke Huan of Chhi that he would be better employed in learning the trade of governing from the people than in poring over Confucian treatises on the subject.[c]

The justified veneration of 'knack' and personal skill or flair was largely a pheno-menon of those long ages when materials and processes were not really fully under control. In a time such as our own, when a whole chemical factory can operate almost automatically, variations in the reactants being sensed by receptor apparatus and continuously adjusted by feedback processes, it is hard even to imagine the lot of the technician who had to carry out empirical procedures of which there was no scientific understanding. Deviations from the normal which would spoil the results desired could only be detected by the man whose observational faculties were extremely alert. Courses of action once found to succeed could only be repeated by the man whose memory was good and whose sleight of hand could rise to the occasion time after time. Materials as variable as wood and potter's clay and the crude uncomprehended metals could only be worked by the man who learnt from decades of experience to know the signs, the 'smell', the physiognomy, of the materials suitable for his pur-poses. There was many a failure and disappointment, but some psychological help was forthcoming from myth and legend, and many of the ancient Taoist metallurgists engaged in rites of purification and ascesis before beginning their operations.[d] This cultivation of intuitive and meditative skill was continued in Zen Buddhism, as in archery and similar arts still today.[e] The craftsman could not express his procedures in logical terms. In fact he could not explain at all; he could only show.

Thus the transmission of the crafts from one generation to another naturally involved a total education of the body and spirit of the learner. Apprenticeship was subjective and personal, not a matter of the intellectual understanding, not at all the appreciation of mathematical functions describing the behaviour of deeply analysed physico-chemical entities. Yet to some extent the skill of the artisans was handed

[a] Cf. Vol. 2, pp. 121 ff.
[b] Cf. Vol. 2, p. 45 above.　　　　　　　　　　[c] Cf. Vol. 2, p. 122 above.
[d] In Sect. 30d below we shall give a striking example of this in a quotation concerning ancient iron-casting. The tradition of religious rite has lasted on in Japan until our own time, as may be seen from Fig. 357, which shows swordsmiths in strange ceremonial dress beginning their temperings after prayer and fasting.
[e] See the first-hand account of Herrigel (1).

[1] 丁庖人　　　　[2] 扁輪匠

down orally in the ubiquitous and invariable practice of 'learning by rote' mnemonic rhymes (*ko chüeh*[1]). This is marked not only in engineering and building literature, which is relatively small, but also among the alchemists and the physicians, whose books are extremely numerous.[a] For the tricky empirical procedures of the medieval chemist nothing was more natural. Even some scholars, despairing of being able to produce accurate maps of the heavens, recommended the learning of the stars by heart.[b] Chêng Chhiao[2] wrote,[c] about +1150:

> Astronomical records are collected in maps rather than in books. Now even if books and manuals are handed down for a hundred (generations) they do not repeat the mistakes and errors, and even those they have can easily be compared and corrected. But maps acquire mistakes at each transmission, and these are perpetuated until it is difficult to find (the stars). Hence reliable maps are rare, and scholars fail to recognise the stars and constellations. However keenly you search in books you cannot get a right picture, and however much you study the charts you cannot rely on what they show. So it is best to chant over every day the *Pu Thien Ko*[3] (Song of the March of the Heavens).[d] Start on a moonless night in the autumn when the heavens are clear like a stretch of water, and chanting a strophe look upwards to study this or that star—after several nights you will have the whole picture of the heavens within your breast.

Of course both this advice, and the mnemonic rhymes of the master-carpenter or the iron-founder, were part of a culture in which the normal method of approaching the study of the classics was learning by heart in school before any explanations were given.[e]

But it must be remembered that in ancient and medieval times (and indeed until the last century) most of the craftsmen remained illiterate. There was therefore not only the basic intellectual difficulty of coining new technical terms,[f] but also serious obstacles to writing them down when they were adopted. Shipbuilding is among the most complex of the mechanical arts, yet one of the deepest living students of this craft in China found in thirty years' experience that hardly any of the best shipwrights whom he met could write.[g] And as we shall see, there are many technical sea terms for which no characters exist at all.[h] Of course all this does not mean that there were no technical manuals in Chinese tradition—in this Section we shall shortly describe an engineering literature of great interest.[i] In the case of shipbuilding, it is true, we know of only one treatise with adequate diagrams, and that was never printed, nor is it very old.[j] But plans, diagrams and models, even to scale, were frequently made

---

[a] Many examples will be found in Sects. 33 and 44.
[b] Cf. Vol. 3, p. 281.
[c] *Thung Chih Lüeh*, ch. 14, p. 1b (in *Thung Chih*, ch. 38), tr. auct. with Lu Gwei-Djen.
[d] A Sui poem; cf. Vol. 3, p. 201.
[e] Even as recently as twenty years ago the bee-like hum of schoolboys chanting the classics was one of the pleasant sounds of the Chinese countryside, as I do well remember.
[f] On this cf. Vol. 2, pp. 43, 260, 491 and elsewhere.
[g] Mr G. R. G. Worcester, in private conversation.
[h] Sect. 29 below.   [i] Cf. pp. 166 ff. below.
[j] It is interesting to remember that in modern engineering parlance 'a Chinese copy' means a copy of a machine or of some component part made by eye, measurement, or tradition, without any diagrams or drawings.

[1] 歌訣   [2] 鄭樵   [3] 步天歌

by the architects, as in the great work of Li Chieh just mentioned, and many other instances are known.[a] Furthermore, when one reaches the higher flights of medieval technology, as in astronomical apparatus, there must have been much paper work done, and in one crucial case, that of the Khaifêng clock-tower of $+1090$, we know that in the initial stages Han Kung-Lien prepared a special monograph on the mathematics and geometry of the clockwork.[b] Still, one can see how great the difficulties were with which inventive mechanical genius had to contend in ancient times. The artisanate was wedded to those customary paths which the experience of ages had created and it must have distrusted innovations almost as much as the Confucian scholars, though for very different reasons.[c] Only now and then in every century did the right men come together for creation, the right combination of hand and brain emerge. Yet in medieval China, seemingly so stable a society in comparison with Europe, it happened at least as often.

Whoever wishes to understand the practice of an eotechnic craft can do nothing better than read the book in which Sturt describes the work of his family firm of Surrey wheelwrights and cart-builders nearly a century ago. This careful autobiography throws a flood of light on the way of life of the skilled workmen just before the coming of modern technology.

With the idea that I was going to learn everything from the beginning I put myself eagerly to boys' jobs, not at all dreaming that, at over twenty, the nerves and muscles are no longer able to put on the cell-growths, and so acquire the habits of perceiving and doing which should have begun at fifteen. Could not Intellect achieve it? In fact, Intellect made but a fumbling imitation of real knowledge, yet hardly deigned to recognise how clumsy in fact it was. Beginning so late in life I know now that I could never have earned my keep as a skilled workman. But with the ambition to begin at the beginning, I set myself, as I have said, to act as boy to any of the men who might want a boy's help....[d]

As an example let us take the traditional English four-wheeled farm wagon.[e] The standard diameter of the front wheels (a dimension probably centuries old) was 4 ft. 2 in. There was need for a swivelling fore-carriage that would give a good 'lock' so that the wagon could turn around in a circle of as small radius as possible.[f] Therefore (a) there was a gentle forward boat-like rise of the floor-timbers of the body, (b) a slight narrowing was effected at the waist, (c) iron shielding was put on the 'sweep' (the back bar of the fore-carriage), and (d) iron 'locking-cleats' protected the side-timbers at the waist. But why were the front wheels not made so small as to go right under the body? Because as it was the axle could clear the ground only by about 2 ft., and while any less clearance might do for a road, it would not do for rough farm tracks and fields. Why then was the body not built higher? Because this would have been very inconvenient for loading and unloading, as well as dangerously liable to overturn. And there was another very good reason why the wheels were made no

---

[a] See Sects. 28 d, 29, below.   [b] See p. 464 below.

[c] On the other hand, when an inventive man did succeed in gathering the right workmen about him, as Ma Chün or Yüwên Khai or Su Sung must have done, they were probably master-craftsmen indeed.

[d] Sturt (1), p. 83.   [e] Cf. Fox (1); Lane (1).

[f] On the history of the bogie or turning-train see Boyer (2).

smaller; the difficulty of fitting the felloes on to the spokes. Naturally the spoke-holes had to be further apart at the periphery of the rim than at its inside circumference, yet this made them hard to put on, and the process was only achieved by the use of a tool called a 'spoke-dog' which strained the spokes together and allowed the felloe to be slipped over their ends.

The nature of this knowledge, says Sturt, should be noted. There was no conscious understanding of the why and the wherefore, only traditional good sense handed down through generations of Surrey wagon-builders. This knowledge

was set out in no book. It was not scientific. I never met a man who professed any other than an empirical acquaintance with waggon-builder's lore. My own case was typical. I knew that the hind-wheels had to be 5 ft. 2 in. high and the fore-wheels 4 ft. 2 in.; that the 'sides' must be cut from the best four-inch heart of oak, and so on. This sort of thing I knew, and in vast detail in course of time; but I seldom knew why. And that is how most other men knew. The lore was a tangled network of country prejudices, whose reasons were known in some respects here, in others there, and so on....The whole body of knowledge was a mystery, a piece of folk knowledge, residing in the folk collectively, but never wholly in every individual. 'However much a man knows,' old Bettesworth used to say, 'there's sure to be somebody as knows more.'[a]

Such is the real background of medieval craftsmanship, in East as well as West. Out of it grew all those great inventions and innovations which preceded the break-through of the Renaissance, when the technique of invention and discovery was itself discovered. It is the scene on which the drama now opening was played.

(4) Tools and Materials

In entering upon the present subject it is first necessary to say something briefly about the tool-chest of the Chinese mechanic (Fig. 353). The task is rendered difficult by the fact that no one, either Chinese or European, has brought together a critical account of Asian tools of all ages, taking into account ethnological relationships and confronting archaeological objects with relevant texts. In fact this has not really been done for our own civilisation either, the literature consisting mainly of speculative books now rather old,[b] with a few brilliant modern contributions on 'instrument typology',[c] and certain borderline studies such as the work of Moles (1) on the history of constructional woodwork. For China, Hommel's account, already mentioned, suffers from total lack of acquaintance with the Chinese literature; but even more regrettable is the fact that he does not give us any of the current technical names (and characters) for the various instruments and devices, though he must have been familiar with all of them.[d] Karlgren's description (13) of Shang dynasty tools and weapons is wholly archaeological in spirit, without any of the practical engineering approach. The

---

[a] Sturt (1), p. 74.

[b] E.g. Noiré (1) and Kapp (1) with the controversies on the 'organ-projection' theory, for which see Horwitz (6).

[c] E.g. Flinders Petrie (1, 2); Curwen (1); Mercer (1). Mr R. A. Salaman's work is eagerly awaited.

[d] Other accounts of the arts and crafts in Japan as well as China, e.g. Kämmerer (1); Yanagi (1); Woltz (1), are not much more informative though some of them reproduce old pictures which may be useful.

PLATE CXXVIII

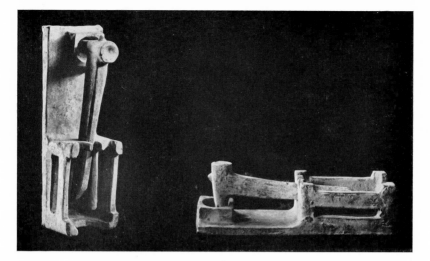

Fig. 358. Han model of man-power tilt-hammer in iridescent green glazed pottery, $10\frac{3}{4}$ in. × $3\frac{1}{2}$ in. × $3\frac{1}{2}$ in. (Nelson Art Gallery, Kansas City).

Fig. 359. The oldest extant drawing of the tilt-hammer, from the *Kêng Chih Thu* of +1210. Edition of +1462.

PLATE CXXIX

Fig. 360. Traditional rope-suspended pile-driver, in use on the Old Silk Road
(orig. photo., 1943).

greatest and most helpful figure, Charles Frémont, is curiously little known,[a] but he combined the historical method and use of sources with actual experiments in subjects such as the physiology of work and strength of materials in a way which has not been done before or since. We shall often refer to him. If these different interests could be combined with an ethnological competence such as Leroi-Gourhan's something valuable would begin to emerge, but the synthetic work has not yet been done, and all I can offer here is a few notes and references which may help to orient our minds.[b]

The most useful viewpoint from which to start is that of Praus (1), who had no difficulty in showing that with the exception of shearing and possibly punching, neolithic man had already discovered all the mechanical principles embodied today in powerful machines for changing the volume and form of matter.

The difference between the far past and the present lies mainly in the application of sources of power far exceeding that of human muscle, but controlled to an ever-increasing extent by mechanical refinements. The few remarks which have to be made may conveniently be arranged according to the classification of mechanical operations in Table 55.

It is curious that in spite of the great social significance of tools and machines, only 10 out of the 214 radicals in the Chinese language represent them,[c] with an additional three radicals for weapons.[d] Possibly this reveals lack of technological interest on the part of Confucian codifiers and lexicographers.

The history of hammers has been considered by Frémont (13), Coghlan (2) and Fischer (3). Much information about the various kinds of hammers (*chui*[1]), mallets (*chhui*[2]), sledgehammers (*ta chhui*[3]), etc., traditionally in use among Chinese artisans, will be found in Hommel.[e] They made great use of the principle of flexible handles for heavy hammers, to add to the force of the impact and reduce the sting on the hands.[f] When the handle of a hammer is attached to a fulcrum at some point along its length, and the head then raised and allowed to fall consecutively either by human power or water power, the instrument (tilt- or trip-hammer; *tui*[4]) has reached the level of a machine, and as such we shall treat it a few pages further on. Frémont could adduce no evidence for the occurrence of this in Europe before the +15th century, yet it goes back well into the Han in China, as is shown by numerous models of tilt-hammers operated by the weight of a man, and found in tombs of the period (Fig. 358). These may be compared with the oldest drawing which we have of the Chinese man-power tilt-hammer, that from the *Kêng Chih Thu* of +1210 (Fig. 359).[g]

---

[a] Partly because he published so much of his work privately, and in obscure periodicals.

[b] We have not had access to the book of Gompertz (1) on the evolution of implements.

[c] Nos. 6, *chüeh*,[5] hook; 17, *khan*,[6] mortar; 18, *tao*,[7] knife; 21, *pi*,[8] spoon or ladle; 69, *chin*,[9] axe; 79, *shu*,[10] axe on a long handle; 127, *lei*,[11] plough; 134, *chiu*,[12] mortar; 137, *chou*,[13] boat; 159, *chhê*,[14] carriage.　　　　　[d] Nos. 62, *ko*,[15] spear; 110, *mao*,[16] lance; 111, *shih*,[17] arrow.

[e] (1) pp. 6, 24, 119, 217, 220, 233, 282.

[f] I myself have often watched astonishing feats of the Szechuanese stone-masons making the rock cuttings and flights of steps in which the province abounds.

[g] Full details of this important work will be given below, p. 166.

[1] 錘　　　[2] 鎚　　　[3] 大鎚　　　[4] 碓　　　[5] 亅　　　[6] 凵　　　[7] 刀

[8] 匕　　　[9] 斤　　　[10] 殳　　　[11] 耒　　　[12] 臼　　　[13] 舟　　　[14] 車

[15] 戈　　　[16] 矛　　　[17] 矢

Table 55. *Classification of mechanical operations for changing the volume and form of matter*

	Percussion	hammer (and anvil) pestle (and mortar) sledgehammer trip-hammer
	Cutting	slashing knife sword slashing + leverage sickle axe adze pushing plane paring chisel gouge pushing + percussion chisel wedge sawing saw hacksaw bandsaw circular saw
	Scraping	gouging turning gouge filing file planing plane all drilling which removes material drill turning lathe and lathe-tools working with abrasive sands working with or without rotary knife grinding
	Detruding (Piercing)	awl needle and pin spear pick-axe
	Shearing (Punching)	scissors, shears punch
	Moulding (Chipless Forming)	pressure + percussion trowel brick-making forging die-work swage-moulding traction, e.g. wire-making extrusion
	Joining	nails dowels soldering gluing welding screws

It is seen used for decorticating rice.[a] A parallel machine is the rope-suspended pile-driver (*ta chhuang chi*[1]). Fig. 360 shows a hand-operated type still common in China being used in repair of a wash-out on the Old Silk Road in Kansu province in 1943. Frémont and Pouderoyen (1) figure closely similar tackle from the Renaissance time in Europe. An important extension of the same technique was the art of deep drilling developed in China in the −1st century (see below, Section 37). Presumably the basal support for hammering action, the anvil (*chen*[2]),[b] goes back far into antiquity;

[a] This technique spread all over the culture-area of Chinese influence, e.g. Indo-China, and appeared also early in India (cf. Grierson (1), p. 195).

[b] Hommel (1), p. 15.

[1] 打椿機   [2] 砧

the Chinese form had no hardy hole until modern times. Logically analogous is the pestle (*chhu*[1]) and mortar (*chiu*[2]), which go back to the Shang time or earlier; Hopkins (11) has drawn attention to the fact that the verb *chhung*[3] (K 1192), to pound in a mortar, was originally a pictograph of the operation.[a]

K 1192

The history of cutting edges in China, and their manifold uses, has yet to be written.[b] Were space and time available, an analysis of all derivatives of the radical *tao*[4] would be of real interest. Besides cutting-edge tools of the workers in wood and metal,[c] there would be of course also the great classes of agricultural implements (such as the sickle, *lien*,[5] the sharp-pointed spade, *fêng*,[6] the ploughshare, *pi*[7] or *chhan*,[8] the hinged chaff-cutter, *cha*,[9] and so on)[d] and of military weapons (such as the sword, *chien*,[10] the crescent-bladed halberd, *yüeh*,[11] etc.).[e] Though *chin*[12] was the ancient term for axe, the word *fu*[13] early replaced it, giving rise to further technical terms, such as *shou fu*[14] for adze.[f] The carpenter's plane was called *pao*,[15] the form of the character perhaps indicating that a knife was enclosed in a box-shaped holder.[g] Though des Noëttes (3) considered that this tool was not known or used in Europe until the +14th century, Feldhaus (1) believed that it existed in Hellenistic times,[h] and a special investigation as to the time of origin of the name and thing in China would be worth while; it may well be earlier. The word *pao*,[15] however, does not occur in Han or pre-Han books, i.e. before about the +2nd century. Chinese planes (cf. Fig. 361) are always worked by pushing away from the worker's body.[i] Outside its box and with the aid of a mallet, the plane knife becomes a chisel (*tso*[16])[j] or gouge (*chhü tso*[17]).

In the technique of sawing, cutting and scraping are combined. The tool goes back to the neolithic and

Fig. 361. Chinese carpenter's plane (drawing from the collection of Mr R. A. Salaman).

[a] Its importance in primitive Chinese life is indicated by the fact that under the character *khan*[18] it appears as the 4th trigram in the Book of Changes (Vol. 2, p. 313 above).

[b] We have already mentioned the strange fact, noted by Andersson (3), that the rectangular knives of the Chinese neolithic have persisted to the present day in the advertising jingle of the itinerant knife-sharpener (Vol. 1, p. 81).

[c] Whose patron or technic deity was Wang Erh, cf. p. 43 above.

[d] Cf. Sect. 41 below, but here may be mentioned the description of Chiang Kang-Hu (1).

[e] Cf. Sect. 30 below, but here may be mentioned the description of Werner (3).

[f] Handles for hatchets, large and small (*kho*[19], *chu*[20]), are among the few tools mentioned by name in the Khao Kung Chi chapter of the *Chou Li* (ch. 12, p. 22a).

[g] Cf. Hommel (1), p. 241.

[h] The plane was not known in Ancient Egypt (Lucas (1), p. 510). The iron planes found at Silchester may be as late as the +4th century, but the bronze ones of Pompeii must be of the +1st (Mercer (1), p. 115). Box planes found at the Saalburg presumably antedate +260 (Schönberger, 1).

[i] Japanese and Korean planes are used in just the opposite way, pulled not pushed, perhaps because their users work on the floor while the Chinese work at a bench.

[j] This word too occurs in the Khao Kung Chi of the *Chou Li* (ch. 11, p. 12a).

[1] 杵	[2] 臼	[3] 舂	[4] 刀	[5] 鎌	[6] 鋒	[7] 鏵
[8] 鑱	[9] 鍘	[10] 劍	[11] 鉞	[12] 斤	[13] 斧	[14] 手斧
[15] 鉋	[16] 鑿	[17] 曲鑿	[18] 坎	[19] 柯	[20] 欘	

was used in all ancient civilisations.[a] In China, the saw (*chü*[1]) achieved many subtle developments and a common form of it is the bow-saw or framed pit-saw for cutting tree-trunks into planks. In such saws the teeth are inclined starting from the middle of the blade in such a way that they point in opposite directions towards the ends; thus the two sawyers, one above the wood and one below, do equal amounts of work.[b] Sometimes logs are sawn hori-zontally, the sawyers walking slowly from end to end on each side (Fig. 363). Full stretching of the bow-saws is effected by tightening the cords on the other side of the central pole of the frame with a toggle stick (Fig. 362).[c] For tree-felling there are crosscut saws. Frame-saws or bow-saws were of course used also in Hellenistic times (Neuburger, 1) and represent an ingenious way of over-coming the dilemma between too thick a blade which would be strong but cut too coarse a kerf, and a blade finer but weaker. Chinese workmen use a great variety of saws and are familiar with many ways of filing and setting the teeth.

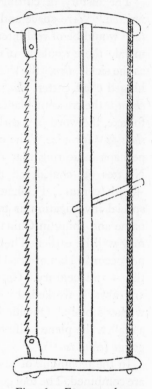

Fig. 362. Frame- or bow-saw with toggle stick, from *Thu Shu Chi Chhêng* (*I shu tien*), ch. 8, p. 27b. Cf. Mercer (1), p. 151.

Since metals can be shaped and cut while in the hot plastic state, the saw plays a much smaller role in metal-working than in carpentry, and the file, which in a sense corresponds to the plane, a correspondingly greater one. Frémont (9, 23) has written the history of the file, which he derives from natural objects primitively available such as the bones of fish in general, and the rough skin of elas-mobranch fishes such as the ray, in particular. Hommel[d] gives a good description of the Chinese file (*tsho*[2]), which differs somewhat from that commonly in use in the West. It has a tang on both ends, one to receive the wooden handle, and the other to carry a longer shaft of wood which fits through a ring-topped spike or eye-bolt (Fig. 364) within which it slides. This guide-rod gives a helpful down-leverage. While occidental artisans hold the article to be filed firmly in a vice and work all round it with the file, the Chinese system is to have the file oscillating in a more or less fixed path, and the article turned around under it as desired for filing (*tsho*[3]). Fig. 365 shows files in use in a +17th-century workshop.

[a] Frémont (21); Lucas (1); Feldhaus (3).

[b] Hommel (1), p. 224; Parsons (1), p. 177. This double-action saw is a combination of the pull-saw characteristic of Ancient Egypt and the push-saw like that now used in Europe. The log is supported on a trestle, and a fresco showing one being cut into planks exactly in the Chinese manner (though with a single-action saw) exists among those of the Campo Santo at Pisa (Lasinio (1); Mercer (1), p. 24). This dates from about +1350.

[c] Or an ingenious bamboo spring may be used. On dating see Ueda Noburu (1).

[d] (1), pp. 17, 38.

[1] 鋸   [2] 鑢   [3] 剉

PLATE CXXX

Fig. 363. Sawyers in the Chungking shipyards in 1944 (Beaton, 1).

PLATE CXXXI

Fig. 364. Apprentices using the Chinese double-tanged file.

PLATE CXXXII

Fig. 365. The double-tanged file in use in a coin-minting workshop of the +17th century
(*Thien Kung Khai Wu*, ch. 8, p. 10*b*).

PLATE CXXXIII

Fig. 366. The strap-drill in use (Hommel, 1), as by shipwrights.

All grindstones and whetstones (*chih*[1], *li*[2]) belong in the same category as files. Grinding (*mo*[3], *yen*[4]) must have reached a high level of skill among the early Chinese because of the importance of jade-working among them.[a] Hommel found no evidence of the use of rotary grindstones,[b] all those that he saw being rectangular blocks worn to a concave upper surface,[c] but the complete absence of the former seems difficult to believe. As we have already seen,[d] the use of rotary tools (the disc-knife, *cha tho*[5]) in jade-working seems to go back to the −3rd century at least, and the edge-runner mill in both its forms[e] can hardly be any younger. The mental connection between the disc-knife and the edge-runner mill (below, p. 195) is seen by the fact that the oldest term for the former which has come down to us[f] is the same as that for the latter, with the addition of 'abrasive' as an adjective (*sha nien*[6]). A special case of a grindstone is that of the inkstone (cf. Vol. 3, pp. 645 ff. above).

Drilling must be considered a branch of scraping or grinding since material is actually removed from the hole made, and on drilling much has been written.[g] This technique brings up the question of the origin of the driving-belt, and takes us therefore immediately into a fundamental engineering problem. Continuous rotary motion can only be attained by the aid of the crank (in one form or another) or the driving-belt, but for many purposes, of which drilling is one, a reciprocating rotary motion is sufficient. It has been said[h] that the Chinese did not make use of continuous rotary motion, but this is a misstatement in view of the antiquity of the crank in China (p. 118 below) and machines such as the square-pallet chain-pump (p. 339 below). If they had not been masters of it, the continuous and carefully adjusted turning of armillary spheres (pp. 481 ff. below) with the aid of water-power would not have been possible. Nevertheless, it does not seem that the artisans themselves applied a continuous drive to their drills (*tsuan*[7]), for even at the present time the reciprocating drive remains in general use, as it is so widely in all eotechnic cultures and civilisations.[i]

The simplest form of reciprocating rotary motion is seen in the shaft-drill, the shaft of which is merely rotated back and forth between the palms of the hands.[j] The ancestor of the driving-belt comes in when a strap or thong is wound round the shaft and each end pulled alternately (similar to the single motion of spinning a top) by one man, while a second holds the drill steady (Fig. 366).[k] Here the belt is not continuous, but it becomes so, in a sense, when its two ends are connected by a piece of wood, as is the case in a bow and its bowstring. If now the bowstring is wound once

---

[a] Cf. Vol. 3, pp. 666 ff., and p. 16 above.　　　　[b] (1), p. 258.
[c] See his fig. 381. Cf. Vol. 4, pt. 1, p. 117 above.　　[d] Sect. 25*f* above.
[e] I.e. that in which the runner or roller describes a circular course, and that in which it moves back and forth in a rectilinear path.
[f] Apart of course from the 'Khun-Wu knife'; cf. above, Vol. 3, p. 667.
[g] Fischer (1); McGuire (1); Hough (1), etc.　　　　[h] Hommel (1), p. 247.
[i] Leroi-Gourhan (1), vol. 1, pp. 55, 70, 101, 170.
[j] McGuire (1), p. 693.
[k] Hommel (1), pp. 331, 332, 338, describing Chinese boat-builders' drills. McGuire (1), pp. 706 ff., 716. The strap-drill or thong-drill is often applied to other uses, such as the lathe, or in India the churn. McGuire (1), p. 742, has detected it in certain Ancient Egyptian symbolic pictures, in which it is not obvious until pointed out.

---

[1] 砥　　[2] 礪　　[3] 磨　　[4] 研　　[5] 剗鉈　　[6] 砂碾　　[7] 鑽

round the drill axis a powerful reciprocating motion will be imparted to the drill each
time the bow is swept back and forth—and this is the bow-drill. In passing, its
possible relation to the bow of the musical instrument known as the fiddle may be
noted. Probably palaeolithic,[a] it is found among many eotechnic peoples still, and
was a well-known tool in Ancient Egypt.[b] It still exists in England in certain trades,
such as those of piano-makers and clock-makers. Here we illustrate a Chinese
carpenter's bow-drill (Fig. 367).[c] Somewhere very long ago, no one knows where or
when, the idea originated of making a hole in the centre of the bow and having it rise
and fall alternately at right angles to the drill axis with a pumping motion, hence the
term pump-drill.[d] Fig. 368 shows a modern Chinese brass-smith's pump-drill.[e]
Seventeenth-century workmen using it are seen in several contemporary illustrations.[f]

As is well known, the drill figured prominently in the very ancient ceremony of
obtaining 'New Fire'.[g] We saw above that in the *Chou Li* a special type of official, the
Ssu Kuan[1] (Director of Fire Ceremonies), was in charge of this.[h] The *Lun Yü* attests
its use for the −5th century,[i] and shows that already there was a series of woods considered suitable for use at the different seasons.[j] Unfortunately, no evidence has
survived, in spite of discussions by scholars of many periods,[k] which would inform us
as to the exact type of instrument used as the fire-drill (*tsuan sui*[2]).

The most spectacular application of drilling in Chinese culture was the art of deep
drilling, practised especially in Szechuan, where there are vast deposits of brine and
natural gas. In Section 37 on the salt industry evidence will be adduced which shows
that by the Early Han period the Chinese had developed very fully the technique of
drilling boreholes two thousand feet deep, and consequently all the techniques
associated with this art. The story forms perhaps the most important part of the
prehistory of petroleum engineering, but we must not anticipate it here.

When the drill is mounted horizontally upon a frame, and is made to bear upon its
working end not a boring tool but the article upon which a chisel or knife is designed
to work, then the lathe has come into existence.[l] Treadles operated by the feet can

---

  a Childe (11).   b Klebs (3), fig. 73, p. 103.
  c McGuire (1), pp. 719, 730; Hommel (1), pp. 246, 247.
  d Holtzapfel & Holtzapfel (1), vol. 4, p. 4; McGuire (1), pp. 733, 735; Hommel (1), pp. 37, 39, 199.
  e This type of drill persists in neotechnic Europe, especially among stone-masons and porcelain-
menders. I remember my father teaching the use of it to me in his workshop. One should note that in its
perfected forms it involves the flywheel.
  f *TKKW*, ch. 10, p. 7*b*, for example.
  g Ethnography in Hough (1), pp. 85 ff. and elsewhere. At the Chhing Ming festival every April in
Hangchow around +1250, the 'New Fire' ceremonies were as dramatic as those of liturgical Christendom; see Gernet (2), p. 208, based on *Mêng Liang Lu*, ch. 2 (p. 148).
  h Vol. 4, pt. 1, p. 87; cf. de Visser (1).   i XVII, xxi, 3.
  j There are many other Han and pre-Han references, such as the *Li Chi*, ch. 12 (Legge (7), vol. 1,
p. 449); the *Kuan Tzu*, ch. 53, etc.
  k Notably by Sung Lien[3] of the Ming in *Ming Wên Tsai*,[4] ch. 37, and Chieh Hsüan[5] in his *Hsüan
Chi I Shu*[6] (Records of Ancient Arts and Techniques). *KCKW*, ch. 4, p. 31*b* is wrong in saying that the
+10th-century *Chung Hua Ku Chin Chu* discusses the fire-drill.
  l Leroi-Gourhan (1), vol. 1, pp. 173, 184, 187; Fischer (2); Montandon (1), p. 488. The best general
history of the lathe in late times is that of Wittmann (1), and now we have Woodbury (4, 5) for the earlier
periods.

  ¹ 司爟   ² 鑽燧   ³ 宋濂   ⁴ 明文在   ⁵ 揭暄   ⁶ 璇璣遺述

PLATE CXXXIV

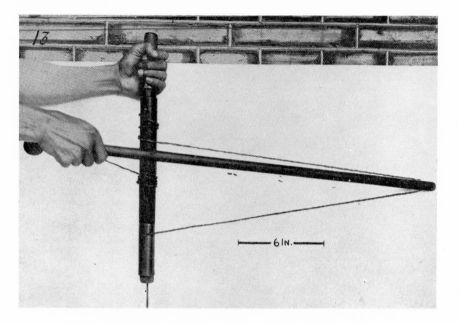

Fig. 367. The bow-drill in use (Hommel, 1), as by carpenters.

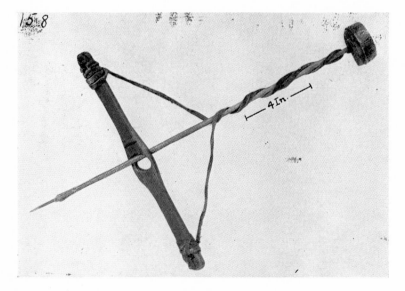

Fig. 368. The pump-drill (Hommel, 1), used by brass-smiths.

easily replace the two hands of the assistant operator, and this simple form of the machine (*chhê chhuang* [1], *hsüan* [2]) is still common among Chinese artisans (Fig. 369).[a] By a simple improvement, one end of the drive belt can be attached to a springy pole above the lathe, thus releasing one of the worker's feet and permitting more concentration on the work—such pole-lathes persisted in general use well into the eighteenth

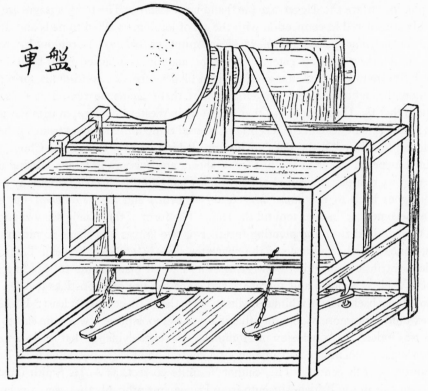

Fig. 369. Alternating-motion treadle lathe, *phan chhê* (Frémont, 5).

century in Europe,[b] and are known from many cultures, such as that of the Algerian Kabyles.[c] The statement that this type of lathe never developed in China[d] should be accepted with caution, since the use of springy bamboo laths was always very characteristic of Chinese technique, as, for example, in one of the two basic types of looms (below, Sect. 31). Little is known as to the antiquity of the lathe in China, but it is

---

[a] Hommel (1), pp. 252, 347. For an Indian example, see Grierson (1), p. 85. Replacement of the belt by connecting-rod, crank or crankshaft, and flywheel gives the 'sewing-machine' drive. Leonardo used this for a lathe (cf. Burstall (1), pp. 122, 141) but it had been current earlier in China for silk-winding machines (Fig. 409) and probably for cotton-gins (Fig. 420). It was one of the steps on the way to the 'steam-engine' assembly for the interconversion of rotary and rectilinear motion (see pp. 380ff.).

[b] Earliest European illustration is probably the MS. Lat. 11560 in the British Museum, f. 84a, of the +13th century.

[c] Holtzapfel & Holtzapfel (1), vol. 4. The pole-lathe is still used in Wales for making wooden bowls and in Buckinghamshire for chair-legs.　　　　[d] Hommel (1), p. 253.

[1] 車牀　　　[2] 鏇

not likely to be later than the Hellenistic time, when Egypt obtained it from the Greeks.[a]

Detrusion, or puncturing where there is no removal of material from the hole, has played an immense part in all man's dealings with matter, but is more typical of primitive than of developed technique. In woodworking there was the awl (*chui* [1]), in agriculture and construction the pickaxe (*ting tzu fu* [2]), in textiles the needle (*chen* [3, 4, 5]) and pin, in warfare the dagger-axe (*ko* [6]) and lance (*mao* [7]). The textile usages are more suitably considered in connection with the art of joining, as allied to nails and dowels.

On shears (*chiao* [8]), scissors (*chien* [9, 10]) and punches (*chhung* [11]) Frémont has written (22); they form a separate category since the material is attacked from both sides at once.[b] Shears in which the tangs of the iron blades formed a continuous spring were common throughout the Roman empire, but there seems no reason at present to suppose that the Han people got them from their Western contemporaries.[c] A pair of such shears is to be seen in the recently recovered Sung tomb frescoes from Yü-hsien [12] in Honan (exhibited at the Palace Museum, Peking, 1952). Sun Tzhu-Chou (*1*) has devoted a special paper to the history of shears and scissors in China, arising out of the recovery of a pair of iron scissors from a Thang tomb.[d] *Chien*,[9] originally a verb, to cut, goes back to the *Shih Ching* (c. −7th cent.) and the *Tso Chuan*, but seems not to become a noun for a tool till the Han.[e] The form of the Thang scissors suggests that they derived from the pivoting together of two knives of a shape common in the Han, hence also the rationale of the character *chiao* [8] ('crossed' knives). There is an obvious intrinsic connection between scissors and all instruments for holding small objects, such as pincers and forceps;[f] and since the use of chopsticks (*chu* [13]; mod. *khuai* [14]) goes back so far in Chinese civilisation (to the −4th century at least; Forke, 15), there would seem more probability that if any travel took place it was from east to west. This was indeed the conclusion of Hommel,[g] who cites evidence that scissors formed of two separate members movable about a central point were not known in Europe before the +10th century. This cannot be quite correct, as A. H. Smith illustrates a Roman pair in the British Museum from Priene, but probably they were uncommon before the late Middle Ages. They were certainly a Venetian import from the Middle East.

[a] Lucas (1), p. 510. Vitruvius (−30) frequently mentions it, as *tornus* (IX, i, 2 and viii, 6; also X, i, 5; vii, 3 and viii, 1).

[b] Traditional types of Chinese instruments in Hommel (1), pp. 17, 21, 38, 200, 217. Southern people, says the *Erh Ya*, called scissors *chi tao*,[15] 'trimming knives'.

[c] As suggested by Leroi-Gourhan (1), vol. 1, p. 276.

[d] A pair in copper, also Thang, had been recovered previously.

[e] *Shih Ming* dictionary, c. +100.

[f] The history of forceps (though without reference to Chinese material) has been compendiously written by Møller-Christensen (1). They go back to the first dynasty of ancient Egypt (c. −3300) as well as to Ur in Mesopotamia; the former seem to be surgical, the latter cosmetic. I am happy to possess a small bronze epilatory forceps of late Chou or Han date, 3·8 cm. long, with legs of equal breadth to the bit and engraved with cloud-scroll designs.

[g] (1), p. 201.

1 錐	2 丁字斧	3 針	4 鍼	5 箴	6 戈	7 矛
8 鉸	9 剪	10 翦	11 揰	12 禹縣	13 箸	14 筷
15 劗刀						

The last basic type of technique is moulding. The means used to fill the mould (*mo* [1]) will vary according to the viscosity of the material involved. Of the casting (*chu* [2])[a] of bronze, iron and other metals, there will be much to say in the Sections on metallurgy (30 and 36), but obviously if the metal is brought to fusion (*yung*, or *jung* [3,4], *hua* [5]) nothing has to be done in the foundry but pour it into the prepared mould. If, however, the material is viscous, as in the case of clay for ceramics, or in brick-making,[b] pressure with or without percussion has to be employed. Some tools, like the trowel (*man* [6]), themselves give shape to what they handle. An outstanding Chinese technique involving moulds is that of house-building with walls of stamped or rammed earth

援鈇線

Fig. 370. Apparatus for wire-drawing, *yuan thieh hsien* (Frémont, 18).

(*terre pisé*);[c] this goes back to the Shang time (−2nd millennium) and will be discussed in Section 28 *d* on building technology. In wire-drawing the moulding is effected by traction of the ductile metal, and it is easily possible to show that no essential change has taken place in the technique since Hellenistic times.[d] Fig. 370 shows Chinese wire-drawing equipment. Hommel gives a good account of the technique as he saw it,[e] with reasons why he thinks that the Chinese were never able, till modern times, to apply it to iron wire.[f] Yet the *Thien Kung Khai Wu* of +1637 gives a clear description of the process,[g] with an illustration of a wire-drawing workshop (see Fig. 332, in Vol. 4, pt. 1).

[a] As opposed to forging, or the making of wrought iron (*tuan* [7,8] *thieh* [9]). Cf. Vol. 5.

[b] Hommel (1), pp. 259 ff.

[c] Hommel (1), pp. 293 ff. Technical terms—'filling up with mud' (*thien ni* [10]), 'ramming earthen walls' (*chhi ni chhiang* [11], *chu ni chhiang* [12]). The process is (in a sense) brick-making *in situ*. Cf. Boehling (1).

[d] Frémont (18); Feldhaus (26). The latter finds the first evidence of the making of iron wire at Nuremberg about +1100.

[e] (1), pp. 25 ff. [f] The iron has to be annealed after every two or three reductions.

[g] Ch. 10, p. 4 *b*, for needle-making. A connection between the origins of iron-wire drawing and those of the magnetic compass needle is to be suspected; cf. our discussion of Sung needles in Vol. 4, pt. 1, pp. 282 ff. above.

[1] 模	[2] 鑄	[3] 熔	[4] 鎔	[5] 化	[6] 鏝	[7] 煅
[8] 鍛	[9] 鐵	[10] 填泥	[11] 砌泥牆	[12] 築泥牆		

The joining of materials together is a branch of technique which may involve a number of the above-mentioned fundamental ones. Projecting structures play the largest part in woodwork—nails (*ting*[1]), dowels[a] (*ho pan ting*[2]), the mortice and tenon (*sun yen*[3], *sun*[4]), etc.—but there is also connecting with glue (*chiao*[5]); a practice which has more importance in metal-working in the form of the analogous techniques of soldering (*khuan*[6], *han*[7,8])[b] and welding.[c] The projecting structure separated from its base is the most important in the arts of sewing and tailoring, needles[d] and pins deriving presumably from the thorns and fish-bones of primitive ages and peoples. The history of the nail has been written by Frémont (4), who figures a Chinese nail-maker's forge taken from a Chhing painting. Nailswages (patterns or templates for nail-making) are discussed by Hommel.[e] Wrought-iron nails were probably always used sparingly by Chinese woodworkers, who achieved the same end by making wood or bamboo pins or dowels. Wrought-iron dowels, however, were considered necessary in Chinese shipbuilding.[f]

Last but not least in importance for the artificer were the tools of measurement.[g] The oldest and simplest were the stretched string or plumb-line (*shêng*[9]), the water-level (*chun*[10]),[h] the measuring-rule (*chhih*[11]), the compasses (*kuei*[12]), the carpenter's square (*chü*[13]), and the balance (*chhüan hêng*[14]) or steelyard (*chhêng*[15]). We have mustered them already in the introduction to the subsection on mensuration (Vol. 4, pt. 1, pp. 15 ff. above). The square and compasses go back so far as to appear in the purest mythological material, for they form the traditional emblems of the organiser gods, Fu-Hsi and his sister-wife Nü-Kua.[i] A tomb-shrine relief showing this has already been reproduced (Vol. 1, Fig. 28). So familiar were the tools of measurement that even Confucian scholars drew upon them for metaphors. For example, the *Hsün Tzu* says:[j]

When the plumb-line with its ink (*shêng mo*[16])[k] is truly laid out, one cannot be deceived as to whether a thing is straight or crooked. When the steelyard (*hêng*[17]) is truly suspended, one cannot be cheated in weight. When the square and compasses are truly applied, one cannot be

---

[a] Hommel (1), p. 257.   [b] Hommel (1), pp. 23 ff.
[c] The last-named is an ancient technique in the Middle East (Maryon & Plenderleith (1), p. 654). But it seems to have no term in Chinese other than 'the greater soldering', or 'joining'.
[d] Leroi-Gourhan (1), vol. 1, p. 276; Hommel (1), p. 199.
[e] (1), pp. 21 ff.
[f] Hommel (1), p. 255.
[g] Cf. on this, the sub-section on survey methods, above, Vol. 3, pp. 569 ff.
[h] The refinement of the bubble in alcohol had to await Melchisedech Thévenot in +1661.
[i] Cf. Granet (1), p. 498, who points out how strange it is that the round thing, the compasses, though associated with the round maleness of heaven, should have been the emblem of the goddess, while the god took the square thing, though associated with the squareness of female Earth. This may conceal the existence of a system of symbolism older than that which came to dominate classical Chinese thought.
[j] Ch. 19, p. 8*b*; tr. Dubs (8), p. 224, mod. auct.
[k] This is a reference to the Chinese carpenter's inked thread, still used today, which, upon being tautened and snapped, makes a straight line across his material. The string passes through the ink-well, which is filled with silk-waste saturated with black ink (Hommel (1), p. 250). We have met with it before (Vol. 2, p. 126). A parallel practice is in use in modern engineering works, where large pieces of steel plate are marked out by the aid of chalked string lines.

[1] 釘	[2] 合板釘	[3] 筍眼	[4] 榫	[5] 膠	[6] 鐝
[7] 釬	[8] 銲	[9] 繩	[10] 準	[11] 尺	[12] 規
[13] 矩	[14] 權衡	[15] 秤	[16] 繩墨	[17] 衡	

mistaken as to squareness and roundness. So when once the *chün-tzu* (the great-souled man) has investigated rightness of conduct (*li*[1]), he cannot be deceived by what is false.

There are many other passages of this kind.[a]

Examples of more complex measurements are frequent in illustrations and early descriptions, for instance, the testing of the strength of cross-bows,[b] the testing of arrows by floating (Vol. 4, pt. 1, p. 39 above) or of specific gravities by the sinking of seeds (Vol. 4, pt. 1, p. 42 above), and several special proof procedures (e.g. wheel trueness) described in the *Chou Li* (Khao Kung Chi).[c] The most striking measuring tool from old China, however, is perhaps the adjustable outside caliper gauge with slot and pin, which looks very like a modern adjustable spanner without the worm. Wu Chhêng-Lo has described and figured a remarkable example[d] dating from the first year of the Hsin dynasty, under Wang Mang, i.e. +9. This is shown in Fig. 371. One side of these sliding calipers is carefully graduated in six inches and tenths, while on the other side there is an inscription in seal characters saying: 'Made on a *kuei-yu* day at new moon of the first month of the first year of the Shih-Chien-Kuo reign-period.' No engineering tool of this kind appears to be known in Europe before the time of Leonardo da Vinci, who sketched something similar.[e]

One cannot quit the subject of tools and their work without alluding to certain characteristic Chinese materials. If this were logically done it would involve a great many questions more suitable for later Sections, e.g. those devoted to building technology (28*d*), botany (38), agricultural arts (42), and so on. In any case, the information on subjects such as the economic lumber trees of China has never been conveniently collected.[f] But a few words may be said of one of the most universally used materials, bamboo,[g] which was not so readily available for other civilisations.

A very large number of species are indigenous to China, probably the commonest being members of the genus *Phyllostachys*.[h] One would have thought that much

---

[a] E.g. *Lü Shih Chhun Chhiu*, ch. 145, tr. R. Wilhelm (3), p. 422.

[b] See Sect. 30*e* below.   [c] See p. 75 below.

[d] (*1*), 1st ed. p. 179, 2nd ed. p. 94. Photographs will be found in Ferguson (3).

[e] Feldhaus (1), col. 1372; but it is a pair of beam-compasses with a screw adjustment. See also Feldhaus (24) and Neduloha (1). According to Davison (8) the sliding-scale calipers were introduced by Vernier in +1631, and the screw micrometer by Gascoigne in +1638, for astronomical purposes. To determine the date and locality of the first European counterpart to the Hsin caliper gauge will require further study.

[f] This is not to deny the merit of the compilations of Chhen Jung (*1*) and Li Shun-Chhing (*1*), but here we need timber technology rather than botany.

[g] Cf. Sowerby (1). The essential basis for the remarkable properties of bamboo has only been revealed by recent work on two-phase materials such as fibre-glass. Composite materials combine substances of high and low tensile strength and modulus of elasticity, so that they can absorb loading stresses which would rupture the weaker component, and at the same time isolate imperfections in the individual units of the stronger component. Bamboo does just this, for in it strong high-modulus fibres of cellulose are combined with a low-modulus plastic matrix, lignin. Hence its multifarious uses in Chinese culture, where also the two-phase structure was dissociated by retting to give the high-tensile-strength fibres alone. We shall find later on (Sect. 30) a parallel to this combination in the pattern-welding of sabres, where wrought iron and steel were forged together (cf. Needham, 32). Nowadays we have high-tensile-strength glass fibres embedded in a matrix of plastic resin, as is described by Slayter (1), with a clear realisation of the historical background of these triumphs of modern technology.

[h] R 755.

[1] 禮

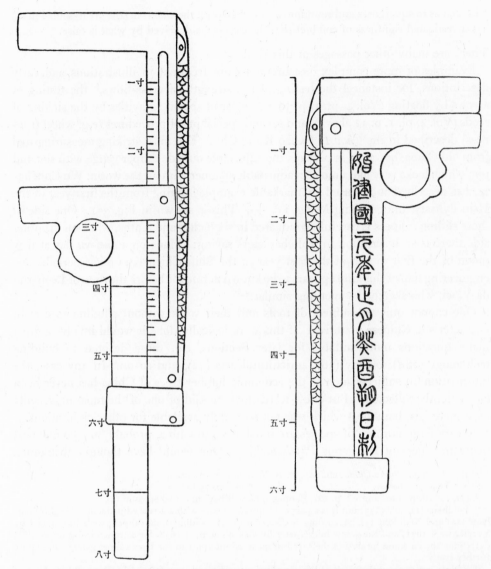

Fig. 371. Adjustable outside caliper gauge with slot and pin, graduated in inches and tenths of an inch; bronze. Self-dated at +9 by the inscription, which reads: 'Made on a *kuei-yu* day at new moon of the first month of the 1st year of the Shih-Chien-Kuo reign-period.' This remarkable metrological tool is thus of the short-lived Hsin dynasty, intermediate between the Earlier and the Later Han. Wu Chhêng-Lo (1); Ferguson (3).

attention would have been devoted to the manifold methods of putting this wonderful material to use, yet the available literature seems very sparse.[a] By way of illustration of some of the techniques, Fig. 372 shows[b] the detail of a litter chair (*chiao*[1], *hua-kan*[2]).

[a] I can refer only to papers by Rundakov (1), and Spörry (1); and to the curious monograph of Spörry & Schröter, which deals only with Japan.
[b] From Mason (1).

[1] 轎   [2] 滑桿

Such constructions make slight use of pegs or lashings, but one bamboo pole is made to hold another tight by chamfering and bending at a right angle, or even doubled back. It is not at all fanciful to point out that furniture, scaffolding and other erections of tubular steel make use at a more neotechnic level of exactly the same principles of structural strength as the ancient applications of the bamboo tube.[a] Yet delicate works of art in bamboo are often described; the *Shan Chü Hsin Hua*[b] of +1360 mentions fastenings in a model of a tortoise which remind one of bolts or rivets. The springiness

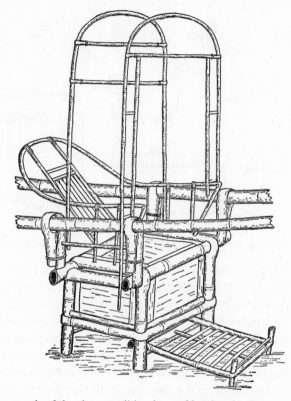

Fig. 372. An example of the clever traditional use of bamboo by Chinese artisans, a litter chair, *chiao tzu* or *hua-kan* (from Mason, 1).

of bamboo laths is also utilised to the full in all kinds of bow and bowstring devices.[c] The remarkable tensile strength of bamboo was not fully realised until recent times, when experiments were made by the Chinese Air Force Research Organisation at Chhêngtu; it was then found that ply-bamboo of formidable qualities could be made by uniting layers of woven laths with aeroplane glue. But all through the centuries this

[a] In +1793 Dr Dinwiddie marvelled much at the dexterity of Chinese artisans erecting mat-sheds and hangars of bamboo, (1), p. 49. And so did I in my turn during the Second World War.
[b] P. 13a; tr. H. Franke (2), no. 29.
[c] I remember being much struck by this when some of my staff rigged up automatic door-closing springs at a moment's notice on some of the doors of our building in Chungking during the Second World War.

property had been empirically utilised to the full in bamboo cables and ropes for many purposes.[a]

In another direction, the bamboo, with internal septae removed, forms a natural pipe, and this fact, as we shall see abundantly throughout the course of this work, exerted a cardinal influence on East Asian invention. In the earliest times it offered itself as a material for flutes and pipe-like instruments of music, instruments which deeply moulded the development of Chinese acoustics through the ages, as we have already seen in Section 26h. Then it was turned to great effect from the Han onwards

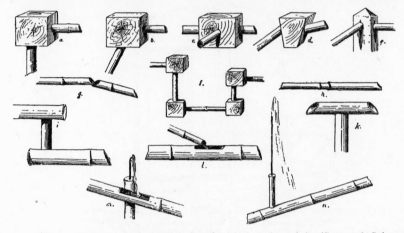

Fig. 373. Ways of using and jointing bamboo for pipes and conduits (Spörry & Schröter).

in the conveyance of brine from the deep borings to the places where evaporation was to take place (cf. Sect. 37). It found use also in piped water installations (see Fig. 373 and p. 129 below). Cut longitudinally it served for light tiles on roofs and every sort of simple channel. But bamboo tubing was used further in alchemy and the beginnings of chemical technology in the form of containers for such purposes as the descensory distillation of mercury and the solubilisation of minerals.[b] It generated the sighting-tube so characteristic of medieval Chinese astronomical instruments,[c] and fulfilled its most fateful destiny by becoming the ancestor of all barrel-guns early in the +12th century (cf. Sect. 30). And lastly let it not be forgotten that bamboo-shoots were found to make excellent eating. 'The bamboo', wrote John Barrow in +1793, concluding a lyrical passage on the subject, 'serves to embellish the garden of the prince, and to cover the cottage of the peasant', to which he slyly added, 'It is the instrument, in the hand of power, that keeps the whole empire in awe'.[d]

[a] Bamboo fibre was long appreciated for making the strong cables required for suspension-bridges; cf. Sect. 28e below. On a journey up the Yangtze in 1908, Esterer (1) made some measurements on the bamboo cables used by the trackers or junk-hauliers. He reckoned a tension of 7362 lb. per sq. in. (av.), which was of the same order as that normally taken by steel wires, yet the breakages were very few. For a similar estimate, see Fugl-Meyer (1). Moreover, while hempen ropes lose some 25% of their strength when wet, the tensile strength of plaited bamboo cables increases about 20% when they are fully saturated with water (Worcester, 1, 2). In the retted state bamboo played its part in the discovery and spread of paper (cf. Sect. 32 below). It also makes good candle-wicks (cf. Vol. 4, pt. 1, p. 80).

[b] See Sect. 33 below, and in the meantime Ho & Needham (3, 4).

[c] Cf. Vol. 3, p. 333.　　　　　　　　　　[d] (1), p. 309.

In all that will follow in this Section, as in several others (e.g. 31 on textile technology), wood and bamboo will appear in greatest prominence as the materials of construction of the machines of ancient and medieval China, only certain essential parts being made of bronze or iron. It may thus seem somewhat paradoxical in an engineering chapter that we shall not find ourselves obliged to take into account the basic principles of metallurgy until we come to the study of military technology. But this does not mean that Chinese culture was particularly war-like; on the contrary its ethos was exactly the opposite. There is good reason to think that iron was used for plough-shares and hoes at a date considerably earlier than that of the first steel weapons (cf. Sects. 30 and 41). The fact is that nowhere in the world was iron used for machinery before an industrial age began to demand long-enduring machines on a large scale. And still more paradoxical is it therefore, that processes of iron and steel metallurgy characteristic of the European industrial age (such as the use of coal in iron smelting, the successful production of cast iron on a large scale, and an ancestral form of the open-hearth process of making steel) were known and used in China from a thousand to fifteen hundred years earlier. But of this we can say no more here.

Glancing over the foregoing paragraphs, one is impressed by the number of unsolved problems. Precisely the simplest components of the engineer's armamentarium, the basic tools, have received the least attention from literary scholars. No one has attempted to analyse the references to tools in Han and pre-Han books. No one has investigated the appearance of so fundamental a machine-tool as the lathe, though some archaeological as well as literary evidence must surely be available. Perhaps the first thing to do for the history of Chinese technology would be to throw more light on the origin and development of the smallest unit operations of engineering achievement and industrial production.

(b) BASIC MECHANICAL PRINCIPLES

We now cross the borderland from tools to machines. It is proposed to deal with this vast subject in the following way; first comes a general discussion of the basic elements of which all machines are built up, including some orientation as to their antiquity in different civilisations. Secondly (pp. 165 ff.), we must consider the fundamental types of machines depicted in the Chinese traditional illustrations, and this will involve necessary bibliographical information, as well as a brief account of the effect of the coming of the Jesuits to China. Thirdly (pp. 243 ff.), our understanding will be enlarged by the literary evidence, after which we shall be in a position to adopt firmer views about the probable times and places of origin of the various engineering techniques than those accepted provisionally at the outset. Between the first and the second stage there will be some account of mechanical toys and automata, partly because they embodied nearly all of the basic mechanical principles, and partly because we have in China little or no iconographic evidence concerning them. At the opening of the third stage there will be an account of vehicles, which can be differentiated rather sharply from all stationary engines. It will then prove convenient to arrange the third stage following a classification of power sources—animal traction, water descent, and wind pressure. The conclusion of the third stage will bring us to the end of what may be regarded as mechanical engineering, but there will still remain the large realm of civil engineering (Sect. 28), so that hydraulics and water conservancy, fortifications, roads and bridges, will follow on. The present Section will reach its conclusion with a few notes on the prehistory of aeronautical engineering.

Let us now begin with a discursive account of the fundamental principles and components of all machines, illustrating it from time to time with examples which will serve to introduce the reader to the world of Chinese technology.

'Machines', said Cuvier in 1816, 'are geometry vivified.' In what follows it will be seen that Chinese civilisation has great engineering achievements to its credit. All the more remarkable, therefore, is it that they came about in an intellectual milieu which included, as we have abundantly seen (in Sect. 19h above), little or no deductive geometry. The only possible conclusion is that the empirical approach was able to overcome the lack of theory, and the only surprise for us is how long it took before the inevitable limitations of this approach were encountered. Moreover, we must not forget the factors of the social environment. In that Chinese society which could never spontaneously generate a Galilean Renaissance there was something at work which encouraged mechanics and engineers to brilliant practical achievements, achievements which neither the men of the Hellenistic world, nor the inhabitants of medieval Europe before the time of Leonardo, fully in possession of their Euclid and their Apollonius, attempted or attained. Such a contrast challenges an explanation, but this is not yet the place for it.

The great difficulty in discussing this subject logically is that all the different devices are so interconnected. Moreover, although it is easy to perceive their possible genetic relations, it is not at all clear in what order they were invented and brought into use. In English there are two classical discussions of the principles and kinematics of simple machines, those of Robert Willis and F. Reuleaux; they are indispensable for the historian of technology and his reader (and all the more so if one should lack personal engineering experience);[a] nor are they any the worse for having been written about

---

[a] Usher (1) drew upon them considerably (cf. 1st ed., pp. 66ff.). They had an interesting predecessor in the work of Augustin de Bétancourt (1808), but the real precursor of such studies, standing at the dawn of modern technology, was John Wilkins with his *Mathematicall Magick; or, Wonders that may be performed by Mechanicall Geometry, concerning Mechanicall Powers and Motions*... (+1648). John Wilkins was probably the only man who ever became both Warden of Wadham and Master of Trinity—certainly the only one who married the sister of the Lord Protector and yet was raised to the episcopate under King Charles the Second. A prominent member of the Invisible College, he became the first secretary of the Royal Society upon its incorporation in +1663. He took a strong interest in Chinese culture and literature, devoting much labour to the elaboration of a universal language with an algebraic-ideographic script (3), which he endeavoured to define precisely. His *Mathematicall Magick* was in two parts, the first, entitled 'Archimedes', dealing with the properties of the simplest fundamental machines; the second, 'Daedalus', examining some of their more elaborate applications. The same pattern was common to a number of books at this time, and we shall find it again in the *Chhi Chhi Thu Shuo* (Diagrams and Explanations of Wonderful Machines), written by Têng Yü-Han (Johann Schreck) and Wang Chêng in +1627, at the other end of the Old World (cf. p. 170 below).

For those who study ancient and medieval machines, books like those of Willis (biography by Hilken, 1) and Reuleaux are sometimes more useful than contemporary works of today. Nevertheless, some of these quite modern books are also most valuable—we may mention those of Hiscox (1, 2); F. D. Jones (1); Klein (1); Dunkerley (1); Steeds (1); Rosenauer & Willis (1).

a hundred years ago. Reuleaux's definition is the more general: 'A machine is a combination of resistant bodies so arranged that by their means the mechanical forces of Nature can be compelled to do work accompanied by certain determinate motions.' That of Willis amplifies the first phrase: 'Every machine will be found to consist of a train of pieces connected together in various ways so that if one be made to move, they all receive a motion, the relation of which to that of the first is governed by the nature of the connection.'

One would not at first sight expect to find this thought in ancient China, but actually it is very old, and common to both the Chinese and the Greeks. The pre-Socratics and other naturalists frequently sought analogies for the world-machine from human machines; Vitruvius (-30) compares the polar axis around which all the stars revolve to a lathe,[a] and Lucretius to a noria,[b] while Wang Chhung ($+80$) compares it to an edge-runner mill or a potter's wheel. The whole Peripatetic philosophy of the 'unmoved mover' implied the idea of a cosmic rotating machine.[c] And the *Ho Kuan Tzu*[1] (Book of the Pheasant-Cap Master)[d] says, in an interesting passage:[e] 'When the embroideries of a State bier move and shake, they do not do so of themselves, but in accordance with the motions of the main carrying-poles; the source is but one.'[f] This passage comes at the end of a paragraph in which good government is said to necessitate the integration (the 'engagement') of ruler and people. He must be their 'prime mover'. In later centuries, too, the machine-analogy played an enormous part in specific sciences such as biology no less than in cosmology; for example, William Harvey's appreciation of the valves of the heart as resembling the valves of a pump ('as by two clacks of a water-bellows to raise water'). Yet this again had a long Chinese ancestry, as we have already seen,[g] in the Taoist conviction that the living body was a self-acting organic automaton, with sometimes one part in charge and sometimes another. But in China no Galilean revolution in mechanics came to awaken physiology from its mystical slumbers.

Heron of Alexandria[h] was apparently the first to classify machine elements; in his view they numbered five: the wheel and axle, the lever, the pulley, the wedge, and

[a] IX, i, 2.   [b] v, 516.

[c] It is met with again in Indian literature of the $+$10th and $+$11th centuries (Somadeva Sūri and Prince Bhōja); see Raghavan (1).

[d] This text is very difficult to date. The passage here quoted would not be likely to be later than the $+$2nd century nor earlier than the $-$3rd.

[e] Ch. 4 (Thien Tsê[2]), p. 10*b*; tr. auct.

[f] *Kai wu chin kang hsi tung chê, chhi yao tsai i yeh.*[3]

[g] Vol. 2, p. 52.

[h] At this place the reader should be reminded that great differences of opinion have existed as to Heron's date. While Sarton (1) and Brunet & Mieli (1) inclined to the view that he flourished about $-$100, there was weighty opinion (Diels (1), Heath (6), vol. 2, p. 306, Heiberg (1), supported by Dr C. Singer, in conversation) in favour of a date in the neighbourhood of $+$200. This may be borne in mind elsewhere in this book, when comparisons with China are under discussion. We ourselves prefer the intermediate date proposed by Neugebauer (6), who identifies the eclipse of $+$62 as having occurred during Heron's working years, so that we think of him as the contemporary of Wang Chhung rather than of Lohsia Hung or Ma Chün. As for Philon and Ctesibius, there is more unanimity that the former's *floruit* was $-$220 and the latter's $-$250. On the whole question Drachmann (2, 9) is indispensable.

[1] 鶡冠子   [2] 天則   [3] 蓋毋錦杠悉動者其要在一也

the endless screw or worm. This was unsatisfactory, but explainable because the treatise in which he did it[a] was concerned only with the lifting of weights.[b] Other early classifications were Indian.[c] In the +7th century Daṇḍin enumerated six types, among which were mobile, stationary, water-works, heat-engines, and mixed devices. Prince Bhōja,[d] about +1050, in his *Samarāṅganā-sūtradhāra*, distinguished between the principle (e.g. rotary or otherwise), the material, the purpose and the form of a machine (*yantra*). His list of merits of a machine is interesting; they included (*a*) proportionateness, (*b*) elegance, (*c*) efficiency for the effect intended, (*d*) lightness, firmness or hardness as the case might be, (*e*) noiselessness when noise was not wanted, (*f*) avoidance of looseness and stiffness, (*g*) smoothness and rhythm in motion, (*h*) controllability in starting and stopping, (*i*) durability. Other classifications could be collected from Arabic sources (though we have met with none in Chinese), but before the Renaissance there were no really analytical treatments of the subject.

Willis (1) divided all machines into classes according to whether the directional relation and the velocity ratio were constant or varying.[e] In each of the three main classes he considered rolling contact, sliding contact, wrapping connection, link-work, and reduplication. Thus in the first class (that where the directional relation and the velocity ratio are both constant), gear-wheels, bevel-wheels, worms, and the rack and pinion would be examples of rolling contact; cams and slots would show sliding contact; pulleys, belts and chain-drives would show wrapping connection; cranks, levers and rods would show link-work; and multiple pulleys or tackle would show reduplication. It is hardly necessary for our purpose to embark on an elaborate classification, and I propose to deal with the machine elements under the following heads: (*a*) levers, hinges and linkwork, (*b*) wheels, gear-wheels, pedals and paddles, (*c*) pulleys, driving-belts and chain-drives, (*d*) crank and eccentric motion, (*e*) screws, worms and helicoidal vanes, (*f*) springs and spring mechanisms, (*g*) conduits, pipes and siphons, (*h*) valves, bellows, pumps and fans.

Reuleaux (1) was probably right in his view that the chief criterion of gradual mechanical perfection was the completeness of the constraint of motion. Excessive play must have been the chief devil in all primitive machines, yet with the materials, tooling and lubricants available there was no other way of making them go round at all. Moreover, Reuleaux pointed out that a very important element in mechanical improvement lay in the substitution of 'pair-closure' or 'chain-closure' for 'force-closure'. An example of the latter would be the resting of a heavy rotary grindstone simply by gravity on bearings with no upper component to keep the axle in place under all

---

[a] Tr. Carra de Vaux (1), from the Arabic version of Qusṭā ibn Lūqā.

[b] The famous definition of Marx (pt. IV, xiii, 1; Paul ed. p. 393) was also partial, since it conceived of every machine as necessarily involving a tool for the production of commodities. This may be true in the broadest sense, if knowledge be included among them.

[c] Cf. Raghavan (1).

[d] He was the ruler of Dhār (Malwa) from +1018 to +1060; see V. Smith (1), p. 189.

[e] Directional relation is constant if while one mechanical component moves in a certain direction, the other also perseveres in its own direction (e.g. a pair of gear-wheels). It varies in such cases as the rocking motion of a saw, or the beam of an early steam-engine. The constancy of the velocity ratio is of course independent of any changes which the actual velocities of the two components may undergo during a given time, for they change at the same rate.

possible circumstances, or the holding of a lathe-tool against the work by the muscular force of the operator. These are certainly the kinds of concepts which would have to be applied in the assessment of engineering achievement in any culture.

(1) LEVERS, HINGES AND LINKWORK

Of the lever and its great early application the balance something has already been said in the physics Section (Vol. 4, pt. 1, pp. 22 ff.). There we saw that the Mohist engineers in the −3rd century must have been acquainted with most if not all of the equilibrium principles stated by Archimedes. In the immediately following centuries, this under-standing of the lever was put to good use in China in the making, almost on a mass-production scale, of crossbow triggers (*nu chi*[1]). These mechanisms, which involved intricate bent levers and catches, were beautiful and delicate bronze castings, and deserve the full description which they will receive below in Section 30 on military technology.[a] On a scale much larger, and with the use of timber, the lever had also been employed from an earlier date in the swape, *shādūf*, or counterweighted bailer bucket (*chieh kao*[2]), which again will be discussed in connection with water-handling machinery (p. 331 below). Lever-presses[b] (beam-presses) were not dominant in China, though the trip-hammer constituted an important application,[c] and heavy loads tended to be hoisted by combinations of levers rather than by pulley tackle.[d] But the most elaborate use of levers in early times in China was undoubtedly in textile machinery, where levers and connecting rods were united with treadles to form complic-ated linkworks. Evidence which will be presented in the appropriate place (Sect. 31 below) shows that the Chinese were far ahead of the West in loom construction—in the −1st century, for example, if not the −4th, they already had the essentials of the draw-loom (*hua chi*[3]) before Europe or perhaps any other civilisation had advanced from the primitive vertical-warp loom to the horizontal-warp loom with its harness of heddles. This precocity is perhaps symbolised by the fact that the Chinese word for loom, *chi*,[4] implies that it is the machine *par excellence*.[e]

The assembly of such a linkwork, however, involves the use of hinges or movable joints. Essentially the hinge (*chiao*[5]) is a pin (*chiao ting*[6]) and two hooks (*kou*[7]), and both these structures were readily available in all antique civilisations.[f] For doors or windows the pin tended to be long and the 'leaves' (the Chinese used the same term *ho yeh*[8]) broad and flat;[g] for links in rods the pin would be short and the hook-tails

---

[a] We have already had in the seismograph (Vol. 3, pp. 628 ff.) an instance of complex link and crank motion, at least if the reconstruction of Wang Chen-To (*1*) be accepted. In any case his discussion of the level of this technique in ancient China is very relevant here.

[b] Beck (1), p. 79; Usher (1), 1st ed. pp. 76, 77. See further pp. 209 ff. below.

[c] Cf. pp. 183, 390 below.   [d] Cf. p. 99 below.

[e] For an appreciation of the role of linkwork in modern machinery see Jones & Horton (1), vol. 1, pp. 391 ff., 418 ff.; vol. 2, pp. 385 ff.; vol. 3, pp. 109 ff., 162 ff., 200 ff., 240 ff.

[f] Sirén (1), vol. 1, pl. 78, gives photographs of a number of Chou specimens.

[g] Modern traditional examples are described in Hommel (1), pp. 292, 300. For ornamented flat bronze plate hinges of early Chou time see White (3), pl. LXXXIX, or Thang Lan (1), pl. 58, fig. 4. The latter album also illustrates some beautiful hollow-casing bronze hinges of the Warring States period

[1] 弩機   [2] 桔槔   [3] 花機   [4] 機   [5] 鉸   [6] 鉸釘   [7] 鉤
[8] 合頁

elongated. Relevant in this connection are the curious bronze hooks on the ends of poles which White (1) described from the Loyang tombs of the −6th century; these seem to have been used for setting up easily dismountable booths or tents.[a] Yuan[b] and Ming[c] books discuss the ancient names for hinges, notably *chin phu*,[1] saying that this was the pin, and that the cylindrical sockets were called *huan niu*[2].[d] One ancient name for the whole hinge was the 'knee-bender', *chhü hsi*[3] or *chhü hsü*.[4]

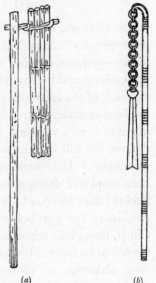

There were prominent uses of links in agriculture and war as well as in textile technology: first the link-flail (cf. below, Sect. 41) called *lien chia*[5],[e] (Fig. 374 *a*) and the war-flail (Fig. 374 *b*); and secondly the component parts of the efficient horse harness (the postillion or chest-trace harness), which though not connected at their junctions by pins (being of leather), nevertheless performed the office of a linkwork system (cf. below, p. 304). This harness was in full use at the beginning of the Han period, that is to say, some ten centuries at least before Europe possessed an efficient harness for horses.

Perhaps more characteristic of European ideas of Chinese civilisation was the collapsible umbrella or parasol (*san*[6],[7]), working by means of the sliding levers still familiar in everyday use. While sun-shades were common in Greek and Roman daily life, and certainly go back to Babylonian times, they were not generally collapsible.[f] We have, however, an indication that the principle of the later collapsible Chinese parasol was used in +21, for in that year Wang Mang had a very large one made as a magic baldachin (*hua kai*[8]) for a ceremonial four-wheeled carriage.[g] The mechanism is said to have been a secret one (*pi chi*[9]), and the +2nd-century commentator, Fu

(a)   *(b)*

Fig. 374. Flails as examples of linkwork and chain connection. (*a*) Farmer's flail from the *Nung Shu* of +1313. (*b*) Iron war-flail from the *Wu Ching Tsung Yao* of +1044. One of the names for the latter, the 'iron crane bird's knee' (*thieh ho hsi*), came to be used as a technical term in +11th-century mechanical engineering for all kinds of combinations of rods and chains in linkwork (cf. p. 461 below).

(pl. 64, fig. 4). Early Han examples are drawn in Kao Chih-Hsi & Liu Lien-Yin (1), p. 651. Certain vessels also had hinges in the Chou and Han. In the museums at Nanking and Canton one can find bronze wine-kettles with hinges excellently made both on the lid and on the birdbeak-shaped spout. Fig. 375 shows an example of a link-hinge from the −10th century.

[a] Thang Lan (1), pl. 63, fig. 3, shows a bronze central canopy holder of Warring States time with free rings surrounding a central boss.

[b] E.g. the *Cho Kêng Lu*.   [c] E.g. the *Shan Thang Ssu Khao* and the *Liu Chhing Jih Chai*.

[d] *Niu* is a knot or button, and the point is that on Chinese garments the button-hole was never *in* the stuff, but formed by looping a silk cord which projected *from* the stuff.

[e] Alternatively, *chia*[10] and *lien chieh*.[11]

[f] Aristophanes, however, has a reference in the *Knights* which might imply collapsibility (Feldhaus (1), col. 945). But it has been suggested that the familiar lever system we use now was a Chinese invention which came to the West later (Feldhaus (2), pp. 45, 46).

[g] *Chhien Han Shu*, ch. 99c, p. 15b, tr. Dubs (2), vol. 3, p. 413; *TPYL*, ch. 702, p. 7a (Pfizmaier (91), p. 288).

[1] 金鋪	[2] 環紐	[3] 屈膝	[4] 屈戌	[5] 連枷	[6] 傘
[7] 繖	[8] 華蓋	[9] 祕機	[10] 架	[11] 捷稽	

PLATE CXXXV

Fig. 375. Link-hinge on a bronze wine-kettle (*ho*) of the − 10th century, self-dated by an inscription naming the Chou High King Mu (photo. CPCRA and BCFA, cf. Thang Lan (*1*), pl. 28). Height 11 in., diameter of mouth 7½ in. From Phu-tu Tshun, Shensi, cf. Watson (*1*), p. 24, pl. 69.

PLATE CXXXVI

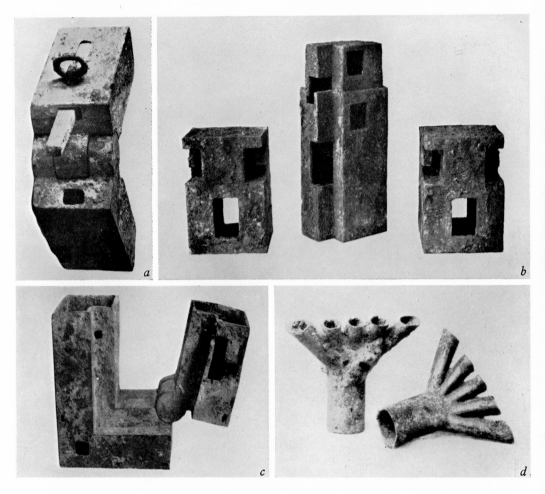

Fig. 376. Bronze castings of complex design from the −6th century (Chou period) excavated at Loyang (White, 1). (*a*) Bronze socketed hinge with locking slide-bolt (pl. 13); (*b*) bronze rebated socketed couplers with holes for tenons and stanchion pins, some using the principle of the bayonet catch (pl. 34); (*c*) bronze rebated hinged corner-fitting socketed to hold wooden members (pl. 11). Cast bronze tubular connecting-joints for assembling the rods or canes of ancient Chinese carriage canopies (Thang Lan, 1); (*d*) two six-way branching holders of the −4th century (pl. 64, fig. 2).

Chhien, adds that his umbrellas all had bendable joints enabling them to be extended or retracted.[a] Collapsible umbrella stays of Wang Mang's time, or shortly after it, have actually been recovered from the tomb of Wang Kuang at Korean Lo-Lang, and are illustrated by Harada & Komai.[b] But the system must go back much earlier, for similar objects of Chou date (−6th century) from Loyang are figured by White.[c] Other pieces of bronze which White described[d] are beautiful castings of very complex design (Fig. 376 b). He calls them 'socketed couplers', though at first sight they look like locks. Bulling (4, 7, 8) believes that some of the designs on the backs of Chou and Han bronze mirrors can only be understood as flattened representations of ceremonial umbrella tops,[e] but her view has not won much support. In any case, after all this, one is hardly surprised, though the discovery is pleasing, to find collapsible umbrellas exactly like those of modern China in the woodcuts of a book printed about +1270 which deals with divination but contains many scenes of daily life.[f]

The question of origins is more interesting than it might appear, for towards the end of the Later Han period, about +160, the folding chair or stool became popular in China.[g] As we shall see later on, it was known first as the *hu chhuang* [1] (barbarian bed), and certainly came from the West, probably from Greek Bactria. But the evidence just given shows that it cannot have been responsible for the appearance of pivoted rods and linkwork in Chinese technology.

Mastery of the art of collapsibility achieved some remarkable successes in the Yuan period which are referred to in a curious entry in the *Shan Chü Hsin Hua* (New Discourses from the Mountain Cabin). Yang Yü tells us that

a lacquer worker of Suchow named Wang, in the Chih-Chêng reign-period (+1341 onwards) made a boat of ox-hide, covered inside and out with lacquer. It was dismountable into parts, and was brought to Shangtu (the Manchurian summer capital of the Yuan emperors), where

---

[a] *Chhi kang chieh yu chhü hsi, kho shang hsia chhü shen yeh.* [2]

[b] (1), vol. 2, pls. XIX, XX.

[c] (1), pls. XIV, XVII. This gives a little colour to the old tradition that the inventor of the collapsible umbrella was Yün shih,[3] the semi-legendary wife of the celebrated −5th-century artisan Kungshu Phan (cf. Li Nien (27), p. 2).

[d] (1), pls. XI, XIII, XXXIV. Some of these have built-in slide-bolts which have to be released before the hinge can open. This is perhaps the place to mention the cast bronze tubular connecting-joints used for assembling the rods or canes of ancient Chinese carriage canopies (cf. Fig. 376 c). At the Chêngchow Archaeological Institute in 1958 I greatly admired some six-way examples of these from the Chhu princely tomb of the Warring States period recently excavated near Hsiao-liu-chuang. Four- and five-way connections for as many ribs are even more numerous in museums. A special paper has been devoted to these interesting objects by Chhang Wên-Chai (2). By the San Kuo period (+3rd century) they were being made of iron, a rather intricate casting job (see Anon. 16). Cf. Anon. (17), pl. 27. The simpler forms of these tubular junctions (three- and many four-way cruciform pieces) appear also in the steppe cultures, as may be seen from the remarkable work of Gallus & Horváth (1) on the pre-Scythic peoples of Hungary, e.g. pls. LII, LIV, LIX, LX. For a similar Hallstatt example see Kossack (1), fig. 3. The junctions in these cultures may well have been parts of horse-gear rather than chariots or carriages.

[e] Naturally a cosmic significance of ceremonial umbrellas is not far to seek; in Sect. 20d the ancient and widely held theory of the heavenly dome (Kai Thien) was described. If the emperor corresponded to the pole-star, no ornament could have been more appropriate for him than a symbolical umbrella. In the *Chou Li* this symbolism is explicit (cf. Biot (1), vol. 2, pp. 475, 488).

[f] *Yen Chhin Tou Shu San Shih Hsiang Shu* (see p. 143 below), ch. 2, pp. 5b, 12a, 14a, 24b and 29b. Cf. Yang Jen-Khai & Tung Yen-Ming (1), vol. 1, pls. 4, 7, 11.   [g] See Stone (1); Ecke (2).

[1] 胡床   [2] 其杠皆有屈膝可上下屈伸也   [3] 雲氏

it was rowed about on the Luan River. No one in Shangtu had ever seen such a thing before, and everyone admired it.

Artisan Wang also constructed an armillary sphere by imperial command. This was collapsible, too, which was very convenient for storage. It was an instrument the skilful construction of which surpassed human imagination. One could really call this man talented. Now (+1360) he is Intendant in charge of one of the (Imperial) Workshops.[a]

Collapsible boats (*chê tieh chhuan*[1]) are heard of again in the +17th century. Liu Hsien-Thing described about +1690 an artisan, Chu Ya-Ling,[2] making such boats with masts and sails yet capable of being folded up and packed in bags for transport.[b] The subject may be followed further in the *Thu Shu Chi Chhêng*.[c]

However original the boats of Artisan Wang may have been, there had been successes considerably earlier in the invention of collapsible or dissectable astronomical apparatus.[d] In the *Yü Hai* we read:[e]

On the 8th day of the sixth month of the 7th year of the Shao-Hsing reign-period (+1137) the commanding general in Szechuan presented to the emperor a 'New Pattern Quick-Method Hemispherical Sky-Dome Plan' (*Chieh fa kai-thien thu hsin shih*[3]). This was the invention of Chang Ta-Chieh[4] of Tzu-chow, the hermit of Tshui-wei-tung (in the mountains), who had made use of old designs of the Thang. He also presented mysterious books from Tshui-wei-tung (namely, the) *Pao Chu Ssu Thien Yü Hsia Pi Shu*[5] (Secret Jade-Box Astronomical Treatise of the Precious Axis), and a *Chin Chien Yao Chüeh*[6] (Essential Principles of the Golden Key). The emperor ordered these things to be (transported) across the water and sent to the temporary court (at Hangchow).

[Comm.][f] The *Jih Li*[7] (Solar Calendar Treatise) records that (Chang) Ta-Chieh, using old Thang designs, invented a 'New Pattern Quick-Method Hemispherical Sky-Dome Plan', such that one could sit down and observe the Tao of the Heavens. It was prepared for the emperor's study throughout one night, and for the (aides-de-)camp to check it over by observation. Thus there was no fatigue of looking upwards because the instrument was placed on a table. Yet if one looked upwards one could see all the heavenly phenomena, though so far away, looking exactly like (the representations) which were (immediately) in front of one's eye. Now 'Quick-Method Hemispherical Sky-Dome Maps' (*Chieh fa kai-thien hua thu*[8]) and a 'Four Cardinal Points Horizon Circle' (*Ssu chêng ti kuei*[9]), on wooden pieces, large and small, with four faces, have also been made. These too were presented to the emperor, who ordered them to be transported across the water to the temporary capital (at Hangchow), and to be handed over to the Bureau of Astronomy for presentation in due form.

The description is not very clear but it indicates in all probability the invention of some kind of hemispherical star maps which could be folded up or dismounted;

---

[a] Pp. 39b ff., tr. H. Franke (2), no. 107, eng. auct.   [b] In the *Kuang-Yang Tsa Chi*.

[c] *Khao kung tien*, ch. 165. Yang Yü's passage is repeated in the *Suchow Fu Chih*[10] (Local History and Topography of Suchow), from which Liu Hsien-Chou (1) quotes it.

[d] The introduction of 'jigsaw maps' in the +5th century will be remembered (Vol. 3, p. 582).

[e] Ch. 1, pp. 35a, b, tr. auct. The abridged version in *Chhou Jen Chuan*, 2nd addendum, ch. 5 (pp. 55 ff.) seems rather garbled.

[f] Chu Pho,[11] whose probable *floruit* is +1150, was the author of the *Jih Li* here quoted by Wang Ying-Lin.

[1] 摺疊船	[2] 朱雅零	[3] 捷法蓋天圖新式	[4] 張大檝
[5] 寶軸司天玉匣秘書		[6] 金鍵要訣	[7] 日曆
[8] 捷法蓋天畫圖		[9] 四正地規	[10] 蘇州府志 [11] 朱朴

portable equipment, in fact, suitable for the imperial impedimenta when the emperor
was in the field with his armies. Chang Ta-Chieh's apparatus may perhaps have been
a dissectable inverted hemispherical mirror marked with the celestial coordinates. In
any case the principle of collapsibility seems to have been in the foreground.

(2) WHEELS AND GEAR-WHEELS, PEDALS AND PADDLES

With the ultimate origin of the wheel we are not directly concerned, for by the time
at which we can begin to speak of the history of technology in China, the Shang period
(c. −1500), the chariot with its wheels had already been introduced from the Meso-

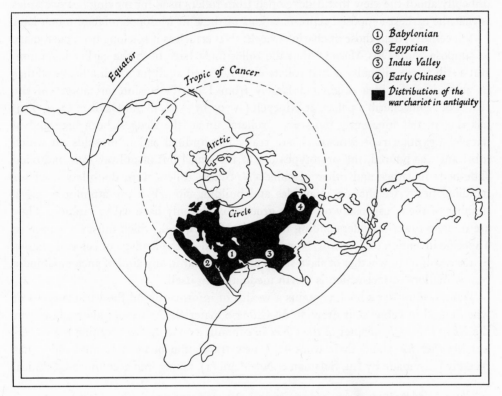

Fig. 377. Polar projection illustrating the distribution of the war-chariot in
antiquity (Bishop, 2).

potamian area. Fig. 377, from Bishop (2), shows the distribution of the war-chariot in
antiquity. There seems to be some doubt as to its point of origin; the traditional view[a]
was that the wheel appeared first in Sumeria about −3000, but Marshall (2) found
models at Harappa in the Mohenjo-daro civilisation (Indus Valley) of similar dating.[b]
The most ancient Sumerian and Chaldean chariots were strange saddle-shaped

[a] Breasted (1); des Noëttes (1); Furon (1).
[b] The earliest examples of spoked wheels belong to northern Mesopotamia, and are datable about
−2000.

structures for 'riding' borne on two wheels, the platform chariot not appearing till after − 2500.[a] This was the form in which it spread both to Egypt and to Shang China, and we can guess something of its nature there from the characters for it on the oracle-bones (cf. below, p. 246). Our ideas about the first origin of the wheel, however, are bound to be somewhat disturbed by the discovery in Mexico of toys with wheels,[b] though it seems quite certain that wheeled vehicles were not used in any Central Amerindian civilisation. If this could be further probed, it might lead to some striking conclusions about social barriers to the application of inventions.

The genetic origin of the wheel has given rise to much discussion[c] and a variety of archaeological evidence has been brought forward, but the problem is not solved. One difficulty about the view that it developed from rollers used for moving heavy statues is that the earliest supposed representations of these in Mesopotamia are much later (− 8th century) than those of chariot-wheels. Nevertheless it has long been customary to suppose, with Otis Mason,[d] that the roller must have been the earlier invention. But a still worse difficulty is that rollers were not used at all for moving heavy statues in antiquity; the idea originated largely from Layard's misinterpretation[e] of the reliefs of Sennacherib's palace at Nineveh (− 705 to − 681).[f] As Davison (3, 4, 7, 9) has shown, the apparently transverse 'rollers' under the sledges, both here and in parallel Egyptian representations, are really longitudinal sliders or skids on which lubricant was poured, the prototypes not of the wheel but of railway rails and oiled slide-beds in lathes and other machinery.[g] Single rollers were doubtless used for handling heavy weights on quaysides and building-sites, but any attempt to use a number of them, especially on rough ground, could only have led to binding. They are no help towards understanding the origin of the wheel, which must be viewed as a specific invention on its own. Nor, incidentally, does the haulage of colossal statues by masses of men, whether of slave status or not, appear in any kind of ancient Chinese representation—a fact which is worth meditating in itself.

Though Forestier's book contains a wealth of information on the various types of wheels used in vehicles, it drew on no Chinese material. As we saw above, however, the *Khao Kung Chi* chapter of the *Chou Li* contains a good deal concerning the wheel-wrights (*lun jen*[1]) and their work.[h] A new translation of the relevant section has recently been made by Lu, Salaman & Needham (1), taking into account the technical

---

[a] But see Childe (10).

[b] Ekholm (1). I have myself seen one of these specimens in the National Museum at Mexico City. In the absence of the horse the vehicles would have required human motive power; in Mexico at any rate.

[c] Forestier (1); Frémont (12); des Noëttes (1); Usher (1), etc.

[d] (2), p. 63.   [e] (1), pp. 24, 26.

[f] Brit. Mus. sculptures 124820, 124822, 124823. On slabs 66 and 67 the 'rollers' still retain stumpy branches untrimmed. Slabs 63 and 64 show the skids end to end under the sledges.

[g] In (4) he computed the number of men required in the al-Bersheh fresco on the assumption of oiled skids, obtaining 179. The number actually shown is 172. On the history of lubrication as related to sliding friction Forbes (21), pp. 159ff., 169ff., may be consulted.

[h] Ch. 11, pp. 7 aff. The classical translation was that of Biot (1), vol. 2, pp. 466ff. from *Chhin Ting Chou Kuan I Su*, ch. 40, pp. 21 bff.

[1] 輪人

studies by Chhing scholars to which reference has already been made.[a] By the Warring States period (−4th century) primitive solid wheels had long given place to very elegantly constructed wheels with hubs (*ku*[1]), spokes (*fu*[2]) and rims composed of felloes (*ya*[3], *jou*[4]).[b] The trueness of the wheel was to be such that it resembled a hanging curtain (*mi*[5]), curving downwards with a beautiful smoothness. In the Han time elm-wood (*Ulmus* spp., *yü mu*[6]) was used for the hubs,[c] rose-wood (*Dalbergia* spp., *than mu*[7])[d] for the spokes, and oak (*Quercus* spp., *chiang mu*[8]) for the felloes.[e] A curious process of unilateral drying (*huo yang*[9]) of the wood for the hubs is given. The hub was drilled through to form an empty space (*sou*[10]) into which the tapering axle was fitted, and between the two a tapering bronze bearing (*chin*[11])[f] was inserted, the exterior being covered by a leather cap (*thao*[12]) to retain lubricant. The thickness of the spokes and the depth of the holes and mortices (*tso*[13]) to receive their tongues (*chü*[14]) and tenons (*tzu*[15]) in hub and rim respectively were very carefully regulated, neither too much nor too little, and the legs (*chiao*[16]) of the spokes were made thinner towards the rim as a measure of 'streamlining' against deep mud.[g] The number of spokes varied widely. A famous text of perhaps the −4th century speaks of chariot-wheels having thirty spokes, and excavations made in 1952 indeed unearthed remains of chariots of this period in which many of the wheels do have this number of spokes.[h] The testing of the completed wheels was elaborate, including the use of geometrical instruments, flotation, weighing, and the measuring of the empty spaces in the assembly by millet grains.[i] The interest of these descriptions[j] is perhaps increased when we reflect that though carriage-wheels were those which most interested the scholar-officials, and therefore those which found a prominent place in the *Chou Li*, the level of Chhi and Han craftsmanship revealed is such that gear-wheels (e.g. for water-mills) would clearly have presented little difficulty to the artisans.[k]

[a] P. 12 above.

[b] The more modern term for the rim of a wheel is *kuo*[17] (derivative from 'suburb'). Since the primary meaning of *ya* is 'tooth', the term probably derived from the gripping or biting of the ground, as in ruts, by the wheel rims. Yet with such a term the transition in thought to actual peripheral teeth, as in gear-wheels, would have been particularly easy. As we shall see, the first traces of these appear during the second half of the Chou period.

[c] As in Europe. In the West it was also customary to use different woods for different wheel components.

[d] See the special study of Schafer (8), and for the Middle East Gershevitch (1).

[e] It is curious that the more springy ash, used for felloes during the last 1500 years in Europe, was not so employed in China, though various kinds of ash-trees are not lacking.

[f] Bronze is assumed in the archaising *Chou Li*, but we know from the *Chi Chiu Phien* of −40 (ch. 3, p. 4a) that bearings of iron were then already long standard.

[g] Cf. p. 259 below concerning the wheels of wheelbarrows.

[h] The text is of course the *Tao Tê Ching*, ch. 11. Duyvendak (18), p. 40, noted the correspondence at once from the photographs of Hsia Nai (1).

[i] The use of this interesting statistical volumetric method in acoustics has already been described (Vol. 4, pt. 1, pp. 199ff. above).

[j] They may be compared with what Ginzrot (1) collected from Greek and Roman sources, in a work of technological history quite extraordinary for the time it was written (1817). Cf. also Mahr (1).

[k] For details of the wooden gear-wheels used in various kinds of traditional Chinese water-raising machinery see pp. 339ff. below. Cf. also p. 11 above.

[1] 轂	[2] 輻	[3] 牙	[4] 輮	[5] 幎	[6] 榆木	[7] 檀木
[8] 櫔木	[9] 火養	[10] 藪	[11] 金	[12] 幬	[13] 鑿	[14] 倨
[15] 笛	[16] 骹	[17] 郭				

'The final development of the wheelwright's art', writes Jope,[a] 'was the construction of a wheel not in one plane but as a flat cone.' This is the technique known as 'dishing'. Such wheels give strength against the sideways thrusts occasioned by the transport of heavy loads over erratic or rutted surfaces. They appear in European illustrations from the +15th century onwards, and one is shown under construction in Jost Amman's woodcut of a wheelwright's shop in +1568. A good sketch of traditional English dished cartwheels, the form of which is often unperceived because so familiar, occurs in Sturt (1) and is here reproduced (Fig. 378). Lu, Salaman &

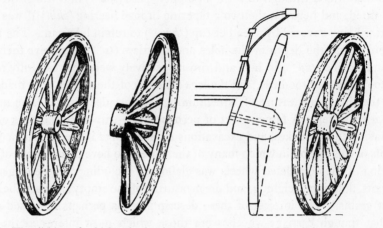

Fig. 378. The dishing of vehicle-wheels: drawings from traditional English examples (Sturt, 1).

Needham (1), however, were able to establish that dishing, far from being a +16th-century occidental perfection, was systematically employed by the wheelwrights of the late Chou and Han.[b] This is demonstrated by several passages in the *Khao Kung Chi*.[c] The text itself speaks of the 'cake-like convexity' (*ping*[1]) of the wheel. Chêng Chung's +1st-century commentary explains the origin of this expression and gives the contemporary technical term for dishing, *lun pi*.[2] The later commentaries expatiate on how this shape was brought about,[d] and Fang Pao,[3] the chief editor of the imperial edition of +1748, draws particular attention to the fact that the Han wheelwrights had the practice and the term. His commentary defines *pi*[2] as shaped like a vessel (*tsêng*[4])

---

[a] (1), p. 552.

[b] Biot (1), vol. 2, p. 468, realised that the wheel was said to be 'légèrement bombé en dedans', but he did not connect this with the technique of dishing as practised in post-Renaissance Europe.

[c] *Chou Li*, ch. 11, pp. 8a, 11b.

[d] The commentators differed a good deal in their explanations. Some thought that the spokes tapered on one side only, others that they were inserted slanting into either hub or felloes. The standard European practice of modern times was to cut the slightly tapering spoke so that it formed a small angle both with its tongue entering the rim spoke-hole and with its tenon entering the hub mortice. Thus while these are right-angle insertions the spoke itself is seen in a cross-section of the entire wheel to connect centre and periphery obliquely. This may also have been done in some ancient Chinese wheels. But since the text of the *Khao Kung Chi* speaks (p. 11a) of treating the spokes (as well as the felloes, p. 12a) with heat, or steam (*jou*[5, 6]), they were almost certainly made to curve, so that both insertions could be at right angles. Cf. the reconstruction in Hayashi Minao (1), p. 217.

[1] 綆   [2] 輪箄   [3] 方苞   [4] 甑   [5] 揉   [6] 樏

PLATE CXXXVII

Fig. 379. The park of vehicles of the Warring States period (−4th or −3rd century) discovered in 1950 during the excavations of the royal tombs at Hui-hsien in Shantung (Anon. (4), pl. 25, fig. 1).

Fig. 380. Detail of two chariots in the Hui-hsien park, showing how much was recoverable from the compacted soil (Anon. (4), pl. 29, fig. 4).

for steaming rice, the bottom of which was made of bamboo, with rounded tapering sides.[a]

In our own time archaeological excavations have brought abundant confirmation of this textual evidence, showing moreover that in Warring States and Han times there was more than one way of producing wheel dish. The investigations of the royal tombs at Hui-hsien in northern Honan, now fully published,[b] revealed a whole park of nineteen vehicles from the −4th or −3rd century.[c] Although the wooden parts had rotted away they had left impressions in the concreted soil so clearly that it was possible to dissect them out down to comparatively small details (Figs. 379 and 380). The reconstruction in this report (Fig. 381) by Hsia Nai, Kuo Pao-Chün *et al.* shows quite straight spokes, and the archaeologists believe that they were all inserted into the hubs in a slanting manner (cf. the models in Figs. 382 and 383). Very different wheels, however, were found in models of Former Han chariots excavated from −1st-century tombs at Chhangsha in Hunan (Fig. 384).[d] Here the curvature of the spokes agrees with the description of the *Chou Li*, but it is so arranged, in shape like a Chinese farmer's hat (*li*[1]), that the concavity of the dish is inward, not outward. The scale drawing in Fig. 386 shows this clearly. Still more remarkable, this inward dishing has been perpetuated in traditional Chinese vehicle wheels down to the present time (Fig. 385).[e]

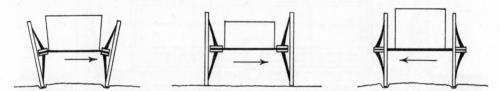

In fact it does not matter much whether the dishing is outwards or inwards. As the adjacent diagram shows, the post-Renaissance type of Western dished wheel, concave outwards and mounted on a downward-pointing axle-bearing on the right of a cart,

[a] This definition may be found already in the *Chi Chiu Phien* dictionary of about −40 (ch. 3, p. 6a). Wang Ying-Lin's Sung commentary compares the shape also to a flat conical fish-basket.

[b] Anon. (4), pp. 47ff., pls. 24 to 31.

[c] Chariot-wheels of the Shang period have not been recovered in a state sufficiently perfect to allow of a conclusion as to whether they were dished or not. But it is highly improbable that they were, for quite recently the excavations of the princely tombs of the State of Northern Kuo (Anon. 27) have brought to light three more parks of chariots. Kuo[2] was extinguished by Chin in −654 (cf. Vol. 1, p. 94, there spelt Kuai), so that these remains date from the first half of the −7th century. None of the wheels shows any sign of dishing, though it is interesting (cf. p. 249) that the hubs are already very elongated. We must surely conclude therefore that the invention of 'dish' was yet another of the technological advances of the Warring States period. Future discoveries may be expected to narrow its date. These need not be expected in China's traditional territory alone, for the remarkable vehicle wheels of the −5th century discovered in the Pazirik tumuli (see Rudenko, 1) may well have originated in China. As the High Altai was not a very suitable region for the use of such carriages, Rudenko suspects that they may have been presents accompanying Chinese princesses married to nomad leaders.

[d] Anon. (11), pp. 139ff., pls. 99 to 102. The essential illustrations from both these sources are reproduced in Lu, Salaman & Needham (1, 2).

[e] How and when the Chinese invention of dishing spread to the rest of the world remains a matter for further investigation.

[1] 笠   [2] 虢

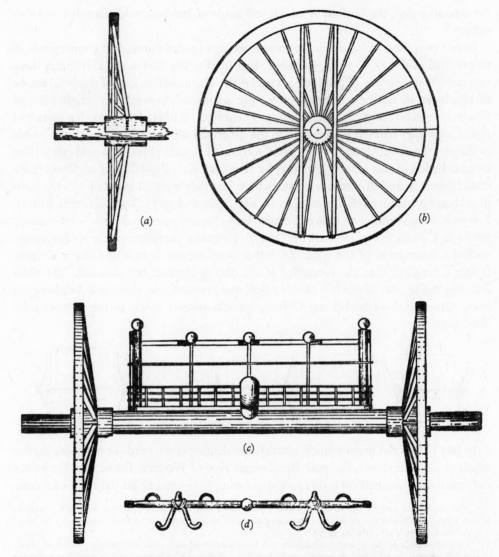

Fig. 381. Reconstruction of the Hui-hsien chariot type, scale drawings (Anon. 4). (a) Cross-section of the wheel, showing dish; (b) elevation of the wheel, showing quasi-diametrical struts; (c) end-on view of assembled chariot; (d) cross-bar with 'yokes'. Scale 2·26 to 50 cm.

was particularly strong against jolts with force directed to the right, since these served only to drive the spokes more securely into the felloes. The same applies to wheels of the Hui-hsien type. But with a wheel of the Chhangsha type on the right of the vehicle, strength was provided against jolts with force directed to the left, for exactly the same reason. In each case, whichever the direction of the strain, the stronger wheel tends to protect the weaker. Of course, for an overhanging body the outward concavity of the Hui-hsien or European type was more convenient than the inward concavity of the Chhangsha type, and it also had the advantage of throwing mud clear.

PLATE CXXXVIII

Fig. 382. Reconstructed model of the Hui-hsien chariot type, viewed from the front
of the right side (Anon. *4*).

Fig. 383. Hui-hsien chariot model ,viewed from above to show the dish of the wheels (Anon. *4*).

PLATE CXXXIX

Fig. 384. Repaired and assembled model chariot from a tomb of the Former Han period
(−1st century) at Chhangsha (Anon. (*11*), pl. 99, fig. 2).

Fig. 385. Two nail-studded cart-wheels with outward dish doing duty as fly-wheels for a manually
operated paternoster pump (orig. photo. Agricultural Machinery Exhibition, Peking, 1958).

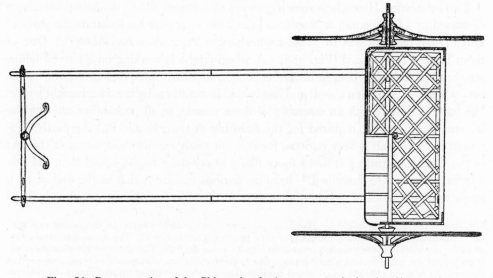

Fig. 386. Reconstruction of the Chhangsha chariot type; a scale drawing (Anon. *11*),
showing the convex or outward dish. Scale 1 to 10 cm.

In all ancient and medieval Chinese vehicles, so far as we know, the axle-bearings
were horizontal. This was also the case very late in Europe, as we find from the
drawings of under-carriages in plates attached to the *Encyclopédie* of Diderot (3).[a]
Such wheels must have had a less enduring life under load than those in which the
spokes meet the ground at right-angles, as in the later European types and the very
elegant Chhangsha design.[b] Indeed some of the Hui-hsien wheels seemed almost to
admit a tendency to weakness, for they were equipped with a curious feature about
which the *Chou Li* is quite silent, namely a pair of quasi-diametral struts (*chia fu*[1])
running from rim to rim on each side of the hub (cf. Fig. 381 *b*). These were almost
certainly inserted into separate felloes, thus adding much to the strength of the wheel,
holding it in dish, and indeed offering a very early example of a kind of truss construction. One finds no further evidence of these structures, either textual or epigraphic,
in China itself, but they are to be met with today in the country carts of Cambodia
(Fig. 387), if in somewhat degenerate form.[c]

[a] See the entry under 'charron' in vol. 3 of the *Planches* (vol. 20 of the whole set). Pl. 3 shows a
timber-tug, pl. 5 a hay-cart, and pl. 6 a small dung-cart; all are on the same principle. Some find it
hard to believe that the artist did not err in drawing these axle-bearings horizontal.

[b] One must remember that the Hui-hsien wheels are some three centuries earlier than the Chhangsha
ones. Moreover, Hui-hsien was in the feudal State of Wei,[2] not the advanced technical culture of Chhi,
whence came in all probability the specifications of the *Khao Kung Chi*.

[c] Cf. Groslier (1), pp. 98ff. These parallel quasi-diametral struts would seem to be connected
with the curious 'six-spoked' wheels ⊕ which still continued to be made in various places until
modern times, and may well derive from extremely primitive wheel forms of the −2nd millennium.
The 'six-spoked' wheels have one diametral component and two quasi-diametral ones. A fine
example from Mercurago in Italy, belonging to the late bronze age (*c.* −1200, contemporary with
the Shang period), has been figured by Munro (1), p. 208, and Childe (11), p. 214; another, rougher,
type, used at any rate until recently by Chinese wheelwrights in Sinkiang, was photographed by

[1] 夾輔   [2] 魏

Two interesting Han reliefs showing scenes in a wheelwright's workshop have been discussed by Lu, Salaman & Needham (2).[a] They may perhaps illustrate the story of Duke Huan of Chhi and the Taoist wheelwright Pien (Pien *lun chiang*[1]).[b] One of them is here reproduced (Fig. 388). A wheelwright is working on a curved felloe, seemingly chiselling a hole in it. Other felloes, of which three or four seem to make up a whole rim, are lying about, and one is held in readiness by the wheelwright's wife. His mate is busied with an assembly of three vessels, in all probability filtering the lacquer, paint or glue required for the finishing of the wheel.[c] But the partly constructed wheel itself is very curious, for it is not easily explicable in terms of the text of the *Chou Li*. Indeed it looks more like a heavy-duty wagon-wheel than the fine chariot-wheels there described.[d] Its most curious feature is that at the end of each spoke there is a block of wood,[e] and as some of the felloes of the rim are already fitted

Hildebrand (1) and reproduced by Harrison (3), p. 72. 'Six-spoked' wheels are frequently met with on −4th-century Greek vase-paintings (Jope (1), pp. 545, 549; Lorimer). The quasi-diametral struts or spokes are morticed into the main cross-strut which carries the hub. This construction clearly derives from that of the most ancient wheels known, the 'tripartite disc' or solid spokeless type, composed of three planks of wood fastened together and cut into discoidal shape, with or without lunate openings on each side of the hub. The three planks are fixed together either by external battens (as in the Hallstatt wheel, *c.* −900, found at Buchau in Germany) or by internal rods or pins running transversely right through the breadth of the planks (as in a −1st-millennium wheel from a Scottish peat-bog near Doune). Both methods are described in detail by Piggott (1), who adds mention of contemporary traditional vehicles still using such solid wheels. For us the most interesting of these, perhaps, are the solid five-plank wheels with quasi-diametral battens carrying the carts which collect marsh grass on the shores of the Poyang Lake in Chiangsi province (Hommel (1), p. 323). As in the case of the Sinkiang carts the axle turns with the wheels. At the other end of the Old World, Portuguese Estremadura still uses similar wheels on ox-carts, as I saw and photographed in 1960. Another obvious development of the quasi-diametral struts was to have four of them ⊕ clasping the hub in its place. This occurs in Egyptian chain-pump gear-wheels (cf. Matchoss & Kutzbach (1), p. 5) and in the broad-tyred drum-wheels of the fish carts of Aveiro in northern Portugal, which go on wet sand. One may thus trace a continuous evolution from the oldest three-plank disc-wheels of Mesopotamian antiquity which date from the end of the −4th or the beginning of the −3rd millennium, and conclude that the invention of radial spokes (occurring just after −2000), no less than that of 'dish', was a major turning-point in the development of the wheelwright's craft. Indeed the hub itself had to be invented, first as a stabiliser and then as a home for spokes. It is interesting that the hieroglyphs of the Mohenjo-daro civilisation include a pictogram of a six-spoked wheel (Wheeler (5), pl. XXIII; McKay (1), pl. XVII). We are indebted to Prof. J. D. Bernal, F.R.S., who was much impressed by this fact, for arousing our interest in 'six-spoked' wheels.

   [a] One was published as the 'Liu village stele' by Chavannes (11), pl. XLIII, seventy years ago; the other is in the Shantung Provincial Museum at Chinan. Both were found near Chia-hsiang in that province, a place which must have been in the old feudal State of Lu.

   [b] Cf. Vol. 2, p. 122.

   [c] Filtration of glue is described in detail by the early +6th-century agricultural encyclopaedia *Chhi Min Yao Shu*, ch. 90; cf. Shih Shêng-Han (1), p. 97. Closely similar arrangements of superimposed vessels on stands clearly in use for filtration, notably in kitchen scenes, may frequently be seen in the reliefs reproduced by Chavannes (9).

   [d] Especially so in the Liu village stele.

   [e] These blocks on the ends of the spokes are so extraordinary that after first becoming acquainted with the Liu village stele I long believed that the subject represented a piece of textile machinery. For it is indeed true that in some types of Chinese spinning wheel (*fang chhê*[2]) the crank- or pedal-operated driving-wheel has no rim at all, the belt being carried on grooved blocks at the ends of the spokes. It then passes over a number of small spindles, often set in a semi-lunar frame, transmitting to them its rotary motion. On these devices see pp. 91 and 104 below, and later Sect. 31. Now that the clearer Chinan relief has come to supplement the Liu village stele, however, there can be no doubt as to what it represents. We may still wonder whether there was any borrowing between the vehicular and the stationary machines, and if so, in which direction.

   [1] 扁輪匠   [2] 紡車

PLATE CXL

Fig. 387. Quasi-diametral struts and outboard axle-bearings on the wheels of a
Cambodian farm-cart (orig. photo., Siem-reap, 1958).

PLATE CXLI

Fig. 388. A wheelwright's shop of the Han period; a stone relief from Chia-hsiang, near Yen-chou (Shantung), found in 1954. Now in the Chinan Museum. To the left the wheelwright is working on a curved felloe, while another is held by his wife. To the right an assistant is filtering lacquer, paint or glue, with one of the feudal gentry standing behind him. Perhaps the relief illustrates the story of Duke Huan of Chhi and the Taoist wheelwright Pien (cf. Vol. 2, p. 122). Lu, Salaman & Needham (2).

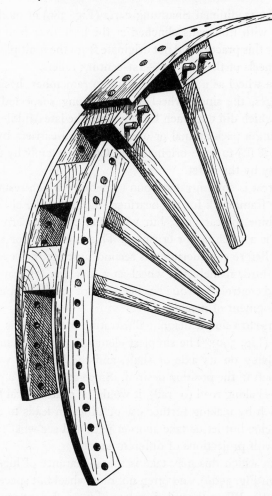

Fig. 389. Conjectural reconstruction of a Han heavy-duty composite cart-wheel.
Lu, Salaman & Needham (2).

on without overlap, they would seem to be curving plates hiding the blocks from the side, indeed what might be called 'curtain-felloes'. With a counterpart on the other side, the whole would then be covered by the strakes of a wooden (or iron) tyre, the intervening spaces being filled in with other blocks, and every component thoroughly pinned together (Fig. 389). So far it has not been possible to find confirmatory textual evidence for the building of such highly composite wheels in early medieval China,[a]

[a] Two hints may be mentioned. In describing Han imperial vehicles, the *Hou Han Shu* (ch. 39, p. 7a), says that the great hunting car has 'double felloes' (*chhung wang*[1]). And a little earlier the same structure has been ascribed, using a different word (*chhung ya*[2]), to four other palace carts or wagons. This would refer to the +1st and +2nd centuries. Secondly, descriptions of the four-wheeled wagons found in tumulus no. 5 at Pazirik in the Altai, which may have been gifts from −5th-century China, state that the felloes of the rims consisted of two halves fastened together longitudinally with wooden pins (Rudenko, 1). This is suspiciously near 'curtain-felloes'.

[1] 重輞   [2] 重牙

but the wheels of the traditional Shantung carts (Fig. 386) of modern times are so excessively studded with iron nails clinched at the back over iron washers that one may wonder whether this practice did not originate from the multiple pinning required by the composite wheels of these + or − 1st-century reliefs.

In thinking of the wheel as a carrier we have to remember, besides the two- and four-wheeled carriages, the single-wheel barrow. Long suspected to have been a Chinese invention, which did not reach Europe until the late Middle Ages, this simple expedient of replacing a pack-animal or one of two hod-carriers by a wheel will be shown below (pp. 258 ff.) to have originated in China certainly by the +3rd century and in all probability by the +1st.

When man's interest is concentrated upon something which rests on an axle, supported by a suitable framework for the bearings (the fu^1 of the old Chinese chariot), the wheel is a machine for carrying; but when the chief concern is with what lies below it, the wheel is a machine for crushing. This is evident for all stages between wheels and rollers. Before considering this second function, however, it will be convenient to study a third, namely the wheel as a machine for transmitting rotatory energy. Mastery and control of torque has been one of the most fundamental features in the gradual development of machinery.

Let us have recourse to a simple diagram illustrating the variations and combinations of axles and wheels (Fig. 390). The simplest elements in the assembly are shown in (a), the wheel revolving on its axle or shaft, and the bearings, whatever they are, which secure that shaft in the position desired. Since adhesion is a principal factor in the passage of a wheel along road (or rail), it would be quite logical to pursue the line of development which by making further use of adhesion leads to all pulleys, windlasses and driving belts, but let us take another path and see what happens when the wheel is furnished with projections of different kinds.

The simplest form which this may take is the appearance of lugs, often not more than four or six (as in Fig. 390b), and often not on a wheel but spaced laterally on the shaft itself (Fig. 390b'). When the axle is rotated by a sufficient source of power, the lugs will alternately depress or raise and then release a set of levers or rods conveniently placed[a]—if their heads are weighted, this is the mechanical pestle, trip-hammer (tui^2) or stamp-mill. Although there seems to be no evidence of this in Europe before the early +12th century (there is a good picture of a vertical stamp-mill in the MS. of the anonymous Hussite engineer),[b] the trip-hammer was both widespread and ancient in China, and water-power was applied to it already in the Han. It seems clearly one

---

[a] This was one of the devices known to the Alexandrian mechanics, as e.g. in the 'windmill' organ-blowing air-pump in Heron's *Pneumatica*; cf. Woodcroft (1), p. 108, Usher (1), 1st ed. p. 92, 2nd ed. p. 140. It also occurs in the puppet-theatre of Heron to activate figures. Usher remarks that these devices were here used for producing motion only, not power; but in Chinese antiquity the trip-hammer was primarily used for practical purposes as a powered engine. Cf. pp. 381, 493.

[b] Sarton (1), vol. 3, p. 1550; Feldhaus (1), col. 915; Berthelot (4); Usher (1), 1st ed. p. 93, 2nd ed. p. 140. The Hussite shows it for human manual motive power (Fig. 619). See further, pp. 113, 394 ff. below.

¹ 轈   ² 碓

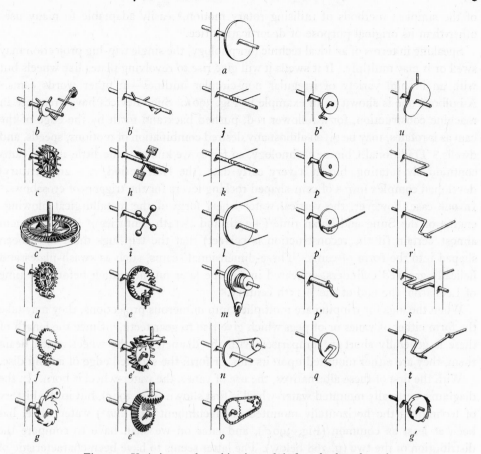

Fig. 390. Variations and combinations of shafts, wheels and cranks.

a	wheel, shaft and bearing
b	lugs on wheel
b'	lugs on shaft
b"	handle mounted on lug, forming crank
c	vanes in vertical water-wheel
c'	vanes in horizontal water-wheel
d	flat teeth on two enmeshing gear-wheels
e	right-angle gearing; pin-wheel and pin-drum or lantern-wheel
e'	right-angle gearing; enmeshing peg teeth
e"	right-angle gearing; bevel-wheels
f	shaped teeth on two enmeshing gear-wheels
g	ratchet-wheel and pawl
g'	crown-wheel
h	vanes or pedals on spokes
i	blocks or bob-weights on spokes
j	discontinuous driving-belt; strap-drill or treadle-lathe
k	discontinuous driving-belt; bow-drill or pole-lathe
l	pulley or windlass
m	differential windlass
n	continuous or endless driving-belt (with mechanical advantage)
o	continuous or endless chain-drive (with mechanical advantage)
p	handle at wheel's edge, forming crank
p'	handle at wheel's edge, fitted with connecting-rod
p"	'oblique' crank handle
q	wheel and crank arm
r	crank arm or eccentric lug
r'	crank fitted with connecting-rod
s	crankshaft and connecting-rod
t	crank or eccentric lug, connecting-rod and piston-rod combination, for the interconversion of rotary and longitudinal rectilinear motion
u	cam an cam-follower

of the simplest methods of utilising rotary motion,[a] easily adaptable to many uses other than its original purpose of decorticating rice.[b]

Speaking in terms of an ideal technic morphology, the single trip-lug projection may swell or it may multiply. If it swells it will give rise to revolving plates like wheels but with an infinite variety of irregular non-circular outlines—in other words cams.[c] A bulbous one is shown as an example in Fig. 390*u*. Such devices have great use in machine construction, for a follower rod, pushed back and forth by the edge of the cam as it rotates, may be given almost any desired combination of motions, speeds, and dwells.[d] Traditional Chinese technology, so far as we know, made little use of cams continuously rotating, but at a very early date (the Han period, *c.* −2nd century) developed complex forms of cam-shaped rocking levers for the triggers of crossbows.[e] In one case, however, the vertical water-wheel form of the metallurgical blowing-engine of the Sung and Yuan time (+13th and +14th centuries),[f] it would seem almost certain (if the reconstruction is correct) that the trip-lugs must have been shaped into the form of cams. 'Three-dimensional' cams, such as swash-plates and helically grooved cylinders, appeared in Europe later on, not much before the time of Leonardo, the end of the +15th century.[g]

When the lugs for tripping are multiplied into numerous projections, they may take the form either of vanes or of pegs which give rise to gear teeth.[h] Since the length of these is generally short in comparison with the diameter of the wheel which bears them, they are either mounted upon its rim, or form the serrated edge of a solid disc.

With the first of these alternatives, the use of vanes, the water-wheel is born. In the diagram a vertically mounted water-wheel is first shown (Fig. 390*c*), but in the history of techniques the horizontally mounted or 'recumbent' (*wo lun*[1]) water-wheel has been at least as common (Fig. 390*c'*), and later on we shall have to compare the distribution of the two (p. 368 below). The latter seems to have been characteristic of Chinese civilisation from the Han to the Thang, after which the other appears also. We shall see further that there is very little difference in the date at which the water-

[a] In the case of the vertical stamp-mill, where the lug acts as a wiper or comma-shaped cam, it was also one of the simplest ways of converting rotary to longitudinal motion. On this subject see particularly pp. 380 ff. below.

[b] Before long we shall find abundant examples of the use of lugs, in the form of 'pinions of one, two, or three', in mechanisms such as the hodometer (p. 281) and the jack-work of mechanical clocks (pp. 455, 462, 485). Trip-mechanisms in modern machinery may be studied in Jones & Horton (1), vol. 1, pp. 118 ff., 148 ff.; vol. 2, pp. 189 ff.; vol. 3, pp. 86 ff.

[c] The history of lugs and cams has been begun by Gille (7).

[d] To gain an idea of their application in contemporary mechanics one may study the first chapter in each of the three volumes of Jones & Horton (1). The range of versatility of cams may be still further increased by mounting them eccentrically, or in yokes (*ibid.* vol. 3, p. 187). See also Willis (1), p. 324.

[e] Cf. Sect. 30*e* below.   [f] See p. 377 below.

[g] We shall return to these devices on pp. 384, 386 below, in connection with the fundamental conversion just referred to.

[h] The close connection between the water-wheel and the gear-wheel is well emphasised by Schuhl (1), p. 7. As Childe (10) has pointed out, some of the oldest Sumerian chariot-wheels are depicted as though they were cogged, and this is seen also on models, the 'cogs' being the heads of the copper nails which studded the felloes. The way thus lay open for the engagement of wheels.

[1] 臥輪

wheel as a power-source makes its appearance in China and the West, so that the question of origin is still unsolved. Here we come upon a basic distinction which it is useful to make depending upon the direction in which the power is transmitted. In using the fall of water as a motive power, force is exerted on the vanes of the wheel and conveyed to the machinery, but force may also be applied to the axle of the wheel and conveyed to the water, with the consequence of forward motion through the medium. This gives us the paddle-wheel boat, which, as will later appear (below, p. 433), is one of the most remarkable Chinese developments of the early Middle Ages. Having failed to find any suitable terms in the literature, I propose to distinguish in what follows between 'ex-aqueous' and 'ad-aqueous' wheels. The difference is analogous in a certain sense to that already mentioned between wheels for carrying and wheels for grinding; the former could be termed 'ex-terrestrial' and the latter 'ad-terrestrial', though here in the first case the medium does not move and in the second the machine need not advance. Much closer is the analogy with air, for the windmill is precisely an 'ex-aerial' wheel, while the aeroplane propeller may be considered an 'ad-aerial' one, though its helical traction and the sufficient forward motion of the aerofoil so given depend of course upon different principles from that of the paddle-wheel in water. In all these cases the prefix 'ad-' will be used when the energy is transmitted from the machine to the medium, and the prefix 'ex-' when the energy is transmitted from the medium to the machine.

Passing now from vanes to pegs and teeth (Fig. 390*d* and *f*), we have before us the history of gear-wheels (*ya lun*[1]), which has frequently been considered.[a] In the West they were essentially a Hellenistic development. The reference to them claimed in Aristotle[b] is in a mainly apocryphal work, the *Problemata Mechanica*, and will not be any earlier than Ctesibius (*c.* −250), Philon of Byzantium (*c.* −220) and Heron of Alexandria (*c.* +60), all of whom used them, or planned the use of them, in a great variety of machines.[c] These men were the contemporaries of Mêng Thien, Liu An and Wang Chhung, but for the Chhin and Former Han period we have little knowledge of the nature of machines constructed with trains of gears; we can only surmise that much must have been going on since a number of specimens of gear-wheels of that time have survived, and since we find so much textual evidence of the use of machinery with toothed wheels in the Later Han, San Kuo and Chin periods. They were needed in water-mills, hodometers, crossbow-arming mechanisms, south-pointing carriages, chain-pumps and mechanised armillary spheres. Tu Shih[d] in +30 and Chang Hêng[e] in +130 were no less familiar with them than Vitruvius in −30.[f] Chang Hêng was famous as a man who could 'make three wheels rotate as if they were one' (*nêng ling*

[a] Frémont (12); Matschoss & Kutzbach (1); Kammerer (1); Feldhaus (4); Woodbury (1, 2); Davison (5).
[b] Feldhaus (1), col. 1339; Neuburger (1), p. 215.
[c] Beck (2); Diels (1), etc.   [d] See p. 370 below.
[e] Cf. his essay *Ying Hsien* (*CSHK*, Hou Han sect., ch. 54, p. 8*b*) in which he says 'Linked wheels may be made to turn of themselves (*tshan lun kho shih tzu chuan*[2])'.
[f] *De Architectura*, IX, viii, 4; X, v and ix, esp. 5.

[1] 牙輪   [2] 參輪可使自轉

san lun tu chuan yeh [1]).[a] Chang Hêng's contemporary, Liu Hsi [2] (*d.* +120), in his synonymic dictionary, the *Shih Ming*, tells us [b] that the human jaws (*i* [3]) were familiarly called 'co-operating wheels' (*fu chhê* [4]) or 'toothed wheels' (*ya chhê* [5]), perhaps on the analogy of the intaking tendency of a mill in which two rollers, connected by gearing, revolve in opposite senses. Similarly, Tu Yü,[6] in his commentary on a passage in the *Tso Chuan*,[c] says that toothed wheels (lit. 'tooth carriers') are meant. The passage is metaphorical; someone is quoting a proverb—'The upper and lower jaw naturally rely upon one another, and when the lips are gone the teeth are cold', but the expression used is *fu chhê hsiang i*,[7] i.e. wheels that give each other mutual support. The point is, not that the writer of the *Tso Chuan* passage, which refers to −654, had gear-wheels in mind, but that they were so common in the Later Han period as to be part of the mental background of commentators.

In recent decades there have been many finds of gear-wheels in tombs dating from the Chhin onwards through the Han, i.e. from about −230. Such discoveries have been multiplied by the great expansion of archaeological research in China during the past dozen years. At least one mould for a bronze toothed wheel has come down to us intact from the Former Han time.[d] First described by Lo Chen-Yü (*3*), and now more accessibly by Wang Chen-To (*3*) and Liu Hsien-Chou (*5*, *6*), this interesting object (Fig. 391 *a*) is made of earthenware, and stamped in three places with a pair of characters only the first of which can be certainly made out ('Eastern') but which are clearly in Han script. The shank into which the axle would fit is square, and the wheel is a ratchet with sixteen slanting teeth ready to receive a pawl.[e] Such an arrangement (Fig. 390*g*) is a special case of gearing needed whenever it is desired that the wheel or roller shall turn only one way and not slip backwards; this particular mould must have been used about −100 for ratchet-wheels forming part of winches, cranes, or arcubal-lista-arming mechanisms.[f] It is unlikely that the Alexandrians had not also thought of this form of gear, but the earliest Western description of it seems to occur in the work of Oribasius (+325 to +400) in connection with surgical instruments.[g]

Another ratchet, this time an actual bronze wheel, was discovered by Chhang Wên-Chai (*1*) in a tomb full of bronze objects at Hsüeh-chia-yai village near Yung-chi in Shansi province.[h] This burial may have occurred in the late Warring States period,

---

[a] This interesting reference to a gear-train was quoted in the +7th century by Li Hsien, the Thang prince who wrote a commentary on the *Hou Han Shu*, from some lost essay of Fu Hsüan (+3rd century); see ch. 89, p. 3*b*, Po-Na edition, ch. 59, p. 4*b*.

[b] Ch. 8 (pp. 107ff.).   [c] Duke Hsi, 5th year (tr. Couvreur (1), vol. 1, p. 253).

[d] It was unfortunately broken during the Second World War, and only about a quarter of it remains in the Shenyang Museum.

[e] The number of teeth shows fairly clearly that the purpose was technological rather than scientific, as e.g. for astronomical apparatus.   [f] Cf. Sect. 30*h* below.

[g] Bk. 49, ch. 346; cf. Feldhaus (1), col. 1043. A development of it which did not, so far as we know, arise spontaneously in China was the crown-wheel (Fig. 390*g'*), which has teeth slanting on one side and straight on the other. This was an essential component of the escapement mechanism of +14th-century European clockwork (cf. pp. 441 ff. below).

[h] The discovery was first made by the cotton farmers Hsieh Shun-Hsiao and Hsieh Po-Hsiao, who knew immediately that the fragments upon which they stumbled were of ancient bronze culture, and promised much more in the vicinity. This incident illustrates the awareness of their heritage by Chinese country folk in 1955. The ratchet wheel is illustrated also by Liu Hsien-Chou (*6*).

[1] 能令三輪獨轉也   [2] 劉熙   [3] 頤   [4] 輔車   [5] 牙車   [6] 杜預   [7] 輔車相依

PLATE CXLII

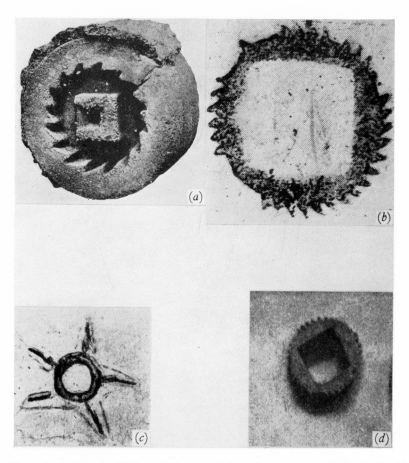

Fig. 391. Gear-wheels from the Chhin and Han periods (−3rd to +3rd centuries). (a) Mould of hard stoneware for a 16-tooth ratchet wheel, c. −100 (Wang Chen-To, 3); (b) bronze 40-tooth ratchet-wheel with large square shank, diam. c. 1 in., c. −200 (Chhang Wên-Chai, 1); (c) small bronze piece, perhaps a pinion of five thorn-teeth, c. −200 (Chhang Wên-Chai, 1); (d) one of a number of bronze 40-tooth gear-wheels with large square shanks, c. −200 (Chhang Wên-Chai, 1).

PLATE CXLIII

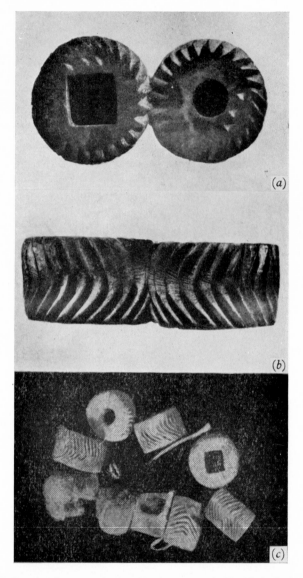

Fig. 392. Objects believed to be gear-wheels with chevron-teeth (double helical gear). (*a*, *b*) Bronze 24-tooth enmeshing gears with round and square shanks, diam. $1\frac{1}{2}$ cm., width 1 cm., *c*. +50 (Liu Hsien-Chou, *6*), from a tomb in Shensi. (*c*) Nine further gear-wheels of the same kind, from Han tombs in Hunan (Shih Shu-Chhing, *1*).

but is perhaps more probably of Chhin or early Han, *c.* −200. Measuring hardly more than an inch in diameter (Fig. 391 *b*) it has forty teeth and the square shank hole is extremely large. The most likely purpose of this ratchet wheel would have been the arming mechanism of a crossbow. The same find also yielded two or three ordinary gear-wheels of bronze, also having forty teeth and almost equally large square shank holes (Fig. 391 *d*).[a] In addition to these there was a small wheel with a round shank (Fig. 391 *c*) which, if not simply an ornament, may have been a pinion of five thorn-teeth. Since such parts needing one or more, but in any case very few, teeth would have been necessary in hodometers (cf. p. 281 below), it may be reasonable to assume that this was something of the kind. But these were not the only objects valuable for technological history in this rich discovery; there were also short tube-like shafts of bronze, bearing gear-teeth either at one end or at both. In addition there were lengths of bronze plate, one with small teeth along one side, another with small teeth on one side and large teeth on the other. Chhang interpreted these as saws, but since by this time iron and steel had long been known and worked,[b] it may be worth suggesting that they were longitudinal racks needed for some piece of bronze machinery.

But the most remarkable and unexpected finds have been those of pairs of gear-wheels with 'chevron-teeth' (*ien tzu chhih lun* [1]), quite analogous to the double helical gear of the twentieth century, as illustrated for instance in Davison (5).[c] Specimens of these bronze objects were first found in 1953 in a tomb of the early Hou Han period at Hung-chhing village near Sian in Shensi (Fig. 392 *a*, *b*),[d] and are dated by Liu Hsien-Chou (6) in the neighbourhood of +50.[e] It will be seen that in the pair photographed one has a square shank while the other's is round. The same difference appeared again in a group figured by Shih Shu-Chhing (*1*) and found five years later in a tomb at Chiang-chia-shan near Hêngyang in Hunan. He illustrates nine in various degrees of perfection (Fig. 392 *c*); made of bronze, they measure 1·5 cm. in diameter, and are 1 cm. wide, so that the working is rather delicate. According to Li Wên-Hsin (*1*), further finds of the same type of gear-wheels have come to light from Han tombs at Chhangsha, further north in the same province. So far their dating has not been given very closely, but they are all certainly Han if not Chhin.[f]

---

[a] There were also fragments of gear-wheels. The even number of teeth may preclude a calendrical–astronomical employment.

[b] Cf. Sect. 30*d* below, and in the meantime, Needham (32).

[c] Or Matchoss & Kutzbach (1), pp. 86 ff. According to Dickinson (5) the invention of single and double helical spur gearing was made by James White in Paris about +1793. He illustrates one of White's own models (fig. 1 on pl. XXXVIII).

[d] See Yen Lei *et al.* (*1*), p. 666 and pl. 4, figs. 4 and 5.

[e] The discoverers themselves, however, prefer a date about −50; see Wu Ju-Tsu & Hu Chhien-Ying (*1*); Anon. (*44*). Cf. Anon. (*43*), p. 79.

[f] It must be said that some Western historians of technology are reluctant to accept these objects as double helical gear-wheels. Prof. R. S. Woodbury of Cambridge, Mass., even suggests (private communication) that they were 'merely rollers for impressing a formal design on clay, or for printing a pattern on paper or textiles'. But in this case it is rather hard to see why they should have been made so carefully of bronze. Experimental notes on their performance when set up on parallel shafts would certainly be of great interest; in the meantime it may be pointed out that they have been accepted as gear-wheels not only by the Chinese archaeologists but by engineers of high competence, notably Prof. Liu Hsien-Chou.

[1] 人字齒輪

Still less well known is the fact that during recent years there have been a number of finds of iron gear-wheels in Han tombs,[a] whether or not of cast iron remains uncertain till full descriptions appear, but presumably so. One of these, a ratchet-wheel of sixteen teeth, about $2\frac{3}{4}$ in. in diameter, is illustrated in a recent survey of important archaeological finds.[b]

Thus during the three centuries and more preceding the time of Chang Hêng gear-wheels large and small were being made for a variety of practical purposes. This spread of date is instructive for comparisons with the Alexandrian engineers, who, broadly speaking, cover the same period, but we should also notice that the distribution of the finds in China is remarkably wide, excluding only the far south and the far north. Interest in gear-wheels and their practical utilisation was therefore not confined to any small region of Han culture. As for the shape of the teeth, all ancient Chinese examples so far seem (if not ratchet-slanted) to approximate to equilateral triangles, exactly as is found in the Anti-Kythera planetarium, probably of the − 1st century.[c] Rounded teeth occur first in a Chinese hodometer specification of + 1027 (see p. 284 below), and with an angle of about 30° in Europe *c*. + 1300.[d] Rounded 'spoke-teeth' are found in the early Western mechanical clocks of the + 14th and + 15th centuries.[e]

It is fortunate indeed that some of the Han gear-wheels were made of bronze and iron, for if all had been made of wood none would have survived to prove knowledge and use. The material employed for them in ancient and medieval China seems to have varied with size and purpose—for water-mills and water-raising machinery they were certainly always of wood, built like wagon-wheels,[f] but for finer machinery such as hodometers and south-pointing carriages[g] they were made from bronze and iron. As Usher has pointed out,[h] the only instance of gear-wheels in power-transmission as opposed to light-duty motions in Western antiquity is the water-mill of Vitruvius.[i] Here an undershot horizontal-axle vertical water-wheel drives a flour-mill by means of right-angle gearing in the form of a pin-wheel and a pin-drum[j] (as in Fig. 390*e*). This right-angle drive was not necessary in Han China if, as seems likely, the earliest water-mills there were driven by horizontal water-wheels,[k] but it appears abundantly in the Chin and later. On the other hand, if we knew more of the mechanism used

<hr/>

[a] See Thang Yün-Ming (*1*); Mêng Hao *et al.* (*1*); Anon. (*45*); Anon. (*46*).

[b] Anon. (*43*), p. 76, fig. 38. This work also gives on p. 79 a useful summary of all the discoveries of gear-wheels, whether of bronze or iron, in Warring States, Chhin and Han tombs or habitations, down to 1961.

[c] This remarkable mechanism is now under study by Prof. D. J. de S. Price, whose monograph on it will be awaited with much interest. See Price (8, 9).

[d] Rounded-off isosceles triangles in geared astrolabes; cf. Price (8).

[e] These had of course been used in wooden millwright's gearing much earlier. Further on the shape of teeth, see pp. 284, 456, 473, 499 below.

[f] Some examples will be shown and discussed on p. 354 below.

[g] Cf. pp. 286 ff. below.   [h] (*1*), 1st ed., pp. 98, 124; 2nd ed., pp. 146, 168.

[i] x, v, 2. See also Beck (1), p. 49.

[j] Essentially two pin-wheels fused together. A Roman example still exists; it is figured by Feldhaus (2), p. 203. One may note how the right-angle pin-wheel generated not only all right-angle gear-wheels, but also the crown-wheel of the verge-and-foliot escapement of clockwork (cf. p. 441 below). In + 1590 the Mogul emperor Akbar had sixteen pin-drums working off one pin-wheel in the machine he invented for boring and cleaning gun-barrels (cf. Blochmann (1), p. 115 and pl. xv).

[k] We shall have much to say on this subject below, pp. 369, 392, 405 ff.

about +120 by Chang Hêng, and his numerous later successors before the invention of the clock escapement early in the +8th century, to bring about the rotation of armillary spheres or celestial globes by water-power,[a] we should very probably have to include this among the earliest examples of power-transmitting gear-work. After the invention of the escapement in +725, there was a great flourishing of gear-wheels in clockwork and jackwork, culminating in the bronze and iron of Su Sung's elaborate masterpiece[b] in +1088, but it is interesting that (as in the +18th century) hard wood for gear-wheels also had its advocates, who appreciated its special lubrication properties and freedom from rust.[c] The pin-drums or lantern-wheels of the Vitruvian water-mill are perhaps the most primitive forms of right-angle gearing—among their successors were the two pin-wheels with enmeshed teeth (Fig. 390 e'), and bevelled gear-wheels (e"). It is generally thought that these last were quite new when Leonardo sketched them,[d] but four hundred years earlier oblique gears of some kind or other had been prominent in Su Sung's clockwork.[e]

From one point of view gear-wheels are wheels which consist only of a hub and a number of short spokes, all rim having vanished. But such spokes can also carry objects of different kinds at their peripheral ends. So far we have thought only of vanes or paddles mounted upon the rims of wheels (as in the water-wheel), but they may also be borne on the ends of free spokes. From the trip-hammer lugs (Fig. 390 b, b'), which were often on the shaft alone, there could be a transition to lugs carrying vanes (h) or weights (i). In the first case we have the radial pedal or radial treadle,[f] a device of enormous importance in the exploitation of human muscular power, and characteristically Chinese. This was the main motor of the square-pallet chain-pump (shui chhê[1]; lung ku chhê[2]; fan chhê[3]), an invention probably of the Later Han. As Haudricourt (1, 2) and Febvre (2) have strongly emphasized, this simple form of treadmill was almost unknown to Mediterranean antiquity, and indeed was never adopted by Western peoples.[g] The crank pedal so familiar on bicycles dates only from

---

[a] See pp. 481 ff. below.

[b] The fullest description is given by Needham, Wang & Price (1), but the essentials are summarised on pp. 446 ff. below.

[c] Notably Wang Fu in +1124 (see p. 499 below).

[d] See Beck (1), p. 100; Matchoss & Kutzbach (1), p. 18.   [e] Cf. below, p. 456.

[f] *Not* the crank pedal, be it noted, *nor* the simple lever fitted at the ends of the thongs of a strap-drill or lathe, which permits the operator's feet to take the place of the hands of a second person. This last is essentially similar to the pedal levers of looms (cf. Fig. 393). We shall shortly meet with a very curious form of treadle which brings about rotary movement without the intervention of a crank (pp. 103, 115, 236 below).

[g] This statement must be qualified by the fact that in some cases Archimedean screws in Roman mines may have been worked by slaves walking on lugs set around their circumference (Bromehead, 7). A little evidence is available also from representations in art. Hudson (1); Treue (1), p. 27; and Forbes (17), p. 677, figure a Pompeii mural painting and a late Egyptian relief which show men treading on Archimedean screws in this way. A terra-cotta model from late Ptolemaic Egypt (Price, 1) confirms the interpretation. The MS. of the Hussite engineer (+1430) also has pictures of men on the outside of stepped wheels applying power to mills (cf. e.g. Matschoss & Kutzbach (1), p. 17). In the absence of the crank, which, as we shall see, makes no appearance before the early Middle Ages, it is rather difficult to see how wheels were rotated in the ancient West, if not by the feet. Yet on the whole a strong impression remains that pedal motions were more widespread and more versatile in their applications in Chinese than in Western culture.

[1] 水車   [2] 龍骨車   [3] 翻車

the middle of the +19th century. Treadmills in Greek and Roman culture there certainly were, but they were large drums into which one or more men could enter and exert force from within.[a] A famous example is the drum-treadmill working a huge

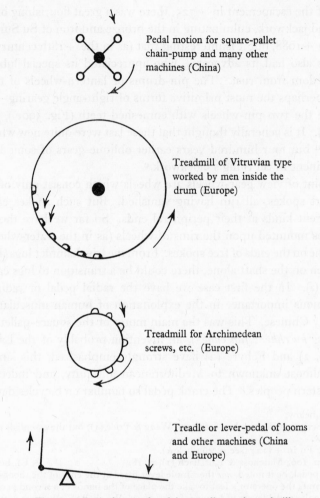

Pedal motion for square-pallet chain-pump and many other machines (China)

Treadmill of Vitruvian type worked by men inside the drum (Europe)

Treadmill for Archimedean screws, etc. (Europe)

Treadle or lever-pedal of looms and other machines (China and Europe)

Fig. 393. Basic forms of pedals, treadles and treadmills.

crane, shown on a relief in the Lateran Museum.[b] Vitruvius describes these in connection with his *tympanum*[c] and noria[d] (*c.* −30). They must have persisted through the Middle Ages, for one is shown in the MS. of the anonymous Hussite engineer

[a] Feldhaus (1), col. 1186.
[b] Feldhaus (1), col. 520; Feldhaus (2), pl. VI; Blümner (1), vol. 3, p. 119.
[c] x, iv, 1 (explained below, p. 360). Sixteenth-century illustrations of Vitruvius show it with a crank, which it assuredly did not have; the text says 'hominibus calcantibus versatur', implying a drum treadmill. Cf. Beck (1), p. 47.
[d] Explained below, p. 356. His man-handled noria (x, lv, 3) is described 'ita cum rota a calcantibus versabitur'; and his water-driven one (x, v, 1) as 'sine operarum calcatura'. Cf. Beck (1), p. 48.

(+1430),[a] and +16th-century representations of them are common.[b] In +1627 they are depicted for the first time in a Chinese book.[c] It is important to realise that these were large and clumsy constructions, quite unsuitable for the use of millions of farmers whose problem was to irrigate their land by small machines and human motive power.[d]

Plates carried on lugs or mounted directly on shafts proved of course to be capable of applications stronger and stranger than the pressure of human feet; they could turn into vanes or sails beaten by the wind—as they first did in early Islamic Persia, spreading thereafter both west and east (cf. pp. 556 ff. below). This was the ex-aerial development, analogous to that of the water-wheel, but the opposite was equally important for it led to all rotary fans and air-compressors, ad-aerial devices in which the Chinese culture-area took the lead (cf. pp. 150 ff. below). And besides these famous uses, we shall find as we go on certain more unusual functions of blocks fitted on the ends of spokes; they may occur as components of heavy-duty composite wagon-wheels, or as carriers for driving-belts where an integral rim is absent (pp. 80, 104).

When the radial lugs carry weights, a bob flywheel results. The most ancient fly-wheel was presumably the neolithic spinning-whorl (Frémont, 12), but it did not find great application before the palaeotechnic period. Another example would be the potter's wheel (*chün*[1]). Second in antiquity would be the heavy discs on bow-drills and pump-drills, attested from so many cultures, as also on the ancient Egyptian crank-drill.[e] In the +11th century such flywheels were applied to pestles for grinding.[f] Radial weights (Fig. 390*i*) found their chief application, perhaps, on the worms of screw-presses from Roman times to the period of printing,[g] and it was doubtless this association with screws which prevented them from being much used in China. True flywheels appear in the +15th century (the Anonymous Hussite), and then steadily acquire their neotechnic importance.

The function of a flywheel can also be performed when the balls or bobs are placed, not directly at the ends of the spokes, but attached to them by short chains which centrifugal force tautens. This device may be oldest in Tibet, where it has been current practice for hand-turned prayer-wheels from time immemorial. About +1490 Francesco di Giorgio sketched such a flywheel,[h] and he had been preceded by the Anonymous Hussite sixty years earlier, who attached swinging weights to the spoke ends.[i] This was just the time when the engineers of the West were interested in ways

---

[a] See, e.g. Berthelot (4); Matchoss & Kutzbach (1), p. 17. One of this date, or earlier, is still in use at an inn between Winchester and West Meon, for raising water. Cf. Sandford (1). Salzman (1), pp. 323 ff., illustrates others.

[b] Cf. Feldhaus (2), pl. XIII; and the relief at the cathedral of Gurk in Carinthia.

[c] *Chhi Chhi Thu Shuo*, ch. 3, p. 10*b* (see below, p. 170).

[d] Haudricourt (1) has underlined the importance of these devices for improving the standard of life. Either one must obtain a more abundant and nutritive product for the same work, or do it at less physiological expense, if labour capacity is fully used, as it was in China.

[e] Below, p. 114.   [f] Below, p. 103.

[g] Cf. Frémont (8) and Drachmann (7). We shall suggest later on (p. 534) a genetic connection with the verge-and-foliot escapement of early Western clockwork.

[h] Lynn White (5), p. 520, (7) p. 116 and fig. 9.   [i] Gille (12).

[1] 均

of overcoming 'dead centre', doubtless because of the increasing use of cranks and crankshafts (cf. pp. 112, 113). It is by no means fanciful to connect this development with the Central Asian domestic slaves who were so numerous in +14th- and +15th-century Italy, and who may have been responsible for many transmissions of technological interest.[a]

It may be remembered here that just as the simple lugs alone may be used for converting rotary to oscillatory motion by tripping, they may also be used, as hand-spikes, for applying the tractive power of men or animals to produce rotary motion.[b] We then have the capstan (if the axle is vertical) or the windlass (if horizontal). But as these generally involve ropes, cords, or belts, they must be left to the next subsection.

Mention was made above of the use of wheels or rollers for crushing or grinding. In Chinese civilisation these techniques were widespread, and go back almost certainly to the Chou. Later we shall examine their various forms, the edge-runner mill with longitudinal travel (*yen chhi*[1]), and the edge-runner mill with circular travel (*nien*[2, 3]).[c] Parallel machines were the roller harrow (*kan*[4]), the hand roller (*kun nien*[5]) and the roller mill with circular travel (*kun nien*[6]). By combining two rollers, with or without gearing, the cotton-gin (*chiao chhê*[7]) and the sugar-cane mill (*ya chê*[8]) were obtained; the ancestors of all steel rolling-mills, mangles, and paper or textile machinery.[d]

This discussion began with rollers, and with them it must also end. For all wheel-carrying shafts must be supported in bearings, and the end of the shaft constitutes a roller. If the bearing is stationary, heat, retardation and wear necessarily arise because of the sliding friction,[e] so that from a time much earlier than is generally supposed, engineers have sought to interpose additional rolling objects between shaft and bearing, thus reducing to a minimum the undesirable effects of frictional drag. Bearings began, no doubt, with the bone or antler hand-holds used by neolithic drillers; in the −2nd millennium Egyptian craftsmen used bowls of soapstone with which to press on their bow-drills.[f] The bearings of the potter's wheel are also of the highest antiquity (the −4th millennium in Mesopotamia); in China small cups of hard porcelain were used for the sockets, in India small concave flint pebbles.[g] Lubricants are no younger; the famous Egyptian wall-painting of c. −1880 at al-Bersheh, which shows 172 men transporting a colossus from the quarry on sledges, has one man pouring oil or grease ahead of the base of the statue on which he rides.[h] As we have already seen, the journals and bearings of vehicle axles were of bronze or iron in Chou and Han China just as

---

[a] We drew attention to this in Vol. 1, p. 189, and the evidence has been much strengthened by the later writings of Lynn White (5, 7).

[b] This distinction recalls that between ad-aqueous and ex-aqueous wheels.

[c] With the rotary disc knife or stone for jade-cutting (*cho yü lun*[9]) we have already dealt (Vol. 3, p. 667 above; cf. also the remark on the rotary grindstone, p. 55 above).

[d] General historical account by Feldhaus (6). Cf. pp. 122, 204 ff. below.

[e] See the excellent lecture of Bowden (1).   [f] Cf. Davison (3, 4).

[g] See Childe (11), esp. p. 198.

[h] Wilkinson (1), vol. 2, frontispiece and pp. 307 ff. Davison (4) has made an interesting calculation showing that the number of hauliers represented is just about that which would have been necessary.

---

¹ 研器   ² 碾   ³ 輾   ⁴ 赶   ⁵ 滚碾   ⁶ 輥碾

⁷ 攪車   ⁸ 軋蔗   ⁹ 琢玉輪

they were in European antiquity.[a] The statement, however, that after the fall of the Roman empire there appear no more metal bearing-surfaces until the clockwork of the +14th century,[b] may possibly be true of Europe but it is certainly wrong for China. The mechanised armillary spheres between the +2nd and the +8th centuries could not have worked at all, even though imperfectly, without metal bearings, and about +720 we even hear of steel bearings being used in the ecliptically mounted observational armillary made by I-Hsing and Liang Ling-Tsan.[c] When we come to the clock-tower of Su Sung in +1088 the details are particularly clear. The main driving-shaft (*thieh shu chu*[1]) of iron has 'cylindrical necks' or journals (*yuan hsiang*[2]) supported on iron crescent-shaped upward-facing bearings (*thieh yang yüeh*[3]). The pointed iron-capped pivot (*tsuan*[4]) of the heavy tiers of jack-wheels is borne in an iron mortar-shaped end-bearing (*thieh shu chiu*[5]). The mounting of the main lever of the linkwork escapement is also lovingly described. At the fulcrum of the lever there is an 'iron shut-in axle' (*thieh kuan chu*[6]) in the form of a horizontal cross-shaft (*hêng kuang*[7]) rocking in 'camel-backs', i.e. upper bearing (*tho fêng*[8]) caps, between end-plates (*thieh hsia*[9]).[d] All this was not the creation of a day—there had been a tradition of three or four centuries already, to say nothing of the 'proto-clocks' going back to the time of Chang Hêng.[e]

Although in these remarkable machines we have not encountered the use of intermediary rolling objects taking off friction, China may be deeply involved in the prehistory of ball-bearings. After the stonemasons of high antiquity we meet first with a systematic use of rollers in the battering-ram and the gate-borer invented by Diades, one of the engineers of Alexander the Great, whom he accompanied in his campaigns (−334 to −323).[f] But here the rollers are still in a straight line, not arranged peripherally around a shaft. True roller-bearings have been claimed for the hubs of the Celtic four-wheeled wagons discovered in a bog at Dejbjerg in Denmark in 1883 and described by Petersen (1) and Klindt-Jensen (1). Here the inside of each hub shows about 32 transverse grooves. Similar internally corrugated hubs have been described from a number of other places and are considered to date from the −1st century, being associated with the La Tène Age III, and supposed to derive from Etruscan origins by way of the Hallstatt culture.[g] The question turns upon what was in the

---

[a] We have also seen that a leather cap retained the lubricant in ancient Chinese vehicle hubs. Vitruvius, x, iv, 1, describes iron journals and bearings for the *tympanum* water-raising machine (cf. p. 360 below). Presently we shall come across a particularly complex assembly of iron bearings in the *trapetum* or olive-crushing mill of classical Greece (p. 202 below; cf. Drachmann, 7).

[b] Davison (4). [c] Vol. 3, p. 350, above.

[d] Nothing has come to light in medieval Chinese descriptions of machinery which indicates what kinds of lubricants were used. We have already seen, however (Vol. 3, pp. 608ff.), that mineral oils from natural seepages were used in lubrication from early times. But doubtless reliance was placed mainly on light and heavy vegetable oils (cf. Hommel (1), p. 328, etc.).

[e] On these see pp. 481 ff. below.

[f] Vitruvius, x, xiii, 3–8, gives a very adequate description.

[g] Finds for example at Langå and Kraghede in Denmark, at Kappel near Buchau in Germany, and at Vurpař in Transylvania (Ardeal).

---

[1] 鐵樞軸 [2] 圓項 [3] 鐵仰月 [4] 鑽 [5] 鐵樞臼
[6] 鐵關軸 [7] 橫桄 [8] 馳峯 [9] 鐵頰

grooves. Although the interpretation of roller-bearings has met with some acceptance among archaeologists,[a] reference to the original paper of Petersen in Danish shows that the pieces of wood which came out from the hub spaces when disinterred were narrow flat strips and not rollers at all. It is therefore quite impossible to regard these Celtic wagons as having been equipped with this invention.[b]

For the earliest roller-bearings we may have to look to China. Among the finds at Hsüeh-chia-yai village in Shansi reported by Chhang Wên-Chai (*1*) there were some remarkable annular bronze objects with internal grooves. These channels were divided into four or eight compartments by small transverse partitions, and each of these was filled with a mass of granular iron rust. Several of these possible 'ball-races' or roller bearings, of different sizes, were found. Since the objects of this tomb must be dated at least as early as the −2nd century, it would seem that the Chinese if not the Celts were taking an interest in the free and smooth run of certain shafts or axles.[c]

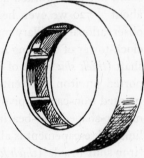

If the rust came from balls or rollers, these objects would certainly be the oldest ball-bearings known.[d] If it did not, pride of place must still go to the strange trunnion bearings of the capstans of the Roman ships built between +44 and +54, and in our own time recovered from the Lake of Nemi south of Rome.[e] These are balls each prolonged at two poles into short shafts which were retained in place by clamps so that they could roll round freely. Eight of them bore the undersurface of the capstan, or so it is assumed for only a fragment containing two remains. Strictly speaking this device was a roller-bearing and not a true ball-bearing, since the spheres could only rotate in one plane, but it has great ancestral interest. Trunnions recur in al-Jazarī (+1206).[f] If we follow further the development of roller-bearings we find that flat rollers similar to those sketched by Leonardo[g] at the end of the +15th century were in use in the +16th, as Agricola testifies.[h] But as we shall later see,[i] certain imperial

[a] E.g. Jope (1), p. 551.

[b] Discussions at the International Congress of the History of Science at Barcelona in 1959 with Dr A. G. Drachmann and Prof. R. S. Woodbury led to the suggestion that the real function of the semi-cylindrical depressions within the hubs was to hold greased leather or perhaps a packing of rags or wool. We are much indebted to Dr Drachmann for letting us know the results of his studies on the actual objects preserved in the National Museum at Copenhagen. I saw them only in 1962.

[c] It was striking to compare this ancient evidence in 1958 with the great movement then being launched in China for establishing a ball-bearing factory in every *hsien* city. Country people could be seen everywhere fitting them in the hubs of cart-wheels of the traditional type.

[d] Continuing his useful role of *advocatus diaboli*, Prof. R. S. Woodbury counsels all due scepticism concerning these objects as ball- or roller-bearings. Even he, however, is willing to hazard a guess (private communication) that the outer bronze annulus was a bearing-case into which were fitted small blocks of iron—to avoid the difficulty of boring accurately so large a hole in an iron plate. We must hope for more light from further discoveries.

[e] See Ucelli di Nemi (1, 2); Dionisi (1); Moretti (1). I had the pleasure of visiting the naval museum on the shores of Lake Nemi in 1955.

[f] Coomaraswamy (2), p. 19.

[g] Cf. Feldhaus (1), col. 600; Beck (1), pp. 324ff.

[h] Hoover & Hoover ed., p. 173. A form of roller-bearing was embodied in clockwork by Eberhardt Baldewin in +1561 (see Lloyd (5), pp. 658, 660).

[i] P. 254 below.

PLATE CXLIV

Fig. 394. Crane pulleys in the Wu Liang tomb-shrine reliefs (restoration by W. Fairbank, 1). The unsuccessful attempt by Chhin Shih Huang Ti to recover the cauldrons of the Chou.

vehicles constructed in China between the +7th and the +11th centuries showed a steadiness hard to account for otherwise. By the +17th century in Europe roller-bearings were a commonplace, freely depicted by Ramelli,[a] and soon fitted under Dutch windmills.[b] Some considerable time after the Nemi capstans, roller-bearings were actually patented by John Garnett in +1787.[c] As for ball-bearings proper, we are told that Cellini used ball mountings for rotating statues in the +16th century[d] and that about +1770 the engineers of the Empress of Russia conveyed heavy blocks of stone considerable distances 'upon cannon-balls rolling between bars of iron'.[e] Probably this was one of the stimuli which led Varlo in 1772 to invent and test true annular ball-races on road vehicles.[f] Thus it had taken nearly two thousand years to attain permanently, after so many brilliant but isolated inventions, those friction-reducing garlands in which our shafts and axles rotate today.

(3) PULLEYS, DRIVING-BELTS AND CHAIN-DRIVES

We now return to a point of departure already mentioned, namely the investigation of what happens when fibrous bodies (sinews, cords, thongs, ropes, etc.), or chains, are wrapped round rotating axles and wheels. Fig. 390 (j) shows the simple arrangement of the strap-drill (p. 55 above) or treadle-lathe, and (k) that of the bow-drill or pole-lathe (p. 57 above)—all such methods produce only alternating rotary motion. To have it continuous, the endless driving-belt is necessary (n),[g] and this was therefore an invention of first-rate importance.[h] Before considering it, however, a few words must be said about the simple pulley (l), i.e. the wheel with furrowed rim able to retain a rope or cord running over it. It has not been possible to identify the people who first realised that friction of a rope passing over a projection would be enormously reduced if a wheel was interposed, but the practice was well known both in Babylonia[i] and ancient Egypt.

Pulleys (*lu-lu* or *lo-lu*[1]) would therefore be expected in China from very early times, and there indeed they were. Fig. 394 shows a famous relief from the Wu Liang tomb-shrines (+147) depicting the attempted recovery of the Chou cauldrons from the river,[j] where two crane pulleys are inferable. A draw-well jar model of Han time

[a] Cf. Matchoss & Kutzbach (1), p. 24; Feldhaus (1), col. 601. The device was a well-head on rollers rotated by man-power and geared so as to raise the well-bucket.
[b] Van Natrus, Polly, van Vuuren & Linperch (1).
[c] See Davison (3). Later on we shall find an even more extraordinary example of patenting age-old devices. Cf. Vol. 3, p. 315 (*a*), and p. 386 below.
[d] Davison (3).   [e] Varlo (1), p. 3; Schmithals & Klemm (1), fig. 99.
[f] He was thus the real pioneer rather than Philip Vaughan in 1794 who usually gets the credit. Vincean wheel-shaped rollers had been applied to vehicle axles, however, by Rowe in +1734.
[g] Unless, of course, direct gearing is used.
[h] Heat rather than light has been generated by the arguments of anthropologists about it. Cf. Lechler (1); Harrison (1).
[i] There are several pictures of it from the Assyrian period between −880 and −850.
[j] We give the restoration of W. Fairbank (1); cf. *CSS*, ch. 4, (pp. 6, 7). These urns or cauldrons were the same as those already mentioned (Vol. 3, p. 503) in the geographical Section. The Chou dynasty was supposed to have inherited them from the remote antiquity of the Hsia, but according to tradition

[1] 轆轤

with its pulley in place is shown in Fig. 395 (Laufer, 3). The *Li Chi*[a] has interesting hints concerning the tackle used for lowering the heavy coffin-lids at important burials, and reports a story about Kungshu Phan (−5th century) on the subject. The term *fêng pei*[1] seems to have meant a kind of four-posted derrick with four pulleys, according to the commentators and to Liu Hsi's *Shih Ming*;[b] in this connection it is interesting that the earliest (Han) pictures of the derricks over the Szechuanese brine wells (cf. Fig. 396 and Sect. 37 below) also show four-membered structures, not pyramidal. Pulleys were so common in the Han that other things were named after them, such as swords with pulley-wheel-shaped hand-guards (*lu-lu chien*[2]),[c] and a variety of date constricted in the middle (*lu-lu tsao*[3]).[d] A pulley hoist is mentioned[e] which enabled the famous calligrapher Wei Chung-Chiang[4] to complete about +230 an inscription high on a tower, and another arrangement of pulleys,[f] powered by a hundred oxen, for fishing a great bell out of a river in +336. Pulleys were in constant demand for palace entertainments—for example, about +915 a whole corps de ballet of two hundred and twenty girls was hauled up a sloping way from a lake in boats.[g] But it is unnecessary to multiply instances.[h]

These heavier uses approximate the pulley to the winch, windlass or capstan. The ordinary well-windlass[i] is shown in nearly all the Chinese books which illustrate machines, beginning with +1313. The main interest in so ancient an instrument is the extent to which it incorporated the crank principle, and to this we shall shortly return (p. 111). The great horizontal winding-drums of the Szechuanese salt industry we shall illustrate later on. The winch was of course also used for tautening when mounted with a ratchet. The following example, taken from the *Thang Yü Lin*,[j] may afford a little light relief. The year is +736.

On the 5th day of the eighth month of the 24th year of the Khai-Yuan reign-period, the (Thang) emperor commanded a rope-walking performance at the imperial tower. First a long

---

they were lost in the Sê River in −335. It was the first emperor, Chhin Shih Huang Ti, who tried to recover them, while on one of his magical perambulations, with the aid of a thousand men, in −219, but he failed (*Shih Chi*, ch. 6, p. 18b; Chavannes (1), vol. 2, p. 154). On the whole subject see Chiang Shao-Yuan (1), pp. 130ff., esp. p. 136.

[a] Ch. 4 (Than Kung), p. 74a (tr. Legge (7), vol. 1, p. 184).

[b] Ch. 20 (p. 317).

[c] Cf. the biography of Chien Pu-I,[5] *Chhien Han Shu*, ch. 71, p. 1b; and the originally pre-Sui book *Ku Yo Fu*[6] (Treasury of Ancient Songs).

[d] *Erh Ya*, under *pien yao tsao*,[7] ch. 14, p. 14a.

[e] *Shih Shuo Hsin Yü*, ch. 21, p. 33a.   [f] *Tzu Chih Thung Chien*, ch. 95 (p. 3008).

[g] *Ju Lin Kung I*,[8] ch. 2, p. 31b. This was organised by Wang Chieh-Ssu,[9] one of the sons of the first ruler of the Chhien Shu State in Szechuan in the Wu Tai period, Wang Chien[10] (d. +918). Bellows worked by smiths churned the lake into artificial waves, and the boats came up through a tunnel in an artificial island hill. Although this was court entertainment, winches and capstans also found great use on the slipways of communications canals all over China (cf. Sect. 29 below).

[h] Pulley-blocks are very rarely shown in Chinese illustrations. They appear, however, in a picture of the Szechuanese salt industry in *TKKW*, ch. 1, p. 72a. But this only in the first edition; in later ones these drawings were replaced by much more elaborate ones and this detail dropped out.

[i] Cf. Ewbank (1), pp. 24ff.; Laufer (3), pp. 72ff. A typical medieval reference is in *Shih Shuo Hsin Yü*, ch. 25, p. 15a. See Fig. 574.

[j] Ch. 5, p. 24a, tr. auct. with Ho Ping-Yü.

[1] 豐碑	[2] 轆轤劍	[3] 轆轤棗	[4] 韋仲將	[5] 雋不疑
[6] 古樂府	[7] 邊腰棗	[8] 儒林公議	[9] 王偕嗣	[10] 王建

PLATE CXLV

Fig. 395. Han pottery jar representing a well-head with pulley and bucket
(Laufer (3), pl. 14).

PLATE CXLVI

Fig. 396. Moulded brick from Chhiung-lai in Szechuan showing the salt industry; Later Han period.
Now in the Hsi-chhêng Museum, Chhêngtu (Rudolph & Wên (1), pl. 91). The derrick over one of the
brine bore-holes is seen on the left, with its winding gear at the top. To the right of its uppermost storey
there is a tank to receive the brine, which flows down through a winding bamboo pipe-line to the row of
evaporating pans seen under a shed on the right. A companion brick (pl. 92), defective in a different
place, shows the descending pipe-line still more clearly, and also three or four other pipes conveying
natural gas to burners under the pans. The best examples of this (see Anon. (22), no. 1, pp. 6, 7) show an
inverted siphon in the pipe-line (cf. p. 128 below). Hunting scenes with cross-bows occupy the upper
right-hand quarter of both pictures. Cf. Figs. 422 and 432 below. The salt industry was already well
developed in Szechuan in −130 and may go back to the −3rd century; cf. Rudolph (4).

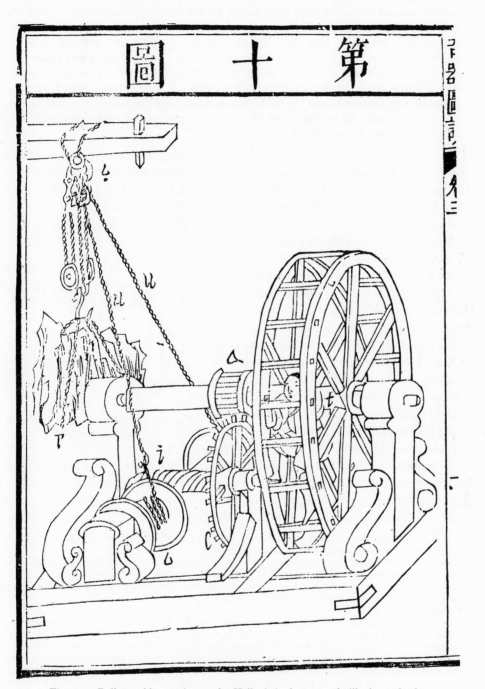

Fig. 397. Pulley tackle, gearing, and a Hellenistic drum-treadmill, shown in the
Chhi Chhi Thu Shuo (+1627).

rope was stretched out so that both ends were at ground level, winding gear was buried there to keep it taut, and there were columns near by from which it was suspended as tight as a lute string. Performing girls then mounted upon it barefoot and paraded to and fro. Looking up, they appeared like flying *hsien* (immortals). Some of them met midway, but just leant sideways a little and passed (without difficulty),[a] others wore shoes yet bowed and extended their bodies with ease; some had painted stilts 6 ft. long, others climbed upon heads and shoulders till there were three or four layers of them, after which they turned somersaults down on to the rope. All moved about, never falling, in accordance with the strict time of drums—it was indeed a wonderful sight.

That form of the principle of mechanical advantage which may be obtained by combining many pulley wheels together in complex tackle was well understood by the Alexandrians, and much of Heron's book, the *Elevator*, is concerned with such devices.[b] They are not described in the books of the main Chinese tradition until the +17th century is reached (cf. Fig. 397),[c] but it would be quite unsafe to assume that they were not known or used by artisans and stonemasons in China before that time.[d] Evidence, whether positive or negative, is lacking.

The history of the related machine known as the 'Chinese windlass' is also obscure. Old textbooks of physics sometimes give it as an example of differential motion, for the axle carries two drums of different diameters (Fig. 398) on which the rope winds and unwinds continuously, thus giving useful work and amplification of effort.[e] Hart (2) mentions one of these in the *Codex Atlanticus* of Leonardo, but no one has yet succeeded in finding any Chinese text which would justify the name which the device bears in the West. Davis certainly speaks [f]as though he had seen it in use in China, but Hommel[g] suspects the possibility of a misunderstanding, since Chinese windlasses (e.g. in coal-mines, gem-mines or wells) commonly have two coils of rope on the same drum, one bucket or seat being lowered as the other comes up. The idea of the com-

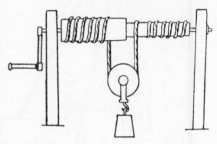

Fig. 398. The differential, or 'Chinese' windlass (Davis, 1).

[a] There were, it would seem, two tightropes across the stage passing over three pulleys, with the main mooring and the winch adjacent to each other on one side.

[b] So also Vitruvius, x, ii. Cf. the anonymous Hussite engineer of +1430; Beck (1), pp. 271 ff. It has usually been assumed, not unreasonably, that pulley tackle was the ancestor of the driving-belt (cf. Beck (1), p. 221), but this derivation is not supported by what the present subsection discloses on the development of technology in China.

[c] *Chhi Chhi Thu Shuo*, ch. 3; see below, p. 215.

[d] A remarkable glimpse of the traditional methods used by Chinese builders for lifting very heavy weights may be found in a note by Tissandier (4), reporting on the erection in 1889 at Peking of some great stone steles like those of the imperial Ming tombs. Pulley tackle was not used at all. The weight was suspended by a stout cable, supported when motionless by two or more purchases round the beams of the scaffolding tower, and raised by the successive actions of a steelyard-like lever exerted through a slip-knot on the main cable, the knot being lowered after each hoist (see Fig. 399). There were five of these levers, each pulled down by five men in the same manner as that of the medieval trebuchet artillery (cf. Sect. 30), but slowly of course, not suddenly. It seems rather probable that such methods were in the classical Chinese tradition of handling heavy loads.

[e] Cf. Ewbank (1), p. 69, who did not think that the device was known in Europe much before 1800.

[f] (1), vol. 3, p. 78.   [g] (1), pp. 2, 118, 119.

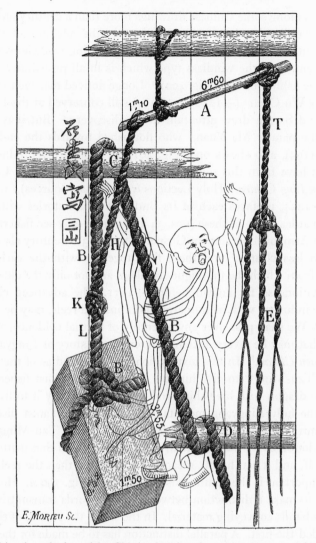

Fig. 399. A traditional method used by Chinese builders for raising heavy weights (Tissandier, 4). A slip-knot hoists the main cable in successive pulls by means of a lever and radiating man-handled ropes like those of pile-drivers and trebuchets. A monk directs the operations. The caption, copied uncertainly by the European draughtsman, seems to say 'Drawn by Shih Sêng-Shih'.

pound drum might of course have been derived from this, and the invention may well have been a local one in China which was never described in literature.[a] Moseley,[b] writing about 1840, says that 'a figure of the capstan with a double axle was seen by

---

[a] A point of interest here is that a differential drum, in the shape of two gear-wheels of unequal diameter fused together, does occur in one Chinese machine, namely the south-pointing carriage as constructed in +1107 (see below, p. 292). It was called a 'tier-wheel' (*tieh lun*[1]). It is very unlikely that this instance was unique.

[b] (1), p. 220.

[1] 疊輪

Dr O. Gregory among some Chinese drawings more than a century old'. In any case, the 'Chinese windlass', whatever its origin,[a] is the ancestor of the well-known Weston Differential Purchase, widely used in engineering workshops today.

A minor invention of the windlass type which is in all probability Chinese is that of the reel on the fishing-rod (see Fig. 400).[b] Lodge noticed that such a reel was shown in a painting by Wu Chen[1] (+1280 to +1354) still conserved at the Freer Gallery in Washington, and Sarton drew attention to his discovery.[c] But it is not the oldest example, for the painter Ma Yuan,[2] who flourished towards the end of the +12th century (c. +1195), also shows one (Fig. 401), in a picture reproduced forty years ago.[d] We even have from this same period printed illustrations in Chinese culture. The *Thien Chu Ling Chhien*[3] (Holy Lections from Indian Sources),[e] printed between +1208 and +1224, prefaces each of its Buddhist moral stories with a wood-block picture, and in at least two of these (nos. 34 and 54) we may see fishermen using rods with reels. An Armenian parchment Gospel of the +13th century also seems to show a reel, though less clearly;[f] and these five, together with the early 17th-century encyclopaedia from which the first figure is taken, are not only the oldest, but also the only representations so far known before +1651. The advanced character of the Chinese textile industry, with its numerous bobbins and reels, may be connected with the invention.[g] We shall pass over any assembly of textual evidence, and draw attention only to what may be a very ancient reference, the story of Lingyang Tzu-Ming[4] in the *Lieh Hsien Chuan*.[h] In this text, which may be dated as of the +3rd or +4th century, Tou Tzu-Ming[5] (to give him his real name), a Taoist fisherman, who once caught a white dragon with his fishing-rod, was rapt away by it to live the life of an immortal on the holy Lingyang mountain. Years later, a Taoist disciple came and asked the country people whether the *tiao chhê*[6] of Tou Tzu-Ming still existed—intending, no doubt, to try his fisherman's luck in turn. All that matters for us is the expression itself, and it can hardly mean anything other than the reel of the rod.

Far more important is the continuous driving-belt, Fig. 390 n. Here at the outset it is necessary to make a distinction between belts or cords transmitting power,[i] and similar endless bands conveying material. In the West, the second of these uses seems to have preceded the first. A parallel distinction has to be made for the more sophisticated and more efficient chain-drive worked by sprocket-wheels and thus overcoming the demon of slip. It will be convenient to deal first with belts and then with chains.

[a] It was, of course, patented in +1788, by James White (Dickinson, 5).

[b] In this illustration the rod is called the 'fishing-rod for turtles (*tiao pieh*[7])'. *San Tshai Thu Hui*, Chhi Yung sect. (+1609), ch. 5, p. 33*b*; copied in *TSCC, I shu tien*, ch. 14, p. 12*b*.

[c] (1), vol. 3, p. 237; followed by Lynn White (7), p. 159.

[d] Anon. (11), Chinese sect., no. 11; Sirén (6), vol. 2, pl. 59, (10) pt. 1, vol. 3, pl. 290. Like Grousset (6), p. 203, he passes this by, oblivious of its technological interest (vol. 2, p. 116).

[e] Recently reproduced photolithographically under the care of Chêng Chen-To.

[f] The especially close links between Armenia and China will be remembered (Vol. 1, p. 224, etc.), and Armenia may well have been the channel of transmission of this technique.

[g] Cf. p. 268 below, and Hommel (1), p. 130.   [h] Ch. 67, tr. & comm. Kaltenmark (2), p. 183.

[i] Liu Hsien-Chou has now devoted a special paper (5) to Chinese inventions in power transmission.

¹ 吳鎮   ² 馬遠   ³ 天竺靈籤   ⁴ 陵陽子明   ⁵ 竇子明
⁶ 釣車   ⁷ 釣鱉

鱉 釣

Fig. 400. Another use of the windlass pulley, the reel on the fishing-rod. Fishing for turtles (*tiao pieh*), from *San Tshai Thu Hui* (+1609), Chhi Yung section, ch. 5, p. 33b.

Feldhaus[a] could adduce no evidence for the driving-belt in Graeco-Roman antiquity, though it seems strange that it should not have occurred to minds such as Heron and Ctesibius. In its purely conveyor form, it may have been present in some of the earliest examples of the 'chain of pots', the water-raising device known as the *sāqīya*,[b] if any of these were ever made with a wheel at the lower end of the band's excursion. But this was no power-transmitter. It is true that in their reconstruction of the military engines described by Biton,[c] Rehm & Schramm (1) showed a driving-belt within the heavy wooden automobile tower (*helepoleōs*, ἑλεπόλεως) invented by Poseidonius.[d] The tower, more than 60 ft. high, and mounted on four wheels, was intended for approaching fortified walls and laying an assault drawbridge upon them; so as it was natural that the motive power should be protected, a man-handled driving-wheel within was made to work upon the wheels. But neither the text nor the Byzantine MS. illustrations say anything of a driving-belt,[e] and from the point of view of engineering probability one might guess that if a chain-drive were not used, directly geared connection with the road-wheels would have been the only alternative. Apart from the early +14th-century textile machinery which we shall discuss in a moment, the earliest picture of a driving-belt which European historians of engineering have so far been able to bring forward[f] is that of a rotary horizontal grindstone in the MS. of the anonymous Hussite engineer (+1430). Even in the 17th and 18th centuries, representations of driving-belts remain rare.[g]

If Asia is to be accorded greater credit for the development of the driving-belt, this may well turn out to be due to the origin there of the only really long-staple textile fibre of antiquity, silk. Now the spinning-wheel (*fang chhê*[1]), in its various forms, was a machine for which the continuous rotary motion assured by a driving-belt was indispensable. But this was (at first sight) an integral part of the technology of short-staple fibres. Naturally enough, its prominence in our own Western culture has long obscured the fact that it had precursors or predecessors in quite another part of the world, designed to handle a quite different type of fibre, for which continuous rotary motion was more important still.

About the origins of the spinning-wheel there has been great uncertainty. We ourselves were long inclined to agree with a widespread view that they should be sought in India,[h] since that culture-area was no doubt the home of cotton culture and cotton

---

[a] (1), col. 1184.   [b] See pp. 352 ff. below.

[c] Biton's *floruit* was *c.* −239; he served Attalus I, king of Pergamon. But his text may well be as late as the +1st century (cf. Sect. 30 below).

[d] Poseidonius was one of the engineers of Alexander the Great, like Diades; his activities must have fallen in the period −334 to −323.

[e] 'What the soldier said is not evidence', as Dr Drachmann remarked in this connection.

[f] E.g. Frémont (12); Feldhaus (1), col. 955; Berthelot (4, 8); Beck (1), p. 282; Singer *et al.* (1); Woodbury (3, 4).

[g] Cf. Beck (1), pp. 408 (Zeising, +1613), 521 (Verantius, +1616), 534 (da Strada, +1629). Crossed belts, to reverse the direction of rotation of the wheels, appear first in the Luttrell Psalter, then in Leonardo (Beck (1), pp. 364, 443) and Cardan (+1550, Beck (1), p. 165). Flat belts and wire cables only in the 19th-century. Another application was the rope railway, which came surprisingly early (+1411), cf. Feldhaus (1), col. 1023, (5).

[h] See, for instance, Horner (1); Born (3) and Schwarz (1).

[1] 紡車

PLATE CXLVII

Fig. 401. 'Angler on a Wintry Lake'; the earliest illustration of the fishing-rod reel, a painting by Ma Yuan, *c.* +1195 (from Sirén, 6).

PLATE CXLVIII

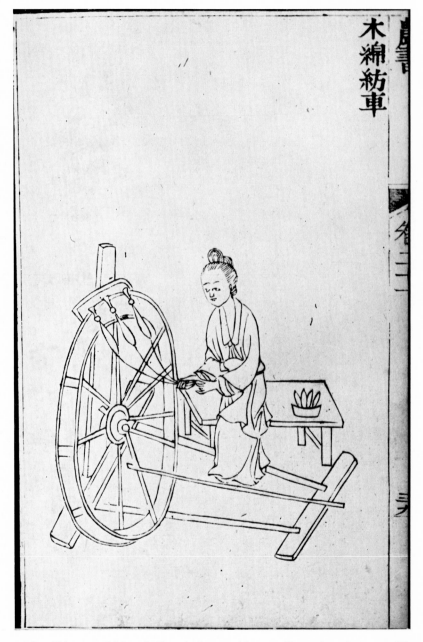

Fig. 402. The Chinese multiple-spindle spinning machine, an illustration of +1313 (*Nung Shu*, ch. 21, p. 29*b*). The three spindles are all rotated by one continuous driving-belt.

PLATE CXLIX

Fig. 403. Contemporary photograph (Hommel, 1) of a three-spindle spinning machine similar to that in the preceding figure. The treadle in these traditional devices is not a crank, but has a direct connection with the wheel at one end and is pivoted on a rough kind of universal joint at the other.

PLATE C L

Fig. 404. The rimless driving-wheel, a form in which the ends of the spokes are connected by thin cords to form a 'cat's cradle' on which the driving-belt is carried. Cotton spinning at a cave-dwelling in Shensi, *c.* 1942.

PLATE CLI

Fig. 405. The oldest representation of the spinning-wheel yet known from any culture; a painting attributed to Chhien Hsüan and in any case datable about +1270 (from Waley, 19). A son is saying farewell to his mother, who turns the crank handle with her right hand and spins the yarn with her left.

PLATE CLII

Fig. 406. The block-spoked driving-wheel, a form in which the ends of the spokes bear grooved blocks on which the driving-belt is carried. Here the five spindles on the semi-lunar mounting are used for spinning hemp, flax, etc. (orig. photo. Tunhuang, 1943). Cf. Figs. 388, 389.

technology. Lynn White (5), however, who has most recently raised the question again in challenging form, believes that this is a false clue and that the spinning-wheel was invented in Europe. The evidence at our disposal inclines us to support his view on India but at the same time impels us to describe the much stronger indications which point to Chinese textile technology as the focus of origin of all such belt-drive machinery. The matter is of great moment, for the spinning-wheel embodied not only power-transmission by 'wrapping connection' but also one of the earliest uses of the flywheel[a] principle as well as a form of mechanical advantage embodied in differential speeds of rotation.

That the spinning-wheel appears in Europe very late has long been known. While +15th-century drawings of it are numerous,[b] the oldest datable Western illustration is usually taken to be that in the Luttrell Psalter[c] of c. +1338, though other well-known ones[d] have come down to us from about the same time. But Chinese icono-graphy has precedence. The multiple-spindle cord-making machine of Leonardo[e] is an almost exact copy of the Chinese multiple-spindle spinning machines illustrated from +1313 onwards. These are usually shown with driving-wheels having integral rims (cf. Fig. 402).[f] This drawing may be understood by comparing it with Fig. 403, a contemporary photograph,[g] where again we see the remarkable direct treadle drive.[h] The presence of three and even five spindles, all driven by one belt in these machines, early in the +14th century, seems to give them a stamp of maturity, and suggests that at that time they had already long been developed. Rimless driving-wheels are also common in Chinese textile technology. In one type the outwardly diverging spokes are connected by thin cords so as to form a bed or 'cat's cradle' (as in the 'spoke-reel') which carries the driving-belt, while in others it passes over grooved blocks set at the end of the spokes. An apparatus of the former type is being used by the old soldier in Fig. 404, sitting outside a Shansi cave-dwelling during the Second World War. The visitor would have been even more impressed if she had been aware that this machine

---

[a] Lynn White (5) sees significance in the flywheels which occur in the *Diversarum Artium Schedula* of Theophilus (early +12th century), who used them on pestles for grinding gold illumination paint (cf. Theobald (1), pp. 14, 19, 174, 191). This does not seem to us very relevant as the potter's wheel is so ancient a thing, and no driving-belts are mentioned by Theophilus.

[b] Gille (14), p. 646, fig. 587, gives one from Gaston Phébus' *Livre de Chasse*. The flyer is present by +1480, as we know from the *Mittelalterliche Hausbuch* (Anon. (15), Essenwein ed. pl. 34a, Bossert & Storck ed. pl. 35); cf. Patterson (1), fig. 168; Feldhaus (1), fig. 710. For Leonardo's automatic flyer worked by a half-gear device (cf. p. 385), see Ucelli di Nemi (3), no. 71; Feldhaus (1), col. 1063, (18), p. 156; Beck (1), p. 108.

[c] Patterson (1), fig. 167; Schwarz (1).

[d] Carus-Wilson (2) illustrates two (pl. 11, b, c) from a British Museum MS. of the *Decretals of Gregory IX* (10 E IV) which she dates about +1320. See also Usher (1), 1st ed. p. 230, 2nd ed. p. 267. Usher took his example to represent a quilling-wheel (see immediately below), I do not know why. Was it not a bad redrawing from the *Decretals*?

[e] See Beck (1), p. 454; Feldhaus (1), col. 1022. There is no reason for insisting that everything Leonardo sketched must have been original. Later (Sect. 31) we shall find evidence which suggests that Chinese textile-machine designs reached Europe in the time of Marco Polo.

[f] For hemp, *Nung Shu*, ch. 22, p. 6a, followed e.g. by *SSTK*, ch. 78, p. 11a. For cotton, *Nung Shu*, ch. 21, p. 29b, followed e.g. by *SSTK*, ch. 77, p. 6a. For doubling cotton, cf. *Nung Shu*, ch. 21, p. 32b, followed e.g. by *SSTK*, ch. 77, p. 19a.

[g] Hommel (1), p. 175.

[h] This dispenses, strictly speaking, with any crank connection; we shall return to the matter on pp. 115, 236 below.

was identical in every particular with that in the picture which qualifies as the oldest representation of a spinning-wheel yet known from any culture. We see it in Fig. 405, part of a painting attributed to Chhien Hsüan [1] and in any case datable about +1270; a dutiful son is taking leave of his mother, who turns the crank handle with her right hand and spins the yarn with her left.[a] The second arrangement is seen in Fig. 406, a photograph which I took at Tunhuang in Kansu in 1943, where this kind of spinning-wheel was used for the twisting of threads of hemp or heavy flax, or for making thick grain-sack yarn from the scrapings of hides.[b] The construction here invites comparison

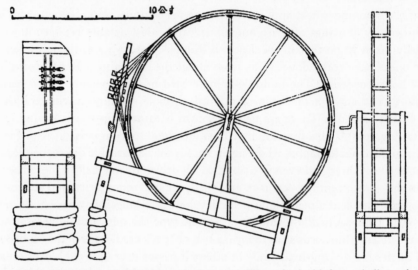

Fig. 407. Scale drawing of a Szechuanese spinning- or quilling-wheel with four spindles each rotated by a separate driving-cord from the main wheel (Than Tan-Chhiung, 1). With multiple driving-cords no semi-lunar mounting for the spindles is necessary. Scale 1½ in. to 1 metre.

with the curious type of composite wheel built in the Han period (p. 81 above), and again raises the problem of a connection between vehicular and stationary wheel build.[c] In comparing these pictures it is interesting to find a prominent semi-lunar spindle mounting[d] in Figs. 402, 403 and 406, where the driving-belt is single. But this disappears when multiple driving-cords are used, one for each spindle, as in the silk-throwing[e] machinery which we shall describe in Section 31, or the hand-crank spinning-wheel of which Fig. 407 is a scale drawing.[f]

[a] See Waley (19), pl. XLVII and p. 240.

[b] The subject of driving-belts is bound up with the cord- and rope-making industry, but lack of space forbids a detailed account of it. Particulars will be found in accessible secondary sources such as Hommel (1), pp. 5ff.; Esterer (1), p. 145; Dickinson (3).

[c] Even popular nomenclature has often been shared by vehicular and stationary wheels; cf. p. 267 below.

[d] This continues unchanged in 18th-century Europe; see Diderot (3), entry under Corderie in *Planches*, vol. 20, pl. II.

[e] To 'throw' silk is to combine together several filaments or fibres into a thicker yarn while at the same time giving it a specific twist generally contrary to that of the component fibres.

[f] Than Tan-Chhiung (1), p. 25.

[1] 錢選

Thus, to sum up, the Chinese illustrations of the spinning-wheel have a clear priority over those of Europe. The literary sources have not yet been fully explored on either side. It is generally agreed[a] that the first European mention occurs in the statutes of a guild at Speyer, where about +1280 it is laid down that 'wheel-spun' yarn could be used for the weft but not for the warp. The presence of the fully developed machine in Chinese paintings and drawings of this time is clear, but we cannot as yet give earlier textual references.[b] During the +13th century, cotton culture was spreading over China for the second time, probably from Sinkiang, but this may not have been any limiting factor for from quite ancient times the Chinese had been using fibres other than silk which needed spinning, notably hemp and ramie. The spinning-wheel may thus have originated at any time after the Han.

Actually, it is not necessary to go outside silk technology at all to look for the birth of the spinning-wheel. The Chinese never wasted anything, and thus they early developed means of dealing with waste silk on cocoons from which the moths had escaped, and coarse silk which could not be wound off from cocoons in the classical way. In fact, if the very beginnings of sericulture were concerned with 'wild' silk-worms, as we must surely assume, this problem would have been the oldest of all. In +1313 Wang Chên shows us[c] the making of floss silk for wadding, etc., from waste cocoons agitated in boiling water;[d] the best of this goes, he says, to make *mien*,[1] the coarsest to make *hsü*.[2] On the following page,[e] he shows us a woman combining together the longer strands end to end by hand with the aid of a 'twisting spindle' (*nien mien chu*[3])—in fact spinning: the production of yarn for rough silk fabric. Although he does not depict any kind of spinning-wheel in use for this purpose, he says that what she is doing is a substitute for it.[f] And then when we turn to recent and contemporary descriptions of the Chinese silk industry we find at once that true spinning-wheels were and are in fact traditionally used for making silk yarn from short lengths directly from the wild cocoons themselves.[g] Thus we may well have here the real point of origin of all spinning-wheels.[h]

Lynn White (5) believes that the immediate precursor of the Western spinning-wheel, from which quite small adjustments would have generated it, was the quilling-

---

[a] Keutgen (1), p. 373; Lynn White (5), p. 518, (7), pp. 119, 173; Gille (14), p. 644.

[b] *TPYL*, ch. 826, pp. 4 aff. unfortunately does not help, as the quotations are all about spinning in general, and describe no techniques. Unfortunately also the *Kêng Chih Thu* (see immediately below) deals only with silk, and with sericulture only in its most developed and standard form.

[c] *Nung Shu*, ch. 21, pp. 23b, 24a, followed e.g. by *NCCS*, ch. 34, pp. 16a, b, 17a, and *SSTK*, ch. 75, pp. 20a, b.

[d] Potash and vegetable detergents, e.g. from the soap bean, were added. Cf. Needham & Lu (1).

[e] *Nung Shu*, ch. 21, pp. 24b, 25a, followed e.g. by *NCCS*, ch. 34, p. 17b.

[f] *Ko tai fang-chi chih kung*.[4] For a contemporary description of the whole process (1880), see F. Kleinwächter in Anon. (45), p. 58.

[g] Cf. J. A. Man for Niu-chuang in Liaoning, F. W. White for the Hankow region, and F. Klein-wächter for Chiangsu and Chekiang, all in Anon. (45), pp. 12, 35, 58.

[h] This would be a refreshing paradox like that of the stern-post rudder, which was invented (as we shall see in Sect. 29) precisely in a culture the ships of which had no stern-posts. Here the seminal environment for the short-staple invention would have been the continuous rotary motion demanded by the long-staple industry.

[1] 綿   [2] 絮   [3] 撚綿軸   [4] 可代紡績之功

wheel for winding yarn on to the bobbins of the weavers' shuttles (*shu*[1], *suo*[2]). This also had its driving-wheel, and a small pulley, mounted on bearings in a framework and connected by an endless belt. But though he could find no evidence for its existence outside Europe, in fact the Chinese pictures and references again long ante-date the European. A fairly clear representation is in the Ypres *Book of Trades*,[a] *c.* +1310, and it is almost certain that the textile wheels in the famous windows of Chartres cathedral (+1240 to +1245)[b] must be interpreted as quilling-machines. There is no earlier sign. At the other end of the Old World we may start with the *Nung Shu* of +1313 which gives a description and two illustrations[c] of the 'weft-machine' or quilling-wheel (*sui chhê*[3] or *wei chhê*[4]). Thence we may go back to the *Kêng Chih Thu*[5] (Pictures of Tilling and Weaving), put together by Lou Shou[6] about +1145. The later editions[d] do not show a quilling-wheel under the rubric 'Weft' (*wei*[7]), but by good fortune that of +1237 clearly does (Fig. 408).[e] Since five reels are placed beside it on the ground, however, it must be in use for twisting and doubling, i.e. 'throwing', the silk on to a small reel, rather than for quilling on to a shuttle bobbin; unless indeed the operations were at that time combined.[f] A modern engraving of a

[a] Gutmann (1); Patterson (1), fig. 183.   [b] Delaporte (1), vol. 2, pl. cxxix, vol. 3, pl. cclxxi.
[c] Ch. 21, pp. 13*b*, 14*a*, *b*. Copied verbally, and iconographically closely, by all the later agricultural encyclopaedias, e.g. *NCCS*, ch. 34, pp. 11*a*, *b*; *SSTK*, ch. 75, pp. 15*a*, *b*.
[d] Of +1462 (Ming) and +1739 (Chhing), reproduced by Franke (11).
[e] The series is reproduced by Pelliot (24), and this is his pl. LVII.
[f] Lynn White (5), p. 518, courageously trying to draw upon Chinese evidence, found that what he could get of it was rather puzzling. The commentary of Franke (11), p. 177, which reproduced the *SSTK* quilling-wheel with crank handle of +1742, is not on any of Lou Shou's poems but on one of the prose passages (p. 134) provided by the committee of literati which embellished the +1739 edition of the *Kêng Chih Thu*. In their elucidation of the picture entitled 'Weft' they mention only the throwing of the silk on to a drum. Thus Franke himself was puzzled too. The trouble is that in the different editions of the 'Pictures of Tilling and Weaving', a number of different operations are shown under a single rubric *Wei*[7] ('Preparing the Weft'). The Sung series (+1145 to +1275) shows as its no. 21 a quilling-wheel in use for doubling and twisting, as we have just noted. No. 21 of the Ming series (+1462), reproduced by Franke (11), pl. LXXXIX, shows something quite different, the winding of silk from the big reels of the reeling-machine (*sao chhê*[8]) on to smaller reels for further use (*lo chhê*[9]). This is often described in the agricultural treatises, cf. *Nung Shu*, ch. 21, pp. 15*b*, 16*a*, *b*, or later *SSTK*, ch. 75, pp. 13*a*, *b*. Then no. 16 of the second Chhing series (+1739), reproduced by Franke (11), pl. XC, on which the literati were commenting, shows twisting and doubling (throwing) being done from four small reels on the ground, not to a reel or bobbin on a quilling-wheel but to a large drum on a special stand. Finally, no. 20 of the Ming series (Franke (11), pl. XCV) shows, we believe, a similar apparatus for doubling and twisting, essentially like that of no. 16 of the second Chhing series. Either its title *Ching*[10] ('Warp') is a +17th-century mistake, or else more probably the machine was used at that time for warp yarn as well as for weft. The *Wei* illustration in the first Chhing edition of +1696 is the same as that of the second save that it still conserves Lou Shou's original poem.

In his important later book, Lynn White (7), esp. pp. 111, 114, unfortunately mishandled the evidence of the *Kêng Chih Thu*, not realising that Franke (11) can only be used in conjunction with the subsequent discoveries of Pelliot (24). We are by no means limited, as he believed, to the witness of a +15th-century edition. About +1275 the painter Chhêng Chhi (see p. 167 below) made a series of illustrations based on those of the printed edition of Wang Kang & Lou Shao in +1237, illustrations which cannot have diverged far from those of +1210, and no doubt resembled fairly closely the original paintings of Lou Shou himself (+1145). Chhêng's scroll was rediscovered in the +18th century and presented to the Chhien-Lung emperor, who caused its contents to be engraved on stone in the Imperial Palace, himself adding a historical introduction (+1769). Although the stones may now have disappeared, rubbings from them have been preserved in the Semallé Scroll, which Pelliot (24) most

[1] 杼   [2] 梭   [3] 繀車   [4] 緯車   [5] 耕織圖   [6] 樓璹
[7] 緯   [8] 繅車   [9] 絡車   [10] 經

PLATE CLIII

Fig. 408. Quilling-wheel in use for silk-throwing and the preparing of the weft; a picture from the *Kêng Chih Thu* (+1210), edition of +1237 (Semallé Scroll). The poem in the background of the scene is one of those by the Chhien-Lung Emperor, and not that of Lou Shou translated in the text (p. 107).

PLATE CLIV

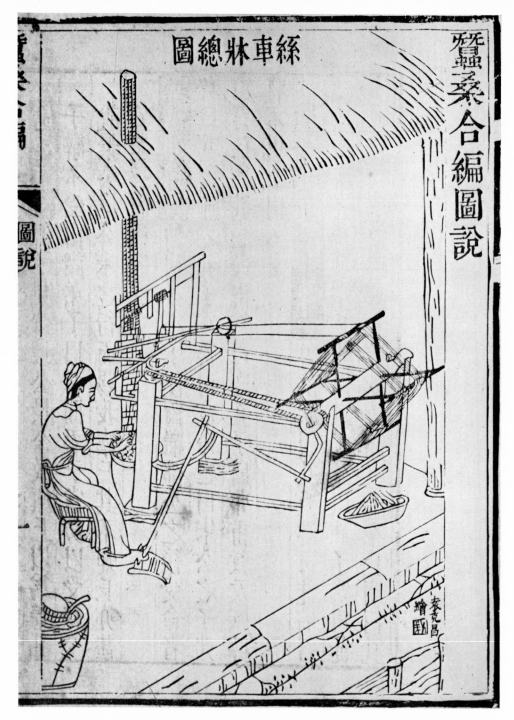

Fig. 409. The classical Chinese silk-reeling machine (*sao chhê*, here *ssu chhê chhuang*), an illustration from the *Tshan Sang Ho Pien* of Sha Shih-An *et al.* (1843). Although this sericultural treatise is so late, it is strictly traditional in character, closely following +17th-century models, and corresponding in every detail of its mechanical descriptions with the literary evidence of Chhin Kuan's late +11th-century *Tshan Shu*, which it faithfully illustrates. The individual fibres of silk are drawn from the cocoons in the heated bath, passing through guiding eyes and over rollers before being laid down on the main reel. There they form broad bands, the fibres oscillating from side to side because they pass under hooks on the simplest form of flyer, the ramping-arm. All the motions of the machine originate from a single treadle action of the operator; this rotates the main reel by means of a crank, but at the same time also a small pulley (at the other end of the frame) which is fitted with an eccentric lug, thus effecting the regular excursion of the ramping-arm. The power is transmitted by a driving-belt. This machine is of importance for the history of technology in several distinct ways (cf. pp. 382 and 404). The inscription in the bottom right-hand corner says 'Drawn by Yuan Kho-Chhang'.

very similar machine in use for doubling is given by de Bavier[a] in his eye-witness description of the Japanese rural silk industry (1874). At the right-hand side of the *Kêng Chih Thu* picture of +1237 we find the original poem of Lou Shou, which runs as follows:[b]

> Steeping the weft, again,[c] is part of the mystery,
> And with hands as cool as the bamboo shoots of spring,
> The country maidens marry two fibres of silk
> And twist them together to one inseparable thread
> Fitted to play its part in a myriad patterns.
> And now, at last, in the late slanting radiance,
> The big wheel's shadow looks like the toad in the moon.[d]
> Yet still, here below, sweet Ah Hsiang speeds her turning,
> And under a deep blue vault the rumbling goes on.[e]

Thus we need have no hesitation in accepting the use of the quilling-wheel for the middle of the +12th century in China, nearly a hundred years before the first European appearance. More remarkable still, texts and reliefs dating from the −1st to the +3rd centuries, consideration of which we must postpone for another part of our discussion (p. 266 below), take the story of the quilling-wheel fully back to the beginning of our era. Such a priority can only be explained by the stimulus of an industry which had to deal with extremely long continuous fibres.

Indeed, if the belt-drive was developed so long ago for winding shuttle-bobbins and for silk-throwing, it is highly probable that it was applied quite early to the most fundamental of all sericultural operations, the reeling or winding-off of the silk filaments from the unbroken cocoons. As we explain elsewhere (p. 2 above and Section 31 below), one can reconstruct completely the classical silk-reeling machine (*sao chhê*[1,2,3]) with its oscillating 'proto-flyer', from a text of approximately +1090, the *Tshan Shu*[4] (Book of Sericulture) of Chhin Kuan,[5] by the aid of +17th-century illustrations[f] (Fig. 409). In this apparatus the main reel on which the silk is wound is powered by a treadle motion, and the ramping-arm of the flyer activated simultaneously by a subsidiary belt-drive.[g] The machine is clearly recognisable in the +1237

---

carefully described and edited. We thus have an authentic record of Sung technology, and if throughout the present work we do not always press for a dating of +1145 as would be not at all unreasonable, we believe that this earliest extant version of the 'Pictures of Tilling and Weaving' does authorise firm conclusions for the beginning of the +13th century (+1210 and +1237). In all our references to it, the implication of the earlier +12th-century date is to be borne in mind.

[a] (1), p. 134 and pl. II, fig. 1. The quilling-wheel is shown in pl. III, figs. 2 and 3.

[b] Tr. auct. with Lu Gwei-Djen. The text agrees with that in the separately transmitted *Kêng Chih Thu Shih*, p. 8a.

[c] This refers to one of the de-gumming processes.

[d] The wheel in this line is clearly the driving-wheel of the quilling or throwing machine. Its outstretched spokes without a solid rim make it look like a spread-eagled toad (*chhan*[6]). Later on we shall see that another term with the same meaning (*hsia-ma*[7]) was often used for the sprocket-wheel of the square-pallet chain-pump; cf. pp. 345 ff.

[e] There is a significance in this noise (see p. 269 below.)

[f] Notably *TKKW*, Ming ed. ch. 2, p. 33a; less well drawn in the later *editio princeps*, ch. 2, p. 20a.

[g] We note elsewhere (p. 404) the historical significance of this duplicity of function.

[1] 繰車   [2] 䌇車   [3] 繆車   [4] 蠶書   [5] 秦觀   [6] 蟾
[7] 蝦蟆

drawing in the *Kêng Chih Thu*.[a] The construction is so important for the history of the driving-belt that readers may already at this stage like to have the actual words of the *Tshan Shu*. It says:[b]

> The pulley (bearing the eccentric lug) is provided with a groove for the reception of the driving-belt, an endless band which responds to the movement of the machine by continuously rotating the pulley (*ku sêng chhi yin i shou huan shêng, shêng ying chhê yün ju huan wu tuan, ku yin i hsüan*[1]).

Although in this apparatus the belt-drive (thus attested for the +11th century) is subsidiary to the main motion, while in the quilling-wheel or spinning-wheel it is essential to it, the machine itself is perhaps more fundamental than either of them.[c] It may be, therefore, that the most ancient use of the power-transmitting endless belt was for the purpose of laying down the fresh silk filaments evenly on the reel. The fact that in the reeling-machine the transmission of power involves no mechanical advantage might also perhaps invite us to think it older. In any case it would be rather surprising if the reeling-machine were not at least as old as the quilling-wheel.

Textile technology is not the only field which gives us driving-belts in medieval China. Our estimate of the date of the first use of an endless rope or cord in that culture may also depend on what exactly the machine was which operated metallurgical bellows by water-power in the +1st century. The 'water-driven reciprocator' (*shui phai*[2]) is a subject of great interest which will be discussed below (pp. 369ff.); here it need only be said that the illustrative tradition starting from +1313 shows a driving-belt working off the main vertical shaft of a 'recumbent' water-wheel, and turning a small pulley with an eccentric lug, thus converting rotary to longitudinal motion for pushing and pulling a piston or fan bellows quite in the manner of the early 19th century. It will be noticed that to the 'connecting-rod' of the reeling-machine, a true piston-rod is here added. It may however be urged that in the Han a simple trip-hammer mechanism was used to depress and release alternately pairs of skin bellows; if so, the driving-belt came later. It was in any case widely used by the second half of the +13th century, which considerably anticipates the heavy industry of Europe. One is tempted to conjecture that the iron-masters and mill-wrights borrowed it from the textile technologists during the Liu Chhao if not in the Han.

Thus far we have been considering 'wrapping connection' as applied to smooth-treaded wheels of the pulley type. But toothed wheels may also participate in this principle, and when gear-teeth were made to enmesh with the hollow spaces in the links of a chain, a new device of great power and potential precision was added to the engineer's stock-in-trade.

---

[a] Pelliot (24), pl. L. Though the artist omitted the driving-belt, it is inescapably implied by the clearly visible ramping-arm and its eccentric lug. The Ming and Chhing drawings (Franke (11), pls. LXXXIII and LXXXIV) get steadily worse, though more and more 'artistic'. Hence Patterson (1), fig. 160, fails to make the machine understandable.

[b] P. 3*a*, tr. auct.

[c] In this paragraph we tread carefully, for it is probable enough that reeling from broken wild cocoons preceded reeling from domesticated whole ones, and this would bring the 'spinning-wheel' back to its most archaic status. But doubtless the ancients of the Shang did their silk 'spinning' by hand.

[1] 鼓生其寅以受環繩 繩應車運如環無端鼓因以旋   [2] 水排

The history of chain-drives (Fig. 390 *o*) is slightly different, as they seem to have been appreciated a little better in Western antiquity than driving-belts. Philon, about −200, described the endless chain of pots so characteristic of later Arabic culture that they bear an Arab name, the *sāqīya*,[a] and so did Vitruvius after him,[b] but this type of chain-pump generally left the band free in the water at the bottom, and therefore did not even bring two wheels tightly together. Even when there were two wheels, power was not transmitted. Philon's own specification, which has been carefully reconstructed by Beck (5),[c] involves a peculiar arrangement of triangular sprocket-wheels to receive the chain at each end of its travel. This design seems so unpractical that many have doubted whether it was ever actually built. Vitruvius (−30) gives no details of his sprocket-wheels, but the *sāqīya* fragments from the great barges of the Lake of Nemi (*c.* +40) clearly show pentagonal ones neatly hollowed out so as to receive the pear-shaped buckets.[d] If Philon's design, then, remained on paper, we might assume a date of about the end of the −2nd century as the time of practical introduction of the device.[e]

Philon of Byzantium is also associated with another use of the endless chain, namely in his remarkable magazine or repeating catapult.[f] According to the reconstructions based on the text,[g] there were two endless chains, one on each side of the stock, moving backwards and forwards round pentagonal cog-wheels, the hind one of which was operated by winch handspikes on the same axle. The string of the torsion catapult could thus be drawn back by a claw attachment fixable at any point to the endless chains, and as this was done, a new arrow fell into the groove from a magazine above. This machine was very ingenious,[h] but there is great doubt as to whether it was ever used on any extensive scale, or even constructed at all. There is certainly no evidence for its use, either textual or archaeological. Moreover, Philon himself minimises its value, saying that there is no point in shooting off so many arrows in the same direction, where they will only be picked up by the enemy and shot back again.[i] In any case the principle is what matters here, and it is interesting to note that while these endless chains transmitted power which was applied to twisting the catapult sinews, they did not transmit power from shaft to shaft, and hence they were not in the direct line of ancestry of the chain-drive proper.

[a] His text is translated by Carra de Vaux (2), pp. 224 ff.; cf. Usher (1), 1st ed. p. 81, 2nd ed. p. 130; see also Ewbank (1), pp. 122 ff. See pp. 352 ff below.

[b] x, iv, 4.

[c] In the interpretation of Drachmann (2), p. 66, the motion of the upper wheel of the bucket-chain was derived from a separate undershot water-wheel by a true chain-drive, but we prefer the reconstruction of Beck, which assumes that the lower wheel of the bucket-chain was a water-wheel itself (as perhaps in some of the later Chinese types, cf. p. 355 below).

[d] See Moretti (1), p. 33; Ucelli di Nemi (2).

[e] At a comparable time in Han China, the noria was probably known (cf. p. 358 below). The *sāqīya* was a later introduction there.

[f] *Mechanica*, IV, 52 ff. The invention was ascribed by him to a predecessor, Dionysius of Alexandria.

[g] Beck (3); Schramm (1); Diels (1); Diels & Schramm (2).

[h] As we shall see later (Sect. 30e), the Chinese crossbow tradition also generated magazine or repeating weapons. These, however, were armed (made ready for action) by lever mechanisms only. Contrasting (as so often) with Greek proposals, the Chinese magazine crossbow was a very practical instrument for certain services.

[i] IV, 59.

This seems to have come with much delay in Europe. Historians of occidental engineering[a] have generally been unable to find any chain-drive in the true sense until the 18th or 19th century. About +1438 Jacopo Mariano Taccola figured an endless hanging chain for manual use like those employed for small hoists in engineering workshops today.[b] About +1490 Leonardo da Vinci made elaborate sketches of hinged-link chains, and used them for purposes such as turning the wheel-lock of a gun.[c] This transmitted the power of a spiral spring, but the chain was not endless. In +1588 Ramelli depicted a chain (again not endless) in oscillatory motion over the driving-wheel of a double-barrel pump (cf. Fig. 613),[d] but he did also describe in three places a true continuous chain-drive,[e] though the wheels so enwrapped seem often to be pulleys rather than sprocket-wheels. Not until about +1770 did de Vaucanson develop an industrially practical chain-drive. This he used for his silk reeling and throwing mills, and towards the end of his life was much occupied with a machine for making its standard links.[f] Hinged links of female shape to fit the male sprockets on the wheels came with Galle in 1832,[g] after which the chain-drive was applied to cars by Aveling in 1863 and to bicycles by J. F. Trefz in 1869.

In China development took place quite otherwise. The most characteristic Chinese water-raising machine, the radial-treadle square-pallet chain-pump, must necessarily have a sprocket-wheel at each end of the flume. This machine, the *fan chhê*,[1] is shown in the oldest of the Chinese books on agricultural machinery which have come down to us (+1145 onwards), and in all later ones.[h] As will in due course be demonstrated,[i] its invention may be securely placed in the Han (c. +1st century), so that it is little later than the Hellenistic *sāqīya*. It was an extremely practical machine, and from the Han period downwards hundreds of thousands of square-pallet chain-pumps dotted the Chinese countryside. Nevertheless, it was not a power-transmitting machine any more than the chain-pumps of the Roman Empire.

a E.g. Uccelli (1); Feldhaus (1); Singer *et al.* (1); Forbes (2).
b Feldhaus (1), col. 562. Again in Ramelli, see Beck (1), p. 210.
c Feldhaus (1), cols. 444ff., 562; Uccelli (1), p. 75; Ucelli di Nemi (3), no. 9.
d This is in several variations, cf. Beck (1), pp. 213ff. One of Ramelli's pump designs (no. 94) has an endless chain, but it only oscillates back and forth to work a pump beam, its driving-wheel being rotated almost to the full extent of travel possible by a lever, connecting-rod and crank deriving from a power-source (see Beck (1), p. 220).
e Cf. Beck (1), p. 221. The nature of the chain and its wrapping connection with the wheels varies curiously. Ramelli's no. 39 has large square links joined together by two small free round links, and the former seem to fit into depressions in the wheels as if between flat cogs. No. 93 has the same kind of chain but shows shallow projections on the floor of the pulley trough fitting between the square links. In no. 126 the chain is round-linked and passes over the wheels without the slightest sign of sprockets, between forked spikes. This arrangement occurs also as a free-hanging hauling chain in nos. 79, 86, 92 and 168; and as the oscillating chain in no. 94 just mentioned, passing in this case over plain flat-troughed pulley-wheels.
f Many of the designs and models of the great French engineer Jacques de Vaucanson (+1709 to +1782) are preserved in the Conservatoire des Arts et Métiers at Paris, and it is said that his chain-making machine is among them. Unfortunately there is no illustration of his chain-drives in his papers on silk technology (1–3), published by the Académie des Sciences.
g Cf. Eude (1), p. 136.
h See below, pp. 339ff.; meanwhile I refer to Hommel (1), p. 50; Ewbank (1), pp. 149ff.
i Sect. 27g.

1 翻車

But it probably inspired the true invention. When we study later on[a] the work of the builders of mechanical clocks in China from the beginning of the +8th century, we shall find that considerable use was made of real chain-drives—'celestial ladders' (*thien thi*[1]). In the great astronomical clock-tower of +1090 the main vertical transmission-shaft proved to be too long, and was soon replaced by modifications in which the power for the armillary sphere on the upper platform was provided by endless chain-drives, successively shorter and therefore more efficient in the various rebuildings.[b] In this masterpiece of Su Sung's,[2] the uppermost shaft rotated a series of three small pinions in a gear-box underneath the armillary (Fig. 410).[c] But it was by no means the earliest, for there is some reason to believe that the mercury-operated clock built by Chang Ssu-Hsün[3] in +978 also contained a chain-drive. Thus at present it would seem that China rather than Europe is responsible for both of the fundamental inventions of the driving-belt and the chain-drive.

(4) CRANK AND ECCENTRIC MOTION

Of all mechanical discoveries that of the crank (*chhü huai*[4])[d] is perhaps highest in importance, since it permits the simplest interconversion of rotary and reciprocating (rectilinear) motion. 'Continuous rotary motion', writes Lynn White in an admirable passage, 'is typical of inorganic matter, while reciprocating motion is the sole form of movement found in living things. The crank connects these two kinds of motion; therefore we who are organic find that crank motion does not come easily to us.... To use a crank, our muscles and tendons must relate themselves to the motion of galaxies

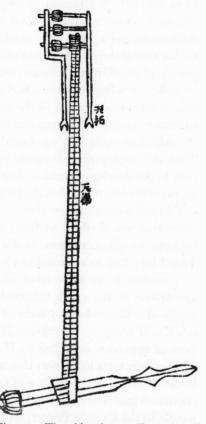

Fig. 410. The oldest known illustration of an endless power-transmitting chain-drive, from Su Sung's *Hsin I Hsiang Fa Yao* (+1090), ch. 3, p. 25 b. This 'celestial ladder' (*thien thi*) was used in the first and second modifications of the great astronomical clock-tower at Khaifêng for coupling the main driving-shaft to the armillary-sphere gear-box. See Fig. 652a and p. 457 below.

---

[a] Sect. 27j.

[b] Cf. below, p. 457, and Needham, Wang & Price (1), pp. 40, 49.

[c] 'The "celestial ladder" is 19·5 ft. long—an iron chain with its links joined together to form an endless circuit, hanging down from an upper chain wheel...and passing round a lower chain wheel on the driving-shaft' (*Hsin I Hsiang Fa Yao*, ch. 3, p. 26a, tr. Needham, Wang & Price (1), p. 47). Exactly what was the nature of this chain we do not know, but the drawing in Su Sung's book gives the impression of double hinged links.

[d] We give, as usual, the Chinese terminology as we proceed (cf. p. 120 below). For the Western terms and their origins, which are particularly interesting, see Lynn White (7), pp. 106ff., 166ff.

[1] 天梯   [2] 蘇頌   [3] 張思訓   [4] 曲柄

and electrons. From this inhuman adventure our race long recoiled.'[a] The principal forms of the device are shown in Fig. 390. The simplest, and perhaps the oldest (or more probably the second oldest), manifestation of the device was born when it occurred to someone that a wheel could be turned by hand more easily if a handle at right angles to its plane were fixed in it at a point near its circumference (Fig. 390p). In due course the direct contact of the worker's arm was replaced by a connecting-rod (shown in p'). Greater leverage was obtained when the handle was mounted on a lug or spoke extending from the wheel's rim, as at (b''), and of course this system may itself have been a modification of the capstan or windlass handspikes (b). As a substitute (whether primitive or degenerative one can hardly say), a piece of wood inserted in the axle at an angle (p'') was found to be capable of performing the office of a crank. Then the crank proper, invisible in (p) and rudimentary in (b''), could manifest itself fully in the developed crank arm or handle as at (q). Conversely, the wheel could also be invisible (cf. p. 89 above), leaving only the crank arm (or eccentric lug, r) which could carry a connecting-rod (r'). Much greater rigidity was acquired by the machine when this was doubled to form a crankshaft (s), and the conversion of rotary to perfectly rectilinear motion, as of a piston, was obtained when the connecting-rod was joined by a link to a second rod as in (t).

According to an impression widely current among historians of technology, the appearance of the crank occurred rather late. Usher says[b] that there is no pre-medieval evidence for any form of crank-handle, and this has been the general view (cf. Haudricourt, 2). Crank-handles (*chhü ping*[1]) indeed appear in many reconstructions of apparatus described by Heron, and others of the Alexandrians,[c] but there is little in their texts to support this, and what little there is may well have been inserted in the +10th century, the time of Qusṭā ibn Lūqā al-Baʿlabakkī, when they were being translated into Arabic.[d] Gille (2) and Forbes (2)[e] find the first evidence of the crank-handle in the Utrecht Psalter, written about +830, where it is seen being used with a rotary grindstone.[f] Trepanation drills with crank-handles have been said to appear in the work of the great +10th-century Spanish Muslim surgeon Abū al-Qāsim al-Zahrāwī (Albucasis).[g] In the first half of the +11th, Theophilus describes the crank-

---

[a] (7), p. 115. I am not sure that the statement is strictly true, as the rotifers may bear witness, but it is broadly so.

[b] (1), 1st ed. p. 119, 2nd ed. p. 160.

[c] Carra de Vaux (1), p. 462; Beck (2), p. 86. Heron's *dioptra* was given a crank handle by Usher (1), 2nd ed. p. 149, but Drachmann (3), p. 127, had already drawn it rightly with little handspikes. Singer (11), p. 87, however, judiciously corrected Singer (2), p. 82. Gille (14), p. 635, seems to err in accepting the genuineness of 'Alexandrian' crank motions. On the archaeological side the critique of Lynn White (7), pp. 105 ff., has disposed of the claim that cranks were fitted to the bilge-clearing chain-pumps of the Lake Nemi ships (c. +40).

[d] My friend Dr A. G. Drachmann, however, feels that certain passages in Oribasius (fl. +362) and even Archimedes, seem to require the assumption of a crank. I hope he will set forth the evidence for this before long.

[e] Pl. 2 and p. 113. Also Lynn White (7), p. 110.

[f] Another MS. illustration, closely similar, occurs in the Psalter of Eadwin of Canterbury (Cambridge, written c. +1150); this is figured by Feldhaus (2), pl. VIII, in colour. Cf. Gille (14), p. 651.

[g] Horwitz (4), but Lynn White (7), p. 170, casts doubt on this. Ibn Sīnā may qualify instead.

[1] 曲柄

handles used in the turning of casting cores.[a] In +1335 Guido da Vigevano illustrated a paddle-boat and a war-carriage or 'tank' propelled by manually turned crankshafts and gear-wheels.[b] By the start of the +15th century the crank had clearly become common,[c] for we see it several times in the Konrad Kyeser MS. (+1405),[d] as also in that of the anonymous Hussite engineer (+1430), where it takes the form of the fully developed (even double-throw) crankshaft.[e] Kyeser uses it for a bucket chain-pump, an Archimedean screw, and a ring of bells; the Hussite for a stamp-mill, a grain mill with right-angle gears, and a paddle-wheel boat. By +1556 the complete two-throw crankshaft is commonplace in Agricola.[f] Gille (2) rightly concludes that no true crankshaft is known before the beginning of the +14th century.

It is not difficult to see how the compound crank or crankshaft was born. The East Anglian Luttrell Psalter of c. +1338 has a rotary grindstone worked by two men using crank-handles placed 180° apart.[g] The cranked windlass for arming or spanning heavy steel crossbows is often considered to date from this period though unimpeachable evidence for it does not antedate +1405 (Kyeser).[h] To originate the form of the crankshaft it was thus only necessary to place two crank-handles in line at the ends of a drum; but mechanical genius must have been required to see the advantages of transmitting torque to or from the centre of the serpentine system, embodying, as it were, the eccentric in the shaft, not only at its two ends.[i] If Guido da Vigevano was not the

---

[a] See Theobald (1), pp. 114, 340, 341.   [b] Cf. Gille (14), p. 651; A. R. Hall (3), (6), p. 726.

[c] Many examples will be found in Hamilton (2); Thomson (1); Gille (14); Feldhaus (20), pp. 223 ff., 236.

[d] Berthelot (5); Usher (1), p. 128. Another example a couple of decades later than this date may be seen in the Prado Museum at Madrid, where a reredos of the Master of Siguenza shows two wheels bearing knives and rotated by very clear crank-handles, all part of a scene of the martyrdom of St Catherine (no. 1336). The same scene, carved a century or so earlier and showing the crank-handles with equal clarity, occurs at the capital of the easternmost column of the south aisle in the beautiful church of Thaxted in Essex.

[e] Berthelot (4); Usher (1); Feldhaus (1), col. 593; Beck (1), pp. 274 ff.; Gille (12), (14), p. 653. This appears again in millwork, excellently drawn, in the *Mittelalterliche Hausbuch* of c. +1480 (Anon. (15), Essenwein ed. pl. 48 a, Bossert & Storck ed. pl. 47).

[f] Hoover & Hoover tr., p. 180.   [g] Millar (1), fol. 78 b.

[h] We regret that we cannot accept the opinion of Mus (2), which Lynn White (7), p. 111, has accredited, that some of the arcuballistae of the Khmer host sculptured on the walls of the Bayon at Angkor Thom in +1185 show arming winches with crank-handles. Not only careful study of the published photographs (references in Sect. 30 e below), but also personal inspection of the reliefs *in situ*, has convinced us that this claim goes too far. It is true, as we shall later show, that rotary arming of crossbows begins in China not in the +14th century but in the +8th (if not earlier)—all the arcuballistae described and figured in the *Wu Ching Tsung Yao* of +1044 (*Chhien Chi*, ch. 13, pp. 6a–12b) have winches—but these are invariably fitted with handspikes rather than cranks. We believe therefore that the Bayon reliefs, which depict artillery introduced by a Chinese expert, Chi Yang-Chün,[1] in +1171, should be interpreted in the same way. For although the hands of the artillerymen are often seen resting on winches, there is no sign of crank-handles at their ends.

[i] Quite soon after the first invention of crankshafts European engineers began to couple them together so that several could be rotated in parallel motion from a single power-source, or vice versa. By +1463 Valturio is drawing a paddle-wheel boat with five crankshafts joined by a single connecting-rod, and his design was printed in +1472 (Lynn White (7), p. 114). About +1490 Francesco di Giorgio Martini sketched a similar pair (*ibid.* fig. 8), and so did Leonardo in his design for a centrifugal pump (Feldhaus (18), p. 47; Beck (1), p. 338). So far as we know, coupled eccentrics of this kind do not occur in China until the treadmill-operated paddle-boats of the 19th century, which may have been influenced by the coupled driving-wheels of locomotives (cf. Fig. 636 and p. 427 below).

[1] 吉楊軍

first to appreciate this, he was of the first generation. But there is more to say about the origin of the crank simple.

Egyptologists have long been acquainted with what may have been a crank drill of primitive brace-and-bit type, which goes back to the Old Kingdom.[a] Fig. 411 (a), taken from Borchardt,[b] shows one such drill, from a relief; including the drilling tool, apparently often of crescent-shaped flint, held in the brace, and the bags of stone above which served as weight and perhaps as fly-wheel. Fig. 411 (b), from de Morgan,[c] and Fig. 411 (c) from Newberry,[d] illustrate the use of the drill for boring holes in stone vessels or hollowing them out, and these drawings are of particular interest since the handle at the top (perhaps a gazelle horn) is oblique (cf. Fig. 390 p''). Klebs reproduced[e] a similar carved relief from a tomb at Saqqara dating from about −2500. He described[f] bored jars of many kinds of hard stones from the Middle Kingdom (basalt, syenite, porphyry, serpentine, etc.). There was even a special hieroglyph for the tool[g] (Fig. 411, d) which seems to depict the brace and bit with a handle more crank-like than it really was.[h] The tool seems to have spread very slowly out of Egypt; Petrie, for example,[i] knew of no Graeco-Roman examples. The earliest European illustrations of the crank-drill or carpenter's brace occur in German, French and Flemish paintings dating from the neighbourhood of +1420 onwards.[j] It is seen also in the +1438 altar panel by the Master of Flemalle in the Mérode Gallery at Brussels.[k] Although the tool was now a crankshaft rather than a crank, the connecting-rod component was of course still the worker's arm, but so also had it been in the case of the ordinary hand-

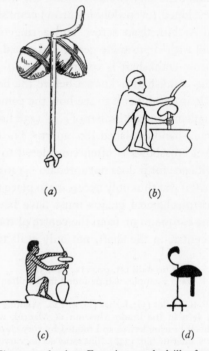

(a)   (b)

(c)   (d)

Fig. 411. Ancient Egyptian crank drills, fore-runners of the brace-and-bit drill, from Old Kingdom reliefs. (a) Oblique-crank flint drill with stone weights, after Borchardt (1); (b) a similar tool in use for drilling out a stone pot, after de Morgan (1); (c) drilling out a vase, after Newberry (1), from the Beni Hasan finds; (d) Middle Kingdom hieroglyph for the crank drill, after Firth & Quibell (1).

[a] Leroi-Gourhan (1), vol. 1, p. 186; Lucas (1), p. 84; Zuber (1).
[b] (2), pp. 142, 143.
[c] Vol. 1, p. 123.
[d] Vol. 1, pl. XI.
[e] (1), p. 82, fig. 66. Copied more accessibly by Childe (15), p. 193.
[f] (2), pp. 107, 115; (3), p. 100.
[g] Firth & Quibell (1), pp. 124, 126.
[h] Assuming that no development had taken place during previous centuries.
[i] (1), p. 39. But he thought he had evidence of an Assyrian one (3), p. 19, pl. XXI. Cf. Aldred (1).
[j] Gille (2, 14); Thomson (1); many examples. Discussion in Lynn White (7), pp. 112 ff.
[k] Mercer (1), p. 206; Heidrich (1), fig. 20.

PLATE CLV

Fig. 412. Oblique crank on a farmer's well-windlass at Kan-nan, Hei-lung-chiang province, Manchuria (orig. photo., 1952). Cf. Hommel (1), pp. 2, 3.

turned grain mill or quern.[a] The first origin of the crank may still have to be sought long before the beginning of our era.[b]

The obliquely pointing position which the handles of some of the ancient Egyptian crank-drills seem to have had is interesting. Assuming for the sake of argument that this almost diagonal substitute for two right angles was a primitive feature, it is striking that it has persisted to the present day in some of the rough windlasses of Chinese miners and farmers (see Fig. 412).[c] It occurs also on well-windlasses in tomb-paintings, notably some of the Jurchen Chin dynasty (+12th century) at Chiang-hsien in Shansi, recently studied by Chang Tê-Kuang (1) and exhibited in 1958 at the Imperial Palace Museum.[d] The oblique 'crank' is also found on some early European hand-querns of about the −1st century (Curwen, 3). The double, or crankshaft, form of this, corresponding to the modern brace-and-bit drill, is the fire-making drill of the South American gauchos;[e] nothing more than an elastic piece of wood bent lightly and twisted like a crank. This, however, though made famous by Darwin and Tylor, is now regarded as of European rather than Amerindian origin.

The most extraordinary kind of oblique crank known to me is the treadle crank found on some Chinese spinning-wheels (Fig. 403) and peculiar to them.[f] In this the treadle is attached to one of the spokes of the wheel nearer the hub than the rim, while its other end is held loosely in the framework; upon being set in motion by the feet as shown in the photograph, one can see that if it were to be copied in an accurate medium such as metal, at least one universal joint would be required. This remarkable

[a] Many consider this the first certain appearance of crank motion, e.g. Lynn White (7), p. 107. But when did it appear? The dating of early European rotary hand-querns is still very uncertain, and the view of Curwen (2) that querns with single vertical peg-handles cannot be placed much before +400 still holds good though nearly three decades of archaeological research have passed since it was formulated. Lynn White discusses the matter in detail. Such a date corresponds to the late Chin period, but as we shall see (p. 189) rotary querns go back in China to the Chhin (−3rd century). A special study of their typology would be necessary to determine the approximate date of appearance of the single vertical peg-handle 'crank' form among them. On the evidence at present available, querns with holes for single vertical peg handles seem to be older in China than in the West by two or three hundred years.

[b] Some reluctance is felt by certain archaeologists in accepting this interpretation, partly perhaps because the crank principle was unappreciated for so many centuries after the first appearance of the ancient Egyptian types. Mr André Haudricourt, for example, would prefer to see in the hieroglyph a representation of the bow-drill (opinion privately communicated). Childe (11) tends to agree with Klebs and von Bissing that the gazelle horn at the top remained stationary, forming a socket or bearing in which the drill could turn. However, as he points out, even if the tool was not one which involved the principle of the crank, it could well mean that the Egyptians had found a new application of rotary motion by −2500. For the potter's wheel had been in use in Mesopotamia for a whole millennium by that time, and in Egypt itself for probably a couple of centuries. Lynn White (7), p. 104, follows Childe; I still think that there is weight in the Egyptian material.

[c] Hommel (1), pp. 2, 3. Possibly querns also (Giglioli, 1). In view of this oblique 'crank-substitute', certain Chinese terms for crank are suggestive—*wan chu*,[1] the 'curved axle'; *chhü chhü*,[2] the 'bent and curved piece'. Cf. 'kurbel' in German.

[d] Also in Chang Tsê-Tuan's painting of +1126, the 'Chhing Ming Shang Ho Thu' (Chêng Chen-To (3), pl. 5), and in the frescoes of the Yung-Lo Kung temple in Shansi (Têng Pai (1), pl. 13), c. +1350.

[e] Frémont (5); McGuire (1), p. 704; Leroi-Gourhan (1), vol. 1, pp. 107, 186.

[f] Cf. pp. 89 above and 236 below.

[1] 彎軸   [2] 屈曲

device seems to be purely Chinese. But it has a certain analogy in the long crank-handles of some medieval European hand-mills.[a]

Historians of occidental technology have had difficulty in dating the common cranked well-hoist, and the most critical recent discussion[b] adduces nothing earlier than the neighbourhood of +1425, a miniature in the *Hausbuch* of the Mendel Foundation in Nuremberg.[c] Chinese medieval well-windlasses therefore take precedence, not only for the oblique crank but also for the regular right-angled variety. The oldest illustration of the latter is in the *Nung Shu*,[d] implying therefore the +13th century, and there can be no doubt about it because the construction is clearly described in the accompanying text.

It has frequently been recognised[e] that the simple quern (*mo*,[1] see Fig. 443), with its upstanding handle near the circumference of the mill-stone, embodies the principle of the crank (cf. Fig. 390 *p*). Elsewhere (pp. 187, 189) we discuss the relative antiquity of this in east and west, but priority cannot yet be determined. In hand-mills throughout the Chinese culture area, a connecting-rod of varying length (besides the worker's arm) was early added to it, certainly by the Sung.[f] Another type (Fig. 444), giving greater mechanical advantage or leverage by increasing the eccentricity,[g] demonstrates Fig. 390 (*b″*). This constitutes, as we saw, a transitional form between the capstan or winch handspikes and the crank. Finally, this type also acquired a connecting-rod of varying length (Fig. 413).[h] In so doing, it gave birth to the eccentric lug system, which was so great a favourite with the medieval Chinese engineers. One finds it, for example, in the +14th-century version of the water-driven metallurgical bellows of the +1st (see on, p. 371), where it turned at higher speed on a small pulley connected with the large driving-wheel by a belt. One finds it also on the silk-reeling machine of the +11th century (above, pp. 2, 108; below, pp. 382, 404), where it operated

---

[a] Bennett & Elton (1), vol. 1, pp. 163, 168. Cf. Leroi-Gourhan (1), vol. 2, p. 162; Jacobi (1); Moritz (1), pl. 15 and pp. 129 ff.

[b] Lynn White (7), p. 167. It is particularly strange that this seemingly obvious application took so long in coming, since the crank had by no means been confined earlier to rotary grindstones. The 'hurdy-gurdy', a fretted stringed instrument, was sounded by a cranked mechanism as early as the beginning of the +10th century (*ibid.* p. 110).

[c] The Foundation is described in some detail by Feldhaus (20), pp. 223 ff. The 'Twelve Brothers of Conrad Mendel's Hospital' were founded by him in +1388; they were lay religious, as if of an order, and continued to work each at his industrial trade. By +1799 when the Hospital was abolished, 799 brothers had enjoyed its shelter. Before +1700 each brother's portrait showed him at his working bench, hence the value of the collection for the history of technology. Feldhaus reproduces a dozen of them.

[d] Ch. 18, p. 26b. The device is indeed badly drawn, in a way which can be appreciated from the more artistic representation in the *Thien Kung Khai Wu*, reproduced in Fig. 574 below. This evidently perpetuated an iconographic tradition which had, as it were, started off on the wrong foot.

[e] As by Haudricourt (1); Horwitz (4, 5); Leroi-Gourhan (1), vol. 1, p. 104; vol. 2, p. 162.

[f] See Hommel (1), fig. 143. This had the advantage that the mill could be operated by several persons instead of only one, and that they would not need to move much from their place.

[g] See also Hommel (1), fig. 156.

[h] It had already done so by the beginning of the iconographic tradition, i.e. in the *Kêng Chih Thu* of +1210 (+1237), see Pelliot (24), pl. xxviii. For a modern photograph see Hommel (1), fig. 159. The early +13th-century (indeed mid +12th-century) evidence for this system in China well antedates the similar systems of the +14th- and +15th-century Italian and German engineers (cf. Gille (2); Lynn White (7), p. 113).

[1] 磨

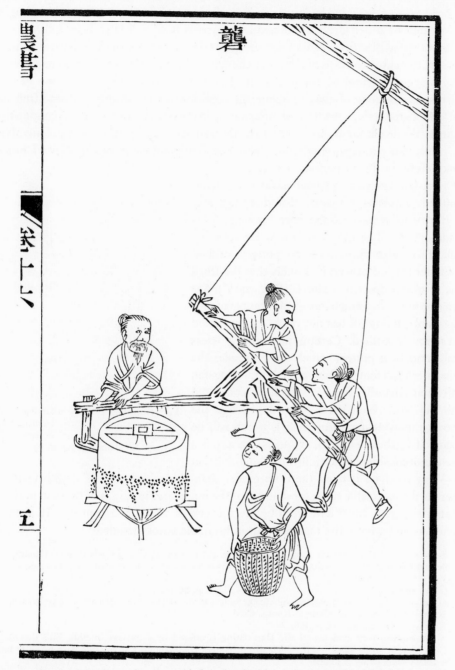

Fig. 413. Crank in the form of an eccentric lug, fitted with connecting-rod and hand-bar to allow of the simultaneous labour of several persons at a grain-hulling mill (*lung*, cf. p. 188 below). *Nung Shu* (+1313), ch. 16, p. 5*a*.

an early type of flyer for laying down the silk in equal layers on the reel, and again was rotated by a belt-drive. This textile apparatus may well have been a direct precursor of the hydraulic blowing-engine, the two being safely referable to the +9th and the +12th centuries respectively, and the latter is a machine of great importance in the history of technology for it embodied, so far as we can see (cf. p. 386 below), the first appearance of that fundamental combination of eccentric, connecting-rod and piston-rod (Fig. 390 *t*) used afterwards in all steam and internal combustion engines. We do not know, however, of any indigenous Chinese machine which involved the crankshaft principle (Fig. 390 *s*), and this seems not to be pictured in a Chinese book before the Jesuit period (+1627).[a]

Crank-handles were, of course, far from being confined to querns in China. Hommel illustrates late forms for rope-making,[b] wire-drawing,[c] silk-winding,[d] etc. The question is how far these go back. From Han tomb models in pottery, studied by Laufer (3) and others, it is likely that the third type of quern described above (Fig. 390 *b″*) was then common. But much more important evidence is available, though it has not yet been assembled and seriously studied. Certain museums[e] possess tomb models of pottery which show, besides the usual quern and foot tilt-hammer, a winnowing-fan with a crank-handle. Fig. 415[f] should be compared with Fig. 414. Now there is no doubt that the rotary winnowing-fan (*fêng chhê*[1]), which will be discussed further below (p. 151 and Sect. 41), is a machine of considerable antiquity in China, and that

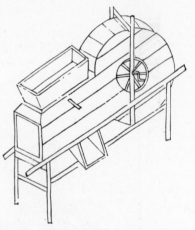

Fig. 414. The Chinese rotary-fan winnowing-machine in its classical form (*Nung Shu*, ch. 16, p. 9*b*), a drawing of +1313. Cf. Hommel (1), p. 77.

the West received it from there no earlier than the middle of the +18th century. There is no reasonable room for doubt that the model in Fig. 415 is of the Han period.[g] That is to say, it must date from before the end of the +2nd century. It is to be saluted as embodying the most ancient indubitable crank-handle.[h]

[a] See *Chhi Chhi Thu Shuo*, ch. 3, pp. 16*b*, 30*b* (Fig. 467). In 1835 Chêng Fu-Kuang (1) illustrated a double crankshaft in the appendix on steamboats which he added to his treatise on optics (see Vol. 4, pt. 1, p. 117 above); Fig. 616.

[b] (1), p. 10.   [c] (1), p. 26.

[d] (1), p. 181. In this technological sphere the oldest term for the crank appears to have been *ni*.[2]

[e] Notably the Nelson Art Gallery, Kansas City.

[f] Cf. Swann (1), pl. 2, opp. p. 378.

[g] Some confirmatory evidence of the Han dating is offered by a passage in Shih Yü's dictionary *Chi Chiu Pien*, p. 50 (−48 to −33), which all subsequent commentators at any rate have taken to refer to the rotary winnowing-machine.

[h] Long after writing this paragraph I was happy to learn from Dr Lynn White that he also regards this Han crank-handle as the most ancient known; cf. his (7), p. 104. Perhaps one might provisionally suggest that the crank-drill of Egypt spread eastwards rather than westwards, the crank then travelling to Western Europe, probably through Byzantium, about the same time as the drawloom (*c.* +5th century). But there is no Byzantine evidence, and that for Islam only very much later.

[1] 風車   [2] 柅

PLATE CLVI

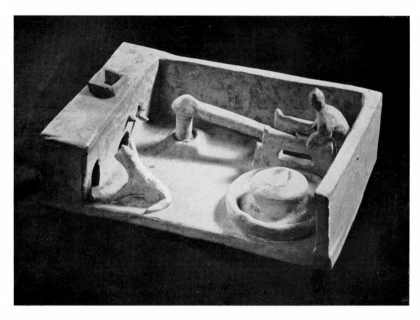

Fig. 415. Farmyard model in iridescent green glazed pottery from a tomb of the Han period (−2nd to +2nd century) in the Nelson Art Gallery, Atkins Museum, Kansas City. On the right a rotary grain-mill and a man working a foot tilt-hammer grain-pounder; on the left a built-in winnowing-machine with hopper and two lower apertures. The crank-handle for working the rotary fan of the winnowing-machine is the oldest representation of the true crank-handle from any civilisation. Model dimensions $8\frac{3}{4}$ in. × 6 in. × $2\frac{1}{2}$ in.

It has been authoritatively said that although 'in China the crank was known, it remained dormant for at least nineteen centuries, its explosive potential for applied mechanics being unrecognised and unexploited'.[a] This is a misapprehension. It is true that Chinese culture did not spontaneously generate modern science and technology, with all the industrialisation that developing capitalism in the West drew forth from it, but within the limits of feudal-bureaucratic society the powers of the crank were widely used and appreciated throughout the Chinese Middle Ages. For three or four hundred years before the time of Marco Polo it was employed in textile machinery for silk-reeling and hemp-spinning, in agriculture for the rotary winnowing-fan and the water-powered flour-sifter, in metallurgy for the hydraulic blowing-engine, and in humbler uses such as the well-windlass. It seems more and more likely that whatever the factors inhibitory to the rise of mechanical and industrial society in China were, technological inventiveness and application was not one of them. Western slowness in developing crank motion offers no less a challenge to historical analysis.

(5) SCREWS, WORMS AND HELICOIDAL VANES

The continuously winding screw-thread, male and female (as in bolt and nut), and the cylindrical worm capable of engaging with an ordinary gear-wheel so that motion may be transmitted between two shafts at right angles, are the most outstanding examples of mechanical systems apparently unknown to Chinese engineers and artisans until the 17th century.[b] On the history of the screw much has been written,[c] and it is quite clear that the principle was very familiar in Hellenistic times. The reputed inventor was Archytas of Tarentum (fl. −365), and all the Alexandrians discuss apparatus involving worms and worm-gearing.[d] Particularly common were the worm- or screw-presses used in the wine and oil industries, shown, for instance, in the wall-

---

[a] Lynn White (7), p. 104, cf. p. 111. We feel, however, that he errs in the other direction in accepting the existence of water-powered blowing-engines in +31 (cf. p. 370 below) as the earliest certain example of crank motion. In the present state of our knowledge we cannot assume that the Han system was the same as that of the Sung and Yuan; it may well have been much simpler, the bellows being operated simply by lugs on a horizontal water-driven shaft. Just when the more complex machinery came in we do not yet know, but by the beginning of the Thang period might be a good guess. As for the various uses of the crank referred to below, some are more clear in dating than others, and the evidence will be found in the appropriate sections. Lynn White was influenced by an opinion of Hommel's, (1), p. 247, but Hommel was essentially an ethnographer who worked in a very limited number of Chinese provinces and set out to describe tools rather than machinery. He neglected the complex technology of silk, visited no iron-works, and saw neither Tzu-liu-ching nor Ching-tê-chen. Moreover, he knew literally nothing of the Chinese scientific and technical literature. And he had of necessity to contradict himself, as in his description of cotton-gins (p. 162). Incidentally, his drawing on that page of a 'Sung dynasty cotton-gin' is incorrect, for he gave the rollers oblique crank-handles though his certain source (Nung Shu, ch. 21, p. 27b) shows right-angled ones very clearly (see Fig. 419).

[b] So foreign to the Chinese technical tradition was the screw that even in 1954, in an article on the bringing of modern agricultural machinery to remote districts, Ma Chi (1) remarked that some of the peasants found difficulty in handling screws and bolts. On the other hand spiral forms in decorative art were quite common in traditional China, as e.g. on temple pillars, and on vessels such as the Buckley bowl (Fig. 295 above).

[c] Frémont (3) concerning worms, (8) on worm-presses, and (19) on screws; Horwitz (6); Treue (1); Singer et al. (1). Feldhaus (1), col. 981, devotes a long entry to it.

[d] Beck (2). A military use occurs in Biton's book on war-machines (cf. p. 102 above), where a counterweighted ladder-bearing swape is raised to the level of city-walls by a screw. Cf. Sect. 30h below.

paintings of Pompeii.[a] Many medieval examples still exist in Europe,[b] but China always used wedge presses. The Archimedean screw for raising water was also well known, and continued in use in most Mediterranean lands after the Arab conquests.[c] It even found its way to India, where it was described by Bhōja (c. +1050) under the name *pātasama-ucchrāya*.[d] Worms were utilised, too, in surgical apparatus, as witness the speculum of Roman physicians,[e] and are mentioned by Paul of Aegina in the +7th century, as well as by most of the great Arabic medical writers. The tapering wood-screw appears in Gallo-Roman times;[f] fibulae and arm-rings of the +3rd to the +6th centuries have screw attachments.[g] There was clearly no break in the European knowledge of screws; in the +13th century Villard de Honnecourt used them for raising weights,[h] and the +15th German engineering MSS. frequently show them.[i] About +1490 metal screws suddenly became common for the fastenings of armour.[j]

A curious problem is raised by the fact that the only people other than Europeans to possess the continuous helical screw were the Eskimos. There has been much discussion[k] as to whether this was an independent invention or due to culture contact with Europeans, but the question is not yet solved.

As Horwitz (6) pointed out, the first picture of a screw (*lo ssu*[1])[l] in Chinese literature occurs in the *San Tshai Thu Hui* encyclopaedia of +1609 (Fig. 416),[m] where it is shown used for the cap of a gun. He himself rightly left open the possibility that the knowledge of the screw in China went back beyond the Jesuits to Arab contacts in the Yuan period or earlier, but for this we are not able to adduce any evidence.

---

[a] See pp. 206ff. below, and Drachmann (7); Blümner (1), vol. 1, p. 188; Forbes (2), pp. 65, 66. Present-day distribution in Benoit (3). Radial balls or bobs acting as a flywheel were added to the worms of coining presses in Leonardo's time (Frémont, 8), but this was certainly not their first appearance, and we shall later suggest (p. 443) that such radial bobs were at the origin of the invention of the verge-and-foliot escapement of Western clockwork.

[b] Such as the wine-press at the Château de Talcy in France, which I examined with Dr G. D. Lu in 1949.

[c] Vitruvius, X, vi. Cf. Ewbank (1), pp. 137ff.

[d] Raghavan (1).

[e] Cf. Milne (1); Treue (1), pp. 108ff.

[f] Frémont (19), p. 17. Its later history has been written by Dickinson (1).

[g] Treue (1), pp. 142ff.

[h] See Lassus & Darcel; Hahnloser (1); Feldhaus & Degen.

[i] E.g. Jacopo Mariano Taccola's *De Re Militari* of c. +1440 (Paris MS. 7329); see Berthelot (4). Also of course the Hussite (Beck (1), pp. 274, 287). The screw jack occurs in +1535; Salzman (1), p. 329. Clamp screws for destroying window-bars are seen in many +16th-century military works (Seselscheiber, +1524; Danner, +1560; Helm, +1570), cf. Treue (1), p. 79.

[j] Treue (1), p. 162. The nut appears to be a still later (+17th-century) development, though in principle going back of course to the female-threaded riders and fixed holders of antiquity. Metal attachments in earlier ages, as e.g. in astrolabes, had been made by the use of rivets held in place by wedge-shaped cotter-pins analogous to the split pins of today. Nevertheless the true nut was known to the later Alexandrians, though sparingly used by them, indeed only in Heron's screw presses. A valuable discussion of this will be found in Drachmann (7), p. 84, who explains the various devices which avoided the need for nuts.

[k] Consult E. Krause (1); Porsild (1); Laufer (20); Treue (1), pp. 137ff. Wardle (1) suggested that the spirally twisted tusk of the narwhal had provided an independent model.

[l] The first character of this name naturally derives from the spiral conch.

[m] Chhi yung sect., ch. 8, p. 6a. And then in *Chhi Chhi Thu Shuo* (+1627), ch. 2, p. 38a, and elsewhere.

---

[1] 螺絲

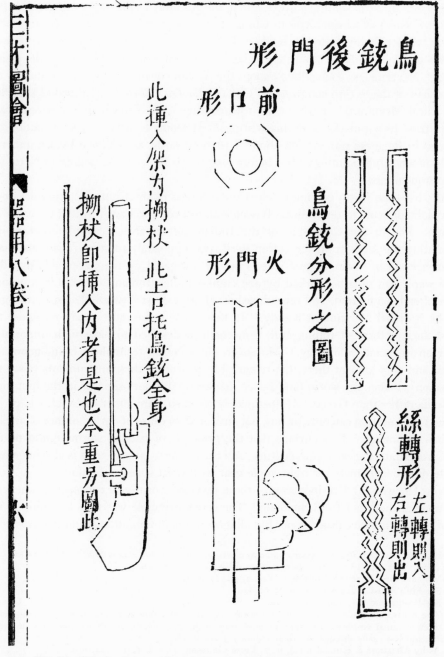

Fig. 416. The first Chinese diagram of a continuous screw or worm and its male and female threads, part of a set of diagrams of the match- or flint-lock musket (*niao chhung*, or 'bird-beaked gun', probably so called from the shape of the lock). From *San Tshai Thu Hui* (+1609), Chhi yung sect., ch. 8, p. 6*a*. During the first half of the +17th century there was a rich proliferation of Chinese literature on fire-arms; this will be discussed in Sect. 30. The screw-cap of the musket is seen on the right of the page. Its upper legend reads 'Diagram of the shapes of the parts of the bird-beaked gun'. Below, on the right: 'The shape of the "silk coil" [i.e. the screw]; if you turn it to the right it comes out, if you turn it to the left it goes in.' In the centre of the page a diagram of the muzzle above, and of the touch-hole underneath. On the left, stock and ramrod. For a brief account of +16th- and +17th-century firearms see Hall (5).

There can be little doubt, however, that the Archimedean water-raising worm (the *lung wei chhê* [1]) was introduced to China at the beginning of the 17th century, when it was repeatedly illustrated in books.[a] But again there is little evidence that it ever spread there, or replaced to any appreciable extent the traditional square-pallet chain-pump. Nevertheless a small book about the Archimedean screw appeared during the latter half of the +18th century, Tai Chen's [2] *Lo Tsu Chhê Chi* [3] (Record of the Class of Helical Machines).[b] Early in the 19th century Chhi Yen-Huai [4] improved it and again tried to popularise it by his writings.[c] It seems to have had much more of a success in Japanese mines from about +1630 onwards, as is shown by a number of well-known scroll-paintings,[d] but before long it was replaced, according to Netto (1), by piston-pumps.

At what time worm gearing reached the frontiers of the Chinese culture-area is at first sight difficult to determine. It seems almost certain, however, that it came with cotton. For in the cotton-gin of the Indian culture-area, a simple machine for separating the seeds from the cotton itself, two oppositely rotating rollers are made to engage not by ordinary gear-wheels but by elongated worms placed side by side.[e] The whole can thus be worked by one crank-handle. Indian examples in museums have often been figured,[f] so I reproduce here (Figs. 417, 418) two Sinhalese cotton-gins in the National Museum at Kandy.[g] In this machine we must surely recognise the most ancient form of rolling-mill, a mechanism destined to have such importance later on in metal technology.[h] Moreover, since cotton technology is indigenous to India and very ancient there, the remarkable gearing in these instruments raises the question whether the worm (and hence the screw) did not originate in the first place in India rather than Greece. Unfortunately, as in so many other cases (cf. e.g. p. 361 below) the question can only be put, not answered, because of our ignorance of ancient Indian technology. It is curious that the principle of parallel worms found no use in Europe until the early 19th century, and the modern use of worms is also in general quite different, namely for right-angle gear with great speed reduction.

Now Horwitz (6) described two of these reversed worm-gear cotton-gins, one from Cambodia, the other from Sinkiang.[i] The screw principle was thus knocking at the door of the Chinese culture-area. But all its equipment for cotton ginning (*kan mien* [5]),

---

[a] E.g. the *Thai Hsi Shui Fa*, incorporated as ch. 19 of *Nung Chêng Chhuan Shu*, as also the *Chhi Chhi Thu Shuo*, ch. 3, pp. 18b, 19b. Again in *Shou Shih Thung Khao*, ch. 38.

[b] We have already referred to this in Vol. 2, p. 516 above.

[c] See Shih Shu-Chhing (1), and p. 528 below.

[d] See Bromehead (7); Treue (1), p. 22.

[e] The worms are often rather long, and always cut at a very oblique angle. It has been suggested that their original employment arose because they were less susceptible to irregularities in tooth-cutting, and less easily disengaged, than ordinary gear-wheels.

[f] As by Matchoss & Kutzbach (1), p. 7; Leroi-Gourhan (1), vol. 1, pp. 104, 110.

[g] For these photographs my grateful thanks are due to the Director Dr P. E. P. Deraniyagala, the Assistant Curator Mr S. T. T. Rajakaruna, and to Mr V. A. Amerasekara. On the Sinhalese cotton-gin (*kapu kapana yantra*) see also Coomaraswamy (5), p. 235, pl. XXIX, 3.

[h] Cf. Feldhaus (6); Usher (2), p. 340.

[i] Of uncertain but comparatively recent date, they are in the Berlin Museum f. Völkerkunde.

---

[1] 龍尾車   [2] 戴震   [3] 贏族車記   [4] 齊彥槐   [5] 赶棉

PLATE CLVII

Fig. 417. A cotton-gin, probably the most ancient form of rolling-mill. In this Sinhalese example the rollers are coupled by means of two enmeshing worms, as generally in the Indian culture-area. National Museum, Kandy (photo. Amerasekara, 1958).

Fig. 418. Another traditional cotton-gin with two enmeshing worms; the single crank has lost its handle. National Museum, Kandy (photo. Amerasekara, 1958).

so far as we know, was made without any gearing, as may be seen from Fig. 419, taken from the *Nung Shu* of +1313, its earliest Chinese illustration,[a] which shows a crank-handle for each roller. This was known as the *chiao chhê*.[1] As it required two operators an improvement took place in the course of time whereby only the lower roller was worked by a hand-crank, while the upper one (smaller and of iron) was worked by a treadle motion assisted by a radial bob flywheel (Fig. 420).[b] This traditional type, the *ya chhê*,[2] may also be seen in Japan, where the flywheel tended to take the form of a single club-shaped weight attached to the shaft.[c] It probably corresponds to what Hsü Kuang-Chhi early in the +17th century referred to as the Chü-yung[3] type, named from a city in Chiangsu but more characteristic of the northern provinces.[d]

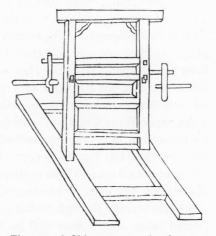

Fig. 419. A Chinese cotton-gin of +1313 (*Nung Shu*, ch. 21, p. 27b). Each roller has a crank-handle.

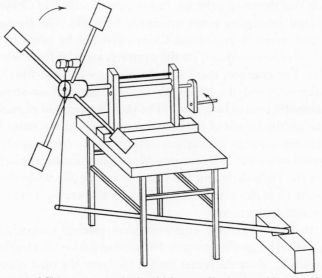

Fig. 420. Another type of Chinese cotton-gin, in which one roller is rotated by a crank-handle while the other is worked by a foot treadle with the aid of an eccentric on a bob fly-wheel (drawing of a traditional example, Liu Hsien-Chou, 5).

[a] Ch. 21, p. 27b, copied in *Nung Chêng Chhüan Shu*, ch. 35, p. 22a.
[b] From Liu Hsien-Chou (5), p. 31. Cf. his (1), pp. 41ff., and Horwitz (6), fig. 12; *TKKW*, ch. 2, p. 26b, first edition, p. 42b.
[c] Chinese and Japanese examples are in the Deutsches Museum at München. The powering of the upper and lower rollers was sometimes interchanged. A small illustration of the Japanese type may be seen in Sargent's translation (1) of the *Nippon Eitai-gura* by Ihara Saikaku (+1685), p. 116. For a description of the Chinese type by a Western traveller in 1793 see Dinwiddie, in Proudfoot (1), p. 75.
[d] *Nung Chêng Chhüan Shu*, ch. 35, p. 22b.

[1] 攪車   [2] 軋車   [3] 句容

From this it would seem therefore that in entering China the cotton-gin left its worm-gear behind. But it does not quite follow that this was a case of refusal to accept a piece of technique foreign to indigenous practices; there may have been machinery already 'in occupation'. The complex story of cotton must be deferred to its proper place,[a] but here we may note that while true cotton was indigenous to India and had been known almost since the Han as *chi-pei*[1] (and other names), the fibres from the silk-cotton or kapok tree[b] (*mu mien*[2]) had been used for textile purposes in China from an early date, principally by the tribal Man peoples of the south. In the Sung and Yuan, when Wang Chên was writing, the phrase *mu mien* was carried over as a name for true cotton, as may be seen indeed in an agricultural encyclopaedia earlier than his, the *Nung Sang Chi Yao*[c] of +1275. During this and the following century true cotton was spreading through China northwards and eastwards. It seems to have been introduced in the first place by two routes: through Burma and Indo-China from about the +6th century, and through Sinkiang in the +13th. Thus worm-gear would have been offered twice. But it may well be that the double-powered rollers of the Chinese machines were already at work from the Han onwards for *mu mien*, and naturally continued in use when true cotton (later generally called *mien hua*[3]) largely supplanted it, becoming indeed one of China's chief textile fibres. In any case the whole story shows once again that the screw principle was not characteristic of Chinese technology.

Perhaps the most interesting point remains to be made. The absence of the continuous screw and worm in traditional China must not be taken to mean that no helicoidal forms were known there; on the contrary, some of these were quite venerable inhabitants. For example, there was the zoetrope or vane-wheel rotated by an ascending air-current,[d] and the helicopter top or horizontal air-screw which itself ascended when rapidly rotated by a cord.[e] The skew-set vanes of all such devices were (like the oblique paddle-boards of horizontal water-mills or the sails of early vertical windmills) essentially separate flat surfaces tangent to the continuously curving helix of a complete screw or worm. It is thus possible to define rather precisely the different achievements of the Hellenistic and the Chinese worlds, for while the former made abundant applications of elongated screw and worm shapes, the latter early developed tangent-plane helicoid structures.[f]

In late +15th-century Europe engineers were placing vane-wheels in kitchen chimneys with shafts and gearing suitable for turning spits—'an elegant automation', as Lynn White puts it, 'since the hotter the fire the faster the roast spins'.[g] Leonardo

---

[a] Section 31 in Vol. 5, pt. 1, below.   [b] *Bombax malabaricum*, R273.   [c] Ch. 2, pp. 11 aff.
[d] Cf. p. 123 in Vol. 4, pt. 1, above.   [e] Cf. p. 583 below.

[f] We are indebted to Mr Rex Wailes and Dr Christopher Zeeman for valuable discussion on this subject. The toy windmill of Heron (cf. p. 556 below) is the only representative of the flat discontinuous tangent type in the West before the time, seemingly in the early Middle Ages, when the boards of horizontal water-mill wheels began to be set at an angle to their shafts. But we have seen how many uses there were for continuous screws and worms in Roman times. Conversely, various kinds of vane-wheels were ancient in China but no unbroken winding helices. Of course, the mill-wheels gave no helical action relative to the fluid and parallel with their axes.

[g] (5), (7), p. 93. Examples in Uccelli (1), pp. 13 ff., figs. 38, 40.

[1] 吉貝   [2] 木棉   [3] 棉花

himself designed one of these.[a] It seems extremely likely that this use of ascending hot-air currents derived from the earlier zoetropes of China (which go back to the Thang if not the Han) and the prayer-wheels of Mongolia and Tibet (cf. p. 566). The transmission could have occurred very easily through the domestic slaves from Central Asia who were brought to Italy in great numbers during the +14th and +15th centuries.[b] As the zoetrope-spits became so common in the great houses of Europe it seems equally likely that they played a part alongside Hellenistic antecedents in stimulating Branca's combinations of air or steam jets and rotor wheels in +1629, a device which soon found its way back to China as we shall presently see (p. 225). Here however the water-mill would have been the dominant influence in his mind, for whether his jet was the hot air rising through a cowl and tube from a furnace[c] or the steam blown forth from a *sufflator* head (cf. p. 226),[d] it was always in line with the plane of rotation of the wheel and not at right angles to it. The zoetrope-spit was more closely allied to the Chinese helicopter top (p. 583) and the Western vertical windmill (p. 555), whence surely descended the screw propeller and the aircraft propeller.

For these reasons it is not impossible that some Chinese inventor may have played a part in the introduction of screw propulsion for ships. In discussing the history of paddle-wheel boats McGregor (1) reported a very peculiar story that a model of a Chinese screw-propeller was brought to Europe and seen by a Col. Beaufoy about +1780, i.e. at a time when screw-propulsion was very much in the air, if not yet in the water.[e] Mark Beaufoy was in fact one of the pioneers of the experimental investigation of ship-model hydrodynamics,[f] and he would certainly have been interested in this. It is not too difficult to track down his own statement. In 1818 Dick (1) proposed the use of a 'spiral oar' or treadmill-operated Archimedean screw for impelling warships through the water, as suggested by his friend Scott of Ormiston,[g] and later in the same year Beaufoy wrote:

A contrivance of this kind I saw between thirty and forty years past, in Switzerland, in the model of a flat-bottomed vessel, brought by Mons. Bosset from the East Indies, but made in China. This model had underneath its bottom a spiral, which was turned when wanted with considerable rapidity by clockwork put in motion by a spring similar to that of a watch. The vessel being placed in a tub of water, the spring wound up and the helm put over, more or less, according as the tub was large or small, the boat continued running in a circle until the clockwork went down.[h]

[a] Feldhaus (1), fig. 98, (2), fig. 428, (18), pp. 86ff.; Uccelli (1), p. 13, fig. 37.
[b] Cf. Vol. 1, p. 189, and Lynn White (5), (7), pp. 93, 116.
[c] *Le Machine*, fig. 2; Uccelli (1), p. 14, fig. 41. Intended for a rolling-mill, through reduction gearing.
[d] *Le Machine*, fig. 25; Uccelli (1), p. 15, fig. 42; Feldhaus (1), fig. 128; Schmithals & Klemm (1), fig. 46. Intended for a vertical stamp-mill, through reduction gearing.
[e] Cf. Gutsche (1). Daniel Bernoulli had proposed the screw-propeller in +1753 and was recommending it as superior to the paddle-wheel when Beaufoy was young.
[f] See his elaborate reports on the behaviour of models (1).
[g] He also suggested the use of steam, and even opined that the screw might be useful for aerial navigation ('an art hitherto much neglected') if used in some way as a non-flapping wing (cf. pp. 585ff. below).
[h] Beaufoy (2), adding an account of disappointing full-scale experiments in the Greenland Dock, and remarks on the screw log, 'invented by the celebrated Dr Hooke'.

The name of the Swiss gentleman gives a clue which might yield more about this curious episode on further investigation; meanwhile we have only to note that the Chinese sculling-oar (the *yuloh*)[a] has affinities with the screw, and that tangent-plane helicoid structures were very much at home in Chinese culture.[b] It is not inconceivable, therefore, that some Chinese artisan thought of making a toy boat move by powering a set of helicopter vanes underneath it. Such a contribution to the main stream of descent of the screw-propeller from Archimedes through Leonardo would not be unworthy of recognition.

(6) SPRINGS AND SPRING MECHANISMS

The elastic properties of bamboo laths have already been mentioned, and it is sure that the Chinese made good use of them from an early time. Springs (*than chi*[1], *than thiao*[2]) were certainly employed in the numerous mechanical toys and automata of which a brief account will shortly be given (p. 156). Other substances, such as horn and sinews, were used in the construction of bows and crossbows, of which we shall treat in the Section on military technology.[c] Springs appear, too, in varieties of the pole lathe, in simple devices such as door-closers,[d] and the traps for wild animals described by Hommel.[e]

Though compound springs made of many leaves had been familiar from late Chou times onwards in the form of crossbows, their application to vehicles as cart-springs never became general, though there are indications that such an invention was made in the +7th century (cf. p. 254 below). In any case, Europe was much more backward in the use of springs with leaves. The crossbow had been known only in the arcuballista artillery forms devised by the Alexandrians, and probably never much used, so it is perhaps not surprising that the use of leaved springs on vehicles did not occur till the end of the +16th century.[f] Spiral springs were known to Leonardo da Vinci, however, and applied to wheel-locks of guns.[g] Spring clocks and watches start about +1480. Of course, springs on vehicles did not become essential until, with the early locomotives, the wheels themselves took on an ad-terrestrial function, and means for ensuring the permanent and simultaneous contact of all four wheels with the substratum became indispensable.

Torsion springs must have been known in China from early times, since they were used in the frame-saw (p. 54 above). Metal springs occur in forceps and padlocks

---

a Cf. Sect. 29*g* below.

b To the zoetrope and helicopter-top here one would be tempted to add the horizontal water-wheel, were it not for the fact that the blades of the Chinese type, unlike other types (cf. pp. 368 ff.), seem never to have been placed at an oblique angle with the main axis. On the other hand, the bamboo buckets of the noria do occupy a tangent-vane helicoidal position (cf. pp. 356 ff.), though not vanes.

c In Vol. 5, pt. 1 (Sect. 30).   d Cf. note (*c*) on p. 63 above.

e (1), p. 126.

f Usher (1), 1st ed. p. 85, 2nd ed. p. 133; Feldhaus (1), cols. 288, 1261.

g Feldhaus (1), col. 445; Uccelli (1), p. 75. We have not encountered any use of spiral springs in traditional Chinese engineering. In view of their obvious affinity with the screw, this is probably what one would expect.

¹ 彈機   ² 彈條

(cf. p. 241 below). Closely related to the spring is the vibrating wire, put to such good use in the cotton bow[a] (*mien than*[1]), for loosening and separating the fibres of the plant material instead of carding; but this was probably an Indian technique which came into China with cotton itself (Frémont, 13).

One of the most remarkable employments of springs in medieval China was for tripping the jack-work of clocks. In a later subsection (pp. 445 ff.) we shall describe the great astronomical clocks erected between the +8th and the +14th centuries, in which time-keeping wheels made figures appear and sounded bells, drums and gongs at the passing hours and quarters. Although it does not appear in the great descriptions, such as that of Su Sung (+1092), the technical term for the springs involved seems to have been *kun than*:[2] 'revolving and snapping springs'. This is found in a statement by Hsüeh Chi-Hsüan,[3] written about +1150 and quoted in the *Hsiao Hsüeh Kan Chu*[4] encyclopaedia.[b] He lists these as one of four different kinds of time-keeper, the others being clepsydras, burning incense sticks, and sundials. A comment says that these springs worked in conjunction with wheels, and when the right hour arrives, the clock 'automatically strikes and sounds the time (*than khou wei shêng*[5])'.[c]

(7) CONDUITS, PIPES AND SIPHONS

In all the foregoing subsections we have been concerned with the utilisation and transmission of mechanical energy. But one of the things which man most desires to do, from the earliest stages of technology, is to transmit liquids and gases from place to place. How was this conduction of fluids accomplished? The whole subject of water supplies[d] and the engineering of artificial canals dug cross-country through earth and rock logically belongs here, but in China (as before in Egypt and Mesopotamia) it was so important that we must reserve it for a special discussion (Section 28*f* below). Conduits or flumes of a more modest nature contrived in wood or split bamboo (*chia tshao*[6]) were always abundantly used in China for small-scale farm irrigation systems (cf. Fig. 421)[e] but also for mining alluvial tin.[f] According to the *Thang Yü Lin*,[g] 'everyone says how good the people of Lungmên are in making hanging channels for water; they lead it up and down as if by magic'.

The history of piping may be followed in numerous writings.[h] Its connection with the important aqueducts so characteristic of civil engineering in European antiquity

---

[a] Hommel (1), p. 163.

[b] Written about +1270 but not printed till +1299. Ch. 1, p. 42*b*.

[c] This is given in the *Tzhu Yuan* entry under *kun than*. We are not sure whether it is the comment of the editors of this encyclopaedia, or part of the quotation from the *Shih-erh Yen Chai Sui Pi*[7] (Miscellaneous Notes from the Twelve-Inkstone Studio), written about 1885 by Wang Chün;[8] with which the entry opens. See further on this Needham, Wang & Price (1), p. 163.

[d] There are valuable papers on this subject by Bromehead (6), Garrison (1), and Grahame Clark (1). Of modern monographs that of Buffet & Evrard (1) is to be preferred to that of Forbes (10).

[e] *Nung Shu* (+1313), ch. 18, p. 21*a*; *TSCC*, *I shu tien*, ch. 5, *hui khao* 3, p. 24*b*.

[f] *Thien Kung Khai Wu*, ch. 14, p. 20*b*.   [g] Ch. 8, p. 24*b*.

[h] Frémont (11); Neuburger (1), pp. 419 ff.; Feldhaus (1), col. 871; Buffet & Evrard (1); Forbes (10); Robins (1).

[1] 棉彈   [2] 輥彈   [3] 薛季宣   [4] 小學紺珠   [5] 彈扣爲聲
[6] 架槽   [7] 十二硯齋隨筆   [8] 汪鋆

Fig. 421. Irrigation flume constructed in wood (*chia tshao*). *Thu Shu Chi Chhêng*, *I shu tien*, ch. 5, p. 24*b*.

is that although the principle of the inverted siphon was well known there was usually no hope of constructing large-diameter pipes capable of withstanding sufficient pressures.[a] The many remarkable aqueducts of Hellenistic times, carried across the landscape on viaducts of stone or brick,[b] have no close parallels in the technique

[a] It was also known in China, as Rickett (1) shows from *Kuan Tzu*, ch. 57, pp. 7b, 8a, but equally hard to achieve on a large scale. For short lifts bamboo piping could be used however, as Han bricks witness (e.g. Anon. (*22*), no. 1, pp. 6, 7).

[b] On this subject see Leger (1); Merckel (1); Bromehead (6); Straub (1); Buffet & Evrard (1); Forbes (10); Winslow (1), etc.

PLATE CLVIII

Fig. 422. Brine conduits in bamboo piping at the Tzu-liu-ching salt-fields, Szechuan, 1944 (Beaton, 1). Cf. Fig. 396 above and Figs. 423, 432 below.

PLATE CLIX

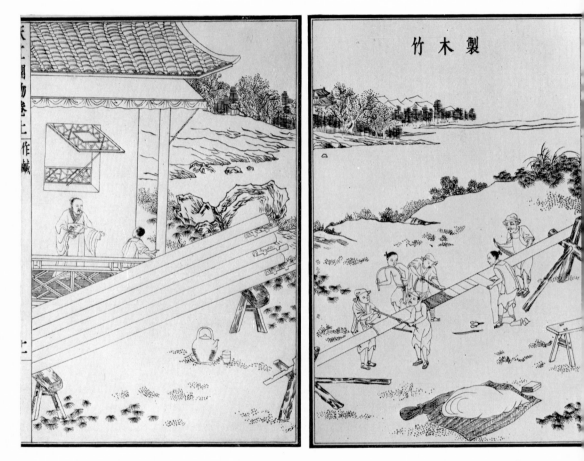

Fig. 423. Connecting a joint in a bamboo drill-haft for the salt-industry in Szechuan, +1637 (*Thien Kung Khai Wu*, ch. 5, pp. 10 *b*, 11 *a*). The bamboo pipe-lines were joined and caulked in a similar way.

of the Chinese. Generally speaking, pipes of hollowed wood were used in the West from the −2nd millennium in Egypt, through the times of Pliny[a] and Conrad of Megenburg (+14th century)[b] to 19th-century London. But copper tubing has been found in Egypt at least as early,[c] and lead pipes, longitudinally soldered, were quite common in Roman cities (cf. Vitruvius).[d] The outstanding case of the use of bronze piping, capable of supporting as much as twenty atmospheres pressure, was the water-supply of Pergamon built by Eumenes II in −180, which involved two inverted siphons of some 60 ft. depth carrying the line across two valleys.[e] But other important works of this type were made in Roman times, notably at Smyrna and Lyons.[f]

We are not able to find any instance of the use of metal piping in truly eotechnic China,[g] but Nature offered there a material which was admirably adapted for the same purpose, and unexpectedly strong, though perishable, namely the stems of bamboo. It may well be that the earliest large-scale use of this took place in the Szechuanese salt-fields, for brine, unlike fresh water, will not permit the growth of algae and consequent rotting of the tubes. Figure 422 shows the appearance of some of these brine 'mains' at Tzu-liu-ching today, and Fig. 423, taken from the *Thien Kung Khai Wu* of +1637, gives an idea of their preparation.[h] The joints are sealed with a mixture of tung-oil and lime. From rubbings of Han bricks which show the salt industry (Fig. 396) it seems certain enough that the bamboo pipe-lines (*lien thung*[1]) were already in full use at that time. For agricultural purposes also bamboo piping was used (cf. a drawing which has been noticed by several Western historians of technology),[i] but it needed frequent replacement. References to piped water-supplies for palaces, houses, farms and villages are not uncommon. But the largest systems of this kind seem to have been due to the great poet-official Su Tung-Pho, who as a Szechuanese knew of the brine pipe-lines in his own province.[j] Under his inspiration, water-mains made of large bamboo trunks were installed at Hangchow in +1089 and at Canton in +1096, caulked with the usual composition and lacquered on the outside.[k] In the latter system there were five parallel mains. Holes were provided at intervals for freeing blockages, and ventilator taps for the removal of trapped air. Significantly, Su Tung-Pho had the help of a Taoist, Têng Shou-An,[2] in planning and executing

[a] XVI, 81. Cf. Buffet & Evrard (1), p. 106.

[b] Machines for boring them out of tree-trunks are pictured in the German technical MSS. of the +15th cent. (Gille, 3), and interested Leonardo as we know from his notebooks.

[c] Cf. Forbes (10), p. 149.

[d] VIII, iv. Cf. Buffet & Evrard (1), pp. 107ff.; Forbes (10), p. 150.

[e] Descriptions will be found in Neuburger (1), and in Buffet & Evrard (1), pp. 55ff., complemented by Forbes (10), pp. 160ff.

[f] Descriptions in Buffet & Evrard (1), pp. 57ff., 80ff.

[g] But from the +15th century onwards pipes of copper and bronze were in use in the Ming palaces (Fan Hsing-Chun (1), pp. 52ff.). [h] Ch. 5, pp. 10b, 11a (not in 1st ed.).

[i] *TSCC, I shu tien*, ch. 5, *hui khao* 3, p. 23b, but the earliest illustration of the kind is in *Nung Shu* (+1313), ch. 18, p. 20a. Reproduced by Feldhaus (1), col. 874; Buffet & Evrard (1), p. 18.

[j] *Tung-Pho Chhüan Chi (Hou chi)*, ch. 4, pp. 6a, b, letters to Wang Ku, nos. 3 and 4, summarised by Lin Yü-Thang (5), pp. 267, 310. Cf. Moule (15), p. 15; and Alley (6), p. 137, translating Tu Fu (+8th cent.).

[k] The Hangchow system was reconstructed in +1270 by the then governor, Chhien Yüeh-Yu, who gives a detailed technical description in his *Hsien-Shun Lin-An Chih*, ch. 33.

[1] 連筒   [2] 鄧守安

these works, which at Canton were supervised by his friend the magistrate Wang Ku.[1] Some of the sections were made of earthenware piping.

As a matter of fact the wide use of piped water-supplies in ancient and medieval China has been greatly underestimated, and it is only now that archaeological excavation has revealed so much concrete evidence of it. Many different kinds of piping (*shui tao*[2]) have come to light. Ancient buildings near the tumulus of the first emperor Chhin Shih Huang Ti in the Wei valley north-east of Lintung[a] have yielded $-$ 3rd-century water-conduits of thick stoneware pentagonal in section (Fig. 424a).[b] Of the same period are the stoneware well-linings (*ching chhüan*[3]) about 3 ft. in diameter

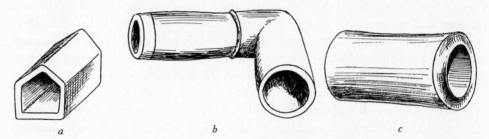

Fig. 424. Ancient and medieval Chinese water-pipes. (a) Section of thick stoneware piping from a Chhin site near Lintung. Length 2 ft., bottom-width 1 ft. 6 in., sides about 1 ft., thickness *c.* 3 in.; (b) straight section and right-angle bend in stoneware or earthenware from a Han site near Hsien-yang. Length 1 ft. 6 in., internal diameter 8 to 9 in.; (c) straight section as found at Thang sites in and near Sian and Loyang, with male and female flanges for jointing. Length 1 ft. 4 in. to 1 ft. 6 in., internal diameter 8 to 9 in. Orig. drawings, 1958, in the Shensi Provincial Museum, Wên Miao, Sian, and the Kuan Kung Museum, Loyang.

and each 1 ft. 6 in. long found at Hsienyang. Also to be seen in the Shensi Provincial Museum at Sian[c] are fine examples of Han earthenware piping, fitting together by male and female flanges[d] and including sharply bending right-angle pieces (Fig. 424b). These pipes continue with little change through the Han and Thang,[e] and are described and illustrated under the name *wa tou*[4] from the Yuan onwards.[f] Once one has seen the things, literary references[g] become very real. At Chhüfou, for example, in the Chou

---

[a] The place of the famous Thang imperial water-gardens, the Hua Chhing Chhih.[5]

[b] There is some uncertainty as to whether the 'spine' was placed upwards or downwards in use. This spined type of conduit seems to be very ancient in China, for other examples, triangular in cross-section, have been found in the Huihsien excavations, where they must date from the $-$ 4th century. Cf. Anon (4), pl. 41; Chhen Kung-Jou (2). A splendid earthenware spout in the shape of a tiger head with wide open mouth, made to fit on to the end of a pipe-line, has been excavated from remains of similar date at the ancient capital of Yen, and is illustrated in Anon (19), fig. 15; Anon (29), pl. 5. See Fig. 425.

[c] This is in the Confucian Temple, of which the Pei Lin is an annexe.

[d] Similar to the ancient earthenware pipes of the Palace of Minos at Knossos in Crete (cf. Buffet & Evrard, p. 33), which however were conical. Other Han water-pipes and well-annuli are preserved in the Nanking Museum.

[e] Thang water-pipes, of about the same dimensions as those of the Han, from the city of Loyang, may be seen in the Kuan Lin Museum there.

[f] *Nung Shu*, ch. 18, p. 28*b*; *TSCC, I shu tien*, ch. 5, *hui khao* 3, p. 29*b*.

[g] Collected, e.g. by Fan Hsing-Chun (1), pp. 50 ff.

[1] 王古   [2] 水道   [3] 井圈   [4] 瓦竇   [5] 華清池

Kung Miao[1] behind the site of the Ling Kuang Tien,[2] a famous temple of Han times, the abundant pottery piping recently discovered[a] gives life to the mention of 'limpid water from mysterious sources coming through underground channels (*yin kou*[3])' in a +2nd-century poem.[b] Again, the *San Fu Huang Thu* (Illustrated Description of the Three Districts in the Capital, Sian), a text of the +3rd century if not the Later Han itself, says[c] that underneath the Shih Chhü Ko[4] palace stones were cut and fitted to make conduits (*lung shih wei chhü*[5]) 'to lead the water in, like the imperial mains (*yü kou*[6]) of today'. This system was made by Hsiao Ho,[7] and must therefore have dated from about −200 at the very beginning of the Former Han.[d] Probably some of these were actual tubes of stone such as the Romans used at Apamoea and elsewhere.[e] Such were the different kinds of piping used in ancient China for water-supply and drainage, further light on which may be expected from archaeological discoveries. Occasionally too pipes were made for air. A rock-cut Taoist temple above the Lintung gardens is kept perpetually cool in summer by a tube bringing cold air from some mountain cleft.[f]

Of the siphon,[g] something has already been said in our discussion on water-clocks (Sect. 20*g*),[h] where we met with two ancient names for it (*yü chhiu*[8] and *kho wu*[9]). A slightly later term used in the San Kuo period[i] was *yin chhung*.[10] Many were the subsequent ways of referring to it, introduced at various dates, for instance *ku chi*,[11] the 'loquacious joker', *hung hsi kuan*,[12] the 'rainbow-shaped sucking pipe',[j] *shui chih*[13] and *tao liu*.[14] Some of these terms deserve much more investigation than we have been able to give them, for as we shall shortly see, they may sometimes have meant much more than the simple siphon, perhaps certain types of syringes or pumps. The siphon was of course widely used in traditional Chinese technique and certainly played as much part in its automata as in those of the Alexandrians, Philon and his successors.[k]

[a] Now in the Shantung Provincial Museum at Chinan.

[b] The *Ling Kuang Tien Fu* by Wang Yen-Shou,[15] of about +140; in *Wên Hsüan*, ch. 11, p. 13*b*, tr. von Zach (6), vol. 1, p. 169.

[c] Ch. 28.

[d] This pavilion was the scene of a celebrated conference in −51 which has already been mentioned (Vol. 1, p. 105). Cf. p. 161 below.

[e] Cf. Buffet & Evrard (1), p. 105.

[f] This is the San Yuan Tung,[16] dedicated to the Taoist Queen of Heaven, Tou Mu,[17] a Chinese version of the Indian goddess Marīci. I do not know how old is the refrigeration device, but I found it working very effectively on a hot day in 1958. For further information on the goddess see Doré (1), vol. 9, pp. 565 ff. She is not to be confused with the Celestial Spouse, patroness of seafarers, whom we have already met (Vol. 3, p. 558), cf. Doré (1), vol. 11, p. 914.

[g] Cf. Feldhaus (1), col. 518. On pipettes cf. Vol. 3, p. 314(*e*) above.

[h] Vol. 3, pp. 320 ff.

[i] Cf. *Piao I Lu*, ch. 1, p. 14*a*. It implies a curving worm.

[j] The term 'rainbow pipe' is particularly interesting because Cambridgeshire countryfolk also use this expression, as my friend Mr Sterland, formerly of the Cambridge Engineering Laboratory, has found.

[k] Cf. Beck (1); Carra de Vaux (2); Woodcroft (1), etc.

1 周公廟	2 靈光殿	3 陰溝	4 石渠閣	5 礱石爲渠
6 御溝	7 蕭何	8 玉虬	9 渴烏	10 陰蟲
11 滑稽	12 虹吸管	13 水巵	14 倒溜	15 王延壽
16 三元洞	17 斗姥			

In the middle of the 13th century, a thousand miles removed from the great centres of technology in that period, east and west, conduits, pipes and siphons were very much in the minds of certain mechanicians at the Mongolian capital of Karakoron. For there, on an island of culture, as it were, surrounded by the vast steppes and rolling hills of Central Asia, the Mongol Khans Küyük and Mangu had in their service a number of west European artisans captured in the wars.[a] Most of what we know about them comes from the account of the Franciscan missionary envoy William of Rubruck.[b] The most eminent was the French artist-craftsman William Boucher,[c] a Parisian goldsmith who had been taken prisoner at Belgrade and who worked at Karakoron from +1246 to +1259. Boucher's most famous achievement was the construction of a great silver fountain in the form of a tree which dispensed four kinds of alcoholic drinks such as kumiss 'automatically' to the imperial guests through the mouths of lions or dragons among its leaves. At the top there was an angel with a trumpet which it could apply to its lips by means of a mechanical arm.[d] All this was accomplished in +1254. But though the silver-work was doubtless of high quality, Boucher's skill as an engineer fell far behind it, for the trumpeting angel had to be moved manually by a hidden slave who blew hard into a long tube at the right moment, and the liquors were not raised up mechanically but poured into long tubes by slaves presumably hidden in the roof of the palace hall. Olschki has much to say[e] of the superiority of medieval European engineering over Chinese 'technical competence' or 'manual craftsmanship', but in fact Boucher's accomplishment (no doubt inhibited by his isolation) was quite inferior,[f] and the mechanicians of China would have had nothing to learn from it.

A series of beautiful Chinese +16th-century vases and jugs bearing representations of fountains upon them has been studied by David (1), who suggests that the motif derived from the magnificent beer-engine constructed by William Boucher three hundred years earlier. This hypothesis is seductive, though not accounting for the disappearance of the tree and the angelic trumpeter, nor for the distinctly Renaissance, even somewhat Italian, character of the designs on the jugs. It raises however a wider issue—the question of the existence of fountains in traditional China. Though David tended to assume that they were unknown there before the time of the Jesuits, it is in fact not difficult to show that evidences of their use may be found from almost every century after the Han. The Jesuit works at the 'Versailles of Peking', the Yuan Ming

---

[a] Cf. Vol. 1, p. 190.   [b] Cf. Vol. 1, pp. 38, 84, 224.

[c] We owe an elaborate study of him to Olschki (4).

[d] It is interesting that William Boucher was an exact contemporary of Villard de Honnecourt, the French engineer so often mentioned in these pages (pp. 229, 404). The latter's apparatus, it will be remembered, included an angel which moved so as to keep pointing at the sun, and an eagle which turned its head mechanically so as to face the deacon when he read the Gospel during the liturgy. They probably worked better than Boucher's devices, but they did not include, as has sometimes been thought, the first mechanical clock escapement (cf. p. 443 below). Though Olschki (4), p. 85, emphasizes the Christian symbolism of the trumpeting angel, the Mongols were closely involved with the Alexander-legend about trumpets blown by the wind, on which see Sinor (2).

[e] (4), p. 61.

[f] See Olschki (4), pp. 64, 88, 93. Boucher was working half a century after al-Jazarī (cf. p. 381) and two and a half after the Chinese naphtha-projector experts (cf. p. 145).

PLATE CLX

Fig. 425. Earthenware tiger-head water-spout to fit the delivery end of a pipe-line, from Chi, the ancient capital of Yen State, −4th or −3rd century (Anon. *19*, *29*, 59).

PLATE CLXI

Fig. 426. The bathing pavilion or Lodge of Artificial Rain in the Royal Gardens at Anurādhapura, Ceylon (orig. photo. 1958). Sheets of water descended through sprays contrived in the entablature of the building, splashing into the pool in the foreground. These gardens were in full use in the +8th century, about the same time as the cool summer retreats of Thang palaces, some descriptions of which are translated in the text.

Yuan, from about +1750 onwards, are too well known to need description, and a reference to Pelliot (27) may suffice.[a] Their predecessors are more important. But first a word on the practical possibilities. Piping being given, a sufficient pressure-head of water was always easy to achieve if the site selected was near a steep declivity, as for example in the case of the Thang water-gardens, the Hua Chhing Chhih, at Lintung, just east of Sian and backing against a lofty outlying foothill of the Chhin-ling Shan. At the same time everything that we shall see later on (pp. 339ff.) in the realm of water-raising appliances goes to show that already in the Han it would have been easy to supply elevated cisterns with fountain-water. But let us turn to some of the descriptions.

Four hundred years before the building of the Yuan Ming Yuan, the last emperor of the Yuan dynasty, Toghan Timur,[b] had surrounded himself with mechanical toys of all kinds, clocks with elaborate jackwork, and fountains of several different sorts. We know about them from the description of Hsiao Hsün,[1] who left a vivid record of the architecture and contents of these Yuan palaces which it had been his duty, as an official, to destroy upon the orders of the first Ming emperor in +1368. This was the *Ku Kung I Lu*.[2] Hsiao Hsün was a Divisional Director of the Ministry of Works, and thus had the opportunity to see in detail the beauty of the buildings which had housed more than a thousand concubines, as well as the arrangements of the workshops in which so many ingenious mechanisms had been made by the emperor himself and his artisans for their delectation. Hsiao Hsün describes[c] dragon-fountains with balls kept dancing on the jets, tiger robots, dragons spouting perfumed mist, and several dragon-headed boats full of mechanical figures—the sort of *shui shih* on which we shall shortly have more to say.[d] It was indeed a shame to destroy them, to 'break down the carved work with axes and hammers', but the demands of a demagogic asceticism were in the ascendant.[e]

Some two hundred years before, Mêng Yuan-Lao, describing in +1148 the glories of Khaifêng, the capital lost to the Chin Tartars, tells us that at a certain temple there were

two statues of the Buddhas Mañjuśrī and Samantabhadra riding on white lions. From the five fingers of each of their outstretched hands, which quivered all the time, streams of water poured in all directions. For this purpose wheels were used[f] to hoist the water up to the

---

[a] Further information will be found in Favier (1), pp. 185ff., 307ff. The chief designer of the hydraulic works was Michel Benoist, assisted by P. M. Cibot. Their biographer, Pfister (1), pp. 814ff., was also under the erroneous impression that fountains were previously unknown in China.

[b] We shall meet with this amateur engineer again in the subsection on clockwork, pp. 507ff. below.

[c] Especially pp. 5aff. Cf. *Ko Ku Yao Lun*, ch. 13, p. 21a. These were probably the mechanical wonders described by Odoric of Pordenone (Yule (2), vol. 2, p. 222) and in Sir John Mandeville's elaboration of him (Letts ed., vol. 1, p. 151). The Byzantine court had had similar delights in the +10th century; cf. Brett (2); F. A. Wright (1), pp. 207, 290ff.

[d] See pp. 160ff. below.

[e] There appears to be an English translation of Hsiao Hsün's book, but we have not seen it (Anon. 42).

[f] From the expression used (*lu-lu*, cf. p. 95), pulleys and buckets would be implied, but the circumstances seem rather to suggest square-pallet chain-pumps or even a pot-chain pump (cf. pp. 339, 352 below).

[1] 蕭洵   [2] 故宮遺錄

top of a high hill behind, where there was a wooden cistern. At the appointed times this was released (through pipes) so that it sprayed like a waterfall.[a]

This must have been quite worth seeing.

Four hundred years earlier still, the great worthies of the Thang had been equally interested in fountains and similar means of cooling halls and pavilions in summer. The *Thang Yü Lin* says:[b]

> After the empress Wu Hou died, the mansions of the princes, princesses and notables in the capital grew daily more magnificent and imposing. During the Thien-Pao reign-period (+742 to +755) the Grand Censor Wang Hung[1] was found guilty of crimes and sentenced to death, so his mansion in Thai-phing-fang was confiscated by the district officials. Several days were not enough for this. In the grounds there was a pavilion called the Lodge of Artificial Rain (Tzu Yü Thing[2]), from the roof of which cascades of water ran down in all directions. If one was there at midsummer one felt as cool as if it were mid-autumn.

This passage does not in itself imply upward-shooting fountains but something perhaps rather more like those lodges or bathing pavilions in Indian lands in which the bathers could sit surrounded by sheets of water descending on all sides.[c] One such remains to this day in the royal gardens at Anurādhapura in Ceylon (Fig. 426).[d] But another text which refers to about the same date indicates true fountains rather clearly. The *Thang Yü Lin* again says:[e]

> When the emperor Ming Huang (Hsüan Tsung) built the Cool Hall (Liang Tien[3]) (about +747), the Remonstrator Chhen Chih-Chieh,[4] submitting a memorial to the throne, admonished most severely against it (on grounds of extravagance). At the request of the emperor, (Kao) Li-Shih[f] summoned him to court. It was when the heat was really extreme. The emperor was in the Cool Hall, and behind his seat the water struck the fan-wheels while cool air played around one's neck and clothes. Chhen Chih-Chieh arrived and was given a seat on a stone chair. A low thunder growled. The sun was hidden from sight. Water rose in the four corners and forming screens fell again with a splash (*ssu yü chi shui chhêng lien fei sa*[5]). The seats were cooled with ice, and Chhen was served with marrow-chilling drinks, so that he began to shiver and his belly was filled with rumblings. Again and again he begged permission to leave, though the emperor never stopped perspiring, and at last Chhen could hardly get as far as the gate before stopping to relieve nature in the most embarrassing way. Next day he recovered his equanimity. But people said that 'when one discusses affairs one should deliberate thoroughly on them first, and not put oneself in the emperor's place'.

For fountains this must suffice. But what were the fan-wheels kept rotating behind the emperor's seat?

---

[a] *Tung Ching Mêng Hua Lu*, ch. 6 (p. 35), tr. auct. with Lu Gwei-Djen.

[b] Ch. 5, p. 33*a*; tr. Shiratori (4), mod. auct. The passage is also in the *Fêng shih Wên Chien Chi*,[6] cit. Liu Hsien-Chou (1), app. p. 135.

[c] The Lodge of Artificial Rain and its counterparts are often mentioned elsewhere, e.g. *Piao I Lu*, ch. 4, p. 3*a*. The *Fêng shih Wên Chien Chi* dates the arrangement at +750.

[d] See Paranavitana (1). I had the great pleasure of visiting this wonderful city, both royal and Buddhist, with Professor Senarat Paranavitana in 1958. The date of the royal gardens would be about contemporary with the Thang. It is interesting that similar arrangements are ascribed to Fu-Lin (Byzantium) in *Chiu Thang Shu*, ch. 198, p. 16*a*; cf. Hirth (1), p. 54. Cf. p. 161 below.

[e] Ch. 4, p. 2*b*; tr. Shiratori (4), mod. auct.

[f] The eunuch most prominent at the court of Hsüan Tsung.

[1] 王鉷   [2] 自雨亭   [3] 涼殿   [4] 陳知節   [5] 四隅積水成簾飛灑
[6] 封氏聞見記

(8) VALVES, BELLOWS, PUMPS AND FANS

'Opening canals for water...or channels for summer floods', says the *Huai Nan Tzu* book, 'are not actions which may be called "contrary to Nature".'[a] But before long the ancient Chinese found that inducing water to go upwards could also be done in ways 'not contrary to Nature'. This will be considered in its place (pp. 330 ff.); here we may study the complex of early inventions connected with the tubular propulsion of liquids (mainly water) and gases (mainly air). Of pushing mechanisms there were many kinds: the flexible animal-skin bag, with or without partial walls of clay, pottery or wood; the piston working in a cylinder; and the rotary fan (or ad-aerial windmill).[b] But the common feature in all machines of this kind except the last was the presence of clack-valves, which needed to be no more than small hinged doors covering the exits and entrances of pipes in the walls of the propulsion chamber.[c]

The progress of these inventions in east and west still presents unsolved problems, but we may approach them in the following way. Universal in China today for all artisanal purposes, and even on a larger scale for minor industries, is the box-bellows (*fêng hsiang*[1]) shown in Fig. 427c. Hommel says rightly[d] that it surpasses in efficiency any other air-pump made before the advent of modern machinery. From the longi-tudinal section (Fig. 427d) it can be seen that the box-bellows is a double-acting force and suction pump;[e] at each stroke, while expelling the air on one side of the piston, it draws in an equal amount of air on the other side. Whenever this bellows first came into general use it provided that fundamental metallurgical necessity, a continuous blast.[f] No less than twelve of the illustrations in the *Thien Kung Khai Wu* (+1637) show its use by metal-workers (cf. Fig. 428, a bronze foundry). In the ordinary Chinese box-bellows, intake valves (*huo mên*[2]) for air are provided at each end of the box, and a single double-acting valve underneath at the junction of the two outlet channels. The cross-section is usually rectangular, allowing for easy construction from wood, and the piston (*huo sai*[3]) is packed with feathers (the ancestors of piston-rings). The

---

[a] The full translation of this passage has already been given in Vol. 2, p. 68.

[b] A careful classification is available in the work of Westcott (1). Cf. Frémont (10, 14).

[c] As Ewbank (1), p. 262, well says, pumps are simply syringes furnished with induction and eduction valves; this permits them to remain permanently mounted in a given relation with their duty. On syringes see immediately below, p. 143.

[d] (1), p. 18 ff.

[e] 'Atmospheric' pumps for liquids are divided primarily into suction pumps and force pumps. Both depend, at least initially, upon the formation of a partial vacuum for filling the cylinder or propulsion chamber, but in suction or lift pumps the liquid then passes through a valve in the piston itself and is simply raised on the upward stroke to a level at which it can discharge through an outlet; while in force pumps the unpierced piston drives out the cylinder's contents on the downward stroke into a rising pipe against as much pressure as the system will stand. Cf. Ewbank (1), pp. 213, 222. The Chinese air-pump, with its solid piston, is so arranged that it sucks and expels on both strokes.

[f] My attention was drawn to this long ago by my metallurgical friend Dr Yeh Chu-Phei. Sir William Chambers, who figured it in +1757, called it a 'perpetual bellows'. There are classical descriptions of it in Lockhart (1), p. 87, and Ewbank (1), pp. 247 ff. Cf. Dinwiddie in Proudfoot (1), p. 74.

---

[1] 風箱　　[2] 活門　　[3] 活塞

common Japanese bellows, though similar, is less ingenious, since the piston carries a valve, and the blast takes place only on the push and not on the pull.[a]

Ewbank much admired the Chinese box-bellows.[b] He pointed out that it was essentially equivalent to the Ctesibian double-barrel force-pump for liquids, with the two cylinders elegantly combined into one.[c] If pipes were connected to the intakes it would become the pump of de la Hire (+1716),[d] and its connection with the principle

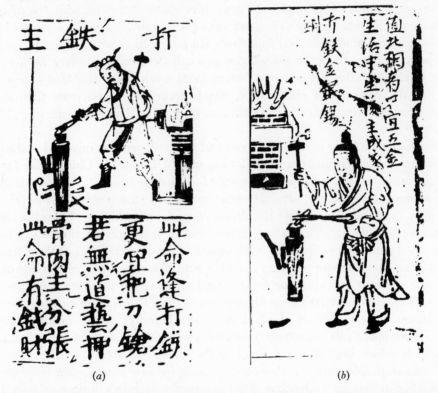

(a)   (b)

Fig. 427. The oldest known illustrations of the Chinese double-acting piston-bellows; two pictures from the *Yen Chhin Tou Shu San Shih Hsiang Shu*, a book on physiognomy and all kinds of divination printed about +1280. (a) A blacksmith at his forge (ch. 2, p. 35a). (b) A silversmith at his forge (ch. 2, p. 36a). The texts are appropriate prognostications.

of the later steam-engine cylinder of James Watt (where steam enters at each end so as to create a vacuum on each side of the piston alternately) is obviously one of close formal resemblance.[e] It could also have served, in principle, as a Boylian air-pump if both the intakes were derived from a sealed chamber. 'The most perfect

---

[a] Hommel (1), p. 20. For Japanese metallurgical piston-bellows see Gowland (1), p. 17, and many other illustrations in his papers.

[b] (1), pp. 247ff., 251.

[c] Ewbank hinted that the Chinese form might have been the origin of the Alexandrian one, but what we know of the possibilities of transmission at that time does not encourage such an idea. On the latter, see further, p. 141 below. The Alexandrian single-barrel piston-bellows associated with musical organs we have already mentioned (Vol. 4, pt. 1, p. 211) and shall consider again, p. 150 below.

[d] Ewbank (1), p. 271. See on, p. 149.   [e] See on, p. 387, and Needham (48).

PLATE CLXII

c

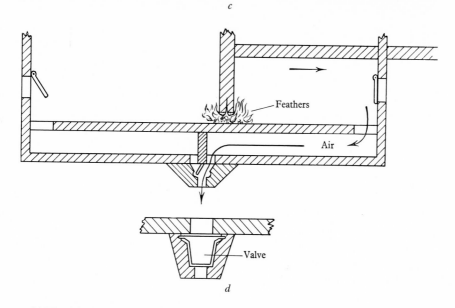

d

Fig. 427. (*c*) The Chinese double-acting piston bellows (Hommel, 1). Beneath, (*d*) is a longitudinal section to show the arrangement which ensures a continuous blast, and a small transverse section of the nozzle to show the mounting of the outlet valve. Air is taken in alternately at each end of the rectangular cylinder through crapaudine valves which open as the piston withdraws and close as the piston approaches. A separate lower compartment closed at the centre conducts the air alternately to each side of a third crapaudine valve within the nozzle. The piston packing consists of feathers, or sometimes of folds of soft paper. In the photograph the outlet is (as commonly) on one side of the box, but in the diagram it is shown, for convenience, at the bottom. The valves with their pivoting pegs resemble in form the characteristic Chinese house-doors (cf. Hommel (1), pp. 292, 298). The swinging outlet valve is seen in the photograph in the neutral position. Example from Kuling, Chiangsi.

PLATE CLXIII

Fig. 428. A battery of double-acting piston bellows in use by bronze-founders, +1637 (*Thien Kung Khai Wu*, ch. 8, p. 6b). According to the caption they are casting a tripod cauldron, and the channels for the molten metal are marked as made of clay.

blowing-machine, and the *chef d'œuvre* of modern modifications of the pump', wrote Ewbank, are the 'facsimiles' of the medieval Chinese piston-bellows.

It is difficult, unfortunately, to bring forward much evidence as to the precise antiquity of this machine, for little research has been devoted to it, and the obvious sources—encyclopaedias such as the *Thai-Phing Yü Lan*—are at first sight unhelpful.[a] But pending the appearance of a history of bellows something like the following may be said. Bellows for metal-working played a very important part in ancient Chinese thought and mythology, as we have already noted in the Section on Taoism.[b] One of the legendary rebels, it will be remembered, was named Huan-Tou,[1,2] 'Peaceable Bellows'. The oldest name for bellows was *tho*,[3,4] the character for which is closely related to *nang*,[5] a skin bag.[c] The most ancient type of bellows in China was therefore no doubt a whole skin with a delivery tuyère fastened in it, working perhaps in pairs. The next development would have been to make the walls of the bellows partly of pottery or wood, with holes in the skin coverings of the pots upon which the feet of the operators took the place of valves—such pots are seen in well-known ancient Egyptian representations,[d] and used by contemporary African peoples.[e] The later terms *ko pai*[6,7] probably refer to this type.[f] Since these skin-covered pots resemble drums it is not surprising that one of the earliest verbs for the act of plying bellows was *ku*,[8] 'to drum', and this is very often found in Chou and Han texts. Reference has already been made[g] to the iron cauldrons which were cast with legal statutes upon them in −512; here the *Tso Chuan* text has the expression '*i ku thieh*'.[9] While some commentators[h] took this to mean a measure of weight which the inhabitants were taxed, others have interpreted it as 'iron blown by the bellows', i.e. cast iron.[i]

Some light on the bellows of the Warring States period is forthcoming from an unexpected source, namely the chapters in the *Mo Tzu* book on military technology. From these it is clear that in the late −4th century it was customary to use toxic smokes made by burning balls of dried mustard[j] and other plants in stoves, the smoke

---

[a] Liu Hsien-Chou (*1*) could find nothing at all, but we shall be able to offer a few hints (p. 139 below). Recently an attempt at a history of blowing machinery has been made by Yang Khuan (*1*); this is valuable though very brief.

[b] See Vol. 2, pp. 115, 117, 119, 330, 483. Cf. Granet (*1, 2, 4*) under 'outre'.

[c] Nevertheless, one may note the presence of the wood radical in *tho*, and the stone one too, suggesting that leather was not the only component. The word *tho* (bellows) occurs also in a doublet as *tho-tho*,[10] the oldest name of the camel. We cannot follow Yang Khuan (*1*) in his argument that this implies that the earliest bellows were camel-back-shaped. As Schafer (*3*) shows, the borrowing was rather in the opposite direction, and for *tho-tho*, probably a loan-word from some Central Asian language, characters were chosen which implied 'the sack-carrier', because of its humps.

[d] Westcott (*1*), pl. XIX; Gowland (*1*), p. 15; Neuburger (*1*), pp. 25, 49; Feldhaus (*1*), col. 369; Ewbank (*1*), p. 238.

[e] Cline (*1*). For ancient European references see Feldhaus (*1*), col. 367ff.; Blümner (*1*), vol. 2, pp. 190ff., vol. 4, e.g. pp. 140ff.

[f] *Pai* is often misprinted in Chinese texts by *pi*,[11] which is a technical term relating to harness.

[g] Vol. 2, p. 522. The text is Duke Chao, 29th year (Couvreur (*1*), vol. 3, p. 456).

[h] As Fu Chhien in the Han.

[i] So Tu Yü in the Chin, and to some extent Khung Ying-Ta in the Thang.

[j] The volatile oil of this plant (as also of the onion and horse-radish) is highly irritant; its active principle is allyl iso-sulphocyanide (cf. Sollmann (*1*), p. 693, or any pharmacological treatise).

[1] 驩兜   [2] 驩兜   [3] 橐   [4] 橐   [5] 囊   [6] 革韛
[7] 革韛   [8] 鼓   [9] 一鼓鐵   [10] 橐駝   [11] 韛

being directed by bellows against troops attacking cities, or blown into the openings of enemy sap tunnels. We learn[a] that 'the bellows are made of ox-hide,[b] with two pots to each furnace, and they are worked by a swape lever tens and hundreds of times (up and down) (*tho i niu phi, lu yu liang fou, i chhiao ku chih pai shih*[1])'. Or 'each stove has four bellows, and when the enemy's tunnel is about to be penetrated, then the oscillating swape levers are furiously worked to blow the bellows fast and fumigate the tunnel (*tsao yung ssu tho, hsüeh chhieh yü i chieh-kao chhung chih, chi ku tho hsün chih*[2])'.[c] The interest of these passages lies in the fact that already by this date a mechanisation of the push-and-pull motion alternating between two cylinders (or pots) seems to have been introduced, exactly as in the double-cylinder force pump

Fig. 429. Annamese single-acting double-cylinder metallurgical bellows (after Schroeder, in Frémont, 14). Here the continuity of the blast is obtained by alternation of excursion of the two manned piston-rods. Copper smelting.

of Heron (+ 1st century)[d] and so many subsequent 'fire-engines' in later Europe. The use of bellows with toxic-smoke projectors must go back to the beginning of the −4th century, for the early chapters of the *Mo Tzu* book also mention them,[e] but without reference to the mechanical swape lever. This invention did away with the necessity of having one person to work each single-acting barrel, such as we see in Schroeder's drawing (Fig. 429) of an Annamese piston-bellows, the continuous blast being obtained by the fact that one piston was rising while the other was descending.

[a] Ch. 62, p. 20*b* (ch. 62, p. 26*a* in Wu Yü-Chiang's reconstructed edition).
[b] In the Han, horse-hide was also used; cf. Yeh Chao-Han (*1*).
[c] Ch. 52, pp. 9*b*, 10*a* (ch. 62, p. 25*a* in Wu Yü-Chiang). Cf. ch. 52, p. 11*a*, 61, p. 17*b*, and 62, p. 20*a*. A reference of nearly the same date, perhaps fifty years later, is in *Han Fei Tzu*, ch. 47, p. 4*b*. For recognition of the importance of the *Mo Tzu* passages credit is due to Yang Khuan (*1*).
[d] *Pneumatica*, ch. 27; Woodcroft (*1*), p. 44; but see Drachmann (*2*), pp. 4 ff., (*9*), p. 155.
[e] Ch. 20, p. 2*b*.

[1] 橐以牛皮鑪有兩甀以橋鼓百十   [2] 竈用四橐穴且遇以桔皋衝之疾鼓橐熏之

From this point it was not a far cry to the conflation of the pistons and cylinders into one, though unfortunately we know nothing of the ingenious conflator.

There is one curious hint that this may have occurred as early as the −4th century. The *Tao Tê Ching* (conservatively of this date) says:

> Heaven and Earth and all that lies between,
> Is like a bellows with its tuyère (*tho yo*[1]);
> Although it is empty it does not collapse (*hsü erh pu chhü*[2]),
> And the more it is worked the more it gives forth (*tung erh yü chhu*[3]).[a]

The statement in the third line could hardly have been made of any skin bellows, but would clearly apply to the piston variety whether the latter was hinged or straight-sliding. Commentators from Wang Pi[b] to Huang I-Chou[c] say that what Lao Tzu was referring to was the *phai tho*,[4] i.e. the 'push-and-pull' bellows. One of them[d] explains further that the *tho* is the outer box or case (*tu*[5]) into which the tuyère (*yo*[6]) is fitted, and that the latter is a tube through which passes the air forced by the 'drumming'. An interesting point is that in the Han period the bellows were worked by hand, if we may judge from the word *han* (or *chhien*)[7], also written *han* (or *chhin*)[8], which the *Shuo Wên* defines as 'the handle of the bellows' but which originally meant the ears of a jar.[e] Nevertheless, some foot-operated types long persisted, such as the large hinged-fan bellows used in the Japanese *tatara*[9] method of iron-smelting.[f] In later literature we find many echoes of the *Tao Tê Ching* passage, as in Chang Hêng's *Hsüan Thu*[10] of +107,[g] and Lu Chi's *Wên Fu*[11] of +302.[h]

Evidence for the existence of the piston air-pump in the Han may perhaps be derived from an interesting passage in the *Huai Nan Tzu* book.[i] Complaining about the decline of primitive simplicity,[j] the writer says that the demands of the metal-workers for charcoal have even led to the destruction of forests. Among the extravagances of the age, 'bellows are violently worked to send the blast through the tuyères in order to melt the bronze and the iron (*ku tho chhui tuo i hsiao thung thieh*[12])'.[k] *Ku tho* could be taken as 'drum-bellows' but the commentator, Kao Yu,[13] who lived about

---

[a] Ch. 5, tr. Hughes (1), p. 147, mod. The version of Waley (4) fails to convey the technical point present in the third line.

[b] Of the San Kuo period (mid +3rd century).

[c] In his *Shih Nang Tho*[14] (Philological Study of Bags and Bellows), +18th century.

[d] Probably Wu Chhêng of the Yuan (early +14th century).

[e] Later dictionaries such as the *Kuang Yün* say 'the handle of a push-and-pull bellows' (+10th century). For a further discussion on this see Yang Khuan(1). Yeh Chao-Han(1) argues that these words signified the valves.

[f] Cf. Gowland (5), pp. 306ff., and Muramatsu Teijirō (1). See Figs. 604, 605. Bloomery 'footeblastes' still existed in the Forest of Dean in +1636 (Schubert (2), p. 187).

[g] *CSHK* (Hou Han sect.), ch. 55, p. 9a.

[h] *Wên Hsüan*, ch. 17, p. 5b; cf. Hughes (7), pp. 106, 171ff., who was led astray by early commentators glossing *yo* as a musical intrument instead of the tuyère.

[i] *C.* −120. Ch. 8, p. 10a.

[j] The context of this standard Taoist threnody has already been expounded in Sect. 10g (Vol. 2, pp. 104ff., 127ff.).

[k] The translation of this important passage by Morgan (1), p. 95, is grossly astray.

[1] 橐籥   [2] 虛而不屈   [3] 動而愈出   [4] 排橐   [5] 櫝   [6] 籥
[7] 鈐   [8] 拑   [9] 蹈鞴   [10] 玄圖   [11] 文賦
[12] 鼓橐吹埵以銷銅鐵   [13] 高誘   [14] 釋囊橐

+200, explains that the word is a verb and means to strike or beat (*chi* [1]). He says further that the bellows (*tho* [2]) is a 'push-and-pull bellows for the smelting furnace (*yeh lu phai tho yeh* [3])'. And he elucidates the unusual word *tuo* by saying that it is a tube of iron conducting the blast of the bellows into the fire (*tho khou thieh tung, ju huo chung chhui huo yeh* [4]). His use of the word *phai* is all the more significant because it was the traditional term for the metallurgical blowing engines operated by water power which came into use early in the +1st century (cf. pp. 369 ff. below). These 'water-reciprocators' (*shui phai* [5]) have been noted by Feldhaus (10) alone among Western historians. Though no contemporary evidence is available as to their design in the Han, all illustrations of them from +1300 onwards show conversion of rotary to longitudinal reciprocating motion, as would be necessary for the piston-bellows.[a]

In any case, we shall shortly show that the piston-bellows was well known even to literary scholars and philosophers in the Sung (+11th and +12th centuries).[b] But the probability of its existence in ancient times in the Chinese culture-area is strengthened by certain anthropological facts which must be mentioned. Among ourselves it is a familiar observation that when one inflates a bicycle tyre the lower end of the pump becomes hot. Indeed the rapid reduction of air to one-fifth of its volume gives a temperature sufficient to ignite tinder.[c] This fact was discovered by the primitive peoples of south-east Asia (especially the Malayan-Indonesian region), who made use of it in one of the most remarkable of all eotechnic devices, the piston fire-lighter (Fig. 431).[d] At the bottom of the instrument the tinder is held in a small cavity. The ethnography of the device was sketched in a classical memoir by Balfour (1), and its presence in Madagascar was one of the pieces of evidence which decisively showed the Malay origin of the indigenous Madagascans.[e] Just how far back in time this application of the piston goes is an extremely difficult question to answer, for there is no way of telling to what extent the primitive populations of such regions have been technologically static over the centuries, but there is certainly every reason for thinking that the invention was autochthonous in south-east Asia.[f] It is indigenous in Yunnan.[g]

---

[a] Most of these clearly show the piston-rod operating a hinged-fan type of bellows like that of the *tatara* (p. 375 below), which might be considered a piston working in a curved cylinder. Perhaps our oldest representation of this is the Hsi-Hsia fresco at Wan-fo-hsia (Yü-lin-khu) published by Tuan Wên-Chieh (1). This must date from between +990 and +1220 (see Fig. 430).

[b] Indeed the +10th is a safe enough estimate (cf. Wang Su, Hsing Lin & Wang Liu, 1).

[c] This fact is at the basis of the most modern theories of the nature of explosion (Bowden & Yoffe). Initiation of explosion by mechanical means is due to the formation of minute hot spots, either because of the adiabatic compression of microscopic pockets of trapped vapour, or, more rarely, friction and viscous heating. Thus the physical chemists became interested in the ancient device of the fire-piston.

[d] Hough (1), pp. 109ff.; Leroi-Gourhan (1), vol. 1, p. 68. We may well believe that it was associated with that other ancestor of all piston-engines and projectile-guns, the blowpipe gun, which also belongs to the culture of the area. Was the piston a tethered projectile, or the projectile a liberated piston?

[e] Madagascan single-acting piston-bellows used in pairs are figured by Ewbank (1), p. 246, and Gille (10). They are similar to the Annamese type in Fig. 429. Some Madagascan forms are double-acting, with two valved pistons on a single piston-rod, and a diaphragm by the central outlet pipe; cf. Ewbank (1), p. 252; and p. 148 below.

[f] It was actually patented in early 19th-century Europe, but this development is considered to have been derivative from south-east Asia.

[g] Medhurst (2).

---

¹ 擊   ² 橐   ³ 冶鑪排橐也   ⁴ 橐口鐵筒入火中吹火也   ⁵ 水排

PLATE CLXIV

Fig. 430. Forge bellows from a Hsi-Hsia fresco at Wan-fo-hsia (Yü-lin-khu), Kansu, dating from the +10th to +13th centuries (Tuan Wên-Chieh, 1). The oldest representation so far known of the fan-piston working in a curved cylinder (cf. p. 375 below), the *tatara* type. For later representations of this see Figs. 602, 604, 605 below.

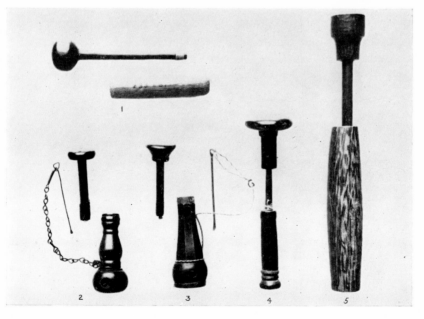

Fig. 431. The fire-piston, an indigenous form of lighter common to South-east Asia and Madagascar, the Malayan culture-area (Hough, 1). Examples from (1) Siam, (2) Lower Siam, (3, 4) the Philippines, (5) Java.

PLATE CLXV

Fig. 432. Part of a scroll-painting of the Chhing period showing (with some artistic license) the method of drawing brine from the deep bore-holes in the Szechuanese salt-fields (from the collection of Mr R. Alley). The long bamboo tube-bucket within the derrick is about to descend into the bore-hole, beside the orifice of which stands a tub ready to receive the brine won and convey it away through a pipe to the evaporating pans. A legend to the right of the tube-bucket tells us that it is 20 ft. long and made of two bamboos connected together. That on the left says that some 60 or 70 bucketfuls are raised each day from bore-holes with abundant brine, some 20 or 30 from those with less. At the base of the bucket there is a valve which opens to let the brine in and closes to retain it on the up-haul. The valve is protected by a ferrule at the end of the tube, the wall of which is slotted to admit the brine just above. The rest of this part of the painting is self-explanatory; the cable passes over broad upper and lower pulleys (*thien kun* and *ti kun*) to the horizontal winding-gear (*phan chhê*) worked by animal power. Cf. Figs. 396 and 422 above.

If then it could be considered part of the stock-in-trade of the ancient Malay-Indonesian-Oceanian component of Chinese culture (cf. Sect. 5b above),[a] the Chinese piston-bellows might well be regarded, at any rate on a working hypothesis, as derived from it. For piston-bellows are rather widespread in the more primitive cultures of East Asia, e.g. the paired form used by the Khas of Laos (Sarraut & Robequin) and the Moi in Annam (cf. Fig. 429). If this were accepted, we might have to suppose that the invention of clack-valves occurred twice: once in the Chinese area, and once in the Mediterranean region[b] when the Hellenistic water-pumps were derived, perhaps, from the ancient Egyptian syringe.

In the fullness of time the fire-piston proved capable of exerting a seminal influence far away from its south-east Asian home.[c] About 1877 Carl Linde, the pioneer of artificial refrigeration, gave a lecture at Munich in the course of which he demonstrated a cigar-lighter made on the fire-piston principle. Among his hearers was Rudolph Diesel, who said in later years that this experience was one of those which had most stimulated him to the invention of the high-compression internal combustion engine now universally known by his name.

The history of piston-bellows is closely related to the history of piston water-pumps. Here the limiting factors were primarily the pressures which any eotechnic system might be likely to withstand. That the Alexandrians developed, and the Romans used, simple bucket suction pumps in which the water is lifted by the piston during its upward stroke, passing through it by a valve from below during its downward stroke, was supposed by Ewbank (particularly for clearing ships' bilges),[d] but Usher regards this as very doubtful.[e] By the time of Agricola, however (mid +16th century), these pumps were in wide use.[f] The force pump (in which the liquid does not pass through the piston, but is driven out by an exit pipe) was, on the other hand, well understood in Hellenistic times, as is shown by the discussion in Vitruvius,[g] who speaks of the cylinder and the piston as *modiolus* and *embolus* respectively. The invention was attributed to Ctesibius,[h] and must have been used quite widely, for a number of such pumps have been found, from the Roman example contrived in a solid wood block and brought to light in a famous excavation at Silchester by Hope & Fox,[i] to the remarkable bronze pumps of Bolsena.[j] Westcott believes that this type of pump

---

[a] Hashimoto (2), p. 4, has suggested that the cyclical characters *ping*[1] and *ting*,[2] the ancient graphs of which are very obscure (K757 and K833), may have been originally pictograms of the parts of the fire-piston.

[b] Frémont (10) suggested that the valve derived in the West from the round Greek and Roman shield which, when suspended, was used to open or shut ventilator holes in roofs or the tops of cupolas. See also Montandon (1), p. 275.

[c] Cf. J. Lehmann (1).   [d] (1), p. 214.

[e] (1), 1st. ed. p. 85, 2nd. ed. p. 134.   [f] Hoover & Hoover ed. pp. 177ff.

[g] x, vii; illustrated in Neuburger (1), p. 229; Usher (1), 1st ed. p. 86, 2nd ed. p. 135; Perrault ed. p. 321.

[h] It must therefore be placed at least as early as the −2nd century; cf. Drachmann (2), pp. 4ff.

[i] Illustrated e.g. in Usher (1), 1st ed. pp. 87, 88, 2nd ed. p. 137.

[j] Some of these have poppet valves like modern internal combustion engines. Cf. A. H. Smith (1), p. 120; Buffet & Evrard (1), p. 75; Singer et al. (1), vol. 2, p. 376.

[1] 丙   [2] 丁

was little used in subsequent centuries, presumably owing to its greater complexity, and it hardly appears again until the time of Cardan (+1550) and Ramelli (+1588).[a]

Generally speaking, piston-pumps for liquids were not a feature of the Chinese eotechnic tradition,[b] and their illustrations in the *Chhi Chhi Thu Shuo*[c] of +1627 may well have been a novelty at the time. Yet there had been one element of traditional art which involved a principle near to that of the suction-lift pump, namely the long bamboo tube-buckets (*chi shui thung*[1]) which were being sent down, from Han times onward, to the brine at the bottom of the bore-holes of the Szechuan salt-field (cf. Figs. 396, 422 and Sect. 37 below). These buckets (cf. Fig. 432) carried a valve at the base by which they were filled, and would have constituted suction-lift pumps if they had fitted tightly to the walls of the bore-hole.[d] But the Chinese aim was different; it must be remembered that the contents had to be raised a distance of 1000 to 2000 ft., not spilled out after a short haul, within the limits of a vacuum which atmospheric pressure could fill. According to Esterer's observations forty years ago, the filling time at the brine was 180 sec., the emptying time at the bore-head 300, the raising time 25½ min. for each load, the dimensions of the buckets 25 m. (*c.* 75 ft.) long by 7·6 cm. (3 in.) diam., and the contents 132 kg. (about 28 gallons). This was a considerable engineering operation.

The relationship of the brine-bucket valves to valves in air-pumps or bellows was perfectly appreciated by Su Tung-Pho[2] in a passage written about +1060. In his description of the Szechuan salt industry, he says:[e]

They also use smaller bamboo tubes which travel up and down in the wells; these cylinders have no (fixed) bottom, and possess an orifice in the top. Pieces of leather several inches in size are attached (to the bottom, forming a valve). As these buckets go in and out of the brine, the air by pushing and sucking makes (the valve) open and close automatically. Each such cylinder brings up several *tou* of brine. All these bore-holes use machinery (hoists). Where profit is to be had, no one fails to know about it.

The *Hou Han Shu*[f] speaks of 'water(-driven) bellows' (*shui pai*[3]). [The expression it actually uses is *shui phai*[4] (cf. p. 370).] This is applied to iron-working in Szechuan, and large ones are used.[g] It seems to me to be the same kind of method as that used in the brine-collecting tube-buckets (*chhü shui thung*[5]) of these salt-wells. Prince Hsien[h] (who made a commentary on the *Hou Han Shu* in the Thang dynasty) did not understand this, and his ideas on the subject were wrong.[i]

---

[a] (1), p. 37. Cf. Beck (1), pp. 176, 214ff., 396ff.; Forbes (2), pp. 142ff.; Usher (2), pp. 330ff.

[b] Feldhaus (1), col. 838, says that piston-pumps were old in Japan, but his statement depended on an account by Netto (1), p. 372, of batteries of lift-pumps used in late 19th-century Japanese mines of traditional type. They much resembled those illustrated in Agricola, and there is no reason to think them independent of him.

[c] Ch. 3, pp. 20*b*, 26*b*, 56*a*ff.

[d] The arrangement was, in fact, described by Tissandier (2) as the 'pompe sans piston, ou pompe chinoise'.

[e] *Tung-Pho Chih Lin*,[6] ch. 6, p. 8*b*, 9*a*; tr. auct.   [f] Ch. 61, p. 3*b*.

[g] One cannot tell from the text whether this should be translated in the past or the present tense; the former is equally likely.   [h] Li Hsien[7] (*fl. c.* +670).

[i] Li Hsien said that *phai*[4] ought to be written *pai*[8] (equiv. *pai*[9]), these two latter characters suggesting bellows of flexible leather.

[1] 汲水筒	[2] 蘇東坡	[3] 水鞲	[4] 水排	[5] 取水筒
[6] 東坡志林	[7] 李賢	[8] 橐	[9] 韛	

This is most valuable evidence on a number of points. Since Su Tung-Pho identifies valves in bellows working like those of the buckets in the shafts, the piston-bellows in some form or other must have been fairly familiar in his time,[a] and though water-driven piston-bellows were probably less common, he speaks as if he had himself seen them.[b] He reproaches Li Hsien for wanting to substitute words meaning 'leather bellows' for the 'push-and-pull' of the text. Then, just over a century later, we find a further reference to piston-bellows in one of the works of the great Neo-Confucian philosopher Chu Hsi. About +1180 he wrote a commentary on the Han alchemical book of Wei Po-Yang,[c] entitling it *Chou I Tshan Thung Chhi Khao I*. Wei had said that four male and female *kua* (the hexagram symbols of the Book of Changes) functioned like the bellows and the tuyère, to which Chu Hsi added the following remark:[d]

> These *kua* are those in which the Yin and the Yang are combined, namely Chen (no. 51), Tui (no. 58), Sun (no. 57) and Kên (no. 52). The bellows (*tho*[1]), the piston-bellows (*pai*[2]), the bag (*nang*[3]) and the tuyère (*yo*[4]) are the tubular spaces (through which they work)....The bellows should sometimes be worked slowly and sometimes rapidly (according to the degree of heating desired), just as the moon waxes and wanes.

And finally one can actually illustrate piston-bellows from the century following Chu Hsi, for a book printed about +1280 gives two small pictures of smiths working at their anvils with unmistakable piston-bellows by their side. This is the *Yen Chhin Tou Shu San Shih Hsiang Shu*[5] (Book of Physiognomical, Astrological and Ornithomantic Divination according to the Three Schools),[e] attributed to Yuan Thien-Kang.[6] We can thus quite safely conclude that the piston-bellows was well known in the Sung. Above (p. 139) reasons were given for thinking that it probably goes back much further, probably long before the Thang. The fact that Yuan Thien-Kang was a diviner of that period (*d. c.* +635) adds further reinforcement to this view.

As has just been said, generally speaking, piston-pumps for liquids were not prominent in Chinese eotechnic practice. But there is sometimes reason to suspect their presence, and sometimes rather extraordinary examples come into the limelight. Let us consider first the simplest ancestor of such pumps, the syringe.[f] In its most primitive form, a tube of bone or metal fixed into a bag of animal origin, it was strictly analogous to the primitive skin-bellows already discussed. There is mention in Hippocrates (*c.* −400) of injections using pig's bladders as containers, and doubtless the earlier techniques of the ancient Egyptian embalmers were similar.[g] Piston syringes seem to begin with the Alexandrians, for Philon alludes to the squirting of rose-water,

---

[a] All the more so as he was a poet, not a technician.
[b] Su was himself a Szechuanese and had spent his youth in that province.
[c] Cf. Sects. 13*g* and 33.   [d] P. 3*a*; tr. auct. On *pai*[2] see C125, D885; *WCTS*, p. 1*b*.
[e] Photolithographically reproduced, Tokyo, 1933; ch. 2, pp. 35*a*, 36*a*. See Fig. 427*a, b*.
[f] Cf. Feldhaus (1), col. 1074. It is strange that such an important ancient device is not indexed in Singer *et al.* (1).
[g] Cf. Wilkinson (1), vol. 2, p. 318, and the elaborate discussion of Lucas (1), pp. 307ff., which unfortunately discloses little concerning the nature of the injection apparatus used at different periods. Cf. also Forbes (12), pp. 190ff.

[1] 橐   [2] 鞴   [3] 囊   [4] 籥   [5] 演禽斗數三世相書   [6] 袁天綱

and a very clear description of a bronze syringe occurs in Heron.[a] Roman examples exist in museums. Celsus, Heron's older contemporary, describes the use of the instrument in aural therapy, and from accounts of Indian surgical equipment it would seem that it developed in that civilisation at least as early.[b] It has a particularly prominent place in India because of its well-known association with a great folk-festival, Holi, when people squirt coloured water and perfumes at each other. In China there is nothing analogous to this, but the instrument is certainly ancient there.[c] From its modern name, *shui chhiang*[1] (water gun), a late appearance might be suspected, but it would be unsafe to assume that because a single-character term for it is lacking the device itself was unknown in antiquity.[d] There were probably a number of other terms which became obsolete. Thus we meet with it in +1044 under the name of *chi thung*.[2] The *Wu Ching Tsung Yao* (Collection of the most important Military Techniques), preoccupied at this point by fire-fighting, says:[e]

> For syringes (*chi thung*) one uses long pieces of (hollow) bamboo, opening a hole in the bottom (septum), and wrapping silk floss round a piston-rod (*shui kan*[3]) inside (to form the piston). Then from the hole water may be shot forth.

This is much more significant than it might seem, for the use of bamboo emphasises once again the cardinal importance of this material for all initiatives concerning tubing in classical Chinese technology.[f] In earlier periods there were probably yet other names. As we have just seen, one of the commonest terms for the siphon proper in ancient and medieval Chinese was *kho wu*,[4] the 'thirsty crow', but though it was common for clepsydra siphons,[g] it is also capable of occurring in contexts where something more than the siphon seems inescapable.[h] For example, we shall find it before long as part of the water-raising machinery built by Pi Lan in +186 for the supply of the city of Loyang (p. 345), and therefore in all probability some simple form of suction-lift pumps. The term *kho wu* would also have been quite appropriate for syringes, which also suck up water, and in considering the interpretation of ancient passages this should be borne in mind. Though hidden by antique terminology, a particular presence may sometimes be implied.

In the +11th century, however, the view is much clearer. The military encyclopaedia just mentioned gives us elsewhere a very remarkable account of a flamethrower

[a] *Pneumatica*, II, 18, tr. Woodcroft (1), p. 80. It was a cylinder within a cylinder, like the modern hypodermic syringe.

[b] See Mukhopadhyaya (1).

[c] For example, in the *Hou Han Shu*, ch. 96B, p. 23b, we read about certain ancient games in which the performers manipulated long artificial fishes and dragons (cf. below, p. 158), and some of these spouted forth water as they danced, 'making a mist that hid the sun'. This must have been something like a battery of our gardener's sprays.

[d] We shall meet presently with just such another problem of terminology (cf. p. 354 below).

[e] (*Chhien chi*), ch. 12, pp. 27a, b; tr. auct. Cf. Sect. 29i.

[f] E.g. the astronomical sighting-tube (Vol. 3, p. 352), the barrels of guns (Sect. 30), and chemical apparatus (Sect. 33). Cf. p. 64 above.

[g] Cf. Vol. 3, p. 323.

[h] The case is analogous indeed with the word siphon, σίφων, as used in Byzantine Greek. As is well known, in connection with Greek fire it meant a pump of some kind, not what we mean by a siphon today.

[1] 水槍   [2] 唧筒   [3] 水桿   [4] 渴烏

for naphtha (Greek fire, in fact), which constituted a liquid piston-pump of ingenious design (Fig. 433). A translation of the passage could not be omitted here.
It runs:[a]

On the right is the naphtha flamethrower (lit. fierce fire oil shooter, *fang mêng huo yü*[1]).
The tank is made of brass (*shu thung*[2]),[b] and supported on four legs. From its upper surface
arise four (vertical) tubes attached to a horizontal cylinder (*chü thung*[3]) above; they are all
connected with the tank. The head and the tail of the cylinder are large, (the middle) is of
narrow (diameter). In the tail end there is a small opening as big as a millet-grain.[c] The head
end has (two) round openings $1\frac{1}{2}$ in. in diameter. At the side of the tank there is a hole with
a (little) tube which is used for filling, and this is fitted with a cover. Inside the cylinder
there is a (piston-)rod packed with silk floss (*tsa ssu chang*[4]), the head of which is wound
round with hemp waste about $\frac{1}{2}$ in. thick. Before and behind, the two communicating tubes[d]
are (alternately) occluded (lit. controlled, *shu*[5]), and (the mechanism) thus determined. The
tail has a horizontal handle (the pump handle), in front of which there is a round cover.
When (the handle is pushed) in (the pistons) close the mouths of the tubes (in turn).[e]

Before use the tank is filled with rather more than three catties of the oil with a spoon
through a filter (*sha lo*[6]); at the same time gunpowder (composition) (*huo yao*[7]) is placed in
the ignition-chamber (*huo lou*[8]) at the head. When the fire is to be started one applies a heated
branding-iron (*lao chui*[9]) (to the ignition-chamber), and the piston-rod (*tsa chang*[10]) is forced
fully into the cylinder—then the man at the back is ordered to draw the piston-rod fully
backwards and work it (back and forth) as vigorously as possible. Whereupon the oil (the
naphtha) comes out through the ignition-chamber and is shot forth as blazing flame.

When filling, use the bowl, the spoon and the filter; for igniting there is the branding-iron;
for maintaining (or renewing) the fire there is the container (*kuan*[11]).[f] The branding-iron is
made sharp like an awl so that it may be used to unblock the tubes if they get stopped up.
There are tongs with which to pick up the glowing fire, and there is a soldering-iron for
stopping-up leaks.

[Comm. If the tank or the tubes get cracked and leak they may be mended by using green
wax. Altogether there are 12 items of equipment, all of brass except the tongs, the branding-
iron and the soldering-iron.]

Another method is to fix a brass gourd-shaped container inside a large tube; below it has
two feet, and inside there are two small feet communicating with them

[Comm. all made of brass],

and there is also the piston (*tsa ssu chang*[4]). The method of shooting is as described above.

If the enemy comes to attack a city, these weapons are placed on the great ramparts, or else
in outworks, so that large numbers of assailants cannot get through.

---

[a] (*Chhien chi*), ch. 12, pp. 66a ff.; tr. auct. with Lu Gwei-Djen & Ho Ping-Yü.

[b] This interpretation is fixed by *TKKW*, ch. 8, p. 4a, ch. 14, p. 7b, etc. and other late Ming sources;
cf. Chang Hung-Chao (*3*), p. 22. The practical use of brass at this date may be noted.

[c] If this is not the hole in the back wall through which the pump-rod passed (and for this purpose it
seems rather too small), we cannot explain it.

[d] Reading *thung*[12] for *thung*.[13]

[e] Like slide-valves.

[f] This must have been a jar in which glowing charcoal was kept, or else perhaps the glowing
composition.

[1] 放猛火油	[2] 熟銅	[3] 巨筒	[4] 拶絲杖	[5] 束	[6] 沙羅
[7] 火藥	[8] 火樓	[9] 烙錐	[10] 拶杖	[11] 罐	
[12] 筒	[13] 銅				

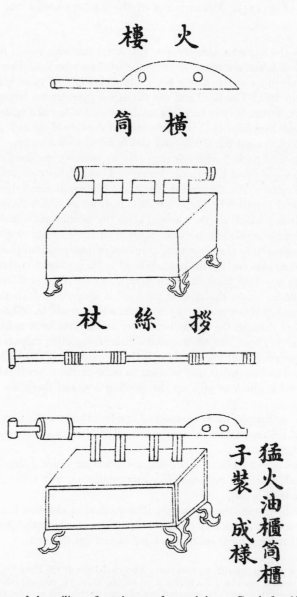

火樓

橫筒

撥絲杖

猛火油櫃筒櫃
子裝 成樣

Fig. 433. Illustrations of the military flamethrower for naphtha or Greek fire (distilled petroleum) from the *Wu Ching Tsung Yao* of +1044 (*Chhien chi*, ch. 12, pp. 66a ff.); a liquid piston-pump of ingenious design, recognisably Chinese in character. From top to bottom, the ignition-chamber (*huo lou*), the horizontal cylinder or barrel (*hêng thung*) connected with the tank below by four vertical tubes, and the double piston-rod with its silk-floss packing (*tsa ssu chang*). The lowest picture, which gives the appearance when fully assembled, is labelled 'Drawing of the complete Fierce Fire Oil (Projector, with its) horizontal barrel and its tank'.

It will readily be allowed that this is a text of great interest, dating as it does from a couple of decades before the time of William the Conqueror.[a] We know indeed that the flamethrower was already in use in the first years of the $+11$th century from a story in which certain officials were laughed at for being more expert with it than with their writing-brushes.[b] The description reads in part rather like a set of army directions concerning a 'Mark II kit', and is none too explicit about the details of the internal mechanism—perhaps these were 'restricted'.[c] We can however be confident that the purpose of the four upright tubes was to enable a continuous jet of flame to be shot forth, just as the double-acting piston-bellows gave a continuous blast of air, and the most obvious way of effecting this was to have a pair of internal nozzles one of which was fed from the rear compartment on the backstroke. According to the reconstruction which we think most probable, here shown in Fig. 434a, b, this meant having two of the tubes secretly connected within the tank. Such a design is very compatible with the directions in the text that the machine was to be started with the piston-rod pushed fully forward, and it also agrees with the statement that the 'two' communicating tubes (i.e. the feed-tubes) are alternately occluded. Only two valves were necessary since the pistons themselves acted like slide-valves on the feeds, but the apparatus was more suitable for a light fluid such as naphtha than it would have been for water, since the feeds were open only at the end of each stroke and the response had to be rapid.[d] That the pistons themselves had no valves is indicated partly by their long and narrow shape, partly by the fact that then only one central feed-tube would have been necessary. Why two pistons instead of a single one (as in the box-bellows) were fitted it is hard to say; possibly for greater rigidity.

Perhaps this apparatus reveals more about the famous Byzantine 'siphon' used for Greek fire than anything yet available from occidental sources. Yet if that gave a continuous jet, it was most probably by the combination of two cylinders in a Ctesibian force-pump system,[e] while the single-cylinder double-acting machine described in the *Wu Ching Tsung Yao* seems characteristically Chinese. In fact it was the principle of the box-bellows used for a liquid, and supplies still further evidence for the dating of the latter. Here we must not enlarge upon the military aspects of our flamethrower.[f] A date very near to $+675$ is still accepted as that of the introduction of Greek fire in the defence of Byzantium by Callinicus, and one must still rely on classical descriptions such as those of Leo Tacticus in the $+8$th and $+9$th centuries. The best opinion, which following Partington (5) we adopt, is that Greek fire was

[a] This flamethrower was, I think, first introduced to Western scholars in the paper of Wang Ling (1). I like to remember that the text describing it was copied out for me nearly twenty years ago by a great scholar, the late Dr Fu Ssu-Nien, long before we possessed a copy.

[b] *Chhing Hsiang Tsa Chi*, ch. 8, p. 6a; we reserve the translation for Sect. 30.

[c] The second apparatus is even more obscurely described, but it gives the impression of having been a smaller portable model.

[d] One is reminded of the 'cut-off' on a modern steam locomotive.

[e] As indeed has been pointed out by Hall (2).

[f] In Sect. 30 on martial technology we shall set fully in their context the various incendiaries of early times, the use of Greek fire by the Chinese as well as the Arabs from the $+10$th century at least, and the development of what we shall term 'proto-gunpowder' as well as the true brisant explosive itself prior to its transmission to the West.

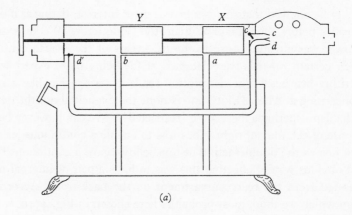

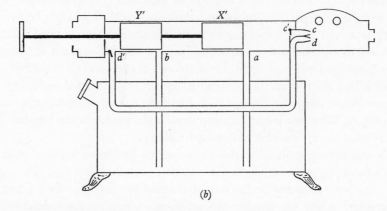

Fig. 434. Reconstruction and explanation of the mechanism of the double-acting double-piston single-cylinder force-pump used in the +11th-century flamethrower. (a) At the end of the fore-stroke; (b) at the end of the back-stroke. The cycle may be visualised as follows:

(1) When the piston-rod is fully pushed in (piston positions X and Y), the naphtha from the forward compartment (or the air, if starting) has been fully expelled through nozzle c. Naphtha enters the rear compartment through feed b, drawn up by the partial vacuum produced by the closure of valve d'.

(2) On the back-stroke feed b is occluded, valve d' opens, and naphtha is expelled through nozzle d. A partial vacuum is produced in the forward compartment by the closure of valve c'.

(3) When the piston-rod is fully withdrawn (piston positions X' and Y'), the naphtha from the rear compartment has been fully expelled through nozzle d. Naphtha enters the forward compartment through feed a, drawn up by the partial vacuum produced by the closure of valve c'.

(4) On the fore-stroke feed a is occluded, valve c' opens, and naphtha is expelled through nozzle c. Thus the cycle is continually repeated until the tank runs dry.

The ignition-chamber contained a low-nitrate gunpowder composition performing the office of a piece of slow-match.

essentially distilled light fractions of petroleum, adjusted as to viscosity and other properties by the addition of small amounts of resins and other substances.[a] Nor can we anticipate here our discussion of gunpowder, though the curiosity of the reader will certainly have been aroused by its use as an igniter in the forward chamber of

[a] Not including saltpetre however.

the projector.[a] It must suffice to say that though the technical term *huo yao* never (in our experience) refers to anything other than mixtures of sulphur, saltpetre and carbonaceous material, the proportion of nitrate in some of the earliest +11th-century compositions was so small that it would have been quite possible to use them as a kind of slow-match[b] in the way described.

If the military engineers of the Sung could produce such elegant pumps to withstand the attacks of enemies such as the Chin Tartars and later the Mongols, why did piston water-pumps seem new in the +17th century in China? Ewbank gave some thought[c] to the reasons why the deep bore-hole buckets and the piston-bellows failed to generate a widespread use of piston water-pumps in the Chinese Middle Ages, concluding that their inhibition was probably due to the very efficiency of the square-pallet chain-pumps. He also maintained,[d] much less plausibly, that the Alexandrian inventions had been influenced by a knowledge of pistons and valves transmitted from Asia. However, it is agreed[e] that the earliest double-acting reciprocating water-pump was that of J. N. de la Hire in +1716, and this assuredly had some connection with the Chinese double-acting air-bellows known by that time to Europeans for nearly two centuries.

From what we have now seen, the converse possibility, namely that the double-acting piston-bellows shown in Chinese illustrations from the middle of the +13th century onwards was an introduction from Europe, is highly unlikely. Its origin there was in fact surprisingly late. Throughout the Middle Ages and the Renaissance period, Europe had depended for its metallurgical blowers either on the trombe[f] or on cuneate skin-and-wood bellows of the type familiar in domestic fireplaces, but larger.[g] These are seen in Mariano (+1440),[h] in Agricola's great work,[i] and abundantly in the *Pirotechnia* of Biringuccio[j] dating from about +1540, where batteries of them are worked in various ways by trip-lugs on water-driven shafts, or by systems of cranks, levers and weights. According to Beckmann's researches,[k] ironfounders about this

---

[a] It will have been noted that the ignition-chamber was fitted with two air-inlets, like a Bunsen burner or a jet engine.

[b] Coarse twine impregnated with saltpetre and slowly burning; used for touching off all kinds of firearms during the first three centuries of gunpowder usage in the West.

[c] (1), p. 250.   [d] (1), p. 268.

[e] Westcott (1), p. 38; Ewbank (1), p. 251.

[f] This was a blower analogous to the ordinary filter-pump; a stream of water descended into a closed space which had an outlet allowing the trapped air to escape. It was particularly associated with the Catalan iron bloom furnaces (on which see Sect. 30*d*) though it had the disadvantage that the air was always damp. The trombe (cf. Percy (2), p. 285; Ewbank (1), p. 476) must certainly derive from the pneumatic-hydraulic apparatus of the Alexandrians, in which water was always driving air out of closed spaces (Boni, 1). Lacking the simple and elegant Chinese double-acting piston-bellows, the Arabs and Byzantines used remarkably complex arrangements to give continuous streams of air (cf. Wiedemann & Hauser (3), and p. 536 below).

[g] The first reference to this kind of bellows is said to be in Ausonius (+4th century), who refers to the valve. The water-driven metallurgical bellows used in German mining centres from about +1200 onwards, described by Johannsen (1), were undoubtedly of this type. We can see a very clear representation of the bellows alone carved in wood on one of the panels in the Norwegian stave church of Hyllestad, dating from the +12th century (Holmqvist (1), pl. LXII, fig. 138).

[h] See Berthelot (4), p. 483; cf. Beck (1), pp. 289, 470.

[i] Hoover & Hoover ed., pp. 208ff., 359ff.,

[j] Cf. Beck (1), pp. 116ff.   [k] (1), vol. 1, pp. 63ff.

time grew tired of the expense and trouble of oiling and maintaining bellows with flexible leather parts, and cuneate all-wood blowers were made in Germany about $+1550$ by Lobsinger, the Schelhorns, and others. They became general during the latter half of the $+17$th century.[a] The development had nothing to do with the cylindrical air-pumps with valves which had been known in Hellenistic and Roman times,[b] and it was not until the middle of the $+18$th century that water-powered blowers of this type were introduced (cf. Fig. 607). By that time, 250 years after the first Portuguese contacts with China, the cylinders were as much Chinese as Ctesibian.[c]

The production of a blast by rotary motion goes back a surprisingly long way in China. This method of propelling air stands of course apart from all other kinds of bellows and pumps in that no valves are necessary, and is logically analogous to a paddle-wheel propelling a liquid up a flume.[d] No doubt the most ancient of all fans were those pieces of any handy flat and relatively rigid material which people caught up to cool themselves in hot summers. Interesting studies of the history of hand fans (*shan*[1], *sha*[2]), ancient in China,[e] have been made;[f] the radial folding fan (*chê shan*[3]), so characteristic of East Asian cultures, seems to have been a Korean invention of the $+11$th century.[g] The alternating action of the punkah (*fêng shan*[4]), though little used, and only in the south, could have suggested the addition of more vanes and a continuous rotation.[h]

When such vanes were first fitted on to a continuously revolving axle in China we do not know, but it was certainly not later than the Han. Nor do we know whether rotary fans of this kind were used first for air-conditioning or for the winnowing process necessary in cereal agriculture. Let us follow these two applications separately.

An important passage in the *Hsi Ching Tsa Chi* (Miscellaneous Records of the Western Capital) refers to the famous inventor Ting Huan[5] whom we often meet

---

a See Schlutter (Schlüter), vol. 1, pl. III *b*, p. 325, vol. 2, pl. VI *g, h, i*, p. 55; de Genssane (1); Ure (1), 1st and 3rd eds. pp. 1127 ff.; Paulinyi (1); Singer *et al.* (1), vol. 4, p. 125. At the great copper works of the Falun mine in Dalarna in Sweden, for instance, as we learn from Lindroth (1), they depended until the end of the $+18$th century wholly on the old leather bellows or the trombe or on bells descending and rising in water (cf. p. 365). By 1800 there were hinged-fan bellows like the Chinese medieval metallurgical bellows (cf. p. 371) and the Japanese *tatara* bellows (cf. p. 375 below and Sect. 30*d*). Piston-bellows came later still.

b Piston-bellows were described by Philon (app. 1), Heron (chs. 76, 77) and Vitruvius (x, viii), but there is no archaeological evidence that they were anything but very exceptional. Intended for organ-blowing (see above, p. 136), they were single-acting, with clack intake-valves, and fed air into a reservoir under constant hydraulic pressure. The design was supposed to go back to Ctesibius (-3rd century); cf. Beck (1), pp. 24 ff.; Drachmann (2), pp. 7 ff., 100, (9), p. 206; Woodcroft (1), pp. 105 ff.

c On this see the evidence presented in Needham (48).

d Such a device actually existed, cf. p. 337 below.

e The earliest mention may be of the -5th century, if the *Wu Ling Tzu*[6] is a genuine work; cf. Forke (13), p. 564.

f See Forke (15); Rhead (1).

g Cf. Giles (12), p. 206, translating from Kuo Jo-Hsü's[7] contemporary *Thu Hua Chien Wên Chih*[8] (Observations on Drawing and Painting). On this book see Hirth (12), p. 109.

h A kind of hinged fan action existed in the metallurgical bellows worked by water-power which we shall presently consider in detail; cf. pp. 369 ff. below.

1 扇   2 箑   3 摺扇   4 風扇   5 丁緩   6 於陵子
7 郭若虛   8 圖畫見聞志

elsewhere in other connections.[a] His *floruit* seems to have been in the neighbourhood of +180. It says:[b]

Ting Huan also made a fan consisting of seven wheels, each ten feet in diameter. They were all connected (*lien*[1])[c] with one another, and set in motion by (the power of) one man. The whole hall became so cool that people would even begin to shiver.

This must have made the Han palaces, with their winding waterways,[d] very pleasant in the heat of a Chinese summer. But air-conditioning was not confined to them, for we continue to meet with it century after century. From a passage quoted not long ago[e] it was clear that rotary fans (*fêng lun*[2]) were operated by water-power in the Cool Halls of Thang palaces, and from numerous mentions in the Sung (e.g. c. +1085 and +1270)[f] the refrigerant effects of artificial draught seem to have been appreciated ever more widely.

Now we can turn to the rotary-fan winnowing-machine. The problem was to find a way of substituting a controllable air-current for the natural breezes on which the farmers of old had depended for separating the chaff from the grain.[g] Metal-workers had had their blast from remote antiquity, but the farmers needed a gentler one, and they solved the problem in a different way. The encyclopaedias of rustic science show two forms of the winnowing-machine, one in which the crank-operated fan is set high up and quite open (Fig. 435), the other in which it is fully enclosed in a cylindrical casing and made to direct its draught over a grain-chute and under a hopper (cf. Fig. 414). Although this latter apparatus looks so like those which we are accustomed to seeing on the farms of our own time, we must remember that its date is only just after +1300. The open form was called the *shan chhê*[3] (and later *yang shan*[4]), the closed form the *yang shan*[4] (later *fêng chhê*[5] and *fêng shan chhê*[6]). The most venerable work, the *Nung Shu*, depicts only the closed form, but *Nung Chêng Chhüan Shu* and *Shou Shih Thung Khao* only the open one; *Thien Kung Khai Wu* has both. From the greater sophistication of the closed type one would be inclined to regard it as the later, while the open type would seem to belong more appropriately to the lifetime of Ting Huan, but as there is some evidence for the former as early as the Han, the latter may be much older than he. I have never seen the open type, nor any contemporary photograph of it, but the closed-box type is still in widespread use.[h] Of course the ancient method of throwing up the grain in baskets still persists in some areas.[i]

---

[a] Already, Vol. 1, pp. 53, 197; Vol. 3, p. 58; Vol. 4, pt. 1, p. 123; and below, pp. 233, 236.

[b] Ch. 1, p. 8*a*, tr. auct. adjuv. Laufer (3), p. 197. Cit. *TPYL*, ch. 702, p. 4*a*; cf. Pfizmaier (91), p. 283.

[c] Note that this word is later in frequent use to describe the enmeshing of gear-wheels. Here also it must surely mean this. A crank might seem to be implied, but a pedal drive is perhaps more probable.

[d] Cf. p. 131 above, p. 161 below.   [e] Cf. p. 134 above.

[f] Kao Chhêng's *Shih Wu Chi Yuan* (Records of the Origins of Affairs and Things), ch. 8, p. 15*b*; Chou Mi's[7] *Wu Lin Chiu Shih*[8] (Institutions and Customs at Hangchow), ch. 3, p. 8*b* (pp. 379ff.), cf. Gernet (2), p. 130.

[g] Should we not see again here that urge for independence of the natural world which led to the magnetic compass for cloudy weather and the mechanical clock for overcast nights? Cf. pp. 465, 541 below.

[h] Many modern representations exist, e.g. Hommel (1), p. 77; Forke (16); Tisdale (2). Scale drawings and cross-sections in Worcester (3), vol. 2, pl. 109 opp. p. 313.

[i] As in medieval Europe (Jope (2), p. 98). I saw countryfolk near Wukung in Shensi winnowing in this way in the summer of 1958.

[1] 連   [2] 風輪   [3] 扇車   [4] 颺扇   [5] 風車   [6] 風扇車   [7] 周密   [8] 武林舊事

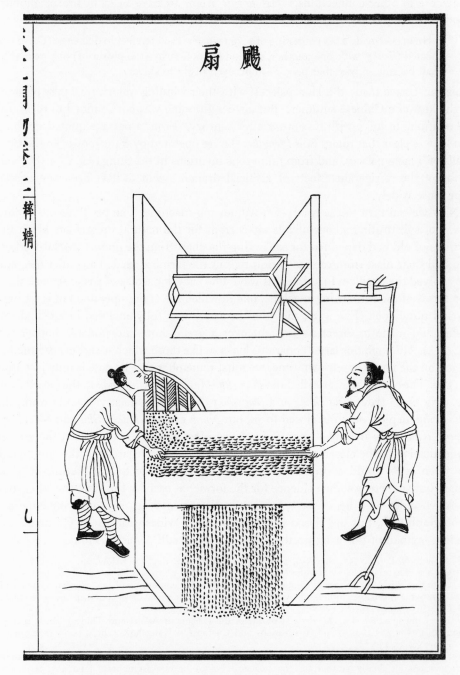

Fig. 435. Open treadle-operated rotary-fan winnowing-machine (*yang shan*), from *Thien Kung Khai Wu* (+1637), ch. 3, p. 9*a*. This form must be at least as ancient as the Han period.

What Wang Chên had to say in +1313 about the rotary-fan winnowing-machine (see Figs. 415, 435) is worth reading.[a]

The rotary winnowing fan (*yang shan*[1]).

According to the *Chi Yün*[2] *yang* means 'flying in the wind', and the *yang* fan is a machine for winnowing grain.[b] To make it one puts at the centre (of a box) a transverse axle fitted with four or six vanes made of thin boards or of bamboo (slips) glued together.[c] There are two types, one with the fan vertically mounted, the other having it horizontally mounted,[d] but both include a driving shaft worked either manually or by means of a treadle, in accordance with which the fan rotates. The mixed grain and chaff from the mortar or the roller-mill is put into the hopper (*kao hsien*[3]), communicating at the bottom with a separator (*pien*[4]),[e] through which the grain falls down as fine as the holes in a *hsi*[5] sieve.[f] As the fan turns, it blows away the husks (*khang*[6]) and bran (*sai*[7]); thus the pure grain is obtained.

Some people raise the fan high up (without enclosing it) and so winnow; this is called the *shan chhê*.[8]

After treading out (*jou*[9]) or beating (with flails)[g] the wheat or other grain, the stalks and husks are all mixed up with it, hence (the need for) winnowing; but these machines are much more efficient than the throwing up in baskets (*chi po*[10]).[h]

As Mei Shêng-Yü[11] the poet says:[i]

> 'There on the threshing-floor stands the wind-maker,
> Not like the feeble round fans of the dog-days,
> But wood-walled and fan-cranked, a cunning contrivance,
> He blows in his tempest all the coarse chaff away,
> Easy the work for those manning the handles—
> No call to wait for the weather, the breezes
> To free the fine grain from its husks, that our fathers
> Needed for tossing their baskets on high.'

Thus at once we find evidence for the existence of the machine early in the +11th century, for the Horatian Mei Yao-Chhen[12] died in +1060. But pressing further backwards, we can establish it in the early part of the +7th also, for that was the time when Yen Shih-Ku[13] was writing his commentary on the *Chi Chiu Phien*[14] (Dictionary

---

[a] *Nung Shu*, ch. 16, pp. 9*b*, 10*a*, tr. auct. with Lu Gwei-Djen; adjuv. Chiang Kang-Hu (1), p. 327.

[b] The *Chi Yün* (Complete Dictionary of the Sounds of Characters) dates from +1037 but was perhaps not completed until +1067.

[c] This approximation to plywood is worth noting, especially in connection with what has been said above, p. 63.

[d] We note that our usage in this book follows the Chinese custom of referring horizontality and verticality to the wheel in question and not to its shaft. Cf. p. 367.

[e] The meaning of this is not quite clear. Perhaps sometimes the machine really contained a sieve, but the reference is more probably to an adjustable trap which controlled the flow of the grain. Indeed the axis of this is seen projecting in Fig. 414, though without the lever attachment which in all the modern models can be hitched at any position on an upright ratchet.

[f] For this see *Nung Shu*, ch. 15, pp. 26*b*, 27*a*.

[g] See *Nung Shu*, ch. 14, pp. 28*b*ff., and above, Fig. 374*a*.

[h] The typical Chinese farm baskets are dustpan-shaped (*Nung Shu*, ch. 15, pp. 24*b*ff.) and therefore convenient for this purpose, especially if they are fitted with short handles (*yang lan*[15]), as shown in *Nung Shu*, ch. 15, pp. 30*a*, *b*.

[i] Wang Chên often quoted him. We have already had an example in Vol. 3, p. 320.

[1] 颺扇   [2] 集韻   [3] 高檻   [4] 圖   [5] 籭   [6] 糠

[7] 粞   [8] 扇車   [9] 蹂   [10] 箕簸   [11] 梅聖俞

[12] 梅堯臣   [13] 顏師古   [14] 急就篇   [15] 颺籃

for Urgent Use), which the Han scholar Shih Yu [1] had put together about −40. His entry on cereal techniques[a] lists 'the tilt-hammer, the grain-mill, the fan-faller, the mortar and the winnowing-basket (*tui, wei, shan thui, chhung, po yang* [2])'. Yen Shih-Ku then explains that the *tui* is for pounding and the *wei* for grinding. He adds some synonyms for rotary grain-mills which we shall study later, and repeats the traditional statement that these were invented by Kungshu Phan.[b] He then goes on to say:

The fan (*shan* [3]) means the rotary winnowing-fan (*shan chhê* [4]), and *thui* [5] explains the principle of the *shan chhê*. Some write this *thui*,[6] but in any case it means 'to fall', that is to say, while (the machine) fans, (the grain) falls through. Other people after pounding toss up in winnowing-baskets, to blow away (*yang* [7]) the chaff. Some write *yang*,[8] but the idea is the same.

These words might thus seem to place the winnowing-machine not only in the early Thang, but also in the latter part of the Early Han (−1st century). Yet how valid was Yen Shih-Ku's interpretation of Shih Yu? Strong support for it can be found in those Han tomb-models already mentioned and figured (Fig. 415), which show what looks extraordinarily like a winnowing-machine with its hopper and crank-handle.[c] There is also the evidence that Ting Huan was using the principle for other purposes in the +2nd century. But certain Han bricks show another device. One man stands behind a pair of uprights some 5 ft. high having long flat fan-boards on them which he seems to be working quickly to and fro, while another man in front of them shakes out grain and chaff from a basket held high above his head.[d] We need to know more about this Han winnowing method, and whether it was oscillatory or rotary, but in any case some later forms of it may well have been the background for Wang Chên's remark about 'horizontally mounted fans' (*wo shan* [9]), of which unfortunately no illustration has come down to us. All in all, however, we need have no hesitation in placing the principle of rotary blowers in the Han time, and perhaps the very early Han (−2nd century).

This contrasts in a remarkable way with the situation in Europe. If Westcott is right in saying[e] that the earliest rotary blowers in Europe are those pictured by Agricola for mine ventilation in the mid +16th century,[f] then it is very hard to believe that the idea did not travel west from China. A striking feature of the Chinese rotary fans is that the air-intake is always shown as central, so that they must be considered the ancestors of all centrifugal compressors. Even the great wind-tunnels

---

[a] P. 50a, b; Wang Ying-Lin's ed. ch. 3, pp. 42b, 43a; tr. auct. with Lu Gwei-Djen.

[b] On both these matters see pp. 188ff. below.

[c] Swann (1), pl. 2, opp. p. 378; Lynn White (7), p. 104; cf. p. 118 above. This Han apparatus seems to be built into a wall of the farmyard rather than constructed separately of wood and bamboo.

[d] The Chhêngtu Municipal Historical Museum has a particularly fine oblong brick showing this scene, but it is not at all uncommon in Han representations of farm techniques, as I noted when visiting many Chinese museums in 1958. The Chhêngtu brick, or a very similar one, is reproduced by Liu Chih-Yuan (1), pl. 7; see Fig. 436.

[e] (1), p. 76.

[f] Hoover & Hoover ed., pp. 204ff. Cf. Frémont (16).

[1] 史游   [2] 碓磑扇隤舂簸揚   [3] 扇   [4] 扇車   [5] 隤   [6] 遺
[7] 揚   [8] 颺   [9] 臥扇

PLATE CLXVI

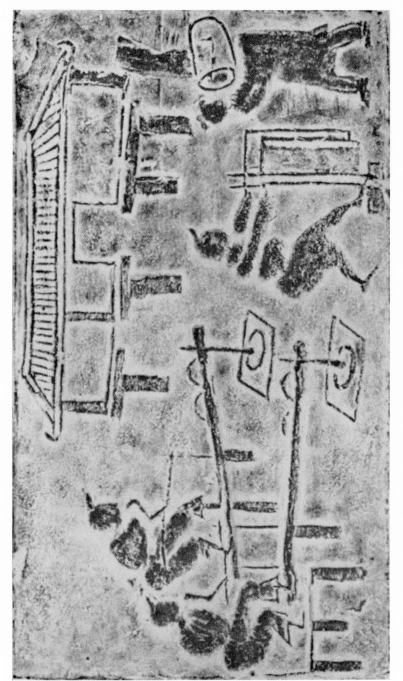

Fig. 436. Grain-pounding with tilt-hammers (left) and winnowing by means of upright fan-boards (right); a Han moulded brick found at Phêng-shan Hsien and now in the Szechuan Provincial Museum, Chhêngtu (from Liu Chih-Yüan, 1). Dimensions 39 × 25 cm. The way in which the fan-boards were used is uncertain, here they look like the usual crapaudine doors worked manually to and fro, but at this time or later they may also have approximated to horizontally mounted rotary fans, set like Persian windmills (cf. p. 557 below), or Buddhist sūtra-repositories (cf. p. 550 below).

of today derive from them. The illustrations of rotary ventilators given by Agricola, however, show both inlet and outlet at the periphery of the casing. These fans therefore must have been very inefficient; indeed it is hard to see how they could have worked at all, unless perhaps the rotor was mounted eccentrically in relation to the casing, which is not suggested either by text or pictures. Nor is there any duct widening so as to take the increasing amount of air collected as it circumscribes the fan.[a] As for the enclosed rotary-fan winnowing-machine, Europe acquired it even later.[b] Special investigations have shown[c] that it was introduced to the West from China early in the +18th century as part of one of a distinct series of waves of agricultural transmissions in alternate directions. But this story we must leave for Sect. 41.

The engineering problems involved in extending rotary and centrifugal principles to the pumping of liquids were not overcome until the end of the +17th century in Europe,[d] and no such types of pump can as yet be reported from eotechnic China.

Since the argument of this subsection has been a little involved, a few words may serve to recapitulate it. It centres round the Chinese single-cylinder double-acting piston air-pump. Historically the most ancient precursor of this must have been the bellows of skin employed by the metallurgical artisans of the Chou period, but at the same time the peoples of east and south-east Asia knew the fire-piston firelighter, and probably the air-pump was born from their union, though the origin of the valves remains obscure. Already in the Warring States (−4th century) oscillating levers were used to work pairs of bellows or pumps, as in Hellenistic Greece, and Han writings (from the −2nd century) afford certain references which point to some kind of push-and-pull bellows which were non-collapsible and therefore probably of piston type. By the Sung (+11th century) the characteristic air-pump is already present in its mature form, implying a prior presence in the Thang. So much for the propulsion of gaseous matter by pistons. Where liquids were concerned the valved buckets used already in the Han for raising brine from deep boreholes approximated to the suction lift pump, and there are strong indications that pumps of this kind, masked now by confusing terminology, were constructed in the Later Han (+2nd century). By the Sung again we have a remarkable use of the piston-pump in the military flamethrower. Yet owing perhaps to the ubiquity of the simpler chain-pumps, the piston-pump for water was uncommon or absent in the Chinese Middle Ages. Finally, rotary blowers make their appearance remarkably early, especially in the practical form of the rotary-fan winnowing-machine, another typical piece of Chinese technology. It seems certain that all European rotary gas-blowers derive from this, but to the Renaissance West must be attributed the extension of the principle to the propulsion of liquids.

[a] I owe these interesting points to Mr Sterland and Mr Stanitz.
[b] Cf. Feldhaus (1), col. 1029. Feldhaus (20), p. 49, went widely astray in saying that it first appeared in China in +1609; evidently he knew of no mention earlier than the *San Tshai Thu Hui*. The models by de Knopperf (+1716) and Evers (+1760) show perfect identity in principle with the classical Chinese design. Chambers figured the Chinese machine in +1757.
[c] Cf. Berg (2); Leser (1), pp. 454, 564ff., (2) p. 449. It seems to have appeared simultaneously in Sweden, Carinthia and Transylvania.
[d] See L. E. Harris (2); Frémont (16); Westcott (1); Beck (1), pp. 225ff.

(c) MECHANICAL TOYS

If there was one field more than any other in which all the basic mechanical principles discussed above were used together, it was that of providing mechanical toys, puppet plays, trick vessels and so on, for the amusement and prestige of successive imperial courts. In the following brief survey we shall come across the names of a number of mechanicians who are known for no other reason than their achievements in this department of 'conspicuous waste', but besides them, many of the best-known engineers of the various periods employed their talents in such services. The subject is closely connected with two bodies of semi-legend, automata on the one hand, and flying machines on the other, both having certain philosophical bearings. We have had something to say already of the former (Vol. 2, p. 53 above), and shall treat lightly of the latter below (p. 568).

The wealth of mechanical toys in the Alexandrian treatises, especially of Heron, is well known.[a] There are figures pouring libations and carrying out all kinds of motions, birds singing and temple doors opening or closing. The treatise on the puppet theatre describes a piece of apparatus which ran along a track under its own power, that of a falling weight.[b] Every possible combination of pipes, siphons, floats, valves, levers, pulleys, gear-wheels, etc., was employed.[c] In later centuries the construction of such ingenious contrivances became a speciality of the Indians and Arabs,[d] who were particularly interested in automata on striking water-clocks[e] and in mechanical cup-bearers.[f] Here they may have had some inspiration from China, for as we shall see, these were notable in Sui and Thang times. The mechanical doves and angels of Villard de Honnecourt in +13th-century Europe were in the same tradition,[g] and as late as +1588 Agostino Ramelli was solemnly illustrating elaborate 'buffets' which delivered different kinds of wine to the accompaniment of singing birds mechanically operated. The 'hydraulic gardens' and puppet plays of the 18th century continued

---

[a] See Beck (1, 4); Prou (1); de Rochas d'Aiglun (1).

[b] This plan had an echo long afterwards in the *Chu Chhi Thu Shuo* of +1627, pp. 11a ff.

[c] Cf. Usher (1), 1st ed. p. 85, 2nd ed. p. 134.

[d] Indian work on *yantra-putrikā*, as they were called, flourished during the +10th and +11th centuries, while the Arab contribution was rather in the +12th, +13th and +14th. Raghavan (1) has described the treatise of the scholar-prince Bhōja (+1018 to +1060) on the subject (the *Samarāṅgaṇa-sūtradhāra*), and the descriptions in the popular epic *Brihatkathā* by Bodhasvāmin (+10th century). It was general in India to credit the Yavanas (the Greeks) with great skill in making such machines.

[e] Cf. pp. 534 ff. below; and Carra de Vaux (3).

[f] The descriptions in Wiedemann & Hauser (2), Coomaraswamy (2) and E. Schroeder (1), to name but a few, centre round a remarkable work, the *Kitāb fī Maʿrifat al-Ḥiyal al-Handasīya* (Book of the Knowledge of Mechanical Contrivances), otherwise known as *Al-Jāmiʿ bain al-ʿIlm waʾl-ʿAml al-Nāfiʿfī Ṣināʿat al-Ḥiyal* (The Work which combines Theory and Practice, and is profitable to the Craft of Ingenious Contrivances). This was written in +1206 at the command of his Sultan by Abūʾl-ʿIzz Ismāʿīl ibn al-Razzāz al-Jazarī, an outstanding engineer and horologist. We shall often have occasion to refer to it hereafter. Famous manuscripts which have come down to us bear dates such as +1341, +1354 and +1486. Automatically refilling wine-cups seems to have been one of the amenities most appreciated at medieval royal courts, both east and west. Raghavan (1) gives details of the earlier Indian achievements of this kind.

[g] Lassus & Darcel (1); Hahnloser (1).

the work,[a] till in our own time we find ourselves surrounded at every stage of life with a thousand 'gadgets' from model railway engines to cigarette-lighters, which would in earlier ages have been the marvellous secrets of imperial courts.

It seems doubtful whether the Chinese mechanical toys were ever inferior to those constructed by the Alexandrians and the Arabs. The theatrical connection is also equally clear in China, as has been pointed out by Sun Chhiai-Ti (1), who derives some elements even of traditional opera from the ancient puppet-plays and shadow-plays.[b] Some believe that the idea of animating such puppets originated from the thought of bringing to life the wooden or clay models of human beings (*yung*[1]) which the Han people placed in their graves as servitors for the dead. It is also urged[c] that many of the enigmatic scenes and designs found on Han tomb-carvings and mirrors represent mechanical toys of various kinds. Palace eunuchs and actors (*huang mên*[2]) functioned as exorcists (*chhang tzu*[3]) in the demon-expelling *no*[4] ceremonies, and wherever actors or acrobats were involved, shamanistic dances and primitive mechanical devices were not likely to be far away.[d] To indigenous practices were perhaps added techniques from abroad; we have already noted[e] the interchanges of acrobats and conjurors which are known to have taken place between Han China and Roman Syria—some Hellenistic mechanical items may have accompanied them.[f] One often finds statements such as that in the *Hou Han Shu*[g] that in +120 the king of Shan[5,h] sent as tribute to the emperor curious drugs, and magicians who could vomit fire and exchange the heads of horses and oxen. Parallel with these Han exhibitions came in the dramatic puppets (*khuai lei*[6]), which later scholars attributed to the

---

[a] Perhaps the most remarkable of these in all Europe may be found at the château and park of Hellbrunn, a few miles south of Salzburg, built by Archbishop Marcus Sitticus of Ems in +1614. There are not only fountains and cascades of all descriptions like those of the beautiful terraces of the Villa d'Este at Tivoli near Rome, but also a great many hydraulic 'practical jokes' such as jets of water which play upon the guests from artificial antlers hung on a wall, or squirt up from the seats of their chairs. Puppets are made to swim round in a basin, spouting water the while, with birds Heronically piping among the stalactites. But the greatest achievement is the puppet theatre of 256 figures actively representing in all detail the life of an 18th-century town, with every tradesman at his work, hydraulically operated. This was built in +1752 by a mining engineer, Lorenz Rosenegger, and must closely resemble the older Chinese works described in the following pages. There is an illustrated account of this puppet theatre by Chapuis (2). An even larger, though probably later, 18th-century example, still in working order, is preserved in the municipal museum of the city of Jindrichkuv Hradec in Czechoslovakia. I have had the pleasure of visiting all three of these interesting places, and record grateful thanks to my hosts Prof. Luciano Petech and Dr Arnošt Kleinzeller. Other 18th- and 19th-century automata are described by von Haringer & Borel (1). The history of individual mechanical figures and automata constitutes a somewhat different story, on which the leading contributions are those of Chapuis & Gelis (1) and Chapuis & Droz (1). This also, however, throws much light on the methods probably used by the old Chinese mechanicians.

[b] For the history of the drama in China, the reader may be referred to works such as those of Arlington (2), Erkes (12) and Buss (1)—but they hardly touch on the present subject.

[c] E.g. by Bulling (1, 6).

[d] Cf. the remarks above (Vol. 1, p. 197; Vol. 2, pp. 53 ff., 121, 132 ff.) on this aspect of Taoism.

[e] Vol. 1, pp. 197 ff. above.

[f] But I know of no evidence which would justify the suggestion of Olschki (4), p. 105, that Ma Chün derived from this tradition, and it is most improbable. Westward transmission also remains open.

[g] Ch. 116, p. 27b.

[h] Some south-western State on the Burma border.

---

[1] 俑   [2] 黃門   [3] 倡子   [4] 儺   [5] 撣   [6] 傀儡

beginning of the dynasty.[a] Early in the +2nd century Chang Hêng the mathematician, astronomer, and engineer, speaks, in his essay on the Western Capital, of the 'plays with artificial fishes and dragons' (*yü lung man-yen chih hsi*[1]), and he may well have advised the workmen how to make them.[b]

An early story, relating to −206, but available to us only in a +6th-century source, the *Hsi Ching Tsa Chi*,[c] deals with a mechanical orchestra of puppets which the first Han emperor found in the treasury of Chhin Shih Huang Ti.[d]

There were also twelve men cast in bronze, each 3 ft. high, sitting upon a mat. Each one held either a lute (*chhin*[2]), a guitar (*chu*[3]), a *shêng*[4] or a *yü*[5] (mouth-organs with free reeds). All were dressed in flowered silks and looked like real men. Under the mat there were two bronze tubes, the upper opening of which was several feet high and protruded behind the mat. One tube was empty and in the other there was a rope as thick as a finger. If someone blew into the empty tube, and a second person (pulled upon) the rope (by means of its) knot, then all the group made music just like real musicians.

No air-pump or bellows seems to have been involved here. It took one person to provide the air-blast by blowing, while another set all the puppets in motion by means of cams, levers, weights, etc., all working off a central drum.

One of the most circumstantial accounts which has come down to us from those times relates to the work of the famous engineer Ma Chün,[6] who flourished in the time of the emperor Ming of the Wei State in the San Kuo period (+227 to +239). In the *San Kuo Chih* we read:[e]

Certain persons offered to the emperor a theatre of puppets, which could be set up in various scenes, but all motionless. The emperor asked whether they could be made to move, and Ma Chün said that they could. The emperor asked whether it would be possible to make the whole thing more ingenious, and again Ma Chün said yes, and accepted the command to do it. He took a large piece of wood and fashioned it into the shape of a wheel which rotated in a horizontal position by the power of unseen water.[f] He furthermore arranged images of singing-girls which played music and danced, and when (a particular) puppet came upon the scene, other wooden men beat drums and blew upon flutes. Ma Chün also made a mountain with wooden images dancing on balls, throwing swords about, hanging upside down on rope ladders, and generally behaving in an assured and easy manner. Government officials were in their offices, pounding and grinding was going on, cocks were fighting, and all was continually changing and moving ingeniously with a hundred variations.

This was the third of his extraordinary accomplishments.

---

[a] *Shih Wu Chi Yuan*, ch. 9, p. 32b.

[b] *Wên Hsüan*, ch. 2, p. 15b; cf. von Zach (6), vol. 1, p. 15. The fullest details on this ceremonial game, which may have been like the dragon processions of today, are to be found in *Hou Han Shu*, ch. 96B, p. 23a, b.

[c] Ch. 3, p. 3a; tr. auct. adjuv. Dubs (2), vol. 1, p. 57. Cf. p. 524.

[d] Besides this there was a hot-air zoetrope (cf. Vol. 4, pt. 1, p. 123), and a mirror which indicated dangerous thoughts. There must certainly have been some substratum for these stories of the −3rd century.

[e] Ch. 29, p. 9a, tr. auct. Here the passage is embodied in the commentary; it is really part of Fu Hsüan's contemporary biography of Ma Chün (*CSHK*, Chin sect., ch. 50, pp. 10aff.), which we have translated above (pp. 39ff.). Parallel mention in *Shih Wu Chi Yuan*, ch. 9, p. 33b.

[f] Note particularly the use of a *horizontal* mounting for the water-wheel (see p. 367 below). The application of such power, even for trivial purposes, marks a step decisively in advance of anything attempted by the Alexandrians.

[1] 魚龍曼衍之戲   [2] 琴   [3] 筑   [4] 笙   [5] 竽   [6] 馬鈞

But Ma Chün was by no means the only mechanician who achieved such successes in his period. Chhü Chih[1] of Hêngyang was famous in the Chin for his wooden dolls' house, with images which opened doors and bowed, and for his 'rats' market', which had figures which automatically closed the doors when the rats wanted to leave.[a] Ko Yu,[2] of Taoist sympathies, was alleged to have made an artificial goat on which he rode away into the mountains,[b] which probably means that he made some ingenious thing.

One would of course expect to find not only mechanical animals like this, but also actual chariots which moved of themselves. Such an automobile was built, so ran the legend, for his mother by Mo Ti in the −4th century, but as the criticism of this involves even more high-flying matters, we will postpone the passage until p. 574 below. What is perhaps surprising is to find self-moving carriages attributed to the Chinese by a serious Muslim writer as late as the neighbourhood of +1115. Among the commercial population in China, says al-Marwazī,

there are men who go about the city selling goods, fruits and so on, and each of them builds himself a cart in which he sits and in which he puts stuffs, goods and whatever he requires in his trade. These carts go by themselves, without any animals (to draw them), and each man sits in his cart, stopping it and setting it in motion just as he desires.[c]

This was one of the things which 'a clever man who had been to China and traded there said…', and from internal evidence it is possible to date this informant's visit between +907 and +923, so that al-Marwazī did not have it at first hand. Perhaps this was an echo not so much of Mohist legends as of what some bona fide traveller had said about Chinese wheelbarrows, still in those times a quite unknown invention in the West (cf. p. 258 below). Or could they have been really 'pedicarts'?

In the +4th century Wang Chia refers to a mechanical man of jade which could turn and move, apparently of itself.[d] But we get a clearer picture of what people were doing from the account in the *Yeh Chung Chi*[3] (Record of Affairs at the Capital of the Later Chao Dynasty) of the masterpiece designed by Hsieh Fei[4] and Wei Mêng-Pien,[5] who worked at the court of the Hunnish emperor Shih Hu between +335 and +345. After describing a wagon-mill (see below, p. 256), it goes on to say:[e]

Hsieh Fei also invented a four-wheeled Sandalwood Car 20 ft. long and more than 10 ft. wide. It carried a golden Buddhist statue, over which nine dragons spouted water. A large wooden figure of a Taoist was made with its hands continually rubbing the front of the Buddha. There were also more than ten wooden Taoists each more than 2 ft. high, all dressed in monastic robes, continually moving round the Buddha. At one point in their circuit each automatically bowed and saluted, at another each threw incense into a censer. All their

---

[a] *TPYL*, ch. 752, p. 2*a*, quoting from a lost book on the Chin period, the *Chin Yang Chhiu*[6] by Sun Shêng.[7]

[b] *TPYL*, ch. 752, p. 2*b*. Cf. p. 272 below. [c] Minorsky (4), pp. 16, 65.

[d] *Shih I Chi*, ch. 3, p. 5*a*.

[e] *TPYL*, ch. 752, p. 3*a*, quotes the passage giving a better text. Tr. auct. adjuv. Pfizmaier (92), p. 152.

[1] 區紙　　　[2] 葛由　　　[3] 鄴中記　　　[4] 解飛　　　[5] 魏猛變

[6] 晉陽秋　　　[7] 孫盛

actions were exactly like those of human beings. When the carriage moved onwards, the wooden men also moved and the dragons spouted their water; when the carriage stopped, all the movements stopped.[a]

We shall meet again with Hsieh Fei and Wei Mêng-Pien in connection with other more important kinds of vehicles.[b]

The automaton cup-bearers and wine-pourers begin to be prominent in the Sui period (early +7th century), under the name of 'hydraulic elegances' (shui shih[1]). The mechanician mostly responsible for this development was Huang Kun,[2] a man in the service of Sui Yang Ti, at the request of whom he wrote a manual, Shui Shih Thu Ching,[3] on the subject. This was edited and enlarged by his friend Tu Pao.[4] According to the accounts,[c] these displays involved the floating of numbers of boats (about 10 ft. long and 6 ft. wide), fitted with mechanical devices and having moving figures on board, along winding stone channels and canals (chhü shui[5]) contrived in palace courtyards and gardens, so passing the guests in turn. All the usual beings were represented, animals and men, immortals (hsien[6]) and singing-girls (chi[7]), playing all kinds of musical instruments, dancing and tumbling, just as in Ma Chün's time.[d]

There were also seven (cup-bearer) boats, 8 ft. long, with wooden men somewhat larger than 2 ft. tall on them. In each boat, one figure held the wine cup standing at the bow, with another beside him in charge of the wine pot, while a third punted at the stern and two others rowed amidships. The boats moved on around the intricate bends of the canal, faster than the 'hydraulic elegances', going round three times when they had gone only once. At each bend, where one of the emperor's guests was seated, he was served with wine in the following way. The 'Wine Boat' stopped automatically when it reached the seat of a guest, and the cup-bearer stretched out its arm with the full cup. When the guest had drunk, the figure received it back and held it for the second one to fill again with wine. Then immediately the boat proceeded, only to repeat the same at the next stop. All these things were performed by machinery set in the water.

Another account gives us the name of one of those responsible for constructing the canals and channels—Thang Hao-Kuei.[8] Such little artificial water-courses[e] have

---

[a] I am not quite clear whether there was religious syncretism here, or whether the Taoists were supposed to be paying homage to a superior religion, or whether even the phrasing does not simply mean followers of the Buddhist Tao. From the present point of view, it does not matter. But the use of some kind of pump for shooting out the water matters very much.

[b] Here may be mentioned certain wooden mechanical toys still traditional in Indo-China—pairs of linked two-wheeled carts, the first having a Buddhist monk automatically beating a drum, the second two women automatically bowing. For a description see Colani (6).

[c] YHSF, ch. 76, pp. 47a ff.; from Ta-Yeh Shih I[9] (Recollections of the Ta-Yeh reign-period) by Yen Shih-Ku,[10] quoted in Thai-Phing Kuang Chi (Chih chhiao sect.), ch. 226, p. 1a (item 1). Also Shih Wu Chi Yuan, ch. 10, p. 25a. Tr. auct. The monograph itself is listed as Shui Shih in the Sui Shu (i wên chih), ch. 34, p. 11b.

[d] Actually a detailed account of the set pieces on these barges has come down to us in what remains of the Shui Shih Thu Ching. There were 72 items, mostly legendary but some historical, e.g. the turtle coming out of the Yellow River with the eight trigrams of Fu-Hsi on its back, Chhin Shih Huang Ti going to meet the ocean spirits, and Chhü Yuan talking with the fisherman. All the performing figures were clothed in rich apparel, and measured more than 2 ft. in height.

[e] Directions for constructing the channels are found in Ying Tsao Fa Shih, ch. 3, p. 10a, ch. 16, p. 13a; and illustrations in ch. 29, pp. 14a ff.

[1] 水飾   [2] 黃袞   [3] 水飾圖經   [4] 杜寶   [5] 曲水   [6] 仙
[7] 妓   [8] 唐豪貴   [9] 大業拾遺   [10] 顏師古

PLATE CLXVII

Fig. 437. An artificial channel for floating wine-cups at royal garden-parties; in the grounds of the Phosŏk-chŏng Pavilion at Kyŏngju in Korea (Chapin, 1). The King of Silla was here in the +10th century continuing a Chinese custom of the +3rd. At the Sui capital by the beginning of the +7th it had given rise to elaborate mechanical toys mounted on boats, which offered wine at the seat of each guest (see text).

Fig. 438. An Arabic version of the mechanical cup-bearers aboard their automatic vessel; from a +1315 MS. of al-Jazarī's book on ingenious devices written at the beginning of the previous century. Freer Gallery, Washington. Beside the cup-bearers there are guests and musicians, all animated by water-power within the boat-shaped hull.

persisted occasionally until today; Fig. 437 illustrates one from the Phosŏk-chŏng¹ Pavilion at Kyŏngju² in Korea (built before +927),[a] and I have seen another in the royal gardens at Anurādhapura, Ceylon, dating from a time corresponding to the Thang. The Sui channels must have been much larger.

When we consider the possible mechanisms of their ingenuities, one is inclined to think that the simplest way in which they could have been accomplished would have been to have the boats connected together and hauled by an endless rope or chain under the water, and the figures operated by power from small paddle-wheels, doubt-less invisible.[b] Later on (p. 417 below) we shall see evidence that the use of such a device at this time (+606 to +616) would be by no means impossible, or even unlikely. The action of the cup-bearer figures at the stopping and starting of the wine-boats might have needed the intervention of projecting trip-pins at each of the guest 'stations', or it could have been arranged that the figures should be motionless so long as the paddle-wheels were revolving, after which springs and weights could come into play. It will be noted that this would be another case of what we are calling in this Section 'ex-aqueous' wheels (pp. 85 above and 412 below),[c] the nearest analogy being with the mills mounted on moored boats and moved by the river current.

Perhaps the custom grew out of an exorcistic ceremony held in Chin times on the 3rd day of the 3rd month, in which cups of wine were floated along little winding channels, 'liu shang chhü shui'.³ About +280 the emperor Wu inquired about the practice and was told by Shu Hsi⁴ that the custom was considered to have originated when Chou Kung established the capital at Loyang.[d] In +353 Wang Hsi-Chih⁵ referred to the ceremony in his famous essay on the Lan Pavilion (*Lan Thing Shih Hsü*⁶).[e] But there was also a festival on the 15th day of the seventh month, when

[a] Chapin (1). Probably one or two centuries before this date, when it is mentioned in the *Samguk Sagi*, ch. 12, p. 5a. Cf. Alley (8) on the hot springs near Chi-hsien.

[b] All we know is that there were 'wheels raising rods and releasing cords (*lun shêng kan chih shêng*⁷)'. Long after this paragraph had been written, there came to my knowledge, through the kindness of Prof. Derek Price, a quite striking parallel among the illustrations of one of the MSS. of al-Jazarī's treatise on automata. This is the copy of +1315 in the Freer Gallery at Washington. Reproduced in Fig. 438, it shows a boat bearing a company of puppet musicians and a cup-bearer as well as a sultan and his friends. Water flowing from a reservoir on the deck tips a bucket periodically, rotating thereby a paddle-wheel which causes the band to sound a fanfare. The water collects in a sump whence it can be returned to the reservoir by means of an Archimedean screw. Here the general design is closely similar to those which we shall study later (p. 535) in the subsection on clockwork, and it would seem that the whole piece was meant to stand on a table rather than to float in a canal. Yet the idea of a boat with a puppet crew and a paddle-wheel is in itself strangely reminiscent of the 'hydraulic elegances' of the Sui court, and one cannot but wonder whether there was not a genetic connection between them during those six hundred years.

[c] One is reminded (as to the 'rowers') of certain mechanical conceits of the European +17th century. In the kitchen of the Hospice de Beaune in France there is a figure which seems energetically to turn the spit, but is in fact set in motion by the falling weight which really works it.

[d] Cf. Bodde (12), p. 31; Hodous (1), p. 102.

[e] *CSHK*, Chin sect., ch. 26, p. 9b. In Sung books, for instance, there is frequent mention of '*liu shang chhü shui*'³—*Hsieh Chhuan Chi*,⁸ ch. 1, p. 36b, and the +1169 *Pei Hsing Jih Lu*,⁹ ch. 1, p. 26a, ch. 2, p. 3a; and *Mêng Liang Lu*, ch. 2, p. 1a.

candles and lights were set afloat.[a] Apparently from the Sui time onwards the channels were constructed indoors and fed with water from clear springs—it must have been a pretty sight to see the mechanical acrobats and Ganymedes floating serenely on their circuitous paths to the accompaniment of musical-box sounds.

In the course of time this interest in boats with mechanical figures spread to the mass of the people, or at least the more affluent of them, leading to a regular trade in model ships. In his *Phêng Chhuang Lei Chi*[1] (Classified Records of the Weed-Grown Window) of +1527, Huang Wei[2] informs us[b] that at Nanking model sailing-boats were beautifully carved with crew and passengers all moving 'by means of a mechanism'. When placed in the water they would sail before the wind, and 'people who liked to busy themselves with miscellaneous affairs' engaged in competitions with them, doubtless on the lovely Hou Hu lake which reflects both the Purple Mountain and the battlements of the far-stretching city walls. By good fortune representations of model boats are preserved in extant Chinese paintings. For instance, Li Sung's[3] scroll (*c.* +1190) entitled *Shui Tien Chao Liang Thu*[4] (Keeping Cool in the Water Pavilion) shows two boys launching their boats in a lake against a background of an elegant beam bridge on piles.[c]

The Sui emperors left a general reputation for active interest in mechanical devices. There remains, for example, an account of automatically opening library doors constructed for Sui Yang Ti (+605 to +616).[d]

In front of the Kuan Wên[5] Hall there was the Library, in which there were fourteen studies, each having windows, doors, couches, cushions, and book-cases, all arranged and ornamented with exceeding great elegance. At every third study[e] there was an open square door (in front of which) silk curtains were suspended, having above two (figures of) flying *hsien*. Outside these doors a kind of trigger-mechanism (*chi fa*[6]) was (contrived) in the ground. When the emperor moved towards the Library he was preceded by certain serving-maids holding perfume-burners, and when they stepped upon the trigger-mechanism, then the flying *hsien* came down and gathered in the curtains and flew up again, while at the same time the door-halves swung backwards and all the doors of the book-cases opened automatically. And when the emperor went out, everything again closed and returned to its original state.

Here we have what might be considered the half-way house between the spontaneously opening temple doors of the Alexandrians and the selenium cells of the Grand Central Station.

People in the Thang continued to feel the fascination of mechanical toys and puppet plays, and some of the latter were very elaborate,[f] such as that which was constructed for the funeral games of a provincial governor in +770. Names of individual mechani-

---

[a] To guide the spirits of those who had been drowned; see Hough (1), p. 254; Bodde (12), p. 62; Arnold (2), p. 490.

[b] Ch. 3, pp. 26*b*, 27*a*.　　　　　　　　[c] It is reproduced in Anon. (37), pl. 67.

[d] *Wên Hsien Thung Khao*, ch. 174 (p. 1506.3), tr. auct. I am indebted to Dr Yuan Tung-Li for bringing this passage to our attention.

[e] Presumably there was a common lobby door to each three studies.

[f] *Thang Yü Lin*, ch. 7, pp. 20*b*, 22*a*; ch. 8, p. 13*a*.

[1] 蓬窗類紀　　　[2] 黃暐　　　[3] 李嵩　　　[4] 水殿招涼圖　　　[5] 觀文殿
[6] 機發

cians have come down to us from this time. There was Yang Wu-Lien,[1] afterwards a general,[a] who made a figure of a monk which stretched out its hand for contributions, saying 'Alms! Alms!', and deposited the contributions in its satchel when they reached a certain weight.[b] This had great success in public on market-days, and collected more than a thousand coins at one time, in aid of what, we are not told.[c] Then there was Wang Chü,[2] who made a wooden otter (*mu tha*[3]) which could catch fish (probably some kind of spring-trap embodying a figure), and Yin Wên-Liang[4] whose wooden cup-bearers and singsong-girls which played the flute were celebrated.[d] About +890 we read of a Japanese called Han Chih-Ho,[5] who achieved renown by his mechanical toys:[e]

A guardsman, Han Chih-Ho, who was Japanese by origin...made a wooden cat which could catch rats and birds. This was carried to the emperor, who amused himself by watching it. Later, Han made a framework which was operated by pedals (*tha chhuang*[6]),[f] and called the 'Dragon Exhibition'. This was several feet in height and beautifully ornamented. At rest there was nothing to be seen, but when it was set in motion, a dragon appeared as large as life with claws, beard, and fangs complete. This was presented to the emperor, and sure enough the dragon rushed about as if it was flying through clouds and rain; but now the emperor was not amused and fearfully ordered the thing to be taken away. Han Chih-Ho threw himself upon his knees and apologised for alarming his imperial master, offering to present some smaller examples of his skill. The emperor laughed and inquired about his lesser techniques. So Han took a wooden box several inches square from his pocket, and turned out from it several hundred 'tiger-flies' (*ying hu tzu*[7]), red in colour, which he said was because they had been fed on cinnabar. Then he separated them into five columns to perform a dance. When the music started they all skipped and turned in time with it, making small sounds like the buzzing of flies. When the music stopped they withdrew one after the other into their box as if they had ranks and precedences. Later on, Han Chih-Ho showed similar toys to the emperor, which could and did hunt flies, like eagles catching sparrows.

The emperor, greatly impressed, bestowed silver and silks on him, but as soon as he had left the palace he gave them all away to other people. A year later he disappeared and no one could ever find him again.

The first part of the account also credits Han Chih-Ho with birds which could fly by means of internal mechanism.[g] There is no need to take these stories *au pied de la lettre*, but we need not doubt that there was some substance behind them.

Indeed, there are close parallels elsewhere. 'Who admires not', wrote Sir Thomas Browne, 'Regiomontanus, his fly, beyond his eagle?' He was referring to the automata

---

[a] To be met with again in Vol. 4, pt. 3, below.

[b] *Tu I Chih*, ch. 1, p. 7*b*; *Chhao Yeh Chhien Tsai*, cit. by Liu Hsien-Chou (*1*).

[c] The weight-and-lever mechanism was exactly what Heron of Alexandria had used for his penny-in-the-slot machine so long before.

[d] *Chhao Yeh Chhien Tsai*, cit. by Liu Hsien-Chou (*1*).

[e] *Tu Yang Tsa Pien*, ch. 2, p. 9*a*, tr. auct. Also *Thai-Phing Kuang Chi*, ch. 227, p. 2*a* (item 3).

[f] This is noteworthy in view of what has been said about pedals (p. 90 above).

[g] *Tu Yang Tsa Pien*, ch. 2, p. 8*b*. They could skim along the surface of the ground at a height of about 3 ft. for one or two hundred paces. Cf. pp. 485, 573 ff. below.

[1] 楊務廉 [2] 王琚 [3] 木獺 [4] 殷文亮 [5] 韓志和
[6] 踏牀 [7] 蠅虎子

described by many writers as having been constructed by the mathematician and astronomer Johannes Müller (+1436 to +1476); one of these was an eagle and the other a fly. Duhem,[a] who carefully examined the traditions, proposed some tentative solutions; the fly, for instance, would beat its wings by means of springs concealed within it, and make the tour of a dinner-table suspended from a hair invisible to the guests, finally approaching the hand of Regiomontanus because of a magnet secretly held by him. Four hundred years earlier, moreover, 'artificial bees' are described in the treatise of Prince Bhōja on automata (*Samarāṅgaṇa-sūtradhāra, c.* +1050).[b] In any case, those who doubt that very small automata can really be made may visit the Museum of Art and History at Geneva, where, in an exhibition installed by Dr Chapuis, actual specimens less than half an inch in height may be seen on display.[c]

In the Sung, glass was added to the stock-in-trade of the artisans who employed themselves in such domains, for we read of a mountain of glass with moving figures, and screens of glass behind which movements went on by the use of water-power, in Chou Mi's[1] *Wu Lin Chiu Shih*[2] (Institutions and Customs at Hangchow)[d] about +1270. Another book by the same writer, the *Chih Ya Thang Tsa Chhao,*[3] describes the performances of Wang Yin-Sêng,[4] who had a horizontally rotating water-wheel like Ma Chün's and could hit any target on it at will with a small bow and arrows. The puppets continued to flourish,[e] and the application of water-power to them in the +13th century is almost certain from their designation: *shui khuai lei.*[5]

By this time they had also become involved in a field rather different from that of simple entertainment; they had been enlisted in the service of horology. To speak of this here would be to anticipate what we shall have to say in the subsection on clock-work below (pp. 455 ff.), but a brief summary may suffice. The ancestors of all Chinese clock-jacks may be identified, perhaps, in the two statuettes of the immortal and the policeman which according to Chang Hêng's specification of +117 were placed on the top of the inflow clepsydra to guide the indicator-rod with their left hands and to point out the graduations with their right.[f] But these did not move. Then in +692, after the time of Huang Kun and Tu Pao, there was the artisan from Haichow, who seems to have been the first to make jacks appear at doors in accordance with the time.[g] When the first escapement came, in +725, I-Hsing and Liang Ling-Tsan arranged for two jacks to strike the hours, standing on the horizon surface of

---

[a] (1), pp. 128 ff.   [b] Raghavan (1), p. 20.
[c] See an article in *The Times*, 17 September 1952.
[d] Ch. 2, pp. 13*a*, 19*a* (pp. 368, 372); ch. 7, p. 8*b*; cf. Gernet (2), p. 204. In Chou Mi's description of the many different kinds of lamps sold at Hangchow twenty or thirty years previously, he says 'another sort (consisted of a) large vase pouring forth a jet of water which rotated a (suitable) mechanism, so that (representations of) all kinds of creatures seemed to move spontaneously'. Later in the same passage the hot-air zoetrope is referred to, so perhaps this was a variety worked by miniature water-power. Possibly it struck the hours in this way as the oil level decreased (cf. p. 535 below and Vol. 3, p. 331).
[e] See *Wu Lin Chiu Shih*, ch. 6, p. 26*b*; ch. 3, p. 1*b*; ch. 7, pp. 13*b*, 15*a*, 16*b*. Cf. *Mêng Liang Lu*, ch. 1, p. 4*a*; ch. 6, p. 3*a*; ch. 13, p. 12*b*; ch. 19, p. 9*a*; ch. 20, p. 13*a*, etc. And *Chhui Chien Lu Wai Chi*[6] (+1260), p. 29*b*.   [f] Cf. Vol. 3, p. 320, and p. 486 below.   [g] Cf. p. 469 below.

[1] 周密   [2] 武林舊事   [3] 志雅堂雜鈔   [4] 王尹生   [5] 水傀儡
[6] 吹劍錄外集

their sphere or globe.[a] Then followed great elaborations. Chang Ssu-Hsün's ingenious clock of +978 had 19 jacks,[b] while the masterpiece of Su Sung and Han Kung-Lien in +1088 had no less than 133.[c] And the tradition was fully maintained by some of the Yuan emperors.[d]

At this point we may suitably leave the parallel European and Chinese traditions of mechanical toys. The latter may or may not have been a little younger in its origin but there was never much to choose between them for ingenuity, and when they came together in the middle of the +13th century, the European tradition did not show up to much advantage. The triumphs of the European 'Gadget Age' were yet to come.

(d) TYPES OF MACHINES DESCRIBED IN CHINESE WORKS

We must now turn our attention to the principal types of useful machines described and illustrated in Chinese books. These books, which, though partly military and naval, centred upon the agricultural and sericultural arts, form families or constellations somewhat analogous to the +15th-century German military engineering manuscripts[e] or the +16th-century French and Italian engineering books.[f] But they cover a longer period, lasting from the middle of the +11th century down to the +18th, just as the Pên Tshao series of pharmaceutical compendia extended over an even wider span of centuries.[g] Later on, when discussing the outlines of Chinese agricultural bibliography,[h] we shall have to refer back to the present paragraphs, in which mention is made only of those works which were part of the iconographic tradition.[i] It will be logical to reserve an account of the farming literature which was not illustrated, and which had little to say about machinery, until the agricultural Section.

After describing, then, the principal books of the agricultural family, we can tabulate the types of machines which appear in them. The present subsection will proceed to deal with the forms of these which were operated by the force of men or animals, reserving the application of water-power and wind-power till a slightly later stage.[j] Besides the series of books already referred to, there appeared, in the Jesuit period, certain small though elaborately illustrated works on machinery, which added a basis of physics and abandoned the agricultural connection. While these might more properly have been discussed separately at the conclusion of this whole Section, there are good reasons for considering them now in relation to the rest.

[a] Cf. p. 474 below.   [b] Cf. p. 470 below.
[c] Cf. Needham, Wang & Price (1), and p. 455 below.
[d] Cf. p. 133 above and p. 507 below.
[e] Described as a group in Sarton (1), vol. 3, pp. 1550ff. For the Chinese military literature see Sect. 30; for the naval see Sect. 29.
[f] Analysed clearly by Beck (1) and Parsons (2).
[g] Cf. Sect. 38.   [h] Sect. 41.
[i] Since this was written, an interesting book on the history of Chinese book illustration has been published by A Ying (1).
[j] Since this was written, a study of Chinese inventions in power-source engineering has been given us by Liu Hsien-Chou (4).

(1) THE NATURE OF THE CHINESE ENGINEERING LITERATURE

It seems that the tradition of agricultural and agricultural-engineering illustration began with admonitory pictures upon the walls of imperial palaces. The emperor Ming Ti[1] of the Chin (who reigned +323 to +325), himself a famous painter, left a series of pictures known as *Pin Shih Chhi Yüeh Thu*[2] (Illustrations for the 'Seventh Month' Ode in the 'Customs of Pin' Section of the *Shih Ching*).[a] There is also evidence, more doubtful, that the emperor Shih Tsung[3] of the Later Chou dynasty (reigned +954 to +959) built a pavilion, the Hui Nung Ko,[4] which was ornamented with scenes of tilling and weaving.[b] In any case some such tradition certainly existed, and it seems even to have had a magical significance, for the Sung encyclopaedist Wang Ying-Lin[5] reports a tradition that in the Thang and Sung on two occasions when the palace frescoes of labour scenes were replaced by mere landscapes, there was trouble among the people and rebellions arose.[c] In due course paintings of this kind were collected into book form, a type of publication the earliest of which was perhaps the *Yüeh Ling Thu*[6] (Illustrations for the Monthly Ordinances (of the Chou Dynasty))[d] by Wang Yai.[7] Nothing is known about it or its author except that it was a Thang production. Its importance is that it was the predecessor of one of the most famous books in all Chinese literature, the *Kêng Chih Thu*[8] (Pictures of Tilling and Weaving).[e]

Such was the artistic and literary importance of this work, as well as its technological interest, that its history and bibliography, which is complex, has given rise not only to many studies in Chinese, but also to two substantial monographs in Western languages, those of Franke (11) and Pelliot (24).[f] The original pictures, each accompanied by a poem, were produced by Lou Shou,[9] an official of the Southern Sung, for presentation to the emperor Kao Tsung, in approximately +1145, i.e. not long after the transfer of the Sung capital to Hangchow.[g] Later, after Lou Shou's death, they were inscribed on stone, and probably also printed, about +1210, by his nephew Lou Yo[10] and grandson Lou Hung.[11] Their value to us lies in the fact that they are the oldest pictures we have of Chinese agricultural, mechanical and textile technology[h]—with the exception of what may be gleaned from carvings in Han tombs and mural paintings of Wei and Thang. Only the military illustrations start earlier (+1044).

---

[a] Pelliot (24), p. 95.　　　　　　　　　　　　　[b] Franke (11), p. 57.
[c] See Pelliot (24), p. 95.　　　　　　　　　　　 [d] Cf. Vol. 3, pp. 195 ff.
[e] A better title would be 'Agriculture and Sericulture', since all the main operations are illustrated, not simply the use of the plough and the loom themselves.
[f] Cf. also Montell (3).
[g] They may have been printed from wood blocks already at this early date, for they soon became very well known.
[h] The original set comprises 21 illustrations of agriculture and 24 of sericulture, making in all 45. There is some doubt as to whether the original pictures were carved in stone. Jäger (4) reported that in 1928 many broken pieces of stone inscribed with Lou Shou's poems, but without drawings, were found buried in the walls of Ningpo, his home city. It is thought that these finds dated from the early years of the +13th century, but the whole question remains unsettled.

[1] 明帝　　　 [2] 豳詩七月圖　　　 [3] 世宗　　　 [4] 繪農閣　　　 [5] 王應麟
[6] 月令圖　　 [7] 王涯　　 [8] 耕織圖　　 [9] 樓璹　　 [10] 樓鑰　　 [11] 樓洪

This raises, of course, the important question of the extent to which the pictures which we have today are faithful copies of the +12th- and +13th-century originals. In +1696 the whole set was redrawn at imperial command by an eminent artist, Chiao Ping-Chên,[1] who followed, as Hirth was the first to point out,[a] the rules of perspective which had been elaborated in the West and introduced by the Jesuits.[b] The Khang-Hsi emperor added a new set of poems, while retaining the old ones of Lou Shou.[c] So highly was the work prized, for its symbolic significance as depicting the foundations of Chinese agrarian culture, that in +1739 the Chhien-Lung emperor ordered the pictures to be copied anew by Chhen Mei,[2] and himself wrote another series of poems for them. The old verses of Lou Shou were now dropped, and a set of prose explanations by seven scholars[d] inserted instead. Very soon afterwards, in +1742, the whole was incorporated into the *Shou Shih Thung Khao* (chs. 52 and 53), Lou Shou's poems being restored and a fourth set, written by the Yung-Chêng emperor between +1723 and +1735, added.

Fortunately, certain sets of pre-Chhing illustrations have been recovered in the present century.[e] Laufer (21) had the good fortune to obtain in Tokyo a Japanese reproduction made in +1676 of an old Chinese edition issued by Sung Tsung-Lu[3] in +1462; and this was published by Franke (11) side by side with the later (+1739) Chhing series. Subsequently, Pelliot (24) went one better by discovering a scroll of rubbings of stone-carvings based on paintings by Chhêng Chhi[4] made during the second half of the +13th century (the Semallé Scroll). These must have derived directly from the edition of +1237 issued by Lou Shao[5] after the death of Wang Kang[6].[f] Discussion is likely to continue on the finer points of these and the many later editions,[g] but there is no great difference between them on the essential technical matters which they represent, and we may therefore feel fairly certain that the older sets are valid for the time of Lou Shou himself.[h]

[a] (9), pp. 57ff.
[b] Chiao Ping-Chên was also an official in the Bureau of Astronomy, hence no doubt his Jesuit contacts.
[c] Cf. Franke (11), pp. 80, 84, 101; Pelliot (24), p. 65; Hedde (1); Julien & Champion (1).
[d] Yu Min-Chung,[7] Tung Pang-Ta,[8] Kuan Pao,[9] Chhiu Yüeh-Hsiu,[10] Wang Chi-Hua,[11] Chiang Ting,[12] and Chhien Wei-Chhêng.[13] See Franke (11), pp. 89, 106, 115; Pelliot (24), p. 78. The pictures were used to ornament ink blocks (Fuchs, 8), and were redrawn by copper-plate engravers in Western countries (Anon., 49).
[e] The Chhing versions are hardly distinguishable but the older ones are very different in style.
[f] Lou Shao was a great-grandson of Lou Shou; and Wang Kang a scholar deeply interested in agriculture who had been an official in Chekiang since +1187, rose to ministerial rank, and had the illustrations carved on wooden blocks.
[g] Cf. Fuchs (8). Various sets of parallel pictures by eminent painters in successive centuries seem also to have existed, and whatever remains of them may yet provide valuable evidence in the history of technology.
[h] Other works of the same kind as Lou Shou's, but which had less good fortune, were also produced in the Sung, e.g. the *Pin Fêng Thu*[14] (Illustrations for the 'Customs of Pin') by Ma Ho-Chih[15] (cf. Pelliot (24), p. 120). In the 18th century, there were several, such as the *Thao Chih Thu Shuo*[16] (Illustrations of the Pottery Industry) of +1743 (cf. David, 2), and the *Mien Hua Thu*[17] (Pictures of Cotton Growing and Weaving) of +1765 (cf. Franke (11), p. 88). It was at one time thought that Han Yen-

[1] 焦秉貞	[2] 陳枚	[3] 宋宗魯	[4] 程棨	[5] 樓杓
[6] 汪綱	[7] 于敏中	[8] 董邦達	[9] 觀保	[10] 裴日修
[11] 王際華	[12] 蔣榑	[13] 錢維城	[14] 豳風圖	[15] 馬和之
[16] 陶冶圖說	[17] 棉花圖			

These bibliographical technicalities, seemingly tedious, have so much importance for the comparative history of technology at the two ends of the Old World that it is necessary to dwell on their meaning for a moment.[a] Chhêng Chhi's paintings must have been made in the close neighbourhood of +1275. The first half of the set was rediscovered and presented to the emperor soon after +1739 while the second half reached the imperial palace in +1769. The Chhien-Lung emperor, realising its importance, immediately had the whole engraved on stone, adding himself a historical introduction and yet another series of poems to go beside Lou Shou's. Whether or not the stones still exist matters the less since Pelliot (24) made available the rubbings of the Semallé Scroll, with all their wealth of fine detail. If then we possess in this way an authentic record of the technology of the Southern Sung, an important contrast with Europe follows, for while the manuscript corpus of the Italian and German engineers informs us about the developments of the +15th and +14th centuries, we are here in presence of reliable Chinese material concerning the +13th and the +12th centuries. Chhêng Chhi must certainly have worked from the block-printed edition of +1237, which derived in turn from a probable printing of +1210, and that in turn was based upon the drawings (quite possibly disseminated as block-prints) of the fountain-head of +1145. One would give a good deal to know whether any technical details changed during this period. We believe it is unlikely that they did, and though we generally date *Kêng Chih Thu* evidence at +1237 or +1210, its probable validity for a century earlier must always be remembered. Here an interesting point emerges, for all through this period (indeed from the +9th century onwards) China had printing while Europe did not. Therefore the criterion of *terminus a quo* has to be applied more strictly to Western MS. illustrations of, say, the +15th century than to Chinese ones of the same kind, if there is evidence of prior typographical transmission in a particular tradition. In other words, we can never assume the existence in the West of an invention (e.g. the crankshaft or the fusee) earlier than its first MS. evidence; but where Chinese iconographic data are known to be based upon still earlier printed editions, we may feel able to give greater probability to the previous existence of the technical detail in question, so far as the limits of the bibliographical facts allow.[b] Descriptive textual evidence (with all its peculiar difficulties of interpretation) is of course a different matter.

One curious feature of the *Kêng Chih Thu* pictures is that those of Chiao Ping-Chên (+1696) are a good deal more like those of Chhêng Chhi (c. +1275) than either of them resemble the Ming edition of +1462 (known from the Japanese version of +1676). As Pelliot says,[c] the only possible explanation for this is that both Chhêng

Chih,[1] the well-known monographer of citrus horticulture (d. +1200), had written a work like the *Kêng Chih Thu*, but this is apparently a mistake (Pelliot (24), p. 100). On the other hand, the famous painter and calligrapher Chao Mêng-Fu[2] (+1254 to +1322) did write 24 poems at the imperial request to accompany pictures, and these (*Thi Kêng Chih Thu Erh-shih-ssu Shou Fêng I Chih Chuan*[3]) are to be found in *TSCC, Shih huo tien*, ch. 39, *i wên* 5, p. 1a. On machines we shall later quote from them.

  [a] Cf. p. 106 above.
  [b] After all, as one of the +13th-century *Kêng Chih Thu* scholars said, the drawings had to go with the poems 'like the lining to the gown'.   [c] (24), p. 93.

  [1] 韓彥直   [2] 趙孟頫   [3] 題耕織圖二十四詩首奉懿旨撰

and Chiao worked directly from extant copies of the edition of +1237. Only a few decades after Chhêng was active, the next fundamental contribution was being prepared, and to it we must now turn.

Whether the illustrations in the accessible editions of the *Nung Shu*[1] (Treatise on Agriculture), produced by Wang Chên[2] in +1313, are equally contemporary with his text is a matter even more important and more difficult to determine. For while so much attention has been paid to the bibliography of the *Kêng Chih Thu*, only a minority of its pictures have information to give us about machines. The *Nung Shu*, on the other hand, shows us no less than 265 diagrams and illustrations of agricultural implements and machines, mostly in a section entitled Nung Chhi Thu Phu[3] (Illustrations of Agricultural Tools). But here again, the authenticity of the illustrative material is strongly vouched for by the significant fact that in no case have we ever found a discrepancy between the text and the pictures. Moreover, the somewhat archaic character of the drawings, quite similar in style to the old series of the *Kêng Chih Thu*, indicates that we may securely take them as valid for Wang Chên's own time.[a] His *Nung Shu* is the greatest, though not the largest, of all works on agriculture and agricultural engineering in China, holding a unique position on account of its date.[b] This is shared to some extent by the *Nung Sang I Shih Tsho Yao*[4] (Essentials of Agriculture, Sericulture, Food and Clothing) written by Lu Ming-Shan[5] in +1314, a work first noticed by Bazin (2). Lu's book was intended to supplement the *Nung Sang Chi Yao*[6] (Fundamentals of Agriculture and Sericulture), which had been compiled by imperial order under the Yuan dynasty already in +1274; but so far as we know, neither of these was ever illustrated.[c] A few years later, in +1318, there appeared a small work by Miao Hao-Chien,[7] the *Tsai Sang Thu Shuo*[8] (Pictures of the Planting of Mulberry-Trees, with Explanations),[d] but it is not conserved.

Throughout the Sung, Yuan and Ming dynasties another class of literature became important, the encyclopaedias for daily use (*jih yung lei shu*[9]). So widespread were these that people generally considered them not worth preserving, and they are now to be found only as rare or unique copies in libraries, as many in Japan and other

---

[a] Cf. Franke (11), p. 46.   [b] And hence its freedom from occidental influences.

[c] For the same reason we pass over here the useful and much earlier *Nung Shu*[10] of Chhen Fu,[11] written about +1149, and still partly conserved. Another work dating from about the same time as Wang Chên's great *Nung Shu* is an 'Everyman's Guide to Agriculture' (*Chung I Pi Yung*[12]) by Wu Tsuan,[13] with a supplement by Chang Fu.[14] This is preserved, according to Chao Wan-Li (1), in the *Yung-Lo Ta Tien*, ch. 13,194, and the publication promised will be eagerly awaited. So far as we know, this work contains no illustrations of rural engineering. Diagrams and specifications are contained, however, in a tractate by Hsüeh Ching-Shih[15] in ch. 18,245 of the same encyclopaedia, which deals with carpentry and wood-work, the making of vehicles, and the construction of looms and other textile machinery. It is entitled *Tzu Jen I Chih*[16] (Traditions of the Joiners' Craft). Although it has been already edited by Chu Chhi-Chhien & Liu Tun-Chên (3), the forthcoming facsimile publications of this will be of still greater interest. Since the *Mu Ching* is lost, and the *Ying Tsao Fa Shih* deals only with building technology, Hsüeh's work, dating as it does from the beginning of the Yuan, gives us the earliest extant text of its kind. We shall highlight it in the Section on textile technology.

[d] Franke (11), p. 60; Pelliot (24), p. 108.

[1] 農書   [2] 王楨   [3] 農器圖譜   [4] 農桑衣食撮要   [5] 魯明善
[6] 農桑輯要   [7] 苗好謙   [8] 栽桑圖說   [9] 日用類書
[10] 農書   [11] 陳旉   [12] 種藝必用   [13] 吳欑   [14] 張福
[15] 薛景石   [16] 梓人遺制

foreign countries as in China itself. Niida (*1*) and Sakai (*1*) have recently described more than twenty of these vade-mecums, printed at dates ranging from about +1350 to +1630. Besides instructions on family customs, popular medicine and hygiene, fortune-telling and the drafting of legal documents, they also give numerous details of farming, sericulture and the arts and crafts. Thus a glance at the *Pien Yung Hsüeh Hai Chhün Yü*[1] (Seas of Knowledge and Mines of Jade; Encyclopaedia for Convenient Use), printed in +1607, finds wood block cuts of the square-pallet chain-pump,[a] the rotary-fan winnowing-machine,[b] the connecting-rod hand-mill,[c] a very simple silk-reeling machine,[d] and some looms.[e] This 'inquire-within-upon-everything' literature thus urgently invites study by historians of technology, and if the Ming editions are not likely to tell us much that we do not already know, those of the Yuan and the Sung may well provide us with important new evidence. Here we come once again upon the effects of the art of printing. Before the popularisation of typography technical monographs or treatises on, say, textiles or iron-working could have had only a very restricted circulation in manuscript, and much must have been lost,[f] but as soon as the democratic medium got into its stride, from the +10th century onwards, technical subjects began to take their place, even if it was often only in popular encyclopaedias, side by side with the topics which had always interested the literati. There is thus unfortunately a built-in obstacle to our quest of technological data for the Thang and earlier, and this is why the evidence of the Tunhuang fresco-paintings and similar sources has such particular importance.

During the three centuries after Wang Chên no further contributions of importance to the literature of rural engineering appeared, but in +1609 the *San Tshai Thu Hui* illustrated encyclopaedia reproduced a good many of the pictures of the *Nung Shu* in somewhat degenerate form.[g] No Western influence is perceptible here, but this was the time of the Jesuits, and shortly afterwards came three books prepared under their inspiration which broke away from the agricultural domination under which Chinese eotechnic mechanical engineering had grown up.[h] The first was the *Thai Hsi Shui Fa*[2] (Hydraulic Machinery of the West) published by Sabatino de Ursis (Hsiung San-Pa[3]) & Hsü Kuang-Chhi[4] in +1612. Then came the *Chhi Chhi Thu Shuo*[5] (Diagrams and Explanations of Wonderful Machines (of the Far West)), by Johann Schreck (Têng Yü-Han[6])[i] & Wang Chêng[7] in +1627; this deals with a great variety of machines,

---

[a] Ch. 20, p. 6*b*; see p. 339 below.   [b] Ch. 20, p. 7*b*; cf. p. 153 above.

[c] Ch. 20, p. 8*a*; see p. 117 above.   [d] Ch. 20, p. 14*a*; cf. p. 382 below.

[e] Ch. 20, pp. 16*b*, 17*a*.

[f] For an example of an important work probably never printed, see Sect. 29*c* below.

[g] Chhi Yung section, ch. 10.

[h] It is a curious fact that though the main Chinese work on architecture, the *Ying Tsao Fa Shih* of +1097 (cf. Sect. 28), contains many illustrations, these are all concerned with measuring-instruments, joinery details, and types of ornament; of cranes and similar apparatus there are none. But this cannot mean that the great buildings of earlier centuries in China were constructed without handling machinery.

[i] In the time of Matteo Ricci the Jesuit order was capable of attracting for its oversea missions some of the best minds of Europe. It was a mobilisation of œcumenical idealism something like that which the League or the United Nations have now and then been able to command in our own time. Johann Schreck (+1576 to +1630) was born a Swiss at Constance and while still very young became known through

[1] 便用學海群玉   [2] 泰西水法   [3] 熊三拔   [4] 徐光啓

[5] 奇器圖說   [6] 鄧玉函   [7] 王徵

including cranes, mills and sawmills, as well as water-raising engines. A shorter companion work, by Wang Chêng alone, the *Chu Chhi Thu Shuo*[1] (Diagrams and Explanations of a number of Machines), appeared in the same year.[a] Owing perhaps to the abundance of the illustrations in these works, they have attracted great attention;[b] an analysis of them will shortly follow.[c]

Between +1625 and +1628, Hsü Kuang-Chhi occupied himself with a new agricultural compendium destined to supersede all earlier works, but it was not published until after his death, when it was edited by Chhen Tzu-Lung[2] as the *Nung Chêng Chhüan Shu*[3] (Complete Treatise on Agriculture), of +1639. This famous book[d] is well illustrated, especially in the irrigation section (chs. 12–17), and later (chs. 18 and 21–3) reproduces nearly all the agricultural machinery of the *Nung Shu* with minor variations. The *Thai Hsi Shui Fa* is reproduced complete (chs. 19, 20), but there is no essential advance along the lines of traditional Chinese engineering beyond Lou Shou and Wang Chên.

Contemporary with the work of Hsü Kuang-Chhi was that of Sung Ying-Hsing,[4] who produced his *Thien Kung Khai Wu*[5] (The Exploitation of the Works of Nature), China's greatest technological classic, in +1637. Though dealing with agriculture and industry rather than engineering in the strict sense, this is perhaps the right place for a brief mention of its contents.[e] It is divided into the following sections:

all the German cultural world for his brilliant attainments in medicine, natural philosophy and mathematics. Passing to Italy in the early years of the 17th century he gained further renown and was, with Galileo, one of the first half-dozen members of the Cesi Academy. A glimpse of his activity among them at this time may be obtained from E. Rosen (1). Already in the north he had won the personal friendship of Kepler. This then was the man who, as Têng Yü-Han, a dedicated exile, found himself in 1626 in far-away Peking, working with a distinguished Chinese scholar, Wang Chêng, on a book which would provide for the first time in Chinese dress the principles of Renaissance mechanics and an account of their applications by the engineers of Europe. This was the *Chhi Chhi Thu Shuo*, which appeared in the following year. For a fuller account see Bernard-Maître (17), and the biography by Gabrieli (1). Schreck latinised his name, punningly, as Terrentius, hence the English form John Terence, best not used. He was elected to the Academy of the Lynxes in 1611, joined the Jesuits in 1612 and reached China in 1621. For the last few years of his life he was engaged in the beginning of the calendar reform which the Jesuits were invited by the imperial court to undertake (cf. Vol. 3, pp. 259, 447).

[a] Pfister (1), p. 157, is wrong in saying that this book was also under joint authorship with Schreck, and that it was concerned with 'les machines indigènes'. On the contrary, the devices were mainly of Wang Chêng's own invention or adaptation, and he was decisively the first 'modern' Chinese engineer, indeed a man of the Renaissance, though so far from its birthplace. A biography of Wang Chêng (+1571 to +1644) is given in Hummel (2), pp. 807ff. The late Prof. Fritz Jäger of Hamburg made himself an authority on Wang Chêng and all his scientific work. I have had the valuable opportunity of consulting the unpublished posthumous papers of Prof. Jäger, thanks to the kindness of the present occupant of the Hamburg chair, Prof. Wolfgang Franke. As this was long after the writing of the present Section, my main satisfaction lay in finding so many conclusions so independently confirmed.

[b] Horwitz (1); Reismüller (1); Feldhaus (2, 11, 12); Goodrich (7); and special studies by Jäger (2), Huang Chieh (2), Shao Li-Tzu (1) and Liu Hsien-Chou (2).

[c] Among the relevant Jesuit publications of this time, we ought to mention also the *Hsi Hsüeh Fan*[6] (A Sketch of European Science and Learning), written in +1623 by Giulio Aleni (Ai Ju-Lüeh[7]) to give an idea of the contents of the books which Nicholas Trigault had brought back for the Pei-Thang Library. See Cordier (8), p. 6; Pfister (1), p. 135; Bernard-Maître (18).

[d] Franke (11), p. 51; Bretschneider (1), vol. 1, p. 82.

[e] There is a special, though short, study of it by Yoshida Mitsukuni (1), and Sung's biography was written by Ting Wên-Chiang (1). Since this Section was first drafted, a remarkable edition of the

[1] 諸器圖說　　[2] 陳子龍　　[3] 農政全書　　[4] 宋應星　　[5] 天工開物
[6] 西學凡　　[7] 艾儒略

     (*a*) Agriculture, irrigation, and hydraulic engineering.
     (*b*) Sericulture and textile technology.
     (*c*) Agriculture and milling processes.
     (*d*) Salt technology.
     (*e*) Sugar technology.
     (*f*) Ceramics industry.
     (*g*) Bronze metallurgy.
     (*h*) Transportation; ships and carts.
     (*i*) Iron metallurgy.
     (*j*) Coal, vitriol, sulphur, and arsenic.
     (*k*) Oil technology.
     (*l*) Paper-making.
     (*m*) Metallurgy of silver, lead, copper, tin, and zinc.
     (*n*) Military technology.
     (*o*) Mercury.
     (*p*) Ink.
     (*q*) Fermented beverages.
     (*r*) Pearls and jade.

The illustrations are the finest of any produced in China on these subjects, and in many cases the only ones, but the text does not always equal them in clarity. Early in the Chhing period the work almost disappeared in China, perhaps because coinage, salt-making and weapon manufacture were government monopolies, but fortunately a copy of the original edition was preserved in Japan, and the best editions we now have are reproductions of that.

Just under a century later, the *Thu Shu Chi Chhêng* encyclopaedia (+1726) continued to reproduce the traditional illustrations.[a] The Yuan series of Wang Chên, modified but not essentially changed, reappears twice,[b] all the *Thien Kung Khai Wu* pictures are also given,[c] as are those of the *Chhi Chhi Thu Shuo* and *Chu Chhi Thu Shuo*.[d] Finally, in +1742, came the *Shou Shih Thung Khao*[1] (Complete investigation of the Works and Days), prepared by imperial order to be the *non plus ultra* of all compendia on agriculture and agricultural engineering. It differs from its predecessors in starting with some geography and meteorology, and including large sections (chs. 19–30, 58–71) on agricultural botany, but the engineeering illustrations deviate in no way from those of the *Nung Chêng Chhüan Shu*. The *Kêng Chih Thu* and the three Jesuit books are all incorporated.

It may be said, therefore, that by the end of the Sung the Chinese tradition of agricultural engineering had reached its fullest development. The *Nung Shu* and the *Thien Kung Khai Wu* are the best sources for it, and though pumps and various kinds

---

Thien Kung Khai Wu has appeared, which includes a Japanese translation and a series of essays on diverse aspects of its contents by Japanese historians of technology; this is the work of Yabuuchi Kiyoshi (*11*). An annotated English translation by Dr and Mrs Sun of Philadelphia is now awaited.
  [a] Generally redrawn with charming artistry.
  [b] *I Shu Tien*, chs. 3–12, and *Khao Kung Tien*, chs. 243–5.
  [c] *Khao Kung Tien*, chs. 174, 178, and 248.
  [d] *Ibid.* ch. 249 (often clearer than the originals, it must be said).

  [1] 授時通考

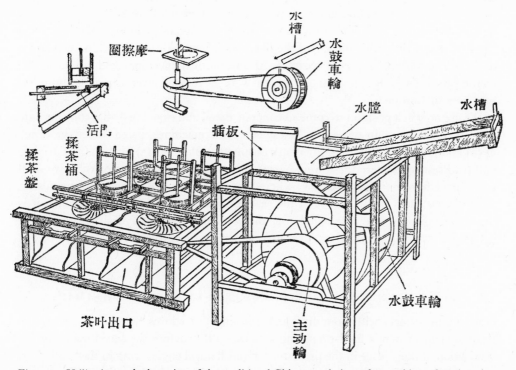

水槽

圈擦摩

水鼓車輪

水膛

水槽

活門

插板

揉茶桶

揉茶盤

茶叶出口

主动轮

水鼓車輪

Fig. 439. Utilisation and adaptation of the traditional Chinese style in craftsmanship and engineering to meet contemporary needs; designs from the National Exhibition of Agricultural Machinery in Peking, 1958. A tea-leaf rolling- and rubbing-mill powered by a small vertical water-wheel (Anon. (*18*), pt. 5, no. 50). The four cheese-like rollers, fixed all together within a rectangular frame, are made to revolve over their gratings by the motion of a discoidal cam at the centre into which the main vertical driving-shaft is fitted eccentrically. Flumes deliver the tea-leaves which drop down through the gratings.

of gearing were introduced by the Jesuits there is no evidence that these innovations were adopted, presumably because social and economic conditions made unnecessary any changes from the classical methods. Hence also there were no essential advances in the books themselves down to the beginning of the 19th century.[a] Indeed, those who, like the writer, have lived in China, know that the early medieval techniques

[a] Strictly speaking, if we think of an iconographic tradition such as this rightly as a whole, we may follow it down to the beginning of the 20th century. The appearance of the *Shou Shih Thung Khao* in no way checked the flow of Chinese literature on agriculture and rural engineering. In +1760, for instance, there appeared the *San Nung Chi*[1] (Records of the Three Departments of Agriculture) by Chang Tsung-Fa,[2] an interesting, very Szechuanese, work unfortunately lacking illustrations. There were many others in the following decades, and all through the nineteenth century works on agriculture and sericulture continued to appear—mention has already been made (p. 2) of some of these, which have important illustrations. Even in the first years of the twentieth, publications on these subjects, such as the *Nung Hsüeh Tsuan Yao*[3] of Fu Tsêng-Hsiang,[4] could be furnished with interesting, and informative, drawings entirely in the traditional style. Moreover, a thorough survey would not neglect certain Japanese contributions. For example, in 1822 Ōkura Nagatsune[5] produced a treatise on agricultural machinery including pile-driving and hydraulic engineering, the *Nōgu Benri-ron*;[6] while in 1836 Tokugawa Naria-ki[7] devoted another entirely to designs for pumps and other kinds of water-raising machinery, the *Ungei Kisan*.[8] The illustrations for both of these were fully traditional.

[1] 三農紀   [2] 張宗法   [3] 農學纂要   [4] 傅增湘   [5] 大藏永常
[6] 農具便利論   [7] 德川齊昭編   [8] 雲霓機纂

(such as the square-paddle chain-pump with radial treadle) are still in full vigour. Their supersession is bound up with the current problems of industrialisation, persistence of wet rice cultivation, and the like, and it may well be that China will adopt designs of modern equipment based on the traditional machines and suited to the country's needs rather than those which were developed over the ages for other purposes in Europe.[a]

We are now in a position to come to concrete detail and examine Table 56, where are assembled all the main types of Chinese eotechnic machines, listed under the heads of (A) Pounding, (B) Grinding, (C) Water raising, (D) Blowing, (E) Sifting, (F) Turning, and (G) Pressing. Where these were operated by the power of water or wind we shall treat of them in more detail below; where they were operated by men or animals we shall discuss them now—except in the cases of water-raising machinery, which it is more convenient to assemble together as the first subsection on hydraulic engineering (p. 330 below), and of blowing apparatus, which has been discussed already (p. 135 above).

(2) Eotechnic Machinery, Powered by Man and Animals

Here we must start with pounding and grinding procedures,[b] probably among the most ancient of man's food-preparing activities. But before we can pursue our investigation of their place in the prehistory of mechanical engineering in East and West a few words may help to recall the reasons for the human need for them. These concerned the problem of de-husking cereal grains. The grain of the cultivated grasses contains, beside the embryo of the future plant, a mass of starch-containing material, the endosperm, which acts as the fostering yolk until starch can be made by the new plant photosynthetically. This through the ages has been man's 'staff of life'. But it is guarded externally by the husk and the aleurone layer, forming when comminuted what we know as chaff and bran; these also contain carbohydrate but in an insoluble

---

[a] These words were first written in 1952. They have proved quite prophetic. When six years later I had the pleasure of attending the National Exhibition of Agricultural Machinery in Peking, I found that great progress had been made in the utilisation and adaptation of the traditional style of engineering and craftsmanship to meet modern needs. I give here three examples: Fig. 439 shows a tea-leaf rolling-mill worked off a vertical water-wheel, Fig. 440 a simple crop-sprayer, and Fig. 441 a hay-cutter with a drive from the wheels; Anon. (18), pt. 5, no. 50; pt. 7, no. 25; pt. 9, no. 19. Some idea of the immense amount of manual labour saved by tea-rolling machinery may be gained from Fortune (1), vol. 2, pp. 235ff.; (4), pp. 197ff., 207ff. All these designs derive patently from Ma Chün and Su Sung rather than from Vitruvius and Leonardo. Of course where the more advanced metallurgical technology is concerned there can be less continuity with the past, yet even in such fields the Chinese technical tradition is exerting its own inspirations. The continuous development from ancient ways has moreover been much encouraged by the emphasis placed at the present time (1959) on industrial decentralisation. We shall describe in Sect. 30 a visit paid in 1958 to a traditional steel-works in Szechuan, where steel has long been made by the highest form of a most ancient process, co-fusion (cf. Needham, 32). Indeed, the fixed idea that *yang fa*[1] (foreign methods) must always necessarily be better than *thu fa*[2] (traditional methods) has been condemned as a 'superstition', and it is not likely that the Chinese will again succumb to it.

[b] In this subsection we deal primarily with grinding machinery for comminution of food or other materials. Grinding machinery for abrasion, shaping or polishing comes into consideration elsewhere, cf. Vol. 3, pp. 667ff., and above pp. 82, 92, 122.

[1] 洋法   [2] 土法

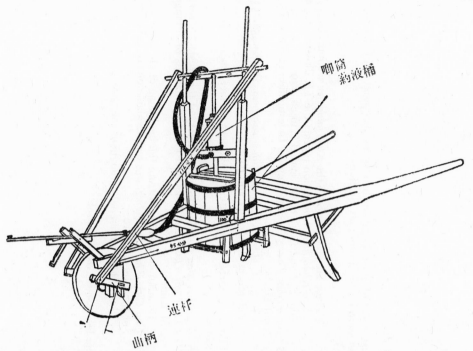

唧筒
約液桶

連杆

凵柄

Fig. 440. Contemporary adaptations; a wheelbarrow-mounted crop-sprayer with a drive from the wheel. Cranks (*chhü ping*) and connecting-rods (*lien kan*) are fitted on each side of the wheel to raise and lower the piston-rod of the pump by means of a cross-head sliding on the upright bars of the frame. As the tank is a tub the machine can be constructed almost entirely of wood. Anon. (*18*), pt. 7, no. 25.

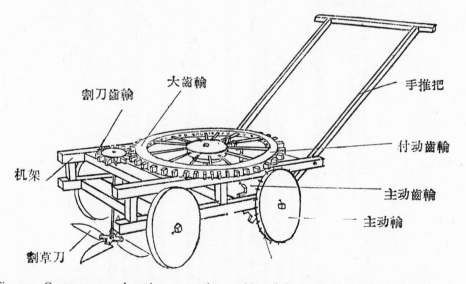

割刀齒輪
大齒輪
手推把

機架
付动齒輪

主动齒輪

主动輪

割草刀

Fig. 441. Contemporary adaptations; a mowing-machine of almost wholly wooden construction. One of the four road-wheels has a gear-wheel on its axle enmeshing at right angles with horizontal gearing above to effect the rapid rotation of hay-cutting knives at the front. Anon. (*18*), pt. 9, no. 19. This road-wheel drive was a very ancient device in Chinese engineering, cf. p. 160 above, and Figs. 528, 533 below.

Key to Table 56

NOTE. This table excludes, in the main, machines of the textile industry and of agricultural husbandry, as also purely military and naval apparatus, which we reserve for discussion in the appropriate Sections.

ABBREVIATIONS

KC *Kêng Chih Thu* ((Poems and) Pictures of Tilling and Weaving) +1145 (+1210)
    Numbers as in Franke (11)

NS *Nung Shu* (Treatise on Agriculture) +1313
    Illustrations numbered consecutively all through

ST *San Tshai Thu Hui* (Universal Encyclopaedia) +1609
    Chhi Yung section, ch. 10

NC *Nung Chêng Chhüan Shu* (Complete Treatise on Agriculture)
    +1628 (+1639)
    References by chapter and page

TK *Thien Kung Khai Wu* (The Exploitation of the Works of Nature)
    +1637
    Illustrations numbered consecutively all through (numbers
    in brackets refer to the first edition)ᵃ

TC *Thu Shu Chi Chhêng* (Imperial Encyclopaedia) +1726
    Illustrations numbered from the beginning of the *I shu tien*
    section.

SS *Shou Shih Thung Khao* (Complete Investigation of the Works
    and Days) +1742
    Ch. 31 ff. The *KC* illustrations reappear in chs. 52, 53.
    References by chapter and page

HF *Thai Hsi Shui Fa* (Hydraulic Machinery of the West) +1612
CC *Chhi Chhi Thu Shuo* (Diagrams and Explanations of Strange
    Machines) +1627
    Illustrations numbered from the beginning of ch. 3
WC *Chu Chhi Thu Shuo* (Diagrams and Explanations of a Variety
    of Machines) +1627
    Illustrations numbered throughout the book

These are all repeated in the *Khao kung tien* section, ch. 244.
The *Khao kung tien* section repeats the *TK* illustrations in
ch. 248 and those of *CC* and *WC* in ch. 249

(The last three books describe Western, rather than Chinese,
machines)

ᵃ The first edition has forty-one illustrations less than the *editio princeps*.

Table 56. *Principal machines described and illustrated in traditional Chinese books*

	Name	Characters	References
(A) POUNDING			
(1) Pestle and mortar, manually operated (man-power)	*chhung chiu*	舂臼	KC/18 NS/116 ST TK/59 (35) TC/129 SS/40.16, 52.38
(2) Tilt-hammer pestle (man-power)	*tha tui*	踏碓	KC/18, 19 NS/117 ST NC/23.2 TK/49, 59 (35), 113 (81) TC/130 SS/40.17, 52.38 NS/118 NC/23.3 TC/131 SS/40.18
(3) Water-counterweighted tilt-hammer ('spoon' tilt-hammer) operated automatically by water fall	*tshao tui*	槽碓	NS/188 ST NC/18.15 TC/41
(4) Battery of trip-hammers actuated by lugs on the shaft of a vertical undershot water-wheel	*chi tui* *shui tui*	機碓 水碓	NS/189 ST NC/18.13 TK/50 (32) TC/40 SS/40.19
(5) Battery of trip-hammers actuated by lugs on a shaft driven by right-angle gearing from a horizontal water-wheel (*wo lun*, 臥輪)	*chi tui* *shui tui*	機碓 水碓	ST
(B) GRINDING			
(a) Apposition of flat or bevelled surfaces			
(1) Quern, manually operated, with short (crank-)handle	*mo* 磨 or *lung* 礱		Not illustrated, though universally used
(2) Quern with long-handled connecting-rod and crank, operated by one or several men	*mo* 磨 or *lung* 礱 *thu lung* 土礱 *mu lung* 木礱		KC/18, 21 NS/119 NC/23.5 TK/46 (29), 112 (81) TC/132 SS/52.44 TK/47 (30) SS/40.10 TK/48 (29) SS/40.11
(3) Rotary grinding-mill operated by two animals, directly yoked	*niu mo* 牛磨 *niu lung* 牛礱		NS/124 ST NC/23.12 TK/51 TC/137 SS/40.25
(4) Rotary grinding-mill operated by two animals by means of a crossed driving-belt from a whim or capstan driving-wheel	*lu mo* 驢磨 *lu lung* 驢礱		NS/120 NC/23.6 TK/53 TC/133 SS/40.13
(5) Multiple rotary grinding-mills, up to eight, directly geared to a driving-wheel operated by animals	*lien mo* 連磨		NS/125
(6) Rotary grinding-mill operated directly by a horizontal water-wheel underneath	*shui mo* 水磨		NS/182 ST NC/18.6 TK/52 TC/33 SS/40.26

Table 56 (continued)

			References
(7)	Rotary grinding-mill operated by a vertical water-wheel and right-angle gearing (Vitruvian mill)	shui mo 水磨 (or shui lung 水礱)	TK/(33) SS/40.12, 40.27
(8)	Multiple rotary grinding-mills, up to nine, operated by a vertical undershot water-wheel, through right-angle gearing	shui chuan lien mo 水轉連磨	NS/186 ST NC/18.7, 18.10 TC/34, 35, 38 SS/40.27, 40.28
(9)	Multiple rotary grinding-mills, directly geared to a driving-wheel operated by a horizontal wind-wheel (windmill)	fêng shan mo 風扇磨 / fêng wei 風磑	} CC/33, 34, 40 WC/4

(b) *Apposition of wheel-edge or roller surfaces*

			References
(1)	Edge-runner mill with longitudinal travel (man-power)	yen nien 研碾	TK/146 (108)
(2)	Rotary disc knife, treadle-operated with alternating motion (man-power) for jade-working (cf. Vol. 3, pp. 667ff.)	cho yü lun 琢玉輪 / cha tho 鍘鉈	} TK/161 (121), 162 (121) TC/KKT/4
(3)	Single edge-runner mill travelling in circular trough (animal-driven)	nien 碾	ST TK/(26)
(4)	Double edge-runner mill travelling in circular trough (driven by two animals)	shih nien 石碾	NS/121 NC/23.8 TK/55 TC/134
(5)	Single or double edge-runner mill operated directly by a horizontal water-wheel underneath	shui nien 水碾 (or nien wei 碾磑) / shuang lun shui wei 雙輪水磑	} NS/184 ST NC/18.8 TK/56 TC/36 SS/40.23
(6)	Double narrow-tread wheel roller (agricultural)	shih tho 石陀	TK/18
(7)	Hand-roller for grinding (man-power)	hsiao kun nien 小滾碾	TK/54 (37) SS/40.22 NS/24, 25 NC/21.12, 21.13 SS/33.12
(8)	Roller harrow (animal-drawn) { Spiked / Not spiked	shih ho tsê 石礰礋 / lu tu 陸碡 / lu tu 碌碡 / kan 趕 / kun nien 輥碾	} KC/5 NS/23 NC/21.11 SS/33.11 TK/41 (34) SS/39.7
(9)	Roller mill (horizontal roller travelling round a circular course), animal-operated		NS/122 ST NC/23.9 TC/135 SS/40.21
(10)	Vertical double-roller sugar-mill, operated by two animals	ya chê chhü chiang 軋蔗取漿	TK/76 (44)
(11)	Horizontal double-roller cotton-gin, manually operated (man-power)	mu mien chiao chhê 木棉攪車 / mien kan 棉趕	} NS/245 NC/35.22 TK/34 (22)
(12)	Combination of grinding-mill, edge-runner mill, and sifter, engageable at will, operated directly by horizontal water-wheel underneath	shui lun san shih 水輪三事	NS/185 ST NC/18.5, 18.9 SS/40.24

(c) WATER RAISING (and LIQUID PROPULSION)

(a) Leverage

Description	Romanization	Characters	References
(1) Double hand-swung bucket	hu tou	戽斗	KC/10, 11 NS/171 NC/17.21 TC/21 SS/37.21, 52.24
(2) Swape (counterweighted bailer; shādūf)	chieh kao	桔槔	KC/13, 14 NS/173 ST NC/17.25 TK/11 (9) TC/23 SS/37.23, 52.30 CC WC SS/38.30
(3) Ditto, modified with flume in beam	ho yin	鶴飲	

(b) Rotary

Description	Romanization	Characters	References
(1) Winch or windlass, with crank, for wells	lu lu	轆轤	NS/174 ST NC/17.29 TK/12 TC/24, 29 SS/37.24
For cranes			CC
For mine-shafts			TK/105 (74), 157 (117), 158 (118)
For salt-well boreholes			TK/66
For pearl-diving			TK/(115)
(2) Capstan winding drum, for salt-well boreholes			TK/68 (42), 71 (42)
(3) Scoop-wheel in flume, hand-operated (man-power), with crank	kua chhê	刮車	NS/172 NC/17.23 TC/22 SS/37.22
(4) Square-pallet chain-pump, hand-operated (man-power), with crank, for short lifts	pa chhê	拔車	TK/14 (8)
(5) Square-pallet chain-pump, operated by radial treadle (man-power)	lung ku chhê / fan chhê / tha chhê / shui chhê	龍骨車 / 翻車 / 踏車 / 水車	KC/13, 14 NS/160 ST NC/17.7 TK/13 (6) TC/13 SS/37.12
(6) Square-pallet chain-pump, driven by one or two animals through right-angle gearing from a driving-wheel	niu chuan fan chhê	牛轉翻車	NS/163 ST NC/17.13 TK/9 (7) TC/16 SS/37.13, 37.16
(7) Square-pallet chain-pump, driven by a horizontal water-wheel through right-angle gearing	shui chuan fan chhê	水轉翻車	NS/164 ST NC/17.11 TK/10 TC/15 SS/37.14
(8) High-lift undershot noria (nā'ūra)	thung chhê	筒車	NS/162 ST NC/17.9 TK/7 (5) TC/14 SS/37.15
(9) High-lift noria for still water, driven by two animals through right-angle gearing from a driving-wheel	wei chuan thung chhê	衛轉筒車	NS/165 ST NC/17.15 TC/17
(10) High-lift pot chain-pump (sāqīya), apparently current-operated	kao chuan thung chhê	高轉筒車	NS/166, 167 ST NC/17.18 TK/8 TC/18 SS/37.17
(11) High-lift pot chain-pump, driven by two animals through right-angle gearing	Ditto		Not illustrated, though undoubtedly used, especially in relatively recent times

Table 56 (*continued*)

			ST
(12) High-lift pot chain-pump, driven by a horizontal water-wheel through right-angle gearing	As c(*b*)/10 above		
(13) Archimedean screw	*lung wei chhê* 龍尾車		HF CC TC/45–9 SS/38.2 ff.
(14) Ctesibian single-acting double-cylinder force-pump	*shui chhung* 水銃 *yü hêng* 玉衡		HF CC TC/50–3 SS/38.14 ff.
(15) Suction-lift pump (cf. pp. 142, 144, 345)	*hêng shêng* 恆升		HF TC/54–7 SS/38.23 ff.
(16) Double-acting single-cylinder double-piston force-pump (flame-thrower)	*fang mêng huo yü* 放猛火油		Illustrated only in the *Wu Ching Tsung Yao* (see pp. 145 ff. above)
(D) BLOWING (AIR PROPULSION)			
(1) Open rotary winnowing-fan, with crank, hand-operated (man-power)	*shan chhê* 扇車 *yang shan* 颺扇		NC/23.11 TK/45 TC/136 SS/40.14
(2) Drum-enclosed winnowing-fan, with crank, hand-operated (man-power)	*yang shan* 颺扇 *fêng chhê* 風車		NS/123 TK/44 (31) SS/40.15
(3) Double-acting single-cylinder single-piston piston-bellows, rectangular section	*fêng shan chhê* 風扇車 *fêng hsiang* 風箱		TK/89, 90, 92, 93, 95, 101, 102, 121, 123, 124, 129, 132
(4) Metallurgical bellows (the 'Water-powered Reciprocator', generally shown with hinged fan-bellows, driven by piston-rod, connecting-rod and eccentric lug from a horizontal water-wheel; a conversion of rotary to longitudinal motion)	*shui phai* 水排		NS/183 ST NC/18.3 TC/32 (see pp. 369 ff.)
(E) SIFTING			
(1) Foot-treadle flour-sifter (man-power)	*ta lo* 打羅 or *mien lo* 麵羅		NC/18.5 TK/60 (36) TC/34 SS/40.30
(2) Flour-sifter driven by piston-rod, connecting-rod and eccentric lug from a horizontal water-wheel (as in D/4 above)	*shui ta lo* 水打羅 *shui chi mien lo* 水擊麵羅		NS/187 ST NC/18.11 TC/39 SS/40.31
(F) TURNING (TRANSMITTING POWER)			
(1) Potter's wheel	*chün* 均 *ta chhüan* 打圈		TK/81 (52), 85 (55), 87 (57)
(2) Vertical undershot water-wheel driving textile machinery (spinning plant)	*shui chuan ta fang chhê* 水轉大紡車		NS/190 & 255 ST NC/18.16 TC/42
(G) PRESSING			
(1) Wedge oil-press	*cha* 榨		NS/126 NC/23.13 TK/111 (80) TC/138 SS/40.29

and indigestible form. Many animals, better provided than man with amylase, and even with cellulose-splitting enzymes contributed by their symbiotic bacteria, can digest all this 'roughage' as well as the inner part of the grain. Human beings, however, deriving but a partial benefit from the bran, can digest only the inner endosperm, and even so, a 'pre-digestion' in the form of cooking is necessary. The simplest and oldest form of this was a toasting or parching of the whole grain which made complete de-husking possible; then the grain was heated with water to about 60° C. so that the starch swelled and gelatinised to form a kind of porridge. But this is something which does not keep. If instead of being heated below boiling temperature the flour is made into a dough with water and baked at about 235° C., durability and consistency is attained, yet the product, even when made very thin, soon stales and is difficult to eat—hence the invention of 'leavening' by means of gas-producing yeasts. But here the gluten proteins of the endosperm have a leading role to play.

Bread cannot be properly leavened and 'risen' unless advantage can be taken of these proteins, for as baking proceeds they are denatured and form an elastic framework which retains the carbon dioxide until perfectly set. If the grain has had to be roasted before hulling, these proteins will have been prematurely denatured, and nothing more than porridge will be possible. Now all cereals other than wheat and rye (including barley, millet and oats) are normally husked in such a way that they cannot be freed from their cover-glumes by ordinary threshing, and this is why they are unsuitable for making aerated bread.[a] But in addition the relative quantity of the proteins in wheat, and also their quality, make this cereal far superior to any other in panification. Furthermore the 'naked' wheats of the species *Triticum vulgare* have proved capable through centuries of domestication of yielding a considerable number of varieties with different desirable properties.

These facts explain the long persistence of pounding processes, variations of the pestle and mortar stemming from the mealing-stone of Palaeolithic man, both in East and West. In classical antiquity maize was of course unknown, sorghum and rice rare and medicinal, oats considered only food for animals, and barley (in the form of kneaded unleavened baps, *maza*, μᾶζα) unfit for baking. Rye was a German-Scandinavian grain which came south but slowly and never very much, while millet was used as a stand-by if the wheat harvest failed. Bread wheat was already supreme in Greek lands by the −4th century, supplanting barley, and naked wheats ousted emmer, the principal husked wheat (*Triticum dicoccum*), in Italy by the −2nd. Thus the Etruscans and the Romans of the early Republic were eaters of porridge (*puls*), made from *far* or *zeia* (ζειά), which was pounded after roasting, and the pounding gave the name *pistor* to the later miller-bakers of Rome. But the Hellenistic world lived, like modern Europe, upon baked bread.[b]

[a] This does not mean that they cannot be used for making flat unleavened cakes like chupatties or *shao ping*, slapped on to the side of a hot iron plate and toasted until crisp, or fried in various ways. Millet is still made into a coarse bread of this kind in north China. In Western antiquity a naked barley was occasionally cultivated; cf. Moritz (3).

[b] On all these questions cf. Zeuner (1) and Moritz (1, 2, 4). Many of us have eaten the porridge of early Republican Rome without knowing it, for in Slavonic lands we have been served with a dark, stiff and rather sour dish called *kasha*, as much of an acquired taste as the rye pumpernickel of Germany.

In China the place which emmer and barley had had in Mediterranean civilisation was taken by millet, both kinds of which, the panicled and the spiked, glutinous and non-glutinous, were indigenous there, and formed the chief cereal of the Yangshao and other Neolithic cultures.[a] But archaeological evidence shows that rice[b] had already penetrated to these cultures before the end of the − 3rd millennium, certainly coming from India. From actual remains in Han tombs the presence of wheat, barley and adlay (Job's tears, *Coix lacrima*) has been established, but certain other cereals, notably kaoliang and buckwheat, are indigenous to China like millet, and were probably cultivated at least as early as the Chou. Exactly when wheat began to be cultivated on a large scale it is hard to say, but it was certainly a crop of the Shang period (*c.* − 1400), having reached China by way of Western and Central Asia, and in subsequent ages it became as characteristic of the Yellow River basin as rice was of that of the Yangtze. Nor do we know when emmer was superseded by the naked wheats, and although throughout later times risen bread was steamed, not baked, in China, one cannot infer from the 'steamers' present from the Neolithic onwards that leavened bread rather than other food was actually steamed in them.[c] We shall probably not go far wrong if we think of the coming of leavened bread in China some time during the Chou period, perhaps in association with the many revolutionary agricultural changes of the early Warring States period. Abundance of bread flour does not depend upon rotary milling, but that invention seems to follow it fairly soon. This brings us back to engineering problems.

Now *kasha* is a porridge made from the coarse grits of emmer (cf. Carleton, 1). Another question which might be raised is that of the unleavened bread (*matza*) of Israel, alone orthodox for temple sacrifice and passover. The prohibition of leaven goes back to the JE stratum of Exodus, and traditionally therefore to the − 8th century, but this may not mean that Israel had naked wheat and therefore risen bread earlier than the peoples of the Western Mediterranean. Perhaps the unleavened bread of later ages was held holy because it was the only form of bread that the patriarchs had known. And the Roman sacrificial bread (*liba*) which was made from emmer might be cited as a parallel. But the truth herein might admit of a wide conjecture. The only part of the world where the staple cereal is still grain roasted before being ground is Tibet. The *tsamba*[1] celebrated by so many travellers is also however a primary element of diet for Mongols, Kazakhs and the Chinese of the western and north-western border areas (cf. Trippner, 1). It is eaten as a paste with warm tea or even cold water, together with butter, vegetables, etc. The chief grain used is *chhing kuo*,[2] a cold-resistant cereal little cultivated in China proper, and seemingly closely related to the spelt or emmer wheats, or to barley; but wheat, oats, naked oats, millet, buckwheat and maize may also be used. References in Chinese literature include *Chhi Min Yao Shu* (*c.* + 540), ch. 10, *San Nung Chi*, ch. 7, p. 10b, and Wu Chhi-Chün (1), vol. 1, ch. 2 (p. 154), vol. 2, ch. 1 (p. 19).

[a] This paragraph is based not only on the fundamental publications of Vavilov (1, 2) and Bishop (2), but also on the kind counsel of Dr Hsia Nai at the Archaeological Institute of Academia Sinica in Peking, and of Dr Chêng Tê-Khun in Cambridge. Cf. Vol. 1, pp. 81 ff., 84.

[b] Rice is something of an exception to the generalisations which we have been making. It does not have to be roasted before being de-husked, and it does contain enough gluten to make a fluffy kind of leavened bread. But all rice-eating peoples have preferred to use it in the form of steamed whole elastic grains suitable as a substratum for meat and vegetable dishes (*mi fan*). This is presumably because it will stand full cooking without disintegrating into porridge form, and as this is so there was no need to go to the trouble of bread-making.

[c] Cf. Vol. 1, p. 82, and Sect. 33 below (in the meantime, Ho & Needham, 3).

[1] 糌粑   [2] 青稞

(i) *Pounding, grinding and milling*

The simplest form of comminution, pounding (impact crushing) by means of the manhandled pestle and mortar, goes back in principle, no doubt, to Mesolithic times and even earlier, but has lasted on for the daily decortication of cereal grains, as well as other purposes, until the present day, not only in China but in a multitude of other places from the Hebrides to Bali.[a] It is fairly frequently represented in ancient Egyptian pictures and Greek vase-paintings. In China the pestle (*chhu*[1])[b] and mortar (*chiu*[2]) was used from antiquity[c] for the hulling of all grains, but especially for the removal of the glumes from rice grains, which are also roughly polished in the process so that they become white and shining by the partial removal of the outer fatty aleurone layer.[d] From Han references[e] we know of mortars made of clay, of wood and of stone, materials to which stoneware was afterwards added. In the Thang southerners used boat-shaped pestles stamping communally in long troughs.[f] Farmers used the pestle for breaking up clods in the fields,[g] as apothecaries and alchemists also for their own purposes, and sometimes the pestle was (and is) suspended from a bamboo bow-spring.[h] In a Chinese cereal-pounding scene of +1210 the pestles are almost indistinguishable from mallets (Fig. 359).[i]

This drawing is noteworthy, however, for the presence of something far more characteristically Chinese, namely the treadle-operated tilt-hammer (*tui*[3]). This was an extremely simple device, using lever and fulcrum to enable the pounding work to be done with the feet and the weight of the whole body, instead of with the hands and

---

[a] It took its place in more advanced technology of course in the stamp-mill familiar to the European Middle Ages for ore-crushing and gunpowder-mixing. As we shall see (p. 394), the vertical stamp-mill was as characteristic of Western practice as the horizontal trip-hammer was of Chinese. Still later descendants of the pestle and mortar are to be seen in the ball mill and the hammer mill (cf. R. H. Anderson (1), pp. 259 ff.).

[b] Also called *thing*[4] or *chhen*.[5]

[c] The Shang graph is a pictograph; cf. K 1067*d* for *chiu*,[6] old, which includes the component. Beside it is *chhung*,[7] to pound, which shows the hands, the pestle and the mortar. *Chhin*,[8] further right, has grain, pestle and hands.

Shang, K 1067*d*	Shang, Hopkins (11)	Chou, K 380

[d] Cf. Hommel (1), p. 101. Daily polishing brought no harm century after century, but when factory milling was introduced in the large eastern Chinese cities a much greater amount of the aleurone layer was removed, and what was left fell prey to moulds in storage. Thus beri-beri was introduced as a disease of industrialisation and caused great ravages until the loss of the B group of vitamins was understood. For this information we are indebted to Prof. B. Platt and Dr Lu Gwei-Djen, who were able to solve the problem in Shanghai in the early thirties of the present century.

[e] E.g. the story of Chhen Hsien[9] in *Chhien Han Shu*, ch. 66, p. 18*a*.

[f] See Liu Hsün's[10] *Ling Piao Lu I*[11] of +895, in *Shuo Fu*, ch. 34, p. 25*a*.

[g] Chou and Han references in Amano (1).

[h] Abel (1), p. 138. European technicians also sprung their pestles from the +15th century onwards (Frémont (13), pp. 22, 23), and Greek examples of this are conjectured (Hock, 1). Cf. Beck (1), p. 521.

[i] Franke (11), pl. XLIV; Pelliot (24), pl. XXIX. This form is also Hebridean (Curwen, 2).

[1] 杵   [2] 臼   [3] 碓   [4] 梃   [5] 挺   [6] 舊   [7] 舂
[8] 秦   [9] 陳咸   [10] 劉恂   [11] 嶺表錄異

arms alone.[a] Illustrated in all the agricultural encyclopaedias (cf. Fig. 359),[b] it is one of the commonest objects of the Chinese countryside. It was used in the same way as the pestle and mortar for decorticating and polishing cereal grains, but also extensively by miners in ore-dressing.[c] In modern times it finds many applications, as in mobile earth-tampers for construction works.[d] As to its antiquity, most Chinese historians would place it without hesitation in the late Chou period, or about the Chhin time, though literary references to it are hard to find before the −1st century. At that date we have the *Chi Chiu Phien* dictionary[e] of −40 and the definitions in the *Fang Yen* of about −15, attributable to Yang Hsiung;[f] but the best statement is that of Huan Than in the *Hsin Lun* of about +20, which we shall give in translation later on.[g]

The pedal tilt-hammer seems not to have been used in other civilisations until much later, if at all. It is not mentioned or illustrated in Europe until a +1537 edition of Hesiod,[h] so it may safely be regarded as derivative there. But it seems to have had European descendants, notably the 'oliver', a sprung treadle-operated tilt-hammer used in forges.[i] If this goes back to the +14th century, as has been supposed, the transmission would have been as medieval as that of the blast-furnace and gunpowder.[j] In any case the mechanised *tui*, the hydraulic trip-hammer, had, as we shall see, much more imposing European progeny in the heavy 'martinet' forge hammers of the 18th century.[k] For this particular machine, animal power was seemingly less suitable, and we do not know of any trip-hammers so worked in China,[l] but when they spread to South America a curious device was invented whereby the ends of the hammer-beams were curved upwards, and depressed successively by the circling whippletree of a plodding animal.[m] Pounding techniques had eventually their own monograph in Chinese literature, though very late, the *Chhu Chiu Ching*[1] by Ong Kuang-Phing.[2]

Grinding procedures were more complex, and led further.[n] In Europe the oldest

---

[a] In one form the lever is so counterweighted that the work can be done by a rocking motion of the body; cf. Hommel (1), p. 100.

[b] There are tomb models and pictorial representations of all ages; e.g. for the Han, Laufer (3), p. 36, fig. 7 and pl. 6. A fine scene on an oblong brick exists in the Chhêngtu Museum, and the Loyang Museum has models. For the Sui a tomb-model is figured in *Wên Wu Tshan Khao Tzu Liao*, 1954 (no. 10), pl. 36. For the Thang see Thang Lan (1), pl. 88, and for the Sung a colourful representation in the frescoes of cave no. 61 at Chhien-fo-tung. Cf. Figs. 358, 436.

[c] Engineering drawings in Louis (1).

[d] Anon (18), pt. 1, no. 61.

[e] P. 50a, b.   [f] Ch. 5, p. 5b.

[g] P. 392 below. *CSHK* (Hou Han sect.), ch. 15, p. 3b; *TPYL*, ch. 762, p. 5a, ch. 829, p. 10a.

[h] Cf. Bennett & Elton (1), vol. 1, p. 94. Lindet (1) supported this interpretation of the passage in Hesiod, but we remain unconvinced. The depiction recurs in the +17th century later, as e.g. in Böckler (1), pl. 10 and relevant text. Actual examples can be seen in the Folklore Museum at Iaşi, Rumania.

[i] Cf. Jenkins (1); Young (1), vol. 2, p. 256, pl. 11, fig. 3.

[j] See Sect. 30d below on the 'Fourteenth-century Cluster' of transmissions.

[k] See on, p. 394.

[l] Except in the important statement of Huan Than about +20, cf. p. 392 below. But no information has come down to us about the construction of these animal-power trip-hammer batteries.

[m] Mengeringhausen & Mengeringhausen (1).

[n] When this account was first written we drew upon the basic works of Lindet (1); Bennett & Elton (1), vol. 1; Curwen (2, 3) and Childe (9). The subsequent publication of the monograph of Moritz (1) marked a great step forward, and with its careful first-hand study of the ancient occidental sources is much preferable to the treatment of Forbes (12). We have not seen Ponomarev (2).

[1] 杵臼經   [2] 翁廣平

instrument combining pressure-crushing and shear force, from the Neolithic onwards, was the grain-rubber, simply a saucer-shaped stone, with a squat and bun-shaped slider movable on its upper surface. Many examples of these are to be seen in museums. Insensibly the grain-rubber developed into the saddle-quern, where a bolster-shaped upper stone rubs or rolls backwards and forwards over a larger longitudinally concave lower stone.[a] This again is often shown in ancient Egyptian pictures, and tomb-models or toys from Egypt and Greece;[b] it may still be seen in use in Mexico (the *metate*) and Africa. The difficulty of continuously feeding in the grain led to the more convenient hopper-rubber (Fig. 442), in which the upper stone has carved out within

Fig. 442. Reciprocating motion in primitive mills. Left, the ancient occidental hopper-rubber. Right, the Olynthian mill, a lever-operated radially oscillating hopper-rubber.

it a hopper with an elongated slit for the grain to pass through. From this there developed the oscillating lever-operated hopper-rubber (Fig. 442) in which one end of the lever was pivoted to a firm bench and the mill was worked back and forth on its lower stone. All these types were incised with parallel grooves in herring-bone or other patterns from the saddle-quern stage onwards.

The transition from these oscillatory forms to the true rotary mill and hand-quern,[c] in which an upper discoidal millstone revolves upon a lower stationary stone, is not at all obvious and indeed presents unsolved mysteries, though the radial motion of the lever-operated hopper-rubber may have inspired the advance.[d] The general principle is shown diagrammatically in Fig. 443; the lower stone is always convex, though it may be only very slightly so, while the upper one is concave with a hole pierced in it through which the grain can fall on to the grinding surfaces. Across this hole is placed a bar, the rynd, which supports the upper stone upon a pin rising from the centre of the lower one. If this pin be continued through a hole in the thickness of the lower stone and attached to a movable lever (the bridge-tree) then a simple method is available for adjusting the exact clearance between the two stones.[e] The

---

[a] It would seem that the longitudinal-travel oscillating edge-runner mill, very characteristic of China, should be a direct descendant of the saddle-quern. We shall be considering edge-runner mills shortly; p. 195 below.

[b] Examples of these are discussed by Moritz (1), pp. 29ff.; some come from Thebes and Rhodes in the −6th and −5th centuries.

[c] Typological analysis by Curwen (2, 3).

[d] This is appreciated by Storck & Teague (1), p. 76, though others find any such transition difficult to envisage.

[e] As in the Scottish quern (Moritz (1), p. 119), and all water-mills and windmills whether in China or the West (Hommel (1), p. 84).

convexity of the lower stone allows the flour to fall out automatically.[a] In some cases the upper stone was so cut as to provide a very large hopper above, as in the famous Roman donkey-mills of Pompeii (Fig. 443).[b] Most hand-querns had a handle placed eccentrically so as to constitute a crank. It may be noted that the rotary mill or quern is morphologically the pestle and mortar turned upside down, the pestle being held stationary below, while the mortar, with a hole made through it, rotates above; this

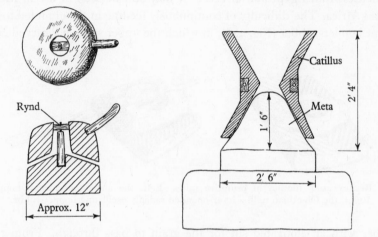

Fig. 443. Rotary mills. Left, the hand-quern or manually operated rotary mill.
Right, the Pompeian mill, an animal-powered rotary mill with large hopper.

may be another clue in the genesis of the invention.[c] There is much to be said in favour of Curwen's opinion (2) that the revolving mill was so great an advance on any previous appliance that it could not have come into being gradually without a decisive act of invention, and the only general precedent it had was the potter's wheel.[d] The problem is to locate the act.

In order to compare the Chinese data with the course of events in Europe it is necessary first to establish a chronological sequence for the inventions just described.

[a] The very great convexity of the Pompeian donkey-mills, says R. H. Anderson (1), proved ancestral to a number of Renaissance devices, though hardly perhaps by direct affiliation. When the *meta* revolved and the *catillus* stood still the machine became the de la Gâche mill of +1722 and generated modern feed-grinders of the sweep type. When it was made to revolve on its side it became the Ramelli mill of +1588, and when part of the *catillus* was removed so as to leave only a concave block, the Böckler mill of +1662. By that time it was not a far cry to the true double roller mill. But that had probably been invented in India many centuries before. The Ramelli and Böckler mills are figured also by Forbes (18), pp. 16, 17.

[b] The famous representations on Vatican marbles have often been reproduced, as by Feldhaus (2), p. 172, and Bennett & Elton (1), vol. 1, p. 186, but there is also an extremely good one on a stone in the Musée Lapidaire at Narbonne; see Espérandieu (1), p. 47. Moritz (1) reproduces several reliefs and other documents.

[c] This exchange of concavity and convexity was recognised in Chinese by the name *tzhu chiu*[1] (female mortar) applied to the upper stone, and *hsiung chiu*[2] (male mortar) to the lower. The upward-projecting pin strengthened the analogy.

[d] The potter's wheel also partook of concavity or convexity according to whether it was of the socketed or pivoted type (Childe, 11). The former was more characteristic of East Asia.

[1] 雌臼   [2] 雄臼

The recent researches of Moritz (1) have rescued the subject from the state of confusion and uncertainty which long prevailed. No verbs implying the 'turning' of mills can be evidenced from any Western culture before the −2nd century, but the contrast between the 'push-and-pull mill' (*mola trusatilis*) and the 'turned mill' (*mola versatilis*) is clear from then onwards.[a] Most misleading was the fact that from the −5th century onwards the word donkey (*onos*, ὄνος) was regularly used to designate the upper stone of the saddle-quern and the hopper-rubber;[b] this led to a conviction that rotary mills powered by animals, originating then, had long preceded querns turned by hand.[c] In fact there is no evidence for rotary mills of any kind in the West before Roman times. Saddle-querns were used in all the ancient occidental cultures, hopper-rubbers following them from about the −6th century,[d] and then the lever-operated hopper-rubbers from the −5th onwards. This last device, which may be called the 'Olynthian mill',[e] must be considered the principal grain-mill of the classical Greek world. But a great puzzle remains in that the hand-quern and the seemingly more developed mass-production 'Pompeian mill' (the donkey-mill or *mola asinaria*) appear at about the same time, though one would expect to find the much smaller manual rotary mill considerably earlier. On archaeological evidence the former goes back only to Pliny's time (+70) but the mentions in Cato[f] fix −160 as a date when it was in fairly common use. Archaeological finds place the latter in pre-Roman Spain about −140 and in the late La Tène culture of the north from about −100, but it cannot be evidenced from literature[g] until −10 unless it was Cato's *mola hispaniensis*, which is probable enough. Plautus, who died in −184, never speaks of turned mills except in the *Asinaria*, which may not be a genuine play (or was perhaps his last, *c.* −185). But he mentions *panis* more frequently than *puls*, and Moritz associates the beginnings of rotary milling with the introduction of commercial bakeries at Rome about −170. The essential invention seems to belong therefore to the close neighbourhood of −200, i.e. shortly after Archimedes' lifetime and the establishment of the Early Han.[h] There is still no reason for disallowing the Latin tradition reported by Varro (−116 to −28) that the rotary mill was invented, so far as he knew, by the Etruscans of Volsinii. Whether we have to think of an origin much further east will appear in what follows.

[a] Pliny, *Nat. Hist.* xxxv, xviii, 135; Cato, *De Re Rust.* x, 4, etc.
[b] Perhaps because of the ear-like handles with which the thing was grasped, perhaps because of a phallic analogy. As a parallel from later technology Moritz cites the term 'donkey-engine', which he expects may give trouble to future historians. We have met with a similar confusion in China already (Vol. 3, p. 317). Instances: Xenophon, *Anab.* I, v, 5; Herodas, VI, 83, etc.
[c] Curwen (2); Childe (9).
[d] These were the hand-mills of Xenophon, *Cyrop.* VI, ii, 31, not querns.
[e] Because so many examples have been found at Olynthus, therefore prior to −348.
[f] *De Re Rust.* x, 4 and XI, 4.
[g] Ps-Virgil, *Moretum*, 24 ff. A dubious reference, cf. Lynn White (7), p. 168.
[h] On the evidence available Moritz has still to leave open the question as to the priority of hand- or animal-mills, but though it may be argued that the former were an adaptation of the latter to meet the needs of troops and travellers, so great an inversion of the usual passage in technological evolution from the simple to the complex and from the small to the large will take a lot of proving. We shall return to the question shortly (p. 192 below).
[i] This we know only because of a quotation in Pliny, *Nat. Hist.* xxxv, xviii, 135. Bailey (1), vol. 2, p. 119, unusually for him, misses the sense doubly in translating it. The view of Curwen (2, 3) that the

There were two words in Chinese for the rotary mill, *mo*[1,2] (alternatively *mo*[3]) and *lung*,[4] depending on the material of which it was made. An ancient dialect word, *chhui*,[5] comprised all these indifferently, and yet another term, *wei*,[6] acquired an even broader significance in that it could include edge-runner mills and roller-mills. The rationale of Chinese eotechnic grinding machinery may be followed not only in original sources but in a number of useful modern studies.[a] The *lung* was essentially for hulling grain and especially for decorticating rice, but the remarkable thing about it was that the Chinese made it from sundried or baked clay (*thu lung*[7]) or wood (*mu lung*[8]).[b] When clay was used it was customary to fix teeth of oak and bamboo into the mill-'stones' while still damp (Fig. 445),[c] to act in the same way as grooves incised in stone. The *mo* on the other hand was essentially a stone grist-mill, for

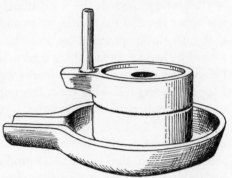

Fig. 444. Typical Chinese rural hand-quern.

comminuting husked grain, rice or naked wheat, to flour (see Figs. 444, 446).[d] In all these various types the clearance was adjusted by the height of the central pin bearing. Rotary mills are illustrated frequently in the Chinese agricultural books,[e] and usually shown equipped with a connecting-rod attached to the crank-handle of such length that it could easily be pushed and pulled by several men (cf. Fig. 413). This tradition of hand-labour inspired the relevant poem of Lou Shou in his *Kêng Chih Thu*:[f]

> Shoulder to shoulder the farmers push and pull,
> Setting the mill in motion with its grinding teeth,
> Making a noise like thunder in the spring.
> As the mill turns the grains fly whirling down,
> Falling in heaps like mountains and rivers,
> Facing each other like high hills.
> They began with a *tou* of grain
> But soon, soon there is plenty to gladden their eyes.

rotary mill was one of the great Greek contributions to civilisation is now quite untenable and the conclusion of Moritz (1), p. 116, that it was invented in some part of the western Mediterranean basin, depended on excluding East Asian evidence from his survey. Volsinii had fallen to the Romans in −280.

   [a] E.g. Laufer (3); Liu Hsien-Chou (1); Hommel (1); Amano Motonosuke (1).
   [b] Cf. Hommel (1), pp. 92, 94. In some forms the baked clay may almost have amounted to stoneware, and it is significant that a variant form of the word *lung*[9] comprises the pottery radical. The technique was an important application of *terre pisé* (cf. Vol. 1, p. 83).
   [c] Cf. Hommel (1), pp. 94, 96.
   [d] *Ibid.* pp. 102 ff.
   [e] E.g. *Nung Shu*, ch. 16, pp. 5a, 6a, 10b, 11a; *Nung Chêng Chhüan Shu*, ch. 23, pp. 5a, 7a, 12a, b; *Shou Shih Thung Khao*, ch. 40, pp. 10a, b, 11a, b, 13a, b, ch. 52, pp. 44b, 45a; *TKKW*, ch. 4, pp. 9b, 10a, b, 12a.
   [f] Tr. auct.

1 磨   2 礳   3 䃺   4 磑   5 䃺   6 磑   7 土磑
8 木磑   9 䃺

PLATE CLXVIII

Fig. 445. Chinese rotary mills; the *lung*, of baked clay or wood, used for hulling grain and decorticating rice. The lower disc or bed-stone is here in process of manufacture. Clay soil is beaten down into the wickerwork form as in *terre pisé* construction (cf. Sect. 28) and the teeth of bamboo (smoked oak for the upper disc), as well as the central pin, are set in place before the drying. Fu-chou, Chiangsi (Hommel, 1).

PLATE CLXIX

Fig. 446. Chinese rotary mills; the *mo*, of stone, used for comminuting husked grain, rice or naked wheat, to flour. Stones outside a mason's workshop in Hopei; his shop-signs hanging on the left.

All such uses were current from the beginning of the Sung.[a] The real problem is what was happening at the beginning of the Han.

It is first of all clear that words implying rotary mills do not occur before the Chhin and Han, though some of them were already in use as verbs with the sense of grinding, smoothing and polishing.[b] This at once suggests that variants of the pestle and mortar were the only grain-handling instruments in use in the Chou and that the rotary mill appeared at the beginning of the −2nd century. At this time in China, curiously also, there is a somewhat imperfect concordance between the literary and the archaeological evidence similar to that which we have encountered in the development of rotary milling in Rome. The point at issue is whether we can feel assured of the existence of rotary mills in the Early Han period. 'Pairs of mills' (*wei i ho*[1]) are mentioned a number of times in the bamboo-slip documents of the Han which have been published by Lao Kan (5), and although few can be exactly dated the whole series runs from −102 to +93. Particularly interesting is the reference to rotary mills in the *Shih Pên*[2] (Book of Origins),[c] a text which must certainly be at least as old as the −2nd century for Ssuma Chhien used it as one of his most important sources in planning and writing the *Shih Chi* (finished in −90).[d] It contains imperial genealogies,[e] explanations of the origin of clan names, and statements regarding inventors legendary and otherwise. What it says is that 'Kungshu (Phan)[3] invented the stone (rotary) mill (*shih wei*[4])'.[f] The *Thu Shu Chi Chhêng* encyclopaedia[g] glosses this in a commentary taken partly from the *Shih Wu Chi Yuan*[h] of +1085:

He made a plaiting of bamboo which he filled with clay (*ni*[5]), to decorticate grain and produce hulled rice; this was called *wei*[6] (actually *lung*[7]). He also chiselled out stones which he placed one on top of the other, to grind hulled rice and wheat to produce flour; this was called *mo*.[8] Both originated in the Chou time.

Kungshu Phan is an old friend of ours (p. 43) and cannot be written off as legendary; he was a famous artisan of the State of Lu, associated with automata and poliorcetics, whose *floruit* must have been within the period −470 to −380. While Laufer (3) thought him much too late for the invention of rotary milling, our hesitations are just

---

[a] This is clear not only from the agricultural encyclopaedias but also from other important sources such as the passage in the *Shih Wu Chi Yuan* which will now be quoted.

[b] For example, some have thought to find an ancient reference to the *lung* in a passage of the *Kuo Yü*, ch. 14 (*Chin Yü*, ch. 8), p. 12a: 'Wên, prince of Chao, built a palace, causing the beams and rafters to be cut and polished (*lung*[7])'—not 'supported on pillar-bases like *lung*'.

[c] Cf. Vol. 1, pp. 51 ff. The text was edited by Sung Chung[9] in the +2nd century.

[d] Cf. Vol. 1, pp. 74 ff.

[e] The information which it gave has been strikingly vindicated by the modern archaeology of the Shang and Chou; cf. Vol. 1, p. 88.

[f] The statement was preserved by a quotation in *Hou Han Shu*, ch. 89, p. 23b. See Chang Shu's edition of the *Shih Pên*, ch. 1, p. 24, Sun Fêng-I's ed., p. 3 and Lei Hsüeh-Chhi's ed., ch. 2 (p. 85). The interpretation of this very early use of the term *wei* as specifically rotary is all the more justified since the hopper-rubber and even the saddle-quern itself seem to have been quite absent from Chinese culture.

[g] *Khao Kung Tien*, ch. 245, *mo wei hui khao*, p. 1a; tr. auct. adjuv. Laufer (3), p. 21.

[h] Ch. 9, p. 3b.

[1] 磑一合   [2] 世本   [3] 公輸般   [4] 石磑   [5] 泥   [6] 磑
[7] 礱   [8] 磨   [9] 宋衷

the opposite, but at any rate he is certainly no more legendary than the Etruscans of Volsinii, and the time at which he lived is not at all improbable for it.[a] After this we have the *Chi Chiu Phien* dictionary[b] of −40, the definitions of the *Fang Yen*[c] about −15, and then the entry of Hsü Shen in the *Shuo Wên Chieh Tzu* of +100, copied by many subsequent lexicographers.[d]

As for the archaeological evidence, we have abundant examples of rotary mills with double hoppers from Han tombs in the form of models (Figs. 447, 414), sometimes alone and sometimes associated with tilt-hammers and other farm equipment.[e] Although only those specimens from Later Han tombs can as yet be exactly dated, there are far too many of them to warrant the view that none are earlier than the +1st century,[f] and Chinese archaeologists do not hesitate to assign them dates as far back as the Chhin. We thus reach the conclusion that everything points to the first half of the −2nd century as the period

Fig. 447. Typical Han tomb-model of rotary hand-quern with a double hopper.

when rotary mills were in general use—and therefore that what was familiar to Cato and Varro was also familiar to Chhao Tsho[1] and Chao Kuo[2] at the other end of the Old World.[g]

We thus have to face a difficult question, diffusion or simultaneous invention? Nor will this be for the last time, for we shall soon see (p. 407) that a very strong case can be made out for an approximately simultaneous appearance of the water-mill itself at the two extremes of East and West. It is a good deal easier, however, to visualise such changes in power source as independent developments, than to accept two quite separate origins for so basic an invention as rotary milling itself—comparable as we

---

a It is quite interesting that the legends of inventors attributed the pestle and mortar to the mythical emperor Huang Ti, or to one of his assistants such as Chhih Chi[3] (*TPYL*, ch. 829, p. 10a, quoting the *Lü Shih Chhun Chhiu*). Cf. Vol. 2, p. 327 above. The various traditions are assembled in Chang Shu's edition of the *Shih Pên*, ch. 1, pp. 14, 15.

b P. 50a, b.

c Ch. 5, p. 5b, cf. p. 184 above.

d References have been collected by Amano (1).

e For example, Laufer (3), pls. 4, 5, pp. 15ff.; fig. 7, p. 36; fig. 10, p. 45, and on the Lo-lang finds Umehara, Oba & Kayamoto (1) and Koizumi & Hamada (1), figured also in Amano (1), p. 36. In 1958 I studied an excellent model in the Loyang Museum. Real millstones from the Han are also known; there is one, for instance, with a diameter of 1 ft. 10 in. in the Sian Museum. Like nearly all the models, it has a double hopper in the upper stone, but unlike them, the lower one is pierced, presumably for a bridge-tree adjustment. For a Northern Wei model see *Wên Wu Tshan Khao Tzu Liao*, 1954 (no. 10), pl. 27. A Thang painting showing a mill worked by two women occurs at Chhien-fo-tung (cave no. 321), and a Sung painting in cave no. 61; this has been reproduced by Chhang Shu-Hung (1), col. pl. 3. A Thang tomb-model with just such double handlebars is figured in Thang Lan (1), pl. 88.

f Dr Chêng Tê-Khun strongly concurs in this opinion.

g Chhao Tsho (d. −154) was a great bureaucrat much concerned with grain production, as well as an engineer and military theoretician. Chao Kuo (fl. −87 to −80) was an outstanding agriculturist expert in all the techniques of his age; he organised the manufacture of cast-iron agricultural implements, and may have been the inventor of a seed-drill plough.

¹ 鼌錯   ² 趙過   ³ 赤冀

must feel it to be with such cultural elements as bronze-founding, the wheel, or cereal agriculture, for which no one has been willing to admit independent beginnings. Forcibly putting this point, but hesitating between Etruscans and Iberians, Moritz[a] retains a suspicion 'that there was a common origin about which we still know nothing'. The dilemma is even more acute as between Chinese and Europeans. We are therefore driven once again to look for some geographically intermediate locality such as Persia or Mesopotamia, whence the fundamental discovery could have spread in both directions.[b]

Now an exception to what was said above about concavity and convexity is constituted by the so-called 'pot
querns' (Fig. 448), in which the upper stone revolves
within the collar of a hollowed cylindrical lower stone.[c]
This design was common enough during the European
Middle Ages, though it seems uncertain whether
primarily for grinding grain. It was never a Chinese type,
but some of the examples from the Middle East may be
very ancient;[d] and though certain forms of these are

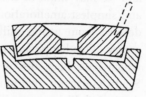

Fig. 448. Pot-quern or paint-mill.

now accepted as the lower bearings of potter's wheels,[e] others may well have been used (as one of their names assumes) for grinding paints, and the existence of larger grain-grinding ones of similar pattern may perhaps be inferred from them.[f] Moreover, these are not the only objects which make a claim to be millstones of high antiquity in the central region; there are querns from Urartu in the neighbourhood of Lake Van, now in the Tiflis Museum, which if not as old as their −8th- or −7th-century attribution, might well be old enough to serve as the archetypes for both the Chinese and the Roman mills.[g] At any rate they call for further investigation.

Here a hitherto unsuggested possibility presents itself. Is it not possible that the Chinese system of making hulling-mills of baked clay, stoneware or wood might have derived from still earlier precedents somewhere in or near the Fertile Crescent or the Indus valley? In this case all evidences of rotary mills in the West, as in China, earlier than the −2nd or −3rd century may well have perished. If only those which were made of durable stone are caught, as it were, by the beam of our archaeological torch, their predecessors might have remained for ever in darkness if the particular agricultural needs of the Chinese had not induced them to continue through the ages this interesting method, assuredly cheap, but also perhaps very ancient. And doubt-less the probable locality of these predecessors was neither Etruria nor Spain nor China, but somewhere in the Middle East. Indeed it is not inconceivable that only

---

[a] (1), p. 116.

[b] Many times already we have been led to assume Babylonian influence spreading in both directions, cf. e.g. Vol. 2, p. 353, Vol. 3, pp. 149, 177, 254, 273, etc. Another instance will occur below, p. 243.

[c] Cf. Bennett & Elton (1), vol. 1, p. 148; Curwen (2), p. 150.

[d] See Childe (9), who understood the need to search for a more eastern focus.

[e] Childe (11).

[f] This is approximately the standpoint of Forbes (12), pp. 141 ff.

[g] Tseretheli (1), reported in Storck & Teague (1), p. 77. In his illuminating review of Russian archaeological discoveries in the Urartu region Piotrovsky (1) makes several mentions of millstones (pp. 35, 45, 63) but unfortunately fails to say whether they were rotary or reciprocating.

the baked clay mills spread, and that stone ones were independently developed both by Kungshu Phan and the Volsinians.

Let us turn now to the application to milling of sources of power other than the muscular strength of man. Here one always feels an inclination to visualise a fixed succession, animal-power coming first, then water-power and eventually wind followed by steam and electricity.[a] But it looks as though any logical scheme of this kind is only partially visible as a real historical sequence, at any rate in the earlier stages, and we may think of it best as a framework for the presentation of facts. The first and simplest amelioration of milling machinery would be the mechanical advantage derived from fitting millstones with gearing, human labour remaining the motive power. This happened quite early in Europe, if one may judge from certain remains which have been discovered at the Saalburg, that Roman fortress on the *limes* in South Germany, including two iron shafts $31\frac{1}{2}$ in. long and bearing at one end pin-drums (cf. Fig. 390 e) with wooden discs and 'teeth' in the form of iron bars.[b] These lantern gear-wheels were undoubtedly intended for the working of mills by right-angled gearing, and as the location precludes a water-mill, human or animal-power was their drive.[c] The date is settled more or less by the fact that the fortress was abandoned about +263. That such an arrangement was not a primary invention is shown by the fact that the right-angled gearing repeats exactly the pattern of the Vitruvian water-mill of about −25, and must almost certainly derive from it. All through the subsequent medieval centuries this manual mill continued to be used, forming indeed a motif in religious art where the four evangelists pour the grain into the hopper, saints man the crank-handles, and bishops receive the issuing hosts in chalice and paten.[d] Geared hand-mills were a sort of *pons asinorum* for the +15th-century German engineers,[e] but in China we find them not until modern times, when they have proved useful in the countryside.[f]

The history of the use of gearing with rotary mills and the history of animal-power for them are intimately connected. From what has already been said we know that in the West mills turned by donkeys go back to the very beginning of firm information about rotary mills themselves, for they date from Cato's time (c. −160).[g] They take precedence over water-mills, therefore, by about a century. As we have seen, too, there is even a tendency to place them earlier than querns and hand-mills. In China, on the other hand, an inversion of the logical order occurs later and seems better

---

[a] An ideal scheme of this kind is set forth at length by Forbes (11), pp. 78 ff.

[b] Illustration in Moritz (1), pl. 14 c.

[c] See Jacobi (1), who experimented successfully with a reconstructed mill of this kind, and Moritz (1), pp. 123 ff.

[d] Material on these mills and pictures has been collected by Nolthenius (1). The manual mill was called in England the Essex mill. I can add a fine example in the Folklore Museum at Iaşi, Rumania.

[e] Cf. Berthelot (4, 5, 6) and Beck (1), pp. 275 ff., 277 ff.

[f] Drawings of manual mills, generally combined with flour-sifters, and sometimes incorporating driving-belts as well as gearing, or instead of it, will be found in Anon. (18), pt. 3, nos. 16, 42, 50, etc.

[g] It is quite strange that neither Clark (1) nor the members of the Newcomen Society who discussed his paper knew of any animal-driven mills with or without gearing before Agricola in the mid +16th century. This kind of machinery survived in Europe until the beginning of the 19th century (Matschoss, 1), and in Arab lands, as in China, until the present time (Brunot, 1). See also Atkinson (1).

established. There water-mills appear to have come first, for while we can find them already at a high state of perfection in +31 (cf. p. 370 below), we get no mention of animal-driven mills before about +175, when Hsü Ching,[1] afterwards an important official, failed to get any post when young and made his living out of mills turned by horses (*ma mo*[2]).[a] Chu I[3] missed this reference when he wrote about +1200 that the poet Yuan Shu[4] (+408 to +453) mentioned donkey-mills in one of his compositions

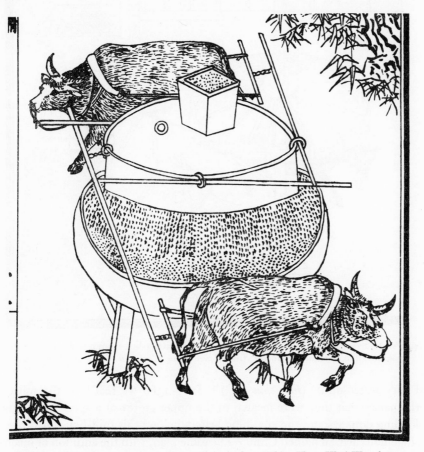

Fig. 449. Ox-driven cereal-grinding mill (*mo*); from *Thien Kung Khai Wu*, ch. 4, p. *12a* (+1637).

in the Liu Sung period, but still the implication was that water-mills had started long before.[b] In the Thang it was customary to use blindfolded mules (*mang lo*[5]).[c] Further evidence for the historical inversion of water-power and animal-power from their logical sequence might be drawn from the curious circumstance that in Wang Chên's time (+1313), and certainly at later periods also, mills driven by animals were

---

[a] *San Kuo Chih*, ch. 38, p. *1a*.   [b] *I Chao Liao Tsa Chi*, ch. 2, p. *49b*.
[c] *Thai-Phing Kuang Chi*, ch. 436, entry under *thui mo lo*.[6]

[1] 許靖   [2] 馬磨   [3] 朱翌   [4] 袁淑   [5] 盲騾   [6] 推磨騾

Fig. 450. Mule-driven cereal-grinding mill with crossed driving-belt from a larger whim wheel giving magnification of angular velocity; from *Thien Kung Khai Wu*, ch. 4, p. 13 *a* (+1637).

called 'dry water-mills' (*han shui mo*[1]).[a] In +1360 we read[b] of the mills of the imperial palace that they were located in the upper storey of a special building, with the plodding donkeys and the gossiping loafers below; this mill-house had been built by an ingenious artisan named Chhü.[2] The same plan, with the addition of mechanised flour-sifters, is now widely found in Chinese rural areas.[c] In the agricultural treatises of the great tradition we find animal-power applied to mills either directly (Fig. 449), or by means of a driving-belt (Fig. 450).[d]

Forces greater than man's own once harnessed, there was no reason why a whole

[a] *Nung Shu*, ch. 16, p. 11 *a*; often repeated elsewhere, e.g. in *Nung Chêng Chhüan Shu*, ch. 23, p. 12 *b*, tr. Laufer (3), p. 23.

[b] *Shan Chü Hsin Hua*, p. 50 *b*, tr. H. Franke (2), p. 126.

[c] Recommended at the National Agricultural Machinery Exhibition of 1958 (see Anon. (*18*), pt. 3, no. 51), it will doubtless last until rural electrification takes over.

[d] In the *Nung Shu* equivalent of this design (ch. 16, pp. 5 *b* ff.) a mechanical advantage of 15 to 1 is stated explicitly in the text.

[1] 旱水磨   [2] 罍

series of mills should not be worked off one driving-shaft. Europe seems to have been very slow to realise this, but historians in China give credit to several people in the +3rd century for the introduction of multiple geared mills. Chi Han,[1] the eminent naturalist (*fl.* +290 to +307), whose book on the strange plants of the south has come down to us,[a] wrote a poetical essay on the 'Eight Mills' (Pa Mo Fu[2]), in which he says:[b]

My cousin, Liu Ching-Hsüan,[3] invented a mill(-house) which showed rare ingenuity and special skill. (It was so arranged) that the weight of eight mills could be turned by only one bullock. So I wrote an essay in which I said that the wooden pieces, square and round, were fashioned (like the universe) by the carpenter's square and the compasses. Below there was quiescence (of the main bearings) just as is symbolised by the *kua* Khun;[4] and above there was motion (of the main driving-wheel) just as is symbolised by the *kua* Chhien.[5,c] Thus the great wheel rotated above in the middle, engaging with the (toothed) wheels of the several mills set around its periphery.

Here doubtless the name of a real inventor has been preserved for us, but he may not have been quite the first, for some sources[d] attribute the design to Tu Yü[6] in a preceding generation (+222 to +284). Whether these multiple mills began in connection with water-power or animal-power is thus not sure, but the 'Eight Mills' tended in later times to become synonymous with multiple installations driven by water-power (*lien chi shui wei*[7]).[e] In a moment we shall see that another early inventor applied the same idea to edge-runner mills. Meanwhile we reproduce the classical picture of a geared animal-driven milling plant (Fig. 451).[f] It is interesting to compare it with similar treadle-driven systems still used in China on co-operative farms.[g]

Reference has already been made to the use of wheel-like objects and rollers for grinding and milling purposes.[h] The simplest is the longitudinal-travel edge-runner mill (*yen nien*;[8] Fig. 452), still commonly used in China, especially by pharmacists and metallurgists,[i] but little known in Europe. Sometimes it is worked by the feet.[j] It would be tempting to regard this as derived directly from the saddle-quern, if the more developed forms of this ancient instrument were known from Shang and Chou China. It might also be thought of in relation to the vertical rotary grindstone turned by a crank, if this had ever been characteristic of Chinese technique (cf. p. 55 above). That it was not so is rather surprising in view of the early appearance there of the rotary disc-knife for jade-cutting—a technique which goes back at least to the beginning of the Han and perhaps to the beginning of the Chou (−13th to −3rd

[a] Cf. Vol. 1, p. 118, and Sect. 38 below.
[b] *CSHK* (Chin sect.), ch. 65, p. 5a, b, from *TPYL*, ch. 762, p. 8a; cited also in *Nung Shu*, ch. 16, p. 12b.
[c] The explanation of this will be found in Vol. 2, pp. 312ff.
[d] E.g. *Wei Shu*, ch. 66, p. 23a, but there is no mention of it in his biography in *Chin Shu*, ch. 34, though many other technical achievements of this remarkable man are mentioned there. The implication of the *Wei Shu* passage is that Tu Yü's multiple mills were water-powered.
[e] Cf. p. 398 and Fig. 622 below.   [f] *Nung Shu*, ch. 16, p. 12a (+1313).
[g] See *Jen Min Jih Pao*, 11 July 1958.   [h] See pp. 82, 92 above.
[i] Cf. Abel (1), p. 177.   [j] Feldhaus (2), p. 66, fig. 77.

[1] 嵇含   [2] 八磨賦   [3] 劉景宣   [4] 坤   [5] 乾   [6] 杜預
[7] 連機水磑   [8] 研磑

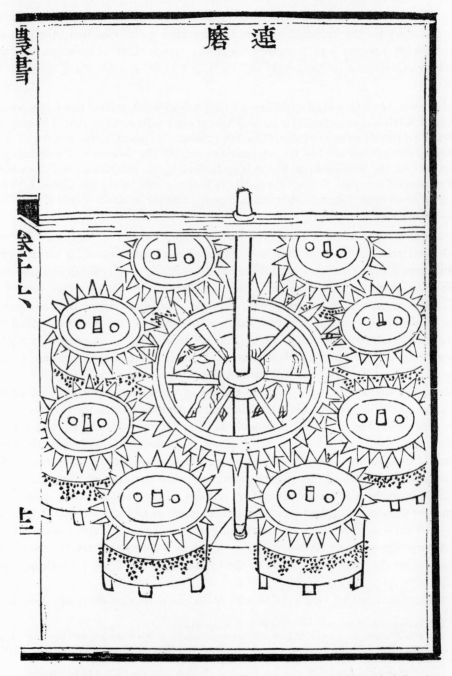

Fig. 451. Geared animal-driven milling-plant (*lien mo*), with eight mills worked directly by the central whim wheel; from *Nung Shu*, ch. 16, p. *12a* (+1313). Cf. Fig. 622.

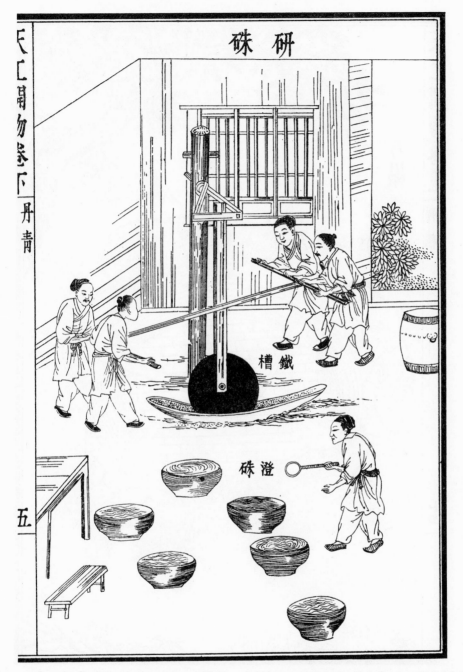

Fig. 452. Longitudinal-travel edge-runner mill (*yen nien*), here grinding cinnabar for vermilion (*chu*), and mainly used in the mineral and pharmaceutical industries; from *Thien Kung Khai Wu*, ch. 16, p. 5*a* (+1637). The trough is labelled as being of iron. Purifying by decantation in the foreground.

Fig. 453. Rotary double edge-runner mill (*shih nien*), for millet, kaoliang, hemp, etc.; from *Thien Kung Khai Wu*, ch. 4, p. 14*a* (+1637). The trailer, for sweeping back the grain into the annular trough, is the only part not shown in the oldest picture of the same device, *Nung Shu*, ch. 16, p. 7*a* (+1313).

PLATE CLXX

Fig. 454. Rotary bogie-wheel edge-runner mill in Hupei.

centuries).[a] Yet in this rotary tool, developed for an industry peculiarly Chinese, we may perhaps recognise the ancestor of all China's edge-runner mills.

A transition exactly analogous to that from grain-rubber to quern took place when the edge-runner wheel was made to revolve in a circular path or trough. This type of mill, the *nien*,[1] is not often illustrated with a single wheel[b] in the Chinese books (cf. Table 56), and the oldest picture which we have of it shows two wheels diametrically opposite each other (Fig. 453). Probably the commonest contemporary variant is that in which the two wheels are arranged bogie-fashion, one immediately behind the other (Fig. 454),[c] yet this the books do not show.[d] The distinction between wheels and rollers is of course not sharp, and the roller-mill (*kun nien*[2] or *Hai-chhing nien*[3]),[e] where a circular path is described, may have been a development from the roller harrow. This apparatus, called variously *shu thu*[4] or *lu tu*[4] (with several alternatives for *lu*[5,6,7]),[f] or if furnished with teeth[g] *shih ho tsê*[8] or *mu ho tsê*,[9] is ancient in China, and there is evidence that the roller harrows of 18th-century Europe were directly derived from it.[h] Such rollers, if used for threshing, would have given rise very naturally to roller-mills since they would be driven round and round the threshing-floor in a circle.[i] The simple hand-roller (Fig. 455) has always been used mainly for millet in China, but the roller-mill (which often carries a hopper at the opposite side of the beam) is used as a method of hulling rice alternative to the *lung*. We give the *Nung Shu* illustration of about +1300 (Fig. 456).[j]

What of the antiquity of these forms of milling in China? Kao Chhêng,[10] in the Sung, thought[k] that the roller-mill began with Tshui Liang[11] about +400, but he was certainly mistaken, since the passage in the *Wei Shu* which he cited (we quote it later, p. 403) refers to the application of water-power to multiple edge-runner mills and roller-mills, and not to their invention. They were undoubtedly familiar in the Han, since they are defined in the *Thung Su Wên*[12] of Fu Chhien[13] about +180.[l] Whether or not they go back, like the rotary grain-mill, to the beginning of the Han, it is hard to say. But there is more than meets the eye in the apparently simple edge-runner mill. In the Section on astronomy we suggested[m] that this device might well have been the

---

[a] See Vol. 3, p. 667.
[b] An example of this kind was photographed, however, by Hommel (1), fig. 133.
[c] Cf. Hommel (1), p. 98.
[d] Engineering drawings are given by Than Tan-Chhiung (1), pp. 179, 230, for china-clay and oil-seed respectively.
[e] *Nung Shu*, ch. 16, p. 9a.   [f] *Ibid.* ch. 12, pp. 14b, 15a.
[g] *Ibid.* ch. 12, pp. 16a, b, 17a.
[h] See Leser (1), pp. 451, 564ff., (2), p. 448. We shall discuss the matter fully in Sect. 41 below.
[i] Chinese agricultural books generally depict the use of the flail rather than the threshing-sledge of occidental antiquity (cf. Curwen (6), p. 99). The classical description of the sledge (*tribulum*) is in Varro (*De Re Rust.* I, lii), but he speaks also of the 'Punic cart', a toothed axle running on low wheels, which seems to have fallen out of use in Europe early. See also Feldhaus (1), col. 221.
[j] Cf. Hommel (1), fig. 149.
[k] *Shih Wu Chi Yuan*, ch. 9, p. 4a.
[l] *TPYL*, ch. 762, p. 8a, gathered into *YHSF*, ch. 61, p. 23a.
[m] Vol. 3, p. 214.

[1] 輾	[2] 輥輾	[3] 海青輾	[4] 摎礦	[5] 碌	[6] 磟
[7] 軈	[8] 石礰礋	[9] 木礰礋	[10] 高承	[11] 崔亮	[12] 通俗文
[13] 服虔					

Fig. 455. Hand-roller (*hsiao nien*) worked by two girls; from *Thien Kung Khai Wu*, ch. 4, p. 13*b* (+1637). The legend on the pedestal says that this instrument is used for ordinary, panicled and glutinous millet as well as kaoliang.

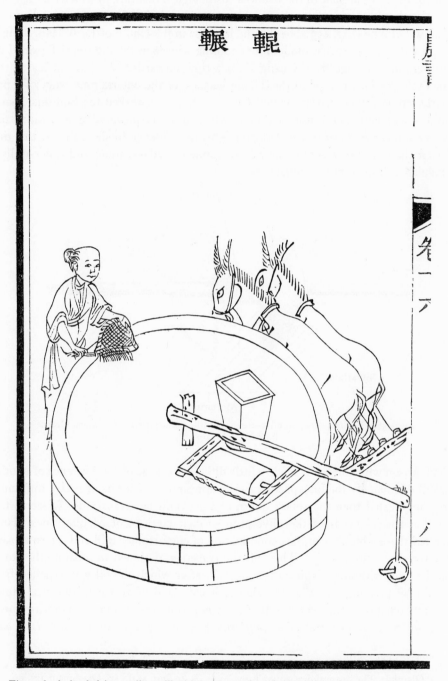

Fig. 456. Animal-driven roller-mill with hopper and road-wheel (*kun nien*); from *Nung Shu*, ch. 16, p. 8*b* (+1313). Note collar-harness and traces.

model for the most archaic of the Chinese cosmological theories, in which case it would have existed at least as early as the Warring States period. Moreover, if the diametric double-wheel type (Fig. 453 above) came into use in the Chhin or Early Han, it might have provided a conceptual model for the idling wheels in the differential gear of the south-pointing carriage from Chang Hêng's time onwards.[a] We have at least proof of the existence of this type in the Thang because of the entertaining story of a particularly strong-armed artisan named Chang Fên[1], who worked for Buddhist monks around +855 and could stop a double-wheeled water-powered edge-runner mill (*shuang lun shui wei*[2]) unaided.[b] Amano (*1*) brings evidence to show that at this time and later the term *wei* was more and more applied to edge-runner and roller-mills as distinguished from rotating millstones.

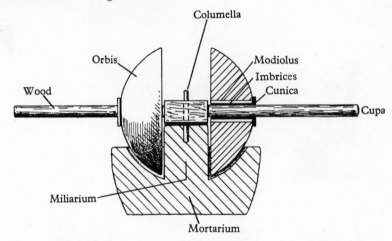

Fig. 457. The *trapetum*, an olive-crushing mill, characteristic of classical Greece (after Drachmann, 7).

Comparison with Europe is made a little difficult because of the existence of another puzzle there, namely the unexpectedly early appearance of forms of rotary mill much more complicated than the rotary quern or the Pompeian grain-mill. It seems that the rather special needs of the olive crop so characteristic of Mediterranean lands generated there a highly complex combination of edge-runner mill and quern as early as the Greek −5th century.[c] This was the *trapetum*, designed to separate the stones from the fruit without crushing them as well, after which the oil was expelled from the pulp by pressing. As Fig. 457 shows, it consisted of two hemispherical stones (*orbis*) revolving on a transverse beam (the *cupa*) which itself rotated on an iron pivot (*columella*) fixed in a central pillar (*miliarium*) of the concave mill-trough (the *mor-*

---

[a] See below, pp. 298ff.   [b] *Yu-Yang Tsa Tsu*, ch. 5, p. 1b.
[c] The *trapetum* must have been fully developed before −348 because a good many examples were found at Olynthus. The classical description is in Cato's *De Re Rust.* xxff.; see Drachmann (7), pp. 137, 142. On these machines the account of Forbes (12), pp. 146ff., is confusing, and for a proper understanding of them the monograph of Drachmann (7) is quite essential. It supersedes Blümner (1), vol. 1, p. 333.

[1] 張芬   [2] 雙輪水䃺

tarium). The meticulous studies of Drachmann (7) and Hörle (1) have elucidated the advanced character of the components, such as the iron bearings (*fistula ferrea*, *imbrices*), collar (*cunica*) and washers (*armillae*), which this machine, with its accurate control of clearance, necessitated. At a comparable period in China probably only the crafts of the wheelwrights, vehicle-builders, and jade-workers would have equalled this.

The other Greek and Roman olive-mill (*mola olearia*) was simpler, though the classical description of it is later,[a] *c.* +50; it consisted of one or two rather narrow rollers or edge-runner wheels describing a travel circle of very small diameter in a trough (Fig. 458). Here the crushing occurred only underneath and not at the sides of the grinding components. Still simpler was the longitudinal trough (*canalis*) in which a millstone (*solea*) set on end was rolled back and forth with its axle gudgeons moving on the walls of the container.[b] This was the only European analogue of the Chinese longitudinal-travel edge-runner mill. Yet another way, the crudest, of crushing olives was to use a toothed board (*tudicula*) something like the threshing-sledge (*tribulum*),[c] and this is interesting because it connects again the techniques of threshing and milling.

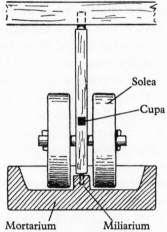

Fig. 458. The *mola olearia*, another type of olive-crushing mill, used in the Hellenistic world (after Drachmann, 7).

In post-classical times the *trapetum* seems to have died out completely in Europe, though it has an analogue, the origin of which is quite unknown, in the oil-mills of India, which consist of a deep annular trough (like the *mortarium*) in which a pestle-shaped crusher is carried round by an animal pulling a radial bar (like the *cupa*).[d] The longitudinal trough was also forgotten, surviving only in local uses.[e] On the other hand the *mola olearia* presumably gave rise in Europe to edge-runner and roller mills of the ordinary type with larger travel circles, and these seem to have been used for many purposes in Europe during and after the Middle Ages.[f] To determine how like these were to the Chinese edge-runner and roller-mills would require a special investigation, but in the meantime one can at least say that when edge-runners appear in the Renaissance engineering books they are of quite Chinese type. Agricola figures

[a] Columella, *De Re Rust.* XII, lii, 6, 7, who preferred it to the *trapetum*. See Drachmann (7), pp. 41 ff., 143; Baroja (5).
[b] Columella, *loc. cit.*, cf. Hörle (1), who noted that Verantius pictured such a mill in +1616 (pl. 24), but by that time Chinese influence might have been at work. Cf. Feldhaus (23).
[c] Cf. Crawford (1).
[d] See Feldhaus (1), col. 719.
[e] Hörle (1) thinks that one can find it, in somewhat crumpled form, in the cider-apple mills of Taunus and Oberwald.
[f] According to Gille (5), p. 116, (7), p. 7, such mills were, and are, traditionally used for olives in Provence and North Africa, for cider-apples and poppyseed-oil in Normandy, for grapes in the Palatinate, and for mustard in the Forez hills west of Lyons. So also wind-driven oil-mills in England (Wailes (3), pl. IX).

none, but Zonca (+1621)[a] and Böckler (+1662) have them.[b] Since the common European name for this type of mill is the 'gunpowder mill' the possibility presents itself that its Western forms may not all derive directly from the Roman *mola olearia* but rather from Chinese influence exerted in the early +14th century at the time of the westward transmission of gunpowder technology.[c] Indeed Zonca shows them being used for this purpose. Further research on the subject would be valuable. As for the ultimate origin of the Chinese edge-runner and roller-mills it seems highly unlikely that they could have derived from a more complex device developed at the other end of the Old World for an industry quite foreign to East Asia, and spontaneous evolution from threshing and harrowing operations, or from the rotary disc-knife used in jade-working, is much more probable.

Though the terminology is not good, the roller-mill must be distinguished from the rolling-mill.[d] While the former consists of one or more rollers travelling continuously in a circular path, the latter utilises the compressing and shearing properties of two rollers adjacent to each other turning in opposite directions. The oldest representatives of these mills, which became so important in the palaeotechnic age of iron[e] for fashioning metal bars and strips, are certainly the cotton-gin (*chiao chhê*[1]) and the sugar-mill (*ya chê*[2]), about which some remarks have already been made (p. 92 and p. 122 above). Neither are likely to have been Chinese in origin, since both the cotton-plant and the sugar-cane are essentially indigenous to South Asia.[f] The chief difference between the two mills is that the gin is always mounted so as to rotate vertically, while the sugar-cane mill generally moves horizontally.[g] The gin cannot have been used in China more than a few centuries before its first illustration in the *Nung Shu* (Fig. 419), and the sugar-mill is not illustrated till the *Thien Kung Khai Wu* (Fig. 459). From this it is not clear how the two rollers engage, but in fact the traditional machine always has sturdy gear-teeth on the rollers themselves, as may be seen from the engineering drawings given by Than Tan-Chhiung.[h]

[a] Pp. 30, 82. Cf. Parsons (2), p. 139; Beck (1), pp. 298, 309; Frémont (2), fig. 97; and H. O. Clark (1) on horse mills.

[b] Cf. pls. 37, 52 and 68. Branca too (+1629), pls. 5, 13.

[c] I remember the late Dr Herbert Chatley making this point in conversation many years ago.

[d] We follow here what is primarily the usage of engineers and metallurgists. Machines with adjacent rollers turning in opposite directions, like the domestic mangle, the cotton-gin, the sugar-cane mill and the rotary printing-press, are termed by flour-millers 'roller-mills', as well as the travelling types which we have just been discussing.

[e] Many studies are available—see Maréchal (1, 2); J. W. Hall (1); Frémont (24); Schubert (2), pp. 304ff.; Usher (2), pp. 342ff. The application of rolling-mills to iron-working for splitting sheets into bars and similar uses dates from the early part of the +16th century, not from the early +18th, as one might be led to think from certain presentations. For non-ferrous metals the principle was used still earlier. Rolling-mills were described by Leonardo (cf. Beck (1), p. 347), but not prominently by any of the German +15th-century engineers. Swedenborg's classical illustration is reproduced by Schmithals & Klemm (1), p. 78.

[f] One suspects that both were distinctively Indian contributions. The first probable mention of the sugar-cane mill in India seems to be of the +1st century; Deerr (1), vol. 1, pp. 44, 55; v. Lippmann (4). Apparently both vertical and horizontal mountings developed there (Deerr, 2).

[g] Cf. Abel (1), p. 200; Hommel (1), p. 113. On vertical and horizontal mountings see pp. 367, 546, etc.

[h] (1), pp. 139ff.

攪車   ² 軋蔗

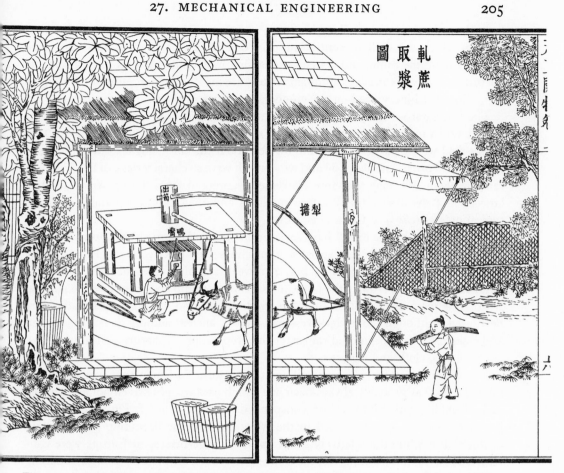

Fig. 459. Animal-driven rolling-mill for expressing the juice of the sugar-cane (*ya chê*); from *Thien Kung Khai Wu*, ch. 6, pp. 6*b*, 7*a* (+1637). The feeding-hole is here called the 'duck-bill', and the main shaft the 'upstanding bamboo-shoot'. Cane has to be fed through thrice.

(ii) *Sifting and pressing*

It only remains now to say a few words about the techniques of sifting and pressing. For sifting powders or bolting flour a treadle-operated machine with rocking motion has long been in use in China. In Fig. 460 the old farmer can be seen throwing his weight from side to side of the oscillating axis.[a] Though we have not found any illustration of the treadle machine (*mien lo*,[1] or in later terminology *shai fên chhi*[2]) before the +17th century (the picture comes from the *Thien Kung Khai Wu*), the *Nung Shu* of +1313 already describes[b] a system of coupling such oscillating sifters to water-wheels (*shui chi mien lo*[3] or *shui ta lo*[4]) with a conversion of rotary to longi-

---

[a] A present-day example appears in Hommel (1), fig. 131, and Feldhaus (2), fig. 76, reproduces one from a scroll-painting of the last century.

[b] *Nung Shu*, ch. 19, pp. 12*a*, 13*a*.

[1] 麪羅   [2] 篩粉器   [3] 水擊麪羅   [4] 水打羅

tudinal motion exactly similar (Fig. 461)[a] to that of the hydraulic metallurgical blowers which we shall presently study in detail.[b] The box-sifter must therefore certainly be of the Sung, and is in all probability much older, but its history remains for further study. Nowadays automatic sifters are incorporated in most of the designs for milling machinery used on the co-operative farms of the Chinese countryside.[c]

The comparative study of pressing plant in East and West presents problems of some interest in the history of technology not hitherto investigated. It happens that in this field we are particularly well informed about the chronological development of presses for those industries of oil and wine which were so characteristic of the Mediterranean region. We may therefore conveniently proceed from the known to the unknown, first discussing the Western methods and then comparing them with the typical presses (cha[1]) of Chinese culture.

The basic Hellenistic texts[d] have revealed in the hands of modern scholars[e] a great deal of information about the successive machines used in Greece and Roman Italy. With the aid of the diagrams in Fig. 462 we may summarise them in Table 57. Thus after the − 1st century the large presses for olive pulp and grapes were equipped with screw mechanisms rather than the winding gear and weights which had before been used. Although the wedge-press was known, ancient authors do not describe it, and it seems to have been used rather for the preparation of pharmaceutical products, essential oils, cloth and papyrus.

The Chinese pattern was quite different from this. In China the most important type of press has probably always been one which uses wedges driven home vertically or horizontally with hammers[f] or a suspended battering-ram (cf. Fig. 463).[g] This it is at any rate which is described in the Nung Shu[h] at the beginning of the + 14th century. But while the relatively small upright wedge-presses of Europe were constructed of a framework of beams, the Chinese horizontal oil-presses, used for obtaining the large variety of vegetable oils characteristic of that culture (e.g. soya-bean oil, sesame-seed oil, rape-seed oil, hemp oil, peanut oil), were (and are) contrived from great tree-trunks slotted and hollowed out. In these is placed the material to be pressed, made into discs ringed with bamboo rope and bound with straw (cf. Fig. 462, 4b); then the blocks are placed in position and the pressure increased from time to time by the wedges. This method takes advantage of the high tensile strength of the

---

[a] We illustrate the best of the traditional drawings from *Shou Shih Thung Khao*, ch. 40, p. 31a.

[b] Pp. 369 ff. below.　　　　[c] Anon. (*18*).

[d] Cato, *De Re Rust.* XII, XIII, XVIII and esp. XIX; Vitruvius, *De Arch.* VI, vi, 3; Heron, *Mechanica*, III, xiii ff.; Pliny, *Nat. Hist.* XVIII, lxxiv, 317. The dating of the inventions depends largely on the passage in Pliny.

[e] Formerly Blümner (1), vol. 1, pp. 337 ff.; Usher (1), 1st ed. pp. 76 ff., 2nd ed. pp. 125 ff. etc. The monograph of Drachmann (7) now supersedes all other treatments, and though Forbes (12), pp. 131 ff. considers a wider range of material he does so less critically. Indeed, without the help of Drachmann his account is barely comprehensible.

[f] In some descriptions a tilt-hammer arrangement was incorporated.

[g] See Hommel (1), pp. 87 ff. Detailed working drawings are given by Than Tan-Chhiung (1), pp. 231 ff. I have had many opportunities of studying this rustic but effective type of press while in China. An example of a similar form is in the Folklore Museum at Iași, Rumania.

[h] Ch. 16, pp. 13b, 14a, b.

[1] 榨

Fig. 460 Treadle-operated sifting or bolting machine (*mien lo*); from *Thien Kung
Khai Wu*, ch. 4, p. 16*b* (+1637).

Fig. 461. Hydraulic sifting or bolting machine (*shui ta lo*), constructed on the same principles as the metallurgical blowing-engines (cf. Figs. 602 ff. below). From *Shou Shih Thung Khao*, ch. 40, p. 31*a* (+1742).

Table 57. *Types of pressing machines used in Greece and Roman Italy*

Type	Description of torculum	Sources	Dating
1a	Indirect lever beam press with windlass and hand-spikes to lower beam (the 'Catonic' press)[a]	Cato (−160) Vitruv. (−25) Pliny (+77)	Already old in the −2nd
1b	Indirect lever beam press with tackle, windlass and handspikes raising weight to hang from beam (the 'Heronic' press)[b]	Heron (+60)	Perhaps also very old
2a	Indirect lever beam press with fixed screw and nut to lower beam[c]	Vitruv. (−25) Heron (+60) Pliny (+77)	Invented c. −30
2b	Indirect lever beam press with rising screw raising weight to hang from beam (the 'Greek' press)[d]	Heron (+60) Pliny (+77)	Perhaps also −1st
2c	Indirect lever beam press with descending screw swallowed in weight and raising it to hang from beam[e]	Heron (+60)	
3a	Direct press with single screw[f]	Heron (+60)	Invented c. +55
3b	Direct press with twin screws[g]	Heron (+60)	
4a	Wedge press[h]	No lit. sources, Pompeian frescoes	In use c. +50

natural wood. It would seem to be ancient, indigenous, and without many parallels in the West.

Indirect lever beam presses were also much used in traditional Chinese technology. One of these, a rope-clutch press common in the paper and tobacco industries (Fig. 464),[i] is closely similar to type 1a in Table 57 and Fig. 462. It differs, however, in having a simple but ingenious disposition of the ropes so that besides hauling down the press beam they act as a brake upon the windlass, thus rendering any pawl and ratchet unnecessary. Again this has all the characteristics of an ancient design. The other main type of Chinese indirect lever beam press is distinctly different from any European form (see Fig. 462, 1c). Since this is used in a typically Chinese industry, the pressing of soya-bean curd, it is unlikely to have been an importation.[j] Here a weight of stone or iron is used, as in type 1b, but instead of being raised by tackle and windlass it is made to depress a lever connected with the beam by an adjustable link-work arrangement, so that when a certain amount of pressure has been applied the weight may be raised and the perforated ratchet-bar re-set so as to continue the pressure as the curd

[a] See Drachmann (7), pp. 50, 145.
[b] *Loc. cit.* pp. 64, 67, 151. This may have developed from one of the oldest forms of all, simply heavy stones tied to the end of the beam.
[c] *Loc. cit.* pp. 53 ff., 146. Cf. Saxl (2), vol. 2, pl. 171b.   [d] *Loc. cit.* pp. 55, 148.
[e] *Loc. cit.* pp. 70, 154.   [f] *Loc. cit.* pp. 56, 158.
[g] *Loc. cit.* pp. 57, 156.   [h] *Loc. cit.* p. 52.
[i] See Hommel (1), pp. 109 ff.
[j] See Hommel (1), pp. 107 ff.; photographs in figs. 164 and 165.

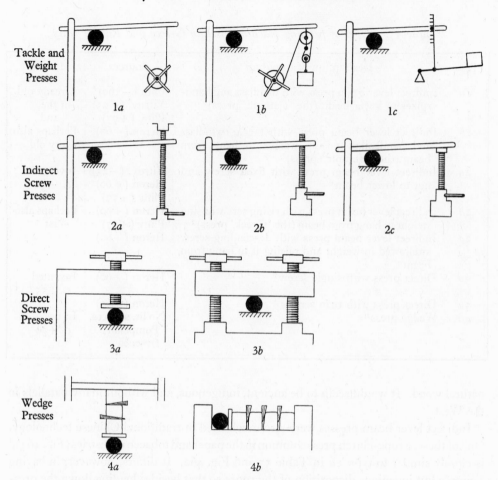

Tackle and Weight Presses

1a   1b   1c

Indirect Screw Presses

2a   2b   2c

Direct Screw Presses

3a   3b

Wedge Presses

4a   4b

Fig. 462. The principal types of pressing plant; table and descriptions in the text.

decreases in bulk owing to expulsion of water. This solution accords with the prominence of linkwork in Chinese eotechnic engineering.[a]

Unfortunately, no studies have been made of the history of these devices.[b] A transmission of the lever beam press to China, perhaps by means of the visits of Roman Syrian merchants,[c] has sometimes been surmised, but there is no evidence for it, and if what we have seen of grain-milling is any criterion, it may perhaps be predicted that we shall find the presses to have been parallel and probably simultaneous developments deriving from the primitive use of heavy stones to weight the end of a beam, or

[a] Cf. pp. 69 ff. above, and the linkwork escapement of the earliest mechanical clocks, p. 460 below.

[b] It is curious that the agricultural encyclopaedias do not illustrate any form of the indirect lever beam press, but that may perhaps be because it was used primarily in certain special trades which came outside their range. On the economic history of oil-pressing a little is known. It seems to have been an important source of income for Buddhist abbeys; cf. Gernet (1); Twitchett (3); Reischauer (2), pp. 71, 212.

[c] Vol. 1, pp. 118, 197 ff. Cf. p. 157 above.

PLATE CLXXI

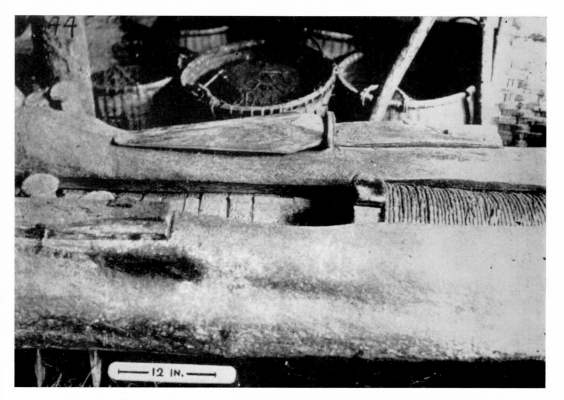

Fig. 463. The most characteristic Chinese oil-press, made from a great tree-trunk slotted and hollowed out; see text. Hommel (1), in Chekiang and Chiangsi.

PLATE CLXXII

Fig. 464. Model of the rope-clutch press as used in the paper and tobacco industries (Mercer Museum, Hommel, 1). The upper press-beam is forced down upon the lower by a rope made fast to a projecting peg in the windlass between the forked ends of the lower beam. As it also passes in two bights round the windlass outside the forked ends, it exerts a strong braking as well as a compressing action, so that the windlass does not tend to fly backward when the hand-spikes are shifted from one hole to another, and a pawl-and-ratchet device is unnecessary. Chien-chhang (Nan-chhêng), Chiangsi, and Tsingtao, Shantung.

wedges to tighten discrete objects. The outstanding difference between the two ends of the Old World was the absence of screw-presses from China, but this is only another manifestation of the fact that this basic mechanism was foreign to that culture.[a]

(3) PALAEOTECHNIC MACHINERY; JESUIT NOVELTY AND REDUNDANCE

So far we have been dealing with machines which were unquestionably and traditionally Chinese. But, as we saw, the palaeotechnic age of coal and iron exerted its first influence on Chinese technology in three engineering books produced in the early years of the +17th century, either by Jesuits or under their influence. These were the *Thai Hsi Shui Fa* (Hydraulic Machinery of the West)[b] of +1612, the *Chhi Chhi Thu Shuo* (Diagrams and Explanations of Wonderful Machines)[c] and the *Chu Chhi Thu Shuo*,[d] both of +1627. It is necessary to examine them with some care, since quite erroneous conclusions have been drawn as to the extent to which they brought to China real novelties in engineering practice.

A number of investigators have devoted their attention to the sources of these books.[e] Working through certain technological illustrations[f] in the *Thu Shu Chi Chhêng* encyclopaedia of +1726, Horwitz noticed that some of them were simply reproductions, rather badly redrawn, as if by artists who did not well understand what they were depicting, of illustrations which he could identify in previous European books. Thus the drum-treadmill operated externally by three men[g] had been copied from Faustus Verantius' *Machinae Novae*[h] of +1615; the endless chain of baskets (bucket conveyor)[i] had been taken from Jacques Besson's *Théâtre des Instruments Mathématiques et Mécaniques*[j] of +1578 (Figs. 465, 466); and the tower windmill working a chain-pump[k] was obviously the same as that shown in Agostino Ramelli's *Diversi e Artificiose Machine*[l] of +1588. But the Chinese encyclopaedia drawings omitted the worm drive of the upper sprocket-wheel of the conveyor,[m] and reproduced the windmill very imperfectly. Later, Reismüller added a fourth identification. He showed that the drum-treadmill operating by means of a two-throw crankshaft two water-pumps with rings or eyes on the ends of their piston-rods (Fig. 467),[n] was a very

---

[a] Cf. pp. 119 ff. above.

[b] By Sabatino de Ursis & Hsü Kuang-Chhi.

[c] By Johann Schreck (Johannes Terrentius) & Wang Chêng.

[d] 'Diagrams and Explanations of a number of Machines' mainly of his own invention or adaptation, by Wang Chêng.

[e] Horwitz (1), Reismüller (1), Jäger (2). For bibliographical details concerning them see Bernard-Maître (18).

[f] All in ch. 249 of the *Khao kung tien* section, into which they had been copied from the older books of the Jesuit period. This chapter was explicitly entitled 'Strange Machines'.

[g] *TSCC, Khao kung tien*, ch. 249, 32nd ill.; from *CCTS*, ch. 3, p. 35 b.

[h] Pl. 23; cf. Beck (1), p. 520.   [i] *TSCC, Khao kung tien*, ch. 249, 8th ill.

[j] Pl. 39.   [k] *TSCC, Khao kung tien*, ch. 249, 21st ill.; from *CCTS*, ch. 3, p. 21 b.

[l] Pl. 73.

[m] The original drawing in the *Chhi Chhi Thu Shuo*, ch. 3, p. 8 b, which we reproduce (Fig. 466), could not be reproached with this, though somewhat crude in comparison with the original. But it gave the worm only in an inset.

[n] *TSCC, Khao kung tien*, ch. 249, 25th ill.; from *CCTS*, ch. 3, p. 26 b, but omitting the fellow in the drum-treadmill.

Fig. 465. An endless-chain bucket-conveyor, as it appeared in the *Chhi Chhi Thu Shuo* of +1627 (ch. 3, p. 8*b*). At the top: 'The Eighth Diagram' (of the Weight Raising section).

PLATE CLXXIII

Fig. 466. The endless-chain bucket-conveyor, from Besson's engineering treatise of +1578.

garbled drawing of a similar machine shown in Vittorio Zonca's *Novo Teatro di Machini e Edificii* (+1607 and +1621);[a] see Fig. 468. Finally Jäger added a number of other identifications.[b]

We are now in a position to say that we have practically all the identifications. Table 58 summarises the contents of the three books. From this it can be seen that a considerable number of the machines and devices which the Jesuits and their collaborators described were specifications which had only just been published in the West, from Cardan and Agricola in the middle of the +16th century, to Besson and Ramelli at its end, with Zonca and Zeising's work only a year or two before the transmission to China took place. But there were also a number of machines which had been known for a long time previously in Europe, since they can be traced to +15th-century MSS. such as that of the Anonymous Hussite, Mariano Taccola, and others. One or two, such as the spring-suspended saw-mill,[c] go back to the +13th century with Villard de Honnecourt. Finally, there were a few, such as the drum-treadmill and the anaphoric water-clock (see pp. 466 ff. below), which were derived directly from Vitruvius—or indeed the Archimedean screws and the Ctesibian force-pumps, which dated still further back. Though specific attributions of the machines are not given, the introductory material has references to Agricola (Kêng-Thien[1]), Simon Stevin (Hsi-Mên[2]), Ramelli (La-Mo-Li[3]), and others. Horwitz and Reismüller were uncertain as to the origin of the numerous mistakes in the Chinese drawings,[d] but inspection of the *Chhi Chhi Thu Shuo* clearly shows that the errors were mostly already contained in the book as it left the hands of Johann Schreck and Wang Chêng.

Feldhaus (11, 12) understood that to be clear of European influence it would be necessary to analyse the content of the *San Tshai Thu Hui* encyclopaedia of +1609. This, and more, has been done in Table 56. But the really relevant question is how far the 'novelties' of the Jesuits were new. Following the lead of Horwitz and Reismüller, Feldhaus affirmed[e] that the pictures of paddle-wheel boats in the Great Encyclopaedia[f] of +1726 were obviously copied from Western sources. But here came in error, for convincing literary evidence will soon be adduced (p. 417 below) establishing the use of treadmill-operated paddle-boats at least as far back as the +8th century in China. Moreover, the demonstration that a certain picture in one of the Jesuit books was directly copied from a Western work does not in itself prove that the idea was new for the Chinese, though it may have been so for the particular

---

[a] P. 110. Cf. Zeising (1), redrawn in Beck (1), fig. 569. The design was an ingenious way of avoiding the necessity for connecting-rods between crankshaft and piston-rods, but not of wide applicability on account of the extreme play. It is seen also in Zonca's book, pp. 103, 107. It stems from di Giorgio (+1475).

[b] In particular, he showed that the systematic description of the principles of mechanics which occupies the first two chapters of the *Chhi Chhi Thu Shuo* was taken from three books, the *Promotus Archimedia* of Marino Ghetaldi of Dubrovnik (+1603), the *Mechanicorum Liber of* Guidobaldo (+1577), and the *Hypomnemata Mathematica* of Simon Stevin (+1586, +1605). But though Schreck had known Galileo well, there is no trace of modern dynamics in the book.

[c] *TSCC, Khao kung tien*, ch. 249, 45th ill.; from *CCTS*, ch. 3, p. 48b.

[d] They realised that the *Thu Shu Chi Chhêng* had copied them from earlier books.

[e] (1), col. 939; (2), p. 71.

[f] *TSCC, Jung chêng tien*, ch. 97, 29th ill. We reproduce it in Fig. 633 below.

[1] 耕田   [2] 西門   [3] 剌墨里

Fig. 467. A water-pump worked by a Hellenistic drum-treadmill, from the *Chhi Chhi Thu Shuo* of +1627 (ch. 3, p. 26*b*). The drawing is so garbled as to be almost incomprehensible. At the top: 'The Eighth Diagram' (of the Water Lifting section).

PLATE CLXXIV

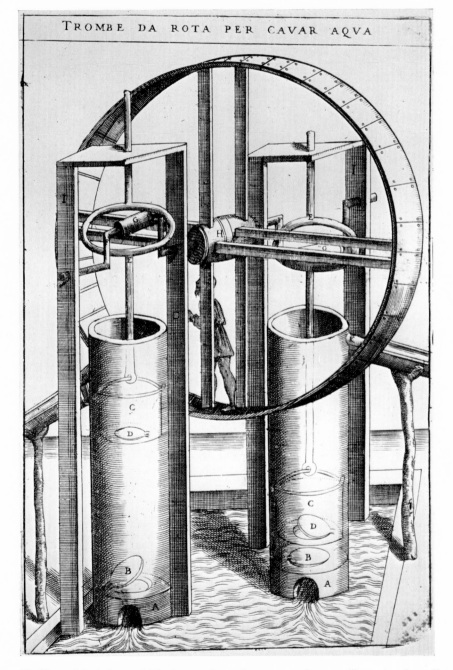

TROMBE DA ROTA PER CAVAR AQVA

Fig. 468. The original from which the previous picture was taken, an illustration from Zonca's engineering treatise of +1607. The machine is a drum-treadmill operating by means of a two-throw crankshaft two lift-pumps which have rings or eyes on their piston-rods. It appears first in the MSS. of Francesco di Giorgio, c. +1475; see Reti (1).

Table 58. *Machines described and illustrated in the Chinese books produced under Jesuit influence*

NOTE. The 16th-century books mentioned here will all be found in the bibliography, but if access to them is difficult, secondary sources which reproduce some of their illustrations have been given below with the following abbreviations: B(1), Beck (1); F (1), Feldhaus (1); F (2), Feldhaus (2); P (2), Parsons (2).

Illustration	Identification in previous European book		Remarks
	Copied from	Identified by	
Thai Hsi Shui Fa (+1612)			
(1) Archimedean screw	Probably specially drawn		Reproduced in *NCCS*, ch. 19, pp. 14*b* ff. *TSCC, I shu tien*, ch. 6, ills. 45–9
(2) Ctesibian force-pump	Ditto		*NCCS*, ch. 19, pp. 31*a* ff. *TSCC, I shu tien*, ch. 7, ills. 50–3
(3) Suction lift-pump	Ditto		*NCCS*, ch. 19, pp. 33*a* ff. *TSCC, I shu tien*, ch. 7, ills. 54–7
(4) Water tanks	Ditto		*NCCS*, ch. 20, pp. 15*b* ff. *TSCC, I shu tien*, ch. 7, ills. 58–62
Chhi Chhi Thu Shuo (+1627). All pictures redrawn for *TSCC, Khao kung tien*, ch. 249			
Chi Chung section (Weight Raising)			
(1 and 2) Steelyard principle			Known in China from the −4th century onwards (see above, Vol. 4, pt. 1, pp. 24 ff.)
(3) Windlass (with handspikes) as crane with three-legged derrick			
(4) Windlass (with crank) as crane with four-legged derrick			Not illustrated in traditional Chinese books, but must have been known and used from Han onwards at least. Cf. Figs. 394, 395, 396
(5) Capstan and pulley as crane with four-legged derrick			
(6) As (4) but doubled			
(7) Crank and driving-belt (using mechanical advantage) for crane			
(8) Conveyor; an endless chain of baskets (ordinary crank seen in use, but worm gear shown)	Besson (+1578), pl. 39 F (1), col. 65	Horwitz (1) Reismüller (1)	Seems to have been used in Sung. See Figs. 465, 466
(9) Conveyor; an endless chain of boxes (ordinary crank turning the upper sprocket-wheel, as in 8)	Ramelli (+1588) P (2), p. 135	auct.	
(10) Vitruvian drum-treadmill as crane with gearing (mechanical advantage) including pulley tackle, worm and two winches	Vitruvius, x, ii (e.g. ed. of Rivius, +1548)		Cf. Schmithals & Klemm (1), pp. 1, 5; Klemm (1), p. 27. See Fig. 397
(11) Capstan as crane drive, with gearing (mechanical advantage) including right-angle gear and one winch	Probably based on Ramelli (+1588) P (2), p. 120		
Yin Chung section (Weight Hauling)			
(12) Haulage over rollers by chain running over sprocket-wheel, operated by crank and right-angle gear (mechanical advantage)			With regard to the use of the endless chains above, and the chain purchase here, the chain drives of Su Sung in the Sung (p. 111 above) should be remembered
(13) Haulage over rollers by ropes; crank and worm moving four winches	Ramelli (+1588) P (2), p. 121	auct.	

Table 58 (*continued*)

Illustration	Identification in previous European book		Remarks
	Copied from	Identified by	
(14) Haulage by ropes and handspike windlasses			
(15) Non-spill buckets mounted on two wheels			
Chuan Chung section (Weight Raising by Turning)			
(16) Well-windlass operated by crank-shaft and gearing			The well-windlass was certainly traditional; the application of gearing may have been new
(17) Well-windlass operated by crank and gearing (mechanical advantage)	Mariano Taccola (+1438) F (1), col. 830	auct.	
Chhü Shui section (Water Lifting)			
(18) Series of three Archimedean screws at right angles, operated by vertical water-wheel and raising from a noria-filled basin			Traditional in Islam and the West
(19) Series of three Archimedean screws parallel with one another, worked off a single rising main shaft driven by a horizontal water-wheel	Cardan (+1550) Ramelli (+1588), ps. 47, 48 B (1), p. 180	auct.	
(20) Suction-lift pump operated by crank (man-power) and rocking beam	Numerous in Agricola (+1556)	auct.	
(21) Vertical (horizontal-axle) wind-wheel on tower-type windmill operating a paternoster pump	Ramelli (+1588)	Horwitz (1) Reismüller (1)	Cf. p. 350
(22) Four groups of men with flume-beamed swapes emptying a caisson or cofferdam	Ramelli (+1588) B (1), p. 229	auct.	These have an ancient Indian and medieval Arabic background; cf. p. 334
(23) Double swape lifted and lowered automatically by a very obliquely cut conical cam rotated by a horizontal water-wheel	Besson (+1578), pl. 46 B (1), p. 201	auct.	See Fig. 471
(24) Hinged-pan swape (man-power)	di Giorgio (+1475) Zonca (+1621), p. 112 F (1), col. 827	auct.	Cf. Reti (1)
(25) Double suction-lift pump with two-throw crankshaft working in rings or eyes on the piston-rods, operated by a drum-treadmill	di Giorgio (+1475) Zeising (+1613) B (1), p. 395 Zonca (+1621), p. 110	Reismüller (1) Feldhaus (1), col. 267 Reti (1)	See Figs 467, 468
(26) Rotary water-pump	Ramelli (+1588) B (1), p. 226	auct.	
Chuan Mo section (Grinding-Mills)			
(27) Inclined treadmill with right-angle gearing	Ramelli (+1588) Zonca (+1621), p. 25 B (1), pp. 211, 297	auct.	See Fig. 470
(28) Drum-treadmill working two mills by right-angle gearing	Vitruvius	auct.	
(29) Two-throw crankshaft (man-power), with bob flywheel, operating mill mounted above on same shaft	*Mittelalterliche Hausbuch* (+1480) F (1), col. 721	auct.	This lacks the flywheel
(30) Falling weight, like a clock weight, driving a mill by gearing			
(31) Field mill (two mills mounted on a four-wheeled wagon), turned by animals	Targone (+1580) Zonca (+1621), p. 88a B (1), p. 310	auct.	Mills mounted on wagons go back to the early Middle Ages in China; cf. pp. 255ff. See Figs. 502, 503

Table 58 (*continued*)

Illustration	Identification in previous European book		Remarks
	Copied from	Identified by	
(32) Treadmill with external steps, worked by three men, to drive two mills	Verantius (+1615), pl. 23 B (1), ch. 22	Horwitz (1) Reismüller (1)	Here Feldhaus (20) is right in saying that the redrawing for *TSCC*, *Khao kung tien*, ch. 249, p. 31a, introduced a serious mistake
(33) Horizontal windmill, carrying four square luffing sails at the ends of its arms, and driving two mills by gearing	Verantius (+1615), pl. 8 F (1), col. 1322 B (1), p. 515	auct.	
(34) Horizontal windmill, carrying four longitudinally streamlined sails and driving two mills by gearing	Verantius (+1615), pl. 9 F (1), col. 1322 B (1), p. 516	auct.	See Fig. 694
(35) Horizontal windmill with square hinged sails attached beneath a ring	Verantius (+1615), pl. 10 F (1), col. 1322 B (1), p. 516	auct.	See Fig. 693
(36) Horizontal windmill with square hinged sails attached above a ring	Verantius (+1615), pl. 10 F (1), col. 1322 B (1), p. 517	auct.	
(37) Horizontal Persian windmill consisting of large vanes in a square tower with side openings	Verantius (+1615), pl. 12 F (1), col. 1324 B (1), p. 517	auct.	See Fig. 691
(38) Horizontal Persian windmill consisting of large vanes in a round tower with side openings	Verantius (+1615), pl. 13 F (1), col. 1325 B (1), p. 517	auct.	
(39) Mill actuated by crankshaft (single throw; man-power), with bob flywheel and right-angle gears	Anonymous Hussite (+1430) F (1), col. 593	auct.	This has a two-throw crankshaft and no flywheel
(40) Multiple mills geared to a horizontal wind-wheel with curved turbine vanes (and a man sitting inside), 'self-explanatory'	Besson (+1578), pl. 50 Verantius (+1615), pl. 11 F (1), col. 1323 B (1), p. 517	auct.	
(41) Variable-height horizontal water-wheel (tide-mill), working two mills, 'self-explanatory'			
Chieh Mu section (Saw-Mills for Wood)			
(42) Saw-mill operated by a crank from an undershot vertical water-wheel; saw vertical; wood moved on by ratchet mechanism	di Giorgio (+1475) Leonardo (c. +1480) Zeising (+1613) B (1), p. 406	auct.	Cf. e.g. Beck (1), p. 323; and p. 383 below
(43) The same, another design		auct.	
(44) Saw-mill operated by crank and rocking beam (man-power), with handspike windlass for moving wood along	Besson (+1578), pl. 13	auct.	
(45) Spring saw-mill (man-power)	Villard de Honnecourt (c. +1240)		Cf. e.g. Klemm (1), pl. 4a; water-power used. See pp. 380, 404 below

Table 58 (*continued*)

Illustration	Identification in previous European book		Remarks
	Copied from	Identified by	
Chieh Shih section (Saw-Mills for Stone) (46) Stone saw-mill (animal-power); saws horizontal; with crank from driving-wheel and right-angle gearing	di Giorgio (+1475) Leonardo (*c.* +1480) Ramelli (+1588) B (1), p. 232 P (2), p. 132	auct.	Cf. e.g. Beck (1), p. 323; and p. 383 below. Ramelli used water power
Chuan Tui section (Trip-Hammers; Stamp-Mills) (47) Stamp-mill, the lugs on the main shaft raising vertical pestles, the shaft turned by two men with cranks	Anonymous Hus- site (+1430) B (1), p. 280	auct.	The only difference here from the water-driven trip-ham- mers used for so many cen- turies in China was that the stamping pestles were moun- ted vertically. Cf. p. 394
Shu Chia section (Revolving Bookcases) (48) Revolving bookcase (gearing shown), wheel vertical	Ramelli (+1588)	Feldhaus (2), pp. 70, 71	See on, pp. 546 ff. and Figs. 679, 680
Shui Jih Kuei section (Water Clocks) (49) Anaphoric water-clock	Vitruvius ix, viii, 8 Diels (1)	auct.	See below, pp. 466 ff. and Fig. 660
Tai Kêng section (Mechanical Ploughing) (50) Plough drawn back and forth across a field by ropes and handspike winches (man-power) (cf. steam-ploughing)	Weimar MS. 328 (+1430) Besson (+1578), pl. 33 F (1), col. 793	auct.	See Fig. 469
Shui Chhung section (Water Shooters) (51) Ctesibian double-cylinder single- acting force pump	Heron Zeising (+1613) B (1), p. 398	auct.	Cf. p. 141 above
(52) The same, on a sled, for fire- fighting			
(53) The same, on wheels, for fire- fighting			
(54) Fire-fighting scene			
Chu Chhi Thu Shuo (+1627) (1) Single-cylinder force-pump			
(2) Agricultural use of flume-beamed swapes	See (22) above		
(3) Crank-operated man-power mill with right-angle gearing	Vitruvius		See p. 192
(4) Mill operated directly by horizontal windmill with vanes or sails of plaited bamboo	See (33–8) above		This recalls the Chinese saltern windmills (cf. below, pp. 558 ff.)
(5) Weight-driven geared mill	Cf. (30) above		
(6) Weight-driven automobile vehicle	Heron Beck (4)		
(7) Combined clock (a Western verge- and-foliot weight-drive system for day hours and a Chinese lead-shot scoop-wheel or striking clepsydra for night-watches), see on, pp. 513 ff.			See Fig. 669
(8) Winch-driven cable plough	See (50) above		
(9) Parts for an arcuballista trigger	See Sect. 30*e*		Old in China (see p. 84 above)

individuals with whom the Jesuits were in contact. Take the case of the endless-chain conveyor or excavator. In view of the antiquity of the radial-treadle square-pallet chain-pump in China, it would almost have been curious if it should never have occurred to anyone there that the same device, modified, could be used in excavations for earthworks. And indeed, in the late +11th-century *Tung Hsien Pi Lu*[1] (Jottings from the Eastern Side-Hall) Wei Thai[2] tells us the following story:[a]

(In the Hsi-Ning reign-period, +1068 to +1077) the city of Linchow had no wells inside the city walls, and it was necessary to draw water from sandy springs outside, where the ground often collapsed. People called it 'scrabbling in the sand'. There was a great desire to extend the city walls to that place (so as to protect the water supply) but the ground was unsafe, and anything built on it was liable to fall down....

So the Acting Commissioner Têng Tzu-Chhiao[3] said to General Lü Kung-Pi,[4] the commander of the forces in Ho-Tung: 'Formerly there used to be a Pa Chu Fa[5, b] (lit. pulling-forth axle method). The sands should be shovelled and pulled out (*chhou*[6]) and the space filled up with powdered charcoal and cement (*chin thu*[7]). City walls can then be built upon this (foundation) without fear of collapse. I should like to use this method and to build new city walls enclosing and protecting the water supply so that Linchow will always be defensible.' Lü Kung-Pi adopted the suggestion and the plan was carried out. And the walls have remained firm to this day, without any subsidences, so that the New Chhin district can be securely defended.

The expression *pa chu fa* here may well have been figurative, signifying simply an exchange method, in which some sort of cement or concrete[c] was substituted for the sand, but if we prefer to take it more literally we must surmise a continuous-chain excavator on the same plan as the chain-pump but using baskets or buckets into which the spoil was shovelled below and ejected above. Perhaps it was something like the endless-belt transporters locally constructed and so much used in China today (cf. Fig. 583).[d]

Before summarising the situation concerning transmission in the +17th century period, let us glance at the sections into which the *Chhi Chhi Thu Shuo* (ch. 3) was

---

[a] Ch. 8, p. 9a, tr. auct.

[b] Note that the term *pa* is the same as that used for the man-power crank-operated square-pallet chain-pump (Fig. 580 below), which would have been the direct model for the excavator which the Jesuits represented as new. See Table 58.

[c] Further light is thrown upon the *chin thu* or cement referred to in this passage by an entry in Fang I-Chih's technical encyclopaedia, the *Wu Li Hsiao Shih* of +1664 (ch. 8, p. 38b). Jotting down some notes on building crafts, he says: 'Sun-dried clay mixed with straw from various plants makes bricks so hard that holes cannot be bored in them. Holien (Pho-Pho's) (fortifications, cf. Sect. 30) were like this—it was one of his numerous techniques. But to mix cement (*chin thu*) the method is to add broken charcoal to broken pottery tiles and stamp down very hard, then it must be baked on all sides; it need not be dried first.' He then tells the story of the Linchow walls in abbreviated form and ends with a few words on wood piling. A study of Chinese traditions regarding mortar, cement, and concrete is much needed, meanwhile it looks as if we have a glimpse here of an early form of the clinkering process essential in the production of the calcium aluminosilicate which gives with gypsum a good setting cement. No doubt if Fang I-Chih had been more explicit we should have heard of the addition of limestone. The further step to concrete by the addition of gravel, slag, etc., would have been much easier.

[d] We shall return to the subject of conveyors and transporters on p. 350 below, in connection with the square-pallet chain-pump.

[1] 東軒筆錄   [2] 魏泰   [3] 鄧子喬   [4] 呂公弼   [5] 拔軸法
[6] 抽   [7] 墐土

divided. The first, Chi Chung[1] (Weight Raising), has eleven items, mostly various forms of pulley tackle.[a] It includes the excavators just mentioned, but the curious thing about it is that the steelyard and the crane-windlass should have been carefully explained—instruments which the Chinese had certainly used since the time of the Warring States. The same applies to the well-windlass in the third section, Chuan Chung[2] (Weight Raising by Turning). The second section, Yin Chung[3] (Weight Hauling), with four items, could have had little novelty except the use of the worm, for Chinese architects had been getting their heavy beams into position for centuries previously. In the fourth section, Chhü Shui[4] (Water Lifting), there are eight items, and here the Archimedean screws and the crankshaft pump were no doubt new. The fifth, Chuan Mo[5] (Grinding-Mills), with fourteen items, is devoted mostly to the application of wind-power, a topic which was not at all new to China (see pp. 558ff. below), though the designs of windmills given can be easily identified in earlier European books. There is also included a mobile mill mounted on a wagon; this was an invention of which Europeans were quite proud, but which had been made also in China, presumably independently, a good many centuries earlier (see pp. 255ff. below). Saw-mills for wood and stone are taken up in the sixth and seventh sections (Chieh Mu[6] and Chieh Shih[7]), with six items. Lastly there comes a vertical stamp-mill of characteristically European, not Chinese, design (Chuan Tui[8]); a revolving bookcase (Shu Chia[9]) about which there will be something to say later (pp. 546ff.) and which was certainly not European only in origin; the Vitruvian anaphoric clock (Shui Jih Kuei[10]); mechanical cable ploughing (Tai Kêng[11]), a plan which can hardly have been much used before the advent of the steam and internal-combustion engines; and four drawings of force pumps for fire-fighting (Shui Chhung[12]), essentially Alexandrian.

It is worth while to take a closer look at the practice of mechanical, or cable, ploughing, for it seems to afford an instance of the adoption in China of the improved gearing arrangements for mechanical advantage introduced by the Jesuits. Fig. 469 shows the illustration given in the *Chhi Chhi Thu Shuo*;[b] it must have been copied from Besson (+1578),[c] but it goes back much earlier than that in Europe, for Feldhaus figures[d] a MS. illustration of the same kind dating from +1430. All these pictures are very unrealistic, for they show only handspike windlasses in use for drawing the plough back and forth across the field, while any effective system must have involved gearing similar to that shown on a previous page in the book of Johann Schreck & Wang Chêng for hauling heavy weights horizontally. Such an arrangement

---

[a] Good illustrations of such tackle, as well as of more complex machinery for building construction, are seen in the *Hwasŏng Sŏngyŏk Ŭigwe*[13] (Records and Machines of the Hwasŏng (Emergency Capital) Construction Service), a Korean book of +1792 described by Chevalier (1). Its editor was Chŏng Yag-yong[14] (+1762 to +1836), a high official of the Yi dynasty, and one of the most original, scientific and progressive thinkers and writers in Korean history; see Henderson (1).

[b] Ch. 3, pp. 54b, 55a.

[c] Pl. 33. Oxen omitted.   [d] (1), col. 793.

[1] 起重	[2] 轉重	[3] 引重	[4] 取水	[5] 轉磨	[6] 解木
[7] 解石	[8] 轉碓	[9] 書架	[10] 水日晷	[11] 代耕	[12] 水銃
[13] 華城城役儀軌		[14] 丁若鏞			

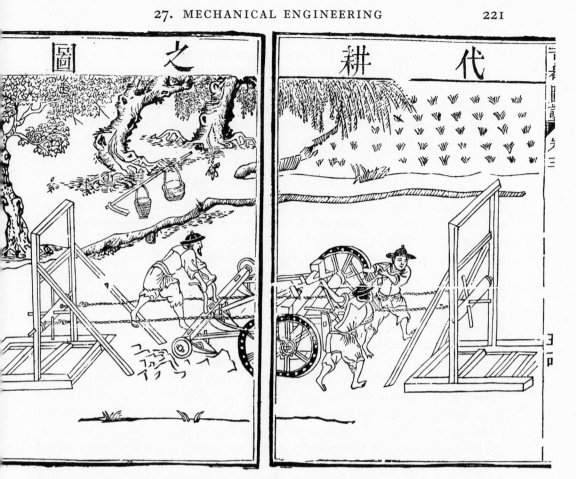

Fig. 469. Mechanical, or cable, ploughing (*tai kêng*); an illustration from the *Chhi Chhi Thu Shuo* (+1627), ch. 3, pp. 54*b*, 55*a*. The idea was old in Europe, but perhaps not practised either there or in China until after the latter part of the +16th century.

is indeed described in +1780 by Li Thiao-Yuan[1] in his book *Nan Yüeh Pi Chi*[2] (Memoirs of the South)[a] as currently in use at that time in Kuangtung.

The 'wooden ox' (*mu niu*[3]) method is a way of ploughing without animals. Two frameworks shaped like the character *jen*[4] are set up, having inside them certain pulleys fixed, and round these are wound long cables some 60 ft. in length, which are attached by iron pulling hooks to the plough. When the method is in use, one man guides the plough while two men sit facing each other on (each of) the frameworks. When the turning is in one direction the plough comes forward; when it is in the other the plough goes back. The work which one man can thus perform is equivalent to the strength of two bullocks. This kind of ploughing is considered very good.

The use of the word 'sitting' here suggests that the men were pedalling rather than turning cranks by hand; if so, this was a very Chinese adaptation. The power stated

[a] Ch. 31, tr. auct. The passage is quoted by Liu Hsien-Chou (*1*), p. 18.

[1] 李調元   [2] 南越筆記   [3] 木牛   [4] 人

also indicates some simple form of gearing. The absence of mechanical advantage in the European +15th-century drawings suggests that cable ploughing was then only an idea, but it probably was actually employed towards the end of the +16th century.[a]

Another clear instance of the acceptance of Western techniques by the Chinese in the Jesuit period was the use of force-pumps mounted on wheeled vehicles for use as fire-engines. These were already being made by Po Yü[1] at Suchow about +1635. Information on fire-fighting organisation in China and Japan has been collected by Horwitz (7).

(i) *A provisional balance-sheet of transmissions*

Let us now try to strike a balance-sheet of the position. First we may list ten machines and devices which were assuredly European introductions: (*a*) Archimedean screw and worm gear, (*b*) the Ctesibian double force-pump, (*c*) the Roman drum-treadmill, (*d*) the tower-type vertical windmill, (*e*) the crank-shaft, (*f*) the inclined treadmill,[b] (*g*) the flume-beamed swape,[c] (*h*) the double swape worked by a rotating conical cam,[d] (*i*) the pan-swape,[e] and (*j*) the rotary water-pump. Of these by far the most important was the fifth, though the days of its employment in external or internal combustion engines[f] were yet far off. The first, the second and the tenth were also important in principle though only the first was fundamentally new, since, as we have seen, the single-cylinder double-acting force-pump had been familiar in the Sung (p. 145 above), and rotary air-compressors (p. 118 above) much earlier. The Archimedean screw for water-raising found considerable use, and even some place in literature, in China and Japan during and after the +17th century (cf. p. 122 above), and the force-pump with its two cylinders spread even more widely as a fire-engine in cities. Of the others, the drum-treadmill was unnecessarily clumsy for a people who had always made a deft use of radial pedals, the vertical windmill was contrary to their engineering traditions, and the rotary pump for liquids was ahead of its time even in the rising industrialisation of the West—the remainder were hardly even practical and never got adopted anywhere.

The +17th-century books next show thirteen machines or devices which it was a perfect work of supererogation to introduce to China. These are (*a*) the steelyard and

---

[a] In the days of steam traction-engines it was of course very common, and it is interesting to find the lineal descendant of the idea, in terms of steel cables and electric winches, participating in current Chinese rural mechanisation (1959); Chhen Po-San (1).

[b] Fig. 470. This was naturally rare in the West because of the oblique-angle gearing which it invited. On the authority of the Director of the Deutsches Museum in München, where a model is shown, it was fairly common in +17th-century Italy—but that was certainly not, as suggested, the earliest use of such gears (cf. p. 456).

[c] I thought for a long time that this apparatus (as illustrated by the Jesuits) was a figment of the imagination of the +16th-century engineers, but it does appear to have been used traditionally in Bulgaria (see Wakarelski, 1). Some forms of it, too, are ancient in India (cf. p. 334).

[d] Fig. 471. I confess to the gravest doubts as to whether this machine was ever built, or even capable of practical use.

[e] This seems to be one of the devices which (if ever constructed) would give the least possible return for work expended.

[f] On these see the classical monograph of Dickinson (4) and papers by Needham (41, 48) and Davison (6).

[1] 薄珏

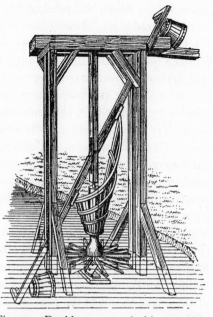

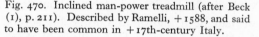

Fig. 470. Inclined man-power treadmill (after Beck
(1), p. 211). Described by Ramelli, +1588, and said
to have been common in +17th-century Italy.

Fig. 471. Double swape worked by a rotating
conical cam (after Beck (1), p. 201). Described
by Besson, +1578.

balance, (b) the windlass, winch and capstan, (c) the crank, (d) the pulley, (e) the
derrick, (f) gear-wheels, including ratchets and right-angle gearing, (g) the siphon,
(h) chain-pumps, both vertical and inclined, (i) driving-belts and chain-drives transmitting power, (j) the mobile wagon-mill, (k) the windmill, (l) the trip-hammer
stamp-mill, (m) pivoted conveniences such as the revolving bookcase. All these the
Chinese had long known and used. The position of the saw-mill is uncertain, for
while it would be surprising if water-power had not been applied to this purpose, as
to so many others, in China, it is not illustrated in any of the traditional books, and
we have as yet found only one textual mention of it, in which the circumstances were
peculiar (cf. p. 424). A second doubtful case is that of the suction lift-pump, since
the Szechuanese brine-well buckets almost amounted to this (cf. p. 142). A third is the
anaphoric water-clock, which was very probably employed in medieval China, though
positive proof of this is difficult. The question is discussed on pp. 466 ff. below. The
introduction of the weight-drive and the spring-drive for clockwork, both distinctively
European contributions, should be added to the list drawn up in the preceding paragraph, but they penetrated China a little later in the +17th century than the period
of the Jesuit publications which we are here discussing.[a]

Finally, we have to consider the machines or devices, not of course shown in the

[a] Historically, of course, the weight-drive no doubt originated from the anaphoric float, whether on
water or sand descending (cf. p. 442). We cannot be quite sure that its free-hanging form was developed
only in Europe. In any case Wang Chêng was quick to see its value, and long before any account
of weight-driven clocks had appeared in Chinese he made use of it in three of his designs (items 5, 6,
and 7).

Jesuit books, which had reached Europe from China at earlier times, or were still to be transferred there. Among these characteristically Chinese inventions we may list: (*a*) the square-pallet chain-pump, (*b*) the double-acting single-cylinder air-pump, (*c*) deep borehole drilling technique,[a] (*d*) the water-power edge-runner (gunpowder) mill, (*e*) water-powered trip-hammers for forges (martinets) as well as grain-pounding, (*f*) rotary fans or air-compressors, notably the winnowing-machine, (*g*) power-transmission by driving-belt, (*h*) power-transmission by chain-drive, (*i*) mechanical clockwork,[b] (*j*) chevron-toothed wheels (double helical gear),[c] (*k*) the horizontal-warp loom,[d] (*l*) the draw-loom, (*m*) silk-reeling machinery, (*n*) silk-twisting and doubling machinery, (*o*) the conversion of rotary to longitudinal reciprocating motion by the combination of eccentric, connecting-rod and piston-rod, embodied in the water-powered metallurgical blowing-engine, (*p*) the wheelbarrow, (*q*) the sailing carriage, (*r*) a form of truss construction embodied in the dishing of vehicle wheels, (*s*) link-work involved in trace-harness, (*t*) collar harness, (*u*) the crossbow and arcuballista, (*v*) the trebuchet or mangonel, (*w*) rocket flight, (*x*) the segmental arch bridge, (*y*) canal lock-gates, and (*z*) numerous inventions in naval construction including the stern-post rudder, water-tight compartments, and fore-and-aft rig. It will be seen that some of these take us out of the present argument and into fields which will be the subject of detailed attention in later Sections. Naturally the Jesuits said nothing about deep-drilling gear or driving-belts since these had been known in Europe for several centuries previously and they perceived that the Chinese were also familiar with them. Conversely they also said nothing about the rotary fan winnowing-machine because Europe did not have it at all; indeed it was probably owing to their own intermediation that it travelled west at the beginning of the +18th century.

There remain a number of further categories of machines and devices about which a word must be said. Some were common to the length and breadth of the Old World, and as yet we know nothing of their real first place of origin—among these one might mention the diolkos or slipway on canals, the most important lock-and-key mechanisms, and the early forms of roller-bearings. Puzzling situations arise when the beginnings of techniques can be dated at closely similar times both in China and the West; the outstanding case of this perhaps is the development of rotary milling at the beginning of the −2nd century, and the application of water-power to milling and other purposes during the −1st. Another category might be constituted by inventions of which the origin is still very obscure, as in the case of the crank, where we can only say that China takes an honourable place in its history. Then there may have been certain Chinese inventions which had died out by the time when the Jesuits came. The applica-

---

[a] Together with its associated hoisting equipment for winning the brine or water, and the arrangements for utilising the natural gas.

[b] Here the essential invention was the escapement, on which see pp. 458 ff. below. In its original home the mechanical clock was powered, as we shall see, by a water-wheel or a sand-driven scoop-wheel, not by a falling weight. It derived from the efforts to secure by water-power the slow rotation of astronomical instruments; efforts which had been going on since the +2nd century. With good reason we retract our statement (Vol. 1, p. 243) that clockwork 'was a distinctively European invention of the early +14th century'.

[c] On this some uncertainty still remains, cf. p. 87.

[d] Unless this reached Europe direct from India.

tion of water-power to the slow rotation of astronomical instruments and thence to mechanical clocks was fairly clearly in a period of decline in + 1600, and another extremer case might be found in the principle of the differential gear, almost certainly used in the south-pointing carriages of the Middle Ages but always confined to a few court technicians and by the end of the + 16th century quite lost. Lastly, certain Jesuit selections remain odd; for example, why they did so much to popularise the Archimedean screw for water-raising, but said so little of the useful screw-presses for oil, wine and perfume typical of the Mediterranean region. In any case, our balance-sheet suggests that when traditional Chinese technology came into confrontation with that of the aspiring Renaissance West, it had very little to be ashamed of so far as fundamental principles were concerned. The Jesuits were right inasmuch as they sensed the coming typhoon of Western mechanisation and scientific industry, for which Chinese society was in no way prepared, but standing at the close of the Middle Ages they greatly over-valued the contributions of the Europe of the past.

(ii) *The steam-turbine in the Forbidden City*

The supreme development of the palaeotechnic period was the steam turbine; the neo-technic age evolved the automobile; and one scene was set in China. About + 1671 Fr Philippe-Marie Grimaldi (Min Ming-Wo[1]) organised an elaborate scientific con-versazione for the young Khang-Hsi emperor, and among the optical and pneu-matical curiosities which were exhibited on that occasion there figured both a model steam-carriage and a model steam-boat. It is best to hear the account in the words of du Halde:[a]

Les Machines Pneumatiques ne picquèrent pas moins la curiosité de l'Empereur. On fit faire d'un bois léger un chariot à quatre roües de la longueur de deux pieds. Au milieu l'on mit un vase d'airain plein de braise et au-dessus un Eolipile, dont le vent donnoit par un petit canal dans une petite roüe à ailes semblables à celles des Moulins à vent. Cette petite roüe en faisoit tourner une seconde avec un essieu, et par leur moyen faisoit marcher le chariot deux heures entières. De peur que le terrain ne lui manquât, on le faisoit marcher en rond en cette manière.

A l'essieu des deux dernières roües, on attacha un timon, et à l'extremité de ce timon un second essieu qui alloit percer le centre d'une autre roüe un peu plus grande que celles du chariot, et selon que cette roüe etoit plus ou moins eloignée du chariot, elle decrivoit un plus grand ou un plus petit cercle.

On appliqua aussi ce principe de mouvement à un petit navire porté sur quatre roües. L'eolipile etoit caché au milieu du navire; et le vent sortant par deux autres petits canaux enfloit ses petites voiles et les faisoit tourner en rond fort long-tems. L'artifice en etoit caché, et l'on entendoit seulement un bruit semblable à celui du vent, ou à celui que l'eau fait autour d'un vaisseau.

Uccelli gives[b] a design of this vehicle reconstructed by Canestrini, but it would seem likely that some form of reduction gearing was incorporated between the steam-

---

[a] (1), vol. 3, p. 270. [b] (1), p. 637.

[1] 関明我

turbine and the front axle, as in another model (Fig. 472).[a] The boat was presumably a four-wheel paddle-boat. According to those who have gone into the matter in detail,[b] the experimental models had been begun by Fr Verbiest already in about +1665. The customary opinion is that the plan derived from the suggestions for steam-turbines published by Branca in +1629[c] but the difference was that none of these would have been at all practicable, while there is no reason to doubt the statement of du Halde that the models of Verbiest and Grimaldi did actually work. It is interesting to note that the turbine principle was not successfully applied to full-scale locomotives till the work of Ljungström in 1922, nor to ship-propulsion till about 1897 with the steam-turbines of Sir Charles Parsons. Liu Hsien-Chou (3), who has most fully summarised the story, and Fang Hao (1), suggest that part of the inspiration of Verbiest and Grimaldi may have been derived from the old Chinese hot-air zoetropes already described (Vol. 4, pt. 1, pp. 122 ff. above), but I am inclined to think that the system of inclined vanes employed in them descended to the helicopter top and hence to the air-screw (see pp. 580 ff. below) rather than to turbine rotors for steam jets, such as the Jesuits used. Even if in some of their designs the rotor was mounted horizontally, the current of vapour had to move (like water over a water-wheel) more or less at right angles to the rotor's axis, while in devices of the zoetrope category the current moves in a direction parallel with this axis. The shape of the boiler in Fig. 473 suggests quite another parallel, and perhaps it was the steam-jet fire-blower which gave inspiration to the Jesuits of the China Mission.

Most people would naturally be inclined to say that in the history of the steam jet Asia would not be involved. But the matter cannot be dismissed quite so easily. From ancient times onwards a jet of steam was used for a purpose other than making a vaned wheel rotate, i.e. for blowing up fires. The story of these steam fire-blowers has recently been fully related by Hildburgh (1),[d] who points out that the famous aeolipile of Heron, rotating by jet propulsion on account of its L-shaped tubes, was only a special form of aeolipile.[e] The commoner sort was simply a kettle or boiler with a pinhole orifice so arranged that a jet of steam could be directed upon a fire. This has several effects: it accelerates combustion because of the draught of air which accompanies the jet, products of combustion are removed, and at the same time the steam is decomposed by the glowing coal to carbon monoxide and hydrogen (water-gas), which immediately burns away.[f] To Philon of Byzantium (c. −210) is ascribed an incense-burner kept glowing by a steam jet,[g] and similar arrangements are mentioned by Vitruvius[h] and

---

[a] Ascribed to Verbiest by Ucelli di Nemi (4), p. 29, who dates it +1681. Uccelli places it as late as +1775 however. Another locomotive model is described on p. 388.

[b] Thwing (1); Rouleau (1); Fang Hao (1).

[c] Cf. p. 125 above. One of Branca's illustrations (fig. 25) is reproduced by Parsons (2), p. 143; Uccelli (1), p. 15, fig. 42; Schmithals & Klemm (1), fig. 46. There may also have been a prior experiment with rudimentary steam-turbine paddle-boats, in +1543; see below, p. 414. We shall return to Branca in connection with the helicopter top, p. 584 below.

[d] Cf. Lynn White (7), pp. 89 ff.

[e] The word derives from the name of the god of the winds, and pila, πίλος, a ball.

[f] Water-gas has quite a wide use in modern technique, and steam-jet blowers are used for industrial boiler furnaces. By varying the mixture with air, a thermostatic control is achieved.

[g] Pneumatica, ch. 57; Drachmann (2), pp. 67, 125; Feldhaus (1), col. 179.   [h] I, vi, 2.

PLATE CLXXV

Fig. 472. Model of the steam-turbine road-carriage constructed by P. M. Grimaldi, together with many other scientific and technical demonstrations, for the entertainment and instruction of the young Khang-Hsi emperor about +1671. A reconstruction in the Museo Naz. della Scienza e della Technica, Milan.

PLATE CLXXVI

Fig. 473. Tibetan steam-jet fire-blower in copper (Museum of Archaeology and
Ethnology, Cambridge). Scale in centimetres and inches.

Heron of Alexandria.[a] Then in the +13th century Albertus Magnus describes[b] a 'sufflator' of this kind in detail, and in the later technological MSS. there are several pictures of bronze busts or heads which direct steam jets from their mouths (Kyeser, +1405; Leonardo da Vinci, c. +1490).[c] The jet 'turbine' follows shortly afterwards (Branca, +1629; Wilkins, for rotating a spit,[d] +1648). In +1545 a hollow bronze statue of a man was found in some ruins at Sondershausen in Germany, and gradually other similar objects appeared; their use was the subject of much dispute until Feldhaus (19) established that these 'Püstriche', or blowers, were fire-blowers corresponding to the description of Albertus Magnus. Examples have been found as far east as South Russia (Ekaterinoslav).

We are now recognisably in presence of the 'pre-natal' form of the steam-engine's boiler, so there is significance in the gradual abandonment of the 'wind-cherub' theme still beloved of cartographers long afterwards. When Cesariano translated Vitruvius into Italian early in the +16th century he depicted the boilers in various elegant, but now no longer anthropomorphic, forms;[e] they were turning, so to say, into the celebrated kettle of James Watt. When Ercker illustrated his 'Treatise on Ores and Assaying' later in the century the steam blowers were just alembics placed alongside the furnaces.[f] But Branca still used an Aeolian head for his steam-jet turbine in +1629 (cf. p. 125 above), and Athanasius Kircher followed suit in +1641.[g] The question was, on or into what should the jet be directed? When della Porta in +1601 demonstrated the emptying of water from a closed vessel by the injection of a jet of steam into it, and the drawing up of water by the vacuum left in a vessel by condensed steam,[h] he was much more truly on the rails which were to lead to the primary break-through of steam-power, i.e. the reciprocating steam-engine rather than the steam-turbine. This was the green light for Ramsay (+1631), d'Acres (+1659), Somerset (+1663) and all their successors.[i] Dircks was surely not wrong in his conviction that the *sufflator* was the chief ancestor of the steam-engine.[j]

But here comes in the unexpected. The human heads and busts were a product of

[a] *Pneumatica*, chs. 74, 75; Drachmann (2), pp. 130ff.; Woodcroft (1), pp. 100ff.—the *miliarium*.

[b] *De Meteoris*, III, ii, 17 (Paris ed. 1890, vol. 4, p. 634).

[c] See Feldhaus (1), col. 843ff., (18), p. 92. Leonardo also appreciated the expansive force of steam, and described an 'Architronito' or steam-cannon, the invention of which he attributed to one Archimedes, probably a Renaissance contemporary (see McCurdy (1), vol. 2, p. 188; Feldhaus (1), col. 394ff.). Though no success attended the efforts of those engineers in later centuries who continued to play with this idea, the Architronito was symbolic in the highest degree if we may look upon the piston as a tethered projectile. Cf. p. 140, and see further Reti (2); Hart (4), pp. 249ff., 295ff.

[d] *Mathematical Magic*, p. 149.

[e] Como, +1521; see Feldhaus (1), col. 26, fig. 10. Cf. Lynn White (7), pp. 90ff.

[f] Prague, +1574 and many later editions; see Sisco & Smith (2), frontispiece and pp. 219, 326ff.

[g] *Magnes*... (Rome, +1641), p. 616.

[h] *Pneumaticorum Libri III* (Naples, +1601), ch. 7, *Magiae Naturalis Libri XX* (Naples, +1589), bk. 19, ch. 3. See Beck (1), pp. 263ff.; Dickinson (4), pp. 4ff.; Wolf (1), p. 544 (confused). The vacuum left by condensed steam had been utilised thaumaturgically in ancient China; cf. Vol. 4, pt. 1, p. 70.

[i] See Dircks (1), pp. 540ff. Lynn White (7), p. 162, provides evidence that something similar was brewing in Bohemia before the end of the +16th century.

[j] Here the accent is on the word steam. One can also put the accent on the word engine, and as we shall presently show (pp. 380ff. below) the Chinese culture-area had a part to play in the development of this much more important than for steam jets or vaned wheels. See further Needham (48).

European fancy,[a] but for the steam-jet fire-blowers there was another focal area, namely the Himalayan region, especially Tibet and Nepal. There they still take the form of bottle-shaped conical copper kettles surmounted by birds' heads, the beaks of which, sometimes quite elongated, point downwards and have the pinhole at the tip (Fig. 473).[b] The fact that an ordinary kettle, which emits steam under very little pressure, does not have the same effect, might plead for a single point of origin of the discovery; if this was Mediterranean, the *sufflator* can hardly have reached the Bactrian region from Egypt in Alexander the Great's time (as Hildburgh suggests), but could have been used there by the later Bactrian Greeks.[c] Alternatively Alexander's veterans may have brought it back. It would certainly have been useful for the fires of dung used by travellers on the Old Silk Road and other desert and mountain regions of Central Asia where wood is scarce and altitude considerable. So far we have no evidence that the Chinese made much use of it, and until this is found we may accept the distribution as being Western and Tibeto-Bactrian. But if so, the remarkable fact presents itself that the Central Asian region of Mongol-Tibetan-Persian culture was that in which the windmill also seems to have its oldest home.[d] And once again there were the Central Asian slaves in Renaissance Italy whose steam jets may at least have reinforced the aeolipiles of Hellenism.[e] Here at any rate is a suggestive juxtaposition of facts, which might prompt us to look for an ancestor of Branca's proposal in the Arabic-Iranian direction, a first combination of steam jet and vaned wheel having occurred somewhere in western Central Asia.[f]

(4) THE 'CARDAN' SUSPENSION

There is one device of which it has not been convenient so far to say anything, either among basic principles or the chief types of Chinese eotechnic machines, namely, that seemingly simple combination of rings whereby an object may be maintained in horizontal equilibrium, the gimbals, or as it is sometimes called, the Cardan suspension.[g]

[a] A rather curious fancy too, for many of the figures are prominently phallic, e.g. the celebrated Jack of Hilton described by Plot in +1686 (p. 433, and pl. XXXIII) and still connected with a jocular tenure.

[b] There are examples in the Cambridge Ethnological Museum, one of which was kindly tested by Dr Bushnell in the writer's presence, with most striking effect. The Himalayan fire-blowers have only one hole, and must be made to 'drink' first by heating the body of the vessel; the European ones have a separate hole for filling.

[c] Cf. Vol. 1, pp. 172 ff., 233 ff. above. There is nothing to prove, of course, that the device was not an ancient one in Central Asia, which travelled westwards in time to stimulate Philon.

[d] Cf. p. 125 above, and pp. 555 ff. below, where it will be shown that the windmill was first harnessed to practical use in Seistan, a region of Parthia close to Greek Bactria (though some seven centuries after Greek Bactria had ceased to exist as such).

[e] Cf. Vol. 1, p. 189, reinforced by Lynn White (5), (7), pp. 93, 116.

[f] Yet another suggestive idea presents itself. The Central Asian fire-blower is shaped very like a still or alembic, as Dr P. C. Mangelsdorf noted when watching one of Dr Bushnell's demonstrations. Could t have had any relation with the history of distilling, whether in India or among the Alexandrian 'alchemists'?

[g] A very original paper was published some forty-five years ago by Laufer (23) on China and the Cardan suspension. At the time of writing of this subsection, however, it was inaccessible to us, and when eventually a transcript of it came to hand among the papers of the late Prof. F. Jäger, I was happy to find that no changes or additions of any consequence were needed.

Most people are familiar with it because they have seen it in use for one of its most widespread Renaissance applications, the mounting of the mariner's compass so that this is independent of the motion of the ship. If three concentric rings are connected together in a series by pivots so that the axes of the pivots are alternately at right angles to each other, and if the central object is weighted and freely movable on the innermost pivot axis, then whatever the position which the outer case comes to occupy, the central object will adjust itself so as to maintain its original position.

This device is known as the Cardan suspension because it was described by Jerome Cardan[a] in his *De Subtilitate* of +1550. The reference occurs in Book 17 (*De Artibus atque artificiosis Rebus*),[b] but Cardan did not claim the invention as his own; he described a chair on which an emperor could sit without being jolted, and he said that the contrivance had previously been used for oil-lamps. Indeed it had been well known in Europe long before his time, though during the +16th and +17th centuries it was attracting particular attention. Leonardo had sketched it about +1500, planning it for use with a compass.[c] Besson (2), in his *Cosmolabe* of +1567, suggested that navigators should sit within large gimbals for taking their observations (Fig. 474); while Branca (+1629) planned a similar horse-drawn ambulance or litter (Fig. 475).[d] During the +17th century, gimbals for ships' compasses came into general use.[e] A gimbal-mounted azimuth compass of the Chinese +18th century is depicted in Fig. 481.

But the suspension had been known already towards the middle of the +13th century, when Villard de Honnecourt was writing his famous notebook. His diagram (Fig. 476),[f] which we may place at about +1237, shows four rings, and is captioned 'Cis engiens est fais p(ar) tel maniere quel p(ar)t q(u)'il tort ades (est) li paelete droite'. Hahnloser points out[g] that the object of the device at that time was for small hand-stoves ('Wärmapfel') which prelates could use during long religious ceremonies in cold cathedrals. Fig. 477 shows a +13th-century example preserved in St Peter's Treasury at Rome. Significantly perhaps, similar portable stoves were known among the Arabs, and Migeon figures a Muslim incense-burner, dated +1271, in the British Museum.[h] The Carrand Collection and similar treasuries have yet other examples, which probably take the use of the device back to the early +12th century both in European and Islamic culture. But there is literary evidence a good deal earlier than this, for the *Mappe Clavicula* (Little Key of Painting), a recipe-book of the +9th

[a] Lived from +1501 to +1576; biographies by Waters (1); Eckman (1); Bellini (1); Ore (1), esp. pp. 120ff., and others.

[b] In the Basel edition of +1560, p. 1028. Earlier editions had given the credit to Juanelo Turriano.

[c] Feldhaus (1), col. 869, (18), p. 120; Gerland (1), p. 250. Sarton (1), vol. 3, p. 716, dates the first use of it for the mariner's compass at +1556. Edward Wright (1) in +1610, translating Çamorano's book of +1581, says that the round box of the compass must be placed 'within two hoopes of latin...fastened within a square box, or a round, so as although the uttermost box be tossed up and downe every way with the motion of the shippe, yet alwaies the superficies and glasse of the inner box may lie level with the Horizon...'.

[d] Pl. 23.   [e] Cf. Breusing (1).

[f] Hahnloser (1), pl. 17.   [g] (1), pp. 45ff.

[h] (1), p. 185. The late Prof. D. S. Rice intended to describe in a special monograph all the Arabic Cardan suspensions of the +13th and +14th centuries.

Fig. 474. The Cardan suspension or gimbals as a nest for navigators while taking their observations, a proposal of Besson, +1567. Cf. Taylor & Richey (1), p. 95.

Fig. 475. The Cardan suspension or gimbals as a cradle for a patient or traveller in a horse-drawn ambulance or litter, a proposal of Branca, +1629.

PLATE CLXXVII

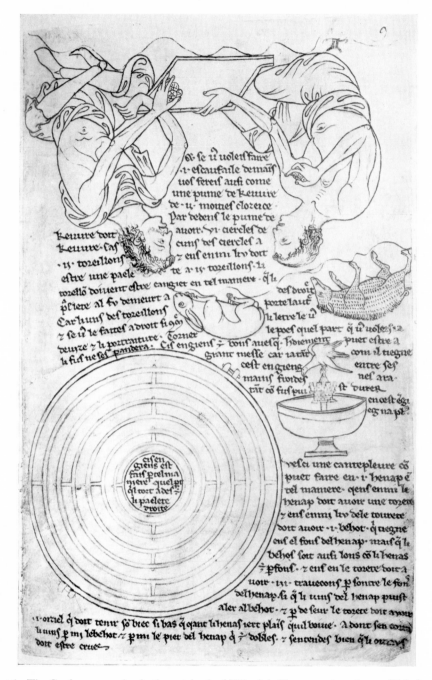

Fig. 476. The Cardan suspension in the notebook of Villard de Honnecourt, *c.* +1237 (Hahnloser, 1).
Within the rings is written 'This engine is made in such a way that the pan remains straight no matter
how it turns'.

PLATE CLXXVIII

Fig. 477. A +13th-century hand-warming stove with Cardan suspension (St Peter's Treasury, Rome; Hahnloser, 1).

Fig. 478. Chinese perfume-burner with Cardan suspension of the Thang period (c. +8th century), in silver-work (Kempe collection, Gyllensvård, 1), diam. c. 2 in. Two rings and three pivot-axes.

PLATE CLXXIX

Fig. 479. A brass Tibetan globe-lamp with Cardan suspension for temple use (author's collection), diam. 10 in. Besides figures of gods and *bodhisattvas* the repoussé work includes germ-letters (symbols of divinities) in medallions. Probably late Chhing (photo. Brunney).

PLATE CLXXX

Fig. 480. Interior of the brass Tibetan globe-lamp showing the suspension of four rings and five pivot-axes (photo. Brunney). A candle-end does duty for the original oil-wick.

century, includes a clear description of a vase surrounded by rings in such a way that rolling motions of the whole sphere are not communicated to it.[a]

Was the invention an Alexandrian one? The 56th chapter of the *Pneumatica* of Philon of Byzantium[b] (*c.* −220)[c] describes an ink-well enclosed in a prismatic box with a hole in each face, any one of which could be used since the ring-suspension within would keep the ink-well the right way up.[d] The statement ends by saying that the design follows an old Jewish pattern for incense-burners. This in itself is suspicious, for it does not sound quite the kind of remark which one of the earlier Alexandrians would have made;[e] moreover, the whole passage is found only in the Arabic MS. translated by Carra de Vaux (2) and not in the Latin MSS. translated by Schmidt (1) and de Rochas d'Aiglun. Furthermore, the description seems out of place among so many devoted purely to pneumatic devices. Sarton, therefore, cautions[f] that it may be an interpolation of later Arabic compilers, perhaps as late as the +13th century.

Strolling one day in a Parisian market (May 1950), I happened to see two brass globe-lamps of Tibetan workmanship,[g] possibly quite recent in date of manufacture, which carried within their fretted casings Cardan suspensions of four rings and five pivot axes.[h] One of the two specimens I was able to buy, and it is seen here in Figs. 479 and 480. It had obviously been intended for the hanging of an oil-lamp in a relatively exposed temple hall or porch. Though I have never encountered devices of this kind in China, I was delighted to find two further examples, incense-burners in beautiful silver-work dating from the Thang period, exhibited in the Carl Kempe Collection of Chinese Gold and Silver at the Victoria and Albert Museum in 1956 (Fig. 478).[i] Then there was the bed-warming brazier which Laufer (23) tells us he found in Sian fifty years ago. It appears moreover that in some provinces lanterns with Cardan suspensions are quite commonly constructed of bamboo, especially those which represent the moon-pearl and are flourished in front of dragons in processions (Fig. 482).[j]

---

[a] See Berthelot (10), p. 64, (11); Thorndike (1), vol. 1, pp. 765 ff.   [b] Cf. B & M, p. 487.

[c] It must be remembered that the dates of all the Alexandrian engineers are still uncertain, and some authorities (such as Sarton (1), vol 1, p. 195) have made him the contemporary of Lohsia Hung (*c.* −120) rather than of Lü Pu-Wei.

[d] Carra de Vaux (2); Beck (5), p. 71; Feldhaus (2), p. 97.

[e] Whatever Philon's exact date was, he certainly lived long before Josephus and Philo Judaeus.

[f] (1), vol. 1, p. 195. Drachmann (2), pp. 67 ff., strongly concurs.

[g] Diameter 10 inches. The repoussé work is typically Tibetan, to judge from S. Hummel (1). Each hemisphere is ornamented with a mounted protector deity and four Bodhisattvas. There are also eight medallions containing four *bījākṣara* or 'germ-letters' (*chung tzu*[1]), used as symbols of divinities, but though the script seems to be the siddham (*hsi-than*[2]) on which van Gulik has given us such an interesting monograph (7), identification has not so far been possible. Siddham was a form of Sanskrit script much cultivated in medieval China and Japan as the vehicle of these symbols and of Tantric Mantrayānic spells (cf. Vol. 2, p. 425).

[h] The number of frames is interesting because modern gyroscopic theory states that five frames are required to give complete freedom of angular motion in all directions of space (Davidson, Saul, Wells & Glenny).

[i] Gyllensvärd (1), nos. 96 and 97; (2) pp. 47 ff, fig. 11, pl. 5 d. These are both about 2 inches in diameter. We reproduce the photograph by the kind permission of Dr Carl Kempe and Dr Bo Gyllensvärd. At least one similar piece is in the Shosoin Treasury at Nara in Japan. Shi Shu-Chhing (2) illustrates it, and adds details of another specimen of the Ming period, together with new textual references.

[j] I am grateful to Mr Wei Tê-Hsin of Sidney Sussex College for information on this; he himself saw such lanterns in his youth in Fukien. Cf. *Hu-chhêng Sui Shih Chhü Ko*, p. 2 a.

[1] 種字   [2] 悉曇

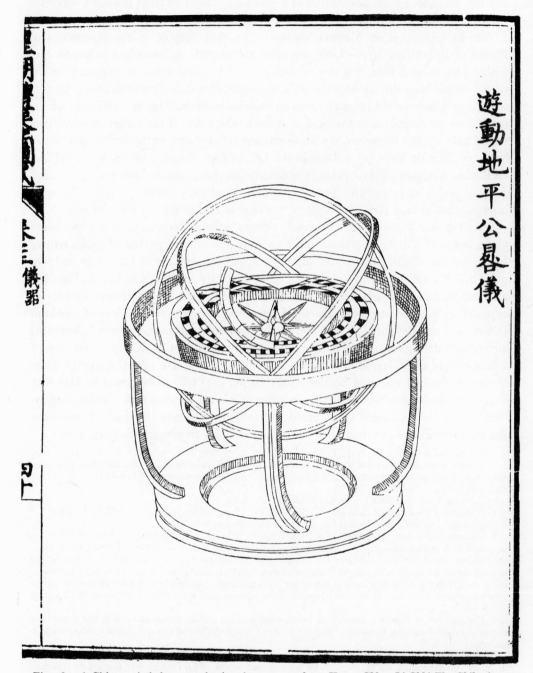

遊動地平公晷儀

Fig. 481. A Chinese gimbal-mounted azimuth compass; from *Huang Chhao Li Chhi Thu Shih*, ch. 3, p. 40*a* (+1759). Such instruments, which combined the magnetic compass with an adjustable gnomon and a graduated rim, enabled navigators to determine precisely the magnetic declination at sea; cf. Taylor (7), pl. 3; Taylor & Richey (1), pp. 31 ff.

That this Sino-Tibetan tradition of gimbals originated from Cardan and the Italian Renaissance becomes very unlikely when we realise that we possess from the +2nd century a Chinese account of the device which is much earlier than anything in Europe or Islam except the dubious passage in Philon.[a] The *Hsi Ching Tsa Chi* says:[b]

In Chhang-an there was a very clever mechanician named Ting Huan[1] (*fl.* +180). He made 'Lamps which are always Full' (*chhang man têng*[2])[c] with many strange ornamentations, such as seven dragons or five phoenixes, all interspersed with different kinds of lotuses. He also made a 'Perfume Burner for use among Cushions' (*wo ju hsiang lu*[3]), otherwise known as the 'Bedclothes Censer' (*pei chung hsiang lu*[4]). Originally (such devices) had been connected with Fang Fêng[5,d] but afterwards the method had been lost until (Ting) Huan again began to make them. He fashioned a contrivance of rings which could revolve in all the four directions,[e] so that the body of the burner remained constantly level (*wei chi huan chuan yün ssu chou, erh lu thi chhang phing*[6]), and could be placed among bedclothes and cushions. For this he gained much renown.

It will be remembered that Ting Huan has been met with before in connection with his zoetrope lamp and air-conditioning fan. The Thang incense-burners mentioned above would thus seem to be descendants of Ting Huan's unspillable censers in a very direct line.

Some colour is given to the statement that Ting Huan only revived an invention which had been current long before, by a curious passage in the *Mei Jen Fu*[7] (Ode on Beautiful Women)[f] written by Ssuma Hsiang-Ju[8] (*d.* −117). This poem, a defence of his continence against the reproaches of a certain prince whose guest he was, gave occasion for a couple of seduction scenes admirably described. The second of these

---

[a] This is still true even if we date the *Hsi Ching Tsa Chi* as late as the +6th century.

[b] Ch. 1, p. 7*b*, tr. auct., adjuv. Laufer (3), p. 196. Partly quoted *TPYL*, ch. 870, p. 1*b*.

[c] Philon of Byzantium also described such lamps in his ch. 20 (Beck (5), p. 68; Drachmann (2), p. 122); the wick-reservoir oil when lowered in level by the burning admitted air to a closed space, thus allowing more fuel to descend. Such ever-filled cruses have a remarkably wide distribution. When in 1958 I visited the Museum of the Ceylon Archaeological Survey at the Sutighara Cetiya, near Dedigama, the great dagoba built at his birthplace by the +12th-century king Parākrama Bāhu I, the curator, Mr Abhaya Devapura, showed me some remarkable hanging oil-lamps in bronze which had been found in the upper relic chamber of the dagoba and were thus contemporary with the king. At the centre of the reservoir of oil with its radiating wicks there is a tank in the form of a well-fashioned elephant, its lingam constituting the pipe through which the reservoir is fed as the level falls below and bubbles of air ascend. I am much indebted for his kind welcome to Mr Devapura, who was inclined to see with me a distant derivation from the Alexandrians (cf. Vol. 1, pp. 176 ff., on the close relations between India and the Hellenistic world in ancient times), though independent invention can never be excluded. It would be interesting to know what method Ting Huan adopted.

[d] This is the name of a spirit anciently worshipped by the Yüeh people, as also of a later State in what is now Chekiang, and of princes who took their names from it. The more usual writing is Fang Fêng.[9] On the other hand the reference may well be to an individual person, and if the second character was a misreading, Fang Fêng[10] might qualify. He was a scholar and military officer of the −1st century who by Wang Mang's time had attained high rank, and his biography (*Chhien Han Shu*, ch. 88, p. 26*a*) says that he was both learned and ingenious. It is not at all easy to decide what the writer had in mind.

[e] This was a stock phrase; the writer certainly meant the three dimensions of space.

[f] *CSHK* (Chhien Han sect.), ch. 22, p. 1*b*. There are translations by van Gulik (8), p. 68, and Margouliès (3), p. 324. The passage was quoted by Chhen Mou-Jen in the *Shu Wu I Ming Su* (Ming), and by *KCCY*, ch. 58, p. 2*a*.

---

[1] 丁緩　　　[2] 常滿燈　　　[3] 臥褥香爐　　　[4] 被中香爐　　　[5] 房風

[6] 爲機環轉運四周而爐體常平　　　　　　[7] 美人賦　　　[8] 司馬相如

[9] 防風　　　[10] 房鳳

takes place in an empty palace or remote imperial rest-house, and among the furniture, hangings, bedclothes, etc., carefully described, we find that 'the metal rings (contained the) burning perfume' (*chin tsa hsün hsiang*[1]).[a] The date of composition would be perhaps in the neighbourhood of −140. Thus it is possible that the invention belongs to the −2nd rather than to the +2nd century.[b]

And indeed Ting Huan's achievement was no isolated instance of an ingenious novelty, for Chinese references to the 'Cardan' suspension can be found in the literature of nearly every following period. The Liang emperor Chien Wên Ti, who reigned in +550, wrote in his poems of hinges on doors and windows as *chin chhü hsü*,[2] 'metal knee-joints'[c] or pivot-and-socket fittings.[d] One must remember that the most typical of such Chinese hinges are in fact pivots working in sockets, leaf-hinges being less common.[e] In +692 an artisan whose name has not come down to us presented to the empress Wu Hou 'wooden warming-stoves which though rolled over and over with their iron cups filled with glowing fuel could never be upset'.[f] So when Thang poets such as Li Shang-Yin[3] (+813 to +858) or Li Ho[4] (*d.* +810) speak of 'perfume-burners with interlocking pivot-and-socket hinges' (*so hsiang chin chhü hsü*[5]), etc., they are probably referring to gimbal suspensions.[d] Then in the Sung there were balls of ivory worked into a series of loose concentric fretted spheres (*kuei kung chhiu*[6]),[g] and since the word *chhê*[7] is sometimes used of the inner components, these may occasionally have been pivoted in the manner of Cardan rings.[h] 'Globe-lamps' (*ta têng chhiu*[8]) appear in the *Ju Shu Chi*[9] (Journey into Szechuan)[i] of +1170, and were used in the defence of Nanking in +1233.[j] Later in the same century Chou Mi mentions[k] *têng chhiu*[8] and *hsiang chhiu*[10] (perfume balls). In all of these the names and descriptions indicate unmistakably the presence of gimbals. Moreover, Sung scholars

---

[a] The background of this is as follows: Ssuma Hsiang-Ju maintained that he was much more chaste than the Confucians and Mohists, for they turned tail and fled at the mere sight of a dancing-girl, while he knew how to withstand temptation and so made no effort to avoid it. His first scene instances this. But at the same time it is, he says, harmful to abstain from all sexual activity, so the second scene describes a happy encounter with a charming girl in a deserted palace. After they had made love his pulse was steadied and his mind fortified. Such a belief was deeply characteristic of Chinese preventive medicine, and the essential content of Ssuma's ode is the skill of when to advance and when to refrain. Nevertheless the form of his second scene may have some connection with the corpus of story about evil spirits taking the form of women in the official dak-bungalows (*thing*[11]) where the bureaucrats rested at night on their journeys and missions (cf. Chiang Shao-Yuan (1), pp. 116ff.); fantasies indeed of wish-fulfilment and Confucian repression.

[b] Dr Chêng Tê-Khun tells me that linked jade rings of Han date which might perhaps be a stylisation of the gimbals are preserved in the collection of Mr Phan at Hongkong.

[c] *Hsü* standing here for *hsi*.[12]   [d] Cf. *Piao I Lu*, ch. 5, p. 4a.

[e] See Hommel (1), pp. 292, 298.

[f] *Mu huo thung thieh chan shêng huo nien chuan pu fan*.[13] This comes from the *Thai-Phing Kuang Chi* (+981), Chi chhiao sect., ch. 226 (item 5), p. 4a, quoting the +8th-century *Chhao Yeh Chhien Tsai*. He was the same Haichow man who made the jack-work (see p. 469 below).

[g] On these balls within balls, cf. Vol. 3, p. 387.

[h] *Ko Ku Yao Lun*, ch. 6, p. 10b.   [i] Ch. 3, p. 14b.

[j] *Kuei Chhien Chih*, ch. 11, p. 4a.

[k] *Wu Lin Chiu Shih*, ch. 2, p. 14a; ch. 9, p. 15a; cf. *Mêng Liang Lu*, ch. 1, p. 4b; ch. 13, p. 8a; ch. 19, p. 13b; ch. 20, p. 4a.

[1] 金錮薰香	[2] 金屈戌	[3] 李商隱	[4] 李賀	[5] 鎮香金屈戌
[6] 鬼功毬	[7] 車	[8] 大燈毬	[9] 入蜀記	[10] 香毬
[11] 亭	[12] 膝	[13] 木火通鐵盞盛火輾轉不翻		

PLATE CLXXXI

Fig. 482. Children engaged in a dragon procession, part of an inlaid lacquer screen in the collection of Dr G. D. Lu (photo. Brunney). In such perambulations, traditionally customary on the 15th day of the first month (cf. Hodous (1), p. 43), a globe-lamp representing the moon-pearl is flourished, as here, in front of the undulating effigy. Gimbals of bamboo within the globe maintain the stability of its light. On the symbolism of the dragon and the pearl cf. Vol. 3, p. 252 and Fig. 95.

themselves distinctly state[a] that they were regarded as identical with what Ting Huan had invented; Ming scholars also.[b] Certain old names, such as *tsan*[1] and *hsün lung*[2] (perfume basket), *yin nang*[3] (silver bag) and *kun chhiu*[4] (rolling sphere) probably also referred to the device, in their opinion. Lastly, the *Hsi Hu Chih*[5] (Topography of the West Lake Region of Hangchow) mentions (in one of its supplements) that *kuan li*[6] (interlocking pivots)[c] were mounted on certain occasions within paper lanterns, and that these were then kicked and rolled along the streets without the lamps inside being extinguished. They were called *kun têng*,[7] rolling lamps. This 18th-century account brings us back, then, to the contemporary folk practices mentioned above.

Developments almost incredible awaited the ring-system of Ting Huan when the suspended perfume-lamp turned into a solid heavy wheel or disc and began to spin on its own axis. At the beginning of this section the gimbals of the mariner's compass were mentioned because this application of the suspension is most commonly known. But it was the gyroscope,[d] the spinning disc with a mounting free in three dimensions, which was destined to replace the very magnetic compass itself, on account of its property, discovered by Foucault a century ago,[e] of orienting itself in the meridian. Hence the gyro-compass,[f] now of such universal application on steel ships and aeroplanes, which indicates true, not magnetic, north. Hence also the gyroscopic stabilisers[g] of ships and especially aircraft, in which the 'automatic pilot' has been the greatest factor in the successful achievement of long-distance and bad-weather flying.[h]

It might be said that in the gimbals the effective power source restoring the original state after displacement is the weight of the object at the centre, and therefore that the power is applied from within outwards. But an invention of even greater importance was made when someone conceived the idea of applying power from the outside, namely the universal joint. Here the outer casing has been changed into a U-piece on the end of a transmitting shaft, corresponding to another U-piece on the end of a receiving shaft, the two being joined by a connecting piece with pivots having axes at right angles to one another. Originally this piece was a ball with pins at right angles. The time of the invention was the end of the 17th century, and it was due either to Schott in +1664[i] or to Robert Hooke ten years later.[j] Its greatest application today is in the transmission shafts of automobiles, for which purpose it was first used by

---

[a] *Shih Wu Chi Yuan* (+1085), ch. 8, p. 8a; also *Wei Lüeh* (c. +1180), cit. *KCCY*, ch. 58, p. 2a, which terms the pivot the 'button' (*niu*[8]) and the socket the 'nostril' (*pi*[9]).

[b] *Liu Chhing Jih Cha* (+1579) which compares the Cardan suspension with the armillary sphere; also *Ming I Khao*[10] by Chou Chhi.[11]

[c] Cf. Vol. 3, p. 314, for a different use of this term in the Sung, when it could mean a float-valve, and p. 485 below for a third meaning, that of trip-lug.

[d] Cf. A. Gray (1) and Schilovsky (1).   [e] Anon. (8).

[f] See T. W. Chalmers' description (1).   [g] Cf. Ross (1).

[h] On general applications of gyrostatics see Davidson, Saul, Wells & Glenny.

[i] Feldhaus (1), col. 870.

[j] Andrade (1), p. 456. Hence it is still called Hooke's joint; cf. Davison (10); Willis (1), p. 272. But a first approximation had been conceived by Besson (+1578), cf. Beck (1), p. 205.

---

[1] 鸞	[2] 薰籠	[3] 銀囊	[4] 滾毬	[5] 西湖志	[6] 關棙
[7] 滾燈	[8] 鈕	[9] 鼻	[10] 名義考	[11] 周祈	

Burstall & Hill in 1825, but it will give service at low speeds when the angle between the two shafts is considerably greater than ever occurs in normal motor-vehicle practice.[a] Few, however, who ride daily in motor-cars, realise that the lineage of so important a device goes back to +2nd-century China.[b]

To sum up, therefore, we are faced in this instance with a situation we shall encounter again (cf. p. 434 below), in which conclusions as to the origin of an invention are rendered a little difficult by the dubious authenticity of the first European reference. If we adopt the cautious view and regard the gimbals of Philon of Byzantium as a late Arabic interpolation, then the credit is Ting Huan's (or Fang Fêng's), and it is not unlikely that the role of the Arabs was to transmit the device from further east. This seems plausible in any case on account of the Thang references from the +7th to the +10th century. But the gimbal suspension was already in Europe by the 9th. The mention of the Jews in the Philon passage may indeed possibly imply that the apparatus was transmitted westwards through Jewish rather than Arab circles. The suggestion has already presented itself[c] that certain important characteristics of Asian thinking were transmitted to the West through Israel, and elsewhere[d] we took an opportunity of discussing in some detail the means by which such transmissions may well have been effected. We also met with another invention possibly transmitted by Hebrew merchants and scholars in the survey instrument known as Jacob's staff.[e]

If one should ask as to the stimulus which gave rise to the invention of the gimbals, the thought inevitably occurs that it was a derivative from the construction of armillary spheres for the astronomers, since their rings also had to be pivoted one within the other. At an earlier stage we caught a glimpse of schools of Chinese artisans making armillary spheres in the +1st century,[f] and I should be ready enough to believe that it was one of these groups to which Ting Huan belonged.[g] At any rate 'Cardan's' suspension is as much Cardan's as 'Pascal's' triangle is Pascal's.

(5) THE LOCKSMITH'S ART

Together with the millwright, the locksmith was certainly one of those medieval artisans who provided some of the skill required at the beginning of the palaeotechnic age of iron and coal. It would therefore be inexcusable, in the present context, if we omitted any reference to the makers of locks and keys. Unfortunately, there have not been even the barest beginnings so far of a history of the locksmith's art in Asia, and we shall therefore have to content ourselves with a very simple sketch, rather to draw attention to the possibilities of the subject for the comparative history of technology than to answer any of the questions which arise.

---

[a] Up to about 25°. For similar devices giving drives up to right angles see Jones & Horton (1), vol. 1, pp. 410ff.; vol. 3, pp. 260ff.

[b] And not only because of Ting Huan, for on pp. 89, 103, 115 above we considered the curious direct treadle drives of medieval Chinese spinning-wheels.

[c] Vol. 2, p. 297.   [d] Vol. 3, pp. 681ff.   [e] Vol. 3, p 575.

[f] Huan Than's account of the old mechanic; Vol. 3, p. 358 above.

[g] The *Hsi Ching Tsa Chi* says that he lived at the capital, which is just where the technical assistants and workers who made instruments for the Bureau of Astronomy and Calendar would be expected to be.

PLATE CLXXXII

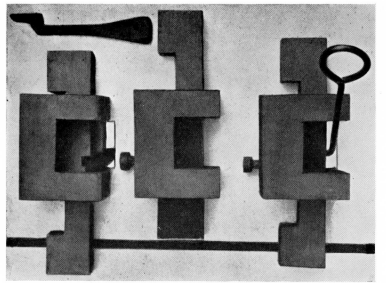

Fig. 484. The evolution of the locksmith's art, continued. The crank-shaped key (*clavicula*) for the Homeric lock; and the simplest form of tumbler (a small pin of wood or metal which fell by its own weight into a mortice in the bolt and had to be lifted up before the bolt would slide back), used in the Laconian lock.

Fig. 483. The evolution of the locksmith's art; models from Frémont (17). The bolt, the limited-excursion bolt, the triple-stapled bolt, and the simplest key-operated bolt.

PLATE CLXXXIII

Fig. 485. Roman keys (British Museum; Smith, 1). (*a*) Anchor-key or T-key, the flukes of which were inserted within the door into mortices in the tumblers to lift them up and allow the withdrawal of the bolt; cf. Fig. 489. (*b*) Plate-key, for raising a latch protected by wards, i.e. labyrinthine obstructions designed to prevent motion by any pattern other than the female tally. Such keys are still in use in Cambridge colleges for 'oaks' or outer doors of sets of rooms. (*c*) Right-angle key with *balanoi* of triangular cross-section, similar to that required for the lock in Fig. 486. (*d*) Rotary key adapted to wards, probably for use with a small padlock and attached to a finger-ring. (*e*) A combination of types (*b*) and (*c*); a plate component to satisfy wards, and four projecting *balanoi* to raise or depress tumblers. Perhaps the rotary principle was used here; if not, and as in a number of other forms, the keyhole had to be of complex form, at least L-shaped.

Although this branch of technique elaborated itself in later times into a thousand fanciful complexities, to say nothing of the ingenuity devoted to modern time-locks, strong-rooms, and the like, the essentials are simple enough. They may be traced in the memoir of Frémont (17)[a] and the accounts of ancient European locks and keys by Diels (1) and A. H. Smith, while Hommel has brought forward valuable observations on the traditional fastenings still in use in China. On distribution there is a well-known monograph by Pitt-Rivers (1). In Fig. 483 one can see that the earliest lock was no more than a bolt, which for convenience was kept attached to the door and made to slide through a block mounted thereon. The first amplification was the addition of a stop and two staples to prevent the bolt from coming out; next an additional staple or lock was added to the wall. Then in order to be able to open the door from outside, a hole was contrived through which the hand could enter, and further refinement diminished this to admit only a mechanical implement, the key. All these types go far back into Egyptian antiquity,[b] and have of course lived on in rustic environments, as archaic genera, in China also, until today.[c]

Diels called[d] this last type the 'Homeric' lock, on account of its mention in the *Odyssey*.[e] The bolt was drawn to, on leaving, by a cord which passed through the door or door-frame and ended in a loop outside. Diels drew attention to a remarkable fact when he pointed out that the kind of key used for opening such a lock was clavicle-shaped (hence its name, *clavicula*), and essentially in the form of a crank (Fig. 484). When inserted through a hole and appropriately turned, it was pressed against a projection or into an indentation on the top or bottom of the bolt, and moved it back, thus opening the door. The use of a crank-shaped object for such purposes and not for assisting rotary motion merits thought,[f] and almost challenges comparison with the wheels on Aztec toys, i.e. in a wheel-less culture.[g]

The first great invention connected with locks was that of tumblers, i.e. small movable pieces of wood or metal which fell by their own weight so as to engage with mortices in the bolt, and which were then raised by suitable projections on the key (Fig. 484). In the beginning they may well have been simple pegs or pins hanging on cords and inserted to prevent accidental opening of the door. Diels called this the 'Laconian' lock, and thought that it was very probably introduced from Egypt to Greece by

[a] A work evidently undertaken with peculiar zest, since Frémont himself came of a family of Parisian locksmiths. Cf. for orientation, the works of Butter (1); Eras (1) and A. A. Hopkins.

[b] Some particularly splendid stone bolts and staples excavated at the ancient Persian temple of Tchoga-Zanbil near Susa, dating from the −13th century, have been described by Ghirshman (3). One of the staples has a side-tube of stone to carry a locking-pin which could be inserted into a cavity in the bolt. The staples are thought to have been attached to the stone or wooden gates with clamps of bronze.

[c] A very ancient type of fastening, which found a home in the highest heavens of literary culture, was the thong and pin system which became universal on those cases (*han*[1]) in which several volumes of a Chinese book are collected.

[d] (1), pp. 40ff.   [e] xxi, 46ff.

[f] See especially pp. 111ff. above.

[g] Cf. p. 74 above. Another example of a crank-shaped object which apparently did not give rise to the crank is that of the bronze Z-stanchions found in Chou tombs at Loyang and figured by White (1), pl. xxvi.

[1] 函

Theodorus of Samos in the − 6th century, as Pliny implies.[a] The projections on the key, which raised the tumblers, were called 'acorns' (balanoi, βάλανοι). There were many ways in which this could be done; the key could enter underneath the bolt, with its projections passing up through the appropriate holes, the key being free to slide with the bolt when it had displaced the tumblers; or the key could actually fit into a hole bored longitudinally in the bolt itself; or the key could enter above the bolt by a separate opening in the lock-case, the tumblers being [-shaped so as to receive it. The first of these types is seen in a specimen of a Roman lock (perhaps + 1st century) now in the British Museum (Fig. 486),[b] and the third is still common in China (Fig. 487). Locks of these types are indeed frequent in many parts of the world, such as Cyprus and Algeria.[c]

One way of raising the tumblers from outside the door was by means of an anchor-shaped key (sometimes called a T-key); this was inserted through a vertical slit, given a quarter turn, and then pulled so as to engage with holes in the tumblers; a leverage action would then raise them. Fig. 485 a shows such an anchor-key from a Roman site,[d] and Fig. 489 shows exactly the same simple instrument in use upon a contemporary traditional Chinese door in Chiangsi. It must be very ancient in China, for two examples of Chou date are shown in White's *Tombs of Old Loyang*.[e] This simple pattern may have given a stimulus for the much greater elaboration seen in Roman plate-keys (Fig. 485 b) which involved 'wards',[f] but these were apparently not used in China. A variation in which the tumblers were made double (the lower portions remaining in the moving bolt), and in various lengths so that only a key with projections exactly of the right lengths would raise all the tumblers to the level of uniform clearance, was known as the 'Jewish' lock[g] and this again has not been described from China.

The rotary principle was developed in late Roman times, and many specimens of turning keys are known from Pompeii and Gaul. Angular displacement now took the place of rectilinear. From the Gallo-Roman latch-raisers which Frémont reproduces,[h]

---

[a] *Hist. Nat.* VII, lvi, 198.

[b] Roman keys are generally right-angled (see Fig. 485 c); this type does not appear to go back beyond the − 2nd century.

[c] According to Pitt-Rivers the first and second of these types are today characteristic of Egypt, the Middle East, Turkey, Persia, etc., while the third centres on Scandinavia and Northern Europe. He did not consider Chinese evidence in this connection.

[d] Cf. Pitt-Rivers, pl. IV, showing specimens from all over Western Europe.

[e] (1), pl. XIII. This gives some colour to the traditional attribution of locksmith inventions to the famous − 5th-century artisan Kungshu Phan, who is said to have replaced older fish-shaped keys, which were unreliable, with keys like 'beetles' (li[1]), which depended on a very perfect fitting of parts, and made all watchmen unnecessary (cf. Liu Ju-Lin, in Li Nien (27), p. 4). This sounds like the invention of tumblers.

[f] This device may be defined as a system of lock in which obstructions (wards) are placed in such a way as to prevent any but the proper key from entering to move latch, bolt or tumblers. The tumbler system, which is a bolt of a bolt, prevents any but the proper key from freeing the bolt. These flat plate-keys, which are maintained in a horizontal position rising and falling with the latch, are still in use in Cambridge colleges, and the door of my own rooms at Caius opened with such a Roman key until 1958 when a Yale lock took its place. One would like to know how far the tradition went back; was it continuous or was it owing to some 18th-century antiquarian Bursar?

[g] Frémont (17), p. 25, fig. 30.

[h] (17), p. 34; also Pitt-Rivers, p. 11 and pls. III and IV.

[1] 蠡

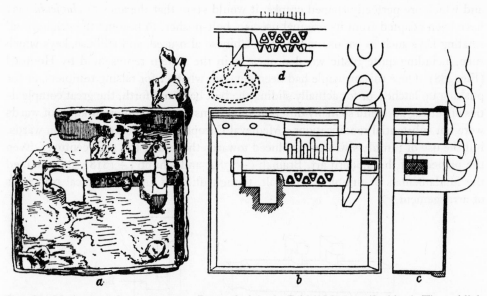

Fig. 486. Mechanism of a +1st-century Roman lock in the British Museum (Smith, 1). The end link of a chain (*c*) is secured by a bolt (*b*) in which there are half a dozen triangular perforations. Appropriate tumblers are held down in these by the action of a spring above, and the right-angled key (*d*) which pushes them up with its projections (*balanoi*) is free to slide with the bolt as it withdraws.

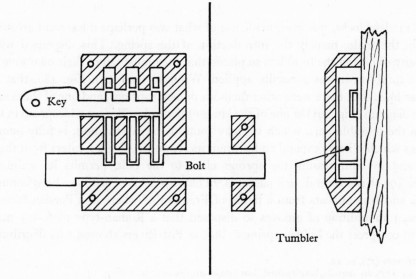

Fig. 487. Mechanism of a type of lock common to Chinese and Roman practice (after Hommel, 1). The key enters above the bolt and parallel with it, lifting the tumblers by mortices made in their sides.

and which are perfectly shaped cranks, it would seem that the ancient *clavicula* may have been adapted from its original use as a bolt-pusher, to become the origin of all rotating keys such as we use today. The Chinese also used, and still use, keys which turn, including quite rustic wooden ones, as in the system represented by Hommel (Fig. 488). Once this principle had been devised, whether for raising tumblers, or for pushing up latches, or for actually sliding the bolt back and forth, the great complexities of cutting keys into so many different shapes answering to different kinds of wards were not substantial modifications. Medieval European locks relied mainly on wards, but the use of tumblers was reintroduced towards the end of the 18th century. Even the pioneers of the modern art, Bramah (+1784) and Chubb (+1818), still retained the fundamentals of tumblers and bolt, though utilising new mechanical principles of arrangement.[a]

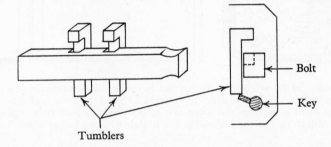

Fig. 488. Mechanism of a common type of Chinese door-lock used for sheds, stores, workshops, etc. (Hommel, 1). The key, which is inserted parallel to the bolt, has two pegs which in their upward path raise L-shaped tumblers notched into the bolt, thereby freeing it for withdrawal. One of the simplest applications of the rotary principle.

All modern locks, however, made use of what was perhaps the second great invention in the trade, namely the introduction of the spring. This dispensed with the necessity of having the tumblers so placed that they must fall of their own weight, for to the tumblers it was generally applied. We have seen it in Fig. 486, that of the Roman lock. But there were other methods of using springs, and one of them appears in the diagram (Fig. 491) of one of the kinds of padlock still most widespread in China. When the movable part, which we may continue to call the bolt, is fully home, the springs which it bears expand and prevent its removal. The key enters from the other end, and by compressing the springs close to the bolt, permits its withdrawal.[b] Figure 490 shows several such padlocks.[c] That this device must have been common to China and Iran appears from a figure of Frémont's,[d] where in a Persian form there has been an addition of grooves so disposed that a Roman-type plate-key must be used to compress the hidden springs. But, as Pitt-Rivers showed,[e] its distribution is

[a] Frémont (17), p. 42.
[b] Such keys are usually longitudinal, but late examples may be rotary.
[c] How far they go back in China is difficult to say. An iron example of the Liao dynasty datable at +959, from Chhih-fêng Ta-ying-tzu in Jehol, is figured in Thang Lan (1), pl. 105. Bronze padlocks of this kind from the Thang are not rare; I noticed a good one in the Sian Museum in 1958.
[d] (17), p. 33, figs. 71 to 73.   [e] (1), pp. 17ff., pls. V, VI, VII.

PLATE CLXXXIV

Fig. 489. Anchor- or T-key in contemporary Chinese use; a lock on a garden gate at Chien-chhang (Nan-chhêng), Chiangsi (Hommel, 1). Here the key is inserted through a vertical slit and turned through 90° to insert the flukes in the mortices; when the bolt is freed it is pushed back by a long pin entering through a second, horizontal, slit. Both key and pin just under 1 ft. long.

PLATE CLXXXV

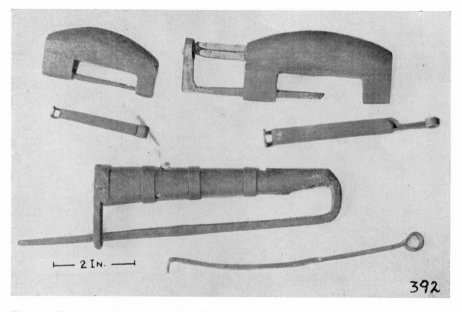

Fig. 490. Examples of the spring padlock shown on the opposite page, in wrought iron and brass, from Anhui and Chiangsi (Hommel, 1). This form was used in ancient and medieval times throughout the length and breadth of the Old World.

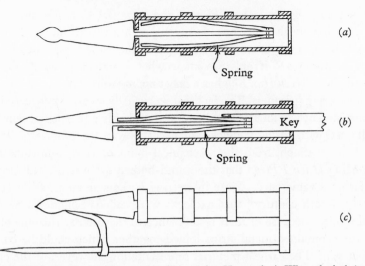

Fig. 491. A type of spring padlock widespread in China (after Hommel, 1). When the bolt is fully home the springs which it bears expand and lock it in position; they are compressed and the bolt freed by the insertion of a forked key at the opposite end of the padlock.

far wider, many specimens being known from Roman Britain,[a] from medieval Sweden and Russia,[b] and from Egypt, Ethiopia, India, Burma and Japan.[c] From his native Wiltshire he described a padlock identical with those which I used to buy in the Chungking markets. In these the bolt is held against the hasp by a helical spring, and the key has a female screw-thread fitting an invisible male screw on the bolt; release is brought about by a backward pull when the screws have fully engaged. Since the screw is essentially European, this must have been a transference from West to East. In any case, the very name of the padlock ('path'- or 'road'-lock, used by pedlars with pad-nags against foot-pads), which shows that it was connected particularly with transportable merchandise, suggests readily how the arts of the locksmith were transmitted along all the trade-routes of the Old World.

So far, so good. But the history of the progress of locksmiths' inventions in China remains extremely obscure, and everything is waiting to be done. One can only offer a few scattered observations, conscious of the first difficulty which needs clearing up, namely the exact significance of the words employed in early times.[d] The *Li Chi*, which

[a] See especially Wright (1); Neville (1); Cuming (1); Fox & Hope (1). I owe these references to Mr H. Courtney Archer, of the Chinese Industrial Cooperatives, who was greatly impressed by the similarity between Roman and Chinese padlocks when visiting the Museum at Cirencester.

[b] Kolchin (1), pp. 128 ff.

[c] Including, near home, a padlock of +1665 attached to the iron Treasurer's chest of the Royal Society. Many of these padlocks were made in the form of fishes (Pitt-Rivers illustrates some, pl. IX), presumably because a fish, which never closes its eyes, would be a good watchman. Indeed, a Thang (or pre-Thang) author, Ting Yung-Hui,[1] says just this, in his *Chih Thien Lu*[2] (cit. *SF*, ch. 74, p. 2b). Cf. *Yen Pei Tsa Chih*,[3] the Liao author of which speaks of 'fish locks'.

[d] The earlier, of course, the more obscure. The *Shu Ching* has a reference in ch. 26 (Chin Thêng), one of the genuine ancient chapters. It concerns the 'metal-bound coffer' in which oaths and oracles were deposited. Omens being auspicious, the Duke of Chou opened it with a *yo*.[4] Commentators

[1] 丁用晦   [2] 芝田錄   [3] 硯北雜志   [4] 籥

may give us Han usage, refers[a] to *chien pi*[1] and *kuan yo*.[2] Commentators seem to think that some spring mechanism was used in the former (presumably for depressing tumblers),[b] and that the latter (lit. tube-pipe) is the same thing as was meant by the word later written *yo*[3]—if only we knew what exactly that was.[c] The section on punishments in the *Hou Han Shu* has a *lang tang so*,[4] apparently a padlock attached to a chain for securing prisoners. The fact that *yo*[2] usually means a flute certainly suggests that the instrument was an elongated key, and probably a female one, such as some of those still used with spring padlocks; or perhaps even the tubular padlock itself. It does not help us much, from the mechanical point of view, unfortunately, to learn from the Ming *Hsien I Pien*[5] that the round-bodied *so*[6] was the old form, while the rectangular ones came later. More interesting is the statement of Tu Kuang-Thing in the early +10th century[d] that padlocks were called 'Solomon's Seal locks' (*wei jui so*[7]), doubtless because of their resemblance to the tubular rhizome of this plant,[e] and that they contained metal strips joined together which could be compressed or extended at will. The earliest pictorial representation, however, of the characteristic Chinese spring padlock occurs in one of the drawings in the *Nung Shu*[8] of +1313, where it is hanging on a granary door.[f]

That some keys fitted over projections, and others into holes, is strongly suggested by one of the biographies in the *Nan Shih*,[g] where a 'key and bolt' (*yo mou*[9]) is mentioned, and the use of a word which primarily means 'male' is interesting in view of the current use of the analogy among locksmiths, electricians and engineers. In +493 we get a glimpse of the Prince of Yü-Lin[10] (Hsiao Chao-Yeh[11]) picking a lock of a city gate with a hooked key (*yo kou*[12]).[h] Similarly, the *I Yuan*[13] says that in the +4th century metal keys (*chin so*[14]) were as long as 2 ft. While it is hard to deduce much from the Thang term for the lock and key (*so hsü*[15]), the *Pei Shih* speaks of 'goose

---

generally interpreted this as a 'pipe with which a treasury is opened', so that Legge (1), p. 154, has 'key' just as Medhurst (1), p. 213, before him, had 'lock'. Barde (3) follows this tradition, but Karlgren (12), pp. 34, 35, departed from it and regarded the *yo* as the bamboo tubes within which the writing-slips were kept. Perhaps here he went astray; in any case we cannot form a very clear conception of this early Chou instrument. Another word, *chüeh*,[16] commonly meaning a buckle, seems to be used for a lock in *Chuang Tzu*, ch. 10, so that Legge (5), vol. 1, p. 281, translated it as 'clasp'. Implying a ring of some sort, it points to the classical Chinese padlock.

[a] Ch. 6, p. 84*a*; Legge (7), vol. 1, p. 299. Parallel passage in *LSCC*, ch. 46 (vol. 1, p. 91); Wilhelm (3), p. 118.

[b] Cf. *Tao Tê Ching*, ch. 27, 'The perfect door has neither bolt nor bar' (Waley (4), p. 177), lit. 'a good door has no shutting bolt yet it cannot be opened (*shan pi wu kuan chien erh pu kho khai*[17])'.

[c] Sirén (1), pls. 91, 92, gives photographs of what he supposes to be keys and locks from the late Chou period, but their construction and use is not at all obvious.

[d] In his *Lu I Chi*.[18]   [e] *Polygonatum officinale*, R688.

[f] Ch. 16, p. 15*b*.

[g] That of Tai Fa-Hsing[19] (+5th century) in ch. 77, p. 4*a*.

[h] *Tzu Chih Thung Chien*, 11th year of the Yung-Ming r.p., ch. 138 (p. 4336). His object was to hold uproarious parties in Hsichow, contrary to the sober course of life prescribed by his father the Crown Prince.

[1] 鍵閉	[2] 管籥	[3] 鑰	[4] 鋃鐺鎖	[5] 賢奕編
[6] 鑮	[7] 萎蕤鑮	[8] 農書	[9] 鑰牡	[10] 鬱林王
[11] 蕭昭業	[12] 鑰鉤	[13] 異苑	[14] 金鑮	[15] 璅須
[16] 繘	[17] 善閉無關鍵而不可開		[18] 錄異記	[19] 戴法興

(neck) keys' (*ku yo*[1]).[a] One wonders whether some of these were not closely related to the *clavicula* and the Gallo-Roman latch-raiser.

Pitt-Rivers was inclined to think[b] that the developed tumbler lock and the spring padlock were Asian inventions which reached Europe about the − 1st century, but the evidence for this is still insufficient. It appears, however, that in the + 9th the Arabs had a high opinion of the products of foreign locksmiths. In his *Examination of Commerce* 'Amr ibn Baḥr al-Jāḥiẓ of Basra (*d.* + 869)[c] listed some of the imports of Iraq as follows:

From China come perfumes, woven silks, plates and dishes (probably porcelain), paper, ink, peacocks, good spirited horses, saddles, felt, cinnamon, and unadulterated 'Greek' rhubarb.[d]

From the Byzantine domains come vessels of gold and silver, *qaisarānī* coins of pure gold, brocaded stuffs, spirited horses, slave-girls, rare utensils in red copper, inviolable locks, and lyres. (Besides these, Byzantium sends) hydraulic engineers, agricultural experts, marble-workers and eunuchs.[e]

This reference is of considerable interest because independent researches in economic history have indicated that it was just during the Thang period in China[f] also that safes or strong-boxes sufficiently unbreakable began to facilitate the development of banking-houses (*kuei fang*[2]). As we have already seen,[g] relations between China and Byzantium (Fu-Lin) were then particularly close.

In any case, it will be clear from what has been said that there was a rather singular community of pattern between locks and keys throughout the Old World. Whether any technical ideas travelled between Asia and Europe, and if so in what direction, remains at present an attractive and important problem for further investigation. Perhaps the fundamental types developed rather early, in Mesopotamian and Egyptian civilisation, and then spread outwards in all directions, to remain essentially unmodified until modern times.

(e) VEHICLES FOR LAND TRANSPORT

Of the great invention of the wheel, which revolutionised all transportation on land, something has already been said.[h] It is obvious, too, that one of the earliest applications of engineering principles was to vehicles, from Sumerian and ancient Egyptian times

[a] *KCCY*, ch. 49, p. 9*a*. They were used for the gates of the inner apartments of the imperial palace.
[b] (1), p. 26.   [c] Cf. al-Jalil (1), p. 111.
[d] Cf. Vol. 1, p. 183.

[e] Tr. Sauvaget (3), p. 11, eng. auct. Unfortunately, Sauvaget omitted a whole line from his translation, so that all the imports mentioned in the passage were attributed to China. This oversight was rectified in the retranslation of Pellat (1), which we have followed. One need not assume that all the Chinese goods necessarily came from China proper, for the Indies and China (Further India) were not always carefully distinguished. The peacocks may well have been Indian and the horse-gear Mongolian. We are grateful to Dr Otto van der Sprenkel for calling our attention in the first instance to this interesting passage.

[f] Kato (1), vol. 1, no. 21; the point suggested by Prof. E. G. Pulleyblank.
[g] Vol. 1, p. 205.   [h] Pp. 73ff. above.

[1] 鵲鑰   [2] 櫃坊

onwards. The story of their diffusion and their developmental modifications, if fully told, would be an epic one, but this is not the place to embark upon it, for it would involve us in the archaeology of times anterior to the earliest Chinese civilisation, and would raise questions of ethnological distribution in those parts of the Old World

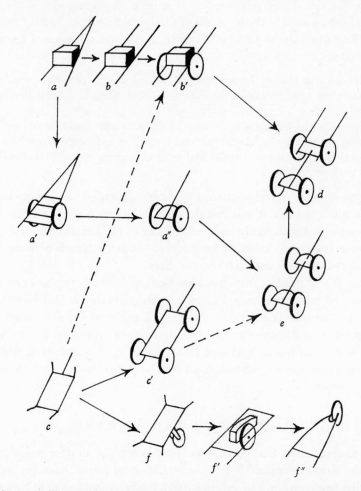

Fig. 492. Diagram to illustrate the evolution of wheeled vehicles. *a*, travois or triangular sled; *a'*, triangular cart; *a''*, pole-chariot; *b*, slide-car or rectangular sled; *b'*, shaft-chariot or true cart; *c*, rectangular draw-sled; *c'*, four-wheeled wagon derived from *c*; *d*, articulated four-wheel wagon (*b'*+*a''*); *e*, articulated four-wheeled wagon (*a''*+*a''*); *f*, four-handled hod or stretcher mounted on a single central wheel (ceremonial litter, cf. p. 271 below); *f'*, wheelbarrow with central wheel; *f''*, wheelbarrow with forward wheel.

furthest removed from China. Thus Fig. 377, taken from Bishop (2), shows the known distribution of the war-chariot in antiquity (towards the end of the -2nd millennium). It may now be considered established that the wheeled car, like the potter's wheel, was invented in Sumeria in the Uruk period (*c.* -3500 to -3200).[a] On the general

[a] Childe (10, 11, 16).

question of the evolution of vehicles it must suffice to refer to classical studies.[a] On ancient and medieval Chinese vehicles and their construction there is also a substantial literature.[b]

According to Haudricourt, a fundamental distinction may be made between vehicles with a central pole to which animals were yoked at the forward end, and vehicles with shafts (prolongations of the side-pieces of the frame) which permitted of more efficient methods of harnessing.[c] There can be no doubt that both types of vehicle derived from sleds or sledges, and that the distinction between single poles and double shafts is more ancient and more important than any which have regard only to the number of wheels. Here it is assumed that the most ancient device was the triangular sled or travois (Fig. 492a), still used in some parts of the world, and that that gave rise to the rectangular sled or slide-car (Fig. 492b);[d] already at this point appears the distinction between the pole and the shafts. Along one line of evolution the triangular sled gave rise to the triangular cart (Fig. 492a'),[e] from which it was an easy transition to the pole-chariot (Fig. 492a''), so familiar from a thousand representations belonging to the antiquity of our own civilisation, and seen indeed on the earliest Mesopotamian saddle-chariots of the −4th millennium.[f] This kind of vehicle probably has also some kind of connection with the plough-beam or pole of the plough. The coupling of two of these chariots together would produce a four-wheeled wagon (Fig. 492e), but this may also have originated from the mounting on wheels of another kind of rectangular sled[g] (Fig. 492c, c'). Meanwhile the slide-car or rectangular sled with shafts had been mounted on wheels to form the shaft-chariot or true cart (Fig. 492b'). This did not make its appearance in Europe until the end of the Roman Empire, i.e. some half a millennium later than in China.[h] When coupled with the pole-chariot, it gave rise in its turn to an articulated four-wheeled wagon (Fig. 492d). The coupling of type e is seen on Bronze-Age rock-carvings in

[a] E.g. Haddon (2), the books of Capot-Rey and Baudry de Saunier, and the brilliant papers of Haudricourt (3); Lane (1); Fox (1) and Childe (10). One of the best recent surveys is that of Haudricourt & Delamarre (1), pp. 155ff. The evolution of the vehicle can also be studied in the rich collection of illustrations assembled by Lefebvre des Noëttes (1). The large work of Ginzrot, though written as long ago as 1817, is still worth consulting. Before the beginning of the −2nd millennium, all wheels were solid (a type which lives on today, as we have seen at p. 80, from China to the Orkneys and Portugal), but spoked wheels spread rapidly after that time.

[b] A thorough study of the vehicles of the Shang and Chou periods is due to von Dewall (1). Representations of vehicles from Han sources have been assembled by Harada & Komai (1), Liu Chih-Yuan (1), Anon. (22), etc., while Hayashi (1, 2) has studied the textual evidence about them. Uchida (1) has written on the vehicles of the nomads of ancient Mongolia. A multitude of further references will find mention as we proceed.

[c] This technique, to which the Chinese made outstanding contributions, will be the subject of a later subsection (pp. 303ff. below).

[d] Still surviving in Ireland (Haddon, 2).

[e] Still surviving in Sardinia and in India.   [f] Des Noëttes (1), figs. 4–10.

[g] In this case the four-wheeled wagon produced would have been of a non-articulated kind. Rectangular sleds of just this type still exist in China (Hommel (1), p. 322). Uchida (1) publishes ancient drawings of Hunnish sleds.

[h] Cf. Fox (1) and the articles in Daremberg & Saglio on the *currus* (vol. 2, p. 1642), the *cisium* (vol. 2, p. 1201) and the *triga* (vol. 9, p. 465). The +3rd-century relief of a boy's chariot (*currus*) with shafts on a Trier sarcophagus (des Noëttes (1), fig. 73, Jope (1), p. 544) is sometimes regarded as the oldest European example. But the *cisium* on the Igel column near Trier, a cart with two mules in shafts driven by two naked youths, is also of the early +3rd century.

Sweden,[a] and that of type *d* is still found in North Italy,[b] while four-wheeled wagons of type *c'* can be noted in La Tène Age bronzes.[c]

The distribution of four-wheeled and two-wheeled vehicles throughout the Old World has been the subject of careful study by Deffontaines (1, 2) and Capot-Rey (1), from which it clearly appears that the former were associated with the steppe country of Northern and Central Asia and Europe, reaching as far west as Eastern France and Northern Italy,[d] and penetrating into Northern India and (formerly) Northern China.[e] Everywhere else[f] two-wheeled carts, derived from two-wheeled chariots, held sway. The conclusion can hardly be avoided that this distribution corresponds to the character of the country, two-wheeled vehicles, turning more easily, being associated with obstructions such as hilly roads, hedges and ditches (in Europe), and irrigation channels (in China).

(1) CHARIOTS IN ANCIENT CHINA

That the chariot which the Chinese originally accepted from the Fertile Crescent in the Shang period was of the pole-and-yoke type (Fig. 492*a''*) seems to be well established from the most ancient forms of the character *chhê*[1] (K 74), one of which I reproduce here. As archaeologists have clearly seen, this must imply the use of the inefficient throat-and-girth harness (cf. below, p. 305).[g] In the Shang and early Chou periods certain horse-trappings found are very similar in design to those of Europe somewhat later (in the Hallstatt period).[h] And the conclusion has been placed beyond all doubt by the discovery in 1950 of a whole park of Warring States chariots in a royal tomb at Liu-li-ko[2] near Hui-hsien,[3] some fifty miles south-west of Anyang (Fig. 379). These chariots were of the −4th or early −3rd century, and though the wood had decayed, its traces in the compacted soil clearly showed the presence of poles and not shafts.[i] Even as late as the Chhin (the latter half of the −3rd century) there are occasional evidences from bronzes[j] of two-

K 74 *c*

---

[a] Berg (1), pl. 26, fig. 2; cf. Childe (10), p. 190. The articulated Dejbjerg wagons of the −1st century already mentioned (p. 93) were also of this type; cf. the drawing in Jope (1), p. 538, and Dvořak (1). These articulated forms were of course the ancestors of the bogie or turning-train, which became prevalent in European wagons from the +15th century onwards (Jope (1), pp. 548, 550). Boyer (2) argues plausibly enough that the principle was never lost in medieval Europe, as some have thought, and Hall (4) finds it in Guido da Vigevano (+1335).

[b] Jaberg & Jud (1), vol. 6, map 1224, fig. 2.

[c] Haudricourt (3).

[d] Cf. the Roman *rheda* (Daremberg & Saglio, vol. 8, p. 862; Jope (1), p. 546). See also Lane (1).

[e] Cf. Moule (3, 8).

[f] I.e. in all the peripheral 'peninsular' and mountainous regions of the Old World.

[g] Chêng Tê-Khun (4); Gibson (4); Hayashi Minao (1), pp. 173 ff.

[h] See, for example, Janse (2).

[i] See Anon. (4), pp. 47 ff., pls. 24–30, 117; Hsia Nai (1). Another remarkable park of vehicles has recently been excavated from the princely tombs of the State of Kuo (cf. Vol. 1, p. 94, there spelt 'Kuai') near the San-mên Gorge of the Yellow River. These date from just before −655. All the twenty chariots had poles, cross-bars and 'yokes' (see Anon. 27). Decisive evidence for still earlier times is provided by the excavations of Ma Tê-Chih, Chou Yung-Chen & Chang Yün-Phêng (1) at Ta-ssu-khung Tshun near Anyang (cf. Watson & Willetts (1), no. 10), the date of this chariot (shown in Fig. 538 below) being soon after −1200.

[j] One of which has been published by Flavigny (1), p. 58 and pl. LVIII, fig. 458. It shows a *quadriga*.

¹ 車   ² 琉璃閣   ³ 輝縣

PLATE CLXXXVI

Fig. 493. A baggage-cart (*chan chhê*) from the I-nan tomb reliefs, *c.* +193
(Tsêng Chao-Yü *et al.* (*1*), pl. 50).

PLATE CLXXXVII

Fig. 494. Chou or Han *quadrigae* in a hunting scene on an inscribed bronze vessel in the
Freer Art Gallery (Harada & Komai, *1*).

wheeled chariots with poles, and the classical writings speak from time to time of *bigae* and *quadrigae*.[a] By the previous century, however, shafted vehicles were coming in.

From the beginning of the Han onwards, all representations (of which a very large number exist, cf. Fig. 493) are unanimous that the chariot had shafts (Fig. 492 *b'*) and was drawn almost always by one horse only, with an efficient harness. This has been further demonstrated by wooden models of Former Han (− 1st-century) date excavated from tombs at Chhangsha in 1950 (Fig. 384).[b] Similar models in pottery and bronze, all recovered from Han tombs, are to be seen in the Chinese museums.[c] Occasionally, moreover (contrary to an often-held opinion),[d] horses were marshalled in tandem to

[a] For example, one of the spurious chapters of the *Shu Ching* (ch. 8, Wu Tzu chih Ko) says that rotten reins are no good for driving six horses (Medhurst (1), p. 123). The word *ssu*[1] meant a team of four. The larger number of horses would not necessarily imply the pole-and-yoke system, for three shafts could project forward from the frame of the vehicle instead of two, thus permitting four horses to be hitched by traces. Such, at least, was des Noëttes' interpretation of a Han tomb-carving collected by Chavannes (des Noëttes (1), fig. 126). Support for this suggestion might be obtained from a remarkable Han bronze vessel in the collection of Jung Kêng, which Sowerby (2) has figured. Two canopied and lattice-sided chariots are drawn respectively by four and five horses abreast, through a landscape of stylised hills and dust-clouds (Fig. 495). In the former case only three lines connect with the vehicle, and

Fig. 495. Han *quadrigae* from an inscribed bronze vessel in Jung Kêng's collection
(after Sowerby, 2)

in the latter case four, but it is impossible to tell from the bronze ornamentation whether they are curved poles or leather traces. Other pictures of chariots with three, four and five horses in just this style, found on the backs of Later Han and Chin bronze mirrors, are figured by Bulling (8), pls. 71, 72, 73 and 74, but their harness is no clearer. Decision is almost equally difficult when we look at a *quadriga* in a hunting scene depicted on a bronze *hsien*[2] of Chou or Han date now in the Freer Art Gallery, and discussed by Umehara (2) and Harada & Komai (vol. 2, pl. VI, 2)—Fig. 494. Here it would seem that there is but one central pole and that the two outermost horses are running in traces. In view of the difficulty of distinguishing traces from reins, however, many more such ancient pictures will have to be analysed, and other evidence will have to come in, before we can feel confident regarding the harness of late Chou and early Han vehicles with more than one horse. We shall return to the subject in connection with harness (cf. pp. 325 ff. below). Meanwhile, see the discussion of Hayashi Minao (1), pp. 211 ff. The arrangement of reins and traces on Chou and Han *quadrigae* has been discussed by Legge (8), vol. 1, p. 192; des Noëttes (1), and Pelliot (46), pp. 265 ff. Later representations of *quadrigae* with central poles also occur, but generally in connection with foreign influence, as for example those of triumphal Buddhist cars in the Tunhuang frescoes. Among a number of these I may cite cave no. 148 of +776 (cf. Chhang Shu-Hung (1), pl. 21).

[b] Anon. (11), pp. 89, 139 ff., pls. 99–102; Hsia Nai (1). We have discussed the wheels of these chariots above (p. 77).

[c] A large pottery model two-wheeled cart from Chhêngtu with a semi-cylindrical awning, and wheels about 2 ft. 6 in. diameter, is in the Chungking Municipal Museum. A smaller bronze one is in the Kansu Provincial Museum at Lanchow, and another bronze, from Nanyang, is figured in *Wên Wu Tshan Khao Tzu Liao*, 1954 (no. 9), pl. 20. Cf. the small bronze model described by Bulling (10).

[d] E.g. Sowerby (3).

[1] 駟   [2] 銑

draw shafted four-wheeled wagons, as is shown on a Han brick published by Rudolph & Wên.[a] This would have been impossible without the efficient breast-strap harness.

For these reasons, Edouard Biot's reconstruction of the chariot of the *Khao Kung Chi* specifications as a *quadriga* with central pole and yokes can no longer be taken to represent the typical chariot of all Chinese antiquity.[b] Nevertheless, the *Chou Li*, for which he provided his illustration, was (as we know) an intentionally archaising book, if indeed the relevant parts of the 'Artificers' Record' contained in it do not themselves go back to the −4th century, i.e. to the time when the pole-and-yoke chariot was beginning to give place to the shaft chariot.[c] It may well be that the Chinese artisans did not make much change in the technical terms when they embarked upon their great invention towards the end of the Chou period, probably at the late Warring States time, of replacing the pole, yoke, and inefficient harness of horse chariots, by the shafts and the efficient harness.[d] A whole monograph could be devoted to a minute examination of the terms which have *chhê*[1] as their radical, drawing upon the *Chou Li* and other late Chou and early Han texts;[e] here it would take us too far even to sketch what could be done, and a few tentative comments must suffice.

At an earlier stage mention has been made of the different wheelwrights and wagon-builders detailed in the *Chou Li* (p. 16), and elsewhere something has been said about the technical terms for the parts of wheels (p. 75). There are so many dimensions given in the text of the *Khao Kung Chi* that quotation would be tedious, but it may be of interest to glance at a sketch of the typical Chhin and Han vehicle based both on textual and archaeological evidence (Fig. 496); it is noteworthy for the great double

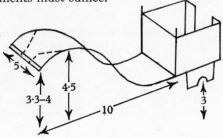

Fig. 496. Sketch of the type of horse-drawn chariot or light cart characteristic of the late Warring States, Chhin and Han periods, based on the descriptions and specifications in the *Chou Li (Khao Kung Chi)*, and on the archaeological evidence. Cf. Figs. 540 ff. below.

[a] (1), p. 33, pl. 84. It closely parallels late medieval European examples, such as the +14th-century drawing in the Luttrell Psalter (Jope (1), p. 548). Another ancient Chinese four-wheeled wagon appears on a Han unglazed pottery urn with reliefs figured by Sirén (1), vol. 2, pl. 101. But they cannot be said to be common.

[b] (1), vol. 2, opp. p. 488. For ancient Indian parallels in chariot and cart construction see Hopkins (1), p. 235, and Gode (3). The latter made use of a curious and little-known translation of parts of the *Chou Li*, now more than a century old—Gingell (1), contemporary with Biot's but seemingly quite independent. In India the pole system predominated, though shafts are found on a copper model from Harappa (Vats (1), p. 105 and pl. XXIII d), cf. Gode (4). Mohenjo-daro models all have poles (Marshall (2), vol. 2, pp. 554ff., vol. 3, pl. CLIV, figs. 7, 10).

[c] The classical archaeological discussions of ancient Chinese chariot- and cart-building carried on in the 18th and 19th centuries all centred on this *Khao Kung Chi*. We shall refer to them in a moment. The most recent study of what it says on the subject is that of Hayashi Minao (2). Cf. Yoshida Mitsukuni (3).

[d] The *Chou Li* text speaks of the curved horse-chariot pole as *chou*[2] and of the straight ox-cart shafts as *yuan*,[3] systematically. But the Han commentary on the opening sentence of the relevant passage says: 'the pole, i.e. the shafts, of the cart' (*chou, chhê yuan yeh*[4]), ch. 11, p. 16 a.

[e] We are glad to learn that Dr Lionello Lanciotti is now working at Rome on a definitive study of ancient Chinese vehicles. An elaborate paper on Shang and Chou chariots by Hayashi Minao (1) appeared just as this section was going to press.

[1] 車   [2] 輈   [3] 轅   [4] 輈車轅也

curve of the shafts. This was the continuation of a tendency already visible in Sumerian and other ancient Mesopotamian pole types,[a] manifesting the connection which there had certainly been. Frequently the shafts narrowed at the top to form a small yoke. The general specification[b] (which was modified for special purposes, such as war-chariots) provided that the axle (*chu*[1], *kho*[2]), of a definite length or gauge (*kuei*[3,4]) should be 3 ft. above the ground. The wheel, whether with spokes (*fu*[5]) fitting into a rim with felloes (*jou*[6], *wang*[7,8]), or solid (*chhuan*[9]),[c] was thus to be some 6 ft. in diameter.[d] At the meeting of wheel and axle there was a hollow outer bearing of bronze or iron (*kung*[10,11], *ti*[12], *kuan*[13]),[e] around which the hub (*chhi*[14], *chih*[15]) projected, strengthened with animal sinews glued together and tightly covered over with leather (*thao*[16]). There could be a metal hub-cap or nave-band (*kuan*[17]), but more usual and more important was the axle-cap (*wei*[18]) which held the wheels in correct position on the axles by means of a linch-pin (*ling*[19], *hsia*[20]) like our cotter-pins.[f] The axle-beam supported the lower frame (*chhun*[21], *fu*[22,23]), which carried the body of the vehicle (*chen*[24]), with its side-pieces (*chê*[25]), vertical members (*chio*[26]), and hand-bars (*shih*[27], *hung*[28]).[g] The pole (*chou*[29]) or shafts (*yuan*[30]) described a curve 'like water pouring out of a vase'. At the forward end the transverse bar (*hêng*[31]) was first attached medially by a connection (*i*[32], *yüeh*[33]) to the pole and carried two 'yokes' (*ê* or *o*[34], *chhiu*[35]), but later came (when the change occurred) to unite the two shafts fixedly together.

The text of the *Khao Kung Chi* makes clear that the standard vehicle specifications were closely connected with those for standard weapons such as spears and lances. It says also that if the pole is not sufficiently curved it will weigh upon the horse, while

[a] Cf. des Noëttes (1), figs. 13, 16; Childe (11), fig. 125 *a*.

[b] Ch. 11, pp. 5 *a*–16 *a* (ch. 40, Biot (1), vol. 2, pp. 463 ff.).

[c] Modern traditional examples may be seen in Hommel (1), p. 323; we have already had occasion to discuss them (p. 80 above).

[d] The Chou foot was about 8 in.; the height of an average man is stated in the text to be some 8 ft.

[e] This generally took the form of an elongated slightly tapering cylinder, but the Lanchow Museum possess an iron object, believed to have been an axle bearing (*chhê chhuan*[36]), which looks like nothing so much as a modern nut, since it is hexagonal externally and circular (though not threaded) internally. Possibly this belonged to a hand-cart or wheelbarrow. Sometimes very elongated hub-bearings appear, as in the four-wheeled wagons found in tumulus no. 5 at Pazirik in the Altai, dated by Rudenko (1) to the −5th century, and possibly of Chinese origin (cf. Frumkin (1); Barnett & Watson). A two-wheeled triumphal car in a Buddhist fresco of about +520 at Mai-chi Shan seems to show something similar; cf. Willetts (2); Wu Tso-Jen (1). So also do many of the *Chin Shih So* reproductions of the +2nd-century Wu Liang reliefs. This device would have minimised wheel unsteadiness. The *Chou Li* always assumes bronze as the metal of axle-bearings, but we know from the *Chi Chiu Phien* dictionary of −40 that iron bearings were then already long standard (ch. 3, p. 4 *a*).

[f] Highly ornamented bronze axle-caps from the late Chou and Han periods are quite common objects in museums. Illustrations may be found in Sirén (1), vol. 1, pl. 64; Harada & Komai (1), vol. 2, pl. xxvff.; Anon. (4), pp. 43, 132, pls. 22, 52, 92, 104; (11), p. 41; (17), pl. 23; (20), p. 106, pls. 47, 53, 68, 74. Fig. 497 is taken from Thang Lan (1), pl. 64. Examples from the early −7th century are given in Anon. (27), pl. 47. The story of Thien Tan (*Shih Chi*, ch. 82, tr. Kierman (1), p. 37) depends on iron axle-caps in −284.

[g] A great variety of bronze chariot-fittings are known, cf. Koop (1), pls. 25, 26, 27, 43 *b*.

1 軸	2 軻	3 軌	4 軏	5 輻	6 輮	7 輞
8 輛	9 輕	10 釭	11 釘	12 軑	13 錧	14 較
15 輒	16 幬	17 輨	18 轊	19 輪	20 轄	21 軨
22 輹	23 轐	24 軫	25 輢	26 輷	27 軾	28 軡
29 輈	30 轅	31 衡	32 輗	33 軏	34 軶	35 輈
36 車串						

if the curves are too sharp it will break. To understand the significance of this curving as it applied to shafts it is necessary to realise that the traces of the breast-strap harness (see below, p. 305) were attached to the centre point of the shafts between the two inflections, thereby exerting a direct and efficient traction on the vehicle. Traces from any additional horses were connected to it by special rings (*chhuan*[1]). The high curving shafts forward of the centre point were really unnecessary, but betray the origin of the arrangement from the earlier pole and yokes, that is to say, from a system in which it was essential that the wooden projection should reach as far forward as the necks of the animals. The ancient Chinese chariot lives on, with straightened shafts, in the common country cart of North China today (Fig. 498), though the semi-cylindrical awning or wagon-roof has long replaced the pavilion- or umbrella-shaped canopies of the fashionable vehicles of the Han.[a] But there are a few Han representations of military or goods carts (*chan chhê*[2]) extraordinarily like the typical vehicles of modern times (Fig. 499).[b]

Though they may have seemed numerous, we have mentioned here only a few of the technical terms which are found in the old texts on chariotry and cart-craft.[c] They have considerable interest and importance, for vehicle-building was one of the oldest occupations of artisans, and a knowledge of the terms which they coined is almost indispensable for the study of much later and more complicated engineering texts concerned with milling, textile or horological machinery.

As regards the gauge of Chinese chariots and carts, we had occasion to notice earlier (Vol. 2, p. 210) the standardisation introduced by Chhin Shih Huang Ti in the −3rd century. So lasting was this tradition in particular parts of the country that when von Richthofen was travelling in China twenty-two centuries later, he found that it was necessary to change the axle-trees of his carts at one point, since in Shensi, Shansi and all North-Western China the gauge was some 20 cm. broader than that of the eastern provinces, and the size of the ruts made a conversion essential. Stores were kept in readiness, however, and the adaptation was quickly effected.[d]

[a] The general history of wheeled transport in Chinese culture, when someone comes to write it, will form a deeply interesting chapter in economic development. In the fairly large literature in Chinese and in Western languages on Chinese economic history we seem to lack studies such as those of Willard (1) on the role of cart and wagon transport in medieval Europe.

[b] Anon. (*22*), pp. 40ff.; Rudolph & Wên (1), pl. 83.

[c] There are of course some difficulties in their interpretation, and authorities may disagree in certain matters. Some of the founders of modern Chinese archaeology devoted much study to the texts concerning vehicles, notably Tai Chen[3] in his *Khao Kung Chi Thu*[4] of +1746 (on which see Kondo, *1*); and Fang Pao[5] in his *Khao Kung Hsi I*[6] two years later. Chiang Yung[7] continued this critical work in his *Chou Li I I Chü Yao*[8] (ch. 7, pp. 10*b*ff.) of +1791. The most complete study perhaps was that of Juan Yuan[9] about 1820, entitled *Khao Kung Chi Chhê Chih Thu Chieh*[10] (Illustrated Analysis of Vehicle Construction in the *Artificers' Record*). Diagrams from this appear in Figs. 500 and 501. The dimensions given in the sketch of the chariot in Fig. 496 are only very approximate, but resemble most closely those given by Juan Yuan (*2*). His contemporary Chhêng Yao-Thien[11] had also worked on the same subject, as in his *Khao Kung Chhuang Wu Hsiao Chi*[12] of about 1805. A comparison of the ideas of these scholars on a particular subject (wheel-dish) will be found in Lu, Salaman & Needham (1).

[d] Von Richthofen (5), vol. 1, p. 546, noted by Horwitz (6), p. 182.

[1] 釧   [2] 棧車   [3] 戴震   [4] 考工記圖   [5] 方苞   [6] 考工析疑
[7] 江永   [8] 周禮疑義舉要   [9] 阮元
[10] 考工記車制圖解   [11] 程瑤田   [12] 考工創物小記

PLATE CLXXXVIII

Fig. 497. Ornamented bronze axle-caps of the Warring States period from a tomb in Shansi (Thang Lan (1), pl. 64).

PLATE CLXXXIX

Fig. 498. Country cart at Lanchow, Kansu (orig. photo., 1943).

Fig. 499. Baggage-cart (*chan chhê*) of the Han period; rubbing of a moulded brick from
Phêng-hsien, in the Chungking Municipal Museum (Rudolph & Wên, 1).

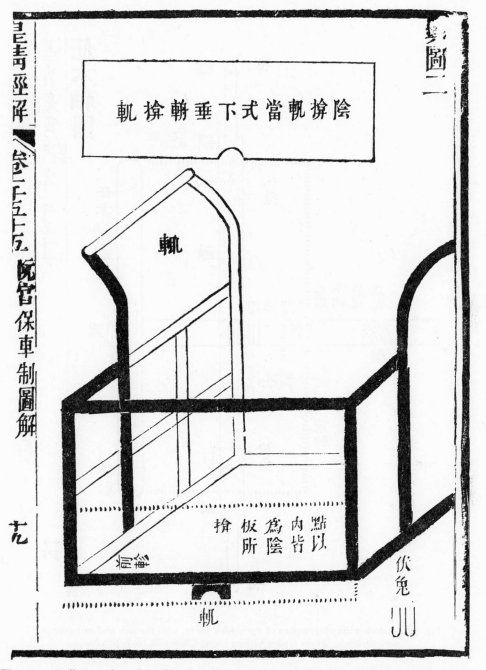

Fig. 500. A diagram from Juan Yuan's archaeological study of vehicle building according to the *Chou Li*, entitled *Khao Kung Chi Chhê Chih Thu Chieh* (1820). This drawing is an isometric projection (cf. Sect. 28*d*) of the body of a Chou or Han chariot. The legends say that the boarding outlined at the top of the picture fits in front of the body over the pole as shown by the dotted line in the main diagram. From *Huang Chhing Ching Chieh*, ch. 1055, p. 19*a*.

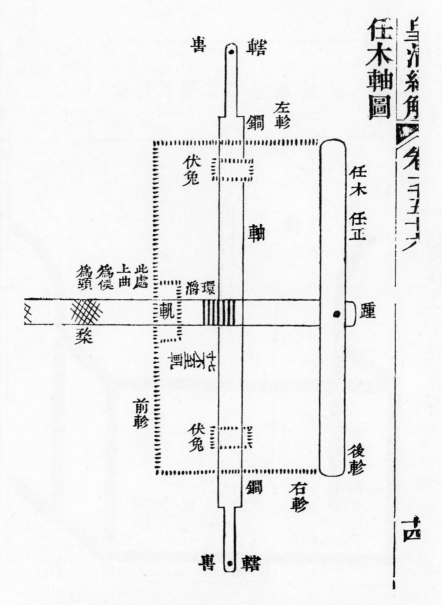

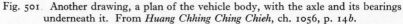

Fig. 501. Another drawing, a plan of the vehicle body, with the axle and its bearings underneath it. From *Huang Chhing Ching Chieh*, ch. 1056, p. 14*b*.

This conservatism is outdone only by the carts of the Indus Valley, which possess the same gauge today as their Bronze Age predecessors, to judge by the ruts at Harappa.[a]

(2) WAGONS, CAMP-MILLS AND HAND-CARTS

In what follows we shall take up in order the Chinese invention of the wheelbarrow and the land-sailing carriage, then certain specialised vehicles of much engineering interest involving gear-trains, the hodometer and the south-pointing carriage; ending finally with a disquisition on the development of the efficient animal harness, a process for which China and Central Asia undoubtedly deserve the thanks of the rest of the world. Before proceeding further, however, there are a few remarks to be made on certain later Chinese vehicles, some of large size,[b] on imperial carriages noted for their lack of vibration, and on camp-mills, i.e machines for grinding and pounding made mobile by being mounted on vehicles and driven by animal-power, either while moving or when brought to rest.

A very large vehicle was built for the emperor Yang of the Sui dynasty about +610. The *Hsü Shih Shuo* says:[c]

> Yüwên Khai[1] built for Sui Yang Ti a 'Mobile Wind-Facing Palace' (*kuan fêng hsing tien*[2]);[d] it carried guards upon its upper deck, and there was room for several hundred persons to circulate in it. Below there were wheels and axles, and when pushed along it moved quite easily as if by the help of spirits. Among those who saw it there was no one who was not amazed.

This was probably far from the first time that large vehicles for special purposes had been constructed. Towers on wheels for attacking city-walls (*chhung*[3], *phêng*[4,5,6]) go back to the treatises on fortification of Mo Tzu's followers (−4th century), as also do chariots armoured in various ways (*chhung*[7]).[e] Much later on, we get an echo of giant vehicles in one of the illustrations of the +17th-century *Thien Kung Khai Wu*, a four-wheeled dray with nine shafts drawn by eight horses, but the use of such vehicles must have been very restricted by the lack of good roads, if not wholly imaginary.[f] Representations of this kind in paintings of late date showing imperial processions are not uncommon,[g] but the impossibility of the technical details of their harness stamps them as fanciful. Nevertheless, judgment should be reserved on what exactly was constructed in late times in China. In Hadji Muḥammad's account of Cathay,

---

[a] Childe (10).   [b] On Han wagon-wheels, cf. pp. 80–1 above.

[c] Ch. 6, p. 11*b*; tr. auct.

[d] The meaning of this name is not obvious; its dictionary translation is 'wind-watching', but it seems very unlikely that the purpose was either meteorological or folkloristic, and I suspect that what we have here is another reference to the use of sails and the wind as motive power, since just this invention had been made in the time of Liang Yuan Ti half a century earlier (see on, p. 278). Cf. Vol. 4, pt. 1, p. 108.

[e] Cf. *Tso Chuan Pu Chu*[8] by Hui Tung[9] (+1718), ch. 6, p. 13*b*.

[f] Ch. 9, pp. 12*b*, 13*a*; cf. *TSCC, Khao kung tien*, ch. 174, pp. 19*a*, 25*b*.

[g] One of this kind, stated to be Yuan in date, is figured by des Noëttes (1), fig. 130, and I saw some earlier than that among the cave-paintings at Tunhuang; there is not much to be got out of them.

[1] 宇文愷   [2] 觀風行殿   [3] 轀   [4] 轒   [5] 轀   [6] 轀
[7] 衝   [8] 左傳補注   [9] 惠棟

delivered to Messer Giov. Battista Ramusio in +1550, there is a curious passage: 'They get their blocks of stone sometimes from a distance of two or three months' journey, conveying them on wagons that have some forty very high wheels with iron tyres; and these shall be drawn by five or six hundred horses or mules.'[a]

Someone should make a study of the monographs on the imperial fleet of vehicles contained in most of the dynastic histories from the *Hou Han Shu* onwards, and traditionally entitled Yü Fu Chih.[1] There is much to be learnt from them about the emperors' cars and carriages,[b] but generally their interest centres on the ornamentation and its symbolism. There was something more to the *yü lu*,[2] however, which attracted great attention in the time of the third Thang emperor, Kao Tsung (+650 to +683), and for long afterwards.[c] Shen Kua tells us about it in the *Mêng Chhi Pi Than* thus:[d]

> In the time of Thang Kao Tsung a large State carriage was made with jade ornaments, and used by him. Three times it was used to carry him to Thai Shan (for the sacred ceremonies on that venerable mountain), and numerous journeys did he also make in it to distant places, but still today (*c.* +1085) it is sound and firm and steady.[e] If a cup of water is placed in the carriage while it is moving, the water does not spill. In the Ching-Li reign-period (+1041 to +1048) all the skilful mechanicians available were called upon to construct another such carriage, but it proved to be too unsteady, so it was left unused. In the Yuan-Fêng reign-period (+1078 to +1085) another carriage was built, but in spite of some marvellous workmanship it fell to pieces on its trials before it was ever presented to the emperor. There is still only the Thang vehicle enduring with all its steadiness and freedom from vibration. But no one has been able to discover the methods by which it was made. Some people maintain that supernal spirits protect the carriage and bear it along; all I can say is that when one is walking behind it one hears certain vague noises coming from within.

This is surely interesting. Shen Kua was not the man to say that a cup of water would not spill in the carriage if in fact it did, and the story reminds one of the prophecy of George Stephenson so much later about the future of railway transportation. Without going so far as to suggest that the principle of the gimbals[f] was made use of (as Branca suggested early in the +17th century), it may have been that the Thang engineer whose name has not been preserved employed roller-bearings or leaf-springs, though he must have enclosed them in such a way that his successors could not ascertain the technique without taking the equipment to pieces. If it was roller-bearings, he long anticipated Leonardo (*c.* +1495), to whom they are attributable in the recent West;[g] and if leaf-springs were used (which in view of their ancient application in the crossbow would seem more probable) the anticipation of European technique

---

[a] Yule (2), vol. 1, p. 294. Cf. van Braam Houckgeest (1), Fr. ed., vol. 1, pp. 108, 222, Eng. ed., vol. 1, p. 143, vol. 2, p. 8.

[b] One notices, for instance, the extra security of double linch-pins.

[c] Cf. *Shih Lin Yen Yü*, ch. 3, p. 15*b*; *Thieh Wei Shan Tshung Than*, ch. 2, p. 2*b*; both Sung books. More or less imaginary illustrations of it occur in the *San Tshai Thu Hui* (Yung chhi sect., ch. 5) and other encyclopaedias (cf. Harada & Komai (1), vol. 2, pl. XVII, 2).

[d] Ch. 19, para. 19 (p. 6*b*); cf. Hu Tao-Ching (1), vol. 2, p. 642; tr. auct.

[e] This would resemble the use today of a State carriage built in +1550 and seems rather unlikely; perhaps what Shen Kua saw was a subsequently rebuilt copy earlier than the +11th century.

[f] This was, of course, old in China; cf. pp. 228ff. Cf. Fig. 475.

[g] Usher (1), 1st ed. p. 188, 2nd ed. p. 227. See also p. 94 above.

[1] 輿服志   [2] 玉輅

would be even more striking, for Feldhaus[a] was unable to adduce any certain example on vehicles earlier than Faustus Verantius' specification of +1595, and des Noëttes[b] could improve on this only by a picture of +1568. Possibly some kind of suspension was used, with chains or leather straps, as in the *chariots branlants* of the European +14th century.[c]

The learned Beckmann, in his history of inventions,[d] devoted a brief section to camp-mills. Mills mounted on wagons, which could follow armies like field ovens and field forges, must have been an obvious military requirement from early times. We have already seen (p. 187 above) that one of the theories of the origin of hand-querns attributes them to the needs of detachments of Roman legionaries. Extension

Fig. 502. Targone's field mill, redrawn by Beck (1), p. 310, from Zonca's engineering treatise of +1607. Two wagon-mounted mills are worked by a horse-whim and gearing while stationary in camp.

of the idea to wheeled transport was only to be expected towards the end of the medieval period when roads improved, and that is what the European records show. In +1607 Zonca drew a picture of a field mill which was worked by animal-power when the wheels of its carriage were pinned down in camp or near billets; he said that it had been invented in +1580 by a military engineer, Pompeo Targone, famous for his part in the siege of La Rochelle and the Huguenot wars.[e] This was often after-wards reproduced, as by Beyer; and more than one draught of a camp-mill geared to the wheels of the carriage, so that it ground in transit, was published during the first decades of the +17th century.[f] In due course the drawing of Targone's camp-mill (Fig. 502) found its way to China, for we see it in the *Chhi Chhi Thu Shuo* of +1627, along with the other machines offered by the Jesuit Johann Schreck (Fig. 503),[g] where the two mills are worked by gearing from a rotating bar and whippletree harnessed to a single horse.

Presumably neither Schreck nor Wang Chêng knew that just about 1300 years

[a] (1), col. 1261. [b] (1), fig. 162; redrawn in Jope (1), p. 549.
[c] On these see Boyer (1). [d] Vol. 2, p. 55.
[e] 1621 ed. p. 88 *a*. But the same invention was made just about the same time by the Mogul emperor Akbar, in India; cf. Blochmann (1), p. 275.
[f] A study of the origin of camp-mills has been made by Feldhaus (8), fig. 72, and (13). All modern mobile agricultural machinery driven from its road-wheels stems from them; cf. Fussell (1); Spencer & Passmore (1); Scott (1); and Dickinson & Titley (1), p. 132. [g] Ch. 3, pp. 33*b*, 34*a*.

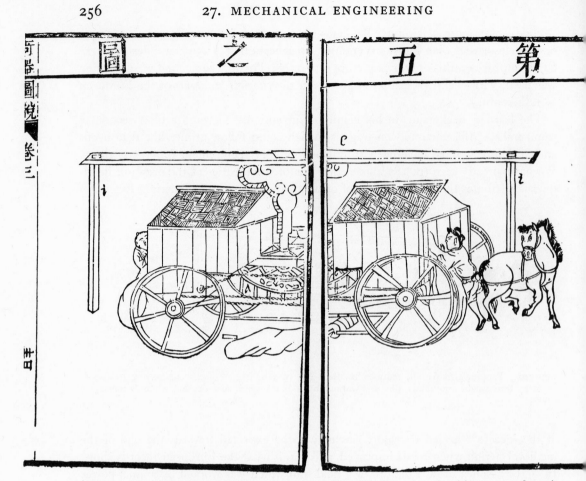

Fig. 503. The field mill as it appeared in the *Chhi Chhi Thu Shuo* of +1627 (ch. 3, pp. 33*b*, 34*a*). At the top: 'The Fifth Diagram' (of the Grinding-Mills Section). The artist added details not in Zonca's picture, suggesting chain or rope drives from the road-wheels (cf. Fig. 465); could he have known of the earlier Chinese automatic camp-mills?

earlier the history of camp-mills had started in China. In the *Yeh Chung Chi*, an account of affairs at the capital of the Hunnish Later Chao dynasty (*c.* +340), Lu Hui, after mentioning that the emperor, Shih Hu, had a south-pointing carriage and a hodometer, says:[a]

He also had pounding-carts (or wagons) (*chhung chhê*[1]), mounted on the body of which there were wooden figures pounding all the while with tilt-hammers as the carts moved. Every ten *li* traversed meant that one *hu* of rice was hulled (on each cart). Moreover, he had mill-carts (camp-mills) (*mo chhê*[2]) with rotating millstones mounted on them, and in these also a *hu* of wheat would be ground every ten *li*. All these vehicles were painted red with bright designs. Each was in the charge of one man, and as it moved along all the skill of the

---

[a] P. 8, but the texts of the quotations in *TPYL*, ch. 752, p. 3*a*; ch. 829, p. 10*a*, are in some places preferable. Tr. auct.

[1] 舂車　　[2] 磨車

construction was displayed. When it stopped the machinery stopped. These mobile mills were made by the Palace Officer Hsieh Fei[a] and the Director of the Imperial Workshops, Wei Mêng-Pien[a]....

Another version[b] has it that each vehicle had one or more hulling tilt-hammers worked off the right wheel or wheels, with rotary millstones worked off the left. We have not met with any later accounts of the use of these machines, but it is hard to believe that they died out completely, even if they had been forgotten by Wang Chêng's time. One should remember that the capital, Yeh,[c] was in northern Honan, and so in the North China plain, doubtless more suitable for the use of such almost nomadic devices than other centres such as Hangchow or Chhêngtu. Indeed, the invention has the air of a cross-fertilisation between Hunnish nomadic needs and Chinese sedentary engineering.

A few words may be added about small vehicles pushed or pulled by men. That wheeled toys existed in the Han we know from carvings on the walls of the tomb of Wang Tê-Yuan,[1] who died in +100; a child in a long-sleeved gown is propelling a two- or four-wheeled object at the end of a stick.[d] Even more remarkable is a scene in one of the Chhien-fo-tung frescoes dating from +851, where a woman is pushing a low cradle-shaped vehicle with four wheels.[e] It seems to contain a person lying down, and as it is not quite long enough for an adult it is probably a perambulator rather than a hearse or an ambulance. As for the shafted hand-cart, it doubtless originated as early as the shafted chariot itself, and the special term for it, *nien*,[2] depicts clearly in bronze forms a cart with pole and yokes pulled by two men.[f] But in its simplest form it must have been much older, for there is a *Shih Ching* reference,[g] and the *Tso Chuan*, describing events of −681, relates how a refugee lord, Nankung Wan,[3] pushed his mother in a hand-cart along the road to safety.[h] In the Han such vehicles seem to have been prominently associated with the internal transport of imperial palaces. Perhaps the reference most interesting to us concerns the Warring States iron-master, Cho shih[4] already mentioned (p. 26).

The Cho family were men of Chao who became rich by iron-smelting. When Chhin defeated Chao (−228 to −222) they were taken captive and deported to Shu (Szechuan). Both husband and wife walked, pushing a cart....[i]

[a] Met with before (pp. 38, 159), and to be met with again (p. 287).
[b] That quoted in the *Shuo Fu*, ch. 73.
[c] Modern Lin-chang,[5] near Anyang in northern Honan, north of the Yellow River.
[d] These reliefs are now preserved in the Sian Museum, where I had the pleasure of studying them in 1958. They had been discovered five years earlier at Sui-tê near Yenan. The style may be seen from Anon. (*21*), pp. 8 ff.
[e] This occurs on the balcony outside cave no. 156.
[f] See Jung Kêng (*3*), p. 726, or Sowerby (*3*).
[g] Mao no. 227, tr. Karlgren (*14*), p. 180. 'Barrow' is not good here (Waley (*1*), p. 130) as it invites confusion with the wheelbarrow proper—on which there hangs a tale.
[h] Duke Chuang, 12th year; tr. Couvreur (*1*), vol. 1, p. 156.
[i] *Shih Chi*, ch. 129, p. 17*a*; tr. auct. adjuv. Swann (*1*), p. 452. Mr Cho was one of the Chhin 'capitalists' to whom we shall return in Sect. 48.

[1] 王得元　　[2] 輦　　[3] 南宮萬　　[4] 卓氏　　[5] 臨漳

They were walking towards the restoration of their fortunes, for in Szechuan Mr Cho found an iron mountain and became much more wealthy than before. Representations of hand-carts on Chhin and Han objects are not very uncommon—one such is shown among the tomb-carvings near Chiating in Western Szechuan (c. +150), reproduced in the collection of Rudolph & Wên.[a]

All through Chinese history such small two-wheeled vehicles were in use. References to them occur in the *Tung Ching Mêng Hua Lu*, a description of life at the old Sung capital Khaifêng, written by Mêng Yuan-Lao about +1140.[b] The 'rickshaw' was their late descendant, but little used in China for some generations until reintroduced from Japan.[c] Several representations are to be seen on the Tunhuang frescoes.[d] Thus cave no. 431, of Northern Wei and Thang date (+5th to +8th centuries) has a very clear picture of a two-wheeled hand-cart pushed by one man. But in cave no. 148, the Thang paintings of which may be dated at +776, there is another man pushing a small cart which might be a single-wheeled wheelbarrow. This brings us to the next phase of our discussion.

(3) THE WHEELBARROW AND THE SAILING-CARRIAGE

Nothing could be more familiar as a vehicle of everyday life than the ordinary wheelbarrow. Perhaps Europeans hardly think of it as a vehicle, since the type used in the West is, as we shall see, very ill-adapted for the carrying of heavy weights, but the Chinese wheelbarrow is still so constructed that as many as six people may ride on it, and it is universally used there for all kinds of freight and passenger transportation.[e] Contrary to what most people would imagine, there is general agreement among historians[f] that the wheelbarrow did not appear in Europe until the late +12th or even the +13th century. Builders of castles and cathedrals must have adopted with alacrity a simple device which cut in half the number of labourers required to haul small loads by substituting a wheel for the front man of the hod or stretcher. But from the beginning, the European design placed the wheel at the furthest forward end of

---

[a] No. 5. In another (no. 2) we see a child being drawn along in a kind of perambulator with large solid wheels and upward-curving shafts which imitate those of full-size chariots. Many of these carvings are in the Man Tung Tzu[1] caves in the river-cliffs, first visited by Baber in 1886, recognised as Han tombs by Torrance (1), and described by W. Franke (2).

[b] As was noticed by Lin Yü-Thang (5), p. 34.

[c] It is with some astonishment that we read in Couling (1), p. 262, Sowerby (3) and Chamberlain (1), p. 236, that this vehicle was the invention of an American missionary in 1870.

[d] I was fortunate enough to have the opportunity of studying them again *in situ* in 1958.

[e] I have a vivid recollection of a long ride in 1943 outside a stretch of the walls of Chhêngtu on one of the wheelbarrows there plying for hire. Matteo Ricci also travelled thus in his day (Trigault, Gallagher tr., p. 317).

[f] De Saunier (1), p. 70; Lynn White (1), p. 147; Forestier (1), pp. 105, 107; Capot-Rey (1), p. 83; des Noëttes (3); Massa (1); Jope (1); Gille (14). One of the earliest representations is in the windows of the Cathedral at Chartres (c. +1220). Another is seen in Bib. Nat. MS. Lat. 9769, a history of the Holy Grail dating from about +1275. From the +15th century there is a fine picture (by an unknown painter) of the St Elizabeth legend, conserved in the Museum at Amsterdam, which shows a curiously concave wheelbarrow in the foreground (about +1475). In Agricola they are frequent.

[1] 蠻洞子

the barrow, so that the weight of the burden was distributed equally between the wheel and the man pushing.

That this was not the case with the Chinese wheelbarrow was appreciated by many Europeans who were in China in the +17th century and afterwards. Thus in +1797 van Braam Houckgeest wrote:[a]

Among the carriages employed in this country is a wheelbarrow, singularly constructed, and employed alike for the conveyance of persons and goods. According as it is more or less heavily loaded, it is directed by one or two persons, the one dragging it after him, while the other pushes it forward by the shafts. The wheel, which is very large in proportion to the barrow, is placed in the centre of the part on which the load is laid, so that the whole weight bears upon the axle, and the barrow men support no part of it, but serve merely to move it forward, and to keep it in equilibrium. The wheel is as it were cased up in a frame made of laths, and covered over with a thin plank, four or five inches wide. On each side of the barrow is a projection, on which the goods are put, or which serves as a seat for the passengers. A Chinese traveller sits on one side, and thus serves to counter-balance his baggage, which is placed on the other. If his baggage be heavier than himself, it is balanced equally on the two sides, and he seats himself on the board over the wheel, the barrow being purposely contrived to suit such occasions.[b]

The sight of this wheelbarrow thus loaded, was entirely new to me. I could not help remarking its singularity, at the same time that I admired the simplicity of the invention. I even think, that in many cases such a barrow would be found much superior to ours.

In addition to this, I should say that the wheel is at least three feet in diameter, that its spokes are short and numerous, and consequently that the felloes are very deep; and that its convexity on the outer side, instead of being very flat, like common wheels, is of a sharp form. This narrowness of the outer edge (rim) of the wheel appeared to me at first sight very unsuitable. It seemed to me that if broader it would have been better adapted to a clayey soil; but I recollected that at Java, the carts drawn by buffaloes have also wheels with narrow felloes, on purpose that in the rainy season they may cut through strong grounds, in which broad wheels would stick fast—as experience taught the learned M. Hooyman, who attempted to employ broad-wheeled carts in the environs of Batavia, but found himself obliged to follow the custom of the country. I am therefore convinced that the Chinese wheel is the best suited to a clayey soil.[c]

By way of illustration to this passage, Fig. 504 shows the general construction of the Chinese wheelbarrow. The European type may have gained something in controllability, but this cannot have been an important factor, since so many thousands of Chinese wheelbarrows carrying heavy loads still to this day pursue their rapid but squealing courses over the Chinese countryside.

What was not known to van Braam Houckgeest, however, nor to the majority of European historians of technology, is that the date of first appearance of this simple vehicle in China is at least as early as the +3rd century. It appears to have been the

---

[a] (1), Fr. ed., pp. 72ff., 86, 108, pl. 3, Eng. ed., vol. 1, pp. 96ff., 114, 144.

[b] The similarity between the Chinese wheelbarrow and an animal bearing pack and pack-saddle was pointed out by Mason (1), p. 436.

[c] The last paragraph here is interesting in view of the attention to 'streamlining' of wheels already noticed in the Chou Li (p. 75 above).

solution of Chuko Liang,[1] the great general of the Kingdom of Shu so noted for his technical interests, to the problem of supplying his armies. The essential passages occur in the *San Kuo Chih*:[a]

In the 9th year of the Chien-Hsing reign-period (+231) (Chuko) Liang again came forth from Chhi Shan, transporting (the army) supplies on 'Wooden Oxen' (*mu niu*[2]).

A page later,[b] the account continues:

In the spring of the 12th year of the same reign-period (+234), knowing that the main army would come out from Yeh-ku (Slanting Valley)[c] he used 'Gliding Horses' (*liu ma*[3]) to transport supplies.

And furthermore:[d]

(Chuko) Liang was a man of great ingenuity. By adding some parts and taking others away, he improved the multiple-bolt arcuballista (*lien nu*[4]). Moreover, the 'Wooden Ox' and the 'Gliding Horse' were both invented by him.

Then the commentary of Phei Sung-Chih[5] (+430) on the history goes on to quote[e] the *Wei Shih Chhun Chhiu*[6] (Spring and Autumn Annals of the (*San Kuo*) Wei Dynasty), by Sun Shêng,[7] as follows:

In the 'Collection of Chuko Liang's Writings' (*Chuko Liang Chi*[8]) there is an account of the method of making the 'Wooden Ox' and the 'Gliding Horse'. The 'Wooden Ox' had a square belly and a curved head, one foot (*chiao*[9])[f] and four legs (*tsu*[10]);[g] its head was compressed into its neck, and its tongue was attached to its belly. It could carry many things, and made thereby the fewer journeys, so it was of the greatest use. It was not suitable for small occasions, but was employed on long journeys; (in one day) it could go several tens of *li* if there was special need, or about twenty *li* if in convoy. The bent part corresponded to the ox's head, the double part[h] corresponded to the ox's limbs (*chiao*[9]), the cross part corresponded to the ox's neck, the revolving part (*chuan chê*[11])[i] corresponded to the ox's foot (*tsu*[10]), the covered part corresponded to the ox's back, the square part corresponded to the ox's belly, the hanging part corresponded to the ox's tongue, the curved part[j] corresponded to the ox's ribs (*lê*[12]), the carved part corresponded to the ox's teeth, the standing part corresponded to the ox's horn, the thin part corresponded to the ox's halter, the handles (*nieh*[13]) corresponded to the traces and the whippletree (*chhiu*[14])[k]—and each half of the axle faced upwards to the double shafts. In the time taken by a man (with a similar burden) to go 6 ft., the Wooden Ox would go 20 ft. It could carry the food supply (of one man) for a whole year,[l] and yet after 20 *li* the porter would not feel tired.[m]

a Ch. 35 (*Shu Shu*, ch. 5), p. 13*a*, tr. auct.
b P. 14*a*.
c See Sect. 28*b* in Vol. 4, pt. 3.
d P. 15*a*.
e P. 16*a*; all tr. auct.
f The wheel.
g Side supports to prevent toppling over.
h The two bearings of the single axle.
i The wheel.
j The housing separating the wheel from the freight.
k This was a technical term also applied to the swing (p. 376 below). In appropriate contexts it could also mean the breeching or crupper straps of harnessed horses (p. 309 below). Cf. p. 328.
l This is quite feasible, assuming a ration of 1 lb. a day, which would mean something of the order of 3 cwt. on the vehicle.
m The whole passage, with variations, is also given, together with the accompanying paragraph on dimensions, in *CSHK* (San Kuo sect.), ch. 59, p. 7*b*.

1 諸葛亮   2 木牛   3 流馬   4 連弩   5 裴松之
6 魏氏春秋   7 孫盛   8 諸葛亮集   9 脚   10 足
11 轉者   12 肋   13 攝   14 鞦

PLATE CXC

Fig. 504. The most characteristic type of Chinese wheelbarrow, with central wheel and housing above it, to carry both heavy loads and passengers. From Tê-an, Chiangsi (Hommel, 1).

It is not at all difficult to discern the wheelbarrow beneath the somewhat picturesque phraseology, which almost seems as if it might have been a kind of code, for after all, the design was military, and could well have been considered 'confidential'.

There follows a long passage in which minute dimensions and measurements are given. It is too tedious to quote, and at first sight so obscure that some have abandoned hope of reconstructing with its aid the exact wheelbarrow of Chuko Liang's time, indeed Goodrich has doubted even the conclusion that a wheelbarrow is involved at all.[a] We can only say that a careful scrutiny of the text has led us to the view that the

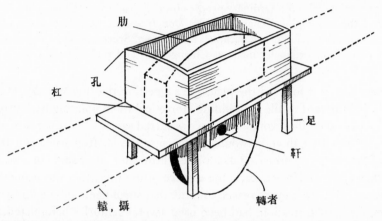

Fig. 505. Reconstruction of Chuko Liang's army service wheelbarrow according to the specifications in the *Chuko Liang Chi* (*c.* +230), preserved by Sun Shêng (*c.* +360) and Phei Sung-Chih (+430). The probable meanings of the technical terms are discussed in the text.

passage is not as garbled as it might seem, and that it points quite clearly to a wheel-barrow of a type which remained traditional thereafter. To make the measurements fit, however, it is necessary to adopt certain identifications; for example, the 'ribs' (*lê*[1]) may plausibly be taken to indicate the internal housing which protects the freight from the large revolving wheel. The box of ribs in an animal is, in a sense, a box within a box. Again, the 'bent part' is the front, the 'cross part' the axle, and the 'double part' the bearings. *Kang*[2] must be taken to mean crossbars, *khung*[3] mortises, and *khan*[4] the small piece of the frame carrying the axle-bearings. The upright sides and ends would be dismountable. All these elements can be seen in the picture of the wheelbarrow given by the *Thien Kung Khai Wu* (Fig. 506), including the internal wheel-housing. There it is called the 'single(-wheel) push-barrow' (*tu thui chhê*[5]), but other terms[b] used at different periods include the nouns *khuang*[6,7,8] and the verb *jung*.[9]

[a] (1), p. 78; tr. auct.

[b] The existence of single characters as names for the wheelbarrow might be interpreted by some as good evidence of its antiquity. Yet these terms are ambiguous, for they also meant the spinning-wheel (or silk-winding machine), in fact any device with a single wheel whether stationary or locomotive. We shall shortly appreciate further this strange association between the wheelbarrow and textile machinery (cf. p. 266).

[1] 肋　　[2] 杠　　[3] 孔　　[4] 軒　　[5] 獨推車　　[6] 軖　　[7] 軭
[8] 軭　　[9] 軴

Sometimes there was a disposition to regard Chuko Liang's devices as mysterious or supernatural, but in the +11th century Kao Chhêng[1] was quite clear that they were wheelbarrows.[a] He wrote:[b]

> Chuko Liang, prime minister of Shu, when he took the field, caused to be made the 'Wooden Ox' and 'Gliding Horse' for the transportation of the army supplies. In Pa and Shu[c] the ways were difficult, and these (vehicles) were more convenient for getting over the hills. The 'Wooden Ox' was the small barrow (*hsiao chhê*[2]) of the present day, and it was so called because it had the shafts projecting in front (so that it was pulled); while the 'Gliding Horse' was the same as that (wheelbarrow) which is pushed by a single person (and so has the shafts projecting behind). Ordinary people nowadays call these 'Chiangchow[3] Barrows'. According to the geographical chapter of the *Hou Han Shu*, there was a Chiangchow in Szechuan. At that time Liu Pei had occupied the whole of Szechuan, so I suspect that Chuko Liang's invention was originally made at Chiangchow, and thus got the name which was continued by later generations.

Kao Chhêng thus put forward the plausible suggestion that the Wooden Ox had the shafts pointing forward, while the Gliding Horse had them pointing backwards.[d] The order of the two inventions would then have occurred exactly according to expectation, since the obvious first thought would be to copy the shaft-chariot, and the transposition of the shafts would have taken place a little later, after some practical experience had been gained. In any case, the essence of the invention was economical, for (as we have seen) there is every reason to think that small hand-carts with two wheels, and shafts, for human traction, had been used several centuries beforehand.

One may pause here to point a moral. In the wheelbarrow we have an outstanding example of those many facts which undermine, and indeed overthrow, the classical European stereotype of China as a civilisation with unlimited man-power incapable of inventing and adopting labour-saving devices. Exactly what the economic situation was in the Han when the wheelbarrow first came widely into use remains for further research to elucidate—it may well be that in various historical periods particular parts of China suffered severe labour-shortages.[e] In any case long priority is here Chinese, and the surprised and grateful barbarians were European.

It has long been customary in China to regard Chuko Liang as the actual inventor of the wheelbarrow. When the present section was first drafted we also inclined to this opinion, but since then evidence has become available which strongly indicates a date rather more than two hundred years earlier as the time of the first invention, i.e. about the end of the Former Han. Of course even in the +3rd century others were concerned besides Chuko Liang. One of the leading technicians of the State of Shu was Phu

---

[a] In later ages, this became traditionally accepted; cf. Lin Chhing's remarks in *Ho Kung Chhi Chü Thu Shuo*, ch. 4, p. 2b.

[b] *Shih Wu Chi Yuan*, ch. 8, p. 2b., tr. auct.   [c] Modern Szechuan.

[d] Though nothing has so far been said concerning the animal traction of wheelbarrows, this became very common in China as we shall shortly see, and the first of these two types would have facilitated it. The man behind had only to keep the vehicle steady. There are of course other possible interpretations of the two designs of Chuko Liang, as we shall suggest in due course.

[e] As we remark elsewhere (pp. 28, 328), we know of no instance in China of the refusal of technical innovation for economic reasons.

[1] 高承   [2] 小車   [3] 江州

PLATE CXCI

Fig. 506 'The one-wheeled push-barrow of the south', from *Thien Kung Khai Wu*, ch. 9, p. 14*b* (+1637).

PLATE CXCII

Fig. 507. A wheelbarrow in a relief of *c.* +150; the tomb of Shen Fu-Chün at Chhü-hsien near Paoning in Szechuan (Segalen, de Voisins & Lartigue, 1).

Fig. 508. A wheelbarrow depicted on a moulded brick taken from a tomb of *c.* +118 at Chhêngtu, Szechuan. Only the right-hand bottom portion of the picture is shown. Rubbing from the Chungking Municipal Museum.

Yuan,[1] whom we shall meet again in connection with iron and steel metallurgy;[a] he held military office in the Western Command under Chuko. A fragment of a letter from Phu Yuan to his commander-in-chief has survived, which says:[b]

I and my workmen now entirely understand your excellent suggestion, and have constructed a 'Wooden Ox' with horizontal timbers joined together and double shafts. (In the time taken by) a man (with a similar burden) to go 6 ft., the 'Wooden Ox' will go 20 ft. A man can carry his whole food supply for a year on it.

Another contemporary who may have been concerned was the naturalist and engineer Li Chuan[2] (*d. c.* +260), who like Chuko Liang (and perhaps under his auspices) improved the crossbow and the arcuballista.[c] In any event, the wheelbarrow was indissolubly associated in later Chinese song and story with the military exploits of the State of Shu.[d]

The fact is that all these men were at best improvers of a single-wheeled carrier which had been in use throughout the Hou Han period. Justification for this view, amounting almost to proof, comes from two sources, epigraphic and (in a more complex way) textual. As has long been known, the Wu Liang tomb-shrine reliefs, dating from *c.* +147, depict the story of Tung Yung[3] and his father. Tung Yung, a young man of uncertain date in the Han, became in later ages one of the Twenty-four Examples of Filial Piety.[e] Losing his mother early he looked after his father, but when the time came he had no funds for the burial, so he borrowed the money on condition that he should be sold into slavery if he could not repay it. From this fate he was redeemed by the remarkable skill of the girl he had married, who after weaving three hundred rolls of silk in a single month revealed that she was in fact the 'Weaving Girl' from on high,[f] and disappeared. According to Tuan Shih,[g] the *Sou Shen Chi*[h] says that Tung Yung tilled the fields and pushed his father about on a kind of barrow called a *lu chhê*.[4] We shall have more to say about this expression, but the Wu Liang relief does indeed show the father sitting on the shafts of a small vehicle (Fig. 509).[i]

---

[a] Sect. 30*d* below. In the meantime see his biography *Phu Yuan Pieh Chuan*,[5] attributed to Chiang Wei,[6] the successor of Chuko Liang as Captain-General of Shu (*d.* +263); *CSHK* (San Kuo sect.), ch. 62, pp. 5*b* ff. (cf. Needham, 32).

[b] Preserved in *Pei Thang Shu Chhao* (+630), ch. 68, reproduced in *CSHK* (San Kuo sect.), ch. 62, p. 6*a*, tr. auct.

[c] *San Kuo Chih*, ch. 42 (*Shu Shu*, ch. 12), p. 9*b*; also *Chhou Jen Chuan*, 2nd addendum, ch. 2 (p. 21).

[d] There is much about Chuko Liang's Army Service Corps in the +14th-century novel by Lo Kuan-Chung, *San Kuo Chih Yen I*, chs. 102 ff., tr. Brewitt-Taylor (1), vol. 2, pp. 446 ff. On this famous book see Lu Hsün (1), pp. 163 ff. He reproduces a lively picture showing the construction of the wooden oxen and gliding horses, as imagined by a late illustrator of this novel.

[e] Cf. Mayers (1), p. 375.

[f] Cf. Vol. 3, p. 276. The spirit of the star Vega.

[g] (*1*), p. 67.

[h] Written by Kan Pao about +348. Cf. Bodde (9, 10).

[i] A closely similar representation, with the structure of the wheelbarrow most distinctly visible, occurs on the back of a bronze mirror figured by Bulling (8), pl. 60. Because of its rim and border pattern she is inclined to date it as late as the +4th century, but since the Tung Yung scene so closely resembles that in Wu Liang's tomb, and there are also other typical Later Han motifs such as the two immortals playing the *liu-po* game (cf. Vol. 3, p. 304), I do not see why it need be placed later than the

---

[1] 蒲元   [2] 李譔   [3] 董永   [4] 鹿車   [5] 蒲元別傳   [6] 姜維

Fig. 509. Tung Yung's father sitting on a wheelbarrow; a scene from the Wu Liang tomb-shrine reliefs of *c.* +147. A rubbing from Jung Kêng (*1*). The inscription at the top says 'Yung's father', and on the left 'Tung Yung was a young man from Chhien-chhêng' (in northern Shantung). Cf. Fig. 510*f*.

In the *Chin Shih So*[a] the Fêng brothers clearly represented the only wheel visible as outside the shafts, and the sides of the cart solid, but as in other cases they were here drawing somewhat on their imagination since the relief itself shows a space between the wheel and the upper parts. Indeed these resemble very closely the typical housing of the large Chinese wheelbarrow wheel, and resting on it is a pot of some kind exactly as objects do today (cf. Fig. 510*f*).

Greater certainty is provided by another carving of about the same date. A wheelbarrow appears sculptured at Chhü-hsien near Paoning on one of the pillars of the

+2nd century. Sirén (10), pls. 22 and 28, figures two pictures of Tung Yung of the +6th century, engraved on the walls of tomb-shrine and sarcophagus, but by this time the vehicles have become distorted chariot forms.

[a] Shih sect., ch. 3, pp. 58, 59. Cf. Jung Kêng (*1*), general rubbing 3, individual rubbings 40, 41.

tomb of Shen Fu-Chün,[1] a Szechuanese notable of about +150. One can see from Fig. 507 that the shafts and bearings are unmistakably outside the single wheel. Segalen, de Voisins & Lartigue, to whom we owe a description of this tomb, themselves interpreted the relief in this way.[a] More recently a Han brick of c. +118 was excavated at Chhêngtu which shows a man pushing a wheelbarrow with a load, in front of a horse-cart with a barrel-vault roof (Fig. 508). Perhaps the load is a dowry box, for this kind of cart is known to have been used by women. Again the housing of the single wheel is clearly to be seen.[b] Thus the pictorial evidence takes us back to the beginning of the +2nd century.

What the texts say is equally important but much more difficult to analyse, and Tuan Shih (1) has done well to raise the question of what exactly the *lu chhê*[2] was. Is the wheelbarrow concealed behind this ancient and soon obsolete term? As we shall see, the problem essentially consists in determining the nature of what at first sight might be translated 'deer-cart', but which in fact can only be correctly translated 'pulley-barrow'. In order to understand this we have to find our way through a maze of definitions and long-disused technical terms in the ancient dialectal dictionaries— an operation unexpectedly rewarding. But first it is necessary to demonstrate the existence and use of this small vehicle and its circumstantial description.

The *Shih Chi* has no mention of it, and it is also seemingly absent from the *Chhien Han Shu*, but from the beginning of the +1st century it is constantly turning up. Pao Hsüan,[3] an upright censor who lost his life under Wang Mang in +3, married when young Huan Shao-Chün,[4] the daughter of a wealthy scholar whose disciple he had been. Since Pao's family was poor he was rather embarrassed when the time came to go home, but the excellent girl 'changed into rough and short clothes, and helped him to push a *lu chhê* back to his village in the country'.[c] This would have been about −30. Fifty years later, when the Red Eyebrows rebellion was threatening the Hsin and preparing the way for the Later Han, a certain official Chao Hsi[5] and his friends were caught in it and surrounded. One of them, Han Chung-Po,[6] had just married a beauty and was much afraid that the rebels might harm her. His companions talked of leaving her behind somewhere on the road, but he angrily rejected the idea and smearing her face with mud, set her on a *lu chhê* which he pushed himself. When they met the 'brigands' he said that she was ill, so they all got safely through.[d] On other occasions, at need, the dead were borne on *lu chhê*. A virtuous official, Tu Lin[7] (d. +47), pushed a *lu chhê* at his brother's funeral,[e] and when a friend of Jen Mo[8] died at

[a] (1), p. 64.
[b] The reproductions in Anon (22), no. 21, pp. 46 ff., and even more in Liu Chih-Yuan (1), no. 55, do not do justice to this moulded brick, and it is by the special kindness of Dr Têng Shao-Chhin, the director of the Chungking Municipal Museum, where the object now is, and of Miss Phan Pi-Ching, that we are able to reproduce a rubbing. The brick was in a fragile state, but the details of the wheelbarrow could be seen perfectly during cleaning.
[c] *Hou Han Shu*, ch. 114, p. 1b, also *Lieh Nü Chuan*, cit. *TPYL*, ch. 775, p. 5a.
[d] *Hou Han Shu*, ch. 56, p. 15a, cit. *TPYL*, ch. 775, p. 4b.
[e] *Hou Han Shu*, ch. 57, p. 7b. *TPYL* cites (loc. [cit.) a parallel passage from the *Tung Kuan Han Chi*.

[1] 沈府君   [2] 鹿車   [3] 鮑宣   [4] 桓少君   [5] 趙熹   [6] 韓仲伯
[7] 杜林   [8] 任末

Loyang, he wheeled his corpse many *li* to the tomb.[a] The independent-minded Fan Jan[1] (*d.* +185), when surrounded by enemies, himself pushed his wife on a *lu chhê* and sent his son out to glean.[b] Many other examples from the Han and somewhat later periods could be given, the term surviving longest naturally enough in poetry.[c] It will suffice to quote what Ying Shao says of the *lu chhê* in his *Fêng Su Thung I* (The Meaning of Popular Traditions and Customs), written in +175.

> The 'pulley-barrow' is narrow and small, and its design follows that of a pulley-wheel (*lu*[2]).[d] Some call it the 'carefree barrow' (*lo chhê*[3]). It may be drawn by an ox or a horse, for which one cuts grass and fodder at stopping-places. Although pushing it is hard work one can lie down when one comes to a rest-house without further worry, hence the name 'carefree barrow'. If you don't have an ox or a horse to attach to it, one man alone can push it wherever he wants to go.[e]

So much for the passing mentions of something which was so familiar that it needed no explanation. When we turn to the ancient lexicographers for further help, we find a vital passage in the *Kuang Ya*[4] (Enlargement of the *Literary Expositor*), a dictionary of dialect synonyms compiled by Chang I[5] in the northern State of Wei during the very same years (+230 to +232) which saw Chuko Liang organising his trains of wheelbarrow porters for the supplies of the rival armies of Shu. What the *Kuang Ya* says is as follows:[f]

> The *sui chhê* is also called the *li-lu* (*sui chhê wei chih li-lu*[6]). The *tao kuei* is also called the *lu chhê* (*tao kuei wei chih lu chhê*[7]).

Which, being rightly interpreted, is as much as to say:

> The silk-winding (or quilling, or twisting and doubling, i.e. throwing machine) is also in some places called the 'pulley-machine'. The 'rut-maker' is also in some places called the 'pulley-barrow'.

Good and sufficient grounds for this understanding were provided by Chhing scholars such as Wang Nien-Sun in his commentary of +1796, and it is well worth while to look over his material. First, the *Shuo Wên* says[g] (+121) that the *sui chhê* is the (stationary) machine used for winding silk on to the bobbin (*fu chhê*[8,9]) in the weaver's shuttle.[h] All the agricultural encyclopaedists, from Wang Chên in +1313 onwards,[i] understand the *sui chhê* of old to be identical with the *wei chhê*,[10] the 'weft-machine' or quilling-wheel, an apparatus similar to the familiar spinning-wheel but used for

---

[a] *Hou Han Shu*, ch. 109B, p. 3b.   [b] *Hon Han Shu*, ch. 111, p. 21a.

[c] See *Phei Wên Yün Fu*, ch. 6 (p. 227.1), ch. 21 (p. 935.2).

[d] For *li-lu*,[11] see immediately below.

[e] *TPYL* preserved this passage, ch. 775, p. 5b. Chung-Fa ed., I Wên, ch. 1 (p. 90). Tr. auct.

[f] *Kuang Ya Su Chêng*, ch. 7B, p. 30a.   [g] Ch. 12, p. 105b.

[h] Few of the sinological dictionaries know this meaning of the word *fu*, but the technical term is clearly defined in the agricultural books. 'What receives the weft is the *fu* (*shou wei yüeh fu*[12])' says Wang Chên.

[i] *Nung Shu*, ch. 21, p. 14b, illustrations on 13b and 14a; followed verbally by e.g. *Nung Chêng Chhüan Shu*, ch. 34, pp. 11a, b, *Shou Shih Thung Khao*, ch. 75, p. 15b, illustr. p. 15a.

[1] 范舟   [2] 鹿   [3] 樂車   [4] 廣雅   [5] 張揖

[6] 繀車謂之麻鹿   [7] 道軌謂之鹿車   [8] 莝車   [9] 莝車

[10] 緯車   [11] 轤轆   [12] 受緯曰莝

winding textile fibres on to the bobbins in the weavers' shuttles.[a] Here of course a large wheel, a driving-belt (*huan shêng* [1])[b] and a small pulley are all required,[c] the value of the invention consisting in the speed given to the little spindle by the mechanical advantage of the disparate sizes of the wheels.

Of this simple machine more later; meanwhile *li-lu* [2] is but a variant of *li-lu*,[3] identical in meaning and nearly homonymous in sound with the most widespread expression for a pulley-wheel, *lu-lu*,[4] already discussed above.[d]

Yang Hsiung's [5] *Fang Yen* [6] (Dictionary of Local Expressions) next confirms that at its much earlier date (− 15) the *sui chhê* was called the *li-lu chhê* [7] in the regions of Chao and Wei—another way of saying 'the pulley-machine'.[e] That it had to have at least one pulley, rapidly rotated by the belt from the driving-wheel, follows from the nature of the case. While in some spinning-wheels (probably much later), five or as many as seven spindles were simultaneously rotated, it was the singleness of the shuttle-bobbin pulley (if not indeed of the large driving-wheel itself) which in early times invited the confusion with the wheelbarrow. The basic linguistic difficulty which we are facing is that the word *chhê* [8] was always ambiguous in that it could mean indifferently a stationary machine or a mobile vehicle. It is as if one should speak of a 'single-wheeler' without giving any clue as to what it did. Moreover, both the components *li* [9] and *lu* [10] could also mean the rut or track left by a vehicle-wheel. We are not at all surprised therefore when the *Fang Yen* goes on to say that in Eastern Chhi and in Hai-Tai the silk-winding machine was called *tao kuei* [11] or 'trace-maker'.[f] In fact the silk fibre being wound on to the bobbin (or the driving-belt endlessly leaving the driving-wheel) was analogised in popular parlance with the track on sand or mud being 'wound off' the vehicle wheel, and what these ancient name-coiners really had in mind was simply the geometrical pattern of a circle and a tangent line.[g] So returning to the *Kuang Ya's* second statement, we see that the *lu chhê*,[12] which we know from a mass of other evidence was sometimes a small vehicle, and had nothing whatever to do with deer, could be either a 'pulley-barrow' or a 'pulley-machine', in the first case a 'rut-maker', in the second a 'trace-maker' (both *tao kuei* [11]), but in all cases a 'single-wheeler'.[h] Hence what Pao Hsüan and his excellent wife were pushing in the last years of the Former Han dynasty was nothing other than that new-fangled device, the wheelbarrow.

Here we could take leave of Wang Nien-Sun and go on our way, but the most ancient origin of the words *li-lu* or *lu-lu* remains intriguing, and this terminological

---

[a] Indeed the agricultural writers all add that the name *sui chhê* persisted as an alternative for *wei chhê*.
[b] The term is Wang Chên's.
[c] On the significance of this simple machine for the history of the driving-belt, cf. p. 106.
[d] P. 95.　　　　　　　　　　[e] See Tai Chen's *Fang Yen Su Chêng*, ch. 5, p. 10*a*.
[f] Ch. 5, p. 10*a*. The *Yü Phien* dictionary of +543 adds *kuei chhê*.[13]
[g] Cf. the colloquialisms of transport: 'filer vite', 'spinning along'; and of the 'march' of events: 'se dérouler'.
[h] Now we can see the full significance of the fact that *khuang* [14, 15, 16] could mean either a spinning-wheel (*fang chhê* [17]) or a wheelbarrow (p. 261 above).

[1] 環繩　　[2] 厤鹿　　[3] 麻鹿　　[4] 轆轤　　[5] 揚雄　　[6] 方言
[7] 轣轆車　[8] 車　　[9] 轣　　[10] 轆　　[11] 道軌　　[12] 鹿車
[13] 軌車　　[14] 軭　　[15] 軖　　[16] 軭　　[17] 紡車

rubbish-heap is worth more turning over. In one of the *Shih Ching* odes [a] a chariot description includes the words *wu wu liang chou* [1]—'the pole is curved like a roof-ridge and has five *wu*'. This last word is usually taken to mean ornamental bands of leather, but the ancient commentary of Mao Hêng (−3rd or −2nd century) explains that the five bands (*shu* [2]) each had a *wu* [3] and that this was in fact a *li-lu*.[4] Here rings to guide traces or reins must be meant, so that the earliest form of the phrase was the simplest possible kind of pulley-block (*c.* −7th century). Another usage occurs in one of the fortification chapters of *Mo Tzu* (*c.* −320), where javelins hurled by the multiple-bolt arcuballista (*lien nu* [5]) are attached to a cord and retrieved by the aid of a reel [b] or windlass (*i li-lu chüan shou* [6]).[c] What follows is more interesting. In another place the *Fang Yen* (−15) says [d] that 'the cord (*chih* [7]) underneath the machine is called in Chhen, Sung, Huai and Chhu a *pi*,[8] while thicker (or longer) ones are called *chi* [9]'. Kuo Pho's commentary (about +300) says that this refers to the *lu chhê*,[10] here evidently in its sense of the silk-winding machine, and the +6th-century *Yü Phien* identifies *chih* [7] as a cord or belt, *so*.[11] Thus in pursuing the ruts of wheelbarrows we come unexpectedly upon fresh evidence that not only the quilling wheel of the +12th century, but also that of the +1st, possessed a belt-drive.

Further confirmation arises from a passage in the *Chou Li* (*Khao Kung Chi*).[e] In the entry for the jade-workers (*Yü jen* [12]) the text says that the *kuei* or sceptre of the emperor has a *pi* in the middle (*Thien tzu kuei chung pi* [13]). Whereupon Chêng Hsüan's +2nd-century commentary says that the word *pi* should be interpreted as identical with the *pi* [14] (i.e. *pi* [8]) of the *lu chhê*, that is to say, 'what has the central function in the design (of the silk-winding or quilling machine) (*i tsu yo chhi chung yang* [15])'. Its object is to prevent any chance of mishap or loss of control. Now the emperor's safety device was a loop of cord passed through the hole in the sceptre in such a way that he could have it round his wrist, thus avoiding any danger of the jade object being inadvertently allowed to slip or fall.[f] To which Wang Nien-Sun adds: 'The *pi* [16] of the sceptre (*kuei* [17]) is an (endless) band (*tsu* [18]) which has the same sort of function as the cord (*pi* [14]) or rope (*so* [11]) of the (silk-quilling) "single-wheeler pulley-machine" (*lu chhê* [10]), because they are both for the control and discipline (*yo shu* [19]) of the thing; hence it is that *pi* [16] is to be understood as *pi*.[14]' The only possible conclusion we can draw is an alternative—if (as is by far the most likely) the shuttle-bobbin silk-winder is meant, then it looks as if we must admit the invention of the driving-belt by the end

[a] Mao no. 128, Hsiao Jung; cf. Legge (8), p. 193; Karlgren (14), p. 82; Waley (1), p. 111.

[b] This arrangement may throw some light on the early invention of the reel of the fisherman's rod in China (cf. p. 100).

[c] Ch. 53, p. 14*b*, tr. Forke (3), p. 608, but the text is very uncertain.

[d] *Fang Yen Su Chêng*, ch. 9, p. 3*b*. We follow Tai Chen here whose elucidation of *c.* +1777 was curiously overlooked by Wang Nien-Sun in 1796.

[e] Ch. 12, p. 1*b* (ch. 42). Biot (1), vol. 2, p. 522, seems not to have understood the commentary.

[f] Cf. e.g. Laufer (8), pp. 87ff.

[1] 五楘梁輈	[2] 束	[3] 楘	[4] 歷錄	[5] 連弩	[6] 以麻鹿卷收
[7] 紩	[8] 畢	[9] 藆	[10] 鹿車	[11] 索	[12] 玉人
[13] 天子圭中必	[14] 繹	[15] 以組約其中央		[16] 必	[17] 圭
[18] 組	[19] 約束				

at least of the Former Han (+ 1st century);[a] but if by any chance a vehicle is intended then again the wheelbarrow shakes out, for the cord or rope could only be the sling so often attached to the shafts and passing over the porter's shoulders (cf. Figs. 506, 512), an aid to stabilisation superfluous for any two-wheel vehicle.

Thus when we return to the statement in the *Kuang Ya* we may feel safe in re-phrasing it as follows: 'The silk-winding (trace-maker) for the bobbins is also called the "pulley-machine"; the (mobile) rut-maker is also called the "pulley-barrow".' We shall probably not go far astray in believing that the semantic unity lay in the circle-and-tangent pattern, and that *li-lu, lu-lu, lo-lu*, etc., were originally onomato-poeic terms deriving from the creaking and squealing of primitive bearings. There were stationary 'single-wheelers' which made a rumbling noise of turning wheels,[b] and there were mobile 'single-wheelers' which groaned and travailed under their loads. Of the two, we may suspect that the silk-winder was much the earlier in view of the great antiquity of silk technology in China (though it may not have had its driving-belt until late in the Former Han); in this case when the 'pulley-barrow' developed it acquired its name from the stationary machine which also seemed to have only one operative wheel.[c]

The foregoing philological long-shots may perhaps have looked like wild fire to some readers. But in fact they have proved to be dead on the target, for since their conclusions were reached, no less than four stone reliefs of Han date depicting quilling-wheels along with looms have been studied and published by Sung Po-Yin & Li Chung-I (1), and Tuan Shih (2), in brief but brilliant contributions to the history of textile technology.[d] To reproduce these reliefs here would anticipate unduly Sect. 31 on textiles in our fifth volume, for which we shall therefore reserve them.

What the difference was between the 'Wooden Ox' and the 'Gliding Horse' of Chuko Liang remains unsolved. As we have seen, Kao Chhêng in the Sung thought that it depended on whether the shafts were before or behind. Certainly the tradi-

---

[a] Hesitations in accepting this may be lessened by the reflection that China was the only ancient culture which enjoyed the availability of a really long-staple textile fibre, namely silk. This would quite naturally have encouraged the development of devices designed to give continuous rotary motion. We indeed believe that not only the quilling-wheel, but also the reeling-machine (*sao chhê*[1,2,3]; cf. pp. 2, 107, 116, 382, 404) was fully in use, if perhaps in the simplest of its forms, by the time of Chhin and Han. Proof follows immediately below.

[b] Cf. the poem of Lou Shou, translated on p. 107 above.

[c] It will be remembered that we have already come across quite a different connection between textile technology and the wheelwright's art in the construction of the heavy-duty composite wheels of the Han (p. 80 above).

[d] Already in 1958 Dr Lu Gwei-Djen and I had been deeply impressed at the Shantung Provincial Museum, Chinan, by one of these representations, a relief from Thêng-hsien excavated five years earlier. Having alluded to this in an account of our studies (Needham, 46), explanatory correspondence followed next year with Mr Calvin Hathaway, in which I interpreted the machines as spinning- rather than quilling-wheels. I mention this here because the former alternative cannot altogether be excluded; hempen cloth was the common wear of the mass of the people in Han times, and hemp fibres had to be spun. Nevertheless the Chinese historians of technology mentioned above take the scenes to represent the working of silk. Even this would not entirely exclude the spinning-wheel (used for broken cocoons, etc.), but it does render its introduction gratuitous since the quilling-wheel was so much the more prominent form of the machine in the sericultural industry.

---

¹ 繰車　　² 𦈕車　　³ 繀車

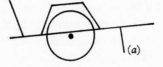

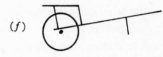

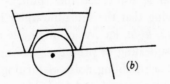

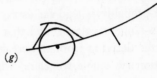

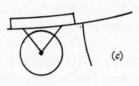

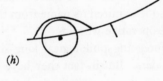

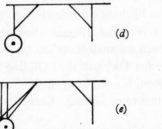

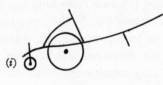

Fig. 510. Types of Chinese wheelbarrows. (a) Wheel central, with housing, taking all the load, a substitute for a pack-animal, the 'pack-horse' type. Chiangsi and many other provinces; also Indo-China. Cf. Figs. 504, 506. (b) The same, with wagon sides. Northern Szechuan, identified with the presumptive design of Chuko Liang (+3rd century). Cf. Fig. 505. (c) Wheel central but centre of gravity high, without housing, axle on sloping struts, flat-car sides, used for moving earth. Shensi, cf. Fig. 511. (d) Small wheel forward, centre of gravity high, axle on vertical stayed struts, stayed support. Szechuan. (e) Two small wheels forward, centre of gravity high, axle on vertical stayed struts, stayed support. Szechuan. (f) Wheel far forward with rectangular housing, straight frame, the 'half-stretcher' type. Shensi and many other provinces. Cf. Figs. 507, 508, 509, 512. This is perhaps the oldest form of the invention, for iconographic evidence takes it back to the +2nd century, and textual evidence to the −1st, though the latter is not decisive as to the form. (g) Intermediate type with curving frame and curved housing. Western Szechuan. Cf. Fig. 513. (h) Intermediate type with curving frame and streamlined housing. Western Szechuan. (i) Intermediate type similar to (g) but with a small auxiliary wheel at the forward end of the frame, useful for clearing obstacles. Hunan and other provinces. Cf. Fig. 517.

tional types of wheelbarrow still current in China today are very numerous, the position of the bearings varying from a central to a very forward point. In the most characteristic Chinese design, as shown in the *Thien Kung Khai Wu* (Fig. 506, Fig. 510a), the wheel seems to have been conceived as a substitute for a pack-animal and takes all the load. One of these types, common at Kuang-yuan (Fig. 510*b*), is regarded by Chinese historians as particularly likely to preserve the pattern used by Chuko Liang.[a] Another (Fig. 511, Fig. 510*c*) is very much in evidence on contemporary public works sites. But the wheel-bearings are also often very far forward; Fig. 512 (cf. Fig. 510*f*) shows a barrow of this kind—particularly interesting since it resembles so closely the barrow on which Tung Yung's father is sitting in the Wu Liang relief (Fig. 509). Often however the wheel-housing is curved rather than rectangular, as in the Szechuanese form shown in Fig. 513 (cf. Fig. 510*g, h*), with the wheel less far forward. And the two main types may even be combined (Fig. 510*i*).[b]

Of the transmission of the technique to Europe about the beginning of the +13th century nothing whatever is known, nor is information easily obtained about the wheelbarrow in the geographically intermediate cultures.[c] But the fact that the wheel in the European types was invariably very far forward, as if to replace one man out of two carrying a hod or stretcher, and therefore taking only half the load, may mean that this was another case of 'stimulus diffusion'.[d] Perhaps Westerners simply heard that something of the kind had been done, and proceeded to imitate it according to their lights without knowing any exact specifications. At the same time it is not possible to exclude Western influence on contemporary Chinese designs, and one might be inclined to see it in the 'half-stretcher' rather than the 'pack-horse' types, though if the Wu Liang tomb-shrine is any indication the former was in China from the beginning, and Chuko Liang's innovation might have been precisely the latter.[e]

Among the most curious designs are those which raise the carrying surface high above one (or even two) small but broad-rimmed wheels (Fig. 510*d, e*) with the aid of oblique stays. Those which we have seen in West China always have the wheels well forward, but in a neighbouring culture the principle of the central wheel is applied to these forms also. This occurs in an interesting vehicle taken note of by Horwitz (6), namely the traditional Korean ceremonial chair-litter, in which the four handles are guided, not carried, by four 'bearers', and the main weight is taken on a central wheel with struts and stays remarkably reminiscent of aeroplane landing-gear (see Fig. 514).[f] This is perhaps an archaic intermediate form between the four-handled hod or stretcher (or the draw-sled) and the wheelbarrow proper.[g] What may be a

[a] This statement is due to a conversation with Prof. Hsü Chung-Shu and other friends at Szechuan University, Chhêngtu, in the summer of 1958.

[b] Cf. Anon. (*18*), pt. 4, no. 14, and below, p. 274.

[c] A note by Lorimer (1) describes the small wheelbarrows with upward-curving frames, like (*g*) and (*h*) in Fig. 510 but without wheel-housings, which are used by the people of Hunza in the Himalayas.

[d] Cf. Vol. 1, pp. 244 ff.

[e] Again such differences might explain his two types. That the 'half-stretcher' type is fully indigenous is also shown by the fact that one clear example of it appears in the famous picture of life in the streets of Khaifêng finished by Chang Tsê-Tuan before +1126 (see on, p. 273). This was painted a century or so before there were any wheelbarrows in Europe at all.

[f] Further references will be found in Osgood (1), p. 141.

[g] Cf. Fig. 492. What have been described as wheeled litters are depicted on Khmer monuments in Indo-China (see des Noëttes (1), p. 102, fig. 113), but they have many wheels and are really wagons with shafts both behind and before, which men are pulling and pushing. On the other hand a vehicle which seems strangely like the Korean litters, but drawn by a horse, is to be seen in the carvings of the temple of Neminātha at Dilwara, Mount Abu, in Rajasthan (Kramrisch (1), pl. 138). This dates from +1231.

literary reference to this occurs in Chang Shun-Min's[1] *Hua Man Chi*,[2] a book of about +1110, where we hear of a *chuan chu chiao tzu*:[3] a litter mounted on a turning wheel.[a] It is noteworthy that even the normal character for litter contains the wheel radical. Indications exist that the wheeled litter may go back to the Former Han, for the biography of Yen Chu,[4] describing the expeditionary force sent to Nan Yüeh in −135, says:[b] 'Chariot-litters (with supplies) passed through the mountains (*yü chiao erh yü ling*[5]).' This puzzled the commentators. Fu Chhien supposed the word *chiao* to refer to mountain paths as narrow as plank bridges, but Chhen-Tsan thought that the vehicles must have been like the 'bamboo-chariots' (*chu yü*[6]) of his own time (*c.* +300) (whatever they were), and Hsiang-Chao believed that the things were in some way 'carried' (*tan*[7]). We probably have here an interesting, though still very obscure, chapter in the early history of the Chinese wheelbarrow.

A few later references to wheelbarrows may be appended. In a previous subsection[c] mention was made of Ko Yu,[8] a semi-legendary personage, Szechuanese of course, variously attributed to Chou or Chin times, who rode away into the mountains on a wooden sheep (*mu yang*[9]) of his own invention.[d] I should not be surprised if this was a piece of folklore which had the real wheelbarrows of Chuko Liang for its basis. It is interesting to see how the range of wheelbarrows had spread far beyond Szechuan, for Lu Yü, who travelled there in +1170, mentions the *tu yuan hsiao chhê*[10] as a feature of Lü-chhêng in Chiangsu.[e] Six years later Tsêng Min-Hsing alluded to the military use of the wheelbarrow in forming protective laagers. He wrote:[f]

In Chiang-hsiang[11] there is a kind of small vehicle called *i têng chhê*,[12] with only one wheel and two shafts....A bamboo basket is fastened on each side by ropes. This device is so efficient that it can take the place of three men; moreover it is safe and steady when passing along dangerous places (cliff paths, etc.). Ways which are as winding as the bowels of a sheep will not defeat it. Not only is it useful for transporting army rations, but at need it can be employed as a defensive obstruction against cavalry. Since the digging of trenches and moats, and the building of forts, take time, the wheelbarrows can be deployed round the perimeter so that the enemy's horses cannot easily pass over. This kind of vehicle can readily go forward and withdraw, and can be used for any purpose. It might well be called a 'mobile fort'.[g]

---

[a] Ch. 8, p. 10*a*.   [b] *Chhien Han Shu*, ch. 64A, p. 4*a*.
[c] Cf. p. 159 above.
[d] His story is told in the *Sou Shen Chi* (tr. Bodde (10), p. 308) and the *Lieh Hsien Chuan* (tr. Kaltenmark (2), p. 95).
[e] *Ju Shu Chi*, ch. 1, p. 8*a*.   [f] *Tu Hsing Tsa Chih*, ch. 9, p. 10*b*, tr. auct.
[g] The significance of this passage for the history of military technology should not be overlooked. In his account of the development of the art of war Oman (1) devotes much attention to the use of laagers as defences against the onslaughts of cavalry. Without giving very much evidence, he attributes the system to the Russians, who called it 'gulai-gorod' or movable city, and it does seem probable that the idea goes back to ancient Scythian or Mongol practices (cf. Lipschutz, 1). Medieval armies frequently protected their baggage-trains by forming laagers (e.g. Edward III at Crécy, and Bela of Hungary at the battle of the Sajo in +1241), but these were incidental. As a primary tactical device it reached its apogee in the epic history of the 'Wagenburg' of the Hussite Wars (+1420 to +1434; Oman (1), vol. 2, pp. 361 ff.). But from the above passage it is clear that the idea was well appreciated—and presumably the manœuvre executed when circumstances called for it—by the Chinese of +1175.

---

[1] 張舜民   [2] 畫墁集   [3] 轉軸轎子   [4] 嚴助   [5] 輿轎而隃嶺
[6] 竹輿   [7] 擔   [8] 葛由   [9] 木羊   [10] 獨轅小車
[11] 江鄉   [12] 一等車

PLATE CXCIII

Fig. 511. Wheelbarrow of type (*c*) on an irrigation project site near Wukung, Shensi
(orig. photo., 1958).

Fig. 512. Wheelbarrow of type (*f*) on the road between Sian and Lintung, Shensi
(orig. photo., 1958).

PLATE CXCIV

Fig. 513. Wheelbarrow of type (*g*) near Kuanhsien, Szechuan (orig. photo., 1958).

Fig. 514. Korean ceremonial chair-litter (Horwitz, 6). The weight is taken by the central wheel with its struts and stays, so that the 'bearers' have simply to guide the vehicle. For its evolutionary position cf. Fig. 492 (*f*).

Not long afterwards Chu I tells us[a] that in the Thang, when Liu Mêng[1] went to pacify the region which is now Ninghsia, he relied on wheelbarrows for the transportation of his supplies. According to the Japanese monk Ennin, however, there was in +845 a Taoist-inspired edict against the use of wheelbarrows (*tu chiao chhê*[2]), which were said to break up the road surface.[b] It certainly cannot long have been in force.

The mention of Chiang-hsiang by Tsêng Min-Hsing reminds us of Kao Chhêng's appellation 'Chiangchow barrows'. This is a little town in the extreme south-west corner of Szechuan, lying in mountainous regions on the road between Hui-li and Hui-tsê in Yunnan which crosses the Yangtze by a ferry near Ta-chhiao almost due north of Kunming. Now in the sources Ko Yu is described as a Chhiang,[3] i.e. a member of one of the tribal peoples indigenous to just this region. We shall probably not go far astray then if we accept it as the original home of the invention in the −1st century, and regard Ko Yu either as the first inventor or perhaps the canonised 'technic deity' of the makers of wheelbarrows.[c]

Animal traction was applied to one-wheel vehicles from an early date in China, as the passage from the *Fêng Su Thung I* is alone sufficient to witness. Perhaps our best illustrations of this may be seen in the famous painting of Chang Tsê-Tuan,[4] finished by +1126, which depicts the popular life of the capital Khaifêng at the time of the Spring Festival (Chhing-Ming Shang Ho Thu[5]).[d] As Figs. 515 and 516 show, many wheelbarrows are moving or stationary in the streets of the city. All but one have the large central wheel and some are very heavily laden; during the loading and unloading they rest on their side-legs. One is being pushed by a single man, and in all cases a porter steadies the vehicle by the shafts behind, while traction is effected either by one man in shafts[e] and one mule or donkey with collar-harness and traces, or by two animals side by side similarly attached. This last system is shown again in a famous picture in the *Thien Kung Khai Wu* (+1637),[f] where in the text we read:[g]

The northern one-wheeled barrow (*tu yuan chhê*[6])[h] is pushed by one man from behind, with (one or more) donkeys pulling it from the front; it is hired by those who dislike riding (on horseback). The travellers sit on opposite sides to balance it,[i] and a mat roof shields them from sun and wind. This kind of conveyance goes as far north as Chhang-an and Chi-ning, and also comes to the capital. When not carrying passengers these barrows will take as much as 4 or 5 *tan* of goods....[j]

[a] *I Chao Liao Tsa Chi*, ch. 2, p. 35a; cf. ch. 1, p. 40a.
[b] See Reischauer (1), p. 247; (2), p. 385.   [c] Cf. Vol. 1, pp. 51ff.
[d] Cf. Chêng Chen-To (3); Hejzlar (1).
[e] Hence the modern northern colloquial expression *erh pa shou chê*.[7]
[f] Ch. 8, pp. 13b, 14a. Photograph in des Noëttes (1), fig. 136.
[g] Ch. 8, p. 10a, tr. auct.
[h] This expression, just now already met with, may indicate that originally the animal-drawn wheelbarrow had a central pole like the ancient pole-chariot. Perhaps the name persisted long after this had been replaced by the more convenient collar-harness and traces. Pictures in some editions of the *Thien Kung Khai Wu* show, however, vestigial shafts. Alternatively the word *yuan* had perhaps become a special technical term for wheel in the barrow world.
[i] A convergent approach to the Irish jaunting-car (cf. Haddon, 3).
[j] About 6 cwt.

[1] 劉蒙   [2] 獨脚車   [3] 羌   [4] 張擇端   [5] 清明上河圖
[6] 獨轅車   [7] 二把手者

The one-wheeled barrow (*tu lun thui chhe*[1]) of the south is (also) pushed by one man (but without animal aid), and carries only 2 *tan*. When it meets pot-holes (in the road) it has to stop; in any case it seldom goes more than 100 *li*....[a]

But perhaps the most original invention was the use of sails, so that wheelbarrows like ships could be borne along by the force of the wind. This admirable device, the *chia fan chhê*,[2] is still widely used in China, notably in Honan and the coastal provinces such as Shantung, and has often been described and illustrated.[b] I reproduce (Fig. 518) a little sketch from Lin-Chhing's[3] *Hung Hsüeh Yin Yuan Thu Chi*[4] (Memoirs of the Events which had to happen in my Life)[c] of 1849, which shows an animal being assisted by the sail; and also (Fig. 519) the careful diagram published by van Braam Houckgeest in +1797, from which one can see that the sails are typical junk fore-and-aft slat-sails with their multiple sheets.[d] The words of van Braam, too, are worth quoting.[e]

Near the southern border of Shantung one finds a kind of wheelbarrow much larger than that which I have been describing, and drawn by a horse or a mule. But judge of my surprise when today I saw a whole fleet of wheelbarrows of the same size. I say, with deliberation, a fleet, for each of them had a sail, mounted on a small mast exactly fixed in a socket arranged at the forward end of the barrow. The sail, made of matting, or more often of cloth, is five or six feet high, and three or four feet broad, with stays, sheets, and halyards, just as on a Chinese ship. The sheets join the shafts of the wheelbarrow and can thus be manipulated by the man in charge.

One had to grant that the apparatus was not a freak, but an arrangement by which, with a favourable wind, the wheelbarrow porters could be greatly assisted. Otherwise such a complicated thing would have been only a bizarre curiosity. I could not help admiring the combination, and was filled with sincere pleasure in seeing twenty or so of these sailing-wheelbarrows setting their course one behind the other.

Unfortunately there are very few references in Chinese literature to sailing wheelbarrows, so that it is as yet not possible to determine the time of their introduction.

The impact which this ingenuity made upon the first European visitors to China in the +16th century, however, can hardly be imagined. In +1585 Gonzales de Mendoza wrote[f] that the Chinese

are great inventers of things, and that they have amongst them many coches and wagons that goe with sailes, and made with such industrie and policie that they do governe them with great ease; this is crediblie informed by many that have seene it; besides that, there be many in the Indies, and in Portugall, that have seene them painted upon clothes, and on their

[a] A modification doubtless due to such difficulties was the addition of a small auxiliary wheel at the front of the vehicle to assist in getting over obstacles (cf. Fig. 517 from Franck (1), p. 632). Cf. Fig. 510*i*, and Anon. (*18*), pt. 4, no. 14.

[b] See e.g. S. W. Williams (1), vol. 2, p. 8; Liu Hsien-Chou (*1*), p. 79; Wagner (1), p. 162; Cable & French (2), p. 80; Dinwiddie in Proudfoot (1), p. 41; Buschan *et al.* (1), vol. 2, pt. 1, pl. XXVI, fig. 2.

[c] Lin-Chhing (+1791 to 1846) was a Manchu of the Wanyen[5] family descended from the emperors of the Jurchen Chin dynasty (+12th century). We shall often meet with him again in connection with hydraulic engineering, a field in which he was notably expert. Cf. his biography in Hummel (2), vol. 1, p. 506.   [d] Cf. Sect. 29*g* below.

[e] (1), Fr. ed., vol. 1, p. 115, pl. 3, Eng. ed., vol. 1, pp. xxv, 152.   [f] (1), p. 32.

[1] 獨輪推車   [2] 加帆車   [3] 麟慶   [4] 鴻雪因緣圖記   [5] 完顏

PLATE CXCV

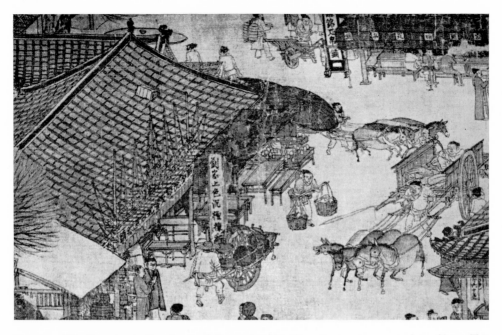

Fig. 515. Wheelbarrows in the streets of the capital, Khaifêng, in +1125; scenes from Chang Tsê-Tuan's painting, Chhing-Ming Shang Ho Thu (Returning up the River to the City at the Spring Festival). At the top, an empty half-stretcher type standing beside a draper's shop; below, a loaded pack-horse type with a very large wheel passing a dyeing establishment.

Fig. 516. Another scene; on the left an empty wheelbarrow being loaded with sack-like objects outside the best hotel; on the right, another one of large wheel size drawn by a mule and guided by two men who pull and push.

PLATE CXCVI

Fig. 517. A train of intermediate-type wheelbarrows with auxiliary wheels at the forward end for surmounting small obstacles (type *i*); on one of the paths through the Nan Ling hills between the Yangtze Valley and Canton (Franck, 1).

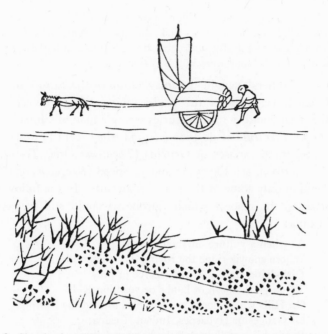

Fig. 518. Sketch of a sailing wheelbarrow, the sail assisting animal traction; from Lin-Chhing's *Hung Hsüeh Yin Yuan Thu Chi* (1849).

Fig. 519. Diagram of a sailing wheelbarrow from van Braam Houckgeest (+1797), showing the batten sail and multiple sheets so characteristic of Chinese nautical practice (cf. Sect. 29g below).

earthen vessell that is brought from thence to be solde: so that it is a signe that their painting hath some foundation.

So also said van Linschoten a dozen years later:[a]

The men of China are great and cunning workemen, as may well bee seene by the Workmanship that commeth from thence. They make and use (waggons or) cartes with sayles (like Boates) and with wheeles so subtilly made, that being in the Fielde they goe and are driven forwards by the Winde as if they were in the Water....

These relations and others[b] caught the imagination of the European mapmakers, so that one finds small vignettes of the land-sailing carriages on almost every atlas published in the +16th and +17th centuries where a map of China is provided.[c] In the *Speculum Orbis Terrarum* of de Jode (+1578) they seem to be lacking, but they figure prominently in the atlases of Ortelius (*Theatrum Orbis Terrarum*, +1584),[d] Mercator (*Atlas*, +1613), see Fig. 520; and J. Speed (*Kingdome of China*, +1626), Fig. 521; to mention only some of the most important.[e] In the following years John Milton immortalised the respect which Europeans could still feel for the strange techniques of China:

> As when a vulture on Imaus bred
> Whose snowie ridge the roving Tartar bounds,
> Dislodging from a Region scarce of prey
> To gorge the flesh of lambs or yeanling Kids
> On Hills where flocks are fed, flies toward the Springs
> Of Ganges or Hydaspes, Indian streams;
> But in his way lights on the barren plaines
> Of Sericana, where Chineses drive
> With Sails and Wind their canie Waggons light....[f]

This was in +1665, but the whole century was fascinated by the story, and puzzled interest in what the Chinese really did continued long after the time of Ortelius. In Campanella's *City of the Sun* (+1623) his Ceylonese Utopians used 'waggons fitted with sails, which are borne along by the wind even when it is contrary, by the marvellous contrivance of wheels within wheels'.[g] In Birch's *History of the Royal Society* we find the following somewhat cryptic entry:[h]

Apr. 1st. 1663; Mr Hooke's paper concerning the Chinese cart with one wheel, mentioned by Martinius in his *Atlas Sinensis*, was read, and discoursed upon, that the said cart was like a wheelbarrow; and the paper was ordered to be filed up.

---

[a] (1), p. 140.

[b] E.g. the despatch of R. Cocks to the East India Company (Calendar of State Papers, Colonial Series, East India, vol. 1) in +1614: '...in the country of Corea...great waggons have been invented to go upon broad flat wheels under sail as ships do, in which they transport their goods....It is said that Ticus Same, called Quabicondono, the deceased Emperor of Japan, did pretend to have convey'd a great army in those sailing waggons to assault the Emperor of China...but was prevented by a Corean nobleman...' (804). The 'emperor' is Hideyoshi, and the Korean perhaps Yi Sunsin (see Sect. 29*i*).

[c] I am much indebted to Mr H. R. Mallett, O.B.E., of the Cambridge University Library, for drawing my attention to this tradition.

[d] The sheet was probably prepared by Ortelius himself; see Bagrow (1); Tooley (1), pp. 106ff.

[e] Thanks to Prof. D. Lach, we add de Bry, +1599, *Indiæ Orientalis*, pt. 2, Icones, pl. xxv.

[f] *Paradise Lost*, III, 431. Cf. Huntley (1).   [g] Cf. Henry Morley (1).

[h] Vol. 1, p. 216.

Borata

*Chinarum gens admodum ingeni-
ofa effe pehibetur, adeo ut currus
excogitarint fabricaverintque,
quos velis ventisque per campos
et loca plana uti navigia per
mare derigere optime norint*

Fig. 520. Vignette of an imagined Chinese land-sailing carriage; from Mercator's *Atlas* of +1613.

Fig. 521. Another vignette of the imagined Chinese land-sailing four-wheeled wagons navigating at will *per terras*; from Speed's *Kingdome of China*, +1626.

Then on 23 November of the same year, Mr Robert Hooke displayed 'a paste-board model of his engine with one wheel, to travel in with ease and speed....'[a] Thomas Hobbes had also been experimenting. In his *Elements of Philosophy* (+1655), which included natural philosophy, he discussed the forces which act upon a sailing-ship beating to windward, and illustrated his argument by trials made on a wooden model of a sailing-carriage.[b] In +1684 the French Academy of Sciences prepared a questionnaire for the Jesuit Philippe Couplet to take back to China with him, in which details of the sailing-carriages were requested.[c] But *The Sacrifice*, Sir Francis Fane's tragedy about Tamerlane, published in London two years later, engagingly ridiculed Chinese antiquity and inventions, including the sailing-carriage.[d] On the other hand Leibniz, when projecting a kind of science museum, suggested that the exhibits should certainly comprise 'le chariot à voiles de Hollande — ou plutost de la Chine'.[e]

What did he mean by the sailing-carriage of Holland? Here the plot thickens. But before elucidating this point we may notice how odd it was that the early European travellers spoke of sailing-carriages rather than sailing wheelbarrows. Whether any of them could actually have seen four-wheeled wagons fitted with sails remains conjectural, for Chinese literature itself is quite silent concerning the use of sails on land in the Ming. Most probably travellers saw sailing wheelbarrows so heavily loaded that their wheels were hidden, and assumed that they must each have at least two.

There are one or two rather important references much earlier, however, indeed in the +6th century. The emperor Yuan of the Liang, in his *Chin Lou Tzu*, wrote:[f] 'Kaotshang Wu-Shu[1] succeeded in making a wind-driven carriage (*fêng chhê*[2]) which could carry thirty men, and in a single day could travel several hundred *li*.' Nothing else is known of this engineer or his sailing-car, but its performance would not be at all impossible, as we shall see in a moment. This would be about +550, and then comes around +610 the 'Mobile Wind-Facing Palace' (*kuan fêng hsing tien*[3]) of Yüwên Khai,

---

[a] Vol. 1, p. 333. It is not distinctly stated that Hooke was occupying himself with a sailing wheelbarrow, but it seems probable, for Martin Martini (+1655, p. 26) after describing Peking wheelbarrows made to carry three seated persons, had added that this was probably the origin of the stories in Europe about the sailing-carriage.

[b] In Molesworth's edition of the English *Works*, vol. 1, p. 340. On the theory of sailing-ships see further in Sect. 29g below. Sailing-carriage models still attract talented inventors; one such was made by the Russian engineer K. E. Tsiolkovsky (1857 to 1936), who later pioneered rocket flight (private communication from Mr D. R. Bentham).

[c] Pinot (2), p. 8. Some travellers, e.g. de Navarrete (+1676), had been sceptical (Cummins (1), vol. 2, p. 212).

[d] For this and for some of the other 17th-century references I am much indebted to the kindness of Mr Jasper Rose of King's College.

[e] *Sämtl. Schriften*, 5e Reihe, vol. 1, p. 564. Among many other mentions in this period we may cite Heylyn (1), p. 357; Anon. (40), p. 31.

[f] Ch. 6, p. 18a, tr. auct. The connection of this emperor (reigned +552 to +554) with such a machine is distinctly interesting as we meet with him in many other places in the present work, for example in connection with geobotanical prospecting (cf. Vol. 3, p. 676). From a special note which Pelliot consecrated to him (26) we learn the significant fact that he was both learned and Taoist. He wrote a commentary (long since lost) on the *Tao Tê Ching*. Once again we see the connection between Taoism and technology, with its necessarily unbiased observation of Nature.

[1] 高蒼梧叔   [2] 風車   [3] 觀風行殿

already mentioned.[a] Allowing for some exaggeration in the account, it seems not unlikely that this was also a land-sailing vehicle.[b]

After that there is a long silence, until just on a thousand years later, when Simon Stevin, the great Dutch mathematician and engineer, constructed a sailing-carriage of which we have many historical details. To Duyvendak (14) we owe an investigation of the story.[c] At a date not exactly determinable, but probably in the autumn of + 1600, Prince Maurice of Nassau invited several ambassadors and distinguished guests, among whom was the learned Grotius, then only 17, to take part in an experiment with a sailing-carriage constructed by Stevin. Two such vehicles had been made, a larger and a smaller, and both succeeded in accomplishing the distance between Scheveningen and Petten along the beach in less than two hours though it took fourteen hours to walk. Grotius composed several Latin poems on the subject, as well as other accounts.[d] Fig. 522, from a contemporary print by de Gheyn, shows the 'fleet' of vehicles.[e] Now Grotius refers to the sailing-carriages of China, and it has been established that he had read van Linschoten, and probably also Mendoza; we may therefore consider it almost sure that Stevin too was acquainted with these works. His successful trials were therefore a direct result of the contact of Europeans with China.[f]

[a] P. 253 above. The normal meaning of the words would of course be 'a temporary palace hall for viewing the scenery and local customs'. But the description which follows authorises the present guess. Another interpretation, adopted by Sickman & Soper (1), pp. 236, 239, makes the thing a rotating pavilion, something like Bernard Shaw's summerhouse on a large scale. There certainly was in several cultures a tradition of an imperial throne or dinner-table set within a baldachin or cupola of astronomical ornament and cosmological significance, all being rotated by animal-power in a basement. Such were Nero's Domus Aurea and the Sassanian Throne of Chosroes, carefully studied by Lehmann (1). Cf. the dome of Quṣair 'Amrah (Vol. 3, p. 389). Yüwên Khai's construction needs further research.

[b] There has been some discussion of ritual sailing-carriages in ancient Egypt, but we think that the case for them has not been made out. In the excavation of the temple of Madīnat Māḍī, built about − 1800, by an Italian expedition in 1936, a kind of wooden frame chassis with four very small solid wheels and numerous holes in its members was discovered. Hellenistic hymns in the temple inscriptions show that there was certainly a legend of kings or gods sailing on the land. Borchardt (3) has also reported the occurrence of a phrase of this kind (hn. t-nt-t!) in much more ancient inscriptions recording a particular temple festival, though the words imply steering a boat rather than running under sail as such. Dittmann (1), followed by Forbes (23, 24), believed that a square sail was set on a mast stepped directly into the central cross-piece of the chassis, but the small size of the wheels and the absence of any steering arrangements seem to us to make this most unlikely. Since however a considerable number of ancient Egyptian representations (many of them given by Dittmann himself) show funeral or other ceremonial cargoes laid in large model boats directly transported upon four-wheeled frames similar to the object discovered, it is surely more probable that the carriage was a mobile base on which such things were mounted for processional purposes. Occasionally, if the wind was in the right quarter and a sail had been set upon the model boat, it might have been found that the task of the oxen or the human pullers was perceptibly eased, and thus the legend might have grown up. However, no picture of a car with a sail directly attached to it, whether with or without hauling animals, has been found, so far as we know, in ancient Egyptian material. Any connection between this custom or legend and the practical successes of China and Europe in later times would be far to seek—but does mankind really ever forget anything? The bipod masts of China (cf. Vol. 4, pt. 3, Sect. 29d) give us pause.

[c] For further details see Forbes (24).

[d] Farraginis Liber, 1; Parallelon Rerumpublicarum, III, ch. 23; translations in Anon. (40), p. 30.

[e] The larger carriage carried a complement of 24 persons, so that Kaotshang Wu-Shu's must have been of the same order of size. Cf. Schmithals & Klemm (1), pl. 39.

[f] Here the reader is referred to Vol. 4, pt. 1, pp. 227 ff., where it was shown that Stevin was connected with a quite different invention also demonstrably based on Chinese antecedents, that of equal temperament in acoustics and music.

Long were the reverberations of that exhilarating day. Stevin's land-ship was copied and recopied[a] through the following couple of hundred years, first by John Wilkins, in his *Mathematical Magick* of +1648 ('That such Chariots are commonly used in the Champain plains of China is frequently affirmed by divers credible Authours'),[b] and then by many writers such as Emerson (+1758)[c] and Hooper (+1774).[d] Wilkins absorbed also a separate tradition, namely the idea of imparting motion to a carriage by means of a windmill and gearing; this went back to the European +14th century, when it had occurred to Guido da Vigevano in +1335, as later to Roberto Valturio about +1460.[e] The Valturian windmill-carriage embodied two vertical windmills but Branca in +1629 had proposed a horizontal windmill, anticipating in form the rotor ship, and it was this which Wilkins popularised. There is no reason to suppose that anything like it was ever built, but during the 18th century quite a number of wagons with sails like Stevin's were used in practice where conditions permitted.[f] Indeed, they continue to this day in many places on the north coasts of Belgium and France, and in California, where the sport of 'land yachts' is widely carried on. I myself saw and rode in them on the beach at La Panne as a small boy just about the time (1907) when Coppens de Houthulst was modernising them by the introduction of the tricycle-wheel system, pneumatic tyres, and light tubular construction. Today they are capable of speeds as high as the fastest express train, especially when they sail upon ice rather than sand. Fig. 523 shows some of these ice-yachts in Kaotshang Wu-Shu's own country, on the Liao River in Manchuria.

It might be thought that the sailing carriage (like the south-pointing carriage, on which see below, pp. 286 ff.) was merely one of the curiosities of history. Even Duyvendak esteemed the practical value of Stevin's 'invention' as nil. But these things have to be placed correctly in the perspective of technological development as a whole. Duyvendak himself pointed out that the sailing-carriage was able to travel with what was, in +1600, almost incredible speed. And here is the root of the matter, for we may say that this transmission from Chinese technology (strange as it may seem) was the first to accustom the European mind to the possibility of high-speed transit on land. A distance of nearly 60 miles was now covered in less than 2 hours, which must have meant that some stretches were travelled over at a speed higher than 30 m.p.h., perhaps in the neighbourhood of 40. When one remembers the excitement caused by modest speeds in the first days of railways, one is not inclined to underestimate the impact on European culture of what was really the first essay at rapid transportation. The Chinese stimulus, if it was no better, cannot be ignored, and the results were overwhelming.

---

a As late as 1834 a set of engravings by B. Rosaspina included one entitled 'Vettura olandese a vela di Simone Stevin Anno 1600'. This has two masts.

b (2), pp. 154 ff.   c Pl. xv.

d Pls. IX, X.

e Feldhaus (4), fig. 11, (1), col. 1274. Paradoxically, da Vigevano's picture of his windmill-driven carriages antedates by about a decade the second earliest remaining illustration of the windmill itself (see on, p. 555).

f References in Feldhaus (1), col. 1270. There is also the curious question of the application of sails to ploughs; on this see Leser (1), pp. 411, 450, (2), p. 451; Chevalier (1), p. 479.

PLATE CXCVII

Fig. 522. The successful sailing-carriages constructed by Simon Stevin (+1600), a print by de Gheyn. These 'land yachts' were inspired by the stories of the sailing-carriages of China, prevalent in Europe during the previous century. These may have transmitted authentic traditions from earlier centuries (cf. pp. 159, 253 and 278), but arose much more probably from the fleets of sailing wheelbarrows encountered by the early Portuguese travellers. Stevin's invention of the sailing-carriage paralleled another also derived from Chinese antecedents, his solution of the acoustic problem of the equal-tempered scale (cf. Vol. 4, pt. 1, pp. 227 ff.).

Fig. 523. Ice-yachts on the Liao River, near Ying-khou in Manchuria, c. 1935.

PLATE CXCVIII

Fig. 524. A drum-carriage of the Han dynasty, used in imperial processions; from the Hsiao Thang Shan tomb-shrines, *c.* +125 (*Chin Shih So* (Shih sect.), ch. 1, p. 133). Rubbing in the collection of Mr E. Bredsdorff from a relief now in the Imperial Palace Museum, Peking. Four musicians are playing on the pan-pipes (*hsiao*, cf. Vol. 4, pt. 1, p. 182 *et passim*, and Fig. 308) below; two others above strike the stand-drum (*chien ku*, cf. Vol. 4, pt. 1, Figs. 301, 303), which has two small clapper-bells (*ling*, cf. Vol. 4, pt. 1, p. 194) hanging from it. There is no doubt that the hodometer or 'automatic mile-measuring drum-chariot' (*chi li ku chhê*) with its puppet figures developed from the vehicle for living musicians, at some time between the beginning of the −1st and the end of the +3rd centuries.

When the foregoing paragraph was first written, its conclusion lacked the support of contemporary 17th-century testimony. But later on, a voice from the past provided it—that of the celebrated Gassendi, writing the life of his friend Fabri de Peiresc. Speaking of the year +1606, Gassendi says:[a]

Also he stept aside to Scheveling, to make triall of the carriage and swiftnesse of the waggon, which some yeers before was made with such Art, that it would run swiftly with sails upon the land, as a ship does in the sea. For he had heard how Grave *Maurice*, after the victory at Nieuport, for triall sake, got up into it, with Don *Francisco Mendoza* taken in the fight, and within two hours was carried to Putten which was 54 miles from Scheveling.[b] He therefore would needs try the same, and was wont to tell us, how he was amazed, when being driven by a very strong gale of wind, yet he perceived it not (for he went as quick as the wind), and when he saw how they flew over the ditches he met with, and skimmed along upon the surface only, of standing waters which were frequently in the way; how men which ran before seemed to run backwards, and how places which seemed an huge way off, were passed by almost in a moment; and some other such like passages.

(4) THE HODOMETER

The idea of a vehicle which would register the distance traversed appealed to mechanicians in more than one ancient civilisation. The hodometer, or 'way-measurer', was quite a simple proposition mechanically. All that was necessary was to make one of the road-wheels drive a system of toothed wheels constituting a reduction-gear train, so that one or more pins revolved slowly, releasing catches at predetermined intervals and striking drums or gongs.

Under the name of *chi li ku chhê*[1] (*li*-recording drum carriage), *ta chang chhê*[2] or *chi tao chhê*[3], this apparatus is mentioned in most of the official dynastic histories from the Chin onwards.[c] None of these references describe the mechanism, except that in the *Sung Shih*, which we shall examine in a moment. Some of them say, however, that at the end of each *li* traversed a wooden figure struck a drum (*ku*[4]) while at the end of each ten *li* another figure struck a gong (*cho*[5]). If the text of the *Ku Chin Chu* of Tshui Pao is to be trusted,[d] this double arrangement existed already in the +3rd century, but since its genuineness is a little uncertain, Wang Chen-To (*3*), who has carefully studied the subject, prefers the view that the more complex machine did not come in till the Sui or Thang. The names of certain engineers who constructed such hodometers with conspicuous success have been preserved, notably Chin Kung-Li[6] in the Thang (+9th century), Su Pi[7] in Wu Tai or Sung (+10th or +11th),

[a] (*1*), p. 104. We are again indebted to Mr Jasper Rose for bringing this account to our notice.
[b] The edition of +1657 misprints 14 here for 54.
[c] *Chin Shu*, ch. 25, p. 3*b*; *Sung Shu*, ch. 18, p. 5*b*; *Nan Chhi Shu*, ch. 17, p. 5*b*; *Sui Shu*, ch. 10, p. 4*a*; *Chiu Thang Shu*, ch. 45, p. 2*b*; *Hsin Thang Shu*, ch. 24, p. 1*b*; *Sung Shih*, ch. 149, p. 16*b*. Cf. also *Hsi Ching Tsa Chi*, ch. 5, p. 2*b* (mid +6th). The second of the names given above was an allusion to music which the legendary emperor Yao was believed to have composed; this agrees with the view that the hodometer developed from a mechanisation of a band borne on a carriage.
[d] Ch. 1, p. 1*a*.

[1] 記里鼓車　　[2] 大章車　　[3] 計道車　　[4] 鼓　　[5] 鐲　　[6] 金公立
[7] 蘇弼

Lu Tao-Lung[1] in +1027 and Wu Tê-Jen in +1107. In +1171 the Chin Tartars captured one or more hodometers belonging to the Sung.[a]

On the above evidence, the invention probably dates at least from the time of Ma Chün[2] (*fl.* +220 to +265). This conclusion is strengthened by the fact that the *Sun Tzu Suan Ching* (Master Sun's Mathematical Manual), which dates from the +3rd to the +5th century, contains a hodometer problem:[b]

> The distance between Chhang-an and Loyang is 900 *li*. Suppose a vehicle the wheel of which covers in one rotation 1 *chang* and 8 *chhih* (18 ft.). How many rotations of the wheel will there be between the two cities?

Apart from the book of Tshui Pao, the oldest description is in the *Chin Shu* (+635) which says:[c]

> The *chi li ku chhê* (mile-measuring drum-carriage) is drawn by four horses. Its shape is like that of the south-pointing carriage.[d] In the middle of it there is a wooden figure of a man holding a drumstick in front of a drum. At the completion of every *li*, the figure strikes a blow upon the drum.

Tshui Pao's own words (*c.* +300) are as follows:

> The *Ta Chang Chhê* was for knowing the distance along a road. It began in the Western Capital. It is also called the mile-measuring carriage (*chi li chhê*). It has two storeys both with a wooden figure. After every *li* traversed the lower figure strikes a drum; after every ten *li* the upper one rings a small bell. The *Shang Fang Ku Shih*[3] (Traditions of the Imperial Workshops) has recorded the method of construction.

An interesting point here is that the invention is clearly attributed to the Early Han period rather than the Later. The book referred to is listed in the Hou Han bibliography, but perished, alas, long ago.

'Drum-carriages' were also known in the Han and perhaps earlier, though they are not positively called 'mile-measuring drum carriages' until the San Kuo. It is probable, therefore, that such carriages in the Early Han were at first simply musical, intended for bands of musicians and drummers in state processions. The earliest seems to be that mentioned in the biography of Tan, prince of Yen Tzhu,[4] where we are only told[e] that flags and banners were carried before and behind it; this would be about −110 for he was the fourth son of Han Wu Ti. About −80 an important official, Han Yen-Shou,[5] also had a drum-chariot in his processions.[f] So did the Shanyu of the Southern Huns,[g] and in +37 horses of particularly fine quality presented by foreigners were harnessed to the Han emperor's drum-chariot.[h] The chapters on the imperial fleet of vehicles in the dynastic histories[i] naturally associate the drum-chariot with *huang mên*[6] people, i.e. eunuchs, palace officials, attendants and familiars,

---

[a] *Chin Shih*, ch. 43, p. 1*b*.   [b] Ch. 3, p. 14*a*, tr. auct.; cf. Mikami (1), p. 25.
[c] Ch. 25, p. 3*b*, tr. Giles (5), vol. 1, p. 223, slightly mod.
[d] See below, pp. 286ff.   [e] *Chhien Han Shu*, ch. 63, p. 9*b*.
[f] *Ibid.* ch. 76, p. 12*b*.   [g] *Hou Han Shu*, ch. 119, p. 5*a*.
[h] *Ibid.* ch. 106, p. 1*a*. Cf. *Chin Lou Tzu*, ch. 1, p. 11*b*; *I Chao Liao Tsa Chi*, ch. 2, p. 23*a*.
[i] *Hou Han Shu*, ch. 39, p. 9*a*; *Chin Shu*, ch. 25, p. 3*b*.

[1] 盧道隆   [2] 馬鈞   [3] 尙方故事   [4] 燕剌王旦   [5] 韓延壽
[6] 黃門

actors, acrobats, conjurers, etc.[a] While at first sight this would seem to strengthen the view that the early Han drum-chariots were purely musical, in fact it almost does the opposite, for we have already seen the intimate connection in ancient times between such entertainers and the makers of mechanical toys.[b] The most reasonable supposition is that the drum-chariot was indeed originally a vehicle for musicians, but that some time in the Early Han (−1st century) the beating of drums and gongs was arranged to work automatically off the road-wheels—Lohsia Hung (c. −110) may perhaps have been involved[c]—and that only then were the possibilities of such an instrument for surveying and charting itineraries realised. One pictorial representation has survived from the Han, namely that in the Hsiao Thang Shan tomb series, dating from about +125 (Fig. 524).[d] Carriages for musicians, whether mechanised or not, survived in imperial processions through many subsequent dynasties.[e]

The question of the date of the origin of the hodometer is important because parallel developments were occurring in Europe, especially in Alexandria, with Heron (c. +60).[f] Before glancing at these, however, let us examine the only extant specification. The *Sung Shih* says:[g]

The Hodometer.
It is painted red, with pictures of flowers and birds on the four sides, and constructed in two storeys, handsomely adorned with carvings. At the completion of every *li*, the wooden figure of a man in the lower storey strikes a drum; at the completion of every ten *li*, the wooden figure in the upper storey strikes a bell. The carriage-pole ends in a phoenix-head, and the carriage is drawn by four horses. The escort was formerly of 18 men, but in the 4th year of the Yung-Hsi reign-period (+987) the emperor Thai Tsung increased it to 30.

In the 5th year of the Thien-Shêng reign-period (+1027) the Chief Chamberlain Lu Tao-Lung[1] presented specifications for the construction of hodometers as follows:

'The vehicle should have a single pole and two wheels. On the body are two storeys, each containing a carved wooden figure holding a drumstick. The road-wheels are each 6 ft. in diameter, and 18 ft. in circumference, one revolution covering 3 paces. According to ancient standards the pace was equal to 6 ft. and 300 paces to a *li*; but now the *li* is reckoned as 360 paces of 5 ft. each.

A vertical wheel (*li lun*[2]) is attached to the left road-wheel; it has a diameter of 1·38 ft. with a circumference of 4·14 ft., and has 18 cogs (*chhih*[3]) 2·3 inches apart.

There is also a lower horizontal wheel (*hsia phing lun*[4]), of diameter 4·14 ft. and circumference 12·42 ft., with 54 cogs, the same distance apart as those on the vertical wheel (2·3 inches). (This engages with the former.)

[a] Cf. Yang Hsiung and the *huang mên*, Vol. 3, p. 358.
[b] Cf. Vol. 1, pp. 197ff.
[c] Astronomer and maker of astronomical instruments; cf. many refs. in Vol. 3.
[d] *Chin Shih So*, Shih sect., ch. 1 (p. 133), interpreted by the Fêng brothers as a cartload of musicians for an imperial cortège. See also Harada & Komai (1), vol. 2, pl. XVIII, 1, and Chavannes (11), pl. XXXVII, who did not himself think that it was a hodometer. Cf. Giles (5), vol. 1, pp. 273ff. The stone is now in the Imperial Palace Museum at Peking.
[e] Much information is collected in *Yü Hai*, ch. 79, pp. 20b, 34a, 41a, 44b, etc.
[f] Vacca (1) and others have conjectured that the Han hodometers were inspired by those of Alexandria. This might depend upon the exact time at which the mechanisation of the 'band-wagons' took place in China, but in any case this process itself, if we are right in assuming it, points to an indigenous development.
[g] Ch. 149, pp. 16bff., tr. Giles (5), vol. 1, p. 224, mod. auct.

[1] 盧道隆   [2] 立輪   [3] 齒   [4] 下平輪

Upon a vertical shaft turning with this wheel, there is fixed a bronze "turning-like-the-wind wheel" (*hsüan fêng lun*[1]) which has (only) 3 cogs, the distance between these being 1·2 inches.[a] (This turns the following one.)

In the middle is a horizontal wheel, 4 ft. in diameter, and 12 ft. circumference, with 100 cogs, the distance between these being the same as on the "turning-like-the-wind wheel" (1·2 inches).

Next, there is fixed (on the same shaft) a small horizontal wheel (*hsiao phing lun*[2]), 3·3 inches in diameter and 1 ft. in circumference, having 10 cogs 1·5 inches apart.

(Engaging with this) there is an upper horizontal wheel (*shang phing lun*[3])[b] having a diameter of 3·3 ft. and a circumference of 10 ft., with 100 cogs, the same distance apart as those of the small horizontal wheel (1·5 inches).

When the middle horizontal wheel has made 1 revolution, the carriage will have gone 1 *li* and the wooden figure in the lower story will strike the drum. When the upper horizontal wheel has made 1 revolution, the carriage will have gone 10 *li* and the figure in the upper storey will strike the bell. The number of wheels used, great and small, is 8 in all, with a total of 285 teeth.

Thus the motion is transmitted as if by the links of a chain, the "dog-teeth" mutually engaging with each other, so that by due revolution everything comes back to its original starting-point (*ti hsiang kou so, chhüan ya hsiang chih, chou erh fu shih*[4]).'

It was ordered that this specification should be handed down (to the appropriate officials) so that the machine might be made.

[The passage concludes by giving a similar specification prepared by Wu Tê-Jen[5] in +1107.]

This description brings out clearly enough the reduction train of gearing and omits only the pegs on the shafts which tripped the wires to operate the puppets. Following its words as closely as possible, model hodometers have been constructed by Bertram Hopkinson[c] (in Giles, 5) and Wang Chen-To (3). That of the latter, elegantly copying the vehicle shown in the Hsiao Thang Shan tomb relief, is shown in Fig. 525. Part of the model's mechanism, opened, is seen in Fig. 526. The canopy of the carriage looks as if it may have been intended to revolve while the vehicle was in motion, and it would have been easy enough to make it do so.

A moment's attention to the last sentence of Lu Tao-Lung's report is worth while. It seems almost the fragment of a poem or essay,[d] but the reference to 'dog-teeth' cogs in +1027 is important, since it shows that the Sung engineers were conscious of the necessity of rounding them off, empirically foreshadowing the kind of shapes used to-day in mathematically defined involute and epicycloid gear teeth. Perhaps a more extended treatment of this problem of form will be discovered in Sung techno-

---

[a] The name for this extremely small wheel shows that the Sung engineers fully understood the principle of velocity ratio, i.e. that the speed of a small wheel will be much faster than that of a large one with which it engages.

[b] One senses here the same lack of development of an adequate technical terminology which has been noted in other connections (cf. Vol. 2, pp. 43, 260, etc.).

[c] Late Professor of Engineering in the University of Cambridge.

[d] There were essays on the hodometer by various authors, such as Yang Wei-Chên[6] of the Yuan (G2415) and Chang Yen-Chen[7] of the Thang, but they are not informative from the engineering point of view (see *TSCC, Khao kung tien*, ch. 175).

[1] 旋風輪   [2] 小平輪   [3] 上平輪   [4] 遞相鉤鑷犬牙相制周而復始
[5] 吳德仁   [6] 楊維楨   [7] 張彥振

PLATE CXCIX

Fig. 525. Working model of a Han hodometer (Wang Chen-To, *3*).

PLATE CC

Fig. 526. Mechanism of the working model showing the gear on the right road-wheel and the reduction gear, including a pinion of three (Wang Chen-To, 3).

logical writings.[a] Historians of engineering in Europe[b] do not seem to know of any attention paid to cog shape before the time of Leonardo da Vinci (c. +1490). The oldest European MS. representation of gear teeth is about +1335, in the treatise of Guido da Vigevano, where the teeth are shown rounded, but this is later than those in Arabic and Chinese books.[c] Empirically, wheelwrights must have known that gear teeth had to be rounded off, and some late +15th-century gear-wheels still extant at the Pile Gate at Dubrovnik show this (Kammerer). The mathematics of the form of gear teeth were first discussed in +1557 by Jerome Cardan.[d] About +1720 empirical rules for shaping teeth were published by Sturm & Leupold,[e] and meanwhile the theoretical geometrical treatment had been begun by P. de la Hire (+1694) in his work on epicycloids.[f]

The classical description of a hodometer in Europe was that of Heron of Alexandria (c. +60),[g] though he did not claim it as a new invention. The more complex of his models recorded the distance travelled by the dropping of balls into receptacles.[h] The machine was also described by Vitruvius,[i] and used in the time of the emperor Commodus (+192), but after that there follows a very long gap, the next appearance of it being at the end of the +15th century in Western Europe.[j] The pattern is therefore the same as that which we have repeatedly met with, i.e. Greek antecedents, paralleled or followed at short distance by Chinese developments which continue throughout the medieval period, and then a reawakening of the subject in Europe. Most famous of the European Renaissance hodometers is doubtless that proposed by Leonardo da Vinci, illustrated by Uccelli[k] and Beck[l] and modelled by Guatelli. The actual use of the instrument in +16th- and +17th-century surveying has already been mentioned.[m] Whether it was ever employed by the Chinese cartographers we do not know.[n]

In general one may say that the hodometers of China were quite worthy ancestors of such remarkable modern machines as the vehicular hodograph (McNish & Tuckerman), which makes a complete plot, with compass-bearings, of the course which it traverses. As Liu Hsien-Chou (6) has pointed out, however, they are also very relevant

---

[a] Cf. above, p. 88, and below, in connection with clockwork, pp. 456, 473, 499.

[b] Such as Woodbury (2); Kammerer (1); Davison (10) or Uccelli (1), p. 71.

[c] Gear teeth in Arabic drawings always have a spiky diagrammatic look (cf. Coomaraswamy, 2). In the *Hsin I Hsiang Fa Yao* of +1090, none unfortunately are shown in profile; in the *Nung Shu* of +1313 they are drawn as if they were truncated steep-sided pyramids.

[d] *De Rerum Varietate*, pp. 263ff.

[e] (1), p. 49, fig. xv.

[f] Further details in Woodbury (2) and Matschoss & Kutzbach (1). An earlier French mathematician and engineer, Desargues (+1593 to +1661), seems to have used epicycloid curves for designing gear teeth, and there is also a claim for Rømer (+1644 to +1710). On the history of gear-cutting machinery, see Woodbury (1).

[g] *Dioptra*, ch. 34, tr. Brunet & Mieli (1), pp. 499, 515; cf. Usher (1), 1st ed., p. 99.

[h] Cf. Diels (1), pp. 64ff.   [i] x, ix, 1–4.

[j] Beckmann (1), vol. 1, pp. 5ff.   [k] (1), p. 49.

[l] (1), p. 424. Particulars of other interesting instruments will be found in Yde-Andersen (1).

[m] Vol. 3, p. 579.

[n] The regular mention of it as part of the fleet of imperial vehicles might be taken to imply that it was primarily a prestige object for processions and parade-grounds, but the strongly quantitative character of medieval Chinese cartography (cf. Sect. 22*d* above) must be remembered.

to the history of clockwork.[a] For besides the method of construction, in which the gear-train driven by the road-wheel included pinions of one or three, the signalisation was auditory by means of jack figures beating on drums or gongs. The hodometer was thus undoubtedly one of the precursors of all horological jack-work.

(5) THE SOUTH-POINTING CARRIAGE

If hodometers were widely spread, another kind of vehicle with gearing was peculiar to the Chinese culture-area. Allusion has already been made to the 'south-pointing carriage' (*chih nan chhê*[1]) in Sect. 26*i* on magnetism, since it was long confused, both by Chinese and Westerners, with the magnetic compass.[b] We know now, however, that it had nothing to do with magnetism, but was a two-wheeled cart with a train of gears so arranged as to keep a figure pointing due south, no matter what excursions the horse-drawn vehicle made from this direction.[c] Yet it is none the less interesting for being mechanical and not magnetic; and if it was probably of little practical use (though we might be unwise to exclude attempted applications in surveying, as for its companion vehicle the hodometer), it has not deserved the summary liquidation which some modern scholars adept in other fields have given it.[d]

The most important passage concerning the history of the south-pointing carriage is that in the *Sung Shu*,[e] written *c.* +500.

> The south-pointing carriage was first constructed by the Duke of Chou (beginning of the −1st millennium) as a means of conducting homewards certain envoys who had arrived from a great distance beyond the frontiers. The country to be traversed was a boundless plain, in which people lost their bearings as to east and west, so (the Duke) caused this vehicle to be made in order that the ambassadors should be able to distinguish north and south.
>
> The *Kuei Ku Tzu* book says that the people of the State of Chêng, when collecting jade, always carried with them a 'south-pointer', and by means of this were never in doubt (as to their position).[f]
>
> During the Chhin and Former Han dynasties, however, nothing more was heard of the vehicle. In the Later Han period, Chang Hêng[2] re-invented it, but owing to the confusion and turmoil at the close of the dynasty it was not preserved.
>
> In the State of Wei (in the San Kuo period) Kaothang Lung[3] and Chhin Lang[4] were both famous scholars; they disputed about the south-pointing carriage before the court, saying that there was no such thing, and that the story was nonsense. But during the Chhing-Lung reign-period (+233 to +237) the emperor Ming Ti commissioned the scholar Ma Chün[5] to

---

[a] See further in Sect. 27*j* below, e.g. p. 494.

[b] Schück & Riedmann (in Schück (1), vol. 2, p. 9) even tried to construct a model with a magnet in the arm of the figure, but without success. The confusion of course still persists in some writers, e.g. Feldhaus (20), p. 55 — in 1954.

[c] The pioneer work on this was done by H. A. Giles (5), vol. 1, pp. 107, 219, 274, and by Hashimoto (3). The subject has been reviewed (down to 1924) by Li Shu-Hua (1).

[d] E.g. J. O. Thomson (1), p. 33: 'the south-pointing chariot may be dismissed as a fable'. It is fair to add that such judgments have been due partly to the fact that the machine got into the legends, and it has been quite justifiable to reject their −3rd millennium datings.

[e] Ch. 18, pp. 4*b*ff., tr. Giles (5), vol. 1, p. 110, mod. auct.

[f] Cf. Vol. 4, pt. 1, p. 269 above, where this story has already been evaluated.

¹ 指南車   ² 張衡   ³ 高堂隆   ⁴ 秦郎   ⁵ 馬鈞

construct one, and he duly succeeded.[a] This again was lost during the troubles attending the establishment of the Chin dynasty.

Later on, Shih Hu[1] (emperor of the Hunnish Later Chao dynasty)[b] had one made by Hsieh Fei;[2] and again Linghu Shêng[3] made one for Yao Hsing[4] (emperor of the (Chiang) Later Chhin dynasty).[c] The latter was obtained by emperor An Ti of the Chin in the 13th year of the I-Hsi reign-period (+417), and it finally came into the hands of emperor Wu Ti of the (Liu) Sung dynasty when he took over the administration of Chhang-an. Its appearance and construction was like that of a drum-carriage (hodometer). A wooden figure of a man was placed at the top, with its arm raised and pointing to the south, (and the mechanism was arranged in such a way that) although the carriage turned round and round, the pointer-arm still indicated the south. In State processions, the south-pointing carriage led the way, accompanied by the imperial bodyguard.[d]

These vehicles, constructed as they had been by barbarian workmen, did not function particularly well. Though called south-pointing carriages, they very often did not point true, and had to negotiate curves step by step, with the help of someone inside to adjust the machinery.

That ingenious man from Fanyang,[e] Tsu Chhung-Chih,[5] frequently said, therefore, that a new (and properly automatic) south-pointing carriage ought to be constructed. So towards the close of the Shêng-Ming reign-period (+477 to +479) the emperor Shun Ti, during the premiership of the Prince of Chhi, commissioned (Tsu) to make one, and when it was completed it was tested by Wang Sêng-Chhien,[6] military governor of Tanyang, and Liu Hsiu,[7] president of the Board of Censors. The workmanship was excellent, and although the carriage was twisted and turned in a hundred directions, the hand never failed to point to the south.[f]

Under the Chin, moreover, there had also been a south-pointing ship.[g]

The pigtailed barbarian[h] Thopa Tao[8] (third emperor of the Northern Wei dynasty)[i] caused a south-pointing carriage to be constructed by an artificer called Kuo Shan-Ming,[9] but after a year it was still not finished. (At the same time) there was a man from Fu-fêng, Ma Yo,[10] who succeeded in making one, but when it was ready he was poisoned by Kuo Shan-Ming.

The first thing needing discussion here is the legendary material. The apparatus came in the course of time to be associated with two fabled events: (a) the battle between Huang Ti and the great rebel Chhih-Yu[11] when the latter made a smoke-screen of fog through which the imperial army had to find its way,[j] and (b) the sending home by Chou Kung of the ambassadors of the Yüeh-Shang[12] people to some place in the far south, for which guidance was necessary. In Klaproth's time it was thought

---

[a] The story of this discussion and achievement occurs in *San Kuo Chih* (*Wei Shu*), ch. 29, p. 9a; also (quoting *Wei Lüeh*) in ch. 3, p. 13b in connection with the construction of the great palace and gardens at Loyang. We have translated the former source on p. 40 above. Cf. *Yü Hai*, ch. 79, p. 28a.

[b] This minor dynasty lasted from +319 to +352 and Shih Hu reigned from +334 to +349.

[c] This minor dynasty lasted from +384 to +417 and Yao Hsing reigned from +394 to +416.

[d] Cf. *Chin Shu*, ch. 25, p. 3b, where a similar order is prescribed.

[e] We have often encountered this eminent mathematician and engineer in foregoing Sections; cf. index entries in Vol. 3.

[f] This was at the very end of the (Liu) Sung dynasty.   [g] On this see Vol. 4, pt. 1, p. 292 above.

[h] The expression used was a pejorative way of referring to northerners in this divided period and arose from the braided plaits of hair which they wore round their heads.

[i] He reigned from +423 to +452.   [j] Cf. Vol. 2, p. 115.

[1] 石虎　　　[2] 解飛　　　[3] 令狐生　　　[4] 姚興　　　[5] 祖沖之
[6] 王僧虔　　[7] 劉休　　　[8] 拓跋燾　　　[9] 郭善明　　[10] 馬岳
[11] 蚩尤　　　[12] 越裳

that Han texts contained these stories,[a] but though the former event is described in the *Shih Chi* [b] and the latter in the *Shang Shu Ta Chuan*[1],[c] there is no mention of the south-pointing carriage. It has been said that there is no reference to it in any Han or pre-Han text, and indeed the only exception seems to be Liu Hsiang's *Hung Fan Wu Hsing Chuan* (*c.* −10)[d] where the allusion is metaphorical. I suspect that this is one of those cases (cf. Vol. 4, pt. 1, pp. 269 ff.) where the word for carriage was slipped in subsequently, the scribe not understanding that the 'south-pointer' reference had been to the lodestone spoon. The two legends,[e] however, appeared in developed form in the *Ku Chin Chu* of Tshui Pao (*c.* +300).[f] This suggests that the south-pointing carriage was really a Later Han or Chin invention, and that the first machine was made, if not by Chang Hêng[g] about +120, then by Ma Chün about +255.

There is certainly no reason to doubt the story about Ma Chün. He was an excellent engineer, and we are fortunate to have much information about him recorded by his friend Fu Hsüan[2] the philosopher; we shall meet with him again when speaking of looms (Sect. 31) and water-raising machinery (p. 346 below). We have already made his acquaintance in connection with the social position of the engineer in medieval China (p. 39 above). The making of the south-pointing carriage is recorded also in the *Wei Lüeh* and other contemporary sources.[h] Kuo Yuan-Shêng,[3] in the *Shu Chêng Chi*[4] (Records of Military Expeditions), a Chin book, says that outside the south gate of the capital there were the Government Workshops (Shang Fang[5]) and that the south-pointing carriage was normally garaged in the north gateway of this factory.[i] He may well have been referring to Ma Chün's machine.[j] It is interesting that Tshui Pao's account ends by saying that the construction is described in a book now long lost, the *Shang Fang Ku Shih* (Traditions of the Imperial Workshops).[k]

[a] The only books then available being late historical compilations. For instance, the *Tzu Chih Thung Chien Kang Mu* (*Chêng Pien*), ch. 15, p. 55*a*.

[b] Ch. 1, p. 3*b* (Chavannes (1), vol. 1, p. 29).   [c] By Fu Shêng[6] (cf. above, Vol. 2, p. 247).

[d] 'Discourse on the Hung Fan Chapter of the *Shu Ching*, in relation to the Five Elements'; passage preserved in *TPYL*, ch. 775, p. 1*b*. There is also a passage in the *Kuei Ku Tzu* book about the story of the guidance of ambassadors (ch. 10, p. 19*b*), repeated in *Yü Hai* (ch. 78, p. 30*b*) and elsewhere, but modern editors regard this as an interpolation from a later commentary. A delightful +18th-century echo of this legend occurs in the memoirs of John Bell of Antermony (1), vol. 2, p. 44, the Scots physician who went to Peking in the train of the Russian ambassador L. V. Ismailov—and he heard it from the lips of the Khang-Hsi emperor himself. On 2 January 1721, at a private audience, '[the Emperor] discoursed of the invention of the load-stone, which, he said, was known in China above two thousand years ago; for it appeared from their records that a certain ambassador from some distant island to the court of China, missing his course in a storm, was cast on the Chinese coast in the utmost distress. The then Emperor, whose name I have forgot, after entertaining him hospitably, sent him back to his own country; and to prevent the like misfortunes in his voyage homeward gave him a compass to direct his course.'

[e] Both translated in Klaproth (1), pp. 74, 78, 80, 82.

[f] The genuineness of this book has often been attacked, but Hashimoto's defence of it seems rather convincing. The passage in question is repeated in Ma Kao's +10th-century *Chung Hua Ku Chin Chu*. Ch. 1 in both cases.

[g] There is no mention of it in Chang Hêng's biography (*Hou Han Shu*, ch. 89).

[h] *San Kuo Chih* (*Wei Shu*), ch. 29, p. 9*a*.   [i] *TPYL*, ch. 775, p. 1*b*.

[j] Cf. *Chin Shu*, ch. 25, p. 3*b*, from which it is clear that the Chin emperors had at least one.

[k] This is listed, as we have seen, in the *Hou Han Shu* bibliography.

[1] 尚書大傳   [2] 傅玄   [3] 郭緣生   [4] 述征記   [5] 尚方
[6] 伏勝

It may be remembered that we referred above (Vol. 4, pt. 1, p. 293) to the fact that at least one of the south-pointing carriages made by the northern barbarian dynasties had lost its machinery when the Sung people captured it, and had to be turned by a man inside. This is from the biography of Tsu Chhung-Chih,[a] the great mathematician, who successfully constructed a new one in +478:

> When emperor Wu of the (Liu) Sung subdued Kuan-chung he obtained the south-pointing carriage of Yao Hsing, but it was only the shell with no machinery inside. Whenever it moved it had to have a man inside to turn (the figure). In the Shêng-Ming reign-period, Kao Ti commissioned Tsu Chhung-Chih to reconstruct it according to the ancient rules. He accordingly made new machinery of bronze, which would turn round about without a hitch and indicate the direction with uniformity.[b] Since Ma Chün's time such a thing had not been.

In the +7th century the invention reached Japan, as we hear of two monks, Chih-Yü [1] and Chih-Yu,[2] constructing such vehicles for the Japanese emperor in +658 and +666.[c] In all probability these ecclesiastical engineers had themselves come from China. The south-pointing carriage was naturally combined sometimes with the hodometer. Later books[d] mention model south-pointing carriages only 15 inches high. Such a one it must have been that Chu Tê-Jun [3] saw at the house of a friend early in the +14th century, and described in his *Ku Yü Thu* (Illustrated Record of Ancient Jade Objects)[e] of +1341. The passage was copied into the *San Tshai Thu Hui* encyclopaedia,[f] where we read:

> The picture on the right (see Fig. 527) is the ornamental (figure) on a (model) south-pointing carriage. By standard measure it was 1·42 ft. high and 7·4 inches long. The (vertical) shaft rose through a hole 3·7 inches in diameter, its own diameter being 3·4 inches. At the top it carried a human figure carved in jade, with one hand constantly indicating the south. The shaft penetrated down through the hole and turned around on its axis. The figure (was represented as) treading upon (an image of) Chhih-Yu. In the Yen-Yu reign-period (+1314 to +1320) I myself saw this (model) at the house of the Senior Academician Yao Mu-An.[4] The colour of the jade was pale reddish-yellow....

This was the figure which so much attracted the attention of occidental sinologists, and which was so often reproduced.[g] What had been a symbol of magic power in the

---

[a] *Nan Chhi Shu*, ch. 52, p. 20*b*; *Nan Shih*, ch. 72, p. 11*b*; tr. Moule (7).

[b] Both the sources add that an engineer 'from the North', So Yü-Lin,[5] claimed to be able to construct a south-pointing carriage at the same time as Tsu Chhung-Chih. He was given similar facilities, but when tests were made in the Lo-Yu Park, his machine worked very badly, so it was laid aside and eventually burnt. It is curious to have to visualise something like the Rainhill 'Rocket' trials taking place more than thirteen centuries earlier so far from Europe. See also *Yü Hai*, ch. 79, p. 49*a*.

[c] *Nihongi*, cit. in the *Wajishi* under relevant dates; see Klaproth (1), p. 93; Li Shu-Hua (1), p. 88; Aston (1), pp. 258, 285. Possibly these were two names for the same monk, but neither appears in the *Kao Sêng Chuan*.

[d] *Shih Wu Kan Chu* (Ming), cit. *KCCY*, ch. 29, p. 24*b*.

[e] Ch. 1, p. 2. [f] Chhi yung sect., ch. 5, p. 10*b*, tr. auct.

[g] See Laufer (8), p. 113; Giles (5), p. 114 (from *TSCC*), etc. Chu Tê-Jun's words were also copied, with his figure, into the *Chin Shih So* (Chin sect., ch. 2, p. 85). Fêng Yün-Phêng added that he had seen at Suchow a figure somewhat similar which might have belonged to a war-chariot. As this was, he said, of bronze, later sinologists, not reading the caption very carefully, assumed that one or more bronze figures for model south-pointing carriages had been seen and described, but this is not the case.

[1] 智踰   [2] 知由   [3] 朱德潤   [4] 姚牧菴   [5] 索馭驎

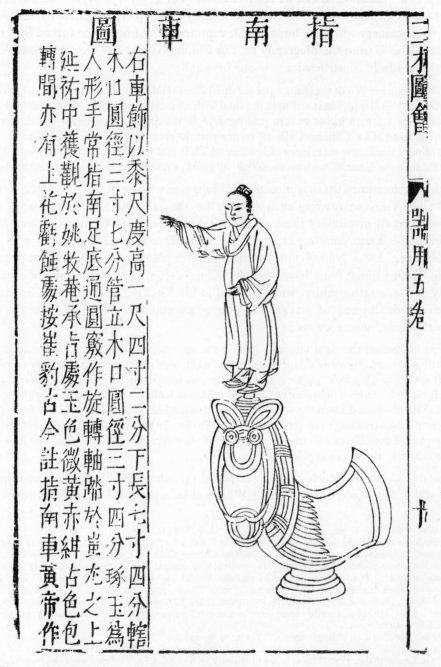

圖 車 南 指

右車飾以黍尺度高一尺四寸二分下長六寸四分轄
木口圓徑三寸七分管立木口圓徑三寸四分琢玉為
人形手常指南足底通圓竅作旋轉軸跗於嵌左之上
延祐中獲觀於姚牧菴承旨廢玉色微黃赤絣占色包
轉間亦有上花欞餙傖按崔豹古今註指南車黃帝作

Fig. 527. Ornamental figure of jade for a model south-pointing carriage, from *San Tshai Thu Hui* (Chhi yung sect.), ch. 5, p. 10 *b* (+1609). The text on the left is translated on p. 289.

+4th century had thus become an entertainment of the learned in the +14th, but it is interesting that as late as the Yuan there were still mechanics who could make it work. After that time no more is heard of it. We must now return to the Sung.

The *Sung Shih* gives a historical account similar to the *Sung Shu*, adding that between +806 and +821 Chin Kung-Li[1] presented a south-pointing carriage and a hodometer to the Thang emperor,[a] and that in +987 the escort was increased from 18 to 30 men. But its main importance is that it gives the only detailed description of the machinery involved, as constructed by two engineers, Yen Su[2] in +1027, and Wu Tê-Jen[3] in +1107. The text says:[b]

In the 5th year of the Thien-Shêng reign-period of the emperor Jen Tsung (+1027), Yen Su, a Divisional Director in the Ministry of Works, made a south-pointing carriage. He memorialised the throne, saying, [after the usual historical introduction]:

'Throughout the Five Dynasties and until the reigning dynasty there has been, so far as I know, no one who has been able to construct such a vehicle. But now I have invented a design myself and have succeeded in completing it.

The method involves using a carriage with a single pole (*yuan*[4]) (for two horses). Above the outside framework of the body of the carriage let there be a cover in two storeys. Set a wooden image of a *hsien* (immortal) at the top, stretching out its arm to indicate the south. Use 9 wheels, great and small, with a total of 120 teeth, i.e.

2 foot-wheels (i.e. road-wheels, on which the carriage runs) 6 ft. high and 18 ft. in circumference,

attached to the foot-wheels, 2 vertical (*ching*[5]) subordinate (*tzu*[6]) wheels, 2·4 ft. in diameter and 7·2 ft. in circumference, each with 24 teeth (*chhih*[7]), the teeth being at intervals of 3 inches apart,

then below the crossbar (*hêng mu*[8]) at the end of the pole (*yuan*[4]) two small vertical wheels [A] 3 inches in diameter and pierced by an iron axle,

to the left 1 small horizontal (*phing*[9]) wheel, 1·2 ft. in diameter, with 12 teeth,

to the right 1 small horizontal wheel, 1·2 ft. in diameter, with 12 teeth,

in the middle 1 large horizontal wheel, of diameter 4·8 ft. and circumference 14·4 ft., with 48 teeth, the teeth at intervals of 3 inches apart.

in the middle a vertical shaft piercing the centre (of the large horizontal wheel) 8 ft. high and 3 inches in diameter; at the top carrying the wooden figure of the *hsien*.

When the carriage moves (southward) let the wooden figure point south. When it turns (and goes) eastwards, the (back end of the) pole is pushed to the right; the subordinate wheel attached to the right road-wheel will turn forward 12 teeth, drawing with it the right small horizontal wheel one revolution (and so) pushing (*chhu*[10]) the central large horizontal wheel to revolve a quarter turn to the left. When it has turned round 12 teeth, the carriage moves eastwards, and the wooden figure stands crosswise and points south. If (instead) it turns (and goes) westwards, the (back end of the) pole is pushed to the left; the subordinate wheel attached to the left road-wheel will turn forward with the road-wheel 12 teeth, drawing with it the left small horizontal wheel one revolution, and pushing the central large horizontal

---

[a] There is more in *Thang Liu Tien*, ch. 17, and *Yü Hai*, ch. 79, pp. 40*b*, 46*b*, 49*b*. Cf. *Sui Shu*, ch. 10, p. 3*b*.

[b] Ch. 149, pp. 14*a* ff., tr. Moule (7), mod. auct., adjuv. Giles (5), vol. 1, p. 219. The Jurchen Chin emperor had both machines in +1171 (*Chin Shih*, ch. 43, p. 1*b*).

[1] 金公立   [2] 燕肅   [3] 吳德仁   [4] 轅   [5] 徑   [6] 子
[7] 齒   [8] 橫木   [9] 平   [10] 觸

wheel to revolve a quarter turn to the right. When it has turned round 12 teeth, the carriage moves due west, but still the wooden figure stands crosswise and points south. If one wishes to travel northwards, the turning round, whether by east or west, is done in the same way.'

It was ordered that the method should be handed down to the (appropriate) officials so that the machine might be made.

In the first year of the Ta-Kuan reign-period (+1107), the Chamberlain Wu Tê-Jen presented specifications of the south-pointing carriage and the carriage with the *li*-recording drum (hodometer). The two vehicles were made, and were first used that year at the great ceremony of the ancestral sacrifice.

The body of the south-pointing carriage was 11·15 ft. (long), 9·5 ft. wide, and 10·9 ft. deep.

(A)   The carriage wheels were 5·7 ft. in diameter, the carriage pole 10·5 ft. long, and the carriage body (*hsiang*[1]) in two storeys, upper and lower. In the middle was placed a partition. Above there stood a figure of a *hsien* holding a rod, on the left and right were tortoises and cranes, one each on either side, and four figures of boys each holding a tassel.

In the upper storey there were at the four corners trip-mechanisms (*kuan-li*[2]),[a] and also 13 horizontal wheels (*wo lun*[3]), each 1·85 ft. in diameter, 5·55 ft. in circumference, with 32 teeth at intervals of 1·8 inches apart. A central shaft, mounted on the partition, pierced downwards.

(C)   In the lower story were 13 wheels. In the middle was the largest horizontal wheel, 3·8 ft. in diameter, 11·4 ft. in circumference, and having 100 teeth at intervals of 1·25 inches apart.

(D)   (On vertical axles) reaching to the top (of the compartment) left and right, were two small horizontal wheels *which could rise and fall*,[b] having an iron weight (attached to) each. Each of these was 1·1 ft. in diameter and 3·3 ft. in circumference, with 17 teeth, at intervals of 1·9 inches apart.

(B)   Again, to left and right, were attached (*fu*[4]) wheels, one on each side, in diameter 1·55 ft., in circumference 4·65 ft., and having 24 teeth, at intervals of 2·1 inches.

(F, G)   Left and right, too, were double gear-wheels (lit. tier-wheels; *tieh lun*[5]), a pair on either side. Each of the lower component gears was 2·1 ft. in diameter and 6·3 ft. in circumference, with 32 teeth, at intervals of 2·1 inches apart. Each of the upper component gears was 1·2 ft. in diameter and 3·6 ft. in circumference, with 32 teeth, at intervals of 1·1 inches apart.

(H)   On each of the road-wheels of the carriage, left and right, was a vertical wheel 2·2 ft. in diameter, 6·6 ft. in circumference, with 32 teeth at intervals of 2·25 inches apart.

(E)   Both to left and right at the back end of the pole there were small wheels *without teeth* (pulleys),[b] from which hung bamboo cords, and both were tied above the left and right (ends of the) axle (of the carriage) respectively.

If the carriage turns to the right, it causes the small pulley to the left of the back end of the pole to let down the left-hand (small horizontal) wheel. If it turns to the left, it causes the small pulley to the right of the back end of the pole to let down the right (small horizontal) wheel.[c] However the carriage moves the *hsien* and the boys stand crosswise and point south.

The carriage is harnessed with two red horses, bearing frontlets of bronze. . . .

ᵃ Presumably to actuate the boys to wave their tassels. On the term *kuan-li* see p. 485 below. Or a hodometer signal may have been incorporated in the design, though nothing is said of drums or gongs.
ᵇ Italics inserted.
ᶜ The sense necessitates the inversion of the words for left and right in the text. The characters could easily be confused. Moreover, the dimensions given throughout the passage cannot all be correct, as they will not fit together.

¹ 箱   ² 關戾   ³ 臥輪   ⁴ 附   ⁵ 疊輪

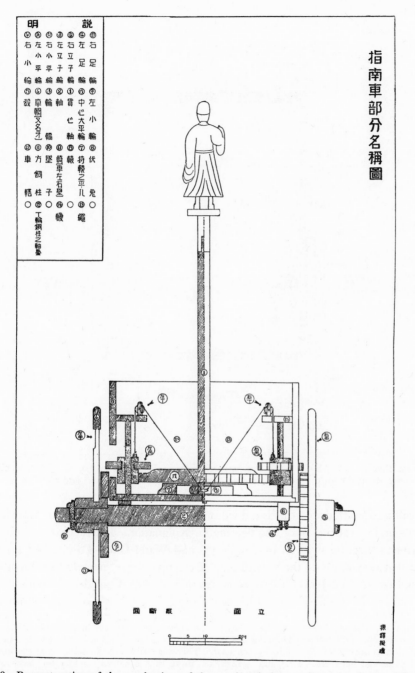

Fig. 528. Reconstruction of the mechanism of the south-pointing carriage according to Moule (7) and Wang Chen-To (3); back elevation. As the rear end of the vehicle's pole ⑧ moves to the left or right, it engages and disengages the suspended gear-wheels to the left and right respectively, thus connecting or disconnecting each road-wheel with the central gear-wheel carrying the pointing figure.

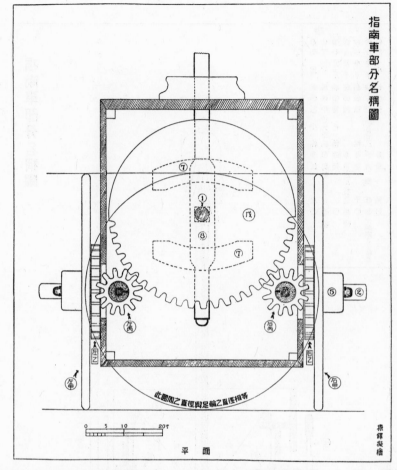

Fig. 529. Plan of the Moule-Wang reconstruction (from Wang Chen-To, 3). The vehicle's pole rotates about the lower bearing of the vertical shaft carrying gear-wheel and pointing figure.

It will be admitted that this is a document distinctly precious for our knowledge of +11th- and +12th-century engineering.[a] Explanations and reconstructions of the mechanism have been given by Moule (7) and Wang Chen-To (3), and I have no doubt that they apply to the second part of the text above, namely the machine constructed by Wu Tê-Jen. This we must examine first.[b] Consider the back elevation (Fig. 528) and the plan (Fig. 529), taken from Wang Chen-To, but embodying the

[a] The *Sung Shih* version may not be the earliest, however, for a passage in almost identical wording occurs in the *Khuei Than Lu*[1] (Thinking of Confucius Asking Questions at Than), a record of governmental matters during the Sung (ch. 13, pp. 1 aff.). This was written by Yo Kho,[2] who was already an important official between +1208 and +1224, i.e. more than a century before the *Sung Shih* was compiled (+1345). Yo Kho gives Wu Tê-Lung[3] instead of Wu Tê-Jen, probably erroneously.

[b] We may leave on one side the 13 wheels in the upper chamber which were purely concerned with the puppet mechanism of the tortoises and cranes, and ensured that the four corner-boys behaved in the same way as the central figure. Cf. here Liu Hsien-Chou (7), p. 104.

[1] 愧郯錄   [2] 岳珂   [3] 吳德隆

PLATE CCI

Fig. 530. Wang Chen-To's working model with the large gear-wheel removed to show the position of the vehicle's pole and the cables which connect its rear end with the left and right rising and falling gear-wheels. These slide freely on their vertical shafts, above which the small pulleys for the cables can be seen.

same solution as that of Moule. The essence of the mechanism was that the subsidiary or inner gear-wheels fixed to the road-wheels engaged with small horizontal toothed wheels which rotated the central large one attached to the same shaft as the main figure. But the small horizontal toothed wheels were never both in gear at the same time; mounted on weights sliding up and down vertical shafts, they were hung on cords passing over pulley-wheels above and attached to the rear end of the carriage-pole below. This is well seen in Wang's model (Fig. 530). Supposing that the carriage, going south, should deviate to the west (i.e. turn to the right), the horses would carry round the pole to the right and hence its rear end would move to the left. This would raise the right small gear-wheel and lower the left one so that it enmeshed with the inner gear of the road-wheel and with the central wheel. As the left road-wheel would be moving while the right one would be out of gear, this train would obviously have the effect of compensating for the change in direction of the carriage and so maintaining approximately the south-pointing direction of the wooden figure. Upon resumption of a straight course, both gear-wheels would again be disengaged.

Wu Tê-Jen's machine was in fact more complicated than this, since he introduced additional components in the gear-train, including (a feature of much interest for the history of technology) two-tier or double-gear wheels.[a] His system can be appreciated from Fig. 531, adapted from Moule (7). The gear-wheel (H) fixed to the road-wheel (A) engaged with another gear-wheel (B) immediately above it, and this in turn engaged and disengaged at a right angle with the vertically movable cogwheel (D), seen below its suspension pulley. This in turn moved (when in action) the lower component (F) of the double gear-wheel, the upper part (G) of which moved the central wheel (C). The object of thus multiplying the cogwheels is not very clear. That these reconstructions do, however, represent Wu Tê-Jen's mechanism, is sufficiently convincing, seeing that he specifically states that two of the wheels could rise and fall, that the cords controlling this were fixed to the rear end of the carriage-pole, and that two of the wheels were pulleys without teeth.

Moule assumed that this reconstruction applied also to the mechanism of Yen Su,[b] and perhaps to that of others such as Ma Chün. But this is not at all sure. Though scrupulously careful to adjust the text as little as possible, Moule had to insert the words '(without teeth)' at the place marked [A] in the exposition of Yen Su.[c] Yet if these wheels were in fact cogwheels the situation would be very different. Now although the solution of Moule and Wang is theoretically feasible, it is mechanically very inelegant, and the amount of play likely to have been involved, together with the difficulty of inducing the wheels to engage and disengage properly, would have made it hardly workable.[d] Moule always dismissed such arguments by saying that the

---

[a] So far as we know, these had appeared first in Heron (+1st century; cf. Beck, 2) or more probably Pappus (c. +300, Tshui Pao's contemporary; cf. Beck (1), p. 29). One of their most important ultimate applications was in the fusee of spring clockwork (cf. pp. 444, 526 below, and Ward, 2).

[b] Another solution for this by Pao Ssu-Ho is given by Liu Hsien-Chou (7), p. 102.

[c] This applies both to the *Sung Shih* and to the *Khuei Than Lu*.

[d] As Dr Frank Malina has pointed out, it could have been improved by the use of wire cable tautly stretched between the gear-wheels underneath the carriage floor to form a continuous system, as also by the use of cam-shaped pulleys, but there is no evidence that such devices were employed.

machine was never intended to serve any practical purpose, but even if, as is probable, it was primarily a symbolic imperial prestige gadget, there is no reason to assume that it never worked properly. Otherwise envoys and tribute-missions, watching it on parade-grounds and in State processions, would not have been impressed. An entirely

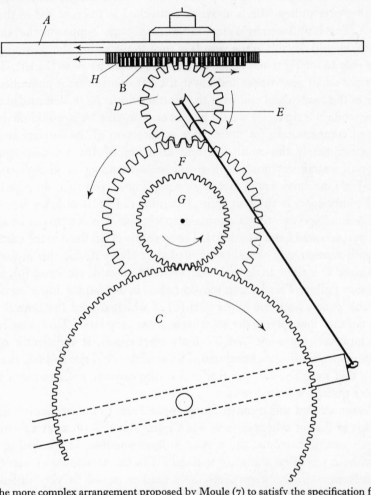

Fig. 531. The more complex arrangement proposed by Moule (7) to satisfy the specification for a south-pointing carriage due to Wu Tê-Jen (+1107) and described in the *Sung Shih*. Translation in the text. Two double gear-wheels are necessary, as explained on p. 295.

different solution was therefore suggested by George Lanchester, himself an eminent engineer, who proposed that the machine (in some cases, at any rate) embodied a simple form of differential gear.[a]

The differential drive is an important element in the modern automobile. As has

[a] He was, however, no sinologist, and owing to lack of competent advice his paper (1), as first presented, embodied acceptance of legendary datings which caused it to be undervalued by the literati. For a lucid and authoritative account of differential motions see Jones & Horton (1), vol. 1, pp. 363ff., 365ff., and for the automobile drive in particular, vol. 1, pp. 379ff.

already been evident, when any wheeled vehicle rounds a corner, the outer wheels travel much further than the inner. The dead axles of carts or wagons permit each wheel to revolve independently, but on a live axle, through which power is transmitted, some device which allows the wheels to move independently while at the same time conveying the torque is essential. The usual type of motor-car differential gear is shown in Fig. 532. The driveshaft A ends in a bevelled pinion B, which drives the broad bevelled gear C, but though concentric with the left road axle, C is not fixed to it. On the contrary, it carries a spider or differential case, D, rigidly fixed to it and carrying bevel gears, E, E' on a transverse arm. The spider is of course also free of the right axle. The drive or torque is transmitted through the stationary teeth of these 'idling wheels' to the bevel gears F, F' on the two half-axles, and E, E' do not move on their own arm or 'stub-shaft' at all, so long as the vehicle is going straight forward. In other words, E, E' and F, F' revolve with the spider and have no motion relative to one another. But if one road-wheel is held back or accelerated relative to its companion,

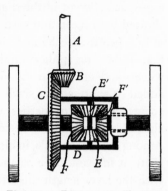

Fig. 532. Diagram to illustrate the principle of the differential drive of motor-vehicles.

it can move independently, drive being maintained, all that happens being that E and E' change their position, running round relative to F and F'.

So much for the 'differentials' of our motor-cars. This nest of gear-wheels used to be called the Starley gear, after the name of the inventor who first applied it to vehicles in 1879. Many have thought that this was the beginning of it, but in an interesting series of papers von Bertele (4, 5) has shown that it was used much earlier, in 18th-century horology.

In order to understand its development in Renaissance Europe one must remember that differential gear, since it embodies right-angle gearing (whether bevelled or not), is a more complicated 'three-dimensional' form of epicyclic gearing.[a] All such gear-trains however differ from conventional gear-wheel assemblies in that the axis of at least one toothed wheel revolves in a circular orbit with respect to the axes of the other wheels. They have proved extremely valuable as constructional elements of many machines (computers, effector mechanisms, etc.)[b] because they permit of numerical relations difficult or impossible to attain in the normal way. Small angular velocities can be easily added or subtracted. Not surprisingly then, the epicyclic system was the first to appear;[c] it is found in a globe clock of about +1575 made by Baldewein (Jost Burgi's predecessor) for William IV at Cassel, and ensures the maintenance

---

[a] We shall shortly encounter a famous case of epicyclic gearing in the 'sun-and-planet' gear fitted by James Watt to his early steam-engines (p. 386 below). For other applications cf. Jones & Horton (1), vol. 1, pp. 175ff., 324ff., 365ff., vol. 3, pp. 305ff.

[b] Many uses of differential gearing, in speed-changing and regulating mechanisms, in control and governing devices for valve gear, turbines, cranes and hoists, etc., will be found described in Jones & Horton (1), vol. 1, pp. 329ff., 376ff., 381ff., 385ff.

[c] One form of it occurs in the Anti-Kythera calendrical computer of about −80; see Price (9). De Dondi used it and Leonardo sketched it; see Price (10).

of a very small velocity difference (366/365).[a] After this it was quite widely used in astronomical clocks, though it seems to have been reinvented several times.[b]

True differential gearing was introduced by Joseph Willlamson about +1720 for correcting for the equation of time.[c] To add this component to the train he made the arm bearing the 'idling wheels' swing back and forth by means of a lever operated by a cam actually cut in the shape of the graph—the 'equation kidney'. Thus the clock showed both mean and solar time.[d] However, since bevel-gears were not yet known in Europe,[e] the mechanism was made with a right-angle crown-wheel. It will be noted that in this first application the main driving force was transmitted through the main wheels and not (as in the automobile) through the idling wheels, which here only added or subtracted a small velocity difference. For a long time after Williamson's achievement the properties of differential gear were little appreciated, and a century elapsed before James White called attention to it in his *Century of Inventions* (1822).[f] Among other proposals, in his dynamometer he inserted a differential in a transmission-shaft, at the same time balancing the arm by a weight which indicated in conjunction with the lever length the torque transferred. This application also differed from the automobile in that all the driving force came through one of the main wheels.[g] Finally the first analytical treatment of epicyclic and differential gearing was given by Robert Willis[h] in 1841.

Thus the history of the differential train would seem to have begun in the early 18th century. The question now opened is whether in its earliest form it does not go back more than a thousand years earlier. If indeed this was really the secret of some at least of the Chinese south-pointing carriage designs, it is interesting to reflect that while in the modern automobile the differential gear fulfils a locomotive and ad-terrestrial function, in the medieval south-pointing carriage its role was precisely the opposite—ex-terrestrial. The close (and natural) connection of the latter with the hodometer, the camp-mill,[i] and the mechanised 'band-wagon'[j] can thus easily be seen. Correspondingly, since the transmission of the driving force was inverted, the small wheels were truly 'idling'.[k]

Lanchester saw that this system could have been used for the mechanism of the south-pointing carriage. The arrangement can readily be understood from the back elevation (Fig. 533) and the photograph of the model (Fig. 534). It is a question of reversing the direction of the drive, which occurs here, as we have noted, not from an engine to the road, but from the road to the south-pointing figure. Just as in the other

  [a] Cf. von Drach (1); von Bertele (5).

  [b] As by George Graham (+1740) and about the same time by the a San Cajetano brothers in Austria and Mudge in England (*c.* +1770). See von Bertele (7).

  [c] Cf. Vol. 3, pp. 313, 329.   [d] Cf. Lloyd (7); von Bertele (4, 10).

  [e] In the subsection on clockwork below (pp. 456, 470) we shall find some evidence that they may have been used already in +11th-century China.

  [f] Cf. Dickinson (5).

  [g] The principle is still used in the Webber differential dynamometer (cf. Jones & Horton (1), vol. 1, p. 366).

  [h] (1), pp. 361 ff., 380, 391, etc. A brief but interesting biography of Willis (1800 to 1875) has been written by Hilken (1).

  [i] Cf. p. 255 above.   [j] Cf. p. 283 above.   [k] Cf. p. 202 above.

plan, the gears A, A_1 fixed to the road-wheels engage with horizontal intermediate wheels, B, B_1 (extended to B_2), but these are now so arranged as to be permanently in gear with an upper and a lower bevelled wheel C_1, C, the upper one of which is concentric with the shaft carrying the pointing figure but is not fixed to it. Between them run two small idling wheels D, D_1 connected by a stub-shaft from the centre of which rises up at right angles the shaft carrying the pointing figure. Relative motions of the road-wheels will now be accurately, and inversely, reflected in the movements of

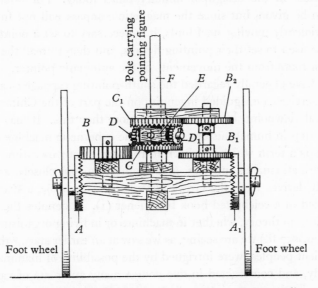

Fig. 533. Reconstruction of the mechanism of the south-pointing carriage according to Lanchester (1); back elevation. Both road-wheels are permanently connected with the pointing figure, which rides on a pair of idling wheels within the differential gear. For further explanation, see text.

the idling wheels and the pointer which they carry. If, for instance, the left-hand road-wheel moves faster than the right, as on a westward turn, B will move while B_1, B_2 will be almost stationary, and C will be turning while C_1 will be almost at rest. Hence D, D_1 will react to precisely the same extent but in the opposite direction. They will then stay put, though C, C_1 will resume their mutually opposite regular motion until further relative change of road-wheel velocities occurs. Fig. 535 shows the pretty behaviour of a model constructed by Lanchester himself.[a]

Does this satisfy Yen Su's specification? If his two small vertical wheels had teeth, it could. But a misprint in the text would have to be assumed, for he should have been recorded as speaking of two large horizontal wheels in the middle, not one. This could be due to a textual error easily made. There should also, however, have been 10 or 11 wheels, and more than 120 teeth. Besides, if Yen Su's machine had used the

[a] Of course in full-scale practice even this device would not have worked well over any considerable distance unless the two road-wheels were made very accurately identical in size, and some measures taken to avoid relative slip. Coales (1) has calculated that a difference of only 1 % between the wheel circumferences would lead to a change of direction of the pointing figure of as much as 90° in a distance only 50 times that between the two wheels.

differential gear, it is not obvious why the motions of the carriage-pole should have been mentioned at all, though the text gives no reason for bringing them in, and no cords or weights are mentioned. In any case, whether or not Yen Su used a form of differential gear, it remains quite possible that Ma Chün or Tsu Chhung-Chih did so, for it is after all simpler as well as more practical and elegant than the clumsy device of engaging and disengaging gear-wheels vertically suspended.[a]

The most astonishing part of the story is that the differential gear is used to operate a 'south-pointer' in the design of military tanks today. For obvious reasons no references can be given, but since the magnetic compass will not function properly within their violently moving steel hulls, it is necessary to set a pointer initially, just as the Chinese used to set their pointing figures, and then to read the direction taken by the tank *en route* from the movements of the automatic pointer.

Sinologists have generally regarded the south-pointing carriage as a kind of playful freak, or an exercise of misguided ingenuity on the part of the Chinese.[b] But from a broad historical viewpoint, it is surely much more than this. It may be said to have been the first step in human history towards the cybernetic machine.

Cybernetics is a term which has been introduced in our own time to cover a vast field in which mathematics, physics, and engineering join closely with biology and physiology. It derives from the Greek *kubernetes*, κυβερνήτης, a steersman, and has been expounded in a celebrated book by Wiener (1). It denotes the field of control and communication theory, whether in machines or in living organisms. Elsewhere in this Section (pp. 157 ff.) we are seeing, as we saw at an earlier stage,[c] that the Chinese, like other ancient peoples, were intrigued by the possibility of inanimate simulacra of the living body, and constructed in due course many automata of various kinds depending on springs, water-power and the like. But it was their own invention of mechanical clockwork[d] which by permitting the motive force to be fully concealed within the body of the automaton had a really far-reaching influence on scientific philosophy. Mechanistic biology, with all its heuristic value, began with the Cartesian estimate of animal bodies as automata. But still it was a far cry from clockwork robots which could do no more than carry out a fixed programme, to living things. Only when it became possible to make machines which could spontaneously revise their proceedings in the light of information as to the degree of approximation to their goals attained, did the realms of mechanics and biology begin to approach closely. Though classical biochemistry and physiology still think of living organisms in terms of power and fuel engineering, advanced science is moving towards the conception both

---

[a] In weighing this probability it may be worth while remembering that the 'Chinese windlass', discussed on p. 98 above, is one of the simplest cases of differential motion, and in this capacity often opens standard accounts of differential gearing (see e.g. Jones & Horton (1), vol. 1, pp. 363 ff.).

[b] Though apparently von Humboldt seriously believed—(2), vol. 1, pp. xxxvi ff.—that it had aided the Chinese to achieve the high level which they attained in their knowledge of the historical geography of Central Asia.

[c] Vol. 2, pp. 53 ff.

[d] See Sect. 27 j below, and in fuller detail Needham, Wang & Price (1). The essential invention here was of course that of the escapement, but it is noteworthy that although the Chinese tower clocks between the +8th and the +14th century were driven by water rather than falling weights or springs, the motive power was always concealed. Cf. p. 540.

PLATE CCII

Fig. 534. Lanchester's working model with the housing removed to show the mechanism. Only one idling wheel was fitted in this example.

PLATE CCIII

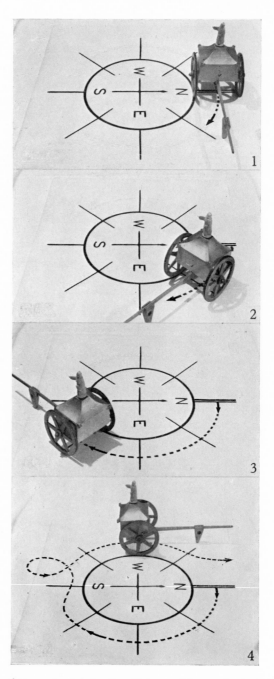

Fig. 535. Action photographs showing the efficiency of the mechanism of the differential-gear reconstruction of the south-pointing carriage (a set from Mr George Lanchester).

of them and of machines in terms of communication engineering. Modern biology has firmly established the existence of a 'homoeostasis' of the internal medium of the body—a remarkable constancy of factors such as osmotic pressure, hydrogen ion concentration, blood-sugar, and the like. So also modern technology has now for a long time past developed all kinds of self-regulating or homoeostatic machines.

Automatic action, in its broadest sense, would cover the coupling of auxiliary motions to the main motion of a machine. We take occasion to discuss elsewhere whether this first appeared in Villard de Honnecourt's water-powered saw-mill (+13th century) or in the Chinese silk-reeling machine (+11th century if not earlier).[a] In any case such auxiliary motions did not establish a self-acting cyclical system. This came with the development of the steam-engine in the +18th century. In its first truly practical form (that of Thomas Newcomen, +1712) it embodied an auxiliary motion in the form of a 'plug-tree' or plug-rod suspended from the beam and actuating the piston of a water-pump which on its down-stroke raised the cold injection water. Naturally the valve gear for admitting steam and cold water into the cylinder of the engine itself played a vital part from the beginning. According to a well-known story (which if not true is *ben trovato*) the automation of this was first effected by a boy attendant, Humphrey Potter, who saved himself trouble by arranging that the beam's rise and fall should spontaneously open and close the valves by suitable cords and catches at the right points in the cycle. It soon became evident that the plug-rod could be made to do this very simply by a series of trip-mechanisms, and indeed this was in all probability part of Newcomen's original design.[b] From that point onwards the road lay clear open to the invention of the slide-valve for single-cylinder double-acting steam-engines by William Murdock in +1799. As explained elsewhere,[c] this principle, introduced by James Watt in +1782, was characteristically Chinese, deriving from the ancient single-cylinder double-acting piston-bellows—with of course the essential difference that the piston no longer acted as a compressor on both strokes but was itself subjected to pressure on both strokes.[d] Here then was a self-acting cyclical system, but it did not establish a self-regulating cyclical system. Watt accomplished this by his now familiar device of the centrifugal governor-balls for regulating the velocity of steam-engines under varying conditions of load (+1787).[e]

This was a true closed-loop 'servo-mechanism', and probably the oldest example of the kind. In the same year Thomas Mead applied it to the control of the gap between the grinding-stones in windmills set by the bridge-tree,[f] for the avoidance of frictional over-heating at high speeds. But in windmills the governor-balls had been preceded by another important automatic control device, the ingenious 'fantail gear' due to Edmund Lee in +1745, which kept the main sails of a post mill auto-

[a] Cf. p. 2 above, p. 404 below.

[b] Cf. Dickinson (4), pp. 30, 41, (6), p. 175; Wolf (2), p. 612, and especially Becker & Titley (1).

[c] Needham (48). Cf. p. 136 above and p. 387 below.

[d] Cf. Dickinson (6), p. 185, (7), pp. 123 ff.; Wolf (2), pp. 622, 624.

[e] The mathematical study of the governor-balls by Clerk-Maxwell in 1868 is regarded as the foundation stone of cybernetics.

[f] Wailes (3), p. 137. Cf. p. 185 above. Governor-balls may have been used in windmills rather earlier than in the steam-engine. For the prior history of the windmill, see pp. 555 ff. below.

matically facing the wind by the use of a small auxiliary windmill mounted at right angles on a horizontally projecting tail substituted for the tail-pole. In tower mills the fantail wind-wheel[a] was geared by a worm to a peripheral rack and pinion at the top of the tower so that it was free to turn until due in the wind direction, when it would stop. By this 'self-becalming' property it ensured that the main sails were always oriented for maximum advantage.[b] Besides these arrangements there were many safety devices in +18th-century mills, giving warning of exhaustion of grain supply, rise in head-water level, etc., precursors, as Coales (1) puts it, of the klaxons and red lights of a modern factory, as also was the south-pointing carriage of old. Some of these took automatic action, e.g. the spring-sails with Venetian-blind shutters introduced by Andrew Meikle in +1772 to spill the wind in gusts and gales. These were later improved upon by the insertion of arrangements permitting adjustment in accordance with the load. By the beginning of the 19th century automatic controls in windmills had been so far developed as to allow of operation by one man alone. The windmill was the prototype of the automatic factory.[c]

Since then we have come to use all kinds of other devices—thermostats,[d] gyro-compass and powered ship-steering systems,[e] computing machines,[f] homing missiles, anti-aircraft fire-control apparatus, and automatically self-adjusting chemical and engineering production equipment.[g] There is now a large literature on automation and on the various types of servo-mechanisms, i.e. power-amplifying motor end-organs which carry out the necessary operations.[h] On the 'sensory' side there is an equal variety of information-producing end-organs—photo-electric cells, chemical meters, current-gauges, thermo-couples, etc. The routing of this information into the machine which must adjust its activities correspondingly is known as 'feedback'. When we desire a motion, says Wiener, to follow a given pattern, the difference between this pattern and the actually performed motion is used as a new input to cause the part regulated to move in such a way as to bring its motion closer to that required by the pattern.[i] When errors or variations from the end desired are automatically neutralised, this is known as negative feedback. Control machinery, like that of the living body, has the task of maintaining performance as close as possible to a predetermined

---

[a] In these wheels the skew angle or 'pitch' of the vanes was large, about 55°, much greater than that of the windmill sails themselves but approximating to that of the Chinese zoetrope and helicopter top, and almost as great as that of the obliquely vaned horizontal water-wheels (cf. pp. 124, 368, 565, 583). For a working drawing see Wailes (3), p. 108.

[b] See Wolf (2), p. 597, and especially Wailes (2, 3).

[c] Other industries were also entering the path to automation at this time, notably textile technology, with the Jacquard punched-card draw-loom (cf. Sect. 31).

[d] History by Ramsey (1).   [e] History by Conway (1).

[f] Here we can only mention the names of the great pioneers; Pascal (+1642), Leibniz (+1671) and Babbage (1833); cf. Wolf (1), pp. 560ff., (2), pp. 654ff. It is particularly interesting that the Jacquard punched cards were the direct genetic ancestors of modern digital computers descending through Babbage to Hollerith.

[g] Cf. Diebold (1).

[h] See for instance the books of Chestnut & Mayer (1); G. H. Farrington (1); Tsien (1) and R. H. Macmillan (1).

[i] Interesting discussions which illuminate the vistas opened up for the exploration of the possibilities of automatic machines, involving new discoveries in information-theory and the like, will be found in Tustin (1); Gabor (1); Cherry (1); Cherry, Hick & McKay (1); and Nagel, Brown, Ridenour et al.

optimum. Experiments are now being carried out by Walter (1) and others in which tentative models of living organisms are being built up from relatively simple circuits connecting receptor and effector units.

This is the background against which it may be said that the south-pointing carriage was the first homoeostatic machine in human history, involving full negative feedback. Of course, the driver has to be included in the loop. But as Coales (1) has acutely pointed out, an attractive carrot held by the pointing figure might have replaced the human driver and closed the loop more automatically.[a] The south-pointing carriage would have been the first cybernetic machine had the actual steering corrected itself, as we could easily make it do today. It is interesting to note, also with Coales, that if the large gear-wheel (C in Fig. 531) were rigidly attached to the pole, then as soon as one of the small gear-wheels engaged, the pole would be restored in a few moments to its original direction. This would of course put intolerable strains on the teeth, and probably lead to prohibitive oscillations, but such an arrangement would have been a purely mechanical self-regulating closed-loop system. There is no textual evidence that this was ever attempted in medieval China.

Yet to produce an instrument which would fully compensate for, and thus consistently indicate, all deviations from a prearranged course, was a real achievement, not only of practice, but also of conception, for the +3rd century.[b] And perhaps it was no coincidence that this simulacrum of a living organism arose in a highly stable civilisation characterised by a highly organic conception of nature,[c] for the tendency to self-regulation is a primary property of living organisms. Wang Chhung in the +1st century, as we have seen,[d] described an animal tropism; the larvae, after being disturbed, swung back on their course. If Chang Hêng and Ma Chün could not make a machine that would accomplish this, at least they made one which would do its best to show where that course lay.

(f) POWER-SOURCES AND THEIR EMPLOYMENT (I), ANIMAL TRACTION

In foregoing sections we have been concerned with the use of animal-power as the prime mover for various kinds of machines, as also for vehicles, and it was convenient to leave vague the exact practical way in which this force was applied. Those who were really concerned with the matter, however, in different historical periods, found themselves face to face with a set of problems which were really of an engineering character.

---

[a] He has given an interesting diagram of the principle of the south-pointing carriage in terms of the conventions of current cybernetic theory.

[b] In the present Section and that on physics (Vol. 4, pt. 1, pp. 229 ff. above) it has always been necessary to distinguish sharply between the south-pointing carriage and the magnetic compass. Yet if the conclusions there reached concerning the knowledge of the directivity of the lodestone spoon in the Han period are correct, it is not without interest that the first constructions of south-pointing carriages with gearing date from the Later Han or San Kuo time. These were homoeostatic *machines* in a sense in which the lodestone was not, for it was a natural object, rather than a contrivance of man. But they may have been built in its image.

[c] Indeed a homoeostatic culture (Needham, 47).   [d] Vol. 4, pt. 1, p. 262 above.

Though we do not usually think of it in this way, any system of harnessing constitutes an elaborate play of linkwork and hinges, in which both the anatomical nature of the draught animal and the structure of the object pulled has to be taken into consideration.

(1) EFFICIENT HARNESS AND ITS HISTORY

That historians of technology have become conscious of the importance of this set of inventions is due to the genius of a French cavalry officer, Lefebvre des Noëttes, who was the first to investigate the harness used in different cultures at different times, as shown by remaining carvings and pictures, and to make actual experiments with reconstructions of the ancient systems used. One is constantly seeing pictures of harnessed animals, in church windows, in manuscripts, on stone monuments, etc., but until one's eyes are opened by the systematic work of a des Noëttes one simply fails to see that quite different methods were employed at different times, and that their gradual improvement involved inventions of great importance. There can be no better way of opening this section than by a brief paraphrase of some of his introductory remarks.[a]

A draught system, he said, is composed of one or more animals, the motive force of which has been applied to traction by means of a special system, the harness, which itself includes several components. A rational harnessing system should permit the complete utilisation of the force of all the animals, and favour their work as a team. The 'modern' harness (i.e. collar-harness), well adapted as it is to the anatomy of the horse, attains this objective, and the force which can be exerted is limited only by the requirement (the load to be pulled) and the quality of the road. The 'antique' harness, on the other hand (which we shall call throat-and-girth harness), could make use only of a small fraction of the possible motive force of each animal, failed to ensure satisfactory collective effort, and in general yielded a very low efficiency. The two are so different that the former cannot possibly be interpreted as a variant of the latter, and cannot, indeed, have grown out of the latter. The throat-and-girth harness was a single clearly recognisable type which remained unchanged from its first appearance in the most ancient illustrations we have, until it finally died out in the Middle Ages in Western Europe. And, moreover, it was the same everywhere, in every ancient realm and culture, equally inefficient. Only one ancient civilisation broke away from this and developed an efficient harness—China.

The throat-and-girth harness is shown[b] in Fig. 536a. It consists of a girth surrounding the belly and the posterior part of the costal region, at the top of which the point of traction is located. Presumably in order to prevent the girth being carried backwards, the ancients combined it with a throat-strap, sometimes narrow, more

a See des Noëttes (1), p. 5 As a prelude to the present section it would be instructive to read the appropriate chapter in Haudricourt & Delamarre (1), pp. 155 ff., which summarises the state of the question, relating harness not only to vehicles but to ploughs. We must continue to bear in mind, however, its relation to machinery powered by animals. The account of harness in Jope (1) needs adjustment, especially in the light of the Chinese contribution.

b From Needham (17), a preliminary presentation of this argument. Cf. Haudricourt (10).

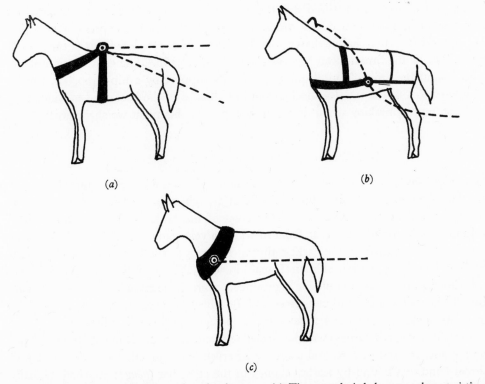

Fig. 536. The three main forms of equine harness. (*a*) Throat-and-girth harness, characteristic of occidental antiquity. Compression of the trachea prevented efficient tractive effort. (*b*) Breast-strap harness, characteristic of ancient and early medieval China. By pressure upon the sternum the line of traction is directly linked with the skeletal system, and full exertion permitted. (*c*) Collar harness, used from late medieval times both in China and the West. As the sternal region is again the bearing point, efficient tractive effort can be exerted.

often broad, which crossed the withers diagonally and surrounded the throat of the animal, thus compressing the sterno-cephalicus muscle and the trachea beneath it.[a] The inevitable result of this was to suffocate the horse as soon as it attempted to put forth a full tractive effort, accompanied by the lowering and advancing of the head. In sharp contrast to this is the 'modern', or collar-harness (Fig. 536 *c*). Here the collar, reinforced and padded, bears directly upon the sternum and the muscles which cover it,[b] thus linking the line of traction intimately with the skeletal system and freeing fully the respiratory channel. The animal is now able to exert its maximum tractive force. The collar-harness was not, however, the only way in which the line of traction could be derived from the chest or sternal region so as to leave the throat free. At some time during the Chhin or early Han in China, or quite probably earlier, in the Warring

[a] When the throat-band rose, as it always tended to do, there was pressure also on the sterno-thyro-hyoideus and even the omo-hyoideus muscles. To appreciate fully the structural relations it is convenient to have at hand a good textbook of veterinary anatomy, such as Sisson & Grossman (1). Cf. especially pp. 25 ff., 269 ff.

[b] The anterior superficial pectoral muscle and the lower parts of the brachio-cephalicus, sterno-cephalicus and cutaneus colli muscles.

States period, someone realised that the animal's shoulders could be surrounded by a trace, which, if suspended by a withers-strap and attached to the point of inflection of the curving cart or chariot shafts, would greatly increase the efficiency of the horse's work. This is the breast-strap harness shown in Fig. 536b.[a] The continuation of the trace round the animal's hind-quarters, and its support by a hip-strap,[b] was not a necessary part of the tractive mechanism, but allowed of the backward movement of the cart, and its braking when descending slopes; to this point we shall return.

(i) *Throat-and-girth harness in Sumer and Shang*

One might say that the throat-and-girth harness of horses was a makeshift alternative for the yoke of the ox.[c] Since the neck of that draught animal projects forwards from the body horizontally, unlike the rising crest of the horse, and since its vertebral column forms a bony contour in front of which a yoke can easily be placed,[d] this was satisfactory enough from the earliest times. Figure 537 shows an ox-cart of age-old type[e] in the streets of Chiu-chhüan (Kansu) in 1958. Abundant illustrations of similar vehicles drawn by oxen could be reproduced, were it necessary, from Han reliefs, Buddhist carvings and Tunhuang frescoes.[f] But the yoke on each side of a central pole, or between shafts, which was suitable for oxen was quite inapplicable to horses,[g] and the throat-and-girth harness was the consequent substitute attachment. In the Shang period and no doubt for several centuries afterwards the junction-point of girth and throat-band was bound by leather thongs to the cross-bar (*hêng*[1]) attached medially to the upward-curving central pole,[h] transmitting thus the tractive force. But for some reason or other, perhaps of symbolism, perhaps of ornament, the useless 'yoke' (*ê* or *o*[2], *chhiu*[3])[i] lived on in vestigial forms with great persistence. Before the beginning of

---

[a] The terms chest-strap or chest-band would of course be more appropriate, since this is not the site of the mammae in the horse, but we retain some usages of common speech.

[b] Alternatively, loin-strap or croup-strap. It was often doubled.

[c] For present purposes we may define the yoke as a forked or curving piece of wood placed over the neck of the bovine animal, or a framework of wood placed around it.

[d] A rather sharp internal 'hump' is formed by the long upward-projecting spinous processes of the 7th cervical vertebra and the 1st to the 5th thoracic vertebrae. Since the curve of the vertebral column so formed is almost as sharp as two right angles (cf. Sisson & Grossman (1), pp. 126ff.), a strong 'ledge' or support is afforded for the yoke; and even when the line of the neck is straight (in some varieties of oxen there is a marked dip or hollow), the tissues under the trapezius muscle are more yielding than those of the corresponding region in the horse.

[e] The ox-yoke itself appears in many variations. A historical and ethnographical summary will be found in Haudricourt & Delamarre (1), pp. 164ff., 180ff. Cf. Huntingford (1).

[f] Cf. Sirén (1), pl. 74; Harada & Komai (1), vol. 2, pl. XXXIV, etc., for unpainted terra-cotta models of Han date, common in museums. A typical Buddhist relief dated +525 is figured by Harada & Komai, vol. 2, pl. XIII (2). At Chhien-fo-tung, many examples from Wei, Sui and Thang, e.g. cave nos. 22, 290.

[g] The thoracic vertebrae of the horse also have spinous processes projecting dorsally but they rise and fall over the scapular region in a flowing curve, which, combined with the powerful suspensory muscles of the neck, make a smooth line along the back and give no 'ledge' of any kind for harness.

[h] Cf. p. 246 above. See particularly the early −7th-century chariot-burials of the State of Kuo (in north-western Honan) described in Anon. (27).

[i] The word *chhiu* seems to have meant the long pieces only, while *ê* or *o* referred to the whole thing; cf. von Dewall (1).

[1] 衡   [2] 軛   [3] 輈

PLATE CCIV

Fig. 537. The age-old ox-yoke harness for bovine animals (orig. photo., 1958, Chiu-chhüan, Kansu). The long upward-projecting spinous processes of the cervical and thoracic vertebrae form a kind of internal hump against which the arched yoke can bear. No such ledge is available in equine animals.

Fig. 538. Burial of a chariot, with horses and charioteer, excavated at Ta-ssu-khung Tshun near Anyang in Honan; Shang period, c. −12th century (Ma Tê-Chih *et al.* (1), cf. Watson (2), pp. 88 ff.). Trenches were cut for the wheels, axle and pole, the wood of which left its traces in the soil. The bronze 'yokes', shaped like narrow wishbones, are still in position by the horses' necks; the bronze hub-caps and the rows of bronze discs used for ornamenting the harness are also still in their places. Two bronze bow-shaped ornaments of uncertain use can be seen within the limits of the chariot body. Photo. CPCRA and BCFA.

PLATE CCV

Fig. 539. One of the bronze 'yokes' for the horses in the Shang chariot burial, probably sheathing for an inner wishbone-shaped lining of curved wood (Ma Tê-Chih *et al.* (*1*), cf. Watson (1), p. 22, pl. 47). These fittings, attached at each end of the cross-bar, were already vestigial, but may have served to protect the bar if the horse reared backwards, as also to guide the reins in their upturned ends. There can be no doubt that at this time throat-and-girth harness was still used in China, as in Babylonia and ancient Egypt. Photo. CPCRA and BCFA.

Fig. 540. A form of 'yoke' still persisting in Han times, seen on a moulded brick excavated near Chhêngtu and now in the Chungking Municipal Museum (Anon. *22*). It is fixed at the centre of the cross-bar connecting the shafts of a light luggage trap, and its ends seem to guide the reins. Dimensions of the brick: 46 × 39 cm.

the Chou it had assumed a narrow V-shaped 'wishbone' form,[a] and its lower ends were turned up, perhaps to act as guides for the reins.[b] Figure 538 shows a Shang chariot burial of about the −12th century excavated at Ta-ssu-khung Tshun near Anyang in Honan.[c] The two bronze-sheathed 'fork-yokes' (Fig. 539) can be seen beside the skulls of the two horses; they must have been suspended from the cross-bar with the reins no doubt resting in their lower 'gutters'. Then at some time about the Warring States period the Chinese abandoned the central pole in favour of two parallel outside bars, and the curving S-shaped shafts of the typical Han chariot,[d] connected above and behind the horse's neck by a cross-bar, were the result. Still, however, the 'fork-yoke' persisted, probably now chiefly because it was often useful as a guide for the reins.[e] This appears to be the arrangement in Fig. 540, a moulded Han brick from Szechuan.[f] After the invention of the breast-strap harness, and the attachment of the traces directly to the point of inflection of the shafts, the cross-bar was of little further use, and it eventually disappeared.[g] So too did the 'fork-yoke'.[h] Here we meet with a question of considerable interest, namely the dates of the two inventions—the shafted vehicle and the breast-strap harness. Did they coincide, or if not what interval separated them, and which came first? Textual researches should have something to say on this problem while we are awaiting a decisive answer from archaeology.

The really astonishing thing about the throat-and-girth harness is the immense spread which it had both in space and time. We find it first in the oldest Chaldean representations[i] from the beginning of the −3rd millennium onwards, in Sumeria[j] and Assyria[k] (−1400 to −800). It was in sole use in Egypt[l] from at least −1500,

---

[a] This was actually made in three pieces, the two long arms being fixed into the holder at the top; cf. von Dewall (1).

[b] There are rather close parallels for the whole arrangement in ancient Egyptian and Assyrian representations (cf. des Noëttes (1), figs. 20, 22, 25, 27).

[c] Ma Tê-Chih, Chou Yung-Chen & Chang Yün-Phêng (1); Watson & Willetts (1), no. 10; Watson (1), p. 22, pls. 46 ff., (2), pp. 88 ff., pl. 11.   [d] Cf. p. 248 above.

[e] But perhaps also as a protection against the breaking of the cross-bar if the horse reared backwards; cf. p. 310 below.

[f] Anon. (22), no. 19, pp. 42, 43; Liu Chih-Yüan (1), no. 50. Even when the shafts were straight, however, as in the chariot models from the −1st century found at Chhangsha (cf. p. 77 above), the 'yokes' were still present, so that the shafts must have pointed upwards at quite an angle unless the cross-bar was very arched (see the illustrations in Anon. 11). Such a highly arched one is indeed shown in Anon. (22), no. 18, p. 41; Liu Chih-Yüan (1), no. 53.

[g] That is to say, when trace harness was used. But along another line of development, it was present at the birth of collar-harness and indeed became the direct ancestor of the hard 'hames' of modern harness. This evolution will be made clear as we proceed.

[h] By the time of the earliest frescoes at Chhien-fo-tung (Tunhuang, see on, p. 319), i.e. the +4th or +5th century, no trace of it is left. But as we shall presently find, the narrow Han 'fork-yoke' may also have been an inspiration for the hard separate 'hames' component of the earliest collar-harness; perhaps it was remembered, or perhaps some derivatives of it lingered on in other parts of the country.

[i] Lejard (1, Ptg.), fig. 2; des Noëttes (1), fig. 17 (the Standard of Ur). In these earliest pictures the girth is sometimes not represented (cf. e.g. the tiger-chariot figured by des Noëttes (1), fig. 16), so that it remains a question whether some considerable time elapsed before the girth was added to the throat-band.

[j] This is implied by the poles of the hobby-horse or saddle-chariots (cf. des Noëttes (1), fig. 4). As is well known, *Equus onager*, not *caballus*, was used before −2000. On Sumerian vehicles see further Childe (11, 16).

[k] See des Noëttes (1), figs. 24, 25 and pp. 21–43.

[l] Winlock (1), figs. 25, 26 (glazed tile, etc., of the 18th dynasty); des Noëttes (1), fig. 38 and pp. 44–61; Breasted (1), fig. 57 (wall-sculptures of Karnak).

where it is shown on all paintings and carvings of chariots and horses, and it was like-wise universal in Minoan[a] and Greek[b] times. Innumerable examples occur in Roman[c] representations of all periods, and the empire of the throat-and-girth system also covered Etruscan,[d] Persian[e] and early Byzantine[f] vehicles without exception. Western Europe[g] knew nothing else until about +600, nor did Islam.[h] Moreover, the south of Asia was almost entirely in reliance on this inefficient harness, for it is seen in most of the pictures of carts which we have from ancient and medieval India,[i] Java,[j] Burma, Siam and other parts of that area. Central Asia, too, has it, e.g. at Bāmiyān.[k] One of its last appearances occurs on a bas-relief of the +14th century at Florence in Italy, where it may be a conscious archaism.[l]

(ii) The first rationalisation; breast-strap harness in Chhu and Han

We may now pass to the efficient breast-strap harness of China with its two traces suspended by the withers-strap (Fig. 536 b). One can only say that it is universal on all Han carvings, reliefs and stamped bricks which show horses and chariots (cf. Chavannes (9, 11), Jung Kêng (1) and indeed every book on Chinese archaeology which reproduces such pictures).[m] The Wu Liang Tzhu tomb-carvings (cf. p. 264 above), of the +2nd century, as depicted in the *Chin Shih So* and on later rubbings, are full of examples of such harness, one of which[n] is shown in Fig. 541. The Hosokawa

[a] Des Noëttes (1), fig. 50 (Inkomi ivory).

[b] Lejard (1, Ptg.), figs. 6, 7 (Attic vase-paintings, −6th century), fig. 10 (−5th-century vase from Milo); Markham (1), fig. 23 (Amphiaros *krater*, −550), fig. 24 (François vase, *c.* −560), fig. 32 (Syra-cusan dekadrachm, −478), fig. 62 (Great Altar at Pergamon, *c.* −170); des Noëttes (1), fig. 58 and pp. 62–75; Zervos (1), the nuptial car of Peleus and Thetis, *c.* −550; Cook (1), pl. 13 (vase, *c.* −570).

[c] Lejard (1, Sculpt.), fig. 50; Espérandieu (1), pp. 12, 13; des Noëttes (1), fig. 72 and pp. 83–8. That this inefficient harness was used for animal-power mills we know from the famous Vatican relief, des Noëttes (1), fig. 74; Moritz (1), pl. 5 b.

[d] Des Noëttes (1), fig. 68, p. 81.

[e] Achaemenid; des Noëttes (1), fig. 64 (Persepolis). Sassanid; des Noëttes (1), fig. 65 (triumphal chariot of Shapur). Pp. 76–80.

[f] Des Noëttes (1), fig. 89 (+9th-century MS.) and pp. 89–92.

[g] *Ibid.* fig. 116 (+6th century, the Trier coffer), fig. 117 (+7th-century MS.), fig. 145 (+11th-century MS.); pp. 104, 105.

[h] *Ibid.* p. 97.

[i] *Ibid.* pp. 99 ff.; fig. 100 (Sanchi, +2nd century), fig. 101 (Graeco-Buddhist copper vase, +1st century); fig. 106 (ivory carving, +15th century). Also the +8th-century frescoes from Polonnaruwa, Ceylon, exhibited at the Musée Cernuschi. The Vedic terms for pole, girth, neck attachment, etc., are discussed by Chakravarti (1), p. 27.

[j] Des Noëttes (1), fig. 112 (Borobodur, +9th century), fig. 114 (statuette, +15th century).

[k] A fresco of the late +5th or early +6th century (Rowland, 1).

[l] Des Noëttes (1), fig. 157, p. 126.

[m] Examples could be multiplied indefinitely. Chavannes (9) gave the reliefs from the Hsiao Thang Shan tomb-shrine of *c.* +129 which had been reproduced a century earlier in *Chin Shih So*, Shih sect., ch. 1 (pp. 106, 110, 126, 133, 135). Jung Kêng (1) gives the Wu Liang tomb-shrine reliefs (*c.* +147) as they are now. The *Chin Shih So* has them in Shih sect., ch. 3 (pp. 52, 69, 83, 85 ff., 104 ff.), etc., and ch. 4 (pp. 46 ff.). Elaborate detail is seen in the more recently discovered I-nan tombs; cf. Tsêng Chao-Yü *et al.* (1), pls. 24, 29, 49, 50. These date from about +193. Many Han bricks from Szechuan are figured in Anon. (22), nos. 18–24, 26 and 27, as also in Liu Chih-Yuan (1), pls. 46–60. Szechuanese carvings as well as bricks may be found in Rudolph & Wên (1), pls. 39, 53, 82, 83, 88, 100. See also des Noëttes (1), figs. 123, 124, 127 and pp. 106–17.

[n] Cf. Shih sect., ch. 3, (p. 105).

PLATE CCVI

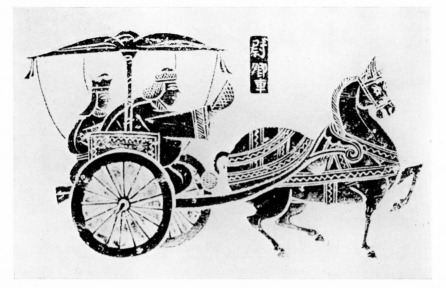

Fig. 541. Han breast-strap harness, a typical representation; the carriage of the Intendant of the Imperial Equipages (Wei Chhing), from the Wu Liang tomb-shrines (c. +147). *Chin Shih So* (Shih sect.), ch. 3 (p. 105). Besides the breast-band or trace suspended by withers-strap (here apparently forked) and hip-strap, a girth is still retained and the old throat-strap has turned into an elongated sling helping to suspend the front of the chest-strap. A breeching is present and there are also ornamental bands over the horse's hind-quarters. Many Han representations are much less elaborate than this, showing only the chest-strap with its suspensions. From a recent rubbing.

Fig. 542. Bronze model of a light luggage trap like that in Fig. 540 but without the roof, Han period. In the Hosokawa Collection, Tokyo (Sirén, 1). Though there is neither breeching nor other parts of the harness, the breast-strap attached to the mid-point of the curving shafts is well seen.

PLATE CCVII

Fig. 543. Breeching and hip-strap hanging from the shafts of a parked chariot; part of the reliefs from the Wu Liang tomb-shrines, c. +147 (Chavannes, 11).

Fig. 544. Another carriage harness relief of the Han period, from the I-nan tomb-shrines c. +193 (Tsêng Chao-Yü et al. 1). Compare with the diagram following.

Collection in Tokyo possesses a splendid Han bronze model of a harnessed horse and cart, reproduced here in Fig. 542, from which the breast-strap and shafts can clearly be seen.[a] The breeching or breech-band, suspended by the hip-strap, seems often to have been left attached to the shafts when the horse was taken out of them, for in a famous picture (Fig. 543) from the Wu Liang tomb-shrines, we see the empty chariot with the breeching between the shafts.[b]

Fig. 545. Elucidation of the I-nan relief to show the forked shafts, the cross-bar, the 'yoke', and the harness components. Technical terms are explained in the text.

Let us examine more closely a Chinese carriage from the end of the +2nd century, the culminating time of Han technique. Figure 544 reproduces a relief from the I-nan tombs[c] and Fig. 545 adds an explanatory drawing. Aided by the commentators such as Fang Pao[d] and Chiang Yung[e] we can at once distinguish the breast-band (chin[1]), the traces (yin[2]),[f] the withers-straps (hsien,[3,4] sometimes called an,[5,6] which more properly meant the saddle of a ridden horse), and the hip-straps and breech-band (chiu[7], chhiu[8]).[g] As often as not, horses harnessed in Han chariots have no girths (pi[9], thing[10]), but those of I-nan do. The shafts of the vehicle (yuan[11], cf. p. 248 above) are now bifurcated, for at or near the point of attachment of the traces a short rod is bound to them so that it serves to keep pointing forwards and downwards the lower extremi-

[a] Sirén (1), vol. 2, pl. 33; Harada & Komai (1), vol. 2, pl. xxiv (3). A similar Han bronze from the Nanyang tombs is illustrated in *Wên Wu Tshan Khao Tzu Liao*, 1954 (no. 9), pl. 20.

[b] Chavannes (11), pls. x and xx; Harada & Komai (1), vol. 2, pl. viii (2); des Noëttes (1), fig. 122. Another representation, showing the breeching hanging on a stand, is seen in the Wu Liang reliefs; Jung Kêng (1), indiv. rubbing no. 55; *Chin Shih So*, Shih sect. ch. 3 (p. 68); Chavannes (9), pl. ccccxcvii and p. 164 (no. 1206), also (11), pl. v.

[c] Tsêng Chao-Yü et al. (1), pl. 50.   [d] In *Khao Kung Hsi I*, ch. 11, p. 5a.

[e] In *Chou Li I I Chü Yao*, ch. 6, p. 9b.

[f] There were a number of synonyms for this term, here omitted. Note that yin[12] alone has the meaning of leading.

[g] The breech-band, breech-strap or breeching corresponds to the crupper of the ridden horse which keeps the saddle from working forward. By a natural extension of meaning the second word, chhiu, could also denominate traces and whippletree (cf. p. 260 above). It participated, moreover, also very naturally, in the term for a swing (cf. p. 376 below).

| [1] 靳 | [2] 靷 | [3] 韅 | [4] 顥 | [5] 鞍 | [6] 窂 | [7] 鞦 |
| [8] 鞧 | [9] 鞃 | [10] 綎 | [11] 轅 | [12] 引 | | |

ties of the 'fork-yoke' (ê¹) which is attached centrally to the cross-bar (hêng²).ᵃ Through special rings (yu huan³) on this run the reins (phei or pi⁴, lê⁵, ti⁶ or thiao⁷), the forward ends of which (chiang⁸) are attached to the bronze cheek-piece levers (piao⁹) at each end of the bit (hsien¹⁰,¹¹). The bridle (lung¹²), or network of straps securing the horse's head-furniture, was also clearly depicted by the carvers of I-nan. Most of these terms have long fallen into disuse and will hardly be found in modern dictionaries.ᵇ When two or more horses were used the inner traces were called yin yin,¹³ and the outer ones yang yin,¹⁴ the inner reins lê⁵ and the outer ones phei.⁴ There were many other terms for ornamental parts of the harness (such as the little jingling bells luan¹⁵) which we must not stay to examine. When the Han harness is well understood, one can appreciate some of the contemporary instructions to drivers; thus, for example, in the Chou Li it is suggested that when mounting gradients the driver should place his weight well forward in the chariot, thus helping to keep the breast-band down.ᶜ Naturally there were various minor variations of the Han trace harness, but they need not delay us here.ᵈ The traces were generally attached to the mid-points of the shafts, but it is clear from some of the reliefs that in the case of heavy baggage-carts, timber-tugs, etc., they were run right back to the body of the vehicle itself, as in modern practice (cf. Fig. 493). This also happened in quadrigae which retained a central pole.ᵉ

So far we have found only one enlightening piece of evidence concerning the first origins of the breast-strap harness in China. It is a form transitional between the throat-and-girth type and the traces (see Fig. 546), depicted with varying degrees of clarity on painted lacquer boxes (lien¹⁶) from the State of Chhu, approximately −4th century in date.ᶠ What seems to be a hard yoke-shaped object, probably of wood, with down-turning ends, has been applied to the horse's chest, and the ends connected to the mid-point of the shafts by traces. Or it may be that two angle-pieces of wood or bone connected by a breast-strap hang at the horse's sides. But the throat-strap has not been quite abandoned, only lengthened so as to support the chest-piece (or

---

ᵃ The function of this system is a little difficult to discern. Perhaps the vestigial yoke lived on in the role of a buffer to prevent the breaking of the light cross-bar when the horse reared back its neck and head. The sockets for the rods at each end of the yoke are clearly visible in Fig. 544. Or perhaps the persistent 'yoke' was part of the trappings and hardly functional at all. It may well have been gilded or at least brightly polished.

ᵇ An elaborate study of some of them will be found in Hayashi Minao (1), pp. 228 ff.

ᶜ Ch. 11, p. 18a (ch. 40, tr. Biot (1), vol. 2, p. 485). The same applied to ox-drawn carts, if the yoke had a throat-strap (yang¹⁷) linking its two ends underneath, as it often did. Balancing on the part of drivers continued to be enjoined until modern times (cf. Philipson (1), p. 52).

ᵈ See Needham & Lu (2). Han carriage-horses were often decked with bands or trappings serving a purely ornamental purpose, especially on the hind-quarters. Cf. Fig. 541.

ᵉ There is a very good example of this in the Hsiao-thang Shan tomb-reliefs (c. +129) immediately following the hodometer (p. 283 above); see Chhang Jen-Chieh (1), pl. 12. Here the Chin Shih So representation (Shih sect., p. 135) is particularly bad, for it made the nearest trace into a shaft.

ᶠ A particularly clear example is figured by Chhang Jen-Chieh (1), pl. 11, who, however, dates it as Han. What seems to be another one, which shows something hanging in the same place as the chest-piece, will be found in Shang Chhêng-Tsu (1), no. 9, pls. 25, 28, 30. This is considered indubitably of Chhu date. One of the best illustrations is given in Anon. (48), pl. 31, an album of Chhu antiquities. All should be studied together.

---

¹ 軶	² 衡	³ 游環	⁴ 轡	⁵ 勒	⁶ 鈞	⁷ 僄
⁸ 韁	⁹ 鑣	¹⁰ 啣	¹¹ 銜	¹² 韁	¹³ 陰靷	
¹⁴ 陽靷	¹⁵ 鸞	¹⁶ 奩	¹⁷ 鞅			

PLATE CCVIII

Fig. 546. Breast-strap harness *in statu nascendi*; a horse and chariot figured on a round lacquer box of the kind made in the State of Chhu, *c.* −4th century, found in a Han tomb near Chhangsha (Chhang Jen-Chieh, *1*). A hard yoke-shaped object surrounds the horse's chest and is connected to the mid-point of the shafts by traces. This is a possible intermediate form between throat-and-girth harness and breast-strap harness.

PLATE CCIX

Fig. 547. Trackers hauling a boat upstream; a fresco painting of the early Thang period from cave no. 322 at Chhien-fo-tung, Tunhuang. Efficient equine harness may have arisen from the practices of human hauliers.

pieces) at the inflection, acting instead of the withers-strap which would suspend the traces; and the girth has also been retained (Fig. 548). This evidence would perhaps suggest that the invention of the shafted chariot was the limiting factor, and that traction from the sternal region was not achieved until the shafts became available. But presently we shall find certain documents perhaps equally venerable which seem to show collars combined with long traces and no shafts.[a] Many archaeological discoveries are still needed; in the meantime the lacquer paintings of Chhu do seem to show us the breast-strap harness in the nascent state.[b]

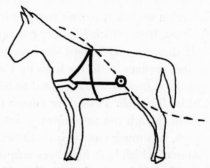

Fig. 548. Diagram to explain the harness system seen in the Chhangsha lacquer painting.

Unless gear of this kind was a completely new invention long afterwards, these hard objects suspended at the side of the horse's shoulder-bones were transmitted through tribes and peoples unknown to turn up again among the Avars (probably from the time when they invaded Hungary from the east in +568) and then among Magyars, Bohemians, Poles and Russians. For burials excavated in these lands dating from between the +7th and the +10th centuries have yielded T-shaped pieces of bone or horn pierced with holes in such a way as to connect conveniently the three harness components—breast-strap, lengthened throat-strap (now as good as a withers-strap), and traces; these last attached either to the vehicle itself or more probably to the tip of the shafts.[c] After the +10th century these objects occur no more, presumably because of the replacement of the breast-strap harness by collar-harness in Central Europe.

What gave the idea for the breast-strap harness we do not know, but there is always the possibility that it derived from what was found most convenient in human haulage.[d]

[a] Cf. Fig. 565 below.

[b] The elongated sling helping to suspend the chest-piece from the top of the girth persisted till the end of the +2nd century, by which time it had long become a purely ornamental band. Or so we may conclude if we rely on the Wu Liang Tzhu reproductions of the Fêng brothers in 1821 (see Fig. 541 and many other places in Chin Shih So, Shih sect., e.g. pp. 126, 127). Unfortunately, we cannot be quite sure about this for many of the harness details have disappeared during the past century, as may be seen from modern rubbings such as those of Jung Kêng (1).

[c] See Gyula (1), pl. LVIII, fig. 4; Arne (1); Zak (1). Lynn White (7), p. 61, calls the arrangement 'a type of rudimentary horse-collar', but this is quite inadmissible; we should rather call it a vestigial remnant of one of the most archaic transitional stages between throat-and-girth harness and breast-strap harness. It is interesting that in the reconstructions of Gyula and Zak the girth is retained just as in the ancient system of Chhu. The Avar harness is also compared by Lynn White (loc. cit.) with reindeer harness, but that is unconvincing too. It is true that Lapp and Siberian reindeer harness is composed of rigid parts as well as flexible parts, straps alternating with pieces of bone in continuous lengths, and that some of the hard parts may be bent in sharp angles, but the single trace to the sledge from the 'collar' runs backwards from its base lengthwise under the body between the legs so that no danger of choking the animal arises; cf. Martin (1); Manker (1). It seems quite impossible that the ancient device of Chhu, a relatively southern State, could have originated from contact with reindeer-using peoples. Of course reindeer-keepers and horse-drivers had important cultural interchanges (cf. Gunda, 1), but in this case it seems unnecessary to bring in a cervid with a quite different type of harness. Presently (p. 316) we shall meet with a group of ornamented metal links in breast-strap harness of Scandinavian regions which may well have been inspired by the bone components of reindeer harness.

[d] On this subject see Mason (1), p. 545, and Haudricourt (8).

The working of boats upstream by large groups of trackers goes back a long way, no doubt, in Chinese history, and men would have been conscious from their own experience that tractive force must be exerted from the sternal and clavicular region in such a way as to permit free respiration. Tracking is represented in the Tunhuang frescoes, from which Fig. 547 may enhance our argument.[a]

If then we may accept the breast-strap harness as an invention of the late Chou period, perhaps it was the basis for the legend of Tsao Fu,[1] the charioteer of King Mu, and the man who was supposed to have been the ancestor of the princes of Chao and Chhin. The *Mu Thien Tzu Chuan* (Account of the Travels of the Emperor Mu),[b] which, though not as ancient a text as it purports to be, is certainly not later than -250, lays much emphasis on the extraordinary speed of Tsao Fu's eight horses, the names of which have long been suspected to be of Turkic origin.[c] Since the book was entombed for nearly six centuries, the mention in the *Shih Chi* is an independent one,[d] and from all the sources the relation of the teams of King Mu to the desert and the steppe comes out clearly; one may conjecture therefore that the reality behind the legend might have been a more efficient harness capable of shifting vehicles when stuck in the sand, as well as of faster and surer travel.[e] At any rate the suggestion is offered to the archaeologists.

(iii) *Comparative estimates*

It is hardly necessary to insist on the advantages of the horse, once properly harnessed, over the ox. While both animals have about the same tractive force, the horse gives some 50% more foot-pounds per second because of its greater speed of movement. Besides this, the horse has greater endurance than the ox, and can work a few more hours each day.

It may be worth while, however, to say a little more about the inefficiency of the equine throat-and-girth harness as compared with the traces and the later collar. The anatomical pattern of course speaks for itself, but des Noëttes made practical experiments in 1910 from which he found that the effective traction of two horses harnessed in the throat-and-girth fashion was limited to about 1100 lb. ($\frac{1}{2}$ ton).[f] Yet many modern vehicles weighed from $\frac{1}{2}$ to 3 tons tare and between $1\frac{1}{2}$ and 9 tons loaded, and were easily drawn by collar harness.[g] A single horse with collar-harness can readily draw a total load of a ton and a half. To this des Noëttes added an examination of the regulations for post vehicles found in the *De Cursu Publico* section of the Theodosian

---

[a] This is an early Thang fresco from cave no. 322. We shall return to the trackers in Section 29*i*.

[b] Cf. Vol. 3, p. 507.   [c] Cf. Chêng Tê-Khun (2), pp. 130, 132; de Saussure (26), pp. 248, 275.

[d] Ch. 5, p. 3*b*, tr. Chavannes (1), vol. 2, p. 8.

[e] This point appeals particularly to me, having had much experience of being stuck in the sand when driving motor-trucks on the edge of the Gobi. It may perhaps be no coincidence that the invention of the collar-harness occurred in just the same part of the world—see on.

[f] Pp. 162 ff. This figure appears to refer to freight loaded, but the carriages he used in his experiments were very light modern ones. Such gigs and traps range from about 550 lb. to about 1000 lb. if four-wheeled. The general argument is therefore unaffected. For these estimates we have to thank Mr and Mrs Clarke of Cambridge.   [g] Pp. 130, 134. Cf. the data of Rankine (1).

[1] 造父

Code (+438).[a] Here the fixed limits of freight varied from as little as 154 lb. for the *birota* to 1082 lb. for the *angaria*, a four-wheeled wagon drawn by several animals.[b] The throat-and-girth harness would not have been able, therefore, to draw modern vehicles, even when empty.

With these figures we may compare the 365 lb. taken by a single Chinese wheel-barrow of the San Kuo period. Moreover, an attentive study of Han illustrations of chariots and carriages, comparing them with those from all other ancient civilisations, clearly shows that the Chinese vehicles were much heavier. While Egyptian, Greek or Roman chariots always appear of minimal size, fit only for two persons at most, with cut-away sides and often drawn by four horses, the Chinese chariots frequently show as many as six passengers (cf. the hodometer or band-cart in Fig. 524).[c] Very frequently too they have heavy upcurving roofs (cf. Fig. 540) and are usually drawn by only one horse. Again, when the Chinese came into contact with Western regions, they distinctly noticed the vehicles of those parts as being remarkably small, and the chapter on Ta-Chhin (Arabia, Roman Syria) in the *Hou Han Shu* says just this.[d]

It is quite easy to add some evidence from Chinese texts which des Noëttes never knew. The *Mo Tzu* book, in a passage which may be taken as of the late −4th century, tells[e] of the comments of Mo Ti on the flying kite constructed by Kungshu Phan:[f]

Kungshu Phan constructed a bird from bamboo and wood, and when completed, it flew. For three days it stayed up in the air, and Kungshu was proud indeed of his skill. But Mo Tzu said to him, 'Your achievement in constructing this bird is not comparable with that of a carpenter in making a linch-pin (*hsia*[1])'.[g] In a few moments he cuts out a piece of wood which, though only three inches long, can carry a load no less than fifty *tan* in weight. Indeed, any achievement which is beneficial to man may be said to be skilful (*chhiao*[2]), while anything not beneficial may be said to be clumsy (*cho*[3]).

In a parallel passage in *Han Fei Tzu*,[h] which would date from the following century, and which will be quoted below in connection with aeronautics, the figure given for the weight of the cart pulled is 30 *tan*. Since the *tan* of the Warring States period is estimated as having been equivalent to 120 lb., Mo Ti's estimate would amount to 3 tons, and that of Han Fei to rather over $1\frac{3}{4}$ tons. Doubtless other figures for loads pulled during the late Chou and Han periods could be found in the texts, but these certainly seem to support the view that the breast-strap harness was little less efficient than the collar-harness itself.[i]

---

[a] VIII, v, 8. Des Noëttes (1), pp. 157ff.

[b] It is likely (Burford, 1) that the *angaria* was drawn by two animals only, often quite probably oxen. If so, this load agrees well enough with the only other specific figure which has come down to us from occidental antiquity, namely 25 talents (1100 lb.) mentioned by Xenophon (*Cyropaedia*, VI, i, 54) in his comments on Cyrus' experiments with siege towers (−4th century).

[c] Three commonly in the Hsiao Thang Shan reliefs of c. +129 (*Chin Shih So*, Shih sect., ch. 1, p. 106).

[d] Ch. 118, p. 10a (tr. Hirth (1), p. 40); cf. *TH*, p. 756.

[e] Ch. 49, p. 10b, tr. Mei Yi-Pao (1), p. 256, mod.   [f] Cf. p. 43 above.

[g] The linch-pin is the piece of wood, analogous to the set-screw, which holds the wheel in place on the axle. It does not, of course, 'carry' the weight, but is one of the indispensable parts of the vehicle which permits it to fulfil its transportative function.   [h] Ch. 11 (ch. 32), p. 2a.

[i] It is interesting to remember that the *Mo Tzu* text is approximately contemporary with the lacquer paintings just discussed which reveal the beginnings of breast-strap harness.

[1] 轄   [2] 巧   [3] 拙

The detailed views of des Noëttes have been the subject of much discussion. For example, Sion (2) urged that the figures in the Theodosian Code may have been minimal ones authorised so as to spare the post-horses, and that heavier freight could have been carried at need, also that ancient horses might have given a better performance than those of des Noëttes on account of training. But no one disputes the general conclusion that the throat-and-girth harness, so long and so widely employed, was four or five times less efficient than the breast-strap or the collar-harness. As for

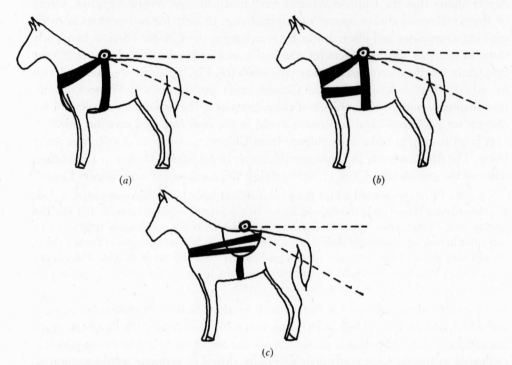

Fig. 549. Attempted ameliorations of throat-and-girth harness. (*a*) The martingale of ancient Egypt, (*b*) the 'false breast-strap' of Assyria and Persia, and (*c*) the saddle breast-strap of Byzantium and Cambodia. None of these solved the problem of efficient equine traction.

the comparison of the two latter, all the practical books concur that trace-harness is never quite as efficient as the collar. Other criticisms of des Noëttes we shall discuss presently (p. 329).

That the throat-and-girth harness was often felt to be unsatisfactory in antiquity is shown by the numerous unsuccessful attempts to improve it. One of the earliest was the addition of a strap analogous to that known as the martingale,[a] which descended from the throat-band, passed between the horse's fore-legs and attached it to the girth (Fig. 549*a*). This was very little use, for the point of traction remaining on the horse's back, the throat-band was always pulled upwards, and in this case it only brought the

---

[a] In Chinese, *yang*.[1]

[1] 鞅

girth forward and galled the elbow region. It can be seen in the prancing horses of Ramses II at Karnak,[a] and has generally been misunderstood by copyists, who have supposed that an extremely loose girth was intended,[b] which would have been no use for anything. A second attempt was what might be called the 'false breast-strap' harness, in which a horizontal band was placed round the chest of the animal, joining the girth at each end (Fig. 549 b). This was not much use either, since the point of pull tended to make the false breast-strap rise and compete with the throat-strap in strangling the horse. It was nevertheless tried again and again, in Assyria (−8th century),[c] by the Sassanid Persians (+4th century),[d] and by the Byzantines (+9th century).[e] A third method was connected with the use of the saddle on draught horses, the throat-strap being attached either further back towards the horse's loins or croup, or lower than the back at the side of the saddle itself (Fig. 549 c). This may be seen in late Byzantine MSS.[f] (+10th to +13th century) as well as at the other end of the Old World in the Khmer culture (sculptured on the +12th-century Angkor Vat),[g] and it found its way through to modern Japanese ploughs[h] and north-west Indian tongas.[i] But still the problem remained unsolved. Except in China.

(iv) *Radiation of the inventions*

In due course the Chinese breast-strap harness of the Han arrived in Europe. Although in the late Roman empire some light vehicles such as the *cisium* began to be fitted with shafts instead of a pole, this does not seem to have involved the adoption of an efficient harness. The famous Gallo-Roman relief at Igel, near Trier (early +3rd century),[j] shows two mules within shafts, but as the girth is clearly visible, and no horizontal traces, it can only be supposed that the throat-strap was retained also; moreover the cart is a very light one, seating only two men. So far as documentary evidence goes, there is no European representation of breast-strap harness before the +8th century.[k] After this it appears more and more frequently, as for instance in the

[a] Des Noëttes (1), fig. 40, p. 47. [b] E.g. Wilkinson (1), vol. 1, pp. 370, 371.

[c] Des Noëttes (1), fig. 21, p. 40 (chariot of Tiglathpileser III).

[d] *Ibid*. (1), fig. 65, p. 48 (Shapur's chariot, relief).

[e] *Ibid*. (1), fig. 90, p. 48 (Cluny coffer). This is certainly not collar harness, as Jope (1), p. 553, by an oversight entitles it.

[f] Des Noëttes (1), figs. 92, 93, pp. 90ff.

[g] McDonald & Loke Wan-Tho (1), pls. 59, 60; des Noëttes (1), fig. 107.

[h] See des Noëttes (1), fig. 137 and p. 119.

[i] Grierson (1), pp. 40, 41; Philipson (1), pp. 55, 59; des Noëttes (1), fig. 181 and pp. 100, 134. The saddle on draught horses continued to be used, of course, after the introduction of collar-harness, as may be seen, for example, on one of the windows at Chartres (*c*. +1215). Jacobeit (2) has also realised the fundamental ineffectiveness of all the expedients described in this paragraph.

[j] Daremberg & Saglio (1), vol. 2, p. 1201. I had the pleasure of studying this famous column *in situ* with my collaborator Dr Lu Gwei-Djen in 1956.

[k] On an Irish monument (see Crawford (1), pls. XLIX, L, and Jope (1), fig. 490). This raises the question of the harness depicted in a relief, claimed as Gallo-Roman, and now in the Musée Calvet at Avignon. Here a four-wheeled wagon is pulled by a number of horses, the nearest one of which (Fig. 550) shows a trace originating from the lowest part of a double throat-band, and apparently joined to the girth at the point where this splits into two, below a saddle-cloth, to pass under the belly of the animal. This relief, which is known to have ornamented the Château of Maraudi in the +16th century, has been fully described by Sautel (1), vol. 2, pp. 227, 229, vol. 3, pls. 57, 59; and is occasionally figured without

Fig. 550   Fig. 551

Fig. 550. The harness of the foreground horse in the Maraudi relief, now in the Musée Calvet, Avignon (Sautel, 1). A trace seems to be combined with a double throat-band, and the girth is forked. Though claimed as Gallo-Roman, the authenticity of this carving is doubtful and the nature of the harness obscure. It may possibly represent a +7th-century attempt to combine throat-and-girth harness with breast-strap harness.

Fig. 551. The nearer of two horses attached to one of the two chariots carved on the lowest architrave of the southern gateway at Sānchi (probably late − 1st century) in India (after Marshall, 3). A prominent throat-band indicates throat-and-girth harness. The traces and withers-strap have no tractive function but seem rather to be intended to keep the horses' long tails out of the driver's way. Possibly the Chinese breast-strap harness was known in India at this time and the artist misunderstood it.

remarkable tapestry found in the Oseberg ship-burial (first half of the +9th century).[a] When Ohthere told King Alfred about +880 of his life in northern Norway and said that 'what little he ploughed he ploughed with horses', the breast-strap harness was doubtless what he used.[b] In the Viking culture of the +9th and +10th centuries the withers-strap of the harness seems to have carried various crest-like metal ornaments, or to have incorporated a curved and ornamented metal component across the horse's back.[c] We see the breast-strap harness again in the drawings illustrating a MS. dated

comment, as in van der Heyden & Scullard (1), fig. 300. The difficulty about it is its date. Des Noëttes (10), noting that the horses have shoes, affirmed that it is a Renaissance 'antique'. Sautel's view is that if not Renaissance, it is of the latest possible 'classical' date. If it could be placed between the +5th and +7th centuries, it might possibly represent an early European attempt to combine the throat-and-girth with the breast-strap harness. But its authenticity is too doubtful to admit of its use in support of any theory. Similar remarks apply to a relief in the Terme Museum at Rome figured by van der Heyden & Scullard (1), fig. 301, and ascribed to a date c. +100. Here evident throat-and-girth harness seems to be accompanied by traces of some kind, perhaps ornamental.

[a] See Holmqvist (1), pl. LX, fig. 134, discussed without technical comment, p. 77. Contrary to Lynn White (7), p. 60, we interpret the picture as showing shafts, not traces, for some are convex upwards and some are drawn clearly as solid. Lynn White gives the impression that Grand (2) deduced the origin of 'modern', or collar, harness from this Norse tapestry, but that is not the case; Grand only felt that the +9th-century Vikings had some form of harness which bore on the shoulders and chest of the horse, and he was led to this view by the (strangely narrow) shafts of the Oseberg wagon (see Brøgger & Schetelig (2), vol. 2, p. 7, pl. 1).

[b] See Sweet (1), p. 18; Ross (1), pp. 20ff.; M. Williams (1), p. 297. We cannot agree with Lynn White (7), p. 61, that the statement of Ohthere implies collar-harness. The name of Ohthere will recall to many Rudyard Kipling's story 'The Knights of the Joyous Venture'.

[c] See Stolpe & Arne (1), pp. 25, 28, 34, 59, pls. XV, fig. 1, XVIII, fig. 1, XXIII, fig. 1, XXIV, fig. 1; Poulsen (1), p. 230, pl. 32, fig. 1; Brønsted (1), p. 144, (2), p. 272, fig. 194; Holmqvist (1), p. 62, pl. XLVI, fig. 107. We know of no justification for the view adopted by Lynn White (7), p. 61, that these objects were intended for use with horse-collars.

PLATE CCX

Fig. 552. Evidence of possible experiments for the improvement of horse harness in the late Roman empire; a boy's shafted *currus* on the +3rd-century sarcophagus of Cornelius Statius, formerly at Trier, now in the Louvre (des Noëttes, 1). The harness is placed too high and fits too tightly round the neck to be a collar, and the shafts meet it too near the mid-point; it is therefore probably a throat-band, occlusive as usual.

Fig. 553. Another Gallo-Roman representation of a harnessed horse; a carved panel in the Landes-museum, Trier (+4th century). The object high on the horse's neck is probably an arched cross-bar joining the shafts, and there seems to be a throat-band fixed to their ends. This would have been no improvement.

PLATE CCXI

Fig. 554. A third late Roman example; horse and cart in the +4th-century mosaics at Ostia, the ancient port of Rome (photo. Richter). Again the harness simulates a collar but is placed too high on the horse's neck, while significantly a girth is present. It is not clear how this throat-and-girth harness was attached to the shafts.

+1023 of the *De Rerum Naturis* of Hrabanus Maurus.[a] The crudeness of the textile makes it a little difficult to distinguish from the collar-harness, but the Bayeux Tapestry seems to show it (+1130),[b] as also contemporary carvings and MSS.[c] After that time it was widely known and used,[d] e.g. on post-coaches in the early 19th century; it found its way to South Africa, whence it was reintroduced into England (Philipson), and in some parts of Italy it is the only harness used. In recent years I have found it common in the Monaco region and Provence, where it is still called 'l'attelage à postillon'. Whether it ever spread to south or south-east Asia is uncertain.[e]

It has long been well known that the other efficient harness, the collar-harness, made its appearance in Western Europe about the beginning of the +10th century, and had been universally adopted by the end of the +12th.[f] Manuscript pictures of about +920 are the first to show it,[g] and afterwards it becomes more and more

---

[a] See Amelli (1); Stephenson (1).

[b] In a ploughing scene. Cf. des Noëttes (1), fig. 146, who regarded it as the oldest European representation of the agricultural use of mules and horses. Cave no. 290 at Chhien-fo-tung, however, has what may possibly be a horse-plough, and this is of the early +6th century. On this see Needham & Lu (2).

[c] Des Noëttes (1), fig. 147 (capital in the crypt of St Denis), fig. 148 (Admont MS. in the Doucet Library).

[d] Cf. a woodcut of +1538 in Justinus' *Œuvres sur les Faictz et Gestes de Troge Pompée...* (Janot, Paris). See also Macek (1), pl. 21 (the +15th-century Jena Codex).

[e] In this connection Haudricourt (4) noted that the Ephthalite Huns took down into India around +480 only the saddle, stirrups, and horseshoes, not the collar-harness. The breast-strap harness seems never to have been adopted in India. Yet the reliefs on the railings of the Bhārhut Stūpa (−2nd century) show what at first sight looks remarkably like breast-strap harness (Marshall (3), pl. XVI, fig. 43). This is seen again in the reliefs at Sānchi (−1st to +2nd century), especially those of the south gateway (Marshall (3), pl. XXIII, fig. 63, lowest architrave; des Noëttes (1), fig. 102; Munshi (1), pl. 19). Close scrutiny reveals, however (Fig. 551), that the horses seem to have throat-and-girth harness as well. What at first sight look as if they might be stumpy chariot-shafts turn out to be the tails of the horses, looped forwards and attached to a horizontal trace suspended by straps from the animals' backs, and giving the illusion of breast-strap harness. We believe that des Noëttes (1), p. 99, was right in saying that these traces had no tractive function. But it did not occur to him that this curious system of keeping the long tails of the horses out of the driver's eye might have been a misunderstanding of the breast-strap harness of late Chou and Han China. Although we do not know very much about contacts between the Chinese and Indian culture-areas at the time, it will be remembered (from Vol. 1, p. 174 above) that Chang Chhien about −130 found Szechuanese products in Bactria which he thought had reached there by the Yunnan-India route, and that there were diplomatic contacts between China and the Kushān empire from at least +90 onwards.

[f] See des Noëttes (1), pp. 123 ff.

[g] Des Noëttes (1), figs. 140, 141, 142 (Latin MS. 8085, Bib. Nat.). A representation in the Trier MS. *Apocalypse*, however, datable *c.* +800, is accepted by its discoverer, Prof. Lynn White, as the oldest picture of collar-harness in Europe, cf. his (7), p. 61 and fig. 3. The absence of any indication of solidity or padding in the band-like shoulder-straps prevents us following him in this. If the two horses were drawing in shafts with a cross-bar yoke, this would not be so serious, for a firm collar could be inferred, but they seem to be in traces. If a band was intended rather than two ropes or straps, then the whole illustration could well be one of trace-harness with the withers-strap slung short (cf. p. 325 below) and so simulating a collar. There are moreover certain Gallo-Roman carvings which demand mention here. The sarcophagus of Cornelius Statius from Trier (now in the Louvre) shows a boy's shafted *currus* with the shafts attached to the mid-point of what might be a collar drawn too high on the neck, or else merely a throat-band (Fig. 552). Cf. des Noëttes (1), fig. 73; Jope (1), p. 544. This is a document of the +3rd century. When visiting the Rheinisches Landesmuseum at Trier Dr Lu and I noticed another representation (Fig. 553) carved on a stone panel of the +4th century. Here again the band to which the shafts are attached is much too high on the neck for a collar, and its exceptionally high relief may have been intended to portray an arching cross-bar joining the shafts. A very similar relief is figured by Espérandieu (2), vol. 5, no. 4035. Again, at the Roman port of Ostia Dr D. Needham drew my attention to a +4th-century mosaic (Fig. 554) showing a four-wheeled wagon (cf. Calza (1), p. 21 and fig. 5; Becatti (1), vol. 4, pl. CVIII, fig. 64). The shafts of this are attached apparently at the centre of a 'collar' which at

common.[a] There is no doubt that over the same period it was largely replacing the old breast-strap harness in China. It is seen in Sung[b] and Yuan[c] paintings (+11th and +14th centuries), and appears, to the exclusion of all other types of harnessing, in the compendia of agricultural engineering already described (p. 166 above).[d] But soon a suspicion arose that the collar-harness goes back a great deal further in China and Central Asia than in Europe. Though des Noëttes himself seems always to have adhered to the view that it arose in the West, he actually figured a photograph[e] by Pelliott (25) of one of the Thang pictures of horses and carts in the Tunhuang cave-temple frescoes (actually from the +9th century), labelling it 'breast-strap and shafts' in spite of the fact that it shows collar-harness very clearly, as we shall shortly see from photographs and drawings specially made later on the spot.[f] He also gave a photograph of a rubbing[g] from Chavannes (9) of what looked like a transitional form between breast-strap and collar harness, from a Northern Wei incised stele of the +6th century.

It is now possible to attain greater precision concerning the time and place of the cardinal inventions and their subsequent arrival in Europe. As has already been indicated,[h] there can be no doubt that the earliest Chinese chariots were of the pole type, with throat-and-girth harness, as is shown by the analysis of the written character as well as the archaeological evidence. This system probably lasted as late as the end of the Warring States period, but was completely superseded by the breast-strap harness during the early Han (i.e. from the −2nd century onwards).[i] In no other part of the world did an efficient equine harness appear so early. For dogs, indeed, the throat-and-girth gear lasted down to the end of the Han, as we know from tomb carvings[j]

---

its upper end even stands out as if padded—but as in all these cases, the position is that of the throat-band and not the collar proper. Moreover a girth is present, though with true collar-harness quite superfluous. Although, therefore, these interesting representations must fail to qualify as collar-harness, one cannot free oneself from the suspicion that some experiments towards better tractive devices were being made in the late Roman empire.

[a] E.g. a +10th-century MS. 135 N.O. in the St Gall Library (des Noëttes (1), fig. 143), an +11th-century MS. 9968 of the Bib. Roy. Brussels (des Noëttes (1), fig. 144), and two +13th-century MSS. (figs. 150, 151). A fine example is found on the 'Kherson' bronze doors of the Church of St Sophia at Novgorod, made by German craftsmen and assembled by Master Abraham about +1150 (Kelly, 1). For a miniature of +1340 see Leix (1), p. 328, and for +15th-century drawings and paintings Born (1), p. 773, (2), p. 782, also Saxl (1). Broadly speaking, the later the representation the clearer it is; some of the early ones place the collar as high on the neck as if it were a throat-band, thus sharing somewhat the uncertainty of the Roman and Gallo-Roman depictions just discussed.

[b] Notably in the famous picture of Chang Tsê-Tuan 'Going up the River at the Spring Festival' (*Chhing Ming Shang Ho Thu*), pls. 8, 11, 16, 18 and 19 in the edition of Chêng Chen-To (3). From Figs. 515, 516 it will be seen that here in +1125 the horse- and mule-collars are of the modern 'combined' type (see on, p. 321). Also Yang Jen-Khai & Tung Yen-Ming (1), pl. 48.

[c] Cf. one in the Brun Collection reproduced by des Noëttes (1), fig. 130.

[d] *Nung Shu* (+1313), ch. 16, pp. 5b, 7a, 8b, 10b; *Thien Kung Khai Wu* (+1637), ch. 1, p. 21b, ch. 3, p. 13a, ch. 9, p. 14a, etc. Needless to say also in +18th-century paintings and modern practice.

[e] (1), fig. 129, cf. p. 111.

[f] Des Noëttes was probably misled because Pelliot's photographs, made fifty years ago, were not always very clear on points of detail, and what are white collars with markedly padded profiles appeared on them only as broad white bands.

[g] (1), fig. 128, cf. p. 111.

[h] See pp. 246 and 306 above.

[i] All these conclusions agree with those reached by Hayashi Minao (*1*) in an elaborate paper on horse vehicle traction from the Shang to the Chhin, which appeared as this section was going to press.

[j] W. Franke (2), fig. 6; Rudolph & Wên (1), pl. 20.

PLATE CCXII

Fig. 555. Han tomb-model of a dog wearing throat-and-girth harness.
Cernuschi Museum, Paris (Hentze, 4).

and figurines[a] one of which is here reproduced (Fig. 555). It still continues in use among some of the tribal peoples of north-eastern Siberia.

That the breast-strap harness made its way in due course to the West across the Eurasian plains we can hardly doubt. Haudricourt (4) offers evidence that it reached Italy at least three centuries before the rest of Europe; he thinks with the Ostrogoths in the early + 5th century. He suggests also that the arrival of the first efficient horse harness in Italy may have had something to do with the precocity of economic awakening there, as compared with other European countries. 'The good fortune of Europe in these matters', he wrote, 'in comparison with the Middle East and India, lay in the fact that it was the natural terminus of those great Asian plains which had been the centres of development of the techniques of locomotion.'

But could this be true of the second efficient horse harness also? That the collar was known in Thang times, i.e. two or three centuries before its first representation in Europe, was borne in upon me with conviction during the month which I spent in 1943 in residence at the cave-temples of Chhien-fo-tung[1] (more properly called Mo-kao-khu[2]) near Tunhuang.[b] In these wonderful cave frescoes the eye of the leisurely visitor with technological awareness is caught by the fact that in nearly all the representations of horses and carts the shafts are attached to the lowest and furthest forward point of the collar-like harness (Fig. 556a). This is radically different from what is seen in drawings and carvings from all other parts of the world,[c] where the attachment, if not in the throat-and-girth position on the horse's back, is to the mid-point of the throat-strap or collar-like band.[d] It bears the clear implication of a pull from the sternal and not the tracheal region. I had already appreciated the significance of this when the first draft of the present subsection was written but at that time the caves were not so well dated as now, and on a return visit in 1958 it was possible to study the representations at Tunhuang in much greater detail.[e]

(v) The second rationalisation; collar-harness in Shu and Wei

Let us begin with the clearest case. In cave no. 156[f] a magnificent panorama depicts the triumphal procession of a Chinese general and provincial governor Chang I-Chhao,[3]

[a] Hentze (4), pl. 31A. Another, about 3 ft. high, taken from a Later Han tomb at Thien-hui-chen, is to be seen in the Chhêngtu Municipal Historical Museum.

[b] With Mr Rewi Alley, Mr Wu Tso-Jen (now President of the Chinese Academy of Fine Arts) and Mr Lo Chi-Mei. It is a pleasure to express our gratitude to the Lama Yi for all his kindness at that time. This is of course the site made famous in the West by Pelliot and Stein long ago.

[c] Except certain early, but subsequent, European pictures, notably a + 12th-century carving in the cathedral of Burgos in Spain (des Noëttes (1), fig. 149), and + 14th- and 15-century Swiss and Austrian MS. illustrations (Wescher (1), p. 2251, (2), p. 2277). Cf. also des Noëttes (1), figs. 155, 159. All these are undoubtedly collar-harness.

[d] Cf. fn. g on p. 317.

[e] I am greatly indebted to Dr Chhang Shu-Hung, Director of the Tunhuang Research Institute, for the warmest of welcomes and for all the Institute's facilities; and to Mrs Chhang (Li Chhêng-Hsien[4]) for her indefatigable help. I also owe much to my colleague Dr Lu Gwei-Djen for discussions at the site.

[f] In Vol. 1, p. 120, I referred to the different enumeration systems which have been used at Chhien-fo-tung. Since then a concordance to them has been published by Hsieh Chih-Liu (1), and we now adhere

¹ 千佛洞   ² 莫高窟   ³ 張議潮   ⁴ 李承仙

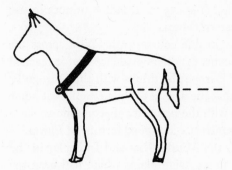

(a) The common element in the Chhien-fo-tung frescoes.

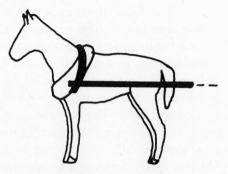

(b) Chhien-fo-tung frescoes; three-component representations.

(c) Contemporary collar-harness of north and north-west China.

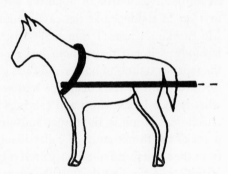

(d) Chhien-fo-tung frescoes; two-component representations.

Fig. 556. Diagrams to explain the evidence of the fresco paintings at the Chhien-fo-tung cave-temples, Tunhuang, Kansu (+5th to +11th centuries); see text.

who recovered the Tunhuang region from the Tibetans in +834. We are fortunately able to date the picture closely for there is good evidence that it was painted in +851.[a] The Commander of the Kuei-I Army and Legate-Plenipotentiary (Kuei-I Chün Chieh-Tu-Shih[1]) is probably celebrating the arrival of an imperial rescript confirming his authority, and he proceeds on his way attended by soldiers, banner-bearers, hunters and musicians. His retinue is followed by a second procession (the more important for us), that of his wife the Lady Sung (Sung Kuo Ho-Nei Chün Fu-jen Sung shih[2]), also composed of riders, musicians and dancers, but including four carts of which three are for the baggage.[b] The main part of the scene is shown in Fig. 557,

solely to the numbering of the Institute, converting other references into this as they arise, whenever possible. The most complete tables are those of Chhen Tsu-Lung (1).

[a] This is the opinion of Dr Chhang Shu-Hung. But there is universal agreement that the date must be in the middle two or three decades of the century.

[b] Previous reproductions of this scene which show the details of harness more or less well have been given by Pelliot (25), pls. XLVII and XLVIII (photos 45142/51, nos. 47 and 48 at Musée Guimet);

[1] 歸義軍節度使   [2] 宋國河內郡夫人宋氏

PLATE CCXIII

Fig. 557. Origins of collar-harness. Part of the procession of the Lady Sung, wife of General Chang I-Chhao, the Exarch of the Tibetan Marches (the Tunhuang region, on the Old Silk Road). A fresco from cave no. 156 at Chhien-fo-tung datable at +851; copy-painting by Yeh Chhien-Yü (1). Besides musicians and dancers (out of the picture to the left), a mounted escort and bearers of two hexagonal reliquaries, there are three baggage-carts (right) and a personal cart with a canopy (left). Though the upper right baggage-cart has been obliterated by the soot of a later fire, all four horses show collar-harness very distinctly. A fifth horse, also wearing collar-harness, can just be made out in the extreme lower right-hand corner of the picture.

Fig. 558. Enlargement of the lowest (foreground) baggage-cart in the procession of the Exarch's consort; a copy-painting by Ho Yi (1948). The curved 'ox-yoke' cross-bar connecting the shafts fits across the front of a well-padded cushion of annular shape passing low across the horse's chest. The collar, in fact, is substituting for the 'hump' of the ox.

PLATE CCXIV

Fig. 559. Typical collar-harness still in use in north and north-west China (orig. photo., 1958, near Chiu-chhüan, Kansu). The collar, still called the cushion (*tien-tzu*), sits free round the neck of the horse or mule, and takes the pressure of a clasping framework (*chia-pan-tzu*), the direct descendant of the ancient yoke, which is attached to the shafts by thongs slipped over pegs at their forward ends. In this example, photographed on the edge of the Gobi Desert, two leading animals were adding their power by means of traces.

and one of the latter, enlarged, in Fig. 558.[a] An attentive study of the pictures *in situ* reveals that in all cases there are three components: (i) the shafts, (ii) a curved piece of wood like the ox yoke or the cross-bar of the Han carriage which connects them together,[b] and (iii) a well-padded collar coming low on the chest (Fig. 556b), and rising behind the cross-bar. It thus becomes immediately evident that in these earliest forms of collar-harness the arrangement was in two parts, and the collar alone was simply an artificial substitute conferring upon the equine animal the equivalent of the internal 'hump' of the ox, i.e. a *point d'appui* against which the yoke could rest.

Our eyes were now opened. Great was our excitement, therefore, to find along the roads in Kansu province and many other parts of north and north-west China that the form of collar-harness still widely used there today perpetuates the double system of the Thang.[c] Figure 559, a photograph taken near Chiu-chhüan, shows it well. The collar (the 'cushion', *tien-tzu*,[1] or the 'hugger', *yung-tzu*[2]) is not itself attached to anything, but there rests upon it a clasping framework (*chia-pan-tzu*[3] or *hsia-pan-tzu*[4]),[d] the descendant of the ancient yoke, which is directly connected with the shafts by thongs slipped over the pegs which they carry at their forward ends (Fig. 556c). In some districts the yokes of oxen assume the shape of a wishbone or a tree-branch forked in a wide angle.[e] In this we may see a transition to the 'yoke' necessary for the horse in the collar-harness invention, and we shall naturally remember also the wishbone-shaped vestigial 'yokes' of the horse-drawn chariots of the Han. Thus the essence of the new device was an 'artificial hump' which would not come off. Only later on were the hard yoke and soft collar combined into one component, as in modern harness, to which the dictionary term for the horse-collar, *hu-chien*,[5] or 'shoulder-protector', applies.[f] And our original question can now be answered, in principle if not yet in date, for many of the earliest European pictures of collar-harness show precisely the Chinese three-component system.[g] Indeed one can actually meet with

Roy (1), p. 123; Gray & Vincent (1), pl. 51A; Chhang Shu-Hung (1), pl. 24; Yeh Chhien-Yü (1), pl. 34. Sirén (10), pls. 66, 66B, offers a very clear copy painting, but the harness of the four carts is not well shown as the artist failed to appreciate the significance of the details. The cart of which we show an enlarged picture is the very one reproduced (and misinterpreted) by des Noëttes (1) in his fig. 129.

[a] For these and for other photographs we are very grateful to Dr Chhang Shu-Hung and the photographic staff of the Tunhuang Research Institute. We are also deeply indebted to Mr Rewi Alley and some of the teachers and students of the Baillie Technical School of the Chinese Industrial Cooperatives at Shantan (now the Technical College of the National Petroleum Administration at Lanchow) for paintings specially made at Tunhuang during and soon after the war. Most of these were made by Mr Ho Yi, a Nakhi from Lichiang in Yunnan, and an able textile designer. Others were made for us by Mr Sorenson, but perished with him in a tragic aeroplane accident in Sinkiang in 1947.

[b] Much credit is due to Mrs Chhang Shu-Hung for appreciating the significance of this vital clue and drawing our attention to it.

[c] Cf. a drawing in Li Chun (1). Photographs in Hommel (1), fig. 490, and des Noëttes (1), fig. 133; clear only if one knows what to look for.

[d] For these names we are indebted to Mr Wang San-Lin of the Chiu-chhüan local government.

[e] For example, around Wukung in Shensi province.

[f] As we have noted above, this fusion was already complete by +1125, the date of Chang Tsê-Tuan's painting. Whether it first occurred in China or the West we do not yet know. Perhaps it was a natural development which happened independently.

[g] Notably in Wescher (2), p. 2277 (+14th century); des Noëttes (1), figs. 150, 154 (+13th and +14th centuries); Lejard (1, Ptg.), fig. 73 (+16th century). The process of fusion can be progressively followed.

[1] 墊子　　　[2] 攤子　　　[3] 梜板子　　　[4] 挾板子　　　[5] 護肩

it still persisting in certain corners of Europe; two years after our study of the North China harness we found it again in the Iberian peninsula at the furthest extreme of the Old World.[a]

Many other pictures of horses and carts are to be seen in the frescoes at Chhien-fo-tung. It was also evident at the first inspection that girths are never visible in them, and traces very rarely, if ever, present. All show the attachment of the shafts to the furthest forward point of the object round the shoulders and chest.[b] Some form of collar-harness was therefore clearly implied even without the crucial evidence of cave no. 156, but in the light thrown by the carts of Chang I-Chhao one feels no hesitation in assuming the existence of the padded 'hump-substitute' when the drawings do not show it. Otherwise the 'horse yoke' could not have worked at all. And indeed while the whole three-component system (Fig. 556b) is seen in at least four or five caves, there are at least five or six more where we find simply the shafts and the yoke (Fig. 556d). Again, by good fortune, the oldest representation of all (cave no. 257), painted under the Northern Wei between +477 and +499,[c] shows collar-harness clearly (Fig. 561).[d] Reins are indicated by a pair of lines, high up on the neck, or just possibly a collar not drawn in its proper place behind the yoke—yet the evidence is decisive, for the yoke itself is well placed, and though no collar appears for it to rest on, it would have been useless without it.[e] The same argument applies to

[a] In 1960 I was able to photograph a number of examples. In the Mancha, as at Almagro, and in Portuguese Estremadura, the *chia-pan-tzu* is an unpainted wishbone-like piece of wood resting against the separate collar; but in Andalusia, between Seville and the Portuguese border, the lower extremities of the *chia-pan-tzu* tend to be linked by a cord or chain exactly as in China. Forms of harness seen in the Algarve, Portugal's southernmost province, approach even closer to the system of the Thang frescoes at Tunhuang, for an arched cross-bar (*cangai*) connects the shafts and is borne on a kind of 'neck-saddle' (*muli*) formed by combining the wishbone-shaped *chia-pan-tzu* (now become functionless and highly ornamented) with the cushion. One sees the same arrangement further north in the Alentejo, but less brightly painted. If traces are used instead of shafts (as for animals turning *sāqīya* chain-pumps), a rough working *chia-pan-tzu* will be placed just behind the ornamental one, replacing the cross-bar. These archaic forms must be studied in the countryside, for ordinary modern collar-harness is more common in town. I do not know whether it is necessary to appeal to the Muslim influence in Spain for an explanation of them, or to assume some special route of transmission from East Asia—perhaps they may be more simply regarded as trans-Pyrenean 'relict fauna', the preservation of stages superseded long since in central European regions. See Fig. 560.

Norway also used to have a type of collar-harness in two parts like the north Chinese *chia-pan-tzu* and *tien-tzu*, if we may judge from an old photograph of a two-wheeled cart from Alfdalen published by Brøgger & Schetelig (2), vol. 2, p. 31.

[b] It is worth noting that Philipson's plates of ideal collar-harness all show that the optimum point where the traction transmitter (shafts, traces, etc.) should be applied to the collar is much lower than its central point when seen from the side, indeed from two-thirds to three-quarters of the semi-circle's circumference.

[c] In the dating of the caves we are now able to take advantage of the valuable study of Mizuno Seiichi (1).

[d] This fresco has often been reproduced; see e.g. Anon. (10), pl. 6; Chhang Shu-Hung (1), colour pl. 1; Chhang Shu Hung (2), cover; Gray & Vincent (1), pl. 6; Sirén (10), pl. 33. It is thought to illustrate the *Ruru Jātaka*.

[e] If one should ask oneself how Fig. 561 differs from Fig. 551 the answer is that at Tunhuang the 'yoke' or cross-bar is at last exactly in the right place on the horse's withers, it could not have stayed there without a collar, and we know the indubitably complete gear of +851 at the same locality. In all the Roman representations on the contrary (Figs. 550 and 552 as well as Fig. 551) there is a band round the throat which is clearly connected with the shafts. Lynn White (7), p. 157, criticising Needham & Lu (2), interprets this Tunhuang fresco-painting as a 'withers yoke or strap between shafts', presumably therefore (if we understand him rightly) as this same Roman system. But a throat-strap would be

PLATE CCXV

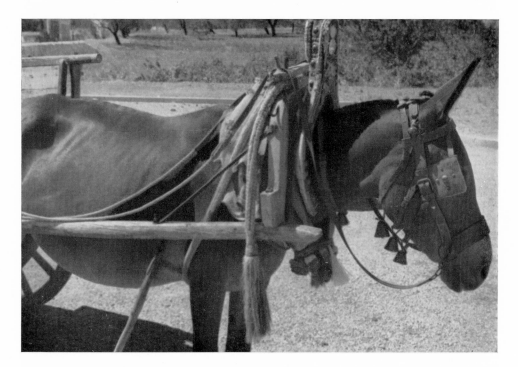

Fig. 560. The Chinese three-component system of equine harness still extant in the Iberian peninsula; an example in southern Portugal (near Tavira, Algarve, orig. photo., 1960). An arched cross-bar connects the shafts just as in the +9th-century frescoes at Chhien-fo-tung, bearing against and upon a free cushion-collar or 'neck-saddle', to which is fixed in front a functionless (but brightly ornamented) wishbone-shaped yoke or *chia-pan-tzu*. Elsewhere, as in the Mancha in Spain, where there is no arched cross-bar, this still performs exactly the same function as in China; and elsewhere again, as in Andalusia near the Portuguese border, it is an approximately rectangular framework of wood and chains or thongs exactly the same as the Chinese pattern. Presumably these Iberian systems are vestiges of the first adoption of the Chinese three-component collar-harness in Europe, before the fusion of the soft cushion and the hard hames into one unit.

PLATE CCXVI

Fig. 561. The oldest representation of equine harness in the Chhien-fo-tung cave-temple frescoes, painted under the Northern Wei between +477 and +499 (cave no. 257; copy-painting in Anon. *10*). A green cart with a brown 'sun-bonnet' roof of the covered-wagon type, striped red and yellow curtains and a green sun-awning, is being drawn by a large white horse. The arching cross-bar is clear, but the artist failed to draw the cushioning collar behind it, without which it would have been useless. Two thin lines higher up the horse's neck may be a collar drawn in the wrong place but are more probably reins, and the breech-strap is indicated by another.

Fig. 562. A slightly later representation, datable between +520 and +524 (cave no. 290; copy-painting in Anon. *10*). The painting, done in the blacks, blues, greens and pale browns characteristic of its period, has a light brown horse pulling a chariot occupied by two monks. The shafts project beyond the horse's chest, implying sternal traction, and one band round the horse's withers does duty both for the yoke and the cushion-collar; we interpret it as the former, and suppose that the artist omitted the latter.

the next oldest pictures, those in cave no. 290, precisely datable between +520 and +524, from which we reproduce Fig. 562. Here only yoke and shafts are shown.[a] From exactly the same reign-period, and with the same ellipsis, comes the cartload of travelling monks in cave no. 428.[b] After this, there are two further two-component frescoes both from the Sui dynasty (c. +600),[c] which bring us back to the certainties of the Thang. As a pendant to this evidence we may mention certain scenes in which animals other than horses are harnessed by yokes within shafts, the padded collar remaining tacit. One of these is of Thang date (probably early +9th century) and shows three carts in line, one drawn by an ox, one by a stag and one by a sheep or large white goat.[d] The artist may have been illustrating a legendary story, but the mechanical principles remain in force.

After this the representations, though generally clearer, parallel or follow in date the presumed earliest appearance of collar-harness in Europe. There are two fine paintings of horses just released from the shafts and still bearing their padded collars; on the shafts both 'yoke' and breeching can be seen.[e] Three-component pictures of horses in their shafts also occur,[f] all before the middle of the +11th century.[g] The conclusion from the Tunhuang evidence is thus quite clear. Collar-harness in its initial form appears explicitly and indubitably in the procession of Chang I-Chhao in +851, nearly a century before the first possibly acceptable documents of Europe. But depictions which can only have been based upon this harness go back to the last quarter of the +5th century and the first quarter of the +6th, so that it would be very reasonable to date the first appearance of it about +475 in the empire of the Northern Wei. And the place would again be significant, for the sands of the borders of the Gobi desert in Kansu and Shensi needed strong tractive apparatus. Where Han trace-

extraordinarily unlikely here, not only because of the considerations already mentioned, but because by +477 the Chinese had not used throat-and-girth harness for nearly a thousand years. The Roman empire on the other hand was only just trying with great difficulty to get away from it. Moreover, no throat-strap is in fact shown.

[a] The horses and carts from this cave may also be viewed in Anon. (10), pls. 25, 26 and 31; and Chhang Shu-Hung (2), p. 6.

[b] This has been reproduced only once, and only incidentally, in Chhang Shu-Hung (1), pl. 5, but a detailed copy-painting will be found in Needham & Lu (2).

[c] Cave no. 419; see Anon. (10), pl. 32; Chhang Shu-Hung (2), p. 10; Sirén (10), pl. 41. It is thought to show the story of Prince Sudhana. Cave no. 420; sketched only in Needham & Lu (2).

[d] Cave no. 159; reproduction in Needham & Lu (2). Another occurs in cave no. 61.

[e] Cave no. 98 of the Wu Tai period (+920 to +960) or very early Sung (before +980); reproduced only in Needham & Lu (2). The donor of this cave was Li Shêng-Thien,[1] King of Khotan (cf. Chhang Shu-Hung (1), pl. 26), a fact which might not be entirely irrelevant to the westward transmission of collar-harness. Also cave no. 146, in which the paintings are of the early Sung, before +1035. Reproduced only in Needham & Lu (2), apart from the old photographs of Pelliot (25), pl. xxvi (photo 45141/18, no. 26 at Musée Guimet).

[f] Cave no. 61 of the Wu Tai period (+920 to +960) or very early Sung (before +980), a general view of which is to be seen in Chhang Shu-Hung (1), pl. 28. This was a cave donated by the Tshao family so important in the history of printing (cf. Sect. 32 below); see Chiang Liang-Fu (1), pls. 28, 29, 30, and Chang Ta-Chhien (1). For the horse and cart see Pelliot (25), pls. ccvi, ccxx (photos 45153/14 no. B/206, and 45153/28 no. B/220 at Musée Guimet), but better in Needham & Lu (2). Also cave no. 146, before +1035; references as in the previous footnote.

[g] Cf. finally cave no. 6 (Wu Tai or early Sung), harness shown in Pelliot (25), pl. cccxxix (photo 45157/3, no. 329Δa/13 at Musée Guimet).

[1] 李聖天

harness would break, the *chia-pan-tzu* could be attached to the shafts by chains, and thus at last the sheer muscular strength available became the only limiting factor.

Yet another group of representations is found on Buddhist steles and reliefs of the Northern and Western Wei periods, in which one sees oxen wearing what looks extremely like collar-harness. I reproduce one of these dated +493 from a rubbing in my own possession (Fig. 563), but many more have been published;[a] they are often more delicately drawn, and the artistic style is characteristic enough. That the animals are oxen is clear because horses (differently depicted) sometimes accompany them in the same scene, and it may well be that the artists intended only to represent a yoke with a loose throat-strap (*yang*[1]) below it. This would be the *joug de garrot*, common later in Europe.[b] They might also have been drawing an inverted U-shaped yoke which came down low on each side of the animal. Perhaps indeed the same yoke could have been used for oxen or horses as the occasion demanded; or perhaps a full horse-collar was sometimes applied to bovine species, as certainly sometimes happened in the West later on.[c] In any case this iconographic group is not at all irrelevant to the early history of collar-harness in China.[d]

A third source of information is constituted by bricks from tombs of the Chhin, Han and San Kuo periods. It would still be premature to fix upon the late +5th century as the assured time of origin of the collar-harness invention, for we may have to recognise it in the +3rd. During the First World War Segalen, de Voisins & Lartigue excavated the tomb of Pao San-Niang[2] at Chaohua, a place just south of Kuangyuan on the Szechuan–Shensi road. This tomb contained a series of magnificent solid moulded bricks (shown here in Fig. 564),[e] each of which has the same picture of a horse and cart. As they are not all equally damaged or imperfect in the same places, it is possible to make out what seems to be a large horse-collar coming low on the chest, apparently amply padded and looking almost like a thick garland. The straight shafts or traces seem to be attached to the side of this with no sign of a yoke. In some cases it is also possible to see a faint girth and perhaps the horizontal line of a breeching. Even the reins appear in some of the bricks. If collar-harness is really present the fact is indeed remarkable, for the date of the tomb is unquestionably in

---

[a] See, for example, two pictures in Sirén (2), pls. 151 B and 169, the former dating from +525 and the latter from +551. Also Binyon (1), pl. VI, a painting purporting to be of +897.

[b] See des Noëttes (1), p. 125 and fig. 153 (a +13th-century ploughing scene).

[c] Cf. des Noëttes (1), fig. 173 (a +13th-century sculpture at Chartres).

[d] Archaeologists have often enough had difficulty in deciding whether a particular piece of harness was intended to be used by oxen or horses. Small bronze objects excavated from +3rd-century tombs in the Rhineland have been taken as models of 'yokes' for horses, to be used with shafts or on both sides of a central pole for throat-and-girth harness (Behrens, 1); and a small wooden yoke of about the same period discovered at Pforzheim by Dauber (1) has been interpreted as a cross-bar between shafts to harness a pony with a throat-strap or *joug de garrot* system of linked hard and flexible pieces (Jacobeit, 1). Lynn White (7), p. 60, accepts the equine relevance of these, but to us it seems more than doubtful.

[e] Segalen *et al.* (1), pl. 58. Reproduced by O. Fischer (1), p. 314, without comment. In their discussion (p. 166) the discoverers did not appreciate the significance of this find for the history of animal traction. A similar series of chariots, not on separate bricks but stamped by the same stamp on a single hollow brick, is seen in Ti Phing-Tzu (1), ch. 2, p. 1b; this is certainly Han, but the harness used is obscure.

[1] 鞅   [2] 鮑三娘

PLATE CCXVII

Fig. 563. Rubbing of a Buddhist stele dated +493 commemorating the monastic profession of Kuo Shih-Jen. An ox-cart in the cortège yokes the animal curiously with what may be a *joug de garrot* or an inverted U-shaped cross-bar suitable either for oxen or for collared horses, or even the horse-collar itself.

PLATE CCXVIII

Fig. 564. Moulded bricks in the tomb of Pao San-Niang at Chaohua, Szechuan, between +220 and +260, depicting something very like collar-harness (Segalen, de Voisins & Lartigue, 1). An amply padded annular object like a large garland surrounds the withers and passes across the sternal region at a low level. In one brick or another, undamaged in different places, shafts or traces and reins or breeching can be made out, together with a distinct girth. If a decorative garland is not simply obscuring normal breast-strap harness here, the invention of collar-harness may have been made in Szechuan.

PLATE CCXIX

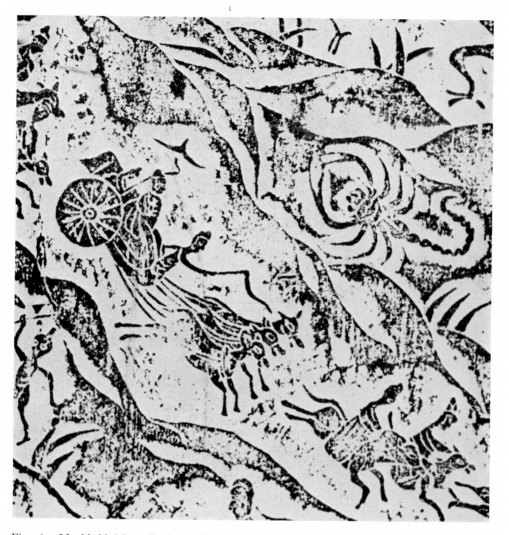

Fig. 565. Moulded brick ascribed to the Han period showing a chariot drawn by three horses in traces all bearing thick annular objects round their withers and chests (collection of Ti Phing-Tzu, *1*). The style of the decoration suggests an earlier date (possibly late Warring States period) and a southern provenance (possibly Chhu, cf. Fig. 546).

PLATE CCXX

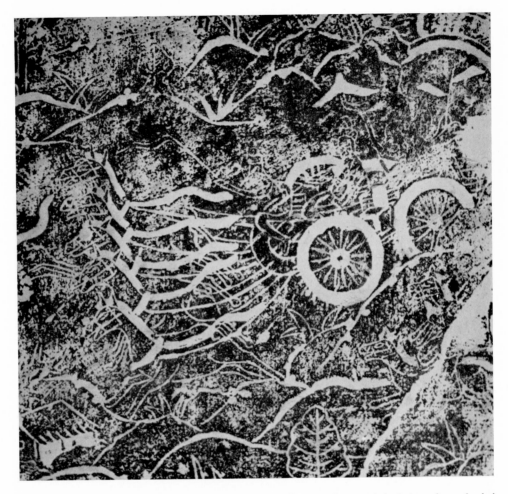

Fig. 566. Moulded brick of Chhin or Early Han date (−3rd or −2nd century) depicting a four-wheeled wagon being drawn rapidly along by five horses each apparently harnessed with a collar and traces. But such an interpretation is subject to reserve; see text. Collection of Ti Phing-Tzu (1).

the San Kuo period, between +221 and +265. Perhaps, then, the muddy clay soil of the Shu kingdom preceded the insidious sand of the Northern Wei, and Thopa Tao was only copying Chuko Liang. Of course the innovation could have first been made several centuries before it was generally adopted.[a] Moreover, the bricks of Pao San-Niang do not stand quite alone, for in a rare collection of reliefs and brick rubbings mostly of unknown provenance published by Ti Phing-Tzu there is one which shows a chariot the three horses of which have thick and distinctly ring-like objects round their chests.[b] Unfortunately, the origin of this hollow moulded brick is uncertain, but it is so curious that we illustrate the relevant part of it in Fig. 565. There is no reason for doubting its genuineness. It cannot be later in date than the Early Han, but it could be as old as the Warring States period, so we can only place it somewhere between the −4th and −1st centuries.[c] Possibly, therefore, the occasional use of collar-harness goes back in China far earlier than its general adoption in the +6th century.

Another hollow moulded brick in Ti Phing-Tzu's collection exemplifies the difficulty in distinguishing clearly sometimes between early examples of collar-harness and trace-harness.[d] As Fig. 566 shows, the style is clearly of the Chhin or Early Han, resembling closely enough the Loyang tiles studied in the well-known book of White (5), and must date the object as of the −3rd or −2nd century. A very unusual four-wheeled wagon is being rapidly drawn along by five steeds, each of which has

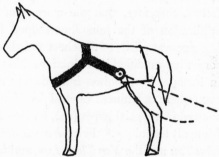

Fig. 567. Diagram to show the effect of a short-slung withers strap, the simulation of a collar in iconographic records.

what seems obviously a collar well down on its chest. But having in mind the deeply incised lines of their backs we dare not take the collar-like objects as padded, and we may prefer to see in them only the breast-strap harness, which was certainly in widespread use by this date.[e] For indeed, if the withers-strap (*hsien*[1]) were slung short, the effect produced by the pull, from the side and at distance, would simulate

[a] Possibly it was this which made Goodrich say (1), p. 54, that the 'breast collar-harness' appeared in +2nd-century China, but we suspect that he was simply referring to the breast-strap harness and not to collar-harness at all. In 1952 the Palace Museum in Peking exhibited a model of a Han *quadriga* chariot in which the horses wore a bizarre type of hard collar formed by a lyre- or wishbone-shaped yoke the lower ends of which were connected by a cross-piece on the chest. Traces were attached at two points on each side. Enquiry failed to reveal authority for this reconstruction, however, and it was decided to withdraw the model pending further study. Galling would be intolerable without the 'cushion'.

[b] (1), ch. 2, p. 6b. Remarkably, there is no sign of shafts, the collars all being attached to traces.

[c] I am greatly indebted to Dr Chêng Tê-Khun for his kind advice on the rubbings of Ti Phing-Tzu. For this particular brick a southern origin seems to be indicated, for besides acrobats, dancers and firewood collectors the scenery is interspersed with elephants, human-headed serpents, scorpions, leopards, etc. Perhaps one should associate it with the region of the State of Chhu. This is the very region which in its lacquer vessels gave us (p. 310 above) precious evidence concerning the invention of the breast-strap harness.   [d] (1), ch. 2, p. 6a.

[e] The lines running backwards to the curved front of the wagon from the shoulders of the horses do look more like traces than shafts, but this would by no means exclude collar-harness.

[1] 顯

that of a true collar (Fig. 567). This is presumably what has happened in the +6th-century Chinese harness shown in Fig. 568, from a Buddhist stele.[a] But a recognition of this fact throws some doubt on the earliest accepted illustrations of collar-harness in Europe; where no padding is shown, it might be safer to regard them as imperfectly fitted breast-strap harness.[b] However, this would not retard the date of collar-harness in Europe much after +1050.

Whenever the padded and stiffened horse-collar was introduced, it was, we are now agreed, a surrogate for the bovine cervical and thoracic vertebrae. But was it an absolutely new thing? To Haudricourt (3, 4, 7) we owe a very interesting hypothesis about its origin, based primarily on philological evidence. Starting with the word 'hames', which in English[c] means the metal skeleton of the modern combined collar,[d] he traced it eastwards through more than twenty eastern European and north Asian languages, revealing thus the fact that in many of the latter it means something apparently quite different, namely the pack-saddle of the Bactrian camel.[e] It would follow, therefore, that the essence of the invention of the collar-harness was the application of the horseshoe-shaped felt-padded wooden ring upon which camel baggage was piled, to the chest and shoulders of the horse, no doubt with dimensions somewhat modified by reduction. No wonder, then, that the invention was not made in Europe. Two points immediately arise. The location of the ingenious novelty, somewhere in Chinese Central Asia,[f] agrees well enough with the fact that, as Laufer showed in a special paper (24), felt-making is a typical piece of nomadic (Hunnish and Mongol) technique.[g] Felt (*chan*[1] or *li*[2]) is mentioned in Han books such as the *Huai Nan Tzu* and the *Yen Thieh Lun*, and felt girdles became quite a fashion about +285. This material, therefore, would certainly not have been a limiting factor in Pao San-Niang's time or even much earlier. Secondly, the camel was used as a pack-animal in the Han period, as we may see from the recent history of the camel in China by Schafer (3). There are references to it in the *Shan Hai Ching*,[h] and in the −2nd and −1st centuries there were government post-camels, Superintendents of Camel Herds, stables for army supply services,[i] and even a Camel Corps on the frontier.[j] There seems no reason therefore why the invention should not have been made as early as the −2nd century, though it may well have been uncommon before the +5th or +6th.

[a] This was the rubbing published by Chavannes (9) and reproduced by des Noëttes (1), fig. 128.

[b] Such a criticism would apply, for instance, to the figs. 140, 141 and 142 of des Noëttes (1).

[c] First recorded use about +1300.   [d] I.e. the fused *chia-pan-tzu* and *tien-tzu*.

[e] The attempt of Jacobeit (2) to derive it from an Indo-European root is quite unconvincing.

[f] Pack-camels are common in the Tunhuang frescoes, e.g. cave no. 61 dating from just before +970; see Anon. (10), pl. 66.

[g] Cf. Olschki (7).   [h] Ch. 3, pp. 33b, 47a, etc.

[i] *Chhien Han Shu*, ch. 96B, p. 10b; an imperial edict of about −90.

[j] Han representations of pack-camels are not infrequent, and later on camel-carts are seen. At Chhien-fo-tung there is a picture of one crossing a bridge (Fig. 569) in a Sui cave (no. 302), c. +600; cf. Hsieh Chih-Liu (1), pl. 11, and *Jen Min Hua Pao*, 1951, 2 (no. 3). In 1958 I had the opportunity of studying contemporary camel-cart harness in Kansu. The shafts were attached to ropes passing both behind and in front of the forward hump, the 'withers-strap' being carried over a small saddle-like leather pad. This one finds also in the camel-carts of Rajasthan, which I have seen with my friend Professor Daya Krishna of Jodhpur, who afterwards kindly sent me photographs.

[1] 氈   [2] 氀

PLATE CCXXI

Fig. 568. Horse and cart on a Buddhist stele of the +6th century (Chavannes, 9). While collar-harness and traces cannot be excluded here, it seems very probable that the harness was of the breast-strap type with the withers-strap slung short. It is not clear whether the shafts end by the horse's tail or further forward. In the latter case the hip-strap indicates breeching.

Fig. 569. Camel-cart in a Sui fresco at Chhien-fo-tung, c. +600 (cave no. 302; copy-painting in Hsieh Chih-Liu, 1). The harnessing is the same as that used in north and north-west China today; see text.

In the documentary evidence assembled by des Noëttes (1) the dates of arrival of the two efficient harnesses in Europe showed a curious inversion. Although in China and Central Asia the general use of breast-strap harness long preceded that of collar-harness, the first appearance of the latter in Europe could be dated about +1000 or a little earlier, while the former was not seen till about +1130. Western monuments of the +8th century (cf. p. 315 above) have now removed this discrepancy. But Haudricourt (3) had independently brought forward philological evidence[a] that this succession was illusory, arising only from the scarcity of pictorial representations. The German word for the traces, *siele*, may be shown to have been borrowed by Slavonic before the dispersion of the Slav peoples, i.e. before the +6th century. Conversely the word for collar, *kummet*, was borrowed by German after this dispersion, probably in the +9th century. It would certainly seem more likely that the times of arrival of the two forms of efficient harness in Europe should have been echelonned in accordance with their original succession in the east. The position may be summarised in the following way:

Form of harness	Des Noëttes (1) (documentary representations)	Newer evidence	Haudricourt (3) (philological evidence)
Throat-and-girth	Very ancient	Very ancient	—
First breast-strap Ch.	+1st	−4th to −2nd	—
Eu.	+12th	+8th	+5th or +6th
First collar-harness Ch.	+17th	−1st (Han bricks)	—
		+3rd (San Kuo bricks)	
		+5th (Tunhuang frescoes)	
Eu.	+10th	—	+9th

Unfortunately, we still know very little indeed about the intermediate stages in the transfer, but Bloch (4) was doubtless justified in emphasising the creatively transmissive role of the Huns in such matters.

It only remains to glance briefly at the relations of these different types of harness to their vehicles. The throat-and-girth harness, with its pole, cross-bar and fork-yokes seen from above, shown in Fig. 570a, is to be contrasted with Fig. 570b, the Han system, where the fork-yoke had not disappeared but had become even more unnecessary because of the full use of breast-strap and traces. It was the predecessor of the simple breast-strap ('postillion') harness of later times (c). But after the Chinese adopted collar-harness, one special feature persisted until the present day, namely the attachment of the hard component of the collar directly to the ends of the shafts (d). This, derived from the ancient ox-yoke method, was only possible because in the Chinese cart the shafts always formed a continuous part of the structure of the vehicle.[b] As Haudricourt (4, 6, 7) has shown, the Russian or Finnish *duga*

[a] Maintained in Haudricourt & Delamarre (1), p. 178.

[b] Haudricourt (7) says that this arrangement was formerly known in the Jura region, and takes it as further evidence of the coming of the collar-harness from China and Central Asia. It was known as 'l'attelage à la Grandvallière'.

(Fig. 570e)[a] originated because if this is not the case, it becomes necessary to keep the ends of the shafts away from one another by a special bar or bow, since they tend to be pulled together in use. The problem is avoided by the attachment of the traces to the vehicle itself, as in some ancient Chinese and in modern European usage (f). This has the further advantage that more than one horse can be attached in file, using the forward ends of the shafts. Meanwhile the hames and the cushion of the collar had fused. Then in the West the yoke, no longer necessary at the level of the horse's

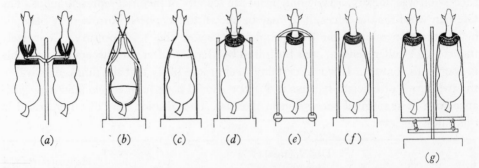

(a)   (b)   (c)   (d)   (e)   (f)

(g)

Fig. 570. Modes of attachment of equine harness to vehicles or machinery. (a) Pole, cross-bar, yokes and throat-and-girth harness; (b) Han breast-strap harness with shafts, yoke vestigial, hip-strap and breeching present; (c) postillion or later breast-strap harness with traces attached to vehicle; (d) traditional Chinese collar-harness with the hard component (chia-pan-tzu), descendant of the yoke and ancestor of the hames, attached directly to the forward ends of the shafts; (e) duga of Russia and Finland, retaining the arched cross-bar because the shafts are not structurally part of the vehicle; (f) modern collar-harness, the traces attached directly to the vehicle; (g) whippletrees for receiving the pull of the traces of collar-harness.

back, moved downwards, either in front of the chest (as in curious forms still persisting in the Landes[b] and South Africa[c]), or underneath the horse's belly (as in the Berber harness),[d] or behind the animal altogether to become eventually the whippletree (Fig. 570g). In the +12th century this was fixed firmly across the pole, but later it was linked to two or more movable whippletrees as shown in the diagram.[e]

(2) ANIMAL POWER AND HUMAN LABOUR

This section cannot end without a reference to the social aspects of efficient equine harness,[f] and the use of animal-power instead of human labour, which have been much

[a] Cf. drawings in Michell (1).   [b] Des Noëttes (1), fig. 182.
[c] Philipson (1), p. 59, pl. xv.   [d] Described by Laoust (1).
[e] Traces and whippletrees were used for oxen in +3rd-century China; we have noted a technical term for them above (p. 260) in the description of the wheelbarrow of Chuko Liang.
[f] In this book there is never opportunity to follow far the social and economic consequences of the inventions which we discuss in it, and in our account of the development of efficient horse harness in the Old World we have taken it as sufficiently obvious that vast effects flowed therefrom. We cannot forbear, however, from mentioning the interesting thesis of Lynn White (7), pp. 67 ff., on the effect of efficient harness in encouraging a proto-urbanisation of rural settlement in Western Europe. From the +11th century onward, though population increased rapidly, the smaller hamlets long inhabited were abandoned, and the peasants agglomerated into larger and larger villages. This was primarily due, he believes, to the replacement of oxen by horses as the basic farm animals, for when the latter could be used it was no longer necessary to live close to the fields tilled, and the larger villages could afford more

debated. Des Noëttes considered his book (1) a contribution to the history of slavery, and so he subtitled it, believing (6, 7) that the invention of the collar-harness had been the chief causative factor in the decline and disappearance of that institution, a laudable detestation of which is very clear in his pages. Historians such as Bloch (3, 5) hastened to point out, however, that the time-relations did not fit this view, since mass slavery vanished from Europe many centuries before the collar-harness first appeared.[a] It is true that we now seem to see the arrival of the breast-strap harness in Europe earlier than was formerly thought, yet the essential point made by Bloch still carries conviction. But it had never affected the converse argument that after all no efficient form of horse harness was invented in those ancient Mediterranean societies which embodied mass slavery and presumably abundant man-power.[b] If the invention of efficient harness was clearly not the cause of the decline of slavery, perhaps the presence of slavery inhibited the invention?

It certainly seems one of the paradoxes of history that in spite of the theoretical brilliance of Stoics and Peripatetics, or Euclid's geometrical insight, or the remarkable ingenuity of the Alexandrian mechanicians—all men who could differentiate, systematise and compile—the ancient Western world never succeeded in solving the problem of harnessing horses efficiently. Perhaps of course it did not try. Here the general ideas of des Noëttes seem to be faulty in further serious ways. Concentrating as he did on equine harness and on its genuine inefficiency in Western antiquity, he encouraged the implication that Greek and Roman culture employed mass slave labour for the movement of heavy weights (in building, shipbuilding, grain-transport, etc.) just as the Egyptians and Babylonians had done centuries earlier. But Fougères (1), and now definitively Burford (1), have been able to demonstrate by studying inscriptions of accounts and other evidence from the −5th century onwards that Greek and Roman engineers moved all their heavy weights (involving loads of up to some 8 tons) quite effectively by the aid of yoked oxen, often in file. Hellenic and Hellenistic civilisation, Burford suggests, did not in fact regard the horse as a draught-animal at all[c]—it was primarily aristocratic, swift and military; and it had always been in short supply. Yet even so, one would have thought that for some purely military purposes

---

amenities both spiritual and material. Such were the roots of that great dominance of city life in late medieval Europe which gave rise to the Renaissance and the rise of capitalism with all their consequences. If this was due to the spread of originally Chinese devices, the intervention of East Asia would have affected the most intimate mechanisms of European social evolution. It is not the last time that we shall meet with this pattern (cf. Sect. 30*i* below).

[a] For his reply see des Noëttes (5). Grand (1) also contributed to the discussion, but his paper has been unavailable to us.

[b] Some of the effects of this weakness (if such they were) produced extraordinary results. The Po valley did not feed Rome with wheat because the expense of haulage with inefficient harness was much greater than bringing wheat by sea from Egypt. What the Po valley grew was barley and fodder, for the meat which could walk by itself to Rome. So at any rate Dr R. J. Forbes, in a stimulating lecture at Cambridge (February 1954).

[c] There are good reasons for accepting this view as more applicable to Greek than to Roman culture. The only Greek inscriptions so far found which clearly refer to draught-horses are those of Callatis on the Black Sea, cf. Tafrali (1); Robert (1); and this place was on the borders of the Greek world. On the other hand it is quite clear that horses and especially donkeys were used in the Roman world, at least from −200 onwards, for working grain-mills (cf. pp. 186 ff. above; the *mola asinaria*, Moritz (1), pp. 62 ff., 74 ff., 97 ff.). Roman reliefs and other representations which depict these mills (cf. des Noëttes (1), fig. 74; Moritz (1), pls. 5*b*, 7*a*, and fig. 9, p. 83) show throat-and-girth harness pretty clearly.

efficient equine harness could have been decidedly useful.[a] In any case, for the Chinese and their Hunnish and Mongol neighbours, ever in the saddle, horses were assuredly not so rare and precious. This is a feature of northern Chinese culture which may well be relevant to the fact that both the efficient forms of harness for traction by horses were invented in its midst.

Here a comparison of Hellenistic slave-owning society with ancient Chinese society inevitably presents itself, but unfortunately the complex problems involved are hardly as yet answerable,[b] and we can only allude to them.[c] It may well be that Chinese society incorporated only domestic slavery, and that the proportion of slaves to the total population (in the Han, for instance) was always small; but the available man-power liable to *corvée* was, at any rate in some periods, abundant enough. Possibly the monsoon climate, with its consequence of strictly fixed seasonal agricultural work, from which Confucian morality and common sense alike forbade the abstraction of the toiling peasant-farmers, led to the use of greater ingenuity in solving the mechanical problems of efficient traction,[d] which might arise at any time of year. In the last resort it was perhaps the nomadic peoples who first faced and solved some of them, aided and stimulated at the borders of culture-contact by the practical genius of the Chinese.[e]

(g) HYDRAULIC ENGINEERING (I), WATER-RAISING MACHINERY

The present subsection, which deals with machinery for raising water from one level to another—an operation of immense importance in any civilisation based on irrigated agriculture—might well have formed part of Sect. 27 *d* above, in which the various fundamental types of machines described and illustrated in Chinese books were discussed.[f]

---

[a] E.g. for bringing catapult artillery more quickly into position, or for moving the impedimenta of a headquarters.

[b] Cf. pp. 23 ff. above. We shall of course deal with these questions as thoroughly as possible in Vol. 7.

[c] Des Noëttes himself was very confused about the Chinese situation. To begin with, he did not recognise the collar-harness as Chinese and Central Asian in origin. Then, while in some places—(1), pp. 110, 111, 122—he saw that the anatomical fitness of the breast-strap rendered it almost as efficient as the collar, he somehow could not bring himself to admit (p. 114) that the ancient Chinese were any better off than the Greeks and Romans. He emphasised, therefore, the great instances of *corvée* labour in Chinese history which were popularly known when he wrote, such as the building of the Great Wall and the digging of the Grand Canal. In order to account for the absence of mass slavery in China, he added that great public works were less important and frequent than in other civilisations, when precisely the opposite is the case. Advancing further into a swamp of errors, he wrote that the Chinese had not used mills because they ate rice, forgetting that rice must be hulled and may be ground, as also that half the Chinese culture-area has always been dependent on cereals other than rice—wheat, millet, buckwheat, etc. Absence of mills (!) he thought helped to account for absence of slavery in China. All these opinions are now but curiosities—yet des Noëttes will always remain a great pioneer.

[d] It has been said that efficient equine harness had to await better understanding of the anatomy of the domestic animals. Now without prejudging the conclusions of Sect. 43, Galen, Herophilus and Erasistratus were probably better anatomists than any of their Chhin and Han contemporaries. Yet so far as the donkeys and horses of Rome were concerned they might never have existed.

[e] For a striking instance of such collaboration cf. p. 257 above.

[f] We have already enlarged upon one piece of water-raising apparatus, the Archimedean screw, in connection with worms and screws in general (Sect. 27 *b*, pp. 120 ff. above). Here it would be out of place since it was not known in the Chinese culture-area until the +17th century.

PLATE CCXXII

Fig. 571. The swape or counterbalanced bucket lever for drawing water from wells; a Han representation from the Hsiao Thang Shan tomb-shrine reliefs, *c.* +125 (Chavannes, 11). Such pictures generally show it, as here, in association with the butcher's shop and kitchen of a great house.

But the significance of the subject is such that it finds its place more naturally as the first part of our account of Chinese hydraulic engineering.[a] It has intimate connections, moreover, with the application and use of water-power, and all that that implies. We shall thus consider in due order (a) the counterweighted bailing bucket, (b) the well-windlass, (c) the scoop-wheel, (d) the square-pallet chain-pump, (e) the vertical pot chain-pump, and (f) the peripheral pot wheel. This enumeration by no means exhausts the variety of water-raising devices known and employed in ancient and medieval times, or traditionally still, but it meets the needs of the Chinese story, and we shall refer to other types of water-lift machinery incidentally as we go along.[b]

(1) THE SWAPE ('SHĀDŪF'; COUNTERBALANCED BAILING BUCKET)

The oldest and simplest mechanism which lightened the human labour of dipping, carrying and emptying buckets,[c] was the swape[d] or well-sweep (chieh kao[1])[e] (often

[a] There is now a brief, but good, account of Chinese traditional types of water-raising machinery by Chhêng Su-Lo (2), but it appeared long after the completion of the present Section.

[b] Similar listings have been given by Forbes (11), pp. 35 ff.; Brittain (1), pp. 219 ff., and others. Unfortunately the terminology of this subject is still in a rather confused state. As useful single names, we still speak of swape or shādūf for the counter-weighted bailing bucket, we call the vertically hanging endless chain of pots the sāqīya, and the wheel with pots around its rim the 'noria'. There is no difficulty about the name for the pallet-carrying endless chain working in a flume, the fan chhê, for it was purely Chinese. But Forbes and Brittain restrict the term 'noria' to peripheral pot wheels which are operated by the current of the water. Furthermore Forbes (cf. (17), pp. 675 ff.) makes sāqīya apply to peripheral pot wheels if worked by men or animals, presumably with treadmills or whims; reserving the term daulāb for endless pot-chains similarly operated. Brittain on the other hand (with a personal background of Egyptian experience) employs the term sāqīya only to designate wheels worked by men or animals using right-angle gearing, whether the buckets are fixed to the wheel or borne on a dipping chain, in other words the ad-aqueous converse of the Vitruvian vertical-wheel ex-aqueous mill. We ourselves prefer to adhere to older authorities, among whom we may cite Ewbank (1), pp. 112 ff., 123 ff., 128 ff., E. W. Lane (1), p. 301, and Chatley (36), p. 159, as explicitly reserving the term sāqīya for the chain of pots; a practice which is borne out as correct by the most exhaustive study of the technical terms in Egyptian Arabic known to us (Littmann, 1). It is true, however, that Arabic dictionaries generally describe the noria as current-operated. The difference really is that Forbes, Brittain et al. seek to apply the terms 'noria', sāqīya, etc., in accordance with the motive power used, while we think that their definition should follow the mechanical structure, i.e. the disposition of the pots. In this we are perhaps influenced by Chinese practice. All such old terminology is still no doubt very loosely used in Arabic countries, and conventions other than that which we here adopt may well be tenable. Various theories about the logical derivation of the forms, and various interpretations of the historical evidence concerning them, are implicit in the presentations of these authors, however; we do not always subscribe to them, and we shall give what opinions we have concerning the evolution of the various devices as we go on. Our grateful thanks are due to Mr Robert Brittain and to Dr D. M. Dunlop for fruitful discussions on this tangled matter of nomenclature. For a comparative practical study see Molenaar (1).

[c] A still more primitive device was the bucket slung on ropes and handled by two men (hu tou[2]). The Kêng Chih Thu illustrates this, and I myself during the Second World War often saw it in use in Kansu province. In China barrel-shaped baskets lined with oilcloth are often used; see Hommel (1), fig. 5. Ball (1) illustrates the system from South India, where it is called the katwa (as in Egypt; Lane) or letha; it will not suffice for more than a 3 ft. lift. Ewbank (1), p. 85, knew it from Egypt. What authority Forbes (11), p. 35, has for applying to it the English word 'swipe' I do not know.

[d] This word is not found in literature before +1773 so applied, but with the meaning of any kind of lever it goes back to +1492. Its alternatives for referring to the well-sweep, i.e. swipe and swype, are recorded from +1600, swip from +1639.

[e] If the arms are of very unequal length, as in Southern Shensi near Meihsien, the machine may be called the 'steelyard swape' (chhêng kao[3]); Fu Chien (1).

[1] 桔槔　　[2] 戽斗　　[3] 秤槔

known by its Arabic name of *shādūf*). This was familiar from early times in Babylonia[a] and ancient Egypt,[b] where it has continued in use till the present day.[c] It may be termed the counterbalanced bailing bucket, and it makes use only of the lever, involving no rotary motion. A long pole is suspended or supported at or near its centre, like a balance-beam, and while one end is weighted with a stone, the other carries the bucket at the end of a rope or a piece of bamboo. In the Old World its distribution became almost universal.[d] The earliest illustrations we have of it in China are those on the mid +2nd century Hsiao-thang Shan[e] and Wu Liang[f] tomb-shrine carvings, often reproduced (cf. Fig. 571). We see it again on the walls of the Chhien-fo-tung cave-temples,[g] and in the *Kêng Chih Thu* of +1210, the *Thien Kung Khai Wu* and other books (Fig. 572). A good contemporary photograph is given by Hommel.[h] But the earliest mention of it in Chinese literature is of the −4th century, in that interesting passage in *Chuang Tzu* which has already been quoted in Sect. 10g above with regard to the paradoxical anti-technology complex of the Taoists. A farmer declines to use the device when Tzu-Kung (Tuanmu Tzhu) suggests that he should do so.[i]

Tzu-Kung had been wandering in the south in Chhu, and was returning to Chin. As he passed a place north of the Han (river), he saw an old man working in a garden. Having dug his channels, he kept on going down into a well, and returning with water in a large jar. This caused him much expenditure of strength for very small results. Tzu-Kung said to him, 'There is a contrivance (*chieh*[1]) by means of which a hundred plots of ground may be irrigated in one day. Little effort will thus accomplish much. Would you, Sir, not like to try it?' The farmer looked up at him and said, 'How does it work?' Tzu-Kung said, 'It is a lever made of wood, heavy behind and light in front. It raises water quickly so that it comes flowing into the ditch, gurgling in a steady foaming stream. Its name is the swape (*kao*[2]).' The farmer's face suddenly changed and he laughed, 'I have heard from my master', he said, 'that those who have cunning devices use cunning in their affairs, and that those who use cunning in their affairs have cunning hearts. Such cunning means the loss of pure simplicity. Such a loss leads to restlessness of

---

[a] Feldhaus (1), col. 827, reproduces a Nineveh bas-relief from Layard, of the −7th century, but this is outdone by Sargonid cylinder-seals of about −2000 (Ward (1), fig. 397; Forbes (11), pp. 16, 31). Cf. Laessøe (1).

[b] The Old Kingdom; Wilkinson (1), vol. 2, p. 4; Grahame Clark (1); Ewbank (1), pp. 94 ff.; Steindorff & Seele (1), p. 183: the New Kingdom, Klebs (3), p. 35, fig. 25.

[c] Illustrations are very frequently seen. Cf. al-Salam (1). A physiological study of work with the *shādūf* was made by Haldane & Henderson (1).

[d] Examples still persist in some parts of Central Europe; cf. Croon (1); Wakarelski (1). During the writing of this Section I met with a whole group of them at Vouvray in Touraine, and later found them common in Hungary especially between Budapest, Tihany and Pécs; as also in Bosnia and the Konavle besides other parts of Yugoslavia.

[e] Chavannes (11), pl. XXXIX.

[f] *Chin Shih So*, Shih sect., ch. 3, (p. 122), ch. 4, (p. 9). Chavannes (11), pls. IV, XIV, XXIII; Chhang Jen-Chieh (1), fig. 17; Laufer (3), p. 68, fig. 16, wrongly captioned Hsiao-thang Shan.

[g] E.g. cave no. 302 dating from the Sui, about +600.

[h] (1), fig. 174.

[i] It is significant that the story is told of Tuanmu Tzhu, the successful merchant and Confucius' wealthiest disciple, who figures also in the chapters on wealthy persons in the *Shih Chi* and the *Chhien Han Shu* (cf. Swann (1), p. 427).

[1] 械   [2] 槹

桔槔

墜石

Fig. 572. A swape or well-sweep (*chieh kao*) from the *Thien Kung Khai Wu* of +1637 (ch. 1, p. 18*a*). *Chui shih* marks the counterweight.

the spirit, and with such men the Tao will not dwell. I knew all about (the swape), but I would be ashamed to use it.[a]

This might be taken as evidence that the swape reached China about the −5th century.

Batteries of swapes raising water in successive levels are often seen in Babylonian and ancient Egyptian representations, described in Arabic MSS., and photographed by contemporary travellers.[b] A later development was to elongate the bucket's spout into a flume, often made from a dug-out palm-tree trunk, this being linked parallel with a counterweighted beam above, and so arranged that it automatically empties itself into the receiving channel on the upward motion.[c] This is the Bengali *dūn*.[d] In India the operation of the device is assisted by a moving counterpoise, i.e. by men who walk back and forth along the upper beam.[e] Finally the bucket, flume and counterpoise were combined into one single unit, or the place of the counterpoise taken by manhandled ropes or gearing.[f] These machines have sometimes been regarded as Muslim inventions because many of them occur in Arabic texts such as the famous book of al-Jazarī (+1206).[g] But in view of their wide distribution and frequency in India, they may more probably have originated there. Significantly they occur in various forms in Indo-China[h] but not further north. In one Arabic type the motion of the flume-beam is produced by the rotation of a lug attached to wheels worked by animal-power, and moving in a slot on the beam, but it is doubtful whether this was ever put into practice.[i] Another design shows a series of quarter-gears rotated by animal power, engaging alternately with a set of lantern-pinions each of which thus raises a flume-beamed swape, and when it has emptied lets it drop back into the water again (cf. Fig. 573).[j] How far such machines were ever common in medieval Islam we do not know. At all events, the flume-beamed swape interested European engineers

---

[a] Ch. 12, tr. auct., adjuv. Legge (5), vol. 1, p. 320; Lin Yü-Thang (1), p. 267. A parallel text, in which the old man rejects the same advice offered by Têng Hsi Tzu[1] instead of Tuanmu Tzhu occurs in the *Shuo Yuan* of Liu Hsiang, written about −20 (ch. 20); tr. H. Wilhelm (2), p. 51. The old man is there called Wu Chang Fu[2] (the Fifty-foot (Garden-plot) Husbandman).

[b] Cf. Jomard (1), p. 780; Ewbank (1), p. 95; Drower (1), fig. 347*a*.

[c] This was really a combination of the ancient swape with another device, the *mote*, consisting only of a scoop-shaped piece of wood suspended at its centre from a kind of a light derrick and used simply to scoop up the water for small lifts; see Ball (1); Ewbank (1), p. 93. All these devices were, and are, in common use in India, but they are not characteristic of China, and were in all probability never known there. In the +16th century this form of *mote* travelled to Holland, whence the name 'Dutch scoop', and to Portugal, where it is called *tranqueira*.

[d] See the descriptions and photographs of Ball (1).

[e] This is true also of large *shādūfs* both in Egypt and India. The *shādūf* will service a lift of from 4 to 10 ft., and while the flume-beamed swape will not lift more than about 3 ft. it will carry much larger amounts of water at each stroke.

[f] Cf. Beck (1), pp. 229, 289, 476. These pivoted lever-and-flume devices may well have had something to do with the +16th-century invention of the *stangenkunst*, on which see below, p. 351, and Beck (1), p. 366; Schmithals & Klemm (1), fig. 37.

[g] See further details on p. 534 below.   [h] Huard & Durand (1), p. 127.

[i] See Wiedemann & Hauser (1).

[j] Coomaraswamy (2), p. 16, pl. 6; Wiedemann & Hauser (1), fig. 15. The apparatus is wrongly described by Forbes (11), p. 47, as a 'mule-driven wheel of pots'. For reasons not at all clear to us, Clutton (1) thinks that it has some logical connection with the pendulum escapement of mechanical clockwork (cf. p. 444 below). The single unit is also shown by al-Jazarī; Wiedemann & Hauser (1), fig. 13.

[1] 鄧析子   [2] 五丈夫

PLATE CCXXIII

Fig. 573. A battery of flume-beamed swapes in al-Jazarī's treatise on mechanical contrivances of + 1206 (Coomaraswamy, 2). An animal-power whim rotates by means of right-angle gearing a set of quarter-gears (drawn in semi-perspective) which engage one after the other with lantern-pinions fixed on the same shafts as flume-beamed swapes. The buckets and flumes, which no longer need counterpoises, are thus periodically raised and emptied into a receiving channel behind the plane of the diagram.

in the +16th century,[a] and was introduced to China by the Jesuits (cf. Table 58 and p. 222 above), though probably never employed there.[b]

The swape is not quite such a jejune instrument as it seems at first sight, for its exact converse, namely the movement of a weight by an alternating water counter-balance, has interesting relations with the development of water-wheels.[c] It was also used for raising beacon-fires (*fêng huo*[1], *fei chü*[2]) on high.[d] And if certain important types of catapults were essentially swapes,[e] so also was that great herald of industrial technology, the Newcomen pumping steam-engine.

(2) THE WELL-WINDLASS

Rotary motion came in with the pulley or drum[f] set at the mouth of the well. At first the rope was simply hauled over it and gathered, then the bucket was counterweighted, and finally the drum was turned by a crank. Han tomb-models (Laufer, 3) show the first of these stages (e.g. Fig. 395).[g] One can see the pulley (*lu-lu*[3]) in its bearings at the top under a small rectangular tiled roof, and the frame is ornamented with the heads of water-bringing dragons. At some later date the drum (*chhang ku*[4]; 'long wheel-hub') was introduced,[h] and the pulley acquired its specific name of *hua chhê*.[5] Figure 574 shows an interesting, though bad, drawing of the drum and crank (*chhü mu*[6]) from the +17th-century *Thien Kung Khai Wu*; but it is not as bad as it looks, for it corresponds remarkably closely with a well-diggers' windlass photographed by Hommel.[i] Both Laufer[j] and Hommel[k] noticed during their travels in China that the drum generally carried two ropes so wound that the counterweight or the unfilled bucket descended as the filled bucket came up. Hommel suggests that this practice may have been at the origin of the 'Chinese windlass' which gains mechanical advantage (discussed above on p. 98).

Other evidence which makes it clear that these simple machines were common in the Han may be found in texts such as *Huai Nan Tzu* where it is recommended not to plant *tzu*-trees near wells, for their roots or branches will impede the movement of the rope and buckets.[l] It is also probable that large capstans worked by animals were used for hauling up the long bamboo buckets of brine in the salt-well boreholes at

---

[a] And, in certain forms, as late as the +18th; cf. de Bélidor (1), vol. 1, pl. 41.

[b] Cf. *Chhi Chhi Thu Shuo*, ch. 3, pp. 22b, 23a, and Ramelli redrawn by Beck (1), fig. 273.

[c] See below, p. 363.

[d] *Piao I Lu*, ch. 7, p. 5a; *Wu Ching Tsung Yao*, ch. 12, pp. 60a, 61a. The converse of the European cucking-stool (Spargo, 1).

[e] The trebuchets; we shall deal with them thoroughly in Sect. 30i.

[f] Cf. pp. 95 ff. above, and Baroja (6).

[g] Such representations from this period are quite common. For other pottery models see e.g. Anon. (4), pl. 35; Thang Lan (1), pl. 88; de Tizac (1), pl. 17. A beautiful bronze model from Chiu-chhüan is in the Kansu Provincial Museum at Lanchow. A depiction in relief is in the I-nan tomb sculptures (Tsêng Chao-Yü *et al.* (1), pl. 48).

[h] It is noteworthy that many of the Han well-pulleys are very broad, looking almost like two cones fitted tip to tip.

[i] (1), fig. 172. [j] (3), p. 72. [k] (1), p. 118.

[l] Cf. Vol. 2, p. 71 above. Ch. 6, p. 10a.

[1] 烽火   [2] 飛炬   [3] 轆轤   [4] 長轂   [5] 滑車   [6] 曲木

Fig. 574. Well-windlass (*lu-lu*) with crank, from *Thien Kung Khai Wu*, ch. 1, p. 18*b* (+1637).
A plan of irrigated plots superimposes itself on the left of the perspective drawing.

this period.[a] Certainly salt derricks with pulleys at the top are seen in several Han representations.[b]

The *nasba* of Iraq or *mote* of India is a modification of the well-bucket used in irrigating on river-banks, in which an animal pulls up a camel-skin, so arranged that when the desired height is reached it discharges through an open limb, this being kept suspended (and therefore closed), when at lower levels, by a subsidiary rope.[c] We have no reason to think that this device was ever used in China, but there is a Chinese reference to it, for the Taoist adept Chhiu Chhu-Chi and his entourage noted such water-raising arrangements at work in the neighbourhood of Samarqand in +1221.[d]

(3) THE SCOOP-WHEEL

The next simplest machine is a hand-operated paddle-wheel sweeping up water into a flume; Fig. 575 shows the illustration given of it in the *Shou Shih Thung Khao*,[e] where, as in the +1313 *Nung Shu*,[f] it is called *kua chhê*.[1] It could be effective only for short lifts. Its simplicity may, however, be deceptive, and it would be unsafe to assume that it was invented before the ex-aqueous water-wheel[g] or the ad-aqueous paddle-wheel[h] themselves. Though ad-aqueous, it is not for the purpose of motion over water, but of transmitting motion to water. A version of the scoop-wheel in which the operator treads upon its perimeter became particularly popular in Japan.[i] Ōkura Nagatsune[2] tells us in his *Nōgu Benri-ron*[3] that these *fumi-guruma*,[4] as they are called, were said to have been invented by two townsmen of Osaka some time between +1661 and +1672, and he gives diagrams and measurements for their construction.[j] Nothing similar has come to light in any Chinese book, but the treadmill scoop-wheel is in fact widely used (or was until recently) in the salterns of East China (Fig. 576). It seems also to have become popular in late +17th- and 18th-century Korea.[k] We have not found any literary references to it. The principle was widely employed in Holland and the English fen country from the +16th century onward, where such 'scoop-wheels' were mounted at the bases of windmills.[l] Westcott[m] gives an elevation

---

[a] Cf. Sect. 37 below.

[b] Notably Szechuanese moulded bricks (Anon. (22), fig. 1; Rudolph & Wên (1), pls. 91, 92; Liu Chih-Yuan (1), pls. 3, 4; Chhang Jen-Chieh (1), fig. 36, etc.). See Fig. 396.

[c] Descriptions will be found, e.g. in Weulersse (3), p. 305; Frémont (12), fig. 37; Ball (1) — and already by Kaempfer in +1712, (1), p. 681 for Persia.

[d] *Chhang-Chhun Chen Jen Hsi Yu Chi*, ch. 1, p. 21a, tr. Waley (10), p. 92.

[e] Ch. 37, p. 22a. [f] Ch. 18, p. 22b.

[g] Probably developed in the −1st century; see below, p. 369.

[h] Probably developed in the +5th century; see below, p. 417.

[i] Illustrations have been published by King (3), p. 265; Anon. (36), p. 97, and others. The principle is the same as that of the Hainan noria; cf. p. 356 below.

[j] The average lift was 2 ft. We are indebted to Dr J. R. McEwan for this reference.

[k] Yi Kwangnin (1), p. 88.

[l] A photograph is given by van Houten (1), p. 136; cf. Wailes (3), pp. 72ff., (7). The first Dutch windmill applied to drainage was built in +1408, but it operated an Archimedean screw; Gille (3); van Houten (1).

[m] (1), pl. v.

[1] 刮車　　　[2] 大藏永常　　　[3] 農具便利論　　　[4] 踏車

Fig. 575. Scoop-wheel (*kua chhê*) or hand-operated paddle-wheel sweeping up water into a flume, from the *Shou Shih Thung Khao* (+1742), ch. 37, p. 22*a*. Effective only for small lifts.

PLATE CCXXIV

Fig. 576. Treadle-operated scoop-wheel (*tha chhê*), perhaps a +17th-century Japanese invention; the user steps on the treads as they come round. About five horizontal windmills (cf. p. 558 below) are in sight in the background of this picture of the Phi-tzu-wo salterns, near Dairen in the Kuantung Peninsula of Liaoning (*c.* 1935).

Fig. 577. Typical Chinese square-pallet chain-pump (*fan chhê*) which raises water in a flume, the pallet-chain passing over a sprocket-wheel powered by two or more men stepping on radial treadles (orig. photo. between Yung-chhang and Yung-chhuan, Szechuan, 1943).

PLATE CCXXV

Fig. 578. Detailed view of the mechanism of the square-pallet chain-pump,
an example in Anhui.

of an 18th-century wheel of this type, which would lift about 1250 gals./min. against a head of about 2 ft. We suspect that the paddle-wheel and flume, or scoop-wheel, was brought into use some time between the Han and the Thang.[a] Possibly it was introduced to Europe as well as Japan and Korea from China, but more historical investigation on it is badly needed.

(4) The Square-Pallet Chain-Pump and the Paternoster Pump

We come now to the most characteristic of Chinese water-raising machines, the square-pallet[b] chain-pump. This device, as it is found today, is seen in Fig. 577; it consists essentially of an endless chain carrying a succession of pallets which draw the water along as they pass upwards through a flume or trough, discharging into the irrigation canal or field at the top. It is known as the *fan chhê*[1] ('turnover wheels'), *shui chhê*[2] ('water machine'), or colloquially, *lung ku chhê*[3] ('dragon backbone machine'). A detailed description of some examples, with dimensions, is given by Hommel,[c] but it may suffice to say that according to the length of the trough (*tshao thung*[4]) it will lift water up to 15 ft., the effective limits being conditioned by leakage and the properties of woodwork. The best inclination is 24° but in practice usually somewhat less. The endless chain is turned by one of four methods: the human hand or foot, animal-power and water-power. Of these the oldest is probably the human foot, since the upper sprocket-wheel can carry on its axle so conveniently the radial treadles already mentioned (p. 89 above)[d]; this is seen in all the agricultural treatises beginning with the *Kêng Chih Thu* (+1210). A nearer view of the sprocket-wheel and pallets, from which a perfect idea of the mechanism can be gained, will be found in Fig. 578.[e] The best Chinese illustration is probably that of the *Thien Kung Khai Wu* (Fig. 579).[f] The same +17th-century work also shows a smaller version of the machine operated by hand (*pa chhê*[5]), using a connecting-rod and eccentric lug or crank on the upper sprocket-wheel axle.[g] Hommel and Ball found examples of these which they were able to photograph.[h] The other widespread way of working the square-pallet chain-pump was by using animal-power, the animal being attached to a whippletree and cogged drive-wheel which engaged at right-angles with a gear-wheel on the sprocket-wheel axle.[i] Figure 581 shows this from the *Thien Kung Khai Wu*,[j] though it appears first in a

---

[a] Though no positive evidence exists that it was not the first water-wheel of all.
[b] The term is not quite satisfactory, for the pallets are usually rectangular rather than square.
[c] (1), pp. 49ff.; cf. J. H. Gray (1), vol. 2, pp. 119ff.; Dinwiddie in Proudfoot (1), p. 75.
[d] Thus the machine may also be called *tha chhê*.[6]
[e] The barrel-shaped treadles here seen remind us of other objects on the end of spokes (cf. pp. 80, 91, 104, above).
[f] Ch. 1, p. 19*a*, first ed. pp. 10*b*, 11*a*.
[g] Ch. 1, p. 19*b*, first ed. p. 12*b*. See Fig. 580.
[h] Cf. Hommel (1), fig. 79. Late +18th-century description by Dinwiddie in Proudfoot (1), p. 75.
[i] Working drawings and description in Worcester (3), vol. 2, p. 312.
[j] Ch. 1, p. 17*a*, first ed. pp. 11*b*, 12*a*.

---

[1] 翻車   [2] 水車   [3] 龍骨車   [4] 槽桶   [5] 拔車   [6] 踏車

Fig. 579. The square-pallet chain-pump in the *Thien Kung Khai Wu* of +1637 (ch. 1, p. 19*a*), here called *tha chhê*.

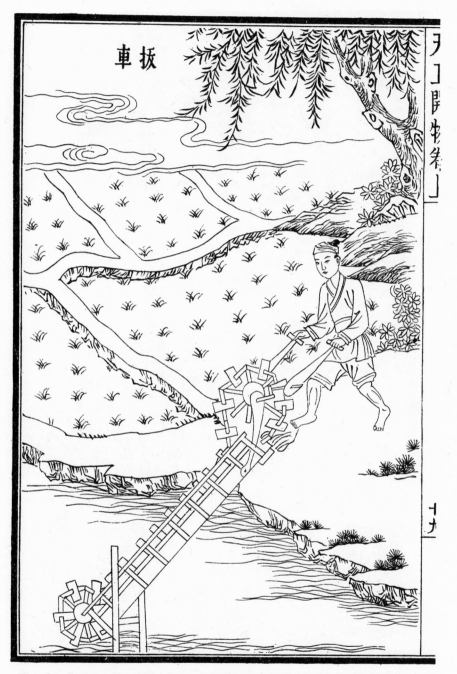

Fig. 580. Square-pallet chain-pump manually operated with cranks and connecting-rods
(*pa chhê*), from *Thien Kung Khai Wu*, ch. 1, p. 19*b* (+1637).

Fig. 581. Square-pallet chain-pump worked by an ox whim and right-angle gearing (*niu chhê*), from *Thien Kung Khai Wu*, ch. 1, p. 17*a* (+1637). The driving axle is labelled 'central shaft', and the legend to the left says that the ox treads a circle wider than the diameter of the horizontal gear-wheel.

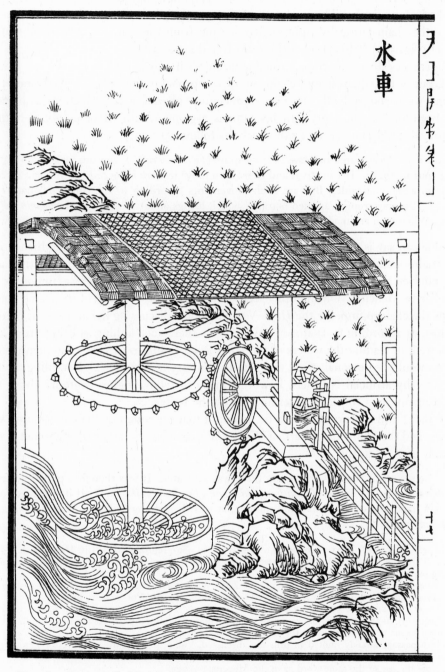

水車

Fig. 582. Square-pallet chain-pump worked by a horizontal water-wheel and right-angle gearing (*shui chhê*), from *Thien Kung Khai Wu*, ch. 1, p. 17*b* (+1637).

painting of *c.* +965,[a] is figured also in +1313,[b] and may be compared with many modern photographs. The last method, apparently less used than the others, was to drive the chain-pump by right-angled gearing from the axle of a horizontal water-wheel (Fig. 582);[c] this also was first illustrated in +1313.[d]

References to specifically identifiable machines in any general literature are always rare, but as the square-pallet chain-pump was probably the favourite water-raiser of China, we are not so badly off. The −4th century gives us a hint which might be near the beginning of the story. In his famous argument with Mencius about human nature, Kao Pu-Hai[1] (Kao Tzu)[e] speaks as follows:[f]

Now if you beat water and cause it to leap up (*po erh yo chih*[2]), you can make it go higher than your forehead. If you stir it and cause it to move (*chi erh hsing chih*[3]),[g] you can force it high up a mountainside. But are such movements according to the (intrinsic) nature of water? It is the strength applied (surely), which brings about these effects (*chhi shih tsê jan yeh*[4]). When men are made to do what is not good, *their* natures are forced in a similar way.

All one can say is that if this is a reference to water-raising machinery (which it has not traditionally been taken to be), the argument fits the chain-pump better than any other device, and certainly much better than the swape, which however was undoubtedly familiar to Mencius. On the other hand, Liu An in the −2nd century seems not to know the chain-pump, for the *Huai Nan Tzu* book, discussing *wei* and *wu wei* (action against, or according to, Nature)[h] says: 'Were there such a thing as using fire to dry up a well, or leading the waters of the Huai River uphill to irrigate a mountain, such things would be personal effort, and actions contrary to Nature....' Retarding the fall of water by artificial channels, however, made to follow higher contours above the main stream, was considered permissible by the writer.

In the +2nd century there can be no doubt that the square-pallet chain-pump was in use, and the following remark of Wang Chhung suggests rather strongly that it was already at work in the +1st.

In the streets of the city of Loyang there was no water. It was therefore pulled up from the River Lo by water-men (*Shui kung chi*[i] *shang Lo chung chih shui*[5]). If it streamed forth quickly (from the cisterns) day and night, that was their doing (*Jih yeh chhih liu, shui kung chih kung yeh*[6]).[j]

---

[a] The Shui Chhê[7] of Kuo Chung-Shu[8] (+937 to +980), reproduced in Liu Hai-Su (*1*), vol. 2, pl. 21. This is in fact the oldest representation that we have of any form of square-pallet chain-pump, antedating by three centuries the *Kêng Chih Thu*.

[b] *Nung Shu*, ch. 18, p. 13*a*.   [c] *Thien Kung Khai Wu*, ch. 1, p. 17*b*, not in first ed.
[d] *Nung Shu*, ch. 18, pp. 14*a*ff.   [e] Mentioned already, Vol. 2, p. 17 above.
[f] *Mêng Tzu*, VI (1), ii, 3; tr. auct.

[g] Legge (3), p. 272, talks of 'damming and leading' here, having contour canals only in mind, but the word *chi* seems to convey more of the sense of active pushing.

[h] In a passage already quoted in full, and discussed, Vol. 2, p. 69, above; *Huai Nan Tzu*, ch. 19, p. 3*b*, tr. Morgan (1), p. 225, mod.

[i] Note the use of the same word as in the Mencius passage.

[j] *Lun Hêng*, ch. 8; tr. Forke (4), vol. 2, p. 382.

¹ 告不害   ² 搏而躍之   ³ 激而行之   ⁴ 其勢則然也
⁵ 水工激上雒中之水   ⁶ 日夜馳流水工之功也   ⁷ 水車
⁸ 郭忠恕

The only other machine capable of such a continuous flow would have been the noria (see on, p. 356), but a series of chain-pumps would fit Wang Chhung's words best, and the probability is that they would have been of the characteristic Chinese square-pallet type. The reference would be to about +80. Just a century later we have the account which has usually been taken as marking the invention itself. It refers to the engineer and master-founder Pi Lan,[1] and occurs in the *Hou Han Shu*'s biography of Chang Jang,[2] a famous eunuch minister (*d.* +189).[a]

He (Chang Jang) further asked Pi Lan to cast bronze statues[b]...and bronze bells...and also to make (lit. to cast) 'Heavenly Pay-off' (Thien Lu[3]) and 'Spread-eagled Toad' (Hsia Ma[4]) (machines)[c] (which would) spout forth water. These were set up to the east of the bridge outside the Phing Mên[5] (Peace Gate) where they revolved (continually, sending) water up to the palaces. He also (asked him to) construct square-pallet chain-pumps (*fan chhê*[6]) and 'siphons' (*kho wu*[7]), which were set up to the west of the bridge (outside the same gate) to spray (*sa*[8]) water along the north–south roads of the city, thus saving the expense incurred by the common people (in sprinkling water on these roads, or carrying water to the people living along them). He also (asked Pi Lan to) mint (lit. cast) bronze coins....

This event was regarded as important enough to merit mention in the imperial annals of the emperor Ling Ti's[9] reign for the year in question (+186).[d] We shall have to return to the passage in connection with the noria (on p. 358), but the construction of the chain-pumps is quite explicitly stated. As for the 'thirsty crows' (i.e. siphons), the term must have been loosely used, for no siphons would have raised water as it is implied that these did. Our explanation is that what Pi Lan constructed included a series of simple suction-lift pumps analogous to the bamboo buckets with valves in their floors, which were probably used in the brine-well boreholes at the time, and which we have already discussed (p. 142 above).[e] All this apparatus was to the west of the bridge, but square-pallet chain-pumps to the east of it also may be concealed by the term *hsia-ma*,[4] about which we shall have more to say presently.

Whatever exactly Pi Lan's water-raising machinery was, we have here at the end of the +2nd century a precious account of a rather advanced water-supply system for an urban area, and the recent discoveries of so many kinds of stoneware piping and conduits from the Chhin and Han[f] give us a clearer picture of how it was organised. Such information about ancient Chinese cities has numerous implications, for example

[a] Ch. 108, p. 24*b*; tr. auct. Chang Jang incurred great unpopularity by instituting taxes for the restoration of the imperial palaces, a task which he entrusted to Pi Lan's colleague the master-architect Sung Tien.[10] It is interesting that without the urge for 'conspicuous waste' these pieces of ancient technology might not have been undertaken and transmitted. In many another instance too, Confucian austerity proved a brake on technological advance (cf. e.g. p. 510 below).
[b] Of men and of the symbolic animals of the four celestial (astronomical) palaces (see Vol. 3, p. 242).
[c] See p. 358 below.   [d] Ch. 8, p. 13*b*.
[e] True siphons might of course have been used in the distribution of the water from cisterns. The figurative meaning of the technical term would also be satisfied by a battery of flume-beamed swapes, such as that in Fig. 573, but for China this would be very peculiar at any period.
[f] Already described, p. 130 above.

| [1] 畢嵐 | [2] 張讓 | [3] 天祿 | [4] 蝦蟆 | [5] 平門 | [6] 翻車 |
| [7] 渴烏 | [8] 灑 | [9] 靈帝 | [10] 宋典 | | |

in connection with public health and hygiene.[a] There is no doubt that Pi Lan's job was done at Loyang, and from later sources such as the *Lo-Yang Chhieh Lan Chi*[1] of +530 it is not difficult to identify the site of his works.[b] Deriving from the Ku Shui[2] stream, the Yang Chhü,[3] a kind of moat, passed round the city walls to the west and south and was bridged outside the Phing Mên (later Phing Chhang Mên[4]) gate,[c] which had been built in +37. The pumps on its east side were evidently reserved for the palaces within the walls, while those on the west serviced the water-pipes of the city streets. As most of the water seems to have gone round the north of the city to enter the Hung Chhih[5] lake, the southern part of the moat was probably very slow-flowing—a fact of some importance for our interpretation of Pi Lan's machines, which thus can hardly have been current-operated.

The other *locus classicus* for the square-pallet chain-pump is in the *San Kuo Chih*[d] and concerns the famous engineer Ma Chün[6] who was active at the court of the emperor Ming Ti[7] of the Wei.

Ma Chün of Fu-fêng was matchless in ingenuity (*chhiao ssu*[8]). According to the essay[e] on him by Fu Hsüan[9]...there was at the capital, within the city, some unused land which could have been made into a park or gardens. But unfortunately no water was available for it. However, Ma Chün constructed square-pallet chain-pumps (*fan chhê*[10]) and had them worked by serving-lads. Whereupon the irrigation water rushed in (at one place) and spouted forth (at another), automatically turning over and over. The skill with which these machines were constructed was a hundred times beyond the ordinary.

This must have been at Loyang between +227 and +239. Later historians, such as the Sung author of the *Shih Wu Chi Yuan*,[f] used to mention Pi Lan and Ma Chün together as the originators of the device. But apart from the evidence already given, it may be that the chain-pump had already acquired its name of 'dragon backbone machine' (*lung ku chhê*[11]) in the Han; in which case some early references to dragon bones, (such as certain letters of Yang Hsiung, for example), dating from the last decades of the −1st century, may have meant the machine and not the fossil. Such, at any rate, was the idea of some Thang commentators.

In thinking over the first invention of the square-pallet chain-pump one has to take into account the fact that the expression *fan chhê*[10] did not always have sole reference to this machine.[g] According to that most ancient of Chinese dictionaries the *Erh Ya* (Warring States material compiled in Chhin and Han),[h] the 'overturning device'

---

[a] Cf. Sect. 44 below. Meanwhile see Needham & Lu Gwei-Djen (1).
[b] See Chou Tsu-Mo's edition, p. 5, and the discussion of Lao Kan (3).
[c] In 1958, on our way to visit the Lungmên defile and its cave-temples south of the city, Dr Lu and I must have passed very near these sites without knowing it, for today they are not very recognisable.
[d] Ch. 29, p. 9a, tr. auct. The account of Ma Chün forms a pendant here to the biography of the musician Tu Khuei.[12]
[e] Translated in full on p. 40 above.
[f] Ch. 45, p. 4b. Cf. *I Lin*, a Thang encyclopaedia.
[g] The point has been raised by Wu Nan-Hsün (1), p. 169.
[h] See Wang Shu-Nan's *Erh Ya Kuo Chu I Tshun Pu Ting*, ch. 7, p. 5a.

[1] 洛陽伽藍記	[2] 榖水	[3] 陽渠	[4] 平昌門	[5] 鴻池
[6] 馬鈞	[7] 明帝	[8] 巧思	[9] 傅玄	[10] 翻車
[11] 龍骨車	[12] 杜夔			

(*fu chhê*[1]) was another name for a kind of bird-trap called *fou*,[2,3] which was more or less synonymous with the *chhung*[4] or *chuo*.[5] On this Kuo Pho, at the beginning of the +4th century, comments that the *fu chhê*[1] was also called the *fan chhê*,[6] and tells us that it had two parallel bars (*yuan*[7]) with a net (*chhuan*[8,9]) at the far end. Here the swape or trebuchet bird-trap is easily recognised; the hunter on seeing birds alight pulls sharply down on a cord attached to the short arm of the lever, thereby lobbing over the long arm and the net attached to it like the sling of a trebuchet catapult.[a] This device goes back to the time of the *Shih Ching*, the early Chou period. Whether it had any connection with the idea of the square-pallet chain-pump is another matter, though Wu Nan-Hsün suggests (not perhaps altogether implausibly) that Pi Lan was inspired partly by the two bars of the swape bird-net for the sides of his flume, and by its overturning motion for the upper action of his endless chain. He recalls also that curious Chou idiom for the human jaws (*ya chhê*[10]) already mentioned,[b] and relates it to the ingesting lower action of the chain. In any case, the two uses of the term *fan chhê*[6] deserve to be recorded.

By the Thang and Sung the square-pallet chain-pump had become a commonplace, mass-produced by thousands of rustic wheelwrights. In +828 its specification was standardised. The *Chiu Thang Shu* says:[c]

In the second year of the Thai-Ho reign period, in the second month...a standard model of the chain-pump (*shui chhê*[11]) was issued from the palace, and the people of Ching-chao Fu[d] were ordered by the emperor to make a considerable number of the machines, for distribution to the people along the Chêng Pai Canal, for irrigation purposes.

The device was also getting into literature—for example, a rhapsody on it was written by Fan Chung-Yen[12] (+989 to +1052).[e] Lü Chhêng[13] tells us that it was customary in his time (the +14th century) for girls to work the square-pallet chain-pumps.[f] Lou Shou, the first author of the *Kêng Chih Thu*, devoted one of his poems to it[g] about

+1145:   The Man of Sung, who pulled up the sprouts, we despise;
           And Chuang (Tzu) of Mêng, who preferred hand-watering with jars.
           These are not so good as (that engine, which works)
           Like a set of birds, each holding the tail of the next in its mouth.
           With this we can change water's flow, and drain a whole lake.
           So dance the rice-plants happily in the caerulean waves
           While the farmer sits on his bamboo mat
           Enjoying the cool of the evening, and the singing and laughing
           Of lads and lasses, under the willows,
           Lit by the setting sun.

[a] See Sect. 30*i* below.   [b] P. 86 above.
[c] Ch. 17, p. 14*a*; tr. auct.   [d] The capital.
[e] *Fan Wên Chêng Kung Chi*, ch. 2; cf. Chhêng Su-Lo (2).
[f] *Lai Hao Thing Shih Chi*[14] (Collected Poems from the Pavilion where the Cranes Come), ch. 3, p. 3*a*.
[g] Tr. auct. Cf. the poem written by Fan Chhêng-Ta in +1186 and translated into English several years ago by Bullett & Tsui (1), stanza 30, p. 23.

[1] 覆車   [2] 罦   [3] 浮   [4] 罿   [5] 罬   [6] 翻車
[7] 轅   [8] 罥   [9] 罥   [10] 牙車   [11] 水車   [12] 范仲淹
[13] 呂誠   [14] 來鶴亭詩集

An early reference to the use of animal-power in working these chain-pumps occurs in Lu Yu's[1] account[a] of his journey to Szechuan in +1170. Fu Lin[2] was using them about the same time for draining the foundations of granary sites.[b] Both these literary mentions are preceded, however, by the early Sung painting already mentioned (p. 344) from about +965.

These machines must always have been used for raising water in civil engineering operations as well as for agricultural irrigation, and thus we see them take their place in Lin Chhing's[3] great Chhing compendium of water-conservancy technology,[c] the *Ho Kung Chhi Chü Thu Shuo*[4] (cf. Sect. 28 below).

The square-pallet chain-pump is probably what was meant in a passage in the travel journal of Chhiu Chhang-Chhun, on his way through Turkestan in +1221 to visit Chinghiz Khan. At Almaligh (near modern Kuldja) the Taoist sage and his party met for the first time with cotton, and observed the work of the local farmers.[d]

They irrigate their fields with canals, but the only method employed by the people of these parts for drawing water (formerly) was to dip in jars and carry them back.[e] When they saw our Chinese water-raising machines, they were delighted with them (*chi chhien Chung yuan chi chhi, hsi*[5]).[f]

'You Thao-hua-shih[6] (Tabghaj, Tabgatch)[g] people are so clever at everything!' they said. Thao-hua-shih is their name for the Han (i.e. the Chinese). Every day the people brought us more and more presents.[h]

Later in the Yuan period, the square-pallet chain-pump spread ever more widely. About +1362, a Korean official, Paek Munbo,[7] urged that they should be adopted in his country, and four centuries later the great progressive Korean scholar Pak Chiwǒn[8] was still advocating the wider use of Chinese water-raising techniques there.[i] In the reign of King Sejong (+1419 to +1450) there was much discussion about the relative

---

[a] *Ju Shu Chi*[9] (Diary of a Journey into Szechuan), ch. 1, p. 6a. This shows, incidentally, that the inventor of the animal-driven chain-pump cannot have been Shan Chün-Liang,[10] at the beginning of the Ming, as suggested by Liu Hsien-Chou (1). The Koreans were also designing ox-whims in +1488 (Yi Kwangnin (1), p. 90).

[b] *Khuei Chhê Chih*, ch. 1, p. 14b.   [c] Ch. 2, pp. 25a, 26a.

[d] *Chhang-Chhun Chen Jen Hsi Yu Chi*, ch. 1, p. 18a, tr. Waley (10), p. 86. Cf. p. 337 above.

[e] Like the old man north of the Han River (Vol. 2, p. 124), cf. pp. 332, 347 above.

[f] Waley rendered this simply as 'our Chinese buckets', but the story then has little point, and the words used imply something more like a machine. Since the noria had then probably long been familiar in Persia, the square-pallet chain-pump is indicated, if not of course the swape.

[g] On this term, which probably derives from the name of the ruling house of the Wei dynasty, Tho-pa, cf. Vol. 1, pp. 169, 186.

[h] This was very kind of them, but they were not so technically backward as the rather patronising tone of the passage might imply. For this was the territory of the Western Liao or Qarā-Khiṭāi State, just deceased, and as we shall shortly see (p. 560 below), its people were almost certainly responsible for transmitting the windmill in the opposite direction, i.e. to China. Cf. also Vol. 4, pt. 1, p. 332.

[i] E.g. in his book on agricultural improvement, *Kwanong Sochho Ansǒl*[11] (+1799). Pak had been greatly impressed by Chinese learning and Chinese technique during his visit as a member of the Korean mission of congratulation on the occasion of the Chhien-Lung emperor's 70th birthday in +1780. In his *Yǒrha Ilgi*[12] he wrote an interesting account of this journey. On his struggle, with like-minded men, to modernise his country, cf. Yang & Henderson (1).

---

[1] 陸游	[2] 傅霖	[3] 麟慶	[4] 河工器具圖說	[5] 及見中原汲器喜
[6] 桃花石	[7] 白文寶		[8] 朴趾源	[9] 入蜀記
[10] 單俊良		[11] 課農小抄案說	[12] 熱河日記	

advantages of square-pallet chain-pumps and norias, the former apparently an intro-duction from China, the latter from Japan. It is clear, at any rate, that the construction of the former was encouraged by the Korean court from about +1400 onwards.[a] In the 19th century such pumps spread widely in Indo-China.[b] Credit for introducing them from the north is often given to the Annamese ambassador Lý Văn-Phúc.[1]

The square-pallet chain-pump, however, was destined to achieve a distribution far wider than that of the Chinese culture-area. Visitors to the Palace of Hampton Court near London may see a remarkable square-pallet chain-pump of pure Chinese pattern, which is said to have been installed there for the removal of sewage in +1516, but which dates more probably from about +1700.[c] As Zimmer (1) points out in his description of it, the size of the pallets (8 × 9 in.) closely resembles that of Chinese practice,[d] but the maker failed to reproduce certain subtleties of the original type, such as greater height than breadth of the pallets, and the arrangement that their grain should be at right angles to the wearing surfaces.

There can indeed be little doubt that this type of chain-pump spread all over the world from China in the +17th century, and it might well be possible to date its travels to within a couple of decades. First Lorini describes something very similar in +1597.[e] European writers on water-raising machinery such as Bate (+1634) and d'Acres (+1660) do not know of it,[f] but Montanus, writing just before +1671, describes what he saw when accompanying one of the Dutch Embassies:[g]

Where there is want of Water, it is convey'd, though a considerable way, out of the Rivers, along digg'd Channels; (by which means all China is made Navigable) and conducted from low to high Places by means of an Engine made of four square Planks holding great store of Water, which with Iron Chains they hale up like Buckets.

The only fault in the observation was that the chains were of wood, not iron. Sinclair, the Scottish physicist and engineer, seems also to refer to these chain-pumps in +1672, when he speaks of 'buckets or plates' for drawing water from mines; as also Meyer and other writers in the following decade. According to Ewbank,[h] chain-pumps were in use on British naval ships for clearing bilge towards the end of the +17th century, having been adopted from Chinese junks.[i]

After +1700, the probable date of the Hampton Court pump, there are numerous descriptions in European engineering literature—Leupold (+1724), van Zyl (+1734), de Bélidor (+1737),[j] Chambers (+1757)[k] and Leonhardt (+1798). At the end of the century Staunton[l] gave a famous cut showing the machine at work, and in +1797 van Braam Houckgeest[m] was so kind as to describe clearly his introduction of it to

---

[a] Much information on this history has been assembled by Yi Kwangnin (1), pp. 84 ff.

[b] Huard & Durand (1), p. 127.

[c] I had the pleasure of studying it, with Miss Su Lin, in 1961.

[d] Details of the typical Chinese dimensions are quoted by Liu Hsien-Chou (1), p. 36, from the Ho Kung Chhi Chü Thu Shuo. [e] Beck (1), p. 247, redraws.

[f] Though of course they describe the sāqīya (cf. p. 352 below).

[g] (1), p. 675; Ogilby tr. [h] (1), pp. 154 ff.

[i] On this subject, see below in Sect. 29i. [j] (1), vol. 1, pls. 36, 37, and p. 360.

[k] (1), p. 13, pl. 18. [l] (2), vol. 2, p. 481.

[m] (1), Fr. ed., vol. 1, p. 57, Eng. ed., vol. 1, p. 74. On him, at length, see Duyvendak (12).

[1] 李文馥

America; a rare example of a statement by a living link in the chain of technique transmission. 'I introduced it', he said, 'into the United States, where it has proved of great utility along the river-banks, on account of the ease with which it is operated.' The Spaniards had taken it to the Philippines, and the Dutch to Batavia, much earlier.[a]

By the Sung period the square-pallet chain-pump may have given rise in China to the endless conveyor chain of containers for excavating sand or earth.[b] In any case, with the Renaissance, the idea arose and spread in the West.[c] Zimmer (2), who has treated of the early history of mechanical handling devices, noted the excavators of Ramelli and Besson (late +16th century), which, as we saw above (p. 211), were offered to the Chinese as a new idea; and demonstrated the ancestry of all conveyor belts and chains. It was to flour-milling that these were first applied, and a description of their earliest forms, due to Oliver Evans (+1756 to +1819), has been given by several authors.[d] The small wooden blades, or 'flights', of the grain-elevator were the lineal descendants of the pallets of Pi Lan and Ma Chün, save that the Archimedean screw principle was sometimes used instead of the endless chain.[e]

Another obvious application of the endless chain was for dredging, as we see in numerous Europeans designs from the +16th century onwards, but this use, with its need for scooping-buckets rather than inter-pallet spaces, derives presumably rather from the *sāqīya* than the *fan chhê*.[f] Here there was a junction of techniques. If the dredger was, and still is, set up slanting like the *fan chhê*, it has buckets like the pot chain-pump or *sāqīya*. The obvious converse logical intermediate device would have pallets (or their equivalent) like the *fan chhê*, but would be set up vertically; and this indeed has long existed under the name of the 'paternoster pump' or 'rag-and-chain pump'. Here the *fan-chhê* flume must have a fourth side. Within a vertical pipe an endless chain brings up balls of metal or lumps of rag and leather which nearly fill the lumen of the pipe and so act as pallets discharging the water at the top. The resemblance of the endless chain to a rosary gave the device its ecclesiastical name. These pumps were popular for mine drainage in +16th-century Europe, and Agricola describes and figures several kinds;[g] but the previous history of the machine is obscure, and no

***

[a] As recently as 1938 it was again carried from China to America because it was found extremely useful in the pumping of crystallising brines; a self-cleaning action prevents clogging. This information was given by Mr Ferris of the Bonneville Salt Company of Salt Lake City to my friend Dr M. R. Bloch of Sdom in Israel.

[b] Cf. p. 219 above.   [c] Cf. Straub (1), p. 237.

[d] E.g. Bennett & Elton (1), vol. 2, p. 194; Storck & Teague (1), p. 165.

[e] The principle persists into the most modern techniques. At the Musée des Travaux Publics in Paris, the French National Railways show a model of a triangular pallet-chain continuous excavator and flumes used for replacing ballast under the permanent way. Moreover, conveyor belts modelled on the *fan chhê* are now playing a great part in the nation-wide canal-building activity of contemporary China. See, for example, a device in Anon. (18), pt. 1, no 48, worked by animal-power; or a news item on the application of Diesel engines (NCNA, 4 February 1960); and Fig. 583.

[f] That is, in most cases. But Zimmer (3) has published a drawing of a Dutch dredger dating from +1562 which clearly shows Chinese pallets in use on a chain, bringing up the silt through a box flume. It may well be that the Chinese pallet-pump was introduced to Europe twice, first in the mid +16th century and then again at the end of the +17th. Cf. Conradis (1). Single-scoop dredgers, in which the bucket is hauled up by a pedal winch, are still used in China (G. R. G. Worcester, unpublished material, no. 102). We shall describe them in Sect. 28f. There are early European analogues.

[g] Hoover & Hoover ed., pp. 190–7.

PLATE CCXXVI

Fig. 583. Belt-conveyors based on the principle of the square-pallet chain-pump at work on a hydraulic engineering project in 1958, the utilisation of the waters of the Thao River in southern Kansu (cf. Anon. 60). A canal some 700 miles long will divert them north of Minhsien, through mountainous country averaging over 6000 ft. above sea level, as far as Chhingyang in the east of the province, so that they irrigate 3·3 million acres instead of running directly into the Yellow River above Lanchow. The canal crosses the head waters of many streams including the Wei River, and passes through many tunnels, one $3\frac{1}{2}$ miles in length.

Fig. 584. Discarded assemblies of cast-iron gear-wheels and sprocket-wheels for paternoster pumps (orig. photo., 1958, at Hsiao-thun-hsiang Agricultural Cooperative, Fêng-thai, Hopei). Steam and electricity are replacing these animal-powered rural water-raisers. In the background, the still older wooden drum-wheel of a *sāqīya* (pot-chain pump).

PLATE CCXXVII

Fig. 585. Triple paternoster pump manually operated with two cranks (orig. photo., Exhibition of Agricultural Machinery, Peking, 1958). The discoidal diaphragms (*phi chhien*) can readily be seen

Fig. 586. Paternoster pump and 'fly-wheel' weighted with two mill-stones, manually operated by eccentric, connecting-rod and hand-bar as in the traditional mills, cf. Fig. 413 (orig. photo., Exhibition of Agricultural Machinery, Peking, 1958).

illustrations of it before the early +15th century are known.[a] It seems to have been foreign to traditional Chinese engineering practice,[b] but this does not mean that it may not have been a European invention deriving from the *fan chhê*, possibly through vague hearsay from intermediate sources.[c] At present there is no evidence. The device had a period of importance, for Agricola described a three-stage paternoster pump in a mine at Schemnitz in the Carpathians worked by 96 horses and raising water at least 660 ft.[d] It may have stimulated some people to think about pistons in cylinders, but as we have seen (p. 141 above), the piston-pump was familiar in Hellenistic and Arabic culture as well as in Agricola's own time, so that the paternoster pump might have derived from the former rather than contributing to it. Provisionally one might almost think of it as a marriage of the old European piston-pump to a rumour from China.

In any case the paternoster pump[e] has today spread over the length and breadth of China as the most efficient and convenient of water-raising machines in the vast movement of rural mechanisation and agricultural improvement. Already in 1952 it was displacing the classical square-pallet chain-pump and even the noria.[f] Six years later I was enabled to study it in many varieties at the National Exhibition of Agricultural Machinery at Peking.[g] Some machines are built entirely of wood, with square diaphragms (*kua shui pan*[1]) working through a vertical square-section wooden pipe, but it is more common to have cast-iron gear and sprocket-wheels (Fig. 584) for animal power, with discoidal 'washers' of leather or rubber (*phi chhien*[2]) running up a metal pipe (Fig. 585). Another photograph (Fig. 586) illustrates the ingenious improvisations now dear to every peasant farmer; a manually-operated connecting-rod and crank like those of the traditional mills (cf. Fig. 413) rotates the right-angle gear with the help of two peripheral weights (old millstones) which form a bob fly-wheel.[h] Elsewhere the small donkey-engines which co-operatives and communes can now afford apply steam-power to the raising of irrigation water. Under the sun of a Chinese summer, an East Anglian contemplating this could not but remember those

[a] A manuscript drawing of a paternoster pump is in the work of Mariano Taccola (+1438), München Codex Lat. 197; see Feldhaus (1), col. 833.

[b] At least, on land, for there is some evidence that the paternoster pump, which became in the 18th and early 19th centuries a favourite device for clearing the holds of ships, was adopted directly from the means employed on Chinese junks (cf. Ewbank (1), pp. 154 ff; Davis (1), vol. 3, p. 82). We shall take up the matter again in Sect. 29*i*.

[c] On the late +14th-century group of important technological transmissions from China to Europe, see p. 544 below.

[d] Not Agricola's home, Chemnitz (Karlmarxstadt) in the Erzgebirge, but Banská Štiavnica in Slovakia, later the site of the first Newcomen steam-engines on the continent (+1721); Voda (1), cf. Nagler (1). Central European mining steam-power may have been delayed by the efficiency of the *Stangenkunst* (cf. p. 334 above), arrangements of rocking pantograph-like levers which could not only transport water directly but transmitted power for pumping from distant prime movers; see Multhauf (4). This arose just after Agricola's time.

[e] Generally called simply *shui chhê*.[3]

[f] My own observations accorded with those of my friend Dr N. W. Pirie, F.R.S., in that year.

[g] Many working drawings and sketches will be found in Anon. (*18*), pt. 1, and many photographs in Anon. (*30*). Among the latter's interesting designs for local construction may be mentioned a vertical water-wheel working a high-lift paternoster pump by a long chain-drive (p. 68).

[h] A working drawing of this will be found in Anon. (*18*), pt. 1, no. 85. Similarly, old cart-wheels are pressed into service as fly-wheels, as we have already seen (Fig. 386).

[1] 刮水板　　[2] 皮錢　　[3] 水車

delightful words borne (for an opposite purpose) by one of the historical pumping-stations in his own gloomy land:

> These fens have often been by Water drown'd;
> Science a remedy in Water found,
> 'The power of Steam', she said, 'shall be employ'd,
> And the Destroyer by itself Destroy'd'.

But soon the *lao hsiang* will be turning a switch.

(5) THE 'SĀQĪYA' (POT CHAIN-PUMP)

To the *sāqīya* ('the cup-bearer girl'?), i.e. the endless chain of pots, let us now turn. It differs doubly from the chain of pallets in the inclined flume, for here the chain hangs normally straight down below the upper wheel, and carries whole pots or buckets which fill at the lower end and discharge at the top.[a] Zimmer (2) introduces his account by reproducing a Babylonian relief of about −700 showing queues of men carrying baskets of earth upwards and descending with them empty.[b] It is not surprising, therefore, that the chain of pots should have been an ancient idea, and there seems little doubt that the description occurring in a text of Philon of Byzantium (*c.* −210) is at any rate partly genuine.[c] Writing about −30, Vitruvius mentions the machine clearly,[d] and we still have the remains of one of the chain-pumps used for emptying the bilges of the great ships of the Lake of Nemi built between +44 and +54.[e] While nothing authorises the view that the ancient Egyptians knew of this device,[f] it spread rapidly in the Hellenistic age all over the Near East, and the *sāqīya* or *daulāb* ('camel-wheel') thus became as characteristic of the Islamic lands as the *fan chhê* was of China.[g] The familiar words 'Or ever the silver cord be loosed, or the golden bowl and the pitcher be broken at the fountain, or the wheel broken at the cistern' must surely refer to the *sāqīya*.[h] One of the most imposing of these machines is that known as Joseph's Well at Cairo, where a borehole descends vertically through solid rock 165 ft. to the lower animal-power chamber, and then further down to reach

---

[a] Cf. Ewbank (1), pp. 122 ff.; Baroja (1, 4); Schiøler (3).

[b] Like many others who have been in China during and since the Second World War, I have seen similar *levées-en-masse* of human labour in the building of airfields, dams and canals. Cf. Sect. 28f below.

[c] *Pneumatica*, ch. 65; cf. Carra de Vaux (2); Beck (5); Drachmann (2), pp. 66, 68.

[d] x, iv, 4; cf. Drachmann (9), p. 151.

[e] See Moretti (1), p. 33. The sprocket-wheels have five cogs moulded to fit buckets pear-shaped in cross-section. A Ctesibian double-cylinder force-pump was also found (*ibid.* p. 34). Fuller details in Ucelli di Nemi (1, 2).

[f] As supposed, for instance, by al-Salam (1), p. 9. Petrie (4), p. 143, says that *sāqīya* pots are common in Roman, but not in ancient Egyptian, dumps. On the other hand, the view of Forbes (11), p. 34, that the so-called 'hanging gardens' of Babylon, *c.* −700, were watered by *sāqīya* working in deep shafts, is not entirely without plausibility. Cf. Schiøler (1).

[g] See E. W. Lane (1), p. 301; Bonaparte (1), vol. 12, pp. 408 ff., and atlas, vol. 2 (Arts et Métiers Section), pls. III, IV and V; Chatley (36), p. 159. A very thorough study of the technical terms in Egyptian Arabic for all the parts of the *sāqīya* is due to Littmann (1). Ball (1) tells us that in 1904 the Dongola province of the Sudan alone had just under 4000 *sāqīya*, each irrigating about 15 acres and supporting 33 people.

[h] Ecclesiastes, xii, 6.

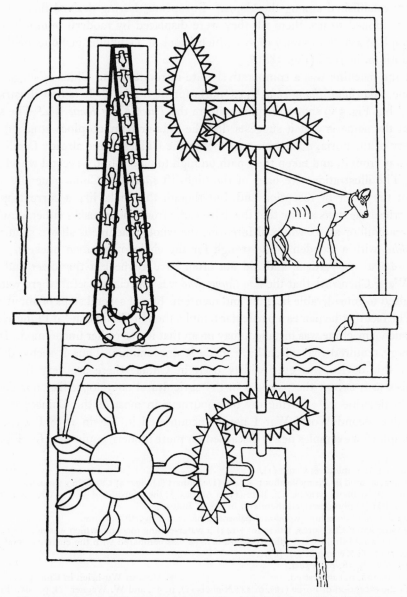

Fig. 587. The *sāqīya* or pot-chain pump, characteristic of Hellenistic and Islamic culture, a diagram in al-Jazarī's book on ingenious mechanical contrivances (+1206), redrawn by Wiedemann & Hauser (1). Alternative forms of power are shown, an animal whim and a vertical water-wheel. Both the gear-wheels and the pot-chain are drawn in semi-perspective, as so often in Arabic pictures of machinery.

a total depth of 297 ft. Water is here raised in two stages by *sāqīya*.[a] Arabic MSS. of the early +13th century, analysed by Wiedemann & Hauser (1) and Schmeller (1), show several forms of pot-chains (see Fig. 587). They passed into Muslim Spain[b]

[a] Norden (1), p. 49, pl. XIX; Ewbank (1), p. 46.
[b] Cf. de Bélidor (1), vol. 1, pl. 39; Schiøler (2). In 1960 I saw many still working in Spain and Portugal.

and were established among the Copts.[a] Europeans, having derived them from the Arabs, continued to use them till they were displaced by modern pumps.[b] I add a photograph of a contemporary *sāqīya* which I took at Yedikülle, just outside the walls of Byzantium, in 1948 (Fig. 588).[c]

That the machine was a comparatively late introduction to China is suggested by the name, *kao chuan thung chhê*,[1] i.e. 'noria for high lifts'.[d] It is first illustrated in +1313,[e] but Fig. 589 shows the picture from the *Thien Kung Khai Wu*.[f] The significance of the name is that it suggests that the pots or bamboo pipes attached to the periphery of the noria, a machine with which the Chinese were already familiar, had flown away from it, and taken their path on high to pass round a second wheel and so return. The illustrations we have of the high-lift *sāqīya* in China generally show a wheel at the lower or reception end, but though this looks like a current-operated paddle-wheel the texts all say that the drive came from above, using either a multiple-pedal treadmill or an ox whim. Moreover, the whole machine is shown slanting like a *fan chhê*, with a wooden guide-trough for the chain of bamboo buckets. Yet the Arabic *sāqīya* was vertical, and did not often have a wheel at the lower end of the chain. Wang Chên says that the *kao thung chhê* was especially useful where water had to be raised to considerable heights, and mentions lifts of several steps of about 100 ft. each. The fact that he names a particular temple, the Hu Chhiu Ssu[2] at Phing-chiang,[g] where such a machine was installed, may mean that it was rather uncommon. Indeed, the energy required may often have made it uneconomic. Thus it seems doubtful whether this device was ever widely used in China; there are very few literary references to it except the passages in the agricultural encyclopaedias, nor do many travellers describe it.[h] During my own journeys in most of the Chinese provinces during the Second World War I never encountered it, but in 1958 I was able to photograph a few examples newly discarded for more modern equipment (cf. Fig. 590).[i]

---

[a] Zimmer (4); Winlock & Crum (1).

[b] One was pictured by Pisanello about +1420 (Degenhart (1), fig. 147), and they were described in the +17th century by many writers, e.g. Bate (1); d'Acres (1); the Marquis of Worcester, see Dircks (1), etc. Wakarelski (1) photographed them still in use in Bulgaria.

[c] The mixture of eotechnic wood and palaeotechnic iron is worth noticing.

[d] The same applies to Korea, where in +1431 a water-raising machine which was not very good for irrigation but would drain water effectively from wells was compared with the current-operated noria from Japan; cf. Yi Kwangnin (*1*), p. 91.

[e] *Nung Shu*, ch. 18, pp. 17*b*ff.

[f] Ch. 1, p. 16*b*, not in first ed.   [g] Modern Wu-hsien in Chiangsu.

[h] With the exception of Forke (16), p. 22; Nichols (1), p. 31, and W. Wagner (1), p. 195. Fu Chien (*1*) found such machines numerous on the south bank of the Wei River in Shensi, and Chatley (36), pl. xxxi (1) illustrates one from the neighbourhood of Tientsin. All these were regular vertical *sāqīya* driven by animals.

[i] The first of these was on Yenthan Island in the Yellow River just below the city of Lanchow; the second at the Fêng-thai Agricultural Cooperative near Peking. The construction of these old gear-wheels is interesting. In the former (which has lost most of its cogs) the rim is more composite than usual because the felloes are clamped by plates on the sides of the wheel, set between spokes which grip the rim externally as well as by the fitting of their tongues into the felloes. In the latter the felloes are joined by wrought-iron dogs, and the spokes are disposed at unequal distances from one another so as to allow for the insetting of the cogs between them alternately in twos and ones.

[1] 高轉筒車   [2] 虎邱寺

PLATE CCXXVIII

Fig. 588. A modernised *sāqīya* at Yedikülle, near Istanbul (orig. photo., 1948). Iron and petrol-cans have replaced the classical wooden drum-wheel and earthen pots, but the right-angle gearing is still much as it would have been in al-Jazarī's time. The donkey whim can be seen underneath the driving shaft.

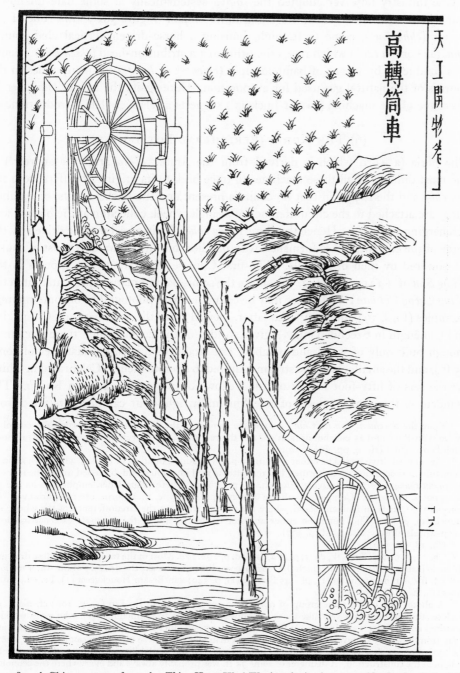

Fig. 589. A Chinese *sāqīya* from the *Thien Kung Khai Wu* (+1637), ch. 1, p. 16*b*. Its foreign origin may be sensed by the inappropriate nomenclature, *kao chuan thung chhê*, 'noria for high lifts'. It is unusual to have a wheel at the lower end, and one suspects operation by paddles, but the texts all make the power come from above.

One industry however adopted the *sāqīya* systematically at some date unknown, namely the salt fields at Tzu-liu-ching in Szechuan. Figure 591 shows[a] one of the towers (*chhê lou*[1]), some 60 ft. high, housing a horse-driven vertical chain-pump (*shui tou*[2] or *tou tzu*[3]) which raises the brine to a distribution-head (*chien wo*[4]) whence it is conducted by means of bamboo pipes (*chien*[5,6] or *hung*[7]) for many miles to the sources of the natural gas used for evaporation. Here again, the lack of character in the name of the machine ('water-buckets') suggests a relatively late introduction.

(6) THE NORIA (PERIPHERAL POT WHEEL)

The noria (a word which is derived from the Arabic *al-nā'ūra*, 'the snorter') is the most difficult of all these machines to trace back to its origin. It differs from the *fan chhê* and the *sāqīya* in that no chain is present and the buckets, pots, or bamboo tubes are attached to the circumference of a single wheel, collecting at the bottom and discharging at the top. Hence the name *thung chhê*.[8] This wheel may be driven by the force of the current, if it is furnished with paddles, but in still water it must of course be powered by men or animals.[b] The first Chinese illustration of it appears in the *Nung Shu* of +1313,[c] but Fig. 592 shows the semi-diagrammatic representation in the *Nung Chêng Chhüan Shu*.[d] These may be compared with photographs of contemporary examples (Figs. 593 and 594);[e] the latter, viewed from the road between Chungking and Chhêngtu in Szechuan (1943), gives an idea of the height which these machines, though built only of wood and bamboo, can attain (in this case a diameter of some 45 ft.), and the smallness of the stream required to operate them.[f] Celebrated in China are the sets of fifty-foot wheels on the Yellow River near Lanchow in Kansu.[g] The grandeur of these masterpieces of eotechnic millwright craft can be appreciated from

---

[a] From the *Szechuan Yen Fa Chih*,[9] a complete account of the technology of the salt fields compiled by Lo Wên-Pin[10] and 25 collaborators in 1882 at the request of the Governor-General of the province, Ting Pao-Chên[11] (ch. 2, pp. 26b, 27a).

[b] The normal drive would be either a multiple-pedal treadmill or an ox or donkey whim. We know of no picture or description of the former, but the latter is well illustrated and discussed under the name of *wei chuan thung chhê*[12] in the *Nung Shu*, ch. 18, pp. 16a ff., where Wang Chên emphasises its use for lakes, pools, moats, etc. Compared with the current-operated type, these forms are rare, however. An unusual type exists on Hainan Island, where the peasant walks on the circumference of the wheel; see Franck (1), p. 321. This might be either ancestor or descendant of the Japanese external treadmill scoop-wheels already discussed (p. 337).

[c] Ch. 18, pp. 11b ff.

[d] Ch. 17, p. 9a; cf. *TKKW*, ch. 1, p. 16a, first ed. pp. 9b, 10a. Most of the traditional Chinese drawings fail to do justice to the great lift which the noria accomplishes, and make the delivery of the water too low.

[e] Cf. for +1794 Staunton (1), pl. 44; Barrow (1), p. 540; van Braam Houckgeest (1), Fr. ed., vol. 1, pp. 55 ff., Eng. ed., vol. 1, p. 72.

[f] A photograph of the same group of norias also appears in Phan Ên-Lin (1), p. 186; cf. his p. 36. A close study of machines very like these in Indo-China is due to Guilleminet (1), who gives working drawings and calculates efficiencies. At Phu'ó'c-lôc a long aqueduct some 50 ft. high conducts the water away from a battery of seven large norias 60 ft. in diameter. All these are of the same order of size as the Syrian norias shortly to be mentioned, but in East Asia the availability of bamboo made it possible to lighten the construction. A diameter of some 75 ft. was probably always and everywhere the economic limit, however.

[g] My first introduction to these was with the geologist Dr E. Beltz, who was also in Lanchow in 1943.

[1] 車樓	[2] 水斗	[3] 斗子	[4] 梘窩	[5] 梘	[6] 筧
[7] 笕	[8] 筒車	[9] 四川鹽法志		[10] 羅文彬	[11] 丁寶楨
[12] 衛轉筒車					

PLATE CCXXIX

Fig. 590. Discarded *sāqīya* gear-wheel and drum-wheel of wood (orig. photo., 1958, at Hsiao-thun-hsiang Agricultural Cooperative, Fêng-thai, Hopei). Cf. Fig. 584, and Lu, Salaman & Needham (2).

PLATE CCXXX

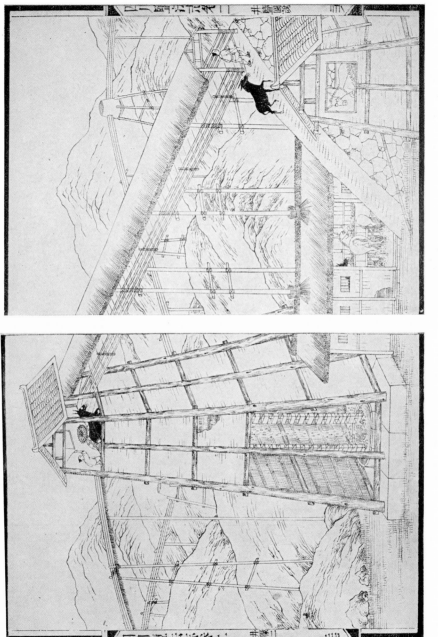

Fig. 591. Industrial use of the *sāqīya* in traditional style; one of the brine-raising towers (*chhê lou*) in the salt-fields at Tzu-liu-ching, Szechuan (*Szechuan Yen Fa Chih* (1882), ch. 2, pp. 26b, 27a). The brine from the bore-holes (cf. Figs. 396, 422, 423 and 432) is collected in tanks such as that seen at the bottom of the tower, and then raised to a height sufficient to allow of downward flow, sometimes for distances of a mile or so, to the evaporating sheds where supplies of natural gas are available. A relief horse is mounting the gallery above the stables on the right, where also another whim can be seen through a window.

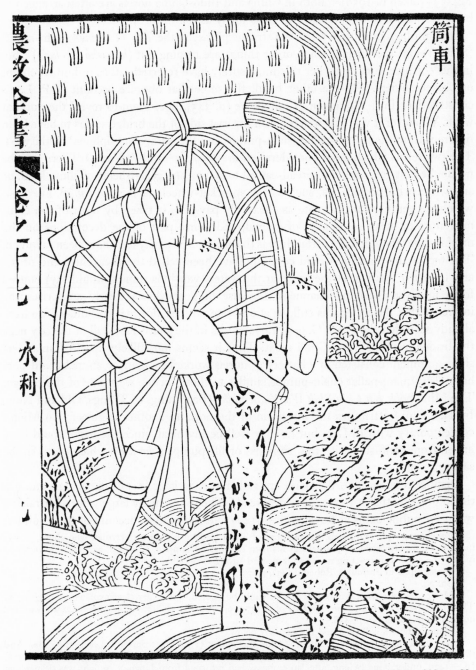

Fig. 592. The noria or peripheral pot wheel (*thung chhê*); a drawing from the *Nung Chêng Chhüan Shu* of +1628 (ch. 17, p. 9*a*). Like most traditional representations of the noria, this fails to do justice to the high lift available; the artist seems to have drawn the delivery flume in plan instead of in elevation.

Fig. 595 (taken in 1958).[a] Both in China and Indo-China norias are often arranged in batteries with a common shaft, up to as many as ten in a row.[b]

It might be tempting to relate the strange phrase in the *Tao Tê Ching*[c] to the noria: 'Some things are loading when other things are tipping out (*huo tsho huo hui*[1])', but in the −4th century or earlier this must surely have referred rather to human chains of basket-carriers. On the other hand, the somewhat obscure account of Pi Lan's constructions in +186, given in full above (p. 345), does seem to reveal the noria at work. Whatever it was that was built on the east side of the bridge seems to have been different from the square-pallet chain-pumps and the pumps or 'siphons' on the west. The two terms which need consideration are *thien lu*[2] and *hsia ma*.[3] There can be no doubt that the second ('spread-eagled toad') means gear-wheels of some kind, or any wheel with projections, such as the sprocket-wheels of the *fan chhê*.[d] The first, which we ventured to translate as 'heavenly pay-off', is taken by most of the commentators to mean some kind of animal, and hence part of the decorations on the machine, but though there certainly was this usage, a pun may here have been involved. The top of the machine would certainly have been called the 'heavenly' end of it,[e] and in fact it was just at the top that the water 'won' (as miners would say) poured forth into the receiving channels. The use of gear-wheels clearly authorises the inference that the water was still, or at any rate not fast-flowing, and indeed we have already seen that the Peace Gate of Loyang was in the south-east wall of the city near the bridge over the Yang Chhü moat. Thus it seems most likely that Pi Lan's east-side equipment consisted of a battery of norias worked by men or animals, and perhaps also square-pallet chain-pumps similar to those on the west side of the bridge. The Ming book *Ku Chin Shih Wu Yuan Shih*[4] (Beginnings of Things, Old and New) by Hsü Chü,[5] attributes the invention of the noria to another engineer, Ko Mien,[6] contemporary with Pi Lan, but he may well be mythical, since the name could have arisen from confusion with the 'siphons' or lift pumps (*kho wu*[7]) which occur in the same passage.

Distinct references to the noria are rather rare in later literature, though one has the impression that it was quite widespread.[f] The use of many large wheels for raising water for public baths at Loyang in +914 is referred to in an account of their builder,

---

[a] Lanchow people still revere the memory of Tuan Hsü,[8] a Kansu countryman who about +1545 made a couple of hazardous journeys at his own expense to Yunnan, Chiangsi, and back, in order to master the art of constructing norias as large and strong as possible.

[b] Photographs in Sarraut & Robequin (1), p. 79.

[c] Ch. 29.

[d] Many examples from different centuries could be quoted, but a very clear explanation is given by Thien I-Hêng in his *Liu Chhing Jih Cha* of +1579; cit. *KCCY*, ch. 48, p. 7b. This reference was first noticed by Schlegel (5), p. 458.

[e] Many instances will be found in Sect. 27*j* below.

[f] It reached Japan apparently about +800; cf. Papinot (1), p. 757; Sarton (1), vol. 1, p. 580. And thence it spread to Korea, as is shown by many quotations assembled by Yi Kwangnin (1), pp. 89ff. In +1429, under King Sejong, after an embassy had returned from Japan, a student was despatched thither to learn the art of making current-operated norias. Between +1431 and +1437 much attention was given to the best forms and designs, and on several occasions standard models were sent out to all the provinces for copying by the local artisans.

[1] 或挈或潟   [2] 天祿   [3] 蝦蟆   [4] 古今事物原始   [5] 徐炬
[6] 葛免   [7] 渴烏   [8] 段續

PLATE CCXXXI

Fig. 593. A group of three norias at work near Chhêngtu, Szechuan (orig. photo., 1943).

PLATE CCXXXII

Fig. 594. A high-lift noria, some 45 ft. in diameter, constructed entirely of bamboo and wood (orig. photo., 1943, from a bridge near Chienyang, Szechuan). The narrowness of the artificial channel which rotates the paddle-wheel as it fills the buckets (bamboo tubes) is noteworthy.

PLATE CCXXXIII

Fig. 595. A battery of high-lift norias (diam. *c.* 50 ft.) in one of the arms of the Yellow River just below Lanchow (orig. photo., 1958). More stoutly built than that in the preceding picture, these can withstand, as here, the great river in spate.

PLATE CCXXXIV

Fig. 596. The oldest representation of a noria; a mosaic at Apamoea in Syria (Mayence, 1). This is of the +2nd century, after Vitruvius and before Pi Lan. But the evidence points to India as the home of the invention.

the monk Chih-Hui;[1] the baths were so large that they could accommodate thousands of people.[a] A few decades later we have the description by an Arab traveller of the water-supply of the city of Shantan (see below, p. 402), which could hardly have been effected without the use of one or more large norias, as in fact it was within living memory. Then about +1170 the philosopher Chu Hsi (cf. above, p. 143) has a reference which must, I think, be interpreted as meaning the noria, although the expression used (*shui chhê*[2]) was more usually applied to the square-pallet chain-pump.[b] He says:[c]

Since the beginning of Heaven and Earth there has been this thing (this machine of Nature) in ceaseless revolution. There is a diurnal rotation, and rotations of months and years. But there is only the one (machine) endlessly turning and rolling. It is like a water-raising machine (*shui chhê*[2]), in which (at a given moment) one (bucket) is upright and another inverted, one is coming up and another going down.

About +1130 a phrase about norias appears in one of the poems of Shen Yü-Chhiu,[3] author of the *Kuei Chhi Chi*[4] (Poems from Tortoise Valley)—*shui hu phien-phien chieh chu yai*,[5] 'the water-buckets gracefully revolve by the river-bank'. Then comes the matter-of-fact description in the *Nung Shu* (+1313), and all later books of the same kind illustrate the noria.

A pleasant literary reference occurs about +1601 in a travel book about Kuangtung by Wang Lin-Hêng.[6] He says, speaking of the southern countryside:[d]

As for the water-raising machines, at the end of every spoke there is a tube (bamboo bucket), upright when rising and inverted when descending, so that as the wheel turns, the water is poured out into a channel. The size of the wheel depends upon the height of the field—even as high as 30 or 40 ft., a field can be watered. Not the slightest human effort is required. This is something like the water-triphammers and water-mills of Chekiang. Such machines as these for raising water would put to shame even the old man who lived north of the Han River long ago;[e] in short the skill of man, when it comes to perfection, can conquer the works of Nature. Who invented (the noria) we do not know, but he ought to be honoured with worship and sacrifices.

---

[a] In this connection see the illuminating study of baths in China by Schafer (7).

[b] This conviction is strengthened by the fact that, as we shall see (p. 361 below), the noria analogy had long been used in Buddhism, and Chu Hsi had certainly devoted much study to that philosophy, though he combated it.

[c] Tr. auct. It has not been possible as yet to locate this text in *Chu Tzu Chhüan Shu* or *Chu Tzu Yü Lei*, but it is taken from the material copied by Wieger (2), p. 188. Its use by recent authors really belongs to the Department of Utter Confusion. The translators of Grousset's last book (6), p. 218, reproduced a text which is in no way what Chu Hsi said, though quoting it as if it were his, and the text is not even a literal translation of Wieger's paraphrase and interjected comments (2), pp. 188 and 191. The muddle existed already in Grousset (5), p. 265. Wieger had indeed spoken of a noria, but evidently confused it with the *sāqīya*, for he included in his 'translation' some words about the full buckets coming up out of the well while the empty ones went back into it. The machine then became a 'chain-pump' in the English translation of Grousset (6), but to add to the bewilderment a footnote is given explaining that the chain-pump is a device for raising water into the rice-fields, portable and operated by treading—though the *fan chhê* has nothing to do with wells and carries no buckets. Of course no references were vouchsafed either by Wieger or Grousset, and the latter inserted several remarks taking the term *li*[7] as laws of Nature, which I hope we have shown (Sect. 18 above) is inadmissible.

[d] *Yüeh Chien Pien*,[8] ch. 3, p. 17*b*, tr. auct.

[e] A reference, of course, to the *Chuang Tzu* story, p. 332 above.

---

[1] 智暉   [2] 水車   [3] 沈與求   [4] 龜谿集   [5] 水戽翩翩接渚涯
[6] 王臨亨   [7] 理   [8] 粵劍編

The noria's history has been studied by Laufer (19), but though the choice of subject did him credit, the paper was not one of his more inspired efforts, since he confused the noria with the *sāqīya* (as so many other writers have done),[a] and even with the *fan chhê*. He also maintained that the noria was not mentioned by Vitruvius. In fact, however, Vitruvius gave about −30 a very clear description of the current-operated noria.[b] That it spread through the Hellenistic world is indicated by the discoveries at Apamoea in Syria, described by Mayence (1), which have revealed a noria in mosaic (Fig. 596). This would be of the early +2nd century, after Vitruvius and before Pi Lan. It was in Syria, too, that the noria was destined to achieve its greatest occidental development in individual size if not in numbers. Though Vitruvius distinctly speaks of buckets (*modioli*), his noria is sometimes described as having had a continuous series of radial boxes round its periphery, as if the very rim of the wheel were hollow and pierced with holes, and this was at any rate a form which it long retained in the Near East.[c] There was thus no doubt a close relation with his *tympanum*,[d] a water-raising wheel completely boxed in and divided into compartments. This, the *tābūt* (i.e. the 'ark') of the Islamic lands, has the disadvantage that it delivers the water only at axle-level and not at the top of the wheel, but it will do for small lifts.[e] Bigland illustrates some norias of the box-rim kind in the Fayum, Egypt;[f] but the most splendid ones ever constructed still exist and labour today at Hama on the Orontes in Syria. Here the largest attain a diameter of nearly 70 ft. and discharge into tall arched aqueducts of stone as well as the more usual trestles.[g] From Mieli we learn that these were originally constructed by Ibn ʿAbd al-Ghanī al-Ḥanafī (+1168 to +1251),[h] and there is much on norias of all kinds in Arabic MSS.[i] The earliest Arabic literary reference is of +884. From Islam follows the usual corollary of Europe—we find the noria again in the late +11th century at work in France,[j] and depicted in +15th-century German MSS.[k] Naturally enough, it was particularly common in Spain.[l] It has persisted in some parts of Europe until today, e.g. the remoter

---

[a] Such as Colin (1), who is otherwise most learned and informative on North African water-raising machinery.

[b] x, v, 1—*sine operarum calcatura*. And Lucretius had mentioned it a few decades before him (*De Rerum Nat.* v, 516), unless this was a reference to water-mills, as some believe. In *De Archit.* x, iv, 3, Vitruvius describes the noria worked by a treadmill. Forbes (11), p. 34, believes that the animal-power noria is first mentioned in Graeco-Egyptian papyri of the −2nd century, but agrees that it was much more frequent in Roman times. We owe to Calderini (1) a special paper which seeks (not altogether successfully) to elucidate the technical terms for water-raising machinery in these papyri. According to Forbes, the current-operated noria first appears in Hellenistic Egypt in +113.

[c] Cf. Ewbank (1), p. 113. In Egypt such norias are called *tambusha* (Hurst, 1).

[d] *De Archit.* x, iv, 1; cf. Drachmann (9), p. 150.

[e] Cf. Lane (1), p. 301. It is always worked by men or animals, never by the force of flowing water. The Romans used it for draining copper-mines in Spain (Baroja (3, 4); Forbes (11), p. 44).

[f] (1), p. 80.

[g] Descriptions in Weulersse (1), pp. 31, 55, (2), fig. 82, pls. 70, 71, (3), p. 256, figs. 50, 51, 52; Dubertret & Weulersse (1), fig. 60; Eddé (1), p. 112; M. O. Williams (1), p. 749.

[h] (1), p. 155; cf. Suter (1), no. 358; Sarton (1), vol. 2, p. 623.

[i] See particularly Wiedemann & Hauser (1); Schmeller (1).

[j] Gille (7), p. 8.

[k] Berthelot (4). Among later references see Alberti (+1512), edition of Paris, +1553, p. 74; Worlidge (+1669); de Bélidor (+1737), vol. 1, pl. 39, vol. 3, pl. 2; Perronet (+1788) reproduced by Straub (1), fig. 35, etc. Eighteen norias were still working between Nürnberg and Forchheim as late as 1956 (Deutsches Museum).   [l] Baroja (3).

sections of Bavaria,[a] and in Bulgaria[b] where the design is very Chinese indeed. And it was not altogether absent in the 19th-century iron age, for steam-driven norias delivering 2000 gallons on each revolution were at work in the eighties for emptying copper residues into Lake Superior.[c]

It seems, then, that we are presented by the noria with a problem similar to that which we shall encounter in the case of the water-mill, namely first appearances separated by rather short time-intervals at the two extremes of the Old World. The problems are of course closely related, since the current-operated noria with paddles is very near the vertical water-wheel for mills. Laufer (19) argued strongly for an origin of the noria in Sogdiana (Central Asian Persia) whence it would have spread in both directions, but he had really no evidence for this,[d] nor is the purely Greek origin urged by Pagliaro (1) or Forbes (11) any more convincing. The question therefore arises whether India might not be the original home of the noria.[e] That the machine was very widespread there in recent times Laufer had no difficulty in showing, but the question is how far it goes back, and that, owing to the well-known difficulty of dating Indian texts, is not easy to say. However, there are references in Pali to a *cakkavaṭṭaka* (turning wheel) which the commentaries explain as *arahatta-ghaṭī-yanta*, or machine with water-pots attached.[f] If this is really the noria and not the *sāqīya*, the mention is interesting, since the date would probably be (on traditional views) about −350.[g] Classical Sanskrit,[h] Jaina Sanskrit and Prakrit,[i] and Buddhist Sanskrit[j] also have references, as well as later Indian languages.[k] Perhaps also the comparisons between the noria and *saṃsāra-cakra*, the wheel of existence, in Buddhist literature, point to an early use of the machine.[l] Indian treatises on engineering will some day

---

[a] King (2).   [b] Wakarelski (1).

[c] See Anon. (44).

[d] The fact that the noria was so widely called the 'Persian Wheel' carries little weight for earlier periods, cf. Morgenstierne (1), pp. 259ff. Ewbank (1), p. 115, maintained that the term 'Persian Wheel' ought to be reserved exclusively for that form of the noria in which the buckets swing on pins, thus losing no water until they are overturned by a catch at the top, but I doubt whether this could be substantiated.

[e] For the information which follows I am much indebted to Sir Harold Bailey.

[f] *Vinaya*, II, 122 (ed. H. Oldenburg).

[g] Another text is *Bhagavad-Gītā*, 18, 61, which speaks of a wheel provided with buckets for irrigation as *yantrārūḍha*; cf. H. Zimmer (2), p. 394. This could perhaps be as old as the −4th century, but may well be as late as the +4th.

[h] Lexicons and *Pañcatantra* have *araghaṭṭa*, +100 to +500. The difficulty of the Pali Canon is that it is no longer considered more reliable than the Sanskrit texts, since it was thoroughly revised towards the end of the +5th century (private communication from Dr E. Conze, cf. 7).

[i] See Jacobi (2), 18, 29, Meyer tr. p. 57; *āraghaṭṭika*,

[j] The Sanskrit-Tibetan dictionary *Mahāvyutpatti* has *arhaṭa-ghaṭī-cakram* (cf. Renou & Filliozat (1), vol. 2, p. 363). Cf. Conze (6) commenting on a difficult passage in the *Abhisamayālaṃkāra* (Ornament of the Intuition of the Doctrine), VII, i, 2, a mid +4th-century text; and its +8th-century commentary by Haribhadra (cf. Renou & Filliozat (1), vol. 2, p. 377). Here again it is hard to exclude decisively the *sāqīya* as the *yantra* referred to.

[k] Cf. J. Bloch (1) for Marāṭhī, p. 393.

[l] See Masson-Oursel (2); Coomaraswamy (4); Foley (1). In the *Divyāvadāna* (−2nd to +2nd century) the Buddha instructs Ānanda to make a *mandala* like a water-wheel so as to show the cycle of rebirths. This sounds like a noria rather than a *sāqīya*. One remembers, too, the ubiquitous 'wheel of fortune' motif in European medieval literature and illustration (cf. Fig. 681), the use of the noria in philosophical analogy by Chu Hsi (p. 359 above), and the place which it took in ancient Chinese cosmological speculation (Vol. 3, p. 489 above). Cf. also Mus (1); Przyłuski (7).

throw greater light on the matter, but as yet hardly any definitive work has been done on them.[a]

Summing up, therefore, the general impression gained is the following. The characteristic Chinese water-raising machine was the square-pallet chain-pump. Originating probably in the +1st century, it knew a worldwide diffusion after the +16th. The characteristic Hellenistic and Arab water-raising machine was the *sāqīya* (vertical chain of pots); this was probably an early Alexandrian invention, and was transmitted to China from Arab lands some time before the +14th century. The noria is the most difficult to place, but provisionally we may adopt the hypothesis that it was invented in India, reaching the Hellenistic world in the −1st century and China in the +2nd.[b]

(h) POWER-SOURCES AND THEIR EMPLOYMENT (II), WATER FLOW AND DESCENT

In the previous subsection we were concerned with man exerting labour to move water about, now we approach the even more epic story of man getting water to labour for him. The invention of the current-operated noria had indeed already done this, though the place and date of its origin remains obscure, and the work accomplished was only the raising of water itself to a higher level. Who was it who first realised that the torque of the axle of the noria could be made to perform tasks beyond the power of unaided human strength, and with greater efficiency and continuity of effort than men or animals could bring? Or was the water-wheel simply a kind of extension of the horizontally rotating millstone? It will not be possible to elucidate all these problems of origin here, but we shall at least be able to contribute to the argument a mass of facts relating to the invention and use of water-wheels in East Asia which have so far not been taken into account by any historian of technology.

The term 'mill' must not be taken here in the narrow sense of the rotary quern only. We shall have to deal with tilt-hammers and trip-hammers, edge-runner mills, saw-mills, air-conditioning fans, textile machinery, and—surprisingly heading most of the others in date—metallurgical bellows operated by water-power.[c]

---

[a] Take for example the *Yaśastilaka Campū*, which contains a description of garden waterworks, written by the South Indian Jain encyclopaedist Somadeva Sūri in +949, and brought to our attention by Raghavan (1, 2). For our present purpose it is rather late, and its vagueness makes interpretation difficult, unless the collaboration of a Sanskritist with a trained engineer could clarify the nature of the machines described. This desideratum applies indeed to most of the past work and future problems of the history of engineering in the Indian culture-area.

[b] Its possible genetic relation to the equally ex-aqueous water-mill wheel will be discussed in detail below (p. 405).

[c] The only paper specifically devoted to water-power in Chinese engineering before the writing of the present section was that of Masai (1); its conclusions were generally in accord with ours. Subsequently the papers of Amano (1) and Liu Hsien-Chou (4) became available. There are no significant differences of opinion.

(i) SPOON TILT-HAMMERS

Let us start with what amounts to a riddle. How could the power of falling water be made use of without any application of continuous rotary motion? The answer is by a device which was the exact opposite of the swape (discussed above, p. 331). Instead of the bailing-bucket being counterweighted to assist the raising of the water, the counterweight was turned into a hammer or pestle and alternately raised and allowed to fall by having a stream of water pouring into the bucket at the other end of the beam[a]. This was the 'spoon tilt-hammer' (*tshao tui*,[1] or trough hammer), illustrated in all the books from the *Nung Shu* of +1313 (see Fig. 597)[b] onwards. Unfortunately, references to it in literature are very few. The *Nung Shu* quotes a poem which is hard to identify, and then much later there is a statement in the Ming dictionary *Chêng Tzu Thung*[2] by Chang Tzu-Lieh:[3]

The mountain people cut out wood in the shape of a spoon, and make it face a mountain torrent to work a water-tilthammer (*shui tui*[4]). Sometimes it works slowly and sometimes quickly, but it doubles the efficiency attainable by man-power. Ordinary people call it the 'spoon tilt-hammer' (*shao tui*[5]).[c]

This would have been written about +1600, though not printed till nearly thirty years later, but the device must have been well known already in +1145, for Lou Shou's poem (given below, p. 393) refers to the water flowing in and out of the slippery spoon (*shih*[6]). Before that we lose sight of it.

No references to it by travellers in China have been found,[d] but Troup (1) studied it carefully in Japan. The *battari*, as it was called, had a trough with a sloping bottom so as to let out the water easily when the counterbalancing effect came into play, and in its most archaic form, such as Lou Shou must have known it, was simply a hollowed-out log. Now between Arima and Sanda in the Settsu province, Troup saw a double trough which swung right round when emptying, and rotated a shaft which worked

[a] A similar balanced lever was incorporated in the coin-in-the-slot machine of Heron of Alexandria (ch. 21, see Woodcroft (1), p. 37; Vowles & Vowles (1), p. 106).

[b] Ch. 19, pp. 13*b*, 14*a*.

[c] Tr. auct. Cit. in *KCCY*, ch. 52, p. 7*a*. Cf. the *Shao-Hsing Fu Chih* (Shui Li sect.) about +1700.

[d] But Dr C. P. Fitzgerald informs me that he saw a battery of spoons operating in 1937 near Erh-yuan (on the road between Tali and Lichiang in Yunnan). He also saw them at Sangan Sa[7] in the Diamond Mountains in north-east Korea. Others have reported them from Korea (e.g. Vowles, 2) and Indo-China (Colani, 5). In Java they take the form of tilting bamboo tubes (*taluktak*) so arranged that after each discharge a kind of chime-stone is made to sound (see Kunst (5) and Crossley-Holland (2), p. 77). Besides the simple duty of warning the farmer of any interruption of flow, this apparatus has been developed by the Javanese into a veritable orchestra, with chime-stones of different pitch and bamboo tubes of different size and emptying-periodicity. So also the Sedang people of Annam make elaborate hammer-struck hydraulic carillons of bamboo (*tang koa*), working automatically in the rice fields for months without ceasing. Somehow the spoons ultimately spread to South America (Baroja (4); Mengeringhausen, J. & M.), presumably from Europe, though references to them are few. According to the Director of the Deutsches Museum at München, where a model is shown, they were used in Switzerland about 1800 for pounding stones and even for saw-mills moved on by ratchets. Their name was *wasser-anke* or *guepfe*.

[1] 槽碓　　[2] 正字通　　[3] 張自烈　　[4] 水碓　　[5] 勺碓　　[6] 匙

[7] 常安寺

Fig. 597. Water-power in its simplest form; the tipping bucket as an intermittent counter-poise. The spoon tilt-hammer (*tshao tui*) from *Nung Shu*, ch. 19, p. 13*b* (+1313).

PLATE CCXXXV

Fig. 598. An +18th-century European use for tipping buckets as intermittent counterpoises; Triewald's blowing-engine for forges and furnaces (+1736). The flume-handled spoons alternately raise and lower bell-shaped cylinders suspended in water, thus sending a continuous blast to the tuyères (Frémont, 2, 10).

vertical stamp-mill pestles by means of lugs (Fig. 599).[a] Here the rotary principle had replaced that of the simple lever.[b] Troup discussed this with the anthropologist Tylor, who asked whether four-spoked examples were known; at the time Troup had not seen any of this kind, but later he found them in the Tonegawa Valley in Jōshū province. The two additional spokes were light, and carried only flat boards with raised rims round the edges, serving merely to assist the main spokes to get round so as to place the troughs under the stream of water. Troup also saw others in which all

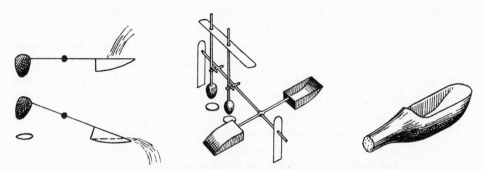

Fig. 599. Possible ancestry of the water-wheel. Left, the simple spoon tilt-hammer (*tshao tui, battari*) for pounding rice, etc.; right, its most archaic form, a hollowed-out log. In the centre, a double-trough example working a vertical stamp-mill (in Settsu province, Japan, after Troup, 1). Since forms with four, six and eight spokes, with pallets and troughs, have been traditional, some connection with the water-wheel is not unlikely.

attempt to mount troughs on the subsidiary spokes had been abandoned, and they carried only flat pallets or paddles. Finally he succeeded in finding and sketching machines with four, six and even eight spokes ending in troughs, all the machines operating stamp-mill pestles by lugs on the shaft.

The relevance of these discoveries for the theory of the origin of the water-mill wheel is obvious enough, but it will be best to postpone its consideration until the conclusion of the argument, when we can view the entire development of water-mills in perspective.

[a] They were said also to exist in several other provinces.
[b] In the +18th century the spoon principle was ingeniously applied by the Swede Martin Triewald to working a blowing-engine for furnaces and forges (Fig. 598). A pair of flume-handled 'spoons' (acting in a converse way to flume-beamed swapes, cf. p. 334) alternately depressed and raised bell-shaped cylinders suspended in a pond with their lower ends open, so that the water, forming a piston, expelled the air through pipes towards the places where the blast was needed (see Frémont (2), p. 158, (10), p. 80). This design dates from +1736, and found much use in undertakings such as the great Falun mine and copper works throughout the rest of the century (cf. Lindroth, 1). Of course the use of water as a piston in this way goes back to the Alexandrians; cf. Philon, *Pneumatica*, ch. 64; Beck (5), p. 75; Drachmann (2), pp. 6ff. And a double bell-blower like that of Triewald, though not worked by water-power, was sketched by Leonardo (Feldhaus (18), p. 46; Beck (1), p. 341). The spoons of the serious-minded Swedes did heavy work, but a facetious Austrian archbishop set them to another task. At the château of Hellbrunn near Salzburg (built in +1614 by Marcus Sitticus of Ems), the numerous water-sports (cf. p. 157) include a face with a periodically protruding tongue—supposedly at the archbishop's critics. The movement is effected by a small stream of water filling a periodically emptying cup, to which the tongue is attached on a pivot. Here we have a connection between the workaday tilt-hammers and the jack-work tipping buckets of the Arab makers of striking water-clocks (cf. pp. 534ff. below). Either the periodic discharge of water, or the work done by the other end of the arm, may be the more important according to the conditions.

(2) WATER-WHEELS IN WEST AND EAST

It is now necessary to sketch very briefly the essentials of what is known about the appearance and spread of the water-wheel in the West. An unsurpassed mine of information is contained in the book of Bennett & Elton,[a] and there are two classical papers which must be mentioned, those of Bloch (2)[b] and Curwen (4). These, and the literature to which they give access, will elaborate, for those who desire it, the basic facts now to be summarised. The oldest known water-mill (a water-wheel working rotary grinding-stones for cereals) in the West was the *hydraletes* ($\dot{v}\delta\rho\alpha\lambda\epsilon\tau\eta s$)[c] described by Strabo[d] about -24 as existing at Cabeira in the Pontus, having formed part of the property which the last Mithridates (Eupator)[e] lost when he was overthrown by Pompey in -65. The first literary reference occurs in a Greek epigram[f] attributed to Antipater of Thessalonica, which might be dated about -30:

> Women who toil at the querns, cease now your grinding;
> Sleep late though the crowing of cocks announces the dawn.
> Your task is now for the nymphs, by command of Demeter,
> And leaping down on the top of the wheel, they turn it,
> Axle and whirling spokes together revolving and causing
> The heavy and hollow Nisyrian stones to grind above.
> So shall we taste the joys of the golden age
> And feast on Demeter's gifts without ransom of labour.

Then comes in -27 the matter-of-fact description in Vitruvius,[g] whose specification has often been illustrated.[h] Pliny's reference (*c.* $+75$) is a more than a little obscure; he says[i] that 'in the greater part of Italy falling pestles are used, and wheels also that water turns as it flows along, and so they mill'. From this it is not clear whether stamp-mill hammers or rotary querns were meant.[j] In the $+3$rd century there were apparently mills along Hadrian's wall[k] and on the tributaries of the Moselle,[l] and in the $+4$th we begin to read of numerous mills, e.g. at Arles[m] and on the hill of Jani-

---

[a] (1), vol. 2. To this Moritz (1) may now be added, and P. N. Wilson (1, 2).
[b] Note additions by Gille (7).   [c] I.e. 'water-grinder'.
[d] XII, 3, 30 (C 556) Jones tr. vol. 5, p. 429.   [e] Biographies by T. Reinach (1) and Duggan (1).
[f] *Greek Anthology*, IX, 418 (Loeb ed., vol. 3, p. 233); tr. auct. adjuv. Curwen (4), etc.
[g] X, V, 2.
[h] Perrault (1), p. 315; Usher (1), 1st ed., p. 124; Bennett & Elton (1), vol. 2, p. 33. Evidence of a water-mill of some kind in Jutland *c.* $\pm$o has been published by Steensberg (1), but it is hard to visualise since nothing was left but the stones of the presumed mill-races. The evidence is however accepted by many archaeologists (Lynn White (7), pp. 81, 160).
  [i] *Nat. Hist.* XVIII, xxiii, 97, reading *ruente* for *ruido*, as d'Arcy Thompson suggested in Vowles (2).
  [j] Cf. Bennett & Elton (1), vol. 1, p. 102; Moritz (1), p. 135.
  [k] Cf. Moritz (1), pp. 136ff.   [l] See below, p. 404.
  [m] Benoit (1); Gille (7); Forbes (19), p. 598. This was the notable factory at Barbegal, which with its sixteen overshot wheels in two descending rows handled 28 tons of flour a day. The date of its construction seems to have been about $+310$. I had the opportunity of paying a personal visit to this site in September 1959, with Dr Dorothy Needham, and could observe the rock cutting some 7 ft. deep which supplied the water at the top of the south-facing slope, to which it had been conveyed by an aqueduct many arches of which still stand. Here again of course nothing is left of the machinery, but the disposition of the masonry indicates vertical Vitruvian water-wheels without possibility of error. To the Rev. Fr François du Roure, Canon of Aix-en-Provence, who resides at the Château de Barbegal, I express warmest thanks for his kind assistance.

culum in Rome, to which most of the water of the aqueduct of Trajan was devoted.[a]
The Justinian Code (+6th century) has much on mills.[b] The water-mill reached
Germany about +770[c] and England again by +838[d] at least. Such is the outline of
its story in the West.

Here, however, we come to an important
distinction. The mill of Vitruvius was quite
unmistakably a vertical undershot wheel
connected by right-angle gearing with the
horizontally placed millstones (Fig. 600),
i.e. a vertical mill.[e] But this is not the
only possible arrangement, and indeed over
large parts of modern Europe, as well as
the Europe we know by historical documents, it is clear that the only mills used
were horizontal mills, i.e. mills in which
the water-wheel was mounted horizontally
in the stream, and connected directly, without gearing, with the upper millstone above
(Fig. 601).[f] Curwen (4) argues that the
mill of Mithridates was of this type, for
nothing is said about gear-wheels, and the
nymphs leaping down on the top of the

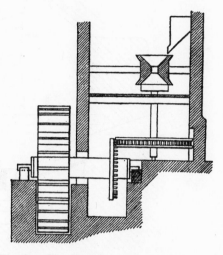

Fig. 600. The Vitruvian mill, a vertical water-
wheel working the mill-stones by right-angle
gearing (Usher, 1).

wheel can hardly have been leaping down on to an overshot vertical wheel since this
type did not appear until the +5th century. One such (built about +470) has been
excavated in the Athenian agora.[g] But the only paintings or mosaics showing mill
water-wheels of the +5th century or earlier are clearly undershot.[h] The action of the
nymphs would agree, however, with the position of the spokes and paddles of a
horizontal wheel, since this was generally arranged so that the water poured down on
to it; a disposition which is frequently seen in the Chinese representations, some of
which are reproduced below.

The distribution of the two types of mill has been acutely discussed by Curwen (4).

[a] Bloch (2), p. 545; Bennett & Elton (1), vol. 2, p. 39.
[b] Bennett & Elton (1), vol. 2, p. 41.
[c] Bloch (2).
[d] Bloch (2); Gille (4); Bennett & Elton (1), vol. 2, p. 97, prefer a charter of +762. Hodgen (1) finds
5624 water-mills in the *Domesday Book* of +1086. The earliest medieval illustration of a water-mill is
French, of the +12th century (Bennett & Elton (1), vol. 2, p. 73).
[e] We retain here the terminology of Bennett & Elton; Curwen unfortunately reversed it, making the
adjective refer to the shaft and not to the wheel. Vertical water-wheels were known to the earlier
Alexandrians, such as Philon, but only for making musical noises by means of the displacement of
air through holes (cf. Schmeller (1); Usher (1), 2nd ed. p. 162; Beck (5), pp. 73 ff.; Drachmann (2),
pp. 64 ff).
[f] For reasons which will shortly be obvious, we cannot adopt the terms 'Greek mill' or 'Norse mill',
as used by Bennett & Elton, for this type. An excellent monograph on it by P. N. Wilson (4) is now
available.
[g] A. W. Parsons (1). A good reconstruction of it is given in Storck & Teague (1) and Forbes (11),
p. 92.
[h] Brett (1); Profumo (1).

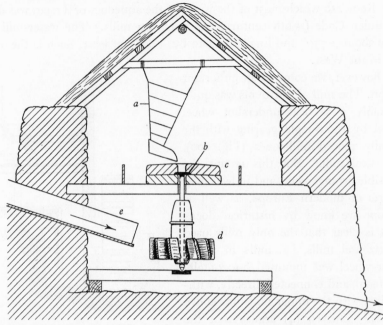

Fig. 601. The horizontal water-wheel, an example from the isle of Lewis (after Curwen, 4). *a*, hopper; *b*, rynd; *c*, stones; *d*, wheel or tirl, with obliquely-set paddles on thick shaft; *e*, chute delivering the water to the side of the wheel behind the shaft, as here viewed.

The early spread of the vertical water-wheel was mainly northwards,[a] and it became in due course characteristic of France, Germany, England and Wales. But the horizontal water-wheel made a peripheral perambulation; it is attested for Lebanon and Syria, Israel,[b] Thessalonica,[c] Yugoslavia,[d] Rumania,[e] Greece, Italy, Provence,[f] Spain, then Ireland,[g] the Shetlands, Scotland,[h] the Faroes,[i] Denmark,[j] Norway and Sweden.[k] What makes the pattern still more interesting is that everywhere east of Syria the horizontal wheel predominated, e.g. in Persia (J. R. Allen, in Goudie, 1), in Turkestan (Cable & French), in the Hunza region of the Himalayas (Morden), and in China. Many travellers have recorded it from there, such as I. L. B. Bishop at the end of the last century;[l] the numerous water-mills of the Chhêngtu plain[m] I have myself observed, and successions of them up to 100 or so in the same small valley have

---

[a] Perhaps because the Goths were impressed with the water-wheels, necessarily vertical, which were mounted on boats in the Tiber in the +6th century (see below, p. 408). Overshot wheels began to predominate in northern Europe from the +14th century onwards.

[b] Avitsur (1); Dalman (1), vol. 3, pp. 245ff., fig. 64.　　　　[c] Curwen (5).

[d] Especially in Bosnia and Hercegovina, where I saw many horizontal water-mills in the summer of 1961; e.g. near Mostar, and at Titograd in Montenegro. Good photographs of the famous mills of Jajce have been published by Pilja (1).

[e] E. Fischer (1), often with spoon-shaped vanes.　　　　[f] De Bélidor (1), vol. 1, pl. 22 and p. 301.

[g] O'Reilly (1); A. T. Lucas (1).　　　　[h] Goudie (1); Dickinson & Straker (1).

[i] K. Williamson (1).　　　　[j] Steensberg (1).

[k] It spread also to the South of France, giving rise to some famous mills at Toulouse (de Bélidor (1), vol. 1, pl. 23; Sicard (1); Bennett & Elton (1), vol. 2, p. 26) and was described by Ramelli in +1588, pl. 115 (W. B. Parsons (2), p. 131).

[l] Personal communication recorded in Bennett & Elton (1), vol. 2, p. 26.

[m] Hosie (4), p. 88. See Figs. 623, 624.

been described, e.g. for Shansi,[a] Kansu[b] and Yunnan.[c] Moreover, it is the commoner form in the Chinese illustrations. Whether or not the mill of Mithridates was really a horizontal mill (as seems likely), the invention was neither Greek nor Roman,[d] and the possibility presents itself that, together with the whole procession of horizontal mills, its origins were further east. In any case, it is distinctly simpler than the vertical mill-wheel, which needed gearing.[e] Whether this is evidence of primitivity or regression we may discuss later. Here we need only add that in the horizontal wheels of most regions, as modern observers have found them, there is either an oblique setting of the paddle-boards (at a 'pitch' angle of some 70°); or else some form of scooping or cupping of the vanes, which gives a similar effect but may remind us of Troup's observations.[f] One would like to know what kind of designs for mills Metrodorus, a Hellenised Persian, took to India in the +4th century, when the water-wheel was regarded there as a new invention.[g]

(3) THE METALLURGICAL BLOWING-ENGINES OF THE HAN AND SUNG

When we ask about the earliest water-wheel in Chinese history[h] we come upon the paradox that it was not used for turning simple millstones, but for the complicated

[a] Miss F. French, personal communication recorded in Curwen (4). Cf. A. Williamson (1) for the border of Chihli and Shansi.

[b] Geil (3), with illustration opp. p. 284; the neighbourhood of Sining.

[c] Dr Dorothy Needham, personal communication, referring to Hsichow, near Tali, on the Erh Hai Lake.

[d] The Pontic kingdom was mainly Persian in culture, only imperfectly Hellenised by the Greek colony city-states on its shores (Rostovtzev & Ormerod (1); Ormerod & Cary (1)). Significance may well attach in the present context to the fact that it was one of the most important mining centres of the Hellenistic world, inheriting the traditions of the iron-working Hittites.

[e] Horizontal water-wheels can be found without difficulty in the designs of the +15th-century engineers. The Hussite has one for a cereal mill; cf. Gille (12); Feldhaus (25). Leonardo applies one with cup-shaped paddles to his rolling mill for making wrought-iron staves of rhomboidal cross-section to be welded into bombard barrels; cf. Beck (1), pp. 430, 432; Feldhaus (25, 26).

[f] Broadly speaking, the paddles are hollow and spoon-like in Turkey, the Lebanon, Greece, Yugoslavia, Provence, Spain and Ireland; flat and upright or oblique in Scandinavia and Scotland; cf. Curwen (4); P. N. Wilson (4). If the oblique flat paddle is to be considered more primitive than the spoon-shaped paddle, then the route of diffusion would have been anti-clockwise round the periphery of the European heart-land rather than clockwise, as usually assumed. Yet one would have to be a very patriotic Irishman to believe that the type of horizontal water-wheel found in the Lebanon derived from Western Celtic regions. It would be more reasonable to envisage two streams of diffusion: an earlier flow of flat paddles anti-clockwise ending in Scotland, and a later clockwise flow of shaped ones ending in Ireland. In any case it is striking that the vanes or paddles of Chinese horizontal water-wheels seem always to have been inserted at right angles to the plane of the wheel—a primitive trait which may indicate great antiquity. Shrouding, i.e. the provision of a rim surrounding the extremities of the blades, appears sporadically over the whole range from China to the Shetlands.

[g] This information, from the Byzantine historian Cedrenos, and from Ammianus Marcellinus, is discussed, with the references, by Bloch (2), p. 540.

[h] We refer to the first specific description. Presently (p. 392) we shall quote a general statement of +20 showing that water-power was then well known for trip-hammer pounding, and that its introduction had taken place some time before. There has been a general tendency to take +31 as the primary date, because the 'water-mill' is associated in the mind with the rotation of driven wheels, as in the Vitruvian grain-mill, and the +14th-century 'water-blowers' also had this (see p. 371 below). But the blowing-engines of the +1st century, as we shall see, may well have used only a shaft with lugs. In this case they involved no principle more advanced than that of the 'water trip-hammers', and the power-producing water-wheel as such must thus be referred to the −1st century, following Huan Than's words (p. 392). The difference in time between its first appearance in Asia Minor and in China is consequently reduced to a minimum.

job of blowing metallurgical bellows. This must mean that there was a tradition of millwrights going back some considerable time before, even though we cannot trace it in literary references. The essential texts run as follows; first, the *Hou Han Shu*:[a]

> In the seventh year of the Chien-Wu reign-period (+31) Tu Shih[1] was posted to be Prefect of Nanyang. He was a generous man and his policies were peaceful; he destroyed evil-doers and established the dignity (of his office). Good at planning, he loved the common people and wished to save their labour.[b] He invented a water-power reciprocator (*shui phai*[2]) for the casting of (iron) agricultural implements.
>
> [Comm.] Those who smelted and cast already had the push-bellows to blow up their charcoal fires, and now they were instructed to use the rushing of the water (*chi shui*[3]) to operate it....[c]
>
> Thus the people got great benefit for little labour. They found the 'water(-powered) bellows' convenient and adopted it widely.

This advanced mechanism comes, therefore, between the dates of Vitruvius and Pliny. The tradition of Tu Shih and his engineers must have persisted in Nanyang, for it was a Nanyang man who spread the knowledge of the technique when he became prominent as an official two centuries later.[d] This we know from the *San Kuo Chih*, which says:[e]

> Han Chi,[4] when Prefect of Lo-ling,[f] was made Superintendent of Metallurgical Production. The old method was to use horse-power for the blowing-engines, and each picul of refined wrought (iron) took the work of a hundred horses. Man-power was also used, but that too was exceedingly expensive. So Han Chi adapted the furnace bellows to the use of ever-flowing water, and an efficiency three times greater than before was attained. During his seven years of office, (iron) implements became very abundant. Upon receiving his report, the emperor rewarded him and gave him the title of Commander of the Metal-Workers (Ssu Chin Tu Wei[5]).

The period referred to must have been a little before +238. Some twenty years later, a new design seems to have been introduced[g] by that ingenious man Tu Yü.[6]

The story continues through the +5th century, for Phi Ling[7] wrote in his *Wu Chhang Chi*[8]:[h]

> The origin of the Pei Chi Lake is that it was (artificially constructed) for the Hsin-Hsing (iron-)smelting and casting works. At the beginning of the Yuan-Chia reign-period (+424 to

---

[a] Ch. 61, p. 3*b*; tr. auct.

[b] I see no reason for taking a superior attitude about this. Well-authenticated cases are known where humanitarian feelings have motivated invention, e.g. Jouffroy's paddle-boat of +1783, for Jouffroy had seen the sufferings of the galley-slaves (Schuhl (1), p. 53).

[c] This commentary is itself a piece of information, for it was written by Li Hsien about +670 and shows that at that time people were quite familiar with the idea of hydraulic blowing-engines. But his further remark, cf. p. 142 above, may mean that fan or piston types had not then replaced flexible leather.

[d] Nanyang had been a centre of the iron and steel industry since the Warring States period. This and all the rest of the siderurgical background to the water-powered blowing-engines will be discussed in full in Sect. 30*d*; in the meantime much is available in Needham (31, 32).

[e] Ch. 24, p. 1*b* (*Wei Shu*); tr. auct. Both these passages are quoted in *TPYL*, ch. 833, and *Nung Shu*, ch. 19, p. 6*b*.

[f] On the North China plain, north of the Yellow River but just within Shantung province.

[g] See *Chin Shu*, ch. 34, p. 9*b*, as interpreted by Yeh Chao-Han (1).

[h] Cit. *TPYL*, ch. 833, p. 3*b*; tr. auct.

[1] 杜詩   [2] 水排   [3] 激水   [4] 韓暨   [5] 司金都尉   [6] 杜預
[7] 皮零   [8] 武昌記

Fig. 602. Metallurgical blowing-engine (*shui phai*) worked by water-power; the oldest extant illustration (+1313), from *Nung Shu*, ch. 19, pp. 5*b*, 6*a*. Full description and discussion in the text. Cf. the similar mechanism in Fig. 461.

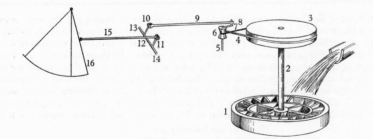

Fig. 603. Interpretation of Wang Chên's horizontal hydraulic reciprocator for iron-works (blast-furnaces, forges, etc.) and other metallurgical purposes (+1313). 1, horizontal water-wheel (*hsia wo lun*); 2, shaft (*li chu*); 3, driving-wheel (*shang wo lun*); 4, driving-belt (*hsien so*); 5, subsidiary shaft; 6, smaller wheel or pulley (*hsüan ku*); 7, eccentric lug or crank (*tiao chih*); 8, crank joint and pin; 9, connecting-rod (*hsing kuang*); 10, 11, rocking roller bell-crank levers (*phan erh*); 12, rocking roller (*wo chu*); 13, 14, bearings; 15, piston-rod (*chih mu*); 16, fan bellows (*mu shan*). We see here a conversion of rotary to longitudinal reciprocating motion in a heavy-duty machine by the classical method later characteristic of the steam-engine, transmission of power taking place, however, in the reverse direction. Thus the great historical significance of this mechanism lies in its morphological paternity of steam power. Both excursions of the bellows are fully mechanised in this design.

$c.+429$) there was a great development of water-power for blowing bellows for metallurgical purposes. But later, Yen Mao,[1] finding that the earthworks of the lake leaked and that it was not much good, destroyed them, substituting man-power bellows (*jen ku phai*[2]), so that it was called the 'treadmill bellows' (*pu yeh*[3]) lake. Now it has got so much out of repair that it cannot be used for smelting, and dries up altogether in winter.

Moreover the *Shui Ching Chu* quotes[a] Tai Tsu's[4] *Hsi Chêng Chi*,[5] a record of the campaign against Yao Hsing, which was written about +420, as saying that the Ku Shui[6] river[b] had been used for the iron industry. At a certain place there were

Fig. 604. A late form of fan bellows (*mu shan*), the *tatara* blower of traditional Japanese iron-works (an +18th-century illustration from Ledebur, 2). See a +5th-century Chinese reference to 'treadmill-bellows', probably something like this, at the top of the page. An +11th-century picture of the fan bellows has been given in Fig. 430 above, very like that in Fig. 602. All these have what are essentially hinged pistons making their excursion in an arc. The usual combination in pairs assured the continuity of the blast. The script accompanying this picture is headed *tetsu no tatara*, 'the foot-bellows for iron'. The right-hand column says: 'ordinary hand-bellows cannot be used for blowing iron, so we use the foot-bellows to melt it till it runs like hot water'. The centre column is a proverb in rhyme to the effect that nothing is so hard that hard work won't soften it. On the left, another verse, somewhat as follows:

> Those of the foundry art
> Right humble men of heart
> So self-forgetful blow,
> Troubles and cares depart,
> Till hard iron treated so
> Doth like hot water flow.

We are indebted to Miss Sakamoto Suzuko for the elucidation of this legend.

[a] Ch. 16, p. 2*b*. We owe this reference to Yang Khuan (*1*).
[b] It runs from the west into the Lo river some distance south-east of the city of Loyang. Cf. pp. 346, 403.

[1] 顏茂   [2] 人鼓排   [3] 步冶   [4] 戴祚   [5] 西征記   [6] 穀水

staithes or quays with machinery for smelting by water-power (*shui yeh*[1]) and an office of the government Iron Authority. Thus we see, says Li Tao-Yuan, that the Ku Shui was used for driving metallurgical blowers. Later on, further mentions may be found in the Thang geography, *Yuan-Ho Chün Hsien Thu Chih* of +814, and probably every century would yield them on investigation.

Machines of this kind are illustrated in nearly all the relevant Chinese books from the time of the *Nung Shu* onwards (+1313), not only for metallurgical bellows, but also for operating flour-sifters and any other machinery requiring a longitudinal motion (cf. Figs. 461, 627b). They have been seen in their traditional form by modern travellers.[a] Their historical importance is due to the fact that from some time in the +1st millennium they embodied the standard conversion of rotary to longitudinal reciprocating motion in heavy-duty machines. For this reason, though the exposition somewhat disturbs our pursuit of the cereal water-mill, it seems necessary to deal with the problem now.

From the time of the earliest illustrations (Fig. 602 shows that given by the *Nung Shu*) this form of conversion of rotary to rectilinear motion is present. At first sight, however, owing to the many mistakes made by the artist, it is almost impossible to make out from this how the machine worked.[b] But by the aid of Wang Chên's *Nung Shu* text, together with the illustrations redrawn in later books, especially the *Thu Shu Chi Chhêng* encyclopaedia[c] and the *Shou Shih Thung Khao* (cf. Fig. 461), a clear understanding of the mechanism can be reached. We may summarise it as follows (see Fig. 603). The motive power came from a horizontal water-wheel, the shaft of which bore at the top a driving-wheel of similar size. Upon this an eccentric could have been mounted directly if the bearings of the wheel had not been built above it. Since that was preferred, a smaller wheel or pulley[d] was mounted alongside in a second frame, and driven by a crossed driving-belt working off the large fly-wheel. This secondary shaft bore, above its bearings, an eccentric lug[e] which connected by

[a] Rocher (1), vol. 2, pp. 196ff., whose description almost exactly parallels that in the *Nung Shu*; Hosie (4), p. 96; A. Williamson (1); Hommel (1), p. 86, in part.

[b] Feldhaus (10) had to stop at this point, but he deserves credit for drawing attention to the importance of the machine, which he knew only from the *San Tshai Thu Hui*.

[c] *I shu tien*, ch. 6, *hui khao* 4, pp. 2a and 8b; the latter works a flour-sifter and is identical with *Shou Shih Thung Khao*, ch. 40, p. 31a.

[d] Throughout this iconographic tradition the artists disguised the small pulley or pinion as something which looks like two or three bubbles. This has led my friend Prof. Aubrey Burstall to conjecture that no driving-belt was present, and that something more like a friction-drive was used, the small wheel being bound round with rope (or with leather stuffed with feathers, etc.) and thus constituting an elastic pinion, rotated by compression against lantern-gear bars on the large driving-wheel. In his interesting working model of the whole machine he has adopted this interpretation. At present we are regretfully unable to do so, for apart from numerous subsidiary arguments we feel that Wang Chên's text implies the presence of a driving-belt as clearly as can be expected in medieval Chinese. On the other hand we can offer no real explanation for the 'bubble' form given to the small pulley or pinion in all the Chinese illustrations of water-powered rotary–longitudinal conversion machinery (cf. the flour-sifter in Fig. 461 and the *TSCC* version of the metallurgical blowing-engine (*I shu tien*, ch. 6, p. 2a, reproduced in Needham (41), fig. 4). See however p. 377 below. One is inclined to wonder why there is no indication of the employment of gearing, as e.g. in Fig. 582, in any of these machines, for the Chinese were adept in its use (cf. pp. 398, 450 ff. below).

[e] In the Section on textile technology we shall see other important examples of just this device, together with clear evidence that it goes back at least to the +11th century. Cf. p. 382 below.

[1] 水冶

means of a rod and working joint with a rocking roller not unlike a bell-crank,[a] this in its turn operating the pole or piston-rod of the bellows itself through another working joint.[b]   Evidence has been given above (pp. 140 ff.) that at least during the Sung, the blowing apparatus in question was some large type of fan- or piston-bellows. Considerable increase in velocity was of course gained by having the eccentric mounted on the small wheel instead of the large one,[c] and this was no doubt purposely intended, in which case the securing of the main drive-wheel by upper instead of lower bearings would have been deliberate.[d]

Now let us compare this with what Wang Chên himself says in a most interesting passage:[e]

According to modern study (+1313!), leather bag bellows (*wei nang*[1]) were used in olden times, but now they always use wooden fan (bellows) (*mu shan*[2]).[f] The design is as follows. A place beside a rushing torrent is selected, and a vertical shaft (*li chu*[3]) is set up in a framework with two horizontal wheels (*wo lun*[4]) so that the lower one is rotated by the force of the water (*yung shui chi chuan*[5]). The upper one is connected by a driving-belt (*hsien so*[6]) to a

[a] The artist failed to bring out the connection of the eccentric with the lug, and inexcusably made the attachment of the connecting-rod's lever as if it was with the piston-rod itself, instead of fitting it into the rocking roller.

[b] Rocking levers on shafts became very popular in Europe from the +16th century onwards for raising water in flumes (cf. Beck (1), pp. 366 ff.) or for transmitting power from waterfalls to hoists and pumps (cf. Lindroth (1), vol. 1, figs. 83–5, 88–91, 248; Jespersen (2), pp. 66 ff.). See further on p. 379 below. This was the *Stangenkunst* already mentioned (p. 334 and p. 351).

[c] The number of strokes of the piston-rod for each revolution of the water-wheel would be much augmented.

[d] We were glad to find, long after writing this, that the general interpretation of Wang Chên's machinery by Yang Khuan (6), p. 57, agrees fully with our own. He adds the interesting point that in the similar crossed-belt horse-whim mill described elsewhere in the *Nung Shu* (ch. 16, pp. 6a, b) the smaller wheel is said to rotate 15 times for each revolution of the driving-wheel. Cf. Fig. 450.

[e] *Nung Shu*, ch. 19, pp. 6b ff.; tr. auct. The details of the machines described by Wang Chên have been studied by several Chinese scholars. Liu Hsien-Chou (1, 4, 7) gives a diagrammatic plan of a similar horse-power bellows, Wang Chia-Chhi (1) and Liu Hsien-Chou (7) show photographs of the model of the hydraulic bellows now in the Imperial Palace Museum which I had the pleasure of seeing in 1958, and Yang Khuan (1, 6) and Li Chhung-Chou (1) have concentrated further on identifying the technical terms in Wang Chên's text. On this their agreement is not complete, and we have adopted the interpretation which seems to us the best.

[f] I think it is certain that Wang Chên is not referring to any kind of rotary fan (as in the winnowing-fan), but is thinking of the piston itself. This may well have been a hinged one, as shown in the traditional drawing, but in any case valves on the intake and output sides of the bellows would have been necessary. The bellows in his illustration, and in all those which derive from it, are somewhat like the cuneate wood bellows of the European +16th century (cf. p. 150 above and Johannsen (2), p. 91), shaped similarly to the wood-and-leather bellows of earlier times (cf. Johannsen (2), pp. 27, 34). An older illustration is seen in a picture of iron-working among the frescoes of the Wan-fo-hsia cave-temples in Kansu (Fig. 430) which are of the Hsi-Hsia period (+986 to +1227); see Tuan Wên-Chieh (1), fig. 18a. But we have found no textual description. Of course descriptions of the straight piston-bellows are also extremely rare. We shall have to return to this subject in Sect. 30d on iron metallurgy, but meanwhile we may draw attention to the fact that the 'hinged piston' fan-bellows of medieval China also have some similarity to the *tatara* bellows of Japan worked by men as treadmills in a see-saw rocking motion (Figs. 604, 605). A very similar form was in operation at the great copper works of the Falun mine in Dalarna (Sweden) in 1816 (Lindroth (1), vol. 2, figs. 76, 77, pp. 162 ff.). The *pu yeh* or treadmill bellows mentioned in the 5th-century passage quoted on p. 372 above may be an early reference to fan-bellows something like *tatara*. For a 'hinged piston' device used by Leonardo, see p. 383 below.

¹ 韋囊   ² 木扇   ³ 立軸   ⁴ 臥輪   ⁵ 用水激轉   ⁶ 絃索

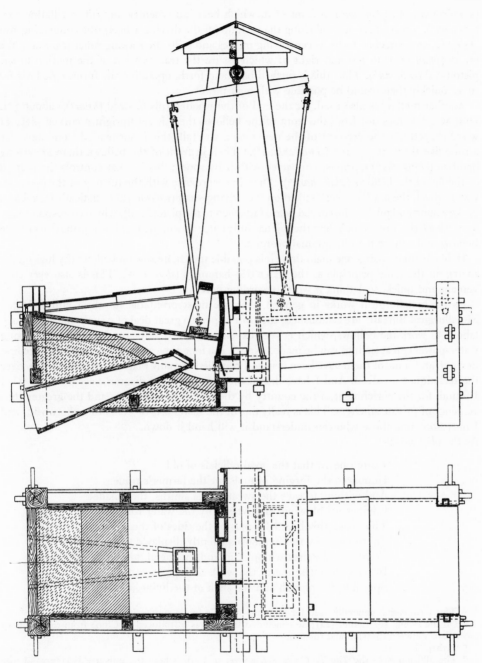

Fig. 605. Elevation and plan of a *tatara* blower (Ledebur, 2). As each fan or hinged piston descends the air is expelled through the central channel (at right angles to the plane of the paper) to the *tuyères*; as it rises air is admitted through the valve at the top of the intake duct (seen in sagittal section, above, left, in the plane of the paper). The two fans, which measure about 5 ft. × 3 ft., are so coupled that as one expels the other takes in, resembling thus the classical double-acting piston-bellows (Figs. 427, 428) and giving a continuous blast. Cf. p. 139.

(smaller) wheel (*hsüan ku*[1]) in front of it, which bears an eccentric lug (lit. oscillating rod, *tiao chih*[2]). Then all as one, following the turning (of the driving-wheel), the connecting-rod (*hsing kuang*[3]) attached to the eccentric lug pushes and pulls the rocking roller (*wo chu*[4]), the levers (*phan erh*[5]) to left and right of which assure the transmission of the motion to the piston-rod (*chih mu*[6]). Thus this is pushed back and forth, operating the furnace bellows far more quickly than would be possible with man-power.

Another method is also used. At the end of the wooden (piston-)rod (*hsün*[7]),[a] about 3 ft. long, which comes out from the front of the bellows, there is set upright a curved piece of wood shaped like the crescent of the new moon, and (all) this is suspended from above by a rope like those of a swing (*chhiu chhien*[8]).[b] Then in front of the bellows there are strong bamboo (*ching chu*[9]) (springs) connected with it by ropes; this is what controls the motion of the fan of the bellows (*phai shan*[10]). Then in accordance with the turning of the (vertical) water-wheel, the lug (*kuai mu*[11]) fixed on the driving-shaft (*wo chu*[4]) automatically (*tzu jan*[12]) presses upon and pushes the curved board (attached to the piston-rod), which correspondingly moves back (lit. inwards). When the lug has fully come down, the bamboo (springs) act on the bellows and restore it to its original position.

In like manner, using one main drive it is possible to actuate several bellows (by lugs on the shaft), on the same principle as the water trip-hammers (*shui tui*[13]). This is also very convenient and quick, so I think it worth recording here.

As for metallurgical works in general, they are most profitable for the country. When Metallurgical Bureaux are established, they often spend a great deal of money and hire much labour to work the bellows, which is very expensive indeed. But by these methods (using water-power) great savings can be made. Now it is a long time since the inventions were first devised, and some of them have been lost, so I travelled to many places to explore and recover the techniques involved. And I have drawn the accompanying diagrams according to what I found, for the enrichment of the country by the official metallurgists and the greater convenience of private smelters. This is really one of the secret arts of benefiting the world, and I only hope that those who can understand it will hand it down.

As the poet says:[c]

'I often heard that the good officials of old
Honoured the forging of tools for the farmer's trade
And wishing to save the sweat of the smiths, they made
Engines for blowing, and wheels by the water roll'd.
This blast, this breath, is part of the tides of the world[d]
As we know by the Symbols anciently display'd.[e]
Come, see the fires whipp'd up and the metal flow
Into its channels; merry the work shall go,
Speeding the plough and the stalk of the living jade.'[f]

---

[a] This term is that generally used for the cross-bar of a frame supporting bells or gongs.

[b] On the swing, see Laufer (22); he concluded that it was introduced to China from the northern barbarians between the Han and the Sui.

[c] Perhaps himself?

[d] The allusion is to the *Tao Tê Ching* (cf. above, p. 139), where the universe is compared to a bellows.

[e] The hexagrams Sun (no. 57) and Li (no. 30); see above, p. 143 and Table 14, Vol. 2.

[f] The translation of this poem is somewhat free.

[1] 旋鼓	[2] 掉枝	[3] 行桃	[4] 臥軸	[5] 攀耳	[6] 直木
[7] 橐	[8] 鞦韆	[9] 勁竹	[10] 排扇	[11] 枴木	[12] 自然
[13] 水碓					

The first paragraph requires no comment, and can be easily understood from the diagram in Fig. 603.[a] The second system, however, though not illustrated in any of the books, is also quite significant, for it describes a type of blowing-engine which used only cams for the conversion of rotary to longitudinal motion. Li Chhung-Chou (1) was the first to see that it depended on a vertically-mounted water-wheel[b] and to realise that Wang Chên was using the term *wo chu* in two quite different senses, first for a horizontal rocking roller like a bell-crank pin and secondly for a horizontal main driving-shaft. His reconstruction was however unsatisfactory and we illustrate (Fig. 606) that of Yang Khuan (9), in which the crescent-shaped boards are pushed back by the lugs as they come round in their arcs, the return excursion being effected by the bamboo springs.[c] The general arrangement was thus the same as that of Villard de Honnecourt's water-powered saw-mill, more or less contemporary in date.

All this, as Wang Chên says, was similar to what we see in Figs. 617 and 618 below, depicting the hydraulic trip-hammer, that simple machine in which a series of tilt-hammers are alternately raised and allowed to fall by catches or lugs rotating with the main shaft. Since it was common in Han times (p. 392) the blowers of Tu Shih

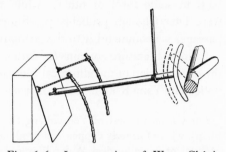

Fig. 606. Interpretation of Wang Chên's vertical hydraulic blowing-engine for iron-works (blast-furnaces, forges, etc.) and other metallurgical purposes (+1313; after Yang Khuan, 9). On the right there is a shaft with a number of projections or lugs (one only is seen in the diagram) rotated by a vertical water-wheel, just as in the common water-powered trip-hammer batteries (Fig. 617). During part of each revolution the lug pushes back a crescent-shaped board at the end of the suspended piston-rod of a fan bellows. The return excursion is effected by bamboo springs, so that only one motion may be said to be mechanised from the main power source.

[a] An interesting point arose, however, while Prof. Burstall's model was under construction. Since the plane of rotation of the small wheel with the eccentric is at right angles to the plane of oscillation of the rocking-levers, and not in the same plane as modern practice would have it, the joints at both ends of the connecting-rod have to be fitted with doubly pear-shaped pins. We do not know how this three-dimensional play could have been overcome otherwise in the +14th century. Possibly it led the original artist astray into the 'bubble' tradition, and one may note in this context the curious pear shape given to the upper bearing of the main driving-wheel in Fig. 602. Cf. pp. 89, 103, 115, Fig. 403; and p. 235 on the universal joint.

[b] In all probability, that is. The machine could have been composed of a horizontal water-wheel and right-angle gearing, but this would have been unnecessarily complicated. It is true, however, that the text does not specify any departure from the horizontal driving-wheel of the first model. Moreover, such gearing is depicted quite often in the Chinese illustrations (cf. Figs. 582, 627 b), and the fondness of medieval Chinese engineers for horizontally-mounted water-wheels (cf. p. xli) must be remembered.

[c] We accept this reconstruction not without some philological misgivings. The solution of Yang Khuan (9) necessitates taking *hsün* to mean the piston-rod instead of the upper bar of a swung frame, and has it suspended by a single rope. Such an arrangement would certainly work, but there are a few points of doubt—among others: (a) the description of the crescent-shaped board might imply 'recumbent' (*yen mu hsing ju chhu yüeh*[1]), (b) a swing normally implies a double rope, (c) the piston-rod is not termed *hsün* in the preceding passage, and in any case the word means primarily a cross-bar, (d) 3 ft. seems rather too short for such a piston-rod, and (e) if an oscillating motion alone was required the piston-rod could have been mounted on or between rollers, without need for any suspension at all. However, alternative reconstructions (see Needham, 48) necessitate departing from the text at least as far, and are less workmanlike from the engineering point of view, so we shall neglect them here.

[1] 偃木形如初月

and Han Chi were surely of this description,[a] but there is always the difficulty that horizontally mounted mill-wheels seem in general to have been more characteristically Chinese than the vertical Vitruvian ones.[b] There is no real answer to this problem as yet. We can at any rate be fairly sure that the bellows of those early times were leather bags in some form or other,[c] while the fan-bellows and straight piston-bellows were later (though probably pre-Sung) introductions. As for the last part of the passage, we commend it to the economic historians.

Chinese literature contains (or rather contained) at least one monograph devoted to the construction of metallurgical blowing-engines. The following curious passage was noticed by Yang Khuan[d] in the *Anyang Hsien Chih*[1] (Topography of Anyang District):

> Forty *li* north-west of Yeh-chhêng there is the place called Copper Mountain (Thung-Shan), where formerly copper was produced. According to the *Shui Yeh Chiu Ching*[2] (Old Manual of Metallurgical Water-Power Technology) water was canalised to blow furnace bellows here in the Later Wei period (+5th and +6th centuries). This was called 'water-power smelting' (*shui yeh*[3]). It was said to have been established (here) by the Director of the Ministries Department Kao Lung-Chih.[4] (The water-wheels were) 1 ft. deep (i.e. broad) and a pace and a half (i.e. 7½ ft.) in diameter.

The appearance of Kao here is not at all surprising. Born in +494, he was an eminent architect, engineer and city builder of the Northern Chhi dynasty, well known for the many hydraulic engineering works which he carried out before his death in +554. He was particularly remembered for the construction of water-mills working many different kinds of machinery.[e] One would give a good deal to recover the manual mentioned by the author of the *Anyang Hsien Chih*.

It is needless to emphasise the outstanding importance of power-driven bellows and forges for the metallurgy of iron, and the remarkably early successes of the Chinese in cast-iron technology cannot be unconnected with the machinery here described.[f] The hydraulic blowers of Tu Shih and his successors, to which there was nothing comparable in the Graeco-Roman world, had indeed a glorious future before them.[g] It is now clear that the use of water-power for metallurgical purposes began

---

[a] We see little value in the suggestion of Brittain (1), p. 255 (in an otherwise admirable book), that these early ironmasters used spoon tilt-hammers, for we know that cast iron was what they were making, and such a system could never have given either the force or the speed of action or the rhythm required even for small blast-furnaces.

[b] Yang Khuan (6), p. 58, adds the further argument that Han Chi's blowing-engines are said to have developed from horse-power whim drives, which would almost certainly have had horizontal driving-wheels.

[c] As depicted in Li Chhung-Chou's drawings. See also Wang Chen-To's interesting interpretation (8) of the Han ironworks relief in Sect. 30d, and meanwhile Needham (32, 48).

[d] (6), p. 65.   [e] Cf. *Pei Chhi Shu*, ch. 18, p. 5b.

[f] Cf. Sect. 30d below, and Needham (31, 32).

[g] In the Chinese texts water-powered forge-hammers are very much less conspicuous than water-powered blowing-engines, and this is probably due precisely to the fact that bloomeries were replaced by blast-furnaces at such an early time. How far hydraulic or martinet hammers were used in, e.g., the forging of co-fusion steel in the Chinese middle ages, is not yet known.

[1] 安陽縣志   [2] 水冶舊經   [3] 水冶   [4] 高隆之

in Europe, notably in Germany, Denmark and France, very much later than in China. Forge-hammers were the first to be mechanised, about the beginning of the + 12th century, and the application of water-power to the bellows for the air-blast followed early in the + 13th, i.e. a century or so before Wang Chên made his researches in the history of technology.[a] Of the specific origin of the plans for the European machines we know nothing, but if anything came westwards overland it would have been the trip-hammer lug first, and then long afterwards the eccentric drive. For the designs in the + 15th-century MSS. show only cuneate leather bellows worked by lugs on the shafts of vertical overshot wheels,[b] and this practice continued through the time of Biringuccio (+ 1540), who figures them in great variety,[c] to the 18th century, when John Wilkinson actually patented in + 1757 a hydraulic blowing-engine[d] essentially similar to that described in the *Nung Shu* of + 1313. The only difference was that while some of Biringuccio's machines included crank motions instead of lugs, Wilkinson's embodied a two-throw crankshaft, as shown in the diagram (Fig. 607).[e] The designs of Ramelli especially (+ 1588) were strangely close to Sung and Yuan Chinese patterns on account of his extensive use of rocking rollers like bell-cranks (Fig. 608).[f] So also in the previous century we find an intimate resemblance in the sketches of Antonio Filarete (*c.* + 1462),[g] and in the subsequent one in the plates of Böckler.[h] One cannot help believing that there was some genetic connection.[i] In any case the Chinese engineers seem to have had a precedence of some ten centuries

[a] This statement depends on records which are very difficult to interpret (cf. Johannsen, 1). Bellows worked by water-power are well established for + 1214 and + 1219 (in the Tyrolese silver and Harz copper districts). Vertical stamp-mills for ore-crushing are clear for + 1135 and + 1175 in Styrian sources, which speak, for instance, of 'unum molendinum et unum stampf'. Mechanised hammers for forging the iron itself first appear probably in + 1116, certainly by + 1249, both sources being French (cf. p. 395). If 'molendinum' here means only the water-wheel then a series of references to 'mills' in connection with iron-working may take the vertical ore stamp-mill back to + 1010; if it means the bellows as well as the water-wheel then these same records may attest trip-lug blowers so operated back to that date. Unfortunately its meaning probably varied at different times and places (cf. Lynn White (7), pp. 83 ff.). Johannsen (14), pp. 92 ff., followed by Dickmann (4), pp. 29 ff., gives examples of water-blowers going as far back as + 738, but there is no precise evidence on their nature; it is most probable that these were trombes, i.e. devices like filter-pumps which needed no moving components (cf. p. 149 above). Trombes probably partnered vertical ore stamp-mills in many places as late as the + 15th century. The first blast-furnaces of Europe did not appear until about + 1380, as we shall see in more detail in Sect. 30*d*. It is quite astonishing to read in Johannsen (14), p. 21, that water-power was never used in Chinese metallurgy until modern times.

[b] For instance, the Anonymous Hussite (+ 1430), Mariano Taccola (+ 1440), the *Mittelalterliche Hausbuch* of + 1480 (Anon. (15), Essenwein ed. pl. 37*a, b*, Bossert & Storck ed. pls. 41, 42). Cf. Berthelot (4); Frémont (14), fig. 30; Beck (1), p. 289; Gille (14), p. 643; Forbes (8), p. 68, (19), pp. 612 ff.

[c] Cf. Beck (1), pp. 116 ff. and Fig. 620.

[d] See Dickinson (2). Cf. de Bélidor (1), vol. 1, pls. 34, 35.

[e] It will be remembered from p. 113 above that the true crankshaft is found in the German engineering MSS. of the + 15th century as well as in Agricola. But a 'piston-rod' as well as a connecting-rod is rarely, if ever, shown with it. Even in Agricola (Hoover & Hoover (1), p. 180) the two are combined.

[f] Especially pl. 137, reproduced in Frémont (2), fig. 142, (10), fig. 88; Forbes (19), p. 613, and elsewhere.

[g] Reproduced and discussed by Johannsen (9).

[h] (1), esp. pl. 78. In spite of what has been said above (p. 150), it is rather striking that down to the end of the + 17th century the illustrations of water-powered blowing-engines almost always show wood-and-leather bellows of the pyramidal cuneate form, rather than all-wood bellows of this shape.

[i] We still know almost nothing of the blowing-engines of intermediate regions of the Old World. Afet Inan (1), p. 41, quotes an interesting account from the travels of Evliya Chelebi (+ 1611 to + 1682), of a bellows at the iron mines of Samakov in Turkey (now Bulgaria) which ten men could not have moved.

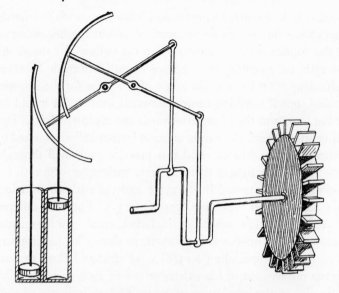

Fig. 607. Schematic semi-perspective diagram of John Wilkinson's hydraulic blowing-engine of +1757 (after Dickinson, 2). A vertical water-wheel works two air-pump pistons by means of a two-throw crankshaft. Apart from the crankshaft (cf. p. 113), this machine was in no essential way different from the blowing-engine described by Wang Chên in +1313.

for the practical trip-hammer principle, and three or four for the combination of eccentric, connecting-rod and piston-rod.[a] Must we not see in the latter the precise, but inverse, pattern of the arrangement of the reciprocating steam-engine?[b]

(4) RECIPROCATING MOTION AND THE STEAM-ENGINE'S LINEAGE

All this brings up in rather acute form the history of the interconversion of rotary and longitudinal reciprocating motion. The oldest examples of this achievement (the bow-drill, the pump-drill and the pole-lathe)[c] all involved non-continuous belting, and in the development of machinery they did not lead far. Next came the use of lugs on a rotating shaft, with springs to ensure return travel. But Gille was stating the half of a half-truth when he said[d] that 'the only way of effecting these conversions known to the Middle Ages involved springs'. He was thinking of course of the hydraulic saw of Villard de Honnecourt (c. +1237), which used the trip-hammer principle but

[a] Assuming that this was not Han and had been developing later through Thang and Sung before Wang Chên's time.
[b] It is true that in the Chinese blowing-engine a kind of 'bell-crank' rocking roller intervenes between the connecting-rod and the piston-rod, but this does not affect the logical situation, and the device might be considered as an interim solution in the search for stability ultimately attained in slides and cross-head. One may also reflect that the beam of the beam-engine was simply a straightened bell-crank, so that Wang Chên's rocking rollers were reincarnated for a time in that form. The only thing lacking in his blowing-engine was the true crankshaft, and that did not appear in such a machine until the time of Wilkinson four centuries later.
[c] Cf. pp. 55 ff. above, as also Childe (11) and Gille (14), p. 645.
[d] (14), p. 652.

PLATE CCXXXVI

Fig. 608. Furnace bellows worked by a vertical water-wheel; a design by Ramelli (+1588). The rocking rollers and bell-cranks (*B*, *C*, *D*), set in motion by the connecting-rod (*F*) and crank (*G*), are very reminiscent of the mechanism of the Chinese blowing-engines of Sung and Yuan times (cf. Figs. 602, 603).

PLATE CCXXXVII

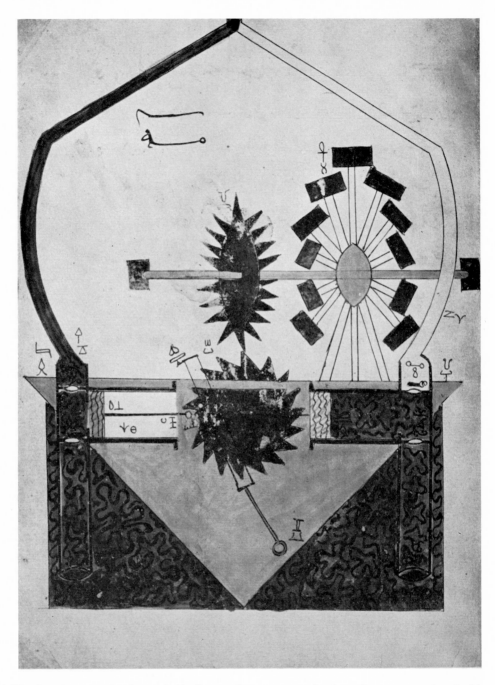

Fig. 609. An avuncular element in the lineage of steam power; the slot-rod water-pump described by al-Jazarī in his treatise on ingenious mechanical contrivances of +1206 (from Coomaraswamy, 2). The upper gear-wheel and the water-wheel are drawn in semi-perspective.

assured the return excursion of the saw by means of a spring pole.[a] More commonly gravity had been used rather than springs—as in the vertical stamp-mills of Western medieval times, or the bellows with counterpoises,[b] or the ancient Chinese trip-hammers and the later European fulling-mills and martinets which so closely resembled them.[c] Gille's formulation, moreover, omitted the two machines of the Middle Ages which lie most directly in the line of ancestry of the steam-engine and the locomotive, namely the connecting-rod system of the Chinese blowing-engines which we have just been examining, and the slot-rod force-pump described by al-Jazarī a century earlier than Wang Chên.

This last machine, described and illustrated (Fig. 609) in his 'Book of the Knowledge of Ingenious Mechanical Contrivances' of +1206, has been discussed by several modern writers.[d] A vertical water-wheel rotated a large gear-wheel on the same shaft, and this engaged with another beneath it[e] (cf. the diagram in Fig. 610). The lower toothed wheel was mounted on a shaft loosely pivoted at one end and free to rotate in an annular space at the other, so that while the wheel turned round upon its geometrical centre, the shaft constituted the equivalent of an eccentric lug as it described its conical path. But instead of being linked directly to a connecting-rod in any way, this eccentric shaft slid up and down within a slotted rod or arm attached to a fixed pivot below and having at each side midway along its length link connections with two piston-rods. Thus as the lower wheel rotated the slot-rod was forced alternately to left and right, moving the pistons successively back and forth.[f] This was a fine piece

[a] As we have just seen, springs were also necessary in the second (vertical water-wheel) design of Wang Chên. They are in no way to be despised as primitive, for they occur in some of the most ingenious modern reciprocating mechanisms. Artificial-respiration pumps, for instance, require a reciprocation cycle of constant velocity. Since this cannot be obtained from the conventional eccentric, connecting-rod and piston-rod combination, a continuously rotating spring arm is made to drop into two notches on a flat wheel connected by steel tape with a bellows-actuating member. The arm is lifted out of the notches by two cam-faced dogs twice in each cycle, and the two return strokes are effected by means of a spring (cf. Jones & Horton (1), vol. 3, pp. 183 ff.). It is interesting that Wang Chên's second type of blower would have had a stroke of approximately constant velocity.

[b] Cf. e.g. Beck (1), p. 119.

[c] Gille (14), p. 635, attributes a 'camshaft' and trip-hammer principle to Heron of Alexandria. What he means is a system of lugs depressing a lever attached to an organ air-pump. This occurs only in the windmill-driven organ blower; *Pneumatica*, ch. 77, cf. Woodcroft (1), p. 108. The principle is there, but the whole design is so hypothetical and unpractical that one has great hesitation in believing it to be the origin of all the trip-lugs of the European Middle Ages. What we say elsewhere (pp. 183 ff., 390 ff.) about the tilt-hammer and trip-hammer in China shows that it was essentially practical and in constant use. It may therefore have a better claim to have been the inspiration of the practical pounding and blowing machinery of Europe.

[d] E.g. Wiedemann & Hauser (1). Coomaraswamy (2) gave the fine reproduction of the drawing which we reproduce, taken from a MS. not later than about +1225 and perhaps contemporary with al-Jazarī himself. Coomaraswamy's explanatory text, however (p. 17), the work of a great art historian rather than an engineer, contains a number of puzzling misnomers. Further details on al-Jazarī and his book are given on p. 534.

[e] According to the description, at right angles to it, and indeed the picture of al-Jazarī is so drawn. Excluding the possibility of the lower gear-wheel being horizontal, it was either cut as a worm, in which case the planes of the wheels would have been at right angles, or (in Prof. Burstall's reconstruction) as a kind of crown-wheel, in which case only the teeth were at right angles to the plane of both wheels. The purpose of this is not at all obvious.

[f] The slotted member is an ingenious device to which engineers of later times often had recourse, and it is still an important component of many machines. Particularly refined use of slot-rods was made by de Dondi in the motions for the planetary dials of his astronomical clock (+1364), a model of which,

of mechanism for the early +13th century, but al-Jazari's device was not as direct an ancestor of the steam-engine combination as the Chinese system which Wang Chên described.[a]

Another machine which must be considered in close connection with the hydraulic blowing-engine is the silk-winding or reeling apparatus already referred to.[b] Though our best illustrations of this come from the early 19th century they agree not only with +17th-century texts but also quite precisely with the *Tshan Shu* (Book of Sericulture) written about +1090. In this machine (Fig. 409) the main winding barrels or reels are worked by a crank-and-pedal motion, but the ramping-arm (the forerunner of the 'flyer')[c] is also operated from the same power-source by means of a driving-belt connecting the main shaft with a pulley at the other end of the frame. This subsidiary wheel then moves the ramping-arm back and forth by means of a lug eccentrically placed. Here then we have the driving-belt just as in Wang Chên's hydraulic blower,[d] as also the smaller wheel with its eccentric, so that the ramping-arm corresponds to the connecting-rod. There is nothing, however, corresponding to the piston-rod, and instead of having any link connection at the further end, the arm is simply held in a ring through which it slides back and forth. Thus this system was still at the stage of the European +15th-century military engineers, whose crankshafts are so often depicted with connecting-rods but never with piston-rods. It resembles too the age-old connecting-rods of the hand-driven millstones (cf. p. 117 above). So the silk-reeler did not have quite all the components of the hydraulic blower. But the fact that it was already a standard piece of mechanism at the end of the +11th century,

destined for the Smithsonian Institution, was exhibited at the Science Museum in London in 1961. In the form of the slotted cross-head or Scots yoke (cf. Jones & Horton (1), vol. 1, pp. 250 ff.) they played a considerable part in steam-engine and pump design by eliminating the connecting-rod and giving the piston-rod a uniform harmonic motion. The slotted member is also valuable for quick-return effects, as in shapers (*ibid.* vol. 1, pp. 300 ff.; vol. 3, pp. 188 ff.), and it occurs in an ingenious arrangement for giving positive accurately-timed reciprocation with a firm lock during the dwell at the end of each stroke (vol. 3, pp. 192 ff.). Slotted rods are also sometimes furnished with an internal rack and periodically thrown by cams from side to side so as to engage with a pinion on a driving-shaft in alternate senses, thus bringing about a reciprocating motion in the direction of their elongated axis. This gives a particularly long stroke (as in windmill pumps; Jones & Horton (1), vol. 1, pp. 260 ff., cf. also the converse case of the Napier motion, where a single rack with two faces perambulates in alternate upper and lower contact with a continuously rotating pinion, pp. 263 ff.). The windmill-pump type of internally racked slot-rod was used as early as +1615 by Solomon de Caus for a similar drive (cf. Beck (1), p. 510). Curved slots with racks give the various varieties of the well-known mangle gear (Jones & Horton (1), vol. 2, pp. 245 ff.). A theoretical treatment of slot systems, both plain and racked, was given by Willis (1), pp. 287 ff., 294, 323, and of course in many later books.

[a] Not as direct, indeed, as I first thought, for we interpreted the text to mean that the lower gear-wheel, normally mounted, carried a simple eccentric lug free to slide up and down in the slot-rod. But Prof. Aubrey Burstall convinced me that a closer adherence to its words revealed a more unusual mechanical principle. We were both pleased to find that Lynn White (7), p. 170, had arrived independently at the same interpretation.

[b] Especially above, pp. 2, 107, 116, and below, p. 404. Photographs in Tisdale (2); cf. Sect. 31.

[c] The function of this, of course, is to assure the even laying down of the fibre or yarn on the reel or bobbin. It survives in textile machinery of the industrial age (Ure (1), 5th ed., vol. 3, pp. 666 ff.).

[d] According to our interpretation of the latter, cf. p. 373 above. We there took note of the rather puzzling fact that the medieval Chinese seem not to have used gearing in their rotary–longitudinal conversion machinery. It was in no way strange to them. The reason escapes us, but we may be sure that the mechanisms such as the silk-reeler and the hydraulic blower worked well enough, for the Chinese were always an intensely practical people, and the texts in question were primarily intended for the dissemination of practical knowledge.

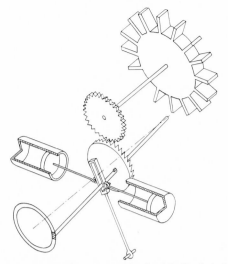

Fig. 610a. Interpretation of the slot-rod force-pump for liquids (Prof. Aubrey Burstall). A gear-wheel rotated by a vertical water-wheel engages with another beneath it. The lower toothed wheel is mounted on a shaft which is loosely pivoted at the right and free to rotate in an annular groove at the left, so that while the wheel turns upon its geometrical centre the shaft describes a conical path and acts as an eccentric lug. Instead of being linked to any connecting-rod, however, it slides up and down within a slotted rod attached to a fixed pivot below and to two piston-rods, one on each side. Thus a continuous flow of liquid proceeds up the discharge pipe seen at the top of the drawing in Fig. 609.

Fig. 610b. Working model of the slot-rod force-pump for liquids (Prof. Aubrey Burstall).

and indeed owing to the antiquity of the silk industry may well have been established practice long before, considerably strengthens the probability that the water-powered blower, with its full 'steam-engine' arrangement for converting rotary to reciprocating longitudinal motion, developed during the Thang and Sung, i.e. during the four or five centuries preceding Wang Chên's description of it. It may very easily, therefore, be older than al-Jazarī's swaying slot-rod.

It is necessary to emphasise that the system of three parts (eccentric, connecting-rod and piston-rod) has not so far been found in any + 14th-century European illustration, and occurs only rarely in the + 15th century. Lynn White, best of guides, gives us nothing earlier than a drawing, now in the Louvre, by Antonio Pisanello, which depicts in very workmanlike fashion a pair of piston-pumps operated by rocking levers raised and lowered by connecting-rods from two cranks fitted 180° apart on the two sides of an overshot water-wheel.[a] It would be reasonable to date this about + 1445 for Pisanello died in + 1456. The same arrangement appears again in the *Mittelalterliche Hausbuch* about + 1480, applied to what looks like a single vertical stamp-mill, but here with only a single crank, connecting-rod and rocking lever.[b] By this time we are fully within the period of activity of Leonardo da Vinci.

When Leonardo faces the problem of interconversion in the late + 15th century, nearly two hundred years after Wang Chên, he shows, as Gille has acutely pointed out,[c] a most curious disinclination to use the eccentric (or crank), connecting-rod and piston-rod combination. In fact he does so only for a mechanical saw.[d] In order to avoid it he has recourse time after time to the most complicated and improbable devices. One celebrated arrangement (Fig. 611) has a lantern pinion on a windlass engaging on each side with right-angle gear in the form of pin-wheels, the motion of

---

[a] No. 2286; see Degenhart (1), fig. 147. The same drawing includes a *sāqīya* driven by an overshot water-wheel through two-stage reduction gearing. We regret that we must exclude what Lynn White (7), p. 113 and fig. 7, proposes as 'the earliest evidence of compound crank and connecting-rod', a MS. drawing of Mariano di Jacopo Taccola which dates between + 1441 and + 1458 (Bayerische Staatsbibliothek, München, Cod. Lat. 197, fo. 82v). We do so not because the two-throw crankshaft is doubly misdrawn, but because it effects the excursion of a lift-pump by what is quite evidently a rope instead of a connecting-rod, gravity alone ensuring the return. This device certainly constitutes an interesting premonition of the three-part system, but no more. By an extraordinary inversion, Lynn White (7), p. 81, misinterprets the construction of the hydraulic blowing-engine of the *Nung Shu* in exactly the same sense. The water-wheel, he says, 'turned a vertical shaft carrying an upper wheel which, by means of an eccentric peg and a cord, worked the bellows of a furnace for smelting iron'. But no such system was used; text and drawing fully concur in making the connecting-rod a rigid bar. Nor have we encountered any Chinese machine which used a cord with an eccentric peg. We are very grateful to Prof. Lynn White for his kindness in bringing these two + 15th-century machines to our notice, and showing us an advance set of the plates of his book.

[b] Anon. (15), Essenwein ed., pls. 24b, 25b; Bossert & Storck ed., pl. 32. Here again there is a misdrawing, for the rocking lever is made to pass through a horizontal instead of a vertical slot. The arrangement does not seem suitable for a stamp-mill, and perhaps a pump is implied, for in the 'Liebesgarten' on the left there is a fountain playing. No text or explanation has survived to help.

[c] (14), p. 654. It will be remembered that the full crankshaft had appeared in Europe during the first half of the + 14th century, and, in the form of the carpenter's brace, during the first half of the + 15th. Cf. pp. 113, 114.

[d] Cf. Beck (1), p. 323; Ucelli di Nemi (3), no. 21; Feldhaus (18), p. 55. An almost exactly similar machine was carved in relief after a drawing by Francesco di Giorgio Martini in + 1474 and later; cf. Feldhaus (20), p. 245, fig. 167; Uccelli (1), p. 64, fig. 200, and especially Reti (1). There is one other device of Leonardo's which employs the principle, a grinding-machine for cylindrical concave glass mirrors (Beck (1), p. 361, fig. 542), but here the 'piston' is hinged (cf. pp. 57, 374).

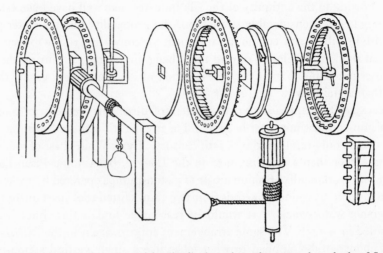

Fig. 611. Interconversion of rotary and longitudinal motion; the unusual methods of Leonardo da Vinci (c. +1490). (i) A lantern-pinion between two pin-wheels with internal ratchets. A lever handle working back and forth in an arc rotates each right-angled gear-wheel alternately in opposite directions by means of the pawl-tooth on its internal drum, but as the lantern-pinion engages with both, it is driven round continuously in one direction only. Thus the alternating rotary motion of the lever is transformed into the continuous longitudinal motion of the windlass-rope (Beck (1), p. 421).

which is restricted by internal ratchets so that they can turn only in opposite directions. A lever working back and forth moves them on by means of internal drums with projecting movable pawls, thus turning the windlass continuously in one direction.[a] A much more brilliant solution was found when Leonardo devised a cylinder with a double helical groove into which a peg on the end of a piston-rod fitted, so that it was driven back and forth as the cylinder continuously rotated (Fig. 612, left). With extraordinary ingenuity he then added a variety of automatically acting gates which prevented misrouting of the peg at the places where the grooves crossed each other.[b] A third design (Fig. 612, centre) used cam-shaped lugs on a rotating shaft acting upon a system of levers hinged by linkwork in such a way that each cam effected successively

[a] Cf. Beck (1), p. 421; Ucelli di Nemi (3), no. 16; Feldhaus (18), p. 68. Double-acting ratchet gearing pushed on by both excursions of a lever is still used in machines of various kinds (cf. Jones & Horton (1), vol. 1, pp. 29, 31; vol. 3, pp. 71 ff.). Wheels with internal ratchets are employed, for instance, in the drives of belt-conveyors (cf. Jones & Horton (1), vol. 1, pp. 53 ff.; vol. 3, p. 70).

[b] Cf. Beck (1), pp. 417 ff.; Ucelli di Nemi (3), no. 122; Willis (1), pp. 157, 321. These cylindrical cams with follower rollers sliding back and forth in the grooves are widely used in contemporary machine construction; cf. Jones & Horton (1), vol. 1, pp. 4, 8; vol. 2, pp. 11 ff., 19 ff., 52 ff., 294; vol. 3, pp. 178, 182. Automatic gates like Leonardo's can be found in wire-making machines and gas-engines; cf. Jones & Horton (1), vol. 1, p. 19; vol. 2, pp. 44 ff. The cylindrical cam finds particular application in the textile industry, as for instance in the 'rotoconer', a machine for winding woollen and worsted yarn on to cones and 'cheeses'. Here the helical groove acts as a traversing device. But no doubt the most familiar example of the use of the grooved cylinder is in the spring screwdrivers sold in every modern tool-shop. The simplest form of the groove system is of course found in the 'face cam', where a pin and roller follows a groove in a rotating plate. Examples of the current use of this may be seen in Jones & Horton (1), vol. 1, pp. 3 ff., vol. 3, pp. 191 ff. It was already used in the +16th century by Ramelli; cf. Beck (1), p. 219. In his design levers were worked up and down by a single channel mounted eccentrically upon a rotating horizontal wheel. I am indebted to Dr Norman Heatley and Dr A. J. W. Haigh of Oxford for discussions on this subject.

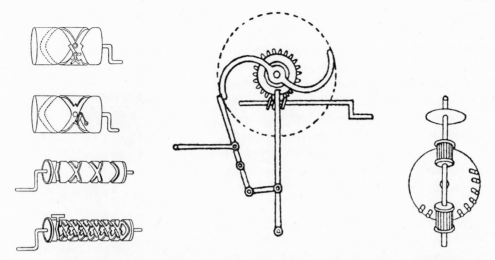

Fig. 612. Other methods of Leonardo (*c.* +1490). On the left, (ii) cylindrical cams with double helical grooves in which a follower pin or roller (seen as a small T-shaped piece in the lowest diagram) moves back and forth along the left- and right-hand threads alternately. Continuous rotary motion is thus converted to reciprocating longitudinal motion. To minimise friction and possible misrouting Leonardo added a variety of automatic gates. At the top, a gate which is closed by the follower itself as it passes; next, a spring-closed gate; below, an arrangement of gates permitting the use of left-hand or right-hand thread exclusively as desired, suitable if the rotary motion were alternating. From Beck (1), pp. 417 ff. In the centre, (iii) curved lugs, levers and linkwork. An S-shaped cam or pair of lugs revolves continuously in one direction (here clockwise, rotated by a worm and crank). The two levers which the lugs displace along a chord almost as long as the radius of the cam's circle are fixed at the lowest joint and at the uppermost but one; on these they can swing. In the drawing the left-hand lug has just pushed the left-hand lever, and hence the piston-rod, nearly to the end of its excursion to the left. In a moment the right-hand lug will begin to push the right-hand lever to the left, thus reversing the excursion of the piston-rod and bringing it back fully to the right. The cycle then repeats, converting continuous rotary to reciprocating longitudinal motion. As the levers must cross over like scissors they cannot be exactly in the same plane. From Beck (1), p. 419. On the right, (iv) the half-gear, one of a wide variety of possible arrangements. A continuously rotating pin-wheel (here seen as a flat disc) furnished with pins or teeth round only half of its circumference, engages alternately with two lantern-pinions on the same shaft, imparting thereby a regularly alternating motion to the latter. Assuming a clockwise rotation, the pins in the diagram are turning the lantern-pinions outwards to the right, but when they reach the uppermost position the lantern-pinions will turn inwards, reversing the motion of the shaft. From Beck (1), p. 321. If the shaft carries a windlass rope or chain, half-gears of this kind will convert continuous rotary to alternating longitudinal motion.

two excursions of the piston-rod.[a] Lastly, Leonardo initiated that scheme which was to be such a favourite of the later Renaissance engineers, the half-gear.[b] In this system (Fig. 612, right) two lantern pinions on the same shaft engage at right angles with a single gear-wheel which has pins set round only half of the circumference of its disc. Thus with the successive contacts the shaft is driven round alternately in each direction. The resultant alternating rotary motion is then easily converted to alternating

---

[a] Cf. Beck (1), p. 419.

[b] And not only with them. Intermittent gears are quite common in modern machinery; cf. Jones & Horton (1), vol. 1, pp. 68 ff., 93 ff., 98 ff.; vol. 2, p. 71; vol. 3, pp. 24, 46 ff., 176. The principle is used in all kinds of ways, with right-angle bevel gears, epicyclic gears, reciprocating racks and so on.

rectilinear motion by two windlass chains or other means (Fig. 613).[a] All these complicated arrangements proved useful in one way or another in the machinery of later centuries,[b] but the simpler and fundamentally important steam-engine system derives from Wang Chên rather than from Leonardo. Why there was this aversion from it in Europe is not at all clear. Serious difficulties were perhaps encountered in assembling the moving parts so that friction and wear were sufficiently overcome,[c] but in this case we should like to know just how and why the technique of the medieval Chinese engineers was more advanced. Better availability of steel bearings might well be the answer.[d]

Perhaps the most extraordinary part of the whole story is that James Watt was driven to the invention of the sun-and-planet gear[e] by the fact that the basic method of converting rotary to rectilinear motion by eccentric and connecting-rod had been patented for steam-engines by James Pickard in +1780.[f] Watt had not patented it himself because he knew that it was old, but probably none of those involved knew anything of the +15th-century German engineers, and certainly no one at that time could have had any suspicion that the Chinese of the Sung period had been intimately and practically acquainted with it. Indeed, on our present information, they were its real inventors.[g]

In considering the ancestry of the component parts of all reciprocating steam-engines and

Fig. 613. A later example of a different use of half-gears for the conversion of continuous rotary to alternating longitudinal motion; one of the pumping systems in the engineering treatise of Ramelli (+1588). Two half-lantern-pinions are mounted in diametrically opposite senses so as to engage with two curved racks (gear-wheel segments) on levers, the other ends of which are attached to pump piston-rods. Moving anti-clockwise, the background half-lantern is about to engage with its rack, the descent of which will raise the corresponding piston-rod and depress the other one because of the pulley and chain connecting the ends of the two racks. Immediately afterwards the foreground half-lantern will engage, reversing the whole process. From Beck (1), p. 215.

---

[a] Cf. Ucelli di Nemi (3), no. 71, where Strobino's reconstruction applies the device to the co-axial textile flyer; and Beck (1), p. 321. For the later applications in Ramelli see Beck (1), pp. 213 ff., 223 ff., and in de Caus, Beck (1), p. 508. Theoretical treatment in Willis (1), p. 293.

[b] There are of course a number of other devices which Leonardo did not mention, e.g. the swashplate (cf. Willis (1), p. 319). But geometrically this is very close to his helical grooves. A form of it was used by Ramelli (Beck (1), p. 217).   [c] As suggested by Gille (14), p. 653.

[d] On this subject see Vol. 3, p. 350, as well as pp. 92 ff. above and pp. 448, 454 below. The Chinese engineers of the Sung and Yuan were perfectly capable of fitting steel pins in cast-iron bushes for their eccentric cranks and linkwork, which would have been just the thing. Their rocking rollers, moreover, involved two working joints between connecting-rod and piston-rod instead of one.

[e] Cf. Reuleaux (1), p. 245; Willis (1), p. 373.

[f] Dickinson (4), pp. 80 ff., (7), pp. 126 ff.; Farey (1), pp. 423 ff.; Lardner (1), pp. 182 ff., F. W. Brewer (1).

[g] Unless of course the system goes back to Tu Shih and his Han engineers in the +1st century, a possibility which is not as yet excluded.

derivative prime movers there is one more thing to remember. Not only, as we found, was the combination of eccentric, connecting-rod and piston-rod apparently first worked out in Sung or Thang (if not in Han) China, achieving there rather than in Europe the most effective of all inventions for the interconversion of rotary and rectilinear motion;[a] but also the double-acting piston and cylinder principle made its first appearance in the Chinese air-pump, sucking and expelling on both strokes (cf. pp. 135, 149), which again was fully developed certainly in the Sung, probably in the Thang, and possibly in the Han. In these inventions the Hellenistic age does not compete, and it is noteworthy that all Chinese datings long precede not only the Renaissance, but also the times of Leonardo da Vinci and even Guido da Vigevano (+14th century).[b] What constituted the fundamental revolution of the European 17th and 18th centuries was the inversion of the direction of motion so that force was transmitted not to but from the piston.[c] One may thus justly conclude (if it is not putting too much strain on our adopted terminology) that the great 'physiological' triumphs of 'ex-pistonian' Europe were built upon a foundation of formal or 'morphological' identity laid by 'ad-pistonian' China.[d] If there was, as seems most probable, a direct genetic connection, this is the place to look for it—not in that strange episode already related[e] when Jesuit mechanicians put a model steam-turbine locomotive through its paces in the palace gardens of the Khang-Hsi emperor.

It only remains to add a word or two about the coming of the steam-engine to China in the 19th century. The steamboat was the carrier.[f] It has generally been thought that the East India Company's steamer *Forbes* was the first to arrive, in 1830, but a passage in the *Hai Kuo Thu Chih*[1] (Illustrated Record of the Maritime Nations), of 1844, shows that the true date was two years before that.[g]

Early in the 8th year of the Tao-Kuang reign-period (April 1828), there suddenly came from Bengal a 'fire-wheel boat' (*huo lun chhuan*[2])'. Now the fire-wheel boat has an empty copper cylinder inside to burn coal, with a machine on the top. When the flames are burning,

[a] In traditional China probably no one noticed the characteristics of the stroke; the irregularity of speed, greatest at mid-stroke, and the momentary dwells in the neighbourhood of each dead centre. Only with the construction of precision steam-driven metal machinery in the European 19th century did such features become evident. Since then much ingenuity has been expended in devising means of obtaining strokes with any desired velocity characteristics (cf. Jones & Horton (1), vol. 1, pp. 249 ff.; vol. 2, pp. 260 ff.; vol. 3, pp. 206 ff.).

[b] We have already paid our homage to the European invention of the crankshaft (p. 113 above), on which Lynn White (7), pp. 112, 114, lays so much emphasis. Important though it was, we feel that the first achievement of the three-part system which forms the structural basis of the reciprocating steam-engine was by far the more revolutionary advance.

[c] Cf. e.g. Dickinson (4) or Usher (1), 2nd ed., pp. 342ff.; and more specifically Farey (1); Galloway (1); Thurston (1); Rolt (1); Dickinson (7) and Dickinson & Titley (1); Cardwell (1).

[d] Curious associations between these two kinematic patterns developed in the +18th century when Newcomen engines were used to pump water which was then used for driving the water-wheels of mines. Cotton-mills, too, needing a particularly steady rotation rate, used water-wheels fed from reservoirs kept supplied by Watt's engines. For an extended discussion of all these matters, see Needham (41, 48).

[e] P. 225 above.

[f] See the monographs by Chhen Chhi-Thien and a brief paper by Worcester (6); but the full story has not yet been given either in Chinese or a Western language.

[g] Ch. 83, p. 4*a*; tr. Chhen Chhi-Thien (2), p. 26.

[1] 海國圖志　　[2] 火輪船

the machine moves automatically, and so do the wheels on both sides of the ship. It can travel at a speed of a thousand *li* in a day and a night. From Bengal to Canton it took only 37 days. According to the foreigners the steamship was invented early in the twenties, but it could not be used for shipping cargo; it is good only for carrying urgent messages.

From this time onwards, steamships became more and more familiar in Chinese waters, and by the decisive assistance which they gave to foreign sea-power during the Opium Wars, caused the greatest consternation among officials and populace alike.

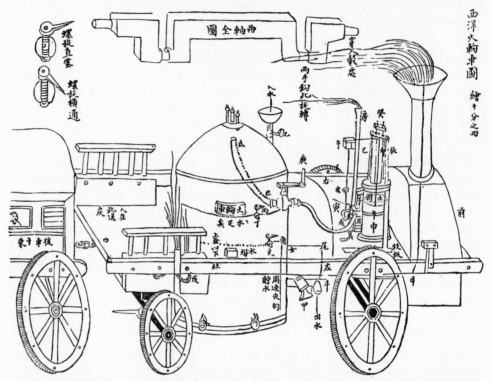

Fig. 614. The first Chinese drawing of a steam locomotive, in Ting Kung-Chhen's *Yen Phao Thu Shuo* (1841), ch. 4, pp. 15 a ff. The title is 'Diagram of a Western Fire-Wheel Carriage'. Many of the markings correspond to *a*, *b*, *c*, etc., but the parts of the boiler are appropriately labelled. Above, centre, the crankshaft (*chhiu chu*), with legends saying that the bearings fit into hubs and that the narrow parts are grasped by the two hands (*liang shou*), i.e. the big ends of the connecting-rods, and turn in them. Above, to the left, two diagrams of screw valve cocks, with a note to say that the position of the handles shows whether open or shut.

It was now realised that a fundamental problem existed, the necessity of mastering all the advances in technology which had come about in European civilisation since the Renaissance; but naturally the growth of a body of Chinese engineers in the modern sense was a slow process. Great credit is due to the earlier pioneers of this movement. One of them was Ting Kung-Chhen,[1] whose *Yen Phao Thu Shuo*[2] (Illustrated Treatise on Gunnery) of 1841 contained diagrams of a model steamboat and steam locomotive (Fig. 614) which he had succeeded in constructing without occidental

---

[1] 丁拱辰   [2] 演礮圖說

assistance.[a] Many years were to pass, however, before the prevailing conservatism was to allow the construction of railways. The building of the Shanghai–Wusung tramway in 1876 and its discontinuance in the following year was an event which long loomed large in Western-language accounts of China. Not till 1881 was the first working line established, as the result of great efforts by the statesman Li Hung-Chang,[1] namely that between Tientsin and the Thangshan coal-mines.

In the meantime, increasing use of steamers had been made by Chinese officials and military men in the fifties and sixties. Chhen Chhi-Thien (2, 3) in his biographies of Tsêng Kuo-Fan[2] and Tso Tsung-Thang,[3] written to elucidate their role in the modernisation of China, has described the efforts of the Chinese engineers of this

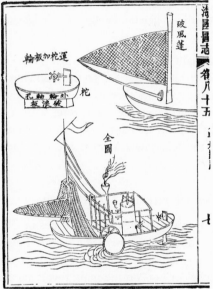
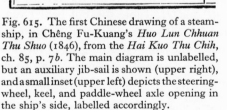

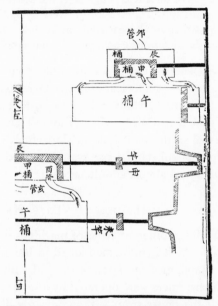

Fig. 615. The first Chinese drawing of a steam-ship, in Chêng Fu-Kuang's *Huo Lun Chhuan Thu Shuo* (1846), from the *Hai Kuo Thu Chih*, ch. 85, p. 7b. The main diagram is unlabelled, but an auxiliary jib-sail is shown (upper right), and a small inset (upper left) depicts the steering-wheel, keel, and paddle-wheel axle opening in the ship's side, labelled accordingly.

Fig. 616. Another of Chêng Fu-Kuang's diagrams (from *Hai Kuo Thu Chih*, ch. 85, p. 14a), showing the first Chinese representation of the slide-valve mechanism of the double-acting reciprocating steam-engine. Comparing this explicitly with the valves of the double-acting piston-bellows, he acutely recognised its parentage.

[a] Ch. 4, pp. 15a ff. Five years later an illustrated description of the mechanism of a paddle-wheel steamer, entitled *Huo Lun Chhuan Thu Shuo*,[4] was given by Chêng Fu-Kuang[5] as an appendix to his book on optics *Ching Ching Ling Chhih*[6] (already mentioned in Vol. 4, pt. 1, p. 117); and this was incorporated into the *Hai Kuo Thu Chih*[7] as ch. 85, with other material, in the following year, 1847 (cf. Fig. 615). This chapter is not by Ting Kung-Chhen,[8] as Lo Jung-Pang (3) says, though he contributed much other matter. Chêng tells us that he had studied pictures and a model, presumably Ting Kung-Chhen's, but had not fully understood the mechanism until he saw diagrams at the house of his friend Ting Shou-Tshun,[9] another of the *Hai Kuo Thu Chih* contributors. According to Lo Jung-Pang (3) the first full-scale Chinese steam paddle-boat was built at Canton in the early forties with the assistance of a European engineer and financed by Phan Shih-Jung.[10] Cf. *I Huan Pei Chhang Chi*, p. 40a.

[1] 李鴻章   [2] 曾國藩   [3] 左宗棠   [4] 火輪船圖説   [5] 鄭復光
[6] 鏡鏡詅癡   [7] 海國圖志   [8] 丁拱辰   [9] 丁守存   [10] 潘世榮

time. The former leader was the founder of the Kiangnan Arsenal near Shanghai, the latter originated the Fuchow Dockyard and the Lanchow woollen mills. Tsêng's temporary arsenal at Anking was the scene of experiments with Chinese-built steam launches in 1862. For 30 July Tsêng Kuo-Fan wrote in his diary:

> Hua Hêng-Fang[1] and Hsü Shou[2] brought here the engine of the fire-wheel boat which had been made by them, for a demonstration. The method is to use fire to make steam, and direct the steam into a cylinder which has three holes. When two of the front holes are closed, the steam goes into the other front hole. The piston automatically goes backward and the wheel turns the upper half circle. When two of the back holes are closed, the steam goes to the other back hole; then the piston automatically moves forward and the wheel completes the rest of the circle. The bigger the fire the greater the quantity of steam. The engine moves forward and backward as if it were flying. This demonstration lasted for an hour. I was so happy that we Chinese could make these ingenious things like the foreigners. No longer will they be able to take advantage of our ignorance.[a]

The scholar-general was evidently trying to explain the action of the valves, but either he failed to get it clear or the text was confused by his secretaries. In the following year the engine served to propel a 29-ft. launch built by another of Tsêng's engineers, Tshai Kuo-Hsiang.[3] Tso Tsung-Thang was at work at the same time with a trial vessel on the West Lake at Hangchow.[b] Only a few years later (1868) the first steamships of substantial size were launched from the Kiangnan Arsenal and the Fuchow Dockyard almost simultaneously.[c]

(5) HYDRAULIC TRIP-HAMMERS IN THE HAN AND CHIN

Of all the different types of machines driven by water-power in early times in China, that which is most mentioned in literature is the trip-hammer (*shui tui*[4]).[d] This was a simple mechanisation of the pedal tilt-hammer mentioned above (p. 183), in which the hammers were operated by a series of catches or lugs on the main revolving shaft. All the books from +1300 onwards illustrate this (cf. Table 56), but here I reproduce (Fig. 617) the picture in the *Thien Kung Khai Wu*.[e] An important point to note is that while the horizontal water-wheel is much the simplest arrangement for rotary millstones, the trip-hammer is best suited by the vertical water-wheel, and so it is

---

[a] Tr. Chhen Chhi-Thien (2), p. 40. Hua Hêng-Fang was a distinguished mathematician.
[b] Chhen Chhi-Thien (3), p. 11.
[c] *Ibid.* p. 47. They were the *Thien Chi*[5] and the *Wan Nien Chhing*[6] respectively. Between 1869 and 1874, the Dockyard built 15 vessels of about 1000 tons each, with engines of up to 250 h.p. After Tso Tsung-Thang had to leave for the North-west, Shen Pao-Chên[7] became Commissioner, with two French engineers, Prosper Giquel (Jih I-Ko[8]) and Paul d'Aiguebelle (Tê Kho-Pei[9]), as technicians in charge. The former became Chief Engineer and wrote an account of the development of the whole enterprise when he retired in 1874 (Giquel, 2).
[d] From the *Thung Su Wên*, written by Fu Chhien about +180, we know of another term for it, *fan chhê*[10] (*YHSF*, ch. 61, p. 23*b*), but this is less commonly met with.
[e] Ch. 4, p. 11*b*, first ed. pp. 61*b*, 62*a*. With this we may compare the scale drawing in Louis (2) of a modern Chinese traditional-type trip-hammer battery used for quartz-crushing. Cf. the descriptions of Barrow (1), p. 565 (+1793); van Braam Houckgeest (1), Fr. ed. vol. 1, pp. 428ff., Eng. ed. vol. 2, pp. 284ff (+1797).

[1] 華蘅芳	[2] 徐壽	[3] 蔡國祥	[4] 水碓	[5] 恬吉
[6] 萬年青	[7] 沈葆楨	[8] 日意格	[9] 德克碑	[10] 輨車

水碓

盖利
用茅

Fig. 617. A battery of hydraulic trip-hammers worked by an undershot vertical water-wheel (*Thien Kung Khai Wu* (+1637), ch. 4, p. 11*b*).

usually figured. One cannot help wondering, however, whether many of the early machines now to be mentioned were not horizontal water-wheels with right-angle gearing. Another point is that in Chinese practice the hammers were always recumbent, and not vertically-acting stamp-mill pestles such as we find in medieval Europe; this permitted of a considerably heavier installation.

One might be tempted to see a reference to the invention of the hydraulic trip-hammer in the story reported in the −3rd century *Lü Shih Chhun Chhiu*[a] about the mother of I Yin, who dreamed that she saw water flowing out of a mortar. However this may be, the earliest explicit statement seems to be in the *Hsin Lun* (New Discourses)[b] of Huan Than (*c.* +20), who remarked:

> Fu Hsi invented the pestle and mortar, which is so useful, and later on it was cleverly improved in such a way that the whole weight of the body could be used for treading on the tilt-hammer (*tui*[1]), thus increasing the efficiency ten times. Afterwards the power of animals—donkeys, mules, oxen and horses—was applied by means of machinery, and water-power too used for pounding, so that the benefit was increased a hundredfold.

Words so general and so assured authorise the conclusion[c] that from at least the time of Wang Mang onwards water-wheels were used more and more for working pounding machinery. Later in the Han dynasty Ma Jung[2] has a poem on the long flute in which he mentions the hammers 'pounding in the water-echoing caves',[d] and in +129 Yü Hsü[3] reported to the emperor that in the lands of the Western Chhiang people[e] water-driven trip-hammers were being introduced along the canals deriving from the mountain streams of the Chhilien Shan.[f]

In the +3rd and +4th centuries references are abundant. Khung Jung[4] (*d.* +208) made a remark in his *Jou Hsing Lun*[5] (Discourse on Mutilative Punishments)[g] which was often quoted afterwards,[h] that the ideas of intelligent men of the day were often better than those of the sages of old, and gave the water trip-hammer as an example.[i] There were men famous for possessing many such machines, even hundreds of them, such as Wang Jung[6] (*d.* +306),[j] whom we have met with already as a computer; Têng Yu[7] (*d.* +326),[k] and Shih Chhung[8] (*d.* +300) who operated them in more than

---

a Ch. 14 (tr. R. Wilhelm (3), p. 179).

b Cit. *TPYL*, ch. 762, p. 5*a*; ch. 829, p. 10*a*; and in *CSHK* (Hou Han sect.), ch. 15, p. 3*b*; tr. auct.; adjuv. Sun & de Francis (1), p. 114. Cf. a quotation from the *Lun Hêng*, cit. *TPYL*, ch. 762, p. 5*a*.

c Drawn e.g. by Yang Lien-Shêng (1), cf. Sun & de Francis (1), p. 114. Yet modern Western historians of technology such as Forbes (11), pp. 87, 110, inform us that the water-wheel 'certainly travelled east' in late Roman and Byzantine times, i.e. from the +3rd and +4th centuries onwards.

d *Chhang Ti Fu*,[9] in *Wên Hsüan*, ch. 18, p. 2*a*.

e Near Liangchow in modern Kansu.

f *Hou Han Shu*, ch. 117, p. 22*a*.

g *CSHK* (Hou Han sect.), ch. 83, p. 11*b*; cit. *KCCY*, ch. 52, p. 6*b*.

h E.g. *Piao I Lu*, ch. 5, p. 1*a*; *TPYL*, ch. 762, p. 7*a*.

i Cf. the disputes in Renaissance Europe between the supporters of the 'Ancients' and the 'Moderns' (p. 6 above). Also Vol. 4, pt. 1, p. 53.

j *Shih Shuo Hsin Yü*, cf. 3B, p. 31*b*; *Chin Shu*, ch. 43, p. 7*a*. Cf. Vol. 3, p. 71.

k *Wang Yin Chin Shu*[10] (Wang Yin's History of the Chin Dynasty), cit. *TPYL*, ch. 762, p. 6*b*.

¹ 碓   ² 馬融   ³ 虞詡   ⁴ 孔融   ⁵ 肉刑論   ⁶ 王戎
⁷ 鄧攸   ⁸ 石崇   ⁹ 長笛賦   ¹⁰ 王隱晉書

PLATE CCXXXIX

Fig. 618. A martinet forge-hammer, from Diderot's encyclopaedia, *c.* +1770. These machines, so important in the European +18th-century iron and steel industry, were worked by water-power in just the same way as their Chinese cereal-pounding ancestors, save that the hammer was generally mounted, as here, parallel with the lug-shaft instead of at right angles with it.

thirty districts.[a] Wei Huan,[1] on the other hand, declined to accept an imperial gift of them,[b] and Wei Shu,[2] afterwards eminent, was so retiring in his youth that he was set to mind them.[c] Poems were written on them, as by Chhu Thao,[3] a friend of Chang Hua's.[d] Tu Yü,[4] a high official and engineer of the +3rd century, established 'combined' trip-hammer batteries (*lien chi tui*[5]),[e] which probably means that several shafts were arranged to work off one large water-wheel. At one time water trip-hammers were not permitted within a certain radius (100 *li*) of the capital,[f] and there seem also to have been imperial ones, which sometimes caused obstruction on the water-ways.[g]

It would be superfluous to adduce more references, which are numerous in Thang and Sung—e.g. some comparisons made by Phei Tu[6] in the +9th[h] and Fang Shao[7] in the +12th.[i] This was the time when Lou Shou wrote his poem for the *Kêng Chih Thu*:[j]

> The graceful moon rides over the wall
> The leaves make a noise, sho-sho, in the breeze.
> All over the country villages at this time of year
> The sound of pounding echoes like mutual question and answer.
> You may enjoy at your will the jade fragrance of cooking rice,
> Or watch the water flowing in and out of the slippery spoon,
> Or listen to the water-turned wheel industriously treading.

During the course of time, moreover, the water trip-hammer was put to many uses other than that of hulling rice. Its use in forges we have already mentioned (pp. 378 ff.), and it is interesting to find a Taoist connection, for the pharmaceutical hermits employed it for crushing mica and other minerals intended as drugs. Li Pai addressed two poems to his wife, who had once been a Taoist nun, when she was visiting a former colleague; one of which goes:[k]

> You, on your visit to the nun Rise-in-the-Air
> Must have reached by now her home in the grey hills,
> Where the streams work the mica-grinding pestles
> And the wind sweeps through the flowers of the rose-bay tree.
> If you find yourself loth to leave this pleasant retreat
> Invite me also to enjoy the sunset glow.

And Pai Chü-I has a poem on 'Visiting Kuo the Taoist and finding him not at home' in which he sees that the trip-hammer for the mica is working on, though unattended.[l]

[a] *Wang Yin Chin Shu*, cit. *TPYL*, ch. 762, p. 6b; *Chin Shu*, ch. 33, p. 13a.
[b] *Wang Yin Chin Shu*, cit. *TPYL*, ch. 762, p. 6b.
[c] *Chin Shu*, ch. 41, p. 1a.
[d] *Shih Shuo Hsin Yü*, ch. 2A, p. 50a; *Chin Shu*, ch. 92, p. 8a.
[e] Quoted by *TPYL*, ch. 762, p. 7a, from the *Chin Chu Kung Tsan*[8] (Eulogia of Distinguished Men of the Chin Dynasty) by Fu Chhang.[9]
[f] *TPYL*, ch. 762, p. 7a, referring to Wang Hun.[10]
[g] *Wang Yin Chin Shu*, cit. *TPYL*, ch. 762, p. 6b, referring to Liu Sung.[11] His biography in *Chin Shu*, ch. 46, p. 1a, says that flooding was caused by the blockage they made, so he recommended that they should be torn down. The emperor gave way and the people were much benefited.
[h] *Thang Yü Lin*, ch. 3, p. 30a.
[i] *Po Chai Pien*[12] (Papers from the Anchored Dwelling), ch. 2, p. 4b.
[j] Tr. auct.　　　[k] Tr. Waley (13), p. 73.　　　[l] Waley (13), p. 74.

[1] 衛瓘　[2] 魏舒　[3] 褚陶　[4] 杜預　[5] 連機碓　[6] 裴度
[7] 方勺　[8] 晉諸公讚　[9] 傅暢　[10] 王渾　[11] 劉頌　[12] 泊宅編

Then in the Ming we have an account of the use of trip-hammers by the paper-makers of Fukien; Wang Shih-Mou[1] in his *Min Pu Su*[2] (Records of Fukien) says that the noise of them is constantly audible to those who travel in boats on the rivers. And the same was said a little later of the perfume-makers of Kuangtung, where the fragrance from the mills was often carried away downstream for miles.[a]

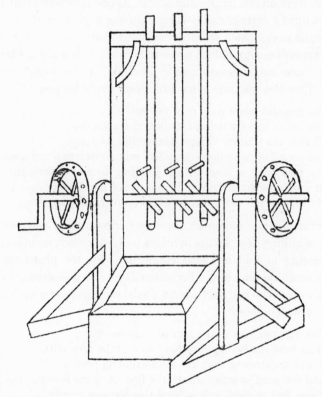

Fig. 619. A vertical pestle stamp-mill, characteristic of medieval Europe; one of the gunpowder mills figured in the MS. of the Anonymous Hussite engineer, *c.* +1430 (from Beck (1), p. 280).

Once more we may note that the water-trip-hammer of Tu Yü and Wang Jung was the lineal ancestor of all heavy mechanical hammers ('martinets') until the introduction of the steam-hammer.[b] Eighteenth-century occidental types of forge-hammer reproduce the design systematically with hardly any modification.[c] It was as characteristically Chinese as the vertical pestle stamp-mill (Fig. 619) was European.[d] While

[a] *Kuangtung Hsin Yü, c.* +1690.
[b] As des Noëttes (3) and Huard (2) have recognised.
[c] For example, the hydraulic hammer of Smeaton about +1780, diagrams of which are reproduced by Wolf (2), p. 639; or that depicted in the *Encyclopédie* of Diderot (Fig. 618), reproduced by Gille (1), p. 94. A photograph of one closely similar to this, still existing at a Yorkshire ironworks, is given by C. R. Andrews (1). Cf. many other pictures in the special monograph of Evrard (2) and in Baroja (4); Johannsen (14), figs. 83, 110, 183.
[d] The advantage of the former type is that the weight of the hammer is taken on the fulcrum and does not fall entirely on the rotating parts.

[1] 王世懋   [2] 閩部疏

the latter appears already in the +15th-century engineering MSS.[a] and in Agricola, the recumbent trip-hammer is not generally seen in Europe until the start of the +17th century, when it is illustrated by Sandrart[b] and Zonca (+1621)[c]—though in China the first printed illustration of it is in the *Nung Shu* of +1313. However, the tilt- or trip-hammer was certainly known in Europe before Zonca's time, because

Fig. 620. Probably the oldest European illustration of the martinet forge-hammer, the cut from Olaus Magnus' *De Gentibus Septentrionalibus*, +1565. Behind the three martinets a water-wheel is seen working the wood-and-leather bellows of an Osmund bloomery furnace. This reminds us that before the coming of the blast-furnace to Europe in the late +14th century water-powered forge-hammers were particularly important for the freeing of the wrought-iron blooms from slag inclusions (cf. Sect. 30d below).

Leonardo often sketched it.[d] Apart from this, the oldest European printed illustration of a martinet forge-hammer is probably the +1565 cut in Olaus Magnus (Fig. 620). Gille (7, 11), following Pereme,[e] takes the forge-hammer of +1116 at Issoudun as the first of such machines in Western Europe, but a more recent study[f] prefers a Catalan

---

[a] Cf. Berthelot (4). Particularly good drawings are in the *Mittelalterliche Hausbuch* (Anon. (15), Essenwein ed. pls. 36b, 38b, Bossert & Storck ed. pls. 40, 43). One may still see these stamp-mills at work in the Maltatal, Carinthia, and there are some in the Folklore Museum at Iaşi, Rumania.

[b] See Frémont (13), p. 93. The Anonymous Hussite, however, has it (+1430); Feldhaus (1), col. 1077; Beck (1), fig. 326. And the *Flores Musicae* of Hugo dem Reutlinger (+1492) shows it (Feldhaus (20), fig. 176) in a scene illustrating the story of Pythagoras (cf. Vol. 4, pt. 1, p. 180).

[c] For fulling-mills, pp. 42, 94. Cf. Beck (1), pp. 299, 313; Uccelli (1), p. 824; Nicolaisen (1).

[d] For file-cutting machinery and other uses as well as for forges; cf. Beck (1), pp. 107, 434 ff., figs. 121, 626, 628, 630, 636, 637, 641, 642, etc.; Uccelli di Nemi (3), nos. 22, 105. Frémont (13), p. 106, reproduces a relevant page from the *Codex Atlanticus*. Date about +1490.

[e] (1), p. 293. Cf. p. 379 above.   [f] Anon. (52), pp. 23 ff.

document of +1190. In all probability these were of the vertical stamp-mill type before Leonardo. Perhaps there may be some significance in the fact that the other great use of the tilt- or trip-hammer operated by water-power was for fulling cloth, since on different grounds (as will be seen in Sect. 31 below) there is reason for thinking that much Chinese textile machine design made its way to Europe about the time of Marco Polo.[a] The fulling-mills at Prato in Italy used trip-hammers traditionally,[b] but it is doubtful whether their design goes back, as suggested, to the beginning of the industry there about +983. Gille (4, 7), following Bloch (2), claims a number of French examples from *c.* +1050 onwards, but the earliest of these are very doubtful cases. Carus-Wilson (1) gives +1185 as the date of the first fulling-mill in England; it belonged to the Yorkshire Templars. Yet again it is most probable that all these pre-Leonardo fulling machines were of the vertical stamp-mill type.[c] Perhaps further evidence will reveal the exact time at which Europe received the recumbent hydraulic trip-hammer from the Chinese culture-area.

(6) WATER-MILLS FROM THE HAN ONWARDS

It is curious that references in early Chinese literature to the mill *par excellence*, the rotary millstones driven by water-power, are much rarer than those to the water-driven trip-hammer. This may arise perhaps from a fluidity of terminology at that time, especially among the scholars whose qualifications were not technical.[d] It would not have been difficult to confuse *chhui*,[1] the equivalent of the grain-mill proper, or even *wei*[2] (cf. p. 188 above) with the word *chui*[3] (hammer) used for *tui*[4] (the tilt- or trip-hammer). Moreover, in the texts already referred to, some books and some editions[e] write *wei*[2] or *mo*[5] instead of *tui*,[4] thus attributing to Tu Yü (+222 to +284) multiple mills worked from a single water-wheel by gearing; and water-mills instead of water-powered trip-hammers to Chhu Thao (*c.* +240 to +280) and Wang Jung (+235 to +306). There seems, in effect, little reason to doubt that water-driven quern mills were working at least as early as the hydraulic blowing-engines of the +1st century, and perhaps some time before them.

The first water-mill illustrations in the main iconographic tradition start from +1313; we reproduce the picture from the *Nung Shu* (Fig. 621) showing the characteristic horizontal water-wheel.[f] Another shows a breast- or over-shot vertical water-

---

[a] Marco Polo was in China, *c.* +1280, just about two centuries before Leonardo's prime.

[b] Uccelli (1), p. 131; Lynn White (7), pp. 83 ff.

[c] On the history of fulling machinery in general see E. K. Scott (1); Pelham (1).

[d] Nevertheless, it must be said that in later ages and among more technical writers the terminology is remarkably clear and self-consistent (cf. pp. 330 ff. above, 449 ff. below). Certain difficulties discussed by Amano (*1*) and Twitchett (3) derive, we believe, from Japanese confusions and aberrations of terms which were rather well defined in China.

[e] E.g. Pai Chü-I's *Liu Thieh* encyclopaedia of *c.* +800, ch. 24. Cf. pp. 195, 393, 399.

[f] Ch. 19, p. 4*a*. The distinct fall of the water upon the wheel should be noted. In his description (ch. 19, p. 4*b*) Wang Chên also gives an account of a mill with a vertical water-wheel and gearing in the Vitruvian style, actuating two pairs of millstones.

---

[1] 磈   [2] 磑   [3] 桘   [4] 碓   [5] 磨

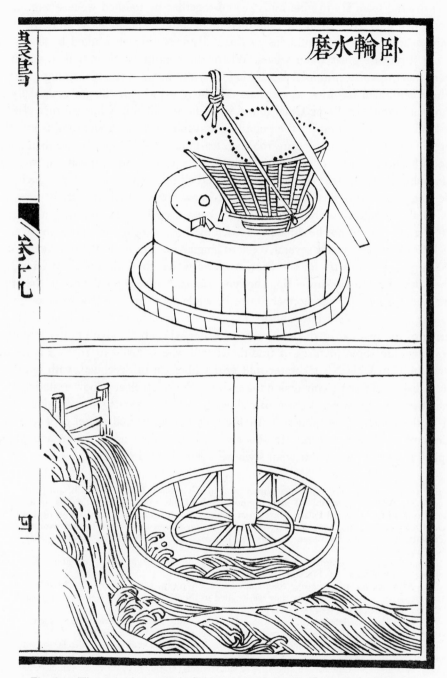

Fig. 621. The most characteristic Chinese form of water-mill, a millstone driven by a horizontal water-wheel (*wo lun shui mo*); from *Nung Shu* (+1313), ch. 19, p. 4*a*.

wheel driving from six to nine mills geared together by toothed wheels (Fig. 622).[a] Wang Chên tells us[b] that in his time there were some installations so large that trip-hammers and edge-runner mills as well as millstones were all worked by shafts and gearing from one great water-wheel. Where the conditions suggested it, the main water-wheel was also equipped as a noria for raising water in times of drought. Some of these combined factories, as we must call them, could mill enough grain daily for a thousand families.[c] Wang Chên found when he travelled in Chiangsi that plant of this kind was widely used for the pounding and rolling of tea-leaves, and from other sources[d] we know that there were 100 such tea-mills in +1083, and more than 260 in +1097. 'Who', asks the poem with which Wang concludes his disquisition, 'were the philosophical artisans (*chê chiang*[1]) who devised all this, with the mastery which they stole from the Author of Change?' Alas, we know none of their names.

A whole technical vocabulary must have been used by these millwrights (*wei po shih*[2]),[e] and it would not be difficult to assemble a series of terms equivalent to those which Curwen (4) gave for several Western languages. For example, the *Nung Chêng Chhüan Shu*[f] gives us *ti*[3] for the framework on which the millstones must rest, *chu mo*[4] or *chhi*[5] for the pin supporting the rynd, *chu mo*[6] or *yen*[7] for the eye in the upper stone and *lou tou*[8] for the hopper above it, while *tsun*[9] or *tshun*,[9] a word originally meaning the metal-shod butt end of a spear, was applied to the gudgeon or lower bearing of the main shaft of the horizontal water-wheel.[g] To give life to these paragraphs we add some pictures of traditional mill-wheels taken in 1958. Figure 623 shows the Chu Chia Nien[10] water-mill near Chhêngtu in Szechuan, with one of its wheels displaced during conversion to service as the small electric power-station of an agricultural co-operative. Figure 624 shows a different variety of horizontal rotor still in the hands of the joiners.[h] In Fig. 625 the horizontal wheel of a mill near Thienshui in Kansu is seen.[i] In this same countryside I easily found and photographed in 1958 mills of Vitruvian style with right-angled gearing.

---

[a] Ch. 19, pp. 10a, b. We reproduce the version in *Shou Shih Thung Khao*, ch. 40, p. 28a. Cf. p. 195 above. Wang Chia-Chhi (1) gives a photograph of the model illustrating this multiple mill which is now in the Imperial Palace Museum at Peking. For a similar but later design by Leonardo, cf. Ucelli di Nemi (3), no. 106. Entirely similar drives for stones in water-mills and windmills continued working well into the present century; cf. Jespersen (1); Wailes (1), fig. 14, (3), fig. 9.

[b] Ch. 19, pp. 12b ff.

[c] Taking the classical estimate of Mencius, I, vii, 24, of eight adults per family this might imply something over five tons. Cf. Fig. 627b, and p. 366.

[d] *Sung Shih*, ch. 94, p. 3b; *Hsing Shui Chin Chien*, ch. 97 (p. 1426); cf. Wang Chia-Chhi (1) and Fig. 439.   [e] As they were called in the Thang period (Gernet (1), pp. 143 ff.).

[f] Ch. 23, p. 12b; cf. Laufer (3), p. 22.

[g] We shall have a case later on (p. 454) where very heavy duty had to be performed by such a vertical end-bearing, with as much accuracy as possible since the purpose was horological.

[h] These are usually fitted with races of their own alongside the main water-wheels, for subsidiary drives.

[i] Very notable here is the dished structure of such wheels; cf. pp. 76 ff. above on vehicle-wheels and the invention of 'dish'. Here of course the purpose would be different; the water continues to do work as it falls towards the centre if the concavity of the wheel is upwards.

[1] 哲匠	[2] 礎博士	[3] 商	[4] 主磨	[5] 臍	[6] 注磨
[7] 眼	[8] 漏斗	[9] 鐏	[10] 朱家碾		

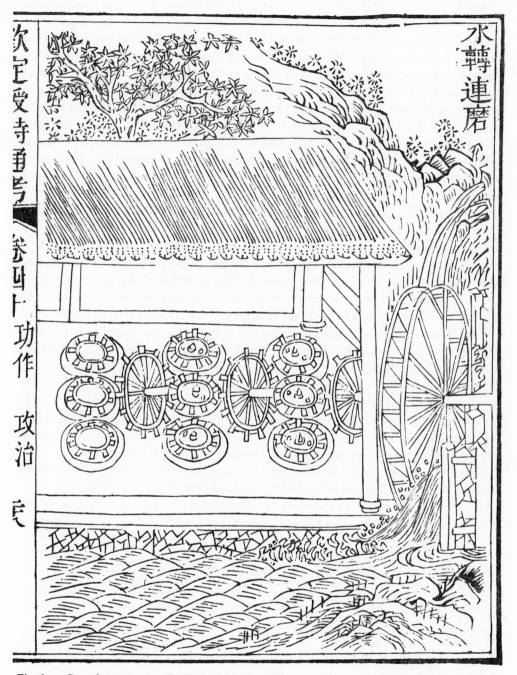

Fig. 622. Geared water-powered milling plant, nine mills worked by an overshot vertical water-wheel and right-angle gearing (*shui chuan lien mo*), an illustration of +1742 (*Shou Shih Thung Khao*, ch. 40, p. 28a). The oldest picture of such a plant is in the *Nung Shu* of +1313. Cf. Fig. 451.

Celebrated water-mills, which the emperor himself came to inspect, were erected by the great +5th-century mathematician and engineer Tsu Chhung-Chih,[1] about +488.[a] About +600, Yang Su,[2] one of the chief technologists of the Sui dynasty, was in control or ownership of thousands of them.[b] As time went on, however, conflict between the interests of water-power users and irrigation controllers greatly increased until from the Sui onwards physiocratic officialdom came into head-on collision with developing mercantile initiative. That the water-mills must not interfere with water-conservancy was explicitly laid down in the *Thang Liu Tien*.[c] The existing fragment of the Thang 'Ordinances of the Department of Waterways (*Shui Pu Shih*[3])',[d] dating from +737, which has been studied and translated by Twitchett (2), has several references to water-mills. Their owners are to construct adequate sluices and to ensure no interruption of traffic (Art. 7),[e] they must remove silt and sandbanks on pain of demolition (Art. 13), and in some cases use of the water-mills is allowed only during certain seasons (Art. 23). At other times, and if the water-supply suffices only for irrigation, the mill is sealed and the millstones impounded by the local government. In this period certain officials distinguished themselves as persecutors of millers (*wei chia*[4])—for example Li Yuan-Hung,[5] who in +721 complained that water-mills (*nien wei*[6]) belonging to wealthy families were jeopardising irrigation, and got authority to destroy them. In +764 Li Hsi-Yün,[7] as aggressive as Don Quixote, demanded that no less than seventy such installations (owned not only by great families but also by Buddhist abbeys) should be demolished, and this was duly done. But the largest destruction of millwrights' work occurred in +778, when eighty plants were torn down,[f] not excluding two water-wheels belonging to the saviour of the empire, the general Kuo Tzu-I[8] who had conquered An Lu-Shan.[g] Our information about these proceedings reveals that the mills were generally the property of imperial concubines and powerful eunuchs or Buddhist abbeys and rich merchants,[h] so the opposition of the Confucian bureaucrats was only one aspect of a perennial antagonism.

A few words may be added about each of these types of ownership. The eminent eunuch of Hsüan Tsung, Kao Li-Shih,[9] was famed about +748 as possessing a mill with five water-wheels which ground 300 bushels of wheat a day.[i] Similar wealth was ascribed to the abbot Hui-Chou[10] about the same time.[j] Already in +612 there

---

a *Nan Chhi Shu*, ch. 52, p. 21a; *Nan Shih*, ch. 72, p. 12a.

b *Pei Shih*, ch. 41, p. 33b; *Sui Shu*, ch. 48, p. 12b; cf. *Tu I Chih*, ch. 1, p. 22a. It is interesting to note in view of what has gone before (p. 26) that Yang Su is said to have employed in his private factories producing fine silk several thousand serving-lads and several thousand girls.

c Ch. 7, p. 9b. But the chief editor, Li Lin-Fu,[11] a high official related to the imperial family, possessed a considerable number of water-mills himself; see *Chiu Thang Shu*, ch. 106, p. 3a.

d A Tunhuang MS., P/2507 in the Bib. Nat. Paris.

e Parallel conflicts between milling and traffic interests occurred in +16th-century England (see P. N. Wilson, 1, 2). Cf. Sect. 28f.

f These events are all recorded in *Thang Hui Yao*, ch. 89 (p. 1622); cf. *Thang Yü Lin*, ch. 1, p. 23a. See further Wittfogel (5); Balazs (5), p. 36.

g See *Chiu Thang Shu*, ch. 120, p. 14a.   h Cf. Pulleyblank (1), p. 29.

i Something of the order of ten tons. *Chiu Thang Shu*, ch. 184, p. 4a. Cf. Fig. 627b.

j *Hsü Kao Sêng Chuan*, ch. 29.

| [1] 祖沖之 | [2] 楊素 | [3] 水部式 | [4] 碨家 | [5] 李元紘 | [6] 磑碨 |
| [7] 李栖筠 | [8] 郭子儀 | [9] 高力士 | [10] 慧冑 | [11] 李林甫 | |

PLATE CCXL

Fig. 623. Contemporary Chinese water-wheels; one of the horizontal wheels of the Chu Chia Nien water-mill near Chhêngtu, Szechuan, awaiting refitting during conversion of the mill to duty as a small electric generating-station (orig. photo., 1958).

Fig. 624. Horizontal rotor still unfinished, Chhêngtu (orig. photo., 1958).

Fig. 625. Horizontal water-wheel in position under a water-mill near Thienshui, Kansu, but not working at the time (orig. photo., 1958). Note shrouding, as in Fig. 623.

were quarrels with monks about water-mill revenues,[a] and two hundred years later Li Chi-Fu,[1] the geographer and meteorologist,[b] came into conflict with them over their claim for tax immunity. Indeed historians are now recognising that mill dues (*wei kho*[2]) were one of the richest sources of income for great abbeys during the Chinese Middle Ages.[c] But enterprising merchants were also fighting the hydraulic bureaucracy. When Wang Fang-I[3] became governor of Suchow in +653 his restoration of order in the city and its neighbourhood included the repair of the moat, and one of the first things he did was to remove some of their water-mills and tax the others heavily to feed the poor.[d] A few years later another official elsewhere, Chhangsun Hsiang,[4] attacked the merchants in the same way.[e] Eventually the whole milling industry had to be co-ordinated officially with the bureaucratic water-control, and we find that in +970 the two offices of Commissioner of Water-Mills (Eastern Region) and Commissioner of Water-Mills (Western Region) were established.[f] Further organisation was called for in +990.

During the Thang the water-mill had radiated to other countries in the Chinese culture-area, to Japan (via Korea) in +610 and +670,[g] and to Tibet about +641.[h] Later on peoples such as the Chhi-tan Tartars were quite familiar with them.[i] Late references we may omit, but Chao Mêng-Fu about +1280 devoted one of his poems on the *Kêng Chih Thu* to water-mills.[j] Early in the +10th century the abundance of water-mills in China caught the attention of an Arab traveller, Abū Dulaf Misʿar ibn al-Muhalhil,[k] who described no less than sixty mills on canals in and around what he believed to be the capital city of Sandābil.[l]

[a] See the story of Wang Wên-Thung[5] in *Kuang Hung Ming Chi*, ch. 6.

[b] Cf. Sections 21 and 22 above (Vol. 3, pp. 490, 520, 544). His denial of their claim is recorded in *Thang Hui Yao*, ch. 89 (p. 1622).

[c] There is abundant material in the monograph of Gernet (1), on which see also Twitchett (3).

[d] *Chiu Thang Shu*, ch. 185A, p. 12b.   [e] *Wên Hsien Thung Khao*, ch. 6 (p. 69.2).

[f] *Shih Wu Chi Yuan*, ch. 7, p. 5a. It would be interesting indeed to make a study of all water-mill legislation in China for comparison with that of European countries, which has been rather extensively worked out by Koehne (1) and others. Perhaps nothing would elucidate more clearly the differences between feudal and feudal-bureaucratic society. While the collection of dues for milling at the 'lord's mill' formed so important a part of European feudal law and dispute, the Chinese bureaucrats were much more concerned with keeping the waterways free from encumbrances which interfered with irrigation and grain-tax traffic. To what extent this official discouragement of mill-work acted as a social factor inhibitory to medieval engineering in China remains to be seen—there were other sides to it as well, e.g. the official interest in the iron and steel industry and the water-power which it needed. This is not the place to enlarge upon the social aspects of water-mills, but these few paragraphs seemed necessary to give some background for the life and work of the artisans concerned.

[g] *Nihongi*; Aston tr. (1), vol. 2, pp. 140, 294. The intermediary of +610 was a Korean monk named Tamjing.[6] He was a remarkable carrier of technological culture, knowledgeable in many arts and crafts besides millwright's work; see Tamura Sennosuke (1), pp. 118 ff. In +840 Ennin still found water-mills remarkable (Reischauer (2), p. 267, (3), p. 156).

[h] So at least Laufer (3), p. 35, but he was only following the general tradition which attributes so many technical introductions to Tibet to the Chinese princess Wên Chhêng Kung Chu[7] who married King Srong-btsan-sgam-po. How much there is in it remains rather obscure.

[i] *Liao Shih*, ch. 48, p. 5b; cf. Wittfogel & Fêng (1), pp. 124, 138; Bretschneider (2), vol. 1, p. 125.

[j] *TSCC*, *Shih huo tien*, ch. 39, *hui khao*, 5, p. 21b.

[k] Cf. Sarton (1), vol. 1, p. 637; al-Jalil (1), p. 178; Minorsky (5), p. 123.

[l] Translation in Ferrand (1), vol. 1, p. 219; cf. von Schlözer (1); Marquart (1); von Rohr-Sauer (1).

[1] 李吉甫   [2] 磑課   [3] 王方翼   [4] 長孫祥   [5] 王文同

[6] 曇徵   [7] 文成公主

This account of Ibn al-Muhalhil will involve us in a momentary digression. Tshen Chung-Mien (1), who has made a special study of the subject, believes that the place Sandābil was the city of Shantan[1] in Kansu province on the Old Silk Road, and that the travellers erred in supposing it to be a capital. His identification is confirmed by the traces of some remarkable water-works which still exist there.[a] At the top of the

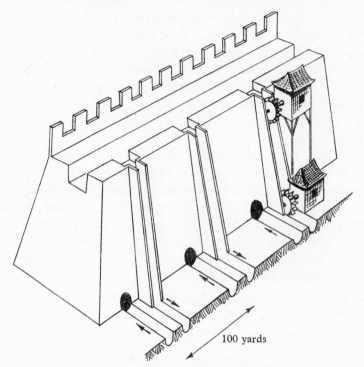

100 yards

Fig. 626. Interpretation of the description of the water-mills of Sandābil (Shantan in Kansu) by Ibn al-Muhalhil, c. +940, in the light of the remains still existing until recently at that Old Silk Road city.

city walls and all along them there runs a wide conduit,[b] giving rise, at intervals of a hundred yards, to a succession of flared brick down-conduits[c] which once conducted the water to dozens of internal water-ways or channels exactly as Ibn al-Muhalhil said about +940. In his time each channel turned two mills,[d] and each street had

[a] Or did so until the recent removal of most of the ancient city-walls. For this information I am indebted to the personal observations of my friend Mr Rewi Alley (June 1952), with whom I first visited Shantan in 1943, and who afterwards made it the centre of that noble undertaking, the Bailie Technical College of the Chinese Industrial Cooperatives. Lacking acquaintance with the site, neither Yule (2), vol. 1, pp. 138, 252, nor any other editor, understood Ibn al-Muhalhil's description.

[b] About 1 ft. 6 in. deep and of an equal width.

[c] About 1 ft. 6 in. wide but only 9 in. deep. These dimensions are due to Mr Courtney Archer, one of Mr Alley's colleagues.

[d] He says that one mill 'let the water flow down underneath it' while the second one in each case 'let the water flow away along the ground'. An attempt has been made to interpret this in the figure; perhaps the upper wheel was overshot and the lower undershot.

[1] 山 丹

two streams, one bringing pure water and the other intended to carry away drainage.[a] So long as the system was in operation, the water was raised to the level of the ramparts by very large norias[b] located at the south-east corner of the city, the emplacements of which remained still visible beside a small deviated river until a year or two ago. The whole system may well have been at work until late in the Chhing dynasty, and it would be interesting to know whether it had a parallel in any other Chinese city. Perhaps Muslim ritual hygiene had exerted a stimulating influence in this frontier region earlier than Ibn al-Muhalhil's time.

There remains little to add to this subsection. The Thang references which we have been considering speak indiscriminately of *shui nien*[1] and *shui wei*[2] or both together, and indeed it was quite natural that the edge-runner mill (cf. Fig. 453 and p. 199 above) should have been powered by water from an early time. The machine is illustrated in the *Thien Kung Khai Wu*,[c] but the oldest picture is of +1313;[d] these may be compared with many modern photographs. I found these mills especially common in Szechuan; generally with only one running stone. Their origin seems fixed rather definitely between +390 and +410, for in the biography of Tshui Liang[3] we read:[e]

When (Tshui) Liang was in Yungchow, he read the biography of Tu Yü, from which he learnt that Tu had devised the 'eight (geared) mills', greatly benefiting his contemporaries thereby. Tshui therefore taught the people (to apply water-power to) edge-runner mills and roller mills ((*shui*) *nien*[4]). After he had attained the position of Grand Counsellor, he memorialised the emperor suggesting that a dam should be built on the Ku Shui east of the Chang-fang bridge (to provide water) for water-powered runner-mills and roller-mills.[f] So these were established in several tens of places, and the profit to the country was ten times greater than ever before.

In +550 the first emperor of the Northern Chhi presented a 'set' of edge-runner mills to the dethroned emperor of Eastern Wei.[g] Although this type of mill shared with others the vicissitudes of government regulation in the Thang period, it persisted virtually unchanged until the present time. Moreover, like others of these simple Chinese designs, it probably made its way to Europe,[h] and in a book such as that of Zonca (+1621) one may see a vertical undershot water-wheel operating an edge-runner mill by right-angle gearing,[i] in no essential way different from that of Tshui

---

[a] In +1221 each of the streets of Samarqand also had two channels of water running along it, according to the *Chhang-Chhun Chen Jen Hsi Yu Chi* (Waley (10), p. 93). The *Sung Shih* (ch. 490, p. 10a) remarks on the water-mills of Turfan; one would like to know of what sort they were and whether they had come from east or west.

[b] See p. 356.

[c] Ch. 4, p. 14b.   [d] *Nung Shu*, ch. 19, p. 8a.

[e] *Wei Shu*, ch. 66, p. 23a; *Wei Lüeh*, ch. 9, p. 18b; *Pei Shih*, ch. 44, p. 19b; tr. auct. adjuv. Laufer (3), p. 33; Yang Lien-Shêng (5), p. 118, (9), p. 130; Wang I-Thung (1). On Tu Yü cf. pp. 195, 393, 396; on the Ku River, pp. 346, 372.

[f] On this place see *Lo-Yang Chhieh Lan Chi*, ch. 4, pp. 9a, 19a, b (pp. 90 ff.). It was seven li west of the city.

[g] *Pei Chhi Shih*, ch. 4, p. 11b. Cf. p. 202 above for a +9th-century mention.

[h] As we have already suggested (p. 204 above), the common European name of 'gunpowder mill' may point to a Chinese origin.

[i] Cf. Beck (1), pp. 298, 309

[1] 水碾   [2] 水磑   [3] 崔亮   [4] 水輾

Liang. Indeed, such mills held their own until the coming of steam and finally electric power in neotechnic times.

Lastly, we must consider a few additional uses of water-power. Saw-mills[a] driven hydraulically have been considered an early development in Europe, one such, on the Moselle, being allegedly referred to as early as *c.* +369 by Ausonius,[b] and a famous one illustrated in his MS. by Villard de Honnecourt (*c.* +1237).[c] Specifications for such machines appear in the *Chhi Chhi Thu Shuo* (+1627), as noted above (pp. 217, 218), but it would not be safe to assume that the idea had never occurred to anyone in China previously, though we have no evidence of it.[d] Water-wheels were there put to all sorts of uses, such as the polishing of stone columns for building, as Hosie saw during his travels in Szechuan more than fifty years ago.[e] Another remarkable employment was that which we have come upon already in the reign of the Thang emperor Ming Huang; whose Liang Tien[1] (Cool Palace) was equipped with water-power about +747 to work air-conditioning fans (*shan chhê*[2]).[f] In view of this, it was perhaps curious that the Chinese did not work out (so far as we know) a rotary blowing system for their hydraulic furnace-bellows.

Particularly remarkable was the use, at least as early as +1313, of water-power for textile machinery. The *Nung Shu* illustrates a spinning-mill (*shui chuan ta fang chhê*[3]) in which we see a vertical undershot water-wheel and a large driving-wheel with a belt-drive on the same shaft working a multiple-bobbin spinning-machine for hemp and ramie, perhaps also for cotton.[g] It is shown in Fig. 627*a*. This should be enough to give pause to any economic historian, especially as Wang Chên clearly says that such installations were common in his time in districts which grew these textile crops.

a Cf. Fischer (4).

b This reference (*Mosella*, v, 362 ff.) has caused a lot of trouble. As Lynn White (3) points out, the location does not fit, for although the poet speaks of marble the only regional stone is blue roofing slate which needs no sawing. Even if marble had been available it would have been sawn, according to ancient practice, by horizontal smooth saws and abrasives, not by vertical toothed saws. Indeed the *Mosella* is now suspected of later provenance, and may well be due to Ermenricus of St Gall, *c.* +850, who perhaps included a description really applying to somewhere else. See further in Lynn White (7), pp. 82 ff.

c Cf. Usher (1), 2nd ed., p. 186; Gille (14), p. 644. In this machine the saw is depressed by lugs on the main shaft of a vertical water-wheel and raised by a spring pole, the feed of the log being brought about by a separate toothed wheel on the main shaft. Lynn White (3), (7), p. 118, considers this 'the earliest instance of a fully automatic industrial machine involving two separate but correlated motions (that of the saw and that of the feed). It therefore marks an epoch in the development of mechanical devices.' But surely the winder-barrels and the ramping-arm of the Chinese silk-winding machine already mentioned (pp. 2, 107 above) would qualify better for this important distinction? The classical description of this occurs, as we have seen, about +1090, and it must be considerably older than that. As we shall see in a moment, water-power was in wide application in China for textile machinery by +1300, and this use therefore clearly existed already in Villard de Honnecourt's time. It is true that we do not know whether such a drive was applied to this particular machine in the +13th and +14th centuries, though it certainly was later on (see immediately below); but power of some sort must always have been applied, and the criterion of separate but correlated motions was fully satisfied in the silk winder-layer.

d Except a very remarkable naval use of saws mounted on boats and powered by paddle-wheels working ex-aqueously (p. 424 below).

e (4), p. 96.   f *Thang Yü Lin*, ch. 4, p. 2*b*, tr. p. 134 above.

g Ch. 19, p. 16*a*; ch. 22, p. 6*b*. A study of the mechanism by Li Chhung-Chou (2) is now available, and Liu Hsien-Chou (7) reproduces the photograph of a model.

1 涼殿   2 扇車   3 水轉大紡車

PLATE CCXLI *a*

Fig. 627 *a*. Water-power applied to textile machinery in the late +13th and early +14th centuries. Left, the 'great spinning-mill' for hemp and ramie, perhaps also for cotton (*Nung Shu*, ch. 22, p. 6*b*). Right, the undershot vertical water-wheel and the large driving-wheel with a belt-drive which powered the plant (*shui chuan ta fang chhê*); Wang Chên tells us that in his time (+1313) such machinery was common in districts which grew these textile crops (*Nung Shu*, ch. 19, p. 16*a*). Cf. Sect. 31 below.

PLATE CCXLI *b*

Fig. 627*b*. Milling plant on a mountain river; scroll-painting by an unknown artist of the Yuan dynasty, *c.* +1300 (Liaoning Provincial Museum). In the tradition of all Chinese painters, the artist worked not from the life, but in tranquil recollection, hence not being a millwright, he confused paddle-wheels with gear-wheels. Nevertheless it is clear that a number of different machines were powered in this mill by two large horizontal water-wheels (centre and right lower compartments). The left upper compartment shows right-angle gearing, probably working a battery of trip-hammers. The centre upper compartment has the main mill-stones and an edge-runner mill in front of the staircase. The right upper compartment has a curious contraption almost certainly to be interpreted as an attempt to draw from memory the crank, connecting-rod and piston-rod combination, i.e. the water-powered reciprocator, working a flour-sifter, perhaps the latticed cupboard seen at the back of it. The left and centre lower compartments show a number of badly drawn gear-wheels, both horizontally and vertically mounted, the exact purpose and connection of which is not clear; but in the right lower compartment the artist has drawn a gear-wheel of tub shape, equally isolated but with admirably designed short pinion teeth, thus revealing to us the well-developed technique of the Yuan millwrights.

Indeed traditional Chinese culture developed a multifarious use of water-power. When the great Korean scholar Pak Chiwǒn visited North China in +1780, he saw it at work, and later wrote in his memoirs: 'When I was passing through San-ho Hsien (about 40 miles east of Peking), I saw water-power used for all kinds of things, blowing air for furnaces and forges, winding off the silk from cocoons, milling cereals—there was nothing for which the rushing force of water to turn wheels was not employed.'[a]

For us all this raises the question of the date of the appearance of the vertical water-wheel in China; there is no illustration of it earlier than +1300, nor any literary text which positively necessitates it, though as has already been said, it would go more naturally with the trip-hammer batteries which are so common from the Han onwards.[b] No solution of the problem seems in view at the present time. In this connection, however, we may note in passing the pleasant sight seen by Troup (1) at a Japanese silk-filature at Tenryu-gawa below Lake Suwa, of a single vertical water-wheel doing double duty as a noria for raising the water needed by the workshop, and as a source of power. This was just what Wang Chên had recommended (cf. p. 398). Finally, we must not forget to mention the most delicate and impressive of all uses of vertical water-wheels, namely their employment, from the +2nd century onwards in China, for assuring the slow rotation of astronomical apparatus. This will be discussed below (pp. 481 ff.). And with a different conception of useful work, the pious Tibetans applied water-power to the continuous rotation of prayer-wheels.[c] After all that this survey has revealed it would be hard to agree with Lynn White that China showed no more imagination than Rome in the application of water-power to industrial purposes.[d]

(7) The Problem of the Inventions and their Spread

When we survey the whole of the above evidence, we may be moved to wonder whether perhaps the horizontal water-wheel and the vertical water-wheel were not two entirely distinct inventions. On this provisional conception, the vertical wheel would have been an adaptation of the (originally Indian?) noria,[e] while the horizontal wheel would have been, as it were, a downward extension of the runner component of the rotary quern. The right-angle gearing required by the Vitruvian type might be considered primarily Alexandrian, for although the Han engineers were quite expert with gear-wheels, the Alexandrians, going back to Ctesibius in Chhin Shih Huang Ti's time, were perhaps chronologically a little ahead of them.[f] On this view, the originally Chinese horizontal water-wheel would have made its appearance in Persian Pontus

a *Yǒn-Am Chip*, ch. 16, cit. Yi Kwangnin (1), p. 102. De Navarrete mentions (+1676) paper-mills (Cummins (1), vol. 1, p. 151).

b In modern, though traditional, China, it is very much in evidence, as Hommel's text and photographs show, (1) pp. 81, 87, 121; both overshot and undershot.

c See Sarat Chandra Das (1), p. 28; Cunningham (1), p. 375; Rockhill (3), p. 232, (4), p. 363; Waddell (1); Simpson (1), pp. 11, 17, 19, etc. Emery & Emery (1) saw water-wheel prayer-cylinders still working thirty years ago near Nan-phing, in the Choni region, an autonomous Tibetan district in southern Kansu above Sung-phan.   d (7), p. 82.

e This view gains something from the fact, pointed out by so many commentators, that Vitruvius treats of mill water-wheels in direct juxtaposition with norias.

f Cf. Drachmann (5). But in the light of the evidence here presented it is impossible to accept the statement of Forbes (11), p. 39, that the water-mill 'came to the East with Buddhism'.

under Mithridates, and then continued its spread, undeterred by the Vitruvian design, around the coasts of Europe to end up finally in Scandinavia as the 'Norse mill'. The fact that Vitruvius makes no mention of the horizontal water-wheel is taken by Bennett & Elton[a] as evidence that it was widely used in the Roman world in his time, but this conclusion has no archaeological basis, and we should prefer to believe that his silence indicates he did not know of it at all. The westward and northward spread of the horizontal wheel must have taken place in the first Christian centuries, by what means exactly remains for further research to elucidate. Somewhere during this process it acquired the oblique setting or 'pitch' of its vanes. The absence of this in Chinese water-wheels perhaps bears witness, as we have said, to their high antiquity.[b] Meanwhile the noria was reaching China, and would have given rise there also to the vertical water-wheel, though perhaps at a later date than in the West; unless, of course, it was the form chiefly used in and before Huan Than's time for the trip-hammers, which always remains possible, indeed probable, for obvious technical reasons.[c]

Let us now return for a moment to the ideas which Troup derived from his study of the Chinese spoon tilt-hammer and stamp-mill, as he saw it at work in Japan. His suggestion was that the spoon tilt-hammer, by a proliferation of its spoons, had been the origin of the overshot vertical water-wheel, while the current-operated noria had given rise to the undershot vertical water-wheel. Unfortunately, the position of the spoon tilt-hammer is a paradoxical one, for while on the one hand it bears all the marks of primitivity (its connection with the age-old swape, and the simple flume), on the other hand there are no early references to it, or at least none have been found so far.[d] Moreover, the theory of Troup has to face the difficulty that the most characteristic ancient Chinese water-wheels seem to have been horizontal ones. A possible danger is that what Troup saw may have been machines built by people who were used to making spoon tilt-hammers, but who had come into contact with builders of true vertical water-wheels; and this suspicion is strengthened by the fact that what the Japanese multiple-spoon 'wheels' worked were vertical stamp-mill pestles and not the characteristic Chinese recumbent hammers. For the present these are questions which will have to remain open.

Since this section was first written, however, new information has come to light which might strengthen considerably a belief in the antiquity of the spoon tilt-hammer in China. I refer to the whole remarkable story of the water-powered mechanical

---

[a] (1), vol. 2, p. 32.

[b] With all this in mind it is impossible to accept the view of Bloch (2), p. 544, that the horizontal water-wheel was a regressive phenomenon. On the other hand the view of Curwen (4) that it is more primitive cannot be said to be proved until we know more of Persian and Chinese water-wheels before the Christian era.

[c] The late tradition (Cedrenos, x, 532 ff.) that water-mills were introduced to the Indians by a Greek, Metrodorus, may not be wrong, but if the noria was really primarily Indian, one would rather expect an indigenous development.

[d] It is curious that many of the Arabic drawings of water-wheels from the early +13th century onwards, seen, for example, in Wiedemann & Hauser (2), depict them as a series of radiating spoons with no wheel rim. What the significance of this may be remains obscure. Moreover, the principle of a counterbalance with water was used occasionally by the Alexandrians (notably Philon, *Pneumatica*, ch. 31; cf. Beck (5), pp. 69 ff.; Drachmann (2), pp. 70 ff.), but only as a component part of other mechanisms and never for doing useful work.

clocks which, with their peculiar linkwork escapements, fill the period between the early +8th century and the time of the first mechanical clocks of Europe (the early +14th); this is told in Section 27*j* below. All the driving-wheels of these time-keepers were fitted with scoops or buckets round the rim, motion being inhibited by the escapement until each 'spoon' successively was full, and Troup, had he known of them, would certainly have been inclined to recognise in them the logical extensions of his eight-spoked spoon-wheels.

As for the comparative dating of water-power as between China and the West, it clearly constitutes a case of approximate simultaneity as puzzling as that of rotary milling itself (cf. p. 190). The Asia Minor date of −70 or so is authorised by subsequent writings of the same century, but the −1st century for China is inescapably indicated by the words of Huan Than (p. 392) written in +20, though in which of its decades a water-wheel was first made to drive the trip-hammers there remains quite obscure. The −1st century is also indicated by the fact that the earliest specific description in a Chinese source refers to metallurgical furnace-blowing, a job more complicated, one would think, than the simple grinding of cereals, and hence implying a longer antecedent development.[a] One must hope that further researches will shed more light, all the more so as (apart from Vitruvius) we know nothing of the mounting, whether vertical or horizontal, of any of the oldest water-wheels.

The vertical water-wheel might seem at first sight to have had the greater future, but in fact it was the other way round, for the horizontal water-wheel was the direct ancestor of one of the most impressive power-sources of the post-Renaissance neotechnic age, namely the hydraulic and steam turbine.[b] Here there would be no space to sketch, even briefly, the history of this machine.[c] In the horizontal water-wheel, as in the vertical undershot wheel, the movement of the rotor results wholly from the impact or impulse of the water acting on the vanes. Vertical overshot wheels, however, are turned mainly by the weight of water rather than its momentum. In Heron's aeolipile, on the other hand, there was neither impulse nor gravitational load, and the movement of the rotor resulted only from the reaction of the steam jets accelerating through the tangential nozzles. But in the turbine rotor water or steam passes through pipes or ducts, or over curved vanes, in such a way that the power derives both from the impulses and the reactions set up between the medium and the curving passages within which it travels. Thus the turbine is essentially a combination of the ancient Chinese water-wheel with the Alexandrian aeolipile. Historically, it springs from the horizontal water-wheels of the South of France (e.g. those at Toulouse already mentioned on p. 368 above), a critical examination of which took place in the +17th century.[d] Already in +1578 Besson had enlarged

<hr />

[a] Here it is natural to ask whether some of Huan Than's powered trip-hammers were not used for forge purposes, like the later 'martinet' hammers (cf. p. 394). The advanced state of Han iron and steel technology would strongly suggest so, but we have as yet found no positive evidence that they did anything but pound grain or decorticate rice. This may well come to light.

[b] The connection was realised by Feldhaus (25) and by Bennett & Elton (1), vol. 2, p. 28; cf. P. N. Wilson (4), p. 19; Lynn White (7), p. 160.

[c] A convenient account of the history of the turbine is given by Usher (1), 1st ed., pp. 345 ff. The standard work is that of Crozet-Fourneyron (1).

[d] Cf. de Bélidor (1), vol. 1, pl. 23 and p. 303; Eude (1), pp. 11 ff.

the horizontal water-wheel into a cone, with spiral ducts or blades on its surface, and this 'tub-wheel' revolved within a closely confining chamber.[a] The Toulouse wheels also had spiral or curved vanes. During the 18th century great attention was given to the study of turbine principles[b], and wheels were made using different proportions of impulse and reaction, but decisive advances awaited the work of Fourneyron, who by 1832 was able to construct turbines of essentially modern type up to 50 h.p. capacity. And what the beginning of the century had done for water, the end of it achieved, at the hands of de Laval and Parsons, for steam. The role which the turbine still plays in the generation of electricity and the propulsion of ships is familiar to everyone. It may long be with us.[c]

By now the reader may well have had enough of mills, but there remain two kinds of mill about which nothing has yet been said, namely mills which were mounted on boats, and mills which told the time. In other words, all paddle-wheel ships and all mechanical clocks are children of the water-mill, and China was the land of their infancy. Let us now consider them in turn.

(i) WHEELS EX-AQUEOUS AND AD-AQUEOUS; SHIP-MILL AND PADDLE-BOAT IN EAST AND WEST

The first story begins, unlike most stories in this book, in Rome. In the year +536 the Goths were besieging the city, and having intercepted the water-supply for the Janiculum mills were on the point of reducing the defenders to starvation, when the Byzantine general, Belisarius, who commanded the garrison, conceived the idea of mounting the mills with their water-wheels on moored boats in the Tiber. So at least Procopius tells us:[d]

When the water of the aqueducts was cut off and the mills stopped, it was not possible to work them with animals for the city was short of food, and provision could scarcely be found for the horses. But Belisarius, an ingenious man, found a remedy for the distress. Below the bridge across the Tiber, which arches to the walls of the Janiculum, he extended ropes, well fastened across the river from bank to bank. To these he moored two boats of equal size, two feet apart, at a spot where the current flowed with greatest velocity under the arches; and placing two millstones in the boats, he suspended the machine by which they are usually turned (i.e. the water-wheel) in the space between. He also contrived, at certain intervals downstream, other machines of the like kind, and these being put in motion by the force of the water, drove as many mills as were necessary to grind food for the city.

It would seem that this system was afterwards very widely used. In Rome itself these floating mills were pictured by Giuliano di San Gallo (+1445 to +1516) in an illustrated MS. studied by Horwitz (10)—from this (Fig. 629) it can be seen that by that

---

    [a] (1), pl. 28. Cf. de Bélidor (1), vol. 1, pl. 19 and p. 302; Beck (1), p. 195; Usher (2), p. 328.
    [b] Segner; von Euler; Smeaton, etc.; cf. Stowers (1); Wilson (3); Bied-Charreton (1).
    [c] As Stowers (2) has written, when the world's supplies of coal and oil have been exhausted and the uranium and thorium reserves are failing too, the winds will presumably still be blowing and the waters as restless as ever in rivers and tides. Far in the future, humanity may need their power.
    [d] De Bello Gothico, I, 15 (v, xix. 19); tr. Dewing (1), vol. 3, p. 191; Bennett & Elton (1), vol. 2, p. 61.

Fig. 628. Ship-mills on the Rhône at Lyons, in a drawing of +1550 (cf. Audin, 1). In the centre, the Pont de la Guillotière, dating from before +1180. Below, in the right-hand corner, a glimpse of the buildings of the Hôtel Dieu, the hospital where François Rabelais (d. +1553) was physician.

time very broad water-wheels were used. On the Dniester (Cash) and Danube (Reichel) they continued till contemporary times. They were at Venice in the +11th century,[a] and in France from the +12th down to the end of the +18th (cf. Fig. 628);[b] in England they made a brief appearance in the +16th only.[c] From the point of view of the diffusion problem it is interesting that in quite modern times these ship-mills were frequently found in Armenia. For parallels with all the European developments numerous references can be found in Arabic literature, beginning with the Banū Mūsā in the +9th century, as Wiedemann (6) has shown.

It has been recognised that ship-mills were not confined to Europe, but historians of technology have overlooked the fact that they were Chinese too. Although we have not found any illustration of them in the likely sources, Wang Chên gave a description in his *Nung Shu* of +1313 which with its single water-wheel working millstones on each of two supporting boats corresponds perfectly with that of the mills of Belisarius.[d] In view of the military supply value of these devices it is interesting to find the description copied in the *Wu Pei Chih*[e] of +1628. The *Thien Kung Khai Wu* (+1637) also speaks of ship-mills, saying that they were particularly common in the south.[f] Trip-hammers were mounted on them as well as millstones, and the wheels could be made to act also as norias whatever the level of the water.

Although we have not so far encountered any Chinese description as early as the time of Belisarius, the employment of ship-mills there can hardly have been much later, for the Ordinances of the Thang Department of Water-ways dating from +737 forbid (Art. 11) ship-mills (*fou wei*[1]) on the river and streams near Loyang as if they were something very well known.[g] Other early references are still scarce, but mills on boats were mentioned by Lu Yu[2] in one of the poems which he wrote on the occasion of his journey to Szechuan in +1170:[h] 'We heard the sound of bells beside the pavilion on its overhanging rock, and could see the mill-boats (*wei chhuan*[3]) moored below in the rushing water.' Then about +1570 Wang Shih-Mou describes in his *Min Pu Su* how the paper-makers of Fukien mounted their trip-hammers on boats each with two water-wheels and furiously pounded away by the aid of the fast-

---

[a] A famous Italian novel, 'Il Mulino del Po', by Bacchelli, is largely concerned with the ship-mills of that river.

[b] Early +14th-century MSS. figure them with castellated superstructures (MS. 11040 in the library of the Dukes of Burgundy at Brussels, a 'History of Alexander the Great'; and MS. Hamilton no. 19 in the Berlin Library). The latter has a legend 'Le fleuve dou Frate [Euphrates] et les Moulins de Babilone'. See la Roncière (1), vol. 2, p. 486; Anthiaume (1), p. 26; Audin (1); Braun & Hogenberg (1).

[c] Much detailed evidence is given by Bennett & Elton (1), vol. 2, ch. 6. The use of floating mills must have been particularly convenient on tidal waters, or in any circumstances where the water-level was likely to vary by flooding or the like. There are several medieval pictures which show mills housed on staging above tall piles (Bennett & Elton (1), vol. 2, pp. 76, 77) and one suspects that these may have had some arrangement whereby the wheel could be raised and lowered according to the level of the water. For Holland and Germany see Nolthenius (2). Cf. p. 217.

[d] Ch. 19, pp. 4b, 5a. These ship-mills, says Wang, were called *huo fa mo*,[4] 'live mills', because they could be moved about from place to place according to the amount of water available or other conditions.

[e] Ch. 138, p. 29b.   [f] Hence the name *Hsin-chün fa*.[5] Ch. 4, p. 2b.

[g] Tr. Twitchett (2), p. 48 from Tunhuang MS. P/2507.

[h] *Chien Nan Shih Kao*,[6] ch. 3.

[1] 浮磑   [2] 陸游   [3] 磑船   [4] 活法磨   [5] 信郡法

[6] 劍南詩稿

PLATE CCXLII

Fig. 629. Ship-mills at work on the River Tiber at Rome, a painting by Giuliano di San Gallo, *c.* +1490 (from Horwitz, 10). Their paddle-wheels are very elongated.

flowing current of their rivers.[a] So also Wang Shih-Chên,[1] writing about his own journey to Szechuan in the early Chhing period:[b]

In Liang-chiang there were many ship-mills (*wei chhuan*[2]), which work on the same principle as the water-raising wheels (*shui chhê*[3]), and are all anchored in the rushing water. The operations of grinding, pounding, and sifting (bolting) are all carried on by the use of water-power. The boats make a noise 'ya-ya, ya-ya' incessantly.

Another visitor was Robert Fortune, who in 1848 travelled through the tea country of northern Fukien. While still in Chekiang province, he found a whole colony of ship-mills near Yenchow.

Leaving the town of Yenchow behind us, our course was now in a north-westerly direction. The stream was very rapid in many parts, so much so that it is used for turning the water-wheels which grind and husk rice and other kinds of grain. The first of these machines which I observed was a few miles above Yenchow. At the first glance I thought it was a steamboat, and was greatly surprised; I really thought that the Chinese had been telling the truth when they used to inform our countrymen in the south that steamboats were common in the interior. As I got nearer I found that the 'steamboat' was a machine of the following description. A large barge or boat was firmly moored by stem and stern near the side of the river, in a part where the stream ran most rapidly. Two wheels, not unlike the paddles of a steamer, were placed at the sides of the boat, and connected with an axle which passed through it. On this axle were fixed a number of short cogs, each of which, as it came round, pressed up a heavy mallet to a certain height, and then allowed it to fall down upon the grain placed in a basin below. These mallets were continually rising and falling, as the axle was driven rapidly round by the outside wheels, which were turned by the stream. The boat was thatched over to afford protection from the rain. As we got further up the river, we found that machines of this description were very common.

Thus in Chekiang, if Fortune's account is right, water-trip-hammers were the chief instrument used. The statements of Lu Yu and Wang Shih-Chên, however, both refer to the Yangtze in eastern Szechuan, and their descriptions, though written so long ago, can be complemented by that of a living observer, Worcester, who has given excellent engineering drawings[c] of the ship-mills still in action around the city of Fouchow (Fou-ling), some sixty miles down-river from Chungking (Fig. 630). In this place they carry four water-wheels on two axles, and are known as *mien fên chhuan*.[4] Those seen by Worcester carried treadle sifters as well, but evidently in former times, or elsewhere, these were connected to the source of power, as shown in books such as the *Nung Shu* (cf. Fig. 461). A minor point of considerable interest recorded by Worcester is that the power-shaft gear-wheel always had 18 teeth, while the quern gear-wheel had only 16. The system may therefore have been a step towards the modern engineering practice of introducing a 'hunting tooth' to ensure even wear. This principle did not come into regular use in marine engineering before the introduction of geared turbines.

---

[a] The passage is quoted by Amano (*1*), p. 31.
[b] *Shu Tao I Chhêng Chi*[5] (Record of the Post-Stages on the Szechuan Circuit), +1672; tr. auct.
[c] (*1*), p. 24.

[1] 王士禎   [2] 碓船   [3] 水車   [4] 麵粉船   [5] 蜀道驛程記

Here we must refer again to the distinction made above between ex-aqueous and ad-aqueous wheels. The water-wheel may be employed to derive work from moving water, or to apply work, with the result of motion, to still water. The ship-mill differs in no way from the ordinary water-mill, since its wheel is ex-aqueous, except in the one essential point that it is mounted upon a structure which has the potentiality of automotive transit over the water if a force intrinsic to it can be brought into operation, i.e. if the wheel can be made ad-aqueous.[a]

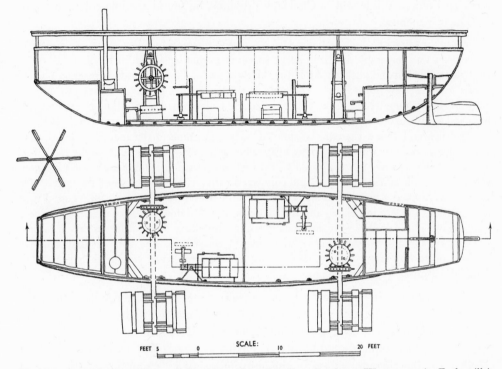

Fig. 630. Scale drawings of one of the ship-mills at Fou-ling, Szechuan (Worcester, 1). Each mill is worked by two stoutly built paddle-wheels and right-angle gearing. Two treadle-sifters (cf. Fig. 460) are seen mounted within the hull.

There is, however, one rather ingenious way in which an ex-aqueous pair of paddle-wheels can bring about motion, namely if a boat be attached by a cable to a point some distance up-stream, and the paddles then made to wind up the cable by the force of the current acting upon them. This idea was present in +15th-century Europe, as we know from several pictures of it in the military-technological MSS., e.g. Mariano Taccola (+1438).[b] Then it was described and figured by Verantius (+1595 and

---

[a] Of course it also differs in that it is quite independent of the level of the water, and one can see from the descriptions that this was a feature which made it particularly attractive to the medieval Chinese millwrights, who had so often to contend with large seasonal variations of water-level. A strange confusion of thought on this subject is manifested in a famous fountain statue of Neptune at Madrid—the conch on which the god stands is provided with paddle-wheels, but it is also drawn by a team of sea-horses.

[b] Berthelot (4); Feldhaus (1), col. 942.

PLATE CCXLIII

Fig. 631. The *liburna* or paddle-wheel ship proposed in the *De Rebus Bellicis* of the late Roman Anonymus; a MS. illustration (Thompson & Flower, 1). Six paddle-wheels are powered by three ox whims within the hull.

Fig. 632. The paddle-wheel boat sketched in the engineering MS. of the Anonymous Hussite (Uccelli, 1), *c.* +1430. Two crankshafts are operated manually.

+1615),[a] and formed the subject of many 18th-century projects and patents,[b] but the rather special conditions required for its successful use doubtless prevented its spread. It was still being taken very seriously in 1825, for an account of that date ascribes it to a Colonel Edward Clark of Philadelphia, and vividly depicts its use below the Trenton Falls on the Delaware, and the Mill Rapids on the Susquehannah.[c] I have no evidence that it was ever used in China, but at Phing-lo in Kuangsi I saw and photographed in 1944 a somewhat similar system of towing, in which the boat was wound upstream by the labour of men handling a capstan on the boat itself.[d]

Yet another use of floating ex-aqueous paddle-wheels is that described by Vitruvius,[e] in which they operate a hodometer to record the distance traversed by the ship. In its modern form,[f] using the screw rather than paddles, this invention has persisted in universal nautical use to the present day. Some Renaissance illustrations of Vitruvius make the device look remarkably like a paddle-boat proper.[g]

At the true ad-aqueous paddle-wheel boat we now arrive.[h] That the idea of working such paddles by the force of men or animals was fully present by the +14th century in Europe is quite undoubted, but how much earlier in that part of the world the idea goes back has been very uncertain, since it depends upon the authenticity of a MS. known as Anonymus *De Rebus Bellicis*.[i] First printed by Gelenius in +1552, this was thought by some to go back to the +6th century, and opinions have greatly differed about it, Neher (1) defending such a date, and Schneider (3) maintaining it to be a forgery of the +14th century. We shall have to come back to this problem. In any case, what the book describes and illustrates is a *liburna* or ship of Roman type, bearing three pairs of paddle-wheels turned by six oxen on the deck (Fig. 631). At the end of the +13th century Roger Bacon mentioned the idea in words rather similar to those of the Anonymus, whose text some think therefore must have been known to him.[j] Then in the various technological MSS. (da Vigevano in +1335; Kyeser's *Bellifortis* of +1405; the anonymous Hussite engineer of +1430 (Fig. 631); Valturio in +1472 with as many as five axles, etc.)[k] there are numerous illustrations of paddle-boats of varying degrees of size and complexity. A little later Giuliano di San Gallo reproduced the Anonymus' picture (Horwitz, 10), and Leonardo himself made a number of sketches with attention to the gearing and crankshafts involved.[l]

[a] Pl. 40, cf. Anon. (5); Beck (1), p. 526.
[b] Cf. Anthiaume (2), p. 135.   [c] Anon. (6).
[d] Groff & Lau (1) have also illustrated this. Worcester (1), p. 52, describes the process in use with the crooked-stern salt junks of Szechuan (cf. Sect. 29d below).
[e] x, ix, 5; cf. Diels (1), p. 67; Torr (1), p. 101; Drachmann (9), pp. 165 ff.
[f] Due to Humphrey Cole (+1577); cf. de Loture & Haffner (1).
[g] E.g. the Florence edition of +1522, p. 182a.
[h] The paper by Horwitz (9) is the best account of its history, though he was not aware of most of the Chinese evidence presented below. There is also a valuable article by McGregor (1) on the patent literature from the +17th century onwards.
[i] See Sarton (1), vol. 1, pp. 416, 430.
[j] See Reinach (2) on *De Secretis Operibus*, c. +1275. Cp. Thompson & Flower (1), p. 119 with Thorndike (1), vol. 2, p. 654.
[k] Berthelot (4, 5); Feldhaus (1), col. 936 ff.
[l] Reproduced and discussed in Horwitz (9); Ucelli di Nemi (3), nos. 4, 80; Feldhaus (18), pp. 124 ff. An armoured paddle-boat was projected in a +16th-century MS. of Byzantine–Turkish origin, Scanderbeg's *Ingenieurkunst und Wunderbuch* (E. Marx, 1); and by Ramelli (+1588), pl. 152.

No practical use of a paddle-wheel boat is recorded in Europe, however, until +1543, when Blasco de Garay constructed such boats for use as tugs, in the harbours of Barcelona and Malaga. Each one was manned by forty men working capstans or treadmills.[a] After +1600 such projects became numerous, and small boats of this kind found fairly wide application.[b] Many distinguished men of science interested themselves in the problem of applying more important power-sources (e.g. Papin and the Marquis of Worcester).[c] In 1819 treadmill paddle-boats were still in regular use on the Loire for fast traffic, and also in America on the Yellowstone expedition.[d] As late as 1885, a 'velocipède nautique' was introduced as a novelty for excursions on French rivers.[e] Credit for the first proposal to apply steam to paddle-wheels goes normally to Jonathan Hulls (+1736),[f] but the first practical successes were those of de Jouffroy d'Abbans in France (+1783)[g] and of Miller, Taylor and Symington in Scotland (+1787). By 1807 Robert Fulton's steamboats began regular service on the Hudson River.[h] Subsequent development is a matter of common knowledge.[i]

When the population of the Chinese coastal cities first saw the steam paddle-boats of the Westerners, an old term *lun chhuan*[1] (wheel-boat) was remembered, and remained in common use for any kind of steamer down to our own times. People in these cities knew little or nothing of their own past, with the exception of a few old-fashioned scholars, for whom much fable was mixed up with fact, and to whom nobody paid any attention.[j] When western historians of technology first saw the picture of a paddle-wheel ship in the *Thu Shu Chi Chhêng* encyclopaedia[k] (+1726), here reproduced as Fig. 633, they did not hesitate to put it down as a garbled reproduction of ideas brought by the Jesuits. Yet the facts were far otherwise. The history of paddle-wheel boats in China goes back to the +8th century at least, and probably to the +5th.

[a] At least, according to the best tradition. Another, deriving from MSS. in the Archivo de Simancas in Castile, and discussed by Arago (2), claims that one of Blasco de Garay's boats had a boiler. Although some rudimentary kind of turbine might not be entirely out of the question (cf. p. 226), it is now considered that the Spanish historian Fernández de Navarrete was misled on this matter by a forged interpolation in the archival sources. Here I am indebted to Captain José-Maria Martinez Hidalgo y Teran, the Director of the Naval Museum installed in the Atarazanas Arsenal at Barcelona, which I had the pleasure of visiting in September 1959. See also Spratt (1, 2). In +1575 there followed the 'Ark of Delft', an armoured structure resting on two hulls and moved by paddle-wheels set between them (la Roncière (1), vol. 2, pp. 486 ff.).

[b] McGregor (1); Anthiaume (2); Reinach (2); cf. Cummins (1), vol. 2, p. 341.

[c] Cf. Galloway (1), pp. 76ff.; Thurston (1), pp. 224ff.; Dircks (1), no. 15.   [d] Marestier (1).

[e] Tissandier (3).

[f] Cf. L. G. Hulls (1); Spratt (1, 2). He may have been anticipated by Denis Papin in +1707.

[g] Cf. Théry (1).   [h] See Spratt (2), pp. 74 ff.

[i] It may be worth remembering that the treadmill-operated paddle-boat is far from dead. It enjoys a genial immortality in the form of those innumerable 'bicycles made for two' which ply along the beaches of the Mediterranean.

[j] I speak of popular knowledge and belief; in fact, as we shall see (p. 431), some of the Chinese naval officials at the time of the Opium Wars were well acquainted with the technological history of their own country, and acted upon it.

[k] *Jung chêng tien*, ch. 97, *shui chan pu, hui khao* 1, p. 30a. The drawing was based on the *Wu Pei Chih* (cf. p. 424) and often subsequently copied, as in Kanazawa Kanemitsu's[2] *Wakan Senyōshū*[3] (Collected Studies on the Ships used by the Chinese and Japanese) of +1761.

[1] 輪船   [2] 金澤兼光   [3] 和漢船用集

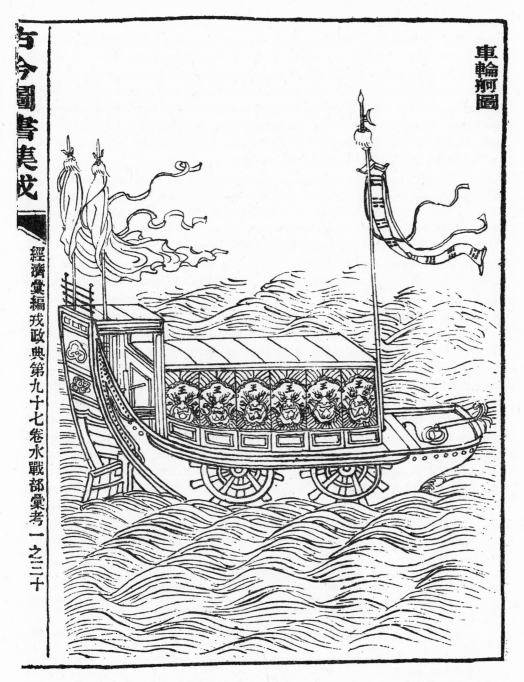

車輪舸圖

Fig. 633. The paddle-wheel warship (*chhê lun ko*) in the *Thu Shu Chi Chhêng* encyclopaedia (+1726), copied from the *Wu Pei Chih* of +1628. The four paddle-wheels are worked by treadmills within the hull.

Naturally the earliest references are less clear than the later. But the original inventor may well have been the famous engineer and mathematician Tsu Chhung-Chih,[1] for in both his biographies [a] there is mention of a 'thousand-league boat' (*chhien li chhuan*[2]), which was tested on the Hsin-Thing[3] river, south of modern Nanking, and proved to be capable of making several hundred *li* in one day without help of wind.[b] This invention was made between +494 and +497, not long before his death in +501. It is possible, however, that protected treadmill paddle-wheel boats may already have been used at the beginning of the century, in a naval action under the command of Wang Chen-O,[4] one of the admirals of the Liu Sung dynasty. In his biographies we find the following passage:[c]

> Wang Chen-O's forces sailed in covered swooping assault craft and small war-junks (*mêng chhung hsiao hsien*[5]).[d] The men propelling the boats were all (hidden) inside the vessels. The Chhiang (barbarians) saw the ships advancing up the Wei (river) but could not see anyone on board making them move. As the northerners had never encountered such boats before, every one of them was sore afraid, and thought that it was the work of spirits.

The passage goes on to describe how after being moored, they cast off upon order being given, and moved away apparently of themselves. Since the barbarians would surely have been familiar with oars and sails, something else is rather strongly suggested. In any case, the date is certain; the action took place in +418.

The following century gives many further indications of paddle-wheel boats. An admiral of the Liang dynasty, Hsü Shih-Phu,[6] in the course of the campaign against the rebel Hou Ching[7] in +552, constructed a number of different kinds of craft to strengthen his fleet. The lists mention[e] *lou chhuan*[8] (boats with several decks), *pho chhuan*[9] (ships with grappling irons),[f] *huo fang*[10] (fire-ships), and *shui chhê*[11] (water-wheel boats). While the usual meaning of this last term is of course the water-raising square-pallet chain-pump,[g] the context here (a list of boats) shows that a vessel of some kind was intended, and the obvious possibility is that paddle-boats were in fact meant. This same expression *shui chhê*[11] for a boat occurs again in the *Ching Chhu Sui Shih Chi*[12] (Annual Folk Customs in the Regions of Ching and Chhu), where it is said that on the 5th day of May the river-people hold races with these 'water-wheel boats'. This book is often ascribed to the Liang period, *c.* +550, and is certainly not later than the beginning of the Thang (*c.* +620). The same boats, it says, were also

---

[a] *Nan Shih*, ch. 72, p. 12*a*; *Nan Chhi Shu*, ch. 52, p. 21*a*. On his work cf. Vol. 3 above, *passim*.

[b] Literary tradition held long subsequently that Tsu's boat had been moved by some kind of internal mechanism. Thus in +1200 Chu I opined that its machinery might have resembled that of the 'wooden oxen' and 'gliding horses' of Chuko Liang (cf. pp. 260 ff. above), which no one any longer understood (*I Chao Liao Tsa Chi*, ch. 2, p. 35*a*). Chu I's naval contemporaries could have enlightened him considerably (cf. p. 422 below).

[c] *Sung Shu*, ch. 45, p. 7*a*; *Nan Shih*, ch. 16, p. 3*a*; *WHTK*, ch. 158, (p. 1380.2); tr. auct. adjuv. Lo Jung-Pang (3).

[d] These terms are significant of the beginnings of ship-armour; see further in Sect. 29*i* below.

[e] *Chhen Shu*, ch. 13, p. 1*b*; *Nan Shih*, ch. 67, p. 11*b*.

[f] Cf. Sect. 29*i* below.   [g] See pp. 339 ff. above.

[1] 祖沖之	[2] 千里船	[3] 新亭江	[4] 王鎮惡	[5] 蒙衝小艦
[6] 徐世譜	[7] 侯景	[8] 樓船	[9] 柏船	[10] 火舫
[11] 水車	[12] 荊楚歲時記			

called *shui ma*[1] (water-horses). An alternative interpretation,[a] equally tenable, would take *shui chhê*[2] as 'water-chariot', and suppose that all the boat names in this entry of Tsung Lin's book on the folk festivals referred simply to the usual many-crewed hand-paddled dragon-boats in which people raced each summer in memory of Chhü Yuan.[b] The classical confusion between *chhê* as 'chariot' and *chhê* as 'machine' renders the problem insoluble pending further evidence. However, another commander in the campaign against Hou Ching, the admiral Wang Sêng-Pien,[3] is said to have had in his fleet 'ships which had two dragons on the sides to enable them to go very fast' (*shuang lung chia hsien hsing shen hsün chi*[4]).[c] Lo Jung-Pang (3) suggests that *shuang lung* may have been a literary emendation for *shuang lun*,[5] 'two wheels', a proposal all the more plausible in that the context says a good deal about portents which appeared to the armies. Some twenty years later, at the siege of Li-yang in +573 when the Northern Chhi State was being invaded by the Chhen, a fourth admiral, Huang Fa-Chhiu,[6] who was also a distinguished military engineer, built and used a number of 'foot-boats' (*pu hsien*[7]) with success.[d] Chang Yin-Lin (2) suggested long ago that these can hardly have been anything else than treadmill-operated paddle-wheel boats.

The circumstantial evidence, therefore, seems distinctly strong that the original invention was made about the time of Tsu Chhung-Chih towards the end of the +5th century if not a little earlier. When we come to the time of Li Kao,[8] prince of the Thang (Tshao Wang Kao[9]),[e] there can be no further doubt. His experiments with paddle-wheel boats were made between +782 and +785, when he was Governor of Hungchow.[f]

Li Kao, always eager about ingenious machines, caused naval vessels (*chan hsien*[10]) to be constructed, each of which had two wheels attached to the side of the boat, and made to revolve by treadmills.[g] These ships moved like the wind, raising waves as if sails were set. As for the method of construction it was simple and robust so that the boats did not wear out.

[a] Preferred, e.g. by Lo Jung-Pang (3).　　　[b] On dragon-boats cf. Sect. 29 *d*.

[c] *Nan Shih*, ch. 63, p. 4 *a*.

[d] *Chhen Shu*, ch. 11, p. 3 *a*; *Nan Shih*, ch. 66, p. 17 *b*. The word *hsien* implies something a good deal larger than a ferry-boat; perhaps a better translation would be 'walked (rather than rowed) war-ships'. Huang Fa-Chhiu designed new types of catapults and a *pho chhê*,[11] doubtless an early form of the 'grappling-irons' (cf. p. 420 below). Worcester (14), pp. 25 ff., reminds us that the long sampans used on the creeks between Hangchow and Shao-hsing are also called 'foot-boats' because of the boatmen's curious method of rowing with their feet while working a steering paddle with their hands. See also Orange (1), p. 496. This method is old (cf. Moule (15), p. 31), but there can be no connection.

[e] We have already met with him in the physics Section, Vol. 4, pt. 1, p. 38 above.

[f] This was worked out by Kuwabara (1); cf. Lo Jung-Pang (3).

[g] When used by Chinese writers, this expression (*tao chih*[12] as here, or *tha chih*[13]) does not, of course, imply the clumsy drum-treadmill of the Romans, which men had to get inside, but the radial-pedal treadmill of the square-pallet chain-pump (cf. p. 339 above). Since as many of these could be fitted as the length of the driving-shaft allowed, and since in any event the pedals took up very little space, this device was much better adapted for use afloat than the ox-driven whim of the Anonymus or the drum-treadmill, though the latter was also proposed for a ship in a +16th-century European MS. described by Horwitz (9). Pedalled paddle-wheel boats may also have incorporated gearing to give magnification of torque and increased speed. All the Chinese texts remark on the speed attained.

[1] 水馬	[2] 水車	[3] 王僧辯	[4] 雙龍挾艦行甚迅疾	
[5] 雙輪	[6] 黃法𣱤	[7] 步艦	[8] 李皋	[9] 曹王皋
[10] 戰艦	[11] 拍車	[12] 蹈之	[13] 踏之	

So far the *Chiu Thang Shu*;[a] the *Hsin Thang Shu* adds that the prince himself taught his artisans how to make these craft, and that their speed was faster than that of a charging horse.[b]

After this, it would be natural to find various echoes of practical paddle-boats in the literature. The story of the pious layman Yuan Tshang-Chi,[1] reported in the late +9th-century *Tu-Yang Tsa Pien*,[c] who was sent home from an island paradise in the sea in a 'wind-riding ship' (*ling fêng ko*[2]), which stirred (*chi*[3]) the water, moving like an arrow, was probably one of them.

It was not until the beginning of the Southern Sung, however, early in the +12th century, that the treadmill paddle-wheel ships really came into their own.[d] After the loss of the capital, Khaifêng, in +1126, the move of the Sung administration to the southern provinces led (among many other things) to the first establishment of a regular Chinese navy based on the maritime expertise of the south. The Yangtze, it was said, must now be China's Great Wall, and battleships must be her watch-towers. At the same time the situation of vigorous defence called forth an upsurge of military and naval inventions, including new gunpowder weapons on the one hand[e] and much more highly developed paddle-wheel boats on the other. Response to this stimulus was quickly apparent, for certain 'flying eight-bladed paddle-wheelers' (*fei lun pa chieh*[4])[f] helped the Sung general Han Shih-Chung[5] to inflict great losses on the army of the Chin Tartars in +1130 when they were trying to extricate themselves from one of their southern campaigns and retreating across the Yangtze to the north.[g] We shall have more to say presently on this battle. It set the pattern for a century afterwards, for cavalry was the typical Jurchen arm, and the Chin State never developed any effective naval power.[h] Engineers now began to report their achievements. Two years after the victory Wang Yen-Hui[6] memorialised as follows:[i]

> To defend the thousand-*li* vastness south of the Great River, it is necessary to have war-ships; to halt the horsemen on the northern plains one must have vehicles.... Boats and vehicles are best if they are light and fast.... I have designed a 'flying tiger war-ship' (*fei hu chan hsien*[7]) with four wheels at the sides. Each wheel, which has eight blades (*chieh*[8]), is rotated by four men. This ship can travel a thousand *li* a day.

This description is of particular interest because it is the oldest we have which mentions four wheels, exactly as depicted in the *Wu Pei Chih* and the *Thu Shu Chi Chhêng*.

---

[a] Ch. 131, p. 5*b*; tr. auct. adjuv. Kuwabara (1).

[b] Ch. 80, p. 11*b*. [c] Ch. 3, p. 3*a*.

[d] A good deal of material on this period was included in our draft, but it has been much extended by the excellent paper of Lo Jung-Pang (3, 1), which reached us just in time.

[e] See further in Sect. 30*k* below.

[f] The phrase could also be interpreted 'paddle-wheel boats and eight-oared boats' or 'boats with paddle-wheels and eight oars'. But the Sung texts in general clearly show the use of various words for oars to mean the blades of the paddle-wheels. Figure 637, too, gives some colour to this.

[g] See *Fêng Chhuang Hsiao Tu*, ch. 2, p. 2*b*; quoted e.g. in *KCCY*, ch. 28, p. 11*b*. Since this is a work of the same century the authority is good.

[h] Unlike the Mongols, as we shall see in Sect. 29*e* below, though they also were originally a nomadic people.

[i] *Yü Hai*, ch. 147, pp. 18*a, b*, an abridgment; tr. Lo Jung-Pang (3), mod.

¹ 元藏機   ² 淩風舸   ³ 激   ⁴ 飛輪八楫   ⁵ 韓世忠
⁶ 王彥恢   ⁷ 飛虎戰艦   ⁸ 檝

As it happened, however, the real proving-ground for the developing invention was not the war against the northerners but internecine strife within the Southern Sung itself. During the disturbances of the Chin invasions an equalitarian peasant revolt had broken out under the leadership of Chung Hsiang,[1] and by +1130 the rebel forces were commanded by a number of gifted captains including Yang Yao[2] and Yang Chhin.[3] Making themselves masters of the Tung-thing Lake, they raided incessantly the cities on its shores. In the following year the Governor of Tingchow, Chhêng Chhang-Yü,[4] embarked upon a big shipbuilding programme to defeat them, including many junks with paddle-wheels designed by a remarkable engineer. From a contemporary Tingchow writer we learn that

one of the soldiers, Kao Hsüan,[5] who had formerly been Chief Carpenter of the Yellow River Naval Guard Force, and of the Pai-pho Vehicular Transport Bureau of the Directorate of Waterways, submitted a specification for wheeled ships which (he claimed) could cope with the enemy....(He first built) an eight-wheel boat as a model, completing it in a few days. Men were ordered to pedal the wheels of this boat up and down the river; it proved speedy and easy to handle whether going forward or backward. It had planks on both sides to protect the wheels so that they themselves were not visible. Seeing the boat move by itself like a dragon, onlookers thought it miraculous.

Gradually the number and size of the wheels were increased until large ships were built which had twenty to twenty-three wheels and which could carry two or three hundred men. The pirate boats, being small, could not withstand them.[a]

This was a truly remarkable piece of technology, the flavour of which we have attempted to capture in a small reconstruction (Fig. 634). It is almost surprising that these craft were not called 'centipede-ships', certainly no other civilisation produced anything like them. But the plan was a very rational one, for in the absence of steam-power and cast-iron wheels it was necessary to distribute the strain over a larger number of paddles.[b] Another source tells us that the biggest wheel-ships (*chhê chhuan*[6])[c] of Chhêng Chhang-Yü were 200–300 ft. long[d] and capable of carrying from seven to eight hundred men.[e]

Very soon these ships were ploughing the waves in the revolutionary service, for a government fleet of twenty-eight sea-going junks and two eight-wheel paddle-boats was stranded in a tidal river, so that all were captured and Kao Hsüan himself taken prisoner. The *Ting Li I Min* continues:[f]

[a] *Ting Li I Min*,[7] a document annotated by Chu Hsi-Tsu (2); tr. Lo Jung-Pang (3), mod. Pai-pho was a place in Honan.
[b] As late as +1786 Patrick Miller fitted five paddle-wheels to his capstan-operated manpower vessel which made 4·3 knots successfully on the Firth of Forth (Spratt (2), p. 43).
[c] This is the first appearance of the most characteristic Sung term for paddle-wheel ships. They are always referred to by the number of *chhê*, and Lo Jung-Pang (3) seems to us to be right in taking this to mean a single wheel and not a pair of wheels, i.e. in the 'machine' sense, not in the 'carriage' sense. Otherwise the recorded numbers of wheels would go beyond belief.
[d] We reserve to Sect. 29i a discussion of the largest sizes attained by Chinese traditional ships.
[e] *Sung Hui Yao Kao*, tsê 145, Shih huo, ch. 50, p. 15b.
[f] In Chu Hsi-Tsu (2), tr. Lo Jung-Pang (3), mod.

[1] 鍾祥　　[2] 楊么　　[3] 楊欽　　[4] 程昌寓　　[5] 高宣　　[6] 車船
[7] 鼎澧逸民

The pirates thus secured the design of the paddle-wheel boat and also the chief designer. He built for Yang Yao a large ship of the Hochow style with several decks and twenty-four wheels, and for Yang Chhin a Ta-tê-shan twenty-two wheeler.... Within two months the pirate bases had over ten many-decked wheel-ships that were stronger and better constructed (than the government ships).

Another contemporary source explains the meaning of the term 'Ta-te-shan'. Li Kuei-Nien[1] wrote:[a]

In the paddle-wheel ships (*chhê chhuan*[2]) men were stationed fore and aft to tread on pedals so that (the vessels) could go forward or backward. The (rebel) ships had names like Ta-tê-shan,[3] Hsiao-tê-shan,[4] Wang-san-chou[5] and Hun-chiang-lung[6] (for their types). They had two or three decks, and some could carry over a thousand men. They were equipped with 'grappling-irons' (*pho-kan*[7])[b] which were like great masts over a hundred feet high. Large rocks were hoisted up to the top of these by means of pulleys and when a government ship came close, they were suddenly let go to smash her. The Hun-chiang-lung, which had a dragon as a figurehead, was the ship that Yang Yao himself used during engagements.

The rebel fleet, which comprised at its height several hundred paddle-wheel ships of all sizes, also used rams. The *Sung Shih* says:[c]

(Yang Yao) launched ships with wheels which churned the water and so moved. They rushed forward as if flying. Moreover they were fitted with rams (*chuang kan*[8]) which damaged and sank the vessels of the imperial navy when they came into collison with them.

Conservative government commanders were nonplussed; in +1135 Wang Hsieh[9] was routed, having declined to follow the advice of his colleagues, recorded in an interesting debate,[d] and make full use of the paddle-wheel ships at his disposal. As time went on, however, some of the 18- and 22-wheel ships were captured from the rebels, and since the government had greater resources it was they in the end who built the largest vessels of the type. This we see from an interesting passage in the *Lao Hsüeh An Pi Chi*,[e] written about +1190, which also shows how the rebels and the government vied with one another in developing naval techniques. After describing the various kinds of 'grappling-irons' that were used, the text goes on to say:

Against the paddle-wheel fighting-ship (*chhê chhuan*[2]) of Yang Yao, the government forces used 'lime bombs' (*hui phao*[10]) thrown from trebuchet catapults.[f] For these they used pottery containers with very thin walls, within which were placed poisonous drugs (or

[a] In his *Chi Yang Yao Pên Mo*,[11] preserved as commentary in Hsiung Kho's[12] *Chung-Hsing Hsiao Chi*,[13] ch. 13, p. 15a; tr. Lo Jung-Pang (3), mod.

[b] On these see Sect. 29i below. The term is unsatisfactory, for while the grappling-irons of the Romans were made to facilitate boarding, some of the Chinese ones were for exactly the opposite purpose, and others were more like spiked hammers, or projectiles launched from above, to smash the enemy ship as it came alongside.

[c] Ch. 365, pp. 8a ff., esp. p. 10b, tr. auct. adjuv. Kuwabara (1). Anon. (18) and Volpicelli (3) deserve credit for having first drawn attention to this passage.

[d] *Chien-Yen I Lai Hsi Nien Yao Lu*,[14] ch. 66 (p. 1116).

[e] Ch. 1, p. 2a, b, tr. auct.   [f] Justification in Sect. 30i.

[1] 李龜年	[2] 車船	[3] 大德山	[4] 小德山	[5] 望三洲
[6] 渾江龍	[7] 拍竿	[8] 撞竿	[9] 王爕	[10] 灰礮
[11] 記楊么本末	[12] 熊克	[13] 中興小紀	[14] 建炎以來繫年要錄	

PLATE CCXLIV

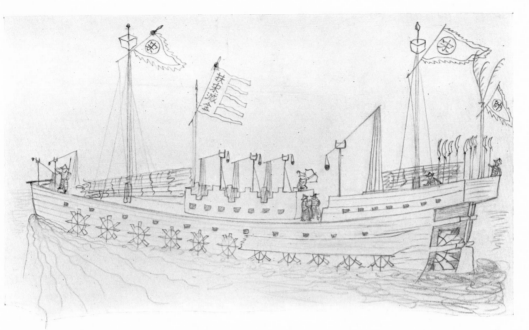

Fig. 634. Reconstruction of one of the multiple paddle-wheel warships of the Sung period (c. +1135) based on the designs of Kao Hsüan (orig. drawing). The largest of these vessels had, as here, twenty-two paddle-wheels (eleven on each side) and a stern-wheel. We have assumed the characteristic junk build, and have added auxiliary sails, a deck-castle, and a number of manned trebuchets (cf. Sect. 30i) for hurling gunpowder bombs, poisonous-smoke containers, etc. (cf. Sect. 30k). The general's pennant flies astern, and a banner hoisted amidships says 'Support the Sung, Destroy the Chin!' Such men-of-war carried crews of two or three hundred sailors and marines.

In the drawing the housing has been removed so as to show the forward six paddle-wheels on the port side.

minerals, probably arsenic), lime, and fragments of scrap-iron (as well as gunpowder).[a] When these were hurled on to the rebel ships during engagements, the lime filled the air like smoke or fog so that their sailors could not open their eyes. The rebels wished to copy this device, but as the right sort of containers could not be found or made within the territories held by them, they failed. So they suffered heavy defeats.

The imperial forces in their turn imitated the (paddle-wheel) ships of the rebels, but made them larger—as much as 360 ft. in length, 41 ft. in the beam, and (with masts) 72½ ft. high. But they had hardly been brought into use before the infantry of (general) Yo Fei[1] decisively conquered the rebels. However, later on, when Wanyen Liang[2] made his invasion (from the north), these paddle-wheel ships were still available, and did excellent service.

These last two remarks need a little explanation. The rebels were finally defeated by the famous commander Yo Fei, who caught the greater part of their fleet in a strange ambush. Having covered the water in an arm of the lake with masses of floating weeds and rotten logs, he lured them in, and when the paddle-wheels were all entangled so that they could not move, his boarding-parties swarmed on to the ships and won a decisive victory. Yang Yao himself was killed.[b] This was in +1135. Scarcely thirty years later, in +1161, the Chin Tartars mounted another expedition against the Sung, and Wanyen Liang, the fourth emperor of the Jurchen Chin (Fei Ti[3]), sought to make a crossing of the Yangtze. This led to the celebrated Battle of Tshai-shih,[4] where after many anxious moments the Sung forces under Yü Yün-Wên[5] gained the day. This encounter is important not only for the variety of gunpowder weapons employed at so early a date,[c] but also because the paddle-wheel warships repeated their achievements on the river accomplished under Han Shih-Chung. Cruising rapidly round Chinshan Island they constantly let off their trebuchet artillery, and struck great fear into the hearts of the Jurchens, who were not much accustomed to any kind of ships and found these almost supernatural. When it was clear that the Yangtze could not be forced, Wanyen Liang was assassinated by his own staff, and the Chin army retreated to the north next day.[d]

All through the Southern Sung there was great activity in building and employing paddle-wheel ships. In +1134 the Deputy Transport Commissioner in Chekiang, Wu Ko,[6] recommended the construction of 9-wheel and 13-wheel warships in the coastal provinces (*chiu chhê shih-san chhê chan chhuan*[7]).[e] In +1183 a Nanking naval commander, Chhen Thang,[8] was specially rewarded for his work in building ninety paddle-wheel and other ships.[f] The imperial court itself took great interest in its automotive vessels, unknown in any contemporary culture. When in +1176 a Nanking

[a] Justification in Sect. 30k.

[b] Sung Shih, ch. 365, pp. 10b, 11a.

[c] A full description will be given in Sect. 30k. On the background of the battle see Cordier (1), vol. 2, p. 169.

[d] Sung Shih, ch. 383, pp. 11a to 13b, esp. 13a; cf. Yü Hai, ch. 147, p. 19a, b; Chung-Hsing Hsiao Chi, ch. 40, pp. 9b, 10b. See also WHTK, ch. 158, (pp. 1381.3ff.) quoting Yang Wan-Li's Hai Chhiu Fu Hou Hsü (in Chhêng-Chai Chi, ch. 44, pp. 6bff.).

[e] Sung Hui Yao Kao, tsê 145, Shih huo, ch. 50, pp. 16b, 17a.

[f] Loc. cit. p. 30a.

[1] 岳飛　　　　[2] 完顏亮　　　[3] 廢帝　　　[4] 采石　　　　[5] 虞允文　　　[6] 吳革
[7] 九車十三車戰船　　　[8] 陳鐙

official, Kuo Kang,[1] advised the conversion of some damaged paddle-wheelers into ordinary junks, the Shun-Hsi emperor replied:

> Our wheel-ships are the fast assault craft (*mêng chhung*[2]) of old. In the campaign of the *hsin-ssu* year (the Battle of Tshai-shih), they brought us victory. How can we afford to have them converted into junks or galleys? Each military district is permitted to design its own ships, but the number of paddle-wheel ships is not to be reduced.[a]

On the contrary, three years later, another edict ordered the equipment of a hundred horse-transport barges with detachable paddle-wheels and casings to protect them.[b] Most interesting was the report of the Yangtze Admiral Shih Chêng-Chih[3] in +1168 that he had constructed very economically a 100-ton warship propelled by a single twelve-bladed wheel (*i chhê shih-erh chiang*[4]).[c]

This solves a problem which must have been puzzling the reader since the first mention of Kao Hsüan's 23-wheel paddle-boat—for evidently that of Shih Chêng-Chih was a stern-wheeler. All the descriptions of ships with odd numbers of wheels must thus refer to vessels with one stern-wheel and a set of pairs of side-wheels. Here we may pause a moment to think a little further about the mechanism of these extraordinary craft.[d] The simplest arrangement which springs to the mind would of course have been to set each pair of paddle-wheels on a single axle and apply the man-power directly to this by means of radial pedals. But one wonders whether in some of these Sung designs the paddle-wheels were not independently mounted on individual bearings amidships; in this case forward action on one side combined with reverse action on the other would have rendered the ship eminently manœuvrable—and some of the descriptions seem to emphasise this. Such an arrangement might have rendered a rudder unnecessary, and that would have been convenient if one of the large stern-wheels was fitted. As for the power available, the number of pedallers carried was sometimes quite large. Thus in +1203 Chhin Shih-Fu[5] built at Chhih-chow two four-wheeled 'sea-hawk' (*hai hu*[6]) warships, covered over on top, armoured with iron plates on the sides, and equipped with spade-shaped rams.[e] The smaller, of 100 tons burden, needed a propulsion crew of 28 men, the larger, of 250 tons,[f] required 42. The largest numbers of pedallers mentioned in any of the sources for a single ship is

---

[a] *Sung Hui Yao Kao*, tsê 145, Shih huo, ch. 50, p. 27b, reproduced in *Yü Hai*, ch. 147, p. 20b; tr. Lo Jung-Pang (3), mod.

[b] *Loc. cit.* p. 28a. Many other decrees are recorded in these sources. One gets the impression from them that it was found undesirable to increase the size of the vessels and the number of paddle-wheels beyond a certain point. Thus in +1135 the order was given to scrap some of the 9-wheel and 13-wheel ships, and to replace them by 5-wheel vessels (*Chien-Yen I Lai Hsi Nien Yao Lu*, ch. 86 (p. 1425) and ch. 89 (p. 1483)). Similarly in +1181 it was ordered that the construction of 8-wheel ships should cease, and that vessels of 7, 6, and 5 wheels should be built (*Sung Hui Yao Kao*, tsê 145, Shih huo, ch. 50, p. 28b).

[c] *Sung Hui Yao Kao*, tsê 145, Shih huo, ch. 50, p. 22a.

[d] No Sung text so far discovered gives details of the power-producing arrangements. They may well have been considered 'restricted information'.

[e] Very full specifications are recorded in *Sung Hui Yao Kao*, tsê 145, Shih huo, ch. 50, pp. 32b ff.

[f] By way of comparison one may remember that this was about the size of Vasco da Gama's flagship, and that Chhin Shih-Fu's smaller type was twice the size of one of Prince Henry's caravels.

[1] 郭剛   [2] 蒙衝   [3] 史正志   [4] 一車十二槳   [5] 秦世輔
[6] 海鶻

200. It is of course impossible to say whether these figures include relays of men, and without knowing more of the mechanism it is hard to compute the horse-power available, but a maximum of 50 h.p. might not be an overestimate, and this would have given a considerable motion to the vessel, certainly quite enough for effective ramming tactics.[a] Moreover the Sung shipwrights may well have found a way to apply the power of the radial pedals of several shafts to the single axle of the paddle-wheel.[b] As is seen in Fig. 637, the 19th-century Chinese paddle-wheel passenger boats worked by tread-mills had coupling-rods and eccentrics powering the stern-wheels from three separate radial-pedal shafts, thus allowing three groups of men to work at the same time. That such an arrangement could have been utilised by the engineers of the Southern Sung will be sufficiently evident from our discussion on pp. 380 ff. above concerning the predecessors of the hydraulic blowing-engines of the late +13th century, and allied machinery such as the silk-winding and reeling apparatus which was already classical in the +11th. Once the basic principle of the eccentric and connecting-rod had been mastered, the use of the latter as a coupling-rod with other eccentrics on other shafts should have been no more difficult (indeed perhaps less so) than the addition of a piston-rod, which brought into existence, as we have seen, the fundamental plan of the reciprocating steam-engine.

So far the accent has been on the naval use of paddle-wheel ships by the Southern Sung, but it seems probable also, as Kuwabara (1) believed, that smaller paddle-boats were used in the great Chinese harbours during the +12th and +13th centuries, especially for towing. Such tugs would have been handy at places like Chhüanchow, where Phu Shou-Kêng[1] was Superintendent of Trading Ships.[c] Certainly small paddle-boats adapted for elegant water-parties and picnics had great success in the +13th century, notably on the West Lake at Hangchow. We hear about these from the author of the *Mêng Liang Lu* (+1275),[d] who says:

> There are also the wheel boats (such as those) belonging to the great house of Chia Chhiu-Ho.[2] On the deck above the cabin there are no men poling or rowing, for these craft move by means of wheels worked by a treadmill, and speed over the water like flying things.

But Neptune remained in conjunction with Mars, and the trumpets of war continued to sound. At the end of the Sung there was a celebrated appearance of paddle-wheel boats during the siege of Hsiang-yang[e] by the Mongols (+1267 to +1273). Two heroic Sung officers, the colonels Chang Shun[3] and Chang Kuei,[4] led a relief convoy

---

[a] As we shall see below, the average speed of the paddle-wheel junks of the 19th century is known to have been about 3½ knots.

[b] The account of Chhin Shih-Fu's ships distinctly suggests this, for although the number of men in the propulsion crew was large there were only two paddle-wheels.

[c] Cf. Vol. 1, p. 180 and Vol. 4, pt. 3. It will be remembered that the first effective use of boats with paddle-wheels in +16th-century Europe was as tugs in harbours. In 1955 it was interesting to read in the daily press that the British Admiralty were placing orders for seven diesel-electric paddle tugs. 'Experience has shown', a correspondent wrote, 'that paddle tugs are more efficient than screw-driven tugs for work in confined basins because of their great manœuvrability and power' (*The Times*, 6 June 1955).

[d] Ch. 12, p. 13b, tr. Moule (5), (15), pp. 30 ff.; Kuwabara (1), mod. auct.

[e] A description of this famous siege will be found below (Sect. 30i).

[1] 蒲壽庚   [2] 賈秋壑   [3] 張順   [4] 張貴

of a hundred paddle-boats laden with supplies, and succeeded in reaching the be-
leaguered city, though both leaders were lost, one on the way in and one on the way
back.[a] In the *Yuan Shih* we read:[b]

> In the third month (of +1272) the outer defences of Fan-chhêng (the secondary city
> across the Han river from Hsiang-yang) fell, after which A-Chu[1] (the Mongol commander)
> built trenches and intensified the investment. Then the Sung colonels Chang Shun and
> Chang Kuei assembled clothing (and other supplies) on a hundred ships and sailed down the
> river to Hsiang-yang. A-Chu attacked them (fiercely, so that) Chang Shun was killed, and
> Chang Kuei only just managed to reach the city (with the convoy). Afterwards they suddenly
> came out with the paddle-boats (*lun chhuan*[2]) and sailed eastwards down the river (towards
> the Sung armies). So A-Chu and his commander Liu Chêng[3] took their respective flotillas
> and waited, burning straw on both banks to illuminate the river so that it looked like day.
> A-Chu pursued Chang Kuei as far as Kuei-mên-kuan and captured him (and others), all the
> rest being killed.

We shall have occasion to return to this remarkable action later on in connection with
gunpowder weapons, for they were used in it by both sides on a considerable scale.[c]

Even now this agonising struggle between the hosts of Sung and Yuan must keep
us for a moment longer, because another incident of the siege involves a strange piece
of technology which may have been an ingenious climax in the use of the ship-mill
principle.[d] Later in the same year (+1272), after the relief convoy had come and
gone, the Sung engineers launched baulks of wood on the river, held together with
iron links, and on these they made a floating bridge so that the garrisons could readily
cross back and forth when reinforcements were needed. But A-Chu came down upon
it with 'mechanical saws' (*chi chü*[4]) devised by the Yuan engineers to cut through the
baulks, and axes to sever the links, after which the bridge was set on fire and totally
destroyed by the marines in the Mongol forces.[e] Although the account is somewhat
uninformative, its obvious interpretation is that saws were mounted on paddle-wheel
boats in such a way that they could be worked from the current when the boats were
held stationary.[f]

Reviewing the evidence assembled, one is struck by the fact that the development of
paddle-boats in medieval China was primarily associated with their value in sea-fights,
especially on lakes and rivers. Indeed the description in the *Thu Shu Chi Chhêng*,
already noticed, refers specifically to paddle-boat ships used for naval purposes. The
illustration is a century older than this encyclopaedia, being derived from the *Wu Pei
Chih*[5] (Record of War Preparations) by Mao Yuan-I[6] (+1628).[g] The accompanying

---

[a] *Sung Shih*, ch. 450, p. 2*a* (translation in Sect. 30*j*).
[b] Ch. 128, p. 2*a*; tr. auct. Cf. Moule (15), p. 73.
[c] See Sect. 30*k*.   [d] *Yuan Shih*, ch. 128, p. 2*b*.
[e] Apparently the same sequence of events recurred in the first month of the following year, when the
engineers and marines of the Yuan were commanded by A-Li-Hai-Ya[7] (p. 7*a*).
[f] Especially as we know that the Sung troops had been using paddle-boats in the same siege that
year, and doubtless some of these had been captured by the Mongols.
[g] Ch. 117, p. 17*a*. Cf. Vol. 3, p. 559.

[1] 阿朮   [2] 輪船   [3] 劉整   [4] 機鋸   [5] 武備志
[6] 茅元儀   [7] 阿里海牙

text is also assuredly much earlier than the +18th century, as in the case of some of the preceding warship illustrations, where the encyclopaedia reproduces almost verbatim an +8th-century text.[a] This one is indeed not so old as that; let us see if we can date it from intrinsic evidence. The passage[b] runs as follows:

The paddle-wheel barque (*chhê lun ko*[1]) is 42 ft. long and 13 ft. broad, having outboard (*wai hsü*[2]) on (each) side a frame (*khuang*[3]) 1 ft. broad, with nothing within (and below) it except four wheels, the bottoms of which reach 1 ft. into the water. Men are ordered to (work a treadmill) so that the wheels turn and the speed is like flying. The flat (part of the deck at the) bow is 8 ft. long, the hold 27 ft. long, and the stern (platform) above the steersman's compartment (*tho lou*[4]) is 7 ft. long. Above the hold, the deckhouse gives through communication fore and aft, with a great beam supporting bulwark boards on each side, each plank being 5 ft. long and 2 ft. broad. Below this are fixed turning pulleys like those which raise hanging windows (*ju tiao chhuang yang*[5]). When approaching the enemy (these loopholes are opened) and (those) inside can let loose *shen phao*[6] (bombs or grenades),[c] *shen chien*[7] (incendiary arrows or rockets or incendiary rockets),[d] and *shen huo*[8] (spattering fire from fire-lances containing rocket composition, or perhaps burning petroleum like Greek Fire).[e] With all this the enemy cannot even see us (because of the protection of the walls of the deckhouse-turret). The enemy being somewhat weakened, our sailors suddenly lift up and fully open the bulwark hatches, (the walls beside the loopholes) acting as a shield, and stand ready within. In addition, raw oxhides are stretched out to protect (*kuo*)[9] the crew (from the enemy's incendiary weapons), while from inside they throw (*phao*)[10] *huo chhiu*[11] (incendiary bombs and toxic smoke bombs), and shoot (*fang*[12]) iron-pointed javelins (*piao chhiang*[13])[f] and use hooks[g] and similar weapons. Thus the enemy ships must (inevitably) be burnt and destroyed.[h]

Now this is certainly not from the Thang, the time of Li Kao, like the other descriptions, but at first sight it might be Sung, for toxic smoke bombs (*huo chhiu*) containing gunpowder are described[i] in +1044 and mentioned in many battle accounts of that period. But this term continued in use long afterwards, and the early Ming date of the text is more forcibly asserted by the recurring use of the word *shen*,[14] 'magical', for three of the types of weapons. This was fashionable for all new inventions of this kind from about +1385 onwards.[j] It seems therefore that we may reasonably date these paddle-boats as +15th century;[k] for if the passage were of Mao Yuan-I's own

[a] See the discussion and translations in Sect. 29*i* below.
[b] Paraphrased very inaccurately in Worcester (3), vol. 2, p. 341. Both here and in another paper (4), he ascribed the illustration and the text to the *San Tshai Thu Hui* encyclopaedia of +1609 (wrongly dating it +1522), but this is a mistake. The derivative statements of Spratt (1, 2) should be corrected accordingly.
[c] Cf. *WPC*, ch. 122, p. 19*b*.   [d] Cf. *WPC*, ch. 129, p. 16*a*.
[e] Cf. *WPC*, ch. 122, p. 23*b*; ch. 126, p. 15*b*; ch. 128, p. 9*b*.
[f] Presumably shot from arcuballistae (cf. Sect. 30*i* below). But the text may simply imply boarding and hand-to-hand fighting as the final phase.
[g] Perhaps the 'grappling-irons', cf. p. 420 above.
[h] Tr. auct.   [i] *WCTY/CC*, ch. 12, p. 67*b*.
[j] See Wang Ling (1), p. 175, and Davis & Ware (1).
[k] If so, curiously contemporary with the European descriptions, such as those of Kyeser and the Hussite engineer and Mariano Taccola (cf. Berthelot, 4, 5), but more detailed.

[1] 車輪舸   [2] 外虛   [3] 框   [4] 舵樓   [5] 如弔窗樣   [6] 神砲
[7] 神箭   [8] 神火   [9] 裹   [10] 拋   [11] 火毬   [12] 放
[13] 鏢鎗   [14] 神

time (early +17th) there would be more mention of barrel-guns and culverins and less emphasis on the earlier types of gunpowder weapons.

Apart from the publication of this description in the *Wu Pei Chih*, this period has not yielded much, but one would probably not have to search very far to find it. The persistence of the naval paddle-boat under the Ming is particularly interesting because it seems to have fallen so much out of the picture during the Yuan. The Mongol dynasty was far from neglecting sea power;[a] on the contrary it emphasised it so much that vessels especially adapted for lake and river combats such as the paddle-wheel ships suffered a decline. Until the coming of an adequate source of power they were clearly unsuitable at sea, and during the period of the great voyages of Chêng Ho,[b]

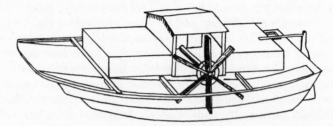

Fig. 635. A +17th-century Japanese drawing of a small manually operated paddle-wheel boat (Purvis, 1).

for example, we hear nothing of them. Only in the +17th century do they regain any prominence, and then a rather theoretical one. A Japanese drawing of a small hand-operated paddle-boat dating from this period (Fig. 635) was reproduced by Purvis (1) in his account of the history of shipping in that country, but he failed to state its source. Whether it was the result of Chinese or Dutch stimulation could not be ascertained without further research, but the former seems much more likely. In any case the paddle-boat continued its millennial life in China, though applied more and more to the humble uses of civilian transportation. For example, we hear of a Suchow craftsman of the Chhing period, Hsü Shih-Ming,[1] who made ferry-boats with paddle-wheels (*chiao tha fei chhê kuo ho*[2]).[c]

Unlike some other medieval Chinese inventions, the treadmill paddle-boats survived in active use down to our own time, especially in the Pearl River at Canton, where they were photographed some thirty years ago (Fig. 636).[d] If this be compared with

---

[a] Cf. Sect. 29*e* below.   [b] Cf. Vol. 3, pp. 556 ff. and Sect. 29*e* below.

[c] Liu Ju-Lin in Li Nien (22), p. 4, from *Suchow Fu Chih*.

[d] We owe two photographs to the kindness of the late Mr P. Paris, who took them in 1929. Our colleague Dr Victor Purcell vividly remembers travelling on the Pearl River paddle-boats between 1921 and 1924 back and forth from Sanshui. Cf. the account of Gibbs (1). When in Canton in the summer of 1958, Dr Lu and I made every effort to find and photograph some of the old treadmill paddle-boats, which had been used as ferries, but in spite of some hours' cruising on the Pearl River and looking into many creeks, we met with no success. It is probable that a recent movement for reconditioning all old pieces of machinery had led to their being broken up and used for other purposes, since steam and motor launches had long replaced them on the river. Nevertheless, our warmest thanks are due to the Canton representatives of Academia Sinica and to the Port Authority for their kindness and eager co-operation in our search.

[1] 徐士明   [2] 脚踏飛車過河

pictures of a model made in the last century (Fig. 637) which Horwitz (9) reproduced, the only obvious difference is that the eccentrics and coupling-rods are not prominent in the photograph, suggesting perhaps that a chain-drive replaced them, but as the places for the three rows of pedallers can clearly be seen, the coupling-rods are probably hidden by the hull bulwarks. It is known that the treadmills had pedals like those of the *fan-chhê* (cf. Fig. 578).[a] Horwitz noted the interesting point that the eccentrics on the model were set 90° apart on the port and starboard sides so as to avoid stoppages at dead centre. In build the boats were constructionally related to the Kwailam junk type.[b] Another fleet of these stern-wheelers plied on the Wusung[1] River between Shanghai and Suchow within living memory. In the nineties of the last century there were about fourteen of these ships carrying some seventy passengers in large and roomy accommodation; pedalled by a group of men six to twenty in number, they did the 100-mile journey in about a day and a night, averaging thus a little under 3½ knots.[c] In the light of all that has gone before, we can now understand the presence of these Mississippi-like stern-wheelers in the Chinese rivers as late as the present century. They had nothing to do either with Robert Fulton or with the Mississippi; they were in the direct line of descent from Shih Chêng-Chi's stern-wheel battleship of +1168.

It is a remarkable fact that when modern Europeans first came to Chinese coastal waters they were quite unable to believe that such paddle-wheel boats could be anything more than an imitation of their own steamboats. A classical encounter[d] took place during the Opium Wars at the Battle of Wusung,[2] where the Huangpho and Wusung rivers enter the Yangtze estuary, in 1842. Here paddle-boats were used on both sides, for the British fleet, advancing to reduce the coastal fortifications, included fourteen such steamers,[e] while on the Chinese side there were five treadmill-powered paddle-wheel junks capable of making 3½ knots. Though fought with courage and skill by their commander Liu Chhang-Chhing,[3] all were captured or destroyed during the battle. They proved of much interest to the British officers, who believed with one accord that they had been copied from the paddle-wheel steamers as something completely new.

> The most remarkable improvement of all [wrote Bernard, patronisingly], and which shewed the rapid stride towards a great change which they (the Chinese) were daily making, as well as the ingenuity of the Chinese character, was the construction of several large *wheeled* vessels, which were afterwards brought forward against us with great confidence at the

[a] G. R. G. Worcester, unpublished material, no. 174, and (14), pp. 88 ff.

[b] See Lovegrove (1). Other services in South China are described by Audemard (5), pp. 64, 66.

[c] Worcester (3), vol. 1, p. 224. Very few of the numerous experimental steamboats before 1810 did better than this speed, and most of them not nearly so well; cf. Spratt (2).

[d] W. D. Bernard (1), vol. 2, pp. 352 ff., accompanied by an engraving of the battle. The paddle-wheel junks are shown in this, but with details too small for reproduction. Other descriptions are given by Worcester (11) and Lo Jung-Pang (3).

[e] Among them was the newly-built *Nemesis* under Cdr. W. H. Hall, the first iron ship (as distinguished from iron-armoured ships, cf. Sect. 29i) ever to take part in a naval engagement. This was twenty years before the famous action between the *Monitor* and the *Merrimac* off Hampton Roads in the American Civil War.

[1] 吳淞   [2] 吳淞   [3] 劉長清

engagement at Woosung, the last naval affair of the war, and were each commanded by a mandarin of high rank, shewing the importance they attached to their new vessels. This, too, was as far north as the Yangtze River, where we had never traded with them; so that the idea must have been suggested to them by the reports they received (from the south) concerning the wonderful power of our steamers or wheeled vessels.[a]

Elsewhere, after again expatiating on the 'extreme ingenuity' of the Chinese, he described the machinery being fitted into a number of hulls which fell into British hands at the capture of Chenhai[1] near Ningpo in 1841.

There were two long shafts [he said], to which were to be attached the paddle-wheels, made of hard wood, about 12 ft. in diameter; there were also some strong wooden cogwheels nearly finished which were intended to be worked by manual labour inside the vessel. They were not yet fitted to the vessels; but the ingenuity of this first attempt of the Chinese, so far north as Chinhai, where they could only have seen our steamers during their occasional visits to Chusan, when that island was before occupied by us, cannot but be admired.[b]

This demonstrates that there were variations in the propulsion mechanism, and that for these naval vessels capstans and hand-spikes connected with the paddle-wheel axles by right-angle mill-gearing were employed, so that the arm was used instead of the leg for providing power. Some of the wheel-junks at the Battle of Wusung were also thus constructed.[c] This system was certainly four or five times as efficient as the direct radial-pedal arrangement, and permitted mechanical advantage by the adjustment of the gear-ratios. One would very much like to know whether the Sung designers ever used it; since they were quite familiar with millwrights' work there is no reason why they should not have done so.[d] Another observer of the Wusung junks, however, Lt Ouchterlony, spoke of 'cranks' working four paddle-wheels on each vessel:

Among the curiosities in the shape of military machines found at Woosung were two junks each fitted with four paddle-wheels about 5 ft. in diameter, worked by two cranks fitted on axles placed athwart in the fore and aft parts of the vessel. They were clumsy enough, but nevertheless useful craft for transporting troops on smooth water, as by making the whole party take their turn, and by working short spells at the cranks, great expedition could be used.[e]

These words recall the assault craft of Wang Chen-O fourteen centuries earlier. Ouchterlony was probably referring to an arrangement of eccentrics and coupling-rods like that in Fig. 637, the power unit being situated in the middle and connected thus to the paddle-wheels both fore and aft.

---

a Bernard (1), vol. 1, p. 280.

b *Ibid.* vol. 2, p. 226. Choushan[2] Island, with its capital, Tinghai,[3] was taken twice by the British during the Opium Wars, first in 1840. For an account of these operations from the Chinese point of view see Anon. (41), pp. 207 ff. For an account of the island by a Western visitor a few years later see Fortune (1), vol. 1, pp. 42, 163, 244, vol. 2, p. 276, (2), pp. 61, 225, 314, etc.

c Cdr. Hall's notes, in Bernard (1), vol. 2, p. 354, quoted by Lo Jung-Pang (3), p. 205.

d It was, of course, the same method as that proposed by the Byzantine Anonymus. It had also been used in successful experiments by Patrick Miller in Scotland in +1786 (Spratt (2), p. 43).

e Ouchterlony (1), p. 298.

1 鎮海   2 舟山   3 定海

PLATE CCXLV

Fig. 636. One of the treadmill paddle-boats of the Pearl River estuary near Canton (photo. Paris, 1929). The large iron stern-wheel, about 9 ft. in diameter, can be seen fitted in the after-gallery under the steersman. Just forward of it, under the awning, there are three sets of handlebars for the pedallers, two rows of whom can be seen at work, resembling those who turn square-paddle chain-pumps (cf. Fig. 579). The funnel belongs to the white Japanese steam-launch behind.

Fig. 637. Stern view of a model of a Cantonese stern-wheel treadmill paddle-boat in the Museum f. Völkerkunde, Vienna (Horwitz, 9). The aft hull bulwarks have been removed so as to show the three eccentrics and coupling rods like those of a steam locomotive; these were set 90° apart on port and starboard sides to prevent dead-centre stoppages.

PLATE CCXLVI

Fig. 638. Design for a treadmill paddle-wheel warship prepared by Chhang Chhing in 1841 during the Opium Wars, including height-adjustable slung rudders fore and aft, an armament of 12 guns, and space for a crew of over a hundred. MS. drawing of Liang Chi-Phing, from Chhen Chhi-Thien (1). Note the standard of the Great Bear (cf. Vol. 3, p. 240, Figs. 90, 103) flying on the left.

Ten years after the battle, J. F. Davis, Governor of Hongkong, reported the talk of the coast as follows:[a]

A native of Chusan had built small vessels on the model of our steamers, with paddle-wheels. It was said that, when ready, he endeavoured to propel them by means of smoke made in the hold; but as they declined altogether to move on such terms, it was subsequently found advisable to turn the wheels by relays of men working with their weight, somewhat on the principle of the treadmill. In this condition they were found by our forces. The Chinese officers tried them, and were satisfied with the rate of their movement, and the use of the guns on board.

After all that we have seen in this chapter any comment on Davis' words would be superfluous. Yet as late as 1950 one of our best authorities could still say of the Battle of Wusung that 'among them (the Chinese war-junks) were several paddle-wheel ships constructed in imitation of steamers'.[b]

Of course there is no need to deny that the Chinese naval technicians were greatly stimulated by the sight of the steam paddle-boats.[c] The 'native of Chusan' may long remain anonymous, but Chhen Chhi-Thien (1) has sought to identify him with Kung Chen-Lin,[1] magistrate of Chia-hsing.[d] Kung was one of the group of followers of the great anti-opium Commissioner Lin Tsê-Hsü,[2] who while cognisant of the past achievements of their own culture were deeply interested in the modern techniques of the West.[e] Kung Chen-Lin afterwards wrote:[f]

In the summer of the *kêng-tzu* year (1840), when the British invaded and occupied Choushan, I was summoned to Ningpo from my post at Ho-chung. There (on the coast) I saw the enemy sails standing like a forest, and among them were ships which stored fire in a

[a] (3), vol. 1, p. 258. I owe this reference to the kindness of my friend Dr Victor Purcell.

[b] Worcester (11). Elsewhere, however, Worcester (4) has stated correctly that the Chinese paddle-boats of the early 19th century were not derived from those of the Europeans. McGregor (1) also wrote of the 'mock steamboats' used by the Chinese.

[c] It would seem that during the Chhing period the Chinese paddle-boats had become purely civilian stern-wheelers on rivers, and that the Sung naval craft had been somewhat forgotten.

[d] Though Kung's birthplace was not in fact Choushan.

[e] Each of these men specialised in one or another of the new engineering techniques. Kung Chen-Lin was a notable inventor who developed (ahead of the West) iron moulds for gun-casting (cf. Sect. 30d), and Wang Chung-Yang also wrote on foundry work. Chêng Fu-Kuang (see p. 389 above) devoted himself to optical instruments and to the steam-engine. The most many-sided was Ting Kung-Chhen,[3] who studied all aspects of gunnery from the making of cannon and projectiles to the positioning of defence batteries and the mounting of artillery within them; he also constructed (as we have seen) a model steamboat and a model locomotive. Others who wrote on gunpowder, as also on mines, bombs, fuses and shells, were Huang Mien,[4] Chhen Chieh-Phing[5] and Ting Shou-Tshun.[6] Then there were those like Hsü Hsiang-Kuang[7] who devoted themselves to the improvement of shipbuilding. Phan Shih-Chhêng,[8] a wealthy merchant and patron of literature, experimented with sea mines, and the first full-scale steamboat was built under the aegis of Phan Shih-Jung.[9] Memoranda by these men and their colleagues occupy chs. 84 to 95 of the 1852 edition of the *Hai Kuo Thu Chih*. Several doctoral dissertations could (and we hope some day will) be written on the Chinese technological pioneers of the Opium War period. The English monographs of Chhen Chhi-Thien, good as they are, make but a beginning, and we have not so far come across anything adequate in Chinese.

[f] In Wei Yuan's[10] *Hai Kuo Thu Chih*[11] (Illustrated Record of Maritime, i.e. Occidental, Nations), ch. 86, p. 2a; tr. Lo Jung-Pang (3), p. 192; Chhen Chhi-Thien (1), p. 35, mod. auct.

[1] 龔振麟　　[2] 林則徐　　[3] 丁拱辰　　[4] 黃冕　　　[5] 陳階平
[6] 丁守存　　[7] 許祥光　　[8] 潘仕成　　[9] 潘世榮　　[10] 魏源
[11] 海國圖志

cylinder and churned the water with wheels. These were surveying the beach, reconnoitring the situation and guiding the other vessels, appearing and disappearing in the waves, and going where they would. People marvelled at their strangeness and wondered at their being powered by fire. But it occurred to me to copy the pattern of these wheel-ships simply replacing steam by man-power. So I asked some artisans to build a small model, and when this was tested on a lake it proved to be quite fast. Hearing of this, the Governor (of Chekiang), Liu Yün-Kho,[1] authorised me to build several full-scale war-junks according to my design; they were ready in a month or so, and proved very manœuvrable at sea.

Here then was a strange 'copy'—much older than its prototype. Similar experiments were going on everywhere. Next year, an official in the Salt Administration at Canton, Chhang Chhing,[2] constructed a treadmill paddle-wheel ship 67 ft. long and 20 ft. in beam, with rudders fore and aft, armed with twelve guns and carrying over a hundred men (Fig. 638).[a] Another designer was Wang Chung-Yang,[3] who began to lay down a number of such craft at Chenhai. According to his own description,[b] these had

two wheels in the fore compartment and two in the aft, each wheel having six blades, the tips of which were level with the bottom of the ship. The hexagonal shafts were enclosed in housing three feet long. Within the ship two men standing shoulder to shoulder (at each handspike of the capstan) and pushing, could make the ship fly over the water. Or one can use treadmills like those of ordinary chain-pumps.... The depth from deck to bottom was rather more than 6 ft., half of which was submerged. If the boat floated too high in the water it was ballasted with stones, for with a draught of 3 ft. the wheels should dip into the water 1 ft.

These were probably the vessels which the British captured at Chenhai when nearly completed. Thus a Chinese technical revival was stimulated by the new steam paddle-boats of the West. But there is every evidence that Kung Chen-Lin and his colleagues knew well of the Thang and Sung antecedents. For example, in a contemporary work, the *Fang Hai Chi Yao*[4] (Selected Material on Coast Defence), Yü Chhang-Hui[5] figured a Sung paddle-boat such as he believed had been used by Yo Fei and Yang Yao (Fig. 639).[c] References to the Sung naval paddle-wheel ships also appear in many official memorials of the time,[d] such as those of Chu Chhêng-Lieh[6] and Chin Ying-Lin.[7] What was really new was the building of steam paddle-wheel ships by the Chinese themselves, and that we have already briefly discussed (pp. 388 ff.) in another connection.

Much more astonishing than all this is the fact that Chinese scholars had been worrying about the history of the invention long before paddle-steamers ever appeared off the Chinese coasts. That remarkable man Fang I-Chih[8] (d. +1671), mathematician,

---

[a] From a MS. drawing of Liang Chi-Phing,[9] reproduced in Chhen Chhi-Thien (1), opp. p. 35. The paddle-wheels were worked by ten men, and the rudders could be quickly raised or lowered according to the direction of motion intended. Such slung rudders were typically Chinese (cf. Sect. 29h), and this was a case where they would have come in very handy. Chhang Chhing's specification is given in *Hai Kuo Thu Chih*, ch. 84, pp. 23a ff. His vessel was finished too late to take part in the war.

[b] *Hai Kuo Thu Chih*, ch. 84, pp. 28b, 29a.   [c] Ch. 15, p. 6b.

[d] Details in Lo Jung-Pang (3).

[1] 劉韻珂   [2] 長慶   [3] 汪仲洋   [4] 防海輯要   [5] 俞昌會

[6] 朱成烈   [7] 金應麟   [8] 方以智   [9] 梁濟平

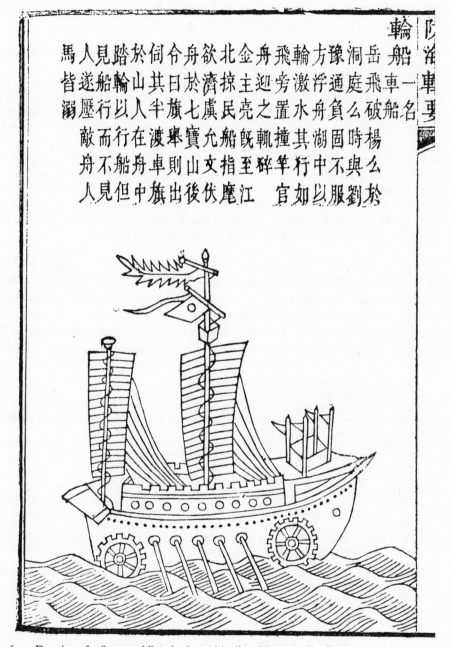

馬人見踏於伺令舟欲北金舟飛輪方豫洞岳　輪　阽
皆遂船輪山其日於濟掠主迎旁激浮通庭飛車　船　車
溺壓行以人半旗七虜民亮之置水舟負么破船名　　要
　敵而行在渡舉寶允船既軌撞其湖固時楊
　舟不船舟車則山文指至碎竿行中不與么
　人見但中旗出後伏麾江　官如以服劉於

Fig. 639. Drawing of a Sung paddle-wheel warship (*lun chhuan*) in Yü Chhang-Hui's work on coastal defence of 1842, the *Fang Hai Chi Yao*, ch. 15, p. 6*b*. The right-hand paragraph of the caption above recalls Yang Yao's use of such warships during his campaign against the government commander Yo Fei early in the +12th century, concluding with a quotation from the *Sung Shih*, ch. 365, translated on p. 420. The second, left-hand, paragraph recalls the defeat of the Chin emperor Wanyen Liang in +1161 by the Sung general Yü Yün-Wên using a fleet of paddle-wheel warships in a surprise attack (cf. p. 421).

scientific encyclopaedist and finally Buddhist monk,[a] has an entry on the subject in his *Wu Li Hsiao Shih* of +1664. In this we read:[b]

Foreign ships (*yang chhuan*[1])[c] have a straight timber beneath (i.e. a keel)[d] and are ballasted so as to be heavy below. They use wheels—but this was also done of old.

Yuan Chhiung[2] (author of the *Fêng Chhuang Hsiao Tu*, cf. p. 418 above) says that Han (Shih-Chung), prince of Chhi,[e] used 'flying' wheel-boats with eight-bladed (paddle-wheels) worked by treadmills (when he surrounded the enemy)[f] at the Battle of Huang-thien-tang (in +1130); they (started and) stopped, and sped back and forth steering to left and right all over the river.[g]

Shih Chhou[3] says[h] that Liu Yü[4] (first emperor of the Liu Sung, +356 to +422) also used them. In the boats there were wheels which were turned (by treadmills), and no one could be seen on deck.

Chang Sui[5] says[i] furthermore that they were employed by Yü Yün-Wên[6] (+1108 to +1174) at the Battle of Tshai-shih (+1161). Within the ships there were treadmills, and as they sailed forward into action their catapults played upon the enemy.

All this material we quickly recognise. The tradition about Han Shih-Chung has already been mentioned. The third paragraph evidently refers to the paddle-boats used by the general Wang Chen-O (p. 416 above) against the northern barbarians. The last records the second great victory of automotive warships on the Yangtze. The really curious thing about Fang I-Chih's remarks is that they were written nearly two hundred years before any European paddle-wheel ships could have been seen in

---

[a] We have a new study of him by Hou Wai-Lu (3).

[b] Ch. 8, p. 23*a*, tr. auct., with Lu Gwei-Djen & Ho Ping-Yü.

[c] There is no doubt that he is talking about European ships because the following entry gives a long and graphic account of one such which made port in Fukien in +1604. He uses the same appellation for it, and adds information on Western shipping obtained from the Jesuit Matteo Ricci.

[d] This was perspicacious, for as we shall see in Sect. 29, one of the cardinal differences between Chinese and European shipbuilding was the absence of any keel in the hull of the junk.

[e] Posthumous title of Han Shih-Chung.

[f] These words come from the original text which Fang I-Chih was abridging.

[g] The circumstances of the battle have been described by Cordier (1), vol. 2, p. 152. The Chin Tartar army retreating northwards was opposed by a great Sung fleet of warships, and its commander Wu-Chu[7] could only hope to get across in small boats, taking advantage of weather when the Sung ships were becalmed. This must have been where Han Shih-Chung's paddle-boats came in. So decisive was the resulting victory that the Chin Tartars never came south of the Yangtze again. The name of the battle is derived from a place near Nanking. Han Shih-Chung was a +12th-century Sir Kenelm Digby; he married a famous courtesan, Liang Hung-Yü,[8] who turned out to be a brave commander in her own right and led a flotilla at this very battle. The passage in the book by their contemporary Yuan Chhiung[2] is delightful; he tells how as a boy he witnessed one of the sea-fights between Sung and Chin from a far vantage-point in the company of one of the imperial physicians who was an elder of his own family, and it would seem that in this combat also paddle-wheel ships took part. One suspects, however, that Han Shih-Chung's paddle-wheel ships were not numerous, for the account of the battle of Huang-thien-tang in the official history makes no mention of them; *Sung Shih*, ch. 364, pp. 7*a*ff., cf. *WHTK*, ch. 158, (p. 1381.3).

[h] We cannot trace this writer, and we are not even sure whether a person is meant, for the words may be the abbreviated title of some book. The source may of course have been a manuscript one.

[i] Chang Sui was a writer of the late Ming whose *Chhien Pai Nien Yen*[9] (Glimpses of a Thousand Years of History) has not been accessible to us.

---

[1] 洋舩	[2] 袁褧	[3] 史紬	[4] 劉裕	[5] 張燧	[6] 虞允文
[7] 兀尤	[8] 梁紅玉	[9] 千百年眼			

Chinese waters. It is possible that there was a misunderstanding and that Fang confused propelling paddle-wheels with the wheels of chain-pumps used for getting rid of bilge-water,[a] but perhaps it is more likely that he had news, either from Western sailors or Jesuit missionaries, of the experiments with treadmill-operated paddle-wheels which had been going on since the time of Blasco de Garay a century or so earlier. This in itself would be an interesting example of culture-contact.

The moment has now come to confront all these pieces of evidence. We have surely no option but to believe that the ship-mills of Belisarius were a direct consequence of the vertical undershot water-wheels of Vitruvius. Once established in the middle of the +6th century, they continued their career in western Europe down to the present time. If we provisionally leave on one side the mysterious *De Rebus Bellicis*, then the origin of the paddle-boat idea (ad-aqueous wheels for the motion of the float) comes very much later in Europe (+15th), in fact no sooner than the idea of making such boats wind themselves up-stream (ex-aqueous wheels for the motion of the float). In China, on the other hand, the paddle-wheel boat proper is remarkably early. If our oldest references are faulty, and Li Kao is the true inventor, then it would be a plausible hypothesis to suggest that wandering Persian merchants in Thang China brought the simple message: 'In the West men have seen boats with wheels', upon which the Chinese assumed (wrongly) that ad-aqueous wheels were meant, and proceeded to construct them.[b] On the other hand, if the real inventor was Tsu Chhung-Chih or Wang Chen-O in the +5th century, they cannot have been inspired by Belisarius, and their work may well have been a spontaneous development of the (Indian?) noria, which would by that time have reached China. This, if substantiated, would give us a possible date for the first introduction of vertical water-wheels of all kinds there.[c] There is, however, one further card in the pack which we have not yet played. It is the *kua chhê*,[1] or scoop-wheel, already mentioned (p. 337 above), used for raising water where very small lifts were required, as from one field to another (six inches or so). This wheel is ad-aqueous, and if it was indeed the first Chinese adaptation of the ex-aqueous noria (for pushing rather than for hoisting water), it might have put the idea in people's heads that something worth while might be performed by applying force *to* a paddle-wheel instead of using it to derive force *from* its medium. Unfortunately, no literary references to the scoop-wheel have been found, apart from its mention in the technical agricultural treatises.[d] As for the origin of ship-mills in China, there is nothing definite to be said; they may have been an independent invention, or a quite separate introduction from Arab contacts, or even possibly a

---

[a] We shall discuss these methods in Sect. 29*i* below.

[b] As already suggested, Vol. 1, p. 246.

[c] We must of course not forget the important statement of Huan Than in +20 (cf. p. 392 above) that water-power was then already used for working trip-hammers. For this the vertical wheel would always have been much more convenient than the horizontal type. The widespread use of the latter in later ages and the general tendency in Chinese engineering to mount wheels horizontally (cf. pp. 369, 406) should not be allowed to prejudice too much our conceptions of what was going on in China in the first five centuries of the +1st millennium.

[d] Even the oldest of these (*Nung Shu*, ch. 18, p. 24*a*) gives no clue as to its history.

[1] 刮車

secondary adaptation deriving from the paddle-wheel naval boats. We lack adequate information about them from pre-Sung times. These considerations show, at any rate, some of the points on which readers of Chinese literature should be alerted.

Did these medieval paddle-boats have anything to do, one may wonder, with the literature on the 'Magic Boat', the self-moving ship on which voyagers saw neither sailors nor machinery? From the summary of Barry (1), we find that such descriptions are common in the Arthurian cycle, and about +1100 in Irish secular tales, but go back in hagiographic texts to about +690. Automatic motion is caused by the presence of the saint or his relics.[a] Three Coptic examples take this even earlier, to the very beginning of the +7th century, and perhaps for this reason Barry sought the origin of the complex in ancient Egyptian predecessors of Charon—yet any Charon is really superfluous. Perhaps we might suggest that the boat with no ferryman and no sailors is a far-away echo of the hidden pedallers of Tsu Chhung-Chih and Wang Chen-O, testifying to the superiority of Chinese technique in the +5th century, in contrast with its reversed position in the +19th.

In the I Ho Yuan (Summer Palace) gardens, near Peking, there stands a famous stone paddle-boat placed there by the last empress of the Chhing, who is said to have used for it funds destined for the building of a real Chinese Navy (Fig. 640). Future generations will surely never destroy this fantasy, which symbolises an honourable circumstance in the history of technology never realised by its builders, namely, the probability that Tsu Chhung-Chih as a practical engineer anticipated Konrad Kyeser by just on a thousand years.[b]

But still we have not reached the dénouement. The problem of the *De Rebus Bellicis* has recently been re-examined by Thompson & Flower and by Mazzarini, with the result that we have now to date it in the +4th century. It must have been written in the close neighbourhood of +370, probably by a Latin of Illyria; and at some point before the time of Charlemagne (perhaps the early +7th century) it was combined with several other Byzantine tracts, including the *Notitia Dignitatum*,[c] to form a kind of corpus. In the +9th or +10th century this was set down in a MS. which afterwards became known as the Speyer Codex, only one leaf of which has survived to provide palaeographers with a dating. The Speyer Codex disappeared about +1602, when it was probably used as binding material, but not before four separate copies had been made of it, the first in +1436 and the fourth in +1542. The whole of the *De Rebus Bellicis* is not at all likely to be a late medieval forgery because no one would then have been able to reproduce the grammar, tricks of style, neologisms and curious quasi-Greek words, or to understand the circumstances and problems for which the Anonymus made his various proposals. Nor is it likely that the paddle-wheel boat (the *liburna*) was inserted in the +14th century, because (*a*) the description

---

[a] We have noticed above (p. 418) one form of this motif, from a late +9th-century Chinese source.

[b] The converse possibility, that Li Kao may have been influenced by a misunderstood report of the mills of Belisarius, has been taken into account above. But even in this case Li Kao's paddle-boat was a practical proposition, anticipating by seven centuries that of Blasco de Garay; quite possibly indirectly generating it.

[c] Somewhat of a parallel to the *Chou Li*.

PLATE CCXLVII

Fig. 640. The celebrated marble ship in the pleasure-gardens of the Summer Palace (I Ho Yuan) near Peking. This Chhing Yen Fang (Ship of the Clear and Quiet Waters), here seen crowded with spring holidaymakers from the contemporary city, was built in 1889 by the Empress-Dowager Tzu-Hsi with misappropriated Navy funds. But its marble paddle-wheels serve to symbolise the leading part played by Chinese engineers from the +5th century onwards in the development of mechanical means of propulsion afloat.

of it comes in the middle of the text and not at either end, while (*b*) it is mentioned by cross-reference at two other places in the text.

For the origins of his idea the Anonymus was no doubt as much indebted to the vertical Vitruvian water-mill as Belisarius later on. But as to its effects, Thompson and all other students of the matter are agreed that there is no contemporary mention of the idea, no evidence that it was anything more than a paper scheme.[a] The memorandum, as he says, was probably intercepted by some civil servant and pigeon-holed without ever reaching the emperor to whom it was addressed, and appears to have stayed in the files for half a millennium after it was written. In these peculiar circumstances it seems extremely unlikely that any word of the invention could have reached Tsu Chhung-Chih or Wang Chen-O a bare century later at the other end of the Old World. Li Kao's paddle-boats, in the latter half of the +8th century, were being built only fifty years or so after the reappearance of the Anonymus' suggestions, so that here again the possibilities of transmission seem small. One cannot help feeling that this constitutes the clearest instance so far unravelled of a strong probability that essentially the same ancient invention was made twice over in different places.[b] Provisionally we can only say that the first specification was Byzantine and the first execution Chinese.

(*j*) CLOCKWORK; SIX HIDDEN CENTURIES

The clock is the earliest and most important of complex scientific machines. Its influence upon the world-outlook of developing modern science was incalculable.[c] No one can doubt that the invention of the mechanical clock was one of the greatest achievements in the history of all science and technology. 'The fundamental solution', wrote von Bertele (1), 'of the problem of securing steady motion by intersecting the progress (of a weight-driven or any powered train) into intervals of equal duration, must be considered as the work of a brain of genius.' The essential engineering task was to devise means of slowing down the rotation of a wheel so that it would keep a constant speed continuously[d] in time with the apparent diurnal revolution of the heavens. The essential invention was the escapement. In what follows we shall show that the first of all escapements arose in China in the middle of a very long line of development of mechanisms for the slow rotation of astronomical models (demonstrational armillary spheres or celestial globes), the primary aim of which was computational rather than time-keeping as such. We shall also show that its first application

[a] It must be admitted also that from the technical point of view a set of ox-driven whims was singularly unsuitable as a nautical power-source. Possibly the plan was tried in one of the European Renaissance experiments. A 'horse-packet' was working successfully at Yarmouth, however, in 1818 (cf. Atkinson (1), pp. 40, 42, 54).

[b] This view is shared by Thompson. Suitable spheres of use might have been an important factor; probably China had more really good navigable rivers, lakes and canals at the times in question than were available to the East Romans.

[c] Cf., for example, Butterfield (1), pp. 8, 44, 59, 111, 120, etc.; Lynn White (7), pp. 124ff.

[d] In the sense that its motion should continue for extended periods without prolonged interruption. In fact, it proved easier to accomplish this by dissecting the motion into very short periods of discontinuity, rather than by any truly continuous braking device.

was to a water-wheel like that of a vertically-mounted water-mill, so that although in later ages mechanical clocks were mostly driven by falling weights or expanding springs, their earliest representatives depended on water-power. The mechanical clock thus owes its existence largely to the art of Chinese millwrights. This story will take some telling. Obviously it differs widely from the account accepted hitherto. How is it that the Chinese contributions to clock-making have been hidden from world history?

It will readily be allowed that few historical events were so rich in consequence as the decision taken by certain southern Chinese officials in +1583 to invite into China some of the Jesuit missionaries who were waiting in Macao. It was the first decisive step in the long process of unification of world science in Eastern Asia, and the better mutual understanding of the great cultures of China and Europe. The two men chiefly concerned were Chhen Jui[1] (+1513 to c. +1585), who was for a short time Viceroy of the two Kuang provinces,[a] and Wang Phan[2] (+1539 to c. +1600), who was Governor of the city of Chao-chhing.[b] They were particularly interested in reports that the Jesuits had, or knew how to make, chiming clocks of modern type, i.e. of metal with spring or weight drives and striking mechanisms. These became known as 'self-sounding bells' (tzu ming chung[3]), by a direct translation of the word 'clock' or cloche, glocke. This is important, for an entirely new name, as this one was, naturally suggested an entirely new thing. The mechanical clocks of the Chinese middle ages had been, as we shall see, extremely cumbrous and probably never very widespread; moreover no special name had distinguished them from non-mechanised astronomical instruments.[c] It was therefore not surprising that the majority of Chinese, even scholars in official positions, now got the impression that the mechanical clock was a new invention of dazzling ingenuity which European intelligence alone could have brought into being. And of course the missionaries (as men of the Renaissance) quite sincerely believed in this higher European science, seeking by analogy to commend the religion of the Europeans as something equally on a higher plane than any indigenous faith.

Both Chhen Jui and Wang Phan obtained the clocks which they desired.[d] The first Jesuit residence, set up at Chao-chhing by Matteo Ricci, had a clock-face on the street with a public self-sounding bell; and this was one of the charges against him when a new governor, Liu Chieh-Chai,[4] closed the mission-house in +1589.[e] But modern horology was irresistible. A magnificent spring clock with chiming bells had been sent

---

a D'Elia (2), vol. 1, p. 164; Trigault (Gallagher tr.), p. 137.
b D'Elia (2), vol. 1, p. 176; Trigault (Gallagher tr.), p. 145.
c Besides, by this time, the clock as a piece of purely time-keeping apparatus had become in the West entirely divorced from its original astronomical connections. As Price (1) has said, 'we must cease to regard the mechanical clock as a time-teller improved beyond all measure by the invention of an escapement. We must regard it instead as a type of astronomical machine which was known and improved throughout the middle ages.' The mechanical clock as a pure time-teller 'was a fallen angel from the world of astronomy'. Price has developed these themes in other papers (4, 8).
d D'Elia (2), vol. 1, p. 166; Trigault (Gallagher tr.), p. 138.
e D'Elia (2), vol. 1, pp. 252, 265; Trigault (Gallagher tr.), pp. 194, 201, 206.

¹ 陳瑞   ² 王泮   ³ 自鳴鐘   ⁴ 劉節齋

PLATE CCXLVIII

Fig. 641. The Jesuits as astronomers in China, a Beauvais tapestry of the late +17th century now at the Musée de l'ancien Évêché, Le Mans (photo. Archives photographiques d'Art et d'Histoire). On an exotic terrace with a pagoda in the background Jesuits in mandarin robes are discussing scientific matters with their Chinese colleagues of the Imperial Bureau of Astronomy, one of whom is examining some heavenly body through a hand-telescope. A bearded father makes some measurements upon a celestial globe like that which Ferdinand Verbiest made for the Peking Observatory in +1673 (cf. Vol. 3, Figs. 158, 176, 191); another dictates some readings to a page, perhaps of sun-spots seen through a reflecting telescope. Behind him stands an ecliptic armillary sphere closely resembling Verbiest's (cf. Vol. 3, Figs. 157, 173, 191, and p. 379). Books bound in Western style remind us of the Pei Thang library of European scientific works established by the Jesuits in Peking (cf. Vol. 3, p. 52), on which see Verhaeren (1).

PLATE CCXLIX

Fig. 642. The muses of the sciences as ladies of the imperial court, an allegorical painting in Western style by Tung Chhi-Chhang (+1555 to +1636) from an album bought at Sian in 1909 by Laufer (28). Left to right: Optics-Acoustics with mirror and flute, Arithmetic-Algebra with a folio volume, Geometry-Mechanics with a pair of compasses and a solid figure, and Astronomy-Geography with globe and map. Tung Chhi-Chhang, who was President of the Ministry of Rites, probably met Matteo Ricci, and may have read with interest Francesco Sambiasi's tractate of +1629 *Shui Hua Erh Ta* (Two Dialogues on Sleep and Allegorical Paintings) prefaced by Li Chih-Tsao (Cordier (8), no. 135).

from Rome as a gift for the emperor, and its peregrinations in the care of the Jesuits seeking to present it form a prominent part of the events of the following years.[a] The impatient inquiries of the Wan-Li emperor in +1600 delivered Ricci and his companions out of the clutches of the eunuch Ma Thang,[1] who had detained them on their way to the capital.[b] After the clock had been installed in the imperial palace, the Jesuits were entrusted with its regulation, and the training of certain eunuchs in clock maintenance and repair.[c] This was the beginning of nearly two centuries of service by the Jesuits, including lay brothers trained as clock-makers, to the Chinese imperial court, where eventually there collected a great variety of clockwork instruments of all kinds.[d] Moreover, wherever they established themselves in provincial cities their mechanical clocks were made known and appreciated.[e] In sum, it is abundantly clear that one of the reasons why the early Jesuit missionaries were so much welcomed by the Chinese was for their interest in clocks and clock-making, hardly less indeed than for their skill as mathematicians and astronomers (cf. Fig. 641).[f]

There can be no doubt that Ricci and his companions regarded efficient mechanical clocks as something absolutely new and unheard-of in China. He says this in his memoirs on several occasions. The clocks for the Cantonese officials in +1583, which struck all the hours automatically, were 'beautiful things never seen nor heard of before in China'.[g] The clock with three bells destined for the emperor, a piece 'which struck all the Chinese dumb with astonishment, was a work the like of which had never been seen, nor heard, nor even imagined, in Chinese history'.[h] The great clock of the Chao-chhing mission house 'with a hand visible from the street which indicated the hours, and a great bell which sounded them, was a thing previously unheard-of'.[i]

---

[a] D'Elia (2), vol. 1, p. 231; vol. 2, pp. 4, 39, 87, 99; Trigault (Gallagher tr.), pp. 180, 296, 320, 348, 355.

[b] D'Elia (2), vol. 2, p. 121; Trigault (Gallagher tr.), p. 369.

[c] D'Elia (2), vol. 2, pp. 126, 128, 159, 313, 471; Trigault (Gallagher tr.), pp. 373, 374, 392, 536.

[d] See the catalogue of Harcourt-Smith (1). In Fig. 643 we illustrate a striking clock or 'self-sounding bell' of European design as it appears in a Chinese work of +1759.

[e] D'Elia (2), vol. 2, p. 382; Trigault (Gallagher tr.), p. 288; cf. de Faria y Sousa (1), pp. 29, 154, 156, 166, 172, 206.

[f] For example, the Khang-Hsi emperor came himself to the Jesuit house in Peking to see the carillon of their turret-clock, which played Chinese tunes each hour. This had been set up by the Portuguese father Gabriel de Magalhaens (+1609 to +1677), and is described in Verbiest's *Astronomia Europaea* ... (+1687), pp. 92 ff. Most of the section entitled 'Horolo-technia' in this book is concerned with the impression made upon the Peking populace by the clock and carillon. It was operated by a spiked drum and trip-wires in some such way as that depicted in +1644 by de Caus (1), pl. 641. The imperial visit was described by Verbiest in a letter of 11 May 1684, transcribed by Bernard-Maître (11): 'Rex venit ad aedes nostras, intravit omnia cubicula, vidit organum et 2 campanas majores horologii suspendendas in turre.' This was the time when the emperor was assiduously studying the natural sciences with the Jesuit experts, and fostering the spread of their knowledge in every possible way. Elsewhere in his book (p. 57), Verbiest was led to write (in a delightful passage which his fellow-Jesuit Bernard-Maître finds rather turgid): 'And now it was as if all the mathematical sciences, led by Astronomy in royal splendour, began together to enter the Imperial Palace—Geometry and Geodesy, Gnomonics and Perspective, Statics and Hydraulics, Music and the Mechanical Arts—all arrayed like the companions of an empress, each handmaiden more lovely than the next.' Cf. Fig. 642.

[g] D'Elia (2), vol. 1, p. 164.   [h] *Ibid.* p. 231.

[i] *Ibid.* p. 252.

[1] 馬堂

Fig. 643. A 'self-sounding bell', i.e. a striking clock of European style such as the Jesuits introduced to China from +1583 onwards (from the *Huang Chhao Li Chhi Thu Shih* of +1759, ch. 3, p. 68a).

Ricci's opinion is thus unmistakable,[a] and no other Jesuit thought differently, so far as we know.

It is true that Ricci and Trigault had something to say of Chinese clocks with driving-wheels which they found on their travels,[b] though they laid little emphasis on them and their descriptions are obscure. It is also true that a number of contemporary

[a] In other cases Ricci was probably quite right in his estimate of the novelty of some of his introductions. He said that the horizontal sundial which he made for the Prince of Chien-An at Nan-chhang was 'something never before seen in China'. This was because the characteristic Chinese form of the sundial was the equatorial, in which the plate is in the equinoctial plane with the pole-pointing gnomon or style at right-angles to it. Chinese geometry had never been adequate for the stereometric projections needed for sundials on vertical or horizontal surfaces. The incident is in D'Elia (2), vol. 1, p. 366.

[b] See pp. 508 ff. below.

Chinese scholars recognised that the clockwork of the Jesuit 'self-sounding bells' was not something fundamentally new in Chinese culture.[a] The former were not anxious (in China) to exalt the achievements of the indigenous past. The latter were insufficiently learned to expound that past as it deserved. This complicated situation had its effects upon the thinking of Europeans later on, and in particular on that of European historians of science. If the Jesuits so firmly believed in the novelty of the mechanical clocks which they introduced to China, who were the later historians of science, penned within the ring fence of the alphabetic languages, to gainsay them? In an interesting passage of Ricci's memoirs (not included in Trigault's book) he wrote:[b]

What with the great facility, commodity and freedom of printing, one can see wherever one goes in Chinese houses how enthusiastically they collect books, much more than is the case with us. And by the same token they print far more books in any year than any other nation. And since they lack our sciences, they make many on other matters, useless and even harmful. But for them there was the greatest novelty in our tidings, telling so much of our own, and of all other nations, what with new laws (of religion), new science and new philosophy, so that much about us came to be printed in their books—partly concerning the arrival of the Fathers and the things we brought with us, pictures, clocks, books, descriptions and mechanical things—partly concerning the laws and sciences that we taught—partly concerning the printing of our books or quotations from them—partly concerning the many epigrams and poems which were composed in our honour. And stories true and false concerning us flew about to such an extent that there will be great memory of us in this kingdom for centuries to come— and what is better, good memory.

Thus novelty was the keynote of the Jesuit experience. Inevitably it moulded the picture of the time formed by subsequent historians, strengthening them in their conclusion that the invention of clockwork was a European one. Towards the end of his discussion of this, Sarton (1) crystallised the accepted view in the following words:[c]

I may add that mechanical clocks were not discovered independently in the Far East. The Chinese (and the Japanese) were of course familiar from early days with sundials, they had obtained some knowledge of clepsydras (from the Roman West?),[d] and they had learned also to measure time with burning tapers or candles; but they never thought of mechanical clocks until some were shown to them by missionaries....

The application of the clock to Far Eastern usage implied some difficulty, because the Chinese (and the Japanese) did not divide the day into twice twelve equal hours, but into two unequal periods (day and night) each divided into six equal divisions....[e]

---

[a] See pp. 524 ff. below.
[b] D'Elia (2), vol. 2, p. 314.   [c] Vol. 3, p. 1546.
[d] Our Section on clepsydras (Vol. 3, pp. 313 ff. above) has shown that the Chinese were indebted to Babylonian and ancient Egyptian rather than Hellenistic influence in this field, while later developments were markedly divergent rather than parallel.
[e] This a complete misapprehension. The statement is true only of the Japanese. As d'Elia (2), vol. 1, p. 128, has correctly stated, the Chinese from time immemorial divided the whole of day and night (from midnight to midnight) into twelve equal double-hours (shih[1]) and one hundred (occasionally 96 or 120) quarters (kho[2]). These were again divided each into 100 (sometimes 60) fên.[3] From an early time it was customary to bisect each double-hour into separate halves, equivalent to our hours, the first being

[1] 時   [2] 刻   [3] 分

Hence the introduction of mechanical clocks necessarily created a deeper disruption of ancient usage in the Far East than in the Christian West.[a] The fact that the Chinese (and the Japanese) did not invent them need not surprise us; it is more remarkable that they adopted them at all. This can be explained only by the necessity of adapting their ways to the Western ways forced upon them.[b]

Here out of eight statements or suggestions only three are right. This should be a warning to all scholars of the provisional nature of their conclusions, and of the danger of preconceived ideas about the comparative contributions of the cultures of East and West. In the first volume of the present work, we ourselves adopted the generally accepted view of the origins of clockwork, and wrote without hesitation:[c] 'The last important introduction (from the West to China) was clockwork, a distinctively European invention of the early +14th century.' The course of research now obliges us to withdraw this statement, once so apparently well-founded, and with it falls one of the greatest bastions of the opinion that a mechanical penchant was always characteristic of occidental, but not of oriental civilisation.

What then was the accepted view, in closer detail? It held that the first successful achievement of slow, regular, and continuous rotation in time with the diurnal revolution of the stars, by means of an escapement acting upon the driving-wheel of a train

known as the *chhu*[1] portion and the second as the *chêng*[2] portion. The first double-hour straddled midnight and the seventh straddled noon. Ginzel (1), vol. 1, pp. 464 ff., knew all this well, and the system had been explained with clarity to the Western world by G. S. Bayer (1) as early as 1735. An adequate account was again given by Ideler (1) in 1837. The twelve equal double-hours probably came from Babylonia, as Bilfinger (2) urged long ago, but the hundred quarters have a strongly indigenous flavour in view of the typical Chinese predilection for decimal systems (cf. Vol. 3, pp. 82 ff.). From the Han onwards the personified spirits of the twelve double-hours were often represented by pottery or terracotta figurines (Fig. 644) having the heads of animals and dressed in dignified robes (cf. Thang Lan (1), pl. 97). Oracle-bone researches now show that 'unequal' hours were indeed current in the Shang and early Chou periods, but they were abandoned about the middle of the −1st millennium, and by the Warring States period a full association of duodenary cyclical characters (and perhaps also the animal cycle) with the twelve equal double-hours had become established. There remained, however, in China down to modern times a concurrent system of five 'unequal' night-watches (*kêng*[3]), each divided into five parts (*chhou*[4]), and varying in length with the seasons. As we shall see below (pp. 455 ff.), many of the designers of clocks from the +11th century onwards made it their aim to mark these intermittent 'unequal' periods as well as the continuously revolving equal double-hours and quarters. Enshoff (1) and Sarreira (1) have maintained that Ricci introduced mechanical clocks keeping 'unequal' or canonical hours to China. There is no foundation whatever for this statement. For further details on the Chinese horary system see Needham, Wang & Price (1), pp. 199 ff. Opportunity may be taken here of amending a mistake on p. 38 of that work. Correcting itself by the text in fig. 14, para. G 8 should read: 'At each fifth of a night-watch a trip-lug is fixed, so that it strikes the gong at every night-watch and division of a night-watch and at sunrise and sunset.'

[a] Again this is an untenable proposition. The 'unequal' or variable-length hours existed only in Japan. And why should more disruption have been created at their supersession there than that which occurred when the first European mechanical clocks obliged the abandonment of the similar canonical hours in the West? On this latter process see Howgrave-Graham (1). On Japanese variable-hour clocks, which embodied many curious devices such as two foliot bars of different lengths, see J. D. Robertson (1); Rambaut (1); Ward (1); Planchon (1), pp. 209 ff.; but especially Takabayashi (1) and Yamaguchi (1).

[b] No Western ways were being forced upon the Chinese until two and a half centuries after Ricci's time. The phrase is appropriate for the period of the Opium Wars and later treaty-port imperialism, but nobody forced the +18th-century emperors to accumulate thousands of clocks in their palaces, nor any of their subjects to use such machines. Disinterested curiosity reigned in good Father Ricci's golden days.

[c] Vol. 1, p. 243.

[1] 初　　　　[2] 正　　　　[3] 更　　　　[4] 籌

PLATE CCL

Fig. 644. Two Presidents of Double-Hours, examples from a set of twelve pottery figurines from a Thang tomb at Han-sên-chai, Sian (photo. CPCRA and BCFA, cf. Thang Lan (*1*), pl. 97). On the left Pig, representing the Hai double-hour (9–11 p.m.); on the right Sheep, representing the Wei double-hour (1–3 p.m.). Cf. Vol. 3, pp. 396, 398, 405; and Needham, Wang & Price (*1*), pp. 199 ff.

PLATE CCLI

Fig. 645. The simplest type of early +14th-century European mechanical clock with weight-drive, crown-wheel, pallets, and verge-and-foliot escapement; an example, probably of later date, from the tower of St Sebald's Church at Nuremberg (Zinner, 4).

of gearing, occurred in Europe very shortly after the beginning of the +14th century.[a] One may take an authoritative statement from Bassermann-Jordan, who wrote:[b] 'We must place the birth of the wheel-clock around +1300. When clocks are mentioned before that time, either sundials or clepsydras are meant, or else there is always something doubtful about the evidence.... The soul of the wheel-clock is the escapement, which hinders the rapid revolution of the wheels. This invention is one of the greatest and cleverest ever made by man, yet the inventor remains unknown and forgotten, commemorated neither by stone nor monument.'[c] It was indeed a turning-point, yet these first mechanical time-keepers were not so much of an innovation as used to be supposed. They descended in fact from a long series of complicated astronomical 'pre-clocks', planetary models, mechanically rotated star-maps, and similar devices designed primarily for exhibition and demonstration rather than for accurate time-measurement.[d] Traces of these remain from Greek, Hellenistic and Arabic times, but the remains are fragmentary and the texts tantalisingly incomplete. We shall make mention of some of them from time to time.

First let us gain a clear idea of the +14th-century mechanisms,[e] and then vivify this skeleton by a few textual references. The simplest form of the early European mechanical clock drew its power from the rotation of a drum brought about by the fall of a suspended weight. This was connected with trains of gearing in great variety, but the movement of the whole was slowed to the required extent by the escapement device known as the verge and foliot. This can best be appreciated by the accompanying illustrations. Figure 645 shows this type of clock in one of its simplest forms, a bracket-clock of about +1380 from St Sebald's church in Nürnberg, described by Zinner (4), and the diagram in Fig. 646 taken from Berthoud (1) gives a clear drawing.[f] The essential parts of the device were the crown-wheel, a toothed wheel with projections like right-angled triangles set perpendicularly to its main plane; the verge or rod standing across this wheel and bearing (at right angles to each other) its two pallets or little plates so as to engage with the crown-wheel; and finally the foliot (crazy dancer), i.e., two weights carried one at each end of a bar set at the top of the verge.[g] The method of action was very simple. The torque of the crown-wheel pushed

---

[a] Besides the reference to Sarton (1), vol. 3, pp. 716 ff., 1540 ff., we may add Beckmann (1), vol. 1, pp. 340 ff.; Usher (1), 2nd ed., pp. 191 ff., 304 ff.; Frémont (7); Baillie (1, 2); Howgrave-Graham (1); Ungerer (2); Bolton (1); Planchon (1); Wins (1). It is noteworthy to find this view shared by Japanese writers, e.g. Takabayashi (1); Yamaguchi (1).

[b] (1), p. 17; cf. p. 170. Similar expressions occur in Bassermann-Jordan (2), pp. 4, 13.

[c] And Michel (15) writes: 'The escapement is perhaps the most extraordinary of all the inventions of man. Nothing in Nature can have served as a model for it, nothing can have suggested it.'

[d] See Price (1).

[e] There is a very large literature on the history of mechanical clocks. The most useful books and papers have just been cited. Cunynghame (1) and J. D. Robertson (1) are small but helpful introductions to the subject. Many well-known works such as Milham (1) and Britten (1) are muddled in presentation or too antiquarian. The books of Saunier (1, 2, 3) are mainly practical but interesting and authoritative. The treatise of Ungerer (1) is primarily devoted to the elaborations of horological jack-work—mechanical figures and displays worked by the striking mechanisms of clocks. Cf. Bolton (1); Bassermann-Jordan (1, 2); Baillie (1, 2); Ungerer (2); Planchon (1, 2); Wins (1); Lloyd (5).

[f] Cf. Beck (1), p. 175.

[g] Time-keeping could be adjusted by varying the distance of the weights from the centre of the foliot bar.

one of the pallets out of the way, giving a swing to the foliot, but this only led to the coming into action of the other pallet and then a swing of the inert weights in the opposite direction. In this way the motion of the wheel was arrested alternately by the two pallets. Thus an oscillatory component received its impulses from the weight drive and at the same time imposed a step-by-step or ticking movement upon the train.

The more familiar examples of these early time-keepers are the large 'turret' clocks made for church towers and the like.[a] Here a further complication is introduced, namely a striking train as well as the going train. The early forms of this are particularly interesting because the verge and pallets could act as a bell-ringing device if the swinging foliot weights were replaced by small hammers applied to a bell; the device would then, when released, run wild.[b] It is very probable that the earliest European mechanical clocks had no pointer or dial-face but simply set off a striking arrangement when each hour, or other prearranged time, arrived.

There can be no doubt that many of the component parts of these clocks were Hellenistic in origin. The falling weight had originally, no doubt, been a falling float, such as we see in the Roman anaphoric clocks, where a dial bearing astronomical markings was made to rotate slowly by a cord attached to a float sinking in a clepsydra.[c] The idea of a dial would have come from the same source. Presently we shall see some evidence that this kind of system may not have been unknown in China, though never dominant there. In the West there was also the automobile puppet theatre minutely described by Heron of Alexandria and reconstructed by Beck (4), where slow motion was obtained by the descent of a heavy weight as grains of sand or cereal escaped through an hour-glass hole at the bottom of the container. Thorndike (7) has drawn attention to a very interesting passage in the commentary of Robertus Anglicus on the *Sphere* of Sacrobosco. Writing in +1271, he said:[d]

Now it is hardly possible for any time-keeping device (*horologium*) to follow the indications of astronomy with absolute accuracy. Yet clock-makers (*artifices horologiorum*) are trying to make a wheel which will accomplish a complete revolution for each one of the equinoctial circles, but they cannot quite perfect their work. If they could, it would be a really accurate clock, and worth more than any astrolabe or other astronomical instrument for reckoning the hours, if one knew how to do this according to the aforesaid method. The way would be this, that a man make a disc of uniform weight in every part so far as could possibly be done. Then a lead weight should be hung from the axis of that wheel, moving it so that it would complete one revolution from sunrise to sunrise....

These words strongly suggest that at this time attempts were being made to construct a practical weight-driven escapement clock, but that success had not yet been

---

[a] On this period see Howgrave-Graham (2); R. A. Brown (1), etc. One of the most famous of these machines is the Dover Castle clock, now in the Science Museum at South Kensington, and often reproduced (as by Ward (1), pl. VII). This is no longer ascribed to its traditional date of +1348, but belongs rather to the end of the century. Yet its plan is of the early type.

[b] Cf. Price (1).

[c] See especially Diels (1), p. 213, and Drachmann (6), (2), pp. 21 ff. Diels' reconstruction is figured by Usher (1), p. 97, 2nd ed. p. 145. The *locus classicus* is Vitruvius, IX, 8. Cf. Kubitschek (1), pp. 209 ff. We shall often return to this device; cf. pp. 466, 503, 517, 536 below.

[d] Basel MS. F. IV. 18, and Bib. Nat. MS. Latin 7392. Tr. Drover (1).

PLATE CCLII

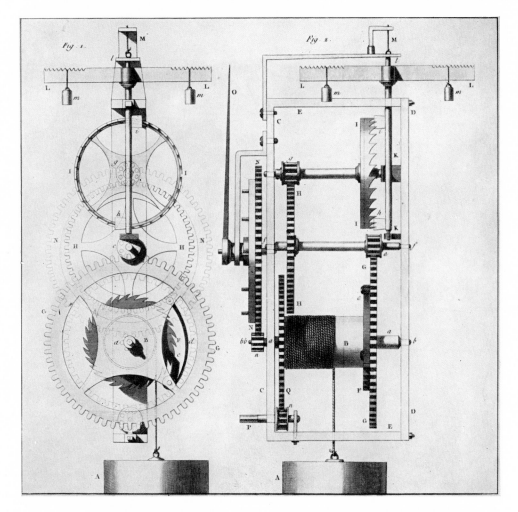

Fig. 646. Drawings from Berthoud (1) to show the essentials of the verge-and-foliot escapement of
early weight-driven mechanical clocks. Left, elevation, right, cross-section.

PLATE CCLIII

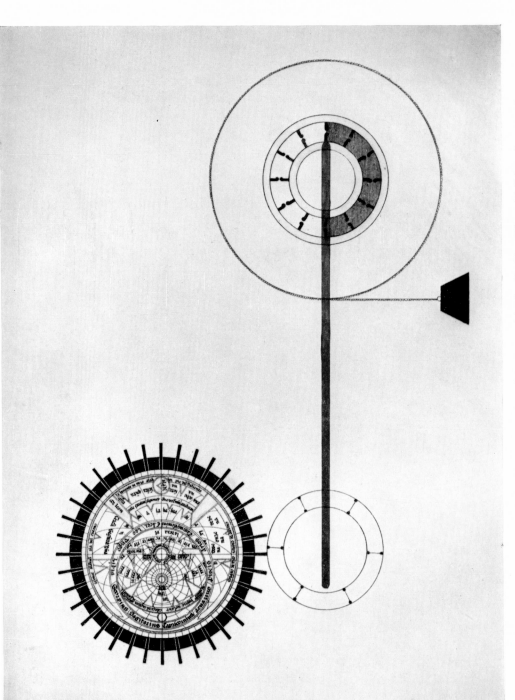

Fig. 647. The mercury leaking-compartment drum 'escapement' in the Castilian *Libros del Saber* of c. +1276.

achieved.[a] However, the mercury clock described in the *Libros del Saber de Astronomia*,[b] compiled for King Alphonso X at Toledo about +1276, could have worked satisfactorily. Here the weight-drive was combined with a hollow drum containing twelve compartments and half filled with mercury, the escapement effect being obtained by the flow viscosity of the mercury in passing through small holes in the walls of the compartments (Fig. 647).[c] Why this system did not spread remains puzzling.[d] Of much significance, however, is the fact that it was used to rotate an astrolabic or anaphoric dial.

As for the verge-and-foliot escapement, Frémont (7) must surely be right in his suggestion that it derived from the radial bob type of fly-wheel.[e] This became associated with the upper ends of the worms of Hellenistic screw-presses, originally used for making wine or oil,[f] when later on, in the +15th century, they began to print books.[g] In the +16th and probably earlier, it was used to assist crank action.[h] The originality lay in its combination with the pallets and crown-wheel so that it oscillated back and forth rather than continuously turning.[i] Now one of the greatest mysteries of the early European clocks was the origin of the escapement principle. For a long time it was thought to appear in a strange design found in the notebook of Villard de Honnecourt about +1237, where a cord carrying weights at each end is wound round two axles, one vertical and one horizontal, finally passing between the spokes of a large wheel on the second axle.[j] It was supposed that the motion was periodically checked, and then released on the recoil. The object of the device was to make a figure of an angel turn and point its finger at the sun.[k] Another design was intended to make an eagle turn its head towards the place where the priest and clerks stood to read the Gospel. But it is now agreed[l] that these mechanisms cannot have been escapements, but simply a means of turning the figures by hand. If so, no predecessor for the first European escapement remains—except the Chinese type shortly to be described.

For three hundred years the verge-and-foliot clock remained unchanged, save for

---

[a] On the general setting of medieval European engineering, cf. des Noëttes (3); Lynn White (1) and C. Stephenson.

[b] See the edition of Rico y Sinobas. Cf. Feldhaus (22).

[c] We say 'escapement effect' because the device was a continuously acting brake and not a true escapement.

[d] Actually in one way it did, giving rise to the falling drum type of water-clock common in the +17th and +18th centuries (cf. Planchon (1); Britten (1), p. 12). This has been discussed above in the Section on clepsydras (Vol. 3, p. 328). One of the earliest references to it is thought to be the account of the *horologium* made by Boethius in +507 for the Burgundian king on behalf of Theodoric, king of the Ostrogoths (Cassiodorus, in *Mon. Germ. Hist. (Auct. Antiquiss.)*, ed. Mommsen, vol. 12, p. 39; also in Migne, *Patrologia Latina*, vol. 69, col. 539 ff.). On the whole subject see now Bedini (3).

[e] Cf. p. 91 above. [f] Drachmann (7), pp. 156, 158; Frémont (8).

[g] Frémont (8). [h] In Agricola, Hoover & Hoover ed., p. 180.

[i] Another technique ancestral to this oscillation was no doubt the alternating rotary motion of the bow-drill and treadle-lathe (cf. pp. 55 ff. above), so universally known and used in antiquity.

[j] Hahnloser ed., pl. 44; Usher (1), pp. 153 ff., 2nd ed., pp. 193 ff.

[k] One cannot but be reminded of the Arabic water-clocks described by al-Jazarī and al-Sāʿātī about +1205, which operated a pointing figure turning on a horizontal clock-face by the fall of a float on the anaphoric clock principle (see Wiedemann & Hauser (4), pp. 135, 156). Villard de Honnecourt certainly also knew water-clocks of some kind (cf. p. 544 below), but we do not know whether, or how, his jacks were connected with them.

[l] Price (1), p. 34; (8), p. 108.

increasing elaboration of striking trains with complex systems of levers and detents. But towards the end of the +16th century the pendulum was coming into wider technological use,[a] and its property of isochronicity began to attract attention.[b] The first application of it to the clock escapement may have been made by Jost Burgi of Prague in +1612, but more credit is due to Galileo and most of all to Huygens.[c] In +1641, after he became blind, a pendulum clock was built for Galileo by his son Vincenzio.[d] But the main features of the pendulum clock, including even the cycloidal arc of swing, were due to Huygens,[e] who constructed his first successful apparatus in +1657, and whose book *Horologium Oscillatorium* achieved its final form in +1673. At first the pendulum was combined with the verge and pallets, as may be seen in Berthoud's diagram (Fig. 648). But about +1680 Wm. Clement devised the familiar anchor escapement, shown by Berthoud in Fig. 649, in which the crown-wheel was replaced by a scape-wheel having teeth in the plane of its rotation.[f] This device has persisted till the present day in many modified forms. Probably its most important modification was the dead-beat escapement introduced by George Graham[g] in +1715; by adjusting the shape of the teeth and pallets this eliminated all that recoil which had been one of the most wasteful features of the early clocks. Then, with progressive solutions of problems such as temperature compensation in the 18th and 19th centuries, we are fully in the modern period, into which it is unnecessary to go further for our present purpose.

In the meantime there had been one other invention of major importance, the application of the spring-drive instead of the falling weight.[h] This permitted the making of portable watches as well as stationary clocks. But it introduced a new difficulty, that of compensating for the variable force exerted as the steel spring ran down; this was overcome by various devices, first an auxiliary spring known as the stackfreed, and then the conical drum known as the fusee. This was a driving barrel of varying diameter, so arranged that the maximal leverage of the cord or chain acting

[a] E.g. by Besson (cf. Frémont, 7); see also Beck (1), pp. 191 ff., 335 ff., 451.

[b] The value of a pendulum escapement had indeed occurred to Leonardo da Vinci, who made a sketch of one, c. +1490, but this had no immediate influence. His priority was pointed out by Reverchon (1).

[c] What Burgi was certainly responsible for was the cross-beat foliot escapement which he incorporated in a number of fine astronomical clocks (see von Bertele, 1, 3, 9). This attained a regularity of ±30 sec./day, i.e. only 3 % of the error previously usual, and nearly as good as many of the early pendulum clocks. Von Bertele does not believe that Burgi ever used the pendulum. Justus Borgen, to give him his proper name (+1552 to +1632), was Automatopaeus to the emperor, and a collaborator of Tycho and Kepler in Prague, where he lived and worked in the Hradčin Castle (cf. Thorndike (1), vol. 7, pp. 614, 623, 630).

[d] At this time J. B. van Helmont was also independently experimenting with the pendulum in relation to time-keeping (*De Tempore*, ch. 50, in *Ortus Medicinae*, +1648); cf. Pagel (8), pp. 398 ff., 409.

[e] Crommelin (1–4, 6); Usher (1), 2nd ed., p. 310.

[f] This invention has often been claimed for Robert Hooke, but the latest examination of the evidence by Lloyd (2) favours Clement. It was important because it allowed the pendulum to swing in an arc of only 3 or 4°, and so reduced the error due to lack of isochronism to a very small value (cf. Ward, 2).

[g] See Lloyd (4). Graham was also the inventor of the cylinder escapement (Ward (1), vol. 2), in which the teeth escape through a slit in a rocking cylinder; he also made the first 'orrery' (on which see p. 474 below, and Vol. 3, p. 339), and the first mercury pendulum.

[h] Usher (1), 2nd ed., p. 305, points out that Peter Henlein of Nürnberg (*fl.* +1510) cannot have been the originator of the spring watch, as has often been supposed. Extant specimens indicate that the invention must have occurred about the last decade of the +15th century, even perhaps as early as +1475.

PLATE CCLIV

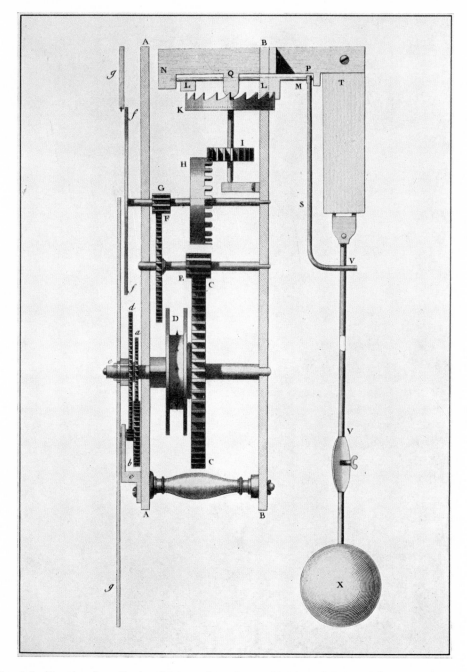

Fig. 648. Drawing from Berthoud (1) to show the combination of the pendulum with the verge
and pallets introduced by Galileo and Huygens in the mid + 17th century. Cross-section.

PLATE CCLV

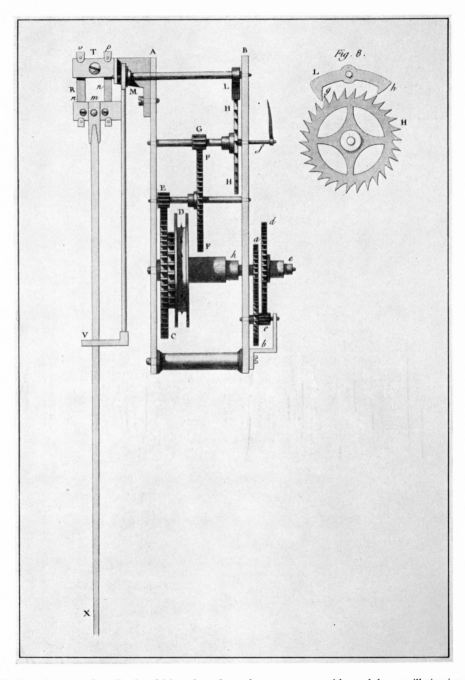

Fig. 649. Drawings from Berthoud (1) to show the anchor escapement with pendulum oscillation introduced by Clement in the late +17th century. Left, cross-section; right, the operation of the anchor
and scape-wheel.

on its largest diameter came into play at the end of the spring's activity when its pull was weakest.[a]

All these conclusions naturally depend upon the evidence of texts as well as of remaining clocks or their parts. It may be considered as quite certain that the earliest type was in use by about +1310 and that all the characteristic features were assembled by +1335.[b] Yet no +14th-century clock has survived immune from later reconstructions so extensive as to render difficult the restoration of the original condition. One of the earliest literary references is due to Dante,[c] who describes quite clearly in a text of +1319 the gear-work of a striking clock. Authentic accounts of clocks occur in the *Chronicles* of G. Fiamma[d] for +1335 and +1344; the former one erected in the tower of a palace chapel at Milan, the latter in another tower at Padua. This was due to Jacopo di Dondi, whose son Giovanni, besides constructing a great clock at Pavia in +1364, wrote a splendid horological treatise[e] which has been carefully studied and summarised by Lloyd (1). His clock was an astronomical masterpiece embodying gear-trains of great complexity to portray accurately the motions of the planets, and to show the ecclesiastical festivals, both fixed and movable.[f] The first mention of the verge and foliot, by Froissart,[g] occurs about the same time, in +1368, and this is also the year of the first modern escapement clock in England.[h]

Such was the picture of the development of the mechanical clock as it stood on the basis of researches in European history alone. The invention of the escapement seemed to have occurred at the beginning of the +14th century with no recognisable antecedents. As Bolton wrote:[i] 'Weight-driven clocks come suddenly into notice at this period in a very advanced stage as regards design, though their workmanship was rough. Their previous evolution must have taken a long time, but there is no reliable record of its stages nor of the men responsible for it.' In 1955, however, a way of solving this problem opened out before us. For the *Hsin I Hsiang Fa Yao*,[1] written in +1092 by a distinguished scientific scholar and civil servant of the Northern Sung dynasty, Su Sung,[2] describes the erection in +1088 of elaborate machinery for effecting the measured slow rotation of an armillary sphere and celestial globe, together

[a] Lloyd (5), pp. 656 ff.; Lynn White (7), p. 128. This device (the 'schnecke' or snail) seems to have been used almost as soon as the spring itself, and Jacob the Czech of Prague (cf. Lloyd, 8) cannot have been, as is often thought, its inventor, about +1525, though doubtless he improved it (cf. Usher (1), 2nd ed., pp. 305, 307, and Lynn White's review of that book). It was known to Leonardo and originated from a mechanism for spanning powerful crossbows (see Feldhaus (18), p. 95, (21) and Berthelot (5, 6) on Kyeser's *Bellifortis* of +1405). Cf. pp. 295 above, 526 below.

[b] Cf. Sarton (1), vol. 3, pp. 1540 ff., Usher (1), 2nd ed., pp. 196 ff. See also Price (8).

[c] *Paradiso*, X, 139 and XXIV, 13.

[d] In Muratori's *Italicarum Rerum Scriptores*, vol. 12. Cf. Lynn White (7), p. 124.

[e] MSS. in Venice, Milan, Padua, Eton and the Bodleian.

[f] It included elliptical gear-wheels, indicators on endless belts of hinged links, and an equatorium similar to that in the Chaucer MS. described and explained by Price (2). A complete working model was made and exhibited at the South Kensington Science Museum in 1961; it is now at the Smithsonian Institution at Washington. On the de Dondi family see Thorndike (1), vol. 3, pp. 386 ff.

[g] In *Li Orologe Amoreus*. We cannot be quite sure, however, that this was driven by a falling weight and not by a water-wheel (cf. below, p. 543).

[h] See Howgrave-Graham (1).

[i] (1), p. 52.

[1] 新儀象法要   [2] 蘇頌

with a profusion of time-keeping jack-work. The book's pithy title might be translated 'New Design for an Astronomical Clock' (lit. 'Essentials of a New Method for (Mechanising the Rotation of) an (Armillary) Sphere and a (Celestial) Globe'). The whole 'Combined Tower' (Ho Thai [1]) could in fact have been nothing more nor less than a great astronomical clock necessitating some form of escapement. And indeed the full translation and study of the elaborate and detailed text[a] not only showed that this was the case, but revealed the considerably earlier origins and development of time-keeping machinery recorded for posterity in Su Sung's remarkable historical introduction. In this way six centuries of Chinese horological engineering, previously hidden, came to light.

(1) Su Tzu-Jung and his Astronomical Clock

To readers of this work Su Sung (Su Tzu-Jung [2]) is by now not an unfamiliar figure. His work has been mentioned frequently in the Section on astronomy[b] in many connections. But he was not only an astronomer and mathematician, he was also a naturalist, for in +1070 or thereabouts he produced, no doubt with a number of assistants, the best work of his age on pharmaceutical botany, zoology and mineralogy, the *Pên Tshao Thu Ching* [3] (Illustrated Pharmacopoeia). Still today this treatise contains precious information on subjects such as iron and steel metallurgy in the +11th century, or the use of drugs such as ephedrine, and we have often referred to it already, notably in the Section on mineralogy.[c] Su Sung was primarily an eminent civil servant, but one of those (by no means few in medieval China) who mastered the scientific and technical knowledge of his time and found opportunities for employing it in the service of the State.

Su Sung (Su Tzu-Jung) was born in +1020 in Fukien, not far from Chhüanchow, the city which Marco Polo was later to know as Zayton. He pursued his career in the bureaucracy with considerable success, associated neither with the Conservative party, though his friends were mostly members of it, nor with that of the Reformers; and he became a specialist in administration and finance.[d] But as was usual in those days, he also received foreign assignments, and in +1077 was despatched as a diplomatic envoy to the Chhi-tan people of the Liao kingdom in the north. Yeh Mêng-Tê,[4] in his *Shih-*

---

[a] Full details will be found in the special monograph of Needham, Wang & Price (1). This work, which revealed for the first time the long history of escapement clocks in China before those of the European +14th century, was first published (2) in January 1956. Only in September of the same year was it found that parallel researches had been proceeding at Peking, with almost exactly the same conclusions. In 1953 and 1954 Liu Hsien-Chou (4, 5) had described Su Sung's gearing, and sketched the mechanism of his escapement, but without noting its historical significance at an 11th-century date. These papers were unknown to us and inaccessible. A further contribution on clockwork as such (6) then followed late in 1956. At the International Congress of the History of Science in Florence in September of that year it was possible—and very pleasurable—to compare results and to reach general agreement among all those concerned.

[b] See Vol. 3, pp. 208, 278, 352, etc.   [c] See Vol. 3, pp. 617, 647, 675, etc.

[d] Biography in *Sung Shih*, ch. 340, pp. 22a ff.; mostly political. Cf. his collected works; *Su Wei Kung Wên Chi*.[5]

¹ 合臺   ² 蘇子容   ³ 本草圖經   ⁴ 葉夢得   ⁵ 蘇魏公文集

Lin Yen Yü,[1] tells us how the opportunity came to him to utilise his astronomical and calendrical knowledge.[a]

When Su Tzu-Jung was taking the provincial examinations (in his youth), it happened that an essay was set on the general principles of the heavens and the earth as manifested in the (structure of the) calendar. He came out top of the list, and ever afterwards he was particularly interested in (astronomy and) calendrical science.

Later on, at the end of the Hsi-Ning reign-period (+1077), (Su Sung) was sent as ambassador to offer congratulations (to the Liao emperor) on the occasion of his birthday, which happened to fall on the winter solstice. (At this time) our (Sung) calendar was ahead of that of the Chhi-tan (Liao) kingdom by one day, and thus the assistant envoy considered that the congratulations should be offered on the earlier of the two days. But the secretary of protocol in the Chhi-tan (Liao) Foreign Office declined to receive them on that day. As the (Liao) barbarians had no restrictions on astronomical and calendrical study,[b] their experts in these matters[c] were generally better (than those of the Sung), and in fact their calendar was correct. Of course, Su Sung was unable to accept it,[d] but calmly and tactfully engaged in wide-ranging discussions on calendrical science, quoting many authorities which puzzled the (Liao) barbarian (astronomers)[c] who all listened with surprise and appreciation. Finally he said that after all, the discrepancy was a small matter, for a difference of only a quarter of an hour would make a difference of one day if the solstice occurred around midnight, and that is considered much only because of convention.[e] The (Liao) barbarians could not reject this argument, and Su Sung was permitted to offer congratulations on the day desired (by his mission).

Upon his return he reported to the emperor Shen Tsung, who was very pleased and said that nothing could have been more embarrassing. When he asked which of the two calendars was right, Su Sung told him the truth, with the result that the officials of the Bureau of Astronomy and Calendar were all punished and fined. Moreover, since later on (foreign) ambassadors might be repeatedly refused reception because someone (at the Sung capital) did not know about the differences in the beginnings of months, and how the Sung envoys had been allowed to have their way, the emperor decreed that national representatives should follow their own choice of date in celebrating festivals, and that (mutual tolerance of calendars) should be observed for the honour of the empire.[f]

At the beginning of the Yuan-Yu reign-period (+1086) the emperor ordered Su Sung to reconstruct the armillary clock (*hun i*[2]), and it exceeded by far all previous instruments in elaboration. A summary of the specifications was handed down to Yuan Wei-Chi[3], Director

---

[a] This book was written about +1130. It contains two versions of the story (ch. 3, p. 14b and ch. 9, pp. 7a ff.) which are conflated here. Tr. auct.

[b] On this aspect of the State support of astronomy in medieval China, see above, Vol. 3, pp. 192 ff.

[c] These men were of course Chinese also, but they had taken service with the northern 'barbarian' dynasty under the aegis of the ruling house and aristocracy, nomadic until a short time before, and still tribal in character.

[d] As a high official of the Sung dynasty, he naturally had to adhere to the Sung calendar. In Chinese custom the promulgation of the calendar by the emperor was a charismatic right and duty corresponding somewhat to the issuing of minted coins, with image and superscription, in occidental countries. Acceptance of the calendar signified acceptance of Chinese imperial authority.

[e] Cf. Maspero (4), p. 258. Owing to an interpolation method then used for plotting the variation of gnomon shadow lengths, Ho Chêng-Thien[4] missed a solstice in +436 because it occurred half an hour before midnight, but in +440 he got the correct day because it occurred three hours after midnight.

[f] This part of the story is reported in very similar terms by Chang Pang-Chi[5] in his *Mo Chuang Man Lu*[6] (Recollections from the Literary Cottage), ch. 2, pp. 13a, b, written about +1131.

[1] 石林燕語　　[2] 渾儀　　[3] 袁惟幾　　[4] 何承天　　[5] 張邦基　　[6] 墨莊漫錄

of Astronomical Observations (Northern Region), (Tung Kuan Chêng[1]). The original model was due to Han Kung-Lien,[2] a first-class clerk[a] in the Ministry of Personnel, who was a very ingenious man. By that time Su Sung had become Vice-President (of the Chancellery Secretariat) and simply gave the ideas to him. He could always carry them out, so that the instrument was wonderfully elaborate and precise. When the (Chin) barbarians captured the capital (Khaifêng[3]) they destroyed the astronomical clock-tower (Ho Thai[4]) and took away with them the armillary clock. Now it is said that the design is no longer known, even to the descendants of Su Sung himself.

These concluding remarks will be elucidated hereafter; all that we need to note now is the high reputation of Su Sung and his assistants in the subsequent generation.

Su's promotion to Vice-Minister took place some twelve years after his embassy and nearly twenty years after the appearance of his pharmacopoeia. It must have been partly the result of the success of the complete working wooden pilot model of the clock, which had been set up in the Imperial Palace at Khaifêng in the previous year (+1088). In +1090 the sphere and globe were cast in bronze, and in +1094 the writing of Su Sung's horological monograph was finished and presented. By this time he was 75, the holder of many honourable titles, and one of the Deputy Tutors of the Heir Apparent. Dying in +1101, he did not see the tragedy of the fall of the capital two decades later and the flight of the Sung empire to the southern provinces.

Our central point of interest now is his description of the power-drive of his sphere, globe and jack-work, together with the escapement which controlled its movement. This is contained in the third chapter of the *Hsin I Hsiang Fa Yao*.[b] But first a word or two on the transmission of the text to us.[c] Though available only in the north at the time of the preceding quotation, it was printed in the south (in Chiangsu) by Shih Yuan-Chih[5] in +1172. A copy of this edition was owned by the late Ming scholar Chhien Tsêng[6] (+1629 to +1699), who reproduced it in a new edition with extreme care. It was printed again later on by Chang Hai-Phêng[7] (+1755 to 1816) and more numerously by Chhien Hsi-Tsu[8] (+1799 to 1844) in the latter year.[d] With a solicitude for the history of science somewhat unexpected in the imperial editors of +1781, they wrote:[e]

The dynasty of your Imperial Majesty now has instruments which in excellence and precision far exceed all those made during the past thousand years. Of course the invention of Su Sung is not to be compared with them. However, we may have something to learn by paying attention to these old matters, for they show that the people of that time were also interested in new inventions.... His book should be considered as something indeed valuable and precious.

---

[a] The text writes mistakenly Chang Shih-Lien.

[b] The first chapter, which is devoted to the armillary sphere and its parts, was largely translated by Maspero (4), pp. 306 ff. The second is short and deals solely with the celestial globe. The whole of the third is given in Needham, Wang & Price (1).

[c] See the account in *Ssu Khu Chhüan Shu Tsung Mu Thi Yao*, ch. 106.

[d] It has been twice reprinted since then.

[e] *Ssu Khu Chhüan Shu Tsung Mu Thi Yao*, ch. 106, tr. auct.

[1] 冬官正   [2] 韓公廉   [3] 開封   [4] 合臺   [5] 施元之
[6] 錢曾   [7] 張海鵬   [8] 錢熙祚

Fig. 650. Pictorial reconstruction of the astronomical clock-tower built by Su Sung and his collaborators at Khaifêng in Honan, then the capital of the empire, in +1090. The clockwork, driven by a water-wheel, and fully enclosed within the tower, rotated an observational armillary sphere on the top platform and a celestial globe in the upper storey. Its time-announcing function was further fulfilled visually and audibly by the performances of numerous jacks mounted on the eight superimposed wheels of a time-keeping shaft and appearing at windows in the pagoda-like structure at the front of the tower. Within the building, some 40 ft. high, the driving-wheel was provided with a special form of escapement, and the water was pumped back into the tanks periodically by manual means. The time-annunciator must have included conversion gearing, since it gave 'unequal' as well as equal time-signals, and the sphere probably also had this (see p. 456). Su Sung's treatise on the clock, the *Hsin I Hsiang Fa Yao*, constitutes a classic of horological engineering. Orig. drawing by John Christiansen. The staircase was actually inside the tower, as in the model of Wang Chen-To (7).

The historical significance of the mechanical rotation of an observational astronomical instrument (a clock-drive) has already been discussed in Vol. 3, pp. 359ff.; cf. also p. 492 below.

Another point of great interest is that it was by no means the only book on astronomical clockwork written during the Sung dynasty. The bibliographical chapter of the *Sung Shih* also records[a] a *Shui Yün Hun Thien Chi Yao*[1] (Essentials of the (Technique of) making Astronomical Apparatus revolve by Water-Power) written by Juan Thai-Fa.[2] But nothing can be ascertained about this author, or his work, or date.

We are now in a position to study the illustrations in Su Sung's book and the working drawings of reconstructions resulting from modern research. Figure 650 shows a pictorial presentation of the general external appearance of the 'Combined Tower' (Ho Thai[3]) or Tower for the Water-Powered Sphere and Globe (Shui Yün I Hsiang Thai[4]).[b] The armillary sphere (*hun i*[5]) is on the platform at the top, the celestial globe (*hun hsiang*[6]) is in the upper chamber of the tower, half sunk in its wooden casing; and below this stands the pagoda-like façade (*mu ko*[7]) with its five superimposed storeys and doors at which the time-announcing figures (jacks) appeared. On the right the housing is partly removed to show the machinery and the water-storage tanks. The scale of the whole is clearly deducible from the internal evidence of the text, and the height must have been between 30 and 40 ft. in all.[c]

Su Sung's general diagram of the works appears in Fig. 651, but its explanation may best be followed in the modern drawing of Fig. 652a.[d] The former sees the structure from the south or front, the latter from the south-east.[e] The great driving-wheel (*shu lun*[8]), 11 ft. in diameter (Fig. 653), carries 36 scoops (*shou shui hu*[9]) on its circumference, into each of which in turn water pours at uniform rate from the constant-level tank (*phing shui hu*[10]).[f] The main driving-shaft of iron (*thieh shu chu*[11]), with its cylindrical necks (*yuan hsiang*[12]) supported on iron crescent-shaped bearings (*thieh*

[a] Ch. 206, p. 10b.

[b] Su Sung's own picture of this has already been given in Vol. 3, Fig. 162, as also in Needham, Wang & Price (1), fig. 5, and in Needham (38), fig. 6. For the pictorial reconstruction we are much indebted to the artist Mr John Christiansen.

[c] Within a particular culture ideas can show extraordinary tenacity. When Khang Yu-Wei wrote the first draft of his magnificent Utopia in 1885, he proposed (Pt. 2, ch. 4), pp. 87 ff. (Thompson tr. p. 103) that there should be 'Time Towers' (*shih piao tha lou*[13]) in every city and along the main roads. These should have elaborate clocks, and orreries with models of the sun, moon, earth and planets, giving graphic representation of the various aspects of time, and explaining eclipses, etc., to the people as well as weaning them away from the old and inconvenient lunar calendar. Su Sung was, as it were, standing behind Khang Yu-Wei's shoulder as he wrote. How delighted both of them would have been with the Planetarium and astronomical museum frequented by the people of Peking today!

[d] A provisional engineering drawing based on further study of the text by Combridge (1), and approximately to scale, is reproduced in Fig. 652b.

[e] Su Sung, as was usual with medieval Chinese engineers, always describes components of machinery by the cardinal points, upper pieces being named 'heavenly' and lower ones 'earthly'. Elsewhere (Needham (38), figs. 9 and 10) I have reproduced two excellent scale drawings by Wang Chen-To (7), who constructed a model of the clock-tower in Peking in 1958. One of these is an approximately isometric projection viewed from the south-east, the other is a cross-section seen from the east. The former therefore corresponds in orientation to our Figs. 650 and 652a. Photographs of the model are reproduced in Anon. (19), fig. 29; Li Jen-I (1) and Combridge (3).

[f] The whole cycle of scoops filled once in every *kho*, and consumed in one double-hour about one and a half tons of water.

[1] 水運渾天機要	[2] 阮泰發	[3] 合臺	[4] 水運儀象臺	[5] 渾儀
[6] 渾象	[7] 木閣	[8] 樞輪	[9] 受水壺	[10] 平水壺
[11] 鐵樞軸	[12] 圓項	[13] 時表塔樓		

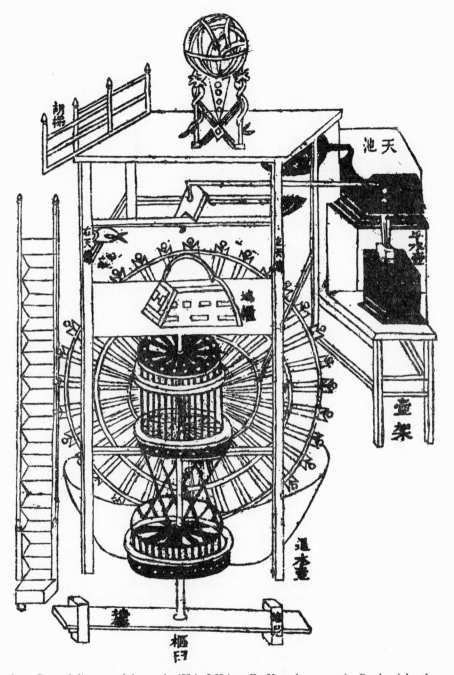

Fig. 651. General diagram of the works (*Hsin I Hsiang Fa Yao*, ch. 3, p. 4*a*). On the right, the upper reservoir tank (*thien chhih*) with the constant-level tank (*phing shui hu*) beneath it. In the centre, foreground, the 'earth horizon' box (*ti kuei*) in which the celestial globe is mounted; below, the time-keeping shaft and wheels supported in the mortar-shaped end-bearing (*shu chiu*). Behind, the main driving-wheel with its spokes and scoops; above, the left and right upper locks (*tso yu thien so*) with the upper balancing lever and upper link, curiously drawn, still higher.

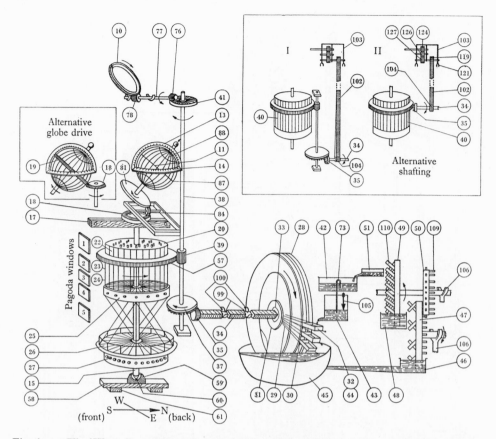

Fig. 652a. The 'Water-Powered Armillary (Sphere) and Celestial (Globe) Tower' (Shui Yün I Hsiang Thai) of Su Sung (+1090); detailed diagrammatic reconstruction of the power and transmission machinery (Needham, Wang & Price, 1). 10, diurnal motion gear-ring (*thien yün huan*) of the armillary sphere; 11, celestial globe (*hun hsiang*); 13, split-ring meridian circle (*thien ching shuang kuei*) of the celestial globe; 14, single-ring horizon circle (*ti hun tan huan*) of the celestial globe; 15, time-keeping shaft (*chi lun chu*); 17, upper bearing beam (*thien shu*); 18, celestial gear-wheel (*thien lun*)—time-keeping shaft wheel no. 1; 19, equatorial gear-ring (*chhih tao ya*) of the celestial globe; 20, wheel for striking the double-hours of the day by bells-and-drums (*chou shih chung ku lun*)—time-keeping shaft wheel no. 2; 22, wheel for striking quarters by bells-and-drums (*shih kho chung ku lun*)—time-keeping shaft wheel no. 3; 23, wheel with jacks reporting the beginnings and middles of double-hours (*shih chhu chêng ssu-chhen lun*)—time-keeping shaft wheel no. 4; 24, wheel with jacks reporting the quarters (*pao kho ssu-chhen lun*)—time-keeping shaft wheel no. 5; 25, night clepsydra (indicator-rod) wheel for striking (the night-watches) on gongs (*yeh lou chin chêng lun*)—time-keeping shaft wheel no. 6; 26, night clepsydra (indicator-rod) wheel with jacks reporting the night-watches and their divisions (*yeh lou kêng chhou ssu-chhen lun*)—time-keeping shaft wheel no. 7; 27, night clepsydra indicator-rod wheel (*yeh lou chien lun*)—time-keeping shaft wheel no. 8; 28, great driving-wheel (*shu lun*); 29, spokes (*fu*); 30, scoop-holders (*hung*); 31, reinforcement rings (*wang*); 32, scoops (*hu*); 33, hub (*ku*); 34, iron driving-shaft (*thieh shu chu*); 35, lower pinion (lit. earth drum, *ti ku*); 37, lower wheel (lit. earth wheel, *ti lun*); 38, vertical transmission-shaft (*thien chu*); 39, middle wheel (*chung lun*); 40, time-keeping wheels (*chi lun*); 41, upper wheel (*shang lun*); 42, upper reservoir tank (*thien chhih*); 43, constant-level tank (*phing shui hu*); 44, water-receiving scoops (*shou shui hu*); 45, sump (*thui shui hu*); 46, lower water-raising tank (*shêng shui hsia hu*); 47, lower noria (*shêng shui hsia lun*); 48, intermediate water-raising tank (*shêng shui shang hu*); 49, upper noria (*shêng shui shang lun*); 50, manual wheel (*ho chhê*); 51, upper flume (*thien ho*); 57, time-keeping gear-wheel (*po ya chi lun*); 58, lower bearing beam (*ti chi*); 59, iron mortar-shaped end-bearing (*thieh shu chiu*); 60, pointed cap of shaft in bearing (*tsuan*); 61, wooden base (*mu chia*); 73, siphon (*kho wu*); 76, rear pinion of armillary sphere drive (*hou thien ku*); 77, armillary sphere drive

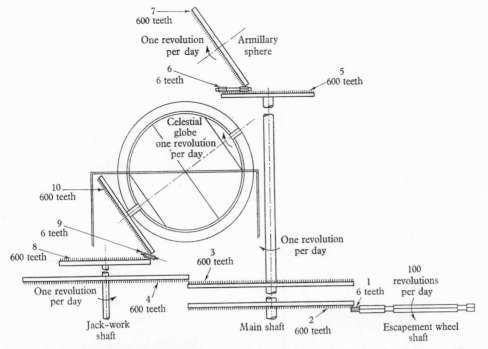

Fig. 652*b*. Provisional scale drawing of the transmission machinery in Su Sung's clock-tower (Combridge, 1). 1, lower pinion (lit. earth drum, *ti ku*); 2, lower wheel (lit. earth wheel, *ti lun*); 3, middle wheel (*chung lun*); 4, time-keeping gear-wheel (*po ya chi lun*); 5, upper wheel (*shang lun*); 6, pinions of armillary-sphere drive (*chhien hou thien ku*); 7, diurnal motion gear-ring (*thien yün huan*); 8, celestial gear-wheel (*thien lun*); 9, celestial idling shaft and pinion (*thien chu*); 10, celestial globe drive-wheel (*hun hsiang thien yün lun*).

Further study (Combridge, 2) suggests that the 600-tooth gear-wheels and the 6-tooth pinions were features of a preliminary experimental design. Su Sung's actual clock-tower almost certainly had solar-sidereal conversion gearing at the top of both shafts, as well as for the 'unequal' time-interval signalisations below (see p. 456).

yang yüeh[1]), ends in a pinion (*ti ku*[2]) which engages (*po*[3]) with a gear-wheel at the lower end of the main vertical transmission-shaft (*thien chu*[4]). This (Fig. 654) drives two components. A suitably placed pinion connects it with the time-keeping gear-wheel (*po ya chi lun*[5]) which rotates the whole of the jack-work borne on the time-keeping shaft (*chi lun chu*[6]). This assembly (seen in the foreground of Fig. 651)

[1] 鐵仰月   [2] 地轂   [3] 撥   [4] 天軸   [5] 撥牙機輪   [6] 機輪軸

layshaft (*thien ku*); 78, front pinion of armillary sphere drive (*chhien thien ku*); 81, celestial globe drive-wheel (*hun hsiang thien yün lun*); 84, celestial idling shaft and pinion (*thien chu*); 87, stars and constellations north and south of the equator (*chung wai kuan hsing*); 88, polar region (lit. Purple Palace Enclosure, *Tzu Wei Yuan*); 99, journal bearings (lit. cylindrical necks, *yuan hsiang*); 100, iron upward-facing crescentic bearings (*thieh yang yüeh*); 102, chain drive (lit. celestial ladder, *thien thi*); 103, gear-box (of celestial ladder) (*thien tho*); 104, lower sprocket wheel (*hsia ku*) of chain drive; 105, water-level marker (lit. arrow of indicator-rod, *chun shui chien*, but probably working automatically to keep the water-level constant); 106, crutched post bearings (*chha shou chu*); 109, handles (*shou pa*); 110, noria buckets (*hu tou*); 119, upper sprocket wheel (*shang ku*) of chain drive; 121, double fork (*shuang chha*) of gear-box; 124, upper pinion (*shang thien ku*) in gear-box; 126, middle pinion (*chung thien ku*) in gear-box; 127, lower pinion (*hsia thien ku*) in gear-box.

Reference numbers and Chinese characters not included here or in the accompanying text, together with further explanations, will be found in Needham, Wang & Price (1).

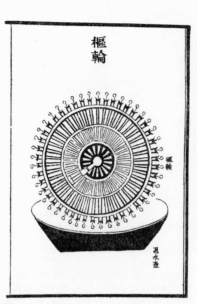

Fig. 653. Driving-wheel (*shu lun*) and
sump (*thui shui hu*), from *Hsin I Hsiang
Fa Yao*, ch. 3, p. 15a.

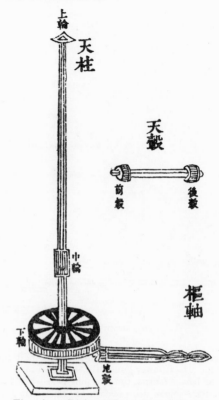

Fig. 654. Right-angle gearing of main
driving-shaft (*shu chu*) and vertical transmission shaft (*thien chu*), from *Hsin I Hsiang
Fa Yao*, ch. 3, p. 16a. Legends in small
characters from above downwards: *shang
lun*, upper (gear-)wheel; *chung lun*, middle
(gear-)wheel; *hsia lun*, lower (gear-)wheel
(= earth wheel); *ti ku*, lower (earth) pinion.
Inset: armillary sphere drive layshaft with
its pinions (*chhien hou thien ku*).

consists of eight superimposed horizontal wheels, seven carrying round the jacks (*ssu
chhen*[1]). Since each of these wheels is from 6 to 8 ft. in diameter, the total weight involved
must have been very considerable, so the base of their shaft is fitted with a pointed
cap (*tsuan*[2]) and supported in an iron mortar-shaped end-bearing (*thieh shu chiu*[3]).[a]

[a] Patterns of machinery repeat themselves in a rather remarkable way. If one looks at the sketch of
the works of the water-mill at Stratford near West Harptree in Somerset, offered by Stowers (1), p. 210,
as a typical example of an 18th-century corn-mill, one is surprised to find Su Sung's design repeated in
every particular—the vertical drive-wheel, the right-angle gearing, and the two vertical transmission
shafts one of which drives the hoist and the other the millstone. Of course no direct connection is
suggested. Another curious parallel exists in the +15th or +16th-century round dovecot tower at
Dunster in Somerset within which a structure of platforms and ladders weighing about ¼ ton and borne
on a massive post of ash some 16 ft. high rotates readily on an un-oiled bearing. This consists of an iron
cone within the base of the post which rests upon an iron pin about 1 in. diameter fixed into a floor beam.
Very relevant to our conceptions of Su Sung's clock-tower are the upright shafts of windmills, which
have been discussed by Wailes (6).

[1] 司辰   [2] 鑽   [3] 鐵樞臼

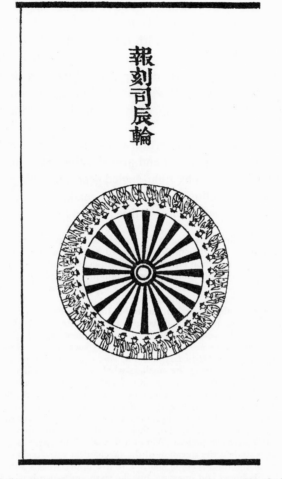

報刻司辰輪

Fig. 655. Wheel with jacks reporting the quarters (*pao kho ssu-chhen lun*), from *Hsin I Hsiang Fa Yao*, ch. 3, p. *12a*. Only 36 out of the 96 jacks are shown.

The jack-work wheels performed a variety of functions,[a] their figures either appearing with placards on which the time was marked, or ringing bells, striking gongs or beating drums as they made their appearances in clothes of different colours at the pagoda doorways.[b] Su Sung's picture of one of these time-keeping wheels is shown in Fig. 655. Their rotation, however, was not the only duty of the time-keeping shaft,

[a] Such as striking all quarters, and the beginnings and middles of the equal double-hours, reporting quarters, reporting the unequal night-watches and their divisions, etc. Full details are given by Needham, Wang & Price (1).

[b] Jack-wheels of this kind seem to go back to the +7th century in China (see p. 469 below), and lasted all through the centuries, as may be seen in many medieval and Renaissance clocks still extant. One such +16th-century clock, by Isaac Habrecht, has been figured by Needham, Wang & Price (1), fig. 27. On the Habrecht family see particularly Bassermann-Jordan (1), pp. 89 ff., 99 ff. A check in Ungerer (1) shows that 15 monumental clocks of major importance in Europe have one or more horizontally rotating jack-wheels (seven in Germany, two in France, and one each in England, Holland, Sweden, Czechoslovakia and Italy.)

for at its upper end it engaged by means of oblique gearing[a] and an intermediate idling pinion with a gear-wheel on the polar axis of the celestial globe (Fig. 652). The angle of these gears corresponded of course with the polar altitude at Khaifêng. Now the text contains a number of notes which record improvements in the clock, probably dating from the last years of the +11th century, and in these an alternative globe drive is given, the uppermost gear-wheel (*thien lun*[1]) rotating an equatorial gear-ring (*chhih tao ya*[2]) on the globe (Fig. 652a).[b] Possibly the original gearing proved difficult to maintain.[c]

We must now return to the main vertical transmission-shaft and the second component which it drives. Its uppermost end provides the power for the rotation of the armillary sphere. This is effected by right-angled gears and oblique gears[a] connected by a short idling shaft. The oblique engagement is made with a toothed ring called the diurnal-motion gear-ring (*thien yün huan*[3]) fitted round the intermediate nest or shell[d]

[a] This can hardly have been bevel gearing in the modern sense, but closer description is impossible, for Su Sung does not explain it in sufficient detail. According to Woodbury (2), bevel gearing appears first in Leonardo and becomes common among the +16th-century engineers (Besson, Ramelli, etc.). But the de Dondi clock of +1364 incorporates a small obliquely cut pinion.

[b] Su Sung's original illustrations of these two forms of globe drive are reproduced in Needham, Wang & Price (1), figs. 21, 28, and in Needham (38), figs. 14, 15.

[c] An alternative and more subtle suggestion, due to Mr John Combridge, proposes that the equatorial drive was really additional to the polar axis drive in some clock-towers, the latter being used for the motion of a sun model round the celestial globe, while the former assured the sidereal rotation, conversion being effected by appropriate gearing. This is strongly supported by the mention of a gear-ring of 478 teeth in the equatorial drive (*Hsin I Hsiang Fa Yao*, ch. 2, p. 2a), probably a scribal error for 487, because $487 \times \frac{3}{4} = 365\frac{1}{4}$, thus being the smallest whole number which will give this year-length conveniently (cf. Needham, Wang & Price (1), p. 36). Hence conversion gearing is to be inferred. The same figure (always with the same inversion) crops up at several places in Su Sung's book, and seems to betray wherever it does so the existence of a sidereal motion as well as the clock's obvious solar motion. As the same number of teeth appears again in one of the descriptions of the drive to the armillary sphere (*Hsin I Hsiang Fa Yao*, ch. 1, p. 17a), there is a distinct probability that the sphere had a sidereal drive to its intermediate component. Work on this is still in progress (cf. Combridge, 2).

Mr Combridge has also studied the question of the mechanical conversion of solar to sidereal time implicit in the structure of the time-keeping shaft. The five upper wheels with their jack-work must have indicated equal double-hours and quarters, but the sixth and seventh wheels announced sunrise, sunset, the variable night-watches and other 'unequal' events. He believes that the eighth wheel was not fixed to the time-keeping shaft but free to rotate on it at an annual rate and driven round by appropriate conversion gearing including 487 teeth on the eighth wheel (cf. Needham, Wang & Price (1), p. 34), 586 teeth on the seventh, and a lay-shaft with 10- and 12-toothed pinions connecting them. The jacks and trip-lugs were mounted on radial arms pivoted like the ribs of a fan (cf. the dusk and dawn culmination diagrams in *Hsin I Hsiang Fa Yao*, ch. 2, pp. 14a ff., one of which is reproduced as fig. 70 by Needham, Wang & Price), and these were controlled automatically by a cylindrical iron cage penetrating the sixth and seventh wheels and free to slide along their noon-to-midnight diameters. The position of the cage in its guiding slots at any season of the year was determined by a horizontal eccentric cam on the upper face of the sidereal eighth wheel consisting of 61 'clepsydra float indicator-rods' of appropriate lengths (cf. Needham, Wang & Price (1), p. 39). Thus the cage travelled slowly to and fro in an annual cycle and regulated the solar signalisations automatically. Such a device is implied rather than described in the text, but its assumption elucidates many hints and details formerly quite obscure (cf. Combridge, 2).

[d] From Sect. 20 on astronomy (Vol. 3, p. 352) it will be remembered that the armillary spheres of this period had three nests or shells of rings. The outer one, the Component of the Six Cardinal Points (*liu ho i*[4]), had the meridian, horizon and equator. The middle or intermediate one, the Component of the Three Arrangers of Time (*san chhen i*[5]), had the solsticial colure, equator and ecliptic. The innermost one, the Component of the Four Displacements (*ssu yu i*[6]), was a polar-mounted declination ring carrying the sighting-tube.

[1] 天輪   [2] 赤道牙   [3] 天運環   [4] 六合儀   [5] 三辰儀   [6] 四游儀

of the armillary sphere not equatorially but along a declination parallel near the southern pole.[a] In this case also the original model proved unsatisfactory and improvements were made as time went on. We know that the main vertical transmission-shaft was made of wood and nearly 20 ft. long. This must soon have showed itself to be mechanically unsound, and in the later variants (probably *c.* +1100) it was first shortened and finally abolished altogether. These designs are shown in the inset in Fig. 652*a*. In the first modification the main vertical transmission-shaft had no other duty than to turn the chief time-keeping gear-wheel, while in the second, the 'earth-wheel' pinion (*ti ku*[1]) connected the main driving-shaft directly with the time-keeping gear-wheel itself, so that no transmission-shaft was necessary. But in both cases the motive power was conveyed to the armillary sphere on the upper platform by means of an endless chain-drive (*thien thi*[2]) rotating three small pinions (*ku*[3]) in a gear-box (*thien tho*[4]); see Fig. 410. In the final design the chain-drive achieved shorter and therefore more efficient form. This feature of the clock may perhaps be considered the most remarkable of all for its time (+11th century), for although an endless belt of a kind had been incorporated in the magazine arcuballista of Philon of Byzantium (−3rd century)[b] there is no evidence that this was ever built, and it certainly did not transmit power continuously. A likelier source for Su Sung's chain-drive may be found in the square-pallet chain-pump so widespread in the Chinese culture area,[c] a device the origin of which we have traced back at least to the +2nd century, and probably to the +1st. Of course this also was for conveying material and not for transmitting power from one shaft to another—hence the originality of Su Sung and his assistants, to whom perhaps indeed all true chain-drives are owing.[d] Such is the interest of this feature that it may be worth while to give the description of it in the words of Su Sung himself:[e]

The chain-drive (lit. celestial ladder) is 19·5 ft. long. The system is as follows: an iron chain with its links joined together to form an endless circuit (*thieh kua lien chou tsa*[5]) hangs down from the upper chain-wheel which is concealed by the tortoise-and-cloud (column supporting the armillary sphere centrally), and passes also round the lower chain-wheel which is mounted on the main driving-shaft. Whenever one link (*kua*[6]) moves, it moves forward one tooth (*chü*[7]) of the diurnal motion gear-ring and rotates the Component of the Three Arrangers of Time, thus following the motion of the heavens.

A brief description of the water-power parts must follow here. Water stored in the upper reservoir (*thien chhih*[8]) is delivered into the constant-level tank (*phing shui*

---

[a] In our reconstruction the length of the idling shaft is exaggerated. The main vertical transmission shaft came right up through the central column under the armillary sphere.

[b] See Beck (3) and Schramm (1). This machine will be discussed further in the Section on military technology in Vol. 5 below.

[c] See above, pp. 339 ff.

[d] Cf. the discussion already given in Sect. 27*b* (pp. 109 ff. above). It may well be, moreover, that Su Sung and Han Kung-Lien were not the first to use the chain-drive in an astronomical clock. As we shall shortly see (p. 471) it probably goes back to one of their predecessors, Chang Ssu-Hsün, about +978.

[e] *Hsin I Hsiang Fa Yao*, ch. 3, p. 26*a*; tr. auct.

[1] 地轂　　　[2] 天梯　　　[3] 轂　　　[4] 天托　　　[5] 鐵括聯周匝　　　[6] 括
[7] 距　　　[8] 天池

hu [1])[a] (Fig. 651) by a siphon (*kho wu* [2]) and so passes to the scoops of the driving-wheel, each of which has a capacity of 0·2 cu.ft.[b] As each scoop in turn descends the water is delivered into a sump (*thui shui hu* [3]). Apparently the clock was never so located as to be able to take advantage of a continuous water supply; instead of this, the water was raised by hand-operated norias (*shêng shui lun* [4]) in two stages[c] to the upper reservoir.

Fig. 656. Upper and lower norias with their tanks and the manual wheel for operating them (*Hsin I Hsiang Fa Yao*, ch. 3, p. 19 *b*).

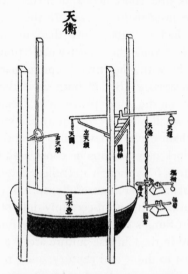

Fig. 657. The 'celestial balance' (*thien hêng*) or escapement mechanism of Su Sung's clockwork (*Hsin I Hsiang Fa Yao*, ch. 3, p. 18 *b*). Legends in small characters, from left to right, upper row: *yu thien so*, right upper lock; *thien kuan*, upper link; *tso thien so*, left upper lock; *kuan chu*, axle or pivot; *thien thiao*, long chain; *thien chhüan*, upper counterweight. Lower row: *thui shui hu*, sump; *ko chha*, checking fork of the lower balancing lever, *shu hêng*; *kuan shê*, coupling tongue; *shu chhüan*, main, i.e. lower, counterweight.

The bearings of these norias (Fig. 656) were supported on crutched columns (*chha shou chu* [5]).

We can now examine what Bassermann-Jordan calls the soul of any time-keeping machine, namely the escapement. All that Su Sung's draftsmen could depict of it for his book is seen in Fig. 657, but fortunately the text is elaborate and for the most

---

[a] On this, see the discussion of the history of the clepsydra in China in the Section on astronomy, Vol. 3, pp. 316 ff. Su Sung's own illustration of these tanks is figured in Needham, Wang & Price (1), fig. 17. They also appear, through a space opened in the tower wall, in Su Sung's general view, Vol. 3, fig. 162; also Needham, Wang & Price (1), fig. 5; Needham (38), fig. 6.

[b] Corresponding to about 12 lb. of water, or rather more than a gallon.

[c] Presumably maintenance mechanics from the Ministry of Works came daily to do this, for the dimensions given allow for rather more than 2 hours running.

[1] 平水壺   [2] 渴烏   [3] 退水壺   [4] 昇水輪   [5] 杈手柱

part clear, enabling the reconstruction of Fig. 658 to be made with some assurance.[a]
The whole mechanism was called the 'celestial balance' (*thien hêng*[1]) and it did indeed
depend upon two steelyards or weighbridges upon which each of the scoops acted in
turn. The first of these, the 'lower balancing lever' (*shu hêng*[2]), prevents the fall of
each scoop until full,[b] by means of a 'checking fork' (*ko chha*[3]). The basic principle is
thus at once revealed, the determination of standard time units by the division of a
constant flow of water into equal parts by a repeated process of accurate and automatic weighing in scoops carried on the driving-wheel. After each weighing
operation the wheel is released so that it can make one step forwards under the
power provided by the combined weight of several previously filled scoops (i.e.
those of the quarter-periphery between '3 o'clock' and '6 o'clock' on the wheel
as seen in Fig. 651). Release takes place as follows: once the weight of water overcomes the counterweight on the lower balancing lever and trips the checking fork,
the scoop swings on its pivot in free fall and smartly trips by means of its projecting
pin a second lever, the 'coupling tongue' (*kuan shê*[4]). This is connected by means of
a chain (*thien thiao*[5]) with another weighbridge, the 'upper balancing lever' (*thien*

---

[a] A beautiful working model of the escapement, using fine sand as the motive fluid, and keeping time
within ± 10 to 20 sec./hour, was constructed by Mr John Combridge of the Engineering Department
of the General Post Office in 1961, and demonstrated to the History of Science Colloquium at Worcester
College, Oxford, in July of that year (Anon. (61); cf. Fig. 659). During its construction (as was foreseen
would be the case, Needham, Wang & Price (1), p. 58), several features were found necessary which
we had not been able to visualise clearly from the study of the text alone but which can now be seen to
be compatible with it. Further models followed, and valuable conclusions drawn from experimental
engineering study have now been published (Combridge, 1). Thus (*a*) the scoops (preferably cylinder-
shaped) must each be counterweighted and free to swing on transverse axles fitted round the rim of the
driving-wheel, within a range delimited by back-stops, (*b*) there must be very little distance between
the checking fork of the lower balancing lever and the coupling tongue which operates the release
through the upper balancing lever, (*c*) the left-hand end of the upper balancing lever must be connected
by a short length of chain with the free end of the right upper lock (cf. Needham, Wang & Price (1),
pp. 33, 57 ff.; the chain can be seen hanging down in Fig. 657), and (*d*) the two upper locks must
immobilise the projecting ends of the spokes. The function of the two locks, formerly uncertain, is
therefore now quite clear. In Prof. Aubrey Burstall's model, the first to be attempted (Burstall, 2),
a damping effect was produced by having the lower part of the driving-wheel dip into the sump, but
the sand-driven models show that this was unnecessary, there is no positive evidence for it in the
text, and the known dimensions of the whole clock-tower do not favour it. It would be interesting
to know, and it should be possible to find out, whether the time-keeping properties of the Chinese
water-wheel linkwork escapement clocks were better or worse than the earliest Western verge-and-foliot
clocks. In an excellent paper Ward (2) has constructed a semi-logarithmic plot of the error in seconds
per day, which ranges from the order of 1000 in +1350 to 0·00001 in 1962. Judging by the performance
of the Combridge models, the daily error in seconds would seem to have been less than 100 as early as
+1080, an accuracy not attained in Ward's graph until the end of the +16th century with the advent of
the cross-beat escapement.

Most recently a reconstruction of the escapement on principles quite different from those of the
Combridge models has been built and published by Burstall, Lansdale & Elliott (1). They regard the
'dragon-head' in Fig. 657 not as the ornamental mouth of the water-stream (cf. Vol. 3, pp. 320 ff.)
but as an interrupter-cam on a separate shaft which has the function of dissociating momentarily the
coupling tongue from the upper balancing lever. In this ingenious system the scoops are fixed, but
since the water leaves them as soon as the escapement has acted, we do not see how the torque on the
driving-wheel could have been sufficient to work all the heavy machinery of the Chinese medieval
clock-towers.

[b] Or nearly full. This is an important point, for although the main part of the time-keeping depended
upon the constancy of flow of the water, as if from a clepsydra, adjustment of the weight on this weighbridge could permit the scoops to descend when less than full, and so the time-keeping could be regulated, within certain limits, by mechanical means.

[1] 天衡   [2] 樞衡   [3] 格叉   [4] 關舌   [5] 天條

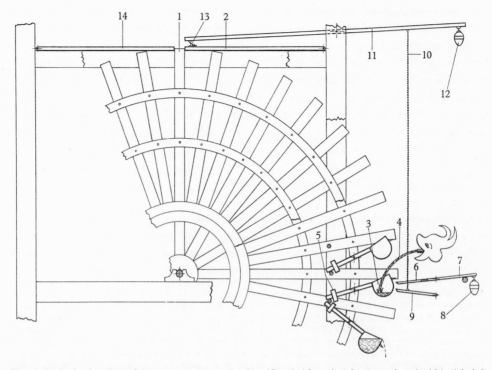

Fig. 658. Scale drawing of the escapement mechanism (Combridge, 1). (1), arrested spoke (*fu*); (2), left upper lock (*tso thien so*), considered as the right in our analysis; (3), scoop (*shou shui hu*) being filled by (4), water jet from constant-level tank; (5), small counterweight; (6), checking fork (*ko chha*) tripped by a projecting pin on the scoop, and forming the near end of (7) the lower balancing lever (*shu hêng*), with (8) its lower counterweight (*shu chhüan*); (9), coupling tongue (*kuan shê*), connected by (10) the long chain (*thien thiao*) with (11) the upper balancing lever (*thien hêng*), which has at its far end (12) the upper counterweight (*thien kuan*), and at its near end, (13) a short length of chain (*thien kuan*) connecting it with (2) the upper lock beneath it; (14), right upper lock (*yu thien so*), considered as the left in our analysis.

The water jet is seen issuing from a dragon mouth (cf. Fig. 657 and p. 504 below).

At the beginning of each 24-second time-interval the driving-wheel is immobilised by the action of the right lock (2) on the spoke (1). As water (4) from the constant-level tank enters the scoop (3), the scoop-holder counterweight (5) is first overcome, and the excess weight of water then rests on the checking fork (6) of the lower balancing lever (7). When the excess overcomes counterweight (8) the lever is suddenly tripped, and the scoop-holder rotates about its pivot so as to fall sharply upon the coupling tongue (9) and trip it in its turn. The long chain (10), which passes freely between the prongs of the checking fork (6), is thus abruptly pulled downwards, depressing the right-hand end of the upper balancing lever (11) with the aid of the upper counterweight (12), normally insufficient to effect this. Momentum is gathered from the loaded scoop for a brief instant while the levers swing, then the short chain of the upper link (13) tightens and jerks the right upper lock (2) out of the way of the spoke. The wheel now makes one quick step clockwise under the driving force of the filled scoops in the lower right-hand quadrant, while the near or left-hand end of the upper balancing lever and the right-hand upper lock fall under their own weight again to arrest the following spoke. Meanwhile the left-hand upper lock (14) has been raised in ratchet fashion as the spoke has passed through, and now falls again behind the next spoke so as to prevent any recoil as the wheel stops. With the return of the linkwork to its original position, the levers (6, 7) and (9) regain their normal places ready for tripping in the following cycle. All the 'tick' processes are accomplished in an instant.

The progress of experimental study, and the ambiguity of the word *kuan*, induced Combridge (2) to propose a useful change in technical terminology. The 'stopping tongue' of Needham, Wang & Price (1) is accordingly here called the 'coupling tongue', and the 'upper stop' becomes the 'upper link'.

hêng [1]), forming a parallel linkage system (the 'iron crane-bird's knee', *thieh ho hsi* [2]).[a]
The upper balancing lever (which gives its name to the whole escapement) carries a
counterweight at its right-hand end, and is fitted at its fulcrum with a crosswise axle
moving in journal bearing caps ('camel-backs', *tho fêng*,[3] between two 'iron cheeks',
thieh hsia [4]). Its left arm, the heavier when the escapement is at rest, ends above the
driving-wheel. From this end hangs the 'upper link' (*thien kuan* [5]), a short length of
chain, joining it to the end of the hinged right 'upper lock' (*thien so* [6]). When the coupling
tongue is tripped with a jerk by the sudden impulse of the falling scoop, it pulls down
the long chain and the right-hand end of the upper balancing lever, thereby raising the
left-hand end, which after an instant 'briefer than a wink' while the short chain
tautens, suddenly withdraws the right upper lock from against the tip of the most
recently arrested spoke. As the gate thus momentarily opens to let the spoke through,
and a new scoop comes under the water delivery jet, the right upper lock descends
again and a left upper lock briskly inserts itself in ratchet manner behind the newly
arrested spoke so as to prevent all recoil. A very elegant part is thus played by the
upper balancing lever, since it acts as an energy-accumulator for the critical operation
of disengaging the right-hand lock abruptly from the heavily-loaded wheel.[b] Thus the
whole cycle repeats.

There were 100 quarters (*kho* [7]) in the Chinese day-and-night period of 12 equal
double-hours (*shih* [8]), hence $8\frac{1}{3}$ quarters in each double-hour, not 8. As the *kho* was
generally divided into 'minutes' (*fên* [9]) sexagesimally,[c] this meant 20 extra *fên* in each
double-hour. The *kho* was thus equivalent to 14 min. 24 sec. of our time.[d] To operate
the jack-work at the correct moments in both the double-hour and the quarter series,
the machinery must have been designed to divide the day both into 24 and 100 equal
parts, and the least common multiple of these numbers, 600, explains why the three
main gear-wheels had just that number of teeth. The unit period was thus 2 min.
24 sec., $\frac{1}{600}$ part of a day. As one of the texts says that each movement of the driving-
wheel corresponded to the passage of six gear-teeth of the time-keeping wheel,[e] it was
at first thought that release occurred only once a quarter, yet it is evident that the time
between two successive releases could not have exceeded the unit period. Closer study
has shown, therefore,[f] that 'movement' here must mean, not one release step or 'tick',

---

[a] This interesting technical term originated from a phrase applied to a weapon, the war-flail. In
Fig. 374 we have shown side by side the well-known farmer's jointed flail (from the *Nung Shu* of +1313,
ch. 14, p. 28*b*) and the war-flail in which a piece of iron is connected to the handle by a chain (from the
Wu Ching Tsung Yao of +1044, ch. 13, p. 14*a*). Since the latter was called the 'crane-bird's knee' as
early as the +3rd century, it was natural enough that the engineers should have borrowed the name for
any arrangement of rods and levers linked by chains. For fuller details see Needham, Wang & Price (1),
p. 56.

[b] The torque-providing weight of water in the scoops of the right-hand lower quadrant not yet
drained out into the sump would have been of the order of 100 lb. Some idea of the total resistance may
be gained from the weight of the clock-tower machinery as a whole, not less than 20 tons of wood, iron
and bronze. Nice adjustment of power to duty must have been necessary—and careful lubrication.

[c] Some of the Chinese calendars, however, gave 100 *fên* to the *kho*; cf. Maspero (4), p. 211.

[d] The Chinese horary systems are discussed by Needham, Wang & Price (1), pp. 199 ff.

[e] *Hsin I Hsiang Fa Yao*, ch. 3, p. 9*a* (Needham, Wang & Price (1), p. 36).

[f] Combridge (1).

[1] 天衡   [2] 鐵鸛膝   [3] 駞峯   [4] 鐵頰   [5] 天關   [6] 天鎖
[7] 刻   [8] 時   [9] 分

but one complete revolution of the driving-wheel, i.e. 36 ticks of 24 sec. each. Thus there were 6 ticks for each time-keeping gear-tooth passage, 36 in every quarter (*kho*), 300 in every double-hour (*shih*), and 3600 every day.[a]

The whole design is strangely reminiscent of the familiar anchor escapement of the late +17th century, since the driving-wheel is also a scape-wheel and the 'pallets' are inserted alternately at two points on its circumference separated by 90° or less, rather than the 180° of the crown-wheel.[b] Although the solution of the problem by chain and linkwork has a certain medieval cumbrousness, the operation is elegant and the performance accurate to an unexpected degree. It certainly far exceeded the inventive capacity of contemporary Europe, that other culture-area where a purely mechanical escapement would later appear. In the Chinese water-wheel linkwork device the action of the arrest and release is brought about not by mechanical oscillation but by the force of gravity exerted periodically as a continuous steady flow of liquid fills containers of limited size.[c] This type of escapement had remained quite unknown to historians of technology until the elucidation of Su Sung's text. Its peculiar interest lies in the fact that it constitutes an intermediate stage or 'missing link' between the time-measuring properties of liquid flow and those of mechanical oscillation. It thus unites, under the significant sign of the millwright's art, the clepsydra and the mechanical clock in one continuous line of evolution.

At one point in the text it is said that one of the jack-wheels intercepts (*chieh*[1]) a gong (*chin chêng*[2]) to strike the night-watches[d] as it turns. All the auditory performances of the jacks must have involved simple contrivances of springs, probably of bamboo. Hence the interest of the statement made in the following century by Hsüeh Chi-Hsüan:[3]

Nowadays time-keeping devices (*kuei lou*[4]) are of four different kinds. There are the clep-sydra (lit. the bronze vessels, *thung hu*[5]), the (burning) incense stick (*hsiang chuan*[6]), the sundial (*kuei piao*[7]), and the revolving and snapping springs (*kun than*[8]).[e]

[a] Vol. 3, pp. 364ff. thus needs correction. The motion now envisaged would have moved on the intermediate component of the armillary sphere on the top platform 0·1° sidereally per tick, so that astronomical measurements with the rotating sidereal coordinate system were possible with precision.

[b] The alternation of the 'pallets' occupies, it will be seen, only a relatively short time in the cycle, taking place during the forward motion of the driving-wheel. Most of the cycle is spent with both 'pallets' inserted, while the scoop is filling. This however was something which the water-wheel link-work escapement had in common with most subsequent escapements, namely that the time it spent in motion was much less than the time spent at rest. In a modern watch, as Michel (15) has pointed out, the wheels are motionless for $\frac{19}{20}$ths of each second, so that they move only 1 hour out of 20, and in precision clocks a much lesser proportion. Hence the long periods which can be worked by horological machinery without serious wear. It is interesting in this connection to note that Su Sung's clock ran successfully from +1092 until +1126, and then again for some years at least after its removal to Peking. So at least the historical records imply.

[c] It is worthy of remark (Combridge, 1) that the total time for a complete revolution of the driving-wheel, comprising 36 ticks, is practically independent of irregularities in the scoops or their counter-weights, or in the net available driving torque. It depends almost wholly on the rate of the fluid flow and the adjustment of the weight on the lower balancing lever.

[d] The 'unequal' night-watches (*kêng*[9]) were five in number, each divided into five parts (*chhou*[10]). They began 12½ *kho* after sunset and ended 12½ *kho* before dawn.

[e] Quoted by Wang Ying-Lin[11] in his scientific encyclopaedia *Hsiao Hsüeh Kan Chu*,[12] written about +1270 but not printed till +1299; ch. 1, p. 42b. Hsüeh Chi-Hsüan lived from +1125 or +1134 to

[1] 戟	[2] 金鉦	[3] 薛季宣	[4] 晷漏	[5] 銅壷	[6] 香篆
[7] 圭表	[8] 輥彈	[9] 更	[10] 籌	[11] 王應麟	[12] 小學紺珠

PLATE CCLVI

Fig. 659. Experimental working model (driven by fine sand) of the Chinese water-wheel linkwork escapement, as demonstrated in July 1961 by Mr John Combridge (Anon. 61). The sand reservoir is at the top on the right. Above the wheel the upper balancing lever and the two upper locks, here working on the scoop-pivot axles, can easily be seen. Below, on the right, a scoop is beginning to fill and will shortly come to rest on the checking fork of the lower balancing lever; just beneath, the coupling tongue trip-lever can be seen connected by its long chain with the upper balancing lever. The former has a suspended weight, the latter is weighted with riders. For a scale working model see Combridge (1).

This last expression seems to be rather a rare one, absent both from the copious engineering vocabulary of Su Sung himself and from the mass of other texts which concern the development of clockwork in medieval China. Nevertheless, it can only refer to the springs which worked the bells and drums as the figures on Su Sung's jack-wheels made their daily rounds.[a] The text is a notable one, for it was adduced in the time of the Jesuits by the few Chinese scholars who knew enough on such subjects in those days to point out that their Renaissance clocks were not the first which had been known in China.[b]

This completes the account of the hydro-mechanical clockwork of Su Sung's great astronomical tower, set up in the form of a working wooden pilot model in the imperial palace at Khaifêng in +1088. It was the time of our Domesday Book and the youth of Abelard. Two years later the metal parts, i.e. the armillary sphere and celestial globe, were duly cast in bronze. The writing of the explanatory monograph must have been well under way in +1092, and it was finally presented to the throne in +1094. Prefixed to it is a remarkable memorial in which Su Sung not only describes the principles of the clock itself, but gives a historical disquisition on all instruments of a similar kind which had existed in previous centuries. This it was which illuminated many other texts not previously comprehensible, permitting the establishment of a history of Chinese clockwork,[c] the outline of which will be found in the following pages. First, however, we must pause a moment to read a little in Su Sung's memorial, for it contains many matters of the greatest interest. He wrote:[d]

When formerly (i.e. after the edict of +1086 ordering the construction of a new clock) I was seeking for help, I met Han Kung-Lien,[1] a minor official in the Ministry of Personnel, who having mastered the *Chiu Chang Suan Shu* (Nine Chapters of Mathematical Art),[e] often used geometry (lit. the methods of right-angled triangles) to investigate the degrees of (motion of the) celestial bodies. Thinking it over, I also became convinced that the ancients used the techniques of the *Chou Pei* (*Suan Ching*)[f] in studying the heavens....

I therefore told (Han Kung-Lien) about the apparatus of Chang Hêng,[2] I-Hsing[3] and Liang Ling-Tsan,[4] and the designs of Chang Ssu-Hsün[5],[g] and asked him whether he could study the matter and prepare similar plans. Han Kung-Lien said that they could be successfully completed, if mathematical rules were followed and the (remains of) the former

+1173. We have not been able to locate the source of the quotation in those of his works which have been available to us.

[a] The hodometer must not be forgotten as one of the ancestors of these trip-mechanisms (cf. pp. 281 ff. above).

[b] Cf. p. 525 below.

[c] For fuller detail, the monograph of Needham, Wang & Price may be consulted.

[d] *Hsin I Hsiang Fa Yao*, ch. 1, pp. 2a ff., tr. auct.

[e] The greatest of the mathematical works of the Han, completed in the +1st century but containing some material as old as the Chhin (−3rd). Cf. the discussions in the mathematical Section, Vol. 3, pp. 24 and 150. For a particular study of one of the most interesting of its methods see Wang & Needham (1).

[f] The oldest of the mathematical works of Chinese antiquity, completed in the −1st century but containing some material as old as the −4th or even the −6th. Cf. the discussions in the mathematical and astronomical Sections, Vol. 3, pp. 19, 199 and 256. Tr. Biot (4).

[g] The work of these men will be explained in the immediately following pages.

¹ 韓公廉　　² 張衡　　³ 一行　　⁴ 梁令瓚　　⁵ 張思訓

machines taken as a basis (*chü suan shu, an chhi hsiang*[1]). Afterwards he wrote a memorandum in one chapter, entitled 'Verification of the Armillary Clock by Geometry (lit. the Right-Angled Triangle Method)' (*Chiu Chang Kou Ku Tshê Yen Hun Thien Shu*[2]), and he also made a wooden model of the mechanism with time-keeping wheels (*mu yang chi lun*[3]). After studying this model I formed the opinion that although it was not in complete agreement with ancient principles, yet it showed great ingenuity, especially with regard to the water-powered driving-wheel, and that it would be desirable to entrust him with the building of it. I therefore recommended to your Imperial Majesty that a (complete) wooden pilot model should first be made and presented to you, and that some officials should be ordered to test its use. If the time-recording (*hou thien*[4]) proved to be correct, then instruments of bronze could be made. On the 16th day of the eighth month in the 2nd year (of the Yuan-Yu reign-period, i.e. +1087) your Imperial Majesty gave an order that my suggestion should be carried out, and that a (special) bureau should be set up, officials appointed and the necessary materials prepared.

I therefore recommended that Wang Yuan-Chih,[5] Professor at the Public College of Shouchow, formerly Acting Registrar of Yuan-wu in Chêngchow prefecture, should be in charge of construction and the receipt and issue of public materials; while Chou Jih-Yen,[6] Director of Astronomical Observations (Southern Region)[a] of the Bureau of Astronomy and Calendar, Yü Thai-Ku,[7] Director of Astronomical Observations (Western Region) of the same Bureau, Chang Chung-Hsüan,[8] Director of Astronomical Observations (Northern Region), and Han Kung-Lien, should be appointed to supervise the construction. (I further recommended) the Assistants in the Bureau, Yuan Wei-Chi,[9] Miao Ching[10] and Chang Tuan,[11] and the Superintendent (Chieh-Chi[12]) Liu Chung-Ching;[13] together with the Students, Hou Yung-Ho[14] and Yü Thang-Chhen,[15] as investigators of the sun's shadow, the clepsydras, and so on. (Lastly, I recommended) the Bureau of Works Foreman Yin Chhing[16] to be Clerk of the Works.

In the fifth month of the 3rd year of the Yuan-Yu reign-period (+1088) a small pilot model was finished, and at your Imperial Majesty's order presented for testing. Afterwards the full-scale machinery was built in wood and completed by the intercalary twelfth month. I (then) begged your Imperial Majesty to send a court official to the Bureau (of Astronomy and Calendar) to explain (the parts to the workmen) in preparation for moving the clock to the palace for presentation....[b] In the tenth month we had sent in a request for instructions regarding the installation, and the Palace Guard Superintendent detailed the Aide-de-Camp Huang Chhing-Tsung[17] (to look after the matter). On the 2nd day of the twelfth month, a letter was sent up asking exactly where (the clock) was to be assembled, and your Imperial Majesty's order came to erect it in the Hall of All Heroes (Chi Ying Tien[18], in the Palace).

With all its vividness of detail, this passage concerning the organisation of one of the greatest technical achievements of the medieval time in any civilisation is certainly worthy of appreciation. Moreover, reading rightly 'between the lines' brings out

---

[a] We came across an instance of this title a few pages above. Though the actual Chinese terms for these regional officials embody the words for spring, summer, autumn and winter, rather than the cardinal points specifically (cf. des Rotours (1), vol. 1, p. 213), it is tolerably certain that the titles referred to the corresponding palaces of the sky (cf. de Saussure, 16a, b), and that the officials themselves were responsible for observing phenomena, usual or unusual, in those sectors.

[b] There is a small lacuna in the text here.

[1] 據算術案器象	[2] 九章鉤股測驗渾天書	[3] 木樣機輪	[4] 候天
[5] 王沈之	[6] 周日嚴	[7] 于太古	[8] 張仲宣	[9] 袁惟幾
[10] 苗景	[11] 張端	[12] 節級	[13] 劉仲景	[14] 侯永和
[15] 于湯臣	[16] 尹清	[17] 黃卿從	[18] 集英殿

several significant points. Han Kung-Lien, that man of brilliant mathematical and mechanical talent, had no post in which he could use it but was found by Su Sung in the minor ranks of his own administrative Ministry.[a] Contrary to common conceptions of medieval working, the new armillary clock was not put together haphazardly by trial and error, but planned in a special memorandum with all the geometrical knowledge that Han could put into it. This certainly makes it easier to understand how the gearing, chain-drives, and other devices were made to carry out successfully their duty of rotating steadily an armillary sphere weighing some 10–20 tons as well as a bronze celestial globe $4\frac{1}{2}$ ft. in diameter. It is also noteworthy that a small wooden model was first made, then a full-scale one was tested against four types of clepsydra[b] as well as star transits, and only after four years were the parts destined for bronze duly cast.

In the last paragraph of his memorial Su Sung wrote:[c]

Thus, as we have seen, the (demonstrational) armillary sphere (*hun thien i*[1]), the bronze observational armillary sphere (*thung hou i*[2]), and the celestial globe (*hun thien hsiang*[3]),[d] are three things different from one another....(Therefore) if we use only one name, all the marvellous uses of (the three) instruments cannot be included in its meaning. Yet since our newly-built machine embodies two instruments but has three uses, it ought to have some (more general) name such as 'Hun Thien'[4] (Cosmic (Engine)). We are humbly awaiting your Imperial Majesty's opinion and bestowal of a suitable name upon it.

And he signed with all his ranks and titles, Imperial Tutor to the Crown Prince, Grand Protector of the Army, Khai-Kuo Marquis of Wukung, etc. Now the two instruments were of course the mechanised observational armillary sphere and the mechanised celestial globe. Clearly the three uses were (*a*) astronomical demonstrations and (*b*) astronomical observations, both with the armillary sphere, together with (*c*) indication on the globe of the positions of all constellations whatever the weather,[e] and their relations to models of the sun, moon, and planets attached to the globe for calendrical verifications.[f] But besides these functions there was also the signalisation of time, both visual and auditory, by elaborate jack-work. Thus Su Sung's request for a new name was of great historical significance. The mechanised astronomical instrument was trembling on the verge of becoming a purely time-keeping machine. Inaudible echo must have answered 'A clock!' But history does not record that the young emperor had any good ideas on nomenclature, and the time-measuring function continued to go unnamed until five hundred years later the Jesuits came with their 'self-sounding bells' to ring in the age of unified world science with its unlimited expansion of appropriate technical terms.

[a] Cf. the discussion on the social milieu of mathematics and astronomy in medieval China in Vol. 3, pp. 152 ff., as also p. 39 above.

[b] *Hsin I Hsiang Fa Yao*, ch. 1, p. 5*a*.   [c] P. 5*b*.

[d] See the discussions in the Section on astronomy (Vol. 3, pp. 383, 343 and 387) for details of the development of these three instruments in China. The demonstrational armillary sphere was fitted with a model earth at the centre instead of the sighting-tube for observations.

[e] There is a perceptible parallel here with the magnetic compass. The Chinese were responsible for two great inventions which rendered man independent of clear weather for his orientation.

[f] The meaning of this has been explained in the Section on astronomy (Vol. 3, p. 361) and the procedure will be referred to again below (p. 487).

[1] 渾天儀   [2] 銅候儀   [3] 渾天象   [4] 渾天

(2) CLOCKWORK IN AND BEFORE THE NORTHERN SUNG

The derivation of mechanical clocks from clepsydras has now been made clear. But the story of the evolution remains to be told.[a] In his memorial Su Sung wrote:[b]

According to your servant's opinion there have been many systems and designs for astronomical instruments during past dynasties all differing from one another in minor respects. But the principle of the use of water-power for the driving mechanism has always been the same. The heavens move without ceasing but so also does water flow (and fall). Thus if the water is made to pour with perfect evenness, then the comparison of the rotary movements (of the heavens and the machine) will show no discrepancy or contradiction; for the unresting follows the unceasing.

This was a nice appreciation of what Europeans were later to think of as the universal writ of the 'law' of gravitation. But Su Sung goes on to give brief descriptions of the predecessors of his own clock, beginning with Chang Hêng's[1] device of the +2nd century. In this we shall follow him, making use of the original texts on which his summary was based. The most convenient plan will be to begin by working backwards, starting from the +11th century and going back as far as the +1st. We shall then be able to make another start from his time in the opposite direction, and following the fate of his own machine, describe the chief events in Chinese clock-making which took place between the +12th century and the arrival of the Jesuits at the end of the +16th.

We must first take up an important mechanical point. It will have been noticed that the description of Su Sung's clock contains nothing resembling a dial. Though the stationary dial-face with a moving pointer is a development associated with the first European mechanical clocks of the +14th century,[c] the rotating dial-face had been used in Hellenistic times.[d] The anaphoric clock, described by Vitruvius[e] about −30, consisted of a bronze disc with a planispheric projection of the stars of the northern hemisphere, and as many as are found between the equator and the tropic of Capricorn which formed the rim of the disc. The circle representing the ecliptic (the zodiac) was provided with 365 small holes, into which was plugged from day to day a little stud representing the sun. The disc was made to rotate by the simple mechanism of a float in a clepsydra attached to a cord terminating in a counterweight and wound

---

[a] It has occasionally been glimpsed, but through a glass darkly. A century ago, a medical missionary in Ningpo, McGowan (4), seeking to advise American clock-makers on the Chinese market, put together some material translated for him from some encyclopaedia, perhaps the *Yü Hai*; but lacking all proper documentation it could not be very convincing. Yet it is still in use (cf. Bedini, 2). More recently the eminent Japanese scholar Shiratori Kurakichi (4) stumbled upon the history of clockwork in the course of his studies on Chinese–Byzantine relations, but his engineering knowledge was insufficient to enable him to translate helpfully the relevant texts which he found.

[b] P. 3a.

[c] Though here again there is a Chinese connection. Was not the magnetic compass needle the first dial-and-pointer reading? Was it not first known in Europe at the end of the +12th century? The reader may refer again to Sect. 26i above.

[d] Allusion has already been made to this in the present subsection, p. 442 above.

[e] IX, viii, 8.

[1] 張衡

round a drum on the horizontal axis bearing the planispheric disc.[a] The disc of the Hellenistic anaphoric clock was separated from the spectator by an immobile network of bronze wires, a vertical wire representing the meridian, three concentric ones the equator and the tropics, and between the tropics other wires indicating the zodiacal months. Across all the circles was an arc denoting the horizon of the place, and there were crosswires dividing the concentric circles into twelve day and twelve night hours, above and below the horizon respectively. Neugebauer (7) and, more completely, Drachmann (6), have shown that this arrangement was the forerunner of the astrolabe of medieval times with its rete. We need not follow further this development since the astrolabe was not known or used in Chinese civilisation. But the question does arise whether any similar kind of mechanised planisphere was employed there. Apart from the Vitruvian description, actual bronze engraved discs from the Roman empire have survived in fragmentary form.[b] No such objects have yet come to light in China, or at least none such has so far been recognised.[c] However, there are certain literary evidences which suggest that simple anaphoric clocks were not unknown there. The following text, for example, dates from the +13th century but refers to devices which were in use in the early part of the +11th.

In his *Tung Thien Chhing Lu*[1] (Clarifications of Strange Things), Chao Hsi-Ku[2] maintains that old bronze objects are good demonifuges and have other strange virtues. He goes on to say:[d]

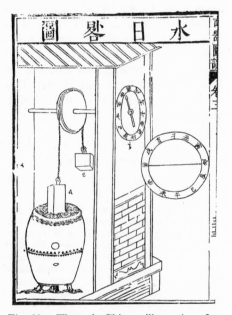

Fig. 660. The only Chinese illustration of an anaphoric clock like those of the Hellenistic and Arabic culture-areas, a drawing from the *Chhi Chhi Thu Shuo* of +1627 (ch. 3, p. 53 b). This *shui jih kuei*, or water-operated sundial, as it was called, was taken by Johann Schreck & Wang Chêng from Fausto Veranzio's engineering treatise of +1616. In the anaphoric clock rotational and other movements are brought about by the rise or fall of a clepsydra float. Anciently the dial itself rotated, but by the +17th century a stationary dial-face with a mobile pointer had come into use.

[a] The reconstruction of Diels (1), p. 213, is reproduced in Usher (1), p. 97, 2nd ed. p. 145; cf. Price (1, 4). A better reconstruction is given in Price (5).

[b] The best known is that found at Salzburg and described by Benndorf, Weiss & Rehm; it dates from about +250. Another, reported by Maxe-Werly (1), comes from the Vosges. Cf. Kibitschek (1), pp. 209 ff. In the Stiftsbibliothek at St Gallen I have seen 9th- and 10th-century drawings of anaphoric clock dials in MSS. of Aratus. Cf. Maddison, Scott & Kent (1).

[c] The only Chinese illustration of an anaphoric clock known to us (Fig. 660) was that given by Johann Schreck & Wang Chêng in their *Chhi Chhi Thu Shuo* of +1627 (ch. 3, pp. 53 b ff.). This was undoubtedly copied from the picture in the *Machinae Novae* of Verantius (+1616), pl. 6, as Jäger found (unpublished notes). The fact that the anaphoric clock was illustrated in 1627 as something new can hardly be taken as proof that it had not been known in earlier centuries. One can only say that it was unfamiliar to Wang Chêng and his circle. Cf. p. 503.

[d] As cit. in *Shuo Fu*, ch. 12, p. 29 b, and *Ko Chih Ku Wei*, ch. 5, p. 11 b. He was writing about +1235.

[1] 洞天清錄   [2] 趙希鵠

Fan Wên Chêng Kung (i.e. Fan Chung-Yen[1])[a] had in his home an ancient mirror. On the back the twelve (double-)hours were marked by hemispherical protuberances like *liu-po* game pieces,[b] and whenever one of these hours arrived, (one of them) shone like the (full) moon.[c] The whole thing rotated without ceasing. There was also a device in another scholar's house called a 'Twelve (Double-)Hour Bell' (Shih-erh shih Chung[2]) which sounded automatically as the hours passed (*ying shih tzu ming*[3]). What magical powers these old bronze objects have!

Evidently Chao Hsi-Ku was not very mechanically minded, yet there is something here calling for explanation about +1020. A second relevant passage comes from the early +10th century and concerns some device which had been in use at the beginning of the +9th. Thao Ku[4] tells us of another 'mirror' in his *Chhing I Lu*[5] (Records of the Unworldly and the Strange):[d]

In the Palace Treasury of the Thang dynasty there was a yellow plate with a circumference of three feet. Around the disc there were designs of animals and other things. In the Yuan-Ho reign-period (+806 to +820) it was occasionally used to see how the symbols changed following the passing of the hours of the day. For instance, at the *chhen* (double-)hour[e] there was a dragon playing among a decoration of flowers and herbs, but when the plate turned to the *ssu* (double-)hour[f] a snake appeared, while when the (double-)hour of *wu* came[g] then the turning plate showed a horse.[h] It was therefore called the 'Twelve (Double-) Hour Plate' (Shih-erh shih Phan[6]). This instrument was handed down (at the end of the dynasty) and was still in existence during the Later Liang dynasty (+907 to +923) of the House of Chu.[i]

Unfortunately, Thao Ku did not distinctly state whether or not the turning was automatic, but this seems probable.

Another account comes from even earlier and seems to show[j] the beginnings

---

[a] An eminent scholar and official in the reign of Jen Tsung, governor of Yenan in Shensi, strongly opposed to Buddhism. See the biography of Fischer (1).

[b] Cf. Vol. 3, pp. 304 ff; Vol. 4, pt. 1, p. 327.

[c] This suggests some kind of lamp shining through a hole in the disc. Lamps lighting with the hours were a feature of many of the Arabic striking water-clocks (cf. Wiedemann & Hauser, 4), such as those described by al-Jazarī in +1206, which also worked on the anaphoric clock principle. Cf. Fig. 33 in Sect. 7h (Vol. 1, pp. 203 ff.). Mention was there made of one of these great clocks still partly extant, at the *madrasah* (college) of Bū'Inānīya at Fez in Morocco (see Michel, 9). Probably the system of the lighting of successive lamps in a series derives from the Jewish ceremony of the 'Festival of Lights' (*Hanukah*), a custom still preserved.

[d] Ch. 2, p. 18b.   [e] 7–9 a.m.

[f] 9–11 a.m.   [g] 11 a.m.–1 p.m.

[h] These were the correct symbolic animals for the successive double-hours.

[i] Tr. auct. An abbreviated version of this text was translated by Schlegel (5), p. 561, and Chavannes (7), p. 65; its possible horological bearing was thus missed.

[j] We word the matter cautiously, for when we first read the passage we took it to refer to some kind of south-pointing carriage (cf. pp. 286 ff.). The difficulty is that the absence of any context gives no clue as to whether *chhê* should be taken as a vehicle or as some other kind of machine with wheels. *Yuan* also normally should mean chariot-shafts. And *chhen* could mean either double-hours or compass-points. However, if the latter were intended, it would be almost certainly either 8 or 24 and not 12. This is perhaps the most decisive ground for the present interpretation.

¹ 范仲淹   ² 十二時鐘   ³ 應時自鳴   ⁴ 陶穀   ⁵ 清異錄
⁶ 十二時盤

of time-keeping jack-work, the figures making their rounds on a horizontally mounted wheel like those of Su Sung's. It is a paragraph of the *Chhao Yeh Chhien Tsai*[1] (Stories of Court Life and Rustic Life) written by Chang Tsu[2] early in the +8th century, and fortunately preserved in *Thai-Phing Kuang Chi*.[a]

> In the Ju-I reign-period (+692) an artisan (*chiang*[3]) from Haichow[b] was presented to the empress[c] Wu Tsê Thien[4]. He made a 'Wheel (for Reporting) the Twelve (Double-)Hours' (Shih-erh chhen Chhê[5]). When the wheel came to the exact south position (*hui yuan chêng nan*[6]), the Wu[7] (south) door opened, and a jack with a horse's head appeared.[d] The wheel revolved round the four directions (as time passed) without the slightest mistake (*ssu fang hui chuan, pu shuang hao li*[8]).

Since this device was made several decades before the first escapement clock, that of I-Hsing (see below, p. 473), and since the effects described would not have required much power in the drive, it is reasonable to assume a sinking-float mechanism. Not long before (about +500) a celestial globe had been rotated in India by means of the anaphoric clock principle—if we may trust the late +15th-century commentary of Parameśvara on the *Āryabhaṭīya*.[e]

But the sinking-float principle seems never to have been prominent in the Chinese culture-area. This was probably because from the outset, as we shall see, it was desired to rotate not only planispheric discs but spherical astronomical instruments, which even if made of wood were quite heavy. The power requirements therefore necessitated the enlistment of the water-wheel—and indeed this was done probably within a century of the first appearance of the water-mill itself. Our immediate hunt is for the first appearance of the escapement.

The most important clock in the Sung dynasty prior to that of Su Sung himself was built by Chang Ssu-Hsün,[9] towards the end of the +10th century. It included sphere and globe, powered by a scoop-bearing driving-wheel and gearing, together with jack-figures to report and sound the hours. Eleven technical terms occur in the description with exactly the same meanings as in Su Sung's text. Chang's clock was a particularly fine and interesting work as it used mercury in the closed circuit instead of water, thus assuring time-keeping in frosty winters. But it must have been somewhat ahead of its

---

[a] Ch. 226 (item 5), p. 3*b*. Tr. auct. Cf. Edwards (1), vol. 1, p. 77.

[b] A good man. He made Cardan suspensions (cf. p. 234) and other ingenious things.

[c] Wu Hou,[10] personal name Wu Chao[11] (+625 to +705), might be described as China's Queen Elizabeth I. In +684, after the death of Kao Tsung, she displaced his successor and ruled the empire alone for the rest of her life. Cf. Fitzgerald (8).

[d] We thought at first that this was the horse-headed god Hayagrīva (van Gulik, 5), but it is the correct animal in the cycle for the noon double-hour corresponding to due south, as witness *Hsiao Hsüeh Kan Chu*, ch. 1, p. 23*a*.

[e] For this reference we are indebted to Dr David Pingree through the kindness of Professor Derek Price. The instrument may well have been a demonstrational armillary sphere rather than a celestial globe. Cf. pp. 481, 539.

---

[1] 朝野僉載　　[2] 張鷟　　[3] 匠　　　　　　　　　[4] 武則天　　[5] 十二辰車
[6] 廻轅正南　　[7] 午　　[8] 四方廻轉不爽毫釐　[9] 張思訓　　[10] 武后
[11] 武照

age, for Su Sung tells us[a] that after Chang's death it soon went out of order and there was no one able to keep it going.[b] The *Sung Shih* says:[c]

At the beginning of the Thai-Phing Hsing-Kuo reign-period (+976) the Szechuanese Chang Ssu-Hsün, a Student in the Bureau of Astronomy, invented an astronomical clock (lit. armillary sphere, *hun i*[1]) and presented the designs to the emperor Thai Tsung, who ordered the artisans of the Imperial Workshops to construct it within the Palace. On a *kuei-mao* day in the first month of the 4th year (+979) the elaborate machine was completed, and the emperor caused it to be placed under the eastern drum-tower of the Wên Ming Hall.

The system of Chang Ssu-Hsün was as follows: they built a tower of three storeys (totalling) more then ten feet in height, within which was concealed all the machinery. It was round (at the top to symbolise) the heavens, and square (at the bottom to symbolise) the earth. Below there was set up the lower wheel (*ti lun*[2]), lower shaft (*ti chu*[3]), and the framework base (*ti tsu*[4]). There were also horizontal wheels (*hêng lun*[5]), (vertical) wheels fixed sideways (*tshê lun*[6]), and slanting wheels (*hsieh lun*,[7] i.e. oblique gearing); bearings for fixing them in place (*ting shen kuan*[8]), a central coupling device (*chung kuan*[9]) and a smaller coupling device (*hsiao kuan*[10]) (i.e. the escapement); with a main transmission shaft (*thien chu*[11]). Seven jacks rang bells on the left, struck a large bell on the right, and beat a drum in the middle to indicate clearly the passing of the quarter(-hours).

Each day and night (each 24 hours) the machinery made one complete revolution, and the seven luminaries moved their positions around the ecliptic. Twelve other wooden jacks were also made to come out at each of the (double-)hours, one after the other, bearing tablets indicating the time. The lengths of the days and nights were determined by the (varying) numbers of the quarters (passing in light or darkness). At the upper part of the machinery there were the top piece (*thien ting*[12]), upper gear(-wheel or -wheels) (*thien ya*[13]), upper linking device (*thien kuan*;[14] another part of the escapement), upper (anti-recoil) ratchet pin (*thien chih*[15]), celestial (ladder?) gear-box (*thien tho*[16]), upper framework beam carrying bearings (*thien shu*[17]), and the upper connecting-chain (*thien thiao*[18]). There were also (on a celestial globe?) the 365 degrees (to show the movements of) the sun, moon and five planets; as well as the Purple Palace (north polar region), the lunar mansions (*hsiu*) in their ranks, and the Great Bear, together with the equator and the ecliptic which indicated how the changes of the advance and regression of heat and cold depend upon the measured motions of the sun.

The motive power of the clock was water, according to the method which had come down from Chang Hêng in the Han dynasty through I-Hsing and Liang Ling-Tsan in the Khai-Yuan reign-period (+713 to +741) (of the Thang). But the bronze and iron (of their clocks) had long gone to rust (*thung thieh chien sê*[19]) and could no longer move automatically. More-

---

[a] *Hsin I Hsiang Fa Yao*, ch. 1, p. 2a.

[b] It is interesting that Chang Ssu-Hsün was a Szechuanese, for that populous province in the west had been the scene just previously, during the late Thang and Five Dynasties periods, of the earliest expansion of another admirable invention, that of block printing. Liu Phien[20] tells us how in +883, on his holiday outings, he used to examine the printed books being sold outside the city walls of Chhêngtu (cf. Carter (1), p. 60). These were mostly on proto-scientific subjects (oneiromancy, geomancy, astrology, planetary astronomy, and speculations of the Yin–Yang school). The province thus seems to have been rather fertile in technological advances at this time. One remembers that it had long been the home of the proto-industrial area of the Tzu-liu-ching brine-field (cf. Sect. 37).

[c] Ch. 48, pp. 3b ff. Parallel text in *Yü Hai*, ch. 4, pp. 29a ff., which is in several places better. Tr. auct.

[1] 渾儀	[2] 地輪	[3] 地軸	[4] 地足	[5] 橫輪	[6] 側輪
[7] 斜輪	[8] 定身關	[9] 中關	[10] 小關	[11] 天柱	[12] 天頂
[13] 天牙	[14] 天關	[15] 天指	[16] 天托	[17] 天束	[18] 天條
[19] 銅鐵漸澁	[20] 柳玭				

over, as during winter the water partly froze and its flow was greatly reduced, the machinery lost its exactness, and there was no constancy between the hot and cold weather. Now, therefore, mercury was employed as a substitute, and there were no more errors....

The images of the sun and moon were also attached high up (to the globe) and according to the old method they had been moved by human hand (each day), but now success was attained in having them move automatically. This was a marvellous thing. (Chang) Ssu-Hsün was considered the equal of the Thang clock-makers and was made Special Assistant in charge of the Armillary Sphere (Engine) (Ssu-Thien Hun I Chhêng[1]).

This passage shows clearly that Chang's clock was very like Su Sung's, with similar drive and similar escapement.[a] The use of mercury was particularly ingenious, however, and it seems certain that a set of planetary models was rotated automatically. Though Su returned to the classical method of moving them by hand (as in the anaphoric clock), later Sung specifications (cf. p. 499 below) also had them on geared wheels, like the orreries and planetaria of later Europe. Perhaps most interesting is the mention of the gear-box, which rather implies the use of a chain-drive like Su Sung's;[b] if so, Chang Ssu-Hsün was an anticipator of Leonardo by five hundred years.

From this we can pass directly to the 'Thang clockmakers'. Who were these men who made, in the +8th century, the most venerable of all escapement clocks? One was a Tantric Buddhist monk, perhaps the most learned and skilled astronomer and mathematician of his time, I-Hsing,[2] the other a scholar, Liang Ling-Tsan,[3] who, like Han Kung-Lien later, occupied a minor administrative post.[c] The technical terms employed in the relevant passages again reveal the essential similarity of the machine to the clock of Su Sung.

These passages are to be found in the official histories of the Thang dynasty[d] and in the *Chi Hsien Chu Chi*[4] (Records of the College of All Sages)[e] written about +750 by Wei Shu.[5] The context of I-Hsing's astronomical clock was his introduction of the

---

[a] Another description of it, derived from that of an eye-witness, exists in the *Fêng Chhuang Hsiao Tu*[6] (Maple-Tree Window Memories), written by Yuan Chhiung[7] and printed soon after +1202; ch. 1, p. 1*b*. The translation will be found in Needham, Wang & Price (1), p. 72.

[b] There is, however, the possibility that the transmission used was a belt-drive, for Su Sung says (*Hsin I Hsiang Fa Yao*, ch. 1, p. 2*a*) that after the death of Chang Ssu-Hsün 'the cords and the mechanism went to rack and ruin (*chi shêng tuan huai*[8])'. But *chi-shêng* might be a single expression meaning 'mechanical cord', i.e. chain. One cannot be quite sure.

[c] In modern terminology, he was a War Office civil servant connected with the Brigade of Guards. Obviously he was another square engineering peg in a round hole. But he left a work entitled *Wu Hsing Erh-shih-pa Hsiu Shen Hsing Thu*[9] (Spirits of the Five Planets and the 28 Hsiu), which was still in existence as a MS. in colours in +1669 in the library of the bibliophile Chhien Tsêng (cf. p. 448 above); Chhü Fêng-Chhi (1), p. 153. A copy is preserved at Osaka; see Sirén (10), vol. 1, pp. 47 ff., pls. 16, 17. Liang's scroll may have been the companion volume of I-Hsing's *Erh-shih-pa Hsiu Pi Ching Yao Chüeh*[10] (*Sung Shih*, ch. 206, p. 9*b*), cf. TW 1309, 1311.

[d] *Hsin Thang Shu*, ch. 31, pp. 1*b* ff. and *Chiu Thang Shu*, ch. 35, pp. 1*a* ff.; the latter is the better text. The version in *Yü Hai*, ch. 4, pp. 21*a* ff. is poor.

[e] The College of All Sages was an Imperial Academy somewhat duplicating the Han-Lin. Both had the duty of drafting edicts and checking their texts. But the College also incorporated all kinds of experts who were available for consultation by the emperor. Entrance to it was by nomination, not by examination; hence the services of scholars and experts not in the strict Confucian tradition could be used. See des Rotours (1), vol. 1, pp. 17, 19.

[1] 司天渾儀丞　　　[2] 一行　　　[3] 梁令瓚　　　[4] 集賢注記　　　[5] 韋述
[6] 楓牕小牘　　　[7] 袁裦　　　[8] 機繩斷壞　　　[9] 五星二十八宿神形圖
[10] 二十八宿秘經要訣

Ptolemaic ecliptically mounted sighting-tube convenient for studying planetary motions on and near the ecliptic.[a] Wei Shu wrote:[b]

> In the 12th year of the Khai-Yuan reign-period (+724) the monk I-Hsing constructed an armillary sphere with an ecliptically mounted sighting-tube in the (Li-Chêng) Library,[c] and when it was finished he presented it (to the emperor). Earlier, he had received an imperial order to reorganise the calendar, and had said that observations were difficult because there was no apparatus with this ecliptic fitting. Just at that time Liang Ling-Tsan made a small model (of the instrument which was wanted) in wood and presented it. The emperor asked I-Hsing to study it, and he reported that it was highly accurate. Therefore a full-scale (sphere) in bronze and iron was made in the Library grounds, taking two years to complete. When it was offered to the throne the emperor praised it exceedingly and asked (Liang) Ling-Tsan and I-Hsing to study (further) Li Shun-Fêng's[1] book, the *Fa Hsiang Chih*[2] (The Miniature Cosmos), so that later on they drew up complete plans of the armillary sphere. And the emperor wrote an inscription in 'eight-tenths' style characters[d] which was carved on the ecliptic ring and which said:
>
>> 'The moon in her waxing and waning is never at fault
>> Her twenty-eight stewards escort her and never go straying
>> Here at last is a trustworthy mirror on earth
>> To show us the skies never-hasting and never-delaying.'
>
> The scholar Lu Chhü-Thai[3] received an imperial order to write an inscription containing the year and month of construction, and the names of the workers, underneath the plate. The observatory used the apparatus for observations, and it is still employed nowadays....
>
> After this, the emperor ordered the casting of bronze for yet another astronomical instrument. The Chief Secretary of the Left Imperial Guard, Liang Ling-Tsan, and his colleague of the Right, Huan Chih-Kuei,[4] took charge of plans for the separate parts, and a great (demonstrational) armillary sphere (*thien hsiang*[5]) was cast 10 ft. in diameter. It showed the lunar mansions (*hsiu*), the equator and all the circumferential degrees. It was made to turn automatically by the force of water acting on a wheel (*chu shui chi lun*[6]). Discussing it, people said that what Chang Hêng (+2nd century) had described in his *Ling Hsien*[7] (Spiritual Constitution of the Universe) could have been no better.
>
> Now it is kept in the College of All Sages at the eastern capital (Loyang). In the courtyard there is the observatory (*yang kuan thai*[8]) where I-Hsing[e] used to make his observations.

This brief account was fortunately expanded in the Thang histories. There we read that in +723 I-Hsing and Liang Ling-Tsan 'and other capable technical

---

a See Vol. 3, pp. 350 ff. This mounting had already been suggested by Li Shun-Fêng, but it was foreign to the equatorial astronomy of China, and was not conserved in later instruments.

b *Yü Hai*, ch. 4, p. 24a—for his own original book is lost.

c Edicts were drafted and checked in the Li-Chêng Library.

d This was a style intermediate between seal characters and *li shu* characters in that proportion.

e A few words on this great man seem necessary here. His secular name was Chang Sui.[9] For the significance of the sect to which he belonged see Sect. 15f (Vol. 2, pp. 425 ff.). Like Matteo Ricci, he astonished people by his feats of memory. Combining knowledge of Indian (and hence indirectly of Hellenistic) mathematics and astronomy with the whole range of the Chinese traditions in these subjects, he was the most outstanding physical scientist of his time. Unfortunately, most of the books he wrote were afterwards lost (with the exception of some doubtful tractates) and only the titles remain. A full-length biography of him, with adequate translations of the available texts which bear on his activities, is urgently required. Cf. Juan Yuan's account of his work in *Chhou Jen Chuan*, chs. 14–16.

¹ 李淳風   ² 法象志   ³ 陸去泰   ⁴ 亙執珪   ⁵ 天像
⁶ 注水激輪   ⁷ 靈憲   ⁸ 仰觀臺   ⁹ 張遂

men (*shu shih* [1])' were commissioned to cast and make new bronze astronomical intruments.

One (of these) was made in the image of the round heavens (*yuan thien chih hsiang* [2])[a] and on it were shown the lunar mansions (*hsiu*) in their order, the equator and the degrees of the heavenly circumference. Water, flowing (into scoops), turned a wheel automatically (*chu shui chi lun, ling chhi tzu chuan* [3]), rotating it (the sphere) one complete revolution in one day and night. Besides this, there were two rings (lit. wheels) fitted round the celestial (sphere) outside, having the sun and moon threaded on them, and these were made to move in circling orbit (*ling tê yün hsing* [4]).[b] Each day as the celestial (sphere) turned one revolution westwards,

[a] This expression is rather unusual. Although Su Sung, three and a half centuries later, believed it to have been a celestial globe, we feel that the general description accords much better with a demonstrational armillary sphere half sunk in a box-like casing the top of which represented the earth's horizon. On the general question of the relation of such spheres with true celestial globes, see Vol. 3, pp. 382 ff. This instrument may have had something to do with the 'celestial dome model' (*kai thien chih chuang* [5]) and the flexible bamboo ruler which I-Hsing employed in connection with his thirty-six charts of the deviations of the moon's path from the ecliptic (*Hsin Thang Shu*, ch. 31, p. 5 a; *Chhou Jen Chuan*, ch. 16, p. 201). Cf. Vol. 3, pp. 357, 392 ff.

[b] According to the account of Chang Ssu-Hsün's clock which we have read above (p. 471), the mechanisation of the movement of the sun and moon models was first effected by him (in +979). But the nuance of the present text indicates automatic motion rather unmistakably. Moreover, there is decisive evidence in the words of a contemporary writer, Chang Yüeh.[6] He says (*Chhüan Thang Wên*, ch. 223, p. 5 a): 'Although the motions (of sun, moon, and globe) all went on simultaneously, their speeds, whether slow or fast, were not equal. Everything returned to its original starting-point, following round in due order and ceaselessly rotating.' Thus I-Hsing and Liang Ling-Tsan rather than Chang Ssu-Hsün, in the +8th century rather than the +10th, were the first to replace the manually adjusted models of earlier times (pegged into holes or threaded on strings) by an 'orrery' system of mechanical motions. A similar instrument was built later on by Wang Fu and his associates about +1124 (cf. p. 499 below). An interesting hypothetical reconstruction (Fig. 661) of the 'orrery' movements from I-Hsing to Wang Fu has been offered by Liu Hsien-Chou (6). As will be seen, this involves both concentric shafting and a number of gear-wheels cut with odd numbers of teeth. Needham, Wang & Price (1) felt some uncertainty about the feasibility of so complex a design in the time of I-Hsing, though not doubting it for Wang Fu and perhaps also Chang Ssu-Hsün. Wiedemann

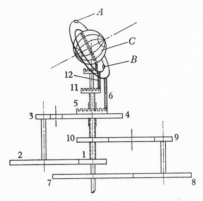

Fig. 661. Reconstruction of a possible mechanism for the 'orrery' movements of medieval Chinese clocks (I-Hsing in +725, Chang Ssu-Hsün in +979, Wang Fu in +1124, etc.). Only the sun and moon movements are shown as well as that for the celestial globe, and the design, due to Liu Hsien-Chou (6), involves both concentric shafting and gear-wheels with odd numbers of teeth. A is the sun model, B the moon, and C the demonstrational armillary sphere or solid celestial globe. The gear-wheels 1 to 6, which work the solar motion, have the following numbers of teeth respectively, 12, 60, 6, 72, 12, 73. The gear-wheels 7 to 12, which work the lunar motion, have 127, 73, 6, 15, 6 and 114 teeth respectively.

(13) and Price (1, 4) have drawn attention to the odd-toothed gearing in Arabic and European astrolabes as one of the most important elements in the prehistory of clockwork. The oldest evidence about these gear-trains is textual, a MS. of al-Bīrūnī written *c.* +1000, but an Arabic geared astrolabe made in +1221 is preserved at Oxford, and a French example of *c.* +1300 at London. This was just about the time when a MS. in the Caius College Library (230/116, pp. 31 ff.), describing methods of cutting gear-wheel teeth, was written at St Alban's Abbey under that great horological abbot, Richard of Wallingford (*c.* +1292 to +1336); see Price (11). There is reason to believe that Su Sung in his clock of +1090 employed at least one odd-toothed gear-wheel, namely that of 487 teeth (cf. p. 456 above). If this is so, no difficulty would have been encountered by Wang Fu and his friends, but whether such wheels could have been made accurately enough in +725 remains a little uncertain. In any case, an orrery movement of some

[1] 術士　　　　[2] 圓天之象　　　[3] 注水激輪令其自轉　　　[4] 令得運行
[5] 蓋天之狀　　　[6] 張說

the sun made its way one degree eastwards, and the moon $13\frac{7}{19}$ths degrees (eastwards). After 29 and a fraction rotations (of the celestial sphere) the sun and moon met. After it made 365 rotations the sun accomplished its complete circuit. And they made a wooden casing the surface of which represented the horizon, since the instrument was half sunk in it. This permitted the exact determination of the times of dawns and dusks, full and new moons, tarrying and hurrying. Moreover there were two wooden jacks standing on the horizon surface, having one a bell and the other a drum in front of it, the bell being struck automatically to indicate the hours, and the drum being beaten automatically to indicate the quarters.

All these motions were brought about (by machinery) within the casing, each depending on wheels and shafts (*lun chu*[1]), hooks, pins and interlocking rods (*kou chien chiao tsho*[2]), coupling devices and locks checking mutually (*kuan so hsiang chhih*[3]) (i.e. the escapement).

Since (the clock) showed good agreement with the Tao of Heaven, everyone at that time praised its ingenuity. When it was all completed (in +725)[a] it was called the 'Water-Driven Spherical Bird's-Eye-View Map of the Heavens' (Shui Yün Hun Thien Fu Shih Thu[4]) or 'Celestial Sphere Model Water-Engine' and was set up in front of the Wu Chhêng Hall (of the Palace) to be seen by the multitude of officials. Candidates in the imperial examinations (in +730) were asked to write an essay on the new armillary (clock).[b]

But not very long afterwards the mechanism of bronze and iron began to corrode and rust, so that the instrument could no longer rotate automatically. It was therefore relegated to the (museum of the) College of All Sages (Chi Hsien Yuan[5]) and went out of use.

Such are the details of the instrument which, so far as we can see, was the first of all escapement clocks.[c] The reference to the checking linkwork is plain, the technical terms used being closely similar to those in the descriptions of the clock of Su Sung. Although the automatic movement of the sun and moon models is not stated with absolute clarity, it is almost certainly implied; the machine had therefore some at least of the features of an orrery or planetarium.[d]

It is not generally known that an occidental orrery was carried to China about a thousand years later, by the embassy of Lord Macartney in +1793. The first planetarium of this modern kind, demonstrating the heliocentric system, was made by George Graham and Thomas Tompion[e] about +1706 for Prince Eugene of Austria[f] in circumstances described not long afterwards by Desaguliers.[g] It was immediately copied, with modifications, for the Earl of Orrery, whence the name still in use for the instrument.[h] A magnificent replica was ordered by the East India Company in

kind there was, and Liu Hsien-Chou may well be right in all that he attributes to I-Hsing, who will thus have even more claim upon our respect and admiration. A solar–sidereal conversion gear-train simpler than that of Liu Hsien-Chou (6, 7) has been suggested by Mr John Combridge (p. 456 above). Liu has published a revised version of his own proposals, with a model, in (7), pp. 109 ff.

[a] This date comes from *Yü Hai*, ch. 4, pp. 25 a, b.

[b] We interpolate this sentence from *Yü Hai*, ch. 4, p. 25 b.

[c] The chief reason for adopting this view at present is that the phraseology of the descriptions is characteristic of those of the succeeding centuries but cannot be traced earlier. Combridge (1, 2) suggests, however, on practical grounds, that the essentials of the water-wheel linkwork escapement go back to Chang Hêng (cf. p. 485), so that the philological criterion may not be as decisive as it seems.

[d] Echoes late and garbled of this clock are often found in 19th-century books, e.g. Planchon (1), p. 247, and still today, e.g. Strickler (1). Generally they derive from the vague account of Gaubil (2), p. 85. Sarton (1), vol. 1, p. 514, termed it an 'uranorama' but was not sure that it depended on water-power.

[e] See Lloyd (4); Gabb & Taylor (1).   [f] Cf. Taylor & Wilson (1).

[g] (1), 1st ed., vol. 1, pp. 430 ff.; 2nd ed., vol. 1, pp. 448 ff.   [h] Cf. Orrery (1); Rice (1).

[1] 輪軸   [2] 鈎鍵交錯   [3] 關鑕相持   [4] 水運渾天俯視圖   [5] 集賢院

+1714, and described by John Harris (1) a few years later. Whether this was from the beginning destined for China we have not ascertained, but the opening of the British Factory at Canton took place in the following year. In any case, an instrument of a similar kind was offered to the Chinese Imperial Court before the end of the century.

G. L. Staunton tells us[a] that in +1792 two Chinese students from the College at Naples came to England to accompany the Macartney mission as interpreters,[b] and began by advising as to the presents to be taken. It was decided that it would be unwise to try to compete with the mechanical toys, and the clocks with elaborate jack-work, which the 'sing-song' trade[c] had by then been pouring into China for close on half a century; some objects of intellectual interest would be more suitable.

> Astronomy being a science [wrote Staunton] peculiarly esteemed in China, and deemed worthy of the attention and occupation of the Government, the latest and most improved instruments for assisting its operations, as well as the most perfect imitation that had yet been made of the celestial movements, could scarcely fail of being acceptable.

Accordingly in the fullness of time, Dr Dinwiddie[d] and Mons. Petitpierre-Boy[e] were to be seen in the Yuan Ming Yuan palace at Peking unpacking, besides the orrery, a celestial and a terrestrial globe, a reflecting telescope, a great burning-lens, a clock showing the lunar phases, and an air-pump.[f] According to Rees[g] this orrery, made by P. M. Hahn of Erfurt and A. de Mylius,[h] had been finished in +1791, a work of many years, and 'allowed to be the most wonderful piece of mechanism ever emanating from human hands'.[i] Thus the high value traditionally placed upon astronomical science in Chinese culture imposed itself on European diplomacy at the end of the +18th century, and a masterpiece of Western astronomical clockwork found its way as a veritable tribute, even if the term was abjured, and the intention partly unconscious, to the land of I-Hsing, Chang Ssu-Hsün and Su Sung.[j] But we must now return to the pioneer orrery of +725.

---

[a] (1), vol. 1, p. 42.

[b] Though the advice of these missionary seminarists may have been good in the first place, they proved inadequate as interpreters when in China partly because they were ignorant of the learning, science and technology of their own culture. This led to results which we shall notice elsewhere.

[c] Cf. p. 522.

[d] The biography of James Dinwiddie (+1746 to 1815) by his grandson (Proudfoot, 1) is well worth reading; thanks are due to Mr J. Cranmer-Byng for directing our attention to it. After his service as Astronomer to the Macartney embassy, Dinwiddie became scientific adviser to the East India Company in Bengal, and later Professor of Natural Philosophy at Fort William College, the forerunner of the University of Calcutta.

[e] On C. H. Petitpierre-Boy (+1769 to c. 1810) see Chapuis, Loup & de Saussure (1), pp. 45 ff. He was a Swiss clock-maker who settled at Macao and then at Manila, married a Javanese and finally was killed by pirates.

[f] See Staunton (1), vol. 1, p. 492, vol. 2, pp. 165, 287; Proudfoot (1), pp. 45 ff.

[g] In Chambers (1), 9th ed. *sub voce.*

[h] Cf. Zinner (8), p. 351. The attribution to Hahn is confirmed by Dinwiddie's own biography (Proudfoot (1), p. 26).

[i] Proudfoot (1), *loc. cit.* A brief Chinese description (*Chang Ku Tshung Pien*, ch. 3, pp. 22a ff.) speaks of it with more moderate enthusiasm.

[j] A Chinese representation of the bringing of astronomical instruments by the Macartney embassy is shown in Fig. 662. It is a silk tapestry preserved in the National Maritime Museum at Greenwich. Being evidently unfamiliar with the objects actually offered, the artist simply depicted one of the

For its background we have to look at the preceding reign. The emperor served by Li Shun-Fêng was Thai Tsung,[a] who ruled with much brilliance from +626 onwards for a quarter of a century. Interested in history and technology as well as in the military arts, he knew how to encourage astronomers, and welcomed Nestorian clergy as well as Taoist priests and Buddhist monks. He entertained cordial diplomatic relations as far west as Byzantium, receiving in +643, for example, an embassy from the Patriarch of Antioch. Such missions may well have brought news of the striking water-clocks at places like Gaza and Antioch.[b] Of course this can have been no more than a 'stimulus diffusion',[c] for there is no reason for thinking that the Byzantine works employed anything more than the sinking-float principle. However, the stimulus would have come just at the right time to encourage Chinese engineers to try to out-do the mechanical toys which formed the striking jack-work of the water-clocks of the Eastern Roman Empire.[d] And indeed the description of I-Hsing's clock does seem to be the first mention of horological escapement-operated jacks in Chinese history. Here he was much better placed than his Greek colleagues, if we are right in our supposition that the water-wheel, providing so much more power than the float, had already long before been characteristic of Chinese astro-mechanical technique.

The emperor for whom I-Hsing worked was Hsüan Tsung,[e] most unfortunate among the rulers of the Thang. Ascending the throne in +712, he prospered for some thirty years, shining as a patron of music, painting and literature. All the greatest of the Thang poets knew his court. In later life, however, growing social and economic strains exposed the country to the military rebellion of An Lu-Shan,[1] a Sogdian general in the Chinese service, from which the dynasty never recovered.[f] Among its incidental results was the death of Hsüan Tsung's famous and beautiful concubine[g] Yang Kuei Fei.[2] Though the Tantrist's clock cannot have been built for her, since she did not join the imperial entourage till +738, her presence evokes considerations of a rather singular kind.

It was, I think, the great medievalist du Cange who first proposed to explain the sudden interest in mechanical clocks at the beginning of the European +14th century

---

+17th-century Jesuit instruments still extant on the old observatory tower at Peking (cf. Vol. 3, pp. 451 ff.). He also gave the mission +16th-century dress, including ruffs. On the Chinese background of the Macartney embassy see Cranmer-Byng (1).

[a] Personal name Li Shih-Min[3] (+597 to +649).

[b] Cf. Vol. 1, pp. 186, 193, 204. See Diels (2). 　　　　　[c] Cf. Vol. 1, p. 244.

[d] We have noted already the Haichow artisan at work in +692 (p. 469). But his device is not said to have made any sound. We have also noted the 'Twelve (Double-)Hour Bell' in the time of Fan Chung-Yen (p. 468). But that was in the third decade of the +11th century, long after the time of I-Hsing. It would seem, therefore, that the chiming jacks of +725 were (so far as we can see at present) the first auditory signals made by any time-keeping device in the Chinese story. As we shall shortly see, all Chinese proto-clocks back to the +2nd century were (according to their descriptions) silent, and the same applies to all the types of clepsydra before the Thang discussed in Sect. 20g above (Vol. 3, p. 319). On the contrary in the West, where pneumatic devices were so popular, clepsydras had been made to blow pipes and whistles from early times—as in Plato's alarm clock (Diels (1), p. 198; Ewbank (1), p. 546). Cf. Athenaeus, IV, 75 ff. This use of siphons and tipping cups continued strongly in the Arab tradition (cf. Wiedemann & Hauser, 4).

[e] Personal name Li Lung-Chi[4] (+685 to +762).

[f] See Pulleyblank (1). 　　　　　　　　　　　[g] Personal name Yang Yü-Huan[5] (d. +756).

[1] 安祿山　　　　[2] 楊貴妃　　　　[3] 李世民　　　　[4] 李隆基　　　　[5] 楊玉環

PLATE CCLVII

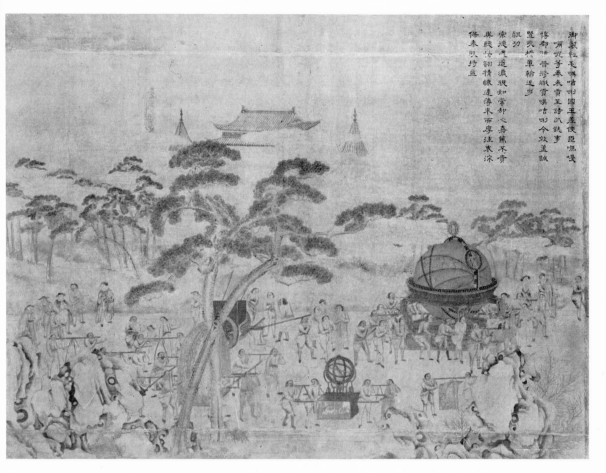

御製紅毛英咭唎國王差使臣嗎嘎
嘶呢等奉表貢至詩以記事
博都咭唎昔償職貢虔
暨爾地單輪遠步
祖功
宗德覃遠覲如常覿心嘉薦不貴
異聽收翻情愫遠傳來而厚往裹深
俾奉恩以拎益

Fig. 662. A Chinese representation of the bringing of astronomical instruments as gifts by the embassy of Lord Macartney in +1793; a silk tapestry in the National Maritime Museum, Greenwich. The artist, doing his best, portrayed a cortège of foreigners in Elizabethan dress with ruffs, bearing forward instruments like those of the Jesuits in the Peking Observatory. The celestial globe is a good representation of that of Verbiest (+1673, cf. Vol. 3, p. 388); the smaller instrument shows the observational armillary described and figured in the *Huang Chhao Li Chhi Thu Shih* of +1759 (ch. 3, p. 32a). This was called a San Chhen I (cf. Vol. 3, p. 351). In the present picture the relative sizes of the globe and the armillary sphere are correctly shown, which suggests that the artist knew them at first hand, but he made both about twice as large as they ought to be with respect to the human figures. No clocks are in evidence, but perhaps horology is represented by a large cage seen passing behind two trees, and doubtless intended to contain the mechanical birds of the sing-song trade (cf. Needham, Wang & Price (1), fig. 55).

In the upper right-hand corner of the tapestry there is an inscription with the following rubric: 'Poem composed by the (Chhien-Lung) Emperor himself on the Occasion of the Arrival of the Ambassador Ma-Kha-Erh-Ni and others, bringing a Memorial and Tribute from the King of the Red-haired (People) of Ying-Chi-Li (England).'

The composition follows:

'Formerly Po-Tu-Ya (Portugal) undertook the task of bringing tribute,
Now Ying-Chi-Li (England) follows suit with all sincerity.
(Her envoys) have out-travelled Shu-Hai and Hêng-Chang of old,
Truly the works and virtues of Our Ancestors have echoed far beyond the seas.
Though the gifts may be commonplace enough Our heart commends the givers;
Strange things indeed We prize not, nor do We listen to boastful claims,
Yet mindful of the great distance journeyed We shall repay a hundredfold,
For in such wise do sage-kings preserve the plenitude of their prosperity.'

Apart from the erudite allusion in the third line to Shu-Hai and Thai-Chang, the legendary surveyors of Yü the Great, who paced out for him the length and breadth of the *oikoumene* (*Huai Nan Tzu*, ch. 4, p. 2a), the poem conveys the imperial attitude with crystal clarity. It is another expression of that same Chinese reaction to the early European industrial age which was immortalised in the celebrated edict on the Macartney embassy (see p. 601 below). Thus was poetic justice executed—temporarily at least—on European claims of superiority by the descendants of the oldest makers of mechanical clockwork.

by the desire of the monks of the great abbeys to know more accurately the time of the night. Thus they could better regulate the hours of their lauds and matins. Whatever may be the plausibility of this suggestion, it was at any rate an attempt to set the invention of the weight-driven verge-and-foliot clocks in a social context, and to find a reason for their rapid adoption. Exactly the same question can be asked about the invention of the water-wheel weighbridge escapement at the beginning of the +8th century in China. What possible need could have been felt at such a period within the Chinese imperial palace—for it was always in close association with the emperor that these masterpieces of medieval engineering arose—for more accurate knowledge of the hours of the night, and for a means of following the march of the constellations even when foul weather rendered them invisible? It is true that time-keeping and stellar depiction were not the only functions of the water-wheel clocks; there was also the use of the mechanised globe as a test of calendrical calculations. This was probably at least as important a determining factor as any other, but we must postpone consideration of it for a few pages. Here let us think only of the bells and gongs automatically giving the time, and the slow rotation of the celestial globe with its map of the heavens.

It will be recalled that the Chinese emperor was a cosmic figure, the analogue here below of the pole star on high.[a] All hierarchies, all officialdom, all works and days, revolved around his solitary eminence.[b] It was therefore entirely natural that from time immemorial the large number of women attending upon him should have been regulated according to the principles of the numinous cosmism which pervaded Chinese court life. Ancient texts give us remarkable insight into the ranks of his consorts and concubines. Though their titles differed considerably during the two millennia which followed the first unification of the empire,[c] the general order comprised one Empress (Hou[1]), three Consorts (Fu Jen[2], Ho Jen[3] or Fei[4]), nine Spouses (etymologically protegées or client ladies; Phin[5], Phin Jen), twenty-seven Beauties (concubines; Shih Fu[6], Mei Jen[7]), and eighty-one Attendant Nymphs (assistant concubines; Nü Yü[8], Yü Chhi[9]). The total adds up to 121, which (certainly by no coincidence) is one third of 365 to the nearest round number. A classical passage in the *Chou Li* (Record of the Rites of the Chou Dynasty)[d] even gives us what might be called a pernoctation rota.

The lower-ranking (women) [it says[e]] come first, the higher-ranking come last. The assistant concubines, 81 in number, share the imperial couch 9 nights in groups of 9. The

---

[a] Cf. Vol. 3, pp. 230ff., 240, 259ff., Fig. 99.

[b] See Creel (3) and Granet (5); Soothill (5). Cf. Vol. 2, p. 287.

[c] Principal references: *Chou Li*, ch. 7; tr. Biot (1), vol. 1, pp. 143, 154, 156; *Li Chi*, ch. 44 (Hun I), p. 42a; tr. Legge (7), vol. 2, p. 432; *Chhien Han Shu*, ch. 99c, p. 23b; tr. Dubs (2), vol. 3, p. 438. These refer to the Han; for the Thang cf. des Rotours (1), pp. 256 ff. Lists of concubine titles are not uncommon; one is given in the anonymous *Chhü Chhao Shih Lei*[10] (Systematic Guide to Court Etiquette), a short book of Sung date (*Shuo Fu*, ch. 34, p. 6b).

[d] Cf. pp. 11ff. above.

[e] Ch. 7 (ch. 2, p. 20b); tr. van Gulik (3), p. 92, Granet (6), p. 26. It occurs in the ancient commentary of Chêng Hsüan, but was not translated by Biot.

| [1] 后 | [2] 夫人 | [3] 和人 | [4] 妃 | [5] 嬪 | [6] 世婦 |
| [7] 美人 | [8] 女御 | [9] 御妻 | [10] 趙朝事類 | | |

concubines, 27 in number, are allotted 3 nights in groups of 9. The 9 spouses and the 3 consorts are allotted 1 night to each group, and the empress also alone 1 night. On the 15th day of every month the sequence is complete, after which it repeats in the reverse order.

Thus it is clear that the women of highest rank approached the emperor at times nearest to the full moon, when the Yin influence would be at its height, and matching the powerful Yang force of the Son of Heaven, would give the highest virtues to children so conceived. The primary purpose of the lower ranks of women was rather to feed the emperor's Yang with their Yin. In the +9th century Pai Hsing-Chien[1] complained that all these rules had fallen into disorder, saying:[a]

Nine ordinary companions every night, and the empress for two nights at the time of the full moon—that was the ancient rule, and the Duennas-Secretarial (Nü Shih[2]) kept a careful record of everything with their vermilion brushes....But alas, nowadays all the three thousand (palace women) compete in confusion....

These secretaries were mentioned already in the *Chou Li*[b] and it is their activities which show us the relevance of these curious matters to the invention of clockwork.

What was at stake was the imperial succession. Chinese ruling houses did not always follow the primogeniture principle, and the eldest son of the empress was not necessarily the heir apparent.[c] Towards the end of a long reign an emperor would have quite a number of princes from which to choose, and in view of the importance of State astrology in China from very ancient times it may be taken as certain that one of the factors in this choice was the nature of the asterisms which had been culminating at the time of the candidate's conception.[d] Hence the importance of the records which had been kept by the Duennas-Secretarial, and the value of an instrument which not only told them the time but from which also the eunuchs could read off the star positions at any desired moment. For example:

The Thai-Yang Shou[3] (star) lies to the west of the Hsiang[4] (the Minister), and is the symbol of the commander-in-chief and the prime minister. It governs the readiness of the country to withstand attack, and the preparation of armaments.[e]

---

[a] *Thien Ti Yin Yang Ta Lo Fu*,[5] p. 56; tr. van Gulik (3).

[b] Ch. 7 (Biot (1), vol. 1, p. 158). For a close parallel at Akbar's court, see Blochmann (1), p. 44.

[c] Theoretically only the sons of the empress could be candidates, but this rule was honoured as often in the breach as the observance. Cf. e.g. Erkes (19), p. 152.

[d] As is well known, ages of individual persons in China are counted not from birth but from conception. In the light of all this, it is rather striking to read in the book of al-Marwazī, written about +1115, when Su Sung's clock was still working in its original position: 'Whenever the King of China wants to enter his women's apartments and to remain alone with the women, the Astrologer goes up to the roof of the house where he is, and observes the stars in order to choose the time propitious for his intercourse with some one of his women' (Minorsky (4), p. 27). This seems to have been taken from the *Zain al-Akhbār* of al-Ghardīzī, written about +1050, and that in turn was based on the lost *Kitāb al-Masālik w'al-Mamālik* of Abū 'Abdallāh al-Jaihānī composed early in the +10th century, i.e. not long before the time of Chang Ssu-Hsün. On the importance of genethliacal astrology in Thang and Sung China cf. Gernet (2), p. 163, based on *Kuei Hsin Tsa Chih (Hou Chih)*, pp. 45a ff., etc., and on the observations of Marco Polo.

[e] *Chin Shu*, ch. 11, p. 9a.

---

[1] 白行簡   [2] 女史   [3] 太陽守   [4] 相   [5] 天地陰陽大樂賦

The two stars of Hsü[1] (Emptiness, the 11th *hsiu*) denote the officials in charge of ancestral worship. They govern northern cities, temples, and all matters pertaining to ritual and prayer. They also govern death and lamentation.[a]

Without doubt the presages of such stars in their transits, to say nothing of comets, novae, and other more unusual phenomena, would have been an important factor in the choice of the successor to the throne. Careful records were thus a fundamental usage in the imperial family. Interest in recording machinery is therefore not at all surprising. But these facts, like the interest in horologia which the Western monks doubtless had, fall short of giving us a reason why the invention of the escapement, as the reply to a demand for more accurate mechanical time-keeping, should have come about particularly at this juncture.

Before quitting the subject of these earliest escapement clocks it is necessary to say a little more about certain inventions which may have paved the way for them. For example, we have found indications of horizontal jack-wheels in the century preceding I-Hsing. Now from the foregoing account it will have been seen that in the absence of oscillators (such as verge-and-foliot or pendulum) an essential part of the water-wheel linkwork escapement was formed by the two weighbridges or steelyards which the scoops had to trip. In order to understand where this component could have been brought in from, we must glance again at the history of clepsydras (*kho lou*[2]) in ancient and medieval China.[b]

The most archaic sort of water-clock, the outflow type, was doubtless a gift to East Asia from the cultural centres of the Fertile Crescent in the -1st or even -2nd millennium.[c] But from about -200 onwards it was replaced almost everywhere in China by the inflow type with an indicator-rod borne on a float (*fou chien lou*[3]).[d] Already in the Han it was well understood that falling pressure-head in the reservoir greatly slowed the time-keeping as the inflow vessel filled. Throughout the centuries two ways of correcting for this were adopted, first the interposition of one or more compensating tanks between the reservoir and the inflow vessel,[e] and then a little later

---

[a] *Chin Shu*, ch. 11, p. 14b; both tr. auct. adjuv. Ho Ping-Yü (1). This astrological text was written only a little more than a century before I-Hsing's time, though certainly containing much ancient material.

[b] For a fuller account see Sect. 20g (Vol. 3, pp. 315 ff.), and Needham, Wang & Price (1), pp. 85 ff.

[c] Mentions of the clepsydra in the *Shih Ching* (Book of Odes), which might date from about the -7th century, are philologically uncertain but not at all historically improbable. What is probably the oldest reference occurs in the *Shih Chi* (ch. 64, p. 1b). In the biography of a general and politician of the -6th century, Ssuma Jang-Chü,[4] who served Prince Ching of the State of Chhi, we are told that while waiting for a rendezvous with another leader, Chuang Chia,[5] Ssuma 'set up a sundial and started the water-clock dripping'. As Ching ruled from -546 to -488, this is a venerable record, contemporary with Confucius, and there seems no reason for regarding it as an anachronism on the part of Ssuma Chhien.

[d] The most complete study of these clepsydras is that of Maspero (4).

[e] This was an admirable means of cumulative regulation, and in late times as many as five tanks above the inflow vessel are known to have been used (cf. de Saussure, 29). In the early $+2$nd century Chang Hêng had at least one compensating tank in the series. There was also at least one in the clepsydra for which Sun Chho[6] wrote an epigrammatic inscription about $+360$ (*Thai-Phing Yü Lan*, ch. 2, p. 13a).

---

[1] 虛　　[2] 刻漏　　[3] 浮箭漏　　[4] 司馬穰苴　　[5] 莊賈　　[6] 孫綽

the insertion of an overflow or constant-level tank in the series.[a] These were the commonest kinds of clepsydras. But there were others which involved weighing the water on some kind of balance, and these have been less studied.

Balance-clepsydras included at least two types, one in which the typical Chinese steelyard (the balance of unequal arms)[b] was applied to the inflow vessel itself, and another in which it weighed the amount of water in the lowest compensation tank. The first sort naturally dispensed with the float and indicator-rod, and was usually made small and portable, and sometimes adapted for the use of mercury instead of water, so that very short intervals of time could conveniently be measured.[c] These were called 'stopwatch clepsydras' (*ma shang kho lou*[1]).[d] The weighing of water in the compensating tank demanded a larger apparatus (*shui chhêng kho lou*[2]), which was used for public and palace clocks throughout the Thang and Sung periods. It permitted the seasonal adjustment of the pressure-head in the compensating tank by having standard positions for the counterweight graduated on the beam, and hence it could control the rate of flow for different lengths of day and night. With this arrangement no overflow tank was required, and the attendants were warned when the clepsydra needed refilling. When one reads the standard description of this clepsydra[e] and other accounts[f] one realises that the men responsible for designing it were two of the great technicians[g] of the Sui dynasty, Kêng Hsün[3] and Yüwên Khai.[4] In other words, the plan was stabilised just about +610, i.e. one hundred and ten years before the work of I-Hsing and Liang Ling-Tsan.[h]

The relevance of these facts is obvious. From the systematic weighing of a water-receiving vessel it was not such a far cry to the weighing of one which both received it and delivered it. That in turn would have pointed the way to the mounting of such

---

[a] The earliest description of this type which we have is from the pen of Yin Khuei[5] (*fl. c.* +550), cf. Maspero (4), p. 193. There is a quite elaborate account by Shen Kua[6] in +1074 (*MCPT*, ch. 8, § 8), cf. Maspero (4), p. 188.

[b] On this, see Vol. 4, pt. 1, pp. 24 ff. above.

[c] As by astronomers in the observation of eclipses, or by others for the timing of races.

[d] They seem to begin with the Taoist Li Lan[7] of the Northern Wei dynasty (*c.* +450). They are often described afterwards, e.g. by Wang Phu[8] in the +12th century. As they depended upon the approach of the steelyard arm to the horizontal they may not be without significance in the history of pointer-readings in general.

[e] *Yü Hai*, ch. 11, pp. 18a ff.; also, greatly abridged, in *Sung Shih*, ch. 76, pp. 3b ff. Full translation in Needham, Wang & Price.

[f] *Sui Shu*, ch. 19, pp. 27b ff.; copied in *Yü Hai*, ch. 11, p. 12a. Full translation in Needham, Wang & Price.

[g] Yüwên Khai was engineer, architect, and Minister of Works under the Sui dynasty for thirty years. On his large sailing carriages see p. 278. He carried out irrigation works and superintended the construction of the Sui Grand Canal (cf. Sect. 28f). He also wrote on the Ming Thang (cf. Vol. 2, p. 287), and made a wooden model of it. Concerning Kêng Hsün see immediately below, p. 482.

[h] Shiratori (4), p. 314, noted the interesting fact that clocks with weights are mentioned by Sulaimān al-Tājir (Sulaimān the Merchant), who was in China in +851 or just before, as being set up on the drum-towers or gate-houses of Chinese cities. The Arabists, from Renaudot (1), p. 25 in 1718, to Sauvaget (2), p. 15, were laudably cautious in interpreting these as weight-driven mechanical clocks, and did not quite know what to do. But now it is clear that what he saw were the large balancing steelyard clepsydras, and his words are particularly interesting because they apply to provincial cities rather than to the capital or the imperial palace.

---

[1] 馬上刻漏   [2] 水秤刻漏   [3] 耿詢   [4] 宇文愷   [5] 殷夔   [6] 沈括
[7] 李蘭   [8] 王普

double-function containers round the rim of a single wheel; in fact to the controlled retention of successive water-receiving and water-delivering scoops so arranged as to pass a weighbridge.[a] Thus it would seem that the weighbridges were directly derived from the steelyard arm which had suspended (and which still for some centuries continued to suspend) the lowest compensating tank of a certain type of clepsydra. It would hardly be possible to demonstrate a clearer succession of stages between the most ancient leaking water-pots and the shock-proof, water-resistant, anti-magnetic, luminous-dial, 24-hour masterpieces which we now wear on our wrists.

(3) THE PRE-HISTORY OF CHINESE CLOCKWORK

We have concluded, then, that what may be called the father and mother of all escapements originated in the first decades of the +8th century. But the history of clockwork which we are sketching cannot stop at this point on its backward course. For between +725 and the beginning of the Christian era there were many other examples of astronomical globes or spheres being slowly rotated by water-power. If we define the escapement as the essence of true clockwork, these earlier devices were not exactly clocks, but they may well be considered predecessors of the clock. Moreover, they were purely astronomical in character and lacked all auditory time-telling components. We shall present briefly the texts concerning them.

Since the meagre power of a sinking float could not have rotated an obliquely mounted sphere or globe, even if made relatively lightly in wood,[b] it is possible that the mechanism consisted of a vertical water-wheel with cups like a noria, doubtless of more simple construction than that of Su Sung. The water-wheel could be attached to a shaft with one trip-lug, quite similar in principle to the water-driven trip-hammer assemblies so common in the Han. Clepsydra drip into the cups would accumulate periodically the torque necessary to turn the lug against the resistance of a leaf-tooth wheel, either itself forming the equatorial ring, or attached to a shaft in the polar axis. Needless to say, the time-keeping properties of such an arrangement would be extremely poor, so poor indeed that it is hard to understand how they could have justified the explicit claims which have come down to us in the texts.[c] Perhaps the linkwork escapement is older than we have dared to suppose.[d]

Examples can be taken from almost every century between the +8th and the +2nd. One of the most outstanding technicians of the +6th century was Kêng Hsün,[1] whom we have already met with on account of his work with clepsydras.[e] A man of matchless technical skill, and witty in argument, he became involved early in life (as

---

[a] Indeed the chain of the 'iron crane bird's knee' might have originated as a manually operated device for bringing the receiving vessels successively under the inflow. If this should turn out to be the case, it would afford a strange parallel to the later automation of the valves of the steam-engine.

[b] And the Chinese texts say frequently enough that bronze instruments were so turned. If the feat was accomplished in India (cf. pp. 469, 539), some excessively light wood or bamboo construction must have been used.

[c] Cf. pp. 483, 485. [d] See the proviso, p. 474 (c) above.

[e] See Sect. 20 (Vol. 3, p. 327), and pp. 36, 480 above.

[1] 耿 詢

we have seen) with a rebellion of southern tribal folk, but when eventually captured was pardoned by the general Wang Shih-Chi[1] on account of his great ingenuity.[a]

After a long time Kêng Hsün met his old friend Kao Chih-Pao,[2] whose knowledge of the heavens had brought him to the position of Astronomer-Royal, and from him (Kêng) Hsün received instruction in astronomy and mathematics. (Kêng) Hsün then conceived the idea of making an armillary sphere (*hun thien i*[3])[b] which should be turned not by human hands but by the power of (falling) water (*pu chia jen li, i shui chuan chih*[4]). When it had been made he set it up in a closed room and asked (Kao) Chih-Pao to stand outside and observe the time (as shown by the) heavens (i.e. the star transits). (His instrument) agreed (with the heavens) like the two halves of a tally. (Wang) Shih-Chi, knowing of this, reported the matter to the emperor Kao Tsu,[c] who made (Kêng) Hsün a government slave[d] and attached him to the Bureau of Astronomy and Calendar.

This account, referring to work which was going on in the neighbourhood of +590, is closely similar to what we are told in all the other cases. Generally no wheel is mentioned, but no float either, and the mechanised instrument is typically set up inside a closed room, with two observers, one inside calling out the indications of the machine, the other outside checking these against the heavenly movements themselves. There may even be no mention of water in the automatic movement, but a water-drive is always to be surmised.

About seventy years earlier the great Taoist physician, alchemist and pharmaceutical naturalist, Thao Hung-Ching[5] (+452 to +536), had done something of the same kind. The reports[e] say that he

made a (demonstrational) armillary sphere (*hun thien hsiang*[6]) more than 3 ft. high, with the earth situated in the middle. The 'heavens' rotated and the 'earth' remained stationary—and it was all moved by a mechanism (*i chi tung chih*[7]). Everything agreed exactly with the (actual) heavens.

Not only that, but he wrote a book about it, the *Thien I Shuo Yao*[8] (Essential Details of Astronomical Instruments);[f] this has long been lost. His mechanised sphere may be dated *c.* +520.

[a] The biography of Kêng Hsün (*Sui Shu*, ch. 78, pp. 7 b ff. and *Pei Shih*, ch. 89, pp. 31 a ff.) shows how very adventurous and uncertain the life of an engineer could be in the +6th century. It has been translated in full by Needham, Wang & Price (1), whence this excerpt.

[b] This passage is quoted in *Hsü Shih Shuo*, ch. 6, p. 11 a, and in *Yü Hai*, ch. 4, p 26 a. The term used suggests that the instrument was a demonstrational armillary sphere (cf. Vol. 3, p. 383), but it may have been a solid globe.

[c] Personal name Yang Chien[9] (+540 to +604), first emperor of the Sui; a forceful monarch but not particularly interested in scientific and technical matters. Kêng Hsün made advisory inclining vessels and many other things for the second emperor (Yang Ti), Yang Kuang[10] (+580 to +618), who greatly encouraged, it will be remembered (p. 162 above) makers of mechanical toys and the like. It is worth noting that Kêng had the idea of filling his hydrostatic trick vessels (cf. Vol. 4, pt. 1, p. 35 above) slowly with clepsydra water—another way of using it other than letting it raise a float or run to waste.

[d] Cf. Wilbur (1), pp. 221 ff.

[e] *Nan Shih*, ch. 76, p. 11 a; copied in *Thai-Phing Yü Lan*, ch. 2, p. 10 a; abbreviated in *Liang Shu*, ch. 51, p. 17 a. Tr. auct.

[f] Listed in the bibliography of *Sui Shu*, ch. 34, p. 15 a.

[1] 王世積   [2] 高智寶   [3] 渾天儀   [4] 不假人力以水轉之   [5] 陶宏景
[6] 渾天象   [7] 以機動之   [8] 天儀說要   [9] 楊堅   [10] 楊廣

A century earlier still, under the Liu Sung dynasty, similar apparatus was set up by Chhien Lo-Chih,[1] an astronomer who was probably the originator of solid celestial globes in China. His work has the special interest that it followed the recovery of the old instruments of Chang Hêng[2] (+78 to +139).

The armillary sphere (*hun i*[3]) made by Chang Hêng was handed down through the (San Kuo) Wei (kingdom) (+221 to +280) and the (Western) Chin (dynasty) (+265 to +317). But after the defeat of China (in +317) at the hands of the (northern Turkic and Hunnish) barbarians, even the later instruments of (Lu) Chi[4] (*fl.* +220 to +245) and (Wang) Fan[5] (+219 to +257) were all lost.[a]

However, in the 14th year of the I-Hsi reign-period (+418) when the emperor An Ti of the (Eastern) Chin (dynasty) was still reigning, Kao Tsu (i.e. Wu Ti, first emperor of the Liu Sung dynasty) (drove north again), captured Chhang-an (the ancient capital) and recovered the old instruments of (Chang) Hêng.[b] Although the forms were still recognisable, the marks of graduation had all gone, and nothing was left of the representations of stars, sun, moon and planets.[c]

Later on, in the 13th year of the Yuan-Chia reign-period (+436) the emperor[d] ordered Chhien Lo-Chih, the Secretary of the Bureau of Astronomy and Calendar, to re-make and cast a (demonstrational) armillary sphere. Its diameter was slightly less than 6·08 ft. and its circumference slightly less than 18·26 ft.[e] The earth was fixed in the centre of the heavens,[f] and there were the two paths of the ecliptic and the equator, the two celestial poles south and north, with the 28 lunar mansions (*hsiu*) depicted round about, as also the Great Bear and the pole star. Each degree corresponded to 0·5 inch. The sun, moon and five planets were strung along the ecliptic. A clepsydra was set up and the whole apparatus made to rotate by its water (*chih li lou kho i shui chuan i*[6]).[g] The transits of stars at dawn and dusk (shown on the instrument) all agreed exactly with the actual movements of the heavens (*yü thien hsiang ying*[7]).[h]

So runs the description in the *Sung Shu*.[i] Chhien Lo-Chih's instruments[j] lasted a long time, for we know[k] that more than a century later they were being cherished by the astronomical bureau of the Liang dynasty (*c.* +555). After the conquest of the

[a] Because the Chinese empire withdrew to Nanking south of the Yangtze, and of course the instruments of Chang Hêng were also left behind in Chhang-an (modern Sian in Shensi).

[b] This passage is also found in *Yü Hai*, ch. 4, p. 15*b*. A rather fuller version of the recovery of the instruments was given in the *I-Hsi Chhi Chü Chu*[8] (Daily Records of the I-Hsi reign-period, +405 to +418), not extant now but quoted in *Thai-Phing Yü Lan*, ch. 2, p. 10*a*. But the date of this source is not much earlier than the *Sung Shu*.

[c] In spite of this phraseology, the instrument was almost certainly a demonstrational armillary sphere.

[d] Now Wên Ti, personal name Liu I-Lung[9] (+407 to 453), third of the Liu Sung.

[e] It is curious that these dimensions indicate a π value very close to 3, since at that time much more sophisticated values were available (see Vol. 3, p. 101).

[f] As has already been said (Vol. 3, p. 389), this is a remarkable development in view of the late (+16th-century) appearance of the same idea in Europe; see Price (3). The model would have rested on a pin in the polar axis, but we are rarely told whether it was a ball or a square flat plate—either figure would have had support from one or other of the ancient Chinese cosmologists (cf. Vol. 3, pp. 211, 217).

[g] Note that this is one of the cases where the instrument was certainly of bronze, and therefore heavy.

[h] Note the philosophical term used—'resonance'—cf. Vol. 2, pp. 285, 304.

[i] Ch. 23, p. 8*b*; parallel text in *Sui Shu*, ch. 19, pp. 17*b* ff.; *KYCC*, ch. 1, p. 23*b*; *Yü Hai*, ch. 4, p. 18*b*.

[j] He also made, *inter alia*, a wooden celestial globe.

[k] *Sui Shu* and *Yü Hai* texts.

[1] 錢樂之   [2] 張衡   [3] 渾儀   [4] 陸績   [5] 王蕃
[6] 置立漏刻以水轉儀   [7] 與天相應   [8] 義熙起居注   [9] 劉義隆

Chhen by the Sui, they were taken to Sian, and then in +605 moved again to the observatory at the Eastern Capital (Loyang). Chhien's 'pre-clock' was therefore almost certainly known to Kêng Hsün and Yüwên Khai, and having withstood so many political upheavals it was probably still available for study by I-Hsing and Liang Ling-Tsan when the College of All Sages in the latter city became the scene of their great invention. It united them with Chang Hêng in no more than two stages.

Another intermediate figure, however, of whom we know was Ko Hêng.[1] In the San Kuo State of Wu (+222 to +280):

> there was also Ko Hêng[a] who was a perfect master of astronomical learning and capable of making ingenious apparatus. He altered the astronomical instrument (*hun thien*[2]) in such a way as to show the earth fixed at the centre of the heavens,[b] and these were made to move round by a mechanism (*i chi tung chih*[3]) while the earth remained stationary.[c] (This demonstrated) the correspondence of the (shadows on the) graduated sundial with the motions (of the heavens) above (*i shang ying kuei tu*[4]). This it was which Chhien Lo-Chih (also) imitated.[d]

We are now within a century of our goal, the work of Chang Hêng[5] in the Later Han period. For he was the first of this line of men who accomplished the continuous slow rotation of astronomical instruments (globes or demonstrational spheres) with the best approximation which they could make to constancy of speed. To readers of this book, Chang Hêng (+78 to +142) is a familiar figure; in Volume 3 there was hardly a Section in which he did not appear (mathematics, astronomy, cartography, etc.).[e] Particularly relevant is the seismograph which he set up at the capital in +132, the first instrument of the kind in any civilisation. The ingenuity of this device, with its inverted pendulum, which continued in use for many centuries afterwards, was so striking that there is nothing inherently improbable in his application of water-power to a drive for an astronomical instrument.

Our two most explicit sources on this matter both date from the Thang, though compiled of course on the basis of ancient documents then existing. One comes from the *Sui Shu*[6] (History of the Sui Dynasty), written about +656 by Wei Chêng[7] and others, the second from Fang Hsüan-Ling's[8] *Chin Shu*[9] (History of the Chin Dynasty) of +635. These two works stand out prominently among the official histories for the length and excellence of their chapters on astronomy and calendrical science.[f] As the texts complement each other, we give them both. The *Sui Shu* says:[g]

---

[a] Perhaps a relative of Ko Hung[10] (+280 to +360) the great alchemist, or his uncle Ko Hsüan[11] (fl. +238 to +255) the Taoist.　　　　　[b] See p. 483 (f), above.

[c] Parallel passage from Sun Shêng's[12] *Chin Yang Chhun Chhiu*[13] quoted in *TPYL*, ch. 2, p. 10a; and, much earlier, in Phei Sung-Chih's commentary in *San Kuo Chih*, ch. 63, p. 5b.

[d] This paragraph is in the *Sui Shu* and *Yü Hai* texts just cited.

[e] See pp. 100, 343, 537. Biographies of him have been written by Chang Yü-Chê (1, 2); Sun Wên-Chhing (3, 4); Li Kuang-Pi & Lai Chia-Tu (1); Lai Chia-Tu (2).

[f] It is particularly to be noted that these Thang records cannot have been projections back to Chang Hêng's time of ideas inspired by I-Hsing's clock, for both were written in the previous century.

[g] Ch. 19, pp. 14b, 15a. The first statement occurs also on p. 2a, and the first and second also on p. 7b, with almost identical wording. Tr. auct.

[1] 葛衡	[2] 渾天	[3] 以機動之	[4] 以上應晷度	[5] 張衡
[6] 隋書	[7] 魏徵	[8] 房玄齡	[9] 晉書　　[10] 葛洪	[11] 葛玄
[12] 孫盛	[13] 晉陽春秋			

...The Astronomer-Royal Chang Hêng again (cast) a bronze instrument (*thung hun i*[1]) on the scale of 0·4 inch to the degree, its circumference being 14·61 ft. It was placed in a closed chamber (*yü mi shih chung*[2]) and rotated by the water of a clepsydra (lit. dripping water) (*i lou shui chuan chih*[3]). One observer watched it behind closed doors and called out to another observer who was looking at the heavens on the observatory platform, saying when such and such a star should be rising, or making its transit, or setting, and everything corresponded like the two halves of a tally.

Then the account in the *Chin Shu* runs:[a]

In the time of the emperor Shun Ti (+126 to +144) Chang Hêng constructed a celestial globe (or more probably a demonstrational armillary sphere; *hun hsiang*[4]), which included the inner and outer circles (*nei wai kuei*[5]),[b] the south and north celestial poles, the ecliptic and the equator, the 24 fortnightly periods, the stars within (i.e. north of) and beyond (i.e. south of) the 28 *hsiu* (equatorial lunar mansions), and the paths of the sun, moon and five planets. The instrument was rotated by the water of a clepsydra (lit. dripping water) (*i lou shui chuan chih*[3]) and was placed inside a (closed) chamber above a hall (*yü tien shang shih nei*[6]). The transits, risings and settings of the heavenly bodies (shown on the instrument in the chamber) corresponded with (lit. resonated with)[c] those in the (actual) heavens (*hsing chung chhu mu yü thien hsiang ying*[7]), following the trip-lug (*yin chhi kuan li*[8]) and the turning of the auspicious wheel (*yu chuan jui lun*[9]).

The general picture is thus quite clear. The procedure was for two observers to compare the indications of the mechanised sphere with the celestial phenomena actually occurring. But as to the machinery employed, only the last sentence gives a clue.[d]

  [a] Ch. 11, p. 5a, tr. auct. adjuv. Ho Ping-Yü (1). Earlier in the same chapter (p. 3b) the *Chin Shu* quotes from some lost work of Ko Hung's,[10] c. +330, which is verbally almost identical with the last two sentences of the *Sui Shu* version. This evidence is much older than the Thang (cf. Vol. 3, p. 218).
  [b] These were probably declination-circles of perpetual apparition and invisibility.
  [c] Note again the use of this philosophically significant technical term (cf. Vol. 2, p. 304).
  [d] The appearance of the word *lun* (wheel) here is of great interest. This is the only text relating to Chang Hêng's device in which it occurs. Perhaps it was called 'auspicious' because it represented, as we shall immediately explain, a great advance in practical technique ancillary to calendar-making. But if *jui* were emendable to *tuan*,[11] so that we could read 'upright wheel', it would make good sense too. Our translation of *kuan-li* as trip-lug is a bold guess, and requires some justification. It must be some technical term, but the lexicographers throw no light on it. That *kuan*[12] was the word used from six to ten centuries later for the links of the mechanical escapement we know, but we hesitate to take back that precise meaning to the present early date. Another use of this form of *kuan-li*[8] occurs in the account of Han Chih-Ho's flying automata about +890, where it refers to the mechanism within their bodies (*Tu Yang Tsa Pien*, ch. 2, p. 8b); cf. p. 163 above. But one finds a much more precise connotation of it in the *Sung Shih*'s description of the south-pointing carriage and the hodometer made by Wu Tê-Jen in +1107 (ch. 149, pp. 15b, 17b), where it can only mean a projecting lug, in fact the pins of pinions of one, two or three. The point is clinched by the fact that the term *thieh po tzu*,[13] 'iron trip tooth', appears alongside it as an alternative expression. For further details on these machines, see pp. 292 and 284 above. Phrases somewhat similar have been met with elsewhere, though their orthography slightly differs. For example, *kuan-li*[14] is the term used by Chou Chhü-Fei[15] in his *Ling Wai Tai Ta*[16] (Questions about what is beyond the Passes) of +1178 for a fish-shaped movable stopper of silver placed inside a bamboo tube through which the tribespeople of the south-west drank wine in their ceremonies (cf. Vol. 3, p. 314). Another form of *kuan-li*[17] was a term used for the interlocking pivots of

[1] 銅渾儀	[2] 於密室中	[3] 以漏水轉之	[4] 渾象
[5] 內外規	[6] 於殿上室內	[7] 星中出沒與天相應	[8] 因其關戾
[9] 又轉瑞輪	[10] 葛洪	[11] 端	[12] 關
[13] 鐵撥子	[14] 關捩	[15] 周去非	[16] 嶺外代答
[17] 關棙			

It is of course unsatisfactory that our chief sources[a] should all date from two to five centuries after Chang Hêng himself.[b] However, contemporary evidence of his achievement does exist, and four pieces of it may be mentioned. First, two later writers at least quote directly from his own books, which in their time were still existing. Thus Su Sung wrote in his memorial (+1092):[c] 'Chang Hêng in his *Hun Thien*[1] (On the Celestial Sphere) says that (one instrument) should be set up in a closed room and rotated by water-power....' And much earlier, about +750, Wei Shu wrote of I-Hsing's clock that when people discussed it they said that what Chang Hêng described in his *Ling Hsien*[2] (Spiritual Constitution of the Universe) could have been no better.[d] Of Chang's two books, finished in +118, we have today only a few fragments and these do not contain the description of the apparatus, but it is tolerably certain that both were fully available until the end of the Thang,[e] and one of them may well have lasted until the collapse of the Northern Sung in +1126, so that Su Sung could have read it.

Besides this, two fragments of Chang Hêng's concerning clepsydra technique were preserved by Thang writers. We have already given them,[f] and need not repeat them, for their content is not their most important feature in the present argument. The first occurs in the *Chhu Hsüeh Chi*[3] (Entry into Learning) encyclopaedia[g] compiled by Hsü Chien[4] in +700; it simply describes the inflow clepsydra with a compensation-tank. The second occurs also in the *Wên Hsüan*[5] anthology, in a commentary[h] by Li Shan[6] of about +660, and mentions the ancestors of jack-work, small statuettes cast on the lids of the inflow vessels, which guided the indicator-rods with their left hand and pointed to the graduations on them with their right. But the important thing is that both Hsü Chien and Li Shan quote these pieces from what appears to be the title of a book, *Lou Shui Chuan Hun Thien I Chih*[7] (Apparatus for Rotating an Armillary Sphere by Clepsydra Water). More probably this title was that of a chapter in Chang Hêng's *Hun I*[8] or *Hun I Thu Chu*[9] (+117), or perhaps of an important appendix of it. In any case, the great collectors of fragments were careful to retain

the rings of Cardan suspensions in lanterns (cf. p. 235). But it also appears in a parallel description of Wu Tê-Jen's mechanical vehicles, namely the *Khuei Than Lu* (ch. 13, pp. 4a, 5b) by Yo Kho, where it is evidently only another way of writing *kuan-li*.[10] Semantically the range of meanings of both words together—connecting, pushing, and inserting something against resistance—covers very well the function of the trip-lug. What it would have had to do here was to push on a geared ring or wheel by one tooth each time that sufficient weight of water had accumulated in one of the scoops on the wheel to force the shaft round against the resistance of the astronomical instrument.

[a] Including the quotation from Ko Hung.

[b] Yet much of what we know of the views of the pre-Socratic philosophers (−6th to −4th centuries) seems to come from quotations in patristic writers as late as the +4th century, and European classical scholars show little hesitation in accepting it.

[c] *Hsin I Hsiang Fa Yao*, ch. 1, p. 3a.   [d] Cf. p. 472 above, for text.

[e] See the bibliographies, *Sui Shu*, ch. 34, p. 15a; *Chiu Thang Shu*, ch. 47, p. 5b; *Hsin Thang Shu*, ch. 59, p. 12b.

[f] In Section 20g (Vol. 3, p. 320).   [g] Ch. 25, pp. 2a, 3a.

[h] Ch. 56, p. 13b. The commentary is on the *Hsin Kho-Lou Ming*[11] (Inscription for a New Clepsydra) written by Lu Chhui[12] in +507.

[1] 渾天	[2] 靈憲	[3] 初學記	[4] 徐堅	[5] 文選	[6] 李善
[7] 漏水轉渾天儀制		[8] 渾儀	[9] 渾儀圖注		[10] 關戾
[11] 新刻漏銘		[12] 陸倕			

this caption when they reproduced the quotations.[a] Again, therefore, the claim to have succeeded in constructing a machine of this kind seems to be Chang Hêng's own, and not something fathered on him by later generations.

The fourth piece of evidence is the most interesting, for though telling us nothing of the mechanism, it reveals the whole principle and procedure for which it was devised. It is a passage in one of the Later Han apocryphal books,[b] the *Shang Shu Wei Khao Ling Yao*[1] (Apocryphal Treatise on the Historical Classic; the Investigation of the Mysterious Brightnesses). Its most probable date is about contemporary with Chang Hêng. It runs as follows:[c]

If the (demonstrational armillary) sphere (*hsüan chi*[2]) indicates a meridian transit when the star (in question) has not yet made it, (the sun's apparent position being correctly indicated), this is called 'hurrying' (*chi*[3]). When 'hurrying' occurs, the sun oversteps his degrees, and the moon does not attain the *hsiu* (mansion) in which she should be. If a star makes its meridian transit when the (demonstrational armillary) sphere has not yet reached that point, (the sun's apparent position being correctly indicated), this is called 'dawdling' (*shu*[4]). When 'dawdling' occurs, the sun does not reach the degree which it ought to have reached, and the moon goes beyond its proper place into the next *hsiu*. But if the stars make their meridian transits at the same moment as the sphere, this is called 'harmony' (*thiao*[5]). Then the wind and rain will come at their proper time, plants and herbs luxuriate, the five cereals give good harvest and all things flourish.[d]

With this it is interesting to compare the comments of Su Sung, who quoted it towards the end of his memorial.[e]

From this we may conclude that those who make astronomical observations with instruments are not only organising a correct calendar so that good government can be carried on (i.e. the administration of agricultural society), but also (in a sense) predicting the good and bad fortune (of the country) and studying the (reasons for) the resulting gains and losses.

Thus he rationalises the common prognosticatory significance attributed by lay folk to the activities of astronomers in medieval China. Not astrological presage but sound calendrical science will make a country prosperous, he says.

Here then is the key to the apparently mysterious arrangement of having one sphere in a closed room and another on the observatory platform. The calendar-making astronomers of Chang Hêng's time were very much concerned with all divergences or discrepancies between the indicated positions of the stars on the one hand and those

---

[a] E.g. in +1267 Wang Ying-Lin in *Yü Hai*, ch. 4, p. 9b, ch. 11, p. 7a. In the 19th century Ma Kuo-Han in *YHSF*, ch. 76, p. 68a, and Yen Kho-Chün in *CSHK* (Hou Han sect.), ch. 55, p. 9a.

[b] On these see Sect. 14f (Vol. 2, pp. 380 ff.)

[c] *Sui Shu*, ch. 19, p. 13b. A preliminary discussion of this passage has already been given in Sect. 20g (Vol. 3, p. 361). Tr. auct. adjuv. Maspero (4).

[d] One late version of the text (*Ku Wei Shu*, ch. 2, p. 2b) made it read in the opposite sense so that the 'hurrying' and 'dawdling' referred to the stars and not to the demonstrational, or rather computational, instrument. This misled Maspero (4), p. 338, into thinking that the whole arrangement was astrological. On the contrary, its calendrical purpose is very clear. We are much indebted to Dr D. Price for elucidating this with us.

[e] *Hsin I Hsiang Fa Yao*, ch. 1, p. 5a.

[1] 尚書緯考靈耀   [2] 璇璣   [3] 急   [4] 舒   [5] 調

of the sun and moon on the other (e.g. the equation of time, lunar perturbations, etc.). If, as is rendered highly probable by several of our texts,[a] the system was to have small objects representing sun, moon and planets attached in some way to the sphere or globe, yet freely movable thereon (for example, beads on threads), then the computer inside the room would adjust their positions in accordance with the predictions of the calendar currently in use. These formulae could then be tested by having the computer say what ought to be happening, whereupon the observer would if necessary correct him. Thus the calendar could be checked. The *Chin Shu* says in two places that it was the computer within who normally spoke first.

When Su Sung built his clock-tower in +1088 his celestial globe within the upper storey and his armillary sphere (still observational though mechanised) on the platform above were perfectly conscious allusions to the precedent established by Chang Hêng so long before. And although a pure time-keeping function now occupied the ground floor, Su Sung retained the graphic models of former centuries. For he tells us that

there are also pearls in different colours denoting the sun, moon and five planets, threaded on silk strings attached at each end by hooks and rings to the south–north axis. Following the waxing and waning, tarrying and hurrying, stopping and retrograding, and all the motions of the seven luminaries, the pearls are made to occupy their corresponding positions.[b] They are rotated day and night following the movements of the heavens. An observer watching the pearls verifies whether the position of a luminary which they indicate agrees with what is observed and measured on the platform (above). If there is no difference (the calendrical formulae are) considered correct; (if there is a difference, the calendar calculations are adjusted).[c]

Thus the whole system was a graphic method of detecting any discrepancies between the motions of the sun, moon and stars, the fundamental, though always incommensurable, regularities on which all calendars had to be based.

The elucidation of this procedure gives us another opportunity of studying the social context of the inventions of the 'pre-clock' (if we may so term Chang Hêng's device) and the escapement clocks. As has already been mentioned,[d] the promulgation of the official calendar was one of the most important acts of the Chinese emperor, and about one hundred of these were issued from the first unification of the empire in the −3rd century until the end of the Chhing dynasty in the +19th. Each bore a specific name consisting of two or three characters, and there is a wealth of information about the dates of their introduction and the astronomers who compiled them. Unfortunately, we lack any masterly survey of the whole in a Western language[e] which would summarise the story, distinguishing between those calendars which simply involved new recensions of existing tables using new saecular terms or radices in the calculations, and those which depended on new measurements so that new constants could be established and new tables made. However, the question may be asked whether there

---

[a] *Sung Shih*, ch. 48, pp. 3*b* ff. concerning Chang Ssu-Hsün; *Chiu Thang Shu*, ch. 35, pp. 1*a* ff.; *Sung Shu*, ch. 23, pp. 8*b* ff. and *Sui Shu*, ch. 19, pp. 17*b* ff. concerning Chhien Lo-Chih.

[b] By manual setting.   [c] *Hsin I Hsiang Fa Yao*, ch. 1, p. 4*b*, tr. auct.

[d] In Sect. 20*c* (Vol. 3, pp. 189 ff.).

[e] The abundant literature, most of which is in Chinese and Japanese, has been referred to in Vol. 3, pp. 390 ff.

was any relation between the horological inventions and the frequency of introduction of new calendars. Were the inventions connected with what one might call periods of 'calendrical uneasiness'?

A preliminary answer is not at all difficult. There are several convenient lists of the Chinese calendars, notably one given by Chu Wên-Hsin (*1*). If we plot the number of new calendars introduced each century between −400 and +1900 we obtain the

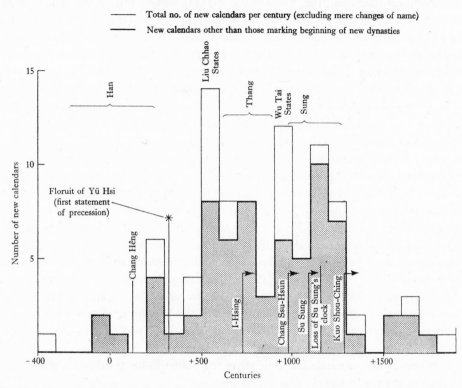

Fig. 663. The relation of calendar-making to the invention of the escapement; plot of the number of new calendars introduced in each century between −400 and +1900. Data from Chu Wên-Hsin (*1*), orig. graph. For interpretation, see text.

graph[a] shown in Fig. 663. The maximum can then be seen to have occurred in the +6th century, when no less than fourteen new calendars were introduced. But a correction is necessary. In order to approach our objective of true astronomical activity it is necessary first to exclude all new calendars which were merely changes of name, and secondly to exclude all those which were introduced to mark the inauguration of new dynasties. Since it was customary for each new ruling house to honour itself in this way, we cannot accept them as evidence of any unavoidable astronomical requirement. We thus have an array of smaller columns. Even these cannot be taken as the true index which we are seeking, for new calendars were often introduced to signalise

[a] Cf. the analogous graph made by Price (6) for the intensity of production of astrolabes in Islam and Christendom between +900 and +1900, or the analysis by Zinner (5) of the categories of printed astronomical books between +1446 and +1630.

new individual accessions within the same dynasty or even changes of reign-period within the same reign. But the corrected columns may be regarded as a rough baro-meter of astronomical activity, for calendrical renewals at these shorter intervals by no means always occurred, and would indeed have been so inconvenient that one may assume that they often resulted from astronomical discussions and controversies.[a]

The general picture seen in Fig. 663 is one of a rise to a double maximum (the +6th to the +8th centuries and the +9th to the +13th), followed by a fall. The obvious interpretation is that the problems were beginning to be posed in the Han, and that they had mostly been solved by the Chhing. Chang Hêng's invention occurs in the preparatory Han period, at a time of few calendars, but one must remember that he was setting in motion a technique which would have to be followed for many years to bring useful results. On the other hand, the indubitable use of the escapement by I-Hsing comes just towards the latter part of the burst of 'calendrical uneasiness' and calendar-making activity of the Liu Chhao and Thang periods. The great clocks of later times all come within the similar period of the Sung.

When the same material is plotted from +200 to +1300 in 25-year periods (Fig. 664), the phases acquire further clarity. There is a relative quiescence to about +500, but from then onwards (apart of course from the spate of inaugurations) there is at least one calendar in each quarter-century and once as many as four. The two or three preceding centuries had been a time of rapidly growing experimentation with astronomical instruments such as armillary spheres.[b] Now also the foreign importations of Buddhism were beginning to take effect, and astronomical ideas were certainly among them. But the calendrical activity of this time belongs to a little-known period and deserves much further study.[c] In any case, the uneasiness stimulated by foreign influences greatly increased in the Thang, and at the beginning of the +8th century the merits of other calendars—Indian, Persian, Sogdian—were hotly debated at the capital.[d] This was the very time at which I-Hsing and Liang Ling-Tsan, as if in answer to desperate demands for some more truly time-keeping machine, made their invention of the water-wheel linkwork escapement.

Towards the end of the Thang things were again quieter, but with the establishment of the Sung discrepancies must have become obvious once more,[e] and we find the clock of Chang Ssu-Hsün (+975) coinciding with a quarter-century in which no less than three calendars were produced. Between +925 and +1225 no quarter-century passed without at least one new calendar, so the clock of Su Sung (+1088) finds its place naturally in the period. The all-time record of six calendars between +1175 and +1200 must almost certainly be a reflection of the harm done to astronomical science

---

[a] Frequently we know that this was the case—as in I-Hsing's own time.

[b] See Table 31 in Vol. 3, and the accompanying discussion.

[c] On the work of men such as Hsintu Fang, Chen Luan, Liu Hsiao-Sun and Chang Mêng-Pin, all good mathematicians as well as astronomers, see Vol. 3, pp. 116, 205, 358, 394.

[d] This was the time when the three families of Indian astronomical mathematicians, to whom belonged, for instance, Chhüthan Hsi-Ta, were established at Chhang-an (see Vol. 3, p. 202). Buddhist scholars were by no means always at one in this period, for I-Hsing was opposed by Chhüthan Hsi-Ta and the Chinese lay astronomer Nankung Yüeh (see Vol. 4, pt. 1, p. 53).

[e] Perhaps the inauguration calendars of the Wu Tai period encouraged calendrical computers.

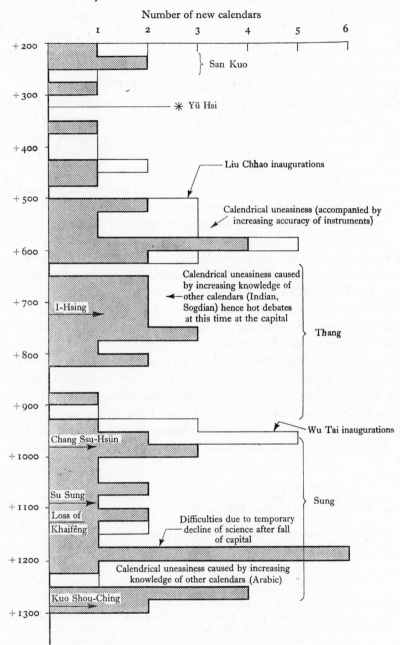

Number of new calendars

Fig. 664. The relation of calendar-making to the invention of the escapement; plot of the number of new calendars introduced in each 25-year period between +200 and +1300. Data from Chu Wên-Hsin"(1), orig. graph. For interpretation, see text.

by the fall of the Sung capital to the Chin Tartars in +1126. We shall have more to say about this shortly. Then in the last part of the +13th century there ensues a burst of activity which is readily explained by Arabic influence analogous to the earlier Indian and Sogdian. Eventually comes the lethargy of the Ming and the beginnings of unified world astronomy with the Jesuits in the early Chhing.

Thus side by side with considerations arising from the private life of the imperial family[a] we can place the needs of calendrical computation. It looks as if the invention of Chang Hêng about +120 derived from the growing doubts which had led Hipparchus to the discovery of the equinoctial precession in −134 and were to lead Yü Hsi to state the same doctrine about +320. But by +700 the more accurate measurement of time had become a burning problem, and the social need of a predominantly agrarian culture for an accurate calendar as well as the intrinsic evolution of astronomical science itself led to the answer found by I-Hsing. It is indeed a strange conclusion that an apparatus so deeply enrooted in Western mechanical industrial civilisation as the clock should have originated in connection with the calendar required by Eastern agricultural people. But other perspectives no less remarkable must be mentioned. In the present work it has often been emphasized that Chinese astronomy was founded and built on a polar and equatorial system while Hellenistic astronomy was primarily ecliptic and planetary.[b] Each had its peculiar advantages and its corresponding triumphs. If Hipparchus was able to state the fact of precession four and a half centuries before Yü Hsi, it was because he was measuring and comparing star positions on ecliptic co-ordinates, and therefore it became evident that their distances from the equinoctial points had changed. But if astronomical instruments were rotated mechanically by Chang Hêng fifteen centuries before the conception of the clock-drive arose in Renaissance Europe,[c] and if in this I-Hsing with his well-documented and successful mechanised time-keeping had priority of

---

[a] Cf. p. 477 above.

[b] See Sect. 20 (Vol. 3, pp. 229, 266 ff.).

[c] As we saw in Vol. 3, pp. 362, 366, the automatic rotation of an observational instrument in Europe was first suggested by Robert Hooke in +1670 and effected by J. D. Cassini using an armillary sphere eight years later. Not until 1824 was an equatorial telescope rotated by clockwork; this was due to Joseph Fraunhofer. The automatic rotation of demonstrational instruments arose rather earlier, but not before the +16th century; cf. H. Werner (1), Crommelin (5). One of the first was built by Juanelo Torriano for Charles V of Spain about +1540 (cf. Morales, 1), but we know very little of it. In his interesting paper on instruments ancestral to the planetarium, Michel (16) illustrates a fine mechanised celestial globe made by Christopher Schissler the younger about +1580, showing automatically the motion of the sun on the ecliptic. From the provisional census published by von Bertele (2, 6) the earliest extant European clockwork-driven demonstrational armillary seems to be that of Josias Habrecht (Fig. 665) dating from +1572. During this period four or five other very splendid ones were made by Jost Burgi, the ingenious colleague of Tycho and Kepler in Prague (see von Bertele, 1, 3). Most of the earlier instruments, such as that of Edward Wright about +1613, were of course constructed on the geocentric system. The oldest extant heliocentric one seems to be the wonderful Gottorp terrarium of +1651 by an unknown maker (Fig. 666), described by von Bertele (2). Another Gottorp instrument, (now in Leningrad), a giant globe 11 ft. in diameter, terrestrial without and celestial within, having seats for observers inside, is of particular interest from the point of view of the Chinese story, since it was rotated by a water-wheel. What escapement mechanism (if any) its builder, Olearius, provided for it about +1660 we should very much like to know, but unfortunately all record of this is lost. The design of Berthoud (Fig. 667) is typical of late 18th-century practice (cf. Vol. 3, Fig. 177, and von Bertele (2), figs. 23a, b, 27a), but is still based on essentially the same principles as that of I-Hsing a thousand years before.

PLATE CCLVIII

Fig. 665. Clockwork-driven demonstrational armillary sphere by Josias Habrecht, made in Strasburg in +1572; probably the oldest extant European instrument of the kind (von Bertele, 2). The sphere, which is constructed on the geocentric system, is rotated by the adjustable train in the casing on the right which works a polar-axis drive. Provision is made for setting the sphere correctly for the latitude of the place. The terrestrial globe is fixed in the centre (cf. the Korean instrument of Vol. 3, Fig. 179, not very different in date), and dials around the north pole indicate sidereal time together with the age and phase of the moon. The dial-faces in the base, like those of the de Dondi clock of +1361, report astronomical positions. Diam. of globe 6¼ in.; location, Rosenborg Castle, Copenhagen.

PLATE CCLIX

Fig. 666. The Copernican terrarium made for the Bishop of Lübeck in +1651 by an unknown master, and later in the possession of the Dukes of Gottorp; probably the oldest extant European heliocentric instrument (von Bertele, 2). The discoidal lid of the vase-shaped case revolves once a year round the vertical axis, which extends beyond the disc and carries a small gilt sphere representing the sun. At some distance from the centre there is another disc through the centre of which rises a shaft carrying a relatively large terrestrial globe surrounded by a number of rings. The globe revolves once a day upon this shaft as it is carried round by the annual motion. A moon model, held in the carrier seen on the left of the earth's disc, demonstrates the lunar phases and periods. Meanwhile the earth and moon models are raised and lowered during their circular motions round the sun and earth respectively by means of a number of elliptical wheels and cams, so as to show the varying inclinations of sun and moon to any point of the earth's orbit. Diam. of discoidal top, 2 ft.; location, Frederiksborg Castle National Museum, Denmark.

PLATE CCLX

Fig. 667. Late +18th-century design for a heliocentric demonstrational armillary sphere by Berthoud (1);
cf. the Jesuit example in the *Huang Chhao Li Chhi Thu Shih* of +1759 (Vol. 3, Fig. 177). The ecliptic
is horizontal and the planetary models rotate in that plane.

nine or ten, it was because Chinese astronomers thought always in terms of equatorial co-ordinates and therefore of declination parallels. Along these tracks all stellar revolution proceeds, but ecliptic latitude and longitude are only an abstract network made by man, a geometrical waste-land, along the lines of which nothing is ever seen to move. Well might Donne say:

> For of Meridians, and Parallels,
> Man hath weav'd out a net, and this net throwne
> Upon the Heavens, and now they are his owne....[a]

In China therefore it was an entirely natural thought to arrange the rotation of a celestial globe or a demonstrational armillary sphere if the plan promised to be useful.[b] What was not perhaps so easy was how to do it.

This subsection may end, then, with a few words to set against their proper background the means which we think Chang Hêng took to solve his problem. To harness the waste dripping of clepsydra water there was one obvious recourse, the art of the millwrights. In +2nd-century China their workmanship was doubtless primitive, but during the previous century the water-powered trip-hammer (*shui tui*[1]) had come into widespread use.[c] Although the most characteristic form of mill-wheel in China later was the horizontally mounted type, the vertical 'Vitruvian' form always persisted,[d] and for the trip-hammer was the more suitable of the two. Moreover, metallurgical blowing-engines powered by water (*shui phai*[2]) had become common in China during the +1st century. If the Chinese vertical water-wheel really derived from the noria,[e] the transition had been made a good while before Chang's time, so that his chief originality lay in arranging for a constant drip into scoops rather than a strong flow and fall on to paddles. The trip-lug on his shaft merely corresponded to those which worked the grain-pounding trip-hammers all around him. However, there was originality also in making it push each time the tooth of a ring or gear-wheel,[f] probably bearing leaf teeth, and controlled by a ratchet.[g] There seems nothing at all in the arrangement which would have been beyond the powers of Han technicians.[h] The trip-lug was their own,[i] though to use it as a pinion of one was rather Alexandrian. You

[a] *An Anatomie of the World; the First Anniversary* (+1611).

[b] How paradoxical it is, therefore, that traditionally the Greeks receive all the credit due to the first inventors of mechanically rotated astronomical instruments. 'It was necessary', writes Michel (16), discussing the limiting factors of this development, 'to await the geometrical genius of the Greeks, with their faculty of "seeing through space", in order to attain the resolution of the mechanics of the heavens as a logical combination of circular motions.' He is thinking, of course, of the Ptolemaic theory, with all its complexities, but in fact the Greek achievement in practice was confined to the rotation of flat discs by sinking floats, and it is unlikely that even the Anti-Kythera planetarium (if powered at all) or that of Archimedes (see p. 534 below) worked on any other principle. There is no doubt that the Greeks were the more successful theorists, but equally none that the Chinese, on the whole, were the better practical instrument-makers.

[c] Cf. pp. 390 ff.   [d] Cf. pp. 405 ff.   [e] Cf. pp. 356 ff.

[f] For the evidence indicating an extensive knowledge of gear-wheels in Han China cf. pp. 85 ff. above.

[g] Cf. Fig. 391 *a, b* and pp. 86 ff.

[h] One remembers the very elaborate puppet theatre constructed less than a century after Chang Hêng's death by the engineer Ma Chün[3] (cf. p. 158). The motive power for that was a water-wheel and there must have been plenty of trip-lugs.   [i] Cf. pp. 82, 381.

[1] 水碓   [2] 水排   [3] 馬鈞

may compare it rather with the peg on the axle of Heron's hodometer [a] in the previous century than with those on the shaft of his organ-blowing windmill.[b] What connections there could have been between the engineers of Alexandria and Han China remains of course a completely unsolved question.[c] It might have been a case of carrying coals to Newcastle, for as we have seen above (pp. 281 ff.) the Chinese hodometers were contemporary, and their acceptable reconstructions involve small pinions of one, two or three teeth.[d]

Presumably Chang Hêng's simple machine was at rest during each period when water was slowly accumulating in one of the scoops. As soon as enough had collected, its weight overcame the resistance of the toothed wheel and armillary sphere, and the trip-lug turned it round by one tooth, then coming to rest against the next. Although we are told optimistically that 'everything agreed like the two halves of a tally' we are bound to assume that the chronometric properties of the device were extremely poor.[e] So much of it would have depended upon play and resistance, the exact size of each scoop, the nature of the bearings of the polar axis, and similar factors. Very probably it was maintained in regular motion only with some difficulty, and perhaps the successive astronomers who made instruments of the same kind were all searching for the right conditions for doing what no one until I-Hsing was really able to do. Or perhaps they made the machine work better than we are inclined to believe. In any case it is hard to exclude Chang Hêng's apparatus absolutely from the horological definition of van Bertele with which this discussion began. The problem of securing slow and regular motion by intersecting the progress of a powered machine into intervals of equal duration was not solved in any Alexandrian design as far as we know, and if Chang Hêng did not solve it himself, he opened the gate to the path which led in the end to its solution.

(4) FROM SU TZU-JUNG TO LI MA-TOU; CLOCKS AND THEIR MAKERS

We now return, as promised, to our focal point, the great clock built by Su Sung (Su Tzu-Jung) in +1088, in order to start again from there and trace the subsequent developments down to the time of the coming of Matteo Ricci (Li Ma-Tou) and his Jesuit horologists. This is a space of almost exactly 500 years, and contrary to all former belief, they were filled with clock-making in China, the success of the mechanicians varying with time and place.

The most obvious question to begin with is what fate befell Su Sung's own clock-tower and the machinery inside it? It so happens that we are rather well-informed

---

[a] B & M, pp. 514 ff. from *Dioptra*, ch. 34. Cf. p. 285 above.

[b] Woodcroft (1), p. 108, from *Pneumatica*, ch. 77. Cf. p. 556.   [c] Cf. Vol. 1, p. 197.

[d] Wang Chen-To (3); Liu Hsien-Chou (6). The latter draws attention to the possible significance of the drum-beating puppets in the hodometer for the jack-work of the later clocks. Nor should the pointing figure of the south-pointing carriage (pp. 286 ff. above) be forgotten in this connection.

[e] There is always the possibility, however, that the accounts of the performance of Chang Hêng's apparatus were quite literal rather than optimistic; cf. the proviso on p. 474 above. A new solution of the problem is offered by Liu Hsien-Chou (7), pp. 98 ff., in terms of reduction-gearing (cf. p. 512).

about this, and the answer is worth giving in detail for it throws quite unexpected light on the political importance attained by clock-making at the end of the Northern Sung. There is no doubt that the bronze components[a] of Su's 'Ho Thai' were cast by imperial order in or just before +1090, after a period of trials superintended by one of the Han-Lin academicians, Hsü Chiang.[1] This scholar, however, had comments to make which led to the construction of an entirely new piece of ancillary apparatus by Su Sung, probably the earliest 'planetarium' which one could actually get inside.

The passage is worth citing.[b]

In the third month of the 4th year of the Yuan-Yu reign-period (+1089) the wooden model (of Su Sung's armillary clock-tower) was completed[c]—previously there had been nothing like it. The Han-Lin Academician Hsü Chiang and others were ordered to examine it. (After due trials), on a *chi-mao* day they said: 'Comparison has been made (of the armillary clock) with the dawns and dusks and the motions of the heavens, and there is perfect agreement, so the order has been given to cast it (i.e. the sphere and the globe) in bronze. Let it be called the Armillary Clock of the Yuan-Yu Reign-Period.'

Some time afterwards (Hsü) Chiang and others also said: 'Now what has always been called the armillary sphere has a spherical outer form, so that the (equatorial) constellation (-positions) in degree(-marks) can be distributed round it. Inside there are the concentric rings and the sighting-tube, which can be used for the observation of the heavens. [Thus we have here two separate instruments, the sphere and the globe, but they could be made into a single instrument.][d] Now what has been erected (contains) the two different things, the [turning][e] sphere and the [turning][e] globe, the former for observing the true numbers of the degrees of the heavenly motion, the latter placed within a closed room, performing the celestial revolution by itself, and checking what is observed by means of the armillary sphere. If these two instruments were to be combined into one, the globe would be (part of) the sphere, and together they would (show) correctly the heavenly motions, performing the function of both. [This also would be a contribution to the (astronomical) equipment of the present dynasty.][e] So we request that an armillary sphere (of this kind) be commissioned.' It was decided to do this.

(Su) Sung understood all these matters well because his own family had possessed a small model (globe or sphere). So he asked (Han) Kung-Lien to make the (necessary) elaborate calculations, and after several years the instrument was completed. It was larger than the height of a man, so that one could enter and sit inside it,[f] its structure being like that of a

---

[a] The sphere and the globe.

[b] It is found, with slight differences of wording, both in *Sung Hui Yao Kao*, tsê 53, Yün li ch. 2, pp. 13 b, 14 a and esp. 23 a, b, a source which may be almost contemporary, and in the *Yü Hai* encyclopaedia (+1267), ch. 4, pp. 46 a ff. The second paragraph of our quotation is a slightly condensed form of the text of *Sung Shih*, ch. 80, pp. 25 a ff. The third is a condensation of the account of Chu Pien (see immediately below) in his *Chhü Wei Chiu Wên*, ch. 8, pp. 10 a ff. Only these last two texts were known to Needham, Wang & Price (1), pp. 115 ff., 117, who therefore misinterpreted Hsü Chiang's suggestion (taking it to be a demand for a new observational armillary sphere without a clock-drive), and discounted Chu Pien's story of the large hollow globe (supposing it to be hearsay or lapse of memory). When the accounts are juxtaposed as they are in the *Yü Hai* and the *Sung Hui Yao Kao*, it becomes clear that what Chu Pien says that Su Sung did was in fact the response to what Hsü Chiang asked for.

[c] Cf. p. 464 above.

[d] The passage enclosed in square brackets is not in the *Sung Shih* version.

[e] The words and passage enclosed in square brackets are not in the *Yü Hai* version.

[f] Reading *chü*[2] for *chü*,[3] with Wang Chen-To (9).

[1] 許將   [2] 踞   [3] 居

lantern or a bird-cage (*kou*[1], *lung*[2]) (with bamboo ribs), and the walls (of silk and paper) pierced with holes according to the positions of the stars. It could be rotated by means of a wheel (*chi lun hsüan chuan chih shih*[3]) so that the star culminations for any particular time before dawn or after dusk[a] could all be seen by looking through the holes. The astronomers and calendrical pundits all flocked to watch its operation, and were quite astonished at it, for nothing like it had previously been achieved.[b]

Thus the auxiliary instrument made at Hsü Chiang's suggestion was a planetarium in the sense that a human observer could enter and study an artificial sky, but not in the sense that the motions of the planets could be mechanically demonstrated. Though the language used resembles that applied so often to water-wheel drives, and *chi* has its nuance, the word *shui* (water) is missing in all sources, so the 'planetarium' may have been set by hand to whatever hour-angle desired; as in the elegant reconstruction recently proposed by Wang Chen-To (*9*), who has the observer sitting in a chair which hangs from the main polar axis. Su Sung doubtless met Hsü Chiang to the extent of providing meridian and horizon rings, but the other great circles would be sufficiently marked on the skin of the world-ball.[c] The year +1092 thus marked the apogee of Su's practical achievements, though he had seven more years to live after finishing his book two years later.

Very soon, however, the masterpieces of Su Sung and Han Kung-Lien were menaced by the controversies of the time. From about +1060 until the fall of the capital in +1126 the public life of the dynasty was torn by violent disputes between the party of the Conservatives and that of the Reformers. Here we cannot enlarge upon the issues which divided them, though perhaps it is not too inaccurate to say that the latter were concerned to strengthen at all costs the bureaucratic character of Chinese society, and were therefore opposed alike to the old-fashioned feudal-minded land-owning gentry and to the merchants or small industrialists.[d] As already mentioned,[e] Su Sung, though probably not an active member of the Conservatives, was identified with them through his friendships. So when in +1094 the Reformers came back into power there was talk of destroying his clock-tower. Chu Pien,[4] in his *Chhü Wei Chiu Wên*[5] (Talks about Bygone Things beside the Winding Wei (River, in Honan)), written about +1140, tells us this.[f]

[a] At any particular time of year, of course.

[b] The *Sung Shih* continues (p. 25*b*): 'In the fourth month of the 7th year of the Yuan-Yu reign-period (+1092) Su Sung was asked to write an inscription for the (combined) armillary sphere and globe', i.e. the 'planetarium' globe. And then, 'in the sixth month, the Armillary Clock (Tower) was completed'.

[c] A few pages ago (p. 492) we noted the similar 'planetarium' globe built by Olearius at Gottorp about +1660. Although that was rotated by water-power while Su Sung's may not have been, it is clear from the context that he could easily have provided it with a mechanical drive of that kind if he had wished to do so. One would hardly have expected to find an ancestor of the Gottorp 'planetarium' globe as much as six hundred years before it.

[d] Cf. Sect. 6*h* (Vol. 1, p. 138). An introductory account will be found in Fitzgerald (*1*), pp. 391 ff., but all the histories of the period in Western languages are most unsatisfactory. However, Williamson (*1*), Ferguson (*4, 7*) and de Bary (*1*) may be read with some advantage.

[e] P. 446 above.

[f] Ch. 8, p. 10*a*, tr. auct.

[1] 篝   [2] 籠   [3] 激輪旋轉之勢   [4] 朱弁   [5] 曲洧舊聞

At the beginning of the Shao-Shêng reign-period (+ 1094), Tshai Pien [1] (Minister of State)[a] suggested that it ought to be destroyed as something which belonged to the previous (Yuan-Yu) reign-period. At that time Chhao Mei-Shu [2] was Assistant Director of the Imperial Library,[b] and as he greatly admired the accuracy and beautiful construction of Su Sung's instruments, he struggled to argue against Tshai Pien, but at first his efforts proved unsuccessful. However, he sought the help of Lin Tzu-Chung,[3] who talked to Chang Tun [4] (Prime Minister)[c] and thus the destruction of the clock was averted. However, after Tshai Ching[5] and his brother[d] came into power nobody dared to say anything to prevent Su Sung's machinery being torn down. How shameful!

The idea was, of course, that each new reign-period must 'make all things new', and politicians of that time could hardly be expected to understand the slow growth of practical scientific knowledge.[e] But in fact the clock remained untouched, and continued to tick over the minutes inexorably until those days of + 1126 when the Chin Tartars were at the gates.

Let us follow its fortunes to the end before returning to the political aspects of horology in the Northern Sung. In this fateful year the capital (Khaifêng) was twice under siege. In September it fell, and both emperors (Hui Tsung and Chhin Tsung) were taken away captive to Peking in the north. For a while the princes and the remains of the court wandered about behind the lines, settling first in one place and then in another, till in + 1129 it was decided to choose Hangchow as the new capital. This was what became the 'flower of cities all' that Marco Polo was to see 140 years later. But the blow at the technological supremacy of the Sung had been extremely severe. One of the most notable things about the sieges of Khaifêng had been the fact that the Chin Tartars exacted as tribute in the periodical armistices whole families of artisans and skilled workmen. They demanded from the city all sorts of craftsmen, including goldsmiths and silversmiths, blacksmiths, weavers and tailors, and even Taoist priests.[f] And there is every reason for thinking that when the capital fell, all the clock-making millwrights and maintenance engineers followed the Chin power and migrated to the north. Probably they accompanied the disassembled clock-tower itself. For the *Chin Shih* says:[g]

After the (Chin) dynasty had captured Pien(-ching) (i.e. Khaifêng) all the astronomical instruments were carried away in carts to Yen (modern Peking or more generally the north-

---

[a] + 1054 to + 1112. A leader-member of the Reformers and son-in-law to Wang An-Shih himself.

[b] This is Chhao Tuan-Yen[6] (b. + 1035), worthy representative of a scholarly family.

[c] + 1031 to + 1101. Originally a friend of the poet Su Tung-Pho, but later one of the leading Reformers.

[d] Tshai Ching (+ 1046 to + 1126) was the elder brother of Tshai Pien, and the two together obtained control of the administration in + 1101. Dying in this year, Su Sung must have been very uncertain as to the fate of his clock.

[e] As Zilsel (4) has shown, the idea of a continuing scientific progress was rare anywhere in the world before the European Renaissance, and then it originated rather from the higher artisanate (cf. Vol. 3, pp. 154, 159) than from academic scholars.

[f] *Sung Shih Chi Shih Pên Mo*, ch. 56. This reference, to which allusion has already been made (p. 20), we owe to the kindness of Dr Wu Shih-Chhang.

[g] Ch. 22, pp. 32b ff. tr. auct. A more extensive excerpt is given in translation by Needham, Wang & Price (1), pp. 131 ff.

¹ 蔡卞   ² 晁美叔   ³ 林子中   ⁴ 章惇   ⁵ 蔡京   ⁶ 晁端彦

eastern regions). The celestial (gear-)wheel (*thien lun*[1]), the equatorial gear-ring (*chhih tao ya*[2]), the time-keeping gear-wheel (*chü po lun*[3]), the celestial globe (*hsüan hsiang*[4]), the bells, drums and quarter-striking jacks (*chung ku ssu-chhen kho pao*[5]), the upper reservoir (*thien chhih*[6]), the scoops, sump and tanks (*shui hu*[7]), etc., all broke or wore out after some years. Only the bronze armillary sphere remained in the observatory (*hou thai*[8]) of the (Chin) Bureau of Astronomy and Calendar. But as Pien(-ching) and Yen were more than a thousand *li* away from one another, the (polar) altitude was quite different, so that it was necessary to alter the arrangement, setting the (south) polar pivot four degrees lower.[a]

In the eighth month of the 6th year of the Ming-Chhang reign-period (+1195) there was a terrible storm with rain and wind, thunder and lightning. The armillary sphere, with its dragon columns, cloud-and-tortoise column,[b] and water-level stand, was struck by lightning. The masonry of the observatory tower was also split asunder so that the sphere fell to the ground and was damaged. The emperor[c] ordered the officials to repair it, and it was replaced upon the tower.

In the Chen-Yu reign-period (+1214 to +1216) (the Chin court and people, pressed by the rising Mongol power) crossed the (Yellow) River (fleeing) southwards. It was proposed that the armillary sphere should be melted down to make things, but the emperor[d] did not have the heart to destroy it. On the other hand its bulk was so large that it would have been difficult to transport by cart, so in the end it was left behind....

Thus the 15-ton armillary sphere of Su Sung eventually fell into the hands of the Mongols towards the beginning of their career of conquest. When Peking became their capital in +1264 it was still available for the astronomical officials, but by then it had suffered from the ravages of time and could no longer conveniently be used.[e] The Mongols now commanded the services of the best and greatest Chinese scientific minds of the age. Su Sung's sphere must have been well known to Kuo Shou-Ching[9] before he made his great invention of the equatorial mounting (the 'simplified instrument' or equinoctial torquetum).[f] Kuo himself was, among so many other things, a maker of clocks. We shall therefore also soon return to him, but first let us take another look at that intriguing period, the end of the Northern Sung.

Historians and sinologists have had much to say about it but few if any have observed that it expired in a blaze of horological exuberance. The *Sung Shih* tells us[g] that in +1124 Wang Fu[10] (a minister of one of the princes) memorialised as follows:

In the 1st year of the Chhung-Ning reign-period (+1102) I chanced to meet at the capital a wandering unworldly scholar (*fang wai chih shih*[11]), who told me that his family name was Wang,[12] and gave me a Taoist book (*su shu*[13]), which discussed the construction of astronomical

---

　　[a] Peking is actually just 5° lat. north of Khaifêng.
　　[b] It will be remembered that this was the central column through which Su Sung had led up his driving-shaft (cf. p. 457).
　　[c] Chang Tsung, personal name Wanyen Ching[14] (accession +1189, d. +1208).
　　[d] Hsüan Tsung, personal name Wanyen Hsün[15] (accession +1213, d. +1224). This decision is a credit to his record.
　　[e] *Yuan Shih*, ch. 48, p. 1b.　　　　　　　　　[f] See Section 20 (Vol. 3, p. 370).
　　[g] Ch. 80, pp. 25b ff.; the passage is fully translated in Needham, Wang & Price (1), pp. 119 ff.

[1] 天輪	[2] 赤道牙	[3] 距撥輪	[4] 懸象	[5] 鐘鼓司辰刻報
[6] 天池	[7] 水壺	[8] 候臺	[9] 郭守敬	[10] 王黼
[11] 方外之士	[12] 王	[13] 素書	[14] 完顏璟	[15] 完顏珣

instruments (*chi hêng*[1]) in detail.[a] So afterwards I asked the emperor to order the Supply Department (Ying Fêng Ssu[2]) to make some models to test what it said, and this they did in the space of two months....

Wang Fu then went on to describe the chief instrument, which seems to have been a demonstrational armillary sphere or a celestial globe, with all the usual features[b] and very elaborately graduated. But it was combined with complicated planetarium machinery which showed automatically not only the positions of the sun and moon but also the phases of the latter. The planets too were shown rising, culminating and setting, moving at varied speeds in advance or retrogradation. The text continues:

A jade balancing mechanism (*yü hêng*[3]) (i.e. the escapement) is erected behind (lit. outside) a curtain, holding and resisting (*chhih o*[4]) the main scoops (*shu tou*[5]). Water pours down, rotating the wheel (*chu shui chi lun*[6]). Lower, there is a cog-wheel (*chi lun*[7]) with 43 (teeth). There are also hooks, pins and interlocking rods one holding another (*kou chien chiao tsho hsiang chhih*[8]). Each (wheel) moves the next without reliance on any human force. The fastest wheel turns round each day through 2928 teeth (*chhih*[9]), the slowest only moves by 1 tooth in every 5 days. Such a great difference is there between the speed of the wheels, yet all of them depend on one single driving mechanism. In precision the engine can be compared with Nature itself (lit. the maker of all things; *tsao wu chê*[10]). As for the rest, it is much the same as the apparatus made (long ago) by I-Hsing. But that old design employed mainly bronze and iron, which corroded and rusted so that the machine ceased to be able to move automatically. The modern plan substitutes hard wood for these parts, as beautiful as jade....

This description is one of the most interesting specifications for a Sung astronomical clock that we have, tantalisingly abridged though it is. The use of hard wood is a particularly striking feature.[c] Wang Fu proceeded to mention other technical respects in which the apparatus excelled its Thang predecessors, especially in the time-telling and striking components, which included a 'candle-dragon' (*chu lung*[11]) which spat forth a pearl from time to time into a bronze lotus,[d] and a jack representing the God of

[a] Here is a third Sung book on clockwork, besides those of Su Sung and Juan Thai-Fa (p. 449).

[b] And some additions, such as some way of showing the Khun-lun Mountains (the cosmic mountain, Mt Meru); an insertion obviously showing Buddhist–Taoist influence.

[c] Owing to the prominence of the early iron turret-clocks it is not generally known that wood was widely used for making clock movements in recent centuries in Europe. Even the greatest clock-makers, such as John and James Harrison, made many clocks of this kind, and one of them, dated +1715, is still going in the Science Museum, London; for description see Lloyd (3) and (10), p. 104 and pls. 127, 128. Lloyd (5) tells us that the Harrisons were of a carpenter family. Wood was used to eliminate friction and avoid oil, with lignum vitae (a naturally oily wood) for the pivots. Oak discs were used for the wheels, the teeth being cut with the grain and screwed in along the rims in groups of from two to five. At the firm of Grauwiler in Basel in the spring of 1956 I saw a 17th-century farmhouse clock with verge-and-foliot escapement, all made of wood except for the weights, pallets, crown-wheel edges, and pins in the lantern pinions. Descriptions of such wooden clocks are often found—for the Black Forest cuckoo clocks and Austria, Kaftan (1); for France, Planchon (2); for New England, Morrell (1). In view of the considerable splashing which must have occurred in the Chinese water-wheel clocks, the use of a light and non-corroding material was especially suitable.

[d] This form of auditory signal is extremely reminiscent of the striking water-clocks of the Byzantine and Arabic culture-areas. The description of Antioch in the *Chiu Thang Shu*, ch. 198, pp. 16a ff., includes an account of such a system (cf. Hirth (1), pp. 53, 213). This would have been written just before +945 and was repeated in later dynastic histories (cf. Vol. 1, pp. 193, 203). We have already

[1] 璣衡	[2] 應奉司	[3] 玉衡	[4] 持扼	[5] 樞斗	[6] 注水激輪
[7] 機輪	[8] 鈎鍵交錯相持		[9] 齒	[10] 造物者	[11] 燭龍

Longevity. He ended by proposing that a special bureau should be set up for constructing several machines of this kind, one to be placed in the Ming Thang,[a] another at the observatory, and three more in other places. A special portable one was to be made to accompany imperial peregrinations. And he added that a book had been written about all this for the benefit of posterity.[b] Eventually construction was ordered, with Wang Fu[c] in charge and Liang Shih-Chhêng[1] as deputy.[d] But only two years afterwards there came the siege of the capital, and one must suppose that all the half-completed pieces, the designs, and the artisans themselves were carried away to the north by the Chin Tartars. Wang Fu, in any case, did not live to know.

There is one very odd thing about his long memorial. Although his proposals were made only thirty years after the building of Su Sung's clock, and although this notable work was still in full function, there is no mention of it whatever. A political reason is of course to be suspected, and this is indeed very probable. Su Sung's clockwork was associated with the Confucian Conservatives—Wang Fu was one of the Taoistic Reformers.

It will be remembered that the Tshai brothers first came to power in +1094, when the young emperor Chê Tsung was able to follow his own inclinations and dismiss the quarrelsome and divided Conservatives. A 'committee of investigation' was at once set up to examine the existing astronomical instruments, including Su Sung's clock, the very existence of which was thenceforward endangered. After Hui Tsung's accession in +1101 the Reformers held power continuously,[e] and times were therefore

suggested that embassies from Byzantium (e.g. in +643, +667, +719 and +720) might have been one of the stimuli acting upon I-Hsing. Cf. the later Arabic water-clocks described in Wiedemann & Hauser (4) which retained the system of dropping balls. The one which the king of Persia presented to Charlemagne in +807 was clearly of this kind (Eginhard's *Annals*, in *Mon. Germ. Histor.* (ed. Pertz), *Scriptorum*, vol. 1, p. 194). The oriental monarch was none other than Hārūn al-Rashīd (cf. Sarton (1), vol. 1, p. 527). On *chu lung* see further *Wei Lüeh*, ch. 6, p. 16*b*.

[a] The cosmological temple of the Chinese emperors (cf. Vol. 2, p. 287). Further on it see Granet (5); Soothill (5); Creel (3).

[b] Was this a fourth Sung book on clockwork? Cf. p. 449 and p. 498.

[c] Wang Fu is one of the strangest characters in all the history of science and technology in China. He is best known as an archaeologist, for his catalogue of the museum of the emperor Hui Tsung was completed in +1111. Entitled the *Po Ku Thu Lu* or *Hsüan-Ho Po Ku Thu Lu*[2] (Illustrated Record of Ancient Objects), it gave an account of bronzes, stone pieces, inscriptions and all kinds of objects in the imperial collection. But Wang Fu was also a geographer and cartographer for he seems to have revised and enlarged an earlier work of Wang Tsêng,[3] the *Chiu Yü Thu Chih*[4] (Illustrated Account of the Nine Regions); cf. *Sung Shih*, ch. 204, p. 5*b*. Anyone who visualised Wang Fu as a shy scholar would, however, be very wide of the mark. He was closely associated with a number of Taoist adepts and certainly acquainted with some of their arts, but this did not prevent him from making an adventurous, and somewhat unscrupulous, career in the official bureaucracy. He rose from secretarial posts to Advisory Censor, and attained ministerial rank in the service of one of the princes. Towards the end of Hui Tsung's reign he advocated the appeasement of the Chin Tartars. Collecting six million strings of cash by tax-farming, he bought five or six empty cities on the borders and represented to the emperor that they had been recaptured from the Chin. Finally, he was executed under Chhin Tsung in +1126 during the disturbances which led to the fall of the capital.

[d] When we first read the passage about these clocks, the name of Liang Shih-Chhêng was not unknown to us. He already occupied a place in the biographical card index, for he was a notable general engineer and builder, who made landscape gardens with botanical and zoological enclosures, fountains, lakes, etc., for the emperor Hui Tsung. We do not know his ultimate fate.

[e] Except between +1107 and +1112, when a particularly alarming comet, and a daylight appearance of the planet Venus (cf. Dubs (2), vol. 3, pp. 349 ff.), caused a temporary restoration of the Conservatives.

[1] 梁師成　　　[2] 宣和博古圖錄　　　[3] 王曾　　　[4] 九域圖志

favourable for the Taoists with whom they were associated. The full working out of the implications of this situation has never, so far as we know, been attempted, but it would be well worth while, for the Taoists had been from the beginning opposed to feudal and feudal-bureaucratic society, and were involved in all the subversive movements which occurred throughout the centuries.[a] At the same time they were the curators and advancers of all kinds of proto-scientific and technological knowledge and practice—in pharmaceutical botany and mineralogy, in astronomy, alchemy, and various forms of engineering.[b] The Reform party of the Sung were bureaucratic scholars who broke away from the typical Confucian ideology and made an alliance with Taoist science and technique.[c] It was highly significant that Wang An-Shih, and again in +1104 Tshai Ching, included mathematics and medicine among the subjects which could be offered in the imperial examinations.

From the beginning of Hui Tsung's reign, therefore, welcome at court was forthcoming for Taoist adepts of all kinds.[d] In +1104 Wei Han-Chin,[1] a thaumaturgist from Chang Ssu-Hsün's province, Szechuan, was entrusted with the casting of nine urns of bronze or iron,[e] in allusion to those legendary vessels which Yü the Great was supposed to have made, bearing upon their outer surfaces some kind of picture-maps of the provinces of the empire.[f] During the first decade of the century, Chu Mien,[2] the son of a Hunanese pharmacist, was despatched by Tshai Ching on a tour of the country with authority to purchase or secure as tribute 'all kinds of valuable articles for the imperial palace. He forced the people to give up their paintings and writings, bronzes and jades, precious stones and ornaments, and every object which would help to adorn the palace or to gratify the luxurious taste of the court.' So at least is the way historians write[g] about his activities, and there is no doubt that Chu Mien himself profited by them, but if we read between the lines, there was a Taoist flavour about this perquisition. The Taoists at court would assuredly have been concerned with all rare drugs, curious gems and other stones, technical secrets[h] and natural products of every sort. And there was Wang Fu ready to catalogue everything which was brought to the imperial museum.

In +1112 two eminent Taoists were presented at court, where they remained for some years close to the emperor. Wang Lao-Chih[3] was an intimate of Tshai Ching's and friendly with Wang Fu.[i] His successor was Wang Tzu-Hsi,[4] also known as a diviner, but of course skilled in many techniques, as men were in those times when magic and science were imperfectly distinguishable. We are told[j] that Wang Tzu-Hsi

---

[a] Cf. Sect. 10 (Vol. 2, pp. 138, 155).   [b] Cf. Sect. 10 (Vol. 2, pp. 34, 57, 130, 161).

[c] The spread of printing and other causes had helped the rise of a new class of less aristocratic literati, more urban and unconnected with the land-owning gentry families; cf. Kracke (2, 3).

[d] There is a distinct similarity between the ambiance here and that of the court of the emperor Rudolf II at Prague some five centuries later.

[e] *Sung Shih*, ch. 462, p. 10*b*.   [f] Cf. Sect. 22*b* (Vol. 3, p. 503).

[g] The words are those of Ferguson (4). Cf. G466.

[h] It will be remembered from p. 292 above that it was in +1107 that Wu Tê-Jen presented his specification, still extant, for the construction of a south-pointing carriage.

[i] He predicted for him initial success but ultimate downfall (*Sung Shih*, ch. 462, p. 11*b*).

[j] *Sung Shih*, ch. 462, p. 12*b*.

[1] 魏漢津   [2] 朱勔   [3] 王老志   [4] 王仔昔

constructed a 'spherical image' (*yuan hsiang*[1]), no doubt a celestial globe, which the emperor kept in a special pavilion. Possibly, therefore, we should recognise in him the unworldly 'Mr Wang' of +1102, who presented the book on clockwork to Wang Fu. During this period, too, the Taoist church received great imperial favour. In +1114 Tshai Ching proposed[a] the construction of a new Ming Thang (cosmological temple) with a square lake, and an imperial Taoist cathedral (Tao Kuan[2]); the first part of this plan was carried out in +1115. A year later another Taoist came into prominence, Lin Ling-Su,[3] whose specialities seem to have been rain-making and Taoist bibliography. The Taoist patrology was incorporated into the imperial library under specially appointed curators. Finally, there was Liu Hun-Khang[4] the geomancer, whose prowess with the magnetic compass led to a striking increase in male progeny among the imperial concubines.

But this entourage of virtuosi was not the kind of court which could have helped even a military emperor to organise the massive resistance needed against the forces of the Chin. With the loss of the northern provinces, the fall of the capital, and the capture of both emperors, the experimental co-operation of Taoists and Reformers came to an end. Had it continued long it might have affected Chinese society profoundly. In any case the curious facts which we have just related, remote though they may perhaps have seemed from the history of clockwork, bring to our notice the existence of two rival schools of horological artisans, one associated with the Conservatives and one with the Reformers. For Han Kung-Lien must have had his group of craftsmen just as Wang Fu (and perhaps Wang Tzu-Hsi) had his. Presumably the political parties vied with one another in the planning and erection of monumental clocks. But this is indeed an unexpected conclusion in view of the usual belief that the clock was something absolutely new to the Chinese in +1600.

We have now clarified some of the social background against which Su Sung's clock was erected, and we have followed its fortunes down to the time of the Mongol power. This pursuit took us to the Chin Tartar State in the north, and so to the Yuan empire. The next question is, what happened in the south, in the area beyond the Yangtze, to which the Sung dynasty withdrew?

Astronomical and engineering science had suffered such a blow that for a considerable time the needs of the imperial observatory could not be met. Although in +1133 astronomers such as Ting Shih-Jen[5] and Li Kung-Chin[6] spoke wistfully about the 'four famous armillary spheres of the former capital',[b] and, with Li Chi-Tsung,[7] made wooden prototype models, casting in bronze was not attempted.[c] About +1136 Yuan Wei-Chi,[8] who had been, it will be remembered, the immediate pupil and

---

[a] *Sung Shih*, ch. 472, p. 6b.

[b] *Ibid.* ch. 81, pp. 13b ff.; *Yü Hai*, ch. 4, pp. 47b ff.; *Hsiao Hsüeh Kan Chu*, ch. 1, p. 10a; *Chhi Tung Yeh Yü*, ch. 15, p. 5b. That Khaifêng had no less than four astronomical observatories is a fact of much interest in itself. The passages are fully translated in Needham, Wang & Price (1).

[c] Much of the difficulty experienced at this time was probably due to the dispersal or death of the necessarily highly qualified artisans.

[1] 圓象   [2] 道觀   [3] 林靈素   [4] 劉混康   [5] 丁師仁
[6] 李公謹   [7] 李繼宗   [8] 袁惟幾

collaborator of Su Sung himself,[a] embarked upon the casting of a small armillary, but it was not successful.[b] There was at this time a demand for an astronomical clock, for in +1144 Su Sung's son, Su Hsi,[1] who had escaped to the south, was called upon to search the family papers, but could produce no designs from which another machine could be made.[c] So the casting of two non-mechanised armillaries was entrusted to Shao O,[2] a palace steward, whose instruments turned out to be usable, if not exactly brilliant.[d] Particularly remarkable is the fact that the great Neo-Confucian philosopher[e] Chu Hsi[3] (+1130 to +1200) interested himself in the clockwork drive and tried hard to reconstruct its mechanism, but without success. In the *Sung Shih* we read:[f]

As for the water-power drive system (*shui yün chih fa*[4]) and the celestial globe, these were no longer available (for use with the armillary sphere). Later on, Chu Hsi had an armillary sphere in his house,[g] and tried hard to investigate the water-drive arrangement (so that it could again be constructed), but without any success. Although some writings of Su Sung were still in existence, they dealt principally with the details of the celestial globe and did not record any working measurements, so it was very difficult to recover his system.

The fact that Chu Hsi took such an active interest in studying the problem of the power-drive for clocks, and also the fact that he proved unable to reconstitute it, are both significant comments on Neo-Confucian philosophy.

The secret of the escapement being thus now temporarily lost, it appears that some of the Southern Sung technicians returned to the simpler float principle of the anaphoric clock. Tsêng Min-Chan[5] (Tsêng Nan-Chung[6]), a scientific worthy of the early +12th century, responsible among other things for the invention or popularisation of the double-pin gnomon equatorial sundial,[h] made water-clocks the description of which suggests this. His son or grandson, Tsêng Min-Hsing,[7] wrote of him about +1176 in the *Tu Hsing Tsa Chih*[8] (Miscellaneous Records of the Lone Watcher), as follows:[i]

The Yü-Chang (i.e. Chiangsi province) water-clock and sundial were invented by Tsêng Nan-Chung, who had mastered astronomy when young. At the beginning of the Hsüan-Ho reign-period (+1119) he passed the *chin-shih* examinations (the third of the degrees) and was posted as magistrate to Nan-chhang (in Chiangsi). When Mr Sun,[9] Auxiliary Academician of the Lung-Thu Pavilion, became commander-in-chief of the army there (and got to know Mr Tsêng) he acquired great respect and affection for him. So when the latter suggested

[a] Cf. pp. 447 and 464 above.   [b] *Chhi Tung Yeh Yü*, ch. 15, p. 5b.

[c] *Sung Shih*, ch. 81, pp. 15a ff.; *Yü Hai*, ch. 4, p. 48b. Cf. *Sung Shih*, ch. 48, pp. 18a ff.

[d] *Loc. cit.* and *Chhi Tung Yeh Yü, loc. cit.*; also *Sung Shih*, ch. 48, p. 18b.

[e] See Sects. 16d, e, f in Vol. 2, and index references.

[f] Ch. 48, p. 19b, tr. auct.

[g] This is interesting evidence that astronomical studies in medieval China were not invariably confined to official and professional circles. For instance Chu Pien wrote: 'In his earlier years Su Sung had had model armillary spheres of small size in his home, and so had gradually come to understand the principles of them...' (*Chhü Wei Chiu Wên*, ch. 8, p. 10a). Cf. p. 495.

[h] See Sect. 20g (Vol. 3, p. 308).   [i] Ch. 2, p. 11a, tr. auct.

[1] 蘇攜   [2] 邵諤   [3] 朱熹   [4] 水運之法   [5] 曾民瞻

[6] 曾南仲   [7] 曾敏行   [8] 獨醒雜志   [9] 孫

the construction of sundials and water-clocks according to new methods, the general was delighted and authorised him to recruit artisans for the purpose.

Thus metal was cast into the shape of vessels, and wooden rods were carved (with graduations). Behind the main vessel four basins and one reservoir were set up.[a] The water in the main vessel (i.e. the inflow receiver) came from the basins (i.e. the compensating tanks) and the water in the basins came from the reservoir. The water poured out through the open mouths of bronze dragon siphons. Beside the indicator-rod stood two wooden figures, the left one in charge of the day quarter-hours and the half (double-)hours of the night. In front of it an iron strip was set up, on which it struck each of these. The right figure was in charge of the (double-)hours of the day, and the night-watches. In front of it there was a bronze gong, which it sounded at each of these times.

(Tsêng Nan-Chung) also made two wooden dials (*mu thu*[1]) with diagrams on them. One was set upon a wooden support for reading (the hours) by the sun's shadow. The other was rotated by water (*yung shui chuan chih*[2]) to imitate the motion of the heavens (*i fa thien yün*[3]). This instrument was highly ingenious and the method very precise, so that it exceeded anything known in former times.

(Tsêng) Nan-Chung often used to observe the phenomena in the heavens by night, and could predict the motions of the stars and constellations, saying that such and such a star (or planet) would pass such and such a degree on such and such a night. One bitter winter, when it was extremely cold, he used to lie on his bed having removed some of the tiles from the roof, the better to watch the heavens. Once he fell asleep so that the frost came down on his body, and afterwards being invaded by cold he died. Alas! his knowledge was not handed down. Only the general constructional designs of his water-clocks and sundials were known to his son. Nowadays water-clocks of this kind are still being made in Chiang-hsiang and some other districts.

Tsêng Nan-Chung died about +1150 and Chu Hsi's efforts would have been about +1170. But, as we have already noted (p. 448 above), Su Sung's book was recovered in full, and first printed, in the south,[b] by Shih Yuan-Chih in +1172. Possibly Chu Hsi's agitation led to this result. In any case it is an interesting comment on the events in the south to find that an elaborate account of the works of Su Sung's clock is contained in the *Chin Shih* but not in the *Sung Shih*. Since these two dynastic histories were both edited by the same scholars, Toktaga[4] the Mongol, and Ouyang Hsüan,[5] in about +1340, the obvious inference is that the engineering description was conserved in the northern Chin archives but not in those of the southern Sung.

This observation restores us to the Mongol period, and to Kuo Shou-Ching, who was mentioned a few pages above. As early as +1262, before Peking became Khubilai Khan's capital, Kuo had made for him a 'Precious Mountain Clock' (Pao Shan Lou[6]), presumably at Shangtu.[c] But in the *Yuan Shih*[d] we have an elaborate description of the illuminated clock (*têng lou*[7]) of the Ta Ming Hall, made almost certainly by Kuo Shou-Ching, for it comes in the midst of a long account of his inventions, though not specifically ascribed to him by name. Since the description deals almost entirely with

[a] Cf. the summary given above, p. 479.   [b] At Wu-hsing in Chiangsu.
[c] *Yuan Shih*, ch. 5, p. 2a.   [d] Ch. 48, pp. 7a, b.

[1] 木圖   [2] 用水轉之   [3] 以法天運   [4] 脫脫   [5] 歐陽玄
[6] 寶山漏   [7] 燈漏

the jack-work, we may omit it here,[a] remarking only that horizontal wheels like Su Sung's were included. The whole mechanism was inside the casing and was driven by water (*chhi chi fa yin yü kuei chung, i shui chi chih*[1]); though no water-wheel is mentioned, adequate power could not otherwise have been obtained. A curious feature concerns the dragons which opened their mouths and rolled their eyes as they chased the 'cloud pearls' which moved up and down;[b] these things were all 'an indication of the evenness of the stream of water, and not just for decoration'. This we cannot explain, but there is no room for doubt that Kuo Shou-Ching was making clocks, more elaborate perhaps, but still in the same tradition as those of I-Hsing and Su Sung. What is interesting, and new, is that the jack-work[c] now completely dominates over the astronomical components. The clock, though not yet endowed with a name of its own, had almost wholly left the world of astronomy towards the end of the +13th century in China. This is a highly significant point, for by about +1310 the same process had occurred in Europe. The exact date of the Ta Ming illuminated clock is unsure,[d] but will not be far from +1276, the year when Kuo Shou-Ching took in hand the restoration of the Peking observatory and its equipment with the best and newest instruments.

These included one piece of apparatus, however, which, though embodying a clockwork drive, presumably water-powered, perpetuated the age-old connection with astronomical uses. It has the added interest that it was seen and described by more than one of the Jesuit missionaries. This late representative of the tradition of Chang Hêng was thus the first of its line to receive the scrutiny of observers from Europe.[e] The standard account of it in the *Yuan Shih*[f] describes a bronze celestial globe of 6 ft. diameter with the usual markings, the moon's path being indicated by a bamboo ring made so as to intersect the ecliptic as desired. The instrument was half sunk in a box-

---

[a] It will be found fully translated in Needham, Wang & Price (1).

[b] On the astronomical symbolism of this motif see Sect. 20 (Vol. 3, p. 252).

[c] The description of this has a distinctly Arabic air, as would be expected, perhaps, when the contacts of Kuo Shou-Ching with Jamāl al-Dīn are remembered (Vol. 3, pp. 372 ff.). It resembles the design of al-Jazarī (cf. Vol. 1, Fig. 33). Both have the signs of the zodiac appearing and disappearing, lamps being lit, balls dropped, etc.

[d] Another account of it, Chhi Lü-Chhien's[2] *Chih Thai Shih Yuan Shih Kuo Kung Hsing Chuang*[3] (Obituary of H.E. the Astronomer-Royal Kuo), gives simply 'Shih Tsu's reign', i.e. officially +1280 to +1294. See *Yuan Wên Lei*,[4] ch. 50, pp. 1 a ff.

[e] It is well worth noting how few men it took to span all the centuries of clockwork drive mechanisms. Ricci saw the globe of Kuo Shou-Ching, but Kuo in the +13th century certainly knew the armillary sphere of Su Sung of two hundred years earlier. There is reason to think that Su Sung and his collaborators in the +11th century had access to the +10th century designs of Chang Ssu-Hsün if not to those of the +8th-century workers (I-Hsing and Liang Ling-Tsan). They in their turn must have been acquainted with what Kêng Hsün had done in +590, if not indeed with the remains of the instruments of Chhien Lo-Chih, which are known to have lasted at least as late as +605. And finally Chhien Lo-Chih's work in +436 was directly inspired by the recovery of the remains of Chang Hêng's apparatus. Thus from Chang Hêng to Matteo Ricci there were only six intermediate men (perhaps only four) spanning fifteen centuries. A strange conservatism can also be detected in the transmission of the astronomical constants involved. Su Sung in the +11th century still used for his machinery values for π and for the year length which had been current when Chang Hêng's work was beginning. Details of this will be found in Needham, Wang & Price (1), p. 78; Combridge (2).

[f] Ch. 48, p. 5 b, tr. Needham, Wang & Price (1), p. 137; adjuv. Wylie (5), Sci. sect. p. 12.

[1] 其機發隱於櫃中以水激之　　　[2] 齊履謙　　　[3] 知太史院事郭公行狀
[4] 元文類

like casing, 'within which there are hidden toothed wheels set in motion by machinery for turning the globe (*chi yün lun ya yin yü kuei chung*[1])'. While a weight drive cannot be entirely excluded as the motive power, this would have been so foreign to the Chinese tradition that a water-wheel is more likely.

In the early spring of the year +1600, during his second visit to Nanking, Matteo Ricci made friends with some members of the staff of the Bureau of Astronomy and Calendar, who came to visit him and to discuss with him the new scientific knowledge which he had brought from Renaissance Europe. We can hear his own words, taken from his journal:[a]

Occasion offered for the Father (Ricci) to go and see the emperor's mathematical instruments, which are set up on a high hill within the city[b] on an open level place surrounded by very beautiful buildings erected of old. Here some of the astronomers take their stand every night to observe whatever may appear in the heavens, whether meteoric fires or comets, and to report them in detail to the emperor. The instruments proved to be all cast of bronze, very carefully worked and gallantly ornamented, so large and elegant that the Father had seen none better in Europe. There they had stood firm against the weather for nearly two hundred and fifty years,[c] and neither rain nor snow had spoiled them.

There were four chief instruments (i.e. the celestial globe, the armillary sphere, a large gnomon, and the equatorial torquetum).[d] The first was a globe (*thung chhiu*[2]), having all the parallels and meridians[e] marked out degree by degree, and rather large in size, for three men with outstretched arms could hardly have encircled it. It was set into a great cube of pure bronze which served as a pedestal for it, and in this box there was a little door through which one could enter to manipulate the works.[f] There was however nothing engraved on the globe, neither stars nor terrestrial features. It therefore seemed to be an unfinished work, unless perhaps it had been left that way so that it might serve either as a celestial or a terrestrial globe.[g]

Another globe was seen and described by one of Ricci's successors towards the end of the same century, Louis Lecomte,[h] but it was so much smaller (only 3 ft. diameter) that it cannot have been the same instrument. When Ferdinand Verbiest[i] refitted the Peking observatory in +1674 he seems not to have destroyed any of the old instruments dating from the Yuan, but later Jesuits did, notably Bernard Kilian Stumpf,[j] who in +1715 (with imperial authorisation, admittedly) melted some of them down to

---

[a] Ed. d'Elia (2), vol. 2, pp. 56ff., who inserted Chinese characters.

[b] Probably the Pei Chi Ko[3] (North Pole Pavilion) hill.

[c] Three hundred and fifty would have been a nearer estimate.

[d] The sphere and torquetum are now preserved on the terrace of the Purple Mountain Observatory at Nanking (cf. Vol. 3, p. 367).

[e] Kuo Shou-Ching's parallels were certainly declination circles parallel with the equator, not parallels in the Western sense aligned with celestial latitude.

[f] Gallagher (1), p. 330, translating Trigault's rewritten version of Ricci's journal, gives this phrase the nuance that the sphere had to be rotated by hand from within the box. But Ricci's words imply adjustment rather than manual labour.

[g] Absence of markings probably meant only that the weather had taken more toll than Ricci allowed for. Terrestrial globes were not in the Chinese tradition, though Jamāl al-Dīn had brought one in +1267, and another was made by the Jesuits in +1623. See Sect. 29*f*.

[h] +1655 to +1728; his account is in (1), p. 65 (+1697).

[i] +1623 to +1688; cf. Vol. 3, pp. 450 ff.   [j] +1655 to +1720; cf. Vol. 3, pp. 380, 452.

[1] 機運輪牙隱於櫃中   [2] 銅毬   [3] 北極閣

make quadrants.[a] This was quite probably the end of Kuo Shou-Ching's globe—an incalculable loss to the history of science.

We are now about to return to our original starting-point, the Jesuits and their experiences in 17th-century China. But first we should take leave of the millennial indigenous tradition of water-powered clockwork. In order to do this, we must enter the private apartments of the imperial palace, where about the middle of the +14th century we find the last emperor of the Yuan dynasty, in extraordinary contrast to the hard-riding desert warriors who were his ancestors, busied—like Louis XVI—in making clocks himself. The following excerpt will show the style of his work:[b]

> The emperor (Shun Ti)[c] himself made (in his workshops, in +1354) a boat 120 ft. long and 20 ft. wide, manned by 24 figures of sailors all in cloth of gold, and holding punt-poles, which sailed about on the lake between the Front and the Back Palaces, (having mechanical arrangements so that the) dragon (figurehead's) head and tail could move about, while it rolled its eyes, opened its mouth and waved its claws.[d]
>
> He himself also constructed a Palace Clock (*kung lou*[1]) 6 to 7 ft. high and half as wide. A wooden casing hid many scoops (*hu*[2]) which made the water circulate up and down within (*yün shui shang hsia*[3]). On the casing there was a 'Hall of the Three Sages of the Western Paradise',[e] at the side of which there was a Jade Girl holding an indicator-rod for the (double-)hours and quarters. When the time arrived this figure rose up on a float, and to left and right appeared two genii in golden armour, one with a bell and the other with a gong. At night these jacks struck the night-watches automatically without the slightest mistake.[f] When the bells and gongs sounded, lions and phoenixes on each side all danced and flew around. East and west of the casing there were 'Palaces of the Sun and Moon', in front of which stood six flying immortals (*fei hsien*[4]). Whenever the noon and midnight (double-)hours arrived, these figures went in procession two by two across a 'Bridge of Salvation' so as to reach the 'Hall of the Three Sages', but afterwards they withdrew and returned to their original positions.
>
> The ingenuity of all this was beyond belief, and people said that surely nothing like it had ever been seen before.

If this judgment was more flattering than accurate, one need not doubt that the imperial jack-work, though quite in the tradition of I-Hsing and Su Sung, was impressive enough. More interesting and notable is the fact that this clock, though almost certainly without dial and pointer, had lost practically all trace of the original astronomical components. Yet in becoming a purely time-keeping machine it had not

[a] Pfister (1), p. 645.

[b] *Yuan Shih*, ch. 43, pp. 13*b* ff.; parallel text in *Hsü Thung Chien Kang Mu*, ch. 27, pp. 11*a* ff. Tr. auct. adjuv. Wieger (1), p. 1735 (a paraphrase). Cf. H. Franke (1).

[c] Also canonised as Hui Tsung; personal name Toghan Timur (Tho-Huan Thieh-Mu-Erh[5]), +1320 to +1370, the tenth and last emperor of his line. Though he reigned for more than thirty years the Mongol empire was in full social and economic decay.

[d] Cf. as a parallel the puppet boat in the book of the Banū Mūsā brothers, who flourished in the +9th century at Baghdad (illustration in Feldhaus (2), p. 236).

[e] Doubtless Confucius, Lao Tzu, and the Buddha.

[f] This implies great ingenuity, for the night-watches were 'unequal', varying in length with the season of the year. The emperor's clock thus in some sense anticipated that made for Chhien-Lung in +1736 by Valentin Chalier (+1697 to +1747), one of the Jesuit horologists. An interesting letter of Chalier about his work has been published by Pelliot (39), tr. Needham, Wang & Price (1), p. 149.

¹ 宮漏　　² 壺　　³ 運水上下　　⁴ 飛仙　　⁵ 妥歡貼睦爾

acquired any new name, and was called merely a *lou*, a 'leaker', like the simplest clepsydras of two thousand years before. It was of course far more, as we can deduce from the water-wheel escapement hinted at by the 'scoops'. By a curious coincidence, this clock was almost contemporary with the wonderful astronomical time-piece of Giovanni de Dondi in Italy (+1364), described by Lloyd (1). This translated astronomical data entirely into terms of dials and pointers, with some indications presented to the onlooker by inscribed links on endless chains, for rotating globes and spheres had been foreign to the European tradition. Shun Ti's clock had already lost them, yet it was in the direct line from the 'pre-clock' of Chang Hêng; de Dondi's profited by the new verge-and-pallet escapement of its time, as also from medieval computing-devices like the equatorium, but it too was in a direct line from the anaphoric dials and pointers of the Hellenistic age.

The final blow to the indigenous tradition might well be dated about +1368, when the new forces of the Ming dynasty captured Peking and ended the Mongol domination. Hsiao Hsün,[1] about twenty years later, in his *Ku Kung I Lu*,[2] left a striking account of the architecture and contents of the Yuan palaces which were destroyed by the first Ming emperor's order.[a] Hsiao himself had been present as a Divisional Director of the Ministry of Works. Although he failed to describe the water-wheel clock with its elaborate jack-work, he expatiated on the dragon-fountains with balls dancing on their jets, the tiger automata, the dragons spouting perfumed mist, and the fleet of dragon-headed mechanical boats. The destruction of all these things, understandable though it was as the act of a dynasty representing the people's resentment at economic exploitation, and able to direct it against quasi-alien overlords, was nevertheless very unfortunate. Much that could have been put to better use probably perished—the Ming, like another revolution later, 'had no use for' clock-makers.[b] No doubt the Chinese horological tradition had become smothered in its own jack-work, and was hopelessly identified with the 'conspicuous waste' of the Mongol court. But its death (if indeed it did quite die at that time)[c] was a circumstance of peculiar historical importance, for it meant that when the Jesuits arrived two hundred and fifty years later there was extremely little to show them that mechanical clocks had ever been known in China.[d]

---

[a] We have already made reference to this in connection with the subject of fountains and automata (p. 133 above).

[b] One should not suppose that this great nationalist uprising was in all respects anti-technological. On the contrary there is much evidence connecting its success with the first adequate large-scale tactical use of metal-barrel cannon, then a new invention.

[c] As we shall see in a moment, it did not die, but underwent a curious and unexpected reincarnation.

[d] Moreover, for reasons not yet at all clear, there was in Ming times a general decline in most of the autochthonous traditions of physical science and technology (with some exceptions such as the ceramics industry). There was thus almost no one who could explain Chinese mathematics, astronomy or other sciences to the Jesuit missionaries. This situation they naturally exploited in several ways. To the Chinese they emphasised as much as possible the superiority of the natural science of Renaissance Europe, because by the aid of it they hoped to convince them of a corresponding superiority of European religion. To the Europeans they praised the ethical and social philosophy of China as much as possible in order to raise the prestige of the Jesuit mission, which was converting not savages, but highly civilised people. The Chinese saw through the first analogical argument very quickly, and determined to adopt

[1] 蕭洵   [2] 故宮遺錄

Their conviction that such machines were something entirely new to the Chinese was, it will be remembered, our very starting-point in this survey.[a] But as to the time-keeping methods in use in China beforehand they spoke in rather mysterious terms. An interesting passage is contained in the admirable description of China which Ricci prefixed to his memoirs,[b] and this was enlarged by Trigault as follows:

They (the Chinese) have very few instruments for measuring time, and those which they do have measure it either by water or by fire. Those which use water are like large clepsydras, and those which use fire are made of a certain odoriferous ash very like that tinder which is used for the touch-wood or slow match of guns and cannon among us.[c] A few other instruments are made with wheels rotated by sand as if by water—but all are mere shadows of our mechanisms, and generally most faulty in time-keeping. As for sundials they know only that one which takes its name from the equator, but they are unable to set it up correctly according to the location[d] (i.e. the latitude).[e]

We are thus obliged to believe that still at the time of Ricci and Trigault some remains were left of the old Chinese driving-wheel clocks.[f] The use of sand seems particularly strange, for the hour-glass is considered to have been an introduction from Europe then of very recent date.[g]

Nevertheless, sand there was, and in tanks.[h] The story of what happened in the Ming has been unravelled by the skill and labour of Yabuuchi (4) and Liu Hsien-Chou (6). We have just seen how the monumental striking water-clock of the last Yuan emperor perished in the wave of Confucian austerity which accompanied the

the 'new' but not the 'Western', the philosophy but not the theology. The Europeans were far more impressed than the Jesuits intended, and in due course Chinese 'morality without supernaturalism' paved the way for deism and revolution. The early Jesuits in China were a most sympathetic group of men and the results of their actions were enormously influential—but almost wholly contrary to what they themselves were consciously working for. Such is the irony of history. The result which concerns us here was that the indigenous scientific and technical tradition of China was very much under-estimated in the West.

[a] P. 437 above.   [b] D'Elia (2), vol. 1, p. 33.

[c] 'Tormentorum nostrorum fomites imitantur.' Needham, Wang & Price (1), p. 155, were puzzled by this phrase, and could only suggest that it might have referred to medical cauterisation. Yet the practice of using tindery incense ash (moxa) for this was characteristically Chinese and was not introduced into European practice until +1674, by Buschoff (cf. Garrison (3), p. 261). What must be the right explanation, slow match for cannon touch-holes, has now been proposed by the Rev. J. Sebes, S.J., and the Rev. Neil Twombly, S.J., of Georgetown; we are indebted to Mr Silvio Bedini for communicating it to us. On the use of incense for measuring time see Vol. 3, p. 330.

[d] This statement was quite erroneous. It could have applied only to the relatively ill-educated provincials with whom the Jesuits were first in contact.

[e] De Christiane Expeditione (Augsburg, 1615), p. 22. The translation of this passage by Gallagher (1), p. 23, seems to depart widely from the Latin.

[f] The Jesuits were of course none too curious in the study of the history of science. Indeed the nuance of 'mere shadows' reminds one of the attitude which some of them adopted towards the ceremonies of Buddhism and Lamaism, often so similar to Christian liturgical practice. Were these not diabolical inventions, horrid mimicries of the true faith?

[g] Wang Chen-To (5), who has gone into this question, concludes that the hour-glass was transmitted to the Chinese and Japanese from Portuguese and Dutch ships towards the end of the +16th century. See Vol. 4, pt. 1, p. 290 above, and Sect. 29f below.

[h] We may note at the outset that in substituting sand for water the Chinese horologists were adopting a medium the escape-rate of which is independent of the pressure-head. They thus circumvented a problem which had dogged the makers of clepsydras for some two thousand years (cf. Vol. 3, pp. 317 ff.) We are grateful to Prof. D. M. Newitt for reminding us of this important point.

rise to power of the nationalist movement of the Ming. Hence it is interesting, though not very surprising, to read in the *Ming Shih*:[a]

After Thai Tsu had conquered the Yuan people, the Astronomer-Royal presented him with a 'rock-crystal clock' (*shui ching kho lou*[1]), within which there were two wooden jacks automatically striking gongs and drums in accordance with the passing hours. But Thai Tsu considered it to be a useless (extravagance) and had it broken up.

Yet if the emperor felt compelled to frown upon anything which savoured of the palatial luxury of alien rulers, he seems to have given positive encouragement to a form of wheel-clock which could be easily made in all provinces and prefectures. The texts which tell us of this new horological design are also in the first of the astronomical chapters of the *Ming Shih*, but they concern a much later date, and the matter arises incidentally. The Chinese astronomers who were the friends and collaborators of the Jesuits are discussing the making of new equipment for the imperial observatories.

In the 7th year of the Chhung-Chên reign-period (+1634), Li Thien-Ching[2] reported that Hsü Kuang-Chhi[3] said that for determining time there had anciently been the clepsydra, but now there was the 'wheels-and-bells' (instrument; i.e. the mechanical escapement clock). Both required human attention, and were therefore not so reliable as taking time from the movements of the heavens themselves. He therefore petitioned that three kinds of instruments, sun-dials, star-dials, and telescopes, should be made. So the emperor authorised him to take charge of the matter.[b]

The historian then goes on:[c]

In the following year (+1635) (Li) Thien-Ching suggested that sand-clocks should be made. At the beginning of the Ming (i.e. about +1360 to +1380) Chan Hsi-Yuan,[4] finding that in bitter winters the water froze and could not flow, replaced it by sand. But this ran through too fast to agree with the heavenly revolution, so to the (main driving) wheel with scoops he added four wheels each having 36 teeth. Later on, Chou Shu-Hsüeh[5] criticised this design because the orifice was too small so that the sand-grains were liable to block it up, and therefore changed the system to one of six wheels (in all), the five wheels each having 30 teeth, at the same time slightly enlarging the orifice. Then the rotation of the machine really agreed with the movement of the heavens. What (Li) Thien-Ching now petitioned for was surely this design deriving from (Chan and Chou).

In this way we come across a new name, hitherto unknown, but of much importance for our story, Chan Hsi-Yuan about +1370. Chou Shu-Hsüeh we have met before;[d] he flourished between 1530 and 1558, leaving a well-deserved reputation as a mathematician, astronomer and cartographer. The date of Chan Hsi-Yuan is clearly confirmed by the fact that the great historian Sung Lien,[6] chief editor of the *Yuan Shih*, wrote an interesting essay[e] on the 'Five-Wheeled Sand-Clock and its Inscription'

---

a Ch. 25, p. 15a; tr. auct.   b Ch. 25, p. 18b; tr. auct.
c Ch. 25, p. 19b; tr. auct.   d Cf. Vol. 3, p. 51.
 e *Wu Lun Sha Lou Ming Hsü*,[7] preserved in *Ming Wên Chhi Shang*[8] (Machines section, ch. 2, pp. 53 ff.), part of *Ku Wên Chhi Shang*,[9] ed. Chhen Jen-Hsi[10] (+1581 to +1636). This essay will be found fully translated in Needham, Wang & Price (1). We are much indebted to Dr Liu Hsien-Chou for a copy of it, as the collection is not available in Cambridge.

¹ 水晶刻漏   ² 李天經   ³ 徐光啓   ⁴ 詹希元   ⁵ 周述學
⁶ 宋濂   ⁷ 五輪沙漏銘序   ⁸ 明文奇賞   ⁹ 古文奇賞   ¹⁰ 陳仁錫

written by him for one of these instruments before his death in +1381. The account of the mechanism given in this text is so complete that Liu Hsien-Chou had no difficulty in making the working drawing shown herewith in Fig. 668. Two things are very clear, first that this type of sand-clock had a scoop-wheel very similar to that in the great water-wheel clocks of Su Sung and others, as well as a certain amount of jack-work just as they did; secondly, that it had a stationary dial-face over which a pointer circulated. The appearance of this feature, so characteristic of all subsequent

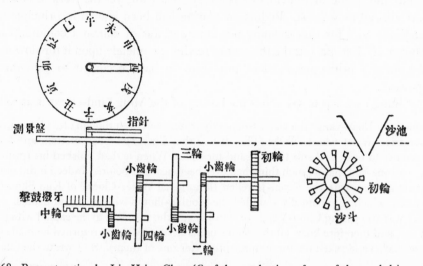

Fig. 668. Reconstruction by Liu Hsien-Chou (6) of the mechanism of one of the sand-driven wheel-clocks of Chan Hsi-Yuan (c. +1370). Its five wheels comprised the driving-wheel (chhu lun) with scoops (sha tou) below the feed (sha chhih), three large gear-wheels (erh, san, ssu lun), and one 'middle wheel' (chung lun) fitted with audible signal trip-lugs (chi ku po ya) and borne on the shaft of the pointer (chih chen) which made the rounds of the dial-face (tshê ching phan). On this the markings for the twelve double-hours can be seen. Four small gear-wheels (hsiao chhih lun) completed the train. The later modification by Chou Shu-Hsüeh (c. +1545) added a fourth large gear-wheel and changed the gear-ratios. The jacks are not shown in the diagram, but they were worked off the middle wheel. Whether the sand-driven clocks of the Yuan and Ming possessed a linkwork escapement of the classical type is uncertain. Though Liu Hsien-Chou believes that they did, his drawing omits it; the alternative is that they kept time purely by reduction-gearing (for a detailed discussion see Needham, Wang & Price (1), pp. 160 ff.).

clocks, in late +14th-century China, is quite curious, since this was just about the time when the dial-face was becoming standardised in Europe too. Perhaps the simplest hypothesis is that both derived independently from the dial of the anaphoric clock, which, as we have seen above (pp. 467, 503), seems to have existed during the earlier Middle Ages in China as well as in the West. Equally important is the evidence of the divorce, now complete, both in China and Europe, between the astronomical functions and the purely time-telling functions of clockwork and proto-clockwork.

The most difficult point to decide about the sand-clocks of the Ming is whether or not they worked with a linkage escapement like the earlier monumental water-wheel clocks from I-Hsing to Toghan Timur. It would have been natural for them to have possessed it, but there is no justification in Sung Lien's text for supposing that they did,[a] and it may be, therefore, that Chan Hsi-Yuan was more original than Sung Lien realised. For he was, perhaps, the inventor of a type of clock with reduction gearing. Computation shows that whether in its original or its modified form, each scoop took only a matter of seconds to fill, and with such an extensive train of gears there could have been no tendency for the driving-wheel to run wild. Of course, with the level of technique available in the 14th century, or even the 16th, the resulting accuracy may well have deserved the strictures it received from the Jesuits, and no doubt the advent of weight- and spring-driven clocks was a very great gain; yet the work of Chan Hsi-Yuan retains all its interest. Reduction gearing had been known in principle to the Alexandrians, and for time-keeping machinery it was of course a commonplace in +14th-century Europe, but the thought of relying completely upon it for slow motion relayed from a prime mover does not seem to have occurred to contemporary Europeans.

What Sung Lien has to say about the history of the Ming sand-clocks is as follows:

Formerly in Luan-yang[1] the water frequently froze,[b] and even though often heated, would still not drive water(-wheel) clocks. Thus it was that Chan Hsi-Yuan of Hsin-an[c] brought his ingenuity to bear, and substituted sand for water. When he had finished his (prototype) clock, everyone said that no such thing had ever been heard of before. Indeed it did not need to fear comparison with the Seven-Jewelled Illuminated Water-Clock[d] of Kuo Shou-Ching (c. +1276), which automatically sounded the (double-)hours with bells and drums.

There was also Chêng Chün-Yung[2] of Phu-yang,[e] who travelled to the capital with (Chan) Hsi-Yuan, and therefore knew all the details of his design. After he returned he made (more of these clocks) and asked me for an inscription (for one of them). So I wrote the following words:

'The "Hoisters of the Water-Pots" were ancient men of State,
And as of old the time they told by Water's constant rate,
But wintry ice spoilt their device—the water turned to land—
Till Chan by Earth did conquer Earth, and moved his wheels with sand;
Which being neither firm nor flood, keeps faith with Heaven's round
And makes the jacks of Master Chêng to beat their rhythmic sound.
So now Stone flows where Water can't, ignoring fire and frost—
Good people all, mark well the dial, and grudge each moment lost.'

And indeed the sand-clocks of the Ming are worth our careful notice too. Their period of dominance was comparatively so recent that the discovery of an actual

---

[a] Here we regret to differ from Liu Hsien-Chou (6). Nevertheless, his conviction that for any kind of accurate time-keeping the Ming sand-clocks must have had some kind of escapement is shared by Mr John Combridge and several colleagues. On the other hand, perhaps they really were as inaccurate as Ricci said (p. 509).

[b] This place was a Chin Tartar city in Hopei province north of Peking, perhaps a temporary capital at the beginning of the Ming dynasty.

[c] A town of this name has long been famous for its magnetic compasses (cf. Vol. 4, pt. 1, p. 294). It is probably the one in Anhui, not its namesake on the Pearl River estuary south of Canton.

[d] Undoubtedly the same as the Illuminated Water-Clock of the Ta Ming Hall (p. 504 above).

[e] A town on the Phuchiang River, a tributary of the Chhien-thang south of Hangchow, in Chekiang province. Southern Chekiang was always associated with the production and working of iron and steel in many small centres.

[1] 灤陽   [2] 鄭君永

specimen in the excavation of a Ming tomb is by no means outside the bounds of possibility.[a]

A late appearance of solid particles fulfilling the function of water in the 'mill-wheel' type of clock brings us to the point where the two traditions, Chinese and Western, joined in a single instrument. By the second decade after the death of Ricci collaboration between Jesuit missionaries and Chinese scholars had become a well-established tradition, and it is to the work of one such pair that we owe the earliest Chinese description of the verge-and-foliot escapement. But in this clock the two traditions fused, for while a mechanism of European type kept equal double-hour time in the front of the cabinet, a scoop-wheel system of Chinese type told the night-watches at the back.

The illustration (Fig. 669), and a brief description, was given by Wang Chêng[1] in his *Chu Chhi Thu Shuo*[2] (Diagrams and Explanations of a number of Machines)[b] of +1627. Perhaps to indicate its double nature, he called it a 'wheel clepsydra' (*lun hu*[3]). A cabinet of two storeys some 2½ ft. high made visible in its central compartment an iron clockwork mechanism driven by a falling weight of lead. One can see the foliots above the gear-train. Wang Chêng described them as follows:

Beside all this there is a cross-shaped (device) having two hitting teeth (*po chhih*[4]) (i.e. pallets) to left and right. All the wheels (of the gear train), impelled by the motion transmitted through them, would move quite fast, if it were not for the hitting teeth in the centre, which slow down the movements, seeming to be pushed to the left yet arresting on the right. This is a most delicate piece of machinery, and the very essence of the 'wheel clepsydra'. But though everything depends on it, it is extremely difficult to describe in writing, and can hardly be represented in a diagram.

Then, since no Chinese clock could be complete without jack-work, the upper storey carried a perambulating figure with a pointer, driven by a belt or chain from the clockwork below, and indicating each of twelve tablets for the double-hours in turn.[c] This figure also tripped a striking mechanism which beat upon the drum in the left-hand compartment, while other arrangements sounded both the drum and a bell in the right-hand compartment at other times.

So far all was purely Western, a manifestation of the knowledge of Wang Chêng's friend the Jesuit Johann Schreck.[d] But inside the cabinet, presumably at the back, there was also 'a night-watches clepsydra (*kêng lou*[5]), with two flumes (*tshao*[6]) and

---

[a] When in Peking in the summer of 1958 I was privileged to be invited to view the tomb of the Wan-Li emperor (the Ting Ling), first of the series now under thorough scientific excavation; cf. Fang Yün (1); Anon. (12).

[b] Pp. 12b ff. tr. auct.

[c] It was a reincarnation of the figure of Helios in the Gaza clock of c. +510, which did just the same thing (cf. Diels, 2).

[d] See pp. 170, 211 above. This collaboration invites the question whether the Ming sand-clocks had any echoes in Europe. A special investigation would be required, but certain +18th-century horologists had much to say of wheels rotated by sand or other granular material—notably F. Perez Pastor, in a book published at Madrid in +1770. From him the trail seems to lead back through Ozanam in France at the beginning of the century to Francisco A. M. Rati in Rome, +1655, and Dominico Martinelli of Spolato, +1654.

[1] 王徵　　　[2] 諸器圖說　　　[3] 輪壺　　　[4] 撥齒　　　[5] 更漏　　　[6] 槽

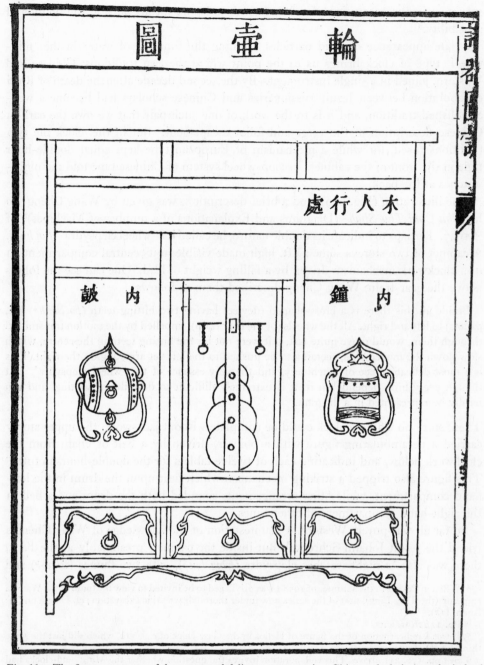

Fig. 669. The first appearance of the verge-and-foliot escapement in a Chinese clock design; the 'wheel clepsydra' (*lun hu*) described by Wang Chêng in his *Chu Chhi Thu Shuo* of +1627 (pp. 12*b* ff.). The foliots can be seen oscillating in the middle compartment of the cabinet above a gear-train, through an opening in the front of the left-hand compartment a drum (*ku*) is visible, and on the opposite side a bell (*chung*). A perambulating jack moved slowly along a gallery above (*mu jen hsing chhu*), pointing to placards of the double-hours one after the other, and tripping a drum-beating mechanism. But strangely, though nothing of this is seen in his diagram, a representative of the classical Chinese horological tradition lurked still at the back of the cabinet, namely a 'clepsydra for the night-watches', which used lead shot. Most probably this was a scoop-wheel of the sand-driven wheel-clock type, with or without a linkwork escapement, and adjustable in some way or other for the 'unequal' intervals throughout the year.

two tubes (*thung*[1]), filled with lead shot (*chhien tan*[2]), and the (rest of the appropriate) machinery'. This rather cryptic statement agrees in no way with the weight-driven clock in the front, but does perhaps suggest that for the unequal night-watches Wang Chêng installed a scoop-wheel clock of the Ming type, using not sand but small lead-shot for filling the buckets.[a] It is unfortunate that he says so little about it; probably it was too familiar to his readers. But one would like to know whether it had a weigh-bridge linkwork escapement or depended on reduction gearing, and also what means he took to adjust the mechanism automatically for the varying night-watch periods at the different times of the year. As in the case of the sand-clocks it is not at all impossible that future archaeological discoveries will reveal it to us.

It seems probable that the bucket-wheel sand-driven clocks did not die out until the close of the +17th century. Some time about +1660 a magistrate in Yunnan, Chi Than-Jan,[3] constructed what was almost certainly an instrument of this kind for a Buddhist temple.[b] Liu Hsien-Thing[4] tells us about it in his *Kuang-Yang Tsa Chi*[5] (Collected Miscellanea), which dates from the last decades of the century:[c]

The Thung-Thien pagoda was really a clock (*tzu ming chung*[6]). It was designed by (Chi) Than-Jan, and built in the form of an Indian or Turkestan Buddhist stūpa (*Hsi-Yü fou-thu*[7]) of three storeys standing on a framework supported on silver blocks. Within the lowest storey of the pagoda there was a bronze wheel revolving in connection with others but not visible from outside. At the front of the middle storey there was an open door showing a '(double-)hour drum' as round as a tub, the circumference of which was divided into 12 placards inscribed in seal characters with the names of the (double-)hours. (This tub-wheel was) rotated by means of the (driving-)wheel below in accordance with the (apparent) diurnal rotation of the heavens. After one revolution of the sun the drum returned to its original starting-point. At each (double-)hour one or other of the placards faced outwards so that it could be fully seen.

There was also a wooden figure which popped out holding placards to announce the quarters, and at the same time striking a bell at the top of the middle storey. Besides this there was a bronze bell hanging in the top storey which was struck by the mechanism. At every quarter the bell sounded one stroke and at each (double-)hour it sounded eight strokes. In front of the bell there was an image of the guardian god Wei-Tho,[8] controlled (by the mechanism from inside), which looked all round to left and to right as if inspecting (what was going on).[d] Above this there was nothing but the roof. The whole design was something hardly ever seen before, and replaced the lotus clepsydra[e] in timing the Buddhist services.

I myself begged Than-Jan to open up the works and let me see them; there were of wheels

---

[a] This interpretation depends upon one's estimate of the size of the shot. If they were substantial lead balls then Wang Chêng may have used an Arabic type of striking clepsydra something like that of King Sejong in Korea a couple of centuries before (cf. p. 517). This also would explain the audible signalisation from a mechanism hidden in the back of the clock.

[b] This non-Christian affiliation, and the absence of any mention of a verge-and-foliot escapement, strongly suggests that the clock did not have a Western weight-drive.

[c] Ch. 3 (p. 141), tr. auct. with Lu Gwei-Djen. Elsewhere (p. 140) the author gives us a brief biography of his friend Chi Than-Jan, from which it appears that he was a worthy local official skilled in medicine as well as mechanics. Both passages are quoted by Liu Hsien-Chou (*1*, suppl.), p. 138.

[d] On Wei-Tho see Doré (1), vol. 7, p. 206.   [e] Cf. Vol. 3, pp. 324 ff.

[1] 筒   [2] 鉛彈   [3] 吉坦然   [4] 劉獻廷   [5] 廣陽雜記
[6] 自鳴鐘   [7] 西域浮屠   [8] 韋馱

large and little more than twenty, all made of yellow bronze, but the workmanship was not very refined and everything depended on the exact shape. After a short period of service it went wrong altogether.

It is a pity that Mr Liu did not tell us more about the motive power, but the tub-wheel with the placards at any rate was clearly horizontal like the jack-wheels of Su Sung. Perhaps his failure to mention any escapement again suggests that reduction gearing was used. In any case the machinery does not seem to have worked too well, and we have the impression of attending the death-bed of a great tradition. Further research may bring to light more details of the work of Chi Than-Jan.

In this brief subsection we have covered some six hundred years. In following the fate of Su Sung's great clock we became involved in the curious politics, ecclesiastical as well as civil, of the court of Hui Tsung early in the +12th century. We saw how its remains came into the possession of the Mongolian rulers and their brilliant men of science such as Kuo Shou-Ching in the +13th, some of whose own water-powered astronomical instruments were handed down to be studied by Jesuits at the end of the +16th century. In one of the Yuan emperors, Toghan Timur, we found a clock-maker of note, apparently the last in China to use water as the motive force; for from the beginning of the Ming (about +1370) Chan Hsi-Yuan and others substituted tanks of sand. This method survived well into the Jesuit period, but by the end of the +17th century sand and scoop-wheels had given place to the standard forms of Europe, first the falling weight and then the spring. We have come a long way, but something yet remains to be said about developments which occurred during the Ming dynasty, from the +15th century onwards.

(5) KOREAN ORRERIES, ASIATICK SING-SONGS, AND THE MECHANISATION OF MT MERU

A few pages ago we were listening to the artisans of the Ming in the Yuan palaces, lamentably 'breaking down the carved work with axes and hammers'. If some of Shun Ti's skilled workmen got away in time, we can hazard a shrewd guess as to where they went. For a few decades later the court of the Yi kingdom of Chosŏn at Seoul in Korea was busied with exactly the same kind of activity, and royal clock-makers were again at work. This merits attention not only for its intrinsic interest but because it was almost the last appearance of water-powered time-keeping instruments.

The fourth king of this dynasty, Sejong[1] (r. +1419 to +1450), gave proof of abilities, and maintained a court, not unworthy to be compared with those of the Caliph al-Ma'mūn[a] or Alfonso X, king of Castile.[b] An enlightened and scholarly ruler, he was the prime inventor of the Korean alphabet in +1446, a remarkable achieve-ment.[c] But he was also fond of science and scientific instruments so that he was able himself to supervise the complete re-equipment of the astronomical observatory of

[a] R. +813 to +833; cf. Hitti (1), p. 310.
[b] R. +1252 to +1284; cf. Sarton (1), vol. 2, p. 834.
[c] Cf. Yi Sangbaek (1); Ledyard (1).

[1] 世宗

the capital, which took seven years from +1432 onwards under the care of Yi Sunji[1] and Chang Yŏngsil[2].[a] Here we are more interested, however, in King Sejong's clocks. It is recorded that when quite young he made a clepsydra together with his father, King Thaejong.[3] Then in +1424 he ordered the construction of 'a kind of clock patterned after a Chinese description'. From its name however, *kêng tien chhi* (*kyŏngjŏmgi*),[4] we might infer that it was some kind of arrangement which indicated the night-watches and their divisions by a series of lamps successively and automatically lit, and balls dropped one after another into ringing bowls, as in the great striking clepsydras of the Arabic world.[b] These of course needed no escapement and depended only on the anaphoric float principle.

Such a surmise is strongly supported by an account of one of the monumental clocks constructed at the Chosŏn capital, preserved in the *Yijo Sillok*[5] (Veritable Records of the Yi Dynasty)[c] and noted long ago, though not elucidated, by Rufus.[d] The 'puppet-clock' in question, which was built by Chang Yŏngsil at Sejong's command in +1434, had three wooden figures of immortals which struck the double-hours on a bell, beat a drum at the beginning of every night-watch, and sounded a gong for every fifth part (*tien*[6] or *chhou*[7]) of each night-watch. The mechanism, minutely described, was as follows. From two inflow clepsydra vessels there rose up two floating indicator-rods, one for the day and night double-hours, the other for the night-watches, dislodging as they did so 12 and 25 small bronze balls respectively from vertical bronze racks fixed above the clepsydra.[e] The double-hour balls fell down one after the other through an appropriate conduit to dislodge 12 iron balls as big as hen's eggs; these then sped down a channel to make one immortal ring its bell, and at the same time poke on a horizontal wheel carrying 12 minor immortals with hour-annunciating placards so arranged that they appeared one after the other at a window. The system for the night-watch bronze balls was a little more complicated in that the tube in which they collected gave on to two separate channels with two rows of waiting iron balls, 5 to make the drum immortal do its work and then to sound the gong for the beginnings of the night-watches, and 20 to mark the passage of fifths of night-watches by sounding the gong alone. An ingenious use of levers sufficed for nearly all the effects. In every case it was arranged that the detent orifices where the small balls tripped the large ones in succession should be closed automatically one after the other. At the end of the night the balls were all collected from the sumps where they had assembled, and

[a] Yi Sunji was an accomplished astronomer much prized by the king, who recalled him urgently from a period of mourning seclusion to undertake this charge. Chang Yŏngsil was originally a government slave but rose to the dignity of Lord Protector of the Army and chief engineer of the Korean court.

[b] Cf. Fig. 33 in Vol. 1, as also pp. 468 above and 534 ff. below.

[c] Sejong sect., ch. 65, pp. 1a ff., photo. edition, vol. 8, pp. 369 ff. The description is largely due to a scholar named Kim Pin[8], who also helped with the re-equipping of the observatory (cf. immediately below). Dr Lu Gwei-Djen collaborated in its study.

[d] (2), p. 31.

[e] As the night-watches and their divisions differed in length according to the season of the year, it must have been necessary to make periodical changes either in the indicator-rods or in the ball-racks for these units of time.

[1] 李純之   [2] 蔣英實   [3] 太宗   [4] 更點器   [5] 李朝實錄
[6] 點   [7] 籌   [8] 金鑌

518 27. MECHANICAL ENGINEERING

replaced at their starting-points, no doubt when the water in the clepsydra vessels was changed. Such was the *tzu chi lou*[1] (automatic striking clepsydra) in the Pao Lou Ko[2] (Poru Kak, Chiming Clepsydra Pavilion). It is thus evident that the clock of +1434 was a striking clepsydra essentially in the Arabic style,[a] though containing a horizontal wheel of circulating jacks like those of Su Sung before and Isaac Habrecht afterwards. One may well wonder what connections these Koreans could have had with the Arabic culture-area;[b] a filiation with a Chinese anaphoric clock tradition as exemplified in Tsêng Min-Chan (cf. p. 503) would perhaps be easier to understand.

Three years later the equipping of the royal observatory was at its height. The *Yijo Sillok*, in a long and informative passage,[c] describes nearly forty astronomical instruments of thirteen types, ranging from small portable sundials oriented by magnetic compasses to modified versions of the 'Simplified Instrument' or equatorial torquetum, originated by Kuo Shou-Ching at the imperial observatory of the Yuan about a century and a half before.[d] Consciously building on his work and somewhat extending it, the Korean astronomers, notably Yi Sunji, Chŏng Chho[3], Chŏng Inji[4] and Yi Chhŏn[5],[e] besides producing instruments of classical type such as the gnomon and the armillary sphere, invented also forms of apparatus hitherto undescribed. Among these were scaphe sundials with stretched wires or threads as styles,[f] a self-orienting armillary dial (*ting nan jih kuei*[6]), and an interesting combination of equatorial sundial and nocturnal known as the *jih hsing ting shih i*[7] (instrument for determining the time by the sun and stars).[g]

Most relevant here however is the fact that two of these pieces of equipment, unlike the automatic clepsydra in the Chiming Pavilion, were mechanised by water-wheel link-escapement clockwork. This is indubitably clear from the descriptions, which use

[a] In the great monograph of Wiedemann & Hauser (4) on Arabic striking clepsydras, there are many descriptions of balls falling through tubes and along channels to activate jacks, cf. pp. 25, 125 ff., 153 ff., 217, 239, 249.

[b] Of course many scholars and artificers of Arabic culture took service under the Mongol government of China in the +14th century, and very probably some went as far as Korea. For a notable case in hydraulic engineering, cf. Sect. 28 *f*, in Vol. 4, pt. 3. 'At the furthest end of Chinese territory', wrote al-Marwazī about +1115, 'lies the land called Sīlā; whoever enters it, whether Muslim or stranger, settles in it and never leaves it, on account of its pleasantness and excellence' (Minorsky (4), p. 27).

[c] Ch. 77, pp. 7*a* ff., photo. edition, vol. 8, pp. 549 ff. A full translation and commentary is in course of publication (Lu, Needham *et al*.). Till now there has been only the brief account in Rufus (2), pp. 30 ff.

[d] This embodied, of course, the invention of the equatorial mounting, a fundamental advance in astronomical equipment. See Vol. 3, pp. 367 ff., 371, and Needham (33).

[e] The most distinguished of these was Chŏng Inji, but he was many things besides an astronomer: a Minister of State, an outstanding writer connected with the introduction of Sejong's *ŏnmun* alphabet in +1446, and the first to compose poetry in the Korean script, as well as eventual director of the historians who produced the *Koryŏ-sa*.[8] Chŏng Chho was an Academician and Intendant of Education; Yi Chhŏn seems to have been in charge of the engineering aspect of the works.

[f] This may require a revaluation of our previous impression that the stretched-thread sundial (Type A) was a Jesuit introduction (cf. Vol. 3, p. 310). It seems here to have developed from a prolongation of the south-pointing style of the equatorial sundial. The forms described are called *hsüan chu jih kuei*[9] and *thien phing jih kuei*.[10]

[g] Cf. Vol. 3, pp. 307, 338. The description of this is largely due to another astronomer, Kim Ton.[11]

[1] 自擊漏   [2] 報漏閣   [3] 鄭招   [4] 鄭麟趾   [5] 李蕆
[6] 定南日晷   [7] 日星定時儀   [8] 高麗史   [9] 懸珠日晷
[10] 天平日晷   [11] 金墩

the classical phraseology to which we have become accustomed. Thus the 3½-ft. diameter celestial globe made of lacquered cloth with a threaded sun model (cf. pp. 471, 488) had 'an ingenious mechanism driven by the force of rushing water yet hidden from sight'.[a] The same applies to another puppet-clock set up in the Pavilion of Respectful Veneration (Hǔmgyǒng Kak or Chhin Ching Ko[1]). Here an artificial golden sun travelled upon its course among clouds from dawn to sunset over (or round) an artificial mountain 7 ft. high, representing perhaps Mt Meru. Meanwhile a jade girl (immortal) beat the double-hours with a wooden stick upon a bell, and simultaneously four out of twelve warrior jacks advanced and retired amidst scenery which changed according to the seasons. Inside there was an 'automatic turning mechanism with a wheel (nei shê chi lun[2]), using clepsydra tank water from (an orifice of) skilfully-worked jade (or other hard stone) to strike upon it (yung hsü lou shui chi chih[3])'.[b] Such words leave no doubt in the mind that between +1434 and +1438 the astronomers and engineers of the Korean court changed over from typically Arabic to typically Chinese horological contrivances, though the latter may well not have been new in the land of Chosǒn. We can hardly fail to recognise here the tradition of I-Hsing, Su Sung and Shun Ti.

As had already been said, the Koreans were probably more interested in science and mechanical technology than any other people on the periphery of the medieval Chinese culture-area.[c] It is thus not perhaps surprising that Korea provides us with a very interesting demonstrational armillary sphere which, though dating from the +17th and +18th centuries, three or four hundred years after King Sejong, still exemplifies the bi-millennial tradition of East Asian astronomical clockwork.[d] Many most characteristic features are to be seen in this piece of apparatus, the mechanical drive of which operates time-telling and striking devices as well as the rotation of the sphere itself.

A photograph of the instrument has already been given in Vol. 3, Fig. 179 (opp. p. 390)[e] in connection with demonstrational armillary spheres. Its length is about 4 ft., its height 3 ft. and its width 1 ft. 9 in.; the armillary sphere being 1 ft. 4 in. in diameter, and the terrestrial globe within it about 3 in. The time of day is shown by discs carried round on arms from a horizontal wheel and displayed through a side window. The sun moves as a model on the ecliptic band (huang tao[4]), marked with

---

[a] Chhi chi shui chi yün chih chhiao tshang yin pu chien.[5]

[b] The wording here is closely reminiscent of that used about Kuo Shou-Ching's water-wheel clock in Yuan Shih, ch. 48, p. 7b. Cf. Needham, Wang & Price (1), p. 136. There follows a reference to inclining vessels (i chhi[6]), on which see Vol. 4, pt. 1, p. 34. This we believe may be an elegant literary allusion to the filling and emptying of the buckets of the scoop-wheel. The description in the Chǔngbo Munhǒn Pigo (see immediately below) is rather fuller (ch. 2, pp. 30a ff.) than that in Yijo Sillok.

[c] Cf. Vol. 3, p. 682.

[d] It is now in the National Museum at Seoul, and currently under renewed study (Chon Sangun (1); Combridge, Ledyard, Lu, Maddison & Needham, 1). Much of the instrument may be original +17th-century work.

[e] From Rufus (2), p. 38 and fig. 26; it was at that time (1936) preserved in the home of Mr Kim Sǒngsu.[7] It is also reproduced in Rufus & Lee (1), and Needham, Wang & Price (1), fig. 59.

[1] 欽敬閣　　　　　[2] 內設機輪　　　[3] 用玉漏水擊之　　　[4] 黃道
[5] 其激水機運之巧藏隱不見　　　[6] 欹器　　　[7] 金性洙

the 24 fortnightly periods. The moon model moves on a the ring representing its path (*pai tao*[1]), marked with the 28 lunar mansions. The mechanism is driven by two weights, one for the wheels and gears of the time-piece, which is regulated by a pendulum (and a verge-and-pallets escapement), the other for the striking device. This sounds a bell by the successive release of iron balls allowed to roll down a trough, after which they are lifted by paddles on a 'noria' to repeat the process. The terrestrial globe shows the influence of some of the +16th-century voyages of exploration.

One stands amazed at the fidelity to ancient tradition manifested in this Korean work. We see (*a*) that the intermediate nest of rings in the armillary sphere is made to rotate, just as in the Thang, and in the clock of +1088, by means of (*b*) a shaft in the polar axis, as was so common among the mechanised spheres of those ages. (*c*) The sphere has a horizon circle fixed externally, as did all Chinese armillary spheres after the time of Chang Hêng. Then there is (*d*) the earth model placed centrally, just as it was in the instruments of Ko Hêng and Liu Chih (+3rd century), and now it is marked (*e*) with the chief continents, like that terrestrial globe brought to Peking from Persia by Jamāl al-Dīn in +1267 for the consideration of Kuo Shou-Ching. We even find (*f*) a special ring for the path of the moon, like that which was incorporated in Li Shun-Fêng's design of +633, as well as in others afterwards. Then within the casing of the mechanism there is a periodical release of balls (*g*) reminiscent of those which fell regularly through Chang Yŏngsil's tubes in +1434, or from the mouth of Wang Fu's (candle dragon) of +1120, or earlier still from the +7th-century steel-yard clepsydras of Antioch, described in the +10th-century *Chiu Thang Shu*. Not only that, but there are the descendants of Su Sung's norias (*h*) ready to raise the balls up again to their starting-point. Finally (*i*) there is a window at which discs, the descendants of Su Sung's time-reporting jacks, present themselves in turn. It would be an instructive thing to have a replica of this whole instrument, together with suitable historical explanations, in every great museum of the history of science and technology in the world.

We are fortunately not without all light concerning its origins, for the *Chŭngbo Munhŏn Pigo*[2] (Complete Serviceable Study of (the History of Korean) Civilisation)[a] embodies a valuable, if somewhat incomplete, survey of the history of Korean scientific and horological instrument-making. After the reign of King Sejong astronomy declined for a time[b] and when in the +17th century it revived again the interest

---

[a] Lit. 'Complete Serviceable Study of the Documentary Evidence of Cultural Achievements (in Korean Civilisation), with Additions and Supplements'. First prepared as the *Tongguk Munhŏn Pigo* in +1770 and thoroughly revised twenty years later. Ch. 2, pp. 22*a* ff., ch. 3, pp. 1*a* ff., studied and translated in collaboration with Dr Lu Gwei-Djen.

[b] In +1467 King Sejo[3] called upon Yu Hŭiik[4] and Kim Nyu[5] to make theodolite-like instruments (*yin ti i*[6]) for surveying, and the next astronomical instrument, a small equatorial torquetum, was not made until +1494, by Yi Kŭkpae[7] under King Sŏngjong.[8] In +1526 all the existing instruments were repaired at the order of King Chungjong[9] by Yi Sun,[10] and in +1548 some new ones were made for King Myŏngjong.[11] Many of all these, including those which had come down from Sejong's observatory, were finally destroyed in the Japanese invasion of +1592, though at least one of the sundial-nocturnals

[1] 白道   [2] 增補文獻備考   [3] 世祖   [4] 俞希益   [5] 金紐
[6] 印地儀   [7] 李克培   [8] 成宗   [9] 中宗   [10] 李純   [11] 明宗

of the court turned more to mechanical time-keeping. Sejong's water-wheel clock with the jade girl immortal in the Pavilion of Respectful Veneration was burnt down and rebuilt twice, first in +1554 and then again in +1614, when Yi Chhung[1] was its restorer.[a] Yi Chhung also restored at the same time the Chiming Clepsydra and its Pavilion.

A new chapter opened when in +1657 King Hyojong[2] ordered Hong Chhŏyun[3] to make an astronomical clock, termed in antique parlance by his edict-writer a *hsüan-chi yü-hêng* (*sŏn-gi okhyŏng*[4]).[b] This instrument turned out to be not much good, so at the instance of the Hongmun Kwan[5],[c] Chhoe Yuji[6] was entrusted with the task and succeeded to perfection.[d] That his clock had a water-wheel linkwork escapement is certain.[e] The court's appetite for armillary clocks now grew rapidly, and under King Hyŏnjong[7] in +1664 two engineers, Yi Minchhŏl[8] and Song Iyŏng,[9] were commissioned to make two further ones. The difference between these two productions is clearly discernible from a memorial by Kim Sŏkchu[10] dated five years later, which by good fortune has been preserved. While Yi Minchhŏl remained faithful to the water-wheel drive and escapement (*hun thien i shui chi chih fa*[11])[f] Song Iyŏng introduced 'Western gear-wheels', i.e. in all probability a weight-drive or spring-drive clockwork mechanism with verge-and-foliots or even pendulum.[g] Thus we find our-

survived. Then a new sphere and globe were made by Yi Hangbok[12] in +1601. After this it becomes difficult to distinguish observational spheres from the armillaries which were incorporated in the clocks, but it seems that one of the former was cast by Yi Minchhŏl and Song Iyŏng in +1669, and another by An Chungthae and Yi Sihwa[13] in +1704. A general account of astronomy and horology under the Yi dynasty (which favoured these sciences from its inception) has been given by Rufus (2), pp. 22 ff., but all its particulars need careful revision.

[a] How long it lasted after that we do not know, but in +1669 Song Chun-gil[14] petitioned for its restoration, so by that time it must have become derelict again. A wooden figure believed to have been part of it was preserved, together with Yi Kŭkpae's equatorial torquetum, at Seoul as late as 1907.

[b] Cf. Vol. 3, pp. 334 ff. and *passim*.

[c] The College for the Cultivation of Literature (Hsiu Wên Kuan[15]), founded in +621, or, to give it the definitive name which it received five years later, the College for the Propagation of Literature (Hung Wên Kuan[5]), was yet another of those erudite government organisations, somewhat parallel to the Han-Lin Academy and the College of All Sages (cf. pp. 471 ff., above), to which Chinese feudal bureaucratism gave rise. The original home of the Library of the Four Categories of Literature, its Scholars and Assistant Scholars (established posts, as we should say) laboured on textual criticism and amendment besides giving regular instruction to students. Eventually it was attached to the Imperial Library. From *Hsin Thang Shu*, ch. 47 (cf. des Rotours (1), vol. 1, pp. 169 ff.) it is interesting to learn that the College had its own librarians, copyists, binders, brush-makers and paper-makers. The corresponding institution in Korea dated only from *c*. +1460, but it seems to have played the same role in astronomy and horology as the College of All Sages did in Thang China, for we constantly hear of its activities in these fields alongside the Royal Observatory.

[d] He was but county magistrate of Kimje,[16] and seems never to have risen higher.

[e] *Chi hêng shui chi tzu yün.*[17]

[f] 'A water-tank was set up above the wooden cover, and the water poured down through the clepsydra spout gradually into the little scoops inside the casing, so that as they became full they made the wheel turn round....'

[g] 'The armillary clock set up by Song Iyŏng was in general design much the same, but instead of using water-tanks he used the gear-wheels of Western clockwork which mutually engaged, being made of predetermined sizes....'

[1] 李冲	[2] 孝宗	[3] 洪處尹	[4] 璇璣玉衡	[5] 弘文館
[6] 崔攸之	[7] 顯宗	[8] 李敏哲	[9] 宋以潁	[10] 金錫胄
[11] 渾天儀水激之法		[12] 李恆福	[13] 李時華	[14] 宋浚吉
[15] 修文館	[16] 金堤	[17] 璣衡水激自運		

selves present at yet another turning-point, for just as Chŏng Chho and his colleagues in +1438 had substituted the water-wheel system for Chang Yŏngsil's anaphoric striking clepsydra, so Song Iyŏng in +1664 went over from the water-wheel system to clockwork of modern type.

Later developments can also to some extent be followed, for we learn that in +1687, under a new king, Sukchong,[1] Yi Minchhŏl was ordered to repair his clock, and this was successfully done.[a] Then in +1732, in King Yŏngjo's[2] reign, although an important Western clock had entered the Seoul Observatory in +1723 as the gift of a Chinese embassy, An Chungthae[3] was instructed to repair the armillary clocks which had been built and maintained in the time of Hyŏnjong and Sukchong. Yŏngjo himself now wrote an interesting essay on the Hall of Right Computation (for Government) entitled *Khuei Chêng Ko Chi*[4], where these time-keeping mechanisms were preserved.[b] After this we hear no more of the armillary clocks, presumably because clocks of ordinary type without astronomical components became more and more common and convenient. Nevertheless it seems highly probable that the wonderful instrument which we have described above is either the original of Song Iyŏng or a later reconstruction of it using most of the original parts, or else possibly a model based on the original design but made by some other instrument-maker during the 18th century. But we must now return to China.

In the 17th and 18th centuries, both before and after the Jesuit dominance in Chinese clock-making had given way to the lay commerce in 'sing-songs',[c] the horological traditions of China had a much greater effect upon European design than has generally been realised. Features emanating from the ancient tradition of I-Hsing and Su Sung were incorporated in European products destined for the 'Asiatick trade', and in Europe itself complex dials and elaborate jack-work became popular in a context of 'Chinoiserie'.

It will be remembered that the re-publication of Su Sung's book occurred in China,

---

[a] 'The design was as follows: inside a large casing there was set up the water (-delivery) tube and the bell arrangements and the stopping mechanism. South of the casing there was set up the armillary sphere, with its Component of the Six Cardinal Points and its Component of the Three Arrangers of Time, just as in the old designs. Of rings there were various sorts, with some features left out and some newly added. The sun and moon each had their rings (for travelling models?). There was no sighting-tube, but in the centre there was placed an earth model of paper with mountains and seas drawn upon it, to represent the (seemingly) level earth. The water(-delivery) tube was connected with the mechanism which rotated the sphere in the north–south polar axis. It was the power of the main axle which rotated the rings also as designated.' And the text goes on to describe the jack figures which sounded and displayed the double-hours and quarters. Then it adds: 'The water was contained in a tank at the top of the casing, and poured down into the (delivery-)tube, so that all the actions and rotations of the machinery, both inside and outside (the casing), were effected by the force of a single stream of water.' All this sounds very like the arrangements of the instrument still preserved, except for the difference of the power-source. Whether the clocks of Chhoe Yuji and of Song Iyŏng were repaired at the same time is not clear.

[b] This ought to be studied and translated in full. It makes no mention of Song Iyŏng's instrument, but it does mention Yi Minchhŏl's, and even more curiously it describes an anaphoric float mechanism, as if someone had reconstructed the old clock of the Chiming Clepsydra Pavilion.

[c] This was the term used in the jargon of the China coast for mechanical toys and clocks with fanciful jack-work. See Greenberg (1); Chang Lien-Chang (1). On the firm of Cox & Beale, which supplied them, see Needham, Wang & Price (1), p. 150.

[1] 肅宗   [2] 英祖   [3] 安重泰   [4] 揆政閣記

under the care of Chhien Tsêng (cf. p. 448 above), about +1665. Within twenty years Christopher Treffler or Trechsler[a] designed and built at Augsburg an astronomical clock called 'Automaton Sphaeridicum'. Vincenzo Coronelli, the great Italian cartographer, saw it and gave a description of it in his *Epitome Cosmographica*[b] ten years later. The instrument was 7 ft. high, and carried a celestial globe with an automatic movement, together with dials and jacks indicating years, months, days, hours and minutes, as well as eclipses for 17 years in advance. Above the globe was mounted an armillary sphere. The whole design, with its crosswise framework base, and the relative positions of globe and sphere, is so reminiscent of Su Sung's clock-tower as to invite the speculation that some knowledge of the latter had percolated to Europe through Jesuit channels in time to inspire Treffler's masterpiece of 1683.[c]

Certain catalogues and inventories of nearly a century later give us a glimpse of the kind of apparatus which was then being made for the Chinese market. James Cox was a clock-maker and jeweller of London, who in 1774 was obliged to sell off his stock in a lottery. His catalogues[d] describe a remarkable 'Chronoscope', the fellow piece of one which had been sent in the 'Triton' Indiaman to Canton in 1769 'and which now adorns the palace of the Emperor of China'. The apparatus, 16 ft. in height, was of course made in gold studded with gems, but embodied many features characteristic of the old Asian horological traditions. Besides a mechanised armillary sphere and celestial globe, there were jack figures which struck bells at the hours and quarters, flying dragons which dropped (real) pearls into the mouths of other creatures waiting to receive them, and an elephant which paraded on a horizontal wheel waving its trunk and tail.[e] The whole thing would have been recognised by Chang Hêng and Su Sung as entirely in their own tradition, yet it was made by men who were unconscious that any such tradition existed.

It is interesting to note, too, that the English were not the only people who offered such works of art to Eastern Asia. Takabayashi reproduces[f] a Japanese sketch of a clock sent from Russia by ship to Nagasaki as a present for the imperial court in 1804. Upon the base which contained the clock-dial or dials there stood an elephant bearing on its back an armillary sphere, presumably mechanised, with a pavilion the spire of which bore—for some inscrutable reason—a wind-wheel.[g] But long before, in +1725,

---

[a] Or perhaps rather his brother John Philipp. Bedini (1, 4) has contributed an admirable study of the Treffler family and their work. See also von Bertele (2).

[b] P. 333; the illustration is reproduced in Stevenson (1), vol. 2, pp. 94 ff. and Needham, Wang & Price (1), fig. 56. On Coronelli and his book see Armão (1), p. 189.

[c] The idea of driving astronomical demonstration models by clockwork was of course not new in Europe. There had been Philipp Irmser's 'Planeten Prunkuhr' of +1556, with its globe, and Jost Burgi's rock-crystal clock of +1610, with its armillary sphere (cf. Lloyd, 6). Cf. pp. 444, 492.

[d] Cox (1, 2). Cf. Lloyd (9).

[e] Perhaps the fullest description is given in the rare and anonymous pamphlet of 1769 (Cox, 3). From the first the work was intended for some Asian destination. It had Turkish and 'Tartar' jacks and figures, 'loorie birds' and 'paddy birds', with other 'rare birds and beasts that are in estimation in the country it is designed for', while the controls were in the form of a star and crescent.

[f] (1), fig. 27 and p. 50.

[g] The elephant motif went back at least to al-Jazarī in +1200, for one of his designs shows an elephant bearing a mahout who strikes a gong, and a rider who, pivoting, directs a pointer at the markings of a dial, the whole being surmounted by a pavilion of singing birds, etc. (Wiedemann & Hauser (4), p. 117).

the Russian embassy of Sava Vladislavitch had carried among its presents something similar.[a]

It hardly falls within the scope of this account to relate the astonishing story of how the Jesuits (and other missionaries) staffed the horological workshops of the imperial Chinese court for just under two centuries.[b] But one subject remains upon which a few words must be said. What did Chinese scholars themselves have to say about the up-to-date clockwork which the Jesuits brought with them? Was there any realisation of its kinship with the earlier inventions described in the Chinese historical literature?

The answer is yes. Already in the +17th century an example can be found. In his scientific and technical encyclopaedia, the *Wu Li Hsiao Shih*, finished in +1664, Fang I-Chih has an entry for 'automatically rotating machines' (*yün chi*[1]).[c] He begins by referring to the Ming sand-clocks of which we have already spoken.

There are certain automatically rotating machines which work neither by water-power nor wind-power. If a small stream is allowed to leave a tank at a higher level (and impinge upon a scoop-wheel on one side, then as it spills out, the other side), becoming lighter, moves upwards, thus turning the machine. Of these some there are which have sand collected in a reservoir which runs down by its own weight and acts as the motive force—thus the motion starts and develops.

He then refers to the 'wooden oxen and running horses' of +3rd-century Szechuan,[d] which he takes to have been something of the same kind,[e] after which he continues:

So the gears of pure iron (*chiai yao*[2]) turn the seven wheels[f], communicating the force precisely, the twelve teeth fitting together, and the large wheel (probably the dial) keeping time with the movement of the Heavens.

Now, he says, all this was just the sort of thing that was done by Chang Hêng, I-Hsing, and Tsu Chhung-Chih. Did not Su Ê[3] record[g] that in the Thang Korean Silla presented a Buddhist set-piece called Wan-Fo Shan[4], with a host of little figures bowing to the sound of bells? And of course there was Chang Ssu-Hsün early in the Sung, with his elaborate mercury-driven jack-work. All this was simply machinery such as Wei Pho[5] and Ma Chün[6] understood.[h]

[a] Cahen (1), p. 93. 'An English striking clock, playing twelve tunes, and embellished with a portrait of Peter the Great, mounted on a crystal globe and valued at Rbls. 700.'

[b] A brief account of this will be found in Needham, Wang & Price (1), based largely on the sketch of Pelliot (39), the book of Chapuis, Loup & de Saussure (1), and miscellaneous notes in Pfister (1). The subject deserves a well-balanced monograph; in the meantime there is an interesting review by Bonnant (1).

[c] Ch. 8, pp. 38*a*, *b*, tr. auct. This passage seems to have escaped notice hitherto.

[d] See pp. 260 ff. above on wheelbarrows, in connection with which they are usually discussed.

[e] Like the puppet theatre of Heron of Alexandria, in fact (cf. p. 442), or the 'automobile' of Wang Chêng (p. 218).

[f] By +1664, then, the sand-driven wheel-clock had become still more complex than it had been in +1635; cf. p. 510 above.

[g] Perhaps in his *Tu-Yang Tsa Pien*,[7] written about +890, but we cannot trace the passage. There is a full account, however, in *Samguk Yusa*, ch. 3, (pp. 141 ff.), which fixes the date at +764 and shows that the action was pneumatic (cf. p. 158 above). For this reference we are indebted to Mr G. Ledyard.

[h] The latter is of course the celebrated engineer of San Kuo and Chin times, but we have not met the former elsewhere.

1 運機   2 錯鎗   3 蘇鶚   4 萬佛山   5 魏朴   6 馬鈞
7 杜陽雜編

And now we have Sun Ju-Li[1] and Sun Ta Niang,[2] with their watches only an inch across (*tshun tzu ming chung*[3]). The workmanship is splendid, but after all what is there really surprising in it?

The passage of Hsüeh Chi-Hsüan quoted above (p. 462) seems to have been more widely known from the 17th to the 19th centuries than the book of Su Sung, though this was carefully copied about +1660 and printed again in 1817 and 1844. The 'revolving and snapping springs' of +1270 were even mentioned in poems. Thus Juan Wên-Ta,[4] in his ode on 'The Chiming Watches of the Red-Haired People' (Hung Mao Shih Chhen Piao[5]) wrote: 'Some say that these are the same sort of *kun-than*[6] which one can read about in Sung books.' Hsüeh's passage was also noted by Wang Jen-Chün,[7] searching about 1895 for evidences of old Chinese inventions to counter the all-inclusive and intimidating claims of occidental science at its most confident. His book, the *Ko Chih Ku Wei* (Scientific Traces in Olden Times), often mentioned by us, was far from being a scholarly production, yet it managed to quote (sometimes even correctly) many valuable texts. Talking of clepsydras,[a] he gives the quotation from Hsüeh, and then goes on:

> The Western striking clock (*Hsi-Yang tzu ming chung*[8]) is derived from the clepsydra. The *Chhou Jen Chuan* (Biographies of Chinese Mathematicians and Scientists) says that the *kun-than* was in fact the same thing as the *tzu ming chung*, and that we had them already before the Sung.

This would thus have been the opinion of the great scholar Juan Yuan, in the last decade of the 18th century.[b] And in another connection Wang Jen-Chün quotes[c] a scholar and official who about 1885 wrote in an account of his embassy to Russia:

> The automatically striking clock was invented by a (Chinese) monk, but the method was lost in China.[d] Western people studied it and developed refined (time-keeping) machines. As for the steam-engine, it really originates from the (monk) I-Hsing of the Thang, who had a way of making bronze wheels turn automatically by the aid of rushing water....[e]

Perhaps Wang Jen-Chün was giving references in his usual random way, for an extremely interesting passage on clockwork by Juan Yuan is to be found, not in the *Chhou Jen Chuan*, but in his collection of miscellaneous pieces of prose and verse, the *Yen Ching Shih Chi*,[9] printed in 1823. Read in the light of all that has gone before, it shows remarkable awareness. Juan Yuan wrote:[f]

[a] Ch. 1, p. 11 *b*.
[b] On this book, see Vol. 1, p. 50, and here, p. 472. In fact, Wên-Ta was his canonisation name.
[c] Ch. 5, p. 28 *a*. The book quoted is Wang Chih-Chhun's[10] *Kuo Chhao Jou Yuan Chi*[11] (Record of the Pacification of a Far Country), ch. 19. A fuller excerpt will be found on p. 599.
[d] I like to recall that when Dr Wang first came to Cambridge in 1946 he drew my attention to this passage, but neither of us was at that time disposed to take it seriously.
[e] Here Wang Chih-Chhun doubtless had primarily in mind the paddle-wheel steamboat, and confused the reciprocating engine with the steam-turbine just then being developed.
[f] Pt. I, sect. 3, ch. 5 (p. 650), tr. auct.

[1] 孫孺理　　　[2] 孫大娘　　　[3] 寸自鳴鐘　　　[4] 阮文達　　　[5] 紅毛時辰表
[6] 輥彈　　　　[7] 王仁俊　　　[8] 西洋自鳴鐘　　[9] 揅經室集
[10] 王之春　　　[11] 國朝柔遠記

The clocks (*tzu ming chung*[1]) (of today) come from the West, but the principle of them is derived from the water-clock (*kho-lou*[2]). The *Hsiao Hsüeh Kan Chu* quotes Hsüeh Chi-Hsüan as saying that there were four sorts of time-keepers (*kuei lou*[3]), the clepsydra, the burning incense, the sundial, and the *kun-than*[4] (revolving and snapping springs).[a] Now what the Yuan people called *kun-than* was just the same as the clock, and already before the Sung there were such devices, but the techniques got lost. The chief merit of the Western apparatus is the science of statics (and mechanics) (*chung hsüeh*[5]), i.e. the knowledge and utilisation of the properties of heaviness and lightness. But in fact all ingenious machinery derives from this, and relies on statics (and mechanics) for its functioning. Depending on the principles (*li*[6]) of this science, clocks are composed of wheels and screws, and from this comes their usefulness. The water-clocks of old had reservoirs of water which gradually diminished as it dripped down, all the while rotating a wheel—was this not because of the weight of the water? Clocks (nowadays) have a coiled piece of iron (*chüan*[7]) (i.e. a spring)[b] pinned inside a bronze drum (*chih thung ku chih chung li chih*[8]),[c] requiring force for its compression, from which state it tends to expand, the energy gradually diminishing—is this not because of the weight used?

Clocks have two drums (*ku*[9]), one for indicating the time,[d] the other for sounding the bell.[e] The time-indicating axle (*thung*[10]) is connected (with the drum) outside by a cord wound round to drive (*to*[11]) the second or 'pagoda' wheel (*tha lun*[12]). Pagoda-wheels (fusees)[f] are shaped like horizontal pagodas (*wo tha*[13]), and the cord is wound round them. The pagoda-wheel drives the third or central gear-train wheel (*chung hsin lun*[14]). The pointers (*chen*[15]) of the time-indicator (i.e. the hands) are on the same axle with this gear-wheel. The (third or) central gear-wheel drives the fourth vertical wheel, and that in turn drives the fifth or toothed (crown-)wheel (*chhih lun*[16]). If this had no (escapement pallets) to control it, if there was nothing to retard the frequency of the sound of the teeth, then the energy of all the wheels could not overcome the force of the iron spring in its drum. Thus obviously there comes a time when the spring has fully expanded. So there is a hanging hammer (*hsüan chui*,[17] i.e. a pendulum) swinging back and forth to control the rate. And the teeth of the toothed (crown-)wheel just correspond (*hsiang ying*[18])[g] (with the escapement pallets) so that there is a retardation transmitted all through from the fourth to the third and so to the second wheels while the spring is gradually released.

The striking-mechanism axle (*thung*[19]) is also connected (with a drum) by a cord to drive a second or pagoda wheel (fusee), and this drives a third or striking-train wheel. The outer axle of this is connected with the striking tooth (a catch), while inside there is set up a lever (*i*[20]) to move the bell hammer. The third or striking-train wheel drives a fourth or bird's-head (*niao thou*[21]) wheel, and this drives a fifth small wheel,[h] and this in turn a sixth or fan wheel

a Cf. p. 462 above.

b This is not the usual modern term for a clock spring, *fa thiao*.[22] In the early 19th century clock-makers used the rather picturesque phrase *kang chhang*,[23] steel intestines (cf. Shih Shu-Chhing, *1*).

c Note the recurrence of the word *li* in this technical explanation; cf. p. 485 above.

d The 'going train'.   e The 'striking train'.

f The fusee is a conical drum with guide grooves for the cord, which compensates for the diminishing force of the uncoiling spring by increasing the torque as the larger diameters come into play. The device was first illustrated and described by Petrus Alemannus in +1477, and sketched by Leonardo. See Lloyd (5), p. 656, (8). Cf. pp. 295, 445 above.

g Note the use here of a philosophical term which we have discussed in Vol. 2, pp. 304 ff.

h The writer seems to have omitted mention of some of the small gear-wheels in the going train.

1 自鳴鐘	2 刻扁	3 晷扁	4 輥彈	5 重學	6 理
7 卷	8 置銅鼓之中捩之		9 鼓	10 筩	11 奪
12 塔輪	13 臥塔	14 中心輪	15 鍼	16 齒輪	17 懸鎚
18 相應	19 筩	20 扗	21 鳥頭	22 發條	23 鋼腸

(*fêng lun*[1]). If there was no fan wheel to modify the force and retard the rotation the bell would be sounded too quickly. The striking-train drum is connected by a mechanism with the time-indicating wheels. When the time comes the catch works and the iron spring is allowed to expand. If the expansion is little the strokes are few, if it is much the strokes are numerous. When the striking is finished the catch holds back the force remaining, and it waits for the next (correct) time.

Alternatively they have two lead weights instead of the iron springs, and then they do not have the two drums, but this also is profiting intelligently from statics (and mechanics). The two drums and all the wheels are mounted within two bronze plates connected together by screws. The motion is all due to the power of rotating wheels. If one studies the pagoda-wheels and the iron springs, one sees that their effect involves also a screw arrangement. But everything depends upon weight overcoming lightness—statics. This is directly derived from the old water-clocks, and was not originated by the Westerners.

Here for the first time we meet with springs and fusees in Chinese dress, even though Juan Yuan seems to have been describing a very simple + 16th-century type of clock.[a] Much more interesting is the unerring penetration with which he discerned the role of practical physics in the senior inventions of the Thang and Sung. To have developed theoretical physics was the great contribution of Renaissance Europe, but Juan Yuan was quite right in believing that the fundamental invention of the mechanical clock had been made without it.

A few years later, another scholar-official, Liang Chang-Chü,[2] who had been looking into some Sung literature, came to just the same conclusions. So in his miscellaneous notes, *Lang Chi Hsü Than*[3] (Further Impressions collected during Official Travels), he wrote:[b]

The *Fêng Chhuang Hsiao Tu*[c] says that in the Thai-Phing Hsing-Kuo reign-period (+976 to +983) there was a Szechuanese Chang Ssu-Hsün[4] who made and presented (to the emperor) an armillary (clock) not at all like those of former times. It had a casing of several storeys totalling more than 10 ft. high, and seven wooden upstanding jacks which could regulate (the time) by striking bells and beating upon drums, beside which there were twelve figures of spirits which came out in rotation bearing placards of the hours. All this was just like the clocks (*tzu ming chung*[5]) of today.

Now in my part of the world, at Fuchow, we have a drum tower on which in former times there was installed a twelve (double-)hour automatic time placard system. According to tradition this was made in the Yuan period by one Chhen Shih-Thang[6] of Fu-ning. It lasted down to the Khang-Hsi reign-period (+1662 to +1722), when the apparatus was taken away by Chou Li-Yuan (Chou Fang-Po). Thus it was Chinese people who made these things, and such skilful technique must surely have been transmitted to the foreigners. Now men in Fukien and Kuangtung and Suchow and other places can all make clocks just as good as theirs. Indeed the (Provincial) Governor, (Chhi) Yen-Huai,[7] himself (recently) constructed out of pure bronze a celestial globe demonstrating the whole universe, using clockwork

---

[a] Cf. Lloyd (5), pp. 651, 656.   [b] Ch. 8, p. 5a, tr. auct.

[c] Written by Yuan Chhiung and completed soon after +1202. The full translation of the original passage will be found in Needham, Wang & Price (1), p. 72. Cf. p. 470 above.

[1] 風輪   [2] 梁章鉅   [3] 浪跡續談   [4] 張思訓   [5] 自鳴鐘
[6] 陳石堂   [7] 齊彥槐

within the horizon to rotate it. It can not only strike the hours and 'quarters' but also shows the passage of the stars and constellations without a hair's-breadth of error. So although Westerners can make such things they could do nothing better than this.

It is good to be introduced here to another man who built, like Kuo Shou-Ching, a water-wheel linkwork escapement clock in the +14th century, but the chief point again is the clearness with which Liang Chang-Chü recognised in the work of Chang and Chhen the ancestry of modern clockwork.

As for the work of Chhi Yen-Huai, we are well informed because of the recent study of Shih Shu-Chhing (1). Chhi was indeed a successful provincial official, and in the twenties of the last century as an amateur clockmaker did build good instruments, one at least of which, a celestial globe (Fig. 670), is still preserved.[a] A longer description of it was given by Chhien Yung[1] some decades later.[b] Like Tai Chen,[c] Chhi interested himself in the improvement of water-raising machinery, especially the Archimedean screw, on the construction of which he wrote a mnemonic rhyming manual, *Lung Wei Chhê Ko*.[2] We may salute in him a modern successor in the direct line from Chang Hêng, I-Hsing and Su Sung, and quite worthy of them.

Thus did the traditional scholars of China attempt to establish, ineffectively perhaps but bravely, the part played by their civilisation in the long epic of human knowledge of Nature and control over natural things, here in the field of time-measurement. Only the discipline of universal history of science could elucidate the truth, and substitute a clear recital for brilliant guesses and unsupported assertions.

The Japanese also thought that mechanical clocks had been invented in China. In the *Nippon Eitai-gura*[3] (Japanese Family Storehouse; or, the Millionaire's Gospel Modernised), an entertaining book on commercial life written about +1685, the author, Ihara Saikaku[4], referred to the matter as follows:[d]

> The Chinese are a self-composed people, and will never be rushed, even to make a living. ...The clock was invented in China. Year after year a man thought about it, with mechanisms ticking by his side day and night, and when he left the task unfinished, his son took over, in a leisurely way, and after him, the grandson. At long last, after three lifetimes, the invention was completed and became a boon to all mankind. But this is hardly the way to make a successful living....

Whether I-Hsing or Chang Ssu-Hsün would have recognised themselves in this description may well be doubtful, but the passage shows at least that at that time Europe was not yet recognised as the fountain of all technical originality. This point is more curious than might at first sight appear, for until now all the oldest Japanese clocks reported have been of an archaic verge-and-foliot type, and it is known that European clock-making was introduced to Japan by the Dutch at the beginning of the 17th

---

[a] For parallel European work of about the same period, and very similar design, see von Bertele (2), fig. 24, illustrating a pair of globes made by Desnos at Paris.

[b] In his *Lü Yuan Tshung Hua*,[5] ch. 12.   [c] Cf. Vol. 2, p. 516.

[d] Tr. Sargent (1), p. 105.

[1] 錢泳   [2] 龍尾車歌   [3] 日本永代藏   [4] 井原西鶴   [5] 履園叢話

PLATE CCLXI

Fig. 670. Astronomical clock in the form of a celestial globe, with concealed works, constructed by Chhi Yen-Huai in 1830 (Shih Shu-Chhing, *1*). Height 1 ft. 4 in. Probable scale 0·01 ft. to a degree. Now in the Anhui Provincial Museum.

PLATE CCLXII

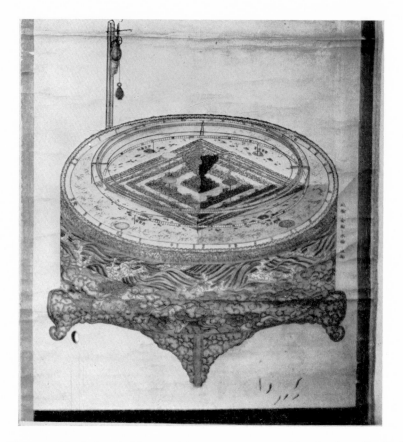

Fig. 671. Uranographic clockwork-driven model or orrery constructed in 1814 by or for the Japanese Buddhist monk Entsū on the principles of the archaic Kai Thien cosmology (cf. Vol. 3, pp. 210 ff.), which he wished to sustain against the world-view of modern science. The earth, square and flat, centres upon Mt Meru, the sacred mountain connecting the earth with the heavens in classical Indian and Buddhist *abhidharma* natural philosophy. It is surrounded by concentric ring-continents separated by annular oceans (cf. Vol. 3, Fig. 242). Above the earth three elevated railways appear to represent the equator and the paths of the sun and moon, and one quarter of the world is seen in darkness, as when the sun is behind the central mountain. The same conception has already been illustrated from a Greek Christian source of the +6th century in Vol. 3, Fig. 218; it seems primarily Indian. To the left a small verge-and-foliot clock can be seen, which probably rotated the planetary models; for different levels of the barrel-shaped case, apparently representing the four elements, also seem to have rotated, perhaps worked by a stronger spring-drive within.

century.[a] What Ihara says might thus suggest that Chinese sand-wheel or water-wheel clocks of Ming or Sung type had in fact been used earlier in Japan also.

Indeed it was in Japan that the design of orreries and planetaria underwent its most curious developments. The first Japanese to construct a celestial globe rotated by clockwork was no doubt Nakane Genkei[1] (+1661 to +1733), one of the founders of modern astronomy in that country.[b] By the time of his death, as we saw at an earlier stage,[c] a copy of the original 'orrery' had been ordered by the East India Company, though it did not reach China until the last decade of the century. Meanwhile similar instruments must have become known in Japan, for from 1810 onwards certain strange cosmological models, rotated by clockwork, began to be reproduced there. Although at first sight these resemble orreries or planetaria in their discoidal character, they were profoundly different for they were based on the most archaic of all the Chinese cosmologies,[d] the Kai Thien shuo,[2] combined with the Indian conception of a central mountain (Mt Meru) connecting the earth with the heavens.[e]

These models were associated with the activities of a remarkable monk, Entsū[3] (+1754 to 1834), whom we have already met[f] as the author of the Bukkoku Rekishō-hen[4] (Astronomy of Buddha's Country), a work in which he attempted to advance the interests of Buddhism by showing the superiority of ancient doctrines over the modern scientific view of the world.[g] Entsū was well versed in the history of Chinese astronomy from Lohsia Hung to Kuo Shou-Ching, but for him its most primitive forms were present verity rather than historical preamble. In 1813 he published in several versions his Shumisen Gimei narabini Jo Wage[5] (Inscription and Preface for the Mt Sumeru Instrument, with Translation), in which he described his cosmological working model. The preface is often reproduced in kakemono scrolls still extant which depict the apparatus (Fig. 671). The heavens are a round dome, here represented as flattened (perhaps sometimes in glass), the earth is square and flat. It centres upon Mt Meru or Sumeru (Khun-lun Shan[6]) which broadens upwards to support the heavens, and surrounding the four inner continents of which Jambūdvīpa is one there are a number of concentric ring-continents separated by annular oceans. Here clearly is the last representative of what we called in the geographical Section the East Asian tradition of religious cosmography,[h] paralleling in many respects the wheel-maps and T-O maps of early medieval Europe.[i] There is ground for thinking that the ultimate origin of all this was Babylonian, but the East Asian tradition (primarily Buddhist and Taoist) certainly

[a] Cf. Needham, Wang & Price (1), p. 168. The full policy of exclusion of foreigners came into effect only from +1641 onwards.

[b] Cf. Hayashi (2), p. 354.   [c] P. 474 above.

[d] Cf. Vol. 3, pp. 210 ff.

[e] We are much indebted to Mr Silvio Bedini for knowledge of these instruments, conveyed to us in the first instance through the kind intermediation of Dr Derek Price, and for permission to reproduce the photograph of a scroll in his collection. We also have to thank Mr Nakayama Shigeru for collaborating with us in the study of Entsū and his work.

[f] See Vol. 3, p. 457. He adopted the philosophical sobriquet of Wu Wai Tzu.[7]

[g] Cf. Mikami (12); Muroga & Unno (1).

[h] Vol. 3, pp. 565 ff.   [i] Vol. 3, pp. 528 ff.

[1] 中根元圭   [2] 蓋天說   [3] 圓通   [4] 佛國曆象編
[5] 須彌山儀銘并序和解   [6] 崑崙山   [7] 無外子

owed much to the old Indian cosmologies dating back to the time of Gautama Buddha.[a] To find it still vigorous in uranoramas of the early 19th century is little short of extraordinary.

Above the flat earth already mentioned one can see in Fig. 671 a single equatorial circle and seemingly two ecliptics, one doubtless for rotating a model sun, the other for a model moon. Constellations characteristic of the four palaces of the heavens are depicted on eight mushroom-like 'umbrellas' at the four quarters; these could hardly be movable. On the left of the instrument a clockwork mechanism with verge-and-foliot escapement and weight-drive may be seen. Not all the scrolls show this,[b] so in some of the models an internal spring motion may have been used. Probably the instrument in Fig. 671 had two drives, the little one on the left serving to move round the sun and moon models, while another, stronger, one inside the drum below, rotated its various storeys. These are marked on the right from top to bottom 'earth-wheel', 'metal-wheel', 'water-wheel' and 'wind-wheel', but whether these had a planetary or an element significance (or both), and how their rotation affected the cosmological model above, remains obscure.[c]

The text of the preface is full of archaic features.[d] Entsū draws justification not only from the Kai Thien classic, the *Chou Pei Suan Ching*,[e] but also from the *I Ching* (Book of Changes).[f] He claims that all his arrangements are in conformity with the teachings of the *Abhidharma Śāstras*, and includes a number of fanciful figures for celestial and terrestrial distances, such as the obliquity of the ecliptic, in *yōjanas*.

The lengths of days and nights [he says] are not equal. This depends upon the fact that the seven environing mountain ranges are of different and varying heights; it is as if one should stand on a step and looking northwards see the ground rising gradually. (In the uranorama) one sees too the four continents all spread out, with square and round, triangular and crescent-shaped, all demonstrating the form of the Great Illusion (*mo yeh*,[1] i.e. Māyā, the universe). Here we cannot exhaust all that can be said, but it will be found in the Indian (Buddhist) calendar writings. This is fully in accordance with what the *Chou Pei* says about numbers and the calendar, and also with the numbers of the Yin and Yang mentioned in the *I Ching*. Sometimes they merely hint, elsewhere they clearly confirm one another. How could one not reverence the divine virtue of the sages?[g]

This is like nothing so much as the belief in the literal inspiration of the first chapters of Genesis.

[a] In itself, as Unno (2) has pointed out, the conception of a central mountain-mass (Mt Meru, Khun-lun Shan) does no discredit to the ancient geographers of the Old World, for it did but recognise the existence of the Tibetan-Himalayan massif. What made the idea archaic as early as the Han was its use as a means of explaining the movements of the heavenly bodies or the alternation of day and night (cf. Fig. 218 in Vol. 3).

[b] E.g. another in Mr Bedini's collection, and an illustration, published but not explained, by Mody (1).

[c] It is said that one of Entsū's models still exists at the Science Museum in Tokyo; if so, a full description would be of much interest.

[d] A draft translation has been made in collaboration with Dr Lu Gwei-Djen, but Entsū's writings are difficult to follow on account of his use of technical terms special to Buddhist astronomy.

[e] Cf. Vol. 3, pp. 19 ff., 210 ff.   [f] Cf. Vol. 2, pp. 304 ff.

[g] Tr. auct. On the *yōjana* see Vol. 4, pt. 1, p. 52.

[1] 摩耶

Entsū seems to have been quite influential in early nineteenth-century Japan. His work was part of a nationalist reaction against the flood of new knowledge (*Rangaku*) emanating from contacts with the Dutch.[a] Other writers too fought back under the banner of Mt Meru, for instance an anonymous calendrical expert (*13*) who produced a *Shumikai Yakuhō-rekki*,[1] and a Buddhist cosmologist, Takai Bankan,[2] whose *Shumisen Zukai*[3] (Illustrated Explanation of Mt Sumeru), issued in 1809, has imaginative pictures of the sacred mountain.

Thus by a strange reversal of roles, that mechanical clockwork which had been brought to East Asia to serve the ends of Christian missionary propaganda found itself enlisted in the service of Buddhist cosmological orthodoxy. But the time for religious interference in scientific matters was fast passing away. It is related that after Entsū's book *Bukkoku Rekishō-hen* appeared, he was criticised most severely by the leading astronomer Inō Tadataka[4] in the following year. Perhaps if we knew more of these discussions we should find in them an early parallel for the famous exchange of T. H. Huxley with Bishop Wilberforce. Yet the bizarre perpetuation of the archaic world-view was still going on in 1848, when further copies of similar scrolls were made.[b] Between 1850 and 1852 the flat-earth models of Entsū's type were combined with complicated clocks in a number of ingenious pieces[c] made by Tanaka Hisashige.[5] But the end was now near for in 1854 came the return of Commodore Perry and the opening of Japan with all that that implied for the unification of world science.

Two pleasant items of incidental intelligence may serve to close this story. In 1809 a Shanghai man, Hsü Chhao-Chün[6] by name, wrote a treatise on clock-making entitled *Tzu Ming Chung Piao Thu Fa*[7] (Illustrated Account of the Manufacture of Self-Sounding Bells).[d] He said in his preface that he came of a family which had been making 'European' clocks for five generations.[e] Perhaps he was a descendant of

[a] Cf. Chesneaux & Needham (*1*); Kuwaki (*1*). The story has not yet been fully told in a Western language. Entsū was competing with such influences as the *Rigaku Hiketsu*[8] (Elements of Physics) by Kamata Ryūkō[9] (1815) and the *Kikai Kanran*[10] (Astronomy and Meteorological Physics) of Aoji Rinsō[11] (1825). China is not without parallels for the anti-*Rangaku* movement (cf. Vol. 3, pp. 454 ff.) but on the whole they were much more progressive scientifically and more concerned with recovering and studying the past in the light of a proper historical perspective.

[b] E.g. some large polychrome wood-block prints under study by Mr Bedini.

[c] See Takabayashi (*1*), figs. 33, 34A, B.

[d] This was included later in the third section of his *Kao Hou Mêng Chhiu* (Important Information on the Universe), already mentioned, Vol. 3, p. 456. There is a parallel Japanese literature. The *Karakuri Zui*[12] (Illustrated Treatise on Horological (lit. Mechanical) Ingenuity), written by Hosokawa Hanzō Yorinao[13] in 1796, has been reproduced in facsimile, with modernised transliteration, by Yamaguchi Ryūji (*1*). The modern names for the escapement are of interest. Chinese clock-makers call it the 'loose arrester' (*chhin tsung chhi*[14]) or the 'barrier-piece' (*chhia tzu*[15]). Dr Liu Hsien-Chou is to be thanked for this information.

[e] Some extant clocks of Hsü Chhao-Chün's time are described by Hsü Wên-Lin & Li Wên-Kuang (*1*). A good deal of discussion has gone on regarding the clocks mentioned in the famous 18th-century novel *Hung Lou Mêng* (Dream of the Red Chamber), completed in +1795. See the papers of Chhen Ting-Hung (*1*), Fang Hao (*2*) and Yen Tun-Chieh (*16*).

[1] 須彌界約法層規　　[2] 高井伴寬　　[3] 須彌山圖解　　[4] 伊能忠敬
[5] 田中久重　　[6] 徐朝俊　　[7] 自鳴鐘表圖法　　[8] 理學秘訣
[9] 鎌田柳泓　　[10] 氣海觀瀾　　[11] 青地林宗　　[12] 機巧圖彙
[13] 細川牛藏賴直　　[14] 擒縱器　　[15] 卡子

Ricci's celebrated friend Hsü Kuang-Chhi [1].[a] And down to the end of the 19th century at least, the clock-makers of that city venerated Matteo Ricci as their tutelary deity or patron saint under the name of Li Ma-Tou Phu-Sa.[2] How amused Su Sung, and especially the Buddhist I-Hsing, would have been!

(6) CLOCKWORK AND INTER-CULTURAL RELATIONS

All that is now needed to draw the material of this story into a coherent whole is a brief summary of the roles of the different culture-areas in the development of mechanical clockwork. Let us examine the provisional scheme which has been embodied in Table 59.

What were the factors leading to the first escapement clock in China? The chief tradition leading to I-Hsing ($+725$) was of course the succession of 'pre-clocks' which had started with Chang Hêng about $+125$. Reason has been given for believing that these applied power to the slow turning movement of computational armillary spheres and celestial globes by means of a water-wheel using clepsydra drip, which intermittently exerted the force of a lug to act on the teeth of a wheel on a polar-axis shaft.[b] Chang Hêng in his turn had composed this arrangement by uniting the armillary rings of his predecessors into the equatorial armillary sphere, and combining it with the principles of the water-mills and hydraulic trip-hammers which had become so widespread in Chinese culture in the previous century.[c] But I-Hsing was indebted to some other sources also, such as the artisans from Ma Chün ($c.\ +300$) onwards who had made horizontal jack-wheels and other mechanical toys worked by water-wheels, and also the Taoist clepsydra experts (e.g. Li Lan, $c.\ +450$) who had made steelyard balances for weighing the water used in time-keeping. In particular it was Kêng Hsün and Yüwên Khai who had arranged to weigh a vessel intermediate in the chain; hence perhaps the idea of weighing a series of them attached to a revolving wheel. As for gearing, we have seen evidence that much experimentation with it was carried on through the centuries by Chinese mechanics.

What were the factors leading to the first escapement clock in Europe ($c.\ +1300$)? The weight-drive descended no doubt from the floats of the Hellenistic anaphoric clocks and mechanical puppet theatres, and was certainly known in its free form in the $+13$th century, as the Moorish drum water-clock of the Alfonsine corpus demonstrates.[d] The use of gearing for simulating time-measurement descended from remote

---

[a] At any rate there were, towards the end of the 18th century, a number of eminent clock-makers of this name, the brothers Hsü I-Ying [3] and Hsü I-Sung,[4] and the son of the latter Hsü Yü.[5] See Tsêng Chao-Yü (1).

[b] As has been pointed out on p. 474 above, it is not yet excluded that the water-wheel linkwork escapement goes back to Chang Hêng himself.

[c] The lug constituted a pinion of one, and the whole 'click-mechanism' was closely similar to some of the machines described by the Alexandrians (e.g. Heron) and Vitruvius. The hodometer, for example. But exactly the same mechanism was used in the Chinese hodometers (p. 282 above), contemporary with them, or almost so.

[d] Rico y Sinobas (1); Feldhaus (22). Cf. Beckmann (1), vol. 1, pp. 349 ff. on the astronomical clock sent by Saladin of Egypt to the emperor Frederick II in $+1232$; the parts moved 'ponderibus et rotis incitatae'. Cf. Schmeller (1).

[1] 徐光啓　　　[2] 利瑪竇菩薩　　　[3] 徐翊漢　　　[4] 徐翊淞　　　[5] 徐玉

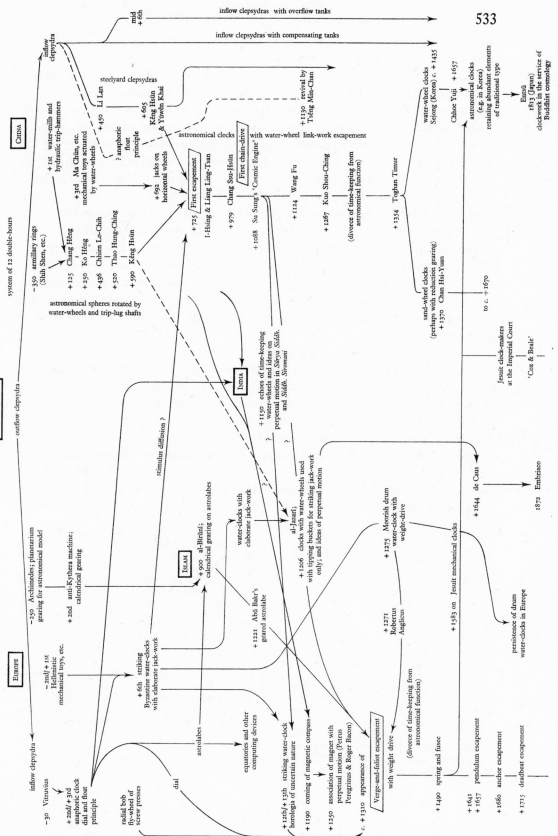

antiquity, for even if we know little of the nature of the planetarium ascribed to Archimedes (*c.* −250),[a] the Anti-Kythera object, with its elaborate gear-wheels, remains[b] to show the extraordinary attainment which Hellenistic (−1st century) technique could reach. Then in the Arab realm there was the application of calendrical gearing to the computational astrolabe. We know about this from a text[c] of Abū al-Raiḥān al-Bīrūnī, dating from about +1000, and there are extant specimens of astrolabes fitted up in this way, notably one at Oxford[d] made by Muḥammad ibn Abū Bakr of Ispahan in +1221. The clock-dial again may be considered a derivative of the astrolabe's face and therefore ultimately of the revolving dial of the anaphoric clock.[e] For jack-work devices there were plenty of precedents in the Byzantine striking water-clocks[f] and their Arabic successors. Only the 'soul' of the mechanical clock, i.e. the escapement itself, had to be provided by the key inventors, whoever they were, of +1300. The form which this actually took, namely the verge-and-foliots, may reasonably be derived from the radial bob fly-wheel, familiar since the early days of Graeco-Roman screw-presses, though now converted by the pallets from discontinuous rotary to regular oscillating motion. But just how original was the basic idea? The preceding six centuries of Chinese escapements suggest that at least a diffusion stimulus travelled from east to west.

To gain a little light on this transmission, if such it was, we must concentrate attention on the years between about +1000 and +1300 and see whether any help can be found from the Islamic and Indian culture-areas. It is noteworthy that arrangements logically equivalent to those of Chang Hêng and his successors are to be found in later Arabic writings. For example, one of the mechanisms described by Ismāʿīl ibn al-Razzāz al-Jazarī in his treatise on striking clepsydras in +1206, as interpreted by Wiedemann & Hauser,[g] consists of a tipping bucket attached to a hinged ratchet which pushes round a gear-wheel by one tooth each time that the bucket fills with water and comes to the emptying point (Fig. 672).[h] This gear-wheel is connected by a cord with what seems to be the plate of an anaphoric clock. And the arrangement possessed a

　　[a] See e.g. Cicero, *De Re Publica*, I, xiv, 21; Ovid, *Fasti*, VI, 277. Most of the relevant passages have been assembled in translated form by Price (8). The best study on it is that of Wiedemann & Hauser (5), but see also Drachmann (2), pp. 36 ff. The only reference to a water-power drive occurs in Pappus (*Opera*, VIII, 2); this dates from the century following Chang Hêng.

　　[b] Rediadis (1); with other references and description in Price (1). This is now regarded as a calendrical analogue computing machine. Internal evidence dates it at −82, and it was probably lost at sea about −65. A detailed monograph on it by Prof. D. J. de S. Price is in preparation; in the meantime see Price (8, 9).

　　[c] Wiedemann (13).

　　[d] Gunther (2), no. 5. Apart from the Anti-Kythera machine, this is almost the earliest example of precision gearing as distinguished from power-transmitting gearing.

　　[e] Drachmann (6); Neugebauer (7); Price (5).

　　[f] Such as the one at Gaza described by Procopius about +510 (Diels, 2). Or, in Sassanid Persia, the time-keeping throne of Khosroes II (*r.* +590 to +628), on which see Christensen (1), p. 461. Presumably all these derived from the parerga of the striking water-clock described by Vitruvius (IX, viii, 5) about −30 and due perhaps to Ctesibius before −200; cf. Drachmann (2), pp. 19 ff., (9), pp. 192 ff.

　　[g] (4), p. 147, elucidating the *Kitāb fī Maʿrifat al-Ḥiyal al-Handasīya* (Book of the Knowledge of Geometrical (i.e. Mechanical) Contrivances), VI, 6. On this work cf. Mieli (1), p. 155; Sarton (1), vol. 2, p. 632. Al-Jazarī's *floruit* was +1180 to +1210 (Suter (1), no. 344).

　　[h] In pondering on the ancestry of the tipping bucket we should not forget what has been said above (p. 363) about the spoon tilt-hammer.

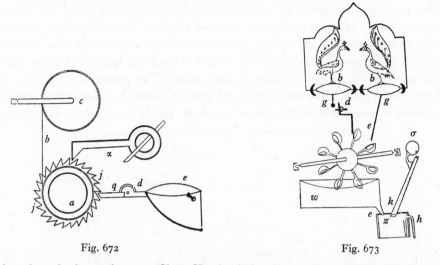

Fig. 672   Fig. 673

Fig. 672. A mechanism analogous to Chang Hêng's trip-lug for rotating a disc or globe by water-power, from al-Jazarī's treatise on striking clepsydras (+1206), after Wiedemann & Hauser (4). The bucket (e), filled by a constant stream of water, tips the hinged lever (d) as it overturns when full, thereby rotating the wheel (a) one step by the action of the linked lever (q) on its ratchet teeth (j, j). A pawl (a) prevents backward rotation. As the wheel turns it winds on the drum (c) by the cord or chain (b). The shafts of the pawl and drum are intended to be at right angles to the plane of the paper.

Fig. 673. Water-wheels in Arabic horological machinery, another device from al-Jazarī's treatise (+1206), after Wiedemann & Hauser (4). A small impulse-turbine or Pelton wheel, rotated periodically by a flow of water from a tipping bucket (not here shown), trips the levers (d, e) thus moving the spindles (g, g) and activating the peacock jacks (b, b). At the same time the water escaping through the sump (w) into the tank (z) causes the pipe (k, σ) to sound, and eventually siphons off through the outflow (h). The shaft of the water-wheel is supposed to be at right-angles to the plane of the paper. Here, as always in Arabic striking clepsydras, the water-wheel is part of the striking, not the going, mechanism.

strange longevity, for it is found in a closely similar form in the book which Isaac de Caus published on mechanical contrivances[a] in +1644. Indeed the principle was still employed by J. B. Embriaco (using two alternate buckets and a pendulum escapement)[b] for a public clock set up in the Pincio Gardens at Rome in 1872. Now al-Jazarī also had water-wheels in his clocks, a fact which can hardly be without significance in relation to earlier Chinese practice. But here they never seem to be used as prime movers actuated by continuous flow for trains of time-keeping gears in consecutive rotation; they simply come into uncontrolled action intermittently whenever the bucket tips out its water, and they operate peacocks and other jack figures, either by trip-levers (Fig. 673)[c] or intermediate gear-wheels (Fig. 674).[d] Where we do find

[a] De Caus (1), pl. 5 and accompanying text, reproduced by Needham, Wang & Price (1), fig. 45. Elsewhere in the same book, he also illustrates (pl. 8) and describes a simple anaphoric clock—rather late in the day. This also is reproduced in Needham et al. (1), fig. 33a.

[b] See Ungerer (1), p. 372. This device had closely similar ancestors in the double-bucket system invented by al-Badī' al-Asṭurlābī (Suter (1), no. 278) who died in +1139; see Wiedemann & Hauser (3); Perrault's clock of +1699; and Triewald's blowing-engine (cf. p. 365 above).

[c] Kitāb, VI, 4 (Wiedemann & Hauser (4), pp. 105, 145).

[d] Kitāb, VI, 2 (Wiedemann & Hauser (4), p. 142). We have already illustrated the external appearance of the jack-work of one of al-Jazarī's great water-clocks in Vol. 1, Fig. 33, opp. p. 204. Ameisenowa (1) offers evidence that the ensemble was intended to represent the 'music of the spheres'.

gearing continuously turned by a water-wheel is in certain machines for delivering a constant supply of air to an organ.[a] In these a vertical shaft, rotated by right-angle gearing from a vertical water-wheel, carries a semilunar cam which raises two valves alternately as it rotates, thus admitting water to two closed spaces in turn so that one can be expelling air while the other is taking it in.[b] This arrangement is referred by al-Jazarī to the Banū Mūsā brothers (+9th century)[c] and to a Byzantine, Apollonius the carpenter, about whom little more is known. But it is not a time-keeper.

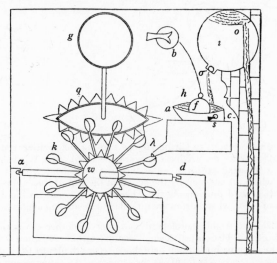

Fig. 674. A third device from al-Jazarī's treatise of +1206 (after Wiedemann & Hauser, 4). A tank (*i*) with an overflow outlet (*o*) delivers water at constant rate through the onyx orifice (*σ*) into the tipping bucket (*a, s, c, h*). Descending periodically through the funnel (*λ*) the water impinges on the cups (*k*) of a small Pelton wheel which is mounted on the shaft (*a, d*) beside a gear-wheel (*w*). Engaging with a larger gear-wheel (*q*) this operates a jack-work disc or sphere (*g*), and further jack machinery is activated by the float (*f*) which is connected by a cord or wire with the small drum (*b*). The turbine shaft is supposed to be at right-angles to the plane of the paper. Again the water-wheel is part of the striking, not the going, mechanism.

There is nothing in the evidence so far available, therefore, which suggests any Arabic influence on the Chinese developments.[d] From the beginning of the +8th century onwards the Chinese clocks were undergoing a steady evolution in such a continuous line that external influence seems very improbable. On the other hand, the Arabic material does indicate the passage westwards of certain Chinese elements.

[a] See Wiedemann & Hauser (3).

[b] The complexity of this method is in striking contrast with the elegant simplicity of the Chinese double-acting piston-bellows (see p. 135).

[c] Suter (1), no. 43.

[d] Even the largest of the clocks described by al-Jazarī, and the Bāb Jairūn clock at Damascus built in +1168 by the father of the physician Riḍwān ibn Rustam al-Khurāsānī al-Sāʿātī (Suter (1), no. 343), who wrote a description of it in +1203 (see Wiedemann & Hauser, 4); though equipped with jack-work and lamp-lighting devices much more complicated than anything previously known in the West, all worked on the anaphoric float principle, with tripping mechanisms acting at intervals. So, too, did the remarkable 'Observatory of the Time and the Hour', a most elaborate water-clock erected in +1325 at Yazd in Persia for Rukn al-Dīn ibn Niẓām al-Ḥusainī by Khalīl ibn Abū Bakr. An excellent description of this has been translated by Sayili (2), pp. 236 ff., from the history of Yazd by Aḥmad ibn Ḥusain, a +16th-century writer. Cf. Sykes (2), p. 421. The remains of striking water-clocks of the +14th century still exist at Fez in Morocco, and have been studied by Price (13).

Here a momentary digression imposes itself. One hardly realises how many of these medieval devices live on in the most ingenious pieces of modern scientific apparatus. Thus the continuous-recording gas calorimeter of C. V. Boys, perfected about thirty-five years ago (see Fig. 675),[a] but still in use today, not only evokes the memory of the water-powered clock of Su Sung, but also embodies components familiar to al-Jazarī and the Alfonsine mechanicians. This apparatus, designed to be independent of everything except the coal-gas supply, has a clock powered by a water-wheel incorporated in a closed-circuit water system. Regulated by an anchor-and-pendulum escapement, this wheel has three functions: (a) controlling the flow of gas through a barostatic meter by powering a differential train which transmits automatically a continuous thermo-barometric adjustment, (b) assuring the cycle of delivery to the calorimeter of successive measured amounts of cold water, (c) driving two recording-paper drums and a clock. The measuring of the water is done by a tipping bucket—just as in the old Arabic designs—so arranged that the jet passes to the water-wheel when the bucket is not receiving. Each half-minute a rod worked by a cam on the water-wheel returns the bucket to the receiving position. Besides this, another Arabic component is to be seen in the segmental compartment drum of the gas meter; and besides the closed-circuit water-wheel, a chain-drive connecting it with the thermo-barometric integrator also recalls the Chinese designs of the +11th century. Paradoxically, the disc-ball-and-cylinder integrator, with its beautifully simple geometrical relationships, though the most Greek part of the whole construction, is yet far beyond what the Greeks themselves ever achieved. On the other hand the differential gear by means of which it modifies the speed of the water-wheel shaft may well have had its origin in the Chinese south-pointing carriage (cf. pp. 296 ff. above). Sir Charles Boys was a man of outstanding mechanical ingenuity[b] and the whole apparatus may have been pure re-invention, yet one can never know what hints or stimuli reached him from the records of the past.

Alternating water-buckets are still to be seen working at the present day in production machinery. Thus they are found in wire-making factories, where an automatic mechanism shunts rapidly-moving white-hot steel rods through different guide-pipes. As long as it is moving past the joint, the steel rod itself acts as a splice, but as soon as it has passed, the moving guide-pipe section mounted on a bell-crank swings over by the weight of a bucket of water which has been slowly filling, and routes the next rod in a different direction, at the same time automatically switching the water-supply to the second, now empty, bucket. In the interval before the arrival of a third rod the process is reversed.[c]

In his systematic censure of vulgar errors (+1646), Sir Thomas Browne took issue with the current representations of St Jerome with a mechanical clock in his study,

[a] The best description is in the book of Hyde & Mills (1), pp. 195 ff., but cf. Boys (1, 2). I am greatly indebted to Major Claud G. Hyde for the photograph here reproduced, and for so kindly showing me the apparatus at Vauxhall. Much credit is due to Mr K. A. Singer, also of the Gas Standards Branch of the Ministry of Fuel and Power, for recognising the antique elements in the Boys calorimeter, and many thanks for calling our attention to it.

[b] See the short biography by Paget (1).   [c] Jones & Horton (1), vol. 1, pp. 413 ff.

for he knew well that this apparatus was of relatively recent origin. At the same time its modernity struck him as very odd, and he wondered whether still further ingenuity might not produce a time-piece which would go of itself. After Dials and Water-glasses

of later years there succeeded new inventions, and horologies composed by trochilick or the artifice of wheels; whereof some are kept in motion by weight, others perform without it. Now as one age instructs another, and time that brings all things to ruin, perfects also every thing; so are these indeed of more general and ready use than any that went before them.... It is I confess no easie wonder how the horometry of Antiquity discovered not this Artifice, how Architas that contrived the moving Dove, or rather the Helicosophie of Archimedes, fell not upon this way. Surely as in many things, so in this particular, the present age hath far surpassed Antiquity; whose ingenuity hath been so bold not only to proceed below the account of minutes, but to attempt perpetual motions, and engines whose revolutions (could their substance answer the design) might out-last the exemplary mobility, and out-measure time itself.[a]

Thus in the mind of Sir Thomas there was a connection between powered clockwork and the idea of perpetual motion. How could this have arisen?

A curious feature about the +12th- and +13th-century Arabic work was the belief

[a] *Pseudodoxia Epidemica*, Bk. v, ch. 18 (Sayle ed., vol. 2, p. 251).

Fig. 675. An example of the continuing utilisation of medieval mechanical and horological devices in modern apparatus, the continuous-recording gas calorimeter of Sir Charles Boys, still in use (photo. Ministry of Fuel and Power, by the courtesy of Major Claud G. Hyde and Mr K. A. Singer). The instrument records, in B.T.U. per cu. ft., the calorific value of coal-gas measured under standard conditions of temperature and pressure at water-vapour saturation. The heat of combustion of continuously burning gas is transferred in a heat-interchanger to continuously flowing water and measured by two expansible thermometers. The movement thus initiated is recorded on a paper band.

Power and timing for most of the components is effected by a water-wheel somewhat like that in Su Sung's clock (+1090), fed from a constant-level tank, but fitted with a pendulum-and-anchor escapement. This water-wheel rotates a meter-shaft by means of a chain-drive like that of Chang Ssu-Hsün (+978). A segmental compartment drum, reminiscent of those of the Alfonsine mechanicians (c. +1275), is fitted concentrically with this shaft and constitutes the meter, gas entering to turn it through a rheostatic strip-valve which adjusts the flow in an automatic manner so as to keep the speed of drum and shaft identical, thus compensating for slight changes in gas pressure. The speed of the drum is also controlled by epicyclic gearing fitted on the shaft between the water-wheel and the meter. This differential train, a descendant of the mechanism of the south-pointing carriage of Yen Su (+1027) and others, is operated by the rise or fall of an air-bell in an annular mercury seal when pressure and temperature conditions deviate from standard, the motion being transmitted through a ball-disc-and-cylinder integrator and noted on a recorder. In standard conditions the ball of polished phosphor-bronze, held in a fork, is stationary in the centre of the disc, but if it is moved to one side or the other by the thermobarometer it turns at a rate proportional to its excursion, thus acting as a bevel gear, rotating the cylinder and varying the speed of the drum through the epicyclic gearing. Meanwhile, a tipping bucket, essentially like those described by al-Jazarī (+1206), fed from the same tank as the water-wheel, doles out fixed amounts of cold water to the heat interchanger at half-minute intervals and is returned automatically to the receiving position by a rod-and-cam device from the water-wheel. A closed circulation of water throughout is effected by a small hot-air engine and pump using the coal-gas supply, which is thus the sole external feed. Hence the apparatus is independent of any failure of water pressure and dispenses with electric current.

Whether Sir Charles Boys was conscious of the antiquity of some of the mechanical devices which he used remains uncertain, but their presence in a modern inventor's stock of ideas is interesting. Older pendulum-regulated water-wheel clocks are known, e.g. that at Dinnet in Aberdeenshire.

Left to right: (a) gas-meter drum within the glass-topped vessel; (b) above, thermo-barometric integrator and recorder, below, air-bell; (c) water-wheel, with delivery pipe above and sump beneath, behind, the rod for re-setting the bucket; (d) above, tipping bucket, centre, heat-interchanger, below, shaft of water-wheel (clock shaft) leading to dial-face, escapement and drive of calorimetric recorder (not seen).

PLATE CCLXIII

(a)　　　　　　　　　　　(b)　　(c)　　　(d)

Fig. 675. See opposite page.

PLATE CCLXIV

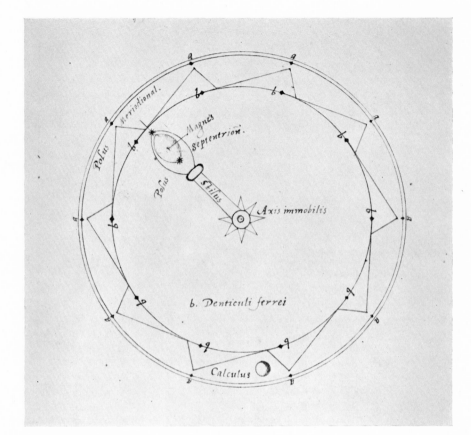

Fig. 676. The magnetic perpetual-motion wheel proposed by Petrus Peregrinus (+1269), a copy of the famous diagram in the +1558 Augsburg edition (*Epistola de Magnete seu Rota Perpetua*) by an anonymous English translator not long afterwards (Caius MS. 174/95). The lodestone (*magnes*) is set on an arm (*stilus*) fixed upon a stationary shaft (*axis immobilis*), while all round the periphery of the wheel there are iron teeth or studs (*denticuli ferrei*). A weight (*calculus*) would, Peter thought, help the motion of the wheel; perhaps he meant it to act as an 'indexer', stopping the wheel periodically at just the right places. 'When one of the teeth comes near the north pole (of the magnet)', he said, 'and owing to the impetus of the wheel passes it, it then approaches the south pole, from which it is driven away rather than attracted....Therefore such a tooth would be constantly attracted and constantly repelled....Being caught between the teeth of a wheel which is continuously revolving, (the weight) seeks the centre of the earth...thereby aiding the motion of the teeth and preventing them from coming to rest in a direct line with the lodestone.'

Although it is difficult to visualise how Peter himself thought that his machine would work, the oneirocritical masters could hardly fail to find in it an adumbration of the electric motor of later times, when electromotive forces would be applied to armatures.

in the possibility of perpetual-motion machines. Soon afterwards, as Sarton has pointed out,[a] this dream began (for the first time) to engross the imagination of European scholar-artisans also.[b] Schmeller (1) has studied a group of Arabic manuscripts[c] dating from about +1200, and containing parts of the text of al-Jazarī's *Kitāb*, which illustrate a number of devices for perpetual motion, sometimes in the form of norias (but neither receiving nor delivering), sometimes in the form of drums containing water-tight chambers of various kinds.[d] Still more remarkable, exactly similar descriptions are found in the *Sūrya Siddhānta* and Bhāskara's *Siddhānta Śiromaṇi*,[e] written about +1150. If the text in the former (ch. 13) were genuine, it would date from the +5th century, but it has long been regarded as an interpolation of about Bhāskara's time.[f] His own perpetual-motion machines were mainly noria-like wheels with mercury in the closed scoops ('quicksilver-holes').[g] The *Sūrya Siddhānta* text first describes a demonstrational armillary sphere and says that it may be sunk in a casing to represent the horizon.[h] It then continues:[i]

16. ...By the application of water ascertainment is made of the revolution of time.

17. One may construct a sphere-instrument combined with quicksilver—this is a mystery; if plainly described, it would be generally intelligible in the world.

18. Therefore let the supreme sphere be constructed according to the instruction of the teacher. In each successive age, this construction, having become lost, is, by the sun(-god's)

19. favour, again revealed to some one or other, at his pleasure....So also, one should construct instruments for the ascertainment of time.

20. When quite alone, one should add quicksilver to the astonishing instrument. By the gnomon, staff, arc, wheel, and shadow-measuring devices of various kinds,

21. according to the instruction of the teacher, a knowledge of time will be gained by the diligent.

22. By water-instruments, by vessels, by the peacock, man, monkey,[j] and by stringed receptacles, one may accurately determine time.

23. Quicksilver-holes, water, and cords, ropes, and oil and water, mercury, and sand are used in these—these applications, too, are difficult.

There is evidently something more here than the simple anaphoric clock device for turning a celestial globe apparently described about +500 in the *Āryabhaṭīya*:[k] 'One should cause a sphere of light wood, equally rounded and of equal weight on all sides, to move in regular time by means of quicksilver, oil and water.' The late +15th-century commentary of Parameśvara explains that the float is a gourd filled with

---

[a] (1), vol. 2, p. 764.   [b] See the excellent history of the subject by Dircks (2).

[c] E.g. Leiden Cod. 499, nos. 1414, 1415; Gotha, Pertsch. cat. no. 1348; Oxford Cod. 954.

[d] Reproduced in Needham, Wang & Price (1), figs. 65, 66.

[e] Ed. Wilkinson & Bapu Deva Sastri; ch. 11, pp. 227 ff.

[f] See Burgess ed., pp. 282, 298. The only reliably old parts of this *Siddhānta* are those included in the *Pañca Siddhāntikā*.

[g] Or in a peripheral channel. One machine had scoops like a water-wheel as well, presumably to get it under way.

[h] Note the significance of this in view of what we know of the earlier Chinese practice (p. 449 above, and Vol. 3, p. 386).

[i] Burgess ed., pp. 305 ff.   [j] Jacks, no doubt, like al-Jazarī's.

[k] Golapāda, 22. We do not know what the apparatus really was; it may equally well have been a demonstrational armillary sphere. This reference is due to Dr David Pingree. Cf. pp. 469, 481.

mercury and descends with the water-level in the clepsydra. Emptying every 24 hours, one complete revolution of the globe is effected. The oil is for lubricating the bearings. It is not surprising that the anaphoric clock principle was transmitted to India, but that had been in a simpler age. The really seminal transmission was that of the belief in the feasibility of perpetual-motion machines, and this must have reached the Arabs from India during the + 12th century, travelling westwards together with the 'Hindu' numerals and place value in one and the same period.[a] By the + 14th century the perpetual-motion mercury wheels of the Indians and Arabs had begun to appear in Latin dress.[b] It may very well not be fanciful to seek the ultimate origin or pre-disposition of the Indian conviction in the profoundly Hindu world-view of endless cyclical change, *kalpas* and *mahākalpas* succeeding one another in self-sufficient and unwearying round. For Hindus as well as Taoists, the universe itself was a perpetual-motion machine.

India thus provides an indubitable association between armillary spheres, time-measuring water-wheels, and devices for perpetual motion. Even Chang Ssu-Hsün's mercury drive seems to be present. One gets a strong impression from some of the Sanskrit texts that the writer was trying to describe water-wheel clocks of Chinese type in veiled language, or else that he knew only vaguely how they worked. Indeed one begins to entertain the belief that the stimulus for the flood of ideas on perpetual-motion devices may have derived from Indian monks or Arabic merchants standing before a clock-tower such as that of Su Sung and marvelling at its regular action.[c]

The *perpetuum mobile* makes its appearance in Europe in the notebooks of Villard de Honnecourt (+ 1237),[d] but much more significantly in the great work of Petrus Peregrinus (+ 1269), *Epistola...de Magnete*.[e] More significantly because the association just mentioned is here enlarged by the addition of the lodestone and its properties.[f] Since, as we have already seen (Vol. 4, pt. 1, p. 246), the first knowledge of the magnetic compass in Europe may be dated close to + 1190, and since there is no remaining doubt that it travelled thither from the Chinese culture-area, the most obvious con-

---

[a] As Lynn White (7), p. 131, has acutely pointed out. Cf. Vol. 3, pp. 10 ff., 15 ff., 146. The first complete account of them was in the *Liber Abaci* of Leonardo Pisano (Fibonacci) in + 1202.

[b] As we know from an anonymous treatise of about + 1340 on natural philosophy (Thorndike (1), vol. 3, p. 578).

[c] More especially as it will be remembered that the water-circulation was a closed one. No stream of water was seen to enter, no mill-race led away. Chinese onlookers without mechanical knowledge might well have spoken to the visitors of magic art. Lynn White (7), p. 130, frowns upon this conjecture, hardly realising, perhaps, the extent of the contact between China and India during the Thang and Sung periods (cf. Vol. 1, pp. 208 ff., 214 ff.).

[d] Cf. Sylvanus Thompson (4); Dircks (2), vol. 2, pp. 1 ff.

[e] I, 10 and II, 3 in particular. The short but important tractate of Pierre de Maricourt or Peter the Stranger, addressed to his friend Siger de Foucaucourt, was first printed just before + 1520 but attributed wrongly to Ramon Lull. The definitive edition which has the value of a MS. was the second, issued at Augsburg in + 1558 by Achilles Gasser. It is entitled *Epistola De Magnete seu Rota Perpetua Motus*. Caius Coll. MS. 174/95 contains a transcript of the Latin introduction of this, with an English translation by an unknown hand of the work itself. A late + 14th-century MS. was published in facsimile by Anon. (46), and the critical Latin edition is that of Hellmann (6). English translations are available by Thompson (5); Mertens (1) and Chapman & Harradon (1).

[f] A glance again at p. 229 above will show the appearance of yet another Chinese invention in Europe at this time—the Cardan suspension. But it was not part of this idea-complex and its connection with the magnetic compass only came about much later.

clusion is that the carriers of the transmission, whoever they were, spoke not only of the magnet, but also of the armillary spheres perpetually in motion urged by their tireless wheels.[a] Of course the idea of perpetual motion may be said to have sprung naturally from the apparent diurnal rotation of the heavens themselves, and no doubt the polar directivity of the lodestone was easily associated with this cosmic movement, but still the coincidence seems more than fortuitous. Peregrinus, writes Taylor (6), conceived of the lodestone as presenting a model or microcosm of the celestial sphere, part corresponding to part, and believed that if correctly poised it would rotate in spontaneous sympathy with the rotating firmament. He therefore proposed to make an armillary sphere with a freely turning axis which held a perfectly shaped and balanced lodestone. Once the axis was elevated so as to correspond exactly with the celestial axis, and the apparatus patiently set in motion by the hand until it took up the motion of the sky, the astronomer would possess an instrument which would serve as a perpetual time-piece and give him indeed all the celestial data he required.[b] Roger Bacon referred to this apparatus three times.[c] Before +1250 he wrote,[d] fresh from observing Petrus Peregrinus at work: 'A faithful and magnificent experimenter is straining to make an armillary sphere out of such a material, and by such a device, that it will revolve naturally with the diurnal heavenly motion.' By +1267 he had revealed[e] that this material was the lodestone—though success had still not been achieved.[f] This would be an instrument to outdo all other astronomical instruments, and worth a king's treasure. Four years later comes the statement of Robertus Anglicus, already quoted (p. 442), which shows that pending the realisation of Peter's plan, men were seeing what could be done with weight-drives. Apparently a Chinese scientific mission to Europe in +1250 could have saved a lot of trouble.

It has been customary in modern science to despise the line of seekers after perpetual motion, and certainly the projectors of the 19th century were tedious enough, but Lynn White (5, 7) has done a good service by pointing out that in correct historical perspective, the idea of perpetual motion had heuristic value. As early as +1235

[a] Dr Derek Price was the first to notice this. Cf. the discussion in Needham, Wang & Price (1), pp. 192 ff.; Price (8), pp. 108 ff. As Lynn White (5) reminds us, Indian influence on Europe was also very apparent during these centuries, not only in the transmission of the Hindu-Arabic numerals (cf. Smith (1), vol. 1, p. 215, vol. 2, pp. 72, 74; Smith & Karpinski (1); and Vol. 3, p. 15) but also in other fields (cf. Thorndike (1), vol. 2, pp. 236 ff.).

[b] '...Arrange the stone on the meridian circle, on its pivots fixed lightly in the poles of the stone, so that it may move in the manner of Armillaries, in such wise that the elevation or depression of its poles may be in accordance with the elevation and depression of the poles of the heavens in the region in which you are. Now if the stone then move according to the motion of the heavens, rejoice that you have arrived at a secret marvel. But if not, let it be ascribed rather to your own want of skill than to a defect of Nature.... By means of this instrument at all events you will be relieved from every kind of clock, for by it you will be able to know the Ascendant at whatever hour you wish, and all the other dispositions of the heavens which astrologers seek after.' 1, 10, tr. Thompson (5).

[c] In the De Secretis, Opus Majus and Opus Minus.   [d] De Secretis.

[e] Opus Minus.

[f] The automatic microcosmic armillary was not the only device for perpetual motion based on magnetism which Petrus Peregrinus proposed. In II, 3 he described a wheel which should revolve by the alternate attraction and repulsion of iron teeth or studs fixed in its periphery relative to a lodestone set on a fixed arm. This, unlike the former, he illustrated by a famous diagram. We reproduce in Fig. 676 the carefully drawn copy of it found in the Caius MS. 174/95, drawn during the lifetime of William Gilbert. Other versions are often figured, as e.g. in Klemm (1), p. 92.

William of Auvergne, then Bishop of Paris, had in his *De Universo Creaturarum* used the phenomena of magnetism to account for the perpetual motion of the celestial spheres.[a] Perhaps this stimulated Petrus Peregrinus. Then Peter was followed by many who clung to the notion that a lodestone, once properly mounted, would revolve unceasingly and automatically. In +1514 Meygret maintained that since it would be turning under the direct influence of the heavens, the usual objections to perpetual motion could not hold. Indeed if one thinks of the idea as an attempt to capture the diurnal rotation of the heavens as a source of terrestrial energy, it seems eminently reasonable, though based on the false assumptions of its time. Not until the 19th century would magnetic fields participate in energy conversions useful to humanity, and not until the age of the first gyro-compasses and artificial satellites would man be able to profit directly from the gravitational forces of the solar system. In +1562 Taisnier described and figured an apparatus which he hoped would give a 'magneticall' perpetual motion,[b] and during the following centuries innumerable projects of a similar kind were brought forward. They signified little; what was important was the influence of Petrus Peregrinus upon William Gilbert. At the end of the +16th century, Gilbert (as Zilsel (3) has pointed out) was much inclined to believe that Peregrinus had been essentially right, because his own experiments with the spherical lodestone or 'terrella' (model earth) had led him to conceive of the earth itself as one vast magnet, in diurnal rotation precisely because of this property. Although the truth of the idea could not be demonstrated, it proved a helpful solvent of some of the classical objections to the Copernican cosmology, which Gilbert indeed himself defended.[c] Even more important, the example of the field of magnetic attraction invisibly extended in space led directly, through Gilbert and Kepler, to the Newtonian concept of universal gravitation.[d] Thus the adopted Indian belief in the possibility of perpetual motion, allied with the transmitted Chinese knowledge of magnetic polarity, deeply influenced modern scientific thought at one of its most crucial early stages (cf. Fig. 677).

Truly they influenced technology too. Lynn White (7) has drawn attention in memorable pages to other features in the philosophical and social history of mechanical clocks. It can hardly be coincidence that their rise in +14th-century Europe paralleled that of the new science of dynamics with its rejection of Aristotle's theory of motion (the continuing push) and its assumption of motive virtue (the transported impetus).[e] Though the term *machina mundi* seems to go back in Christendom to Dionysius the Areopagite, it was no +18th-century deist, but one of the pioneers of the new medieval physics, Nicholas d'Oresme (*d*. +1382) who first used the metaphor

---

[a] Cf. Duhem (3), vol. 3, p. 259; Sarton (1), vol. 2, p. 588; Thorndike (1), vol. 2, pp. 338 ff. This was just the time when Villard de Honnecourt was working, though his *perpetuum mobile* was not magnetical. Though hardly better able than Huai Nan Tzu to distinguish magic from science, William was quite a sceptic in his way, and did not believe that a magic charm could stop a water-mill (Thorndike, *loc. cit.* p. 351). Cf. p. 202 above. The world-outlook of power technology was raising its head.

[b] Cf. Dircks (2), vol. 1, pp. 6, 18 ff., vol. 2, pp. 33 ff.; Thorndike (1), vol. 5, pp. 580 ff.

[c] Cf. Wolf (1), pp. 293 ff.

[d] Cf. Vol. 4, pt. 1, pp. 60, 236, 334; Butterfield (1), pp. 56, 79, 126 ff.; Needham (45).

[e] Cf. Vol. 4, pt. 1, pp. 57 ff.; Dugas (1, 2); Clagett (1, 2).

PLATE CCLXV

Fig. 677. European medieval man peers out through the firmament to see the rotational machinery of the solid celestial spheres, a German woodcut of the +16th century (Zinner, 9). The gap between the rims of the sky and the earth is an ancient motif of folklore cosmology (cf. Vol. 3, p. 215). Here we are reminded of the way in which three strands of ancient Asian knowledge and belief helped to destroy the prison of the solid crystalline celestial spheres in which Europeans had lived from Ptolemy's time (+2nd century) to the Renaissance. Chinese knowledge of magnetic polarity and directivity, transmitted westwards by the end of the +12th century (the time of Alexander Neckam, cf. Vol. 4, pt. 1, pp. 245 ff.) and described in part by Petrus Peregrinus (+1269), led to the experiments of William Gilbert (c. +1590). Indian belief in the possibility of perpetual motion, which had reached Europe through Islam by about +1200, now flowered in the conception that the earth itself was a spherical magnet in continuous rotation. The Copernican hypothesis was thus supported from an unexpected quarter, and Gilbert explicitly upheld it. Not only the earth but also the sun, moon and planets, should be considered 'magnetical bodies', from which 'the virtue magnetical is poured out on every side around in an orbe'. Hence Kepler's efforts to explain the motions of the planets magnetically; and, by the analogy of attraction, Newton's concept of the universal law of gravity. By Newton's time a third strand had exerted great influence in Europe, namely the idea traditional in Chinese astronomy and cosmology that the planets and stars were luminaries of unknown nature floating stationary or moving with various speeds in infinite empty space (cf. Vol. 3, pp 438 ff.).

PLATE CCLXVI

Fig. 678. A monastic water-driven wheel-clock of +13th-century Europe, an illumination in a biblical MS. (Bodleian Library, 270b, fol. 183v.). The incident portrayed is the moving back of the sun 10° by the clock for King Hezekiah during his illness at the word of the prophet Isaiah (2 Kings xx. 5–11 and Isaiah xxxviii. 8). Moved by the −8th-century king's goodness and sorrow, the Lord added fifteen years to his life and promised deliverance from the Assyrians; 'and this shall be a sign unto thee from the Lord, that the Lord will do this thing that he hath spoken—behold, I will bring again the shadow of the degrees, which is gone down in the sundial of Ahaz, ten degrees backward. So the sun returned ten degrees, by which degrees it was gone down.' The mechanism of the clock limned by the monk about +1285 is hard to make out, but a water-wheel seems to be present, with spout, wires, cords, and bells. The picture thus suggests that water-wheel linkwork escapement clocks of the Chinese type were used in Europe before the advent of the weight-driven wheel-clock (Drover, 1). Certainly this horo-logium is not at all like a sundial, in spite of the biblical text which the monk was illustrating.

Renaissance artisans amused themselves in making sundials which would give Hezekiah's sign. The American Philosophical Society in Philadelphia possesses an instrument of this sort made by Christopher Schissler of Augsburg in +1578. When the flattened bowl of the scaphe is filled with water, refraction causes the sun's shadow to move just 10° backwards.

of the universe as a vast clock set in motion with all harmony by God.[a] Thus complex celestial uniformity as well as particular temporal intervention would now show forth the Creator's wisdom and power, so that in abbeys and cathedrals the miracle-play had to make room for the astronomical clock. As Lynn White says, there was in these impressive pieces neither pious deception (as in Hellenistic temples), nor over-awing mystery (as in Byzantine palaces), but frank admiration of mechanical potentiality and skill (as, he might have added, in Chinese academies and courts). When the city-halls of burghers began to be adorned with such instruments the real divergence of the Western from all other cultures manifested itself significantly indeed—but that is another story.[b]

By the middle of the +13th century [in the West, wrote Lynn White], a considerable group of active minds, stimulated not only by the technological successes of recent generations but also led on by the will-o'-the-wisp of perpetual motion, were beginning to generalise the concept of mechanical power. They were coming to think of the cosmos as a vast reservoir of energies to be tapped and used according to human intentions. They were power-conscious to the point of fantasy. But without such fantasy, such soaring imagination, the power technology of the Western world would not have been developed.[c]

Now it is clear also that without the earlier discoveries and speculations of Chinese and Indian naturalists there might well have been neither the fantasy nor the power. Out of this background stepped not only Roger Bacon and Peter the Stranger but Francis Bacon too. What the world has not yet recognised is that in fact they all talked like Taoists—in the echoing words of the Han and Thang:

Man's might can conquer the changes of Nature, make thunder in winter and ice in summer, make the dead walk and the dry wood blossom, confine a spirit in a bean...open doors in paintings and make images speak....[d]

and again: 'The sage commands Nature and is not commanded by Nature.'[e]

Although much can be said about the technological thought of the +12th and +13th centuries in Europe, the somewhat humiliating fact remains that the actual nature of the 'horologes' then used, mostly in abbeys and cathedrals, constitutes one of the darkest patches of our ignorance. But Howgrave-Graham (1) has distinguished a cluster of records from +1284 onwards which suggest the rise of a new invention at this time, or the unwonted popularity or development of some existing device. The monumental clocks concerned were almost surely not driven by falling weights. Of the clock in the cathedral at Wells in western England, for example, the Welsh poet Dafydd ap Gwilym (+1343 to +1400) wrote in curious terms—it had 'orifices' in it, as well as wheels, weights, ropes, hammers, and also 'heads' and 'tongues'; words strangely reminiscent of Su Sung's 'coupling tongue'. As late as +1353 there was

---

[a] See Thorndike (1), vol. 3, p. 405, vol. 4, p. 169.
[b] Cf. our account of the social and economic background of the history of science and technology in East and West in Vol. 7. Meanwhile see Vol. 3, pp. 154ff.
[c] (7), p. 133.
[d] Vol. 2, p. 444 above, from the *Kuan Yin Tzu* book, ch. 7, p. 2b. Cf. also p. xlii above.
[e] Vol. 2, p. 60 above, from the *Kuan Tzu* book, ch. 37, p. 8a. *Shêng jen tshai wu pu wei wu shih.*[1]

[1] 聖人裁物不爲物使

'a cistern in the clock' of the great tower of Windsor Castle, set up for Edward III by a *magister horologii* and two other Lombards.[a] To Drover (1) we owe the study of a remarkable manuscript[b] dating from about +1285, which has a picture apparently showing a water-wheel clock (Fig. 678). The wheel is distinctly reminiscent of the driving-wheel of Su Sung; water can be seen pouring from an animal head into a sump below, and there are other objects, including a row of five bells, more difficult to make out.[c] Drover has also analysed the accounts of some of the mysterious *horologia*. For instance, the *Chronicle* of Jocelyn of Brakelond tells[d] how in +1198 when a fire broke out in the abbey church of Bury St Edmunds in East Anglia, the monks ran to the clock to get water. Inscriptions in the abbey of Villers in Belgium, dating from +1268, which have been studied by Sheridan (1), also indicate without doubt a water-clock. Yet Cologne in +1235 had an 'Uhrengasse' which was inhabited by smiths.[e] The issue therefore is whether the clocks of the late +12th and early +13th century in Europe were water-wheel clocks or not. We must hope that further discoveries may throw light on this question. If the existence of water-wheel clocks could be proved, it would suggest a transmission at the time of the Crusades, such as seems to have occurred with the invention of the windmill (see pp. 564 ff. below), rather than at the later time of Marco Polo and Petrus Peregrinus.[f] In that case the inventors of the verge-and-foliot would have had about a century or so to try out their improvements on the old hydro-mechanical escapement. But in some ways it is easier to imagine that the transmission was of the stimulus diffusion type only, and that firm conviction of the prior successful solution of the problem elsewhere led European scholar-artisans to solve it themselves in a different manner.[g] We really do not know.

[a] Brown (1).   [b] Oxford Bodl. Cod. 270b.
[c] A row of bells in a curiously similar position occurs in another representation of a clock, carved in a relief of Tubal-cain on the façade of the cathedral at Orvieto about +1320, where it is combined with a set of right-angle gear-wheels and what might conceivably be a weight-drive; see J. White (1). Compare with this one of the Kyeser MSS. illustrations (+1405) reproduced by Gille (12), pt. 2, fig. 12. We are not enamoured of the suggestion of Lynn White (7), p. 120, that the Bodleian picture of Hezekiah's clock represents an Alfonsine compartment-drum system, but how it worked remains (in the *pai hua*) anybody's guess.
[d] Butler ed., p. 106.   [e] Bassermann-Jordan (1), p. 17.
[f] These two periods approximately coincide, one may remember, with the two foci of reception by the West of Greek and Arabic astronomical knowledge, the translation of the Toledan Tables in +1087 and that of the Alfonsine Tables in +1274. The corpus of auxiliary texts to these restored the sphere and globe to medieval Europe.
[g] In meditating about the nature and form of this transmission or stimulus, it is desirable to have in mind others which are more or less certain and which occurred at times not far different. If we may choose the date +1300 to +1320 as a focal one for the appearance of the first working mechanical clocks in Europe, we may remember that +1325 was equally focal for gunpowder, the original home of which had been +10th-century China. Evidence for this statement will be found in Sect. 30. Towards +1380 we find the first blast-furnaces producing cast iron in Europe—but in China this technique had been developed from the −4th century onwards and by the Sung period had reached a high level of excellence. Evidence for this statement will be found in Sect. 30d (cf. Needham (31, 32) for shorter surveys). Towards +1375, and also in the Rhineland, comes the first European block-printing, an art which had been current in China since the +9th century (cf. Carter, 1). Still closer to clockwork in time are the great segmental arch bridges of Europe, the first being about +1340, though in China structures of this kind had first appeared in the +7th century (see the discussion in Sect. 28e). The +14th century thus presents itself as a time of adoption by Europeans of a number of important techniques which had already been known and used for a long time in the Chinese culture-area. One may indeed believe that Europeans did not know exactly where they came from, but it is asking too much

In any case, whether the Chinese escapement came to Europe in person, or only as a rumour, I-Hsing's great contribution was to introduce a truly chronometric principle into the mechanical as opposed to the clepsydric part of the clock. At first this was not great, for it is clear that in the Chinese clocks the major part of the time-keeping was effected by the constancy of the flow of the water. The mechanism could only intervene in so far as changing the weight on the weighbridge would permit the scoops to fall before they were quite full. This is why we must recognise in it the missing link between the purely hydrodynamic clepsydra and the purely mechanical clock. For when the inventions of early +14th-century Europe had been made, the verge-and-foliots took over the greater part of the time-keeping duty. This was still not completely embodied in the escapement since any considerable change in the weight hanging from the drum would affect the fastness or slowness of the clock.[a] Not until the introduction of the pendulum in the 17th century was an approach to a truly isochronous mechanism made. That this had taken two millennia is not surprising when one considers the leisurely growth of human technology as a whole before the Renaissance. But the Chinese contribution was vital. Its recognition[b] enables us henceforward to estimate at their true value statements such as the following—so often found. 'The Chinese', writes Lübke (1), 'never made any discoveries comparable with those of Europeans in the technique of clock-making. The clocks (collected in such great numbers in the Forbidden City) have of course nothing to do with time-measurement in Old China.' And, says Planchon (1), with superb (though unconscious) irony, 'the Chinese have never produced any mechanical clockwork properly so called —in this field they have only been bad imitators'.

The profound influence of clockwork on the world-outlook of developing modern science was noted at the beginning of this subsection. But emphasis has also rightly been placed[c] on the importance of the trade of clock- and watch-maker for the growth of scientific instrument-making in Renaissance Europe. These craftsmen became for science what the millwright was for industry—a fruitful source of ingenuity and workmanship. The millwrights had been there all through the Middle Ages, and the clockmakers from the beginning of the +14th century. Their presence was certainly one of the important roots of Renaissance science, pure and applied, for a supply of artisans was ready to generate makers of machines and instruments as soon as these things were demanded and devised. By now it is abundantly clear that China also

to maintain that they were all new and independent discoveries. There is evidence of a somewhat similar wave of adoptions towards the end of the +12th century, when within a few decades of the year +1190 Europeans came to know and use the magnetic compass, the stern-post rudder, and the wind-mill. It is curious and rather striking that these two periods are also closely characterised by an influx of eastern knowledge in the fields of astronomy and cosmology. For the Toledan Tables and their auxiliary texts became known in Europe by about +1200, while the Alfonsine Tables and their corresponding texts were more quickly appreciated, exerting their influence shortly after +1300. The conclusion is almost inescapable that clockwork and all that it imples fits perfectly into this picture as an importation of the second wave.

[a] Cf. von Bertele (1).

[b] Not likely, perhaps, to be readily accorded by all Western scholars. It will be urged that the European oscillating escapement was the only true, the only important, escapement. So already R. S. Woodbury, in his review of Price (8).

[c] E.g. by Bernal in his Beard Lectures, (1), p. 235.

had such artisans, in skill and ingenuity at least as eminent. If therefore China had no Renaissance and no development of modern science and technology, the presence of artisans was evidently not in itself enough. And though clock-making in China seems never to have become a mass industry before the time of the Jesuits (as it did in +15th- and +16th-century Europe),[a] the building of mill-work and water-raising machinery of all kinds was spread throughout the length and breadth of the empire. The manifold activity of skilled millwrights was therefore not enough either.[b] Yet 'it was the work of the millwright', says Bernal,[c] 'that gave rise to the first genuinely European invention, that of the clock....' Although in the light of the knowledge here set forth, the second part of the sentence can no longer be sustained, the remarkable insight shown in the first pays a debt which we all owe to the engineers of the Middle Ages.

(k) VERTICAL AND HORIZONTAL MOUNTINGS; THE REVOLVING BOOK-CASE IN EAST AND WEST

In the foregoing subsections we have become very conscious of the distinction necessary in the history of technology between wheels mounted vertically and wheels mounted horizontally.[d] The noria and the Vitruvian power-wheel, with their relations the ship-mill and the paddle-wheel, have a distribution different in time and space from the horizontal power-wheel.[e] In the material which we must now examine, we shall see that horizontal wheels are again characteristic of China and indeed of all Central Asia and Iran, both as regards certain wheels turned by hand in libraries, and as regards wheels rotated by a new source of power, the wind; while conversely the Vitruvian system continued to exert its influence in the West.[f]

The revolving book-case[g] is not perhaps a very inspiring subject from the modern technological point of view, for it led to nothing further, though in its heyday, as one of the numinous aspects of great Buddhist temples, it deeply affected many Chinese Confucian scholars not easily led astray by their emotions. Its interest for us lies

[a] The significance of the trade for the development of industry in general may be seen from the fact that Marx (1), pp. 334 ff., gave watch-making as the classical example of heterogeneous manufacture— a machine industry before the rise of factory production. William Petty also used the watch to illustrate the division of labour.

[b] Some preliminary thoughts on the great question of the factors which led to the rise of modern science and technology in Europe alone have already been given in Vol. 3, pp. 154 ff. A fuller discussion will follow in the last volume (Sects. 46 ff.).

[c] (1), p. 234.

[d] What social significance the distinction between vertical-axis capstans and horizontal-axis winches may have, appeared just as I was writing this, in a news item stating that the trawlers of a certain country were viewed with suspicion by some because they were fitted, like warships, with capstans instead of winches.

[e] It is fairly obvious that both vertical and horizontal ex-aqueous wheels had their disadvantages; the former often necessitated right-angle gearing, while the latter required a very substantial lower bearing since it had to take all the weight.

[f] It should be noted that we adhere to the definitions that a 'vertical' system means a system with vertical wheel and horizontal axle, while a 'horizontal' system means a system with horizontal wheel and vertical axle.

[g] All students of this subject are greatly indebted to Goodrich (7), who has written an exhaustive monograph on it, here laid under contribution.

rather in the fact that it illustrates perfectly the preference of Western engineers for vertical, and of Chinese engineers for horizontal mountings. But it is also interesting as furnishing another case, parallel with the chain-excavator referred to above (p. 219), in which the Jesuits introduced to their Chinese friends, as something new, a device which had been known in China nearly a thousand years previously.

The *Chhi Chhi Thu Shuo* of +1627 pictured[a] a revolving bookcase (Fig. 679) by the aid of which a scholar could consult many books without moving from his seat; simple gearing devices ensured that the books should remain right side up as the structure revolved. Giles[b] and Feldhaus[c] both sensed that this was a redrawing of a European design; the former rebutted a 'claim, without any grounds', on the part of the Chinese, to a 'purely western invention, the revolving reading-desk';[d] and the latter, more constructively and correctly, identified the original as occurring in the work of Ramelli (+1588);[e] Fig. 680. Here then was another transmission of light and ingenuity from west to east.

Nevertheless, as Goodrich (7) says, prototypes of the revolving bookstand existed in China even in the very province of Wang Chêng himself, Johann Schreck's collaborator in the making of the *Chhi Chhi Thu Shuo* (cf. Table 58 above), and one cannot but wonder whether they had not been in some way ancestral to the European design. From the +6th century onwards it had been customary to erect large (sometimes very large) book-cases in Buddhist temples so that the whole of the *Tripiṭaka* or scriptures could be contained in them, housed in appropriate drawers. During the course of centuries whatever practical use they may once have had faded into the background, and their chief significance became ritual, persons wishing to acquire merit being able to turn the whole structure round and thus perform an act symbolic of the Buddhas and Bodhisattvas 'turning the Wheel of the Law'.[f]

According to Nanjio,[g] who was one of the first to give attention to these *lun tsang*[1] or 'revolving repositories', their traditional inventor was Fu Hsi[2] in +544, and statues of him and his two sons generally appear on them. The ascription is not quite certain, however, since contemporary biographies of him do not describe the invention, and

---

[a] Ch. 3, p. 52a. [b] (5), vol. 1, p. 108.

[c] (2), p. 70. Both these scholars knew only the *TSCC* reproductions (*Khao kung tien*, ch. 249, p. 48a).

[d] No one has been able to identify the author of the claim to which Giles took exception.

[e] P. 317, pl. 188. In the Chinese redrawing the books in the revolving machine have been made into Chinese books, but those in the bookcase at the back of the picture still stand up on end, and it was this which aroused Giles' suspicion.

[f] It would take us too far off our course to pursue questions of the symbolism of the wheel in comparative religion; the reader is referred to works such as those of Simpson (1); d'Alviella (1); Przyluski (4), etc. Though the wheel as a symbol was far less important in Europe than in Asia, it may well be that the Roman drum-treadmill was stylised into the 'Wheel of Fortune', cf. Fig. 681, from an early edition of Petrarch. Cf. Feldhaus (20), p. 282. That in ancient India the noria was taken as a 'type' or model of the wheel of existence and of rebirths, we have already seen (p. 361 above). This, like the Taoist doctrines of cyclical recurrence (cf. Vol. 2, p. 75), was in the dimension of time. But in space also there was a wheel, a world-wheel, the hub of which was Mt Meru (cf. p. 529), and hence the just and virtuous world-monarch was called *cakravartī-rāja*, the 'wheel-king' (*chuan lun wang*[3]) of universal peace (cf. Zimmer (2), pp. 127 ff.). Equally universal, the Buddha 'set in motion the wheel of the sacred doctrine'. Hence the device on India's national flag today.

[g] (1), p. xxv.

[1] 輪藏 [2] 傅翕 [3] 轉輪王

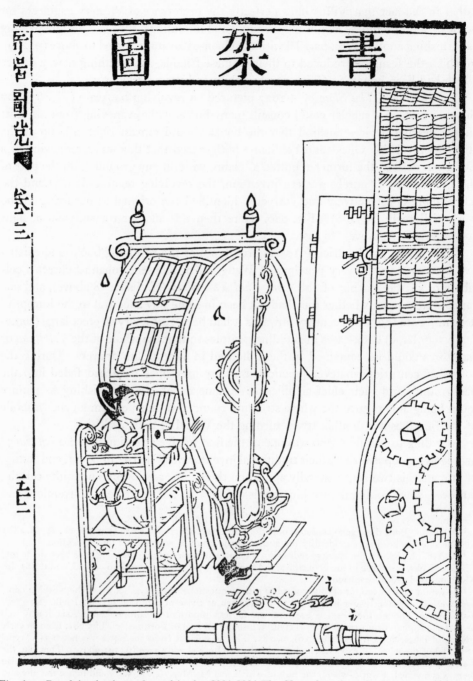

Fig. 679. Revolving bookcase figured in the *Chhi Chhi Thu Shuo* of +1627, ch. 3, p. 52*a*. Although the books in it have vertical rulings for Chinese type, those on the shelves are bound in European style.

PLATE CCLXVII

Fig. 680. The source of the illustration used by Johann Schreck & Wang Chêng, the revolving book-case depicted in Ramelli's engineering treatise of +1588. A simple arrangement of gearing keeps the successive book-rests at the same angle as they are rotated. From this it can be seen that the Chinese draftsman of Fig. 679 drew the gear-wheel system erroneously.

PLATE CCLXVIII

Fig. 681. The Wheel of Fortune, a device from the woodcut title-page of the first translation of Petrarch's *De Remediis utriusque Fortunae libri duo* into a living language, Czech, printed at Prague, +1501. Six men cling to a wheel with a crank-handle while the poet indites above. Possibly this conception originated from the Vitruvian drum-treadmill. Crank-handles began to appear on these 'wheels of destiny' as early as the +12th century, i.e. after the grindstone and the hurdy-gurdy, but before the crossbow crannequin and the well-windlass (cf. pp. 111 ff. above, and Lynn White (7), p. 110). But the symbolism of the wheel was much more important for religious and philosophical thought in Asia than in Europe.

the first one which does is as late as +1056. Nevertheless, it may be taken fairly safely that the practice dates from about the +6th century, whether or not it really derives from Fu Hsi, who was a famous scholar of the time, and known for his syncretistic tendencies.[a] Literary references to the revolving repositories begin from +823, this being the date of a stele now in the Pei Lin at Sian, which bears an inscription concerning such a machine erected in a temple some five *li* outside that city. There are others from the ninth and tenth centuries,[b] but the evidence becomes really abundant in the eleventh, perhaps because of the printing of the Buddhist Canon shortly before (+971 to +983). Yeh Mêng-Tê[1] wrote about +1100:[c]

> When I was young I saw several four-sided revolving repositories. Recently in all centres large and small, and even in remote mountain fastnesses and tiny villages, in six or seven out of ten temples, one can hear the sound of the wheels of the revolving cases turning.

In the Sung nine examples were particularly famous, and then after a break during the Yuan, we have fourteen descriptions of the building of such structures under the Ming, the latest in +1650—significantly some time after the designs in the *Chhi Chhi Thu Shuo*, but completely disregarding them.

For in fact the Buddhist revolving repositories were all as horizontal, technically, as the horizontal water-wheels. One still existing[d] has been carefully described by Liang Ssu-Chhêng (*1*), from whom we take Figs. 682 and 683, showing the way the octagonal revolving book-case is mounted in the temple, as well as its external appearance. This example is the Lung-Hsing[2] Temple at Chêngting[3] in Hopei, and probably dates only from the Sung, with later repairs, though the temple itself goes back to +586. From the Sung, too, we have the equally interesting diagrams and descriptions in the great treatise on architecture, the *Ying Tsao Fa Shih* of Li Chieh (+1103). Here the 'Turning Treasury of the Sūtras' (*chuan lun ching tsang*[4]) is discussed in two places[e] and a figure is also given (Fig. 684).[f] Its height was to be 20 ft. and its diameter 16 ft., and this is just about the size of some of the existing specimens,[g] but others were certainly much larger, approximating to the 70 ft. of the prayer-cylinders at the Yung Ho Kung in Peking.[h] Some took the strength of ten men to revolve.

For the purposes of the Buddhists, Ramelli's gear-wheels would of course not have been necessary, yet it does not follow that gearing was not used for other purposes,

---

[a] Cf. Vol. 2, pp. 409 ff. above.

[b] All are minutely discussed in Goodrich (7). Here we need only mention the example at Wu-thai Shan[5] seen by the Japanese monk Ennin in +840 (cf. Reischauer (2), p. 247, (3), p. 196).

[c] *Chien Khang Chi*[6] (Record of Nanking Affairs), ch. 4, p. 6b; tr. Goodrich (7).

[d] There are, of course, numerous others, and numerous eye-witness accounts by foreign travellers, collected by Goodrich (7). Boerschmann (9) has given a first-hand description of the fine example at the Tha-Yuan Ssu[7] temple on Wu-thai Shan.[5] Its axle-length is 41 ft. and the octagonal case rises nearly 34 ft. above the ground floor of the hall which houses it. The rotation is accomplished by four men working on radial arms in a cellar. The 18 storeys give 144 boxes for statues or books.

[e] Ch. 11, pp. 1a ff.; ch. 23, pp. 1a ff.   [f] Ch. 32, p. 21b.

[g] Goodrich (7), p. 150.

[h] Cumming (1), p. 394. This is the most famous Lamaist temple in China.

---

¹ 葉夢得   ² 隆興寺   ³ 正定   ⁴ 轉輪經藏   ⁵ 五台山
⁶ 建康集   ⁷ 塔院寺

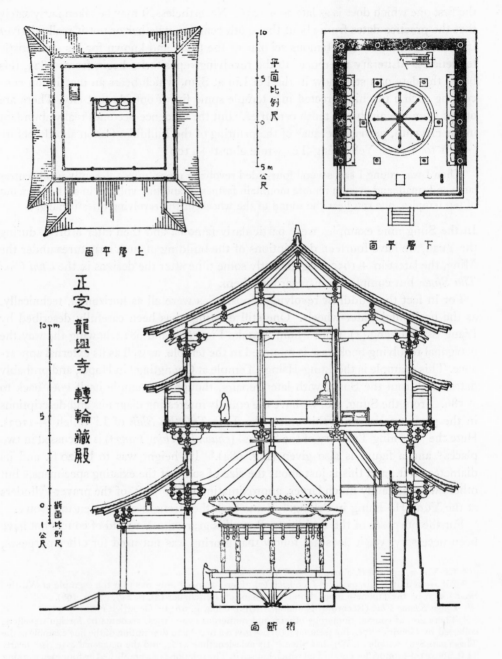

Fig. 682. Scale drawings of the Repository Hall of the Lung-Hsing Temple at Chêngting in Hopei (Liang Ssu-Chhêng, *1*). Above, left, upper-storey plan, right, ground-floor plan; below, section in elevation.

PLATE CCLXIX

Fig. 683. One of the revolving repositories intended to contain the Buddhist patrology, the *Ta Tsang* or *Tripiṭaka*. Detail of the octagonal carved and decorated bookcase in the Lung-Hsing Temple at Chêngting in Hopei (Liang Ssu-Chhêng, *1*). Though the temple dates from +586, the time of the originator of such bookcases, the existing repository is probably not earlier than the +12th century.

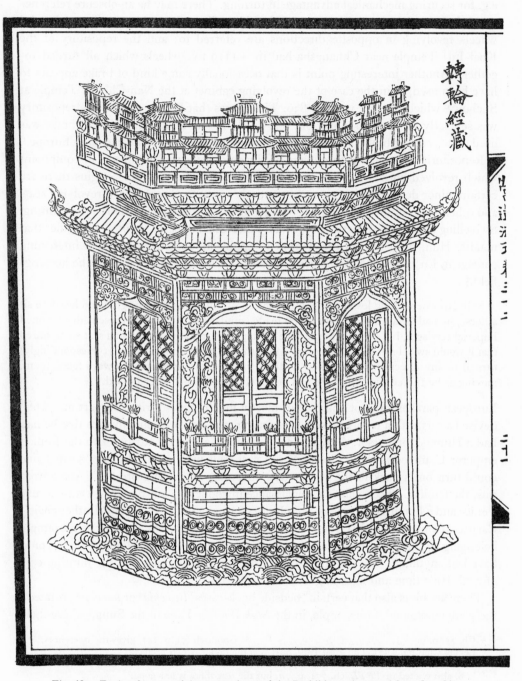

轉輪經藏

Fig. 684. Design for a revolving respository of the Buddhist scriptures (*chuan lun ching tsang*), from the *Ying Tsao Fa Shih* of +1103, by Li Chieh (ch. 32, p. 21*b*).

e.g. for securing mechanical advantage in turning. There may be an obscure reference to this in Yang Shen's + 16th century *Tan Chhien Tsung Lu*,[a] where 'male and female' wheels revolving in opposite directions are referred to; and the repository of the Khai-Fu[1] Temple near Chhangsha had in +1119 five wheels which all turned together.[b] Another interesting point is that occasionally some kind of brake appears to have been used, as in the case of the revolving cabinet at the Nan-Chhan[2] Temple at Suchow,[c] which was put up in +836. This shows that the purpose of the repository was scholarly rather than pious, but one would like to know what kind of brake was used, since curved brake-bands first appear only in the time of Leonardo in Europe.[d]

Repositories for the scriptures were not by any means the only pieces of equipment which revolved horizontally in medieval China, and in considering origins there are certain minor devices which we cannot entirely neglect. One thinks of revolving seats and other pieces of furniture, and also of the first cases used for movable-type printing. Swivelling chairs appear very early, in the Later Chao dynasty about +345 under that notable Hun, Shih Hu,[3] a real patron of engineers and inventors.[e] An interesting passage in Lu Hui's *Yeh Chung Chi* shows us this mechanised nomad in the hunting-field.[f]

Shih Hu when young loved going out to the chase, but as later on he grew too heavy to sit a horse, he had a hunting-car (*lieh nien*[4]) made, similar in shape to the present-day manned imperial carriages, but actually borne (like a litter) by twenty men. It had a seat so mounted that it would swivel round mechanically, and at the top a curved roof which (correspondingly) turned in any direction. Thus when he took aim at birds or beasts it always faced them, moving as he turned his body. He was an excellent shot and never missed.

European parallels are not absent, but mostly much later.[g] Prince Rupert in +1680 devised a cart with a revolving seat for hunting, certainly without any idea that he had had a Hunnish precursor. A tradition of uncertain authority attributes to the Roman emperor Commodus (*r.* +180 to +192) a chariot with a swivel-chair 'so that one could turn one's back to the sun or take advantage of passing breezes'. Apart from this, the familiar office-chair is a Renaissance introduction, starting in + 15th-century Venice and a hundred years later popular in Germany under the name 'Luther chair'. Carpaccio's picture of St Jerome (+1505) shows one, and an 18th-century example belonging to Thomas Jefferson is preserved at Philadelphia.[h] None of these could have had anything to do with the Chinese revolving book-cases, but the hunting-cars of Shih Hu's time and later might well be relevant.

There are hints also that certain 'bedside book-cases' (*wo chia*[5] or *lan chia*[6]) rotated; these are mentioned, for example, in the *Shih Wu Chi Yuan* of the Sung,[i] where their

---

　　[a] Ch. 15, p. 9.　　　　　　　　　　　　　[b] Goodrich (7), p. 141, gives the references.
　　[c] Lévi & Chavannes (2). The great Pai Chü-I was the donor, when prefect.
　　[d] Usher (1), 1st ed. p. 130, 2nd ed. p. 174; Beck (1), fig. 421.　　　　　　[e] Cf. pp. 159 and 256.
　　[f] P. 8, also quoted in *TPYL*, ch. 774, p. 7a, but the text there is less good.
　　[g] Cf. Feldhaus (1), col. 1100. Loewe (3) is now even showing, from Han limes strip-documents, and indeed from *Mo Tzu*, that ancient Chinese forts often had revolving arcuballistae and rotating crossbow turrets; cf. Sect. 30h.　　　　　[h] Cf. Giedion (1), pp. 289 ff.　　　　　[i] Ch. 40, p. 7b.

　[1] 開福寺　　　　[2] 南禪寺　　　　[3] 石虎　　　　[4] 獵輦　　　　[5] 欹架　　　　[6] 孄架

invention is ascribed to the early +3rd century. What certainly did rotate, however, was the *huo tzu pan yün lun*[1] or wheel-shaped case for systematic storage of a fount of movable type, introduced by its first inventor, Pi Shêng,[2] about +1045. We reproduce here (Fig. 685) the famous illustration in the *Nung Shu*[a] of +1313, but a proper

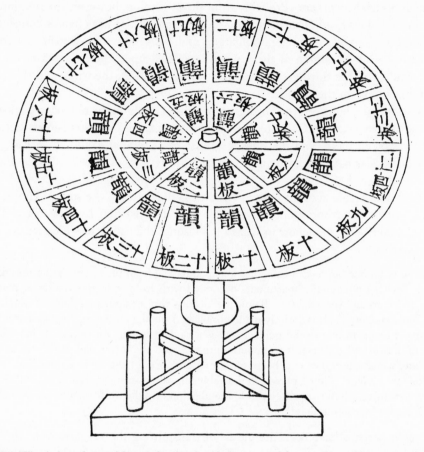

Fig. 685. Wheel-shaped rotatable case for the classified storage of founts of movable type invented by Pi Shêng about +1045, from Wang Chên's *Nung Shu* (+1313), ch. 22, p. 22*b*. All the characters were stored according to their *yün* or rhyme in 24 compartments, 8 inner and 16 outer, thus coming easily to the compositor's hand.

account of this first of all compositors' work must naturally be postponed until Section 32. As in the case of the chairs[b] the horizontal mounting imposed itself.

The original reason for the invention of the revolving book-cases may well have been connected, as Goodrich suggests, with the great burden of translation work assumed by the Chinese Buddhists during the early centuries of our era. Wheel- or

---

[a] Ch. 22, p. 22*b*.   [b] And the jack-wheels; cf. p. 455 above.

[1] 活字板韻輪   [2] 畢昇

cylinder-libraries never arose in India[a] or Central Asia, nor among Chinese Confucians or Taoists. No one, says Goodrich, can have read the story of those years without being impressed by the self-sacrificing, often self-effacing, activity of hundreds of scholars and copyists in the labour of translation of the *sūtras*; and a central store of books which could conveniently be rotated would have left the sides of the halls free for the desks of the workers. Probably from the beginning, however, the rotation was a piece of religious symbolism as much as a convenience. It has obvious connections with the ceremony of circumperambulation, and with the prayer-wheels of Lamaist Buddhism.[b] But discussion of them would encroach upon the question of the discovery and use of the third great eotechnic source of power, the wind, and this is the very subject next on the agenda.

To recapitulate, however, the burden of these brief notes on the revolving book-case, it is clear that its origin in China took place perhaps a thousand years before Ramelli's design was taken there, and the question must be reiterated, could these European Renaissance ideas have received any stimulus from reports of travellers to the East of Asia?

At least one Arabic account exists. In +1420 Shāh Rukh, the son of Tīmūr, sent an embassy to the Ming emperor, and the narrative written by Ghiyāth al-Dīn-i Naqqāsh describes at Ganchow in Kansu a 'kiosque' the nature of which we can now well recognise.[c]

> In another temple there is an octagonal kiosque, having from the top to the bottom fifteen stories. Each story contains apartments decorated with lacquer in the Cathayan manner, with ante-rooms and verandahs.... Below the kiosque you see figures of demons which bear it on their shoulders.... It is entirely made of polished wood, and this again gilded so admirably that it seems to be of solid gold. There is a vault below it. An iron shaft fixed in the centre of the kiosque traverses it from bottom to top, and the lower end of this works in an iron plate, whilst the upper end bears on strong supports in the roof of the edifice which contains this pavilion. Thus a person in the vault can with a trifling exertion cause this great kiosque to revolve. All the carpenters, smiths, and painters in the world would learn something in their trades by coming here!

This might perhaps have been the means of transmitting the suggestion to the West, but until someone can bring evidence of revolving book-cases in occidental libraries, or plans for them, with dates, our question must remain unanswered.[d] There was a long space of time during which such a suggestion could have made its way westwards.

---

[a] So far as I know. But we must not forget the existence in India of 'merry-go-rounds' (*rathadolā*) for the delectation of the royal beauties, described by Prince Bhōja in his engineering work of about +1050 (see Raghavan, 1). Such machines would have involved bearings and gearing quite analogous to those of the rotating libraries.

[b] An essentially scholarly origin of the revolving book-cases might account for the fact that water-power was never applied to them in China. Horizontal water-wheels, however, were certainly applied to rotate prayer-cylinders in Tibet, for several examples were seen and sketched by Simpson (1), pp. 11, 19, 20; Emery & Emery (1), etc.

[c] Tr. Rehatsek (1); Quatremère (3), the latter cit. in Yule (2), vol. 1, p 277, cf. p. 179.

[d] Someone should look into the question of the development of horizontally rotating reading-desks or lecterns for the heavy parchment books of the European Middle Ages. Giedion (1), pp. 285 ff., has something to say on this, and figures such a desk of +1485, but the matter needs much more study.

The fact that Ramelli's was a vertical type, and that all the Chinese ones, from Fu Hsi onwards, were horizontal, would simply have been characteristic of the two engineering traditions.

(*l*) POWER-SOURCES AND THEIR EMPLOYMENT (III), WIND FORCE; THE WINDMILL IN EAST AND WEST

The *Chronicle* of Jocelyn de Brakelond, which deals with the affairs of the Abbey of Bury St Edmunds in East Anglia and constitutes one of the most famous of monastic annals, tells how Dean Herbert built an illegal windmill in +1191 which competed with the mills of the Abbey and was pulled down upon the orders of the Abbot Samson.[a] It was long believed[b] that this gave us the earliest certain instance of a windmill anywhere in Western Europe, but Lynn White (3) has collected five firm records slightly earlier.[c] It is at any rate certain that from this time onwards windmills spread rapidly, coming into use throughout occidental countries in the +13th century. No illustration, however, survives from before about +1270, the date of the so-called 'Windmill Psalter', probably written at Canterbury.[d] After the time of the well-known engraved brass at King's Lynn, Norfolk (+1349), representations become numerous.[e] The Western vertically-mounted windmill was from the beginning a kind of reversed or ex-aerial propeller, and though no doubt essentially an empirical development, derived topologically from the Archimedean screw and not from the Vitruvian water-mill. It was thus deeply occidental, but it involved one new mechanical problem, the orientation of the main driving-shaft (the 'windshaft') so as to present the sails (or 'wheel') in a position at right angles to the direction of the wind.[f] Two distinct types appeared, smaller mill-housings revolving round a central post or pivot permanently fixed on (or in) the ground ('post-mills'), and larger mills consisting of a tower of brick or masonry roofed by a movable cap[g] carrying the sails and the shaft ('tower-mills'). All the earlier mills were of the post type, supported by four diagonal slanting legs (the 'quarterbars'); one of them, for example, is pictured by the anonymous Hussite engineer (+1430).[h]

The rather sudden appearance of these windmills in Europe still presents a considerable mystery. We can find only one forerunner in Western antiquity—the so-called 'anemurion' (ἀνεμούριον) described in the *Pneumatica* of Heron of Alexandria

[a] 'Herbertus decanus levavit molendinum ad ventum super Hauberdun'; Rolls ed., vol. 1, p. 263, Butler tr., p. 59. The chronicle was much popularised by the version of Carlyle.

[b] Bloch (2); Bennett & Elton (1), vol. 2, pp. 224 ff.; Usher (1); Horwitz (11); Vowles (1, 2), etc.

[c] In Normandy and Provence, +1180 (Delisle (1), p. 514, (2); Giraud (1), vol. 2, p. 208); in Yorkshire and Suffolk, +1185 and +1187 (Lees (1), pp. 131, 135, cxlvi); and in Crusader Syria, +1190 (Ambroise, Hubert tr. ll. 3227 ff.). Gille (4, 11) further claims a date of +1162 for Arles. On the general question see Delisle (2) and Ponomarev (1), but the latter's methodology seems to me unsound.

[d] See Wailes (3), p. 149, (4).   [e] Cf. Wailes (1), p. 3, (4), pp. 190 ff.

[f] We shall say no more of this here as the matter has already been discussed on p. 301 above.

[g] Formerly called 'turret', by the literati.

[h] Beck (1), p. 277. For the later history of windmills in Europe see Bennett & Elton (1), vol. 2; Wailes (1, 3, 4, 5). The importance of the role of millwrights in early industrialisation has often been pointed out (cf. p. 545 above).

(+1st century),[a] but no one knows quite what to make of it. According to interpretations traditional since the +16th century, this apparatus consisted of a vertically mounted windmill driving a shaft with trip-lugs which operated a piston air-pump attached to a musical organ. Fortunately an ancient or medieval MS. rendering of the original diagram has in this case survived,[b] answering to the usual reference letters in the Greek text, and from this we can see that the only accretion was the musical organ. All that the title says is 'How to make an instrument that sounds a pipe when the wind blows'. Heron describes the disc of the wind-wheel as carrying 'plates like those of the so-called weather-cocks or wind-vanes (anemuria)', and the ancient diagram shows about twenty-eight of these set rather closely together around the large hub. No hint in the text points either to a horizontal mounting or to any sort of shield-walls, so we are almost bound to conclude that vanes on a vertical wheel were set slanting in screw-fashion, unless indeed the arrangement simply imitated the paddles of a vertical water-wheel. The text does in any case distinctly say that the wind-wheel could be turned about to the position most suitable for the wind to spin it; a mobility that was allowed for in the post-Renaissance reconstruction (quite conformably with Heron's words) by having the four lugs play upon a plate at the end of the pump lever.

As no perceptible echo was caused by this device or toy in the Hellenistic world, and no other writer mentions it, there has been a tendency to reject it altogether as a later interpolation.[c] But this it cannot be, for every one of the three Greek recensions[d] contains the passage, and this book never passed through Arabic. Yet while we must accept the genuineness of Heron's wind-motor there is no reason for thinking that it was more than a toy carried in the hand, or that it led to anything further.[e] The true origins of wind-power seem to lie elsewhere, though Heron may perhaps have been remembered long afterwards at a critical time in the West.[f]

The history of windmills really begins in Islamic culture, and in Iran. According to a story due to 'Alī al-Ṭabarī (c. +850) and repeated by Ibn 'Alī al-Mas'ūdī (c. +947)[g] and other writers, the second orthodox Caliph, 'Umar ibn al-Khaṭṭāb, was murdered in +644 by a captured Persian technician Abū Lu'lu'a, who claimed to be able to construct mills driven by the power of the wind, and was bitter about the high taxes

---

[a] I, 43, W. Schmidt ed. (2), p. 205 and fig. 44; ch. 77 in Woodcroft (1), p. 108. Schmidt's illustration, reproduced by Forbes (19), p. 614, is the better. We are greatly indebted to Dr A. G. Drachmann of Copenhagen for his excellent advice in this difficult matter before the appearance of his special study (8).

[b] Schmidt ed. (2), p. xl, fig. 44a. It seems to be the same in all texts which have it.

[c] Thus Forbes (19), p. 615, with unwonted scepticism (especially where European developments are in question), says that the Heronic windmill drawings 'must be dismissed as interpretations by Christian or Muslim scribes who knew the windmill proper'.

[d] Including the pre +6th-century Pseudo-Heron.

[e] Cf. the presentations of Horwitz (11) and Vowles (1, 2). Dr Drachmann suggests that it is just possible that the windmills of the Aegean might derive directly from Heron. These are low towers with eight or twelve arms mounted vertically on the shaft, each carrying a triangular lateen sail. The type is found as far west as the coasts of the Iberian peninsula. If they really go back to Hellenistic times, the literature of the first Christian millennium is singularly silent concerning them. Pictures in Baroja (2); Stillwell (1); Cobbett (1).

[f] On the manuscript and typographical transmission of Heron's *Pneumatica* see Boas (1).

[g] Tr. de Meynard & de Courteille (1), vol. 4, pp. 226 ff.; Wiedemann (7); Jacob (1), p. 89.

to which he had been subjected.[a] More certain, perhaps, is the mention of windmills in the works of the Banū Mūsā brothers (+850 to +870),[b] while a century later several reliable authors are speaking of the remarkable windmills of Seistan (e.g. Abū Isḥāq al-Iṣṭakhrī and Abū al-Qāsim ibn Ḥauqal).[c] Seistan is an arid sandy region, renowned for the continuous blowing of high winds; it is situated in the area where Persia marches with Afghanistan and Baluchistan, and where the Helmand River runs down

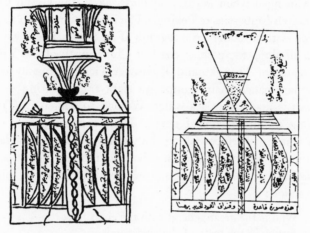

Fig. 686. Diagrams of windmills from the cosmographical treatise of al-Dimashqī, *c.* +1300 (MSS. at Leiden and Berlin, from Horwitz, 11). The horizontal rotors with their vanes below, the millstones in an upper storey.

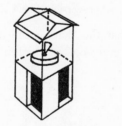

Diagrammatic perspective view

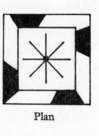

Plan

to its inland lakes. A very detailed description of these windmills occurs in the *Nukhbat al-Dahr* (Cosmography) of Abū 'Abdallāh al-Anṣārī al-Ṣūfī al-Dimashqī[d] about +1300. From this it is clear that the Iranian windmills were horizontal in type, and enclosed in shield-walls so that the wind entered only on one side, in turbine-fashion, moreover the querns themselves were in an upper storey and the vanes or

[a] Cf. Jacob (2), p. 60; W. Muir (1), p. 187; Lynn White (7), p. 86.

[b] See Wiedemann (6).

[c] Excerpted in Barbier de Meynard (1), p. 301. Cf. the mention in the *Kitāb al-Rujārī* of 'Abdallāh al-Idrīsī (+1154); Jaubert (1), pp. 442 ff.

[d] See Mieli (1), p. 275. Tr. Mehren (1), p. 246.

sails below (cf. Fig. 686).[a] A little earlier, the encyclopaedist Abū Yaḥyā al-Qazwīnī had also enlarged upon the windmills of Seistan.[b] They still grind on today, and many modern travellers have visited them[c] (cf. Fig. 687). While retaining their shield-walls, however, and sometimes adding further curtain-walls, their wheels have become greatly broadened so as to form tall upstanding rotors, and the millstones have been placed beneath.

The first European to see windmills in China was Jan Nieuhoff, who met with them at Paoying[1] in Chiangsu when journeying north along the Grand Canal in +1656 with one of the Dutch Embassies to Peking.[d] The illustration which he gave (Fig. 688) may be compared with a modern photograph (Fig. 689).[e] These windmills are still extensively used all along the eastern coast of China north of the Yangtze, and particularly in the region of Thangku and Taku near Tientsin, mainly as prime movers for operating square-pallet chain-pumps by right-angle gearing in the numerous salterns where salt is made from sea-water. We owe to Chhen Li (1) a recent study of them including minute technical detail. Their construction is of considerable interest, for the vanes or surfaces taking the wind-pressure do not radiate from the central axis, but are in fact true junk slat-sails[f] mounted on eight masts forming the periphery of a skeleton drum.[g] Chhen Li gives a local riddle which helps us to understand the construction:

> Who is the great general with the eight faces,
> Strong in the teeth of the wild winds?
> He has eight masts that follow the wind and turn,
> Wearing a hat at the top, and standing on a needle below,
> His two ends can revolve at your wish
> And make the waters come or go wherever you like.[h]

Looking at Chhen's diagrams (inset) we see that the 'hat' is the upper bearing of the central axle, and the 'needle' is the pin or gudgeon on which it revolves below. But the ingenuity of the whole contrivance is seen in the fact that it dispenses entirely with the shield-walls used in Persia. It was able to do this by adopting the fore-and-aft rig of Chinese junks for the windmill sails, as may be understood by the diagram on the next page. In position *A*, the sail is held taut against the wind by its 'sheet'

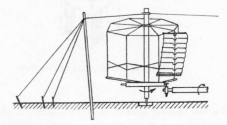

rope (*phêng lan shêng*[2]), but when it reaches position *C*, the sail blows right outwards

---

[a] From al-Dimashqī MSS. at Leiden and Berlin, copied by Horwitz (11).

[b] See Mieli (1), p. 150.

[c] E.g. Tate (1), vol. 3, p. 251; Kennion (1), pl. 6; le Strange (3), pp. 337, 409, 411; Maillart (1), p. 120 and opp. p. 86; Bagnold (1), pp. 144 ff.; Sykes (2), p. 397.

[d] Cf. Petech (4).   [e] King (3), fig. 177.

[f] Each 6 × 10 ft. giving 480 sq. ft. available surface. See further in Sect. 29g on Chinese sails in general.

[g] Diam. 15 ft.   [h] Tr. auct.

[1] 寶應   [2] 蓬攬繩

('luffs'), coming back into the eye of the wind in position E, freely hanging and opposing no wind-resistance. By position G, the sheet has tightened again, preparing the sail to receive the full force of the wind. Chhen Li found that effective wind-pressure was exerted for considerably more than 180° of the cycle, since in position D the sail does a certain amount of work when it is 'sailing into the wind'.[a] The whole system constitutes an invention of great interest and practical importance since thousands of these simple machines are at work at the present day.[b]

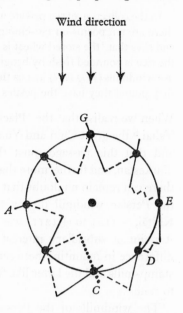

Unfortunately, there are very few literary references in Chinese to this or any other type of windmill, though perhaps some might be found in the local topographies of the provinces where they occur, and this search has not yet been made.[c] Several decades before Nieuhoff's visit, Sung Ying-Hsing recorded in his *Thien Kung Khai Wu*[d] that these windmills were commonly used at Yang-chün[1] and elsewhere,[e] which would take us back to the beginning of the +17th century. But the only really important reference so far found relates to the borders of Sinkiang; it occurs in the *Shu Chai*

---

[a] It also starts to work in position G just before the half-point of the cycle is reached, and it goes on doing work in position C for a short time after the luff has taken place.

[b] Modern designs based on them will be found in Anon. (*30*), pp. 72 ff. Figure 690 shows experiments in progress at the Peking Agricultural Inventions Exhibition in 1958.

[c] Bathe (1), p. 4, has accepted, and Lynn White (7), p. 86, has disseminated, a story that windmills were used in China on the Grand Canal for hauling vessels from one level to another by means of rollers on inclined planes. But the authority is not good. It occurs only in certain editions (1837, 1842) of the 'Book of Trades', a career-guide to the crafts and professions which came out in many forms during the 19th century (cf. Anon. 62–5) and deserves a special study. The passage, so far as I have found, appears only when the article on 'The Engineer' is present. Thus in Whittock *et al.* (1), p. 194, we read: 'China, in her institutions hostile to art, has nevertheless encouraged the making of canals; and their convenience having aided in supplying a ready transit of her commodities, she has, more from cunning than a wish to develop the powers of the human mind, intersected her country with them. The canal which runs from Canton (*sic*) to Peking is in length upwards of 800 miles and was executed about 700 years since; it has no locks, tunnels or aqueducts, and when stopped by mountains or other impediments, a rolling bridge is resorted to, and sometimes inclined planes. These rolling bridges consist of a number of cylindrical rollers which turn easily on pivots, and are sometimes put in motion by a windmill, so that the same machinery serves a double purpose, that of working the mill, and drawing up the vessels. In this manner they draw their vessels from the canals on one side of the mountain to the other.' We have encountered no Chinese text or illustration which would confirm this use of windmills. Besides, the passage seems rather confused, and indeed no less suspect than unkind. The usual method of working junks over the double inclined planes or *diolkoi* (to which we shall return in Sect. 28*f*) was by means of multiple 'capstanes' (cf. Lecomte's +17th-century account, (1), pp. 104 ff.). For a later description see Barrow (1), p. 512; for a well-known cut, Davis (1), vol. 1, p. 138; and for a recent photograph, Cressey (1), fig. 138. But one can never exclude the possibility of further evidence, especially since at least one water-wheel was used in +17th-century Holland for hauling boats over the double inclined planes which themselves go back there to the +12th; cf. Tew (1).

[d] Ch. 1, p. 6*a*, first ed. p. 9*a*.   [e] I.e. the coastal lands of Chiangsu.

[1] 揚郡

Lao Hsüeh Tshung Than[1] (Collected Talks of the Learned Old Man of the Shu Studio), a book of the Yuan or late Sung time by Shêng Jo-Tzu.[2] He says:[a]

In the collection of the private works of the 'Placid Retired Scholar' (Chan Jan Chü Shih[3]), there are ten poems on Ho-chung Fu.[4] One of these describes the scenery of that place ... and says that 'the stored wheat is milled by the rushing wind (*chhung fêng mo chiu mai*[5]) and the rice is pounded fresh by hanging pestles (*hsüan tui chhu hsin kêng*[6]). The westerners there use windmills (*fêng mo*[7]) just as the people of the south use water-mills (*shui mo*[8]). And when they pound they have the pestles hanging vertically.'

When we realise that the 'Placid Retired Scholar' was none other than Yehlü Chhu-Tshai,[b] the great Chin and Yuan statesman and patron of astronomers and engineers, and that this passage must therefore refer to the year +1219, when he visited Turkestan, and furthermore that Ho-chung Fu was the place which we call Samarqand, there can remain no doubt that the northern Chinese must have been acquainted with the Persian windmill during the +13th century. The Western Liao State (Qarā-Khiṭāi, +1124 to +1211) was thus almost certainly the focus of this transmission.[c] A point of subsidiary interest is that Yehlü Chhu-Tshai distinctly mentioned the difference in mounting between Western and Chinese trip-hammers (the former like stamp-mills and the latter like forge tilt-hammers) which we have already had occasion to note.[d]

The windmills of the Persian culture-area were described more fully some two hundred years later by other Chinese visitors. In the *Chhih Pei Ou Than* (Chance Conversations North of Chhih-chou) Wang Shih-Chên says:[e]

In the western countries called Herat (Ha-Lieh[9]) and Samarqand (Sa-Ma-Erh-Han[10]) there are many windmills. Walls (of bricks) are built into the form of a house, having at the top openings facing the four directions, outside which screens can further be set up so as to catch the wind. Inside the chamber (below) a wooden axle is fitted, with sails (lit. wind-riding boards) attached to it above and the millstones (driven by) it underneath. Whatever the quarter from which the wind blows, the axle always goes on turning, and the more it blows the more work is done. This was what Yehlü Wên-Chêng was referring to in his poem, where he says....

And he goes on to quote the two lines we have just cited. He then adds, however,

The (people of those parts) also have a (kind of punkah winnowing)-fan. Under an awning (at the side of the mill) they hang high up a cloth which has a lot of hair along its lower edge, and facing it they have a cord which pulls (and waves) it automatically (following the rotation of the mill). Thus when there is wind they do not need to pull it by hand. See Chhen Chhêng's *Hsi Yü Lu* (Record of the Western Countries).

---

[a] Ch. 1, p. 5a, b, tr. auct.   [b] +1190 to +1244; cf. above, Vol. 1, p. 140.
[c] Cf. Wittfogel & Fêng (1), p. 661. It had then just fallen to the Mongols. A number of other transmissions have also been attributed to the Hsi Liao people; cf. Vol. 1, p. 133, Vol. 3, pp. 118, 457, and Vol. 4, pt. 1, p. 332 above.
[d] Pp. 184, 394 above.   [e] Ch. 23, p. 10b. The work dates from +1691.

[1] 庶齋老學叢談   [2] 盛如梓   [3] 湛然居士   [4] 河中府
[5] 衝風磨舊麥   [6] 懸碓杵新粳   [7] 風磨   [8] 水磨
[9] 哈烈   [10] 撒馬兒罕

PLATE CCLXX

Fig. 687. Horizontal windmills in Seistan (Vowles, 1). Here the rotors are tall and elongated within their shield-walls and curtain-walls while the grinding stones are housed beneath.

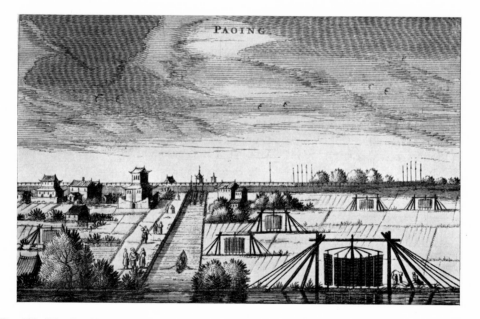

Fig. 688. The first European picture of Chinese horizontal windmills, from Nieuhoff's account of one of the Dutch embassies, +1656. The place is Paoying in Chiangsu.

PLATE CCLXXI

Fig. 689. Typical Chinese horizontal windmill working a square-pallet chain-pump in the salterns at Taku, Hopei (King, 3). The fore-and-aft mat-and-batten type sails luff at a certain point in the cycle and oppose no resistance as they come back into the eye of the wind (see diagram on p. 559).

Fig. 690. Modern design for a pump windmill based on the traditional luffing-sail type but of more compact and economical construction (orig. photo. Agricultural Machinery Exhibition, Peking, 1958).

Here no doubt a simple lug and lever was employed to activate the subsidiary mechanism. What is more interesting is the reference, and indeed the original passage is easy to find in the *Hsi Yü Fan Kuo Chih*[1] (Records of the Strange Countries of the West),[a] a short tractate written by Chhen Chhêng.[2] Together with a coadjutor, Li Ta,[3] Chhen had paid a visit to Samarqand and Herat, among other places, in +1414, on a diplomatic mission to 'show the flag' analogous to those more famous ones which took Chêng Ho down to the South Seas.[b] They penetrated more than 12,000 *li* west of Chia-yü-kuan[c] and wrote a joint diary still preserved, the *Hsi Yü Hsing Chhêng Chi*.[4] Thus we know of a second occasion when windmill designs could have been carried east.

The most probable supposition, therefore, seems to be that the horizontal windmill was introduced to China either by Central Asian, Chhi-tan or Arab–Persian merchants overland, or by Arab–Indian sailors or merchants through the ports,[d] some time during the Sung or Yuan periods.[e] This transmission may well have been only the message that wind-power could be used with a horizontal rotor, whereupon Chinese nautical technicians proceeded to construct for their friends in the salt industry the 'fore-and-aft' windmill as we now have it. This explains at any rate why the distribution remained coastal; inland, at any rate away from the great rivers, there were no skilled sail-makers.[f]

By the +16th century the Persian horizontal windmills had become well known in Europe, and designs based on them figured largely in the engineering book of Verantius (+1615, but probably written *c.* +1595).[g] Ironically enough, two of these[h] quickly found their way back East in the Jesuit book *Chhi Chhi Thu Shuo* of +1627 (Fig. 691), though the Chinese had probably long been using a much more practical type. Another form penetrated to the New World, and found employment there for driving sugar-mills in the West Indies, e.g. on St Kitts, where Labat saw it in +1696 (Fig. 692). This must surely have been a westward transmission from Iberian culture, originally derived from Muslim Spain.[i] The principle of luffing sails instead of shield-

---

[a] Pp. 7*b*, 8*a*. It also says that the water-mills were 'much the same as those in China'. Does this not imply horizontal water-wheels?

[b] Cf. Vol. 1, p. 143; Vol. 3, pp. 556 ff.   [c] Cf. Vol. 1, pp. 143, 169 and Fig. 14.

[d] Hence the interest of the statement made by al-Idrīsī in +1154 that there were windmills for grain-grinding on islands near Malaya (tr. Ferrand (1), vol. 1, pp. 172, 194; cf. Gerini (1), p. 535). Perhaps Arabs had set them up there. Horwitz (11) gives details of small windmills for various uses now found in Sumatra, the Celebes, etc., which might derive from these.

[e] That the entry of the windmill took place no earlier is perhaps suggested by the fact that it never acquired a specific character, nor even a specific phrase, but has always been known simply as *fêng chhê*,[5] in confusion with the rotary winnowing-fan (cf. p. 118 above), which is certainly much older.

[f] It may of course be remembered also that some parts of China are remarkably windless, notably Szechuan.

[g] Besson also produced a design in +1578, cf. Wailes (4). The first appearance in Europe, so far as we know, had been in the MS. of Mariano Jacopo Taccola *c.* +1445 (cf. Uccelli (1), p. 10, fig. 28; Lynn White, 5, 7). They never competed successfully against the longer-established vertical windmills, though there was a good deal of argument about their respective merits. In +1659 d'Acres (1) was a great contemner of 'horizontal sailes'. Marin (1) describes an extant example in the Ligurian alps.

[h] Nos. 37 and 38 in Table 58 above. Cf. Beck (1), p. 517, figs. 788 and 789.

[i] Windmills at Tarragona under Muslim rule 'established by the men of former times' are referred to in the *Kitāb al-Rawḍ* of Ibn 'Abd al-Mun 'im al-Ḥimyarī, written in +1262. See Lévi-Provençal (1), p. 153.

[1] 西域番國志   [2] 陳誠   [3] 李暹   [4] 西域行程記   [5] 風車

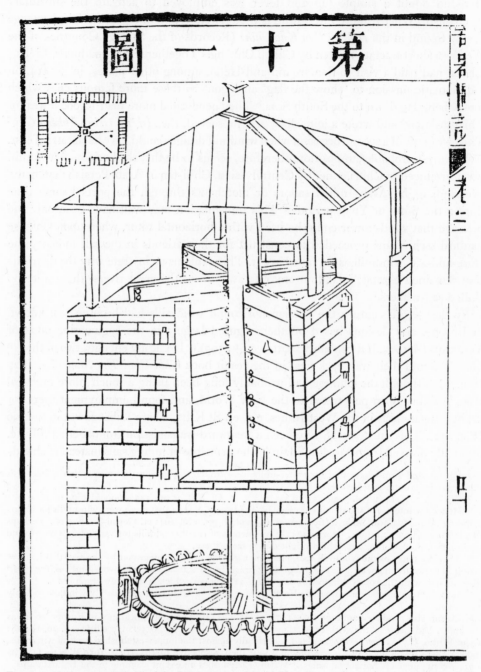

Fig. 691. Horizontal windmill of the Persian type with shield-walls, put forward as a new design in the *Chhi Chhi Thu Shuo* of +1627 (ch. 3, p. 40*b*), because present in the engineering treatise of Verantius (+1615). At the top: 'The Eleventh Diagram' (of the Milling section).

PLATE CCLXXII

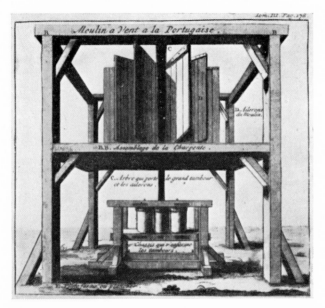

Fig. 692. West Indian horizontal windmill working a sugar-mill (cf. Fig. 459), a drawing in Labat, +1696. Identical in design with the windmills of Seistan, this must surely have been a transmission from Muslim Spain.

walls also spread far and fast in the +16th century for G. H. Rivius included such a design[a] in the book on architectural engineering which he published at Nürnberg in +1547. The idea was also taken up in due course by Verantius, who hinged his sails in various ways, or made them permanently open on one side and streamlined on the other;[b] and again the industrious Johann Schreck presented these to his Chinese readers as original novelties from the creative West (Figs. 693 and 694).[c] Such a

Fig. 693. Verantius' windmill with suspended luffing flaps, reproduced in the *Chhi Chhi Thu Shuo* (ch. 3, p. 39*a*). Redrawing by Beck (1).

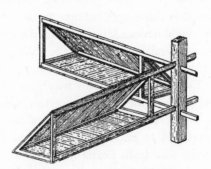

Fig. 694. Verantius' streamlined 'anemometer' sails designed to oppose minimum wind-resistance in the return phase, also copied in the *Chhi Chhi Thu Shuo* (ch. 3, pp. 37*b*, 38*a*). Redrawing by Beck (1).

reintroduction in less elegant form was one of those superfluities on which we have already remarked,[d] and in those parts of the country where the 'maritime' windmill was used, it continued to be built as before without the least attention being paid to Verantius' designs.[e] Finally, some account of the Chinese pattern more detailed than Nieuhoff's may have reached Europe in the 18th century, for Bennett & Elton allude[f]

[a] Bk. 3, p. 41, see Beck (1), p. 184, fig. 204. This was quite a rapid move, connected doubtless with the first Chinese contacts of Portuguese travellers. As early as +1509 they had met junk captains and their crews in Malacca, and from +1517 onwards, the date of the ill-fated embassy of Tomé Pires, Portuguese ships were calling at Chinese ports and trading along the coasts, though illegally after +1522. Cf. Vol. 2, p. 525 and Sect. 29*e* below.

[b] Somewhat analogous to a modern anemometer. This principle had long been known in East Asia, as we shall see immediately below. If the old print of Hohentwiel castle in +1641 reproduced by Uccelli (1), p. 10, fig. 27, is worthy of credence, some horizontal windmills of this kind were working during the +17th century. And they are still to be found in certain parts of the world, notably Quebec Province in Canada, where they are used for pumping water (see Bathe (1); de la Rue (1), pl. XIII*a*).

[c] *Chhi Chhi Thu Shuo*, nos. 33 to 36, and 40; see Table 58 above.

[d] Cf. p. 223 above. Yet Laufer (3), p. 19, concluded, by some strange process of reasoning, that the *Chhi Chhi Thu Shuo* description 'proves' that windmills were previously unknown in China.

[e] At the present time, however, some experimental designs resemble those of Verantius. Fig. 696 shows a water-raising windmill of anemometer type under construction beside the Old Silk Road near Anhsi in Kansu (1958).

[f] (1), vol. 2, p. 326; cf. Bathe (1). Stephen Hooper was a leading protagonist of the horizontal windmill from about +1785 onwards and constructed several successful examples (see Wailes (3), pp. 84 ff., pls. VI, VII).

to a number of new inventions then patented there for making the returning sails of horizontal windmills present their edges only to the wind. But in Europe this type of windmill was never to any extent adopted.

Jesuit influence may have left its mark in China in another way, however. In certain of the eastern Chinese provinces wind-power is harnessed for water-raising by means of a curiously constructed windmill the axis of which is set neither vertically nor horizontally but obliquely (Fig. 695). These are particularly common between Shanghai and Hangchow.[a] Now from the +16th century onwards in Holland very similar small windmills have been in use, their axles being continuous with the inclined shafts of Archimedean screws; as may be seen in the photographs given by van Houten (1).[b] Most probably, therefore, the oblique windmill was introduced to China in the +17th century as part of a compact piece of equipment which included the Archimedean screw.[c] When the latter then failed to supersede the traditional types of water-raising machines the oblique windmills continued in certain districts, being harnessed by appropriate gearing to square-pallet chain-pumps.

Two related questions may now be asked, though they cannot be answered: first the origin of the European vertical windmills of the +12th century, and secondly that of the Persian horizontal ones of the +7th. With regard to the former, Vowles (1) developed an elaborate argument to show that there could have been a transmission of the idea from Persia northwards and westwards by means of Scandinavian and Russian traders along the Baltic–Orient routes. Against this there is no positive evidence, but in such circumstances one would have expected to find windmills in Russia and Scandinavia earlier than in western Europe, and that has not been demonstrated. In fact Muslim Spain is a more likely intermediary region.

But whatever the route which the idea may have taken, the Western windmill was from the beginning so different from the Persian that a high degree of inventive novelty must have entered into it. The message from the east may simply have been that over there people found wind-wheels possible and useful; whereupon the artisans of northern Europe contrived them according to their lights. A persistent tradition, indeed, has maintained that the idea of windmills was brought back by the first crusaders (+1090 to +1170).[d] A further question then presents itself—what exactly were those lights? What were the specific technical influences which led the first Western millwrights to their solution of the vertical mounting? It is too simple to say that the European windmill derived from the vertical Vitruvian water-wheel, for from the outset its axis was placed (quite unlike that of the latter) parallel to the lines

[a] Personal communication from Dr N. W. Pirie, F.R.S., 1952. Cf. Chang Han-Ying (1) for Anhui.
[b] A somewhat similar machine is described and illustrated in de Bélidor (1), vol. 2, pl. 2.
[c] It will be remembered from Sect. 17d, and above, pp. 122, 528, that Tai Chen wrote a short monograph on this device (*Lo Tsu Chhê Chi*) in the latter part of the +18th century. It appears prominently also in the *Thai Hsi Shui Fa* incorporated in the *Nung Chêng Chhüan Shu*.
[d] Cf. Bennett & Elton (1), vol. 2, p. 230; Lopez (1), p. 613. Ambroise says that the German crusaders built in +1190 'the first windmill that had been made in Syria', but that does not mean that the idea had not come from other Islamic lands. In western Europe as late as +1408 a windmill was called a 'moulin turquois à vent' (Delisle, 2). Here is a parallel with the 'torquetum' discussed in Vol. 3, pp. 370, 378 (cf. Sayili (2), p. 385).

PLATE CCLXXIII

Fig. 695. Oblique-axis windmill characteristic of Chiangsu and Chekiang (photo. Pictorial Press). Although these windmills are fitted with typically Chinese mat-and-batten type sails, and work, as here, square-pallet chain-pumps for irrigation, they probably derive from Dutch originals of the +17th century with occidental sails which turned Archimedean screws fitted directly on the windshaft.

Fig. 696. Contemporary Chinese design for an experimental water-raising windmill of streamlined 'anemometer' type, under construction beside the Old Silk Road near Anhsi, Kansu (orig. photo., 1958).

PLATE CCLXXIV

Fig. 697. Model by Greville Bathe of the 'Jumbo' windmill or 'go-devil' used extensively on the plains of Kansas and Nebraska *c.* 1890 (photo. by the courtesy of Mr Rex Wailes). Such wind-wheels, though vertical, form a rare exception to the principle that Western windmills (unlike Vitruvian water-wheels) have always been placed with their axes parallel to the lines of flow of the medium.

of flow.[a] In fact, the Western windmills, necessarily facing into the wind, were always 'air-screws' in reverse.[b] But their effective ancestor can hardly have been the continuous or Archimedean screw; for their sails were for many centuries perfectly flat surfaces, though placed of course at a certain angle (traditionally 17°)[c] to the plane of the circle in which they rotated, borne at the forward end of the windshaft.[d] They thus constituted, like the vanes on Chinese zoetropes or helicopter tops,[e] planes tangent to a continuously curved screw. Now the only European machine which could with reasonable probability have given the inspiration required in the middle of the +12th century was (paradoxically) the horizontal water-mill. We know at least from many descriptions of traditional examples of these during the past two centuries that in many cases (e.g. in Norway, Shetland and the Faroes) their vanes were inserted slightly obliquely so as to catch the full force of the water as it fell upon them.[f] Elsewhere (from Turkey and Serbia to Ireland) spoon- or ladle-like shapes were used, again approximating to tangent planes of a continuous screw.[g] All that is lacking is proof that these inclined-vane forms go back into the +12th century. Could this be established we should be well justified in believing what must still remain a surmise, that the Western windmill was the solution of men who knew both the vertical and the horizontal water-mill,[h] and who were convinced that somehow or other the Saracens

[a] The only exception to this occurs in the curious so-called 'Jumbo' mills of late 19th-century America, which were like water-mills with their lower halves concealed within casings (Fig. 697). The system was said to be cheap and effective, but it never spread.

[b] It was early empirically found that the best arrangement was to have the windshaft up-canted somewhat (10° to 15°), thus balancing its weight against that of the sails and assisting them to clear housings below.

[c] Analogous to the 'pitch' of a screw. This is the figure given in Jousse's treatise just after +1700, and Mr R. Wailes tells us that it has been verified in surviving +18th-century structures, notably in Massachusetts. The angle shown in Ramelli's famous drawing of +1588, however, seems no more than some 10° (cf. Wailes (5), p. 91; Vowles & Vowles (1), p. 125). About +1740, however, de Bélidor (1), vol. 2, pl. 2 and p. 38, seems to recommend much more—about 35°. During the +18th century windmill sails were given a twist ('weathered') so that the angle of setting changed along their long axis. In this way they came to approximate more closely to segments of a continuous screw form. The shape found empirically to be most satisfactory involved a skew angle of some +22° at the root and about −3° at the tip; wind-tunnel tests have confirmed the experience of millers. Probably the best shape was arrived at by adjustment of the structure so that the sail-cloth did not flap. For an illustration see Wailes (1), fig. 41, and frontispiece.

[d] The sails of windmills were not fixed to any wheel or hub in the ordinary sense. Their main members ('whips' if radial, 'stocks' if diametral) were first morticed right through the squared end of the windshaft, and then after cast iron came in, socketed in box-shaped 'poll ends' or 'canister heads' at the end of the windshaft, or fixed to cast iron 'crosses' thereon; cf. Wailes (1), p. 22.

[e] Cf. Vol. 4, pt. 1, p. 123, and p. 583 below. It is very unlikely that European artisans of the +12th century knew of these devices.

[f] Cf. p. 369 above. See the reviews of Curwen (4) and Wilson (2, 4), and original descriptions by Landt (1); Hibbert (1) and others. So far as we can judge by personal observations, travellers' descriptions, and a study of the agricultural encyclopaedias, the vanes in the wheels of Chinese horizontal water-mills were not mounted obliquely (cf. Fig. 625). Skew-set water-wheels can indeed be seen in China nowadays (cf. Fig. 624) but it is doubtful how far they are traditional.

[g] Cf. p. 369 above. Again see Curwen (4); A. T. Lucas (1), and Wilson (2, 4), and for an original account particularly McAdam (1). In all the types of water-wheels the comparable skew angle or 'pitch' was of course more like 70°.

[h] If this is the correct theory of the origin of the Western windmill, the whole-wheel drive had a remarkable resurrection in the 19th century when several English tower-mills were equipped with 'annular sails' (cf. Wailes (1), p. 33, fig. 69, (3), p. 99, pl. XVI, and the appendix by F. C. Johansen on p. 185). This arrangement still lives on vigorously in the common wind-pumps of the present

had already successfully mastered the power of the wind. Finally, Alexandrian tradition can never be excluded; though Heron's 'anemurion' was but a toy wind-wheel, the mill of Dean Herbert repeated its vertical pattern. The medieval European solution was a distinct advance on any other form because it ensured that the whole area of the rotor surface received the pressure of the wind all the time; no part was shielded by walls, and the cycle of rotation contained no inactive luffing segment.

Last comes the question of the origin of the Persian windmills from the +7th century onwards. Vowles emphasised, of course, the Arabic translations of the Alexandrian mechanicians, and one could also imagine, with Forbes,[a] a direct influence of the horizontal water-mill, whether from its Pontic or its Chinese home. But in that windy corner of Iran, adjacent as it is to the Tibetan massif, we are perhaps justified in suspecting quite a different origin, namely the tradition which gave rise to wind-driven prayer-wheels in Mongol–Tibetan culture. This possibility has been raised by Horwitz (11) and connects with the revolving book-cases of which we have already treated.[b] That prayer-cylinders were commonly rotated in Tibet by horizontal water-wheels has also been mentioned,[c] but the use of wind-power has been reported by many travellers.[d] Although too often very imprecise in their accounts, the information which they give reveals several different forms. First, what Pallas saw in the late +18th century among the Khoshot, Kalmuk and Kirghiz tribes between the Aral, the Volga and Lake Sarpa above Astrakhan, was of great interest, for on and around the tombs of their Khans (Fig. 698) were generally five small prayer windmills looking for all the world like modern anemometers.[e] Closely similar apparatus was studied by Rockhill (3) about 1890 at Shang-chia, a Mongol settlement on the Tsaidam Plateau (cf. Fig. 699). Besides these streamlined cup-vane types, Horwitz (11) and Laufer (19) have described Tibetan prayer-drums with curved wind-vanes (Fig. 700) preserved in Western museums.[f] While both these forms rotated horizontally, using wind-force from the side and embodying a principle of streamlining, a third form, more closely related to the screw, took advantage of an uprising air-current. The Abbé Huc on his journey of 1844 noticed many prayer-drums fixed on the tops of Mongol yurts and rotated by the rising stream of hot air.[g] Later travellers confirmed his observations often enough.[h] This at once reminds us of the cinematographic toys or zoetropes which at an earlier stage we could trace back to the Thang, even, it may be, to the

day. Another point of interest is that in Russia post-windmills exist in which the millstones are placed above the main driving-shaft, 'in effect, water-mills on stilts with windmill sails substituted for the water-wheels' (Wailes (4), p. 624, (5), p. 98). This is taken as a kind of intermediate form between the water-mill and the developed windmill. Cf. p. 564.

[a] (19), p. 617.   [b] Pp. 546 ff. above.

[c] P. 405 above; cf. Sarat Chandra Das (1), p. 28; Cunningham (1), p. 375; Rockhill (3), p. 232, (4), p. 363; Waddell (2), p. 149, and Simpson (1), pp. 11, 17, 19.

[d] Notably Bonvalot (1), vol. 2, p. 143; Rockhill (3), p. 147; Waddell (2), p. 149.

[e] (1), vol. 1, p. xiv and pl. 6; vol. 2, pp. 304, 335 and pl. 16. Here we may find an echo of Troup's theory of the origin of the water-wheel; cf. pp. 406 ff. above. An actual example, from the Chahar Mongol region, is to be seen in the East Asian ethnological collections in the National Museum at Copenhagen.

[f] The lineal descendants of these we may see every day in the Savonius S-rotor which provides those twirling ventilators familiar on the roofs of motor-vans and refrigerated railway wagons.

[g] (1), vol. 1, p. 202.   [h] E.g. Gilmour (1), p. 165; Rockhill (4), p. 363.

PLATE CCLXXV

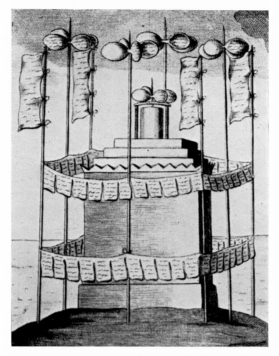

Fig. 698. Prayer-flags and four-cup 'anemometer' prayer-wheels on and around the tomb of a Khan of the Khoshot, Kalmuk and Kirghiz tribes near Astrakhan, as seen by Pallas in +1776.

Fig. 699

Fig. 700

Fig. 699. Another example of four-cup 'anemometer' prayer-wheels, seen at the Mongol settlement of Shang-chia on the Tsaidam plateau in 1890 (Rockhill, 3).

Fig. 700. The Savonius S-rotor forestalled in a particular type of Tibetan wind-driven prayer-drum, an example in tin (Museum f. Völkerkunde, Vienna). Cf. Horwitz (11) and Laufer (19).

Chhin and Han;[a] and foreshadows the even more curious invention of the helicopter top, presently to be described.[b]

Perhaps, then, the windmills of Seistan were primarily of Tibetan or Mongol inspiration? Here the difficulty is that prayer-cylinders designed for the automatic repetition of the famous mantra are unlikely to have antedated the reign of K'ri-srong-lde-brtsan (+755 to +797) during which Buddhism conquered Tibet.[c] The statement often made that Fa-Hsien found prayer-wheels in Central Asia on his pilgrimage in +400 rests purely on a mistranslation, and none of the other Buddhist travellers mentions them.[d] Of course this does not prove that the Central Asian peoples were not using wind-driven gadgets of religious significance in pre-Buddhist times. Certainly the use of pennants to flutter in the wind and attract the attention of the gods is something very old, probably much older than the prayer-wheel, and common perhaps to all the shamanic systems of north and central Asia.[e] And on the other side of the chronological balance-sheet, we must recall that though the oldest reference to Persian windmills seems to be datable at +644, they figure more prominently in Arabic writings of the +9th and +10th centuries. By then the greeting to the jewel in the lotus might have had time to work its benevolent technological effects for suffering humanity. To sum the matter up, a Mongol–Tibetan Shamanist and Buddhist ancestry must be regarded as at least as probable as the more conventional Graeco-Arabic one.[f]

Thus the Chinese windmill is a characteristic contribution of its own, derived, we may certainly say, from the typically Asian horizontal windmills of Iran, but embodying devices borrowed from nautical technique so ingeniously as to make it almost a new invention. The European vertical windmill, on the other hand, equally original, though probably also due to a stimulus from Iran, seems more likely to have derived from the Archimedean screw and the vertical as well as the horizontal water-wheel.[g] The origins

[a] Cf. Vol. 4, pt. 1, p. 123 above.　　　　　[b] Cf. p. 583 below.

[c] The religion had made hardly any impression on Tibet before +630, during the reign of the great promoter of culture Srong-btsan-sgam-po (+617 to +650). Not until the +11th century was Lamaism fully established, and the prayer-wheels might have been as late as that.

[d] Many have fallen into this trap, e.g. Cunningham (1), p. 375; Horwitz (11), p. 99; and Forbes (19), p. 615 (who even says they were wind-driven prayer-wheels) as recently as 1956. In ch. 5 of his Fo Kuo Chi,[1] completed in +416, Fa-Hsien[2] is speaking of a small country called Chieh-chha,[3] somewhere in the Himalayas (possibly Ladak or Kashgar) and ends by saying 'As for the excellent rules and customs which the monks of these places have, there is not room to tell of them all (sha-mên fa yung chuan chuan shêng, pu kho chü chi[4])'. The first translator, Rémusat (1), in 1836, went astray at this point, and took the words to mean: 'The monks, conformably to the law, make use of wheels.' Those who followed him, Beal (1) in 1869 and Legge (4) in 1886, gave the right sense, as was subsequently pointed out by Rockhill (3), p. 334; Simpson (1), p. 34; Goodrich (7), p. 154, and others. All modern translators, such as Li Yung-Hsi (1), agree. Cunningham appealed to certain Indo-Scythian coins as confirmation of his belief in the great antiquity of the prayer-wheel, but one may agree with the remark of Rockhill, that as the mace-like objects held in the hand might be almost anything, prayer-wheels were not impossible. Thus the mistake of a single hour is not corrected in a hundred autumns. Unfortunately, too, the puzzle of the origin of prayer-wheels remains with us.

[e] I have myself often seen them in the Sino-Tibetan borderlands.

[f] In this Lynn White (5) concurs. Elaborating later (7), pp. 86, 116, he suggests that the horizontal windmill in Europe was a direct transmission by the Central Asian slaves of +14th- and +15th-century Italy (cf. pp. 92, 125, and Vol. 1, p. 189).

[g] This does not mean that it never had any connections with nautical technology. Spanish and Portuguese windmills near the coasts have long used pieces of sailcloth resembling lateen sails; pictures

[1] 佛國記　　　[2] 法顯　　　[3] 竭叉　　　[4] 沙門法用轉轉勝不可具記

of the Persian horizontal windmill must remain for the present undetermined, but they seem likely to have had just as much to do with the Mongol–Tibetan wind-driven prayer-mechanisms as with the horizontal water-wheel or the toys of ancient Alexandria.

(m) THE PREHISTORY OF AERONAUTICAL ENGINEERING

In the foregoing section we have been concerned with wheels or rotors which, according to our terminology, might be called 'ex-aerial', that is to say, wheels or windmill sails so designed as to utilise the force of the winds for doing work. In earlier sections we also saw that 'ad-aerial' wheels or fans were known to ancient and medieval China, whether for use in winnowing, or for cooling palace halls in summer weather.[a] Although these employments did not involve the motion of any vehicle, which is one of the greatest uses of ad-aerial rotors today, we shall shortly see, not only that some precursors of the aeroplane propeller existed in China, but that one of them played a cardinal part in the development of modern aerodynamic thinking. More, the rise of this new science in the 19th century depended fundamentally upon the study of an apparatus which had not been known in Europe before the 16th, and which was partly Chinese in origin, namely the kite. The kite's stretched fabric is of course not a shaped aerofoil like the aeroplane wing,[b] but the most essential difference between them is the fact that the lift of the kite is provided by the fortuitous air-stream of the wind, while that of the aeroplane's wing is made mechanically by its propellers. Pilots of today who call aeroplanes 'kites' perhaps hardly realise how fitting historically is this slang term.[c]

'Lieh Tzu could ride upon the wind. Cool and skilfully sailing, he would go on for fifteen days before returning. . . .'[d] These words (from the Section on Taoism) we shall not have forgotten. They give us, indeed, just the proper starting-point for the present

---

in Mengeringhausen & Mengeringhausen (1); Forbes (19); Baroja (2). But the mills in the interior provinces, against which Don Quixote fought, resemble French tower-mills closely. In 1960 I had the opportunity of close examination of a number of the small Portuguese windmills, especially at São Braz de Alportel and Santiago de Cacem. I was struck by the fact that the sails used are true lateens, with a small luff edge; their bellying towards the leach gives the necessary helicoidal structure. The eight arms, connected at their periphery with rope or wire, are set staggered into the windshaft at four points.

[a] Pp. 118, 134, 151 above.

[b] At least not normally. Yet some Chinese kites were astonishingly made with cambered surfaces (see Fig. 705). Examples are figured by Chanute (1); Tissandier (5) and Gibbs-Smith (4), p. 36, but with no indication of date. It would be very interesting to know how far back in Chinese history this technique originated. Cf. Wei Yuan-Tai (1).

[c] In the years before the First World War, some of the pioneers of flying, especially S. F. Cody, habitually used the term 'power-kite' for their experimental aeroplanes. In 1907 he fitted a 12-h.p. engine to one of his modified man-lifting kites (cf. p. 590 below), and flew it without a pilot. The Farman aeroplanes of 1910 were often called 'box-kite' aircraft, although by that time the cellular type of construction was being rapidly abandoned. It had been very prominent, however, in the Voisin-Farman biplanes of 1908, and the Voisin-Archdeacon float glider of 1905; in both these the main wings were divided into kite-like cells by vertical partitions, and there was a similar tail-unit aft. Cf. Gibbs-Smith (1), pp. 57, 73, and pls. VIII (e) and IX (d).

[d] Chuang Tzu, ch. 1; cf. Vol. 2, p. 66 above.

PLATE CCLXXVI

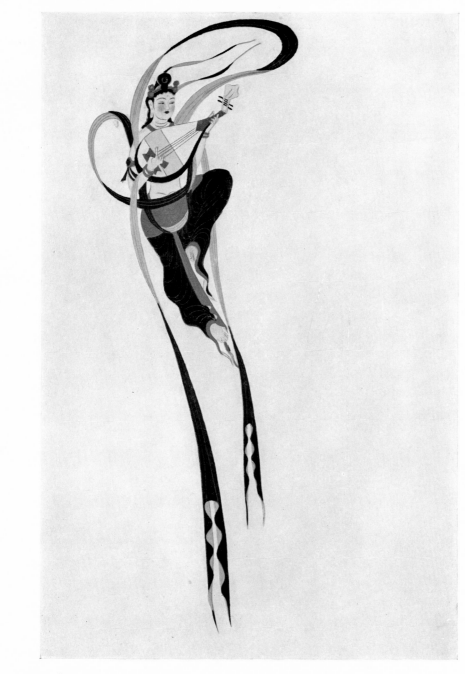

Fig. 701. The myth and magic of flight, one of the innumerable Buddhist *apsaras* (angels) depicted in the fresco paintings of the Chhien-fo-tung cave-temples at Tunhuang. A representation of the Thang period (*c.* +8th century) from cave no. 44, copied by Shih Wei-Hsiang. Weightless, she strikes the *phi-pha* lute (cf. Vol. 4, pt. 1, p. 130) to celebrate the Buddha's enlightenment. When thus he attained to the definition of the Four Holy Truths and the Noble Eightfold Way,

'So glad the world was—though it wist not why—
That over desolate wastes went swoooning songs
Of mirth, the voice of bodiless Prets and Bhuts
Foreseeing Buddh; and Devas in the air
Cried, 'It is finished, finished!'... (*Light of Asia*, p. 110).

review.[a] For though the Taoists elevated the conception of 'making excursion on the winds' and 'riding on the immensity of the universe' to philosophical heights, the idea itself had grown out of that primitive Asian shamanism which was one of the roots of their school. As has been abundantly shown, both for the most characteristic forms in Central and Northern Asia,[b] and for less typical varieties further afield,[c] the shamanic rites of possession and ecstasy almost invariably involve an imagined journey through the skies or the heavens, an ascension, a magic flight.[d] Hence the frequency of this theme among the poets of the Warring States and the Han.[e] The obsession of flight is constantly found in Chinese Taoism; the *hsien*,[1] or perfected adepts and immortals, were represented in Han times as having feathers or dressed in feathers;[f] emperors and alchemists alike ascended to the heavens when they died, as we see from innumerable Taoist biographies;[g] and down to modern times a literary term for Taoist priests was *yü kho*[2]—'feathered guests'. The entry of Buddhism to China only intensified the traditions of flying beings, for the *gandharvas* and *apsaras* of Indian mythology,[h] transformed outwardly by the Graeco-Bactrian focus through which they passed, winged their way to the east as the *fei thien*,[3] where they gave rise, in the Wei frescoes of Tunhuang, to some of the most exquisite and beautiful representations in all Chinese art.[i] One sees them in nearly every cave at Chhien-fo-tung; an example is given in Fig. 701. Meanwhile at the other end of the Old World, a parallel line of development had produced the angels of Hebrew and Christian tradition, and presumably also the broomstick-riding witches of medieval legend.[j] We may take it that

[a] Reference may here be made to previous treatments of the Chinese contributions to the development of aeronautical science. The note of Giles (9) was superseded by the remarkable contribution of Laufer (4), whose main failing was a tendency to take the legendary material too seriously. Parallel accounts for the occident are those of Feldhaus (14) and Hennig (2). The history of aviation itself constitutes now a large literature, some helpful items in which will be cited later on, including the works of Brown (1); Hodgson (2) and Davy (1). Perhaps the most brilliant study available is that of Duhem (1, 2), but it ends at the time of Montgolfier. It is thus fortunately now complemented by the book of Gibbs-Smith (1), for whose researches we were happy to make available a copy of the present section in draft form; his excellent work carries on the epic mainly from Montgolfier's time. Henceforward the reader may find it advantageous to have at hand some fairly up-to-date introduction to aeronautical science, such as that of Surgeoner (1) or Sutton (1). Abundant illustrations in Dollfus & Bouché (1).

[b] The most recent summary of the material is the interesting book of Eliade (3). For the Buriat Mongols, Tungus, Yakuts, Ostjaks, etc., see pp. 175, 211 ff.

[c] Métraux, on the South American Indians (1).

[d] Eliade (3), p. 415.

[e] Instances will be given in a moment. Sometimes on a bronze mirror we see a pair of scholars riding through nothingness in an aerial car drawn by birds or dragons—as for example that figured by Bulling (8), pl. 66, and dated by her in the close neighbourhood of +70.

[f] Cf. Vol. 2, p. 141 above. A typical story about feathered fairy girls who married mortal farmers occurs in the *Hsüan Chung Chi* (+6th century), *YHSF*, ch. 76, p. 8a.

[g] Laufer (4), p. 27; de Harlez (4); L. Giles (6); Kaltenmark (2), pp. 15, 23, 125, 127.

[h] Cf. Eliade (3), pp. 362 ff.

[i] The reader is referred to the excellent monograph of Nagahiro Toshio (1) on the subject.

[j] Cf. Laufer (4), p. 9. The devils of Christendom, it seems, had membranous bat wings only from the mid +13th century onwards; earlier occidental representations give them the bird wings of fallen angels. Baltrušaitis (1), pp. 151 ff., has brought forward evidence suggesting that this development was due to Chinese influence. There are certainly close iconographic similarities between the flying devils of later Europe and the demons of earlier Taoist and Buddhist China.

[1] 仙   [2] 羽客   [3] 飛天

the whole complex goes back to imaginations of winged and flying genii in ancient Mesopotamia and Egypt.[a] Its only connection with the present discussion is that it put ideas into people's heads; ideas (for example) of aerial cars and their makers.

(i) Legendary Material

Legends of self-propelled aerial cars, as opposed to flying vehicles drawn by winged animals and to unassisted personal flight in the style of Daedalus and Icarus, go back quite a long way in China, where they were associated with a mythical foreign person or people called Chi-Kung.[1] In the text of the *Shan Hai Ching* (Classic of the Mountains and Rivers), which may represent Early Han ideas, these people appear as three-eyed hermaphrodites, but there is no mention of their aircraft.[b] This appears suddenly in the works of two +3rd century contemporaries, Chang Hua and Huangfu Mi. The former, in his *Po Wu Chih* (Record of the Investigation of Things), says:[c]

> The Chi-Kung people were good at making mechanical devices (*shih kang*[2]) for killing birds. They could also make aerial carriages (*fei chhê*[3]) which, with a fair wind, travelled great distances. In the time of the emperor Thang[4,d] a westerly wind carried such a car as far as Yüchow, whereupon Thang had the car taken to pieces, not wishing his own people to see it. Ten years later there came an easterly wind (of sufficient strength), and then the car was reassembled and the visitors were sent back to their own country, which lies 40,000 *li* beyond the Jade Gate.[e]

Exactly the same story occurs in the *Ti Wang Shih Chi* (Stories of the Ancient Monarchs) of Huangfu Mi, who took Chi Kung to be a person, however, rather than a people.[f] It is then echoed time after time, e.g. in the +5th century by Shen Yo in his commentary on the Bamboo Books[g] and in the +6th by *Chin Lou Tzu*[h] and the *Shu I Chi*,[i] naturally with variations.[j] Long before the Sung it had become a literary commonplace.

Some interest attaches to the iconographic tradition associated with the Chi-Kung story. The oldest picture we have of the flying car is in the rare encyclopaedia *I Yü Thu Chih* (Illustrated Record of Strange Countries), compiled some time after +1392 and printed in +1489.[k] It shows a rectangular chariot with two occupants and one curious wheel (Fig. 702) which appears to be toothed. If Giles (9) and Laufer (4) were right in interpreting this wheel (from the general drawing of the picture) as

---

[a] No doubt the psychologists have something to say of its origins.
[b] Ch. 7, p. 3*a*.   [c] Ch. 2, p. 1*b*, tr. Giles (9), mod. auct.
[d] Legendary founder of the Shang dynasty, therefore mid −2nd millennium.
[e] Yümên Kuan on the Old Silk Road.
[f] A third early source is the Chin *Kua Ti Thu*,[5] cited in *Yü Hai*, ch. 78, p. 20*a*.
[g] *Chu Shu Chi Nien*, ch. 1, p. 21*a*.   [h] Ch. 5, p. 22*b*.
[i] Ch. 2, p. 13*b*.
[j] Also in the *Hsüan Chung Chi*, cit. *TPYL*, ch. 752, p. 3*a* (missed by Ma Kuo-Han).
[k] See Moule (4); Sarton (1), vol. 3, p. 1627; cf. Vol. 3, p. 513.

---

[1] 奇肱   [2] 拭扛   [3] 飛車   [4] 湯   [5] 括地圖

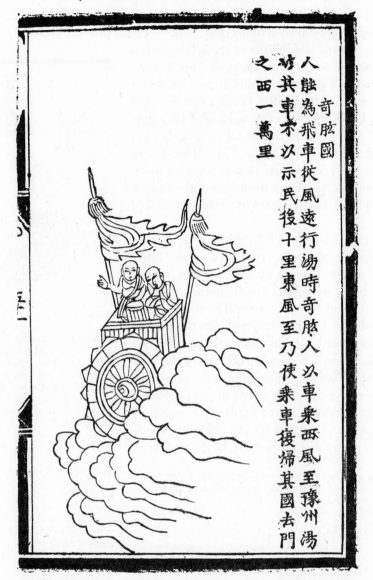

之西一萬里 奇肱國 人齕為飛車從風遠行湯時奇肱人以車乘西風至豫州湯 誌其車不以示民後十里東風至乃使乘車復歸其國去門

Fig. 702. The oldest printed picture of the aerial car of the mythical Chi-Kung people, a page from the *I Yü Thu Chih* encyclopaedia, *c.* +1430, first printed +1489. The inscription is approximately identical with the passage of the *Po Wu Chih* translated in the text. The one wheel visible appears to be toothed; did the artist think of it as facing the air like a vertical wind-wheel?

meant to be placed at right angles to the direction in which the car is flying through rolling clouds,[a] then it adumbrates a propeller. If we knew only this picture[b] such an identification would not perhaps be very convincing, but a variant occurs in some

[a] From the direction of the flags it would perhaps seem more likely that the wheel was imagined as one of a pair of aerial cart- or paddle-wheels.

[b] Which found its final form in the *Thu Shu Chi Chhêng, Pien i tien*, ch. 45; thence copied by Duhem (2), fig. 2, and others. Here two wheels were drawn.

editions of the *Shan Hai Ching*[a] which shows a clearly recognisable attempt by the artist to depict screw-bladed rotors (Fig. 703).[b] We shall produce evidence a few pages further on that the possibilities of the helicopter top or 'Chinese top' for powered flight were appreciated as early as the +4th century, and it may be, therefore, that some of the medieval artists who depicted the car of Chi-Kung were able to imagine the applicability of such rotors to horizontal motion. It may, for instance, be significant that Thao Hung-Ching in the late +5th century refers to a 'wheeled flying car' (*fei lun chhê*[1]) in which the Prince of the Eastern Sea (one of the Taoist hierarchy) made a round of visits.[c]

Flying cars drawn by birds, griffons, or dragons were a separate tradition. It started in the Han[d] and was strongly taken up by the Buddhists, several examples being present in the Tunhuang frescoes (Wei and Thang),[e] cf. Fig. 704a. The theme is also Indian[f] and may well be connected with the concept of 'vehicles of the Gods' in that mythology, Garuḍa and the like, as well as with the chariots of the solar and planetary spirits which appear in occidental mythology also, e.g. Phaethon. Again, its origins are probably in Babylonia, as the Etana myth suggests.[g]

What began as mythology was naturally transmuted into poetry as time went by. In the Section on

Fig. 703. Another pictorial version of the aerial car of the Chi-Kung people, from the *Shan Hai Ching Kuang Chu* (text of the −2nd century or earlier, +17th-century commentary). The inscription follows the *Shan Hai Ching* (ch. 7, pp. 3a ff.) fairly closely at first, saying: 'The people of the Chi-Kung country have each one arm and three eyes, and they are partly male and partly female. They are able to construct flying carriages which can follow the wind and travel great distances. Their land lies north of the I-Pei (country of the One-Armed Men). The skill of the Chi-Kung people is truly marvellous; by studying the winds they created and built flying wheels, with which they can ride along the paths of the whirlwinds. They visited us in the time of the emperor Thang.' The artist has here drawn the aerial car with two wheels, but both seem to be intended to represent screw-bladed rotors.

---

[a] Notably the *Shan Hai Ching Kuang Chu*[2] of +1667 edited by Wu Jen-Chhen.[3] It has been reproduced by Bazin (1); Duhem (2) and others.

[b] Other editions simply have an ordinary chariot with a dragon in an ox-yoke.

[c] In *Chen Kao* (TT 1004).

[d] Besides Bulling (8) just mentioned, Chavannes (9) reproduced a relief dated +87 (car with birds) and Harada & Komai (1) give a picture of a cloud chariot (*yün chhê*[4]) said to be by the Chin painter Ku Khai-Chih.[5]

[e] Especially caves nos. 296–311, with an important picture in no. 305.

[f] Laufer (4), pp. 44 ff.

[g] Laufer (4), pp. 58 ff. It is a curious coincidence that just about the same time that Chang Hua and Huangfu Mi were writing about the aerial cars of the Chi-Kung people in China, the corpus of legends about Alexander the Great which bears the name of Pseudo-Callisthenes was coming to completion in Alexandria and Byzantium. One of the many features of technical interest in this 'Alexander-Romance', as it is called, is the story that the great king embarked upon an aerial ascent in a car drawn by two or more large birds or gryphons. After the +9th century the story became very popular in Western Europe and is often found on façades and misericords, as e.g. in the choir at Lincoln. For further information see Cary (1), pp. 9 ff., 38, 59, 134 ff. and 296 ff.; Millet (1); M. D. Anderson (1).

---

[1] 飛輪車　　　[2] 山海經廣注　　　[3] 吳任臣　　　[4] 雲車　　　[5] 顧愷之

PLATE CCLXXVII a

Fig. 704a. A bird-drawn flying car in the Chhien-fo-tung frescoes, Hsi and Ho as the charioteers of the sun (cf. Vol. 3, p. 188).
Painting of the Western Wei period, c. +545, from cave no. 285, copied by Ho Hsi-Liang & Fan Wên-Tsao.

PLATE CCLXXVII *b*

Fig. 704 *b*. Kite-flying at Haikuan (Allom & Wright).

astronomy[a] we had occasion to refer to the accounts by numerous writers, ancient and modern, of imaginary flights through the sky to the moon or the sun. Lucian of Samosata (*c.* +160) is paralleled by Chang Hêng (+135), in his *Ssu Hsüan Fu*,[1,b] and indeed by Chhü Yuan earlier in the great *Li Sao* (*c.* −295),[c] though they are romantic while he is allegorical.

(2) THAUMATURGICAL ARTISANS

From the writers and artists we must now pass to the thaumaturgical artisans. In the end, someone actually does something. The invention of a wooden kite (*mu yuan*[2]) is ascribed in various ancient texts to Mo Ti[3] (the founder of the Mohist school; *d.* −380),[d] and to his contemporary Kungshu Phan,[4] the famous engineer of the State of Lu.[e] Whether it was in the shape of a bird is not clear. The character *yuan* continued to mean the bird which we call a kite (*Milvus lineatus* and related species),[f] and when applied to the flying device was usually qualified by the adjectival word *chih*,[5] paper. The *Han Fei Tzu* book, written about −255, says:[g]

> Mo Tzu made a wooden kite (*mu yuan*[2]) which took three years to complete. It could indeed fly, but after one day's trial it was wrecked. His disciples said 'What skill the Master has to be able to make a wooden kite fly!' But he answered 'It is not as clever as making a wooden ox-yoke peg (*i*[6]). They only use a short piece of wood, eight-tenths of a foot in length, costing less than a day's labour, yet it can pull 30 *tan*,[h] travelling far, taking great strain, and lasting many years. Yet I have worked three years to make this kite which has been ruined after one day's use.' Hui Tzu[i] heard of it and said: 'Mo Tzu is indeed ingenious, but perhaps he knows more about making yoke-pegs than about making wooden kites.'

This last remark may be taken as a hit at Mo Ti's utilitarianism. A closely similar passage occurs in the *Mo Tzu* book itself,[j] where Kungshu Phan is said to have constructed a bird from bamboo and wood, which stayed aloft for three days without coming down. Mo Tzu then engages him in a similar conversation about utilitarianism. In later times everybody knew these stories, which are repeated in *Pao Phu Tzu*, where Ko Hung, talking (*c.* +300) of people who made artificial things as good as real ones, speaks of Kungshu Phan's kites swaying and somersaulting (*mu yuan chih phien fan*[7]);[k] in the +6th-century *Shu I Chi* (with elaborations);[l] in the +12th-century *Hsü Po Wu Chih*;[m] and in the Ming *Hung Shu*[8] (Book of the Wild Geese).[n] The third

---

[a] Vol. 3, p. 440.
[b] *CSHK* (Hou Han sect.), ch. 52, p. 1*b*, also *Wên Hsüan*, ch. 15, p. 1*a* (tr. von Zach (6), vol. 1, p. 217).
[c] B. K. Lim (1), pp. 92 ff., better Hawkes (1), pp. 28 ff.
[d] Vol. 2, cf. pp. 165 ff. above.   [e] Cf. p. 43 above.
[f] R314.
[g] Ch. 11 (ch. 32), p. 2*a*, tr. auct. adjuv. Liao Wên-Kuei (1), vol. 2, p. 34.
[h] A weight equivalent to nearly 2 tons. Cf. the parallel passage, p. 313 above.
[i] The paradoxer, cf. Vol. 2, pp. 189 ff. above.
[j] Ch. 49, p. 9*b* (tr. Mei Yi-Pao (1), p. 256). See p. 313.
[k] *Nei Phien*, ch. 8, p. 10*b*.   [l] Ch. 2, p. 6*b*.   [m] Ch. 10, p. 7*b*.
[n] By Liu Chung-Ta.[9] Quoted in Wu Nan-Hsün (1), p. 168.

[1] 思玄賦   [2] 木鳶   [3] 墨翟   [4] 公輸般   [5] 紙   [6] 輗
[7] 木鳶之翩翻   [8] 鴻書   [9] 劉仲達

of these had no doubt that the devices of Mo Ti and Kungshu Phan were kites, such as were flown by Sung children; and the fourth repeats what was probably a tradition (though other statements of it have not come to hand) that Kungshu Phan flew wooden man-lifting kites over the city of Sung during a siege, either for observation or as vantage-points for archers. If this should be considered unlikely for the −4th century, we shall nevertheless see in a moment that the military use of kites goes back a long way in Chinese history.[a]

It is interesting to find that Wang Chhung sought to discredit the traditions about Mo Ti and Kungshu Phan. About +83 the great sceptic wrote, in his *Lun Hêng*:[b]

> The books of the literati talk about the great skill of Kungshu Phan and Mo Ti, saying that they carved from wood kites which flew for three days without coming down. That they made wooden kites which would fly is quite possible but the report that these did not alight for three days must be exaggerated. If such a thing had the shape of a bird, how could it fly for three days without resting? If it could soar, why only for three days? It might indeed have been equipped with some mechanism by which it was set in motion and continued to fly, so that it did not descend; in this case the story should say that it flew continuously, not only for three days. There is another report, too, that Kungshu Phan lost his own mother because of his skill. He constructed for her a wooden chariot and horses with a wooden driver, and when it was ready and she had taken her place inside, it sped away and never returned; thus she was lost to him. Since the mechanism of the wooden kite was (presumably) equally well constructed, it also should have continued to fly without stopping. But if the mechanism would function but a short while, and therefore the kite could not keep flying for more than three days; so also the wooden carriage should have come to a stop on the road within three days' journey, instead of carrying the mother completely away. The two stories (in fact contradict each other and) must be wide of the truth.

Wang Chhung, therefore, though in captious mood, did not disbelieve in the possibility of artificial flight itself.

Another attempt at this seems to have been due to a younger contemporary of Wang Chhung, namely Chang Hêng the great astronomer and engineer (+78 to +139). The main information we have about it comes from a book called *Wên Shih Chuan*[1] (Records of the Scholars) by Chang Yin[2].[c] The passage, quoted twice in the *Thai-Phing Yü Lan*,[d] says that a wooden bird (*mu niao*[3]) was made, with wings and pinions (*yü ho*[4]), having in its belly a mechanism which enabled it to fly several *li* (*fu chung yu shih chi, nêng fei shu li*[5]). We are inclined to think that while the devices of Mo Ti and Kungshu Phan were kites, probably shaped roughly like birds, the invention of Chang Hêng could have involved the air-screw of the helicopter top, though the only motive power available to him for such a purpose would have been springs. We certainly need not take too seriously the statement about the distance flown. In Chang

---

[a] And, more surprisingly, the use of man-lifting kites also.

[b] Ch. 26, tr. Forke (4), vol. 1, p. 498; mod. Leslie et auct.

[c] If he was the same Chang Yin who lived *c.* +170 to +190 his information would be rather likely to be accurate. But the book is not mentioned before the Sui bibliography.

[d] Ch. 752, p. 2*b*; ch. 914, p. 8*a*.

[1] 文士傳   [2] 張隱   [3] 木鳥   [4] 羽翮   [5] 腸中有施機能飛數里

Hêng's own writings there are references to the machine. In his essay on the use of leisure in retirement (*Ying Hsien*[1]),[a] he says (+126):

> Certain base scholars used to report evil of me to the emperor, but I decided not to worry about such affairs, or to learn their 'unique arts' (of civil service intrigue). Yet linked wheels may be made to turn of themselves, so that even an object of carved wood may be made to fly all alone in the air (*mu tiao yu nêng tu fei*[2]). With drooping feathers I have returned to my own home; why should I not adjust my mechanisms and put them in working order (so that I may fly still higher than before)?

Here then he seems to mention his own mechanical interests, using them as an analogy for his own situation out of office.[b]

The chief Western parallel to all this is the 'flying dove' of Archytas of Tarentum, more or less of a Pythagorean,[c] whose *floruit* was in the neighbourhood of −380, making him a contemporary of Mo Ti and Kungshu Phan. Unfortunately, there is little reference to his model aircraft earlier than Aulus Gellius (*fl.* +130 to +180), the contemporary of Chang Hêng and Chang Yin, who quotes from Favorinus of Arelates,[d] an older man (*fl.* +150), the report that it flew by means of some expanding vapour contained within it.[e] According to other accounts[f] a weight and pulley were involved, and while the object could fly it could not rise again after falling. This might suggest that a launching mechanism was used, after which the model went forward in gliding flight assisted by whatever power-source is implied by the reference to compressed air or steam. The invention seems much more in the Alexandrian manner than of the time of Archytas, and it is distinctly curious, in view of the medieval expertise of China regarding gunpowder rockets,[g] that the jet-propulsion principle seems to be hinted at in the Greek and not in the Chinese sources. All the great Alexandrian mechanicians, however, were concerned with pneumatic devices—Ctesibius with pumps,[h] Heron with hydrostatic systems, organs, steam-jets and wind-power,[i] and

---

[a] *CSHK* (Hou Han sect.), ch. 54, p. 8*b*; *Hou Han Shu*, ch. 89, p. 3*b*, with commentary; tr. auct.

[b] Other references to flying birds of wood also occur in the same essay, p. 10*a*. It will be remembered that mention was made earlier of a lost writing of Chang Hêng's, the *Fei Niao Thu*,[3] which can be dated at +114. In the Section on Geography (Vol. 3, pp. 538, 576) we considered the possibility that this work had something to do with map-making, though as the last character is uncertain, the real subject may have been calendrical science. A third possibility is that we should interpret the title 'Diagrams of the Mechanism of the Flying Bird'. As for the problem of what this was, the ornithopter type (flapping-wing aircraft) should not be forgotten. Leonardo's favourite obsession, it lasted as late as Lilienthal in the eighties of the last century, and many model forms of it have successfully flown (Gibbs-Smith (1), pp. 19, 21). But perhaps it will never be much use until the complexity of the wing-mechanisms approaches that of the wings of birds themselves (*ibid.* pp. 267 ff.).

[c] On him in general see Freeman (1), p. 234; Sarton (1), vol. 1, p. 116; Neuburger (1), p. 231.

[d] On him see B & M, p. 578.

[e] *Noctes Atticae*, x, 12, ix ff. 'Nam et plerique nobilium Graecorum et Favorinus philosophus memoriarum veterum exsequentissimus affirmatissime scripserunt, simulacrum columbae e ligno ab Archyta ratione quadam disciplinaque mechanica factum volasse; ita erat scilicet libramentis suspensum et aura spiritus inclusa atque occulta concitum.'

[f] Cf. Laufer (4), p. 64.

[g] And even winged rockets, as we shall duly see in Sect. 30 below.

[h] Cf., for example, Sarton (1), vol. 1, p. 184; Drachmann (2, 9).

[i] Sarton (1), vol. 1, p. 208; Drachmann (2, 9).

[1] 應閒   [2] 木雕猶能獨飛   [3] 飛鳥圖

Ctesibius and Philon[a] with catapults (ballistae) involving the use of compressed air.[b]

Conceivably, therefore, the account in Aulus Gellius may refer to a light model with glider wings launched from an inclined platform by a weight, and containing a space with a narrow backward-pointing outlet, through which a jet of steam could issue, as in Heron's aeolipile.[c] Alternatively the model was perhaps hung from a pole on a 'whirling arm' and driven round on the end of it.[d] Thus if there is anything in these speculations, Mo Ti, Kungshu Phan and Chang Hêng may have experimented, before the +2nd century was out, with two of the great components of modern aeronautical science, the kite-wing and the air-screw;[e] while Archytas or the Alexandrians may possibly have used the jet principle.

(3) THE KITE AND ITS ORIGINS

Let us now examine more closely the chief material basis for Chinese aeronautical stories, the kite of wood, bamboo, and paper. Its use in Asia would seem to be exceedingly old, since anthropologists have found it in a wide distribution radiating south and east of China through Indo-China, Indonesia, Melanesia and Polynesia (Chadwick, 1). In some parts of this area kite-flying was practised as a religious function connected with gods and mythical heroes. Often tabooed to women, the kite frequently carried, as in China, attachments such as strings or pipes to make musical or humming noises in the air.[f] An important practical application was found for it in a method of fishing, to remove the hook and bait far from the sinister shadows of the boat and the fisherman.[g] In China a game was played with kites.[h]

As to the origin of the kite, Waley (15) suggested that perhaps it derived from an ancient Chinese method of shooting off an arrow with a line attached to it, so that both arrow and prey could be recovered by hauling it in—as represented in the character i[1].[i] Raglan (1) declined to accept this for the reasons that kite-flying consists not in standing still and pulling a solid object towards one, but in towing it to start it from

---

[a] Sarton (1), vol. 1, p. 195.

[b] In the pneumatic catapult (aerotonon) the cord was attached to levers themselves fixed at right-angles to pistons fitting tightly into bronze cylinders. The act of stretching the cord compressed the air in the cylinders, and this, upon the pulling of the trigger mechanism, assured the return of the cord to its original position and the despatch of the arrow or other projectile. There is no evidence, however, that the machine was anything but a military curiosity, or even that it was ever built at all. Cf. Schramm (1); Beck (3); Neuburger (1), p. 224.

[c] Duhem (1), p. 125, agrees with these general conclusions.

[d] This suggestion is due to Mr Gibbs-Smith.

[e] Allusion has been made already to the philosophical context and significance of their efforts at automation, cf. Vol. 2, pp. 53 ff. Later Chinese work with flying models has been mentioned on p. 163 above.　　　　　　　　　　　　　　　　　　[f] See immediately below, p. 578.

[g] This Indonesian-Melanesian device has been monographed by Balfour (2); Anell (1) and Plischke (1); see also Montandon (1), p. 244.

[h] The cord near the kite being covered with crushed glass or porcelain glued on, the players seek to get their kites to windward of their rivals so that the cords are cut through, and the losing kite comes fluttering to earth. This is an autumn sport; Laufer (4), p. 32; Wei Yuan-Tai (1). Cf. Fig. 704b.

[i] We shall shortly find an unexpected, but more acceptable, function for this method in the history of technology, Sect. 28e below.

[1] 夷

the ground,[a] while if a hunting method had been at the beginning of the story, the religious associations of the kite might be harder to explain. He himself suggested as the kite's ancestor the bull-roarer, which produces a loud humming noise on the end of the string with which it is whirled round, and still exists as an important ritual object in many primitive cultures. But the origin of the kite lies so far back in Asian history that all such theories can be but speculative.

Perhaps we have already seen an ancient Chinese representation of a kite without noticing it; on the −4th-century Hui-hsien bowl (Fig. 299 in Vol. 4, pt. 1, at the bottom on the left.)

We may well be prepared to regard the devices of Mo Ti and Kungshu Phan as the earliest references to kites, though Laufer was disinclined to do so. The scholars of the Sung, notably Kao Chhêng[b] and Chou Ta-Kuan[c] (+13th century), recorded a story that the Han general Han Hsin[1] (d. −196) flew a kite over a palace to measure the distance which his sappers would have to dig in order to make a tunnel through which the troops could enter. They used the expression 'paper kite' though paper was not available until three centuries after his time; moreover, the story has not been found so far in any contemporary source, but some foundation for it is by no means impossible. In the Thang book *Tu I Chih* we find the following:[d]

In the Thai-Chhing reign-period (+547 to +549) of Liang Wu Ti, Hou Ching[2] rebelled, and besieged Thai-chhêng (Nanking), isolating it from loyal forces far and near. Chien Wên[3] (i.e. Hsiao Kang,[4] later emperor for one year, +550) and the crown prince (Hsiao) Ta-Chhi[5] decided to use many kites (*fu yuan*[6]) flying in the sky to communicate knowledge of the emergency to the army leaders at a distance. The officers of Hou Ching told him that there was magic afoot, or that messages were being sent, and ordered archers to shoot at the kites. At first they all seemed to fall but then they changed into birds which flew away and disappeared.

This probably means that kites were used for signalling, and despatches sent out by carrier pigeons.[e] Then in +781 a loyal general of the Thang, Chang Phei,[7] besieged in Lin-ming, signalled with kites to inform his fellow commanders of his predicament, and the city was eventually relieved.[f] Again in the +13th century kites were used in war between the Chin Tartars and the Mongols. At the famous siege of Khaifêng in +1232, when the army of the Jurchen Chin was shut up in their capital by Ögötäi's forces,

The besieged sent up paper kites with writing on them, and when these came over the northern (i.e. the Mongol) lines, the strings were cut (so that they fell among) the Chin prisoners (there). (The messages) incited them (to revolt and escape). People who saw this said: 'Only a few days ago the Chin (commanders) were using red paper lanterns (for

---

[a] Besides, the feathering of an arrow would be far too small to give it any lift. A Chinese farmer's hat on the end of a string would be a much more likely ancestor.

[b] *Shih Wu Chi Yuan*, ch. 8, p. 27a.

[c] *Chhêng Chai Tsa Chi*[8] (c. +1295), a reference first noted by St Julien (4) and L. C. Anderson. Quoted *KCCY*, ch. 60, p. 8b.

[d] Ch. 2, p. 4a; also quoted in *KCCY*. Tr. auct.

[e] See below, Sect. 30.   [f] Cf. *Chiu Thang Shu*, ch. 187B, p. 13a (biography).

[1] 韓信   [2] 候景   [3] 簡文   [4] 蕭綱   [5] 大器   [6] 縛鳶

[7] 張伾   [8] 誠齋雜記

signalling) and now they are making use of paper kites. If the generals think they can defeat the enemy by such methods they will find it very difficult.'[a]

Thus we have an early instance of a 'leaflet raid', for the messages were simply propaganda urging the captured Chin soldiers to rise and fight their way back to their own side.[b] These examples may be enough to show the continual military uses which were found for kites in China, and this perhaps lends additional plausibility to the original association with Mo Ti, the interest of whose disciples in military technique has been emphasised more than once elsewhere in this book.[c]

That kite-flying as a pastime also goes back a long way is evident too. One sees pictures of it in the Tunhuang frescoes from the Wei period onward.[d] Literary descriptions occur in +10th-century books such as the *Tiao Chi Li Than*[1] (Talks at Fisherman's Rock),[e] and frequently in the Sung and Ming.[f] Though the practice of fitting Aeolian harps on kites may have started in the Thang or before, it is closely associated with the name of a famous maker of kites[g] in the +10th century, Li Yeh.[2] Those of bamboo with one thin bamboo string are called *fêng chêng*,[3] 'wind-psalteries', or *fêng chhin*,[4] 'wind-zithers'; those with seven silk strings fixed across a gourd-shaped framework are called *yao chhin*,[5] 'hawk-lutes'.[h] Sung references to this practice are numerous.[i] In a cognate custom, whistles (*ko ling*[6] or *shao tzu*[7]) of bamboo, gourd or horn, are fixed to the tail-feathers of pigeons.[j] This is certain for the Sung period and not likely to have started later than the Thang.[k]

---

[a] *Chin Shih*, ch. 113, p. 18b, tr. auct.

[b] The slightly modified quotation in the *Thung Chien Kang Mu*, pt. 3, ch. 19, p. 50a, b, was translated more than a century ago by St Julien for Reinaud & Favé (2), p. 288. Misunderstanding the nature of the writings carried by the kites, he supposed them to be magic charms, and added a patronising footnote about similar cases of Chinese credulity in the Opium Wars. But an independent and contemporary source, the *Kuei Chhien Chih*, written in +1235 by Liu Chhi, which gives a fuller account (ch. 11, p. 4a), makes it quite clear that the messages called upon the prisoners to return, adding that they were all promised promotion if they got back, 'running between the arrows and the stones'. Thus once again Chinese common sense was turned by sinologists into non-sense. Feldhaus, for his part, (1), col. 654, confused two incidents of the siege, and placed lanterns and kites wrongly in the context of the Mongol hot-air balloon dragon-standards (cf. p. 597 below).

[c] Above, Vol. 2, pp. 165 ff.; below, Sect. 30.

[d] Notably in caves nos. 332 and 148, both Thang, the latter dated +698.

[e] P. 38b.   [f] *Wu Lin Chiu Shih*, ch. 6, p. 15b; *Li Hai Chi*, p. 4b.

[g] See *Wu Tai Shih Chi*, ch. 30, p. 11b.

[h] Moule (10), pp. 105, 111; cf. also the special article of Hsü Chia-Chen (1).

[i] For example, *Tu Hsing Tsa Chih* (+1176), ch. 1, p. 9b, or a reference in a poem by Fan Chhêng-Ta c. +1180 (*Shih Hu Tzhu*, p. 13a). The *Wu Lin Chiu Shih* by Chou Mi (about +1270) lists Aeolian harps for kites as being on sale in Southern Sung Hangchow, and mentions the names of two men, Chou San[8] and Lü Phien-Thou,[9] who were renowned for making them (ch. 6, pp. 15b, 30b). They are also discussed in the Ming book *Hsün Chhu Lu*[10] (Enquiries and Suggestions about Popular Customs) by Chhen I.[11] Cf. Wu Nan-Hsün (1), p. 168.

[j] Moule (10), p. 67. These can fill the air above a Chinese city with delicious sounds, as I know from personal experience, well remembering that strange sky-music which one used to hear in the lanes of the Kweichow town of Anshun—as well as other places in China. Cf. further Laufer (4), p. 72, (26); Bodde (12), p. 22; Wang Shih-Hsiang (1).

[k] Moule's estimate of date was too cautious; he would have been delighted to find the pigeon-whistles on sale in his own Hangchow in the Southern Sung, for they are indeed listed (as *po-ko ling*[12]) among the special commodities in *Wu Lin Chiu Shih*, ch. 6, p. 15b.

[1] 釣磯立談	[2] 李鄴	[3] 風箏	[4] 風琴	[5] 鷂琴
[6] 鴿鈴	[7] 哨子	[8] 周三	[9] 呂偏頭	[10] 詢芻錄
[11] 陳沂	[12] 鵓鴿鈴			

A kite is supported upon the wind by a combination of three forces: its weight, the resistance of the air, and the compensating tension of the string. According to the strength of the wind the kite moves in a great circle of which the string is the radius, rising when the wind freshens and falling to a vertical position in which it can no longer remain suspended when the wind falls. The operator can then, by holding the string taut and running, keep the kite airborne during the calm period by providing sufficient flow under its plane surfaces, just as the artificial airstream lifts the true powered aeroplane. In the 18th century, some of the greatest European mathematicians devoted attention to the theory of kites (e.g. Newton, Desaguliers, d'Alembert, Euler).[a] But already, during the centuries of existence of the kite in China,

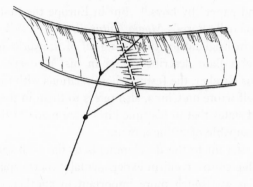

Fig. 705. Chinese cambered-wing kite (after Tissandier, 5).

several interesting refinements had been introduced, such as the addition of a second cord permitting the control of the angle of attack according to the wind strength.[b] Surfaces were also (as we have seen) made concavo-convex,[c] but unfortunately we do not yet know whether this practice began earlier than the end of the + 18th century, at which time the first suggestions of curved aerofoils were made in Europe.[d] Fig. 705 shows a typical Chinese cambered-wing kite. Here a point which should not be overlooked is the historical relation between the kite, the aeroplane and the sailing-carriage. The Chinese origin of the latter has already been described;[e] and the kite might almost be considered as a detached sail of the sailing-carriage.[f] At various times efforts were indeed made, not without success, to tow land-vehicles by means of kites, the most famous being that of Pocock in 1827.[g]

[a] Duhem (1), p. 199.　　　　　　　　　[b] *Ibid.* p. 195.
[c] See Chanute (1) and especially Tissandier (5) reproduced by Gibbs-Smith (4), fig. 11, p. 36. 'Monoplane' kites were generally not less than 3 ft. broad and often 11 ft. long. The two ends are often curved like plant leaves, and bird-shaped kites may have long rigid tails. Similar principles governed kite-making in Japan, to judge from the study of Tissandier (6). Cf. Wei Yuan-Tai (1); le Cornu (1); Needham (42).
[d] See below, p. 581.　　　　　　　　　[e] See above, p. 274.
[f] Duhem (1), p. 180. Did not della Porta call it a 'flying sayle'?
[g] *Ibid.* p. 193; Gibbs-Smith (1), pp. 12, 162. He had a successful run on the road between Bristol and Marlborough.

(4) THE HELICOPTER TOP; KO HUNG AND GEORGE CAYLEY ON THE 'HARD WIND' AND 'ROTARY WAFTS'

We now approach the most important part of this discourse, the examination of the role which ancient and medieval Chinese aeronautical devices played as part of the basis for the vast modern development of aerodynamics and aviation. That the kite was unknown in Europe until the end of the +16th century, when it was brought back by the early travellers, is well appreciated. 'It makes its début', said Laufer, 'as a Chinese contrivance, and not as a heritage of classical antiquity.'[a] This does not mean that the kite was unknown in the Islamic world; it was probably not new there in the +9th century when Abū 'Uthmān al-Jāḥiẓ described the flying of kites 'made of Chinese carton and paper' by boys.[b] But in Europe the first description of kites occurs in Giambattista della Porta's *Magia Naturalis*[c] of +1589. A few decades later they were employed in England for the letting off of fireworks in the air, as it is for this purpose that John Bate describes them in his *Mysteryes of Nature and Art* (+1634).[d] Athanasius Kircher, the Jesuit, whose relations with the China Jesuits were close, and who himself wrote on China, also refers to them in the *Ars Magna Lucis et Umbra* (+1646), and states that in his time kites were made in Rome of such dimensions that they were capable of lifting a man.

All this is highly relevant to the developments of the 19th century. The study of kites did indeed in due course confirm experimentally their capacity to carry human aeronauts aloft, but it was much more important in another way, because closely connected with the search for suitable glider and aeroplane wings. In 1804 Sir George Cayley constructed a successful model aircraft with plane (flat) kite wings and a tail rudder-elevator consisting of two plane kites intersecting at right angles.[e] This was 'the first true aeroplane in history'.[f] Plane surfaces attached to whirling arms were also used by him in the same year for his fundamental physical experiments on air-

---

[a] (4), p. 37. So also Plischke (2) in a later careful re-examination of the matter.

[b] In the Book of Animals (*Kitāb al-Ḥayawān*); Laufer (4), p. 37; Hitti (1), p. 382.

[c] Bk. 20, ch. 10 (English ed. +1658, p. 409), a 'flying Sayle'. In the +16th-century descriptions it is not always easy to distinguish the true kite from the hot-air dragon-balloon which had been popular earlier (cf. p. 597 below), but Johann Schmidlap (+1560) and Johann Mathesius (+1562) were probably talking about the latter and not the former. Della Porta was quickly followed by Jacob Wecker (+1592) and Daniel Schwenter (+1636). The latter talks of whistles which could be fitted on the kite as well as fireworks, a distinctly Asian (if not specifically Chinese) trait.

[d] Bate gives a picture of kite-flying, reproduced in Duhem (2), fig. 65; Gibbs-Smith (1), pl. 1 (*a*); but the first European illustration was that of Hellenius (+1618), in an engraving of Middelburg in Holland. As Plischke (2) points out, the fact that the kite appears first in Holland and England suggests that it was brought to Europe by Dutch or English merchants; if the transmission had been Portuguese we should expect it to have occurred half a century earlier. Neither della Porta nor Bate used the word 'kite', and Gibbs-Smith (1), p. 163, finds it 'inexplicable' that this should have been chosen out of all other possible bird names. But in fact the English name 'kite' is just a direct translation of the old Chinese term. The words used in other European languages, e.g. *Drache, cerf-volant*, etc., could also have been derived from the different animal forms so often given by the Chinese to their kites. Thus the terminology points rather clearly to China as the source of the transmission.

[e] See Cayley (1), p. 26. For his biography (+1773 to 1857) see Hodgson (1, 3); Pritchard (1). He is not to be confused with the Cambridge mathematician, Arthur Cayley (1821–95).

[f] Gibbs-Smith (1), pp. 10, 162, 190, and pl. 11 (*b*), (8); Needham (42).

resistance, angle of incidence, and other aerodynamic phenomena.[a] But the study of the wings of birds had long been proceeding in parallel, and Cayley himself had realised as early as 1799 that 'to make a surface support a given weight by the application of power to the resistance of air' was the basic problem. Any aeroplane must have, he clearly saw, a main supporting wing,[b] with a tail-unit to exercise control; this we know from a dated silver medallion engraved by him, and from contemporary drawings.[c] His pioneer design, in which these features were incorporated, is regarded as 'the first illustration in history of an aeroplane of modern type'. Moreover, it embodied the discovery of the aerofoil.[d] Although Cayley understood that a cambered wing gave better lift, he did not feel compelled to build it into his full-size machines, for in his model gliders he relied on the production of a curved surface by the airflow itself, acting as it did on his fabric wings which had spars only along the leading and trailing edges. But many of his drawings show the camber very clearly. Then, within fifty years, the conviction grew that one must imitate the cross-section of avian wings [e] rather than the stretched paper of the kite, and double-surfaced wings with different upper and lower profiles were introduced.[f] In this way streamlined sections were attained which combined the advantages of the convex upper surface with those of a nearly flat lower one, the former reducing air-pressure by accelerating the speed of flow and so creating an upward suction; the latter obviating concavity turbulence and compressing the air by decreasing its flow speed.[g] In later years Cayley himself carried the aerofoil principle further in successful models (1818, 1849 and 1853) and full-scale passenger-carrying monoplane gliders (1849 and 1853), which however continued to show traces of the plane kite both in their wings and in their rudder-elevator tails.[h] He had already for many years mastered lateral stability by means of the dihedral angle at which he set his wings, and longitudinal stability by means of the tail-plane which he fitted. The majority of those who experimented with model flying-machines in the second half of the century adopted curved aerofoil shapes of one kind or another for their wings.[i]

But still the paper bird of China had not exerted its full influence upon aeroplane

[a] See Cayley (1), pp. 22 ff. and frontispiece.

[b] That is to say, in ordinary speech, for a monoplane has strictly two wings, port and starboard, and a biplane two sets of wings.

[c] Gibbs-Smith (1), pp. 10, 189, and title-page. In this Cayley had been preceded by certain imaginative writers, notably Restif de la Bretonne (+1781).

[d] A key point here is whether Cayley or any of the other 18th- or 19th-century pioneers knew of the cambered wings of some of the Chinese kites. Kite-makers in China had doubtless been led to this development by their preoccupation with the imitation of animal, especially bird, forms (cf. Wei Yuan-Tai, 1), and not from formulated aerodynamic considerations. Yet the influence, could it be established, would be of great interest, and a search in the literature of early aviation, both manuscript and printed, might be very rewarding. There were many who experimented with large cambered-wing kites, e.g. Maillot (1).

[e] Cf. Cayley (1), p. 52.

[f] Henson (1842) seems to take the credit for this, followed by Pénaud (1876) and many others.

[g] Cf. Surgeoner (1), vol. 1, pp. 32 ff.; Gibbs-Smith (1), pp. 256 ff., 263 ff. Here we have a nice illustration of the way in which the aeroplane wing embodies, in a very real sense, both the bird-wing and the paper-kite plane. The later developments are sketched in Brooks (1).

[h] See Cayley (2, 3) and Gibbs-Smith (1), pp. 10, 190 ff. and pl. 11 (d). Cayley's actual design has now been discovered by Gibbs-Smith (6).

[i] E.g. Henson (1847), Wenham (1866) and Lilienthal (1891).

design, for in 1893 the Australian Lawrence Hargrave invented the box-kite for greater stability and lift,[a] normally two cells connected by booms to form a tandem frame, and it was this which inspired most of the biplane builders of the first decade of the present century.[b] Thus although the plane (flat) kite was not by any means the only influence on glider design, there was justification enough for the use of the term 'power-kite' for some of the aeroplanes of this period, and for its legacy in common speech today.[c] Meanwhile from Cayley's time onwards the study of suitable power-sources was steadily progressing,[d] so that at last the tow-rope and falling gradient could be replaced by energy generated within the aircraft itself. For this the invention of the air-screw or propeller was an absolute essential, and in a moment we must turn to consider what its origins were.

Before proceeding to this, however, let us leap backward in time some fifteen centuries, and pause to notice a very remarkable passage, on aerodynamics one might almost say, written by the great Taoist adept and alchemist Ko Hung about +320. This is what we find in the *Pao Phu Tzu*:[e]

Someone asked the Master about the principles (*tao*[1]) of mounting to dangerous heights and travelling into the vast inane. The Master said....[f] 'Some have made flying cars (*fei chhê*[2]) with wood from the inner part of the jujube tree, using ox leather (straps) fastened to returning blades so as to set the machine in motion (*huan chien i yin chhi chi*[3]).[g] Others have had the idea of making five snakes, six dragons and three oxen, to meet the "hard wind" (*kang (fêng*[4]))[h] and ride on it, not stopping until they have risen to a height of forty *li*.[i] That region is called Thai Chhing,[5] (the purest of empty space). There the *chhi*[6] is extremely hard, so much so that it can overcome (the strength of) human beings. As the Teacher[j] says: "The kite (bird) flies higher and higher spirally, and then only needs to stretch its two wings, beating the air no more, in order to go forward by itself. This is because it starts gliding (lit. riding) on the 'hard wind' (*kang chhi*[7]). Take dragons, for example; when they first rise they go up using the clouds as steps, and after they have attained a height of forty *li* then they

---

[a] Cf. Gibbs-Smith (1), pp. 30, 73, 162, 318 and pl. v (*c*); Needham (42).

[b] Cf. fn. (*c*) on p. 568 above.

[c] Cf. Vivian & Marsh (1), p. 190. As late as 1910 expositions of flying theory, such as that of Ferris (1), generally started by considering kites. The book of Chanute (1) gives the fullest account of the influence of kites on glider design.

[d] He himself experimented with a gunpowder motor, (1), p. 42, in 1807.

[e] *Nei Phien*, ch. 15, pp. 12a ff.; *Tao Tsang* ed., pp. 13a ff.; cit. *TPYL*, ch. 15, p. 4b (abridged). Tr. auct.

[f] The first sentences of Ko Hung's reply concern the use of drugs which will make the body ethereally light, and he also says a few words about stilts.

[g] Some texts read *huan*,[8] 'disposed in a ring', but the technical sense is not much altered thereby.

[h] This phrase might also be translated 'rushing wind' or 'violent wind', or alternatively 'wind from very high in the sky, from the four stars of the box of the Great Bear'. The idea of speed, it is important to note, is contained implicitly in the expression 'hard wind'.

[i] The distance which Ko Hung happens to mention would be equivalent to about 65,000 ft. Today meteorological balloons commonly reach this height, and manned balloons have been as high as 100,000 ft. About 40,000 ft. is an operational level for stratosphere flying. Rockets, of course, have reached far out beyond Ko Hung's wildest dreams, but this Section was first written before the sputnik and cosmic-vehicle age had dawned.

[j] A reference to Chuang Tzu?

---

[1] 道   [2] 飛車   [3] 還劍以引其機   [4] 罡風   [5] 太清
[6] 氣   [7] 剄蒸   [8] 霽

rush forward effortlessly (lit. automatically) (gliding)." This account comes from the adepts (*hsien jen*[1]), and is handed down to ordinary people, but they are not likely to understand it.'

For the beginning of the +4th century this is truly an astonishing passage, and if it can be equalled by any Greek parallel, one would be glad to know the reference. There can be no doubt that the first plan which Ko Hung proposes for flight is the helicopter top; 'returning (or revolving) blades' can hardly mean anything else, especially in close association with a belt or strap. This kind of toy was termed in 18th-century Europe the 'Chinese top',[a] though it seems to have been known in the West already in late medieval times.[b] In +1784 it attracted the attention of Launoy and Bienvenu in France[c] who made a bow-drill device drive two contra-rotating propellers consisting of silk-covered frames (Fig. 706).[d] In +1792 it stimulated Sir George Cayley, who may truly be called the father of modern aeronautics, to his first experiments on what he afterwards called 'rotary wafts' or 'elevating fliers'. He tells us so himself in the paper which he contributed to *Nicholson's Journal*[e] in 1809, and he too used a bow-drill spring to work two feather air-screws which kept the top mounting into the air.[f] The ordinary 'Chinese top' was simply an axis bearing radiating blades set at an angle, and given a powerful rotation by the pulling of a cord previously wound round the stem, hence Ko Hung's reference to leather straps. One of its commonest names in China was the 'bamboo dragonfly' (*chu chhing-thing*[2]).[g] Figure 707 is a diagram drawn by Cayley himself which he sent to a French engineer, Dupuis-Delcourt,[h] in 1853, saying that while the original toy would rise no higher than 20 or 25 ft., his improved models would 'mount upward of 90 ft. into the air'.[i] This then was the direct ancestor of the helicopter rotor and the godfather of the aeroplane propeller.

There can be little doubt that the helicopter top was connected in its origin with the hot-air zoetrope (*tsou ma têng*[3]), which has already been discussed in the physics Section (Vol. 4, pt. 1, p. 123), and with the Mongol prayer-wheel operated by a chimney

[a] The first appearance of this toy in China is a very obscure question. Search should be made among Chinese paintings of the genre which depicts pedlars selling children's toys (cf. p. 586).
[b] Gibbs-Smith (7) convincingly adduces a painting of *c.* +1460 at the Musée de l'ancien Eveché, Le Mans, and a +16th-century stained-glass panel in the Victoria and Albert Museum. Feldhaus (20), p. 285, fig. 191, adds a remarkable helicopter top with no less than three superimposed airscrews painted in a picture by Pieter Breughel the elder, *c.* +1560, and now in the Kunsthistorische Museum at Vienna. Gibbs-Smith (9) accepts Chinese provenance before Leonardo's time.
[c] It is exceedingly unlikely that they knew of Leonardo da Vinci's helical screw helicopter design; Duhem (1), pp. 279ff., (2), fig. 42*b*; Ucelli di Nemi (3), no. 3; Feldhaus (18), p. 149; Beck (1), p. 351, etc.
[d] Cf. Duhem (1), p. 282, (2), p. 232, fig. 156*a*; Gibbs-Smith (1), pp. 171 ff.
[e] Cayley (2), in Hubbard & Ledeboer (1), vol. 1.
[f] Cf. Duhem (1), p. 282, (2), p. 232, fig. 156*b*; Gibbs-Smith (1), p. 189, pl. 11 (*a*).
[g] Many Chinese scientific friends, e.g. the entomologist Dr Chu Hung-Fu, and our colleague Dr Lu Gwei-Djen, well remember playing with them as children. See Tatin (1) for a typical 19th-century European type, powered by Pénaud's rubber motor.
[h] For publication in a new aeronautical journal.
[i] Hubbard & Ledeboer (1), vol. 1, opp. p. x; Hodgson (2), fig. 135. One of Cayley's more extraordinary designs was that for a 'convertiplane', i.e. an aircraft with helicopter screws for vertical ascent which would close to form wings when the desired altitude had been reached, horizontal motion being assured by two ordinary propellers (1843); cf. Gibbs-Smith (1), pp. 11, 143, 190, 193 ff. and pls. 11 (*c*), xxi (*f*). Not till 102 years later were successful vertical take-off machines of this kind built, notably the McDonnell xv (1) and the Fairey Rotodyne.

[1] 仙人　　[2] 竹蜻蜓　　[3] 走馬燈

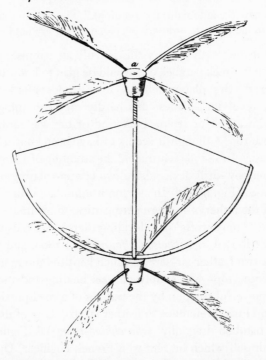

Fig. 706. The 'bamboo dragonfly' (*chu chhing-thing*) or Chinese helicopter top, studied by Launoy & Bienvenu in +1784 and by Cayley in +1792 (drawing by Cayley in 1809). A bow-drill spring rotates two feather air-screws (*a*, *b*) which carry the top high up into the air.

air-current, which we mentioned only a few pages above (p. 566). The use of a similar horizontal vane-wheel to work a roasting-spit, often found in the great kitchens of Europe,[a] was apparently not known in East Asia. All these were essentially rotors with vanes (*yeh lun*[1]), moving in relation to currents of air parallel with their axes,

[a] One of these was sketched by Leonardo about +1485; cf. Duhem (2), fig. 42*a*; Beck (1), fig. 605; Uccelli (1), fig. 37. As we noted on p. 124 above, this was an elegant automation since the hotter the fire the faster the roast would spin. In the +16th century it was frequently described (cf. Beck (1), fig. 375; Uccelli (1), figs. 38, 40). In +1629 Branca proposed (pl. 2) to use the ascending air-current from a forge to work a small rolling-mill by means of reduction gearing, but he employed a vertical paddle-wheel, not a zoetrope vane-wheel. Conversely, there was a horizontal wheel in his famous Aeolian stamp-mill (pl. 25), but again it was a paddle-wheel 'turbine' and not a vane-wheel. Still, wind-wheels were certainly 'in the air' at this time. Branca's illustrations are often reproduced (e.g. Uccelli (1), figs. 41, 42; Schmithals & Klemm (1), fig. 46; Vowles & Vowles (1), p. 126). Lynn White (5, 7) well appreciates the connection of the spit vane-wheel with the marine screw and the aeroplane propeller. He is inclined to place it in a group of Sino-Tibetan inventions adopted by Europe in the +15th century, associating it with the horizontal windmill (p. 567 above), the ball-and-chain flywheel device (p. 91 above), and also with such cultural borrowings as the 'Dance of Death' motif (cf. Baltrušaitis, 1). If Gibbs-Smith (7) is right in his identifications, the helicopter top must be added to the list. Lynn White further makes the interesting suggestion that this 'cluster of transmissions' may be connected with the slave-trade which brought thousands of Tartar domestic servants to Italy in medieval times, and which reached its height by the middle of the +15th century (cf. Vol. 1, p. 189; further references in Lynn White, 5). On p. 544 above and in Sect. 30*d* we define a 'Twelfth-century Cluster' and a 'Fourteenth-century Cluster' of transmissions. Both were more important than this fifteenth-century one, but the second and third may well be due, at least in part, to the remarkable social movement of which Lynn White reminds us.

[1] 葉輪

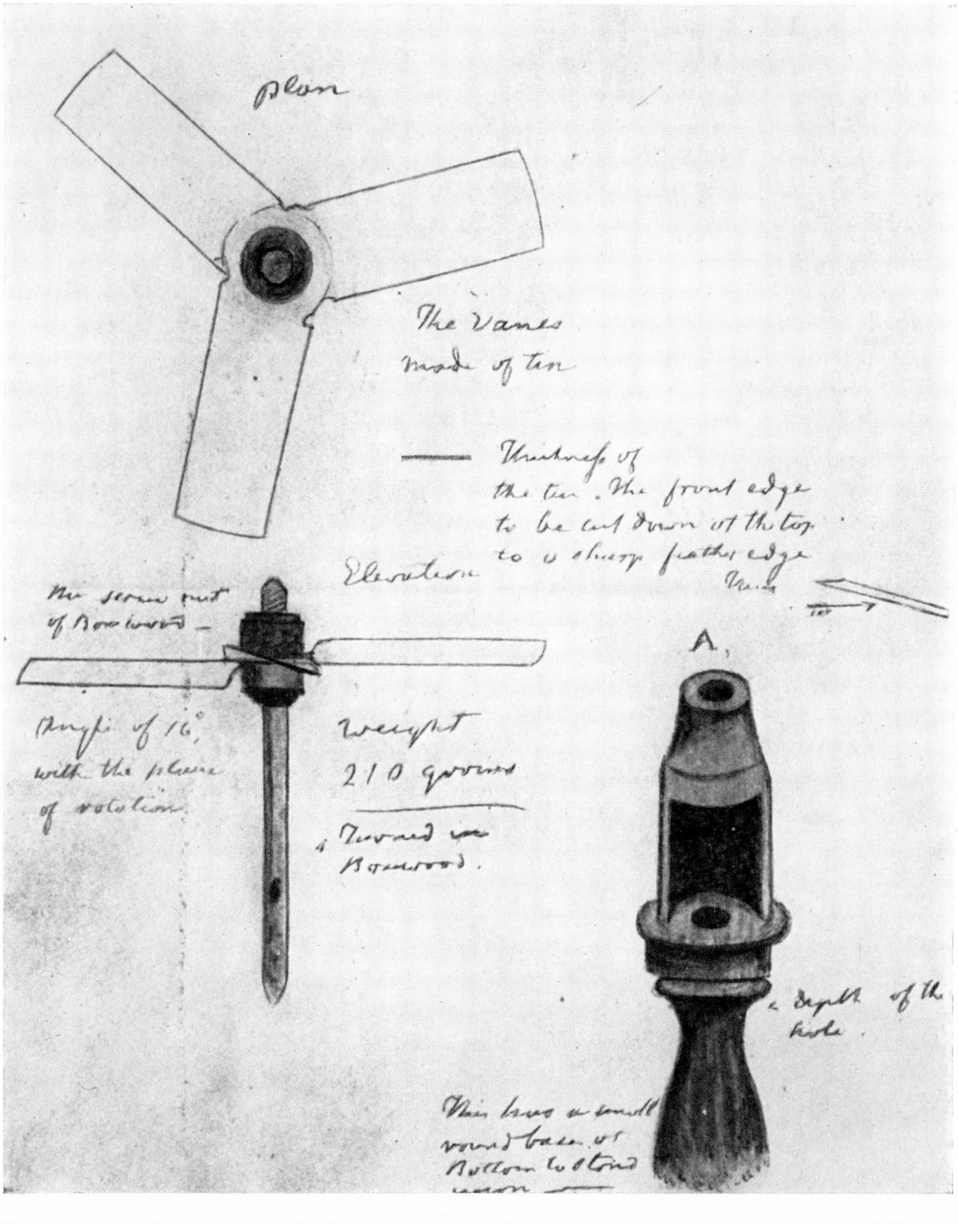

PLATE CCLXXVIII

Fig. 707. Page of drawings sent by Cayley to Dupuis-Delcourt in 1853 illustrating an improved Chinese helicopter top which would mount more than 90 ft. into the air. From Hubbard & Ledeboer (1). This was the direct ancestor of the helicopter rotor and the godfather of the aeroplane propeller.

the helicopter top ad-aerially because of the motion imparted to it by the cord, the zoetrope and the prayer-wheel ex-aerially because of the ascending current of hot air. But since the aeroplane propeller had to be vertically mounted, not only to bring about motor transportation (as the marine screw propels the ship), but to assure the airborne character of the flying-machine itself by driving the wings forward and so providing the necessary airstream lift, it was likely to spring from the European rather than the Chinese engineering tradition. For time after time in the present book we have shown that Chinese technicians preferred horizontal mountings and Westerners vertical ones.

The role of the vertical windmill in the generation of the aeroplane propeller has been particularly well appreciated by Gibbs-Smith.[a] Some twenty years before the practical work of Launoy and Cayley, an obscure French mathematician, Alexis Paucton (1), revived (it seems quite independently)[b] the idea of Leonardo for a helicopter screw, but he added to his proposed aeronef or 'ptérophore' an air-screw for horizontal motion—both being of the continuous Archimedean or marine variety (1768).[c] Significantly, he entitled his book a contribution to the 'theory of windmills'. But it seems that Vallet was the first to try out a vertical ad-aerial air-screw in practice when in 1784 he attempted, fruitlessly, to move a river-boat by a hand-operated propeller.[d] The same year however saw a successful ascent by Blanchard and Sheldon at Chelsea in a balloon equipped with a single hand-driven propeller, most significantly called a 'moulinet'.[e] The effect it produced was of course minimal, and Blanchard later suggested that steam power would some day be employed to drive it. Also in 1784 Meusnier[f] proposed that an elongated balloon should be fitted with three propellers in series, thus anticipating the dirigible.[g] But this took a long time to develop, and it was 1843 before the propellers of Monck Mason's clockwork-powered model airship carried it the length of a London hall.[h] Though full-size dirigibles became feasible by the end of the ensuing decade,[i] another development in this same year was even more significant for the future, namely the design of W. S. Henson for an 'aerial steam carriage' which fitted propulsive air-screws to a fixed-wing aeroplane.[j]

The transition from the spatial position of the helicopter rotor to that of the aeroplane propeller had however already been made (at the ex-aerial level) in China. Liu Thung,[1] in his early +17th-century *Ti Ching Ching Wu Lüeh*[2] (Descriptions of

[a] (1), pp. 3, 5, 170 ff.
[b] His inspiration was derived explicitly from the Archimedean screw.
[c] See Duhem (1), pp. 280 ff. Paucton also suggested the use of screw propulsion for ships.
[d] Gibbs-Smith (1), pp. 8, 170 ff.
[e] *Ibid.* pp. 170, 336 ff.; illustration in (3). Vallet also experimented with a propeller attached to the gondola of a balloon. The year 1784 was a kind of *annus mirabilis*, for as we have already seen Launoy & Bienvenu were at the same time working on their helicopter top. Moreover the previous year had seen Montgolfier's invention of the first practical balloon.
[f] He could not have had a more appropriate surname.
[g] See Uccelli (1), p. 893, fig. 16; Duhem (2), p. 224, fig. 151; Gibbs-Smith (1), p. 171, (3).
[h] Gibbs-Smith (1), p. 337, (3).
[i] Notably Giffard's airship of 1852.
[j] This impressive design is frequently reproduced; see Brooks (2), fig. 214; Schmithals & Klemm (1), fig. 120; Gibbs-Smith (1), pl. 11 (e); Needham (42).

[1] 劉侗   [2] 帝京景物署

Things and Customs at the Imperial Capital),[a] says that after kite-competitions were
forbidden, many-coloured wind-wheels (*fêng chhê*[1]) were made,[b] which when set up
facing the wind (*ying fêng*[2]), or rapidly carried in the hand, whirled round showing
their red and green colours confusingly. The arms of these wind-wheels, which were
of course vertically mounted,[c] were also used (in strange echo of Heron) to do work as
lugs by depressing a lever and beating a drum (see inset). There is room for some
surprise, at first sight, that such things as the helicopter

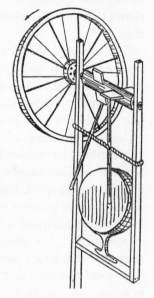

top and the zoetrope, with blades or vanes set screw-
fashion, should have originated in China, a culture to
which the screw and worm were, as we have seen (p. 124
above) essentially foreign. Evidently the setting of vanes
at a skew angle so as to constitute flat surfaces tangent to
the curves of a worm, and thus give or receive motion, did
not involve the invention of the continuously curving
forms of the screw.[d]

In quite a different manner, moreover, Chinese technology had already prepared the way for those vertically
mounted rotary roarers which would one day send the wings
of aircraft tearing through the heavens. At an earlier stage
(pp. 150 ff.) we saw how advanced the medieval Chinese
technicians were in their construction of rotary fans,
notably the winnowing-fan used in agriculture, but also
air-conditioning fans for palace halls. All these were
vertically mounted just as propellers would one day be,
and although they gave radial rather than axial flow, the rotary blowers of China
preceded those of Europe by some fifteen centuries.

Perhaps the most extraordinary prefiguration occurred when the Chinese toy-
makers proceeded to fit Liu Thung's wind-wheels to children's kites. Early in the
present century these were quite common at Nanking.[e] Such ex-aerial wheels rotating
just for joy on cambered wings exactly at the time when the aeroplane, with all its

---

[a] Quoted by Liu Hsien-Chou (*1*), pp. 68 ff.

[b] Note the ambiguity of the term; cf. fn. (e) on p. 561 above.

[c] 'Pinwheels' set up to make a noise in the wind are fairly common in south-east Asia, e.g. Bali
(cf. Bateson & Mead, 1). Other such playthings can be seen in the paintings of Hieronymus Bosch
(*c.* +1500); Gibbs-Smith (1), p. 171. It would be important to establish the time of first appearance
of toy wind-wheels in China. This could perhaps be done by studying one of the subjects which the
old Chinese artists delighted to paint, namely the pedlars of children's toys. Here a preliminary recon-
naissance found none in the painting of this kind by Su Han-Chhen[3] (*c.* +1115 to +1170) reproduced
by Cohn (2), pl. 77. But there is another by Wang Chen-Phêng,[4] dated +1310, which shows at least
two, perhaps four; two look like small vertically-mounted anemometers, and one seems to be of the
folded-paper type. This picture will be found in Sirén (10), vol. 6, pl. 47.

[d] Nor did that wonderful invention of the aboriginal Australians, the boomerang, closely related
though it is to the helicopter rotor and the gyroscope (cf. Duhem (1), pp. 269 ff.). Of course, such a
failure is easier to understand in a culture which remained at so primitive a level of general technology.

[e] Personal recollection of Dr Lu Gwei-Djen. We shall see presently that they may have stimulated
Chinese artists' conceptions of aerial voyages in the novels which they illustrated.

---

¹ 風車   ² 迎風   ³ 蘇漢臣   ⁴ 王振鵬

potentialities for good and evil, was being born, might almost symbolise contrasting conceptions of civilisation.[a]

Let us now return to Ko Hung. His words about the series of different kinds of animals would be incomprehensible if we did not know well the perennial Chinese tradition of making kites in the shapes of animals.[b] I have no doubt that what he was referring to were man-lifting kites, and though as yet we have no evidence that Ko Hung or any of his contemporaries constructed such large instruments, there would have been really nothing to prevent it. For people expert in kite-flying the possibility was obvious.[c] And it happens that we do possess from a time not long after that of Ko Hung himself, a remarkable account of this very thing.

The setting was the reign of a cruel and tyrannical emperor in a short-lived dynasty, Kao Yang,[1] Wên Hsüan Ti of the Northern Chhi, who ruled from +550 to +559. Though we may not need to believe all that we are told of his excesses, there is no doubt that humane Confucian government was at a discount in his time. One of his more peculiar methods of punishment was to make his prisoners participate in dangerous experiments on flight. Thus in the chapter in the *Sui Shu* on the history of law we find the following:[d]

> On one occasion the emperor visited the Tower of the Golden Phoenix[e] to receive Buddhist ordination. He caused many prisoners condemned to death to be brought forward, had them harnessed with great bamboo mats (*chhü chhu*[2]) as wings, and ordered them to fly down to the ground (from the top of the tower). This was called a 'liberation of living creatures'.[f] All the prisoners died, but the emperor contemplated the spectacle with enjoyment and much laughter.[g]

This was by no means the first time in Chinese history that trials had been made of wing-beating or ornithopter flight. As far back as the beginning of the +1st century there had been a well-authenticated attempt to imitate avian motions, though the name of the inventor has not been preserved.[h] In +19, Wang Mang, the only Hsin emperor, pressed by the nomadic warriors on the north-west frontiers, mobilised all who professed to be the masters of strange arts, and had them put to practical test.

---

[a] Some time after writing these lines, I was interested to find a remark by Wu Nan-Hsün (*1*), p. 169, that in China the kite began by being applied to war-like purposes and ended as a children's toy. With this as a text for a sociological discourse, much could be said.

[b] Chadwick (1) and others have reproduced 18th-century pictures of Chinese kite-flying which show this. Wei Yuan-Tai (1) illustrates contemporary examples. Cf. Fig. 704 *b*.

[c] It has been said that man-lifting kites are shown in medieval Chinese paintings (Duhem (1), p. 201), but this we have not been able to confirm.

[d] Ch. 25, p. 10 *b*, tr. Balazs (8), p. 56, eng. auct.

[e] This was one of three at the north-west of the capital (Yeh, near modern Lin-chang, north of the Yellow River). It was the tallest of them, about 100 ft. in height.

[f] A Buddhist practice for acquiring merit; birds and fishes were let go after being caught. But pre-Buddhist also, and *Lieh Tzu* cannot be dated thereby (cf. Bodde, 19).

[g] The commentary in *TCTC*, ch. 167 (p. 5189) says that any prisoner who descended successfully was to be pardoned. We have already come across another instance of the use of prisoners in experiments (Vol. 2, p. 441). That was for testing alchemical elixirs about +400. We shall return to the matter in Sect. 33; meanwhile see Ho Ping-Yü & Needham (4).

[h] The incident has already been referred to, Vol. 1, p. 110 above.

¹ 高洋　　² 蘧篨

One man said that he could fly a thousand *li* in a day, and spy out the (movements of the) Huns. (Wang) Mang tested him without delay. He took (as it were) the pinions of a great bird for his two wings (*ta niao ho wei liang i*[1]), his head and whole body were covered over with feathers,[a] and all was interconnected by means of (certain) rings and knots (*huan niu*[2]).[b] He flew a distance of several hundred paces, and then fell to the ground. (Wang) Mang saw that the methods could not be used, but wishing to gain prestige from these (inventors) he ordered that they should be given military appointments and presented with chariots and horses, while waiting for the army to set forth.[c]

This attempt, then, forms an early Chinese link in the long line connecting the Daedalus legend with Meerwein[d] and even with Lilienthal. Ko Hung must certainly have known about it. One feels some difficulty in drawing any sharp distinction between the wing-beating 'tower-jumpers', as modern historians of aeronautics like to call them, and the eventually successful gliders, for some of the former (no doubt fortuitously) glided long enough before landing to survive. So although the first true glider flights were those of Cayley's passengers in 1852 and 1853, the idea has very ancient origins. After all, birds themselves glide on motionless planes as well as beating their wings in their ascent. We must therefore place Wang Mang's pioneer, as well as the back-room experts of Kao Yang, in that line of temerarious bird-men[e] which runs through the Saxon monk Aethelmaer in the +11th century[f] to the numerous experimenters of the +17th and +18th[g] of which Meerwein was one of the most intelligent. The wild youth of the ornithopter endeavour ended with the Montgolfier period; whether it has a future, who can say?

But there was something more interesting in the wicked emperor's proceedings of +559 than the crude imitation of birds. The *Tzu Chih Thung Chien* (Comprehensive Mirror of History for Aid in Government), drawing upon other contemporary official sources, says:[h]

Kao Yang made Thopa Huang-Thou[3] (Yuan Huang-Thou[4]) and other prisoners take off from the Tower of the Golden Phoenix attached to paper (kites in the form of) owls (*ko chhêng chih chhih i fei*[5]). Yuan Huang-Thou was the only one who succeeded in flying as far

---

[a] There was probably some magic in this; cf. p. 569 and what has been said about the feathered immortals in Vol. 2, p. 141. We can be sure that Wang Mang's inventor was a Taoist.

[b] Hinges and pivots are also implied.

[c] *Chhien Han Shu*, ch. 99c, p. 5b, tr. auct. adjuv. Dubs (2), vol. 3, p. 382. We have to thank Prof. Liu Hsien-Chou for drawing our attention again to this passage.

[d] C. F. Meerwein, architect to the prince of Baden, built and tested a flying-machine in 1781; see Duhem (1), p. 231, (2), p. 204, fig. 137; Gibbs-Smith (1), pp. 311 ff. It was really a glider in which the pilot lay prone under an elongated ovoid wing with sharp ends and upward convexity. The surface necessary to support a man was correctly calculated, and the dihedral angle of the two halves of the monoplane could be altered only slightly. This was the moment when beating wings gave place to adjustable but essentially motionless wings.

[e] Cf. Duhem (1), pp. 110, 150, 167, 210, 222, 229, 231; Hodgson (2); Gibbs-Smith (1), pp. 3 ff., 6 ff., 12.

[f] Better known as Eilmer (or, erroneously, Oliver) of Malmesbury; he was an old man in +1066; cf. Sarton (1), vol. 1, p. 720; Lynn White (6).

[g] E.g. Guidotti (+1628), Besnier (+1678), de Bacqueville (+1742), etc.

[h] Ch. 167 (p. 5189), tr. Balazs (8), p. 132, eng. auct.

[1] 大鳥翮爲兩翼   [2] 環紐   [3] 拓跋黃頭   [4] 元黃頭
[5] 各乘紙鴟以飛

as the Purple Way, and there he came to earth. But then he was handed over to the President of the Censorate, Pi I-Yün,[1] who had him starved to death.

Here the context was the destruction of the Thopa and Yuan families which had been the ruling houses of the Northern, Eastern and Western Wei dynasties. In this, the last, year of the tyrant's reign, there had been a massacre of no less than 721 surviving members of these families, and Yuan Huang-Thou was himself a prince of the Wei.[a] For us of course the technical aspects are the main interest, and it is quite remarkable that in these experiments kites were used. Since the imperial road called the Purple Way was 5 *li* (about 2½ km.) north-west of the city, the prince succeeded in 'riding on the hard wind' for a considerable distance. Moreover, the circumstances show that what was going on was not quite simply a cruel emperor's sport with prisoners, for the cables of the kites must have required man-handling on the ground with considerable skill, and with the intention of keeping the kites flying as long and as far as possible.

Thus we have one circumstantial account of man-lifting kites within a couple of centuries of Ko Hung's time, and others are probably still buried in the texts. By the time when Marco Polo was in China (*c.* +1285) man-lifting kites were in common use, according to his description,[b] as a means of divination whereby sea-captains might know whether their intended voyages would be prosperous or not.

And so we will tell you [he says] how when any ship must go on a voyage, they prove whether her business will go well or ill. The men of the ship will have a hurdle, that is a grating, of withies, and at each corner and side of this framework will be tied a cord, so that there be eight cords, and they will all be tied at the other end to a long rope. Next they will find some fool or drunkard and they will bind him on the hurdle, since no one in his right mind or with his wits about him would expose himself to that peril. And this is done when a strong wind prevails. Then the framework being set up opposite the wind, the wind lifts it and carries it up into the sky, while the men hold on by the long rope. And if while it is in the air the hurdle leans towards the way of the wind, they pull the rope to them a little so that it is set again upright, after which they let out some more rope and it rises higher. And if again it tips, once more they pull in the rope until the frame is upright and climbing, and then they yield rope again, so that in this manner it would rise so high that it could not be seen, if only the rope were long enough. The augury they interpret thus; if the hurdle going straight up makes for the sky, they say that the ship for which the test has been made will have a quick and prosperous voyage, whereupon all the merchants run together for the sake of sailing and going with her. But if the hurdle has not been able to go up, no merchant will be willing to enter the ship for which the test has been made, because they say that she could not finish her voyage and would be oppressed by many ills. And so that ship stays in port that year.

[a] Cf. *TH*, p. 1235.

[b] Moule & Pelliot ed., vol. 1, pp. 356 ff. The account survived only in the 'Z' MS. version, one of the most complete but also one of the rarest. Delighted at discovering this passage in the winter of 1960, I did not know that Lynn White (6) had also come upon it, but it was good to have his confirmation of my interpretation. It was probably, as he says, because this text had so little circulation that the idea of the man-carrying kite was not followed up in Europe for nearly three centuries. Only from +1589 onwards did della Porta mention it in his section on the 'flying Sayle' in *Magia Naturalis*.

[1] 畢義雲

Surely this was one of the strangest sights to be seen in the fabled + 13th-century ports of Zayton and Khanfu.

The wonders of modern aviation have thrown kites so much into the background that it is generally quite forgotten that they could ever supply sufficient lift to carry human beings into the air.[a] Yet this development played its part in the history of aviation. A number of tentative trials of this kind took place from the time of Pocock about 1825 onwards (Simmonds; Biot; Cordner; Wise, etc.),[b] but full success was not attained until the work of B. F. S. Baden-Powell in 1894.[c] Here a turning-point was the invention of the Australian Hargrave in the nineties, who (as we have seen, p. 582) devised the box-kite of rectangular cells, and thereby produced one of the precursors of the biplane.[d] By 1906 it was possible for a man to remain for an hour at a height of 2600 ft. suspended by a train of kites.[e] The significance of this was great. Only a few years before, Alexander Graham Bell had written: 'a properly constructed flying-machine should be capable of being flown as a kite, and conversely, a properly constructed kite should be capable of use as a flying-machine when driven by its own propellers'.

Lastly, what is to be said of Ko Hung's 'hard wind'? From the examples he gives of the gliding and soaring of birds, it is obviously nothing else than the property of 'air-lift', the bearing or rising of the inclined aerofoil subjected to the forces of an airstream, whether natural or artificial. It will not be forgotten that we have met with the 'hard wind' of the Taoists before (Vol. 3, pp. 222 ff. above), in the astronomical Section, where its role as a natural cause of planetary or stellar motion came in for remark. It was there suggested that someone had observed the high resistance of a strong current of air from the orifice of a metallurgical tuyère. But Ko Hung applies the concept very clearly to gliding flight, as indeed had Chuang Chou before him, when he wrote about the wings of the giant *phêng*[1] bird being airborne upon the density (*chi hou*[2]) of the wind beneath.[f] Ko Hung ends by attributing to him the idea that flying things rise up 'using the clouds as steps', which may be more than a poetic metaphor, hinting as it does at the existence of those ascending air-currents which modern glider pilots have learnt so well to utilise. Something of these could probably have been observed in the behaviour of smoke, and particularly of the mists and clouds

---

[a] Nothing need be said here of that other great service which the Chinese kite performed for modern science in the hands of Benjamin Franklin when in + 1752 he identified the electricity of the lightning flash with that of the Leiden jar. Three years earlier Alexander Wilson had used a battery of kites to carry thermometers to a height of 3000 ft. to determine the temperature of the clouds (Pledge (1), p. 317).

[b] Descriptions in Hodgson (2); Vivian & Marsh (1), p. 56; Duhem (1), p. 201; Gibbs-Smith (1), pp. 12, 16, 46, 162.

[c] Gibbs-Smith (1), pp. 34, 162. Great improvements were made by S. F. Cody from 1901 onwards (cf. p. 568).

[d] Cf. Gibbs-Smith (1), pp. 337 ff.

[e] Vivian & Marsh (1), p. 189. Figure 708 shows a kite-train bearing aloft a military observer at the Rheims meeting of 1909. On this cf. Gibbs-Smith (1), p. 247; Broke-Smith (1).

[f] See Vol. 2, p. 81 above. After all, the augurs had been watching bird flight for a very long time (Vol. 2, p. 56 above). But they had been doing so in the West too.

[1] 鵬   [2] 積厚

PLATE CCLXXIX

Fig. 708. A train of kites bearing aloft a military observer (Lt. Bassel) at the Rheims
aeronautical meeting of 1909 (Vivian & Marsh, 1).

on the lofty mountain heights which the Taoists delighted to frequent.[a] And so we end our study of one of the most remarkable (indeed prophetic) ancient texts on the prehistory of aviation which any literature can show.

(5) THE BIRTH OF AERODYNAMICS

Let us now attempt to place all this in correct perspective with regard to the growth of aeronautical science and practice. Leaving aside for the moment the development of aerostatic machines (the balloon was really a product of 18th-century pneumatic chemistry), and of jet-propulsion (for we must treat of the Chinese invention of rockets elsewhere, in Section 30), we may concentrate attention upon wings or aerofoils, and air-screws. Man's attention (at least in the West) was attracted first by the beating of the bird wing, its gliding properties being neglected; hence Leonardo's main interest was, as we have seen, in flying-machines on the flapping or ornithopter principle.[b] The decisive contribution of Alfonso Borelli, in his *De Motu Animalium* of +1681, was to show that human muscles were anatomically and physiologically incapable of providing motive power for unaided winged flight with the materials then available, and there the matter rested. But the idea of beating wings was very tenacious. The first conception of powered flight still envisaged them,[c] and so did the first dissociation of the bearing function from the propulsive function.[d] It was George Cayley at the beginning of the 19th century who broke completely with the old obsession, and became the precursor of Lilienthal and the Wrights rather than the successor of Aethelmaer and Leonardo.[e] As we know from the Cayley papers,[f] he was the first to analyse the aerodynamic properties of the atmosphere by physical means, the first to lay down the scientific principles of heavier-than-air flight, the first to experiment with a captive plane at various angles of incidence,[g] the first to make model and full-size glider aeroplanes with rudders and elevators and to test them in free flight, the first to discuss streamlining and the 'centre of pressure' of a surface in an air-stream, the first to realise that curved wings give a better lift than flat ones, and to recognise the existence of a low-pressure region above them, the first to suggest multiple superimposed wings, and the first to state that the lift of a plane varies as the square of the relative air-speed multiplied by the density. All this was done in a single decade, between 1799 and 1810. How great a pioneer Cayley was may be appreciated

[a] Cf. the observation of Cotte in +1785 that certain cloud layers might move in directions quite different from that shown by the weathercock at ground level (Duhem (1), p. 188). See also Scorer (1) on lee waves in the atmosphere.

[b] Cf. pp. 575, 583 and Hart (1, 2, 4).

[c] John Wilkins, in his *Mathematical Magick* of +1648, suggested the use of a steam-engine, Bk. 2, chs. 6–8.

[d] Tito Livio Burattini, a Venetian engineer in the service of the King of Poland, built in +1647 at Warsaw a model aircraft which had four fixed glider wings as well as others which were made to beat; cf. Duhem (1), pp. 161 ff.

[e] We speak here of Leonardo da Vinci theoretically, for it is certain that his projects could have exercised no influence on the development of aviation before the latter part of the 19th century; see Gibbs-Smith (1), pp. 187 ff.

[f] Edited by Hodgson (3). Cf. Gibbs-Smith's summary of his achievements, (1), pp. 10, 189, (8).

[g] The 'whirling-arm' he used for this was an 18th-century device with which Robins had studied ballistics (+1746) and Smeaton windmill sails (+1759).

even further when we remember his studies on the air-screw, arising from the Chinese top, already mentioned; his anticipation of the internal combustion engine; and his practical and rational proposals for applying power to balloons. Thus the dirigible, adumbrated by Meusnier and others in +1784, became a reality with Giffard in 1852, and in due course handed over its propellers to their more onerous duties in heavier-than-air flying-machines.

Cayley was not indeed the first to see the importance of the Chinese top, for Launoy and Bienvenu in +1784 had already successfully experimented with it, and Paucton even earlier, in +1768, had proposed both the helicopter rotor and the vertical air-screw propeller. Although he himself worked from first principles, his position in history is a singularly focal one, the junction-point of Western vertical and Chinese horizontal mountings, as they contributed to modern aeronautical engineering. The first model aircraft (in the modern sense) to fly on the helicopter principle, however, was that of W. H. Phillips in 1842; the same year in which Henson began to patent a design basically similar (as we have seen, p. 585) to that of a modern twin-engined monoplane.[a] The first 'modern' powered model aeroplane was built on this pattern in the same decade by Henson & Stringfellow, but it could hardly sustain itself and made only slow descending power-glides. Thus it was 1857 before the first model was made which would take off under its own power, fly freely for some distance and land safely. This was the achievement of Félix du Temple de la Croix, employing first clockwork and later steam.[b] After his work many experiments with models were made, and with them it was possible to study such important phenomena as stalling.[c] Aerofoil design, with the realisation that the upper surface of the wing must be convex, was advanced in the work of Wenham (1866), Pénaud (1876) and others, while the last decades of the century saw widespread empirical work with full-scale gliders.[d] Finally in 1903 came the first successful full-scale flying by the Wright brothers, using the internal combustion engine and an aeroplane in all fundamental respects identical

[a] Gibbs-Smith (1), p. 13; the model of W. H. Phillips was steam-driven and the blades were rotated by jets from their tips—a remarkable anticipation of current practice.

[b] Gibbs-Smith (1), pp. 15, 314 ff.; in 1874 du Temple's full-scale steam-powered aeroplane just succeeded in being airborne for a short distance after taking off down an inclined ramp.

[c] As the angle of attack of a wing is increased, the lift coefficient rises from zero at the 'no-lift angle' to a maximum (about 16°) at which 'stalling' occurs and the aircraft ceases to fly, rolls violently, and enters a falling spin. The cause of this is the separation of the airstream from the upper surface of the wing, causing turbulence near the trailing edge, consequent drag, sudden reduction of lift, and backward motion of its line of action. The result is a tendency to pitch forwards and dive uncontrollably, and rolling occurs because the stall usually occurs on one wing first. Movable flaps which can be lowered at the trailing edges of the wings increase the lift at the slow speeds necessary for landing. By inserting slots in the wing (now usually a single slot along the wing's leading edge, with a slat of roughly aerofoil cross-section forward of it), Handley-Page found that the stalling angle could be raised to 26° or more. The reason is that the smooth stream of air leaving the trailing edge of the slat prevents the separation of the flow from the upper surface of the wing and the consequent reduction of lift. The great practical value of this device is that it allows of a greater lift coefficient and therefore much slower landing speeds. Although the turbulence which occurs in stalling is incidental, there is reason to believe that the remarkable invention of the slotted wing may have been suggested by the fenestrated rudders of Chinese ships, which will be described below in Sect. 29h (private communication from Prof. E. V. Telfer). Sir Frederick Handley-Page tells me, however, that he cannot now remember such a stimulus.

[d] Otto Lilenthal, P. S. Pilcher, O. Chanute and others, whose deeds are recounted in Gibbs-Smith (1), pp. 28 ff.

PLATE CCLXXX

Fig. 709. Pictures from the *Ching Hua Yuan*, a novel by Li Ju-Chen written about 1815, adumbrating practical aircraft. The illustrations were provided for the edition of 1832 by Hsieh Yeh-Mei. A scene from ch. 66 shows an aerial car with four road-wheels and two screw-bladed rotors set between them on each side. The inscription says: 'The Emperor takes a flying car to consult with the Crown Prince on the publication of the Yellow Rescript (of successful candidates); while the Empress conducts the examination of the talented beauties.'

PLATE CCLXXXI

Fig. 710. Another scene from the *Ching Hua Yuan*, ch. 94, shows three flying cars each with four screw-bladed rotors taking the place of the road-wheels, and engaging with gear-wheels which seem to be connected with some hidden power mechanism. Drawing by Hsieh Yeh-Mei, 1832. The inscription says: 'The Prince of Wên-Yen, obeying orders, returns to his own country (by air); and the Erudite Girl, thinking of her parents, departs (by air) for the Mountain of the Immortals.' A veritable airport.

with those of today.[a] This incorporated both the devices, the ad-aerial screw rotor and the kite wing, which Ko Hung had spoken of sixteen centuries earlier. The kite was married to the windmill.[b]

Such was the key combination. Though the idea of it was implicit in Cayley's work, it remained (to use a singularly appropriate expression) 'in the air' during the first decades of the 19th century, crystallising only in Henson's famous design of 1842–3, and in the models of Henson & Stringfellow (1847) and Félix du Temple de la Croix (1857). It would be interesting indeed if Chinese imagination participated in this crucial period, and the facts are well worth examination. Duhem[c] & Huard[d] have both pointed out that a Chinese novel which appeared about this time was illustrated by pictures of imaginary flying-machines which combined propeller blades with kite surfaces resembling those of a biplane. This novel was difficult to identify from their descriptions,[e] but it is in fact the *Ching Hua Yuan*[1] (Flowers in a Mirror), written by Li Ju-Chen[2] (+1763 to 1830) between 1810 and 1820.[f] Published eight years later, it was reprinted in 1832, when Hsieh Yeh-Mei[3] added 108 pictures, and it is in two of these that the aircraft are shown. The first (for ch. 66, see Fig. 709) shows a car amidst clouds, open save for an awning; it has four wheels as if for land travel, but between them on each side there is a screw-bladed rotor in a position analogous to that of the wheel on a paddle-wheel boat. The second (for ch. 94, see Fig. 710) is more interesting, for it shows three flying cars, each with four screw-bladed rotors taking the place of the ordinary land wheels, and, most curiously, between each of these propellers a large gear-wheel which seems to connect them with a power-source. This clearly shows that the artist had in mind mechanically driven ad-aerial (not ex-aerial) wheels. Duhem is probably claiming too much when he says[g] that the rectangular bodies of the machines and the parallel awnings above them are like the lifting surfaces of a biplane. After all, even the box kite was not known in China or anywhere else at that time. But there may have been significance in something which probably was familiarly known there, namely, those children's kites fitted with toy wind-wheels which rotated as the kite flew;[h] and whether or not his ideas derived from them, it seems that Hsieh Yeh-Mei did really conceive of wheels acting in some way on the air and nor merely rotated by it. So ended the classical contributions of Chinese culture to aeronautics.

[a] See Gibbs-Smith (1), pp. 35 ff., 224 ff.; Brooks (2).
[b] So Gibbs-Smith (2).
[c] (1), pp. 12, 268, (2), pp. 200–4.   [d] (2), p. 29.
[e] They gave it the title of 'Une Fleur dans la Neige', and one of them wrote the author's name 'Jin Ho Yuen'. I am indebted to Dr Huard, Mr P. van der Loon, and Mr Cyril Burch, for helping me to clear up the mystery. The true title refers to evanescent beauty, and the novel was of permanent social significance since it dealt with questions such as the emancipation of women, though superficially an account of fantastic adventures and strange countries; see Hummel (2), p. 473; Lu Hsün (1), pp. 329 ff; Adkins (1). The aircraft are mentioned in chs. 66, 70, 82, 85, 86, 87, 91, 94 and 95.
[f] No. 4228/4231 in the catalogue of Courant (3); for translations cf. Davidson (1), p. 3.
[g] (2), p. 200.   [h] Cf. p. 586 above.

[1] 鏡花緣   [2] 李汝珍   [3] 謝葉梅

(6) THE PARACHUTE IN EAST AND WEST

What remains in this Section is of comparatively minor importance, yet not without interest. On account of its great simplicity one would suppose that the parachute idea, analogous to that of the sea-anchor, would be quite old in many civilisations. Feldhaus, however, finds no instance in Europe earlier than the description by Leonardo about +1500 in the Codex Atlanticus,[a] which was followed before long (perhaps quite independently) by the 'Homo volans' of Faustus Verantius (c. +1595, first published +1615, pl. 38). Historians doubt whether it was ever tried in practice before Blanchard and perhaps Montgolfier used it for animals about +1778, and the personal descents of Lenormand and Garnerin made some years later.[b]

In China, however, there are much older references. In the *Shih Chi*, completed by −90, Ssuma Chhien related a story[c] about the legendary emperor Shun.[1] His father Ku Sou[2] wanted to kill him, and finding him at the top of a granary tower, set fire to it, but Shun escaped safely by attaching a number of large conical straw hats together and jumping down. The +8th-century commentator Ssuma Chên[3] understood this clearly in the sense of the parachute principle, saying that the hats acted like the great wings of a bird to make his body light and bring him safely to the ground.[d] A much later, but much more circumstantial, reference occurs in the *Thing Shih*[4] (Lacquer Table History),[e] written by Yo Kho[5] in +1214. The grandson of the great general Yo Fei is describing what he saw in Canton as a young man when his father was governor in +1192. After an interesting description of the manners and customs of the foreign community of Arab merchants established there, he speaks of their mosques and of a 'grey cloud-piercing minaret like a pointed silver pen'. Inside this there was a winding spiral staircase for the muezzin, with round look-out openings at every several tens of steps, from which the Arabs watched and prayed for their ships arriving in the spring. Yo Kho goes on:[f]

On the very top there is a huge golden cock instead of the usual (Buddhist) wheels (on pagodas), but it is now short of one leg. The Cantonese people used to say that this defect dated from the time of the former governor Lei Tshung[6] (c. +1180), when some robber came and stole it away, leaving no trace behind him. They said that one day in the market there was a poor man selling something made of pure gold. When someone picked it up and asked him how he had got it, he said 'The foreigners used to be so strict that no one could enter their establishments, but I hid above a beam for three nights and got inside the minaret with some dried food to sustain me during the daytime. Then at night I used a steel (saw) to cut it off, and hid it within my clothes, but I could not get more than one leg.' Again they

---

[a] (1), col. 279; (18), p. 141. Cf. Ucelli di Nemi (3), no. 84.

[b] Hodgson (2); Gibbs-Smith (1), pp. 165 ff.   [c] Ch. 1, p. 22b, tr. Chavannes (1), vol. 1, p. 74.

[d] Indian contemporaries of Ssuma Chên also describe a descent from a tower by means of 'umbrellas'; *Prabhavakacarita*, IX, 87–9, cf. Raghavan (1). The story concerns the two nephews of the Buddhist commentator Haribhadra, whom we have encountered already in connection with water-raising machinery (p. 361 above).

[e] The passage was noted by Kuwabara (1), pt. 2, p. 30.

[f] Ch. 11, p. 6b, tr. auct. with Lu Gwei-Djen.

¹ 舜   ² 瞽叟   ³ 司馬貞   ⁴ 桯史   ⁵ 岳珂   ⁶ 雷澡

asked him how he got away, to which he replied 'I descended by holding on to two umbrellas (*yü kai*[1]) without handles. After I jumped into the air the high wind kept them fully open, making them like wings for me, and so I reached the ground without any injury.' Although the robber only stole one leg, down to the present time they have never been able to repair it.

Perhaps the thief had heard some story-teller relating the tale of Shun and his father; in any case it is remarkable that his own words should have been preserved.[a] All these indications must mean that the idea was current in China, but the observation of air-resistance to an outstretched fabric is so simple that it may well have originated in many places independently. Indeed it would follow merely from the use of ship sails. If the parachute principle was not developed in China as it was in later Europe, this was because it was naturally ancillary to aviation itself, a typical piece of post-Renaissance technology.

Surprisingly, however, we have unusually concrete evidence that the invention was in fact at least once a transmission to the west. The Ambassador of Louis XIV in Siam, Simon de la Loubère, who was there in +1687 and +1688, described in his *Historical Relation* the exploits of Chinese and Siamese acrobats, saying:[b]

> There dyed one, some Years since, who leap'd from the Hoop, supporting himself only by two Umbrella's, the hands of which were firmly fix'd to his Girdle; the Wind carry'd him accidentally sometimes to the Ground, sometimes on Trees or Houses, and sometimes into the River. He so exceedingly diverted the King of Siam, that this Prince had made him a great Lord; he had lodged him in the Palace, and had given him a great Title; or, as they say, a great Name.

Now the researches of Duhem[c] have revealed that this passage was read in the following century by L. S. Lenormand, who was stimulated by it to make practical trials in +1783, from the tops of trees and buildings, which were quite successful.[d] It was Lenormand who gave the parachute its present name and recommended it to Mont-golfier, who fully appreciated its importance. This led to the descent of Garnerin from a balloon in +1797. There are not many cases in which so clear a line of trans-mission is detectable.

(7) THE BALLOON IN EAST AND WEST

The balloon may be said to be related topologically to the parachute, for a sufficient constriction of the latter's orifice turns it into the former. But physically they are quite different, for in one case the descent of a curved fabric surface is delayed by the drag of the aerial medium, while in the other, its ascent is facilitated by the presence of a medium lighter than air confined beneath it. As we have already indicated, the

[a] If any reader should doubt the possibility of his feat, similar accounts collected by Kerlus (1) may prove convincing. The facts could easily be tested.

[b] (1), p. 47.

[c] (1), pp. 237, 248, 263, (2), pp. 232 ff.; cf. Huard (2), p. 29.

[d] It is certain that Lenormand did not know of the suggestions of Leonardo, and that he was stimu-lated by the Siamese relation; it is almost certain that he was ignorant of Veranzio's proposal.

[1] 雨蓋

balloon or aerostat, the 'cloud captured in a bag', was a product of the pneumatic chemistry of the European 18th century.[a] Remarkably enough, the first two aerial voyages ever made by man were accomplished in a single year, +1783, by Pilâtre de Rozier and the Marquis d'Arlandes in a Montgolfier hot-air balloon, and then in the following month by J. A. C. Charles and his mechanic Robert in a hydrogen balloon. The simpler of these forms, using nothing but hot air, could have originated very much earlier than this, and in fact in model form it did.

Easter merry-makers in +17th-century Europe had an entertaining trick of making empty eggshells rise in the air literally 'under their own steam'. This is reported in many books, for example Jacques de Fonteny's poem *L'Oeuf de Pasques* of +1616, which describes it as a traditional custom.[b] The procedure was simple enough, requiring only a little deftness; the contents of an egg being emptied through a small hole and the shell very carefully dried, the right amount of dew (pure water) was introduced and the hole closed with wax. Then in the hot sun the egg would move uneasily, grow light, and rise up into the air, floating a moment before falling.[c] How ancient this trick was in Europe we do not know, but we were quite astonished to read of it in the book of Duhem because we had already come across a similar model of the lighter-than-air flying-machine in a Chinese text, not of the +17th but of the −2nd. This is the *Huai-Nan Wan Pi Shu*[1] (The Ten Thousand Infallible Arts of the Prince of Huai-Nan), that compendium of ancient Taoist techniques which we have occasion to refer to so often in this work.[d] Liu An's book of secrets, if not exactly now as he himself knew it, must certainly be a Han compilation. The text says, in its usual concise way: 'Eggs can be made to fly in the air by the aid of burning tinder.' And an ancient commentary incorporated in the text explains: 'Take an egg and remove the contents from the shell, then ignite a little mugwort tinder (inside the hole)[e] so as to cause a strong air-current. The egg will of itself rise in the air and fly away.'[f] Thus the method of Liu An was more akin to that of the Montgolfier brothers than to that of the eggs raised by steam, since nothing but hot air was employed. The discovery of this text puts a rather different complexion on the relations of China and Europe in the prehistory of aerostatic flight.

When the present Section was first drafted we doubted whether China had had any part in this, but we are now inclined to think that the Han tradition was never lost. China was likely to be the home of hot-air balloons for several different reasons. Paper was available, as nowhere else in the world, from the Han period onwards, and

---

[a] Cf. Duhem (1), pp. 332, 369, 370 ff., 418 ff., 437, 442 ff.; Gibbs-Smith (4), pp. 53 ff., 64 ff., 74 ff.

[b] The literary references are given in Duhem (1), p. 401. Illustration in Fludd (2), p. 186.

[c] The water-vapour formed by the evaporation of the dew expels the air through the pores in the shell, and ultimately when the wax melts, through the hole. The hot steam inside the egg is just sufficiently buoyant to raise the shell for a short time into the air before being dissipated itself by the same channels.

[d] Cf. Sections 26 (Vol. 4, pt. 1, pp. 69, 91, 279, 316 above), 30, 33, 34, 40 and 44. On the book's bibliography, see Kaltenmark (2), p. 32, and on its comparative position in the literature of China and Europe, Ho Ping-Yü & Needham (2).

[e] Normally used for making incense sticks and moxibustion cones (cf. Vol. 1, Fig. 29).

[f] *TPYL*, ch. 736, p. 8b and ch. 928, p. 6b, tr. auct.

[1] 淮南萬畢術

the development of the classical globular paper lanterns would have encouraged experimentation. When their upper openings were too small and the source of light and heat unusually strong, they must sometimes have shown a tendency to rise and float free of support. And indeed it is not difficult to find instances of the popular survival of hot-air balloons as an ancient sport in the Chinese culture-area. Goullart, for instance, gives a graphic description of seasonal customs involving this in the Lichiang region of Yunnan province.[a] He tells us that in July, the critical month before the rainy season, the rice was already planted and the people did not have much to do, so in the evenings, besides dancing, the young men and the Nakhi girls flew hot-air balloons made of rough oiled paper pasted over a bamboo framework. With bunches of burning pine splinters underneath, these would sail up into the night air, some floating in the distance like red stars for several minutes before bursting into flame and falling far away. Duhem, again, reports similar pastimes in Cambodia.[b] Medieval descriptions are still needed to fill the gap in continuity, but it is probable that further search will reveal them. The Han evidence and the ethnological evidence together make a *prima facie* case for a perennial Chinese tradition, and indeed it is very unlikely that the tribal people and peasants of north-western Yunnan derive their proceedings from the France of Montgolfier.[c]

It might even be urged that a practice originally Chinese was brought to the knowledge of Europeans at the time of the Mongol invasions. Much evidence has been collected from eastern European chronicles[d] that hot-air balloons shaped like dragons were used for signalling or as standards by the Mongol army at the Battle of Liegnitz in +1241, and this is accepted as assured by many writers.[e] Certainly many of the early +15th-century German works on military technology, such as the MSS. of Konrad Kyeser's *Bellifortis*, show drawings of horsemen holding what appear to be flying dragons in the air on the end of cords.[f] He states that they contained oil-lamps as well as combustibles to give an effect of vomiting forth fire. It is rather difficult to evaluate these descriptions and pictures, which in some respect recall kites rather than hot-air balloons, and the subject needs further study, but we are inclined to believe that a considerable aerostatic element was involved.[g] Whatever the arrangements actually were, they seem to have continued into the +16th century, for an account of the entry of Charles V into München in +1530, with an accompanying contemporary woodcut, attests the appearance of a similar flying or floating and fire-breathing dragon.[h]

---

[a] (1), p. 178. Mr Wei Tê-Hsin remembers like ploys in Fukien.   [b] (1), p. 409.

[c] This presents a case somewhat parallel to that of the Cardan suspension seen above, p. 233. Gimbal lamps in their nests of pivoted rings made during the last two centuries on the frontiers of Tibet derive much more probably from the Chinese tradition beginning with Ting Huan in the +2nd century than from the Italy of Jerome Cardan.

[d] See Hennig (1); Feldhaus (8, 15, 20).

[e] See Feldhaus (1), cols. 653 ff.; Forbes (4a); Plischke (2). Large paper lanterns for signalling are often referred to in Chinese accounts of military operations, as at the siege of Khaifêng by the Mongols in +1232 (*TCKM*, pt. 3, ch. 19, p. 50a; cf. p. 577 above), but we have not met with any statement that they floated in the air.

[f] See Berthelot (5); Feldhaus (1), *loc. cit.*, and others.

[g] Duhem (1), for example, pp. 404, 411, 412, is sceptical, but reproduces the illustrations, (2), pp. 36 ff., figs. 17, 18. Cf. the dragon standards of Romans and Parthians (Feldhaus (1), col. 198).

[h] Bassermann-Jordan (1), pp. 64 ff.

If some of these creatures had an orifice behind as well as at the front, they were perhaps precursors of the wind-sock. And here again there is an East Asian background, for Tissandier (6), describing some of the kites of Japan, mentions and depicts a huge hollow paper fish with a large open mouth and a smaller opening at the rear, which was used for decoration and as a sensitive wind-vane or weathercock. As he was writing in 1886, no reverse influence from occidental aviation technology could have been in play, and if the wind-sock was found in Japan it had almost certainly been in China earlier.

Perhaps these phenomena may have some relation to the ideas of Albert of Saxony (+1316 to +1390),[a] who imagined that things might float at the surface between the sphere of air and that of fire, just as they floated at the surface between water and air. Could he have known of the Mongol hot-air dragon-standards? In any event, his suggestion stimulated the Jesuit Caspar Schott in +1658, who was the first to speak of the possibility of aerostatic flight.[b] Schott also discussed the flying eggs of de Fonteny and others. Then in +1709 came the activities of the Brazilian priest Fr Gusmão, who succeeded in burning the curtains of the King of Portugal's audience hall with a model Montgolfière.[c] Barely eight decades had then to pass before men were truly in the air. Thus by tracing a tenuous thread, illuminated only very fitfully, we come back to a view often formerly held on less secure evidence, namely that China did play a considerable part in the prehistory of the balloon and the airship.

So far as we know, traditional China carried out no aerostatic experiments of Montgolfian scale and daring.[d] But the great interest taken by Chinese scholars, however, in the news which came from France in the penultimate decade of the +18th century, might bear witness to the existence of certain traditions. In one of his most curious letters, written from Peking on 15 November +1784 (just a year after the first Paris ascents), the Jesuit J. J. Amiot (4) described the interest which the literati were showing, and said that they were disposed to reconsider in a Montgolfian light ancient stories long dismissed as fables. Amiot himself, no disparager of ancient China's sages, wondered whether perhaps Huang Ti or Shen Nung might not have known of some fluid lighter than air long since forgotten. 'This suggestion', he added, 'I send for what it may be worth.' But there was excitement not only in the capital. The southerner Wang Ta-Hai travelled abroad in the East Indies between +1783 and +1790, making notes which he collected in his *Hai Tao I Chih Chai Lüeh* prefaced in

[a] Cf. Sarton (1), vol. 3, pp. 1428 ff.

[b] Duhem (1), pp. 334 ff., 339.

[c] The epic, or tragi-comedy, of Fr Gusmão is a very involved story, and may be read in Duhem (1), pp. 417 ff. or Gibbs-Smith (4), pp. 53 ff., (5).

[d] There is a peculiar story, generally dismissed out of hand, which may yet prove to have some basis at present unknown. Giles (9) tells us that a certain Fr Besson is said to have written in +1694 that a balloon ascended from Peking at the accession of the emperor 'Fo Kien' in +1306. But no new reign began in that year, and no emperor bore that name. Pfister (1), in his encyclopaedia of the Jesuit mission in China, knows no such Father as Besson, but there was a Joseph Besson (+1607 to +1691) who was a missionary in Syria, so the rumour may have been started by him. Curious 18th-century prints in Chinoiserie style of this aerostat, pictured as an elongated dirigible with nine gondolas, continue to appear in popular articles. Cf. Duhem (1), pp. 85, 376, 403, 409, but even he brings no solution for this puzzle; nor Feldhaus (2), p. 54.

the following year. In this he gave an account, necessarily at second hand, of the new balloon or 'sky-ship' (*thien chhuan*[1]):[a]

This boat is short and small, resembling a dome-shaped pavilion and capable of containing ten men. Attached to it there is a pair of bellows, or air-pump, of exquisite workmanship, in shape like a globe; several people work this with all their might, whereupon the ship flies up high into the heavens. There it is borne about by the winds, but if they wish to navigate it they spread sails and make use of quadrants to measure distances. When they reach their destination they take in their sails and let the ship descend. It has been reported that these ships have been burnt and injured by the sun's rays, while persons venturing in them have been scorched to death, so people hardly dare to go on using them.

This sounds like an echo of a hydrogen balloon, with a second conversation about Daedalus and Icarus occupying the same line. But it testifies to the living interest shown by Chinese scholars and travellers, conscious of their own past, in the opening phase of man's conquest of air and space.

(n) CONCLUSION

About the year 1911 an old gentleman taking a stroll in Peking had his attention drawn to an aeroplane flying overhead, but with perfect sang-froid remarked 'Ah, a man in a kite!'[b] Chinese reactions to modern technology did not stop there, however, and quite a number of authors, though lacking that balanced judgment which an exhaustive acquaintance with sources both eastern and western alone could give, did not fail to maintain the emergence of occidental technology from oriental origins. Wang Chih-Chhun,[2] for instance, wrote:[c]

The useful arts and techniques originated from the earliest generations; thus geometry was invented by Jan Tzu (Jan Chhiu,[3] one of the disciples of Confucius), but later on the Chinese lost his books and Western people studied them, so that they became skilled in mathematics. So also the automatically striking clock was invented by a (Chinese) monk, but the method was lost in China. Western people studied it and developed refined (time-keeping) machines. As for the steam-engine, it really originates from (the monk) I-Hsing of the Thang, who had a way of making bronze wheels turn automatically by the aid of rushing water—all that was added was the use of steam and the change of name. As for fire-arms, they originated in the fighting at Tshai-shih[4] in the time of Yü Yün-Wên[5] (of the

---

[a] Ch. 5, tr. Anon. (37), p. 55. We say necessarily at second hand, for it is quite unlikely that any Montgolfian balloons had been seen in Asia before 1800. Possibly the first were those used by the French in their invasion of Viet-Nam (Tonkin) in 1884 (cf. Lê Thánh-Khôi (1), p. 378). Several interesting contemporary Chinese prints showing these observation-balloons have been published by Tissandier (1). The French balloon corps had been started in 1877, and the same writer gives details of the corresponding Chinese unit trained by French instructors at the request of Li Hung-Chang in 1888. This was the embryo of the Chinese Air Force of today. French balloon observers were used again during the troubles of 1900, and an album of their photographs of Peking exists (Anon. 43).

[b] For this story I am indebted to Mrs Ingle, formerly of Chhilu University.

[c] In his *Kuo Chhao Jou Yuan Chi*[6] (Record of the Pacification of a Far Country); an account of his ambassadorship to Russia, ch. 19; quoted *KCKW*, ch. 5, p. 28*a*, tr. auct. Cf. p. 525.

[1] 天船　　[2] 王之春　　[3] 冉求　　[4] 采石　　[5] 虞允文
[6] 國朝柔遠記

Sung); when he defeated the enemy by the aid of certain firearms called *phi-li(-phao)*.[1,a] Thus the wonderful techniques of Western people are all based on the remains of ancient (Chinese) inventions. How could they be cleverer than Chinese people?

Wang Chih-Chhun's reaction (about +1885) was no doubt considered chauvinistic, and when such statements reached the ears of the 'gentlemen of the world' at that time, they were laughed out of court. But the progress of sober history has swung the pendulum the other way, and there now appears more solid basis for such protests, exaggerated though they may have been, than seemed at the time. Whoever has read through the foregoing pages of this Section will allow that the balance shows a clear technological superiority on the Chinese side down to about +1400.

Historians are coming to recognise this. When des Noëttes (3) drew up his table of medieval acquisitions, he stated clearly that many of them had come from the East of Asia. We can no longer accept, wrote Carl Stephenson, the easy formula of an Italian Renaissance in which all the great scientific and technological advances arose from the rediscovery of the Greek and Latin classics. Rather they depended upon the expertise of the medieval artisans and the stimuli which these men had received from the peoples of Asia. Medieval technology, wrote Lynn White, consisted not simply of the equipment inherited from the Roman–Hellenistic world modified by the ingenuity of Western Europeans; it embodied also vitally important elements derived from the northern barbarians, the Byzantine and Near East, and from the Far East. Haudricourt (1) is more precise—'The history of technology', he says, 'is still in its childhood, but nevertheless one can affirm that the astonishing industrial development of Europe was conditioned by a previously continuous inflow of Asian novelties, Indo-Iranian in antiquity, Chinese in the middle ages.' And Febvre (2) adds: 'Europe was an Asian cross-roads'.[b]

The lateness of the date at which the Europeans still thought they had a good deal to learn from the Chinese may come as something of a surprise. Greenberg (1), approaching the matter from the point of view of economic history, has pointed out that 'until the epoch of machine production, when technical supremacy enabled the West to fashion the whole world into a single economy, it was the East which was the more advanced in most of the industrial arts'. He then throws light on the genesis of the Opium Wars by showing, from many hitherto unpublished documents, that while the West was eager to get tea, silk, textiles, porcelain, lacquer, and the like, there was nothing from Europe which the Chinese at that time (late +18th and early +19th centuries) wanted in exchange.[c] The East India Company resorted to the drug

a See above, p. 421, and below, Vol. 5, pt. 1 (Sect. 30). The battle occurred in +1161.

b Cf. also Lopez (1); Chatley (23); Middleton Smith (1); Bodde (13); Huard (2). Sometimes recognition is given to medieval technological innovations with a rather misleading emphasis. Thus Crombie (1), e.g. pp. 16 ff., is so concerned to praise the scientific achievements of the European Middle Ages that in enumerating inventions he fails to add that the majority of them were not made in Europe at all. It is true that he partly repairs this elsewhere (2).

c This fact found classical expression in the famous edict of the Chhien-Lung emperor on the Macartney mission in +1793, translated most recently by Cranmer-Byng (1), p. 134: 'We have never

1 霹靂炮

traffic in order to avoid the drain of bullion from the West. So high was the prestige of Chinese 'know-how' in those days that in 1847 St Julien (4) wrote:

Il est permis de penser que pour satisfaire aux besoins des arts et servir les progrès de la civilisation, la génie des Européens trouvera par lui-même, pendant bien des siécles encore, après des essais et des efforts longtemps continués, une multitude d'inventions utiles ou bienfaisantes, que les Chinois avaient trouvées avant eux, mais qui gisent cachées dans leurs livres, et y resteront inconnues, tant qu'un gouvernement libéral et éclairé ne fera pas entreprendre, à ses frais ou sous ses auspices, soit le dépouillement, soit la traduction, des ouvrages où des procédés scientifiques et industriels, applicable à notre état social et à nos besoins, sont consignés et nettement décrits.[a]

And one cannot discount his words entirely as those of a sinologist in search of financial support, for half a century afterwards the silk industry of Lyons was still organising full-scale missions[b] to study the traditional procedures of the Chinese sericultural provinces.

For the details of the transmissions, in so far as it is possible to formulate them at the present stage, the reader is referred to pp. 222 ff., 544, 584 above. The only important basic machine which the Chinese did not have was the continuous screw, and for this the great development of pedals and treadles (unfamiliar in Europe) was no small compensation. Some techniques, such as the water-wheel, evolved in parallel in the two civilisations. Perhaps the most extraordinary reflection which occurs to us is that while China (allowing something for probable losses of texts) has nothing to show which equals the systematic treatment of mechanics and pneumatics by the Alexandrians (Ctesibius, Philon, Heron), those men were living in a civilisation which was incapable of utilising horse power to draw weights of more than half a ton, and which persisted in the universal employment of the primitive vertical loom. Some have thought, indeed, that the more advanced character of oriental technology was dimly realised by the Greeks, and it has even been suggested that the fable of Atlantis

valued ingenious articles', it said, 'nor do we have the slightest need of your country's manufactures.' There is general agreement with Greenberg's view; cf. Lattimore (7) and Gallagher & Purcell (1). Purcell regards the words of this edict as the key to the understanding of Sino-Western relations throughout the +18th century (p. 580), and emphasises also the converse fact that the traditional industries of China served an important export market. Chinese ramie, silk and other textiles flowed into Mexico via Manila; Burma and South-east Asia obtained ironware in quantity from China; and great quantities of Chinese porcelain went to Borneo and all Indonesia. Indeed the marked success of Chinese mercantile settlers led to serious persecutions, notably in Dutch Indonesia in +1740, and in the Spanish Philippines in +1709, +1755 and +1763 (Gallagher & Purcell, pp. 587, 591).

[a] This conviction was very general at the time. The following year saw the appearance of Rondot's review (1) of the manufactures of China. Two years later the Mission Press at Shanghai issued a translation of excerpts on sericulture from the *Nung Chêng Chhüan Shu* of +1639 (Anon. 39), illustrated with Chinese diagrams of the early 19th century (cf. pp. 2, 107, 382 above). Julien's study of the history and fabrication of Chinese porcelain (7) followed in 1856, and the general account of ancient and modern industries from Chinese sources by Julien & Champion (1) in 1869.

[b] See Anon. (23). So also, as late as 1896, the Blackburn Chamber of Commerce sent a mission to China; cf. Bourne, Neville & Bell (1). Though most of these later groups were primarily interested in the opening up of trade (and even perhaps in 'spheres of influence' or other forms of partition), their problems were intimately related to the study of indigenous industrial techniques. In 1884 Tissandier (7) could write: 'Que de procédés de l'Extrême-Orient déroutent encore nos chimistes! La fabrication de la laque, celle de l'encre de la Chine, du vermillon, la métallurgie et la confection des tam-tams, la fabrication de certains papiers, de plusieurs espèces de vernis, sont encore pratiqués dans le Céleste Empire et au Japon, à l'aide des procédés que jusqu'ici nous cherchons en vain.'

in Plato's dialogue *Critias*,[a] the island of Atlantis, whose people were such great architects and builders of irrigation-canals and bridges, has reference to the civilisations of Asia. If so, it is probably rather to Mesopotamia than further East, since at the time when it was written (*c.* −360, not long after the death of Mo Ti) Chinese superiority in technique had hardly gained the lead which it did some centuries later. Yet very large public works had been constructed in China in the −6th and −5th centuries.[b]

Let the last words come from the realm of Islam, for the Arabs were very well qualified to be impartial judges of engineers both European and Chinese. From Ruy Gonzales Clavijo, Spanish ambassador to Timur Lang (Tamerlane, +1336 to +1404), we hear that at Samarqand 'the craftsmen of Cathay are reputed to be the most skilful by far beyond those of any other nation; and the saying is that they alone have two eyes, that the Franks may indeed have one, while the Muslims are but a blind folk'.[c] Truer perhaps, and better, was the remark of Abū 'Uthmān 'Amr ibn Baḥr al-Jāḥiẓ (*d.* +869)—'Wisdom hath alighted on three things; the brain of the Franks, the hands of the Chinese, and the tongue of the Arabs.'[d]

---

[a] Jowett (1), vol. 3, pp. 429 ff., 519 ff.　　[b] Cf. Chi Chhao-Ting (1), p. 66.

[c] Cit. Olschki (4), p. 98; Yule (2), vol. 1, pp. 174, 264. We do not know who originated this saying, or where, but it was repeated over and over again. Abū Manṣūr Tha'ālibī, writing in about +1030, reports it, and so does Sharaf al-Zamān Ṭāhir al-Marwazī just under a century later, in the following words: 'The people of China are the most skilful of men in handicrafts—no other nation approaches them. The people of Rūm (the Eastern Roman Empire) are proficient too, but they do not reach the level of the Chinese. The latter say that all men are blind in craftsmanship, except the men of Rūm, who being one-eyed, know but half the business' (cf. Minorsky (4), pp. 14, 65). A curious Sufic mystical parable in the *Mathnawī* of Jalāl al-Dīn Rūmī (*d.* +1273) involves a contest in painting between Greeks and Chinese at a Sultan's court (I, ll. 3467 ff. Nicholson tr. vol. 2, p. 189). Then the adage about the blindness comes again in the *Fleurs des Histoires d'Orient* written by Prince Haython of Armenia in +1307 (Bk. 1, ch. 1; Yule, *ibid.* p. 258), and before long it was copied from somewhere by Sir John Mandeville (*c.* +1362), ch. 23 (Letts ed., vol. 1, p. 151). After Clavijo it reappears in Nicolò Conti (*c.* +1440; Yule, *ibid.* p. 175) and in Josafat Barbaro (*c.* +1470; Yule, *ibid.* p. 178). Its estimate of the Chinese craftsmen was echoed by a host of travellers; for example the technicians of the Macartney embassy in +1793 such as Dinwiddie (1), p. 49; Barrow (1), p. 306; cf. Cranmer-Byng (2), p. 264.

[d] Hitti (1), pp. 90, 382, quoting *Majmū'at Rasā'il*, pp. 41 ff.; cf. al-Jalil (1), pp. 109 ff.

BIBLIOGRAPHIES

A CHINESE AND JAPANESE BOOKS BEFORE +1800
B CHINESE AND JAPANESE BOOKS AND JOURNAL ARTICLES SINCE +1800
C BOOKS AND JOURNAL ARTICLES IN WESTERN LANGUAGES

In Bibliographies A and B there are two modifications of the Roman alphabetical
sequence: transliterated *Chh-* comes after all other entries under *Ch-*, and transliterated *Hs-* comes after all other entries under *H-*. Thus *Chhen* comes after *Chung*
and *Hsi* comes after *Huai*. This system applies only to the first words of the titles.
Moreover, where *Chh-* and *Hs-* occur in words used in Bibliography C, i.e. in a
Western language context, the normal sequence of the Roman alphabet is observed.

When obsolete or unusual romanisations of Chinese words occur in entries in
Bibliography C, they are followed, wherever possible, by the romanisations adopted
as standard in the present work. If inserted in the title, these are enclosed in square
brackets; if they follow it, in round brackets. When Chinese words or phrases occur
romanised according to the Wade–Giles system or related systems, they are assimilated to the system here adopted without indication of any change. Additional notes
are added in round brackets. The reference numbers do not necessarily begin
with (1), nor are they necessarily consecutive, because only those references required
for this volume of the series are given.

Korean and Vietnamese books and papers are included in Bibliographies A and B.

ABBREVIATIONS

See also p. xxxvii

AA	Artibus Asiae
AAA	Archaeologia
AAE	Archivio p. Anthropol. e Etnol.
AAN	American Anthropologist
AAS	Arts asiatiques (continuation of Revue des Arts Asiatiques)
A/AIHS	Archives Internationales d'Histoire des Sciences (continuation of Archeion)
AB	Art Bulletin (New York)
ABA	Ars Buddhica (Tokyo)
ABAW/PH	Abhandlungen d. bayr. Akad. Wiss. München (Phil.-Hist. Klasse)
ABORI	Annals of the Bhandarkar Oriental Research Institute (Poona)
ACA	Acta Archaeologica (Copenhagen)
ACLS	American Council of Learned Societies
ACP	Annales de Chimie et Physique
ACSS	Annual of the China Society of Singapore
ADI	Anzeiger f. d. Drahtindustrie
ADVS	Advancement of Science (British Assoc. London)
AEST	Annales de l'Est (Fac. des Lettres, Univ. Nancy)
AGHST	Agricultural History (Washington, D.C.)
AGMN	Archiv. f. d. Gesch. d. Medizin u. d. Naturwissenschaften (Sudhoff's)
AGMNT	Archiv. f. d. Gesch. d. Math., d. Naturwiss. u. d. Technik (cont. as QSGNM)
AGMW	Abhandlungen z. Geschichte d. Math. Wissenschaft
AGNT	Archiv f. d. Gesch. d. Naturwiss. u. d. Technik (cont. as AGMNT)
AH	Asian Horizon
AHAW/PH	Abhandlungen d. Heidelberger Akad. Wiss. (Phil.-Hist. Klasse)
AHES	Annales d'Hist. Écon. et Sociale
AHES/AHS	Annales d'Hist. Sociale
AHES/MHS	Mélanges d'Hist. Sociale
AHOR	Antiquarian Horology
AHR	American Historical Review
AHSNM	Acta Historica Scientiarum Naturalium et Medicinalium (edidit Bibliotheca Universitatis Hauniensis, Copenhagen)
AJ	Asiatic Journal and Monthly Register for British and Foreign India, China and Australia

AJSC	American Journal of Science and Arts (Silliman's)
AKML	Abhandlungen f. d. Kunde des Morgenlandes
AL	Annuaire du Bureau des Longitudes (Paris)
AM	Asia Major
AMA	American Antiquity
AMK	Alte u. Moderne Kunst (Österr. Zeitschr. f. Kunst, Kunsthandwerk u. Wohnkultur)
AN	Anthropos
ANM	Annales des Mines
ANS	Annals of Science
ANTJ	Antiquaries Journal
AO	Acta Orientalia
APAW/PH	Abhandlungen d. preuss. Akad. Wiss. Berlin (Phil.-Hist. Klasse)
AP/HJ	Historical Journal, National Peiping Academy
AQ	Antiquity
AQC	Antique Collector
ARAB	Arabica
ARAE	Archaeologiai Ertesitö (Budapest)
ARASI	Annual Report of the Archaeol. Survey of India
ARJ	Archaeological Journal
ARLC/DO	Annual Reports of the Librarian of Congress (Division of Orientalia)
ARO	Archiv Orientalni (Prague)
ARSI	Annual Reports of the Smithsonian Institution
ARUSNM	Annual Reports of the U.S. National Museum
AS/BIHP	Bulletin of the Institute of History and Philology, Academia Sinica
AS/CJA	Chinese Journal of Archaeology (Academia Sinica)
ASAE	Annales du Service des Antiquités de l'Égypte
ASEA	Asiatische Studien; Études Asiatiques
ASGB	Aeronautical Society of Great Britain
ASPN	Archives des Sciences Physiques et Naturelles (Geneva)
ASRAB	Annales de la Soc. (Roy.) d'Archéol. (Brussels)
ASURG	Annals of Surgery
ASWSP	Archiv. f. Sozialwissenschaft u. Sozialpolitik
AT	Atlantis
AX	Ambix

BABEL	Babel; Revue Internationale de la Traduction	BSNAF	Bulletin de la Société Nationale des Antiquaires de France
BAFAS	Bulletin de l'Assoc. Française pour l'Avancement des Sciences	BSRCA	Bulletin of the Society for Research in [the History of] Chinese Architecture
BAVH	Bulletin des Amis du Vieux Hué (Indo-China)	BTG	Blätter f. Technikgeschichte (Vienna)
BBSHS	Bulletin of the British Society for the History of Science	BUA	Bulletin de l'Université de l'Aurore (Shanghai)
BCS	Bulletin of Chinese Studies (Chhêngtu)	BUM	Burlington Magazine
BEFEO	Bulletin de l'École Française de l'Extrême Orient (Hanoi)	BUSNM	Bulletin of the U.S. National Museum
BFMRS	Biographical Memoirs of Fellows of the Royal Society	BVB	Bayerische Vorgeschichtsblätter
BGSC	Bulletin of the Chinese Geological Survey	BVSAW/PH	Berichte ü. d. Verhandl. d. Sächs. Akad. Wiss. (Leipzig)
BGTI	Beiträge z. Gesch. d. Technik u. Industrie (cont. as Technik Geschichte—see BGTI/TG)	CAF	Congrès Archéologiques de France
		CAMR	Cambridge Review
		CAR	Catholic Review
BGTI/TG	Technik Geschichte	CEN	Centaurus
BI/LRPC	Bulletin d'Information de la Légation de la République Pop. de Chine (Berne)	CET	Ciel et Terre
		CHER	Chhing-Hua Engineering Reports (Kung Chhêng Hsüeh Pao)
BIHM	Bulletin of the (Johns Hopkins) Institute of the History of Medicine (continued as Bulletin of the History of Medicine)	CHESJ	Chhing-Hua Engineering Society Journal (Kung Chhêng Hsüeh Hui Hui Khan)
		CHI	Cambridge History of India
BLSOAS	Bulletin of the London School of Oriental and African Studies	CHIM	Chimica (Italy)
		CHJ	Chhing-Hua Hsüeh Pao (Chhing-Hua (Ts'ing-Hua) University Journal of Chinese Studies)
BM	Bibliotheca Mathematica		
BMFEA	Bulletin of the Museum of Far Eastern Antiquities (Stockholm)	CHJ/T	Chhing-Hua (Ts'ing-Hua) Journal of Chinese Studies (New Series, publ. Thaiwan)
BMFJ	Bulletin de la Maison Franco-Japonaise (Tokyo)		
BMRAH	Bulletin des Musées Royaux d'Art et d'Histoire (Brussels)	CIBA/T	Ciba Review (Textile Technology)
BNAWCC	Bulletin of the Nat. Assoc. Watch and Clock Collectors (U.S.A.)	CJ	China Journal of Science and Arts
BNGBB	Berichte d. naturforsch. Gesellschaft Bamberg	CLHP	Chin-Ling Hsüeh Pao (Nanking University semi-annual Journal)
BNI	Bijdragen tot de taal-, land- en volken-kunde v. Nederlandsch Indië	CLIT	Chinese Literature
		CLTC	Chen Li Tsa Chih (Truth Miscellany)
		CME	Chartered Mechanical Engineer
BNYAM	Bulletin of the New York Academy of Medicine	CMIS	Chinese Miscellany
		CON	Connoisseur
BRGK	Berichte d. Römisch-Germanischen Kommission (d. deutsches Archäol. Inst.)	CP	Classical Philology
		CQ	Classical Quarterly
		CR	China Review (Hongkong and Shanghai)
BRO	Brotéria (Lisbon)		
BSAF	Bulletin de la Société Astronomique de France	CRAAF	Comptes Rendus hebdomadaires de l'Académie d'Agriculture de France
BSEIC	Bulletin de la Société des Études Indochinoises	CRAS	Comptes Rendus hebdomadaires de l'Acad. des Sciences (Paris)
BSEIN	Bulletin de la Soc. d'Encouragement pour l'Industrie Nationale	CREC	China Reconstructs
		CRRR	Chinese Repository
BSG	Bulletin de la Société de Géographie (continued as La Géographie)	CTE	China Trade and Engineering
		CUL	Cambridge University Library
		CUP	Cambridge University Press
BSISS	Bulletin de la Société Internationale des Sciences Sociales	D	Discovery

DAE	*Daedalus (Journ. Amer. Acad. Arts and Sciences)*		*bandes d. Geschichtslehrer Deutschlands)*
DEI	*De Ingenieur ('s-Gravenhage)*	*H*	*History*
DHT	*Documents pour l'histoire des Techniques (Cahiers du Centre de Documentation d'Hist. des Tech.; Conservatoire Nat. des Arts et Métiers, Paris)*	*HCCC*	*Huang Chhing Ching Chieh* (Yen Chieh, ed.)
		HE	*Hesperia (Journ. Amer. Sch. Class. Stud. Athens)*
DI	*Die Islam*	*HH*	*Han Hiue (Han Hsüeh); Bulletin du Centre d'Études Sinologiques de Pékin*
DKM	*Die Katholischen Missionen* (Bonn)		
DNAT	*Die Natur* (Halle a/d Saale)	*HJAS*	*Harvard Journal of Asiatic Studies*
DNATU	*Die Natuur* (Amsterdam)		
DP	*Dissertationes Pannonicae ex Instituto Numismatico et Archaeologico Universitatis de Petro Pázmány Nominatae Budapestinensis Provenientes*	*HMSO*	Her Majesty's Stationery Office
		HORJ	*Horological Journal*
		HP	*Hespéris (Archives Berbères et Bulletin de l'Institut des Hautes Études Marocaines)*
DTP	*Revista de Dialectología y Tradiciones Populares* (Madrid)	*I*	*L'Ingegnere*
DUZ	*Deutsche Uhrmacher Zeitung*	*IAE*	*Internationales Archiv f. Ethnographie*
EAN	*Edgar Allen News*	*IAQ*	*Indian Antiquary*
EDJ	*Edinburgh Journal* (Chambers')	*IAUT*	*Revista del Instituto de Antropología de la Universidad Nacional de Tucuman*
EHOR	*Eastern Horizon* (Hongkong)		
EHR	*Economic History Review*		
EMJ	*Engineering and Mining Journal*	*IB*	*Ingeniør og Bygningsvaesen* (Copenhagen)
EN	*Engineer*		
END	*Endeavour*	*ICS*	*L'Italia che Scrive*
ENG	*Engineering*	*ILN*	*Illustrated London News*
ENPJ	*Edinburgh New Philos. Journ.*	*IM*	*Imago Mundi; Yearbook of Early Cartography*
EPJ	*Edinburgh Philosophical Journal*		
ERE	*Encyclopaedia of Religion and Ethics*	*ISIS*	*Isis*
		ISL	*Islam*
ETH	*Ethnos*		
EWD	*Electrical World* (New York)	*JA*	*Journal Asiatique*
		JAFL	*Journal of American Folklore*
FEQ	*Far Eastern Quarterly* (continued as *Journal of Asian Studies*)	*JAN*	*Janus*
		JAOS	*Journal of the American Oriental Society*
FI	*France-Illustration*		
FK	*Földrajzi Közlemények* (Hungarian Journal of Geography)	*JAS*	*Journal of Asian Studies* (continuation of *Far Eastern Quarterly, FEQ*)
FL	*Folklore*		
FLV	*Folk-Liv*	*JBAA*	*Journal of the British Archaeological Association*
FMNHP/AS	*Field Museum of Natural History* (Chicago) *Publications*, Anthropological Series	*JCE*	*Journal of Chemical Education*
		JCUS	*Journal of Cuneiform Studies*
		JEH	*Journal of Economic History*
GB	*Globus*	*JESHO*	*Journal of the Economic and Social History of the Orient*
GBA	*Gazette des Beaux-Arts*		
GC	*Génie Civil*	*JGJRI*	*Journal of the Ganganatha Jha Research Institute*
GERM	*Germania*		
GGM	*Geographical Magazine*	*JGLGA*	*Jahrbuch d. Gesellschaft f. löthringen Geschichte u. Altertumskunde*
GJ	*Geographical Journal*		
GR	*Geographical Review*		
GRO	*Geografia Revuo* (Esperanto; Beograd)	*JHI*	*Journal of the History of Ideas*
GRSCI	*Graphic Science*	*JHMAS*	*Journal of the History of Medicine and Allied Sciences*
GTIG	*Geschichtsblätter f. Technik, Industrie u. Gewerbe*		
		JHOAI	*Jahreshefte d. österreich. Archäol. Institut* (Vienna)
GWU	*Geschichte in Wissenschaft u. Unterricht (Zeitschr. d. Ver-*	*JHS*	*Journal of Hellenic Studies*

JIN	*Journal of the Institute of Navigation* (U.K.)	*KVHAAH*	*Kungl. Vitterhets Historie och Antikvitets Akademiens Handlingar* (Oslo)
JISI	*Journal of the Iron and Steel Institute* (U.K.)		
JJHS	*Japanese Journal of the History of Science* (*Kagaku-shi Kenkyū*)	*LFI*	*La Fonderia Italiana*
		LN	*La Nature*
JKHSW	*Jahrbuch d. Kunsthistorischen Sammlungen in Wien*	*LRCW*	*Locomotive, Railway Carriage and Wagon Review*
JNPCA	*Journal of Natural Philos., Chem. and the Arts* (Nicholson's), (later united with *PMG*)	*LSH*	*La Suisse Horlogère*
		LSYC	*Li Shih Yen Chiu* (Peking) (*Journal of Historical Research*)
JOSHK	*Journal of Oriental Studies* (Hongkong Univ.)	*MA*	Man
JPIME	*Journal and Proceedings, Institution of Mechanical Engineers*	*MAI/NEM*	*Mémoires de l'Académie des Inscriptions et Belles-Lettres*, Paris (*Notices et Extraits des MSS.*)
JRAES	*Journal of the Royal Aeronautical Society* (formerly *Aeronautical Journal*)	*MANU*	*Manuscripta*
		MAS	*Mémoires de l'Académie des Sciences* (Paris)
JRAI	*Journal of the Royal Anthropological Institute*	*MC*	*Métaux et Civilisations* (continued as *Techniques et Civilisations*)
JRAS	*Journal of the Royal Asiatic Society*		
JRAS/C	*Journal of the Ceylon Branch of the Royal Asiatic Society*	*MC/TC*	*Techniques et Civilisations* (formerly *Métaux et Civilisations*)
JRAS/KB	*Journal* (or *Transactions*) *of the Korea Branch of the Royal Asiatic Society*	*MCB*	*Mélanges Chinois et Bouddhiques*
		MCHSAMUC	*Memoires concernant l'Histoire, les Sciences, les Arts, les Mœurs et les Usages, des Chinois, par les Missionnaires de Pékin* (Paris 1776–)
JRAS/NCB	*Journal* (or *Transactions*) *of the North China Branch of the Royal Asiatic Society*		
JRCAS	*Journal of the Royal Central Asian Society*	*MDGNVO*	*Mitteilungen d. deutsch. Gesellsch. f. Natur. u. Volkskunde Ostasiens*
JRSA	*Journal of the Royal Society of Arts*	*MDIK*	*Mitteilungen d. deutschen Inst. f. ägypt. Altertumskunde in Kairo*
JRSAI	*Journal of the Royal Society of Antiquaries of Ireland*	*MENG*	*Mechanical Engineering* (New York)
JS	*Journal des Sçavans* (1665–1778) and *Journal des Savants* (1816–)	*MF*	*Mercure de France*
		MGNM	*Mitteilungen a. d. germanisches National-Museum*
JSHB	*Journal Suisse d'Horlogérie et Bijouterie* (later *J. Suisse d'Horlog.*)	*MGSC*	*Memoirs of the Chinese Geological Survey*
		MH	*Medical History*
JWCBRS	*Journal of the West China Border Research Society*	*MIARA*	*Machines et Inventions approuvées par l'Académie* (des Sciences de Paris)
JWCI	*Journal of the Warburg and Courtauld Institutes*	*MIT*	Massachusetts Institute of Technology
JWH	*Journal of World History* (UNESCO)	*MLN*	*Modern Language Notes*
		MMG	*Mechanics' Magazine*
		MMI	*Mariner's Mirror*
KDVS/AKM	*Kgl. Danske Videnskabernes Selskab* (Archaeol.-Kunsthist. Medd.)	*MN*	*Monumenta Nipponica*
		MP	*Il Marco Polo*
KDVS/HFM	*Kgl. Danske Videnskabernes Selskab* (Hist.-Filol. Medd.)	*MRAI/DS*	*Mémoires presentés par divers savants à l'Académie Royale des Inscriptions et Belles-Lettres* (Paris)
KHS	*Kho Hsüeh* (*Science*)		
KHTP	*Kho Hsüeh Thung Pao* (*Science Correspondent*)	*MRASP*	*Mémoires de l'Académie Royale des Sciences* (Paris)
KK	*Kokka*	*MRDTB*	*Memoirs of the Research Dept. of Tōyō Bunko* (Tokyo)
KKTH	*Khao Ku Thung Hsün* (*Archaeological Correspondent*)	*MS*	*Monumenta Serica*
KMJP	*Kuang Ming Jih Pao*	*MSAF*	*Mémoires de la Société* (Nat.) *des Antiquaires de France*
KSHP	*Kuo Sui Hsüeh Pao*		

MSFO	Mémoires de la Soc. Finno-Ougrienne	PHFC	Proceedings of the Hampshire Field Club and Archaeol. Soc.
MSOS	Mitteilungen d. Seminar f. orientalischen Sprachen (Berlin)	PHY	Physis (Florence)
		PIGE	Proc. Inst. Gas Engineers (U.K.)
MSTRM	Mainstream (New York)	PIME	Proceedings of the Institute of Mechanical Engineers (U.K.)
MT	Metallurgia		
MW	Middle Way (Journal of the Buddhist Society, U.K.)	PKR	Peking Review
		PL	Philologus, Zeitschrift f. d. klass. Altertums
N	Nature	PMJP	Mitteilungen aus Justus Perthes' Geographische Anstalt (Petermann's)
NALC	Nova Acta; Abhandl. d. Kaiserl. Leop.-Carol. Deutsch. Akad. Naturf. (Halle)		
		PNHB	Peking Natural History Bulletin
NATV	Naturens Verden (Copenhagen)	PO	Poona Orientalist
NAVC	Naval Chronicle	POPA	Popular Astronomy
NAW	Nieuwe Archief voor Wiskunde	POPS	Popular Science (U.S.A.)
NBAC	Nuovo Bull. di Archeologia Cristiana (Rome)	PPHS	Proceedings of the Prehistoric Society
NCR	New China Review	PPS	Proceedings of the Physical Society
NGM	National Geographic Magazine		
NGWG/PH	Nachrichten v. d. k. Gesellsch. (Akademie) d. Wiss. z. Göttingen (Phil.-Hist. Klasse)	PRH	Praehistoria (Prague)
		PRIA	Proceedings of the Royal Irish Academy
NJKA	Neue Jahrbücher f. d. klass. Altertum, Geschichte, deutsch. Literatur u. f. Pädagogik	PRS	Proceedings of the Royal Society (before division into series A and B)
NKKZ	Nihon Kagaku Koten Zensho (Collection of works concerning the History of Science and Technology in Japan)	PRSA	Proceedings of the Royal Society (Ser. A)
		PRSM	Proceedings of the Royal Society of Medicine
NO	New Orient (Prague)	PSA	Proceedings of the Society of Antiquaries
NQ	Notes and Queries		
NQCJ	Notes and Queries on China and Japan	PSAS	Proceedings of the Society of Antiquaries of Scotland
NRRS	Notes and Records of the Royal Society	PTRS	Philosophical Transactions of the Royal Society
NS	New Scientist		
NSN	New Statesman and Nation (London)	QSGNM	Quellen u. Studien z. Gesch. d. Naturwiss. u. d. Medizin (continuation of Archiv f. Gesch. d. Math., d. Naturwiss. u. d. Technik, formerly Archiv f. d. Gesch. d. Naturwiss. u. d. Technik)
NUK	Natur u. Kultur (München)		
NV	The Navy		
OAR	Ostasiatische Rundschau		
OAS	Ostasiatische Studien		
OAV	Orientalistisches Archiv (Leipzig)	RA	Revue Archéologique
OAZ	Ostasiatische Zeitschrift	RBA	Revue de Botanique Appliquée
OB	Orientalistische Bibliographie	RBMQR	Rotol and British Messier Quarterly Review
OC	Open Court		
OE	Oriens Extremus (Hamburg)	RBS	Revue Bibliographique de Sinologie
OLL	Ostasiatischer Lloyd	RDI	Rivista d'Ingegneria
OLZ	Orientalische Literatur-Zeitung	REJ	Royal Engineers Journal
OR	Oriens	RG	Revista de Guimarães
ORA	Oriental Art	RGHE	Revue de Géographie Humaine et d'Ethnologie
ORT	Orient		
OSIS	Osiris	RHS	Revue d'Histoire des Sciences (Centre Internationale de Synthèse, Paris)
OUP	Oxford University Press		
PA	Pacific Affairs	RHSID	Revue d'Histoire de la Sidérurgie (Nancy)
PC	People's China		
PET	Petroleum	RPLHA	Revue de Philol., Litt. et Hist. Ancienne
PFL	Pennsylvania Folklife		

RQS	*Revue des Questions Scientifiques* (Brussels)		*TAPS*	*Transactions of the American Philosophical Society* (cf. *MAPS*)
RR	*Review of Religion*			
RRIL	*Rendiconti d. R. Istituto Lombardo*		*TAS/J*	*Transactions of the Asiatic Society of Japan*
RSH	*Revue de Synthèse Historique*		*TASCE*	*Transactions of the American Society of Civil Engineers*
RSO	*Rivista di Studi Orientali*			
RTS	Religious Tract Society		*TBK*	*Tōyō Bunka Kenkyūjo Kiyō* (Memoirs of the Institute of Oriental Culture, Univ. of Tokyo)
RUB	*Revue de l'Univ. de Bruxelles*			
RUSCP	*Reports of the U.S. Commissioner of Patents*			
			TBS	*Tōyō no Bunka to Shakai* (Oriental Culture and Society)
S	*Sinologica* (Basel)			
SA	*Sinica* (originally *Chinesische Blätter f. Wissenschaft u. Kunst*)		*TCKM*	*Thung Chien Kang Mu* (Chu Hsi et al. ed.)
			TCULT	*Technology and Culture*
SAE	*Saeculum*		*TFTC*	*Tung Fang Tsa Chih* (*Eastern Miscellany*)
SAM	*Scientific American*			
SAN	*Shell Aviation News*		*TG/K*	*Tōhō Gakuhō, Kyōto* (*Kyoto Journal of Oriental Studies*)
SB	*Shizen to Bunka*			
SBE	*Sacred Books of the East* series		*TG/T*	*Tōhō Gakuhō, Tōkyō* (*Tokyo Journal of Oriental Studies*)
SDPVT	*Sbornik pro Dějiny Přírodnich Věd a Techniky* (*Acta Historiae Rerum Naturalium necnon Technicarum*) (Prague)		*TGAS*	*Transactions of the Glasgow Archaeological Society*
			TH	*Thien Hsia Monthly* (Shanghai)
SE	*Stahl und Eisen*		*TIIC*	*Transactions of the Indian Institute of Culture* (Basavangudi, Bangalore)
SH	*Shih Huo* (*Journal Hist. of Economics*)			
SIS	*Sino-Indian Studies* (Santiniketan)		*TIM*	*Time Magazine*
			TIYT	*Trudy Instituta Istorii Yestestvoznania i Tekhniki*
SL	*Shui Li* (*Hydraulic Engineering*)			
SLAVA	*Slavia Antiqua* (Poznań)		*TJSL*	*Transactions (and Proceedings) of the Japan Society of London*
SN	*Scientific News*			
SN	*Shirin* (*Journal of History*) (Kyōto)		*TK*	*Tōyōshi Kenkyū* (*Researches in Oriental History*)
SO	*Sociologia* (Brazil)		*TM*	*Terrestrial Magnetism and Atmospheric Electricity* (continued as *Journ. Geophysical Research*)
SP	*Speculum*			
SPAW	*Sitzungsberichte d. preuss. Akad. d. Wissenschaft*			
SPCK	Society for the Promotion of Christian Knowledge		*TNS*	*Transactions of the Newcomen Society*
SPMSE	*Sitzungsberichte d. physik. med. Soc. Erlangen*		*TP*	*T'oung Pao* (*Archives concernant l'Histoire, les Langues, la Géographie, l'Ethnographie et les Arts de l'Asie Orientale*, Leiden)
SPRDS	*Scientific Proceedings of the Royal Dublin Society*			
SR	*Sperry Review*			
SSC	*Bulletin Annuel de la Société Suisse de Chronométrie*		*TR*	*Technology Review*
			TRDG	*Tenri Daigaku Gakuhō* (*Bulletin of Tenri University*)
SSE	*Studia Serica* (*West China Union University Literary and Historical Journal*)		*TWHP*	*Thien Wên Hsüeh Pao* (*Acta Astronomica Sinica*)
SWAW/PH	*Sitzungsberichte d. k. Akad. d. Wissenschaften Wien* (Vienna) (Phil.-Hist. Klasse)		*TYG*	*Tōyō Gakuhō* (*Reports of the Oriental Society of Tokyo*)
SWYK	*Shuo Wên Yüeh Khan* (*Philological Monthly*)		*UJA*	*Ulster Journal of Archaeology*
			UMK	*Uhrmacherkunst* (*Verbandszeitung d. deutsch. Uhrmacher*)
SYR	*Syria*			
T	*Technica* (*Bull. du Comité Belge de Bijouterie et de l'Horlogérie*)		*VAG*	*Vierteljahrsschrift d. astronomischen Gesellschaft*
TAP	*Annals of Philosophy* (Thomson's)		*VDI*	Verein d. deutschen Ingenieure
			VS	*Variétés Sinologiques*

WCC	*Bulletin of the Nat. Assoc. Watch and Clock Collectors* (U.S.A.)	ZAES	*Zeitschrift f. Aegyptische Sprache u. Altertumskunde*
WEAR	*Wear* (a journal of friction and lubrication studies)	ZAGA	*Zeitschrift f. Agrargeschichte u. Agrarsoziologie*
WHJP	*Wuhan University Daily*	ZDMG	*Zeitschrift d. deutsch. Morgenländischen Gesellschaft*
WP	*Water Power*		
WWTK	*Wên Wu Tshan Khao Tzu Liao (Reference Materials for History and Archaeology)*	ZDVTS	*Z Dejín Vied a Techniky na Slovensku* (Bratislava)
		ZGT	*Zeitschrift f. d. gesamte Turbinenwesen*
YAHS	*Yenching Shih Hsüeh Nien Pao (Yenching University Annual of Historical Studies)*	ZHWK	*Zeitschrift f. historische Wappenkunde* (continued as *Zeitschr. f. hist. Wappen- und Kostumkunde*)
YCHP	*Yenching Hsüeh Pao (Yenching University Journal of Chinese Studies)*	ZMP	*Zeitschrift f. Math. u. Physik*
		ZOIAV	*Zeitschrift d. österr. Ingenieur u. Architekten Vereines*
Z	*Zalmoxis; Revue des Études Religieuses*	ZTF	*Zeitschrift f. techn. Fortschritt*
		ZVDI	*Zeitschrift d. Vereines deutsch. Ingenieure*

A. CHINESE BOOKS BEFORE +1800

Each entry gives particulars in the following order:

(a) title, alphabetically arranged, with characters;

(b) alternative title, if any;

(c) translation of title;

(d) cross-reference to closely related book, if any;

(e) dynasty;

(f) date as accurate as possible;

(g) name of author or editor, with characters;

(h) title of other book, if the text of the work now exists only incorporated therein; or, in special cases, references to sinological studies of it;

(i) references to translations, if any, given by the name of the translator in Bibliography C;

(j) notice of any index or concordance to the book if such a work exists;

(k) reference to the number of the book in the *Tao Tsang* catalogue of Wieger (6), if applicable;

(l) reference to the number of the book in the *San Tsang* (Tripiṭaka) catalogues of Nanjio (1) and Takakusu & Watanabe, if applicable.

Words which assist in the translation of titles are added in round brackets.

Alternative titles or explanatory additions to the titles are added in square brackets.

It will be remembered (p. 603 above) that in Chinese indexes words beginning *Chh-* are all listed together after *Ch-*, and *Hs-* after *H-*, but that this applies to initial words of titles only.

Where there are any differences between the entries in these bibliographies and those in Vol. 1–3, the information here given is to be taken as more correct.

References to the editions used in the present work, and to the *tshung-shu* collections in which books are available, will be given in the final volume.

ABBREVIATIONS

C/Han	Former Han.
E/Wei	Eastern Wei.
H/Chhin	Later Chhin.
H/Han	Later Han.
H/Shu	Later Shu (Wu Tai).
H/Thang	Later Thang (Wu Tai).
J/Chin	Jurchen Chin.
L/Sung	Liu Sung.
N/Chou	Northern Chou.
N/Chhi	Northern Chhi.
N/Sung	Northern Sung (before the removal of the capital to Hangchow).
N/Wei	Northern Wei.
S/Chhi	Southern Chhi.
S/Sung	Southern Sung (after the removal of the capital to Hangchow).
S/Thang	Southern Thang.
W/Wei	Western Wei.

An-yang Hsien Chih 安陽縣志.
  Local History and Topography of Anyang District (Honan).
  Chhing, +1693.
  Ma Kuo-Chêng 馬國正 *et al.*
  Enlarged +1738 by Chhen Hsi-Lu 陳錫輅.
  Enlarged 1819 by Kuei Thai 貴泰.

Bukkoku Rekishō-hen.
  See Entsū (1).

Chan Kuo Tshê 戰國策.
  Records of the Warring States.
  Chhin.
  Writer unknown.

Chang Ku Tshung Pien 掌故叢編.
  Collected Historical Documents.
  Chhing, various dates.
  Ed. Palace Museum, Peking, 1928–30.

Chen-Chung Hung-Pao Yuan-Pi Shu 枕中鴻寶苑祕書.
  The Infinite Treasure of the Garden of Secrets; (Confidential) Pillow-Book (of the Prince of Huai-Nan).
  See *Huai-Nan Wang Wan Pi Shu.*
  Cf. Kaltenmark (2), p. 32.

Chen Kao 眞誥.
  True Reports.
  Liang, early +6th century (but the earliest material contained in it is dated +365).
  Thao Hung-Ching 陶弘景.

Chêng Tzu Thung 正字通.
  Dictionary of Characters.
  Ming, +1627.
  Chang Tzu-Lieh 張自烈.

Chi Chiu Phien (or *Chang*) 急就篇 (章).
  Dictionary for Urgent Use.
  C/Han, −48 to −32.
  Shih Yu 史游.
  With +13th-century commentary by Wang Ying-Lin 王應麟.

Chi Yang Yao Pên Mo 記楊么本末.
  The History of (the Rebellion of) Yang Yao from Beginning to End.
  Sung, *c.* +1140.
  Li Kuei-Nien (b) 李龜年.
  Preserved only in fragments.

Chi Yün 集韻.
  Complete Dictionary of the Sounds of Characters [cf. *Chhieh Yün* and *Kuang Yün*].
  Sung, +1037.
  Compiled by Ting Tu 丁度 *et al.*
  Possibly completed in +1067 by Ssuma Kuang 司馬光.

Chien Khang Chi 建康集.
  Record of Nanking Affairs.
  Sung, *c.* +1110.
  Yeh Mêng-Tê 葉夢得.

Chien Nan Shih Kao 劍南詩稿.
  Collected Poems from Szechuan (the country south of Chien-Mên-Kuan).
  Sung, *c.* +1170.
  Lu Yu 陸游.

Chien-Yen I Lai Hsi Nien Yao Lu 建炎以來繫
年要錄.
A Chronicle of the Most Important Events
since the Chien-Yen Reign-Period
(+1127 to +1130) [of the Southern Sung].
Sung, c. +1220.
Li Hsin-Chhuan 李心傳.

Chih Thien Lu 芝田錄.
The Field of Magic Mushrooms.
Thang or before.
Ting Yung-Hui 丁用晦.

Chih Ya Thang Tsa Chhao 志雅堂雜鈔.
Miscellaneous Records of the 'Striving for
Elegance' Library.
Sung, c. +1270.
Chou Mi 周密.

Chin Chhi Chü Chu 晉起居注.
Daily Court Records of the Emperors of
the Chin Dynasty.
Pre-Sui.
Liu Tao-Hui 劉道會.

Chin Chu Kung Tsan 晉諸公讚.
Eulogia of Distinguished Men of the Chin
Dynasty.
Pre-Sui.
Fu Chhang 傅暢.

Chin Lou Tzu 金樓子.
Book of the Golden Hall Master.
Liang, c. +550.
Hsiao[蕭繹
(Liang Yuan Ti 梁元帝).

Chin Shih 金史.
History of the Chin (Jurchen) Dynasty
[+1115 to +1234].
Yuan, c. +1345.
Tho-Tho (Toktaga) 脫脫 & Ouyang Hsüan
歐陽玄.
Yin-Tê Index, no. 35.

Chin Shih So.
See Fêng Yün-Phêng & FêngYün-Yuan (1).

Chin Shu 晉書.
History of the Chin Dynasty [+265 to
+419].
Thang, +635.
Fang Hsüan-Ling 房玄齡.
A few chs. tr. Pfizmaier (54–57); the astro-
nomical chs. tr. Ho Ping-Yü (1). For
translations of passages see the index of
Frankel (1).

Ching Chhu Sui Shih Chi 荊楚歲時記.
Annual Folk Customs of the States of Ching
and Chhu [i.e. of the districts correspond-
ing to those ancient States; Hupei,
Hunan and Chiangsi].
Prob. Liang, c. +550, but perhaps partly
Sui, c. +610.
Tsung Lin 宗懍.
See des Rotours (1), p. cii.

Ching Ching Ling Chhih.
See Chêng Fu-Kuang (1).

Ching Hua Yuan.
See Li Ju-Chen (1).

Ching Shih Ta Tien.
See *Yuan Ching Shih Ta Tien.*

Chiu Chang Suan Ching.
See *Chiu Chang Suan Shu.*

Chiu Chang Suan Shu 九章算術.
Nine Chapters on the Mathematical Art.
H/Han, +1st century (containing much
material from C/Han and perhaps Chhin).
Writer unknown.

Chiu Thang Shu 舊唐書.
Old History of the Thang Dynasty [+618
to +906].
Wu Tai, +945.
Liu Hsü 劉昫.
Cf. des Rotours (2), p. 64.
For translations of passages see the index of
Frankel (1).

Chiu Wu Tai Shih 舊五代史.
Old History of the Five Dynasties [+907 to
+959].
Sung, +974.
Hsüeh Chü-Chêng 薛居正.
For translations of passages see the index of
Frankel (1).

Cho Kêng Lu 輟耕錄.
[sometimes *Nan Tshun Cho Kêng Lu*].
Talks (at South Village) while the Plough is
Resting.
Yuan, +1366.
Thao Tsung-I 陶宗儀.

Chou I.
See *I Ching.*

Chou I Tshan Thung Chhi Khao I.
See *Tshan Thung Chhi Khao I.*

Chou Kuan I Su 周官義疏.
Collected Commentaries and Text of the
*Record of the Institutions (lit. Rites) of
(the) Chou (Dynasty)* (imperially com-
missioned).
Chhing, +1748.
Ed. Fang Pao 方苞 et al.

Chou Li 周禮.
Record of the Institutions (lit. Rites) of
(the) Chou (Dynasty) [descriptions of all
government official posts and their duties].
C/Han, perhaps containing some material
from late Chou.
Compilers unknown.
Tr. E. Biot (1).

Chou Li Chêng I 周禮正義.
Amended Text of the *Record of the Institutions (lit. Rites) of (the) Chou (Dynasty)*
with Discussions (including the H/Han
commentary of Chêng Hsüan 鄭玄).
C/Han, perhaps containing some material
from late Chou.
Compilers unknown.
Ed. Sun I-Jang (1899) 孫詒讓.

Chou Li I I Chü Yao 周禮疑義舉要.
Discussion of the Most Important Doubtful
Matters in the *Record of the Institutions
(lit. Rites) of (the) Chou (Dynasty)*.

Chou Li I I Chü Yao (*cont.*)
Chhing, +1791.
Chiang Yung 江永.

Chou Pei Suan Ching 周髀算經.
The Arithmetical Classic of the Gnomon
and the Circular Paths (of Heaven).
Chou, Chhin and Han. Text stabilised about
the −1st century, but including parts
which must be of the late Warring States
period (*c.* −4th century) and some even
pre-Confucian (−6th century).
Writers unknown.

Chou Shu 周書.
History of the (Northern) Chou Dynasty
[+557 to +581].
Thang, +625.
Linghu Tê-Fên 令狐德棻.
For translations of passages see the index of
Frankel (*1*).

Chu Chhi Thu Shuo 諸器圖說.
Diagrams and Explanations of a number of
Machines [mainly of his own invention
or adaptation].
Ming, +1627.
Wang Chêng 王徵.

Chu Shu Chi Nien 竹書紀年.
The Bamboo Books [annals, fragments of a
chronicle of the State of Wei, from high
antiquity to −298].
Chou, −295 and before, such parts as are
genuine. (Found in the tomb of An Li
Wang, a prince of the Wei State *r.* −276
to −245; in +281.)
Writers unknown.
See van der Loon (*1*).
Tr. E. Biot (*3*).
Reconstruction of the genuine parts by
Chu Yu-Tsêng and Wang Kuo-Wei; see
Fan Hsiang-Yung (*1*).

Chu Tzu Chhüan Shu 朱子全書.
Collected Works of Master Chu (Hsi).
Sung (ed. Ming), *editio princeps* +1713.
Chu Hsi 朱熹.
Ed. Li Kuang-Ti 李光地 (Chhing).
Partial trs. Bruce (*1*); le Gall (*1*).

Chu Tzu Yü Lei 朱子語類.
Classified Conversations of Master Chu (Hsi).
Sung *c.* +1270.
Chu Hsi 朱熹.
Ed. Li Ching-Tê 黎靖德 (Sung).

Chuang Tzu 莊子.
[= *Nan Hua Chen Ching.*]
The Book of Master Chuang.
Chou, *c.* −290.
Chuang Chou 莊周.
Tr. Legge (*5*); Fêng Yu-Lan (*5*); Lin Yü-
Thang (*1*).
Yin-Tê Index no. (suppl.) 20.

Chuang Tzu Pu Chêng
The Text of *Chuang Tzu*, Annotated and
Corrected.
See Liu Wên-Tien (*1*).

Chung-Hsing Hsiao Chi 中興小紀.
Brief Records of Chung-hsing (mod.
Chiang-ling on the Yangtze in Hupei).
Sung, *c.* +1150.
Hsiung Kho 熊克.

Chung Hua Ku Chin Chu 中華古今注.
Commentary on Things Old and New in
China.
Wu Tai (H/Thang), +923 to +926.
Ma Kao 馬縞.
See des Rotours (*1*), p. xcix.

Chung I Pi Yung 種藝必用.
Everyman's Guide to Agriculture (lit.
What one must Know and Do in the Art
of Crop-Raising).
Sung, *c.* +1250.
Wu I (or Tsuan) 吳懌 (欑).
With Supplement (*Pu I*) by Chang Fu 張福.
In *Yung-Lo Ta Tien*, ch. 13,194.
Ed. Hu Tao-Ching.

Chŭngbo Munhŏn Pigo 增補文獻備考.
Complete Serviceable Study of (the History
of Korean) Civilisation [lit. Complete
Serviceable Study of the Documentary
Evidence of Cultural Achievements (in
Korean Civilisation), with Additions and
Supplements].
Revision and enlargement of the *Tongguk
Munhŏn Pigo*.
Korea, +1770, revised in +1790 by Yi
Manun 李萬運, and finally in 1907.

Chhang-Chhun Chen Jen Hsi Yu Chi 長春眞
人西遊記.
The Western Journey of the Taoist (Chhiu)
Chhang-Chhun.
Yuan, +1228.
Li Chih-Chhang 李志常.

Chhao Yeh Chhien Tsai 朝野僉載.
Stories of Court Life and Rustic Life [or,
Anecdotes from Court and Countryside].
Thang, +8th century, but much remodelled
in Sung.
Chang Tso 張鷟.

Chhen Shu 陳書.
History of the Chhen Dynasty [+556 to
+580].
Thang, +630.
Yao Ssu-Lien 姚思廉, and his father
Yao Chha 姚察.
A few chs. tr. Pfizmaier (59). For transla-
tions of passages, see the index of
Frankel (*1*).

Chhêng Chai Tsa Chi 誠齋雜記.
Miscellanea of the Sincerity Studio.
Sung, *c.* +1295.
Chou Ta-Kuan 周達觀.

Chhi Chhi Mu Lüeh 奇器目略.
Enumeration of Strange Machines.
Chhing, +1683.
Tai Jung 戴榕.

Chhi Chhi Thu Shuo 奇器圖說.
[= *Yuan Hsi Chhi Chhi Thu Shuo Lu Tsui.*]

Chhi Chhi Thu Shuo (*cont.*)
Diagrams and Explanations of Wonderful Machines.
Ming, +1627.
Têng Yü-Han (Johann Schreck) 鄧玉函 & Wang Chêng 王徵.

Chhi Min Yao Shu 齊民要術.
Important Arts for the People's Welfare [lit. Equality].
N/Wei (and E/Wei or W/Wei), between +533 and +544.
Chia Ssu-Hsieh 賈思勰.
See des Rotours (1), p.c; Shih Shêng-Han (1).

Chhi Tung Yeh Yü 齊東野語.
Rustic Talks in Eastern Chhi.
Sung, c. +1290.
Chou Mi 周密.

Chhien Han Shu 前漢書.
History of the Former Han Dynasty, [−206 to +24].
H/Han (begun about +65), c. +100.
Pan Ku 班固, and (after his death in +92) his sister Pan Chao 班昭.
Partial trs. Dubs (2), Pfizmaier (32–34, 37–51), Wylie (2, 3, 10), Swann (1).
Yin-Tê Index, no. 36.

Chhien Pai Nien Yen 千百年眼.
Glimpses of a Thousand Years of History.
Ming, +16th century.
Chang Sui 張燧.

Chhih Pei Ou Than 池北偶談.
Chance Conversations North of Chhih (-chow).
Chhing, +1691.
Wang Shih-Chên 王士禎.

Chhin-Ting Chou Kuan I Su.
See *Chou Kuan I Su.*

Chhin-Ting Hsü Wên Hsien Thung Khao 欽定續文獻通考.
Imperially Commissioned Continuation of the *Comprehensive Study of* (*the History of*) *Civilisation* (cf. *Wên Hsien Thung Khao* and *Hsü Wên Hsien Thung Khao*).
Chhing, ordered +1747, pr. +1772 (+1784).
Ed. Chhi Shao-Nan 齊召南, Hsi Huang 嵇璜 *et al.*
This parallels, but does not replace, Wang Chhi's *Hsü Wên Hsien Thung Khao.*

Chhin-Ting Ku Chin Thu Shu Chi Chhêng 欽定古今圖書集成.
See *Thu Shu Chi Chhêng.*

Chhin-Ting Shou Shih Thung Khao 欽定授時通考.
See *Shou Shih Thung Khao.*

Chhin-Ting Shu Ching Thu Shuo 欽定書經圖說.
The *Historical Classic* with Illustrations.
Chhing (edition by imperial order, 1905).
Ed. Sun Chia-Nai 孫家鼐 *et al.*

Chhin-Ting Ssu Khu Chhüan Shu, etc.
See *Ssu Khu Chhüan Shu.*

Chhing Hsiang Tsa Chi 青箱雜記.
Miscellaneous Records on Green Bamboo Tablets.
Sung, early +11th century.
Wu Chhu-Hou 吳處厚.

Chhing I Lu 清異錄.
Records of the Unworldly and the Strange.
Wu Tai, c. +950.
Thao Ku 陶穀.

Chhou Jen Chuan 疇人傳.
Biographies of Mathematicians and Astronomers.
Chhing, +1799.
Juan Yuan 阮元.
With continuations by Lo Shih-Lin 羅士琳, Chu Kho-Pao 諸可寶, and Huang Chung-Chün 黃鍾駿.
In *HCCC,* chs. 159 ff.

Chhü Chhao Shih Lei 趨朝事類.
Systematic Guide to Court Etiquette.
Sung.
Writer unknown.

Chhu Chiu Ching.
See Ong Kuang-Phing (1).

Chhu Hsüeh Chi 初學記.
Entry into Learning [encyclopaedia].
Thang, +700.
Hsü Chien 徐堅.

Chhu Tzhu 楚辭.
Elegies of Chhu (State) [or, Songs of the South].
Chou, c. −300 (with Han additions).
Chhü Yuan 屈原 (& Chia I 賈誼, Yen Chi 嚴忌, Sung Yü 宋玉, Huainan Hsiao-Shan 淮南小山 *et al.*).
Partial tr. Waley (23); tr. Hawkes (1).

Chhu Tzhu Pu Chu 楚辭補註.
Supplementary Annotations to the *Elegies of Chhu.*
Sung, c. +1140.
Ed. Hung Hsing-Tsu 洪興祖.

Chhü Wei Chiu Wên 曲洧舊聞.
Talks about Bygone Things beside the Winding Wei (River in Honan).
Sung, c. +1130.
Chu Pien 朱弁.

Chhüan Thang Wên.
See Tung Kao (1).

Chhui Chien Lu Wai Chi 吹劍錄外集.
The Chhui Chien Miscellany.
Sung, c. +1260.
Yü Wên-Pao 俞文豹.

Chhun Chhiu 春秋.
Spring and Autumn Annals [i.e. Records of Springs and Autumns].
Chou; a chronicle of the State of Lu kept between −722 and −481.
Writers unknown.
Cf. *Tso Chuan*; *Kungyang Chuan*; *Kuliang Chuan.*
See Wu Khang (1); Wu Shih-Chhang (1); van der Loon (1).
Tr. Couvreur (1); Legge (11).

Chhun Chhiu Fan Lu 春秋繁露.
String of Pearls on the *Spring and Autumn Annals*.
C/Han, *c.* −135.
Tung Chung-Shu 董仲舒.
See Wu Khang (1).
Partial trs. Wieger (2); Hughes (1); d'Hormon (ed.).
Chung-Fa Index, no. 4.

Chhung Hsü Chen Ching 冲虛眞經.
See *Lieh Tzu*.

Erh Ya 爾雅.
Literary Expositor [dictionary].
Chou material, stabilised in Chhin or C/Han.
Compiler unknown.
Enlarged and commented on *c.* +300 by Kuo Pho 郭璞.
Yin-Tê Index no. (suppl.) 18.

Fa-Hsien Hsing Chuan. 法顯行傳.
See *Fo Kuo Chi*.

Fan Wên Chêng Kung Wên Chi 范文正公文集.
Collected Works of Fan Chung-Yen.
Sung, *c.* +1060.
Fan Chung-Yen 范仲淹.

Fang Hai Chi Yao.
See Yü Chhang-Hui (1).

Fang Yen 方言.
Dictionary of Local Expressions.
C/Han, *c.* −15 (but much interpolated later).
Yang Hsiung 揚雄.

Fang Yen Su Chêng 方言疏證.
Correct Text of the *Dictionary of Local Expressions*, with Annotations and Amplifications.
Chhing, +1777.
Tai Chen 戴震.

Fêng Chhuang Hsiao Tu 楓䆫小牘.
Maple-Tree Window Memories.
Sung, late +12th century.
Yuan Chhiung 袁褧.
Completed by a later writer soon after +1202.

Fêng Shih Wên Chien Chi 封氏聞見記.
Things Seen and Heard by Mr Fêng.
Thang, late +8th century.
Fêng Yen 封演.
Cf. des Rotours (2), p. 104.

Fêng Su Thung I 風俗通義.
The Meaning of Popular Traditions and Customs.
H/Han, +175.
Ying Shao 應劭.
Chung-Fa Index, no. 3.

Fo Kuo Chi 佛國記.
[= *Fa-Hsien Chuan* or *Fa-Hsien Hsing Chuan*.]
Records of Buddhist Countries [also called Travels of Fa-Hsien].
Chin, +416.
Fa-Hsien (monk) 法顯.

Tr. Rémusat (1), Beal (1), Legge (4), H. A. Giles (3), Li Yung-Hsi (1).

Fu Tzu 傅子
Book of Master Fu.
Chin, +3rd century.
Fu Hsüan 傅玄.

Hai Kuo Thu Chih.
See Wei Yuan & Lin Tsê-Hsü (1).

Hai Tao I Chih Chai Lüeh 海島逸誌摘略.
Brief Selection of Lost Records of the Isles of the Sea [or, a Desultory Account of the Malayan Archipelago].
Chhing, between +1783 and +1790, preface +1791.
Wang Ta-Hai 王大海.
Tr. Anon. (37).

Han Fei Tzu 韓非子.
The Book of Master Han Fei.
Chou, early −3rd century.
Han Fei 韓非.
Tr. Liao Wên-Kuei (1).

Han Wei Tshung-Shu 漢魏叢書.
Collection of Books of the Han and Wei Dynasties [first only 38, later increased to 96].
Ming, +1592.
Ed. Thu Lung 屠隆.

Ho Kuan Tzu 鶡冠子.
Book of the Pheasant-Cap Master.
A very composite text, stabilised by +629, as is shown by one of the MSS. found at Tunhuang. Much of it must be Chou (−4th century) and most is not later than Han (+2nd century), but there are later interpolations including a +4th- or +5th-century commentary which has become part of the text and accounts for about a seventh of it (Haloun (5), p. 88). It contains also a lost 'Book of the Art of War'.
Attrib. Ho Kuan Tzu 鶡冠子.
TT/1161.

Ho Kung Chhi Chü Thu Shuo.
See Lin Chhing (2).

Hou Chou Shu 後周書.
See *Chou Shu*.

Hou Han Shu 後漢書.
History of the Later Han Dynasty [+25 to +220].
L/Sung, +450.
Fan Yeh 范曄.
The monograph chapters by Ssuma Piao 司馬彪 (d. +305).
A few chs. tr. Chavannes (6, 16); Pfizmaier (52, 53).
Yin-Tê Index, no. 41.

Hua-chhêng Chhêng I I Kuei.
See *Hwasŏng Sŏngyŏk Ŭigwe*.

Hua Man Chi 畫墁集.
Painted Walls.
Sung, *c.* +1110.
Chang Shun-Min 張舜民.

Hua Yang Kuo Chih 華陽國志.
  Record of the Country South of Mount
  Hua [historical geography of Szechuan
  down to +138].
  Chin, +347.
  Chhang Chhü 常璩.
Huai Nan Hung Lieh Chieh 淮南鴻烈解.
  See *Huai Nan Tzu*.
Huai Nan Tzu 淮南子.
  [=*Huai Nan Hung Lieh Chieh*.]
  The Book of (the Prince of) Huai-Nan [com-
  pendium of natural philosophy].
  C/Han, c. −120.
  Written by the group of scholars gathered
  by Liu An (prince of Huai-Nan) 劉安.
  Partial trs. Morgan (1); Erkes (1); Hughes
  (1); Chatley (1); Wieger (2); Wallacker (1).
  Chung-Fa Index, no. 5.
  TT/1170.
Huai-Nan (Wang) Wan Pi Shu 淮南(王)萬畢術.
  [Prob. = *Chen-Chung Hung-Pao Yuan-Pi
  Shu* and variants.]
  The Ten Thousand Infallible Arts of (the
  Prince of) Huai-Nan [Taoist alchemical
  and technical recipes].
  C/Han, −2nd century.
  No longer a separate book but fragments
  contained in *TPYL*, ch. 736 and elsewhere.
  Reconstituted texts by Yeh Tê-Hui in *Kuan
  Ku Thang So Chu Shu*, and Sun Fêng-I
  in *Wên Ching Thang Tshung-Shu*.
  Attrib. Liu An 劉安.
  See Kaltenmark (2), p. 32.
  It is probable that the terms *Chen-Chung*
  枕中 Confidential Pillow-Book; *Hung-
  Pao* 鴻寶 Infinite Treasure; *Wan-Pi*
  萬畢 Ten Thousand Infallible; and
  Yuan-Pi 苑祕 Garden of Secrets; were
  originally titles of parts of a *Huai-Nan
  Wang Shu* 淮南王書 (Writings of the
  Prince of Huai-Nan) forming the Chung
  Phien 中篇 (and perhaps also the Wai
  Shu 外書) of which the present *Huai
  Nan Tzu* book (q.v.) was the Nei Shu 內書.
Huang Chhing Ching Chieh 皇清經解.
  Collection of (more than 180) Monographs
  on Classical Subjects written during the
  Chhing Dynasty.
  See Yen Chieh (1) (ed.).
Hun I 渾儀.
  [=*Hun Thien* or *Hun I Thu Chu*.]
  On the Armillary Sphere.
  H/Han, +117.
  Chang Hêng 張衡.
  Fragment in *YHSF*, ch. 76.
Hun I Thu Chu 渾儀圖注
  The Armillary Sphere, with Illustrations and
  Commentary.
  See *Hun I*.
Hun Thien 渾天.
  The Celestial Sphere (Instrument).
  See *Hun I*.

Hun Thien Hsiang Shuo (or *Chu*) 渾天象說(注).
  Discourse on Uranographic Models.
  San Kuo, c. +260.
  Wang Fan 王蕃.
  In *CSHK* (San Kuo sect.), ch. 72, pp. 1a ff.
Hung Fan Wu Hsing Chuan 洪範五行傳.
  Discourse on the Hung Fan chapter of the
  Historical Classic in relation to the Five
  Elements.
  C/Han, c. −10.
  Liu Hsiang 劉向
Hung Hsüeh Yin-Yuan Thu Chi.
  See Lin Chhing (1).
Hung Lou Mêng 紅樓夢.
  The Dream of the Red Chamber [novel].
  Chhing, +1792.
  First part by Tshao Chan 曹霑 (d. +1763),
  second part by Kao Ê 高鶚.
  Tr. Wang Chi-Chen.
Hung Ming Chi 弘明集.
  Collected Essays on Buddhism. (Cf. *Kuang
  Hung Ming Chi*.)
  S/Chhi, c. +500.
  Sêng-Yu 僧祐.
Hung Shu 鴻書.
  Book of the Wild Geese.
  Ming.
  Liu Chung-Ta 劉仲達.
Hwasŏng Sŏngyŏk Ŭigwe 華城城役儀軌.
  Records and Machines of the Hwasŏng Con-
  struction Service [for the Emergency
  Capital at Suwŏn].
  Korea, +1792, presented +1796, pr. 1801.
  Chŏng Yag-yong 丁若鏞.
  See Chevalier (1); Henderson (1).
Hsi Chêng Chi 西征記.
  Narrative of the Western Expedition (of
  Liu Yü against the H/Chhin State of Yao
  Hsing).
  Chin & L/Sung, c. +420.
  Tai Tsu 戴祚.
Hsi Ching Tsa Chi 西京雜記.
  Miscellaneous Records of the Western
  Capital.
  Liang or Chhen, mid +6th century.
  Attrib. Liu Hsin 劉歆 (C/Han) or
  Ko Hung 葛洪 (Chin), but probably
  Wu Chün 吳均.
Hsi Hsüeh Fan 西學凡.
  A Sketch of European Science and Learning
  [written to give an idea of the contents of
  the 7000 books which Nicholas Trigault
  had brought back for the Pei-Thang
  Library].
  Ming, +1623.
  Ai Ju-Lüeh (Giulio Aleni) 艾儒略.
Hsi Hu Chih 西湖志.
  History of the West Lake Region (Hangchow).
  Chhing, +1734.
  Li Ê (ed.) 李鶚.
Hsi Yu Chi.
  See *Chhang-Chhun Chen Jen Hsi Yu Chi*.

Hsi Yü Fan Kuo Chih 西域番國志.
Records of the Strange Countries of the West.
Ming, *c.* +1417.
Chhen Chhêng 陳誠.

Hsi Yü Hsing Chhêng Chi 西域行程記.
Diary of a Diplomatic Mission to the
Western Countries (Samarqand, Herat,
etc.).
Ming, +1414.
Chhen Chhêng 陳誠 & Li Ta 李暹.

Hsi Yu Lu 西遊錄.
Record of a Journey to the West.
Yuan, +1225.
Yehlü Chhu-Tshai 耶律楚材.

Hsiao Hsüeh Kan Chu 小學紺珠.
Useful Observations on Elementary Know-
ledge.
Sung, *c.* +1270, but not pr. till +1299.
Wang Ying-Lin 王應麟.

Hsieh-Chhuan Chi 斜川集.
Collected Poems of (Su) Hsieh-Chhuan
(Su Kuo).
Sung.
Su Kuo 蘇過.

Hsien I Pien 賢奕編.
Leisurely Notes.
Ming.
Liu Yuan-Chhing 劉元卿.

Hsien-Shun Lin-An Chih 咸淳臨安志.
Hsien-Shun reign-period Topographical
Records of the Hangchow District.
Sung, +1274.
Chhien Yüeh-Yu 潛說友.

Hsin I Hsiang Fa Yao 新儀象法要.
New Design for an Astronomical Clock (lit.
Essentials of a New Device (for making
an) Armillary Sphere and a Celestial
Globe (revolve)) [including a chain of
gears for keeping time and striking the
hours, the motive power being a water-
wheel checked by an escapement].
Sung, +1094.
Su Sung 蘇頌.

Hsin Kho-Lou Ming 新刻漏銘.
Inscription for a New Clepsydra.
Liang, +507.
Lu Chhui 陸倕.

Hsin Lun 新論.
New Discussions.
H/Han, *c.* +20.
Huan Than 桓譚.
Cf. Pokora (9).

Hsin Thang Shu 新唐書.
New History of the Thang Dynasty [+618
to +906].
Sung, +1061.
Ouyang Hsiu 歐陽修 & Sung Chhi 宋祁.
Cf. des Rotours (2), p. 56.
Partial trs. des Rotours (1, 2); Pfizmaier
(66–74). For translations of passages see
the index of Frankel (1).
Yin-Tê Index, no. 16.

Hsin Wu Tai Shih 新五代史.
New History of the Five Dynasties [+907
to +959].
Sung, *c.* +1070.
Ouyang Hsiu 歐陽修.
For translations of passages see the index of
Frankel (1).

Hsü Kao Sêng Chuan 續高僧傳.
Further Biographies of Eminent (Buddhist)
Monks (cf. *Kao Sêng Chuan* and *Sung Kao
Sêng Chuan*).
Thang, +660.
Tao-Hsüan 道宣.
TW/2060.

Hsü Po Wu Chih 續博物志.
Supplement to the *Record of the Investiga-
tion of Things* (cf. *Po Wu Chih*).
Sung, mid +12th century.
Li Shih 李石.

Hsü Shih Shuo 續世說.
Continuation of the *New Discourses on the
Talk of the Times* (cf. *Shih Shuo Hsin Yü*).
Sung, *c.* +1157.
Khung Phing-Chung 孔平仲.

Hsü Thung Chien Kang Mu.
See *Thung Chien Kang Mu Hsü Pien* and
Thung Chien Kang Mu San Pien.

Hsü Wên Hsien Thung Khao 續文獻通考.
Continuation of the *Comprehensive Study of
(the History of) Civilisation* (cf. *Wên
Hsien Thung Khao* and *Chhin-Ting Hsü
Wên Hsien Thung Khao*).
Ming, finished +1586, pr. +1603.
Ed. Wang Chhi 王圻.
This covers the Liao, J/Chin, Yuan and
Ming dynasties, adding some new
material for the end of S/Sung from
+1224 onwards.

Hsüan-Chi I Shu 璇璣遺述.
Records of Ancient Arts and Techniques
(lit. of the Circumpolar Constellation
Template).
Chhing.
Chieh Hsüan 揭暄.

Hsüan Chung Chi 玄中記.
Mysterious Matters.
Date uncertain, pre-Sung, perhaps +6th-
century.
Mr Kuo 郭氏.
YHSF, ch. 76, pp. 28a ff.

Hsüan-Ho Po Ku Thu 宣和博古圖.
[= *Po Ku Thu Lu.*]
Hsüan-Ho reign-period Illustrated Record
of Ancient Objects [catalogue of the
archaeological museum of the emperor
Hui Tsung].
Sung, +1111 to +1125.
Wang Fu 王黼 or 敝, *et al.*

Hsüan Thu 玄圖.
Delineations of the Great Mystery (the
Universe) [fragment only].
H/Han, +107.

Hsüan Thu (cont.)
  Chang Hêng 張衡.
  Preserved only in quotations in *TPYL* and
    elsewhere.
Hsün Chhu Lu 詢芻錄.
  Enquiries and Suggestions (concerning
    Popular Customs and Usages).
  Ming.
  Chhen I 陳沂.
Hsün Tzu 荀子.
  The Book of Master Hsün.
  Chou, *c.* −240.
  Hsün Chhing 荀卿.
  Tr. Dubs (7).

I Chao Liao Tsa Chi 猗覺寮雜記.
  Miscellaneous Records from the I-Chao
    Cottage.
  Sung, *c.* +1200.
  Chu I 朱翌.
I Ching 易經.
  The Classic of Changes [Book of Changes].
  Chou with C/Han additions.
  Compilers unknown.
  See Li Ching-Chhih (*1, 2*); Wu Shih-
    Chhang (*1*).
  Tr. R. Wilhelm (2); Legge (9); de Harlez
    (*1*).
  Yin-Tê Index, no. (suppl.) 10.
I Lin 意林.
  Forest of Ideas [philosophical encyclo-
    paedia].
  Thang.
  Ma Tsung 馬總.
  TT/1244.
I Yü Thu Chih 異域圖志.
  Illustrated Record of Strange Countries.
  Ming, *c.* +1420 (written between +1392
    and +1430); pr. +1489.
  Compiler unknown, perhaps Chu Chhüan
    朱橚.
  Cf. Moule (4); Sarton (1), vol. 3, p. 1627.
  A copy is in the Cambridge University
    Library.
I Yuan 異苑.
  Garden of Strange Things.
  L/Sung, *c.* +460.
  Liu Ching-Shu 劉敬叔.

Jehol Jih Chi.
  See *Yŏrha Ilgi.*
Jou Hsing Lun 肉刑論.
  Discourse on Mutilative Punishments.
  H/Han, *c.* +200.
  Khung Jung 孔融.
Ju Lin Kung I 儒林公議.
  Public-Spirited Sayings of Confucian
    Scholars.
  Sung.
  Writer's name lost.
Ju Shu Chi 入蜀記.
  Journey into Szechuan.

Sung, +1170.
Lu Yu 陸游.

Kao Hou Mêng Chhiu 高厚蒙求.
  Later issued as *Kao Hou Mêng Chhiu Tsê
    Lüeh* 摘略.
  Important Information on the Universe
    [astronomy and celestial and terrestrial
    cartography].
  Chhing, *c.* +1799, repr. 1807–29, repr. 1842.
  Hsü Chhao-Chün 徐朝俊.
Kao Sêng Chuan 高僧傳.
  Biographies of Outstanding (Buddhist)
    Monks [especially those noted for learning
    and philosophical eminence].
  Liang, between +519 and +554.
  Hui-Chiao 慧皎.
  TW/2059.
Karakuri Zui 機巧圖彙.
  Illustrated Treatise on Horological (lit.
    Mechanical) Ingenuity.
  Japan, +1796.
  Hosokawa Hanzō Yorinao 細川半藏賴直.
  Reproduced in facsimile, with modernised
    transliteration of the text, in Yamaguchi
    Ryūji (*1*).
Kêng Chih Thu 耕織圖.
  Pictures of Tilling and Weaving.
  Sung, presented in MS., +1145, and perhaps
    first printed from wood blocks at that
    time; engraved on stone, +1210, and
    probably then printed from wood blocks.
  Lou Shou 樓璹.
  The illustrations published by Franke (11)
    are those of +1462 and +1739; Pelliot
    (24) published a set based on an edition of
    +1237. The original illustrations are lost,
    but cannot have differed much from these
    last, which include the poems of Lou Shou.
    The first Chhing edition was in +1696.
Kêng Chih Thu Shih 耕織圖詩.
  Poems for the *Pictures of Tilling and Weaving*.
  Sung, *c.* +1145.
  Lou Shou 樓璹.
Khang-Hsi Tzu Tien 康熙字典.
  Imperial Dictionary of the Khang-Hsi
    reign-period.
  Chhing, +1716.
  Ed. Chang Yü-Shu 張玉書.
Khao Kung Chi 考工記.
  The Artificers' Record [a section of the
    Chou Li, q.v.].
  Chou and Han, perhaps originally an official
    document of Chhi State, incorporated
    c. −140.
  Compiler unknown.
  Tr. E. Biot (1).
  Cf. Kuo Mo-Jo (*1*); Yang Lien-Shêng (7).
Khao Kung Chi Chhê Chih Thu Chieh.
  See Juan Yuan (2).
Khao Kung (Chi) Chhuang Wu Hsiao Chi.
  See Chhêng Yao-Thien (2).

Khao Kung (Chi) Hsi I 考工析疑.
An Examination of Doubtful Matters in the
Artificers' Record (of the *Chou Li*).
Chhing, +1748.
Fang Pao 方苞.

Khao Kung Chi Thu 考工記圖.
Illustrations for the *Artificers' Record* (of
the *Chou Li*) (with a critical archaeo-
logical analysis).
Chhing, +1746.
Tai Chen 戴震.
In *HCCC*, chs. 563, 564; reprinted
Shanghai 1955.
See Kondō (*1*).

Khuei Chhê Chih 暌車志.
A Cartload of Queer Phenomena [The
Khuei kua section in the *I Ching* has a
reference to a 'wagon full of ghosts', and
this is also a constellation].
Sung.
Kuo Thuan 郭彖.

Khuei Than Lu 愧郯錄.
Thinking of Confucius Asking Questions
at Than.
Sung, +1210 to +1220.
Yo Kho 岳珂.

Kikai Kanran.
See Aoji Rinsō (*1*).

Ko Chih Ching Yuan 格致鏡原.
Mirror of Scientific and Technological
Origins.
Chhing, +1735.
Chhen Yuang-Lung 陳元龍.

Ko Chih Ku Wei.
See Wang Jen-Chün (*1*).

Ko Ku Yao Lun 格古要論.
Handbook of Archaeology.
Ming, +1387, enlarged and reissued
+1459.
Tshao Chao 曹昭.

Ku Chin Chu 古今註.
Commentary on Things Old and New.
Chin, *c.* +300.
Tshui Pao 崔豹.
See des Rotours (*1*), p. xcviii.

Ku Chin Shih Wu Yuan Shih 古今事物原始.
Beginnings of Things, Old and New.
Ming.
Hsü Chü 徐炬.

Ku Kung I Lu 故宮遺錄.
Description of the Palaces (of the Yuan
Emperors).
Ming, +1368.
Hsiao Hsün 蕭洵.

Ku Wei Shu 古微書.
Old Mysterious Books [a collection of the
apocryphal Chhan-Wei treatises].
Date uncertain, in part C/Han.
Ed. Sun Chio 孫瑴 (Ming).

Ku Wên Chhi Shang 古文奇賞.
Collection of the Best Essays of Former
Times.

Ming, *c.* +1630.
Ed. Chhen Jen-Hsi 陳仁錫.

Ku Yo Fu 古樂府.
Treasury of Ancient Songs.
Nucleus pre-Sui.
First authors and compiler unknown.
Ed. and first printed in Yuan, *c.* +1345, by
Tso Kho-Ming 左克明.

Ku Yü Thu 古玉圖.
Illustrated Description of Ancient Jade
Objects.
Yuan, +1341.
Chu Tê-Jun 朱德潤.

Kuan Tzu 管子.
The Book of Master Kuan.
Chou and C/Han. Perhaps mainly compiled in the Chi-Hsia Academy (late
−4th century) in part from older
materials.
Attrib. Kuan Chung 管仲.
Partial trs. Haloun (2, 5); Than Po-Fu
et al. (*1*).

Kuang Hung Ming Chi 廣弘明集.
Further Collection of Essays on Buddhism
(cf. *Hung Ming Chi*).
Thang, *c.* +660.
Tao-Hsüan 道宣.

Kuang Ya 廣雅.
Enlargement of the *Erh Ya; Literary
Expositor* [dictionary].
San Kuo (Wei), +230.
Chang I 張揖.

Kuang Ya Su Chêng 廣雅疏證.
Correct Text of the *Enlargement of the
Erh Ya*, with Annotations and Amplifi-
cations.
Chhing, +1796.
Wang Nien-Sun 王念孫.

Kuang-Yang Tsa Chi 廣陽雜記.
Collected Miscellanea of Master Kuang-
Yang (Liu Hsien-Thing).
Chhing, *c.* +1695.
Liu Hsien-Thing 劉獻廷.

Kuang Yün 廣韻.
Enlargement of the *Chhieh Yün; Dictionary
of the Sounds of Characters.*
Sung.
(A completion by later Thang and Sung
scholars, given its present name in +1011.)
Lu Fa-Yen 陸法言 *et al.*

Kuangtung Hsin Yü 廣東新語.
New Talks about Kuangtung Province.
Chhing, *c.* +1690.
Chhü Ta-Chün 屈大均.

Kuei Chhi Chi 龜谿集.
Poems from Tortoise Valley.
Sung, *c.* +1130.
Shen Yü-Chhiu 沈與求.

Kuei Chhien Chih 歸潛志.
On Returning to a Life of Obscurity.
J/Chin, +1235.
Liu Chhi 劉祁.

Kuei Ku Tzu 鬼谷子.
　Book of the Devil Valley Master.
　Chou, −4th century (perhaps partly Han or later).
　Writer unknown; possibly Su Chhin 蘇秦 or some other member of the School of Politicians (Tsung-Hêng Chia).

Kung Pu Chhang Khu Hsü Chih 工部廠庫須知.
　What should be known (to officials) about the Factories, Workshops and Storehouses of the Ministry of Works.
　Ming, +1615.
　Ho Shih-Chin 何士晉.

Kung Shih Tiao Cho Chêng Shih Lu Pan Mu Ching Chiang Chia Ching 工師雕斲正式魯班木經匠家鏡.
　The Timberwork Manual and Artisans' Mirror of Lu Pan, Patron of all Carvers, Joiners and Wood-workers.
　Date unknown, contents traditional and certainly partly medieval.
　Repr. 1870 and many other dates.
　Apparent author Ssuchêng Wu-Jung 司正午榮.
　Editors Chang Yen 章嚴 and Chou Yen 周言.

Kuo Chhao Jou Yuan Chi.
　See Wang Chih-Chhun (1).

Kuo Yü 國語.
　Discourses on the (ancient feudal) States.
　Late Chou, Chhin and C/Han, containing early material from ancient written records.
　Writers unknown.

Kwanong Sochho Ansŏl 課農小抄案說.
　Proposals for Agricultural Improvements.
　Korea, +1799.
　Pak Chiwŏn 朴趾源.

Lai Hao Thing Shih Chi 來鶴亭詩集.
　Collected Poems from the Pavilion where the Cranes Come.
　Yuan, +14th century.
　Lü Chhêng 呂誠.

Lang Chi Hsü Than.
　See Liang Chang-Chü (2).

Lao Hsüeh An Pi Chi 老學庵筆記.
　Notes from the Hall of Learned Old Age.
　Sung, c. +1190.
　Lu Yu 陸游.

Li Chi 禮記.
　[=*Hsiao Tai Li Chi.*]
　Record of Rites [compiled by Tai the Younger].
　(Cf. *Ta Tai Li Chi.*)
　Ascr. C/Han, c. −70 to −50, but really H/Han, between +80 and +105, though the earliest pieces included may date from the time of the *Analects* (c. −465 to −450).
　Attrib. ed. Tai Shêng 戴聖.

Actual ed. Tshao Pao 曹褒.
　Trs. Legge (7); Couvreur (3); R. Wilhelm (6).
　Yin-Tê Index, no. 27.

Li Chi Chu Su 禮記注疏.
　Record of Rites, with assembled Commentaries.
　Text C/Han, commentaries of all periods.
　Ed. Juan Yuan (1816) 阮元.

Li Hai Chi 蠡海集.
　The Beetle and the Sea [title taken from the proverb that the beetle's eye view cannot encompass the wide sea—a biological book].
　Ming, late +14th century.
　Wang Khuei 王逵.

Li Sao 離騷.
　Elegy on Encountering Sorrow [ode].
　Chou (Chhu), c. −295.
　Chhü Yuan 屈原.
　Tr. Hawkes (1).

Li Shih A-Pi-Than Lun 立世阿毗曇論.
　Lokasthiti Abhidharma Śāstra; Philosophical Treatise on the Preservation of the World [astronomical].
　India, tr. into Chinese +558.
　Writer unknown.

Liang Shu 梁書.
　History of the Liang Dynasty [+502 to +556].
　Thang, +629.
　Yao Chha 姚察 and his son Yao Ssu-Lien 姚思廉.
　For translations of passages see the index of Frankel (1).

Liao Shih 遼史.
　History of the Liao (Chhi-tan) Dynasty [+916 to +1125].
　Yuan, c. +1350.
　Tho-Tho (Toktaga) 脫脫 & Ouyang Hsüan 歐陽玄.
　Partial tr. Wittfogel, Fêng Chia-Shêng et al.
　Yin-Tê Index, no. 35.

Lieh Hsien Chuan 列仙傳.
　Lives of Famous Hsien (cf. *Shen Hsien Chuan*).
　Chin, +3rd or +4th century, though certain parts date from about −35 and shortly after +167.
　Attrib. Liu Hsiang 劉向.
　Tr. Kaltenmark (2).

Lieh Nü Chuan 列女傳.
　Lives of Celebrated Women.
　Date uncertain, nucleus probably Han.
　Attrib. Liu Hsiang 劉向.

Lieh Tzu 列子.
　[=*Chhung Hsü Chen Ching.*]
　The Book of Master Lieh.
　Chou and C/Han, −5th to −1st centuries.
　Ancient fragments of miscellaneous origin finally cemented together with much new material about +380.

Lieh Tzu (*cont.*)
Attrib. Lieh Yü-Khou 列禦寇.
Tr. R. Wilhelm (4); L. Giles (4); Wieger
(7); Graham (6).
TT/663.
Ling Hsien 靈憲.
The Spiritual Constitution (or Mysterious
Organisation) of the Universe [cosmo-
logical and astronomical].
H/Han, +118.
Chang Hêng 張衡.
In *YHSF*, ch. 76.
Ling Piao Lu I 嶺表錄異.
Southern Ways of Men and Things [on the
special characteristics and natural history
of Kuangtung].
Thang, *c.* +895.
Liu Hsün 劉恂.
Ling Thai Pi Yuan 靈臺秘苑.
The Secret Garden of the Observatory
[astronomy, including a star list, and
State astrology].
N/Chou, *c.* +580.
Revised in Sung by Wang An-Li 王安禮
and much probably added in Ming; two
different books circulate under this title
and authorship.
Yü Chi-Tshai 庾季才.
Ling Wai Tai Ta 嶺外代答.
Information on What is Beyond the Passes
(lit. a book in lieu of individual replies to
questions from friends).
Sung, +1178.
Chou Chhü-Fei 周去非.
Liu Chhing Jih Cha 留青日札.
Diary on Bamboo Tablets.
Ming, +1579.
Thien I-Hêng 田藝衡.
Liu Thieh 六帖.
The Six Cards [encyclopaedia].
Thang, *c.* +800.
Pai Chü-I 白居易.
Enlarged in Sung by Khung Chuan 孔傳.
Lo Tsu Chhê Chi 贏族車記.
Record of the Class of Helical Machines.
Chhing, late +18th century.
Tai Chen 戴震.
Lo-Yang Chhieh Lan Chi 洛陽伽藍記
(or *Loyang Ka-Lan Chi*, *sêng ka-lan* trans-
literating *sanghārāma*).
Description of the Buddhist Temples and
Monasteries at Loyang.
N/Wei, *c.* +547.
Yang Hsüan-Chih 楊衒之.
Lou Shui Chuan Hun Thien I Chih 漏水轉渾
天儀制.
Method for making an Armillary Sphere
revolve by means of water from a Clep-
sydra [perhaps only a part of the *Hun I*, q.v.,
to which it is attached in *YHSF*, ch. 76].
H/Han, +117.
Chang Hêng 張衡.

Lu I Chi 錄異記.
Strange Matters.
Sung, +10th century.
Tu Kuang-Thing 杜光庭.
Lu Pan Ching 魯班經.
The Carpenter's Classic, or Manual of Lu
Pan (Kungshu Phan).
Date unknown.
Writer unknown.
Lu Pan Ching 魯班經.
See *Kung Shih Tiao Cho Chêng Shih Lu
Pan Mu Ching Chiang Chia Ching*.
Lü Shih Chhun Chhiu 呂氏春秋.
Master Lü's Spring and Autumn Annals
[compendium of natural philosophy].
Chou (Chhin), −239.
Written by the group of scholars gathered
by Lü Pu-Wei 呂不韋.
Tr. R. Wilhelm (3).
Chung-Fa Index, no. 2.
Lun Hêng 論衡.
Discourses Weighed in the Balance.
H/Han, +82 or +83.
Wang Chhung 王充.
Tr. Forke (4); cf. Leslie (3).
Chung-Fa Index, no. 1.
Lun Yü 論語.
Conversations and Discourses (of Con-
fucius) [perhaps Discussed Sayings,
Normative Sayings, or Selected Sayings];
Analects.
Chou (Lu), *c.* −465 to −450.
Compiled by disciples of Confucius (chs.
16, 17, 18 and 20 are later interpolations).
Tr. Legge (2); Lyall (2); Waley (5); Ku
Hung-Ming (1).
Yin-Tê Index no. (suppl.) 16.
Lung Wei Chhê Ko
See Chhi Yen-Huai (*1*).

Mei Jen Fu 美人賦.
Ode on Beautiful Women.
C/Han, *c.* −140.
Ssuma Hsiang-Jo 司馬相如.
Mêng Chhi Pi Than 夢溪筆談.
Dream Pool Essays.
Sung, +1086; last supplement dated +1091.
Shen Kua 沈括.
Ed. Hu Tao-Ching (*1*); cf. Holzman (1).
Mêng Hua Lu.
See *Tung Ching Mêng Hua Lu.*|
Mêng Liang Lu 夢粱錄.
The Past seems a Dream [description of
the capital, Hangchow].
Sung, +1275.
Wu Tzu-Mu 吳自牧.
Mêng Tzu 孟子.
The Book of Master Mêng (Mencius).
Chou, *c.* −290.
Mêng Kho 孟軻.
Tr. Legge (3); Lyall (1).
Yin-Tê Index no. (suppl.) 17.

Mien Hua Thu 棉花圖.
Pictures of Cotton Growing and Weaving.
Chhing, +1765.
Fang Kuan-Chhêng 方觀承.

Min Pu Su 閩部疏.
Records of Fukien.
Ming, c. +1580.
Wang Shih-Mou 王世懋.

Ming I Khao 名義考.
Studies of Names and Things.
Ming.
Chou Chhi 周祈.

Ming Shih 明史.
History of the Ming Dynasty [+1368 to
+1643].
Chhing, begun +1646, completed +1736,
first pr. +1739.
Chang Thing-Yü 張廷玉 *et al.*

Ming Wên Tsai 明文在.
Extant Literature of the Ming Dynasty.
Chhing.
Ed. Hsüeh Hsi 薛熙.

Mo Ching 墨經.
See *Mo Tzu.*

Mo Chuang Man Lu 墨莊漫錄.
Recollections from the Literary Cottage.
Sung, c. +1131.
Chang Pang-Chi 張邦基.

Mo Tzu (incl. *Mo Ching*) 墨子.
The Book of Master Mo.
Chou, −4th century.
Mo Ti (and disciples) 墨翟.
Tr. Mei Yi-Pao (1); Forke (3).
Yin-Tê Index, no. (suppl.) 21.
TT/1162.

Mu Ching 木經.
See *Kung Shih Tiao Cho Chêng Shih Lu
Pan Mu Ching Chiang Chia Ching.*

Mu Thien Tzu Chuan 穆天子傳.
Account of the Travels of the Emperor Mu.
Chou, before −245. (Found in the tomb of
An Li Wang, a prince of the Wei State,
r. −276 to −245; in +281.)
Writer unknown.
Tr. Eitel (1); Chêng Tê-Khun (2).

Munhŏn Pigo.
See *Tongguk Munhŏn Pigo.*

Nan Chhi Shu 南齊書.
History of the Southern Chhi Dynasty
[+479 to +501].
Liang, +520.
Hsiao Tzu-Hsien 蕭子顯.
For translations of passages see the index
of Frankel (1).

Nan Hua Chen Ching 南華眞經.
See *Chuang Tzu.*

Nan Shih 南史.
History of the Southern Dynasties [Nan
Pei Chhao period, +420 to +589].
Thang, c. +670.
Li Yen-Shou 李延壽.

For translations of passages see the index
of Frankel (1).

Nan Tshun Cho Kêng Lu 南村輟耕錄.
See *Cho Kêng Lu.*

Nan Yüeh Pi Chi 南越筆記.
Memoirs of the South.
Chhing, +1780.
Li Thiao-Yuan 李調元.

Nihongi 日本紀.
Chronicles of Japan [from the earliest times
to +697].
Japan, +720.
Writers unknown.
Tr. Aston (1).

Nippon Eitai-gura 日本永代藏.
The Japanese Family Storehouse; or, the
Millionaire's Gospel Modernised [dis-
courses on commercial life].
Japan, +1685.
Ihara Saikaku 井原西鶴.
Tr. G. W. Sargent (1).

Nōgu Benri-ron.
See Ōkura Nagatsune (1).

Nung Chêng Chhüan Shu 農政全書.
Complete Treatise on Agriculture.
Ming. Composed +1625 to +1628;
printed +1639.
Hsü Kuang-Chhi 徐光啓.
Ed. Chhen Tzu-Lung 陳子龍.

Nung Hsüeh Tsuan Yao.
See Fu Tsêng-Hsiang (1).

Nung Sang Chi Yao 農桑輯要.
Fundamentals of Agriculture and Seri-
culture [Imperially Commissioned].
Yuan, +1273.
Preface by Wang Phan 王磐.

Nung Sang I Shih Tsho Yao 農桑衣食撮要.
Essentials of Agriculture, Sericulture, Food
and Clothing.
Yuan +1314 (again pr. +1330).
Lu Ming-Shan (Uighur) 魯明善.

Nung Shu 農書.
Treatise on Agriculture.
Yuan, +1313.
Wang Chên 王禎.

Pai Chhuan Hsüeh Hai 百川學海.
The Hundred Rivers Sea of Learning [a
collection of separate books; the first
tshung-shu].
Sung, late +12th or early +13th century.
Compiled and edited by Tso Kuei 左圭.

Pai Hai 稗海.
The Sea of Wild Weeds [a *tshung-shu* col-
lection of 74 books].
Ming.
Compiled and edited by Shang Chün
商濬.

Pai Hu Thung Tê Lun 白虎通德論.
Comprehensive Discussions at the White
Tiger Lodge.
H/Han, c. +80.

Pai Hu Thung Tê Lun (cont.)
  Pan Ku 班固.
  Tr. Tsêng Chu-Sên (1).
Pai Khung Liu Thieh.
  See *Liu Thieh.*
Pao Phu Tzu 抱樸(朴)子.
  Book of the Preservation-of-Solidarity
    Master.
  Chin, early +4th century.
  Ko Hung 葛洪.
  Partial trs. Feifel (1, 2); Wu & Davis (2); etc.
  TT/1171-1173.
Pei Chhi Shu 北齊書.
  History of the Northern Chhi Dynasty
    [+550 to +577].
  Thang, +640.
  Li Tê-Lin 李德林, and his son Li Pai-Yao
    李百藥.
  A few chs. tr. Pfizmaier (60).
  For translations of passages see the index
    of Frankel (1).
Pei Chou Shu 北周書.
  See *Chou Shu.*
Pei Hsing Jih Lu 北行日錄.
  Diary of a Journey to the North.
  Sung, +1169.
  Lou Yo 樓鑰.
Pei Shih 北史.
  History of the Northern Dynasties [Nan
    Pei Chhao period, +386 to +581].
  Thang, c. +670.
  Li Yen-Shou 李延壽.
  For translations of passages see the index
    of Frankel (1).
Pei Thang Shu Chhao 北堂書鈔.
  Book Records of the Northern Hall
    [encyclopaedia].
  Thang, c. +630.
  Yü Shih-Nan 虞世南.
Pên Tshao Thu Ching 本草圖經.
  The Illustrated Pharmacopoeia.
  Sung, c. +1070 (presented +1062).
  Su Sung 蘇頌.
  Now contained only as quotations in the
    Thu Ching Yen I Pên Tshao (*TT*/761)
    and later pharmacopoeias.
Pên Tshao Yen I 本草衍義.
  The Meaning of the Pharmacopoeia Eluci-
    dated.
  Sung, +1116.
  Khou Tsung-Shih 寇宗奭.
  Partly contained also in the *Thu Ching Yen I
    Pên Tshao* (*TT*/761) and as quotations in
    later pharmacopoeias.
Phei Wên Yün Fu 佩文韻府.
  Encyclopaedia of Phrases and Allusions
    arranged according to Rhyme.
  Chhing, +1711.
  Ed. Chang Yü-Shu 張玉書 *et al.*
Phêng Chhuang Lei Chi 蓬窗類紀.
  Classified Records of the Weed-Grown
    Window.

Ming, +1527.
  Huang Wei 黃暐.
Phu Yuan Pieh Chuan 蒲元別傳.
  Biography of Phu Yuan (the celebrated
    iron-master and swordsmith of Shu).
  San Kuo (Shu), c. +255.
  Attrib. Chiang Wei 姜維.
  Only in fragments, *TPYL*, *CSHK*, etc.
Piao I Lu 表異錄.
  Notices of Strange Things.
  Ming.
  Wang Chih-Chien 王志堅.
Pien Yung Hsüeh Hai Chhün Yü 便用學海
    群玉.
  Seas of Knowledge and Mines of Jade;
    Encyclopaedia for Convenient Use.
  Ming, +1607.
  Compiler unknown.
  Ed. Wu Wei Tzu 武緯子.
Po Chhai Pien 泊宅編.
  Papers from the Anchored Dwelling.
  Sung, c. +1117.
  Fang Shao 方勺.
Po Ku Thu Lu 博古圖錄.
  See *Hsüan-Ho Po Ku Thu.*
Po Wu Chih 博物志.
  Record of the Investigation of Things (cf.
    Hsü Po Wu Chih).
  Chin, c. +290 (begun about +270).
  Chang Hua 張華.
Pu Thien Ko 步天歌.
  Song of the March of the Heavens [astro-
    nomical].
  Sui, end +6th century.
  Wang Hsi-Ming 王希明.
  Tr. Soothill (5).

Rigaku Hiketsu.
  See Kamata Ryūkō (1).

Samguk Yusa 三國遺事.
  Remains of the Three Kingdoms.
  Korea, c. +1280.
  Iryŏn (monk) 一然.
San Fu Huang Thu 三輔皇圖.
  Illustrated Description of the Three
    Districts in the Capital (Chhang-an,
    Sian).
  Chin, late +3rd century, or perhaps
    H/Han.
  Attrib. Miao Chhang-Yen 苗昌言.
San Kuo Chih 三國志.
  History of the Three Kingdoms [+220 to
    +280].
  Chin, c. +290.
  Chhen Shou 陳壽.
  Yin-Tê Index, no. 33.
  For translations of passages see the index of
    Frankel (1).
San Kuo Chih Yen I 三國志演義.
  The Romance of the Three Kingdoms
    [novel].

San Kuo Chih Yen I (*cont.*)
Yuan, *c.* +1370, first known edition +1494.
Lo Kuan-Chung 羅貫中.
Text revised and considerably changed
c. +1690 by Mao Tsung-Kang
毛宗崗.
Tr. Brewitt-Taylor (1).

San Nung Chi 三農紀.
Records of the Three Departments of
Agriculture.
Chhing, +1760.
Chang Tsung-Fa 張宗法.

San Tshai Thu Hui 三才圖會.
Universal Encyclopaedia.
Ming, +1609.
Wang Chhi 王圻.

Shan Chü Hsin Hua 山居新話.
Conversations in the Mountain Retreat on
Recent Events.
Yuan, +1360.
Yang Yü 楊瑀.
Tr. H. Franke (2).

Shan Hai Ching 山海經.
Classic of the Mountains and Rivers.
Chou and C/Han.
Writers unknown.
Partial tr. de Rosny (1).
Chung-Fa Index, no. 9.

Shan Thang Ssu Khao 山堂肆考.
Books seen in the Mountain Hall Library.
Ming, +1595.
Phêng Ta-I 彭大翼.

Shang Shu Ta Chuan 尚書大傳.
Great Commentary on the *Shang Shu*
chapters of the *Historical Classic*.
C/Han, −2nd century.
Fu Shêng 伏勝.

Shang Shu Wei Khao Ling Yao 尚書緯考靈曜.
Apocryphal Treatise on the *Shang Shu*
chapters of the *Historical Classic*; Investiga-
tion of the Mysterious Brightnesses.
C/Han, −1st century.
Writer unknown.
Now contained in *Ku Wei Shu*, chs. 1 and
2.

Shao-Hsing Fu Chih 紹興府志.
Local History and Topography of Shao-
hsing (Chekiang).
Chhing, +1719.
Ed. Yü Chhing-Hsiu 俞卿修.

Shih Chi 史記.
Historical Records [or perhaps better:
Memoirs of the Historiographer(-Royal);
down to −99].
C/Han, *c.* −90 [first pr. *c.* +1000].
Ssuma Chhien 司馬遷, and his father
Ssuma Than 司馬談.
Cf. Burton Watson (2). Partial trs.
Chavannes (1); Burton Watson (1);
Pfizmaier (13–36); Hirth (2); Wu Khang
(1); Swann (1), etc.
Yin-Tê Index, no. 40.

Shih Ching 詩經.
Book of Odes [ancient folksongs].
Chou, −9th to −5th centuries.
Writers and compilers unknown.
Tr. Legge (8); Waley (1); Karlgren (14)

Shih-erh Yen Chai Sui Pi.
See Wang Chün (1).

Shih Hu Tzhu 石湖詞.
Songs of the Lakeside Poet.
Sung, *c.* +1180.
Fan Chhêng-Ta 范成大.

Shih I Chi 拾遺記.
Memoirs on Neglected Matters.
Chin, *c.* +370.
Wang Chia 王嘉.
Cf. Eichhorn (5).

Shih-Lin Yen Yu 石林燕語.
Informal Conversations of (Yeh) Shih-Lin
(Yeh Mêng-Tê).
Sung, +1136.
Yeh Mêng-Tê 葉夢得.
See des Rotours (1), p. cix.

Shih Ming 釋名.
Explanation of Names [dictionary].
H/Han, *c.* +100.
Liu Hsi 劉熙.

Shih Nang Tho 釋囊橐.
Philological Study of Bags and Bellows.
Chhing, +18th century.
Huang I-Chou 黃以周.

Shih Pên 世本.
Book of Origins [imperial genealogies,
family names, and legendary inventors].
C/Han (incorporating Chou material).
Ed. Sung Chung (H/Han) 宋衷.

Shih Shuo Hsin Yü 世說新語.
New Discourses on the Talk of the Times
[notes of minor incidents from Han to
Chin]. Cf. *Hsü Shih Shuo.*
L/Sung, +5th century.
Liu I-Chhing 劉義慶.
Commentary by Liu Hsün 劉峻 (Liang).

Shih Wu Chih Yuan 事物紀原.
Records of the Origins of Affairs and
Things.
Sung, *c.* +1085.
Kao Chhêng 高承.

Shou Shih Thung Khao 授時通考.
Complete Investigation of the Works and
Days [Imperially Commissioned; a
treatise on agriculture, horticulture and all
related technologies].
Chhing, +1742.
Ed. O-Erh-Thai (Ortai) 鄂爾泰, with Chang
Thing-Yü 張廷玉, Chiang Phu 蔣溥,
et al.

Shu Chai Lao Hsüeh Tshung Than 庶齋老學
叢談.
Collected Talks of the Learned Old Man of
the Shu Studio.
Late Sung or Yuan.
Shêng Jo-Tzu 盛如梓.

Shu Chêng Chi 述征記.
Records of Military Expeditions.
Chin, +3rd or +4th century.
Kuo Yuan-Sêng 郭緣生.

Shu Chin Phu 蜀錦譜.
Monograph on the Silk Brocade Industry of Szechuan.
Yuan, +14th century.
Fei Chu 費著.

Shu Ching 書經.
Historical Classic [or, Book of Documents].
The 29 'Chin Wên' chapters mainly Chou (a few pieces possibly Shang); the 21 'Ku Wên' chapters a 'forgery' by Mei Tsê 梅賾, c. +320, using fragments of genuine antiquity. Of the former, 13 are considered to go back to the −10th century, 10 to the −8th, and 6 not before the −5th. Some scholars accept only 16 or 17 as pre-Confucian.
Writers unknown.
See Wu Shih-Chhang (1); Creel (4).
Tr. Medhurst (1); Legge (1, 10); Karlgren (12).

Shu Hsü Chih Nan 書叙指南.
The Literary South-Pointer [guide to style in letter-writing, and to technical terms].
Sung, +1126.
Jen Kuang 任廣.

Shu I Chi 述異記.
Records of Strange Things.
Liang, early +6th century.
Jen Fang 任昉.
See des Rotours (1), p. ci.

Shu Shu 蜀書.
See *San Kuo Chih*.

Shu Tao I Chhêng Chi 蜀道驛程記.
Record of the Post Stages on the Szechuan Circuit.
Chhing, +1672.
Wang Shih-Chên 王士禎.

Shu Wu I Ming Su 庶物異名疏.
Disquisition on Strange Names for Common Things.
Ming.
Chhen Mou-Jen 陳懋仁.

Shui Ching 水經.
The Waterways Classic [geographical account of rivers and canals].
Ascr. C/Han, prob. San Kuo.
Attrib. Sang Chhin 桑欽.

Shui Ching Chu 水經注.
Commentary on the *Waterways Classic* [geographical account greatly extended].
N/Wei, late +5th or early +6th century.
Li Tao-Yuan 酈道元.

Shui Pu Shih 水部式.
Ordinances of the Department of Waterways.
Thang, +737.
Writer unknown.
Tunhuang MS., P/2507, Bib. Nat. Paris.
Tr. Twitchett (2).

Shumikai Yakuhō-rekki.
See Anon. (13).

Shumisen Gimei narabini Jo Wage.
See Entsū (2).

Shumisen Zukai.
See Takai Bankan (1).

Shuo Fu 說郛.
Florilegium of (Unofficial) Literature.
Yuan, c. +1368.
Ed. Thao Tsung-I 陶宗儀.
See Ching Phei-Yuan (1).

Shuo Wên.
See *Shuo Wên Chieh Tzu.*

Shuo Wên Chieh Tzu 說文解字.
Analytical Dictionary of Characters.
H/Han, +121.
Hsü Shen 許慎.

Shuo Yuan 說苑.
Garden of Discourses.
Han, c. −20.
Liu Hsiang 劉向.

Sou Shen Chi 搜神記.
Reports on Spiritual Manifestations.
Chin, c. +348.
Kan Pao 干寶.
Partial tr. Bodde (9).

Ssu Chou Chih.
See Lin Tsê-Hsü (1).

Ssu Hsüan Fu 思玄賦.
Thought the Transcender [ode on an imaginary journey beyond the sun].
H/Han, +135.
Chang Hêng 張衡

Ssu Khu Chhüan Shu 四庫全書.
Complete Library of the Four Categories (of Literature). [Chhing Imperial MS. Collection.]
A vast MS. collection commissioned by the Chhien-Lung emperor in +1772. For ten years some 360 scholars, headed by Chi Yün 紀昀, were employed in collating the texts of the 3461 books regarded as the most noteworthy and valuable. 6793 books of lesser interest were described in the analytical catalogue but not embodied in the collection. Each of the finished sets comprised more than 36,000 *pên*. Of the 7 MS. sets 3 still exist in China, and a selection has been printed as a *tshung-shu.*
See Mayers (1); Têng & Biggerstaff (1), pp. 27 ff.

Ssu Khu Chhüan Shu Chien Ming Mu Lu 四庫全書簡明目錄.
Abridged Analytical Catalogue of the *Complete Library of the Four Categories* (of *Literature*) (made by imperial order).
Chhing, +1782.
There are two versions of this: (a) ed. Chi Yün 紀昀, which contains mention of nearly all the books in the *Thi Yao*; (b) ed. Yü Min-Chung 于敏中, which

Ssu Khu Chhüan Shu Chien Ming Mu Lu (*cont.*)
contains entries only for the books which
were copied into the imperial MS. sets.

Ssu Khu Chhüan Shu Tsung Mu Thi Yao 四庫
全書總目提要.
Analytical Catalogue of the *Complete Library
of the Four Categories* (*of Literature*) (made
by imperial order).
Chhing, +1782.
Ed. Chi Yün 紀昀.
Indexes by Yang Chia-Lo; Yü & Gillis.
Yin-Tê Index, no. 7.

Su-Chou Fu Chih 蘇州府志.
Local History and Topography of Suchow
[Chiangsu].
Chhing, +1748.
Ed. Yarshan 雅爾哈善.

Su Wei Kung Wên Chi 蘇魏公文集.
Collected Works of Su Sung.
Sung.
Su Sung 蘇頌.

Sui Shu 隋書.
History of the Sui Dynasty [+581 to +617].
Thang, +636 (annals and biographies);
+656 (monographs and bibliography).
Wei Chêng 魏徵 *et al.*
Partial trs. Pfizmaier (61–65); Balazs (7, 8);
Ware (1).
For translations of passages see the index
of Frankel (1).

Sun Tzu Suan Ching 孫子算經.
Master Sun's Mathematical Manual.
San Kuo, Chin or L/Sung.
Master Sun (full name unknown) 孫子.

Sung Hui Yao Kao 宋會要稿
Drafts for the *History of the Administrative
Statutes of the Sung Dynasty*.
Sung.
Collected by Hsü Sung (1809) 徐松.
From the *Yung-Lo Ta Tien*.

Sung Shih 宋史.
History of the Sung Dynasty [+960 to
+1279].
Yuan, *c.* +1345.
Tho-Tho (Toktaga) 脫脫 & Ouyang
Hsüan 歐陽玄.
Yin-Tê Index, no. 34.

Sung Shih Chi Shih Pên Mo 宋史紀事本末.
The Rise and Fall of the Sung Dynasty.
Ming.
Fêng Chhi 馮琦 & Chhen Pang-Chan
陳邦瞻.

Sung Shu 宋書.
History of the (Liu) Sung Dynasty [+420
to +478].
S/Chhi, +500.
Shen Yo 沈約.
A few chs. tr. Pfizmaier (58).
For translations of passages see the index
of Frankel (1).

Szechuan Yen Fa Chih.
See Lo Wên-Pin *et al.* (1).

Ta Chhing Li Chhao Shih Lu 大清歷朝實錄.
Veritable Records of the Chhing Dynasty.
Chhing, +1644 onwards.
Printed from the copy in the Ancestral
Temple of the Manchu Dynasty,
Chhang-chhun, 1937.

Ta Yuan Chan Chi Kung Wu Chi 大元氈罽
工物記.
Record of the Government Weaving Mills.
Yuan.
Writer unknown.
From the *Yuan Ching Shih Ta Tien.*

Tan Chhien Tsung Lu 丹鉛總錄.
Red Lead Record.
Ming, +1542.
Yang Shen 楊慎.

Tao Tê Ching 道德經.
Canon of the Tao and its Virtue.
Chou, before −300.
Attrib. Li Erh (Lao Tzu) 李耳(老子).
Tr. Waley (4); Chhu Ta-Kao (2); Lin Yü-
Thang (1); Wieger (7); Duyvendak (18);
and very many others.

Tao Tsang 道藏.
The Taoist Patrology [containing 1464
Taoist works].
All periods, but first collected in the Thang
about +730, then again about +870 and
definitively in +1019. First printed in the
Sung (+1111 to +1117). Also printed in
J/Chin (+1186 to +1191), Yuan
(+1244), and Ming (+1445, +1598 and
+1607).
Writers numerous.
Indexes by Wieger (6), on which see Pelliot's
review; and Ong Tu-Chien (Yin-Tê Index
no. 25).

Tetsuzan Hitsuyō Kiji 鐵山必要記事.
Record of the Essentials of Iron Technology.
Japan, +1684.
Shimohara Shigenaka 下原重仲.
Repr. *NKKZ*, vol. 10.

Thai Hsi Shui Fa 泰西水法.
Hydraulic Machinery of the West.
Ming, +1612.
Hsiung San-Pa (Sabatino de Ursis) 熊三拔
& Hsü Kuang-Chhi 徐光啓.

Thai-Phing Kuang Chi 太平廣記.
Miscellaneous Records collected in the
Thai-Phing reign-period.
Sung, +981.
Ed. Li Fang 李昉.

Thai-Phing Yü Lan 太平御覽.
Thai-Phing reign-period Imperial Encyclo-
paedia (lit. the Emperor's Daily Readings).
Sung, +983.
Ed. Li Fang 李昉.
Some chs. tr. Pfizmaier (84–106).
Yin-Tê Index, no. 23.

Thang Hui Yao 唐會要.
History of the Administrative Statutes of
the Thang Dynasty.

Thang Hui Yao (cont.)
Sung, +961.
Wang Phu 王溥.
Cf. des Rotours (2), p. 92.
Thang Liu Tien 唐六典.
Institutes of the Thang Dynasty (lit. Administrative Regulations of the Six Ministries of the Thang).
Thang, +738 or +739.
Ed. Li Lin-Fu 李林甫.
Cf. des Rotours (2), p. 99.
Thang Shu.
See *Chiu Thang Shu* and *Hsin Thang Shu.*
Thang Yü Lin 唐語林.
Miscellanea of the Thang Dynasty.
Sung, collected *c.* +1107.
Wang Tang 王讜.
Cf. des Rotours (2), p. 109.
Thao Chih Thu Shuo 陶冶圖說.
[= *Thao Yeh Thu* and *Thao Yeh Thu Shuo.*]
Illustrations of the Pottery Industry, with Explanations.
Chhing, +1743.
Thang Ying 唐英.
Tr. Julien (7), pp. 115 ff.; Bushell (4), pp. 7 ff.; Sayer (1), pp. 4 ff.
Thao Yeh Thu 陶業圖.
See *Thao Chih Thu Shuo.*
Thao Yeh Thu Shuo 陶冶圖說.
See *Thao Chih Thu Shuo.*
Thieh Wei Shan Tshung Than 鐵圍山叢談.
Collected Conversations at Iron-Fence Mountain.
Sung, *c.* +1115.
Tshai Thao 蔡絛.
Thien-Chu Ling Chhien 天竺靈籤.
Holy Lections from Indian Sources.
Sung, +1208 to +1224.
Writer unknown.
Thien Kung Khai Wu 天工開物.
The Exploitation of the Works of Nature.
Ming, +1637.
Sung Ying-Hsing 宋應星.
Thing Shih 桯史.
Lacquer Table History [notes jotted down on a lacquer table each day by the author after his day's work as an official, and then copied by a secretary before being erased and leaving the table free for further use].
Sung, +1214.
Yo Kho 岳珂.
See des Rotours (1), p. cxi.
Thu Ching Pên Tshao 圖經本草.
See *Pên Tshao Thu Ching.*
The name belonged originally to a work prepared in the Thang (*c.* +658) which by the 11th century had become lost. Su Sung's *Pên Tshao Thu Ching* was prepared as a replacement. The name *Thu Ching Pên Tshao* has often been applied to Su Sung's work, but wrongly.

Thu Ching Yen I Pên Tshao 圖經衍義本草.
The Illustrated and Elucidated Pharmacopoeia.
Largely a conflation of the *Pên Tshao Yen I* and the *Pên Tshao Thu Ching* but with many additional quotations.
Sung, *c.* +1120.
Khou Tsung-Shih 寇宗奭.
TT/761.
Thu Shu Chi Chhêng 圖書集成.
Imperial Encyclopaedia.
Chhing, +1726.
Ed. Chhen Mêng-Lei 陳夢雷 *et al.*
Index by L. Giles (2).
Thung Chien Chhien Pien 通鑑前編.
History of Ancient China (down to the point at which the *Comprehensive Mirror of History* begins).
J/Chin, *c.* +1275.
Chin Li-Hsiang 金履祥.
Thung Chien Kang Mu 通鑑綱目.
Short View of the *Comprehensive Mirror (of History, for Aid in Government)* [the *Tzu Chih Thung Chien* condensed].
Sung (begun +1172), +1189.
Chu Hsi 朱熹 (and his school).
With later continuations, *Thung Chien Kang Mu Hsü Pien* and *Thung Chien Kang Mu San Pien.*
Definitive edition, with all commentaries, etc., *c.* +1630, ed. Chhen Jen-Hsi 陳仁錫.
Partial tr. Wieger (1).
Thung Chien Kang Mu Hsü Pien 通鑑綱目續編.
Continuation of the *Short View of the Comprehensive Mirror (of History, for Aid in Government)* [covering the Sung and Yuan periods].
Ming, +1476, pr. after +1500.
Ed. Shang Lu 商輅.
Thung Chien Kang Mu San Pien 通鑑綱目三編.
Continuation of the *Short View of the Comprehensive Mirror (of History, for Aid in Government)* [covering the Ming period].
Chhing, +1746.
Ed. Shen Tê-Chhien 沈德潛 & Chhi Shao-Nan 齊召南.
Thung Chih 通志.
Historical Collections.
Sung, *c.* +1150.
Chêng Chhiao 鄭樵.
Cf. des Rotours (2), p. 85.
Thung Chih Lüeh 通志略.
Compendium of Information [part of *Thung Chih*, q.v].
Thung I Lu 通藝錄.
Records of Old Arts and Techniques.
Chhing, end +18th century.
Chhêng Yao-Thien 程瑤田.

Thung Su Wên 通俗文.
Commonly Used Synonyms.
H/Han, +180.
Fu Chhien 服虔.
YHSF, ch. 61.
Ti Ching Ching Wu Lüeh 帝京景物畧.
Descriptions of Things and Customs at the
Imperial Capital (Nanking).
Ming, c. +1638.
Liu Thung 劉侗.
Ti Wang Shih Chi 帝王世紀.
Stories of the Ancient Monarchs.
San Kuo or Chin, c. +270.
Huangfu Mi 皇甫謐.
Tiao Chi Li Than 釣磯立談.
Talks at Fisherman's Rock.
Wu Tai (S/Thang) & Sung, begun c. +935.
Shih Hsü-Pai 史虛白.
Ting Li I Min 鼎澧逸民.
Recollections of Tingchow.
Sung, c. +1150.
Writer unknown.
Cf. Chu Hsi-Tsu (2).
Tongguk Munhŏn Pigo. 東國文獻備考.
See *Chŭngbo Munhŏn Pigo*.
Tshan Shu 蠶書.
Book of Sericulture.
Sung, c. +1090.
Chhin Kuan 秦觀.
Tshan Thung Chhi 參同契
The Kinship of the Three; or, The
Accordance (of the *Book of Changes*) with
the Phenomena of Composite Things
[alchemy].
H/Han, +142.
Wei Po-Yang 魏伯陽.
Comm. by Yin Chhang-Sêng 陰長生.
Tr. Wu & Davis (1).
TT/990.
Tshan Thung Chhi Fa Hui 參同契發揮.
Elucidations of the *Kinship of the Three*
[alchemy].
Yuan, +1284.
Yü Yen 俞琰.
TT/996.
Tshan Thung Chhi Fên Chang Chu Chieh 參同
契分章註解.
The *Kinship of the Three* divided into
Chapters, with Commentary and Analysis.
Yuan, c. +1330.
Chhen Chih-Hsü 陳致虛 (Shang Yang
Tzu 上陽子).
TTCY pên 93.
Tshan Thung Chhi Khao I 參同契考異.
A Study of the *Kinship of the Three*.
Sung, +1197.
Chu Hsi 朱熹 (originally using pseudonym
Tsou Hsin 鄒訢).
TT/992.
Tso Chuan 左傳.
Master Tsochhiu's Tradition (or Enlarge-
ment) of the *Chhun Chhiu* (Spring and
Autumn Annals) [dealing with the period
−722 to −453].
Late Chou, compiled from ancient written
and oral traditions of several States
between −430 and −250, but with
additions and changes by Confucian
scholars of the Chhin and Han, especially
Liu Hsin. Greatest of the three com-
mentaries on the *Chhun Chhiu*, the others
being the *Kungyang Chuan* and the
Kuliang Chuan, but unlike them,
probably originally itself an independent
book of history.
Attrib. Tsochhiu Ming 左邱明.
See Karlgren (8); Maspero (1); Chhi Ssu-
Ho (1); Wu Khang (1); Wu Shih-
Chhang (1); van der Loon (1); Eberhard,
Müller & Henseling.
Tr. Couvreur (1); Legge (11); Pfizmaier
(1–12).
Index by Fraser & Lockhart (1).
Tso Chuan Pu Chu 左傳補注.
Commentary on *Master Tsochhiu's Enlarge-
ment of the Chhun Chhiu*.
Chhing, +1718.
Hui Tung 惠棟.
Tu Hsing Tsa Chih 獨醒雜志.
Miscellaneous Records of the Lone Watcher.
Sung, +1176.
Tsêng Min-Hsing 曾敏行.
Tu I Chih 獨異志.
Things Uniquely Strange.
Thang.
Li Jung 李冗.
Tu-Yang Tsa Pien 杜陽雜編.
The Tu-yang Miscellany.
Thang, end +9th century.
Su Ê 蘇鶚.
Tung Ching Mêng Hua Lu 東京夢華錄.
Dreams of the Glories of the Eastern
Capital (Khaifêng).
S/Sung, +1148 (referring to the two decades
which ended with the fall of the capital
of N/Sung in +1126 and the completion
of the move to Hangchow in +1135),
first pr. +1187.
Mêng Yuan-Lao 孟元老.
Tung Hsien Pi Lu 東軒筆錄.
Jottings from the Eastern Side-Hall.
Sung, end +11th century.
Wei Thai 魏泰.
Tung-Pho Chhüan Chi (or *Chhi Chi*) 東坡全集
(七集).
The Complete (or Seven) Collections of
(Su) Tung-Pho [i.e. Collected Works].
Sung, down to +1101, but put together later.
Su Tung-Pho 蘇東坡.
Tung-Pho Chih Lin 東坡志林.
Journal and Miscellany of (Su) Tung-Pho
[compiled while in exile in Hainan].
Sung, +1097 to +1101.
Su Tung-Pho 蘇東坡.

Tung Thien Chhing Lu (*Chi*) 洞天清錄 (集).
Clarifications of Strange Things
[Taoist].
Sung, *c.* +1240.
Chao Hsi-Ku 逍希鵠.

Tzu Chih Thung Chien 資治通鑑.
Comprehensive Mirror (of History) for Aid
in Government [−403 to +959].
Sung, begun +1065, completed +1084.
Ssuma Kuang 司馬光.
Cf. des Rotours (2), p. 74; Pulleyblank (7).
A few chs. tr. Fang Chih-Thung (1).

Tzu Jen I Chih 梓人遺制.
Traditions of the Joiners' Craft.
Yuan, +1263.
Hsüeh Ching-Shih 薛景石.
In *Yung-Lo Ta Tien*, ch. 18,245.
Ed. Chu Chhi-Chhien & Liu Tun-Chên.

Tzu-Ming-Chung Piao Thu Fa.
See Hsü Chhao-Chün (2).

Ungei Kisan.
See Tokugawa Nariaki (1).

Wajishi 和事始.
The Origins of Affairs in Japan.
Japan, +1696.
Kaibara Ekiken 貝原益軒.

Wakan Sanzai Zue 和漢三才圖會.
The Chinese and Japanese Universal
Encyclopaedia (based on the *San Tshai
Thu Hui*).
Japan, +1712.
Terashima Ryōan 寺島良安.

Wan Pi Shu 萬畢書.
See *Huai-Nan* (*Wang*) *Wan Pi Shu*.

Wang Yin Chin Shu 王隱晉書.
Wang Yin's History of the Chin Dynasty.
Pre-Sui.
Wang Yin 王隱.

Wei Lüeh 魏略.
Memorable Things of the Wei Kingdom
(San Kuo).
San Kuo (Wei) or Chin, +3rd or +4th
century.
Yü Huan 魚豢.

Wei Lüeh 緯署.
Compendium of Non-Classical Matters.
Sung, +12th century (end).
Kao Ssu-Sun 高似孫.

Wei Shih Chhun Chhiu 魏氏 (or 世) 春秋.
Spring and Autumn Annals of the (San
Kuo) Wei Dynasty.
Chin, *c.* +360.
Sun Shêng 孫盛.

Wei Shu 魏書.
See *San Kuo Chih*.

Wei Shu 魏書.
History of the (Northern) Wei Dynasty
[+386 to +550, including the Eastern
Wei successor State].
N/Chhi, +554, revised +572.

Wei Shou 魏收.
See Ware (3).
One ch. tr. Ware (1, 4).
For translations of passages, see the index of
Frankel (1).

Wei Wu Chhun Chhiu 魏武春秋.
= *Wei Shih Chhun Chhiu*, q.v.

Wên Fu 文賦.
Rhapsody on the Art of Letters.
Chin, +302.
Lu Chi 陸機.
Cf. Hughes (7).

Wên Hsien Pei Khao 文獻備考.
See *Chŭngbo Munhŏn Pigo*.

Wên Hsien Thung Khao 文獻通考.
Comprehensive Study of (the History of)
Civilisation (lit. Complete Study of the
Documentary Evidence of Cultural
Achievements (in Chinese Civilisation)).
Sung, begun perhaps as early as +1270
and finished before +1317, printed
+1322.
Ma Tuan-Lin 馬端臨.
Cf. des Rotours (2), p. 87.
A few chs. tr. Julien (2); St Denys (1).

Wên Hsüan 文選.
General Anthology of Prose and Verse.
Liang, +530.
Ed. Hsiao Thung (prince of the Liang) 蕭統.
Tr. von Zach (6).

Wên Shih Chuan 文士傳.
Records of the Scholars.
Chin.
Chang Yin 張隱.

Wu Ching Tsung Yao 武經總要.
Collection of the most important Military
Techniques [compiled by Imperial
Order].
Sung, +1040 (+1044).
Ed. Tsêng Kung-Liang 曾公亮.

Wu Li Hsiao Shih 物理小識.
Small Encyclopaedia of the Principles of
Things.
Chhing, +1664.
Fang I-Chih 方以智.
Cf. Hou Wai-Lu (3, 4).

Wu Lin Chiu Shih 武林舊事.
Institutions and Customs of the Old Capital
(Hangchow).
Sung, *c.* +1270 (but referring to events
from about +1165 onwards).
Chou Mi 周密.

Wu Ling Tzu 於陵子.
Book of Master Wu Ling.
Ascr. Chou, −4th or −5th century.
Chhen Thien-Tzu 陳田子.

Wu Pei Chih 武備志.
Treatise on Armament Technology.
Ming, +1628.
Mao Yuan-I 茅元儀.

Wu Shu 吳書.
See *San Kuo Chih*.

Wu Tai Shih Chi.
See *Hsin Wu Tai Shih.*
Wu Yüeh Chhun Chhiu 吳越春秋.
Spring and Autumn Annals of the States of
Wu and Yüeh.
H/Han.
Chao Yeh 趙曄.

Yeh Chung Chi 鄴中記.
Record of Affairs at the Capital of the Later
Chao Dynasty.
Chin.
Lu Hui 陸翽.
Yen Chhin Tou Shu San Shih Hsiang Shu 演
禽斗數三世相書.
Book of Physiognomical, Astrological and
Ornithomantic Divination according to
the Three Schools.
Ascr. Thang, first printed Sung, late +13th
century.
Attrib. Yuan Thien-Kang 袁天綱.
Yen Ching Shih Chi.
See Juan Yuan (*3*).
Yen Pei Tsa Chih 硯北雜志.
Miscellaneous Notes from North of Yen.
Liao.
Lu Yu-Jen 陸友仁.
Yen Phao Thu Shuo.
See Ting Kung-Chhen (*1*).
Yen Thieh Lun 鹽鐵論.
Discourses on Salt and Iron [record of the
debate of −81 on State control of com-
merce and industry].
C/Han, *c.* −80.
Huan Khuan 桓寬.
Partial tr. Gale (*1*); Gale, Boodberg & Lin.
Yen-Yen Chi.
See *Yŏn-Am Chip.*
Yijo Sillok 李朝實錄.
Veritable Records of the Yi Dynasty
[Chosŏn kingdom, Korea].
Korea, +1392 to 1910.
Officially compiled.
Ying Hsien 應閒.
Essay on the Use of Leisure in Retirement.
H/Han, +126.
Chang Hêng 張衡.
Ying Tsao Fa Shih 營造法式.
Treatise on Architectural Methods.
Sung, +1097; printed +1103; reprinted
+1145.
Li Chieh 李誡.
Yo Shu 樂書.
Book of Acoustics and Music.
N/Wei, *c.* +525, or E/Wei, *c.* +540.
Hsintu Fang 信都芳.
In *YHSF*, ch. 31, pp. 19*a* ff.
Yo Shu Chu Thu Fa 樂書註圖法.
Commentary and Illustrations for the *Book
of Acoustics and Music.*
N/Wei, *c.* +525 or E/Wei, *c.* +540.
Hsintu Fang 信都芳.

Partially preserved in *Yo Shu Yao Lu,*
ch. 6, pp. 18*a* ff.
Yo Shu Yao Lu 樂書要錄.
Record of the Essentials in the Books on
Music (and Acoustics).
Thang, *c.* +670.
Wu Huang Hou 武皇后 (the Empress
Wu, later known as Wu Tsê Thien),
probably written while emperor Kao
Tsung was still reigning.
Incomplete, and only preserved because
copied by Kibi no Makibi 吉備眞備
between +716 and +735.
Yŏn-Am Chip 燕巖集.
Collected Writings of (Pak) Yŏn-am (Pak
Chiwŏn).
Korea, +1770 onwards, ed. 1901.
Pak Chiwŏn 朴趾源.
Yŏrha Ilgi 熱河日記.
Diary of a Mission to Jê-Ho [a Korean
embassy to congratulate the Chhien-
Lung emperor on his seventieth birthday].
Korea, +1780.
Pak Chiwŏn 朴趾源.
Yü Hai 玉海.
Ocean of Jade [encyclopaedia].
Sung, +1267 (first printed Yuan, +1351).
Wang Ying-Lin 王應麟.
Cf. des Rotours (*2*), p. 96.
Yü Phien 玉篇.
Jade Page Dictionary.
Liang, +543.
Ku Yeh-Wang 顧野王.
Extended and edited in the Thang (+674)
by Sun Chhiang 孫強.
Yu-Yang Tsa Tsu 酉陽雜俎.
Miscellany of the Yu-yang Mountain (Cave)
[in S.E. Szechuan].
Thang, +863.
Tuan Chhêng-Shih 段成式.
See des Rotours (*1*), p. civ.
Yuan Chien Lei Han 淵鑑類函.
The Deep Mirror of Classified Knowledge
[literary encyclopaedia; a conflation of
Thang encyclopaedias].
Chhing, +1710.
Ed. Chang Ying 張英 *et al.*
Yuan Ching Shih Ta Tien 元經世大典.
Institutions of the Yuan Dynasty.
Yuan, +1329 to +1331.
Partly reconstructed and ed. Wên Thing-
Shih (1916) 文廷式.
Cf. Hummel (*2*), p. 855.
Yuan-Ho Chün Hsien Thu Chih 元和郡縣圖志.
Yuan-Ho reign-period General Geography.
Thang, +814.
Li Chi-Fu 李吉甫.
Cf. des Rotours (*2*), p. 102.
Yuan Hsi Chhi Chhi Thu Shuo Lu Tsui 遠西
奇器說錄最.
Collected Diagrams and Explanations of
Wonderful Machines from the Far West.

Yuan Hsi Chhi Chhi Thu Shuo Lu Tsui (cont.)
 See *Chhi Chhi Thu Shuo.*
Yuan Shih 元史.
 History of the Yuan (Mongol) Dynasty
  [+1206 to +1367].
 Ming, *c.* +1370.
 Sung Lien 宋濂 *et al.*
 Yin-Tê Index, no. 35.
Yuan Tai Hua Su Chi 元代畫塑記.
 Record of the Government Atelier of
  Painting and Sculpture.
 Yuan.
 Writer unknown.
 From the *Yuan Ching Shih Ta Tien.*
Yuan Wên Lei 元文類.
 Classified Collection of Yuan Literature.

 Yuan, *c.* +1350.
 Ed. Su Thien-Chio 蘇天爵.
Yüeh Chien Pien 粵劍編.
 A Description of Kuangtung [travel diary].
 Ming, shortly after +1601.
 Wang Lin-Hêng 王臨亨.
Yung-Lo Ta Tien 永樂大典.
 Great Encyclopaedia of the Yung-Lo reign-
  period [only in manuscript].
 Amounting to 22,877 chapters in 11,095
  pên, only about 370 being still extant.
 Ming, +1407.
 Ed. Hsieh Chin 解縉.
 See Yuan Thung-Li (*1*).

B. CHINESE AND JAPANESE BOOKS AND
JOURNAL ARTICLES SINCE +1800

A Ying (1) 阿英.
Chung-Kuo Lien Huan Thu Hua Shih Hua
中國連環圖畫史話.
A History of Chinese Book Illustration.
Ku-Tien I-Shu, Peking, 1957.

Amano Motonosuke (1) 天野元之助.
Chūgoku no Usu no Rekishi 中國のうす
の歴史.
On the History of Mortars and Mills in
China.
SB, 1952, **3**, 21.

Anon (4).
Hui-hsien Fa Chüeh Pao-Kao 輝縣發掘報告.
Report of the Excavations at Hui-hsien.
Academia Sinica Archaeological Research
Institute Pub.
Science Press, Peking, 1956.

Anon. (10).
Tunhuang Pi Hua Chi 敦煌壁畫集.
Album of Coloured Reproductions of the
fresco-paintings at the Tunhuang cave-
temples.
Peking, 1957.

Anon. (11).
Chhangsha Fa Chüeh Pao-Kao 長沙發掘
報告.
Report on the Excavations (of Tombs of
the Chhu State, of the Warring States
period, and of the Han Dynasties) at
Chhangsha.
Acad. Sinica Archaeol. Inst.
Kho-Hsüeh, Peking, 1957.

Anon (12).
Ti-Hsia Kung Tien—Ting Ling 地下宮殿—
定陵.
A Palace underground—the Ting Ling (Tomb
of the Wan-Li emperor of the Ming).
Wên-Wu, Peking, 1958.

Anon. (13).
Shumikai Yakuhō-rekki 須彌界約法層規.
Mount Sumeru Calendar Calculations.
Japan, c. 1820.

Anon. (14) (ed.).
*Chung-Kuo ti Nu-Li Chih yü Fêng-Chien
chih Fen-Chhi Wên-Thi Lun Wên Hsüan
Chi* 中國的奴隸制與封建之分期問
題論文選集.
Selected Essays on the Question of the
Periodisation of Slavery and Feudalism
in China.
Peking, 1956.

Anon. (16).
*Loyang Shih-liu Kung Chhü Tshao Wei Mu
Chhing-Li* 洛陽十六工區曹魏墓清理.
A Tomb of the (San Kuo) Wei period in the
16th District at Loyang [iron tubular
connecting joints for carriage canopies].
KKTH, 1958 (no. 7), 51.

Anon. (17).
Shou-hsien Tshai Hou Mu Chhu Thu I Wu
壽縣蔡侯墓出土遺物.
Objects Excavated from the Tomb of the
Duke of Tshai at Shou-hsien.
Acad. Sinica Archaeol. Inst.
Peking, 1956.

Anon. (18).
*Chhüan Kuo Nung Chü Chan-Lan Hui Thui-
Chien Chan Phin* 全國農具展覽會推
薦展品.
Catalogue of Recommended Designs at the
National Exhibition of Agricultural
Machinery.
Peking, 1958.

Anon (19).
*Chung-Kuo Li-Shih Po-Wu-Kuan Yü-Chan
Shuo-Ming* 中國歷史博物館預展說
明.
Guide to the Pre-View of the Exhibition at
(the opening of) the (Peking) Museum of
Chinese History (with abridged English
translation inserted).
Wên-Wu, Peking, 1959.

Anon. (20).
Loyang Chung-chou Lu 洛陽中州路.
Antiquities (of the Neolithic, Chou and
Han periods) discovered during the
Rebuilding of Chungchow Street at Loyang.
Kho-Hsüeh, Peking, 1959.

Anon (21).
Shih Kho Hsüan Chi 石刻選集.
A Selection of Stone Carvings (from all
periods).
Wên-Wu, Peking, 1957.

Anon. (22).
*Szechuan Han Hua Hsiang Chuan Hsüan
Chi* 四川漢畫像磚選集.
A Selection of Bricks with Stamped Reliefs
from Szechuan.
Wên-Wu, Peking, 1957.

Anon. (27).
Shang-Tshun-Ling Kuo Kuo Mu Ti
上村嶺虢國墓地.
The Cemetery (and Princely Tombs) of
the State of (Northern) Kuo at Shang-
tshun-ling (near Shen-hsien in the San-
mên Gorge Dam Area of the Yellow
River).
Institute of Archaeology, Academia Sinica,
Peking, 1959 (Yellow River Excavations
Report no. 3).

Anon. (29).
Chung-Kuo Li-Shih Po Wu Kuan 中國歷史博物館.
The (New) National Museum of History (at Peking).
WWTK, 1959 (no. 10), 11.

Anon. (30).
Kai Liang Thi Shui Kung Chü 改良提水工具.
Improved Techniques in Water-Raising Machinery.
Shui-Li Tien-Li, Peking, 1958.

Anon. (37).
Chung-Kuo Chien Chu 中國建築.
Chinese Architecture (and Bridge-building) [album].
Dept. of the History of Architecture, Academia Sinica, and the Architectural School of Chhing-hua University.
Wên-Wu, Peking, 1957.

Anon. (43).
Hsin Chung-Kuo-ti Khao-Ku Shou-Huo 新中國的考古收獲.
Successes of Archaeology in New China.
Wên-Wu, Peking, 1961.
Names of the 22 writers in this collective work are given on p. 135.

Anon. (44).
Shensi Chhang-an Hung-chhing-tshun Chhin Han Mu ti-erh-tzhu Fa-Chüeh Chien Chi 陝西長安洪慶村秦漢墓第二次發掘簡記.
Preliminary Report of the Second Series of Excavations of Chhin and Han Tombs at Hung-chhing-tshun near Sian in Shensi [including details of late C/Han bronze gearwheels].
KKTH, 1959 (no. 12), 662.

Anon. (45).
Nan-yang Han-Tai Thieh-Kung-Chhang Fa-Chüeh Chien Pao 南陽漢代鐵工廠發掘簡報.
Preliminary Report of the Excavation of a Han Iron-works at Nanyang [including details of iron gearwheels of the C/Han period].
WWTK, 1960 (no. 1).

Anon. (46).
Fu-chien Chhung-an-chhêng Tshun Han Chhêng I-Chih Shih Chüeh 福建崇安城村漢城遺址試掘.
Trial Excavations of the Remains of a Han City at Chhung-an City Village in Fukien [including a discovery of iron gearwheels of the late C/Han period].
KKTH, 1960 (no. 10), 1.

Anon. (48).
Chhu Wên Wu Chan-Lan Thu Lu 迖文物展寬圖錄.
Album of Antiquities from the State of Chhu (Exhibition).
Peking, 1954.

Aoji Rinsō (1) 青地林宗.
Kikai Kanran 氣海觀瀾.
A Survey of the Ocean of Pneuma [astronomy and meteorological physics].
Japan, 1825; enlarged in 1851.
In *NKKZ*, vol. 6.

Chang Hung-Chao (1) 章鴻釗.
Shih Ya 石雅.
Lapidarium Sinicum; a Study of the Rocks, Fossils and Minerals as known in Chinese Literature.
Chinese Geol. Survey, Peiping: 1st ed. 1921, 2nd ed. 1927.
MGSC (ser. B), no. 2 (with Engl. summary).
Crit. P. Demiéville, *BEFEO*, 1924, **24**, 276.

Chang Hung-Chao (3) 章鴻釗.
Chung-Kuo Yung Hsin ti Chhi-Yuan 中國用鋅的起源.
Origins and Development of Zinc Technology in China.
KHS, 1923, **8** (no. 3), reprinted in Wang Chin (2), p. 21.

Chang Ta-Chhien (1) 張大千.
Tunhuang Mo-Kao-Khu Pi Hua 敦煌莫高窟壁畫.
Copy Paintings of the Frescoes of the Mo-Kao-Khu (Chhien-fo-tung) Cave-Temples near Tunhuang [album].
Peking, 1947.

Chang Tê-Kuang (1) 張德光.
Shansi Chiang-hsien Phei-chia-pao Ku Mu Chhing-Li Chien-Pao 山西絳縣裴家堡古墓清理簡報.
Brief Report on the Old Tombs (of J/Chin Dynasty period, + 12th cent.) at Phei-chia-pao near Chiang-hsien in Shansi (province).
KKTH, 1955, **1** (no. 4), 58.

Chang Tzu-Mu (1) 張自牧.
Ying Hai Lun 瀛海論.
Discourse on the Boundless Sea (i.e. Nature). [An attempt to show that many post-Renaissance scientific discoveries had been anticipated in ancient and medieval China.]
c. 1885.

Chang Yin-Lin (1) 張蔭麟.
Ming Chhing chih Chi Hsi Hsüeh Shu ju Chung-Kuo Khao-lüeh 明清之際西學術入中國考略.
History of the Penetration of Western Science and Technology into China in the late Ming and early Chhing Periods.
CHJ, 1924, **1** (no. 1), 38.

Chang Yin-Lin (2) 張蔭麟.
Chung-Kuo Li-Shih shang chih 'Chhi Chhi' chi chhi Tso-chê 中國歷史上之「奇器」及其作者.
Scientific Inventions and Inventors in Chinese History.
YCHP, 1928, **1** (no. 3), 359.
Reprinted in (8), pp. 64ff.

Chang Yin-Lin (4) 張蔭麟.
Chi Yuan Hou Erh Shih Chi Chien Wo Kuo ti-I Wei Ta Kho-Hsüeh Chia; Chang Hêng 紀元後二世紀間我國第一位大科學家; 張衡.
Chang Hêng; Our first Great Scientist (+2nd century).
TFTC, 1925, 21 (no. 23), 89.
Reprinted in (8), pp. 323 ff.

Chang Yin-Lin (5) (tr.) 張蔭麟.
Sung Yen Su Wu Tê-Jen Chih-Nan-Chhê Tsao-Fa Khao 宋燕肅吳德仁指南車造法考.
Investigation of the Method of Construction of the South-Pointing Carriage used by the Sung engineers Yen Su and Wu Tê-Jen [tr. of Moule (7)].
CHJ, 1925, 2 (no. 1).

Chang Yin-Lin (6) 張蔭麟.
Sung Lu Tao-Lung Wu Tê-Jen Chi-Li-Ku-Chhê chih Tsao-Fa 宋盧道隆吳德仁記里鼓車造法.
The Method of Construction of the Hodometer used by the Sung engineers Lu Tao-Lung and Wu Tê-Jen.
CHJ, 1926, 2 (no. 2), 635.
Reprinted in (8), pp. 124 ff.

Chang Yin-Lin (7) 張蔭麟.
Sung Chhu Szechuan Wang Hsiao-Po Li Shun chih Luan (I Shih-pai chih Chün-Chhan Yün-Tung) 宋初四川王小波李順之亂(一失敗之均產運動)
The Revolt of Wang Hsiao-Po and Li Shun in Szechuan [+993 to +995]; an unsuccessful Communist Movement.
CHJ, 1937, 12 (no. 2), 315.
Reprinted in (8), pp. 146 ff.

Chang Yin-Lin (8) 張蔭麟.
Chang Yin-Lin Wên Chi 張蔭麟文集.
Collected Works of Chang Yin-Lin [posthumous].
Ed. Lun Wei-Liang 倫偉良.
With introductory essays by Wang Huan-Piao 王煥鑣, Chang Chhi-Yün 張其昀 & Hsieh Yu-Wei 謝幼偉.
Chung-Hua Tshung-Shu, Thaipei and Hongkong, 1956.

Chêng Chen-To (3) (ed.) 鄭振鐸.
Sung Chang Tsê-Tuan 'Chhing-Ming Shang Ho Thu' Chüan 宋張擇端「清明上河圖」卷.
[Album of] Reproductions of the Painting 'Going up the River to the Capital at the Spring Festival' finished by Chang Tsê-Tuan in +1126, with Introduction.
Wên-Wu Ching-Hua, 1959 (no. 1), and separately, Wên-Wu, Peking, 1959.

Chêng Fu-Kuang (1) 鄭復光.
Ching Ching Ling Chhih 鏡鏡詅癡.
Treatise on Optics by an Untalented Scholar.
Contains as appendix Huo Lun Chhuan Thu Shuo 火輪船圖說, On Steam Paddle-boat Machinery, with Illustrations.
Pref. 1846, pr. 1847.

Chêng Thien-Thing (1) 鄭天挺.
Kuan-yü Hsü I-Khuei 'Chih Kung Tui' 關於徐一夔「織工對」.
On Hsü I-Khuei's Conversations with the Weavers.
LSYC, 1958 (no. 1), 65.

Chiang I-Jen (1) 蔣逸人.
Kuan-yü Sung Tai i Wang Hsiao-Po Li Shun chi Chang Yü têng wei Shou ti Nung Min Chhi-I ti chi-ko Wên-Thi 關於宋代以王小波李順及張餘等為首的農民起義的幾個問題.
The Question (of the Nature of) the Peasant Uprising led by Wang Hsiao-Po, Li Shun & Chang Yü in the Sung Dynasty.
LSYC, 1958 (no. 5), 47.

Chiang Liang-Fu (1) 姜亮夫.
Tunhuang; Wei Ta-ti Wên-Hua Pao-Tsang 敦煌; 偉大的文化寶藏.
(The Chhien-fo-tung Cave-Temples near) Tunhuang—a Great National Treasure.
Ku-Tien Wên-Hsüeh, Shanghai, 1956.
Abstr. J. Gernet, RBS, 1959 2, no. 26.

Chiang Shao-Yuan (1) 江紹原.
Chung-Kuo Ku Tai Lü-Hsing chih Yen-Chiu 中國古代旅行之研究.
Le Voyage dans la Chine Ancienne.
Shanghai, 1934.
Fr. transl. by Fan Jen.
Comm. Mixte des Œuvres Franco-Chinoises, Shanghai, 1937.

Chŏn Sangun (1) 金相運.
Senki Gyokkō (Temmon Tokei) ni tsuite 璇璣玉衡(天文時計)につして.
On a Clockwork Armillary Sphere [of the Yi Dynasty].
JJHS, 1962 (2nd ser.), 1 (no. 3), 137.

Chou Chhing-Chu (1) 周清澍.
Wo Kuo Ku Tai Wei-Ta-ti Kho-Hsüeh Chia; Tsu Chhung-Chih 我國古代偉大的科學家; 祖冲之.
A Great Chinese Scientist; Tsu Chhung-Chih (mathematician, engineer, etc.).
Essay in Li Kuang-Pi & Chhien Chün-Yeh (q.v.), p. 270.
Peking, 1955.

Chu Chhi-Chhien 朱啓鈐 & Liang Chhi-Hsiung 梁啓雄 (1 to 6).
Chê Chiang Lu [parts 1 to 6] 哲匠錄.
Biographies of [Chinese] Engineers, Architects, Technologists and Master-Craftsmen.
BSRCA, 1932, 3 (no. 1), 123; 1932, 3 (no. 2), 125; 1932, 3 (no. 3), 91; 1933, 4 (no. 1), 82; 1933, 4 (no. 2), 60; 1934, 4 (nos. 3 and 4), 219.

Chu Chhi-Chhien 朱啓鈐, Liang Chhi-Hsiung 梁啓雄 & Liu Ju-Lin 劉儒林 (1).

Chu-Chhi Chhien (*cont.*)

　Chê Chiang Lu [part 7]　哲匠錄.
　Biographies of [Chinese] Engineers, Archi-
　　tects, Technologists and Master-Crafts-
　　men (continued).
　BSRCA, 1934, **5** (no. 2), 74.

Chu Chhi-Chhien　朱啓鈐 & Liu Tun-Chên
　劉敦楨 (*1, 2*).
　Chê Chiang Lu [parts 8 and 9]　哲匠錄.
　Biographies of [Chinese] Engineers, Archi-
　　tects, Technologists and Master-Crafts-
　　men (continued).
　BSRCA, 1935, **6** (no. 2), 114; 1936, **6**
　　(no. 3), 148.

Chu Chhi-Chhien　朱啓鈐 & Liu Tun-Chên
　劉敦楨 (*3*) (ed.).
　'*Tzu Jen I Chih*'　梓人遺制.
　The 'Traditions of the Joiner's Craft' (a
　　short treatise by Hsüeh Ching-Shih of
　　the Yuan Dynasty (+1263), preserved in
　　ch. 18,245 of the *Yung-Lo Ta Tien*)
　　[edited with notes].
　BSRCA, 1932, **3** (no. 4), 135.

Chü Chhing-Yuan (*1*)　鞠清遠.
　Han Tai ti Kuan-Fu Kung-Yeh　漢代的
　官府工業.
　Government Industry in the Han Period.
　SH, 1934, **1** (no. 1), 1.

Chü Chhing-Yuan (*2*)　鞠清遠.
　Yuan Tai Hsi-Kuan Chiang-Hu Yen-Chiu
　元代係官匠戶研究.
　A Study of the Government Artisans of the
　　Yuan Dynasty.
　SH, 1935, **1** (no. 9), 367.
　Tr. Sun & de Francis (1), p. 234.

Chu Hsi-Tsu (*2*)　朱希祖.
　Yang Yao Shih Chih Khao Chêng　楊么事
　迹考證.
　A Study of the Records of (the Rebellion
　　of) Yang Yao.
　Shanghai, 1935.

Chu Wên-Hsin (*1*)　朱文鑫.
　Li Fa Thung Chih　曆法通志.
　History of Chinese Calendrical Science.
　Com. Press, Shanghai, 1934.

Chhang Jen-Chieh (*1*)　常任俠.
　Han Tai Hui Hua Hsüan Chi　漢代繪畫選集.
　Selection of Reproductions of Han Drawings
　　and Paintings (including Stone Reliefs,
　　Moulded Bricks, Lacquer, etc.).
　Chao-Hua I-Shu, Peking, 1955.

[Chhang Shu-Hung] (*1*) (ed.)　常書鴻.
　Tunhuang Mo-kao-khu　敦煌莫高窟.
　(The Cave-Temples at) Mo-kao-khu
　　[Chhien-fo-tung] near Tunhuang.
　Kansu Jen-min, Lanchow, 1957.

[Chhang Shu-Hung] (*2*) (ed.)　常書鴻.
　Chung-Kuo Tunhuang I Shu Chan　中國敦
　煌藝術展.
　Catalogue of the Tokyo Exhibition of Tun-
　　huang (Chhien-fo-tung Cave-Temples) Art.
　Tokyo, 1958.

Chhang Wên-Chai (*1*)　暢文齋.
　*Shansi Yung-chi Hsien Hsüeh-chia-yai Fa-
　hsien-ti i Phi Thung-Chhi*　山西永濟縣
　薛家崖發現的一批銅器.
　On a group of Bronze Objects (of Chhin or
　　Former Han period) discovered at Hsüeh-
　　chia-yai near Yung-chi Hsien in Shansi
　　(Province) [gear-wheels, ball-bearings (?),
　　etc.].
　WWTK, 1955 (no. 8), 40.

Chhang Wên-Chai (*2*)　暢文齋.
　*Wo tui Chi-Chien Chhu Thu Chhi Wu ti
　Jen-Shih*　我對幾件出土器物的認識.
　On the Identification of Various Objects
　　excavated from Tombs [tubular con-
　　necting joints for carriage canopies,
　　etc.].
　WWTK, 1958 (no. 8), 39.

Chhen Jung (*1*)　陳嶸.
　Chung-Kuo Shu Mu Fên Lei Hsüeh　中國
　樹木分類學.
　Illustrated Manual of the Systematic Botany
　　of Chinese Trees and Shrubs.
　Agricultural Association of China Series.
　Nanking, 1937.

Chhen Kung-Jou (*1*)　陳公柔.
　*Shih-sang Li Chi-hsi Li Chung so chi-tsai-ti
　Sang-Tsang Chih-Tu*　士喪禮既夕禮中
　所記載的喪葬制度.
　On the Funeral Institutions recorded in the
　　I Li (Personal Conduct Ritual) [archaeo-
　　logical evidence that they correspond to
　　the practices of the early Warring States
　　period].
　AS/CJA, 1956, **4**, 67.

Chhen Kung-Jou (*2*) *et al.*　陳公柔.
　*Lo-yang Chien-pin Tung Chou Chhêng Chih
　Fa-Chüeh Pao-Kao*　洛陽澗濱東周城
　址發掘報告.
　Report on the Excavations of an Eastern
　　[Later] Chou City at Chien-pin near
　　Loyang.
　AS/CJA, 1959 (no. 2), 15.

Chhen Li (*1*)　陳立.
　*Wei Shen-mo Fêng Li Mei-yu tsai Hua-Pei
　Phu Pien li Yung? Po-Hai Hai-Phin
　Fêng-Chhê Thiao-Chha Pao-Kao*　爲什
　麼風力沒有在華北普遍利用? 渤海海
　濱風車調查報告.
　Why has the Windmill not been more
　　widely used in North China? A Report
　　on the Construction of the Windmills (of
　　the Salterns) used on the coast of the
　　Gulf of Chihli (near Tientsin).
　KHTP, 1951, **2** (no. 3), 266.

Chhen Shih-Chhi (*1*)　陳詩啓.
　Ming Tai ti Kung-Chiang Chih-Tu　明代
　的工匠制度.
　Regulations concerning Artisans in Ming
　　Times.
　LSYC, 1955 (no. 6), 61.
　Abstr. *RBS*, 1955, **1**, no. 155.

Chhen Ting-Hung (1) 陳定閎.
Kuan-yü 'Hung Lou Mêng' chung chih
Chung chi chhi Tha 關於「紅樓夢」中
之鐘及其他.
On the Clocks in the novel Dream of the
Red Chamber and related matters.
TFTC, 1945, **40** (no. 21), 42.

Chhêng Su-Lo (2) 程溯洛.
Chung-Kuo Shui-Chhê Li-Shih Ti Fa-Chan
中國水車歷史底發展.
An Account of the History and Develop-
ment of the Chinese Square-Pallet Chain-
Pump.
Essay in Li Kuang-Pi & Chhien Chün-Yeh
(q.v.), p. 170.
Peking, 1955.

Chhêng Su-Lo (3) 程溯洛.
Chung-Kuo Chung Chih Mien-Hua Hsiao
Shih 中國種植棉花小史.
A Brief History of the Cultivation of Cotton
in China.
Essay in Li Kuang-Pi & Chhien Chün-Yeh
(q.v.), p. 216.
Peking, 1955.

Chhêng Yao-Thien (2) 程瑤田.
Khao Kung Chhuang Wu Hsiao Chi 考工
創物小記.
Brief Notes on the Specifications (for the
Manufacture of Objects) in the Artificers'
Record (of the Chou Li).
Peking, c. 1805.
In HCCC, chs. 536–539.

Chhi Ssu-Ho (1) 齊思和.
Huang Ti chih Chih Chhi Ku Shih 黃帝之
制器故事.
Stories of the Inventions of Huang Ti [and
his Ministers].
YAHS, 1934, **2** (no. 1), 21.

Chhi Yen-Huai (1) 齊彥槐.
Lung Wei Chhê Ko 龍尾車歌.
Mnemonic Rhyming Manual on the Archimedean Screw.
c. 1820.

Chhien Lin-Chao (1) 錢臨照.
Shih 'Mo Ching' chung Kuang-hsüeh Li-hsüeh
chu Thiao 釋墨經中光學力學諸條.
Expositions of the Optics and Mechanics in
the Mohist Canons.
In Li Shih-Tsêng hsien-sêng Liu-shih Sui chi
nien Lun Wên Chi 李石曾先生六十歲
紀念論文集. Studies presented to Mr Li
Shih-Tsêng on his 60th Birthday.
Nat. Peiping Academy, Kunming, 1940.

Chhien Wei-Chhang (1) 錢偉長.
Wo Kuo Li Shih shang-ti Kho-Hsüeh Fa-
Ming 我國歷史上的科學發明.
Scientific Discoveries and Inventions in
Chinese History.
Chinese Youth Publishing Soc., Peking,
1953, 2nd edn. 1954.

Chhien Yung (1) 錢泳.
Lü Yuan Tshung Hua 履園叢話.
Collected Garden Stroll Conversations.
1870.

Chhü Fêng-Chhi (1) (ed.) 瞿鳳起.
Yü Shan Chhien Tsun-Wang Tshang Shu Mu
Lu Hui Pien 虞山錢遵王藏書目錄彙
編.
Classified Conflation of the (Three) Biblio-
graphies of the (early Chhing) Bibliophile
Chhien Tsêng.
Ku-Tien Wên-Hsüeh, Peking, 1958.

Chhüan Han-Shêng (2) 全漢昇.
Chhing Chi ti Chiang-Nan Chih-Tsao Chü
清季的江南製造局.
The Chiang-nan (Kiangnan) Arsenal at the
End of the Chhing Dynasty.
AS/BIHP (Thaiwan ed.), 1951, **23**, 145.

Entsū (1) 圓通.
Bukkoku Rekishō-hen 佛國曆象編.
On the Astronomy and Calendrical Science
of Buddha's Country [actually India and
China].
Kyōto, 1810, 1815.

Entsū (2) 圓通.
Shumisen Gimei narabini Jo Wage 須彌山
儀銘并序和解.
Inscription for the Mt Sumeru Instrument
[cosmographical model] and Japanese
Translation of the Preface.
Japan, 1813.

Fan Hsiang-Yung (1) 范祥雍.
Ku-Pên 'Chu Shu Chi Nien' Chi-Chiao
Ting-Pu 古本竹書紀年輯校訂補.
Revised Edition of the Genuine Fragments
of the Bamboo Books [annals], (correcting
and supplementing the earlier recon-
structions of Chu Yu-Tsêng and Wang
Kuo-Wei).
Hsin Chih Shih, Shanghai, 1956.
Abstr. RBS, 1959, **2**, no. 60.

Fan Hsing-Chun (1) 范行準.
Chung-Kuo Yü Fang I-Hsüeh Ssu-Hsiang
Shih 中國預防醫學思想史.
History of the Conceptions of Hygiene and
Preventive Medicine in China.
Jen-Min Wei-Sêng, Peking, 1953, 1954.

Fang Hao (1) 方豪.
Chêng Chhi Chi yü Huo Chhê, Lun Chhuan,
Fa-Ming yü Chung-Kuo 蒸汽機與火車
輪船發明於中國.
The Invention of the Steam Engine, the
Locomotive and the Paddle-Boat in China.
TFTC, 1943, **39** (no. 3), 45.

Fang Hao (2) 方豪.
'Hung Lou Mêng' Hsin Khao 紅樓夢新考.
A New Study of the novel Dream of the
Red Chamber.
SWYK, 1943, **4**, 921.

Fêng Yün-Phêng 馮雲鵬 & Fêng Yün-Yuan
馮雲鵷.
Chin Shih So 金石索.

Feng Yün-Phêng (cont.)
  Collection of Carvings, Reliefs and Inscriptions.
  (This was the first modern publication of the Han tomb-shrine reliefs.)
  1821.
Fu Chien (1) 傅健.
  Shensi Mei-hsien Chhü Yen chih Thiao-Chha 陝西郿縣渠堰之調查.
  An Enquiry into the Canals and Dams of Mei-hsien (south of the Wei River) in Shensi Province.
  SL, 1934, 7, 239.
Fu Tsêng-Hsiang (1) 傅增湘.
  Nung Hsüeh Tsuan Yao 農學纂要.
  Essentials of Agricultural Science.
  1901.
Furushima Toshio (1) 古島敏雄.
  Nihon Nōgyō Gijutsu-shi 日本農業技術史.
  History of Agricultural Technology in Japan.
  2 vols., Tokyo, 1959.

Hamada Kōsaku 濱田耕作 & Umehara Sueji 梅原末治 (1).
  Keishū Kinkan-Tsuka to Sono Ihō 慶州金冠塚と其遺寶.
  A Royal Sepulture, 'Kinkan-Tsuka' or 'Gold-Crown Tomb', at Kyŏngju, Korea, and its Treasures.
  Sp. Rep. Serv. Antiq. Govt.-Gen., Chōsen, 1924, no. 3, 2 vols. text, 1 vol. plates.
Harada Yoshito 原田淑人 & Komai Kazuchika 駒井和愛 (1).
  Shina Koki Zukō 支那古器圖考.
  Chinese Antiquities (Pt. 1, Arms and Armour; Pt. 2 Vessels [ships] and Vehicles).
  Academy of Oriental Culture, Tokyo, 1937.
Hashimoto Masukichi (2) 橋本增吉.
  Shina Kodai Rekihōshi Kenkyū 支那古代曆法史研究.
  Über die astronomische Zeiteinteilung im alten China.
  Tokyo, 1943.
  (Tōyō Bunko Ronsō, no. 29 東洋文庫論叢.)
Hashimoto Masukichi (3) 橋本增吉.
  Shinan Sha Kō 指南車考.
  An Investigation of the South-Pointing Carriage.
  TYG, 1918, 8, 249, 325; 1924, 14, 412; 1925, 15, 219.
Hayashi Minao (1) 林巳奈夫.
  Chūgoku Senshin-jidai no Basha 中國先秦時代の馬車.
  Chariots and Horses in the Shang and Chou Periods.
  TG/K, 1959, 29, 155.
Hayashi Minao (2) 林巳奈夫.
  Shūrei-Kōkōki no Shasei 周禮考工記の車制.

Chariot Construction in the Artificers' Record of the Chou Li.
  TG/K, 1959, 30, 275.
Ho Chi-Mei 何寄梅 & Jui Kuang-Thing 芮光庭.
  Kung-Chü ti Ku-Shih 工具的故事.
  The Story of Tools and Machines.
  Pei-ching Shu-Tien, Peking and Shanghai, 1954.
Hou Wai-Lu (3) 侯外廬.
  Fang I-Chih—Chung-Kuo ti Pai Kho Chhüan Shu Phai Ta Chê-Hsüeh Chia 方以智—中國的百科全書派大哲學家.
  Fang I-Chih—China's Great Encyclopaedist Philosopher.
  LSYC, 1957 (no. 6), 1; 1957 (no. 7), 1.
Hu Tao-Ching (1) 胡道靜.
  'Mêng Chhi Pi Than' Chiao Chêng 夢溪筆談校證.
  Complete Annotated and Collated Edition of the Dream Pool Essays (of Shen Kua, +1086).
  2 vols.
  Shanghai Pub. Co., Shanghai, 1956.
  Crit. rev. Nguyen Tran-Huan, RHS, 1957, 10, 182.
Huang Chieh (2) 黃節.
  Wang Chêng Chuan 王徵傳.
  Life of Wang Chêng.
  KSHP, 3, no. 6, p. 7a.
Hsieh Chih-Liu (1) 謝稚柳.
  Tunhuang I-Shu Hsü Lu 敦煌藝術敘錄.
  Catalogue of the Subjects of the Frescoes of (the Chhien-fo-tung Cave-Temples near) Tunhuang.
  Shanghai, 1955.
Hsü Chhao-Chün (1).
  See Kao Hou Mêng Chhiu.
Hsü Chhao-Chün (2) 徐朝俊.
  Tzu-Ming-Chung Piao Thu Fa 自鳴鐘表圖法 or Shuo 說.
  Illustrated Account of the Manufacture of Mechanical Clocks.
  Shanghai, 1809.
  A MS. copy of this date is in the Peking National Library.
Hsü Chia-Chen (1) 徐家珍.
  Fêng Chêng Hsiao Chi 風箏小記.
  A Note on Aeolian Whistles (attached to Kites).
  WWTK, 1959 (no. 2), 27.
Hsü Wên-Lin 徐文鄰 & Li Wên-Kuang (1) 李文光.
  Than Chhing Tai ti Chung Piao Chih Tsao 談清代的鐘表制造.
  On the Design of Some Chinese Clocks of the Chhing period.
  WWTK, 1959 (no. 2), 34.

Jeon Sangwoon (1).
  See Chŏn Sangun (1).

Juan Yuan (*1*).
    See *Chhou Jen Chuan*.
Juan Yuan (*2*). 阮元.
    Khao Kung Chi Chhê Chih Thu Chieh 考
        工記車制圖解.
    Illustrated Analysis of Vehicle Construction
        in the *Artificers' Record* (of the *Chou Li*).
    Peking, *c.* 1820.
    In *HCCC*, chs. 1055, 1056.
Juan Yuan (*3*). 阮元.
    Yen Ching Shih Chi 揅經室集.
    Collected Writings from the Yen-Ching
        Studio.
    1st part 1823; 2nd part 1830; 3rd part 1844.
Jung Kêng (*1*). 容庚.
    Han Wu Liang Tzhu Hua Hsiang Khao Shih
        漢武梁祠畫像考釋.
    Investigations on the Carved Reliefs of the
        Wu Liang Tomb-shrines of the [Later]
        Han Dynasty.
    2 vols.
    Yenching Univ. Archaeol. Soc.
    Peking, 1936.
Jung Kêng (*3*). 容庚.
    Chin Wên Pien 金文編.
    Bronze Forms of Characters.
    Peking, 1925, repr. 1959.

Kamata Ryūkō (*1*). 鎌田柳泓.
    Rigaku Hiketsu 理學秘訣.
    Elements of Physics.
    Japan, 1815.
    In *NKKZ*, vol. 6.
Kao Chih-Hsi 高至喜, Liu Lien-Yin 劉廉銀
        et al. (*1*).
    *Chhangsha Shih Tung-pei Chiao Ku Mu
        Tsang Fa-Chüeh Chien-Pao* 長沙市東
        北郊古墓葬發掘簡報.
    Short Report on the Excavations of Tombs
        (of Warring States and Later Periods) in
        the North-eastern Suburbs of Chhangsha.
    KKTH, 1959 (no. 12), 649.
Katō Shigeshi (*1*). 加藤繁.
    Shina Keizai-shi Kōshō 支那經濟史考證.
    Studies in Chinese Economic History.
    2 vols.
    Tōyō Bunko, Tokyo, 1952, 1953.
    (Tōyō Bunko Pubs. Ser. A, nos. 33 and 34.)
Khang Yu-Wei (*1*). 康有爲.
    Ta Thung Shu 大同書.
    Book of the Great Togetherness [an
        Utopia].
    Conceived and first drafted +1884 and 1885,
        first printed (in part) Shanghai, 1913;
        San Francisco, 1929; first complete
        publication Chung-hua, Shanghai, 1935;
        repr. Peking, 1956.
    Abridged tr. Thompson (*1*).
Koizumi Akio (*1*). 小泉顯夫.
    Rakurō Saikyōzuka 樂浪彩篋冢.
    The Tomb of the Painted Basket, [and two
        other tombs] of Lo-lang.

Koseki Chōsa Hōkoku (Archaeol. Res.
    Rep.), no. 1.
    Soc. Stud. Korean Antiq., Seoul, 1934.
    (With Engl. summary.)
Kondō Mitsuo (*1*). 近藤光男.
    Tai Shin no Kōkōkizu ni tsuite 戴震の考
        工記圖について.
    On Tai Chen and his *Khao Kung Chi Thu*.
    TG/T, 1955, **11**, 1; abstr. *RBS*, 1955, **1**,
        no. 452.
Kuo Mo-Jo (*1*). 郭沫若.
    Shih Phi Phan Shu 十批判書.
    Ten Critical Essays.
    Chün-i, Chungking, 1945.
Kuo Mo-Jo (*2*). 郭沫若.
    Ku Tai Shih-Hui chih Yen-Chin 古代社
        會之研究.
    Studies in Ancient Chinese Society.
    Shanghai, *c.* 1927.
Kuo Mo-Jo (*6*). 郭沫若.
    Nu Li Chih Shih-Tai 奴隸制時代.
    On the Period of Slave-Owning Society
        (in Ancient China).
    Jen-Min, Peking, 1954.
Kuwabara Jitsuzō (*1*). 桑原騭藏.
    *Sōmatsu no Teikyo-Shihaku Saiiki-jin Ho
        Ju-Kō no Jiseki* 宋末の提舉市舶西域
        人蒲壽庚の事蹟.
    On Phu Shou-Kêng, a man of the Western
        Regions, who was Superintendent of
        Merchant Shipping at Chhüanchow to-
        wards the end of the Sung Dynasty.
    Shanghai, 1923; Iwanami-Shoten, Tokyo,
        1935.
    Rev. P. Pelliot, *TP*, 1927, **25**, 205.
    Chinese tr., *Phu Shou-Kêng Khao*, by Chhen
        Yü-Ching (Chung-Hua, Peking, 1954).

Lai Chia-Tu (*1*). 賴家度.
    '*Thien Kung Khai Wu*' *chi chhi Chu chê*;
        Sung Ying-Hsing 天工開物及其著者;
        宋應星.
    The *Exploitation of the Works of Nature* and
        its Author; Sung Ying-Hsing.
    Essay in Li Kuang-Pi & Chhien Chün-Yeh
        (q.v.), p. 338.
    Peking, 1955.
Lai Chia-Tu (*2*). 賴家度.
    Chang Hêng 張衡.
    Chang Hêng; a Biography.
    Shanghai People's Pub. Co., Shanghai, 1956.
Lao Kan (*3*). 勞榦.
    Pei Wei Loyang Chhêng Thu ti Fu-Yuan
        北魏洛陽城圖的復原.
    Reconstruction of the Plan of [the Capital]
        Loyang, as it was during the Northern
        Wei period [according to the *Lo-yang
        Chhieh Lan Chi*].
    AS/BIHP, 1948, **20**, 299.
Lao Kan (*4*). 勞榦.
    Liang Han Hu-Chi yü Ti-Li chih Kuan-Hsi
        兩漢戶籍與地理之關係.

Lao Kan (4) (cont.)
Population and Geography in the Two Han Dynasties.
AS/BIHP, 1935, 5 (no. 2), 179.
Tr. Sun & de Francis (1), p. 83.

Lao Kan (5) 勞榦.
Chü-Yen Han Chien Khao Shih 居延漢簡考釋.
A Study of the Han Bamboo Tablets from Chü-yen (Edsin Gol).
Peking, 1949.

Li Chhung-Chou (1) 李崇州.
Ku-Tai Kho-Hsüeh Fa-Ming Shui-Li Yeh Thieh Ku Fêng Chi 'Shui Phai' chi chhi Fu-Yuan 古代科學發明水力冶鐵鼓風機「水排」及其復原.
Reconstruction of the 'Water-powered Reciprocator' or Hydraulic Blowing Engine for Iron-works, an ancient discovery in applied science.
WWTK, 1959 (no. 5), 45.

Li Chhung-Chou (2) 李崇州.
Shih-Chieh shang Tsui-Tsao-ti Shui Fang Chi Chhê—'Shui Chuan Ta Fang Chhê' 世界上最早的水力紡績車;「水轉大紡車」.
The World's Oldest Water-Powered Spinning Machinery—the 'Great Water-Driven Spinning Machine' (described by Wang Chên in the Nung Shu of +1313).
WWTK, 1959 (no. 12), 29.

Li Ju-Chen (1) 李汝珍.
Ching Hua Yuan 鏡花緣.
Flowers in a Mirror [novel].
1828, repr. 1832 and often later; punctuated edition 1923 with introduction by Hu Shih.
See Hummel (2), p. 473.

Li Kuang-Pi 李光璧 & Chhien Chün-Yeh 錢君曄 (ed.) (1).
Chung-Kuo Kho-Hsüeh Chi-Shu Fa-Ming ho Kho-Hsüeh Chi-Shu Jen Wu Lun Chi 中國科學技術發明和科學技術人物論集.
Essays on Chinese Discoveries and Inventions in Science and Technology, and on the Men who made them.
San-Lien Shu-Tien, Peking, 1955.

Li Kuang-Pi 李光璧 & Lai Chia-Tu 賴家度 (1).
Han Tai ti Wei Ta Kho-Hsüeh Chia; Chang Hêng 漢代的偉大科學家;張衡.
A Great Scientist of the Han Dynasty; Chang Hêng (astronomer, mathematician, seismologist, etc.).
Essay in Li Kuang-Pi & Chhien Chün-Yeh (q.v.), p. 249.
Peking, 1955.

Li Nien (27) (ed.) 李儼.
Chung-Kuo Ku-Tai Kho-Hsüeh Chia 中國古代科學家.

(Twenty-nine) Chinese Scientists of Former Times [biographical essays by twenty-four scholars].
Kho-Hsüeh, Peking, 1959.

Li Wên-Hsin (1) 李文信.
Jang Khao-Ku Kho-Hsüeh tsai Tsu Kuo Shih-Hui Chu-I Chien-Shê Kao Chhao Chung Chuang Ta 讓考古科學在祖國社會主義建設高潮中壯大.
The Relation of Archaeology to the Strengthening of the Socialist Movement in China.
WWTK, 1956 (no. 3).

Liang Chang-Chü (1) 梁章鉅.
Lang Chi Tshung Than 浪跡叢談.
Impressions Collected during Official Travels.
c. 1843.

Liang Chang-Chü (2) 梁章鉅.
Lang Chi Hsü Than 浪跡續談.
Further Impressions Collected during Official Travels.
c. 1846.

Liang Ssu-Chhêng (1) 梁思成.
Chêng-Ting Tiao Chha Chi Lüeh 正定調查紀略.
The Ancient Architecture at Chêngting (Hopei); [on the revolving library at the Lung-hsing Temple at Chêngting.]
BSRCA, 1933, 4 (no. 2), 1.

Lin Chhing (1) 麟慶.
Hung Hsüeh Yin-Yuan Thu Chi 鴻雪因緣圖記.
Illustrated Record of Memories of the Events which had to happen in My Life.
1849.
Cf. Hummel (2), p. 507.

Lin Chhing (2) 麟慶.
Ho Kung Chhi Chü Thu Shuo 河工器具圖說.
Illustrations and Explanations of the Techniques of Water Conservancy and Civil Engineering.
1836.
Cf. Hummel (2), p. 507.

Lin Tsê-Hsü (1) 林則徐.
Ssu Chou Chih 四洲志.
Information on the Four Continents [especially on the West].
c. 1840.
One of the chief sources of Wei Yuan & Lin Tsê-Hsü (q.v.).

Liu Chih-Yuan (1) (ed.) 劉志遠.
Szechuan Shêng Po-Wu-Kuan Yen-Chiu Thu Lu 四川省博物館研究圖錄.
Illustrated Studies on the (Reliefs, Bricks, and other Objects in the) Szechuan Provincial Museums (Chungking and Chhêngtu).
Ku-Tien I-Shu, Peking, 1958.

Liu Hai-Su (1) 劉海粟.
Ming Hua Ta Kuan 名畫大觀.

Liu Hai-Su (1) (cont.)
Album of Celebrated Paintings.
4 vols.
Chung-Hua, Shanghai, 1935.

Liu Hsien-Chou (1) 劉仙洲.
Chung-Kuo Chi-Chieh Kung-Chhêng Shih-
Liao 中國機械工程史料.
Materials for the History of Engineering in
China.
CHESJ, 1935, 3, and 4 (no. 2), 27. Re-
printed Chhing-hua Univ. Press, Peiping,
1935.
With supplement, CHER, 1948, 3, 135.

Liu Hsien-Chou (2) 劉仙洲.
Wang Chêng yü Wo Kuo ti I Pu Chi-Chieh
Kung-Chhêng-Hsüeh 王徵與我國第一
部機械工程學.
Wang Chêng and the First Book on
[Modern] Mechanical Engineering in
China.
CHER, 1943, 1; CLTC, 1943, 1 (no. 2), 215.

Liu Hsien-Chou (3) 劉仙洲.
Chung-Kuo tsai Jê Chi Li-Shih shang ti Ti-
Wei 中國在熱機歷史上的地位.
The Position of China in the History of
Heat Engines.
TFTC, 1943, 39 (no. 18), 35.

Liu Hsien-Chou (4) 劉仙洲
Chung-Kuo tsai Yuan Tung Li fang-mien ti
Fa-Ming 中國在原動力方面的發明.
Chinese Inventions in Power-Source
Engineering.
CHER, 1953 (n.s.), 1 (no. 1), 3.

Liu Hsien-Chou (5) 劉仙洲.
Chung-Kuo tsai Chhuan Tung Chi chien
fang-mien ti Fa-Ming 中國在傳動機件
方面的發明.
Chinese Inventions in Power Transmission.
CHER, 1954 (n.s.), 2 (no. 1), 1; with
addendum 1954 (n.s.), 2 (no. 2), 219.

Liu Hsien-Chou (6) 劉仙洲.
Chung-Kuo tsai Chi Shih Chhi fang-mien ti
Fa Ming 中國在計時器方面的發明.
Chinese Inventions in Horological Engi-
neering.
CHER, 1956 (n.s.), 4, 1; TWHP, 1956, 4
(no. 2), 219.
Eng. tr. (1) in Actes du VIIIe Congrès
Internat. d'Hist. des Sci., Florence, 1956,
vol. 1, p. 329.

Liu Hsien-Chou (7) 劉仙洲.
Chung-Kuo Chi-Chieh Kung-Chhêng Fa-
Ming Shih 中國機械工程發明史.
A History of Chinese Engineering Inven-
tions.
Kho-Hsüeh, Peking, 1962.
Rev. Wang Hsü-Yün, KMJP, 20 June 1962.

Liu Wên-Tien (1) 劉文典.
Chuang Tzu Pu Chêng 莊子補正.
Emended Text of The Book of Master
Chuang.
Com. Press., Shanghai, 1947.

Liu Ying (1) 劉穎.
Chih Nan Chhê Hsin Shih 指南車新釋.
New Suggestions about the South-Pointing
Carriage [tr. and rev. of Lanchester, 1].
WHJP (Scientific and Industrial Supple-
ment), 1947, 19 and 26 Feb.

Liu Yung-Chhêng (1) 劉永成.
Chieh Shih Chi-Ko Yu Kuan Hang Hui ti
Pei-Wên 解釋幾個有關行會碑文.
Notes concerning Handicraft Guilds on
Stone Inscriptions found in Suchow and
its Vicinity [Chhing period].
LSYC, 1958 (no. 9), 63.

Liu Yung-Chhêng (2) 劉永成.
Tui Su-Chou 'Chih Tsao Ching Chih Chi'
Pei-Wên ti Khan-Fa 對蘇州織造經制
記碑文的看法.
A Stone Inscription on Management in the
Silk-Weaving Industry in Suchow
[+1647].
LSYC, 1958 (no. 4), 87.

Liu Yung-Chhêng (3) 劉永成.
Chhien-Lung Su-Chou Yuan Chhang Wu San
Hsien 'I-Ting Chih-Fang Thiao-I Chang
Chhêng' Pei 乾隆蘇州元長吳三縣議
定紙坊條議章程碑.
Notes on the Regulations for Paper-Mills
agreed by the Three Districts of Yuan
(-ho), Chhang(-chou) and Wu during
the Chhien-Lung reign-period [+1736
to +1795].
LSYC, 1958 (no. 2), 85.

Lo Chen-Yü (3) 羅振玉.
Hsüeh Thang so Tshang Ku Chhi Wu Thu
Shuo 雪堂所藏古器物圖說.
Illustrated Description of Ancient Objects
Preserved in the Snow Studio. c. 1910.

Lo Jung-Pang (1) 羅榮邦.
Chung-Kuo chih Chhê Lun Chhuan 中國
之車輪船.
(The History of) the Paddle-Wheel Boat in
China.
CHJ/T, 1960 (n.s.), 2 (no. 1), 213.
Eng. tr. Lo Jung-Pang (3).

Lo Wên-Pin et al. (1) 羅文彬.
Szechuan Yen Fa Chih 四川鹽法志.
Memorials of the Salt Industry of Szechuan.
Compiled officially at the request of the
Governor-General of the province, Ting
Pao-Chen, 1882.

Ma Kuo-Han (1) (ed.) 馬國翰.
Yü Han Shan Fang Chi I Shu 玉函山房
輯佚書.
The Jade-Box Mountain Studio Collection
of (Reconstituted) Lost Books.
1853.

Ma Tê-Chih 馬得志, Chou Yung-Chen 周
永珍 & Chang Yün-Phêng 張雲鵬 (1).
I-chiu-wu-shih-san Nien Anyang Ta-Ssu-
Khung Tshun Fa-Chüeh Pao-Kao 一九
五三年安陽大司空村發掘報告.

Ma Tê-Chih *et al.* (*cont.*)
  Report on the Excavations at Ta-Ssu-
    Khung Tshun (Village) near Anyang,
    1953.
  AS/CJA, 1955, **9**, 25.
Masai Tsuneo (*1*) 增井經夫.
  Shina no Suisha 支那の水車.
  On Water-Mills in China (including the
    Hydraulic Blowing Engine).
  In *Tōyōshi Shusetsu* 東洋史集説.
  Dr Kato Presentation Volume, *Katō Hakase
    Kanreki Kinen* 加藤博士還暦記念.
  Tokyo, 1938.
Mêng Hao 孟浩, Chhen Hui 陳慧 & Liu
    Lai-Chhêng 劉來城 (*1*).
  *Ho-pei Wu-an Wu-chi Ku Chhêng Fa-
    Chüeh Chi* 河北武安午汲古城發掘記.
  An Account of Excavations at the old City
    of Wu-chi at Wu-an in Hopei [including
    details of iron gearwheels of H/Han date].
  KKTH, 1957 (no. 4), 43.
Mizuno Seiichi (*1*) 水野清一.
  Tonkō Sekkutsu nōto 敦煌石窟ノト.
  Notes on the Tunhuang Cave-Temples
    (Chhien-fo-tung).
  ABA, 1958, **34**, 8.
Muramatsu Teijirō (*1*) 村松貞次郎.
  Saikin no 'Tatara' Jijō 最近のタタラ
    事情.
  The 'Tatara' Method of Iron-Smelting in
    Recent Times.
  JJHS, 1953 (no. 26), 30.

Nagahiro Toshio (*1*) 長廣敏雄.
  Hiten no Geijutsu 飛天の藝術.
  A Study of Hiten [Fei Thien] or Flying
    Angels.
  Asahi Shimbunsha, Tokyo, 1949.
Nagasawa Kikuya (*1*) 長澤規矩也.
  Shina Gakujutsu Bungeishi 支那學術文藝
    史.
  History of Chinese Scholarship and
    Literature.
  Tokyo, 1938.
  Tr. Feifel, see Nagasawa (1).
Niida Noboru (*1*) 仁井田陞.
  *Gen-Min-jidai no Mura no Kiyaku to
    Kosaku Shōsho Nadō (Nichiyō Hyakka-
    zensho no Rui nijūshu no uchi Kara)*
    元明時代の村の規約と小作證書な
    ど(日用百科全書の類二十種の中か
    ら).
  Village Rules and Tenant Bills of the Yuan
    and Ming Periods as seen in Twenty
    Ordinary (Popular) Encyclopaedias of the
    Time.
  TBK, 1956, **8**, 123.
Ninagawa Noritane (*1*) 自壹至五.
  Kanko Zusetsu 觀古圖説.
  Account of Traditional Japanese Arts and
    Industries.
  Tokyo, 1883.

Ōkura Nagatsune (*1*) 大藏永常.
  Nōgu Benri-ron 農具便利論.
  Advantageous Specifications for Agri-
    cultural Tools and Machinery.
  Japan, 1822.
  In *NKKZ*, vol. 11.
Ong Kuang-Phing (*1*) 翁廣平.
  Chhu Chiu Ching 杵臼經.
  Manual of Pounding Techniques.
  c. 1830.
Phan Chieh-Tzu (*1*) 潘絜兹.
  Tunhuang Mo-kao-khu I-Shu 敦煌莫高
    窟藝術.
  The Art of the Cave-Temples at Mo-kao-
    khu [Chhien-fo-tung] near Tunhuang.
  Jen-min, Shanghai, 1957.
Phêng Tsê-I (*1*) (ed.) 彭澤益.
  *Chung-Kuo Chin-Tai Shou-Kung Yeh Shih
    Tzu-Liao 1840–1949* 中國近代手工
    業史資料 1840–1949.
  Materials for the Study of the Chinese
    Artisanate and the Growth of Industrial
    Production between 1840 and 1949.
  4 vols.
  San-Lien, Peking, 1957.
  Abstr. *RBS*, 1962, **3**, no. 343.

Sakai Tadao (*1*) 酒井忠夫.
  Mindai no Nichiyō Ruisho to Shōmin Kyōiku
    明代の日用類書と庶民教育.
  The 'Daily Use Encyclopaedias' (Jih Yung
    Lei Shu) of the Ming Dynasty and the
    Education of the Common People.
  Art. in *Kinsei Chūgaku Kyōikushi Kenkyū*
    近世中國敎育史研究.
  Studies on the History of Education in
    China in Recent Centuries, pp. 25–
    155.
  Ed. Hayashi Tomoharu 林友春.
  Tokyo, 1958.
Sha Shih-An 沙式菴, Lu I-Mei 陸伊湄
    & Wei Mo-Shen 魏默深 (*1*).
  Tshan Sang Ho Pien 蠶桑合編.
  Collected Notes on Sericulture.
  1843, 2nd ed. 1845.
  Cf. Liu Ho & Roux (1), p. 23.
Shang Chhêng-Tsu (*1*) 商承祚.
  *Chhangsha Chhu Thu Chhu Chhi Chhi Thu
    Lu* 長沙出土楚漆器圖錄.
  Album of Plates and Description of Lacquer
    Objects from the State of Chhu exca-
    vated at Chhangsha (Hunan).
  Shanghai, 1955.
Shao Li-Tzu (*1*) 邵力子.
  *Chi Nien Wang Chêng Shih Shih San Pai
    Chou Nien* 紀念王徵逝世三百周年.
  In Memory of the 300th Anniversary of the
    Death of Wang Chêng.
  CLTC, 1944, **1** (no. 2), 210.
Shen Ping-Chhêng (*1*) 沈秉成.
  Tshan Sang Chi Yao 蠶桑輯要.

Shen Ping-Chhêng (1) (cont.)
Essentials of Sericulture.
1869.

Shigezawa Toshirō (1) 重澤俊郎.
'Shūrei' no Shisōshiteki Kōsatsu 周禮の思
想史的考察.
A Study of the Leading Ideas (on State,
Society and Industry) in the *Record of
Institutions of the Chou Dynasty*.
TBS, 1955, 4, 42; 1956, 5, 31.

Shih Shu-Chhing (1) 史樹青.
Chhi Yen-Huai so-Chih-ti Thien-Wên Chung
齊彥槐所制的天文鐘.
On the Astronomical Clock constructed by
Chhi Yen-Huai.
WWTK, 1958 (no. 7), 37.

Shih Shu-Chhing (2) 史樹青.
Ku-Tai Kho-Chi Shih Wu Ssu Khao 古代
科技事物四考.
Four Notes on Ancient Scientific Technology: (*a*) Ceramic objects for medical
heat-treatment; (*b*) Mercury silvering of
bronze mirrors; (*c*) Cardan Suspension
perfume burners; (*d*) Dyeing stoves.
WWTK, 1962, (no. 3), 47.

Ssuchêng Wu-Jung (1) 司正午榮.
See *Kung Shih Tiao Cho Chêng Shih Lu
Pan Mu Ching Chiang Chia Ching.*

Sudō Yoshiyuki (2) 周藤吉之.
Nan-Sō no Nōsho to sono Seikaku 南宋の
農書とその性格; 特に王禎「農書」の
成立と關聯して.
Chinese Books on Farming in the Southern
Sung; with special reference to the
Compilation of the 'Treatise on Agri-
culture' by Wang Chên.
TBK, 1958, 14, 133.

Sun Chhiai-Ti (1) 孫楷第.
*Chin-Tai Hsi Chhü Yuan Chhu Sung Khuai-
lei Hsi Ying Hsi Khao* 近代戲曲原出
宋傀儡戲影戲考.
A Study of the Origins of Modern Opera
from Sung Puppet-Plays and Shadow-
Plays.
Fujen Univ., Peking, 1950.

Sun I-Jang (3) 孫詒讓.
See *Chou Li Chêng I.*

Sun Tzhu-Chou (1) 孫次舟.
*Sung-hsien Thang Mu so chhu Thieh Chien
Thung Chhih chi Mu Chih chih Khao Shih*
當縣唐墓所出鐵翦銅尺及墓誌之
考釋.
Study of a Pair of Iron Scissors and a
Copper Measuring-Rod from a Thang
Tomb at Sunghsien (Honan).
BCS, 1941, 1, 61; abstr. *HH*, 1949, 2, 389.

Sun Wên-Chhing (3) 孫文青.
Chang Hêng Nien Phu 張衡年譜.
Chronological Biography of Chang Hêng.
CLHP, 1933, 3, 331.

Sun Wên-Chhing (4) 孫文青.
Chang Hêng Nien Phu 張衡年譜.

Life of Chang Hêng [enlargement of (3)].
Com. Press, Shanghai, 1935; 2nd ed.
Chungking, 1944; 3rd ed. Shanghai,
1956.

Sung Po-Yin 宋伯胤 & Li Chung-I 黎忠義
(1).
*Tshung Han Hua-Hsiang-Shih Than-So
Han-Tai Chih-Chi Kou-Tsao* 從漢畫像
石探索漢代織機構造.
The Looms (and other Textile Machinery)
of the Han Period as revealed by (recently
discovered) Han Stone Reliefs.
WWTK, 1962 (no. 3), 25.

Suzuki Osamu (1) 鈴木治.
Ka Ran Kō 和鑾考.
A Study of the Ancient Chariot or Harness
Bells *ho* and *luan*.
TRDG, 1957, 8 (no. 3), 65; Abstr. *RBS*,
1962, 3, no. 477.

Takabayashi Hyōe (1) 高林兵衞.
Tokei Hattatsu Shi 時計發達史.
A History of Time-Measurement.
Tōyō Shuppansha, Tokyo, 1924.

Takai Bankan (1) 高井伴寬.
Shumisen Zukai 須彌山圖解.
Illustrated Explanation of Mount Sumeru
[religious cosmography].
Japan, 1809.

Takakusu Junjirō 高楠順次郎 & Watanabe
Kaigyoku 渡邊海旭 (1) (ed.).
*Ta Chêng Hsin Hsiu Ta Tsang Ching
(Taishō Shinshū Daizōkyō).* 大正新修
大藏經.
The Chinese Buddhist Tripiṭaka.
Tokyo, 1924 to 1929. 55 vols.
Catalogue by the same, Fascicule Annexe to
Hōbōgirin, Maison Franco-Japonaise,
Tokyo, 1931.

Têng Kuang-Ming (1) 鄧廣銘.
Yo Fei Chuan 岳飛傳.
Biography of Yo Fei (great Sung general).
San-Lien, Peking, 1955.

Têng Pai (1) (ed.) 鄧白.
Yung-Lo Kung Pi Hua 永樂宮壁畫.
Album of (Twenty) Coloured Plates selected
from the Frescoes of the Yung-Lo (Taoist)
Temple (in Shansi), *c.* +1350.
Jen-Min I-Shu, Shanghai, 1959.

Than Tan-Chhiung (1) 譚旦問.
Chung Hua Min Chien Kung I Thu Shuo
中華民間工藝圖說.
An Illustrated Account of the Industrial
Arts as traditionally practised among the
Chinese People.
Chung-Hua Tshung-Shu, Thaipei and
Hongkong, 1956.

Thang Lan (1) (ed.) 唐蘭.
*Wu Shêng Chhu Thu Chhung-Yao Wên-Wu
Chan-Lan Thu Lu* 五省出土重要文物
展覽圖錄.
Album of the Exhibition of Important

Thang Lan (*1*) (*cont.*)
   Archaeological Objects excavated in Five
   Provinces (Shensi, Chiangsu, Jehol,
   Anhui and Shansi).
   Wên-Wu, Peking, 1958.
Thang Yün-Ming (*1*) 唐雲明.
   Pao-ting Tung-Pi Chiu Chhêng Thiao-Chha
   保定東壁舊城調查.
   A Study of the Old City beside the Eastern
   Wall of Paoting [including details of iron
   gearwheels of H/Han date].
   WWTK, 1959 (no. 9), 82.
Ti Phing-Tzu (*1*) 狄平子.
   Han Hua (1st series) 漢畫. (It is doubtful
   if any more was ever published.)
   Collection of Drawings [Inscribed Bronzes
   and Stones, Moulded Bricks, etc.] of the
   Han period.
   Shanghai (probably issued by a curio
   dealer), 19?
Ting Kung-Chhen (*1*) 丁拱辰.
   Yen Phao Thu Shuo 演礮圖説.
   Illustrated Treatise on Gunnery [and many
   aspects of engineering].
   Chhüanchow, 1841, repr. 1843.
Ting Wên-Chiang (*1*) 丁文江.
   Biography of Sung Ying-Hsing 宋應星
   (author of the *Exploitation of the Works
   of Nature*).
   In the *Hsi Yung Hsüan Tshung-Shu* 喜詠
   軒叢書, ed. Thao Hsiang 陶湘.
   1929.
Tokugawa Nariaki (*1*) 德川齊昭編.
   Ungei Kisan 雲霓機纂.
   On Pumping Machinery.
   Japan, 1836.
   In *NKKZ*, vol. 11.
[Tsêng Chao-Yü] (*1*) (ed.) 曾昭燏.
   *Kuan-yü Wo Kuo Tsao Chung Li-Shih-ti
   San-ko Tshai-Liao* 關于我國造鐘歷
   史的三個材料.
   Three Contributions to the History of the
   Clockmaking Industry in China.
   Nanking Museum, 1958 (mimeographed).
   (Issued in connection with an exhibition of
   the history of clockwork held that year in
   Shanghai and Nanking.)
Tsêng Chao-Yü 曾昭燏, Chiang Pao-Kêng
   蔣寶庚 & Li Chung-I 黎忠義 (*1*).
   *I-nan Ku Hua Hsiang Shih Mu Fa-Chüeh
   Pao-Kao* 沂南古畫像石墓發掘報告.
   Report on the Excavation of an Ancient
   [Han] Tomb with Sculptured Reliefs at
   I-nan [in Shantung] (*c.* +193).
   Nanking Museum, Shantung Provincial
   Dept. of Antiquities, and Ministry of
   Culture, Shanghai, 1956.
Tshen Chung-Mien (*1*) 岑仲勉.
   *Wu Chhuan ti Chung-Kuo Ku Wang Chhêng
   yü chhi Shui-Li Li Yung* (*fu Shan-Tan Ta
   Fo Ku Chi*) 誤傳的中國古王城與其
   水力利用(附山丹大佛古蹟).

Medieval Mistakes about the Capital of
   China [Sandabil] and the Use of Water-
   Power there (with an appendix on the
   Remains of the Great Buddha (Temple)
   at Shantan).
   TFTC, 1945, **41** (no. 17), 39.
Tuan Shih (*1*) 段拭.
   Han Hua 漢畫.
   Painting and Bas-Reliefs in the Han
   Period.
   Ku-Tien I-Shu, Peking, 1958.
Tuan Shih (*2*) 段拭.
   *Chiangsu Thung-shan Hung-lou Tung Han
   Mu Chhu Thu Fang-Chih Hua-Hsiang-
   Shih* 江蘇銅山洪樓東漢墓出土紡
   織畫像石.
   A Stone Relief of the Later Han Dynasty
   with a Weaving Scene, from Hung-lou
   near Thung-shan in Chiangsu province.
   WWTK, 1962 (no. 3), 31.
Tuan Wên-Chieh (*1*) (ed.) 段文杰.
   Yü Lin Khu 榆林窟.
   The Frescoes of Yü-lin-khu [i.e. Wan-fo-
   hsia, a series of cave-temples in Kansu].
   Tunhuang Research Institute, Chung-Kuo
   Ku-Tien I-Shu, Peking, 1957.
Tun Li-Chhen (*1*) 敦禮臣.
   Yenching Sui Shih Chi 燕京歲時記.
   Annual Customs and Festivals of Peking.
   Peking, 1900.
   Tr. Bodde (12).
Tung Kao (*1*) (ed.) 董誥 *et al.*
   Chhüan Thang Wên 全唐文.
   Collected Literature of the Thang Dynasty.
   1814.
   Cf. des Rotours (2), p. 97.

Uchida Gimpū (*1*) 內田吟風.
   Jōdai Mōko ni okeru Sharyō Kōtsū 上代蒙
   古に於ける車輛交通.
   Vehicles in Ancient Mongolia.
   TK, 1940, **5**, 24.
Ueda Noburu (*1*) 上田舒.
   Tōa ni okeru Nokogiri no Keifu 東亞にぞ
   ける鋸の系譜.
   Ancient and Traditional Saws of China and
   Japan.
   KKZ, 1957, **42**, 169
   Abstr. *RBS*, 1962, **3**, no. 941.
Umehara Sueji (*2*) 梅原末治.
   *Beikoku 'Furiya' Bijutsu-kan Shozō no Zōgan
   shuryō bun dōsen* 米國フリヤ美術館
   所藏の象嵌狩獵文銅洗.
   An Inlaid Hunting Motif on a Chinese
   bronze vessel preserved in the Freer
   Gallery of Art.
   In *Kuwabara Hakase Kanreki Kinen Tōyō
   Shironsō* 桑原博士還曆記念東洋史
   論叢.
   Papers on Far Eastern History presented to
   Dr Kuwabara on his Anniversary, p. 17.
   Tokyo, 1934.

Umehara Sueji 梅原末治, Oba Tsunekichi
小場恒吉 & Kayamoto Kamejirō 榧本
龜次郎 (1).
Rakurō Ōkō-bo 樂浪王光墓.
The Tomb of Wang Kuang at Lo-lang, Korea.
Soc. for the Study of Korean Antiquities;
Chōsen Koseki Kenkyū-Kwai 朝鮮古
蹟研究會.
Detailed Reports of Archaeological Research,
no. 2.
Seoul (Keijo), 1935, 2 vols.

Unno Kazutaka (2) 海野一隆.
*Konron Shisuisetsu no Chiri Shisōshi teki
Kōsatsu* 崑崙四水説の地理思想史
的考察.
A Study of the Legend of Mount Khun-
Lun and its Four Rivers in relation to
the History of Geographical Thought.
SN, 1958, **41**, 379.

Unno Kazutaka (3) 海野一隆.
Chūgoku Bukkyō ni okeru Sekai Kubun setsu
中國佛教にわける世界區分説.
The Geographical Partition of the World in
Chinese Buddhism [wheel-maps centered
on Khun-Lun Shan = Mt Meru].
In *Tanaka Shūsaku Kyōju Kokikinen
Chirigaku Ronbunshū* 田中秀作教授
古稀記念地理學論文集.
Commemorative Volume for the 70th
Birthday of Prof. Tanaka Shūsaku, p. 106.
Kyoto, 1956.

Wang Chen-To (1) 王振鐸.
*Han Chang Hêng Hou-Fêng Ti-Tung I Tsao-
Fa chih Thui-Tshê* 漢張衡候風地動儀
造法之推測.
A Conjecture as to the Construction of the
Seismograph of Chang Hêng in the Han
Dynasty.
YCHP, 1936, **20**, 577.

Wang Chen-To (3) 王振鐸.
*Chih-Nan-Chhê Chi-Li-Ku-Chhê chih Khao
Chêng chi Mo Chih* 指南車記里鼓車
之考證及模製.
Investigations and Reproduction in Model
Form of the South-Pointing Carriage and
the Hodometer (*Li*-Measuring Drum
Carriage).
AP/HJ, 1937, **3**, 1.

Wang Chen-To (5) 王振鐸.
*Ssu-Nan Chih-Nan Chen yü Lo-Ching Phan
(Hsia)* 司南指南針與羅經盤 (下).
Discovery and Application of Magnetic
Phenomena in China, III (Origin and
Development of the Chinese Compass Dial).
AS/CJA, 1951, **5** (n.s., **1**), 101.

Wang Chen-To (7) 王振鐸.
*Chieh Khai-liao Wo Kuo 'Thien Wên
Chung' ti Pi-Mi* 揭開了我國「天文鐘」
的秘密.
The Secret of Our (Medieval) 'Astro-
nomical Clocks' revealed.

WWTK, 1958 (no. 9), 5.

Wang Chen-To (8) 王振鐸.
*Han Tai Yeh Thieh Ku-Fêng-Chi ti Fu
Yuan* 漢代冶鐵鼓風機的復原.
On the Reconstruction of the Metallurgical
Iron-Casting Bellows of the Han period.
WWTK, 1959 (no. 5), 43.

Wang Chen-To (9) 王振鐸.
Chung-Kuo Tsui-Tsao-ti Chia-Thien-I
中國最早的假天儀.
The Earliest Chinese Planetarium.
WWTK, 1962 (no. 3), 11.

Wang Chia-Chhi (1) 王家琦.
*Shui Chuan Lien Mo, Shui Phai ho Yang
Ma* 水轉連磨水排和秧馬.
On Water-powered Multiple Geared Mills,
Water-powered Reciprocators (Hydraulic
Metallurgical Blowing-Engines), and
Rice-planting Boats.
WWTK, 1958 (no. 7), 34.

Wang Chih-Chhun (1) 王之春.
Kuo Chhao Jou Yuan Chi 國朝柔遠記.
Record of the Pacification of a Far Country
[his ambassadorship to Russia].
c. 1885.

Wang Chin (2) (ed.) 王璡.
*Chung-Kuo Ku-Tai Chin-Shu Hua-Hsüeh
chi Chin Tan Shu* 中國古代金屬化學
及金丹術.
Alchemy and the Development of Metal-
lurgical Chemistry in Ancient and
Medieval China [collected essays].
Chung-Kuo Kho-Hsüeh Thu-Shu I-Chhi
Kung-ssu, Shanghai, 1955.

Wang Chün (1) 汪鋆.
Shih-erh Yen Chai Sui Pi 十二硯齋隨筆.
Miscellaneous Notes from the Twelve-
Inkstone Studio.
c. 1885

Wang Chung-Lao (1) 王仲犖.
*Kuan-yü Chung-Kuo Nu-Li Shê-Hui ti
Wa-Chieh chi Fêng-Chien Kuan-Hsi ti
Hsing-Chhêng Wên-Thi* 關於中國奴
隸社會的瓦解及封建關係的形
成問題.
The Question of the Breakdown of Slave
Society in China and the Formation of
the Feudal Relationship.
Wuhan, 1957.

Wang Fang-Chung (1) 王方中.
*Sung Tai Min Ying Shou-Kung-Yeh-ti Shê-
Hui Ching-Chi Hsing-Chih* 宋代民營手
工業的社會經濟性質.
Private Handicraft Industry during the
Sung Dynasty.
LSYC, 1959 (no. 2), 39.

Wang Hsien-Chhien (1) (ed.) 王先謙.
Huang Chhing Ching Chieh Hsü Pien 皇清
經解續編.
Continuation of the *Collection of Monographs on Classical Subjects written during
the Chhing Dynasty.*

Wang Hsien-Chhien (1) (cont.)
1888.
See Yen Chieh (1).
Wang Hsien-Chhien (2) 王先謙.
Chuang Tzu Chi Chieh 莊子集解.
Collected Commentaries on the Book of
Master Chuang.
1909.
Wang Jen-Chün (1) 王仁俊.
Ko Chih Ku Wei 格致古微.
Scientific Traces in Olden Times.
1896.
Wang Nien-Sun (2).
See Kuang Ya Su Chêng.
Wang Su 王蘇, Hsing Lin 邢琳 & Wang Liu
王劉(1).
Wo Kuo tsai Kang-Thieh Yeh Lien Kung
Yeh shang ti wei-ta Chhuang-Tsao 我國
在鋼鐵冶煉工業上的偉大創造.
The Great Contributions of China to the
Technology of Iron and Steel.
WWTK, 1959 (no. 1), 26.
Wei Yuan 魏源 & Lin Tsê-Hsü 林則徐 (1).
Hai Kuo Thu Chih 海國圖志.
Illustrated Record of the Maritime [Occi-
dental] Nations.
1844; enlarged ed. 1847; further enlarged
ed. 1852.
For the problem of authorship see Chhen
Chhi-Thien (1).
Wu Chhêng-Lo (1) 吳承洛.
Chung Hsi Kho-Hsüeh I-Shu Wên-Hua Li-
Shih Pien Nien Tui Chao 中西科學藝
術文化歷史編年對照.
Comparative Tables of Scientific Techno-
logical and Scholarly Achievements in
China and Europe.
KHS, 1925, 10, 1.
Wu Chhêng-Lo (2) 吳承洛.
Chung-Kuo Tu Liang Hêng Shih 中國度
量衡史.
History of Chinese Metrology [weights and
measures].
Com. Press, Shanghai, 1937; 2nd ed.
Shanghai, 1957.
Wu Chhi-Chün (1) 吳其濬.
Chih Wu Ming Shih Thu Khao 植物名實
圖考.
Illustrated Investigation of the Names and
Natures of Plants.
1848.
Wu Ju-Tsu 吳汝祚 & Hu Chhien-Ying
胡謙盈 (1).
Pao-chi ho Hsi-an Fu Chin Khao Ku Fa-
Chüeh Chien Pao 寶雞和西安附近考
古發掘簡報.
Preliminary Report of Archaeological Ex-
cavations near Paochi and Sian [in-
cluding bronze gearwheels of the late
C/Han period found at Hung-chhing-
tshun].
KKTH, 1955 (no. 2), 33.

Wu Nan-Hsün (1) 吳南薰.
Chung-Kuo Wu-Li-Hsüeh Shih 中國物理
學史.
A History of Physics in China (preliminary
draft, based on courses of lectures).
Pr. pr. Dept. of Physics, Wuhan Univ.,
Wuhan, 1954.
Wu Yü-Chiang (1) (ed.) 吳毓江.
Mo Tzu Chiao Chu 墨子校注.
The Collected Commentaries on the Book
of Master Mo (including the Mohist
Canon).
Tu-li, Chungking, 1944.

Yabuuchi Kiyoshi (4) 藪內清.
Chūgoku no Tokei 中國の時計.
[Ancient] Chinese Time-Keepers.
JJHS, 1951 (no. 19), 19.
Yabuuchi Kiyoshi (11) (ed.) 藪內清.
'Tenkō Kaibutsu' no Kenkyū 天工開物の
研究.
A Study of the Thien Kung Khai Wu (The
Exploitation of the Works of Nature,
+1637).
Tokyo, 1953.
Chinese translation of the eleven critical
essays, by Su Hsiang-Yü et al.; Chung-
Hua Tshung-Shu, Thaipei and Hong-
kong, 1956.
Yamaguchi Ryūji (1) 山口隆二.
Nihon no Tokei; Tokugawa Jidai no Wadokei
no Kenkyū 日本の時計;德川時代の
和時計の研究.
Time-Measurement in Japan; Studies on
the Clocks of the Tokugawa Period.
Nihon Hyōronsha, Tokyo, 1942.
Yang Chung-I (1) 楊中一.
Thang Tai ti Chien-Min 唐代的賤民.
The Lowest Ranks of Society in the Thang
Dynasty period.
SH, 1935, 1 (no. 4), 124.
Tr. Sun & de Francis (1), p. 185.
Yang Chung-I (2) 楊中一.
Pu-Chhü Yen-Ko Lüeh Khao 部曲沿革
略考.
A Study of the Evolution of the Status of
'Dependent Retainers'.
SH, 1935, 1 (no. 3), 97.
Tr. Sun & de Francis (1), p. 142.
Yang Jen-Khai 楊仁愷 & Tung Yen-Ming
董彥明 (1) (ed.)
Liaoning Shêng Po-Wu-Kuan Tshang Hua
Chi 遼寧省博物館藏畫集.
Album of Pictures illustrating the Collection
of Paintings in the Liaoning Provincial
Museum.
Wên-Wu, Peking, 1962.
Yang Khuan (1) 楊寬.
Chung-Kuo Ku-Tai Yeh-Thieh Ku-Fêng Lu
ho Shui-Li Yeh-Thieh Ku-Fêng Lu ti Fa-
Ming 中國古代冶鐵鼓風爐和水力
冶鐵鼓風爐的發明.

Yang Khuan (*1*) (*cont.*)
On the Blast Furnaces used for making
Cast Iron in Ancient China, and the
Invention of Hydraulic Blowing-Engines
for them.
Essay in Li Kuang-Pi & Chhien Chün-
Yeh (q.v.), p. 71.
Peking, 1955.

Yang Khuan (*4*) 楊寬.
Chung-Kuo Li-Tai Chhih Tu Khao 中國
歷代尺度考.
A Study of the Chinese Foot-Measure
through the Ages.
Com. Press, Shanghai, 1938; revised and
amplified ed. 1955.

Yang Khuan (*6*) 楊寬.
*Chung-Kuo Ku-Tai Yeh-Thieh Chi-Shu ti
Fa-Ming ho Fa-Chan* 中國古代冶鐵技
術的發明和發展.
The Origins, Inventions, and Development
of Iron [and Steel] Technology in Ancient
and Medieval China.
Jen-Min, Shanghai, 1956.

Yang Khuan (*9*) 楊寬.
*Kuan-Yü Shui-Li Yeh-Thieh Ku-Fêng-Chi
'Shui Phai' Fu-Yuan ti Thao-Lun* 關於
水力冶鐵鼓風機水排復原的討論.
Queries on the Reconstruction of the
(Vertical Water-Wheel) Hydraulic
Blowing-Engines in Ironworks (of the
Sung and Yuan periods).
WWTK, 1959 (no. 7), 48.

Yang Lien-Shêng (*1*) 楊聯陞.
Tung Han ti Hao Tsu 東漢的豪族.
The Great Families of the Eastern Han.
CHJ, 1936, **11** (no. 4), 1007.
Tr. Sun & de Francis (*1*), p. 103.

Yeh Chao-Han (*1*) 葉照涵.
*Han Tai Shih Kho Yeh-Thieh Ku-Fêng Lu
Thu* 漢代石刻冶鐵鼓風爐圖.
A Han Relief showing a Blast Furnace for
making Cast Iron.
WWTK, 1959 (no. 1), 2 & 20.

Yeh Chhien-Yü (*1*) 葉淺予.
Tunhuang Pi Hua 敦煌壁畫.
The Wall-Paintings at Tunhuang (the
Chhien-fo-tung Cave-Temples).
Chao-Hua I-Shu, Peking, 1957.

Yen Chieh (*1*) (ed.) 嚴杰.
Huang Chhing Ching Chieh 皇清經解.
Collection of [more than 180] Monographs
on Classical Subjects written during the
Chhing Dynasty.
1829; 2nd ed. Kêng Shen Pu Khan, 1860.
Cf. Wang Hsien-Chhien (*1*).

Yen Kho-Chün (*1*) (ed.) 嚴可均.
*Chhüan Shang-Ku San-Tai Chhin Han San-
Kuo Liu Chhao Wên* 全上古三代秦漢
三國六朝文.
Complete Collection of Prose Literature
(including Fragments) from Remote
Antiquity through the Chhin and Han

Dynasties, the Three Kingdoms and the
Six Dynasties.
Finished 1836; published 1887–93.

Yen Lei *et al.* (*1*) 閻磊.
*Shensi Chhang-an Hung-chhing-tshun Chhin
Han Mu ti-erh-tzhu Fa-Chüeh Chien-Chi*
陝西長安洪慶村秦漢墓第二次發掘
簡記.
Short Catalogue of Objects found in the
Second Excavation of Chhin and Han
Tombs at Hung-chhing-tshun near Sian.
KKTH, 1959 (no. 12), 662.

Yen Tun-Chieh (*16*) 嚴敦傑.
*Pa 'Hung Lou Mêng Hsin Khao' nei Hsi-
Yang Shih Kho yü Chung-Kuo Shih Kho
Pi-Chiao* 跋紅樓夢新考內西洋時刻
與中國時刻之比較.
A Comparison of Chinese and Western
Horary Systems by way of a Postscript
to [Fang Hao's] 'New Study of the novel
Dream of the Red Chamber'.
TFTC, **40** (no. 16), 27.

Yen Yü (*1*) 燕羽.
*Chung-Kuo Li-Shih shang ti Kho [-Hsüeh]
Chi[-Shu] Jen Wu* 中國歷史上的科
技人物.
Lives of [twenty-two] Scientists and Tech-
nologists eminent in Chinese History [in-
cluding e.g. Chang Hêng, Tsu Chhung-
Chih, Yü Yün-Wên and Yüwên Khai].
Chün-Lien, Shanghai, 1951.

Yen Yü (*4*) 燕羽.
*Chung-Kuo Ku-Tai kuan-yü Shen Ching
Tsuan Chüeh Chi-Chieh ti Fa-Ming* 中國
古代關於深井鑽掘機械的發明.
The Invention of Deep Drilling Technique
for Wells and Boreholes in Ancient
China.
Essay in Li Kuang-Pi & Chhien Chün-Yeh
(q.v.), p. 186.
Peking, 1955.

Yi Kwangnin (*1*) 李光麟.
Yijo Suri Sa Yŏn-gu 李朝水利史研究.
History of Irrigation during the (Korean)
Yi Dynasty.
Korean Research Centre, Seoul, 1961
(Korean Studies series, no. 8).

Yonezawa Yoshio (*1*) 米澤嘉圃.
*Kandai ni okeru Kyūtei Sakuga Kikō no
Hattatsu* 漢代に於ける宮廷作畫機
構の發達.
On the Development of the Organisation
of Drawloom Weaving in Palace Work-
shops during the Han Dynasty.
KK, 1938, nos. 571, 574–7.
Abstr. *MS*, 1942, **7**, 364.

Yonezawa Yoshio (*2*) 米澤嘉圃.
Gishin Nambokuchō Jidai no Shōhō 魏晉南
北朝時代の尚方.
The Imperial Workshops in the Wei, Chin
and Northern and Southern Dynasty
periods.

Yonezawa Yoshio (2) (cont.)
  TG/T, 1939, **10**, 303.
  Abstr. *MS*, 1942, **7**, 364.
Yoshida Mitsukuni (1) 吉田光邦.
  '*Tenkō Kaibutsu' ni tsuite* 天工開物につ
    いて.
  On the *Thien Kung Khai Wu* (Exploitation
    of the Works of Nature), [a +17th-
    century technological treatise].
  JJHS, 1951, **18**, 12.

Yoshida Mitsukuni (3) 吉田光邦.
  Shūrei-Kōkōki no Ikkōsatsu 周禮考工記
    一考察.
  Notes on some Aspects of Technology in
    the 'Artificers' Record' of the *Chou Li*.
  TG/K, 1959, **30**, 167.
Yü Chhang-Hui (1) 俞昌會.
  Fang Hai Chi Yao 防海輯要.
  Essentials of Coast Defence.
  1822.

C. BOOKS AND JOURNAL ARTICLES IN
WESTERN LANGUAGES

ABEL, CLARKE (1). *Narrative of a Journey to the Interior of China....* Longman, London, 1818.

D'ACRES, R. (1). *The Art of Water Drawing* (water raising machinery). Brome, London, 1659 and 1660. (Facsimile reproduction, Extra Publication no. 2 of the Newcomen Society, Heffer, Cambridge, 1930.)

ADKINS, E. C. S. (1). '*Ching Hua Yuan*; China's *Gulliver's Travels.*' *ACSS*, 1954, 34.

AFET INAN (1). *Aperçu Général sur l'Histoire Économique de l'Empire Turc-Ottoman.* Maarif Matbaasi, Istanbul, 1941. (Publ. de la Soc. d'Hist. Turque, Ser. VII, no. 6.)

AGRICOLA, GEORGIUS (GEORGE BAUER) (1). *De Re Metallica.* Basel, 1556. See Hoover & Hoover. Account of the engineering in *De Re Metallica* in Beck (1), ch. 8.

ALBERTI, L. B. (1). *De Re Aedificatoria.* Rembolt & Hornken, Paris, 1512. Ital. tr. *I Dieci libri di Architettura.* Venice, 1546. Fr. tr. Paris, 1553. Eng. tr. London, 1726. See Olschki (5).

ALLEY, REWI (6) (tr.). *Tu Fu; Selected Poems.* Foreign Languages Press, Peking, 1962. (Selection by Fêng Chih.)

ALLEY, REWI (8). 'Thangshan and the Eastern [Manchu Dynasty Imperial] Tombs [near Chi-hsien in Hopei].' *EHOR*, 1963, **2** (no. 12), 39.

D'ALVIELLA, GOBLET (1). 'Moulins à Prières, Roues Magiques, et Circumambulations.' *RUB*, 1897. Also art. 'Prayer-Wheels.' *ERE*, vol. 10, 394, 500.

AMEISENOWA, Z. (1). 'Some Neglected Representations of the Harmony of the Universe.' *GBA*, 1958, 349. (Commemoration Volume for Hans Tietze.)

AMELLI, A. M. (1) (ed). *Miniature Sacre e Profane dell'Anno 1023 illustranti l'Enciclopedia Medioevale de Rabano Mauro.* Montecassino, 1896.

AMIOT, J. J. M. See de Rochemonteix, P. C. (1).

AMIOT, J. J. M. (4). Letter on barometric pressure, static electricity and magnetism, and stories of flying men or spirits ('aérambules') in Chinese Literature, 15 Nov. 1784. *MCHSAMUC*, 1786, **11**, 569.

ANDERSON, L. C. (1). 'Kites in China.' *NCR*, 1921, **3**, 73.

ANDERSON, M. D. (1). *The Choir-Stalls of Lincoln Minster.* Friends of Lincoln Cathedral, Lincoln, 1951.

ANDERSON, R. H. (1). 'The Technical Ancestry of Grain-Milling Devices.' *MENG*, 1935, **57**, 611; *AGHST*, 1938, **12**, 256.

ANDERSSON, J. G. (3). 'An Early Chinese Culture.' *BGSC*, 1923, **5**, 1.

ANDRADE, E. N. DA C. (1). 'Robert Hooke' (Wilkins Lecture). *PRSA*, 1950, **201**, 439. (The quotation concerning fossils is taken from the advance notice, Dec. 1949.) Also *N*, 1953, **171**, 365.

ANDRADE, E. N. DA C. (3). 'The Early History of the Vacuum Pump.' *END*, 1957, **16** (no. 61), 29.

ANDREWS, C. R. (1). *The Story of Wortley Ironworks; a Record of Eight Centuries of Yorkshire Industry.* S. Yorkshire Times, Mexborough, 1950.

ANELL, B. (1). *Contribution to the History of Fishing in the Southern Seas.* Almquist & Wiksell, Stockholm, 1955.

ANON. (5). Illustration (+1616) of a paddle-wheel boat winding itself upstream on a towrope attached to a fixed point, with commentary. *GTIG*, 1927, **11**, 293.

ANON. (6). 'Plan for ascending Rapids in Rivers.' *MMG*, 1825, **3**, 225.

ANON. (8). 'Leon Foucault [and the gyroscope].' *SR*, 1951, **3**, 16.

ANON. (11). *A Gallery of Japanese and Chinese Paintings.* Kokka, Tokyo, 1908.

ANON. (13). *Streyd-Buch von Pixen, Kriegsrüstung, Sturmzeuch und Feuerwerckh* (artillerist's manual). MS. Kaiserhaus, Vienna cod. 5135, early +15th century. See Sarton (1), vol. 3, p. 1551.

ANON. (14). Technical drawings of military engineering by the Anonymous Engineer of the Hussite Wars, c. 1430. MSS. Cod. Lat. 197, Hofbibliothek, Munich; 328 Weimar. See Sarton (1), vol. 3, p. 1551.

ANON. (15). *Das Mittelalterliche Hausbuch.* Album of an arquebus-maker, c. 1480, containing various engineering drawings. MS. Wolfsee Castle. Ed. A. von Essenwein, Frankfurt 1887; H .T. Bossert & W. F. Storck, Leipzig, 1912; Joh. Graf. v. Waldburg-Wolfegg, Munich, 1957. See Sarton (1), vol. 3, p. 1553.

ANON. (18). 'Miller of Dalswinton anticipated by the Chinese' (on Yang Yao's paddle-boat in the Sung). *CR*, 1882, **11**, 201; *NQCJ*, **26**, 128.

ANON. (23). *La Mission Lyonnaise d'Exploration Commerciale en Chine, 1895–1897.* Chambre de Commerce, Lyon, 1898.

ANON. (30). (Antiquus Idiota, ps.) 'Sui and Ancient Chinese Swords.' *CJ,* 1928, **8**, 299.

ANON. (32) (ed.). *New China* (album of photographs). Foreign Languages Press, Peking, 1953.

ANON. (36) (ed.). 'Changing Japan seen through the Camera.' Album of photographs, with foreword by G. Alsot. Asahi Shimbun, Tokyo & Osaka, 1933.

ANON. (37) (tr.). 'The Chinaman Abroad; or, a Desultory Account of the Malayan Archipelago, particularly of Java, by Ong-Tae-Hae' (Wang Ta-Hai's *Hai Tao I Chih Chai Lüeh* of +1791). *CMIS,* 1849 (no. 2), 1.

ANON. (38). *Handy Technical Dictionary in Eight Languages* (English, French, German, Italian, Polish, Portuguese, Russian and Spanish). Disce Publications, and K.L.R. Publishers, London, 1952.

ANON. (39). 'Dissertation on the Silk Manufacture and the Cultivation of the Mulberry; translated from the Works of Tseu-Kwang-K'he [Hsü Kuang-Chhi], called also Paul Siu, a Colao, or Minister of State in China.' (Translation of excerpts of chs. 31 to 34 of the *Nung Chêng Chhüan Shu* of +1639). *CMIS,* 1849 (no. 3); also sep. pub.; Mission Press, Shanghai, 1849.

ANON. (40). *Humane Industry; or, a History of Most Manual Arts, deducing the Original, Progress, and Improvement of them; furnished with Variety of Instances and Examples, shewing forth the Excellency of Humane Wit.* Herringman, London, 1661.

ANON. (41). *An Outline History of China.* Foreign Languages Press, Peking, 1958.

ANON. (42). *Hsiao's Record of the Imperial Palaces at Khanbaliq.* No place, no date, but pr. in China.

ANON. (43). *La Chine à Terre et en Ballon; Reproduction de Photographies des officiers du Génie du Corps Expéditionnaire, 1900–1901* (album). Berger-Levrault, Paris n.d. (1902?).

ANON. (44). 'Les Roues élévatoires dans les Mines de Cuivre.' *LN,* 1884, **12** (pt. 2), 344.

ANON. (45). *Silk; Replies from Commissioners of Customs to the Inspector-General's Circular no. 103, Second Series; to which is added 'Manchurian Tussore Silk' by N. Shaw.* Shanghai, 1917. (China, Maritime Customs, II, Special series, no. 3.)

ANON. (46). *The Epistle of Petrus Peregrinus on the Magnet, reproduced from a MS written by an English hand about +1390.* Quaritch, London, 1900.

ANON. (49). *The Rice Manufacture in China; from the Originals brought from China.* (Album of 24 copper-plate engravings, with engraved title-page, drawn by A.H.) Bowles, Bowles & Sayer, London, *c.* +1780. (Copied from the *Kêng Chih Thu* in one of its Chhing editions, Khang-Hsi rather than Chhien-Lung, all plates reversed.)

ANON. (52). 'Le Martinet — Esquisse d'une Morphologie du Martinet; les Origines du Moulin à Fer; Trois Documents, etc.' *RHSID,* 1960, **1** (no. 3), 7.

ANON. (59). *Labour and Struggle; Glimpses of Chinese History.* Museum of Chinese History, Peking. (Supplement to *CREC,* Apr. 1960.)

ANON. (60). 'A Canal through the Mountains' (the utilisation of the water of the Thao River in Southern Kansu). *PKR,* 1958 (no. 24), 17.

ANON. (61). 'A Model of Su Sung's Escapement.' *HORJ,* 1961, 481. 'Heavenly Clockwork—a Sequel' (by B.L.H.). *AHOR,* 1962, 297.

ANON. (62). *Book of Trades; or, Library of the Useful Arts.* 3 vols. Tabart, London, 1804, 1811.

ANON. (63). *Book of English Trades, and Library of Useful Arts.* Rivington, London, 1835.

ANON. (64). *Book of Trades; or, Circle of the Useful Arts.* Griffin, Glasgow, 1835; 10th ed. Griffin, London, 1852.

ANON. (65). *Book of Trades.* SPCK, London, n.d. (1862).

ANON. (71). 'The Economy of the Chinese illustrated by a Notice of the Tinkers, with a description of the Bellows.' *CRRR,* 1836, **4**, 37.

ANON. (72). *Water-Conservancy in New China* (album of photographs with bilingual captions and text). For the Ministry of Water-Conservancy; People's Art Pub. Ho., Shanghai, 1956.

ANTHIAUME, A. (2). *Le Navire, sa Propulsion en France, et principalement chez les Normands.* Dumont, Paris, 1924.

ARAGO, D. F. (2). 'Notice historique sur les Machines à Vapeur.' *AL,* 1829, repr. 1830 and 1837, 221 (230). Repr. in *Œuvres,* vol. 5, pp. 1 ff. Gide & Baudry, Paris; Weigel, Leipzig, 1855.

ARLINGTON, L. C. (2). *The Chinese Drama from the Earliest Times until Today....* Kelly & Walsh, Shanghai, 1930.

ARMÃO, E. (1). *Vincenzo Coronelli, Cenni sull'Uomo e la sua Vita, Catalogo Ragionato delle sue Opere, Lettere....* Bibliopolis, Florence, 1944. (Biblioteca di Bibliografia Italiana, no. 17.)

D'ARNAL, E. SCIPION, ABBÉ (1). *Mémoire sur les Moulins à Feu établis à Nîmes.* Nîmes, 1783.

ARNE, T. J. (1). 'Skandinavisches Holzkammergräber aus der Wikingerzeit in der Ukraine.' *ACA,* 1931, **2**, 285.

ARNOLD, BROTHER. See Mertens, J. C. (1).

ARNOLD, Sir EDWIN (1). *The Light of Asia; or, the Great Renunciation, being the Life and Teaching of Gautama [Buddha], (as told by an Indian Buddhist).* London, 1879, frequently reprinted.

ARNOLD, Sir EDWIN (2). *Seas and Lands.* Longmans Green, London, 1894.

ASTON, W. G. (tr.) (1). '*Nihongi*', *Chronicles of Japan from the Earliest Times to +697.* Kegan Paul, London, 1896; repr. Allen & Unwin, London, 1956.

ATKINSON, F. (1). 'The Horse as a Source of Rotary Power.' *TNS*, 1962, **33**, 31.

AUDEMARD, L. (5). *Les Jonques Chinoises; IV, Description des Jonques.* Museum voor Land- en Volkenkunde & Maritiem Museum Prins Hendrik, Rotterdam, 1962.

AUDIN, MARIUS (1). *Les Vieux Moulins du Rhône* (ship-mills at Lyon). Lyon, n.d. See also *Le Confluent du Rhône et de la Saône; les Emplacements qu'il a occupés depuis les périodes géologiques jusqu'à nos jours, les Transformations qu'il a subies, et ses derniers Avatars.* Cumin & Masson, Lyon, 1919.

AVITSUR, S. (1). *On the History of the Exploitation of Water-Power in Eretz Israel.* Avehsalom Institute for Homeland Studies, Tel-Aviv, 1960.

BACCHELLI, R. (1). *Il Mulino del Po.* Eng. tr. F. Frenaye. Hutchinson, London, 1952.

BADDELEY, J. F. (2). *Russia, Mongolia, China; being some Record of the Relations between them from the Beginning of the +17th century to the Death of the Tsar Alexei Mikhailovitch (+1602 to +1676), rendered mainly in the form of Narratives dictated or written by the Envoys sent by the Russian Tsars or their Voevodas in Siberia to the Kalmuk and Mongol Khans and Princes, and to the Emperors of China; with Introductions Historical and Geographical, also a series of Maps showing the Progress of Geographical Knowledge in regard to Northern Asia during the +16th, +17th and early +18th Centuries; the Texts taken more especially from Manuscripts in the Moscow Foreign Office Archives....* 2 vols. Macmillan, London, 1919.

BAGNOLD, R. (1). *Libyan Sands.* Hodder & Stoughton, London, 1935.

BAGROW, L. (1). 'Ortelii Catalogus Cartographorum.' *PMJP*, 1929, Ergänzungsband **43**, no. 199; 1930, Ergänzungsband **45**, no. 210.

BAILEY, K. C. (1). *The Elder Pliny's Chapters on Chemical Subjects.* 2 vols. Arnold, London, 1929 and 1932.

BAILLIE, G. H. (1). *Clocks and Watches; an historical Bibliography.* NAG Press, London, 1951.

BAILLIE, G. H. (2). *Watches.* Methuen, London, 1929.

BALAZS, E. (=S.) (5). 'Beiträge z. Wirtschaftsgeschichte d. T'ang-Zeit.' *MSOS*, 1931, **24**, 1; 1932, **25**, 93; 1933, **26**, 166; ref. F. Otte, *OAZ*, 1933, **9** (19), 40.

BALAZS, E. (=S.) (7) (tr.). 'Le Traité Économique du *Souei-Chou* [*Sui Shu*].' *TP*, 1953, **42**, 113. Sep. pub. as *Études sur la Société et l'Économie de la Chine Médiévale*, no. 1. Brill, Leiden, 1953.

BALAZS, E. (=S.) (8). 'Le Traité Juridique du *Souei-Chou* [*Sui Shu*].' *TP*, 1954, **42**, 113. Sep. pub. as *Études sur la Société et l'Économie de la Chine Médiévale*, no. 2. Brill, Leiden, 1954. (Bibliothèque de l'Inst. des Hautes Études Chinoises, no. 9.)

BALFOUR, H. (1). 'The Fire Piston.' Art. in *Anthropological Essays presented to Edw. Burnett Tylor in honour of his 70th Birthday*, 1907.

BALFOUR, H. (2). 'Kite-Fishing.' In *Essays and Studies presented to William Ridgeway...on his 60th Birthday.* Ed. E. C. Quiggin, Cambridge, 1913. Pp. 583 ff.

BALL, E. B. (1). 'The Influence of the Mechanical Mind on the Development of Irrigation throughout the Ages.' *JPIME*, 1940, **142**, 407.

BALTRUŠAITIS, J. (1). *Le Moyen Âge Fantastique; Antiquités et Exotismes dans l'Art Gothique.* Colin, Paris, 1955.

BARBOTIN, A. (1). *Les Industries de l'Indochine française; Notions élémentaires de Sciences appliquées.* Impr. d'Extrême-Orient, Hanoi, 1917.

BARDE, R. (3). 'Les Éléments et les Nombres.' Unpublished MS.

BARNETT, R. D. & WATSON, W. (1). 'The World's Oldest Persian Carpet, preserved for 2400 Years in perpetual ice in Central Siberia; astonishing new Discoveries from the Scythian Tombs of Pazirik.' *ILN*, 1953, **223**, 69.

BAROJA, J. CARO (1). 'Sobre la Historia de la Noria de Tiro [i.e. Sāqīya].' *DTP*, 1955, **11**, 15–79.

BAROJA, J. CARO (2). 'Disertación sobre los Molinos de Viento.' *DTP*, 1952, **8**, 212.

BAROJA, J. CARO (3). 'Norias, Azudas, Aceñas.' *DTP*, 1954, **10**, 29–160.

BAROJA, J. CARO (4). 'Sobre Maquinarias de Tradición Antigua y Medieval.' *DTP*, 1956, **12**, 114.

BAROJA, J. CARO (5). 'En la Campiño de Córdoba (Observaciones de 1949).' Oil-mills and presses. *DTP*, 1956, **12**, 270.

BAROJA, J. CARO (6). 'Sobre Cigüeñales [crank-winches] y otros Ingenios para elevar Agua.' *RG*, 1955, **56**, 25.

BARROW, John (1). *Travels in China.* London, 1804. German tr. 1804; French tr. 1805; Dutch tr. 1809.

BARRY, P. (1). 'The Magic Boat.' *JAFL*, 1915, **28**, 195.

DE BARY, W. T. (1). 'A Re-appraisal of Neo-Confucianism.' *AAN*, 1953, **55** (no. 5), pt. 2, 81. (American Anthropological Association Memoirs, no. 75.)

VON BASSERMANN-JORDAN, E. (1). *Alte Uhren und ihre Meister*. Diebener, Leipzig, 1926.

VON BASSERMANN-JORDAN, E. (2). *Die Geschichte d. Rädeuhr unter bes. Berücksichtigung d. Uhren d. bayrischen National-Museums*. Keller, Frankfurt-am-Main, 1905.

VON BASSERMANN-JORDAN, E. (3) (ed.). *Die Geschichte d. Zeitmessung u. d. Uhren*. de Gruyter, Berlin & Leipzig, 1920–5. Only three parts published: L. Borchardt, *Die Altägyptische Zeitmessung*, 1920; K. Schoy, *Gnomonik d. Araber*, 1923; J. Drecker, *Theorie d. Sonnenuhren*, 1925.

VON BASSERMANN-JORDAN, E. (4). *Uhren*. 4th ed. completely revised by H. von Bertele. Klinkhardt & Biermann, Braunschweig, 1961. (Bibliothek f. Kunst- u. Antiquitäten-freunde, no. 7.)

BATE, JOHN (1). *The Mysteryes of Nature and Art: conteined in foure severall Tretises, the first of Water Workes, the second of Fyer Workes, the third of Drawing, Colouring, Painting and Engraving, the fourth of Divers Experiments, as wel serviceable as delightful; partly collected, and partly of the author's peculiar practice and invention*. Harper, Mab, Jackson & Church, London, 1634, 1635. Bibliography in John Ferguson (2).

BATESON, G. & MEAD, M. (1). *Balinese Character, a Photographic Analysis*. Special Publications of the New York Academy of Sciences, 1942, no. 2.

BATHE, G. (1). *Horizontal Windmills*. Pr. pr. Philadelphia, 1948.

DE BAVIER, E. (1). *La Sériculture, le Commerce des Soies et des Graines, et l'Industrie de la Soie, au Japon*. Georg, Lyon, 1874; Dumolard, Milan, 1874; Baillière & Tindall, London, 1874.

BAYER, GOTTLIEB SIEGFRIED (1). *De Horis Sinicis et Cyclo Horario Commentationes, accedit eiusdem Auctoris Parergon Sinicum de Calendariis Sinicis, ubi etiam quaedam in Doctrina Temporum Sinica emendantur*. Academy of Sciences, St Petersburg, 1735.

BAZIN, M. (1). 'Notice du *Chan Hai King* [*Shan Hai Ching*]; Cosmographie Fabuleuse attribuée au Grand Yu.' *JA*, 1839 (3ᵉ sér.), **8**, 354.

BAZIN, M. (2). 'Le Siècle des Youen [Yuan], ou Tableau Historique de la Littérature Chinoise depuis l'Avènement des Empereurs Mongols jusqu'à la Restauration des Ming.' *JA*, 1850 (4ᵉ sér.), **15**, 1 & 101.

BEAL, S. (1) (tr.). *Travels of Fah-Hian [Fa-Hsien] and Sung-Yün, Buddhist Pilgrims from China to India (+400 and +518)*. Trübner, London, 1869. Incorporated in Beal (2).

BEAL, S. (2) (tr.). *Si Yu Ki [Hsi Yü Chi], Buddhist Records of the Western World, transl. from the Chinese of Hiuen Tsiang [Hsüan-Chuang]*. 2 vols. Trübner, London, 1884. 2nd ed. 1906. Repr. Susil Gupta, Calcutta, 1957, 4 vols., as *Chinese Accounts of India, translated from the Chinese of Hiuen Tsiang*.

BEAL, S. (3) (tr.). *The Life of Hiuen Tsiang [Hsüan-Chuang] by the Shaman [Śramana] Hwui Li [Hui-Li], with an Introduction containing an account of the Works of I-Tsing [I-Ching]*. Trübner, London, 1888; Kegan Paul, London, 1911.

BEATON, C. (1). *Chinese Album* (photographs). Batsford, London, 1945.

BEAUFOY, MARK (1). *Nautical and Hydraulic Experiments, with numerous scientific Miscellanies*. Pr. pr. London, 1834. 3 vols. announced but C.U.L. has only the first.

BEAUFOY, [MARK], Col. (2). 'On the Spiral Oar; Observations on the Spiral as a Motive Power to impel ships through the Water, with Remarks when applied to measure the Velocity of Water and Wind.' *TAP*, 1818, **12**, 246.

BECATTI, G. (1). *Scavi di Ostia*. 4 vols. Libreria dello Stato, Rome, 1960.

BECK, T. (1). *Beiträge z. Geschichte d. Maschinenbaues*. Springer, Berlin, 1900.

BECK, T. (2). 'Herons (des älteren) Mechanik.' *BGTI*, 1909, **1**, 84.

BECK, T. (3). 'Der altgriechische u. altrömische Geschützbau nach Heron dem älteren, Philon, Vitruv und Ammianus Marcellinus.' *BGTI*, 1911, **3**, 163.

BECK, T. (4). 'Herons (des älteren) Automatentheater.' *BGTI*, 1909, **1**, 182.

BECK, T. (5). 'Philon von Byzanz (etwa 260–200 v. Chr.).' *BGTI*, 1910, **2**, 64.

BECKER, C. O. & TITLEY, A. (1). 'The Valve Gear of Newcomen's Engine.' *TNS*, 1930, **10**, 6.

BECKMANN, J. (1). *A History of Inventions, Discoveries and Origins*. 1st German ed., 5 vols., 1786 to 1805. 4th ed., 2 vols. tr. by W. Johnston, Bohn, London, 1846. Enlarged ed., 2 vols. Bell & Daldy, London, 1872. Bibl. in John Ferguson (2).

BEDINI, S. (1). 'Johann Philipp Treffler, Clockmaker of Augsburg.' *BNAWCC*, 1956, **7**, 361, 415, 481.

BEDINI, S. (2). 'Chinese Mechanical Clocks.' *BNAWCC*, 1956, **7** (no. 4), 211.

BEDINI, S. (3). 'The Compartmented Cylindrical Clepsydra.' *TCULT*, 1962, **3**, 115.

BEDINI, S. (4). 'Agent for the Archduke; another chapter in the Story of Johann Philipp Treffler, Clockmaker of Augsburg.' *PHY*, 1961, **3**, 137.

BEHRENS, G. (1). 'Die sogennante Mithras-Symbole.' *GERM*, 1939, **23**, 56.

DE BÉLIDOR, B. F. (1). *Architecture Hydraulique; ou l'Art de Conduire, d'Elever et de Ménager les Eaux, pour les différens Besoins de la Vie.* 4 vols. Jombert, Paris, 1737–53.

BELL, ALEXANDER GRAHAM (1). 'The Tetrahedral Principle in Kite Structure.' *NGM*, 1903, **14**, 219. 'Aerial Locomotion.' *NGM*, 1907, **18**, 1. See also Grosvenor, G. H. (1).

BELL OF ANTERMONY, JOHN (1). *Travels from St Petersburg in Russia to Diverse Parts of Asia.* Vol. 1, *A Journey to Ispahan in Persia,* 1715 to 1718; *Part of a Journey to Pekin in China, through Siberia,* 1719 to 1721. Vol. 2, *Continuation of the Journey between Mosco and Pekin; to which is added, a translation of the Journal of Mr de Lange, Resident of Russia at the Court of Pekin,* 1721 and 1722, etc., etc. Foulis, Glasgow, 1763.

BELLINI, A. (1). *Gerolamo Cardano e il suo Tempo (sec. XVI).* Hoeppli, Milan, 1947.

BENNDORF, O., WEISS, E. & REHM, A. 'Zur Salzburger Bronzescheibe mit Sternbildern.' *JHOAI*, 1903, **6**, 32.

BENNETT, R. & ELTON, J. (1). *History of Corn Milling.* 4 vols. Simpkin Marshall, London, 1898. (i, *Headstones, Slave and Cattle Mills;* ii, *Watermills and Windmills;* iii, *Feudal Laws and Customs of Mills;* iv, *Some Famous Feudal Mills.*)

BENOIT, F. (1). 'L'Usine de Meunerie Hydraulique de Barbegal (Arles) à l'Époque Romaine.' *AHES/AHS*, 1939, **1**, 183; *RA*, 1940 (6e sér.), **15**, 18.

BENOIT, F. (3). 'Moulins à Graines et à Olives de la Méditerranée; Essai de Stratigraphie.' In *Travaux du 1er Congrès International de Folklore* (Paris, 1937). Arbault, Tours, 1938. (Pub. du Département et du Musée des Arts et Traditions Populaires.)

BERG, G. (1). *Sledges and Wheeled Vehicles.* Stockholm, 1935. (Nordiska Museets Handlingar, no. 4.)

BERG, G. (2). 'Den Svenska Sadesharpan och dem Kinesiska' (on the coming of the rotary winnowing-fan from China to Europe). Art. in *Nordiskt Folkminne; Studien tillagnade C. W. von Sydow.* Stockholm, 1928.

BERLIN, I. (1). *Two Concepts of Liberty.* OUP, Oxford, 1959.

BERNAL, J. D. (1). *Science in History.* Watts, London, 1954. (Beard Lectures at Ruskin College, Oxford.)

BERNARD, W. D. (1). *Narrative of the Voyages and Services of the 'Nemesis' from 1840 to 1843, and of the Combined Naval and Military Operations in China; comprising a complete account of the Colony of Hongkong, and Remarks on the Character of the Chinese, from the Notes of Cdr. W. H. Hall R.N., with personal observations.* 2 vols. Colburn, London, 1844.

BERNARD-MAÎTRE, H (1). *Matteo Ricci's Scientific Contribution to China,* tr. by E. T. C. Werner. Vetch, Peiping, 1935. Orig. pub. as *L'Apport Scientifique du Père Matthieu Ricci à la Chine,* Hsienhsien, Tientsin, 1935 (rev. Chang Yü-Chê, *TH*, 1936, **3**, 538).

BERNARD-MAÎTRE, H. (9). 'Deux Chinois du 18e siècle à l'École des Physiocrates Français.' *BUA*, 1949 (3e sér), **10**, 151.

BERNARD-MAÎTRE, H. (11). 'Ferdinand Verbiest, Continuateur de l'Œuvre Scientifique d'Adam Schall.' *MS*, 1940, **5**, 103.

BERNARD-MAÎTRE, H. (17). 'La Première Académie des Lincei et la Chine.' *MP*, 1941, 65.

BERNARD-MAÎTRE, H. (18). 'Les Adaptations Chinoises d'Ouvrages Européens; Bibliographie chronologique depuis la venue des Portugais à Canton jusqu'à la Mission Française de Pékin.' *MS*, 1945, **10**, 1–57, 309–88.

VON BERTELE, H. (1). 'Precision Time-Keeping in the pre-Huygens Era.' *HORJ*, 1953, **95**, 794. Germ. tr. *BTG*, 1954. Fr. tr. *JSHB*, 1954 (no. 9–10), 391, (no. 11–12) 463, 1955 (no. 3–4), 167.

VON BERTELE, H. (2). 'Clockwork Globes and Orreries.' *HORJ*, 1958, **100**, 800. Sep. pub. Antiq. Horol. Soc. London, 1958.

VON BERTELE, H. (3). 'Nuevos Documentos sobre la Obra de un Relojero Suizo Genial Jost Burgi, "el Segundo Arquímedes", +1552 to +1632.' *JSHB* (Spanish edition), 1956 (no. 11–12); 1957 (no. 1–2).

VON BERTELE, H. (4). 'The Origin of the Differential Gear and its Connection with Equation Clocks.' *TNS*, 1960, **30**, 145. (Paper read Oct. 1956.)

VON BERTELE, H. (5). 'Zur Geschichte der Äquationsuhren-Entwicklung.' *BTG*, 1956, **19**, 78.

VON BERTELE, H. (6). 'Globes and Spheres; Globen und Sphären; Globes et Sphères.' *Swiss Watch & Jewelry Journal*, Scriptar, Lausanne, 1961. (First part of a lecture at the Royal Society of Arts, before the British Horological Institute and the Antiquarian Horol. Soc. 28 Nov. 1957.)

VON BERTELE, H. (7). 'Das "Rädergebäude" des David a San Cajetano.' *AMK*, 1957, **2** (nos. 7–8), 23.

VON BERTELE, H. (8). 'Jost Burgis Beitrag zur Formentwicklung der Uhren.' *JKHSW*, 1955, **51**, 160.

VON BERTELE, H. (9). 'Jost Burgi's Pupils and Followers.' *CON*, 1955, **135**, 96.

VON BERTELE, H. (10). 'The Development of Equation Clocks; a phase in the History of Hand-setting Procedure.' *LSH*, 1959 (no. 3), 39 (no. 4), 15; 1960 (no. 1), 17 (no. 4), 37; 1961 (no. 1), 25.

VON BERTELE, H. See von Bassermann-Jordan (4).

BERTHELOT, M. (4). 'Pour l'Histoire des Arts Mécaniques et de l'Artillerie vers la Fin du Moyen Âge (1).' *ACP*, 1891 (6ᵉ sér.), **24**, 433. (Descr. of Latin MS. Munich, no. 197, the Anonymous Hussite engineer (German), *c.* +1430; of Ital. MS. Munich, no. 197, Marianus Jacobus Taccola of Siena, *c.* +1440; of *De Machinis*, Marcianus, no. XIX, 5, *c.* +1449; and of *De Re Militari*, Paris, no. 7239: Paulus Sanctinus, *c.* +1450, the MS. from Istanbul.)

BERTHELOT, M. (5). 'Histoire des Machines de Guerre et des Arts Mécaniques au Moyen Âge; (II) Le Livre d'un Ingénieur Militaire à la Fin du 14ème siècle.' *ACP*, 1900 (7ᵉ sér.), **19**, 289. (Descr. of MS. *Bellifortis*, Göttingen, no. 63, Phil., K. Kyeser +1395 to +1405; and of Paris MS., no. 11015, Latin, Guido da Vigevano, *c.* +1335.)

BERTHELOT, M. (6). 'Le Livre d'un Ingénieur Militaire à la Fin du 14ème Siècle.' *JS*, 1900; 1 and 85. (Konrad Kyeser and his *Bellifortis*.)

BERTHELOT, M. (7). 'Sur le Traité *De Rebus Bellicis* qui accompagne le *Notitia Dignitatum* dans les Manuscrits.' *JS*, 1900, 171.

BERTHELOT, M. (8). 'Les Manuscrits de Léonard da Vinci et les Machines de Guerre.' *JS*, 1902, 116. (Argument that L. da Vinci knew the drawings in the +4th-century Anonymous *De Rebus Bellicis*, and also many inventions and drawings of them by the +14th- and early +15th-century military engineers.)

BERTHELOT, M. (10). *La Chimie au Moyen Âge;* vol. 1, *Essai sur la Transmission de la Science Antique au Moyen Âge* (Latin texts). Impr. Nat. Paris, 1893.

BERTHELOT, M. (11). 'Sur le suspension dit de Cardan.' *CRAS*, 1890, **111**, 940.

BERTHOUD, F. (1). *Histoire de la Mésure du Temps par les Horloges.* 2 vols. Impr. de la République, Paris, An 10 (1802).

BERTUCCIOLI, G. (1). 'Un Sinologo Scomparso; Giovanni Vacca, 1872–1953.' *ICS*, 1953, **36** (no 4/5), 1.

BESSNITZER, U. (1). *Der Gezewg mit seiner Zugehörunge* (on weapons and machines), 1489. MS. Cod. Palat. Germ. 130, Univ. Heidelberg. See Sarton (1), vol. 3, p. 1553.

BESSON, JACQUES, (1). *Théatre des Instruments Mathématiques.* Lyon, 1578 (written before 1569). Latin ed. Lyon, 1582. Account in Beck (1), ch. 10. An earlier version, *Instrumentorum et Machinarum quas Jacobus Bessonus...excogitavit,* is dated 1569 but bears no indication of locality, perhaps Orléans, as the author died there in that year.

BESSON, JACQUES (2). *Le Cosmolabe, ou Instrument Universel concernant toutes Observations qui peuvent faire les Sciences Mathématiques, tant au Ciel, en la Terre, comme en la Mer.* de Roville, Paris, 1567.

DE BÉTANCOURT, AUGUSTIN (1). *Essai sur la Composition des Machines,* tr. from Spanish by Lanz. Paris, 1808. Engl. tr. *Analytical Essay on the Construction of Machines.* London, 1810.

BEYER, J. M. (1). *Theatrum Machinarum Molarium; Schauplatz der Mühlenbaukunst.* Leipzig, 1735; Dresden, 1767.

BIED-CHARRETON, R. (1). 'L'Utilisation de l'Énergie Hydraulique; ses Origines, ses Grandes Étapes.' *RHS*, 1955, **8**, 53.

BIGLAND, E. (1). *Journey to Egypt.* Jarrolds, London, 1948.

BILFINGER, G. (2). *Die Babylonische Doppelstunde; eine chronologische Untersuchung.* Wildt, Stuttgart, 1888.

BINYON, L. (1). *Chinese Paintings in English Collections.* Van Oest, Paris & Brussels, 1927. (Eng. tr. of the French text in Ars Asiatica series, no. 9.)

BIOT, E. (1) (tr.). *Le Tcheou-Li ou Rites des Tcheou* [*Chou*]. 3 vols. Imp. Nat., Paris, 1851. (Photographically reproduced Wêntienko, Peiping, 1930.)

BIOT, E. (4) (tr.). 'Traduction et Examen d'un ancien Ouvrage intitulé *Tcheou-Pei,* littéralement "Style ou signal dans une circonférence".' *JA*, 1841 (3ᵉ sér.), **11**, 593; 1842 (3ᵉ sér.), **13**, 198 (emendations). (Commentary by J. B. Biot, *JS*, 1842, 449.)

BIOT, E. (17). 'Notice sur Quelques Procédés Industriels connus en Chine au XVIᵉ siècle.' *JA*, 1835 (2ᵉ sér.), **16**, 130.

BIOT, E. (20). 'Mémoire sur la Condition des Esclaves et des Serviteurs gagés en Chine.' *JA*, 1837 (3ᵉ sér.), **3**, 246.

BIRCH, T. (1). *History of the Royal Society of London.* Millar, London, 1756.

BIRINGUCCIO, VANUCCIO (1). *Pirotechnia.* Venice, 1540, 1559. Eng. tr. C. S. Smith & M. T. Gnudi, Amer. Inst. Mining Engineers, New York, 1942. Account in Beck (1), ch. 7. See Sarton (1), vol. 3, pp. 1554, 1555. Bibliography in John Ferguson (2).

BISHOP, C. W. (1). 'Chronology of Ancient China.' *JAOS*, 1932, **52**, 232.

BISHOP, C. W. (2). 'Beginnings of Civilisation in Eastern Asia.' *AQ*, 1940, **14**, 301; *JAOS*, 1939, Suppl. no. 4, p. 35.

BLOCH, J. (1). *La Formation de la Langue Marathe.* Champion, Paris, 1920. (Bib. de l'École des Hautes Études Orientales, no. 215.)

BLOCH, MARC. See Febvre, L. (5).

BLOCH, MARC (2). 'Avènement et Conquêtes du Moulin à Eau.' *AHES*, 1935, **7**, 538.

BLOCH, MARC (3). 'Les Inventions Médiaevales.' *AHES*, 1935, **7**, 634; 1936, **8**, 513.

BLOCH, MARC (4). 'Les Techniques, l'Histoire, et la Vie; Note sur un Grand Problème d'Influences' (introduction to Haudricourt, 4). *AHES*, 1936, **8**, 513.

BLOCH, MARC (5). 'Technique et Évolution Sociale; à propos de l'Histoire de l'Attelage et de celle de l'Esclavage.' *RSH*, 1926, **41**, 91. Criticism of des Noëttes (1) who replied (5). Bloch's reply *RSH*, 1927, **43**, 87.

BLOCHMANN, H. F. (1) (tr.). *The 'Ā'īn-i Akbarī' (Administration of the Mogul Emperor Akbar) of Abū'l Fazl 'Allāmī*. Rouse, Calcutta, 1873. (Bibliotheca Indica, *NS*, nos. 149, 158, 163, 194, 227, 247 and 287.)

BLÜMNER, H. (1). *Technologie und Terminologie der Gewerbe und Künste bei Griechern und Römern*. 4 vols. Teubner, Leipzig & Berlin, 1912.

BOAS, M. (1). 'Heron's *Pneumatica*; a Study of its Transmission and Influence.' *ISIS*, 1949, **40**, 38.

BÖCKLER, G. A. (1). *Theatrum Machinarum Novum*. Schmitz & Fürst, Nuremberg, 1662, 1673, 1686. See Beck (1), ch. 23.

BODDE, D. (9). 'Some Chinese Tales of the Supernatural; Kan Pao and his *Sou Shen Chi*.' *HJAS*, 1942, **6**, 338.

BODDE, D. (10). 'Again Some Chinese Tales of the Supernatural; Further Remarks on Kan Pao and his *Sou Shen Chi*.' *JAOS*, 1942, **62**, 305.

BODDE, D. (12). *Annual Customs and Festivals in Peking, as recorded in the 'Yenching Sui Shih Chi'* [by Tun Li-Chhen]. Vetch, Peiping, 1936. (Revs. J. J. L. Duyvendak, *TP*, 1937, **33**, 102; A Waley, *FL*, 1936, **47**, 402.)

BODDE, D. (13). *China's Gifts to the West*. Amer. Council on Education, Washington, 1942. (Asiatic Studies in American Education, no. 1.)

BODDE, D. (17). 'The Chinese Cosmic Magic known as "Watching for the Ethers".' Art. in *Studia Serica Bernhard Karlgren Dedicata*. Ed. E. Glahn. Copenhagen, 1959. P. 14.

BODDE, D. (19). '*Lieh Tzu* and the Doves; a Problem of Dating.' *AM*, 1960, **7**, 25.

BOECKLER, G. A. See Böckler, G. A.

BOEHLING, H. B. H. (1). 'Chinesische Stampfbauten' (making of pisé-de-terre walls). *S*, 1951, **3**, 16.

BOERSCHMANN, E. (9). 'Die grosse Gebetmühle im Kloster Ta-Yüan-Si [Tha Yuan Ssu] auf dem Wu-Tai-Schan.' *SA* (Forke Festschrift Sonderausgabe), 1937, 35.

BOLTON, L. (1). *Time Measurement*. Bell, London, 1924.

BONAPARTE, NAPOLEON (1) (ed.). *Description de l'Égypte, ou Receuil des Observations et des Recherches qui ont été faites en Égypte pendant l'Expédition de l'Armée Française*. 9 vols., with Atlas in 2 vols. Paris 1809–22.

BONI, B. (1). 'Sull'origine italiana delle Trombe idroeoliche.' *LFI*, 1958, **7** (no. 5), 161.

BONI, B. (2). 'Aldo Mieli, 1879–1950.' *CHIM*, 1956 (no. 2).

BONNANT, G. (1). 'The Introduction of Western Horology into China.' *LSH* (Internat. ed.), 1960 **75** (no. 1), 28. Sep. pub., Geneva, 1960.

BONVALOT, G. (1). *Across Thibet*. 2 vols. Tr. C. B. Pitman. Cassell, London, 1891.

Book of Trades
   c. +1310 (Ypres MS.). See Gutmann (1).
   1804, 1811. See Anon. (62).
   1835. See Anon. (63).
   1835, 1852. See Anon. (64).
   1837, 1842. See Whittock, N. *et al*. (1).
   1862. See Anon. (65).

BORCHARDT, L. (2). *Das Grabdenkmal des Königs Ne-User-Re*. Hinrichs, Liepzig, 1907.

BORCHARDT, L. (3). 'Schiffahrt auf dem Lande.' *ASAE*, 1939, **39**, 377.

BORN, W. (1). 'Man's Labour throughout the Year' (in relation to the crafts and the signs of the zodiac). *CIBA/T*, 1939 (no. 22), 771.

BORN, W. (2). 'Craftsmen as Children of the Planets.' *CIBA/T*, 1939 (no. 22), 779.

BORN, W. (3). 'The Spinning-Wheel.' *CIBA/T*, 1939, **3** (no. 28), 982.

BOSSERT, H. T. & STORCK, W. F. (1). *Das mittelälterliches Hausbuch*... (Anon.15). Seemann, Leipzig, 1912.

BOURNE, F. S. A., NEVILLE, A. H. & BELL, H. (1). *Report of the Mission to China of the Blackburn Chamber of Commerce, 1896–97*. Ed. W. H. Burnett. 2 vols. Blackburn, 1898.

BOURNE, J. (1). *A Treatise on the Screw Propeller, Screw Vessels, and Screw Engines*. London, 1867.

BOWDEN, F. P. (1). *Recent Studies of Metallic Friction* (Thomas Hawksley Lecture, Institution of Mechanical Engineers). Preprint; London, 1954.

BOWDEN, F. P. & YOFFE, A. D. (1). *The Initiation and Growth of Explosion in Liquids and Solids*. Cambridge, 1952. (Cambridge Monographs on Physics.)

Boxer, C. R. (1) (ed.). *South China in the Sixteenth Century; being the Narratives of Galeote Pereira, Fr. Gaspar da Cruz, O.P., and Fr. Martin de Rada, O.E.S.A. (1550–1575).* Hakluyt Society, London, 1953. (Hakluyt Society Pubs. 2nd series, no. 106.)

Boyer, M. N. (1). 'Mediaeval Suspended Carriages.' *SP*, 1959, **34**, 359.

Boyer, M. N. (2). 'Mediaeval Pivoted Axles.' *TCULT*, 1960, **1**, 128.

Boys, C. V. (1). 'A Recording and Integrating Gas Calorimeter.' *PIGE*, 1922.

Boys, C. V. (2). 'My Recent Progress in Gas Calorimetry.' *PPS*, 1936, **48**, 881.

van Braam Houckgeest, A. E. (1). *An Authentic Account of the Embassy of the Dutch East-India Company to the Court of the Emperor of China in the years 1794 and 1795 (subsequent to that of the Earl of Macartney), containing a Description of Several Parts of the Chinese Empire unknown to Europeans; taken from the Journal of André Everard van Braam, Chief of the Direction of that Company, and Second in the Embassy.* Tr. L. E. Moreau de St Méry. 2 vols., map, but no index and no plates; Phillips, London, 1798. French ed. 2 vols., with map, index and several plates; Philadelphia, 1797. The two volumes of the English edition correspond to vol. 1 of the French edition only.

de Brakelond, Jocelyn (1). *Memorials of St Edmund's Abbey.* Ed. T. Arnold. Rolls Series, London, 1890. See also Butler, H. E.

Branca, Giovanni (1). *Le Machine.* Rome, 1629. Account in Beck (1), ch. 24.

Braun, G. & Hogenberg, F. (1). *Civitates Orbis Terrarum....* 1577 to 1588.

Breasted, J. H. (1). *The Conquest of Civilisation.* Harper, New York, 1926.

Bretschneider, E. (1). *Botanicon Sinicum; Notes on Chinese Botany from Native and Western sources.* 3 vols. Trübner, London, 1882 (printed in Japan). (Repr. from *JRAS/NCB*, 1881, **16**.)

Bretschneider, E. (2). *Mediaeval Researches from Eastern Asiatic Sources; Fragments towards the Knowledge of the Geography and History of Central and Western Asia from the Thirteenth to the Seventeenth Century.* 2 vols. Trübner, London, 1888.

Brett, G. (1). 'A Byzantine Water-Mill.' *AQ*, 1939, **13**, 354.

Brett, G. (2). 'The Automata in the Byzantine "Throne of Solomon".' *SP*, 1954, **29**, 477.

Breusing, A. (1). *Die nautischen Instrumente bis zur Erfindung des Spiegelsextanten.* Bremen, 1890.

Brewer, F. W. (1). 'Notes on the History of the Engine Crank and its Application to Locomotives.' *LRCW*, 1932, **38**, 373.

Brewitt-Taylor, C. H. (1) (tr.). '*San Kuo*', or the Romance of the Three Kingdoms. Kelly & Walsh, Shanghai, 1926. Reissued as *Lo Kuan-Chung's 'Romance of the Three Kingdoms', 'San Kuo Chih Yen I'*. Tuttle, Rutland, Vermont, and Tokyo, 1959.

Brinckmann, J. (1). *Kunst und Handwerk in Japan.* Wagner, Berlin, 1889.

Brittain, R. (1). *Rivers, Man and Myths.* Doubleday, New York, 1958; Longmans, London, 1958.

Britten, F. J. (1). *Old Clocks and Watches, and their Makers.* 6th ed. Spon, London, 1932. New ed. edited G. H. Baillie, C. Chitton & C. A. Ilbert, 1954.

Brøgger, A. W. & Schetelig, H. (2). *Osebergfundet.* 4 vols. Univ. Oldsaksamling, Oslo, 1928.

Broke-Smith, Brig. P. W. L. (1). 'The History of Early British Military Aeronautics.' *REJ*, 1952, **66**, 1, 105, 208. Sep. pub. pr. pr.

Bromehead, C. E. N. (6). 'The Early History of Water Supply.' *GJ*, 1942, **99**, 142 and 183.

Bromehead, C. E. N. (7). 'Ancient Mining Processes as illustrated by a Japanese Scroll [*c.* +1640].' *AQ*, 1942, **16**, 193.

Brøndsted, J. (1). 'Danish Inhumation Graves of the Viking Age.' *ACA*, 1936, **7**, 81.

Brøndsted, J. (2). *Early English Ornament; the Sources, Development and Relation to Foreign Styles of pre-Norman Ornamental Art in England.* Hachette, London, 1924; Levin & Munksgaard, Copenhagen, 1924.

Brooks, P. W. (1). 'The Development of the Aeroplane' (Cantor Lecture). *JRSA*, 1959, **107**, 97.

Brooks, P. W. (2). 'Aeronautics [in the late Nineteenth Century].' Art. in *A History of Technology.* Ed. C. Singer *et al.* Vol. 5, p. 391. Oxford, 1958.

Brown, C. L. M. (1). *The Conquest of the Air; an Historical Survey.* Oxford, 1927.

Brown, R. A. (1). 'King Edward's Clocks' (King Edward III of England; Windsor, Westminster, Queensborough and Langley, +1351 to +1377). *ANTJ*, 1959, **39**, 283.

Brunet, P. & Mieli, A. (1). *L'Histoire des Sciences (Antiquité).* Payot, Paris, 1935.

Brunot, L. (1). 'Le Moulin à Manège à Rabat-Salé.' In *Memorial Henri Basset; Nouvelles Études Nord-Africaines et Orientales*, p. 91. (Pub. de l'Institut des Hautes Études Marocaines.) Geuthner, Paris, 1928.

de Bry, Theodor (1). *Indiae Orientalis.* Frankfurt, 1599. Part of *Collectiones Peregrinationum in Indiam Orientalem et Occidentalem*, 1590 to 1634.

Buffet, B. & Evrard, R. (1). *L'Eau Potable à travers les Âges.* Solédi, Liége, 1950.

Bullett, Gerald & Tsui Chi (1). *The Golden Year of Fan Chhêng-Ta.* Cambridge, 1946.

BULLING, A. (1). 'Descriptive Representations in the Art of the Chhin and Han Period.' Inaug. Diss., Cambridge, 1949.

BULLING, A. (4). '[Umbrella Motifs in the] Decoration of Chou and Han [Metal] Mirrors.' Communication to the 23rd International Congress of Orientalists. Cambridge, Sept. 1954.

BULLING, A. (6). 'The Meaning of China's most Ancient Art; an Interpretation of Pottery Patterns from Kansu (Ma-chhang and Pan-shan) and their Development in the Shang, Chou and Han periods.' Brill, Leiden, 1952 (Revs. H. G. Creel, AA, 1953, 16, 320; J. Needham, A/AIHS, 1954, 7, 71).

BULLING, A. (7). 'The Decoration of Some Mirrors of the Chou and Han Periods.' AA, 1955, 18, 20.

BULLING, A. (8). The Decoration of Mirrors of the Han Period; a Chronology. Artibus Asiae, Ascona, 1960 (AA, Suppl. no. 20). (Rev. S. Cammann, JAOS, 1961, 81, 331.)

BULLING, A. (10). 'A Bronze Cart [of Chou or Han].' ORA, 1955 (n.s.), 1, 127.

BURFORD, A. (1). 'Heavy Transport in Classical Antiquity.' EHR, 1960 (2nd. ser.), 13, 1.

BURGESS, E. (1) (tr.). Sūrya Siddhānta; translation of a Textbook of Hindu Astronomy, with notes and an appendix. Ed. P. Gangooly, introd. by P. Sengupta. Calcutta, 1860. (Reprinted 1935.)

BURKITT, M. C. (1). The Old Stone Age. Cambridge (2nd ed.), 1949.

BURKITT, M. C. (2). Our Early [Neolithic] Ancestors. Cambridge, 1926.

BURKITT, M. C. (3). Prehistory. Cambridge, 1925.

BURLINGAME, R. (1). March of the Iron Men; a Social History of Union through Invention. Scribner, New York, 1939.

BURLINGAME, R. (2). Engines of Democracy; Inventions and Society in Mature America. Scribner, New York, 1940.

BURLINGAME, R. (3). Backgrounds of Power. Scribner, New York, 1950.

BURSTALL, A. (1). A History of Mechanical Engineering. Faber & Faber, London, 1963.

BURSTALL, A. F. (2). 'Experimental Working Models of the Chinese Double-acting Piston Bellows and of a (simplified) Chinese Water-wheel Escapement.' In Needham (41), with illustrations.

BURSTALL, A. F., LANSDALE, W. E. & ELLIOTT, P. (1). 'A Working Model of the Mechanical Escapement in Su Sung's Astronomical Clock Tower.' N, 1963, 199, 1242.

BURY, J. B. (1). The Idea of Progress. Macmillan, London, 1920.

BUSCHAN, G., BYHAN, A., VOLZ, W., HABERLANDT, A. & M., & HEINE-GELDERN, R. (1). Illustrierte Völkerkunde. 2 vols. in 3. Stuttgart, 1923.

BUSS, K. (1). Studies in the Chinese Drama. Four Seas, Boston, Mass., 1927 (lim. ed.). 2nd Ed., Cope & Smith, New York, 1930.

BUTLER, H. E. (1) (tr.). The Chronicle of Jocelin of Brakelond. Nelson, London, 1951.

BUTTER, F. J. (1). Locks and Lockmaking. Pitman, London, 1926.

BUTTERFIELD, H. (1). The Origins of Modern Science, +1300 to 1800. Bell, London, 1949.

CABLE, M. & FRENCH, F. (2). China, her Life and her People. Univ. of London Press, London, 1946.

CAHEN, G. (1). Some Early Russo-Chinese Relations. Tr. and ed. W. S. Ridge. National Review, Shanghai, 1914. Repr. Peking, 1940. Orig. ed. Histoire des Relations de la Russie avec la Chine sous Pierre le Grand (+1689 à +1730). 1912.

CALDERINI, A. (1). 'Macchine Idrofore secondo i Papiri Greci.' RRIL, 1920 (ser. II), 53, 620.

CALZA, G. (1). Ostia. Libreria dello Stato, Rome, 1950. (Ministero della Pubblica Istruzione; Itinéraires des Musées et Monuments d'Italie.)

CAPOT-REY, R. (1). Géographie de la Circulation sur les Continents. Gallimard, Paris, 1946.

CARDAN, JEROME (1). De Subtilitate. Nuremberg, 1550 (ed. Sponius); Basel, 1560. Account in Beck (1), ch. 9. Bibliography in John Ferguson (2).

CARDAN, JEROME (2). De Rerum Varietate, Basle, 1557.

CARDWELL, D. S. L. (1). Steam Power in the +18th Century; a Case Study in the Application of Science. Sheed & Ward, London, 1963. (Newman History & Philosophy of Science Series, no. 12.)

CARLETON, M. A. (1). Emmer; a Grain for Semi-arid Regions. U.S. Dept. of Agriculture Farmers' Bulletin no. 139. Washington, D.C., 1901.

CARLYLE, THOMAS (1). Past and Present (contains the account of the Chronicles of the Abbey of Bury St Edmunds). London, 1843.

CARTER, T. F. (1). The Invention of Printing in China and its Spread Westward. Columbia Univ. Press, New York, 1925, revised ed. 1931. 2nd ed. revised by L. Carrington Goodrich. Ronald, New York, 1955.

CARUS-WILSON, E. M. (1). 'An Industrial Revolution of the 13th century' (the fulling mill in England). EHR, 1941, 11, 39.

CARUS-WILSON, E. M. (2). 'The Woollen Industry [in the Middle Ages].' Ch. 6 in Cambridge Economic History of Europe. Ed. M. Postan & E. E. Rich. Cambridge, 1952. Vol. 2, p. 355.

CARY, G. (1). *The Medieval Alexander.* Ed. D. J. A. Ross. Cambridge, 1956. (A study of the origins and versions of the Alexander-Romance; important for medieval ideas on flying-machine and diving-bell or bathyscaphe.)

CARY, M. (1). 'Maës, qui et Titianus.' *CQ*, 1956, **6** (n.s.), 130.

CASH, J. A. (1). 'Behind the News in Rumania, Bessarabia and Bukovina.' *GGM*, 1936, **3**, 143; 1940, **10**, 143.

DE CAUS, ISAAC (1), Ingénieur et Architecte, Natif de Dieppe. *Nouvelle Invention de lever l'Eau plus Hault que sa Source, avec quelques Machines movantes par le moyen de l'eau et un discours de la conduite d'ycelle.* London, 1644. Eng. tr. by John Leak, Moxon, London, 1659.

DE CAUS, SOLOMON (1). *Les Raisons des Forces mouvantes avec diverses Machines et plusieurs Desseins de Grottes et de Fontaines.* Frankfurt, 1615. Account in Beck (1), ch. 21.

CAYLEY, Sir GEORGE (1). *Aeronautical and Miscellaneous Notebook, c. 1799–1826.* Ed. J. E. Hodgson. Heffer, Cambridge, 1933. (Newcomen Society Extra Publications, no. 3.)

CAYLEY, Sir GEORGE (2). *On Aerial Navigation.* Collected (but abridged) papers from *JNPCA*, 1809, **24**, 164 and 1810, **25**. Ed. Hubbard & Ledeboer, ASGB, London, 1910. (Aeronautical Classics series, no. 1.)

CAYLEY, Sir GEORGE (3). 'Retrospect of the Progress of Aerial Navigation, and Demonstration of the Principles by which it must be governed.' *MMG*, 1843, **38**, 263.

CHADWICK, N. K. (1). 'The Kite; a Study in Polynesian Tradition.' *JRAI*, 1931, **61**, 457.

CHAKRAVARTI, P. C. (1). 'The Art of War in Ancient India.' Univ. of Dacca Press, Ramna, Dacca, 1941. (Univ. of Dacca Bulletin, no. 21.)

CHALMERS, T. W. (1). *The Gyroscopic Compass.* Constable, London, 1920.

CHAMBERLAIN, B. H. (1). *Things Japanese.* Murray, London. 2nd ed. 1891; 3rd ed. 1898.

CHAMBERS, EPHRAIM (1). *Cyclopaedia; or, An Universal Dictionary of Arts and Sciences.* 2 vols. London, 1728; 2nd ed. 1738. Italian tr. Venice, 1749. 7th ed. 1752 with suppl. by G. L. Scott; 8th ed., 5 vols., 1788 reorganised by Abraham Rees; enlarged 1819.

CHAMBERS, Sir WM. (1). *Designs of Chinese Buildings, Furniture, Dresses, Machines and Utensils; to which is annexed, A Description of their Temples, Houses, Gardens, etc.* London, 1757.

CHANG HAN-YING (1). 'New View of Water Conservancy.' *CREC*, 1959, **8** (no. 8), 2.

CHANG LIEN-CHANG (1). 'The [Chinese] Clock and Watch Industry.' *CREC*, 1962, **11** (no. 2), 31.

CHANG YÜ-CHÊ (1). 'Chang Hêng, a Chinese Contemporary of Ptolemy.' *POPA*, 1945, **53**, 1.

CHANG YÜ-CHÊ (2). 'Chang Hêng, Astronomer.' *PC*, 1956 (no. 1), 31.

CHANUTE, O. (1). *Progress in Flying Machines.* New York, 1894, 1899.

CHAO WAN-LI (1). 'The Yung-Lo Encyclopaedia.' *CLIT*, 1959 (no. 6), 142.

CHAPIN, H. B. (1). 'Kyongju, ancient Capital of Silla' and 'Korea in Pictures.' *AH*, 1948, **1** (no. 4), 36.

CHAPMAN, S. & HARRADON, H. D. (1). 'Archaeologica Geomagnetica; Some Early Contributions to the History of Geomagnetism; I, The Letter of Petrus Peregrinus de Maricourt to Sygerus de Foucancourt, soldier, concerning the Magnet (+1269).' *TM*, 1943, **48**, 1, 3.

CHAPUIS, A. (1). 'Relations de l'Horlogerie Suisse avec la Chine; la Montre Chinoise.' In Chapuis, Loup & de Saussure q.v.

CHAPUIS, A. (2). 'Les Jeux d'Eau et les Automates Hydrauliques du Parc d'Hellbrunn près Salzburg.' *LSH*, 1952, **67** (no. 4), 39.

CHAPUIS, A. & DROZ, E. (1). *Les Automates; Figures Artificielles d'Hommes et Animaux; Histoire et Technique.* Neuchâtel, 1950.

CHAPUIS, A. & GELIS, E. (1). *Le Monde des Automates; Étude Historique et Technique.* 2 vols. Priv. publ., Paris, 1928.

CHAPUIS, A., LOUP, G. & DE SAUSSURE, L. (1). *La Montre 'Chinoise'.* Attinger, Neuchâtel, n.d. (1919) (rev. P. Pelliot, *TP*, 1921, **20**, 61).

CHATLEY, H. (2). 'The Development of Mechanisms in Ancient China.' *TNS*, 1942, **22**, 117. (Long abstr. without illustr., *ENG*, 1942, **153**, 175.)

CHATLEY, H. (23). *The Origin and Diffusion of Chinese Culture.* China Soc. London, Oct. 1947.

CHATLEY, H. (36). 'Far Eastern Engineering.' *TNS*, 1954, **29**, 151. With discussion by J. Needham, A. Stowers, A. W. Skempton, S. B. Hamilton *et al.*

CHAVANNES, E. (1). *Les Mémoires Historiques de Se-Ma Ts'ien [Ssuma Chhien].* 5 vols. Leroux, Paris, 1895–1905. (Photographically reproduced, in China, without imprint and undated.)

    1895   vol. 1 tr. *Shih Chi*, chs. 1, 2, 3, 4.

    1897   vol. 2 tr. *Shih Chi*, chs. 5, 6, 7, 8, 9, 10, 11, 12.

    1898   vol. 3 (i) tr. *Shih Chi*, chs. 13, 14, 15, 16, 17, 18, 19, 20, 21, 22.

                vol. 3 (ii) tr. *Shih Chi*, chs. 23, 24, 25, 26, 27, 28, 29, 30.

    1901   vol. 4 tr. *Shih Chi*, chs. 31, 32, 33, 34, 35, 36, 37, 38, 39, 40, 41, 42.

    1905   vol. 5 tr. *Shih Chi*, chs. 43, 44, 45, 46, 47.

CHAVANNES, E. (7). 'Le Cycle Turc des Douze Animaux.' *TP*, 1906, **7**, 51.

CHAVANNES, E. (9). *Mission Archéologique dans la Chine Septentrionale.* 2 vols. and portfolios of plates. Leroux, Paris, 1909–15. (Publ. de l'École Franç. d'Extr. Orient, no. 13.)

CHAVANNES, E. (11). *La Sculpture sur Pierre en Chine aux Temps des deux dynasties Han.* Leroux, Paris, 1893.

CHEN, GIDEON. See Chhen Chhi-Thien.

CHÊNG CHEN-TO. See Průsek (3).

CHÊNG TÊ-KHUN (2) (tr.). *Travels of the Emperor Mu.* JRAS/NCB, 1933, **64**, 124; 1934, **65**, 128.

CHÊNG TÊ-KHUN (4). 'An Introduction to Chinese Civilisation' (mainly prehistory). ORT, 1950: Aug. p. 28, 'Early Inhabitants'; Sept. p. 28, 'The Beginnings of Culture'; Oct. p. 29, 'The Building of Culture'.

CHÊNG TÊ-KHUN (5) (ed.). *Illustrated Catalogue of an Exhibition of Chinese Paintings from the Mu-Fei Collection* (held in connection with the 23rd International Congress of Orientalists). Fitzwilliam Museum, Cambridge, 1954.

CHÊNG TÊ-KHUN (9). *Archaeology in China*: vol. 1, *Prehistoric China*; vol. 2, *Shang China*; vol. 3, *Chou China*; vol. 4, *Han China*. Heffer, Cambridge, 1958– .

CHERRY, E. C. (1). 'A History of the Theory of Information.' In *Symposium on Information Theory.* Min. of Supply, London, 1950 (mimeographed), p. 22.

CHERRY, E. C., HICK, W. E. & McKAY, D. M. (1). Symposium on Cybernetics. ADVS, 1954, **10**, 393; N, 1953, **172**, 648.

CHERRY, T. M. (1). 'Anthony George Maldon Michell, 1870–1959.' BMFRS, 1962, **8**, 91.

CHESNEAUX, J. & NEEDHAM, J. (1). 'Les Sciences en Extrême-Orient du 16ème au 18ème Siècle.' In *Histoire Générale des Sciences*, vol. 2, p. 681. Ed. R. Taton. Presses Universitaires de France, Paris, 1958.

CHESTNUT, H. & MAYER, R. W. (1). *Servomechanisms and Regulating System Design.* Wiley, New York, 1951; Chapman & Hall, London, 1951 (General Electric monograph series, no. 1). (Rev. J. Greig, N, 1953, **172**, 91.)

CHEVALIER, H. (1). *Cérémonial de l'Achèvement des Travaux de Hoa-Syeng (Corée), 1800, traduction et resumé* (an illustrated Korean work on city fortifications described). TP, 1898, **9**, 394.

CHEVALIER, H. (2). 'La Charrue en Asie.' GC, 1899, **36**, 26; 1901, **38**, 346.

CHHEN CHHI-THIEN (1). *Lin Tsê-Hsü; Pioneer Promoter of the Adoption of Western Means of Maritime Defence in China.* Dept. of Economics, Yenching Univ., Vetch (French Bookstore), Peiping, 1934. ([Studies in] Modern Industrial Technique in China, no. 1.)

CHHEN CHHI-THIEN (2). *Tsêng Kuo-Fan; Pioneer Promoter of the Steamship in China.* Dept. of Economics, Yenching Univ., Vetch (French Bookstore), Peiping, 1935. ([Studies in] Modern Industrial Technique in China, no. 2.)

CHHEN CHHI-THIEN (3). *Tso Tsung-Thang; Pioneer Promoter of the Modern Dockyard and the Woollen Mill in China.* Dept. of Economics, Yenching Univ., Vetch (French Bookstore), Peiping, 1938. ([Studies in] Modern Industrial Technique in China, no. 3.)

CHHEN PO-SAN (1). 'Cable-tow Traction for Farm Tools.' PKR, 1959 (no. 8), 23.

CHHEN TSU-LUNG (1). 'Table de Concordance des Numérotages des Grottes de Touen-Hoang [Tunhuang].' JA, 1962, **250**, 257.

CHHU TA-KAO (2) (tr.). *Tao Tê Ching, a new translation.* Buddhist Lodge, London, 1937.

CHI CHHAO-TING (1). *Key Economic Areas in Chinese History, as revealed in the Development of Public Works for Water-Control.* Allen & Unwin, London, 1936.

CHIANG FÊNG-WEI (1) (ed.). *Gems of Chinese Literature.* Progress Press, Chungking, 1942.

CHIANG KANG-HU (1). *On Chinese Studies.* Com. Press, Shanghai, 1934. (On printing, p. 252; agriculture and farm implements, pp. 295, 304; dentistry, p. 331.)

CHIANG SHAO-YUAN (1). *Le Voyage dans la Chine Ancienne, considéré principalement sous son Aspect Magique et Religieux.* Commission Mixte des Œuvres Franco-Chinoises (Office de Publications), Shanghai, 1937. Transl. from Chinese by Fan Jen.

CHILDE, V. GORDON (1). *The Bronze Age.* Cambridge, 1930.

CHILDE, V. GORDON (9). 'Rotary Querns on the Continent and in the Mediterranean Basin.' AQ, 1943, **17**, 19.

CHILDE, V. GORDON (10). 'The First Waggons and Carts—from the Tigris to the Severn.' PPHS, 1951, 177.

CHILDE, V. GORDON (11). 'Rotary Motion [down to −1000]'. Art. in *History of Technology*, vol. 1, p. 187. Ed. C. Singer, E. J. Holmyard & A. R. Hall. Oxford, 1954.

CHILDE, V. GORDON (13). 'A Prehistorian's Interpretation of Diffusion.' In *Independence, Convergence and Borrowing, in Institutions, Thought and Art*, p. 3. Harvard Tercentenary Publication, Harvard Univ. Press, Cambridge, Mass., 1937.

CHILDE, V. GORDON (16). 'Wheeled Vehicles [in Early Times to the Fall of the Ancient Empires].' Art. in *A History of Technology*, ed. C. Singer *et al.*, vol. 1, p. 716. Oxford, 1954.

CHRISTENSEN, A. (1). *l'Iran sous les Sassanides*. Levin & Munksgaard, Copenhagen, and Geuthner, Paris, 1936 (2nd ed. 1942, Copenhagen only). (Ann. Mus. Guimet, Bibl. d'Études, no. 48.)

CHURCHILL, A. & CHURCHILL, J. (1) (ed.). *A Collection of Voyages and Travels....* Churchill, London, 1704; 2nd ed. 1732–52.

[CIBOT, P. M.] (4). 'Du Kong-Pou [Kung Pu], ou du Tribunal des Ouvrages Publics.' *MCHSAMUC*, 1782, **8**, 278.

CLARK, GRAHAME (1). 'Water in Antiquity.' *AQ*, 1944, **18**, 1.

CLARK, H. O. (1). 'Notes on Horse-Mills.' *TNS*, 1928, **8**, 33.

CLAVIJO, RUY GONZALES (1). *Embassy to Tamerlane*. London, 1928.

CLERK-MAXWELL, J. (1). 'On the Motion of Governor-Balls in Steam Engines.' *PRS*, 1868, **16**, 270.

CLINE, W. (1). *Mining and Metallurgy in Negro Africa*. Banta, Menasha, Wisconsin, 1937 (mimeographed). (General Studies in Anthropology, no. 5, Iron.)

CLUTTON, C. (1). 'The First Mechanical Escapement?' *AHOR*, 1962, **3**, 332.

COALES, J. (1). 'The Historical and Scientific Background of Automation.' *ENG*, 1956, **182**, 363.

COBBETT, L. (1). 'Mediterranean Windmills.' *AQ*, 1939, **13**, 458.

COGHLAN, H. H. (2). 'The Prehistory of the Hammer.' *TNS*, 1946, **25**, 181.

COHN, W. (2). *Chinese Painting*. Phaidon, London, 1948; 2nd ed. 1951.

COLANI, M. (5). 'Ethnographie Comparée; V, Pièces et Coutumes Extrême-Orientale ou Indonésienne en Indochine.' *BEFEO*, 1938, **38**, 212.

COLANI, M. (6). 'Ethnographie Comparée; VI, Pièces paraissant être d'Origine Indochinoise.' *BEFEO*, 1938, **38**, 225. 'VII, Documents ethnographiques divers.' *BEFEO*, 1938, **38**, 233.

COLIN, G. S. (1). 'La Noria Marocaine et les Machines Hydrauliques dans le Monde Arabe.' *HP*, 1932, **14**, 22.

COLLADON, M. & CHAMPIONNIÈRE, M. (1). 'Note sur les Machines à Vapeur de Savery.' *ACP*, 1835, **59**, 24.

COMBRIDGE, J. H. (1). 'The Celestial Balance; a Practical Reconstruction.' *HORJ*, 1962, **104**, 82. Repr. Antiq. Horol. Soc., London, 1962.

COMBRIDGE, J. H. (2). 'The Chinese Water-Balance Escapement.' *N*, 1964, **204**, 1175

COMBRIDGE, J. H. (3). 'The Chinese Water-Clock.' *HORJ*, 1963, **105**, 347.

CONRADIS, H. (1). 'Alte Baggermaschinen.' *BGTI/TG*, 1937, **26**, 51.

DE CONTI, NICOLÒ (1). In *The Most Noble and Famous Travels of Marco Polo, together with the Travels of Nicolò de Conti, edited from the Elizabethan translation of J. Frampton (1579), etc.*, by N. M. Penzer. Argonaut, London, 1929. Bibliography by Cordier (5).

CONWAY, H. G. (1). 'Some Notes on the Origins of Mechanical Servo-Mechanisms.' *TNS*, 1954, **29**, 55.

CONZE, E. (6). 'Marginal Notes to the *Abhisamayālaṅkāra*.' *SIS*, 1957, **5** (no. 3), 1.

CONZE, E. (7). 'Recent Progress in Buddhist Studies.' *MW*, 1959, **34** (no. 1), 6.

COOK, R. M. (1). *Greek Painted Pottery*. Methuen, London, 1960.

COOMARASWAMY, A. K. (2). 'The Treatise of al-Jazarī on Automata [+1206]; Leaves from a MS. of the *Kitāb fī Maʿarifat al-Ḥiyal al-Handasīya* in the Museum of Fine Arts, Boston, and elsewhere.' Mus. of Fine Arts, Boston, 1924. (Communications to the Trustees, no. 6.)

COOMARASWAMY, A. K. (4). 'The Persian Wheel [Noria].' *JAOS*, 1931, **51**, 283.

COOMARASWAMY, A. K. (5). *Mediaeval Sinhalese Art*. 2nd ed. Pantheon, New York, 1956.

CORDIER, H. (1). *Histoire Générale de la Chine*. 4 vols. Geuthner, Paris, 1920.

CORDIER, H. (5). 'Deux Voyageurs dans l'Extrême-Orient au 15e et 16e Siècles.' *TP*, 1899, **10**, 380 (de Conti and Varthema; bibliographical only).

CORDIER, H. (8). *Essai d'une Bibliographie des Ouvrages publiés en Chine par les Européens au 17e et au 18e siècle*. Leroux, Paris, 1883.

LE CORNU, J. (1). *Les Cerfs-Volants*. Nony, Paris, 1902.

CORONELLI, VINCENZO (1). *Epitome Cosmographica*. Poletti, Venice, 1693.

COULING, S. (1). *Encyclopaedia Sinica*. Kelly & Walsh, Shanghai; Oxford and London, 1917.

COURANT, M. (3). *Catalogue des Livres Chinois, Coréens, Japonais, etc. dans le Bibliothèque Nationale, Département des Manuscrits*. Leroux, Paris, 1900–12.

COUVREUR, F. S. (1) (tr.). '*Tch'ouen Ts'iou*' [*Chhun Chhiu*] *et* '*Tso Tchouan*' [*Tso Chuan*]; *Texte Chinois avec Traduction Française*. 3 vols. Mission Press, Hochienfu, 1914.

COUVREUR, F. S. (2). *Dictionnaire Classique de la Langue Chinoise*. Mission Press, Hsienhsien, 1890 (photographically reproduced Vetch, Peiping, 1947).

COX, JAMES (1). *A Descriptive Catalogue of the several Superb and Magnificent Pieces of Mechanism and Jewellery exhibited in Mr Cox's Museum at Spring Gardens, Charing Cross*. Cox, London, 1772.

COX, JAMES (2). *A Descriptive Inventory of the several Exquisite and Magnificent Pieces of Mechanism and Jewellery comprised in the Schedule annexed to an Act of Parliament made in the 13th year of*

H.M. King George III, for enabling Mr James Cox of the City of London, Jeweller, to dispose of his Museum by way of Lottery. Cox, London, 1774.

[Cox, JAMES] (3). *A Description of a Most Magnificent Piece of Mechanism and Art.* London, 1769.

CRANMER-BYNG, J. L. (1). 'Lord Macartney's Embassy to Peking in +1793 from Official Chinese Documents.' *JOSHK*, 1958, **4**, 117.

CRANMER-BYNG, J. L. (2) (ed.). *An Embassy to China; being the Journal kept by Lord Macartney during his Embassy to the Emperor Chhien-Lung, +1793 and +1794.* Longmans, London, 1962. Macartney (1, 2); Gillan (1).

CRAWFORD, H. S. (1). *Handbook of Carved Ornament from Irish Monuments of the Christian Period.* Royal Society of Antiquaries of Ireland, Dublin, 1926.

C[RAWFORD], O. G. S. (1). 'A Primitive Threshing-Machine' (the *tribulum*). *AQ*, 1935, **9**, 335.

CREEL, H. G. (3). *Sinism; a Study of the Evolution of the Chinese World-View.* Open Court, Chicago, 1929. (Rectifications of this by the author will be found in (4), p. 86.)

CREEL, H. G. (4). *Confucius; the Man and the Myth.* Day, New York, 1949; Kegan Paul, London, 1951. Reviewed D. Bodde, *JAOS*, 1950, **70**, 199.

CRITOBULOS OF IMBROS (1). *De Rebus gestis Mechemetis* (on gun-founding), *c.* 1467. See Sarton (1), vol. 3, p. 1553.

CROMBIE, A. C. (1). *Robert Grosseteste and the Origins of Experimental Science.* Oxford, 1953.

CROMBIE, A. C. (2). *Augustine to Galileo; the History of Science, +400 to +1650.* Falcon, London, 1952.

CROMMELIN, C. A. (1). 'Les Horloges Publiques ou de Clocher et l'application du Pendule à ces Horloges.' *JSHB*, 1952 (no. 5/6).

CROMMELIN, C. A. (2). 'The Clocks of Christian Huygens.' *END*, 1950, **9**, 64.

CROMMELIN, C. A. (3). 'La Contribution de la Hollande à l'Horlogerie.' *SSC*, 1949, **2**, 1. (Mainly on the clocks of Huygens.)

CROMMELIN, C. A. (4). *Descriptive Catalogue of the Huygens Collection in the Rijksmuseum voor de Geschiedenis der Natuurwetenschappen.* Nat. Mus. of the Hist. of Sci. Leiden, 1949.

CROMMELIN, C. A. (5). 'Planetaria, a Historical Survey.' *AHOR*, 1955, **1**, 70.

CROMMELIN, C. A. (6). *Huygens' Pendulum Experiments, Successful and Unsuccessful.* Lecture to a joint meeting of the Antiquarian Horological Society and the British Horol. Institute, Dec. 1956. Repr. from H. Alan Lloyd (6). (Antiq. Horol. Soc. Pubs. no. 2.)

CRONE, E., DIJKSTERHUIS, E. J. & FORBES, R. J. (1) (ed.). *The Principal Works of Simon Stevin.* Amsterdam, 1955– .

CROON, L. (1). 'Technische Kulturdenkmale in Niedersachsen.' *BGTI/TG*, 1933, **22**, 138.

CROSSLEY-HOLLAND, P. C. (2). 'Non-Western Music.' Ch. 1 in *The Pelican History of Music*, ed. A. Robertson & D. Stevens. Vol. 1, *Ancient Forms to Polyphony.* Penguin, London, 1960.

CROWTHER, J. G. (1). *The Social Relations of Science.* Macmillan, London, 1941.

CROZET-FOURNEYRON, M. (1). *L'Invention de la Turbine.* Paris, 1925.

CUMING, H. S. (1). 'The History of Keys.' *JBAA*, 1856, **12**, 117.

CUMMING, C. F. G. (1). *Wanderings in China.* London, 1888.

CUMMINS, J. S. (1) (ed. & tr.). *The Travels and Controversies of Friar Domingo de Navarrete (+1618 to +1686).* 2 vols. Cambridge, 1962. (Hakluyt Society, 2nd series, nos. 118, 119.)

CUNNINGHAM, A. (1). *Ladak; Physical, Statistical and Historical, with Notices of the Surrounding Countries.* Allen, London, 1854.

CUNYNGHAME, Sir H. H. (1). *Time and Clocks.* Constable, London, 1906.

CURWEN, E. C. (1). 'Implements and their Wooden Handles.' *AQ*, 1947, **21**, 155.

CURWEN, E. C. (2). 'Querns.' *AQ*, 1937, **11**, 133.

CURWEN, E. C. (3). 'More about Querns.' *AQ*, 1941, **15**, 15.

CURWEN, E. C. (4). 'The Problem of Early Water[-wheel] Mills.' *AQ*, 1944, **18**, 130.

CURWEN, E. C. (5). 'A Vertical Water-Mill near Salonika.' *AQ*, 1945, **19**, 211.

CURWEN, E. C. (6). *Plough and Pasture.* Cobbett, London, 1946. (Past and Present, Studies in the History of Civilisation, no. 4.) Re-issued in Curwen & Hatt (1).

CURWEN, E. C. & HATT, G. (1). *Plough and Pasture; the Early History of Farming:* Pt. I, *Prehistoric Farming of Europe and the Near East;* Pt. II, *Farming of Non-European Peoples.* Schuman, New York, 1953. (Life of Science Library, no. 27.)

CUVIER, G., BARON (1). *Réflexions sur la Marche des Sciences et sur leur Rapport avec la Société.* Paris, 1816.

D'ACRES. See d'Acres.

DALMAN, G. (1). *Arbeit und Sitte in Palästina.*
  Vol. 1: *Jahreslauf und Tageslauf* (in two parts).
  Vol. 2: *Der Ackerbau.*
  Vol. 3: *Von der Ernte zum Mehl (Ernten, Dreschen, Worfeln, Sieben, Verwahren, Mahlen).*

Vol. 4: *Brot, Öl und Wein.*
Vol. 5: *Webstoff, Spinnen, Weben, Kleidung.*
    Bertelsmann, Gütensloh, 1928– . (Schriften d. deutschen Palästina-Institut, nos. 3, 5, 6, 7, 8;
    Beiträge z. Forderung christlicher Theologie, ser. 2, Sammlung wissenschaftlichen Monographien,
    nos. 14, 17, 27, 29, 33, 36.)
DANIELL, T. & DANIELL, W. (1). *Oriental Scenery.* London, 1814.
DAREMBERG, C. & SAGLIO, E. (1). *Dictionnaire des Antiquités Grecques et Romains.* Hachette, Paris, 1875.
DARMSTÄDTER, L. (1) (with the collaboration of R. du Bois-Reymond & C. Schäfer). *Handbuch zur
    Geschichte d. Naturwissenschaften u. d. Technik.* Springer, Berlin, 1908.
DAS, SARAT CHANDRA (1). *Journey to Lhasa and Central Tibet,* ed. W. W. Rockhill. Murray, London,
    1902.
DAUBER, A. (1). 'Römische Holzfunde aus Pforzheim.' *GERM,* 1944, **28**, 227.
DAUMAS, M. (1). 'Les Instruments Scientifiques aux 17e et 18e Siècles.' Presses Univ. de France,
    Paris, 1953.
DAUMAS, M. (3). 'Le Brevet du Pyréolophore des Frères Niepce (1806).' *DHT,* 1961, **1**, 23.
DAVEY, H. (1) (with appendices by W. G. Norris, Sir Frederick Bramwell, H. W. Pearson, J. H.
    Crabtree, W. E. Hipkins, Messrs Thornewill & Warham, W. B. Collis & H. S. Dunn). 'The
    Newcomen Engine.' *PIME,* 1903, 655.
DAVID, Sir PERCIVAL (1). 'The Magic Fountain in Chinese Ceramic Art; an Exercise in Illustrational
    Representation.' *BMFEA,* 1952, **24**, 1.
DAVID, Sir PERCIVAL (2). 'The *Thao Shuo* and "Illustrations of Pottery Manufacture"; a critical study
    and a review reviewed.' *AA,* 1949, **12**, 165.
DAVIDS, T. W. RHYS, Mrs (Mrs C. A. F. RHYS DAVIDS). See Foley, C. A.
DAVIDSON, MARTHA (1). *A List of Published Translations from the Chinese into English, French and
    German.* Pt. 1. *Literature exclusive of Poetry.* Edwards, for Amer. Council of Learned Societies,
    Ann Arbor, Michigan, 1952.
DAVIDSON, M., SAUL, G. C., WELLS, J. A. & GLENNY, A. P. (1). *The Gyroscope and its Applications.*
    Hutchinson, London, n.d. (1946).
DAVIS, J. F. (1). *The Chinese; a General Description of China and its Inhabitants.* 1st ed. 1836, 2 vols.
    Knight, London, 1844, 3 vols. 1847, 2 vols. French tr. by A. Pichard, Paris, 1837, 2 vols. Germ.
    trs. by M. Wesenfeld, Magdeburg, 1843, 2 vols., and by M. Drugulin, Stuttgart, 1847, 4 vols.
DAVIS, J. F. (3). *China during the War and since the Peace.* 2 vols. Longman, London, 1852.
DAVIS, TENNEY L. & WARE, J. R. (1). 'Early Chinese Military Pyrotechnics' (analysis of *Wu Pei Chih,*
    chs. 119 to 134). *JCE,* 1947, **24**, 522.
DAVISON, C. ST. C. (1). *Historic Books on Machines.* HMSO (Science Museum), London, 1953. (Book
    Exhibitions, no. 2.)
DAVISON, C. ST. C. (3). 'Bearings since the Stone Age.' *ENG,* 1957, **183**, 2.
DAVISON, C. ST. C. (4). 'Wear Prevention in Early History.' *WEAR,* 1957, **1**, 155.
DAVISON, C. ST C. (5). 'A Short History of Gears from Archimedes to the Present Day.' *ENG,* 1956,
    181, 132.
DAVISON, C. ST C. (6). 'The Internal Combustion Engine; some Early Stages in its Development.'
    ENG, 1956, **182**, 258.
DAVISON, C. ST C. (8). 'The Evolution of the Workshop Micrometer.' *EN,* 1960, 196 (29 July).
DAVISON, C. ST C. (9). 'Transporting Sixty-ton Statues in Early Assyria and Egypt.' *TCULT,* 1961,
    2, 11.
DAVISON, C. ST C. (10). 'Geared Power Transmission.' *CME,* 1962, **9**, 140.
DAVY, M. J. B. (1). *Aeronautics; Heavier-than-Air Aircraft. A brief Outline of the History and Develop-
    ment of Mechanical Flight with reference to the National Aeronautical Collection.* Pt. 1, *Historical
    Survey.* HMSO (Science Museum), London, 1949. (Pt. II is the Catalogue.)
DEERR, N. (1). *The History of Sugar.* 2 vols. Chapman & Hall, London, 1949. Crit. rev. J. R. Parting-
    ton, *A/AIHS,* 1950, **3**, 964.
DEERR, N. (2). 'The Evolution of the Sugar-Cane Mill.' *TNS,* 1940, **21**, 1; 'The Early Use of Steam-
    Power in the Cane-Sugar Industry.' *TNS,* 1940, **21**, 11.
DEFFONTAINES, P. (1). 'Note sur la Répartition des Types de Voitures.' In *Mélanges de Géographie et
    d'Orientalisme offerts à E. F. Gautier.* Arbault, Tours, 1932.
DEFFONTAINES, P. (2). 'Sur la Répartition Geographique des Voitures à Deux Roues et à Quatre Roues.'
    In *Travaux du 1er Congrès International de Folklore, Paris, 1937.* Arbault, Tours, 1938, p. 117.
    (Pub. du. Département et du Musée National des Arts et Traditions Populaires.)
DEGENHART, B. (1). *[Antonio] Pisanello.* Schroll, Vienna; Chiantore, Turin, 1941.
D[EHERGNE], J. (3). 'Bibliographie de Quelques Industries Chinoises; Techniques Artisanales et
    Histoire Ancienne.' *BUA,* 1949 (3ᵉ sér.), **10**, 198. (1. General Works; 2. Textiles; 3. Metallurgy,
    Weapons, Gunpowder; 4. Paper and Printing; 5. Ceramics.)

DELAPORTE, Y. (1). *Les Vitraux de la Cathédrale de Chartres.* 1 vol. and 3 vols. of plates. Houvet, Chartres, 1926.

DELISLE, L. (1). *Études sur la Condition de la Classe Agricole et l'État de l'Agriculture en Normandie au Moyen Âge.* Paris, 1851.

DELISLE, L. (2). 'On the Origin of Windmills in Normandy and England.' *JBAA*, 1851, **6**, 403.

DEONNA, W. (1). 'Le Mobilier Délien.' In *Exploration Archéologique de Délos*, ed. T. Homolle *et al.* de Boccard, Paris, 1938, vol. 18.

DESAGULIERS, J. T. (1). *A Course of Experimental Philosophy*, 2 vols. Innys, Longman, Shewell & Hitch, London, 1734; 2nd ed. Innys, Longman, Shewell & Hitch with Senex, London, 1745.

VON DEWALL, MAGDALENE (1). *Pferd und Wagen im frühen China; Aufschlüsse zur Kulturgeschichte aus der 'Shih Ching' Dichtung und Bodenfunden der Shang- und frühen Chou-Zeit.* Saarbrücken, 1962. (Saarbrücker Beiträge z. Altertumskunde, no. 1.)

DEWING, H. B. (tr.) (1). *Procopius' 'History of the Wars'.* Heinemann, London, 1919–54. (Loeb Classical Library.)

DIBNER, B. (2) (ed.). *The 'New Discoveries [Nova Reperta]'; the Sciences, Inventions and Discoveries of the Middle Ages and the Renaissance as represented in 24 Engravings issued in the early 1580's by Stradanus [Jan van der Straet].* Album of 19 plates and one title-page, also containing the 3 plates and one title-page of *America Retectio*. Burndy Library, Norwalk, Conn., 1953.

DICK, T. L. (1). 'On a Spiral Oar.' *TAP*, 1818, **11**, 438.

DICKINSON, H. W. See Singer, C. (12).

DICKINSON, H. W. (1). 'The Origin and Manufacture of Wood Screws.' *TNS*, 1942, **22**, 79.

DICKINSON, H. W. (2). 'John Wilkinson [engineer].' *BGTI*, 1911, **3**, 215.

DICKINSON, H. W. (3). 'A Condensed History of Rope-Making.' *TNS*, 1942, **23**, 71.

DICKINSON, H. W. (4). *A Short History of the Steam-Engine.* Cambridge, 1939. 2nd. edn. ed. A. E. Musson, Cass, London, 1963.

DICKINSON, H. W. (5). 'James White and his *New Century of Inventions.' TNS*, 1950, **27**, 175.

DICKINSON, H. W. (6). 'The Steam-Engine to 1830.' Art. in *A History of Technology*, ed. C. Singer *et al.*, vol. 3, p. 168. Oxford, 1958.

DICKINSON, H. W. (7). 'James Watt, Craftsman and Engineer.' Cambridge, 1936.

DICKINSON, H. W. & STRAKER, E. (1). 'The Shetland Water-Mill.' *TNS*, 1933, **13**, 89.

DICKINSON, H. W. & TITLEY, A. (1). *Richard Trevithick, the Engineer and the Man.* Cambridge, 1934.

DICKMANN, H. (4). *Aus der Geschichte der deutschen Eisen- und Stahlerzeugung.* Stahleisen MBH, Düsseldorf, 1959. (Monographien ü Stahlverwendung, no. 1.)

DIDEROT, D. (2) (ed.) (with d'Alembert, J.). *Encyclopédie ou Dictionnaire raisonné des Sciences, des Arts et des Métiers, par une Société des Gens de Lettres*, 17 vols. Briasson, David, le Breton & Durand, Paris, 1753 to 1765. Supplement, ed. Panckoucke, 4 vols. Rey, Amsterdam, 1776–7. General index, 2 vols., 1780.

DIDEROT, D. (3) (ed.) (with d'Alembert, J.). *Recueil de Planches sur les Sciences, les Arts libéraux et les Arts mécaniques, avec leur Explication*, 12 vols. Briasson, David, le Breton & Durand, Paris, 1763 to 1777. The illustrations for the Encyclopaedia, making in all a set of 35 vols. A selection of 485 plates republished with introduction and notes by C. C. Gillispie, Dover, New York; Constable, London, 1959.

DIEBOLD, JOHN (1). *Automation; the Advent of the Automatic Factory.* New York, 1952.

DIELS, H. (1). *Antike Technik.* Teubner, Leipzig and Berlin, 1914; enlarged 2nd ed., 1920. (rev. B. Laufer, *AAN*, 1917, **19**, 71).

DIELS, H. (2). 'Über die von Prokop beschriebene Kunstuhr von Gaza; mit einem Anhang enthaltende Text und Übersetzung d. ἔκφρασις ὡρολόγιου des Prokopios von Gaza.' *APAW/PH*, 1917 (no. 7).

DIELS, H. & SCHRAMM, E. (1) (ed. & tr.). 'Herons *Belopoiika.' APAW/PH*, 1918 (no. 2).

DIELS, H. & SCHRAMM, E. (2) (ed. & tr.). 'Philons *Belopoiika* [Bk. IV of the *Mechanica*].' *APAW/PH*, 1918 (no. 16).

DIELS-FREEMAN: FREEMAN, K. (1). *Ancilla to the Pre-Socratic Philosophers; a complete translation of the Fragments in Diels' 'Fragmente der Vorsokratiker'.* Blackwell, Oxford, 1948.

DIJKSTERHUIS, E. J. (1). *Simon Stevin.* The Hague, 1943.

DIONISI, F. (1). *Les Navires de Nemi.* Galilei, Rome, 1941.

DIRCKS, H. (1). *The Life, Times, and Scientific Labours of the Second Marquis of Worcester, to which is added a Reprint of his 'Century of Inventions' (+1663), with a Commentary thereon.* Quaritch, London, 1865.

DIRCKS, H. (2). *Perpetuum Mobile; or, a History of the Search for Self-Motive Power from the 13th to the 19th century.* Spon, London. Vol. 1, 1861; vol. 2, 1870.

DITTMANN, K. H. (1). 'Der Segelwagen von Medīnet Māḍī.' *MDIK*, 1941, **10**, 60.

DOBRZENSKY, J. J. V. (1). *Nova, et amaenior, de Admirando Fontium Genio, Philosophia.* Ferrara, 1657 or 1659.

DODWELL, C. R. (1) (ed. & tr.). *Theophilus 'De Diversis Artibus'*. Nelson, London, 1961.

DOLLFUS, C. & BOUCHÉ, H. (1). *Histoire de l'Aéronautique; Texte et Documentation*. l'Illustration, Paris, 1932; repr. 1938, 1942.

DOOLITTLE, J. (1). *A Vocabulary and Handbook of the Chinese Language*. Fuchow, 1872.

DORÉ, H. (1). *Recherches sur les Superstitions en Chine*. 15 vols. T'u-Se-Wei Press, Shanghai, 1914–29

Pt. I, vol. 1, pp. 1–146: 'Superstitious' practices, birth, marriage and death customs (*VS*, no. 32).
Pt. I, vol. 2, pp. 147–216: talismans, exorcisms and charms (*VS*, no. 33).
Pt. I, vol. 3, pp. 217–322: divination methods (*VS*, no. 34).
Pt. I, vol. 4, pp. 323–488: seasonal festivals and miscellaneous magic (*VS*, no. 35).
Pt. I, vol. 5, sep. pagination: analysis of Taoist talismans (*VS*, no. 36).
Pt. II, vol. 6, pp. 1–196: Pantheon (*VS*, no. 39).
Pt. II, vol. 7, pp. 197–298: Pantheon (*VS*, no. 41).
Pt. II, vol. 8, pp. 299–462: Pantheon (*VS*, no. 42).
Pt. II, vol. 9, pp. 463–680: Pantheon, Taoist (*VS*, no. 44).
Pt. II, vol. 10, pp. 681–859: Taoist celestial bureaucracy (*VS*, no. 45).
Pt. II, vol. 11, pp. 860–1052: city-gods, field-gods, trade-gods (*VS*, no. 46).
Pt. II, vol. 12, pp. 1053–1286: miscellaneous spirits, stellar deities (*VS*, no. 48).
Pt. III, vol. 13, pp. 1–263: popular Confucianism, sages of the Wên miao (*VS*, no. 49).
Pt. III, vol. 14, pp. 264–606: popular Confucianism, historical figures (*VS*, no. 51).
Pt. III, vol. 15, sep. pagination: popular Buddhism, life of Gautama (*VS*, no. 57).

VON DRACH, A. (1). *Die zu Marburg im Mathematisch-Physikalischen Institut befindliche Globusuhr Wilhelms IV von Hessen als Kunstwerk und astronomisches Instrument*. Elwert, Marburg, 1894.

DRACHMANN, A. G. (1). 'Hero's and Pseudo-Hero's Adjustable Siphons.' *JHS*, 1932, **52**, 116.

DRACHMANN, A. G. (2). 'Ktesibios, Philon and Heron; a Study in Ancient Pneumatics.' *AHSNM*, 1948, **4**, 1–197.

DRACHMANN, A. G. (3). 'Heron and Ptolemaios.' *CEN*, 1950, **1**, 117.

DRACHMANN, A. G. (5). 'On the alleged Second Ktesibios.' *CEN*, 1951, **2**, 1.

DRACHMANN, A. G. (6). 'The Plane Astrolabe and the Anaphoric Clock.' *CEN*, 1954, **3**, 183.

DRACHMANN, A. G. (7). 'Ancient Oil Mills and Presses.' *KDVS/AKM*, 1932, **1** (no. 1). Sep. publ. Levin & Munksgaard, Copenhagen, 1932.

DRACHMANN, A. G. (8). 'Heron's Windmill.' *CEN*, 1961, **7**, 145.

DRACHMANN, A. G. (9). *The Mechanical Technology of Greek and Roman Antiquity; a Study of the Literary Sources*. Munksgaard, Copenhagen, 1963.

DROVER, C. B. (1). 'A Mediaeval Monastic Water-Clock.' *AHOR*, 1954, **1**, 54.

DROWER, M. S. (1). 'Water-Supply, Irrigation and Agriculture [from Early Times to the End of the Ancient Empires].' Art in *A History of Technology*, ed. C. Singer et al. Oxford, 1954, vol. 1, p. 520.

DUBERTRET, L. & WEULERSSE, J. (1). *Manuel de Géographie; Syrie, Liban et Proche Orient: I, La Péninsule Arabique*. Imp. Cath. Beirut, 1940.

DUBS, H. H. (2) (tr., with the assistance of Phan Lo-Chi and Jen Thai). *'History of the Former Han Dynasty', by Pan Ku; a Critical Translation with Annotations*. 3 vols. Waverly, Baltimore, 1938–.

DUBS, H. H. (8) (tr.). *The Works of Hsün Tzu*. Probsthain, London, 1928.

DUCASSÉ, P. (1). *Histoire des Techniques*. Presses Univ. de France, Paris, 1945.

DUGAS, R. (1). *Histoire de la Mécanique*. Griffon (La Baconnière), Neuchâtel, 1950. Crit. rev. P. Costabel, *A/AIHS*, 1951, **4**, 783.

DUGAS, R. (2). *La Mécanique au XVIIème Siècle; des Antécédents Scholastiques à la Pensée Classique*. Griffon, Neuchâtel, 1954. Crit. rev. C. Truesdell, *ISIS*, 1956, **47**, 449.

DUGGAN, A. (1). *King of Pontus; the story of Mithridates Eupator who alone challenged the Might of Rome*. Coward & McCann, New York, 1959.

DUHEM, J. (1). *Histoire des Idées Aéronautiques avant Montgolfier*. Inaug. Diss., Paris. Sorlot, Paris, 1943.

DUHEM, J. (2). *Musée Aéronautique avant Montgolfier*. Sorlot, Paris, 1943. (This is essentially the volume of plates accompanying Duhem (1), but it contains lengthy notes on each illustration.)

DUHEM, P. (1). *Études sur Léonard de Vinci*. 3 vols. Hermann, Paris.
Vols. 1, 2: 'Ceux qu'il a Lus et Ceux qui l'ont Lu.' 1906, 1909.
Vol. 3: 'Les Précurseurs Parisiens de Galilée.' 1913.
Pt. 1: Albert of Saxony, Bernardino Baldi, Themon, Cardan, Palissy, etc.
Pt. 2: Nicholas of Cusa, Albertus Magnus, Vincent of Beauvais, Ristoro d'Arezzo, etc.
Pt. 3: Buridan, Soto, Nicholas d'Oresme, etc.

DUHEM, P. (3). *Le Système du Monde; Histoire des Doctrines Cosmologiques de Platon à Copernic*. 5 vols. Paris, 1913–17.

DUNBAR, G. S. (1). 'Henry Chapman Mercer [1856 to 1930], Pennsylvania Folklife Pioneer [and Rudolf P. Hommel, 1887 to 1950].' *PFL*, 1961, **12** (no. 2), 48.

DUNKERLEY, S. (1). *Mechanism.* Longmans Green, London, 1905.

DUYVENDAK, J. J. L. (12). 'The Last Dutch Embassy to the Chinese Court' (+1794 to +1795). *TP*, 1938, **34**, 1, 223; 1939, **35**, 329.

DUYVENDAK, J. J. L. (14). 'Simon Stevin's "Sailing-Chariot"' (and its Chinese antecedents). *TP*, 1942, **36**, 401.

DUYVENDAK, J. J. L. (18) (tr.). '*Tao Tê Ching*', the Book of the Way and its Virtue. Murray, London, 1954 (Wisdom of the East series). Crit. revs. P. Demiéville, *TP*, 1954, **43**, 95; D. Bodde, *JAOS*, 1954, **74**, 211.

DVOŘAK, A. (1). 'Knížeci Pohřby na Vozech ze starši Doby Železné.' *PRH*, 1938, **1** (1).

EBERHARD, W. (21). *Conquerors and Rulers; Social Forces in Mediaeval China* (theory of gentry society). Brill, Leiden, 1952. Crit. E. Balazs, *ASEA*, 1953, **7**, 162; E. G. Pulleyblank, *BLSOAS*, 1953, **15**, 588.

ECKE, G. V. (2). 'Wandlungen des Faltstuhls; Bemerkungen z. Geschichte d. Eurasischen Stuhlform.' *MS*, 1944, **9**, 34.

ECKMAN, J. (1). 'Jerome Cardan.' *BIHM*, 1946, suppl. no. 7.

EDDÉ, J. (1). *Géographie; Liban-Syrie.* Imp. Cath., Beirut, 1941.

EDWARDS, E. D. (1). *Chinese Prose Literature of the Thang Period.* 2 vols. Probsthain, London, 1937.

EGGERS, G. (1). 'Wasserversorgungstechnik im Altertum.' *BGTI/TG*, 1936, **25**, 1.

EICHHORN, W. (4). 'Zur Vorgeschichte des Aufstandes von Wang Hsiao-Po und Li Shun in Szechuan (+993 bis +995).' *ZDMG*, 1955, **105**, 192.

EICHHORN, W. (5). 'Wang Chia's *Shih I Chi*.' *ZDMG*, 1952, **102** (N.F. **27**), 130.

EKHOLM, G. F. (1). 'Wheeled Toys in Mexico.' *AMA*, 1946, **11**, 222.

D'ELIA, PASQUALE (2) (ed.). *Fonti Ricciane; Storia dell'Introduzione del Cristianesimo in Cina.* 3 vols. Libreria dello Stato, Rome, 1942–9. Cf. Trigault (1); Ricci (1).

ELIADE, MIRCEA (3). *Le Chamanisme, et les Techniques Archaïques de l'Extase.* Payot, Paris, 1951.

EMERSON, W. (1). *The Principles of Mechanics, explaining and demonstrating the general Laws of Motion, the Laws of Gravity...Projectiles...Pendulums...Strength and Stress of Timber, etc.* Robinson, London, 1758, 1773, 1782.

EMERY, H. C. & EMERY, C. [or EMORY] (1). 'Stages from Choni to Sung-Pan.' *CJ*, 1924, **2**, 539; 1925, **3**, 25.

ENSHOFF, P. D. (1). 'Pater Ricci's Uhren.' *DKM*, 1937, **65**, 190. 'Hatte China die ungleichen Stunden?' Typescript paper in the Jäger Collection.

ERAS, V. J. M. (1). *Locks [and Keys].* Dordrecht.

ERCKER, L. (1). *Beschreibung allerfurnemisten Mineralischen Erzt und Berckwercks arten....* Prague, 1574, with ten later editions. See Sisco & Smith (2).

ERKES, E. (12). 'Das chinesische Theater vor d. Thang-Zeit.' *AM*, 1935, **10**, 229.

ERKES, E. (19). *Geschichte Chinas, von den Anfängen bis zum Eindringen des ausländischen Kapitals.* Akademie-Verlag, Berlin, 1956.

ERKES, E. (20). 'Ursprung und Bedeutung der Sklaverei in China.' *AA*, 1937, **6**, 294.

ERKES, E. (21). 'Das Problem der Sklaverei in China.' *BVSAW/PH*, 1954, **100** (no. 1).

ESPÉRANDIEU, E. (1). *Souvenir du Musée Lapidaire de Narbonne.* Commission Archéologique, Narbonne, n.d.

ESPÉRANDIEU, E. (2). *Receuil Général des Bas-Reliefs, Statues et Bustes de la Gaule Romaine.* Imp. Nat. Paris, 1908, 1913.

ESPINAS, A. (1). *Les Origines de la Technologie.* Alcan, Paris, 1897.

VON ESSENWEIN, A. (1). *Mittelälterliches Hausbuch; Bilderhandschrift des 15 Jahrh...* (Anon. 15). Keller, Frankfurt-am-Main, 1887.

ESTERER, M. (1). *Chinas natürliche Ordnung und die Maschine.* Cotta, Stuttgart and Berlin, 1929. (Wege d. Technik series.)

EUDE, E. (1). *Histoire documentaire de la Mécanique Française, d'après le Musée Centennal de la Mécanique à l'Exposition Universelle de 1900.* Dunod, Paris, 1902.

EVLIYA CHELEBI (1). *Narrative of Travels in Europe, Asia and Africa, in the +17th century, by Evliya Effendi,* tr. J. von Hammer. 3 vols. London, 1846–50.

ÉVRARD, R. (2). *Forges Anciennes* (a study of martinet hammers in Belgium and Luxembourg). Editions Solédi, Liége, 1956.

EWBANK, T. (1). *A Descriptive and Historical Account of Hydraulic and other Machines for Raising Water, Ancient and Modern....* Scribner, New York, 1842. (Best ed. the 16th, 1870.)

VON EYBE ZUM HARTENSTEIN, L. (the younger) (1). *Kriegsbuch.* 1500. MS. 1390, Univ. Erlangen. See Sarton (1), vol. 3, p. 1564.

FAIRBANK, WILMA (1). 'A Structural Key to Han Mural Art.' *HJAS*, 1942, **7**, 52.

FANE, Sir Francis (1). *The Sacrifice, a Tragedy.* London, 1686.

FANG YÜN (1). 'Ming Tomb Discoveries.' *PKR*, 1958, **1** (no. 32), 17.

FAREY, J. (1). *A Treatise on the Steam Engine, Historical, Practical and Descriptive.* Longman, Rees, Orme, Brown & Green, London, 1827.

DE FARIA Y SOUSA, MANUEL (1). *Imperio de la China y Cultura Evangelica en el por los Religiosos de la Compañia de Jesus....* Officina Herreriana, Lisbon, 1731.

FARRINGTON, G. H. (1). *Fundamentals of Automatic Control.* Chapman & Hall, London, 1951. (Crit. J. Greig, *N*, 1953, **172**, 91.)

FAVIER, A. (1). *Pékin; Histoire et Description.* Peking, 1897. For Soc. de St Augustin, Desclée & de Brouwer, Lille, 1900.

FEBVRE, L. (2). 'Civilisation, Moteurs et Mouvements.' *AHES/MHS*, 1942, **2**, 56. Discussion of Haudricourt (1).

FEBVRE, L. (5). 'Marc Bloch fusillé.' *AHES/MHS*, 1944, **6**, 5. 'De l'Histoire au Martyre; Marc Bloch, 1886 à 1944.' *AHES/MHS*, 1945, **7**, 1.

FELDHAUS, F. M. (1). *Die Technik der Vorzeit, der Geschichtlichen Zeit, und der Naturvölker* (encyclopaedia). Engelmann, Leipzig and Berlin, 1914.

FELDHAUS, F. M. (2). *Die Technik d. Antike u. d. Mittelalter.* Athenaion, Potsdam, 1931. (Crit. H. T. Horwitz, *ZHWK*, 1933, **13** (N.F. **4**), 170.)

FELDHAUS, F. M. (3). *Die Säge, ein Rückblick auf vier Jahrtausende.* Dominicus, Berlin, 1921.

FELDHAUS, F. M. (4). *Die geschichtlichen Entwicklung d. Zahnrades.* Stolzenberg, Berlin-Reinickendorf, 1911.

FELDHAUS, F. M. (5). *Zur Geschichte d. Drahtseilschwebebahnen.* Zillessen, Berlin-Friedenau, 1911. (Monogr. z. Gesch. d. Technik, no. 1.)

FELDHAUS, F. M. (6). *Geschichte d. Kugel-, Walzen- und Rollenlager.* Fichtel & Sachs, Schweinfurth, 1914.

FELDHAUS, F. M. (7). *Buch der Erfindungen.* Oestergaard, Berlin, 1908. Short version of Feldhaus (2).

FELDHAUS, F. M. (8). *Ruhmesblätter d. Technik von der Urerfindungen bis zur Gegenwart.* 2 vols. Brandstetter, Leipzig, 1924. Similar to Feldhaus (9).

FELDHAUS, F. M. (9). *Kulturgeschichte d. Technik.* 2 vols. Salle, Berlin, 1928. Superseded by Feldhaus (2).

F[ELDHAUS], F. M. (10). 'Gebläse in China.' *GTIG*, 1927, **11**, 310.

FELDHAUS, F. M. (11). 'Technisher Inhalt der chinesischen Encyclopädie von +1609.' *GTIG*, 1915, **2**, 56.

FELDHAUS, F. M. (12). 'Die europäischen Maschinen im alten China.' *ZTF*, 1916, no. 7, 177.

FELDHAUS, F. M. (13). 'Darstellungen d. ersten Feldmühle.' *GTIG*, 1917, **4**, 28.

FELDHAUS, F. M. (14). *Altmeister des Segelfluges.* Schultz, Berlin-Lichterfelde, 1927.

FELDHAUS, F. M. (15). 'Erdöl in der Luftschiff-fahrt.' *PET*, 1952, **5**, 633.

FELDHAUS, F. M. (18). *Leonardo der Techniker u. Erfinder.* Diederichs, Jena, 1913.

FELDHAUS, F. M. (19). 'Ü. Zweck u. Entstehungszeit d. sog. Püstriche.' *MGNM*, 1908, 140.

FELDHAUS, F. M. (20). *Die Maschine im Leben der Völker; ein Überblick von der Urzeit bis zur Renaissance.* Birkhäuser, Basel and Stuttgart, 1954.

FELDHAUS, F. M. (21). 'Über den Ursprung von Federzug und Schnecke.' *DUZ*, 1930, **54**, 720.

FELDHAUS, F. M. (22). 'Die Uhren des Königs Alfons X von Spanien.' *DUZ*, 1930, **54**, 608.

FELDHAUS, F. M. (23). *Geschichte d. Ölgewinnung.* Festschr. d. Verbandes d. deutschen Ölmühlen, 1925.

FELDHAUS, F. M. (24). *Geschichte des technischen Zeichnens*, 2nd ed. rev. and enlarged with the assistance of E. Schruff. Kuhlmann AG, Wilhelmshafen, 1959. Crit. rev. R. S. Hartenberg, *TCULT*, 1961, **2**, 45. Also appeared serially in Eng. tr. in *GRSCI*, 1960–.

FELDHAUS, F. M. (25). 'Beiträge z. älteren Geschichte der Turbinen.' *ZGT*, 1908, **5**, 569.

FELDHAUS, F. M. (26). 'Beiträge z. Geschichte des Drahtziehens.' *ADI*, 1910, 137, 159 & 181.

FELDHAUS, F. M., BIEDENKAPP, G., KOLLMANN, J., LUX, J. U. & REITZ, A. (1). *Der Ingenieur, seine kulturelle, gesellschaftliche und sociale Bedeutung, mit einem historischen Überblick ü. des Ingenieurwesen.* Franck, Stuttgart, 1910.

FELDHAUS, F. M. & DEGEN, A. (1). 'Villard aus Honnecourt, ein Technikus d. 13 Jahrhunderts.' *ZOIAV*, 1906, **58** (no. 30), 429.

FERCKEL, C. (1). 'On the *De Secretis Mulierum*.' *AGMN*, 1914, **7**, 47.

FERGUSON, JOHN (2). *Bibliographical Notes on Histories of Inventions and Books of Secrets.* 2 vols. Glasgow, 1898; repr. Holland Press, London, 1959. (Papers collected from *TGAS*.)

FERGUSON, J. C. (2). *Survey of Chinese Art.* Commercial Press, Shanghai, 1940.

FERGUSON, J. C. (3). (*a*) 'The Chinese Foot Measure.' *MS*, 1941, **6**, 357. (*b*) *Chou Dynasty Foot Measure.* Privately printed, Peiping, 1933. (See also a note on a graduated rule of *c.* +1117, 'A Jade Foot Measure', *TH*, 1937, **4**, 391.)

FERGUSON, J. C. (4). 'The Southern Migration of the Sung Dynasty.' *JRAS/NCB*, 1924, **55**, 14.

FERGUSON, J. C. (6). 'Transportation in Early China.' *CJ*, 1929, **10**, 227.

FERGUSON, J. C. (7). 'Political Parties of the Northern Sung Dynasty.' *JRAS/NCB*, 1927, **58**, 35.

FERRAND, G. (1). *Relations de Voyages et Textes Géographiques Arabes, Persans et Turcs relatifs à l'Extrême Orient, du 8ᵉ au 18ᵉ siècles, traduits, revus et annotés etc.* 2 vols. Leroux, Paris, 1913.

FERRIS, R. (1). *How it Flies; or, the Conquest of the Air; the History of Man's Endeavours to Fly and of the Inventions by which he has succeeded.* Nelson, New York, 1910.

FIRTH, C. M. & QUIBELL, J. E. (1). *The Step Pyramid [of Saqqarah].* Service des Antiquités de l'Égypte, Cairo, 1935.

FISCHER, E. (1). 'Sind die Rümanen ein Balkanvolk?' *OAV*, 1910, **1**, 70 (72).

FISCHER, H. (1). 'Beiträge z. Geschichte d. Werkzeugmaschinen, I' (boring tools and machines). *BGTI*, 1912, **4**, 274.

FISCHER, H. (2). 'Beiträge z. Geschichte d. Werkzeugmaschinen, II' (lathes). *BGTI*, 1913, **5**, 73.

FISCHER, H. (3). 'Beiträge z. Geschichte d. Werkzeugmaschinen, III' (hammers and triphammers). *BGTI*, 1915, **6**, 1.

FISCHER, H. (4). 'Beiträge z. Geschichte d. Holzbearbeitungsmaschinen.' *BGTI*, 1911, **3**, 61.

FISCHER, OTTO (1). *Die Kunst Indiens, Chinas und Japans.* Propylaea, Berlin, 1928.

FITZGERALD, C. P. (1). *China; a Short Cultural History.* Cresset Press, London, 1935.

FITZGERALD, C. P. (8). *The Empress Wu.* Cresset Press, London, 1956.

FITZGERALD, KEANE (1). 'An Attempt to improve the Manner of working the Ventilators [of Mines] by the Help of the Fire-Engine [i.e. Newcomen's atmospheric steam-engine].' *PTRS*, 1758, **50**, 727.

FLAVIGNY, R. C. (1). *Le Dessin de l'Asie Occidentale Ancienne et les Conventions qui le régissent.* Maisonneuve, Paris, 1940.

FLEMING, A. P. M. & BROCKLEHURST, H. J. (1). *A History of Engineering.* Black, London, 1925.

FLINDERS PETRIE, W. M. See Petrie, W. M. Flinders.

FLUDD, ROBERT (2). *Utriusque Cosmi Majoris scilicet et Minoris Metaphysica, Physica atque Technica Historia.* Galler, Oppenheim, 1617. A subsequent title-page, heading a later part (1618), generally bound with the first, reads *Naturae Simia seu Technica Macrocosmi Historia.*

FOLEY, C. A. (Mrs T. W. Rhys Davids) (1). 'On Buddha's instruction to Ānanda to make a Mandala like a water-wheel to show the cycle of rebirths'; a comment to Waddell (3). *JRAS*, 1894 (N.S.), **26**, 389.

FOLEY, C. A. (Mrs T. W. Rhys Davids) (2). 'Economic Conditions according to Early Buddhist Literature.' *CHI*, vol. 1, ch. 8.

DA FONTANA, GIOVANNI (1). *Bellicorum instrumentorum liber cum figuris et ficticiis literis conscriptus.* MS. Cod. icon. 242 Hofbibliothek, München, 1410–20. See Sarton (1), vol. 3, p. 1551.

FORBES, R. J. (1). *Bibliographia Antiqua; Philosophia Naturalis.* Nederl. Inst. v. h. Nabije Oosten, Leiden.

 Vol. I. Mining and Geology. 1940.

 Vol. II. Metallurgy. 1942.

 Vol. III. Building Materials, 1944.

 Vol. IV. Pottery, Faience, Glass, Glazes, Beads. 1944.

 Vol. V. Paints, Pigments, Varnishes, Inks and their Applications. 1949.

 Vol. VI. Leather Manufacture and Applications. 1949.

 Vol. VII. Fibrous Materials, Preparation and Industries. 1949.

 Vol. VIII. Paper, Papyrus and other Writing Materials. 1949.

 Vol. IX. Man and Nature. 1949.

 Vol. X. Science and Technology. 1950.

 Suppl. I. General. 1952.

FORBES, R. J. (2). *Man the Maker; a History of Technology and Engineering.* Schuman, New York, 1950. (Crit. rev. H. W. Dickinson & B. Gille, *A/AIHS*, 1951, **4**, 551.)

[FORBES, R. J.] (4a). *Histoire des Bitumes, des Époques les plus Reculées jusqu'à l'an 1800.* Shell, Leiden, n.d.

FORBES, R. J. (5). 'The Ancients and the Machine.' *A/AIHS*, 1949, **2**, 919.

FORBES, R. J. (8). 'Metallurgy [in the Mediterranean Civilisations and the Middle Ages].' In *A History of Technology*, ed. C. Singer *et al.*, vol. 2, p. 41. Oxford, 1956.

FORBES, R. J. (10). *Studies in Ancient Technology.* Vol. 1, *Bitumen and Petroleum in Antiquity; The Origin of Alchemy; Water Supply.* Brill, Leiden, 1955. (Crit. Lynn White, *ISIS*, 1957, **48**, 77.)

FORBES, R. J. (11). *Studies in Ancient Technology.* Vol. 2, *Irrigation and Drainage; Power; Land Transport and Road-Building; The Coming of the Camel.* Brill, Leiden, 1955. (Crit. Lynn White, *ISIS*, 1957, **48**, 77.)

FORBES, R. J. (12). *Studies in Ancient Technology*. Vol. 3, *Cosmetics and Perfumes in Antiquity; Food, Alcoholic Beverages, Vinegar; Food in Classical Antiquity; Fermented Beverages, −500 to +1500; Crushing; Salts, Preservation Processes, Mummification; Paints, Pigments, Inks and Varnishes.* Brill, Leiden, 1955. (Crit. Lynn White, *ISIS*, 1957, **48**, 77.)

FORBES, R. J. (13). *Studies in Ancient Technology*. Vol. 4, *Fibres and Fabrics of Antiquity; Washing, Bleaching, Fulling and Felting; Dyes and Dyeing; Spinning; Sewing, Basketry and Weaving; Looms; Weavers.* Brill, Leiden, 1956. (Crit. Lynn White, *MANU*, 1958, **2**, 50.)

FORBES, R. J. (14). *Studies in Ancient Technology*. Vol. 5, *Leather in Antiquity; Sugar and its Substitutes in Antiquity; Glass.* Brill, Leiden, 1957.

FORBES, R. J. (15). *Studies in Ancient Technology*. Vol. 6, *Heat and Heating; Refrigeration, the art of cooling and producing cold; Lights and Lamps.* Brill, Leiden, 1958.

FORBES, R. J. (17). 'Hydraulic Engineering and Sanitation [in the Mediterranean Civilisations and the Middle Ages].' Art. in *A History of Technology*, ed. C. Singer *et al.*, vol. 2, p. 663. Oxford, 1956.

FORBES, R. J. (18). 'Food and Drink [from the Renaissance to the Industrial Revolution].' Art. in *A History of Technology*, ed. C. Singer *et al.*, vol. 3, p. 1. Oxford, 1957.

FORBES, R. J. (19). 'Power [in the Mediterranean Civilisations and the Middle Ages].' Art. in *A History of Technology*, ed. C. Singer *et al.*, vol. 2, p. 589. Oxford, 1956.

FORBES, R. J. (21). *More Studies in Early Petroleum History.* Brill, Leiden, 1959.

FORBES, R. J. (23). 'Een Oud-Egyptisch Voorganger van Simon Stevin.' *DNATU*, 1941, **61**, 160.

FORBES, R. J. (24). 'The Sailing Chariot.' Art. in *The Principal Works of Simon Stevin*, vol. 5, 1963. See Crone *et al.* (1).

FORESTIER, G. (1). *La Roue.* Berger-Levrault, Paris and Nancy, 1900.

FORKE, A. (3) (tr.). *Me Ti [Mo Ti] des Sozialethikers und seine Schüler philosophische Werke.* Berlin, 1922. (*MSOS*, Beibände, **23–25**.)

FORKE, A. (4) (tr.). *'Lun-Hêng', Philosophical Essays of Wang Chhung.* Vol. 1, 1907. Kelly & Walsh, Shanghai; Luzac, London; Harrassowitz, Leipzig. Vol. 2, 1911 (with the addition of Reimer, Berlin). Photolitho re-issue, Paragon, New York, 1962. (*MSOS*, Beibände, **10** and **14**.) Crit. P. Pelliot, *JA*, 1912 (10ᵉ sér.), **20**, 156.

FORKE, A. (9). *Geschichte d. neueren chinesischen Philosophie* (i.e. from the beginning of the Sung to modern times). de Gruyter, Hamburg, 1938. (Hansische Univ. Abhdl. a. d. Geb. d. Auslands-kunde, no. 46 (ser. B, no. 25).)

FORKE, A. (12). *Geschichte d. mittelälterlichen chinesischen Philosophie* (i.e. from the beginning of the Former Han to the end of the Wu Tai). de Gruyter, Hamburg, 1934. (Hamburg. Univ. Abhdl. a. d. Geb. d. Auslandskunde, no. 41 (ser. B, no. 21).)

FORKE, A. (13). *Geschichte d. alten chinesischen Philosophie* (i.e. from antiquity to the beginning of the Former Han). de Gruyter, Hamburg, 1927. (Hamburg. Univ. Abhdl. a. d. Geb. d. Auslandskunde, no. 25 (ser. B, no. 14).)

FORKE, A. (15). 'On Some Implements mentioned by Wang Chhung' (1. Fans, 2. Chopsticks, 3. Burning Glasses and Moon Mirrors). Appendix III to Forke (4).

FORKE, A. (16). 'Von Peking nach Ch'ang-An und Loyang.' *MSOS*, 1898, **1**, 1.

FORKE, A. (17). 'Der Festungskrieg im alten China.' *OAZ*, 1919, **8**, 103. (Repr. from Forke (3), pp. 99 ff.)

FORSTER, L. (1). 'Translation; an Introduction.' Art. in *Aspects of Translation*, ed. A. H. Smith, p. 1. Secker & Warburg, London, 1958. (University College, London, Communication Research Centre; Studies in Communication, no. 2.)

FORTI, UMBERTO (1). *Storia della Tecnica Italiana alle Origini della Vita Moderna.* Sansoni, Florence, 1940.

FORTI, UMBERTO (2). *Storia della Tecnica dal Medioevo al Rinascimento.* Sansoni, Florence, 1957.

FORTUNE, R. (1). *Two Visits to the Tea Countries of China, and the British Tea Plantations in the Himalayas, with a Narrative of Adventures, and a Full Description of the Culture of the Tea Plant, the Agriculture, Horticulture and Botany of China.* 2 vols. Murray, London, 1853.

FORTUNE, R. (2). *Three Years' Wanderings in the Northern Provinces of China, including a Visit to the Tea, Silk and Cotton Countries; with an Account of the Agriculture and Horticulture of the Chinese, New Plants, etc.* Murray, London, 1847. Abridged as vol. 1 of Fortune (1).

FOUGÈRES, G. (1). Review of the first version of L. des Noëttes (1). *JS*, 1924, **109**, 321.

FOX, C. (1). 'Sleds, Carts and Waggons.' *AQ*, 1931, **5**, 185.

FOX, G. E. & HOPE, W. H. ST JOHN (1). 'Excavations on the Site of the Roman City of Silchester in Hampshire in 1900.' *AAA*, 1901, **57**, 247.

FOX, LANE. See Pitt-Rivers.

FRANCK, H. A. (1). *Roving through Southern China.* Century, New York, 1925.

FRANKE, H. (1). 'Some Remarks on the Interpretation of Chinese Dynastic Histories' (with special reference to the Yuan). *OR*, 1950, **3**, 113.

FRANKE, H. (2) (tr.). 'Beiträge z. Kulturgeschichte Chinas unter der Mongolenherrschaft' (complete translation and annotation of the *Shan Chü Hsin Hua* by Yang Yü, +1360). *AKML*, 1956, **32**, 1–160 (rev. J. Prüsek, *ARO*, 1959, **27**, 476).

FRANKE, H. (3). 'Volksaufstände in d. Geschichte Chinas.' *GWI*, 1951, **1**, 31.

FRANKE, H. (13). 'Neuere Arbeiten zur Soziologie Chinas.' *SAE*, 1951, **2**, 306.

FRANKE, H. (15). 'Kulturgeschichtliches über die chinesische Tusche.' *ABAW/PH*, 1962, N.F. **54**, 1–158.

FRANKE, O. (11) (intr. & tr.). *Kêng Tschi T'u [Kêng Chih Thu]; Ackerbau und Seidegewinnung in China, ein kaiserliches Lehr- und Mahn-Buch.* Friederichsen, Hamburg, 1913. (Abhandl. d. Hamburgischen Kolonialinstituts, vol. 11; Ser. B, Völkerkunde, Kulturgesch. u. Sprachen, vol. 8.)

FRANKE, W. (2). 'Die Han-Zeitlichen Felsengräber bei Chiating (West Szechuan).' *SSE*, 1948, **7**, 19.

FRANKEL, H. H. (1). *Catalogue of Translations from the Chinese Dynastic Histories for the Period +220 to +960.* Univ. Calif. Press, Berkeley and Los Angeles, 1957. (Inst. Internat. Studies, Univ. of California, East Asia Studies, Chinese Dynastic Histories Translations, Suppl. no. 1.)

FRANKLIN, BENJAMIN (1). 'Maritime Observations' (a letter to Mr Alphonsus le Roy dated Aug. 1785). *TAPS*, 1786, **2**, 294 (p. 301); abstracted in *NAVC*, 1803, **9**, 32.

FRASER, E. D. H. & LOCKHART, J. H. S. (1). *Index to the 'Tso Chuan'.* Oxford, 1930.

FREEMAN, J. R. (1). 'Flood Problems in China.' *TASCE*, 1922, **85**, 1405.

FREEMAN, K. (1). See Diels-Freeman.

FREEMAN, K. (2). *The Pre-Socratic Philosophers, a companion to Diels, 'Fragmente der Vorsokratiker'.* Blackwell, Oxford, 1946.

FRÉMONT, C. See Sauvage, E. (1).

FRÉMONT, C. (1). *Études Expérimentales de Technologie Industrielle, No. 10: Évolution des Méthodes et des Appareils employés pour l'Essai des Matériaux de Construction, d'après les Documents du Temps* (Renaissance onwards). (Internat. Congr. Strength of Materials, Paris, 1900.) Dunod, Paris, 1900.

FRÉMONT, C. (2). *Études Expérimentales de Technologie Industrielle, No. 15: Évolution de la Fonderie de Cuivre, d'après les Documents du Temps.* Renouard, Paris, 1903.

FRÉMONT, C. (3). *Études Expérimentales de Technologie Industrielle: No. 37; Machine à Mesurer le Rendement des Vis; Origines de la Vis et des Engrenages.* For École Nat. Sup. des Mines, Dunod & Piriat, Paris, 1910. Completed in Frémont (19)

FRÉMONT, C. (4). *Études Expérimentales de Technologie Industrielle, No. 40: Le Clou.* BSEIN, 1912 and sep. Paris, 1912.

FRÉMONT, C. (5). *Études Expérimentales de Technologie Industrielle, No. 44: Origine et Évolution des Outils.* BSEIN, 1913 and sep. Paris, 1913. (Brace and bit, potters' wheel, lathe, rolling-mill, etc.)

FRÉMONT, C. (6). *Études Expérimentales de Technologie Industrielle, No. 45: Origine et Évolution des Outils Préhistoriques.* Paris, 1913. Completed in Frémont (20).

FRÉMONT, C. (7). *Études Expérimentales de Technologie Industrielle, No. 47: Origine de l'Horloge à Poids.* Paris, 1915. (Incl. springs and spiral springs; pendulums.)

FRÉMONT, C. (8). *Études Expérimentales de Technologie Industrielle, No. 48: Le Balancier à Vis pour Estampage.* Paris, 1916. (Worm press for coining and printing.)

FRÉMONT, C. (9). *Études Expérimentales de Technologie Industrielle, No. 49: La Lime.* Paris, 1916. Completed in Frémont (23).

FRÉMONT, C. (10). *Études Expérimentales de Technologie Industrielle, No. 50: Origines et Évolution de la Soufflerie.* Paris, 1917. Cf. Frémont (14).

FRÉMONT, C. (11). *Études Expérimentales de Technologie Industrielle, No. 54: Origine et Évolution du Tuyau.* Paris, 1920. Cf. Frémont (14).

FRÉMONT, C. (12). *Études Expérimentales de Technologie Industrielle, No. 57: Origine de la Poulie, du Treuil, de l'Engrenage, de la Roue de Voiture, etc.; Étude sur le Frottement des Cordes et sur les Palans.* BSEIN, 1921 and sep. Paris, 1921.

FRÉMONT, C. (13). *Études Expérimentales de Technologie Industrielle, No. 64: Le Marteau, le Choc, le Marteau Pneumatique.* Paris, 1923. (Hammers and vibrators.)

FRÉMONT, C. (14). *Études Expérimentales de Technologie Industrielle, No. 66: La Forge Maréchale.* Paris, 1923.

FRÉMONT, C. (15). *Études Expérimentales de Technologie Industrielle, No. 67: Origine et Début de l'Évolution de la Chaudière à Vapeur.* Paris, 1923. (From the still.)

FRÉMONT, C. (16). *Études Expérimentales de Technologie Industrielle, No. 68: Origine et Évolution des Pompes Centrifuges.* Paris, 1923.

FRÉMONT, C. (17). *Études Expérimentales de Technologie Industrielle, No. 70: La Serrure, Origine et Évolution.* Paris, 1924.

FRÉMONT, C. (18). *Études Expérimentales de Technologie Industrielle, No. 72: l'Essai de Traction des Métaux.* Paris, 1927.

FRÉMONT, C. (19). *Études Expérimentales de Technologie Industrielle, No. 75: La Vis.* Paris, 1928.

FRÉMONT, C. (20). *Études Expérimentales de Technologie Industrielle, No. 76: Les Outils, leur Origine, leur Évolution.* Paris, 1928. (Mostly prehistoric: scraper, chisel, hatchet, gouge, bore, etc.)

FRÉMONT, C. (21). *Études Expérimentales de Technologie Industrielle, No. 77: La Scie.* Paris, 1928.

FRÉMONT, C. (22). *Études Expérimentales de Technologie Industrielle, No. 81: Le Cisaillement et le Poinçonnage des Métaux.* Paris, 1929.

FRÉMONT, C. (23). *Études Expérimentales de Technologie Industrielle, No. 82: La Lime.* Paris, 1930.

FRÉMONT, C. (24). *Études Expérimentales de Technologie Industrielle, No. 27: Origine du Laminoir* (rolling-mill). Paris, 1907.

FRIESE, H. (1). 'Zum Aufsteig von Handwerkern ins Beamtentum während der Ming-Zeit.' *OE*, 1959, **6**, 160.

FRUMKIN, G. (1). 'Archéologie Soviétique en Asie.' *ASEA*, 1957, **11**, 73.

FUCHS, W. (8). 'Zum *Kêng Chih Thu* der Mandju-Zeit und die japanische Ausgabe von 1808.' *OAS*, 1959, 67. An enlarged version of 'Rare Chhing Editions of the *Kêng Chih Thu*'. *SSE*, 1947, **6**, 149.

FURON, R. (1). *Manuel de Préhistoire Générale.* Payot, Paris, 1943.

FUSSELL, G. E. (1). *The Farmer's Tools, 1500 to 1900; the History of British Farm Implements, Tools and Machinery before the Tractor came.* Melrose, London, 1952.

GABB, G. H. & TAYLOR, F. SHERWOOD (1). 'An Early Orrery by Thomas Tompion and George Graham, recently acquired by the Museum of the History of Science, Oxford.' *CON*, 1948 (Sept.), 24 & 55.

GABOR, D. (1). 'Communication Theory, Past, Present and Prospective.' In *Symposium on Information Theory*, p. 2. Min. of Supply, London, 1950 (mimeographed).

GABRIELI, G. (1). *Giovanni Schreck Linceo, Gesuita e Missionario in Cina e le sue Lettere dall'Asia.* Rome, 1937.

GALE, E. M. (1) (tr.). *Discourses on Salt and Iron ('Yen Thieh Lun'), a Debate on State Control of Commerce and Industry in Ancient China, chapters 1–19.* Brill, Leiden, 1931. (Sinica Leidensia, no. 2.) (Crit. P. Pelliot, *TP*, 1932, **29**, 127.)

GALE, E. M., BOODBERG, P. A. & LIN, T. C. (1) (tr.). 'Discourses on Salt and Iron (*Yen Thieh Lun*), Chapters 20–28.' *JRAS/NCB*, 1934, **65**, 73.

GALLAGHER, J. & PURCELL, V. (1). 'Economic Relations in Africa and the Far East [from 1713 to 1763].' Ch. 24 in *The New Cambridge Modern History*: Vol. 7, *The Old Régime*, ed. J. O. Lindsay, p. 566.

GALLAGHER, L. J. (1) (tr.). *China in the 16th Century; the Journals of Matthew Ricci, 1583–1610.* Random House, New York, 1953. (A complete translation, preceded by inadequate bibliographical details, of Nicholas Trigault's *De Christiana Expeditione apud Sinas* (1615). Based on an earlier publication: *The China that Was; China as discovered by the Jesuits at the close of the 16th Century: from the Latin of Nicholas Trigault.* Milwaukee, 1942.) Identifications of Chinese names in Yang Lien-Shêng (4). (Crit. J. R. Ware, *ISIS*, 1954, **45**, 395.)

GALLOWAY, R. L. (1). *The Steam-Engine and its Inventors; a Historical Sketch.* Macmillan, London, 1881.

GALLUS, S. & HORVÁTH, T. (1). 'Un Peuple Cavalier pré-Scythique en Hongrie; Trouvailles archéologiques du premier Âge du Fer et leurs Relations avec l'Eurasie' (−8th to −7th centuries). 2 vols. *DP*, 1939 (ser. 2), no. 9.

GARRISON, F. H. (1). 'History of Drainage, Irrigation, Sewage-Disposal, and Water-Supply.' *BNYAM*, 1929, **5**, 887.

GARRISON, F. H. (3). *An Introduction to the History of Medicine.* Saunders, Philadelphia, 1913; 4th ed. 1929.

GASSENDI, P. (1). *The Mirror of True Nobility and Gentry, being the Life of N. C. Fabricius, Lord of Peiresk.* Tr. W. Rand. London, 1657.

GAUBIL, A. (2). *Histoire Abrégée de l'Astronomie Chinoise.* (With Appendices 1, Des Cycles des Chinois; 2, Dissertation sur l'Éclipse Solaire rapportée dans le *Chou-King* [*Shu Ching*]; 3, Dissertation sur l'Éclipse du Soleil rapporteé dans le *Chi-King* [*Shih Ching*]; 4, Dissertation sur la première Éclipse du Soleil rapportée dans le *Tchun-Tsieou* [*Chhun Chhiu*]; 5, Dissertation sur l'Éclipse du Soleil, observée en Chine l'an trente-et-unième de Jésus-Christ; 6, Pour l'Intelligence de la Table du *Yue-Ling* [*Yüeh Ling*]; 7, Sur les Koua; 8, Sur le Lo-Chou (recognition of Lo Shu as magic square).) In *Observations Mathématiques, Astronomiques, Géographiques, Chronologiques et Physiques, tirées des anciens Livres Chinois ou faites nouvellement aux Indes, à la Chine, et ailleurs, par les Pères de la Compagnie de Jésus*, ed. E. Souciet. Rollin, Paris, 1732, vol. 2.

GEIL, W. E. (3). *The Great Wall of China.* Murray, London, 1909.

DE GENSSANE, M. (1). *Traité de la Fonte des Mines par le Feu du Charbon de Terre.* Paris, 1770.

DE GENSSANE, M. (2). 'Machine pour Eléver l'Eau par le Moyen du Feu, simplifiée par M. de G...' (an automatised Savery steam pumping system using a form of 'cataract' or tipping bucket). *MIARA*, 1744, **7**, 227.

GERINI, G. E. (1). *Researches on Ptolemy's Geography of Eastern Asia (Further India and Indo-Malay Peninsula).* Royal Asiatic Society and Royal Geographical Society, London, 1909. (Asiatic Society Monographs, no. 1.)

GERLAND, E. & TRAUMÜLLER, F. (1). *Geschichte d. physikalischen Experimentierkunst.* Engelmann, Leipzig, 1899.

GERNET, J. (1). *Les Aspects Économiques du Bouddhisme dans la Société Chinoise du 5ᵉ au 10ᵉ siècles.* Maisonneuve, Paris, 1956. (Publications de l'École Française d'Extrême-Orient, Hanoi & Saigon.) (Revs. D. C. Twitchett, *BLSOAS*, 1957, **19**, 526; A. F. Wright, *JAS*, 1957, **16**, 408.)

GERNET, J. (2). *La Vie Quotidienne en Chine à la Veille de l'Invasion Mongole (+1250 à +1276).* Hachette, Paris, 1959.

GERSHEVITCH, I. (1). 'Sissoo at Susa.' *BLSOAS*, 1957, **19**, 317.

GHIRSHMAN, R. (3). 'Tchoga-Zanbil près Suse; Rapport Préliminaire de la 6ᵉ Campagne.' *AAS*, 1957, **4**, 113.

GIACOMELLI, R. (1). 'I Modelli delle Macchine Volanti di Leonardo da Vinci.' *I*, 1931, **5** (no. 2).

GIBBS, C. D. I. (1). 'The River-Life of Canton.' *NV*, 1930, p. 73.

GIBBS-SMITH, C. H. (1). *The Aeroplane; an Historical Survey of its Origins and Development.* HMSO for Science Museum, London, 1960.

GIBBS-SMITH, C. H. (2). 'The Birth of the Aeroplane' (Cantor Lecture). *JRSA*, 1959, **107**, 78.

GIBBS-SMITH, C. H. (3). 'The Origins of the Aircraft Propeller.' *RBMQR*, 1959, April, June & September.

GIBBS-SMITH, C. H. (4). *A History of Flying.* Batsford, London, 1953.

GIBBS-SMITH, C. H. (5). 'The Work of Fr. Gusmão.' *JRSA*, 1949, **97**, 822.

GIBBS-SMITH, C. H. (6). 'The First Manned Aeroplane' (Sir George Cayley's kite-shaped glider of 1853; its nature and appearance). *The Times*, 13 June 1960, p. 11. (The announcement refers to a paper in *MMG* for 15 Sept. 1852, and reproduces its drawings.)

GIBBS-SMITH, C. H. (7). 'Origins of the Helicopter.' *NS*, 1962, **14**, 229.

GIBBS-SMITH, C. H. (8). 'Sir George Cayley, "Father of Aerial Navigation" (1773 to 1857).' *NRRS*, 1962, **17**, 36.

GIBBS-SMITH, C. H. (9). 'Note on Leonardo's Helicopter Model.' In Hart (4) as Appendix.

GIBBS-SMITH, C. H. See Hart, I. B. (4).

GIBSON, H. E. (4). 'Communications in China during the Shang Period.' *CJ*, 1937, **26**, 228.

GIEDION, S. (1). *Mechanisation takes Command; a Contribution to anonymous History.* Oxford, 1948.

GIGLIOLI, E. H. (1). 'Chinese querns.' *AAE*, 1898, **28**, 376.

GILES, H. A. (1). *A Chinese Biographical Dictionary.* 2 vols. Kelly & Walsh, Shanghai, 1898; Quaritch, London, 1898. Supplementary Index by J. V. Gillis & Yü Ping-Yüeh, Peiping, 1936. Account must be taken of the numerous emendations published by von Zach (4) and Pelliot (34), but many mistakes remain. Cf. Pelliot (35).

GILES, H. A. (5). *Adversaria Sinica:*
1st series, no. 1, pp. 1–25. Kelly & Walsh, Shanghai, 1905.
      no. 2, pp. 27–54. Kelly & Walsh, Shanghai, 1906.
      no. 3, pp. 55–86. Kelly & Walsh, Shanghai, 1906.
      no. 4, pp. 87–118. Kelly & Walsh, Shanghai, 1906.
      no. 5, pp. 119–144. Kelly & Walsh, Shanghai, 1906.
      no. 6, pp. 145–188. Kelly & Walsh, Shanghai, 1908.
      no. 7, pp. 189–228. Kelly & Walsh, Shanghai, 1909.
      no. 8, pp. 229–276. Kelly & Walsh, Shanghai, 1910.
      no. 9, pp. 277–324. Kelly & Walsh, Shanghai, 1911.
      no. 10, pp. 326–396. Kelly & Walsh, Shanghai, 1913.
      no. 11, pp. 397–438 (with index). Kelly & Walsh, Shanghai, 1914.
2nd series no. 1, pp. 1–60. Kelly & Walsh, Shanghai, 1915.

GILES, H. A. (9). 'Spuren d. Luftfahrt im alten China.' *GTIG*, 1917, **4**, 79. (A translation by A. Schück of Giles (5), 1st ser., pp. 229 ff., with comments by F. M. Feldhaus.)

GILES, H. A. (12). *Gems of Chinese Literature; Prose.* 2nd ed. Kelly & Walsh, Shanghai, 1923. (i) For texts see Lockhart (1); (ii) Abridged edition, without acknowledgement of authorship but with the inclusion of the Chinese texts of the pieces selected, ed. Chiang Fêng-Wei (1), Chungking, 1942.

GILES, L. See Yetts, W. P. (18).

GILES, L. (6). *A Gallery of Chinese Immortals ('hsien'), selected biographies translated from Chinese sources (Lieh Hsien Chuan, Shen Hsien Chuan, etc.).* Murray, London, 1948.

GILFILLAN, S. C. (2). *The Sociology of Invention.* Follett, Chicago, 1935.

GILLE, B. (1). 'Notes d'Histoire de la Technique Métallurgique; I, Les Progrès du Moyen Âge; Le Moulin à Fer et le Haut Fourneau.' *MC*, 1946, **1**, 89.

GILLE, B. (2). 'La Naissance du Système Bielle-Manivelle.' *MC/TC*, 1952, **2**, 42.

GILLE, B. (3). 'Léonard de Vinci et son Temps.' *MC/TC*, 1952, **2**, 69.

GILLE, B. (4). 'Le Machinisme au Moyen Âge.' *A/AIHS*, 1953, **6**, 281.

GILLE, B. (5). 'Contributions à une Histoire de la Civilisation Technique; I, L'Antiquité Classique.' *MC/TC*, 1953, **2**, 109.

GILLE, B. (6). 'Contributions à une Histoire de la Civilisation Technique; II, L'évolution des Techniques au 16ᵉ siècle. *MC/TC*, 1953, **2**, 119.

GILLE, B. (7). 'Le Moulin à Eau, une révolution technique médiévale.' *MC/TC*, 1954, **3**, 1.

GILLE, B. (9). *La Technique Sidérurgique et son Évolution; Catalogue de l'Exposition du Musée Historique Lorrain à l'occasion du Colloque International 'Le Fer à travers les Âges'.* Palais Ducal, Nancy, 1955.

GILLE, B. (10). 'Études sur les Métallurgies Primitives; l'ancienne Métallurgie du Fer à Madagascar.' *MC/TC*, 1955, **4**, 144.

GILLE, B. (11). 'Les Développements Technologiques en Europe de +1100 à +1400.' *JWH*, 1956, **1**, 63.

GILLE, B. (12). 'Études sur les Manuscrits d'Ingénieurs du +15ᵉ siècle.' *MC/TC*, 1956, **5**, 77, 216.

GILLE, B. (13). 'Les Problèmes Techniques au +17ᵉ siècle.' *MC/TC*, 1954, **3**, 177.

GILLE, B. (14). 'Machines [in the Mediterranean Civilisations and the Middle Ages].' Art. in *A History of Technology*, ed. C. Singer *et al.*, vol. 2, p. 629. Oxford, 1956.

GILMOUR, J. (1). *Among the Mongols.* RTS, London, 1892.

GINGELL, W. R. (1) (tr.). *The Ceremonial Usages of the Ancient Chinese, B.C. 1121, as described in the 'Institutes of the Chou Dynasty strung as Pearls'.* A translation of the abridgement *Chou Li Kuan Chu.* London, 1852. (Preface written at Fuchow, 1849.)

GINZEL, F. K. (1). *Handbuch d. mathematischen und technischen Chronologie, das Zeitrechnungswesen d. Völker.* 3 vols. Hinrichs, Leipzig, 1906.

GINZROT, J. C. (1). *Die Wagen und Fahrwerke d. Griechen u. Römer und anderer altern Völker, nebst d. Bespannung, Zäumung u. Verzierung ihrer Zug-, Reit-, und Last-Thiere.* 2 vols. Lentner, Munich, 1817.

GIQUEL, P. (1). 'Mechanical and Nautical Terms in French, Chinese and English.' In Doolittle, J. (1), vol. 2, p. 634.

GIQUEL, P. (2). *The Foochow Arsenal and its Results, from the commencement in 1867 to the end of the Foreign Directorate, on the 16th Feb. 1874,* tr. from French by H. Lang. Shanghai Evening Courier, Shanghai, 1874.

GIRAUD, C. (1). *Essai sur l'Histoire du Droit français au Moyen-Âge.* 2 vols. Paris, 1846.

GODE, P. K. (3). 'Carriage Manufacture in the Vedic Period [in India] and in Ancient China.' *ABORI*, 1947, **27**, 288. Repr. in *Studies in Indian Cultural History*, vol. 2 (Gode Studies, vol. 5), p. 129.

GODE, P. K. (4). 'The Indian Bullock-Cart; its Prehistoric and Vedic Ancestors' *PO*, 1941, **5**, 144. Repr. in *Studies in Indian Cultural History*, vol. 2 (Gode studies, vol. 5), p. 123.

GOMPERTZ, MAURICE (1). *The Master Craftsmen; the Story of the Evolution of Implements.* Nelson, London, 1933.

GOODRICH, L. CARRINGTON (1). *Short History of the Chinese People.* Harper, New York, 1943.

GOODRICH, L. CARRINGTON (7). 'The Revolving Book-Case in China.' *HJAS*, 1942, **7**, 130.

GOODRICH, L. CARRINGTON & FÊNG CHIA-SHÊNG (1). 'The Early Development of Firearms in China.' *ISIS*, 1946, **36**, 114, with important addendum *ISIS*, 1946, **36**, 250.

GOTO, S. (1). 'Le Goût Scientifique de Khang-Hsi, Empereur de Chine.' *BMFJ*, 1933, **4**, 117.

GOUDIE, G. (1). 'On the Horizontal Water-Mills of Shetland.' *PSAS*, 1886, **20**, 257.

GOULLART, P. (1). *Forgotten Kingdom* (the Lichiang districts of Yunnan). Murray, London, 1955.

GOWLAND, W. (1). 'Copper and its Alloys in Prehistoric Times.' *JRAI*, 1906, **36**, 11.

GOWLAND, W. (5). 'The Early Metallurgy of Copper, Tin and Iron in Europe as illustrated by ancient Remains, and primitive Processes surviving in Japan.' *AAA*, 1899, **56**, 267.

GRAND, R. (1). 'La Force Motrice Animale à travers les Âges et son Influence sur l'Évolution Sociale.' *BSISS*, 1926.

GRAND, R. (2). 'Utilisation de la Force Motrice Animale; Vues sur l'Origine de l'Attelage Moderne.' *CRAAF*, 1947, **33**, 706. Abstr. *BSNAF*, 1947, 259.

GRANET, M. (1). *Danses et Légendes de la Chine Ancienne.* 2 vols. Alcan, Paris, 1926.

GRANET, M. (2). *Fêtes et Chansons Anciennes de la Chine.* Alcan, Paris, 1926; 2nd ed. Leroux, Paris, 1929.

GRANET, M. (4). *La Religion des Chinois.* Gauthier-Villars, Paris, 1922.

GRANET, M. (5). *La Pensée Chinoise.* Albin Michel, Paris, 1934. (Évol. de l'Hum. series, no. 25 bis.)

GRANET, M. (6). *Études Sociologiques sur la Chine.* Presses Univ. de France, Paris, 1953.

GRANGER, F. (ed. & tr.) (1). *Vitruvius on Architecture*. 2 vols. Heinemann, London, 1934. (Loeb Classics edn.)

GRAY, A. (1). *A Treatise on Gyrostatics and Rotational Motion*. Macmillan, London, 1918.

GRAY, B. & VINCENT, J. B. (1). *Buddhist Cave-Paintings at Tunhuang*. Faber & Faber, London, 1959.

GRAY, J. H. (1). *China, a History of the Laws, Manners and Customs of the People*. 2 vols., ed. W. G. Gregor. Macmillan, London, 1878.

GREENBERG, M. (1). *British Trade and the Opening of China, 1800–1842*. Cambridge, 1951.

GRIERSON, Sir G. A. (1). *Bihar Peasant Life*. Patna, 1888; reprinted Bihar Govt., Patna, 1926.

GROFF, G. W. & LAU, T. C. (1). 'Landscaped Kuangsi; China's Province of Pictorial Art.' *NGM*, 1937, **72**, 700.

GROSLIER G. (1). *Recherches sur les Cambodgiens*. Challamel, Paris, 1921.

GROSVENOR, G. H. (1). 'Dr [Alexander Graham] Bell's Man-Lifting Kite.' *NGM*, 1908, **19**, 35.

GROUSSET, R. (5). *Histoire de la Chine*. Fayard, Paris, 1942.

GROUSSET, R. (6). *The Rise and Splendour of the Chinese Empire*. Tr. A. Watson-Gandy & T. Gordon. Bles, London, 1952.

GUATELLI, R. A. (1). Models of Leonardo da Vinci's machines. *TIM*, 1939, 29 May; 1951, **58**, 15 Oct., 29.

GUIDOBALDO (Guido Ubaldo del Monte) (1). *Mechanicorum Liber*. Pisa, 1577.

GUILLEMINET, P. (1). 'Une Industrie Annamite; les Norias de Quảng-Ngải.' *BAVH*, 1926, **13** (no. 2), 97–216.

VAN GULIK, R. H. (3). '*Pi Hsi Thu Khao*'; *Erotic Colour-Prints of the Ming Period, with an Essay on Chinese Sex Life from the Han to the Chhing Dynasty (−206 to +1644)*. 3 vols in case. Privately printed, Tokyo, 1951 (50 copies only, distributed to the most important Libraries of the World). (Crit. W. L. Hsü, *MN*, 1952, **8**, 455; H. Franke, *ZDMG*, 1956, (N.F.) **30**, 380.)

VAN GULIK, R. H. (5). 'Hayagrīva; the Mantrayānic Aspect of the Horse-Cult in China and Japan.' *IAE*, 1935, **33** (Suppl.).

VAN GULIK, R. H. (7). *Siddham; an Essay on the History of Sanskrit Studies in China and Japan*. Internat. Acad. Indian Culture, Nagpur, 1956. (Sarasvati-Vihara series, no. 36.)

VAN GULIK, R. H. (8). *Sexual Life in Ancient China; a Preliminary Survey of Chinese Sex and Society from ca. −1500 to +1644*. Brill, Leiden, 1961.

GUNDA, B. (1). 'l'Importance anthropo-géographique du Contact des Éleveurs de Renne et Éleveurs de Cheval.' *FK*, 1940, **67**, 308.

GUNTHER, R. T. (2). *The Astrolabes of the World*. 2 vols. Oxford, 1932.

GUTMANN, A. L. (1). 'Cloth-Making in Flanders.' *CIBA/T*, 1938, **2** (no. 14), 466.

GUTSCHE, F. (1). 'Die Entwicklung d. Schiffsschraube.' *BGTI/TG*, 1937, **26**, 37.

GYLLENSVÄRD, BO (1). *Chinese Gold and Silver[-Work] in the Carl Kempe Collection*. Stockholm, 1953; Smithsonian Institution, Washington, D.C., 1954.

GYLLENSVÄRD, BO (2). 'Thang Gold and Silver.' *BMFEA*, 1957, **29**, 1–230.

GYULA, L. (1). 'Beiträge z. Volkskunde der Avaren, III' (in German and Magyar). *ARAE*, 1942 (3rd ser.), **3**, 334 & 341.

HADDAD, SAMI I. & KHAIRALLAH, AMIN A. (1). 'A Forgotten Chapter in the History of the Circulation of the Blood.' *ASURG*, 1936, **104**, 1.

HADDON, A. C. (1). *The Study of Man*. Murray, London, 1908.

HADDON, A. C. (2). 'The Evolution of the Cart.' In Haddon (1), p. 161.

HADDON, A. C. (3). 'The Origin of the Irish Jaunting-Car.' In Haddon (1), p. 200.

HAHN, P. M. & DE MYLIUS, A. (1). *Description of a Planetarium, or Astronomical Machine, which exhibits the most remarkable Phaenomena, Motions, and Revolutions of the Universe*. London, 1791.

HAHNLOSER, H. R. (1) (ed.). *The Album of Villard de Honnecourt*. Schroll, Vienna, 1935.

HALDANE, J. S. & HENDERSON, YANDELL (1). 'The Rate of Work done with an Egyptian Shadouf.' *N*, 1926, **118**, 308.

DU HALDE, J. B. (1). *Description Géographique, Historique, Chronologique, Politique et Physique de l'Empire de la Chine et de la Tartarie Chinoise*. 4 vols. Paris, 1735, 1739; The Hague, 1736. Eng. tr. R. Brookes, London, 1736, 1741. Germ. tr. Rostock, 1748.

HALL, A. R. (2). 'A Note on Military Pyrotechnics [in the Middle Ages].' Art. in *A History of Technology*, ed. C. Singer *et al.*, vol. 2, p. 374. Oxford, 1956.

HALL, A. R. (3). *The Military Inventions of Guido da Vigevano*. Proc. VIIIth International Congress of the History of Science, p. 966. Florence, 1956.

HALL, A. R. (4). 'More on Mediaeval Pivoted Axles.' *TCULT*, 1961, **2**, 17.

HALL, A. R. (5). 'Military Technology [from the Renaissance to the Industrial Revolution].' Art. in *A History of Technology*, ed. C. Singer *et al.*, vol. 3, p. 347. Oxford, 1957.

HALL, A. R. (6). 'Military Technology [in the Mediterranean Civilisations and the Middle Ages].' Art. in *A History of Technology*, ed. C. Singer *et al.*, vol. 2, p. 695. Oxford, 1956.

HALL, J. W. (1). 'The Making and Rolling of Iron.' *TNS*, 1928, **8**, 40.

HAMILTON, S. B. (2). 'Bridges [from the Renaissance to the Industrial Revolution].' Art. in *A History of Technology*, ed. C. Singer *et al.*, vol. 3, p. 417. Oxford, 1957.

HANDLEY-PAGE, F. (1). 'The Handley-Page [Slotted] Wing [for Aircraft].' *JRAES*, 1921, **25**, 263.

HARADA, YOSHITO & KOMAI, KAZUCHIKA (1). *Chinese Antiquities*. Pt. 1, *Arms and Armour*; Pt. 2, *Vessels [Ships] and Vehicles*. Academy of Oriental Culture, Tokyo Institute, Tokyo, 1937.

HARCOURT-SMITH, S. (1). *A Catalogue of various Clocks, Watches, Automata and other miscellaneous objects of European workmanship, dating from the 18th and early 19th centuries, in the Palace Museum and the Wu Ying Tien, Peking*. Palace Museum, Peiping, 1933.

VON HARINGER, J. & BOREL, V. (1). 'Über mechanische Spielereien.' *AT*, 1947, **19**, 96.

DE HARLEZ, C. (4) (tr.). *Livres des Esprits et des Immortels* (transl. of *Shen Hsien Chuan*). Hayez, Brussels, 1893.

HARRIS, JOHN (1). *Astronomical Dialogues*. London, 1719.

HARRIS, JOHN (2). *Lexicon Technicum; or, an Universal English Dictionary of Arts and Sciences*. London, 1704–10. Bibliography in John Ferguson (2).

HARRIS, L. E. (2). 'Some Factors in the Early Development of the Centrifugal Pump, 1689 to 1851.' *TNS*, 1952, **28**, 187.

HARRISON, H. S. (1). 'The Origin of the Driving-Belt.' *MA*, 1947, **47**, 114.

HARRISON, H. S. (2). 'Opportunism and the Factors of Invention.' *AAN*, 1930, **32**, 106.

HARRISON, H. S. (3). 'Discovery, Invention and Diffusion [from Early Times to the Fall of the Ancient Empires].' Art. in *A History of Technology*, ed. C. Singer *et al.*, vol. 1, p. 58. Oxford, 1954.

HART, I. B. (1). 'Leonardo da Vinci as a Pioneer of Aviation.' *JRAES*, 1923, **27**, 244. 'Leonardo da Vinci's MS. on the Flight of Birds.' *JRAES*, 1923, **27**, 289.

HART, I. B. (2). *The Mechanical Investigations of Leonardo da Vinci*. Chapman & Hall, London; Open Court, Chicago, 1925.

HART, I. B. (3). 'The Scientific Basis for Leonardo da Vinci's Work in Technology.' *TNS*, 1953, **28**, 105.

HART, I. B. (4). *The World of Leonardo da Vinci, Man of Science, Engineer, and Dreamer of Flight* (with a note on Leonardo's Helicopter Model, by C. H. Gibbs-Smith). McDonald, London, 1961. Rev. K. T. Steinitz, *TCULT*, 1963, **4**, 84.

HAUDRICOURT, A. G. (1). 'Les Moteurs Animés en Agriculture; Esquisse de l'Histoire de leur Emploi à travers les Âges.' *RBA*, 1940, **20**, 759.

HAUDRICOURT, A. G. (2). 'Contribution à l'Étude du Moteur Humain.' *AHES/AHS*, 1940, **2**, 131.

HAUDRICOURT, A. G. (3). 'Contribution à la Géographie et à l'Ethnologie de la Voiture.' *RGHE*, 1948, **1**, 54. (Contains the theory of the relationship of the camel packsaddle to the collar harness of the horse.)

HAUDRICOURT, A. G. (4). 'De l'Origine de l'Attelage Moderne.' *AHES*, 1936, **8**, 515.

HAUDRICOURT, A. G. (6). 'L'Origine de la Duga.' *AHES/AHS*, 1940, **2**, 34.

HAUDRICOURT, A. G. (7). 'Lumières sur l'Attelage Moderne.' *AHES/AHS*, 1945, **8**, 117.

HAUDRICOURT, A. G. (8). 'Relations entre Gestes Habituels, Forme des Vêtements et Manière de Porter les Charges.' *RGHE*, 1949, **2**, 58.

HAUDRICOURT, A. G. (10). 'Les Premières Étapes de l'Utilisation de l'Énergie Naturelle.' Art. in *Les Origines de la Civilisation Technique*, vol. 1 of *Histoire Générale des Techniques*, ed. M. Daumas. Presses Univ. de France, Paris, 1962.

HAUDRICOURT, A. G. & DELAMARRE, M. J. B. (1). *L'Homme et la Charrue à travers le Monde*. Gallimard. Paris, 1955.

HAUSER, F. (1). *Über das 'Kitāb al-Ḥiyal', das Werk über die sinnreichen Anordnungen der Banū Musā*. Mencke, Erlangen, 1922. (Abhdl. z. Gesch. d. Naturwiss. u. d. Med. no. 1.)

DE HAUTEFEUILLE, JEAN (1). *Pendule Perpetuelle, avec un nouveau Balancier, et la Manière d'élever l'Eau par le moyen de la Poudre à Canon*. Paris, 1678.

HAWKES, D. (1) (tr.). '*Chhu Tzhu*'; *the Songs of the South—an Ancient Chinese Anthology*. Oxford, 1959. (Rev. J. Needham, *NSN*, 18 July 1959.)

HAWTHORNE, J. G. & SMITH, C. S. (1) (ed. & tr.). '*On Diverse Arts*', *the Treatise of Theophilus*. Univ. of Chicago Press, Chicago, 1963.

HAYASHI, TSURUICHI (2). 'Brief History of Japanese Mathematics.' *NAW*, 1905, **6**, 296 (65 pp.); 1907, **7**, 105 (58 pp.). (Crit. Y. Mikami, *NAW*, 1911, **9**, 373.)

HAZARD, B. H., HOYT, J., KIM HA-TAI, SMITH, W. W. & MARCUS, R. (1). *Korean Studies Guide*. Univ. of Calif. Press, Berkeley and Los Angeles, 1954.

HEATH, Sir THOMAS (6). *A History of Greek Mathematics*. 2 vols. Oxford, 1921.

HEDDE, I. (1). '*Kang Tchi Tou [Kêng Chih Thu]*'; *Description de l'Agriculture et du Tissage en Chine*. Bouchard-Huzard, Paris, 1850.

HEIBERG, J. L. (1). 'Geschichte d. Mathematik u. Naturwissenschaften im Altertum.' Art. in *Handbuch d. Altertumswissenschaft*, vol. 5 (1), 2. Beck, Munich, 1925.

HEIDRICH, E. (1). *Alt Niederländische Malerei*. Diederichs, Jena, 1910.

HEJZLAR, J. (1). 'The Return of a Legendary Work of Art; the most famous Scroll in the Peking Palace Museum, "On the River during the Spring Festival" by Chang Tsê-Tuan (+1125).' *NO*, 1962, 3 (no. 1), 17.

HELLMANN, G. (6) (ed.). *Rara Magnetica, +1269 to +1599*. Berlin, 1898. (Neudrucke von Schriften und Karten über Meteorologie und Erdmagnetismus, no. 10.)

HENDERSON, G. (1). 'Chŏng Ta-San; a Study in Korea's Intellectual History.' *JAS*, 1957, 16, 377.

HENNIG, R. (1). 'Beiträge z. Frühgeschichte d. Aeronautik.' *BGTI*, 1918, 8, 100.

HENNIG, R. (2). 'Zur Vorgeschichte d. Luftfahrt.' *BGTI*, 1928, 18, 87.

HENTZE, C. (4). *Chinese Tomb Figures; a Study in the Beliefs and Folklore of Ancient China*. Goldston, London, 1928.

HERON OF ALEXANDRIA (1). *Spiritalium Liber*. Urbino, 1575. *Mechanici*, Venice, 1572.

HERRIGEL, E. (1). *Zen in the Art of Archery*. Tr. from German by R. F. C. Hull with introdn. by D. T. Suzuki. Routledge & Kegan Paul, London, 1953.

HERRMANN, A. (1). *Historical and Commercial Atlas of China*. Harvard-Yenching Institute, Cambridge, Mass., 1935.

VAN DER HEYDEN, A. A. M. & SCULLARD, H. H. (1). *Atlas of the Classical World*. Nelson, London, 1959.

HEYLYN, P. (1). *Microcosmos; or, a Little Description of the Great World*. Lichfield & Short, Oxford, 1621.

HIBBERT-WARE, S. (1). *A Description of the Shetland Islands*. Edinburgh, 1822.

HILDBURGH, W. L. (1). 'Aeolipiles as Fire-Blowers.' *AAA*, 1951, 94, 27.

HILDEBRAND, J. R. (1). 'The World's Greatest Overland Explorer' (Marco Polo). *NGM*, 1928, 54, 505 (544).

HILKEN, T. J. N. (1). 'The Rev. Robert Willis, Engineer and Archaeologist (1800 to 1875).' Unpublished paper.

DE LA HIRE, J. N. (1). 'Mémoire pour la Construction d'une Pompe qui fournit continuellement de l'Eau dans le Reservoir.' *MRASP*, 1716, 322.

HIRTH, F. (1). *China and the Roman Orient*. Kelly & Walsh, Shanghai; G. Hirth, Leipzig and Munich, 1885. (Photographically reproduced in China with no imprint, 1939.)

HIRTH, F. (9). *Über fremde Einflüsse in der chinesischen Kunst*. G. Hirth, Munich and Leipzig, 1896.

HIRTH, F. (12). 'Scraps from a Collector's Notebook.' *TP*, 1905, 6, 373 (biographies of Chinese painters and archaeologists). Subsequently reprinted in book form, Stechert, New York, 1924.

HISCOX, G. D. (1). *Mechanical Movements, Powers, Devices and Appliances used in Constructive and Operative Machinery and the Mechanical Arts*... (an illustrated glossary). Henley, New York, 1899.

HISCOX, G. D. (2). *Mechanical Appliances, Mechanical Movements, and Novelties of Construction*. Henley, New York, 1923.

HITTI, P. K. (1). *History of the Arabs*. 4th ed. Macmillan, London, 1949; 6th ed. 1956.

HO PENG-YOKE. See Ho Ping-Yü.

HO PING-YÜ (1). *The Astronomical Chapters of the 'Chin Shu'*. Inaug. Diss., Singapore, 1957.

HO PING-YÜ & NEEDHAM, JOSEPH (2). 'Theories of Categories in Early Mediaeval Chinese Alchemy' (with transl. of the *Tshan Thung Chhi Wu Hsiang Lei Pi Yao*, c. +7th cent.). *JWCI*, 1959, 22, 173.

HO PING-YÜ & NEEDHAM, JOSEPH (3). 'The Laboratory Equipment of the Early Mediaeval Chinese Alchemists.' *AX*, 1959, 7, 57.

HO PING-YÜ & NEEDHAM, JOSEPH (4). 'Elixir Poisoning in Mediaeval China.' *JAN*, 1959, 48, 221.

HO SHAN (1). 'New Inventions in Irrigation Pumps [Gas Explosion Pumps invented by Tai Kuei-Jui, Phêng Ting-I and others].' *PKR*, 1958, 1 (no. 23), 14.

HOCK, G. (1). 'Ein Beitrag zur vorgeschichtlichen Technik.' Art. in *Schumacher Festschrift*, p. 80. Wilckens, Mainz, 1930; also in Deonna (1), pp. 125, 126.

HODGEN, M. T. (1). 'Domesday Water-Mills.' *AQ*, 1939, 13, 261.

HODGSON, J. E. (1). 'Notes on Sir George Cayley as a Pioneer of Aeronautics.' *TNS*, 1922, 3, 69.

HODGSON, J. E. (2). *The History of Aeronautics in Great Britain from the Earliest Times to the latter half of the Nineteenth Century*. Oxford, 1924.

HODGSON, J. E. (3). Introduction to Cayley (1). *Aeronautical and Miscellaneous Notebook (c. 1799–1826) of Sir George Cayley, with an Appendix comprising a List of the Cayley Papers*. Heffer, Cambridge, 1933. (Newcomen Society Extra Publication no. 3.)

HOLMQVIST, W. (1). 'Germanic Art during the +1st Millennium.' *KVHAAH*, 1955, 90, 1–89.

HOLTZAPFEL, C. & HOLTZAPFEL, J. J. (1). *Turning and Mechanical Manipulations*. Holtzapfel, London, 1852–94.
  I Materials (C.H.), 1852.
  II Cutting Tools (C.H.), 1856.
  III Abrasive and Miscellaneous Processes (C.H. rev. J.J.H.), 1894.
  IV Hand, or Simple, Turning (J.J.H.), 1881.
  V Ornamental, or Complex, Turning (J.J.H.), 1884.
HOMMEL, R. P. See Dunbar, G. S. (1).
HOMMEL, R. P. (1). *China at Work; an illustrated Record of the Primitive Industries of China's Masses, whose Life is Toil, and thus an Account of Chinese Civilisation*. Bucks County Historical Society, Doylestown, Pa., 1937; John Day, New York, 1937.
DE HONNECOURT, VILLARD. See Lassus & Darcel; Hahnloser.
HOOPER, W. (1). *Rational Recreations, in which the Principles of Numbers and Natural Philosophy are clearly and copiously elucidated, by a Series of easy, entertaining and interesting Experiments, among which are all those commonly performed with the Cards*. 4 vols. Davis, London, 1774, 1787.
HOOVER, H. C. & HOOVER, L. H. (1) (tr.). *Georgius Agricola 'De Re Metallica', translated from the 1st Latin edition of 1556, with biographical introduction, annotations and appendices upon the development of mining methods, metallurgical processes, geology, mineralogy and mining law from the earliest times to the 16th century*. 1st ed. Mining Magazine, London, 1912; 2nd ed. Dover, New York, 1950.
HOPE, W. H. ST J. & FOX, G. (1). 'Excavations on the Site of the Roman City of Silchester' (water-pumps). *AAA*, 1896, **55**, 215 (232).
HOPKINS, A. A. (1). *The Lure of the Lock*. New York, 1928.
HOPKINS, E. W. (1). 'The Social and Military Position of the Ruling Class in India, as represented by the Sanskrit Epic.' *JAOS*, 1889, **13**, 57–372 (with index) (military techniques, pp. 181–329).
HOPKINS, L. C. See Yetts, W. P. (12).
HOPKINS, L. C. (11). 'Pictographic Reconnaissances, VII.' *JRAS*, 1926, 461.
HÖRLE, J. (1). Articles *trapetum* and *tudicula*. In Pauly-Wissowa, vol. VIA, pt. ii, pp. 2187 ff.; vol. VIIA, pt. 1, p. 774.
HORNER, J. (1). *The Linen Trade of Europe during the Spinning-Wheel Period*. McCaw, Belfast, 1920.
HORVÁTH, ÁRPÁD (1). 'A Tűzesgép' ([A History of] the Steam Engine), (in Magyar). Táncsics Kőnyv-kiadó, Budapest, 1963.
HORWITZ, H. T. (1). 'Ein Beitrag zu den Beziehungen zwischen ostasiatischer und europäischer Technik.' *ZOIAV*, 1913, **65**, 390.
HORWITZ, H. T. (2). 'Ü. ein neueres deutschen Reichspatent (1912) und eine Konstruction v. Heron v. Alexandrien.' *AGNT*, 1917, **8**, 134.
HORWITZ, H. T. (3). 'Technische Darstellungen aus alten Miniaturwerken.' *BGTI*, 1920, **10**, 175.
HORWITZ, H. T. (4). 'Die Entwicklung der Drehbewegung' (history of the crank). *BGTI*, 1920, **10**, 177.
HORWITZ, H. T. (5). 'Die Drehbewegung in ihrer Bedeutung f. d. Entwicklung d. materiellen Kultur' (history of the crank). *AN*, 1933, **28**, 721; 1934, **29**, 99.
HORWITZ, H. T. (6). 'Beiträge z. aussereuropäischen u. vorgeschichtlichen Technik.' *BGTI*, 1916, **7**, 169.
HORWITZ, H. T. (7). 'Beiträge z. Geschichte d. aussereuropäischen Technik.' *BGTI*, 1926, **16**, 290.
HORWITZ, H. T. (9). 'Zur Geschichte des Schaufelradtriebes.' *ZOIAV*, 1930, **82**, 309, 356.
HORWITZ, H. T. (10). 'Giuliano di San Gallo' (+1445 to +1516, his illustrated MS.). *BGTI*, 1926, **16**, 200.
HORWITZ, H. T. (11). 'Über das Aufkommen, die erste Entwicklung und die Verbreitung von Wind-rädern.' *BGTI/TG*, 1933, **22**, 93.
HOSIE, A. (4). *Three Years in Western Szechuan; a Narrative of Three Journeys in Szechuan, Kweichow and Yunnan*. Philip, London, 1890.
HOUCKGEEST, A. E. VAN BRAAM. See van Braam Houckgeest.
HOUGH, W. (1). 'Fire as an Agent in Human Culture.' *BUSNM*, 1926, no. 139.
VAN HOUTEN, J. H. (1). 'Protection contre la Mer et Assèchements en Hollande.' *MC/TC*, 1953, **2**, 133,
DE HOUTHULST, WILLY COPPENS (1). 'Le Sport du Char à Voile.' *FI*, 1950, **6** (no. 250), 117. Cf. R. Miller, *POPS*, May 1950, 132.
HOWGRAVE-GRAHAM, R. P. (1). 'Some Clocks and Jacks, with Notes on the History of Horology.' *AAA*, 1927, **77**, 257.
HOWGRAVE-GRAHAM, R. P. (2). 'New Light on Ancient Turret-Clocks.' *TNS*, 1955, **29**, 137.
HSIA NAI (1). 'New Archaeological Discoveries.' *CREC*, 1952, **1** (no. 4), 13.
HSÜ CHI-HÊNG (1). 'The Man who built the first Chinese Railway [Chan Thien-Yu].' *CREC*, 1955. **4** (no. 7), 26.

HUARD, P. (2). 'Sciences et Techniques de l'Eurasie.' *BSEIC*, 1950, **25** (no. 2), 1. This paper, though correcting a number of errors in Huard (1), still contains many mistakes and should be used only with care; nevertheless it is valuable on account of several original points.

HUARD, P. & DURAND, M. (1). *Connaissance du Viêt-Nam.* École Française d'Extr. Orient, Hanoi, 1954; Imprimerie Nationale, Paris, 1954.

HUBBARD, T. O'B. & LEDEBOER, J. H. (1). *Aeronautical Classics, edited for the Council of the Aeronautical Society of Great Britain.* ASGB, London, 1910–11.

I   *Aerial Navigation*, by Sir George Cayley (1809).
II   *Aerial Locomotion*, by F. H. Wenham (1866).
III *The Art of Flying*, by T. Walker (1810).
IV   *The Aerial Ship*, by Francesco Lana (1670).
V   *Gliding*, by P. S. Pilcher (1897); and *The Aeronautical work of John Stringfellow* (b. 1799).
VI   *The Flight of Birds*, by Alfonso Borelli (1680).

HUBERT, M. J. (1) (tr.). *Ambroise's 'Estoire de la Guerre Sainte'.* New York, 1941.

HUC, R. E. (1). *Souvenirs d'un Voyage dans la Tartarie et le Thibet pendant les Années 1844, 1845 & 1846* [with J. Gabet], revised ed., 2 vols. Lazaristes, Peiping, 1924. Abridged ed. *Souvenirs d'un Voyage dans la Tartarie, le Thibet et la Chine...*, ed. H. d'Ardenne de Tizac, 2 vols. Plon, Paris, 1925. Eng. transl., by W. Hazlitt, *Travels in Tartary, Thibet and China during the years 1844 to 1846.* Nat. Ill. Lib. London, n.d. (1851–2). Also ed. P. Pelliot, 2 vols. Kegan Paul, London, 1928.

HUC, R. E. (2). *The Chinese Empire; forming a Sequel to 'Recollections of a Journey through Tartary and Thibet'.* 2 vols. Longmans, London, 1855, 1859.

HUDSON, D. S. (1). 'Some Archaic Mining Apparatus.' *MT*, 1947, **35**, 157.

HUDSON, G. F. (1). *Europe and China; A Survey of their Relations from the Earliest Times to 1800.* Arnold, London, 1931 (rev. E. H. Minns, *AQ*, 1933, **7**, 104).

HUGHES, A. J. (1). *History of Air Navigation.* Allen & Unwin, London, 1946.

HUGHES, E. R. (1). *Chinese Philosophy in Classical Times.* Dent, London, 1942. (Everyman Library, no. 973.)

HUGHES, E. R. (7) (tr.). *The Art of Letters, Lu Chi's 'Wên Fu', A.D. 302; a Translation and Comparative Study.* Pantheon, New York, 1951. (Bollingen Series, no. 29.)

HUGHES, E. R. (9). *Two Chinese Poets [Pan Ku and Chang Hêng]; Vignettes of Han Life and Thought.* Princeton Univ. Press, Princeton, N.J., 1960.

HULLS, JONATHAN (1). *New Machine for propelling Ships.* Privately printed, London, 1737.

HULLS, L. G. (1). 'The Possible Influence of Early Eighteenth Century Scientific Literature on Jonathan Hulls, a Pioneer of Steam Navigation.' *BBSHS*, 1951, **1**, 105.

HUMMEL, A. W. (2) (ed.). *Eminent Chinese of the Chhing Period.* 2 vols. Library of Congress, Washington, 1944.

HUMMEL, S. (1). *Tibetisches Kunsthandwerk in Metall.* Harrassowitz, Leipzig, 1954.

HUNTINGFORD, G. W. B. (1). 'Prehistoric Ox-Yoking.' *AQ*, 1934, **8**, 456.

HUNTLEY, F. L. (1). 'Milton, Mendoza and the Chinese Land-Ship.' *MLN*, 1954, **69**, 404.

HUSSITE ENGINEER, the anonymous. See Anon. (14).

HYDE, C. G. & MILLS, F. E. (1). *Gas Calorimetry.* Benn, London, 1932.

IDELER, L. (1). *Über die Zeitrechnung d. Chinesen. SPAW*, 1837, 199. Sep. publ. Berlin, 1839.

JABERG, K. & JUD, J. (1). *Sprach- und Sach-Atlas Italiens und d. Südschweiz.* Bern, 1934.

JACOB, G. (1). *Arabisches Beduinenleben.* Berlin, 1897.

JACOB, G. (2). *Der Einflüss d. Morgenlandes auf das Abendland, vornehmlich während des Mittelalters.* Hanover, 1924.

JACOBEIT, W. (1). 'Zur Rekonstruktion der Anschirrweise am Pforzheimer Joch.' *GERM*, 1952, **30**, 205.

JACOBEIT, W. (2). 'Zur Geschichte der Pferdespannung.' *ZAGA*, 1954, **2**, 17.

JACOBI, H. (1). Art. on milling in *Der obergermanisch-räetische Limes des Römerreiches.* Abt. B, vol. 2 (1), no. 11 (Lieferung 56). Also in *Saalburg Jahrbuch*, vol. 3, pp. 21, 75 ff., 85, 92, figs. 41, 42, 46 & pl. 17.

JACOBI, HERMANN (2). *Ausgewählte Erzählungen in Mahārāshṭrī....* Leipzig, 1886. Eng. tr. J. J. Meyer, *Hindu Tales.* London, 1909.

JÄGER, F. (2). 'Das Buch von den wunderbaren Maschinen [Chhi Chhi Thu Shuo and Chu Chhi Thu Shuo]; ein Kapitel aus der Geschichte der abendländisch-chinesischen Kulturbeziehungen.' *AM* (N.F.), 1944, **1**, 78.

JÄGER, F. (4). 'Der angebliche Steindruck des Kêng-Tschi-T'u [Kêng Chih Thu] von Jahre +1210.' *OAZ*, 1933, **9** (19), 1.

ABD AL-JALIL, J. M. (1). *Brève Histoire de la Littérature Arabe.* Maisonneuve, Paris, 1943; 2nd ed. 1947.

JANSE, O. R. T. (2). 'Tubes et Boutons Cruciformes trouvés en Eurasie.' *BMFEA*, 1932, **4**, 187.

JAUBERT, P. A. (1) (tr.). *Géographie d'Edrisi* (the *Kitāb al-Rujārī* of al-Idrīsī, +1154). Imp. Roy. Paris, 1836. (Recueil de Voyages et de Mémoires publié par la Société de Géographie, no. 5.)

JENKINS, RHYS (1). 'The Oliver [treadle-operated sprung tilthammer for forges]; Iron-Making in the Fourteenth Century.' *TNS*, 1931, **12**, 9.

JENKINS, RHYS (2) (ed.). *R. d'Acres' 'Art of Water-Drawing', published by Henry Brome, at the Gun in Ivie Lane, London, 1659 and 1660.* Newcomen Society (Heffer), Cambridge, 1930. (Newcomen Society Extra Pubs. no. 2.)

JESPERSEN, A. (1). *A Preliminary Analysis of the Development of the Gearing in Watermills.* Privately printed, Virum, Denmark, 1953.

JESPERSEN, A. (2). *The Lady Isabella Waterwheel of the Great Laxey Mining Company, Isle of Man, 1854–1954; a Chapter in the History of Early British Engineering.* Privately printed, Virum, Denmark, 1954.

JOHANNSEN, O. (1). 'Die erste Anwendung der Wasserkraft im Hüttenwesen.' *SE*, 1916, **36**, 1226.

JOHANNSEN, O. (2). *Geschichte des Eisens*, 2nd ed. Verlag Stahleisen MBH, Düsseldorf, 1925. See Johannsen (14).

JOHANNSEN, O. (9). 'Filarete's Angaben über Eisenhütten; ein Beitrag z. Geschichte des Hochöfens und das Eisengusses in 15 Jahrh.' *SE*, 1911, **31**, 1960 & 2027. (On the *Trattato di Architettura* of Antonio Averlino Filarete, *c.* +1462.)

JOHANNSEN, O. (14). *Geschichte des Eisens*. 3rd ed. Verlag Stahleisen MBH, Düsseldorf, 1953. (Johannsen (2) completely rewritten and tripled in size.)

JOMARD, M. (1). Description of shādūf batteries in Egypt. In Bonaparte (1); *Grande Description de l'Égypte* (État Moderne section), vol. 2. Memoirs, pt. 2.

JONES, F. D. (1). *Ingenious Mechanisms.* Industrial Press, New York, 1930. See Jones & Horton (1).

JONES, F. D. & HORTON, H. L. (1). *Ingenious Mechanisms for Designers and Inventors; Mechanisms and mechanical Movements selected from Automatic Machines and various other forms of Mechanical Apparatus as outstanding examples of Ingenious Design, embodying Ideas or Principles applicable in designing Machines or Devices requiring Automatic Features or Mechanical Control.* 3 vols. Industrial Press, New York, 1951–2. Machinery Pub. Co., Brighton, 1951–2.

JOPE, E. M. (1). 'Vehicles and Harness [in the Mediterranean Civilisations and the Middle Ages].' Art. in *A History of Technology*, ed. C. Singer *et al.*, vol. 2, p. 537. Oxford, 1956.

JOPE, E. M. (2). 'Agricultural Implements [in the Mediterranean Civilisations and the Middle Ages].' Art. in *A History of Technology*, ed. C. Singer *et al.*, vol. 2, p. 81. Oxford, 1956.

JOWETT, B. (1) (tr.). *The Dialogues of Plato.* Oxford, 1892.

JULIEN, STANISLAS (4). 'Notes sur l'Emploi Militaire des Cerfs-Volants, et sur les Bateaux et Vaisseaux en Fer et en Cuivre, tirées des Livres Chinois.' *CRAS*, 1847, **24**, 1070.

JULIEN, STANISLAS (7) (tr.). *Histoire et Fabrication de la Porcelaine Chinoise* (partial transl. of the *Ching Tê Chen Thao Lu* with notes and additions by A. Salvétat, Chemist, and an appended memoir on Japanese Porcelain tr. by J. Hoffmann from the Japanese). Mallet-Bachelier, Paris, 1856.

JULIEN, STANISLAS (8). Translations from *TCKM* relative to +13th-century sieges in China (in Reinaud & Favé, 2). *JA*, 1849 (4ᵉ sér.), **14**, 284 ff.

JULIEN, STANISLAS & CHAMPION, P. (1). *Industries Anciennes et Modernes de l'Empire Chinois, d'après des Notices traduites du Chinois....* (Paraphrased précis accounts based largely on *Thien Kung Khai Wu.*) Lacroix, Paris, 1869.

KAEMPFER, ENGELBERT (1). *Amoenitatum Exoticarum Fasciculi V; quibus Continentur Variae Relationes, Observationes et Descriptiones Rerum Persicarum et Ulterioris Asiae, multa attentione, in peregrinationibus per universum orientem, collectae....* Meyer, Lemgoviae, 1712.

KAFTAN, R. (1). *Illustrierter Führer durch das Uhren-Museum der Stadt Wien...Geschichte d. Rädeuhr.* Deutscher Verlag f. Jugend u. Volk, Vienna, 1929.

KALTENMARK, M. (2) (tr.). *Le 'Lie Sien Tchouan' [Lieh Hsien Chuan]; Biographies Légendaires des Immortels Taoistes de l'Antiquité.* Centre d'Études Sinologiques Franco-Chinois (Univ. Paris), Peking, 1953. (Crit. P. Demiéville, *TP*, 1954, **43**, 104.)

KÄMMERER, E. A. (1). *Trades and Crafts of Old Japan.* Tuttle, Tokyo, 1961. Reproduction of an early Tokugawa album of 50 plates.

KAMMERER, O. (1). 'Die Entwicklung der Zahnräder.' *BGTI*, 1912, **4**, 242.

KAPP, ERNST (1). *Grundlinien einer Philosophie d. Technik.* Braunschweig, 1877.

KARLGREN, B. (1). *Grammata Serica; Script and Phonetics in Chinese and Sino-Japanese. BMFEA*, 1940, **12**, 1. (Photographically reproduced as separate volume, Peiping, 1941.) Revised edition, *Grammata Serica Recensa*, Stockholm, 1957.

KARLGREN, B. (12) (tr.). 'The Book of Documents' (*Shu Ching*). *BMFEA*, 1950, **22**, 1.

KARLGREN, B. (13). 'Weapons and Tools of the Yin [Shang] Dynasty.' *BMFEA*, 1945, **17**, 101.

KARLGREN, B. (14) (tr.). *The Book of Odes; Chinese Text, Transcription and Translation*. Museum of Far Eastern Antiquities, Stockholm, 1950. (A reprint of text and translation only from his papers in *BMFEA*, **16** and **17**; the glosses will be found in **14**, **16**, and **18**.)

KATO, SHIGESHI (1). *Studies in Chinese Economic History*. (English summaries of *Shina Keizai-shi Kōshō*. 2 vols. Toyo Bunko Pubs. Ser. A, nos. 33 and 34; in pamphlet form.) Toyo Bunko, Tokyo, 1953.

KENNION, R. L. (1). *By Mountain, Lake and Plain*. London, 1911.

KERLUS, G. (1). 'Curiosités physiologiques; les Sauteurs.' *LN*, 1884, **12** (pt. 2), 213, 282.

KEUTGEN, F. (1). *Urkunden zur städtischen Verfassungsgeschichte*. Berlin, 1901.

KIANG KANG-HU. See Chiang Kang-Hu.

KIERMAN, F. A. (1). *Four Late Warring States Biographies [from the 'Shih Chi']*. Harrassowitz, Wiesbaden, 1962. (Far Eastern & Russian Instit., Univ. of Washington, Seattle, Studies on Asia.)

KING, F. H. (2). *Irrigation in Humid Climates*. U.S. Dept. of Ag. Farmers' Bull. no. 46. Govt. Printg. Off., Washington, 1896.

KING, F. H. (3). *Farmers of Forty Centuries; or, Permanent Agriculture in China, Korea and Japan*. Cape, London, 1927.

KIRCHER, A. (3). *Magnes, sive de Arte Magnetica*. Rome, 1641.

KLAPROTH, J. (1). *Lettre à M. le Baron A. de Humboldt, sur l'Invention de la Boussole*. Dondey-Dupré, Paris, 1834. Germ. tr. A. Wittstein, Leipzig, 1884; résumés, P. de Larenaudière, *BSG*, 1834, Oct.; Anon. *AJ*, 1834 (2nd ser.), **15**, 105.

KLEBS, L. (1). 'Die Reliefs des alten Reiches (2980–2475 v. Chr.); Material zur ägyptischen Kultur-geschichte.' *AHAW/PH*, 1915, no. 3.

KLEBS, L. (2). 'Die Reliefs und Malereien des mittleren Reiches (7–17 Dynastie, c. 2475–1580 v. Chr.); Material zur ägyptischen Kulturgeschichte.' *AHAW/PH*, 1922, no. 6.

KLEBS, L. (3). 'Die Reliefs und Malereien des neuen Reiches (18–20 Dynastie, c. 1580–1100 v. Chr.); Material zur ägyptischen Kulturgeschichte.' Pt. I. 'Szenen aus dem Leben des Volkes.' *AHAW/PH*, 1934, no. 9.

KLEIN, A. W. (1). *Kinematics of Machinery*. McGraw Hill, New York, 1917.

KLEINGUNTHER, A. (1). Πρῶτος Εὑρετής (Greek lists of inventors, and technic deities). *PL*, 1933, Suppl. **26**, 26.

KLEMM, F. (1). *Technik; eine Geschichte ihrer Probleme*. Alber, Freiburg and München, 1954. (Orbis Academicus series, ed. F. Wagner & R. Brodführer.) Engl. tr. by Dorothea W. Singer, *A History of Western Technology*. Allen & Unwin, London, 1959.

KLINDT-JENSEN, O. (1). 'Foreign Influences in Denmark's Early Iron Age.' *ACA*, 1950, **20**, 1–231 (claim for roller-bearings on a –1st-century cart). See correspondence between J. T. Emmerton and C. St C. Davison in *ENG*, 1957, **183**, 292, 326.

KOEHNE, C. (1). 'Die Mühle im Rechte d. Völker.' *BGTI*, 1913, **5**, 27.

KONEN, H. (1). *Physikalischen Plaudereien; Gegenwartsprobleme und ihre technische Bedeutung*. Buch-gemeinde, Bonn, 1941.

KOOP, A. J. (1). *Early Chinese Bronzes*. Benn, London, 1924.

KOSSACK, G. (1). *Zur Hallstattzeit in Bayern*. *BVB*, 1954 (no. 20), 1.

KRACKE, E. A. (1). *Civil Service in Early Sung China* (+960 to +1067), *with particular emphasis on the development of controlled sponsorship to foster administrative responsibility*. Harvard Univ. Press, Cambridge, Mass., 1953 (Harvard-Yenching Institute Monograph Series, no. 13) (revs. L. Petech, *RSO*, 1954, **29**, 278; J. Průsek, *OLZ*, 1955, **50**, 158).

KRACKE, E. A. (2). 'Sung Society; Change within Tradition.' *FEQ*, 1954, **14**, 479.

KRACKE, E. A. (3). 'Family versus Merit in Chinese Civil Service Examinations under the Empire' (analysis of the lists of successful candidates in +1148 and +1256). *HJAS*, 1947, **10**, 103.

KRACKE, E. A. (4). *Translation of Sung Civil Service Titles*. École Prat. des Hautes Études, Paris, 1957. (Matériaux pour le Manuel de l'Histoire des Song; Sung Project, no. 2.)

KRAMRISCH, S. (1). *The Art of India; Traditions of Indian Sculpture, Painting and Architecture*. Phaidon, London, 1955.

KRAUSE, E. (1). 'Die Schraube, eine Eskimo Erfindung?' *GB*, 1901, **79**, 8 & 125.

KROEBER, A. L. (1). *Anthropology*. Harcourt Brace, New York, 1948.

KUBITSCHEK, W. (1). 'Grundriss d. Antiken Zeitrechnung.' Beck, Munich, 1928. (Handbuch d. Altertumswissenchaft, ed. W. Otto, Abt. I, Teil 7.)

KUDRAVTSEV, P. S. (1). *Istoria Physiki* (in Russian). 2 vols. Ministry of Education, Moscow, 1956.

KUDRAVTSEV, P. S. & KONFEDERATOV, I. R. (1). *Istoria Physiki i Tekhniki*. Ministry of Education, Moscow, 1960.

KUNST, J. (5). *Music in Java*. 2 vols. The Hague, 1949.

KUWABARA, JITSUZO (1). 'On Phu Shou-Kêng, a man of the Western Regions, who was the Superintendent of the Trading Ships' Office in Chhüan-Chou towards the end of the Sung Dynasty, together with a general sketch of the Trade of the Arabs in China during the Thang and Sung eras.' *MRDTB*, 1928, **2**, 1; 1935, **7**, 1 (revs. P. Pelliot, *TP*, 1929, **26**, 364; S. E[lisséev], *HJAS*, 1936, **1**, 265). Chinese translation by Chhen Yü-Ching, Chunghua, Peking, 1954.

KUWAKI, AYAO (1). 'The Physical Sciences in Japan, from the time of the first contact with the Occident until the time of the Meiji Restoration.' In *Scientific Japan, Past and Present*. Ed. Shinjo Shinzo, p. 243. IIIrd Pan-Pacific Science Congress, Tokyo, 1926.

KYESER, KONRAD (1). *Bellifortis* (the earliest of the +15th-century illustrated handbooks of military engineering, begun +1396, completed +1410). MS. Göttingen Cod. Phil. 63 and others. See Sarton (1) vol. 3, p. 1550; Berthelot (5), (6).

LABAT, J. B. (1). *Nouveau Voyage aux Isles de l'Amerique*. Paris, 1722.

LAESSØE, J. (1). 'Reflections on Modern and Ancient Oriental Water-works.' *JCUS*, 1953, **7**, 5.

LAMOTTE, E. (1) (tr.). *Mahāprajñāpāramitā Sūtra; Le Traité [Mādhyamika] de la Grande Vertu de Sagesse, de Nāgārjuna*. 3 vols. Louvain, 1944 (rev. P. Demiéville, *JA*, 1950, **238**, 375.)

LANCHESTER, G. (1). *The Yellow Emperor's South-Pointing Chariot* (with a note by A. C. Moule). China Society, London, 1947. Chinese tr. and rev. by Liu Ying, q.v.

LANDT, JØRGEN (1). *A Description of the Feroe Islands*. Tr. from Danish, London, 1810.

LANE, E. W. (1). *An Account of the Manners and Customs of the Modern Egyptians (1833 to 1835)*. Ward Lock, London, 3rd ed. 1842; repr. 1890.

LANE, R. H. (1). 'Waggons and their Ancestors.' *AQ*, 1935, **9**, 140.

LANE-FOX. See Pitt-Rivers.

LANG, O. (1). *Chinese Family and Society*. Yale Univ. Press, New Haven, Conn. 1946.

LAOUST, E. (1). *Mots et Choses Berbères; Notes de Linguistique et d'Ethnographie (Dialectes du Maroc)*. Paris, 1920.

LARDNER, DIONYSIUS (1). *The Steam Engine Explained and Illustrated; with an Account of its Invention and Progressive Improvement, and its Application to Navigation and Railways; including also a Memoir of Watt*. 7th ed. Taylor & Walton, London, 1840.

LASINIO, C. (1). *Pitturi del Campo Santo di Pisa*. Florence, 1813.

LASSUS, J. B. A. & DARCEL, A. (1) (ed.). *The Album of Villard de Honnecourt*. Paris, 1858. Facsimile additions by J. Quicherat, and Eng. tr. by R. Willis. London, 1859.

LATHAM, B. (1). *Timber, its Development and Distribution; an Historical Survey*. Harrap, London, 1957 (rev. R. J. Forbes, *A/AIHS*, 1958, **11**, 97).

LATTIMORE, O. (7). 'The Industrial Impact on China, 1800 to 1950.' Art. in *Proc. 1st Internat. Conference of Economic History, Stockholm, 1960*, p. 103. Mouton, The Hague, 1960.

LAUFER, B. (1). *Sino-Iranica; Chinese Contributions to the History of Civilisation in Ancient Iran*. *FMNHP/AS*, 1919, **15**, no. 3 (Pub. no. 201) (rev. and crit. Chang Hung-Chao, *MGSC*, 1925 (ser. B), no. 5).

LAUFER, B. (3). *Chinese Pottery of the Han Dynasty*. (Pub. of the East Asiatic Cttee. of the Amer. Mus. Nat. Hist.) Brill, Leiden, 1909. (Photolitho re-issue, Tientsin, 1940.)

LAUFER, B. (4). 'The Prehistory of Aviation.' *FMNHP/AS*, 1928, **18**, no. 1 (Pub. no. 253). Cf. 'Mitt. u. d. angeblicher Kenntnis d. Luftschiff-fahrt bei d. alten Chinesen.' *OLL*, 1904, **17**; *OB*, 1904, no. 1489, p. 78; also *OC*, 1931, **45**, 493.

LAUFER, B. (8). *Jade; a Study in Chinese Archaeology and Religion*. *FMNHP/AS*, 1912, **10**, 1–370. Repub. in book form, Perkins, Westwood & Hawley, South Pasadena, 1946 (rev. P. Pelliot, *TP*, 1912, **13**, 434).

LAUFER, B. (10). 'The Beginnings of Porcelain in China.' *FMNHP/AS*, 1917, **15**, no. 2 (Pub. no. 192) (includes description of +2nd-century cast-iron funerary cooking-stove).

LAUFER, B. (19). 'The Noria or Persian Wheel.' Art. in *Oriental Studies in honour of Cursetji Erachji Pavry*, ed. A. V. W. Jackson, p. 238. Oxford, 1933.

LAUFER, B. (20). 'The Eskimo Screw as a Culture-Historical Problem.' *AAN*, 1915, **17**, 396.

LAUFER, B. (21). 'The Recovery of a Lost Book' (the Japanese version of the +1462 edition of the *Kêng Chih Thu*). *TP*, 1912, **13**, 97.

LAUFER, B. (22). 'The Swing in China.' *MSFO*, 1934, **67**, 212.

LAUFER, B. (23). 'Cardan's Suspension in China.' In *Anthropological Essays presented to Wm Henry Holmes in honour of his 70th Birthday*, ed. F. W. Hodge, p. 288. Washington, 1916.

LAUFER, B. (24). 'The Early History of Felt.' *AAN*, 1930, **32**, 1.

LAUFER, B. (25). 'The Bird-Chariot in China and Europe.' In *Anthropological Papers written in honour of Franz Boas and presented to him on the 25th anniversary of his doctorate*, ed. B. Laufer, pp. 410 ff. Stechert, New York, 1926.

LAUFER, B. (26). 'Chinese Pigeon Whistles.' *SAM*, 1908, 394.

LAUNOY, M. & BIENVENU, M. (1). *Instruction sur la nouvelle Machine inventée par Messieurs Launoy et Bienvenu, avec laquelle un Corps monte dans l'Atmosphère et est susceptible d'être dirigé.* Paris, 1784.

LAYARD, A. H. (1). *Discoveries among the Ruins of Nineveh and Babylon.* London, 1845.

LÊ THÁNH-KHÔI (1). *Le Viet-Nam; Histoire et Civilisation.* Éditions de Minuit, Paris, 1955.

LECHLER, G. (1). 'The Origin of the Driving-Belt.' *MA*, 1947, **47**, 52.

LECOMTE, LOUIS (1). *Nouveaux Mémoires sur l'État présent de la Chine.* Anisson, Paris, 1696. (Eng. tr. *Memoirs and Observations Topographical, Physical, Mathematical, Mechanical, Natural, Civil and Ecclesiastical, made in a late journey through the Empire of China, and published in several letters, particularly upon the Chinese Pottery and Varnishing, the Silk and other Manufactures, the Pearl Fishing, the History of Plants and Animals, etc. translated from the Paris edition, etc.* 2nd ed. London, 1698. Germ. tr. Frankfurt, 1699–1700.)

LEDEBUR, A. (2). 'Über den japanischen Eisenhüttenbetrieb.' *SE*, 1901, **21**, 842.

LEDYARD, G. (1). *The Invention of the Korean Alphabet by King Yi Sejong* (r. +1419 to +1450). In preparation.

LEE. See Li.

LEE SANG-BECK. See Yi Sangbaek.

LEEMANS, W. F. (1). 'Some Marginal Remarks on Ancient Technology' (essay-review of Forbes 10, 11, 12, 13, 14, 15). *JESHO*, 1960, **3**, 217.

LEES, B. A. (1). *Records of the Templars.* London, 1935.

LEGER, A. (1). *Les Travaux Publics aux Temps des Romains.* Paris, 1875.

LEGGE, J. See Ride, Lindsay (1).

LEGGE, J. (1) (tr.). *The Texts of Confucianism, translated:* Pt. I. *The 'Shu Ching', the religious portions of the 'Shih Ching', the 'Hsiao Ching'.* Oxford, 1879. (*SBE*, no. 3; reprinted in various eds. Com. Press, Shanghai.) For the full version of the *Shu Ching* see Legge (10).

LEGGE, J. (2) (tr.). *The Chinese Classics, etc.:* Vol. 1. *Confucian Analects, The Great Learning, and the Doctrine of the Mean.* Legge, Hongkong, 1861; Trübner, London, 1861. Photolitho re-issue, Hongkong Univ., Hongkong, 1960.

LEGGE, J. (3) (tr.). *The Chinese Classics, etc.:* Vol. 2. *The Works of Mencius.* Legge, Hongkong, 1861; Trübner, London, 1861. Photolitho re-issue, with Notes by A. Waley in supplementary volume, Hongkong Univ., Hongkong, 1960.

LEGGE, J. (4) (tr.). *A Record of Buddhistic Kingdoms; being an account by the Chinese monk Fa-Hsien of his Travels in India and Ceylon* (+399 to +414) *in search of the Buddhist Books of Discipline.* Oxford, 1886.

LEGGE, J. (5) (tr.). *The Texts of Taoism.* (Contains (a) *Tao Tê Ching*, (b) *Chuang Tzu*, (c) *Thai Shang Kan Ying Phien*, (d) *Chhing Ching Ching*, (e) *Yin Fu Ching*, (f) *Jih Yung Ching.*) 2 vols. Oxford, 1891; photolitho reprint, 1927. (*SBE*, nos. 39 and 40.)

LEGGE, J. (7) (tr.). *The Texts of Confucianism:* Pt. III. *The 'Li Chi'.* 2 vols. Oxford, 1885; reprint, 1926. (*SBE*, nos. 27 and 28.)

LEGGE, J. (8) (tr.). *The Chinese Classics, etc.:* Vol. 4, Pts. 1 and 2. *'Shih Ching'; The Book of Poetry.* 1. The First Part of the *Shih Ching*; or, the Lessons from the States; and the Prolegomena. 2. The Second, Third and Fourth Parts of the *Shih Ching*; or the Minor Odes of the Kingdom, the Greater Odes of the Kingdom; the Sacrificial Odes and Praise-Songs; and the Indexes. Lane Crawford, Hongkong, 1871; Trübner, London, 1871. Repr., without notes, Com. Press, Shanghai, n.d. Photolitho re-issue, Hongkong Univ., Hongkong, 1960.

LEGGE, J. (9) (tr.). *The Texts of Confucianism.* Pt. II. *The 'Yi King'* [*I Ching*]. Oxford, 1882, 1899. (*SBE*, no. 16.)

LEGGE, J. (10) (tr.). *The Chinese Classics, etc.:* Vol. 3, Pts. 1 and 2. *The 'Shoo King'* (*Shu Ching*). Legge, Hongkong, 1865; Trübner, London, 1865. Photolitho re-issue, Hongkong Univ., Hongkong, 1960.

LEGGE, J. (11). *The Chinese Classics, etc.:* Vol. 5, Pts. 1 and 2. *The 'Ch'un Ts'ew' with the 'Tso Chuen'* (*Chhun Chhiu* and *Tso Chuan*). Lane Crawford, Hongkong, 1872; Trübner, London, 1872. Photolitho re-issue, Hongkong Univ., Hongkong, 1960.

LEHMANN, J. (1). *Rudolf Diesel and Burmeister and Wain.* Copenhagen, 1938.

LEHMANN, K. (1). 'The Dome of Heaven.' *AB*, 1945, **27**, 1.

LEIX, A. (1). 'Mediaeval Dye Markets in Europe.' *CIBA/T*, 1938 (no. 10), 324.

LEJARD, A. (1). *Le Cheval dans l'Art.* Gründ, Paris, 1948.

LEONARD, J. N. (1). *Tools of Tomorrow.* New York, 1935.

LEONHARDI, F. G. (1). *Beschreibung zweyer chinesische Maschinen z. Bewässerung ihrer Garten.* Leipzig, 1798.

LEROI-GOURHAN, ANDRÉ (1). *Évolution et Techniques.* Vol. 1. *L'Homme et la Matière*, 1943; Vol. 2. *Milieu et Techniques*, 1945. Albin Michel, Paris.

LEROI-GOURHAN, ANDRÉ (2). 'Les Armes' (classification of shock and projectile weapons). In *Encyclopédie Française*, 1936, vol. 7, pp. 7·12–1 ff.

LESER, P. (1). *Entstehung und Verbreitung des Pfluges*. Aschendorff, Münster, 1931. (Anthropos Bibliothek, no. 3.)

LESER, P. (2). 'Westöstlichen Landwirtschaft; Kulturbeziehungen zwischen Europa, dem vord. Orient u. d. Fern-Osten, aufgezeigt an landwirtschaftlichen Geräten und Arbeitsvorgängen.' In *P. W. Schmidt Festschrift*, ed. W. Koppers, pp. 416 ff. Vienna, 1928.

LETTS, M. (1) (ed. & tr.). *Mandeville's 'Travels'; Texts and Translations*. 2 vols. Hakluyt Society, London 1953. (Pubs. 2nd series, nos. 101, 102.)

LEUPOLD, J. (1). *Theatrum Machinarum Generale*. Leipzig, 1724. *Theatrum Machinarum Molarium*. Deer, Leipzig, 1735.

LÉVI, S. & CHAVANNES, E. (2). 'Quelques Titres Énigmatiques dans la Hiérarchie Ecclésiastique du Bouddhisme Indien.' *JA*, 1915 (11ᵉ sér.), **6**, 307.

LÉVI-PROVENÇAL, E. (1) (tr.). *La Péninsule Ibérique au Moyen-Âge* (translation of the *Kitāb al-Rawḍ* of Ibn 'Abd al-Mun'im al-Ḥimyarī, +1262). Brill, Leiden, 1938.

LI CHUN (1). 'A Pair of Skinny Horses.' *CREC*, 1960, **9** (no. 1), 36.

LI JEN-I (LI JEN-YI) (1). 'The Chinese Clock-Making Industry.' *HORJ*, 1963, **105**, 329; followed by letter from J. H. Combridge, p. 347.

LI SHU-HUA (1). 'Origine de la Boussole, I. Le Char Montre-Sud.' *ISIS*, 1954, **45**, 78. Engl. tr., slightly enlarged, and with Chinese characters added. *CHJ/T*, 1956, **1** (no. 1), 63. Sep. pub. in Chinese and English, I-Wên Pub. Co., Thaipei, 1959.

LI SHUN-CHHING (Lee Shun-Ching) (1). *Forest Botany of China*. Com. Press, Shanghai, 1935.

LI YUNG-HSI (1) (tr.). *'A Record of the Buddhist Countries,' by Fa-Hsien*. Chinese Buddhist Association, Peking, 1957.

LIAO WÊN-KUEI (1) (tr.). *The Complete Works of Han Fei Tzu, a Classic of Chinese Legalism*. 2 vols. Probsthain, London, 1939, 1959.

LILLEY, S. (3). *Men, Machines and History*. Cobbett, London, 1948.

LIM BOON-KÊNG. See Lin Wên-Chhing.

LIN WÊN-CHHING (1) (tr.). *The 'Li Sao'; an Elegy on Encountering Sorrows, by Chhü Yüan of the State of Chhu* (ca. *338 to 288 B.C.*).... Com. Press, Shanghai, 1935.

LIN YÜ-THANG (1) (tr.). *The Wisdom of Lao Tzu [and Chuang Tzu], translated, edited, and with an introduction and notes*. Random House, New York, 1948.

LIN YÜ-THANG (5). *The Gay Genius; Life and Times of Su Tung-Pho*. Heinemann, London, 1948.

LINDET, L. (1). 'Les Origines du Moulin à Grains.' *RA*, 1899 (3ᵉ sér.), **35**, 413; 1900, **36**, 17

LINDROTH, S. (1). *Gruvbrytning och Kopparhantering vid Stora Kopparberget intill 1800-talets Början* (in Swedish). 2 vols. Almqvist & Wiksell, Upsala, 1955. (Skrifter utgivn av Storakopparbergs Bergslags Aktiebolag—a corporation founded in +1288.) A résumé of this work, with map, appeared in *Sweden Illustrated*, 1954.

VAN LINSCHOTEN, JAN HUYGHEN (1). *Itinerario, Voyage ofte Schipvaert van J. H. van L. naer oost ofte Portugaels Indien* (+1579 to +1592). Amsterdam, 1596, 1598; ed. Kern, The Hague, 1910. Repr. Warnsinck-Delprat, 5 vols., The Hague, 1955. Eng. tr. by W. Phillip, *John Huighen Van Linschoten his discours of Voyages into ye Easte and West Indies*. Wolfe, London, 1598. Ed. A. C. Burnell & P. A. Tiele, London, 1885. (Hakluyt Soc. Pub., 1st ser., nos. 70, 71.) (Information about the Chinese coast dating from *c*. +1550 to +1588 collected at Goa *c*. +1583 to +1589.)

LIPSCHUTZ, A. (1). *'Ayil y Ayllu.'* Art. in *Miscellanea Paul Rivet*, p. 339. Mexico City, 1958.

VON LIPPMANN, E. O. (4). *Geschichte des Zuckers, seiner Darstellung und Verwendung, seit den ältesten Zeiten bis zum Beginne der Rübenzuckerfabrikation; ein Beitrag zur Kulturgeschichte*. Hesse, Leipzig, 1890.

LITTMANN, E. (1). 'Die Sāqiya.' *ZAES*, 1940, **76**, 45.

LIU HO & ROUX, C. (1). *Aperçu Bibliographique sur les anciens Traités Chinois de Botanique, d'Agriculture, de Sériculture et de Fungiculture*. Bosc & Riou, Lyons, 1927.

LIU HSIEN-CHOU (1). *On Chinese Inventions of Time-Keeping Apparatus*. Actes du VIIIᵉ Congrès International d'Histoire des Sciences, Florence, 1956, p. 329. (Eng. tr. of Liu Hsien-Chou, 6.)

LLOYD, H. ALAN (1). *Giovanni de Dondi's Horological Masterpiece of +1364*. Privately printed, no place, no date (London, 1954).

LLOYD, H. ALAN (2). 'The Anchor Escapement.' *HORJ*, 1952 (Suppl.), April.

LLOYD, H. ALAN (3). 'John Harrison, 1693–1776.' *LSH*, 1953, Oct.

LLOYD, H. ALAN (4). 'George Graham, Horologist and Astronomer.' *HORJ*, 1951, **93**, 708.

LLOYD, H. ALAN (5). 'Mechanical Time-Keepers [+1500 to +1750].' Art. in *A History of Technology*, ed. C. Singer, E. J. Holmyard & A. R. Hall. Vol. 3, p. 648. Oxford, 1957.

LLOYD, H. ALAN (6) (pref. & ed.). *Catalogue of the Tercentenary Exhibition of the Pendulum Clock of Christiaan Huygens* (at the Science Museum, South Kensington, London, 1956, 1957). (Pubs. Antiq. Horol. Soc. no. 2.) London, 1956.

LLOYD, H. ALAN (7). 'Notes on very early English Equation Clocks.' *HORJ*, 1943, **85**, 314.

LLOYD, H. ALAN (8). 'The Origin of the Fusee.' *ANTJ*, 1951, **31**, 188.

LLOYD, H. ALAN (9). 'English Clocks for the Chinese Market.' *AQC*, 1951, **22** (no. 1), 25.

LLOYD, H. ALAN (10). *Some Outstanding Clocks over Seven Hundred Years*. Leonard Hill, London, 1960.

LO JUNG-PANG (3). 'China's Paddle-Wheel Boats; the Mechanised Craft used in the Opium War and their Historical Background.' *CHJ/T*, 1960 (N.S.), **2** (no. 1), 189. Abridged Chinese tr. Lo Jung-Pang (*1*).

LOCKHART, J. H. S. (1). '*Han Wên Tshui Chen*'; *the Texts of the Translations made by H. A. Giles in 'Gems of Chinese Literature'*. Shanghai, 1927.

LOCKHART, W. (1). 'The Medical Missionary in China; a Narrative of Twenty Years' Experience.' Hurst & Blackett, London, 1861.

LOEWE, M. (3). 'The Han Documents from Chü-yen.' Inaug. Diss. London, 1963.

LOPEZ, R. S. (1). 'Les Influences Orientales et l'Éveil Économique de l'Occident.' *JWH*, 1954, **1**, 594.

LORIMER, H. L. (1). 'The Country Cart of Ancient Greece.' *JHS*, 1903, **23**, 132.

LORIMER, E. O. (1). 'Primitive Wheelbarrows.' *AQ*, 1936, **10**, 463. Cf. 'The Burusho of Hunza.' *AQ*, 1938, **12**, 5.

LORINI, BUONAIUTO (1). *Delle Fortificationi*. Venice, 1597. Account in Beck (1), ch. 12.

DE LOTURE, R. & HAFFNER, L. (1). *La Navigation à travers les Âges; Évolution de la Technique Nautique et de ses Applications*. Payot, Paris, 1952.

DE LA LOUBÈRE, S. (1). *A New Historical Relation of the Kingdom of Siam by Monsieur de la Loubère, Envoy-Extraordinary from the French King to the King of Siam, in the years 1687 and 1688, wherein a full and curious Account is given of the Chinese Way of Arithmetick and Mathematick Learning.* Tr. A. P., Gen[t?] R.S.S. [i.e. F.R.S.]. Horne Saunders & Bennet, London, 1693 (from the Fr. ed., Paris, 1691).

LOUIS, H. (1). 'A Chinese System of Gold Milling.' *EMJ*, 1891, 640.

LOUIS, H. (2). 'A Chinese System of Gold Mining.' *EMJ*, 1892, 629.

LOVEGROVE, H. (1). 'Junks of the Canton River and the West River System.' *MMI*, 1932, **18**, 241.

LU GWEI-DJEN, SALAMAN, R. A. & NEEDHAM, JOSEPH (1). 'The Wheelwright's Art in Ancient China; I, The Invention of "Dishing".' *PHY*, 1959, **1**, 103.

LU GWEI-DJEN, SALAMAN, R. A. & NEEDHAM, JOSEPH (2). 'The Wheelwright's Art in Ancient China; II, Scenes in the Workshop.' *PHY*, 1959, **1**, 196.

LU HSÜN (1). *A Brief History of Chinese Fiction*, tr. Yang Hsien-Yi & Gladys Yang. Foreign Languages Press, Peking, 1959.

LÜBKE, A. (1). 'Altchinesische Uhren.' *DUZ*, 1931, **55**, 197. 'Chinesische Zeitmesskunde.' *NUK*, 1931, **28**, 45; *UMK*, 1936, **61**, 324; *OAR*, 1930, **11**, 586.

LÜBKE, A. (2). *Der Himmel der Chinesen*. Voigtländer, Leipzig, c. 1935.

LUCAS, A. (1). *Ancient Egyptian Materials and Industries*. Arnold, London (3rd ed.), 1948.

LUCAS, A. T. (1). 'The Horizontal Water-Mill in Ireland.' *JRSAI*, 1953, **83**, 1.

McADAM, R. (1). 'A Mill-Wheel found in the Bog of Moycraig, near Ballymoney, Co. Antrim.' *UJA*, 1856, **4**, 6.

McCURDY, E. (1). *The Notebooks of Leonardo da Vinci, Arranged, Rendered into English, and Introduced by....* 2 vols. Cape, London, 1938.

McCURDY, G. G. (1). *Human Origins*. 2 vols. New York, 1924.

McDONALD, MALCOLM & LOKE WAN-THO (1). *Angkor*. Cape, London, 1958.

McGOWAN, D. J. See Wang Chi-Min (2), biography no. 5.

McGOWAN, D. J. (1). 'Methods of Keeping Time known among the Chinese.' *CRRR*, 1891, **20**, 426. (Reprinted *ARSI*, 1891 (1893), 607.)

McGOWAN, D. J. (4). 'On Chinese Horology, with Suggestions on the Form of Clocks adapted for the Chinese Market.' *RUSCP*, 1851 (1852), **32**, 335. (Reprinted *AJSC*, 1852 (2nd ser.), **13**, 241; *EDJ*, 1853.)

McGREGOR, J. (1). 'On the Paddle-Wheel and Screw Propeller, from the Earliest Times.' *JRSA*, 1858, **6**, 335.

McGUIRE, J. D. (1). 'A Study of the Primitive Methods of Drilling.' *ARUSNM*, 1894, 625.

McMILLAN, R. H. (1). *An Introduction to the Theory of Control in Mechanical Engineering*. Cambridge, 1951 (rev. J. Greig, *N*, 1953, **172**, 91).

McNISH, A. G. & TUCKERMAN, B. (1). 'The Vehicular Odograph.' *TM*, 1947, **52**, 39.

MA CHI (1). 'Making new Farm Implements.' *CREC*, 1954, **3** (no. 1), 30.

MACARTNEY, GEORGE (Lord Macartney) (1). *Journal kept during his Embassy to the Chhien-Lung Emperor (+1793 and +1794)*, ed. J. L. Cranmer-Byng (2). Longmans, London, 1962.

MACARTNEY, GEORGE (Lord Macartney) (2). *Observations on China*, ed. J. L. Cranmer-Byng (2). Longmans, London, 1962.

Macek, J. (1). *The Hussite Movement in Bohemia*. Orbis, Prague, 1958.

Maddison, F. (1). 'A +15th-Century Spherical Astrolabe.' *PHY*, 1962, **4**, 101.

Maddison, F., Scott, B. & Kent, A. (1). 'An Early Medieval Water-Clock' (a Catalan monastic anaphoric horologe with weight-driven striking mechanism, *ca.* +1000). *AHOR*, 1962, **3** (no. 12).

Magaillans. See Magalhaens.

de Magalhaens, Gabriel (1). *A New History of China, containing a Description of the Most Considerable Particulars of that Vast Empire.* Newborough, London, 1688. Tr. from *Nouvelle Relation de la Chine.* Barbin, Paris, 1688. The work was written in 1668.

Mahr, O. (1). 'Zur Geschichte des Wagenrades.' *BGTI/TG*, 1934, **23**, 51.

Maillart, Ella (1). *The Cruel Way.* London, 1947.

[Maillot, M.] (1). 'Le Cerf-Volant; Théorie du Cerf-Volant—un Cerf-Volant gigantesque.' *LN*, 1886, **14** (pt. 2), 269.

Mandeville, Sir John (+1362). See Letts, M. (1).

Manker, E. (1). *De Svenska Fjällapparna.* STF, Stockholm, 1947. (Handböcker om det Svenska Fjället, no. 4.)

de Manoury d'Ectot, Marquis (1). 'Rapport sur une Nouvelle Machine à Feu presentée à l'Académie et executée aux Abattoirs de Grenelle.' *ACP*, 1821, **18**, 133.

Maréchal, J. (1). *Histoire de la Métallurgie du Fer dans la Vallée de la Vesdre.* Éditions Wallonie, 1942.

Maréchal, J. (2). 'Évolution de la Fabrication de la Fonte en Europe et ses Relations avec la Méthode Wallonne d'Affinage.' *MC/TC*, 1955, **4**, 129. Abridged in *Actes du Colloque International 'Le Fer à travers les Âges'*, Nancy, Oct. 1955, p. 517. (*AEST*, 1956, Mémoire no. 16.)

Marestier, J. B. (1). *Mémoire sur les Bateaux à Vapeur des États-Unis.* Paris, 1824.

Margouliés, G. (3). *Anthologie Raisonnée de la Littérature Chinoise.* Payot, Paris, 1948.

Mariano, Jacopo. See Taccola.

de Maricourt, Pierre. See Peregrinus, Petrus.

Marin, G. (1). 'Horizontala Ventmuelilo (Ligurio).' *GRO*, 1963, **5** (no. 5), 14.

Markham, S. D. (1). *The Horse in Greek Art.* Johns Hopkins Univ. Press, Baltimore, 1943. (Johns Hopkins Univ. Studies in Archaeol. no. 35.)

Marquart, J. (1). *Osteuropäische und Ostasiatische Streifzüge.* Leipzig, 1903. (Das Itinerar des Mis'ar ben al-Muhalhil nach der chinesischen Hauptstadt, pp. 74 ff.)

Marshall, Sir John (2). Excavations at Mohenjo-Daro. *ARASI*, 1926–7.

Marshall, Sir John (3). 'Monuments of Ancient India.' *CHI*, vol. 1, ch. 26.

Martin, F. (1). *Sibirische Sammlung.* Stockholm, 1897.

de la Martinière, Breton (1). *China, its Customs, Arts, Manufactures, etc. edited principally from the Originals in the Cabinet of the late Mons. Bertin* [+1719 to +1792], *with Observations Explanatory, Historical and Literary...* (tr. from the French). 3rd ed., Stockdale, London, 1812, 1813; repr. 1824.

Marx, E. (1). 'Bericht u. ein Dokument mittelälterliche Technik' (Scanderbeg's *Ingenieurkunst u. Wunderbuch*, +16th). *BGTI*, 1926, **16**, 317.

Maryon, H. & Plenderleith, H. J. (1). 'Fine Metal-Work [in Early Times before the Fall of the Ancient Empires].' In *A History of Technology*, ed. C. Singer *et al.*, vol. 1, p. 623. Oxford, 1954.

Mason, O. T. (1). 'Primitive Travel and Transportation.' *ARUSNM*, 1894, 237.

Mason, O. T. (2). *The Origins of Invention; a study of Industry among Primitive Peoples.* Scott, London, 1895.

Maspero, H. See Merlin, A. (1).

Maspero, H. (4). 'Les Instruments Astronomiques des Chinois au temps des Han.' *MCB*, 1939, **6**, 183.

Massa, J. M. (1). 'La Brouette.' *MC/TC*, 1952, **2**, 93.

Masson-Oursel, P. (2). 'La Noria, prototype du Saṃsāra.' In *Mélanges R. Linossier*, vol. 2, p. 419. 1932.

Matschoss, C. (1). 'Die Maschinen d. deutsch. Bergs- und Hüttenwesens vor 100 Jahren.' *BGTI*, 1909, **1**, 1.

Matschoss, C. & Kutzbach, K. (1). *Geschichte des Zahnrades.* VDI-Verlag, Berlin, 1940 (*nebst Bemerkungen zur Entwicklung der Verzahnung*, by K. Kutzbach). (Published to celebrate the 25th anniversary of the Zahnradfabrik Friedrichshafen Aktiengesellschaft, and distributed by that firm.)

Maxe-Werly, L. (1). 'Notes sur des Objets antiques découverts à Gondrecourt (Meuse) et à Grant (Vosges).' *MSAF*, 1887, **48**, 170.

Mayence, F. (1). 'La Troisième Campagne de Fouilles à Apamée' (the +2nd century noria in mosaic). *BMRAH*, 1933 (3ᵉ sér.), **5**, 1.

Mayers, W. F. (1). *Chinese Reader's Manual.* Presbyterian Press, Shanghai, 1874; reprinted, 1924.

Mayers, W. F. (2). 'Bibliography of the Chinese Imperial Collections of Literature' (i.e. *Yung-Lo Ta Tien; Thu Shu Chi Chhêng; Yuan Chien Lei Han; Phei Wên Yuan Fu; Phien Tzu Lei Pien; Ssu Khu Chhüan Shu*). *CR*, 1878, **6**, 213, 285.

MAYOR, R. J. G. (2). 'Slaves and Slavery [in Ancient Greece].' In *A Companion to Greek Studies*, ed. L. Whibley, pp. 416, 420. Cambridge, 1905.

MEDHURST, W. H. (1) (tr.). *The 'Shoo King' [Shu Ching], or Historical Classic* (Ch. and Eng.). Mission Press, Shanghai, 1846.

MEDHURST, W. H. (2). 'The Fire-Piston in Yunnan.' *CR*, 1876, **5**, 202.

MEHREN, A. F. (1). *Manuel de la Cosmographie du Moyen Âge, trad. de l'Arabe 'Nokhbet ed-Dahr fi Adjaib-il-birr wal-Bah'r' de Shems ed-Din Abou Abdallah Mohammed de Damas....* Copenhagen, 1847.

MEI YI-PAO (1) (tr.). *The Ethical and Political Works of Mo Tzu*. Probsthain, London, 1929.

VON MEMMINGEN, ABRAHAM (attrib.) (1). *Feuerwerksbuch*. Written *c.* 1422, pr. Augsburg, 1529. See Sarton (1), vol. 3, p. 1551.

DE MENDOZA, JUAN GONZALES (1). *Historia de las Cosas mas notables, Ritos y Costumbres del Gran Reyno de la China, sabidas assi por los libros de los mesmos Chinas, como por relacion de religiosos y oltras personas que an estado en el dicho Reyno*. Rome, 1585 (in Spanish). Eng. tr. Robert Parke, *The Historie of the Great & Mightie Kingdome of China and the Situation thereof; Togither with the Great Riches, Huge Citties, Politike Gouvernement and Rare Inventions in the same* [undertaken 'at the earnest request and encouragement of my worshipfull friend Master Richard Hakluyt, late of Oxforde']. London, 1588 (1589). Reprinted in Spanish, Medina del Campo, 1595, Madrid, 1944; Antwerp, 1596 and 1655; Ital. tr. Venice (3 editions), 1586; Fr. tr. Paris, 1588, 1589 and 1600; Germ. and Latin tr. Frankfurt, 1589. New ed. G. T. Staunton, London 1853. (Hakluyt Soc. Pubs. 1st ser., nos. 14, 15.)

MENGERINGHAUSEN, J. & MENGERINGHAUSEN, M. (1). 'Technische Kulturdenkmale in Ausland.' *BGTI/TG*, 1933, **22**, 137.

MERCER, H. C. See DUNBAR, G. S. (1).

MERCER, H. C. (1). *Ancient Carpenter's Tools illustrated and explained, together with the Implements of the Lumberman, Joiner, and Cabinet-Maker, in use in the Eighteenth Century*. Bucks County Historical Society, Doylestown, Pennsylvania, 1929.

MERCKEL, C. (1). *Die Ingenieur-Technik im Altertum*. Springer, Berlin, 1899.

MERCZ, MARTIN (1). 'Kunst aus Büchsen zu schiessen' (gunnery). 1471. MS. Liechtenstein. See Sarton (1), vol. 3, p. 1553.

MERLIN, A. (1). *Notice sur la Vie et les Travaux de Monsieur Henri Maspero [1883 à 1945], Membre de l'Académie*. Institut de France, Académie des Inscriptions et Belles-Lettres, Séance Publique Annuelle, 23 Nov. 1951.

[MERTENS, J. C.] (Brother Arnold) (1) (tr.). 'The Letter of Petrus Peregrinus on the Magnet.' *EWD*, 1904, **43** (no. 13), 598; and sep. with an introduction by M. F. O'Reilly (Brother Potamian), New York, 1904.

MÉTRAUX, A. (1). 'Le Chamanisme chez les Indiens du Gran Chaco,' *SO*, 1944, **7**, 3: Le Chamanisme chez les Indiens de l'Amérique du Sud Tropicale,' 208; 'Le Chamanisme Araucan,' *IAUT*, 1942, **2**, 309.

MEYER, E. (1). 'Zur Geschichte d. Anwendungen der Festigkeitslehre im Maschinenbau....' *BGTI*, 1909, **1**, 108.

MEYERHOF, M. (1). 'Ibn al-Nafīs und seine Theorie d. Lungenkreislaufs.' *QSGNM*, 1935, **4**, 37.

MEYERHOF, M. (2). 'Ibn al-Nafīs (+13th century) and his Theory of the Lesser Circulation.' *ISIS*, 1935, **23**, 100.

MEYGRET, AMADEUS (1). *'Questiones...in Libros de Coelo et Mundo Aristotelis*. Paris, 1514.

DE MEYNARD, C. BARBIER (1). *Dictionnaire Géographique, Historique et Littéraire de la Perse et des Contrées Adjacentes....* Imp. Imp. Paris, 1861.

DE MEYNARD, C. BARBIER & DE COURTEILLE, P. (1) (tr.). *Les Prairies d'Or* (the *Murūj al-Dhabab* of al-Mas'ūdī, +947). 9 vols. Paris, 1861-77.

MICHEL, H. (9). 'Un Service de l'Heure Millénaire.' *CET*, 1952, **68**, 1.

MICHEL, H. (15). 'A propos des premières Montres; Over de eerste Horloges.' *T*, 1956, March, 129.

MICHEL, H. (16). 'Les Ancêtres du Planetarium.' *CET*, 1955, **71** (no. 3-4), 1.

MICHELL, T. (1). *Russian Pictures drawn with Pen and Pencil*. Murray, London, 1889.

MIELI, ALDO. See Boni, B. (2).

MIELI, ALDO (1). *La Science Arabe, et son Rôle dans l'Évolution Scientifique Mondiale*. Brill, Leiden, 1938.

MIGEON, G. (1). *Manuel d'Art Mussulman; Arts plastiques et industriels*. 2 vols. Paris, 1927.

MIKAMI, Y. (1). *The Development of Mathematics in China and Japan*. Teubner, Leipzig, 1913 (Abhdl. z. Gesch. d. math. Wissenschaften mit Einschluss ihrer Anwendungen, no. 30.) Photolitho re-issue, Chelsea, New York, n.d. (1961). Revs. H. Bosmans, *RQS*, 1913, **74**, 64; J. Needham, *OLZ*, 1962.

MIKAMI, Y. (12). 'A Japanese Buddhist View of European Astronomy' (the *Bukkoku Rekishō-hen* of Entsū, +1810). *NAW*, 1912, **10**, 233.

MILESCU, NICOLAIE (SPǍTARUL) (1). *Descrierea Chinei* (in Rumanian, originally written in Russian, *c.* 1676), with preface by C. Bărbulescu. Ed. Stat pentru Lit. şi Artă, Bucarest, 1958. This work in 58 chs. is, with the exception of chs. 3, 4, 5, 10 and 20, essentially a Russian translation and adaptation of Martin Martini's text accompanying the maps in his *Atlas Sinensis* (Amsterdam, 1655). Milescu prepared it in the course of his diplomatic mission (1675 to 1677) as the Ambassador of the Tsar of Russia to the Emperor of China. See Baddeley (2).

MILESCU, NICOLAIE (SPǍTARUL) (2). *Jurnal de Cǎlǎtorie în China* (in Rumanian, originally written in Russian, 1677, as the report to the Tsar from his Ambassador to the Chinese Emperor), with preface by C. Bărbulescu. Ed. Stat pentru Lit şi Artă, Bucarest, 1956; repr. with a new preface by C. Bărbulescu, Ed. pentru Lit, Bucarest, 1962. Eng. tr. Baddeley (2), vol. 2, pp. 242 ff.

MILHAM, W. I. (1). *Time and Timekeepers; History, Construction, Care, and Accuracy, of Clocks and Watches*. Macmillan, New York, 1923.

MILLAR, E. G. (1). *The Luttrell Psalter* [*East Anglian, c.* +*1340*]. British Museum, London, 1932.

MILLET, G. (1). 'L'Ascension d'Alexandre.' *SYR*, 1923, **4**, 85.

MILNE, J. S. (1). *Surgical Instruments in Greek and Roman Times*. Oxford, 1907.

MINAKATA, K. (3). 'Flying Machines in the Far East.' *NQ*, 1909 (10th ser.), **11**, 425.

MINORSKY, V. F. (4) (ed. & tr.). *Sharaf al-Zamān Ṭāhir al-Marwazī on China, the Turks and India* (*c.* +1120). Royal Asiatic Soc. London, 1942 (Forlong Fund series, no. 22).

MINORSKY, V. F. (5). *Abū Dulaf Misʿar ibn Muhalhil's Travels in Iran*. Univ. Press, Cairo, 1955.

MITTELÄLTERLICHES HAUSBUCH. See Anon. (15).

MODY, N. H. N. (1). *A Collection of Japanese Clocks*. (Lim. ed.) Kegan Paul, London & Kobe, 1933; Tokyo, 1932.

MOINET, L. (1). *Nouveau Traité générale astronomique et civil d'Horlogerie théorique et pratique....* Paris, 1848 and 1875.

MOLENAAR, A. (1). *Water Lifting Devices for Irrigation*. FAO, Rome, 1956. (Agricultural Development Paper, no. 60.)

MOLES, ANTOINE (1). *Histoire des Charpentiers*. Gründ, Paris, 1949.

MØLLER-CHRISTENSEN, V. (1). *The History of the Forceps; an Investigation on the Occurrence, Evolution and Use of the Forceps from Prehistoric Times to the Present Day*. Levin & Munksgaard, Copenhagen, 1938; Oxford University Press, London, 1938.

MÖNCH, P. (1). 'Buch der Stryt und Buchssen' (military engineering). 1496. MS. Cod. Palat. Germ. 126, Univ. Heidelberg. See Sarton (1), vol. 3, p. 1554.

MONTANDON, G. (1). *L'Ologénèse Culturelle; Traité d'Ethnologie Cyclo-Culturelle et d'Ergologie Systématique*. Payot, Paris, 1934.

MONTANUS, A. (1). *Atlas Chinensis; being a Second Part of a Relation of Remarkable Passages in two Embassies from the East India Company of the United Provinces to the Viceroy Singlamong and General Taising Lipovi, and to Konchi* [*Khang-Hsi*], *Emperor of China and East Tartary...*, tr. J. Ogilby. Johnson, London, 1671.

DAL MONTE, GUIDOBALDO. See Guidobaldo (e Marchionibus Montis).

MONTELL, G. (3). 'The *Kêng Chih Thu* (Illustrations of Tilling and Weaving).' *ETH*, 1940 (nos. 3–4), **5**, 165.

MORALES, A. (1). *Las Antigüedades de las Ciudades de España*. Alcalá, 1575.

MORDEN, W. J. (1). 'By Coolie and Caravan across Central Asia.' *NGM*, 1927, **52**, 369.

MORETTI, G. (1). *Il Museo delle Nave Romane di Nemi*. Libreria dello Stato, Rome, 1940.

MORGAN, E. (1) (tr.). *Tao the Great Luminant; Essays from 'Huai Nan Tzu'*, with introductory articles, notes and analyses. Kelly & Walsh, Shanghai, n.d. (1933?).

MORGAN, M. H. (1). *Vitruvius; the Ten Books on Architecture*. Harvard Univ. Press, Cambridge, Mass., 1914.

MORGENSTIERNE, G. (1). 'Iranian Notes.' *AO*, 1922, **1**, 245.

MORITZ, L. A. (1). *Grain-Mills and Flour in Classical Antiquity*. Oxford, 1958.

MORITZ, L. A. (2). ''Αλφιτα, a Note.' *CQ*, 1949, **43**, 113.

MORITZ, L. A. (3). 'Husked and "Naked" Grain.' *CQ*, 1955, **49** (N.S. **5**), 129.

MORITZ, L. A. (4). 'Corn.' *CQ*, 1955, **49** (N.S. **5**), 135.

MORLEY, HENRY (1) (ed.). *Ideal Commonwealths*. London, 1885 (includes Thomasso Campanella's 'City of the Sun').

MORRELL, R. (1). 'Talks on [Wooden] Clocks.' *WCC*, 1951, **4**, 437.

MOSELEY, H. (1). *Mechanical Principles of Engineering and Architecture*. Brown, Green & Longmans, London, 1855 (2nd ed.).

MOULE, A. C. (3). 'The Bore on the Ch'ien-T'ang River in China.' *TP*, 1923, **22**, 135 (includes much material on tides and tidal theory).

MOULE, A. C. (4). 'An Introduction to the *I Yü Thu Chih*.' *TP*, 1930, **27**, 179.

MOULE, A. C. (5). 'The Wonder of the Capital' (the Sung books *Tu Chhêng Chi Shêng* and *Mêng Liang Lu* about Hangchow). *NCR*, 1921, **3**, 12, 356.

MOULE, A. C. (7). 'The Chinese South-Pointing Carriage.' *TP*, 1924, **23**, 83. Chinese tr. by Chang Yin-Lin (5).

MOULE, A. C. (8). 'Carriages in Marco Polo's Quinsay.' *TP*, 1925, **24**, 66.

MOULE, A. C. (10). 'A List of the Musical and other Sound-producing Instruments of the Chinese.' *JRAS/NCB*, 1908, **39**, 1–162.

MOULE, A. C. (15). *Quinsai, with other Notes on Marco Polo*. Cambridge, 1957.

MOULE, A. C. & PELLIOT, P. (1) (tr. and annot.). *Marco Polo (+1254 to 1325); The Description of the World*. 2 vols. Routledge, London, 1938. Further notes by P. Pelliot (posthumously pub.). 2 vols. Impr. Nat. Paris, 1960.

MOULE, A. C. & YETTS, W. P. (1). *The Rulers of China, −221 to +1949; Chronological Tables compiled by A. C. Moule, with an Introductory Section on the Earlier Rulers, ca. −2100 to −249 by W. P. Yetts*. Routledge & Kegan Paul, London, 1957.

MUIR, Sir WILLIAM (1). *The Caliphate; its Rise, Decline and Fall*. Grant, Edinburgh, 1915. Revised ed. T. H. Weir, 1924.

MUKHOPADHYAYA, G. (1). *The Surgical Instruments of the Hindus, with a Comparative Study of the Surgical Instruments of the Greek, Roman, Arab and Modern European Surgeons*. 2 vols. Calcutta Univ., Calcutta, 1913.

MULTHAUF, R. P. (4). 'Mine Pumping in Agricola's Time and Later.' *BUSNM*, 1959 (no. 218), 113. (Contributions from the Museum of History & Technol, no. 7.)

MUMFORD, LEWIS (1). *Technics and Civilisation*. Routledge, London, 1934.

MUMFORD, LEWIS (2). *The Culture of Cities*. Secker & Warburg, London, 1938; often reprinted.

MUMFORD, LEWIS (3). *The Condition of Man*. Secker & Warburg, London, 1944.

MUMFORD, LEWIS (4). 'An Appraisal of Lewis Mumford's *Technics and Civilisation* (1934).' *DAE*, 1959, **88**, 527.

MUNRO, R. (1). *Lake-Dwellings of Europe*. Cassell, London, 1890.

MUNSHI, K. M. (1). *The Saga of Indian Sculpture*. Bharatiya Vidya Bhavan, Bombay, 1957.

MUROGA, NOBUO & UNNO, TAZUTAKA (1). 'The Buddhist World-Map in Japan and its Contact with European Maps.' *IM*, 1962, **16**, 49.

MUS, P. (1). 'La Notion de Temps Réversible dans la Mythologie Bouddhique.' AEPHE/SSR, 1939, 1.

NAGASAWA, K. (1). *Geschichte der chinesischen Literatur, und ihrer gedanklichen Grundlage*. Transl. from the Japanese by E. Feifel. Fu-jen Univ. Press, Peiping, 1945.

NAGEL, E., BROWN, G. S., RIDENOUR, L. N. *et al.* (1). 'Automatic Control, Control Systems, Automatic Chemical Plant, Role of the Computer, Information Theory, etc.' *SAM*, 1952, **187** (no. 3), 44 ff.

NAGLER, J. (1). 'Die Erste "Curieuse Feuer-Maschine" in Österreich; eine Grossleistung von Joseph Emanuel Fischer von Erlach.' *AMK*, 1957, **2** (no. 7–8), 26.

NANJIO, B. (1). *A Catalogue of the Chinese Translations of the Buddhist Tripiṭaka*. Oxford, 1883. (See Ross, E. D.)

VAN NATRUS, L., POLLY, J., VAN VUUREN, C. & LINPERCH, P. *Groot Volkomen Mollenbock*. Amsterdam, 1736.

DE NAVARRETE, DOMINGO. See Cummins (1).

NEDULOHA, A. (1). 'Kulturgeschichte des technischen Zeichens.' *BTG*, 1957, **19**; 1958, **20**; 1959, **21**. Sep. pub. Springer, Vienna, 1960. Crit. L. R. Shelby, *TCULT*, 1963, **4**, 217.

NEEDHAM, JOSEPH (2). *A History of Embryology*. Cambridge, 1934. Revised ed. Cambridge, 1959; Abelard-Schuman, New York, 1959.

NEEDHAM, JOSEPH (16). 'Central Asia and the History of Science and Technology.' *JRCAS*, 1949, **36**, 135.

NEEDHAM, JOSEPH (17). *Science and Society in Ancient China*. Watts, London, 1947. (Conway Memorial Lecture, South Place Ethical Society.) Revised ed. *MSTRM*, 1960, **13** (no. 7), 7.

NEEDHAM, JOSEPH (20). 'Science in Chungking.' *N*, 1943, **152**, 64. 'The Chungking Industrial and Mining Exhibition'. *N*, 1944, **153**, 672. Reprinted in Needham & Needham (1).

NEEDHAM, JOSEPH (21). 'Science in Western Szechuan, I. Physico-Chemical Sciences and Technology.' *N*, 1943, **152**, 343. Reprinted in Needham & Needham (1).

NEEDHAM, JOSEPH (23). 'Science and Technology in the North West of China.' *N*, 1944, **153**, 238. Reprinted in Needham & Needham (1).

NEEDHAM, JOSEPH (25). 'Science and Technology in China's Far South East.' *N*, 1946, **157**, 175. Reprinted in Needham & Needham (1).

NEEDHAM, JOSEPH (31). 'Remarks on the History of Iron and Steel Technology in China' (with French translation; 'Remarques relatives à l'Histoire de la Sidérurgie Chinoise'). In *Actes du Colloque International 'Le Fer à travers les Âges'*, pp. 93, 103. Nancy, Oct. 1955. (*AEST*, 1956, Mémoire no. 16.)

NEEDHAM, JOSEPH (32). *The Development of Iron and Steel Technology in China.* Newcomen Soc. London, 1958. (Second Biennial Dickinson Memorial Lecture, Newcomen Society.) Repr. Heffer, Cambridge, 1964. French tr. (unrevised, with omissions and additions in the illustrations), *RHSID*, 1961, **2**, 187, 235, 1962, **3**, 1, 61, and sep. pub.

NEEDHAM, JOSEPH (33). 'The Peking Observatory in A.D. 1280 and the Development of the Equatorial Mounting.' Art. in *Vistas of Astronomy* (Stratton Presentation Volume), ed. A. Beer, vol. 1, p. 67. Pergamon, London, 1955.

NEEDHAM, JOSEPH (34). 'The Translation of Old Chinese Scientific and Technical Texts.' Art. in *Aspects of Translation*, ed. A. H. Smith, p. 65. Secker & Warburg, London, 1958 (Studies in Communication, no. 2); and *BABEL*, 1958, **4** (no. 1), 8.

NEEDHAM, JOSEPH (38). 'The Missing Link in Horological History; a Chinese Contribution.' *PRSA*, 1959, **250**, 147. (Wilkins Lecture, Royal Society.) Abstract, with illustrations, in *NS*, 1958, **10**.

NEEDHAM, JOSEPH (41). *Classical Chinese Contributions to Mechanical Engineering.* Univ. of Durham, Newcastle, 1961. (Earl Grey Lecture.)

NEEDHAM, JOSEPH (42). 'Aeronautics in Ancient China.' *SAN*, 1961 (no. 279), 2; (no. 280), 15.

NEEDHAM, JOSEPH (45). 'Poverties and Triumphs of the Chinese Scientific Tradition.' Art. in *Scientific Change*, ed. A. C. Crombie; Heinemann, London, 1963. Symposium on the History of Science, Oxford, 1961.

NEEDHAM, JOSEPH (46). 'An Archaeological Study-Tour in China, 1958.' *AQ*, 1959, **33**, 113.

NEEDHAM, JOSEPH (47). 'Science and China's Influence on the West.' Art. in *The Legacy of China*, ed. R. Dawson, p. 234. Oxford, 1964.

NEEDHAM, JOSEPH (48). 'The Prenatal History of the Steam-Engine.' (Newcomen Centenary Lecture.) *TNS*, 1964.

NEEDHAM, JOSEPH. See Chesneaux, J. & Needham J.

NEEDHAM, JOSEPH & LU GWEI-DJEN (1). 'Hygiene and Preventive Medicine in Ancient China'. *JHMAS*, 1962, **17**, 429. Abridgement in *HEJ*, 1959, **17**, 170.

NEEDHAM, JOSEPH & LU GWEI-DJEN (2). 'Efficient Equine Harness; the Chinese Inventions.' *PHY*, 1960, **2**, 121.

NEEDHAM, JOSEPH, WANG LING & PRICE, DEREK J. DE S. (1). *Heavenly Clockwork; the Great Astronomical Clocks of Medieval China.* Cambridge, 1960. (Antiquarian Horological Society Monographs, no. 1.) Prelim. pub. *AHOR*, 1956, **1**, 153.

NEEDHAM, JOSEPH, WANG LING & PRICE, DEREK J. DE S. (2). 'Chinese Astronomical Clockwork.' *N*, 1956, **177**, 600. Chinese tr. by Hsi Tsê-Tsung, *KHTP*, 1956 (no. 6), 100.

NEEDHAM, JOSEPH, WANG LING & PRICE, DEREK J. DE S. (3). 'Chinese Astronomical Clockwork.' *Actes du VIIIe Congrès International d'Histoire des Sciences*, p. 325. Florence, 1956.

NEHER, R. (1). 'Anonymus *De Rebus Bellicis*.' Inaug. Diss., Tübingen, 1911.

NETTO, C. (1). 'Ü. japanisches Berg- u. Hütten-Wesen.' *MDGNVO*, 1879, **2**, 368.

NEUBURGER, A. (1). *The Technical Arts and Sciences of the Ancients.* Tr. from the Germ. ed., *Die Technik des Altertums*, Voigtländer, Leipzig, 1919, by H. L. Brose. Methuen, London, 1930 (with a drastically abbreviated index and the total omission of the bibliographies appended to each chapter, the general bibliography, and the table of sources of the illustrations).

NEUGEBAUER, O. (6). 'Über eine Methode zur Distanzbestimmung Alexandria-Rom bei Heron.' *KDVS/HFM*, 1939, **26**, no. 2 (p. 21) and no. 7.

NEUGEBAUER, O. (7). 'The Early History of the Astrolabe.' *ISIS*, 1949, **40**, 240.

NEVILLE, R. C. (1). 'Description of a remarkable Deposit of Roman Antiquities of Iron discovered at Great Chesterford in Essex, in 1854.' *ARJ*, 1856, **13**, 1.

NEWBOULD, G. T. (1). 'The Atmospheric Engine at Parkgate.' *TNS*, 1935, **15**, 225.

NICHOLS, F. H. (1). *Through Hidden Shensi.* New York, 1902.

NICHOLSON, R. A. (2) (tr.). *The 'Mathnawī' of Jalāl al-Dīn Rūmī.* 8 vols. Brill, Leiden, 1925; Luzac, London, 1925 (Gibbs Memorial series, N.S., no. 4).

NICOLAISEN, N. A. MØLLER (1). *Tycho Brahes Papirmølle [+1589] paa Hven.* Gyldendal, Copenhagen, 1946.

NIEUHOFF, J. (1). *L'Ambassade [1655-1657] de la Compagnie Orientale des Provinces Unies vers l'Empereur de la Chine, ou Grand Cam de Tartarie, faite par les Sieurs Pierre de Goyer & Jacob de Keyser; Illustrée d'une tres-exacte Description des Villes, Bourgs, Villages, Ports de Mers, et autres Lieux plus considerables de la Chine; Enrichie d'un grand nombre de Tailles douces, le tout receuilli par Mr Jean Nieuhoff. . .* (title of Pt. II: *Description Generale de l'Empire de la Chine, ou il est traité succinctement du Gouvernement, de la Religion, des Mœurs, des Sciences et Arts des Chinois, comme aussi des Animaux, des Poissons, des Arbres et Plantes, qui ornent leurs Campagnes et leurs Rivieres; y joint un court Recit des dernieres Guerres qu'ils ont eu contre les Tartares*). de Meurs, Leiden, 1665.

NIEUWHOFF. See Nieuhoff.

DES NOËTTES, R. J. E. C. LEFEBVRE (1). *L'Attelage et le Cheval de Selle à travers les Âges; Contribution à l'Histoire de l'Esclavage.* Picard, Paris, 1931. 2 vols. (1 vol. text, 1 vol. plates). (The definitive version of *La Force Animale à travers les Âges.* Berger-Levrault, Nancy, 1924.) Abstracts *LN*, 1927 (pt. 1).

DES NOËTTES, R. J. E. C. LEFEBVRE (3). 'La "Nuit" du Moyen Âge et son Inventaire.' *MF*, 1932, **235**, 572.

DES NOËTTES, R. J. E. C. LEFEBVRE (5). 'La Force Motrice Animale et le Rôle des Inventions Techniques.' *RSH*, 1927, **43**, 83.

DES NOËTTES, R. J. E. C. LEFEBVRE (6). 'La Conquête de la Force Motrice Animale et la Question de l'Esclavage.' *BAFAS*, 1927, **56** (no. 70), 25.

DES NOËTTES, R. J. E. C. LEFEBVRE (7). 'L'Esclavage antique devant l'Histoire.' *MF*, 1933, **241**, 567.

DES NOËTTES, R. J. E. C. LEFEBVRE (10). 'Le Char de Vaison.' *CAF*, 1923, **86**, 375.

NOIRÉ, L. (1). *Das Werkzeug u. seine Bedeutung f. d. Entwicklungsgeschichte d. Menschheit.* Diemer. Mainz, 1880.

NOLTHENIUS, A. T. (1). 'Les Moulins à Main au Moyen-Âge.' *MC/TC*, 1955, **4**, 149.

NOLTHENIUS, A. T. (2). 'Schipmolens.' *DEI*, 1955, **67**, 420.

NORBURY, J. (1). 'A Note on Knitting and Knitted Fabrics [before the Industrial Revolution].' Art. in *A History of Technology*, ed. C. Singer et al., vol. 3, p. 181. Oxford, 1957.

NORDEN, F. L. (1). *Voyage d'Égypte et de Nubie.* Impr. de la Maison Royale des Orphelins, Copenhagen, 1755.

O'MALLEY, C. D. (1). 'A Latin Translation (+1547) of Ibn al-Nafîs, related to the Problem of the Circulation of the Blood.' *JHMAS*, 1957, **12**, 248. Abstract in *Actes du VIIIᵉ Congrès International d'Histoire des Sciences*, p. 716. Florence, 1956.

O'REILLY, J. P. (1). 'Some Further Notes on Ancient Horizontal Water-Mills, Native and Foreign.' *PRIA*, 1902, **24**, Sect. C, 55.

OLAUS MAGNUS (1). Abp. of Upsala. *Historia de Gentibus Septentrionalibus, earumque diversis Statibus, Conditionibus....* Rome, 1555; Antwerp, 1558; Basel, 1567. Abridgement by C. S. Graphaeus, Antwerp, c. 1565. Eng. tr., Streater, Mosely & Sawbridge, London, 1658.

OLIVER, R. P. (1). 'A Note on the *De Rebus Bellicis*.' *CP*, 1955, **50**, 113 (on the interpretation of the steel-spring arcuballistae described by the Anonymus of +370).

OLSCHKI, L. (4). *Guillaume Boucher; a French Artist at the Court of the Khans.* Johns Hopkins Univ. Press, Baltimore, 1946 (rev. H. Franke, *OR*, 1950, **3**, 135).

OLSCHKI, L. (5). *Geschichte d. neusprachlichen wissenschaftlichen Literatur, I. Die Literatur d. Technik u. d. angewandte Wissenschaften vom Mittelalter bis zur Renaissance.* Winter, Heidelberg, 1918; Olschki, Florence, 1919.

OLSCHKI, L. (7). *The Myth of Felt.* Univ. of California Press, Los Angeles, Calif., 1949.

OMAN, C. W. C. (1). *A History of the Art of War in the Middle Ages.* 1st ed. 1 vol. 1898; 2nd ed. 2 vols. 1924 (much enlarged); vol. 1, +378 to +1278; vol. 2, +1278 to +1485. Methuen, London (the original publication had been a prize essay printed at Oxford in 1885; this was reprinted in 1953 by the Cornell Univ. Press, Ithaca, N.Y., with editorial notes and additions by J. H. Beeler).

ORANGE, J. (1). *The Chater Collection; Pictures relating to China, Hongkong, Macao, 1655 to 1860, with Historical and Descriptive Letterpress....*' Butterworth, London, 1924.

ORE, OYSTEIN (1). *Cardano, the Gambling Scholar.* Princeton Univ. Press, Princeton, N.J., 1953.

ORMEROD, H. A. & CARY, M. (1). 'Rome and the East.' In *CAH*, vol. 9, p. 350.

ORRERY, the Countess of Cork and, (1) (ed.). *The Orrery Papers.* 2 vols. Duckworth, London, 1903.

DA ORTA, GARCIA (1). *Coloquios dos Simples e Drogas he cousas medicinais da India compostos pello Doutor Garcia da Orta.* de Endem, Goa, 1563. Latin epitome by Charles de l'Escluze, Plantin, Antwerp, 1567. Eng. tr. *Colloquies on the Simples and Drugs of India*, with the annotations of the Conde de Ficalho, 1895, by Sir Clements Markham. Sotheran, London, 1913.

OSGOOD, C. (1). *The Koreans and their Culture.* Ronald, New York, 1951.

OUCHTERLONY, J. (1). *The Chinese War; an Account of all the Operations of the British Forces from its Commencement to the Treaty of Nanking.* Saunders & Otley, London, 1844.

PAGEL, W. (8). 'J. B. van Helmont's *De Tempore*, and Biological Time.' *OSIS*, 1949, **8**, 346.

PAGET, Sir RICHARD (1). 'Sir Charles Boys' (C. V. Boys; obituary). *PPS*, 1944, **56**, 397.

PAGLIARO, A. (1). 'Pahlavī *katas*, "canale", Gr. Καδος.' *RSO*, 1937, **17**, 72.

PAK, C. See Read, B. E. & Pak Kyebyŏng.

PALLAS, P. S. (1). *Sammlungen historischen Nachrichten ü. d. mongolischen Völkerschaften.* St Petersburg, 1776. Fleischer, Frankfurt and Leipzig, 1779.

PANCIROLI, GUIDO (1). *Rerum Memorabilium sive Deperditarum pars prior (et secundus) Commentariis illustrata et locis prope innumeris postremum aucta ab Henrico Salmuth.* Amberg, 1599 and 1607;

Schonvetter Vid. et Haered. Frankfurt, 1617, 1646, 1660. Eng. tr. *The History of many Memorable Things lost, which were in Use among the Ancients; and an Account of many Excellent Things found, now in Use among the Moderns, both Natural and Artificial... now done into English.... To this English edition is added, first, a Supplement to the Chapter of Printing, shewing the Time of its Beginning, and the first Book printed in each City before the Year 1500. Secondly, what the Moderns have found, the Ancients never knew; extracted from Dr Sprat's History of the Royal Society, the Writings of the Honourable Mr Boyle, the Royal-Academy at Paris, etc.* London, 1715, 1727. French tr. Lyon, 1608. Bibliography in John Ferguson (2).

PANIKKAR, K. M. (1). *Asia and Western Dominance.* Allen & Unwin, London, 1953.

PAPINOT, E. (1). *Historical and Geographical Dictionary of Japan.* Overbeck, Ann Arbor, Mich. 1948. Photographically reproduced from the original edition, Kelly & Walsh, Yokohama, 1910. Eng. tr. of the French original, Sanseidō, Tokyo, and Kelly & Walsh, Yokohama, 1906.

PARANAVITANA, S. (1). 'The Magul Uyana of Ancient Anurādhapura' (royal parks and gardens). *JRAS/C*, 1944, **36**, 194.

PARIS, J. A. (1). *Philosophy in Sport made Science in Earnest: being an Attempt to illustrate the first principles of Natural Philosophy by the aid of popular Toys and Sports of Youth.* Illustrated by George Cruickshank. London, 1827; 6th ed. 1846.

PARSON, A. W. (1). 'A Roman Water-Mill in the Athenian Agora.' *HE*, 1936, **5**, 70.

PARSONS, J. B. (1). 'The Culmination of a Chinese Peasant Rebellion; Chang Hsien-Chung in Szechuan, +1644 to +1646.' *JAS*, 1957, **16**, 387.

PARSONS, W. B. (1). *An American Engineer in China.* McClure-Philips, New York, 1900.

PARSONS, W. B. (2). *Engineers and Engineering in the Renaissance.* Williams & Wilkins, Baltimore, 1939.

PARTINGTON, J. R. (5). *A History of Greek Fire and Gunpowder.* Heffer, Cambridge, 1960.

PASTOR, F. PEREZ (1). *Tratado de los Reloxes Elementares, o el Modo de Hacer Reloxes con el Agua, la Tierra, el Ayre, y el Fuego; y en que, con la mayor facilidad, y poquisima Costa, se aprende á añadirles los mas prodigiosos Movimientos de los Astros, y Planetas, como de diversas Figuras, el Canto de las Aves, y otras Invenciones....* Lozano, Madrid, 1770.

PATTERSON, R. (1). 'Spinning and Weaving [in the Mediterranean Civilisations and the Middle Ages].' Art. in *A History of Technology*, ed. C. Singer *et al.*, vol. 2, p. 191. Oxford, 1956.

PAUCTON, M. (1). *Théorie de la Vis d'Archimède, de laquelle on déduit celle des Moulins, conçue d'une Nouvelle Manière* (a helicopter suggested). Butard, Paris, 1768.

PAULINYI, Á. (1). *Príspevok k Technologickému Vývinu Hroneckých Železiarní v Prvej Polovici 19 Storočia* (*Beitrag zur Gesch. d. technischen Entwicklung der Rohnitzer Eisenwerke in der ersten Hälfte des 19 Jahrhunderts*) (in Czech with German summary). *SDPVT*, 1962, **7**, 159.

PELHAM, R. A. (1). *Fulling Mills; a Study in the Application of Water-Power to the Woollen Industry.* Society for the Protection of Ancient Buildings, London, n.d. (1955). (Wind and Watermill Section, S.P.A.B. booklet no. 5.)

PELLAT, C. (1). 'Jāḥiẓiana; I, Le *Kitāb al-Tabaṣṣur bi-l-Tijāra* (De la Clairvoyance en Matière Commerciale) attribué à Jāḥiẓ.' *ARAB*, 1954, **1**, 153.

PELLIOT, P. (24). 'À Propos du *Kêng Tche T'ou [Kêng Chih Thu]*.' In *Mémoires concernant l'Asie Orientale*, ed. Senart, Barth, Chavannes & Cordier, vol. 1. Leroux, Paris, 1913.

PELLIOT, P. (25). *Les Grottes de Touen-Hoang [Tunhuang]; Peintures et Sculptures Bouddhiques des Époques des Wei, des T'ang et des Song [Sung].* Mission Pelliot en Asie Centrale, 6 portfolios of plates. Paris, 1920–24.

PELLIOT, P. (26). Note on Liang Yuan Ti and his writings. *TP*, 1912, **13**, 402, n. 3.

PELLIOT, P. (27). *Les Influences Européennes sur l'Art Chinois au 17ᵉ et au 18ᵉ siècle.* Imp. Nat., Paris, 1948. (Conférence faite au Musée Guimet, Feb. 1927.)

PELLIOT, P. (39). 'L'Horlogerie en Chine' (a review of Chapuis, Loup & de Saussure, q.v.). *TP*, 1921, **20**, 61.

PELLIOT, P. (46). Notes on stirrups and other horse-trappings in China in a review of the first version of Lefebvre des Noëttes (1). *TP*, 1926, **24**, 256.

PENZER, N. M. (1) (ed.). *The Most Noble and Famous Travels of Marco Polo; together with the Travels of Nicolo de Conti, edited from the Elizabethan translation of J. Frampton (1579), etc.* Argonaut, London, 1929.

PERCY, J. (2). *Metallurgy; Iron and Steel.* Murray, London, 1864.

PEREGRINUS, PETRUS (Pierre de Maricourt) (1). *Epistola de Magnete seu Rota Perpetua Motus.* 1269. First pr. by Achilles Gasser, Augsburg, 1558 (a MS. copy of this, with Engl. tr. by an unknown hand, is in Gonv. and Caius Coll. MS. 174/95). Second pr. in Taisnier (1). See Thompson, S. P. (5); Hellmann, G. (6); Anon. (46); [Mertens, J. C.] (1); Chapman & Harradon (1).

PEREME, A. (1). *Recherches historiques et archéologiques sur la Ville d'Issoudun.* Paris, 1847.

PERRAULT, CLAUDE (tr.) (1). *Abregé des Dix Livres d'Architecture de Vitruve.* Paris, 1674, 1684; Amsterdam, 1681 (1691).

PERRAULT, CLAUDE (2). 'Horloge à Pendule qui va par le Moyen de l'Eau, inventée par M. P. . . .' *MIARA*, 1699, **1**, 39, 41.

PERRONET, J. R. (1). *Description des Projets et de la Construction des Ponts de Neuilli, de Mantes, d'Orléans, etc.* Paris, 1788.

PETECH, L. (4). 'L'Ambasciata Olandese del 1655/1657 nei Documenti Cinesi.' *RSO*, 1950, **25**, 77.

PETERSEN, H. (1). *Vognfundene i Dejbjerg Praestegaardsmose ved Ringkjøping.* Copenhagen, 1888.

PETRIE, W. M. FLINDERS (1). *Tools and Weapons, illustrated by the Egyptian Collections in University College, London, and 2000 outlines from other Sources.* Constable, London, 1917.

PETRIE, W. M. FLINDERS (2). *Arts and Crafts of Ancient Egypt.* Edinburgh, 1910.

PETRIE, W. M. FLINDERS (3). *Six Temples at Thebes.* Quaritch, London, 1897.

PETRIE, W. M. FLINDERS (4). *The Wisdom of the Egyptians.* London, 1940.

PFISTER, L. (1). *Notices Biographiques et Bibliographiques sur les Jésuites de l'Ancienne Mission de Chine* (+1552 *to* 1773). 2 vols. Mission Press, Shanghai, 1932 (*VS*, no. 59).

PFIZMAIER, A. (91) (tr.). 'Denkwürdigkeiten v. chinesischen Werkzeugen und Geräthen.' *SWAW/PH*, 1872, **72**, 247, 265, 272, 275, 295, 308, 313, 315. (Tr. chs. 701 (screens), 702 (fans), 703 (whisks, sceptres, censers, etc.), 707 (pillows), 711 (boxes and baskets), 713 (chests), 714 (combs and brushes), 717 (mirrors), *Thai-Phing Yü Lan.*)

PFIZMAIER, A. (92) (tr.). 'Kunstfertigkeiten u. Künste d. alten Chinesen.' *SWAW/PH*, 1871, **69**, 147, 164, 178, 202, 208. (Tr. chs. 736, 737 (magic), 750, 751 (painting) and 752 (inventions and automata), *Thai-Phing Yü Lan.*)

PFIZMAIER, A. (93) (tr.). 'Zur Geschichte d. Erfindung u. d. Gebrauches d. chinesischen Schrift-gattungen.' *SWAW/PH*, 1872, **70**, 10, 28, 46. (Tr. chs. 747, 748, 749, *Thai-Phing Yü Lan.*)

PHAN ÊN-LIN (ed.) (1). *Scenic Beauties in Southwest China* (album of photographs). China Travel Service, Shanghai, 1939.

PHILIPSON, J. (1). *Harness; as it has been, as it is, and as it should be. . .with Remarks on Traction and the Use of the Cape Cart.* Reid, Newcastle-on-Tyne, 1882.

PIGGOTT, S. (1). 'A Tripartite Disc Wheel from Blair Drummond, Perthshire.' *PSAS*, 1957, **90**, 238.

PILJA, DUŠAN (1). *The District of Jajce; a Review of the Old and the New in Bosnia and Hercegovina.* Narodna Prosvjeta, Sarajevo, 1959.

PINOT, V. (2). *Documents Inédits relatifs à la Connaissance de la Chine en France de 1685 à 1740.* Geuthner, Paris, 1932.

PIOTROVSKY, B. B. (1). 'Ourartou.' Art. in *Ourartou, Neapolis des Scythes, Kharezm*, ed. C. Virolleaud, p. 13. Maisonneuvve, Paris, 1954.

PIPPARD, A. J. S. & BAKER, J. F. (1). *Analysis of Engineering Structures.* Arnold, London, 1936.

PITT-RIVERS, A. H. LANE-FOX (1). *On the Development and Distribution of Primitive Locks and Keys.* Chatto & Windus, London, 1883.

PLANCHON, M. (1). *L'Horloge; son Histoire rétrospective, pittoresque et artistique.* Laurens, Paris, 1899; 2nd ed. 1912.

PLANCHON, M. (2). *Catalogue; Rapport, Horlogerie.* Classe 96 of the Musée Retrospectif, Exposition Universelle Internationale. Paris, 1900.

PLEDGE, H. T. (1). *Science since* +1500. HMSO, London, 1939.

PLISCHKE, H. (1). *Der Fischdrachen* (fishing-kite). Leipzig, 1922.

PLISCHKE, H. (2). 'Alter und Herkunft des europäischen Flächendrachens' (kite). *NGWG/PH* (Mittl. u. neueren Gesch.), 1936, **2**, 1.

POGGENDORFF, J. C. (1). *Geschichte d. Physik.* Barth, Leipzig, 1879.

DE POIROT, LOUIS (1) (tr.). '[Préface ou Introduction], [by the Yung-Chêng Emperor, his successor] [aux] Instructions Sublimes et Familières de Cheng-Tzu-Quogen-Hoang-Ti [Shêng Tsu Jen Huang Ti, i.e. the Khang-Hsi Emperor].' *MHSAMUC*, 1783, **9**, 65–281. (Ital. tr. by Louis de Poirot, S.J., from the Manchu text, done into French by Mme la Comtesse de M**.)

POKORA, T. (9). 'The Life of Huan Than.' *ARO*, 1963, **31**, 1.

POLE, W. (1). *A Treatise on the Cornish Pumping Engine.* London, 1844.

PONOMAREV, N. A. (1). 'On the Times and Places of the Appearance of the First Windmills' (in Russian). Contribution to symposium: *Istoria Mashinost-Roenia* (History of Machine Con-struction). *TIYT*, 1960, **29**, 352. (MS. Engl. tr. by A. L. Nasvytis & J. D. Stanitz.)

PONOMAREV, N. A. (2). *The History of Flour- and Grain-Milling Technology.* Acad. Sci. USSR, Moscow, 1955.

PORSILD, M. P. (1). 'The Screw Principle in Eskimo Technique.' *AAN*, 1915, **17**, 1.

DELLA PORTA, J. B. (Giambattista) (2). *Pneumaticorum Libri III.* Naples, 1601. Ital. tr. *I Tre Libri de' Spiritali.* Naples, 1606.

POUDEROYEN, A. (1). 'Les Polders aux Pays-Bas, leur assèchement, leur drainage; les Moulins de Pompage, d'Industrie, et les moulins à boue.' *MC/TC*, 1953, **2**, 142; 1954, **3**, 16, 39.

Poulsen, P. (1). 'Der Stand der Forschung ü. d. Kultur d. Wikingerzeit.' *BRGK*, 1932, **22**, 182.
[Powell, Thomas] (1). *Humane Industry; or, a History of most Manual Arts, deducing the Original, Progress, and Improvement of them; furnished with Variety of Instances and Examples, shewing forth the Excellency of Humane Wit.* Herringman, London, 1661. Bibliography in John Ferguson (2).
Praus, A. A. (1). 'Mechanical Principles involved in Primitive Tools and those of the Machine Age.' *ISIS*, 1948, **38**, 157.
Price, D. J. de S. (1) 'Clockwork before the Clock.' *HORJ*, 1955, **97**, 810; 1956, **98**, 31.
Price, D. J. de S. (2) (ed.). The '*Equatorie of the Planetis*' [*probably written by Geoffrey Chaucer*]. With a linguistic analysis by R. M. Wilson. Cambridge, 1955 (rev. H. Spencer-Jones, *JIN*, 1955, **8**, 344).
Price, D. J. de S. (3). 'A Collection of Armillary Spheres and other Antique Scientific Instruments.' *ANS*, 1954, **10**, 172.
Price, D. J. de S. (4). 'The Prehistory of the Clock.' *D*, 1956, **17**, 153.
Price, D. J. de S. (5). 'Precision Instruments: to +1500.' Art. in *A History of Technology*, ed. C. Singer, E. J. Holmyard & A. R. Hall, vol. 3, p. 582. 1957.
Price, D. J. de S. (6). 'An International Check-list of Astrolabes.' *A/AIHS*, 1955, **8**, 243 & 363.
Price, D. J. de S. (8). 'On the Origin of Clockwork, Perpetual Motion Devices, and the [Magnetic] Compass.' *BUSNM*, 1959 (no. 218), 81. (Contributions from the Museum of History & Technol. no. 6.) Crit. R. S. Woodbury, *TCULT*, 1960, **1**, 270.
Price, D. J. de S. (9). 'An Ancient Greek Computer' (the Anti-Kythera calendrical analogue computing machine). *SAM*, 1959, **200** (no. 6), 60.
Price, D. J. de S. (10). 'Leonardo da Vinci and the Clock of Giovanni de Dondi.' *AHOR*, 1958. **2**, 127, 222.
Price, D. J. de S. (11). 'Two Mediaeval Texts on Astronomical Clocks.' *AHOR*, 1956, **1** (no. 10), 156.
Price, D. J. de S. (13). 'Mechanical Water-Clocks of the +14th Century at Fez, Morocco.' Communication to the Xth Internat. Congress of the History of Science, Ithaca, N.Y. 1962. Abstracts vol., p. 64.
Price, F. G. H. (1). 'Note on a Curious [Terra-cotta] Model of an Archimedean Screw, probably of the Late Ptolemaic Period, found in Lower Egypt.' *PSA*, 1897 (2nd ser.), **16**, 277.
Pritchard, J. L. (1). *Sir George Cayley, the Inventor of the Aeroplane.* Parrish, London, 1961. Rev. J. C. Hunsaker, *TCULT*, 1963, **4**, 88.
Procopius (1). *De Bello Gothico.* See Dewing, H. B. (1), vols. 3, 4, 5.
Profumo, A. (1). Note on a late Roman painting of a water-mill. *NBAC*, 1917, **23**, 108.
Prou, V. (1). 'Les Théâtres d'Automates en Grèce au 2ᵉ siècle avant l'Ère Chrétienne, d'après les Αυτοματοποιικα d'Héron d'Alexandrie.' *MRAI/DS*, 1884 (1ᵉ sér.), **9**, 117.
Proudfoot, W. J. (1). *Biographical Memoir of James Dinwiddie, LL.D., Astronomer in the British Embassy to China (1792–4), afterwards Professor of Natural Philosophy in the College of Fort William, Bengal; embracing some account of his Travels in China and Residence in India, compiled from his Notes and Correspondence by his grandson....* Howell, Liverpool, 1868.
Prušek, J. (3). 'Chêng Chen-To; in memoriam.' *ARO*, 1959, **27**, 177.
Przyłuski, J. (4). 'Le Culte de l'Étendard chez les Scythes et dans l'Inde.' *Z*, 1938, **1**, 13.
Przyłuski, J. (7). 'La Roue de la Vie à Ajanta.' *JA*, 1920 (11ᵉ sér.), **16**, 313.
Pulleyblank, E. G. (1). *The Background of the Rebellion of An Lu-Shan.* Oxford, 1954. (London Oriental Series, no. 4.)
Pulleyblank, E. G. (2). 'The Date of the Staël-Holstein Scroll.' *AM*, 1954, (N.S.) **4**, 90.
Pulleyblank, E. G. (6). 'The Origins and Nature of Chattel Slavery in China.' *JESHO*, 1958, **1**, 185.
Purvis, F. P. (1). 'Ship Construction in Japan.' *TAS/J*, 1919, **47**, 1; 'Japanese Ships of the Past and Present.' *TJSL*, 1925, **23**, 51.

Quatremère, E. M. (3) (tr.). 'Notice de l'Ouvrage Persan qui a pour Titre *Matla Assaadeïn oumadjma-albahreïn*, et qui contient l'Histoire des deux Sultans Schah-rokh et Abou-Saïd. (The account by Ghiyāth al-Dīn-i Naqqāsh of the embassy from Shāh Rukh to the Ming emperor.) *MAI/NEM*, 1843, **14**, pt 1, 1–514 (387).

Raghavan, V. (1). 'Yantras or Mechanical Contrivances in Ancient India.' *TIIC*, 1952 (no. 10) 1–31.
Raghavan, V. (2). 'Gleanings from Somadeva Sūri's *Yaśastilaka Campu*.' *JGJRI*, 1944, **1**, 378, 467.
Raglan, Lord (1). *How Came Civilisation?* Methuen, London, 1939.
Raistrick, A. (1). *Dynasty of Iron Founders; the Darbys and Coalbrookdale.* Longmans Green, London, 1953.
Rambaut, A. (1). 'Note on some Japanese Clocks lately purchased for the Science and Art Museum.' *SPRDS*, 1889, (N.S.) **6**, 332.
Ramelli, Agostino (1). *Le Diversi e Artificiose Machine del Capitano A.R.* Paris, 1588. Account in Beck (1), ch. 11.

RAMSEY, A. R. J. (1). 'The Thermostat; an Outline of its History.' *TNS*, 1946, **25**, 53.

RANKINE, W. J. McQ. (1). *A Manual of the Steam Engine and other Prime Movers*. Griffin & Bohn, London, 1861.

READ, BERNARD E. (1) (with LIU JU-CHHIANG). *Chinese Medicinal Plants from the 'Pên Ts'ao Kang Mu' A.D. 1596...a Botanical, Chemical and Pharmacological Reference List*. (Publication of the Peking Nat. Hist. Bull.). French Bookstore, Peiping, 1936 (chs. 12–37 of *Pên Tshao Kang Mu*) (rev. W. T. Swingle, *ARLC/DO*, 1937, 191).

READ, BERNARD E. (2) (with LI YÜ-THIEN). *Chinese Materia Medica; Animal Drugs*.

		Serial nos.	Corresp. with chaps. of *Pên Tshao Kang Mu*
Pt. I	Domestic Animals	322–349	50
II	Wild Animals	350–387	51 *A* and *B*
III	Rodentia	388–399	51 *B*
IV	Monkeys and Supernatural Beings	400–407	51 *B*
V	Man as a Medicine	408–444	52

PNHB, 1931, **5** (no. 4), 37–80; **6** (no. 1), 1–102. (Sep. issued, French Bookstore, Peiping, 1931.)

READ, BERNARD E. (3) (with LI YÜ-THIEN). *Chinese Materia Medica; Avian Drugs*.   245–321   47, 48, 49

Pt. VI Birds

PNHB, 1932, **6** (no. 4), 1–101. (Sep. issued, French Bookstore, Peiping, 1932.)

READ, BERNARD E. (4) (with LI YÜ-THIEN). *Chinese Materia Medica; Dragon and Snake Drugs*.

Pt. VII Reptiles   102–127   43

PNHB, 1934, **8** (no. 4), 297–357. (Sep. issued, French Bookstore, Peiping, 1934.)

READ, BERNARD E. (5) (with YU CHING-MEI). *Chinese Materia Medica; Turtle and Shellfish Drugs*.

Pt. VIII Reptiles and Invertebrates   199–244   45, 46

PNHB (Suppl.), 1939, 1–136. (Sep. issued, French Bookstore, Peiping, 1937.)

READ, BERNARD E. (6) (with YU CHING-MEI). *Chinese Materia Medica; Fish Drugs*.

Pt. IX Fishes (incl. some amphibia, octopoda and crustacea)   128–198   44

PNHB(Suppl.), 1939. (Sep. issued, French Bookstore, Peiping, n.d. prob. 1939.)

READ, BERNARD E. (7) (with YU CHING-MEI). *Chinese Materia Medica; Insect Drugs*.

Pt. X Insects (incl. arachnidae etc.)   1–101   39, 40, 41, 42

PNHB (Suppl.), 1941. (Sep. issued, Lynn, Peiping, 1941).

READ, BERNARD E. (8). *Famine Foods listed in the 'Chiu Huang Pên Tshao'*. Lester Institute, Shanghai, 1946.

READ, BERNARD E. & PAK C. (PAK KYEBYŎNG) (1). *A Compendium of Minerals and Stones used in Chinese Medicine, from the 'Pên Ts'ao Kang Mu'*. *PNHB*, 1928, **3** (no. 2), i–vii, 1–120. (Revised and enlarged, issued separately, French Bookstore, Peiping, 1936 (2nd ed.).) Serial nos. 1–135, corresp. with chs. of *Pên Tshao Kang Mu*, 8, 9, 10, 11.

REDIADIS, P. (1). Account of the Anti-Kythera machine (+2nd century). In J. N. Svoronos (or Sboronos), *Das Athener Nationalmuseum*. 3 vols. text, 3 vols. plates. Beck & Barth, Athens, 1908–37. Textband 1, *Die Funde von Anti-Kythera*.... Greek ed. *To en Athēnais Ethnikōn Mouseion*, Beck & Barth, Athens, 1903–11.

REHATSEK, E. (1) (tr.). 'An Embassy to Khata or China A.D. 1419; from the Appendix to the *Ruzat al-Safa* of Muhammed Khavend Shah or Mirkhond, translated from the Persian....' *IAQ*, 1873, 75, (the embassy from Shāh Rukh, son of Tīmūr, to the Ming emperor; narrative written by Ghiyāth al-Dīn-i Naqqāsh).

REHM, A. & SCHRAMM, E. (1) (ed. & tr.). 'Biton's Bau von Belagerungsmaschinen und Geschützen (griechisch und deutsch).' *ABAW/PH*, 1929 (N.F.), no. 2.

REICHEL, E. (1). 'Aus der Geschichte d. Wasserkraftmaschinen' (ship-mills on the Danube). *BGTI*, 1928, **18**, 57.

REIN, J. J. (1). *Industries of Japan; together with an Account of its Agriculture, Forestry, Arts and Commerce.* Hodder & Stoughton, London, 1889.

REINACH, S. (2). 'Un Homme à Projets du Bas-Empire.' *RA*, 1922 (5e sér.), **16**, 205 (text, translation and commentary of the Anonymus *De Rebus Bellicis, c.* +370).

REINACH, T. (1). *Mithridate Eupator, Roi de Pont.* Paris, 1890.

REINAUD, J. T. & FAVÉ, I. (2). 'Du Feu Grégeois, des Feux de Guerre, et des Origines de la Poudre à Canon chez les Arabes, les Persans et les Chinois.' *JA*, 1849 (4e sér.), **14**, 257.

REISCHAUER, E. O. (1). 'Notes on Thang Dynasty Sea-Routes.' *HJAS*, 1940, **5**, 142.

REISCHAUER, E. O. (2) (tr.). *Ennin's Diary; the Record of a Pilgrimage to China in Search of the Law* (the *Nittō Guhō Junrei Gyōki*). Ronald Press, New York, 1955.

REISCHAUER, E. O. (3). *Ennin's Travels in Thang China.* Ronald Press, New York, 1955.

REISMÜLLER, G. (1). 'Europäische und chinesische Technik.' *GTIG*, 1914, **1**, 2.

RÉMUSAT, J. P. A. (1) (tr.). '*Foe Koue Ki* [*Fo Kuo Chi*]', ou *Relation des Royaumes Bouddhiques; Voyage dans la Tartarie, dans l'Afghanistan et dans l'Inde, exécuté, à la Fin du 4e siècle, par Chy Fa-Hian* [*Shih Fa-Hsien*]. Impr. Roy. Paris, 1836. Eng. tr. *The Pilgrimage of Fa-Hian* [*Fa-Hsien*]; *from the French edition of the 'Foe Koue Ki' of Rémusat, Klaproth and Landresse, with additional notes and illustrations.* Calcutta, 1848. (Fa-Hsien's *Fo Kuo Chi.*)

[RENAUDOT, EUSEBIUS] (1) (tr.). *Anciennes Relations des Indes et de la Chine de deux Voyageurs Mahometans, qui y allèrent dans le Neuvième Siècle, traduites d'Arabe, avec des Remarques sur les principaux Endroits de ces Relations.* (With four Appendices, as follows: (i) Eclaircissement touchant la Prédication de la Religion Chrestienne à la Chine; (ii) Eclaircissement touchant l'Entrée des Mahometans dans la Chine; (iii) Eclaircissement touchant les Juifs qui ont esté trouvez à la Chine; (iv) Eclaircissement sur les Sciences des Chinois.) Coignard, Paris, 1718. Eng. tr. London, 1733. The title of Renaudot's book, which was presented partly to counter the claims of the pro-Chinese party in religious and learned circles (the Jesuits, Golius, Vossius etc., see Pinot (1), pp. 109, 160, 229, 237), was misleading. The two documents translated were: (*a*) The account of Sulaimān al-Tājir (Sulaiman the Merchant), written by an anonymous author in +851, (*b*) The completion *Silsilat al-Tawārīkh* of +920 by Abū Zayd al-Ḥasan al-Shīrāfī, based on the account of Ibn Wahb al-Baṣri, who was in China in +876 (see Mieli (1), pp. 13, 79, 81, 115, 302; al-Jalil (1), p. 138; Hitti (1), pp. 343, 383; Yule (2), vol. 1, pp. 125–33). Cf. Reinaud (1); Sauvaget (2).

RENOU, L. & FILLIOZAT, J. (1). *L'Inde Classique; Manuel des Études Indiennes.* Vol. 1, with the collaboration of P. Meile, A. M. Esnoul & L. Silburn. Payot, Paris, 1947. Vol. 2, with the collaboration of P. Demiéville, O. Lacombe & P. Meile. École Française d'Extrême Orient, Hanoi, 1953; Impr. Nationale, Paris, 1953.

RETI, LADISLAO (1). *Francesco di Giorgio Martini's* (+1439 to +1502) *Treatise on Engineering* (*Trattato di Architettura*), *and its Plagiarists.* Communication to the Xth International Congress of the History of Science, Ithaca, N.Y., 1962. Abstracts Vol., p. 36. Full pub. *TCULT*, 1963, **4**, 287.

RETI, LADISLAO (2). 'Leonardo da Vinci nella Storia della Macchina a Vapore.' *RDI*, 1957, 21.

REULEAUX, F. (1). *Kinematics of Machinery; Outlines of a Theory of Machines* (tr. A. B. W. Kennedy from *Theoretische Kinematik*, Wieweg, Braunschweig, 1875). London, 1876. French. tr. by A. Debize: *Cinématique; Principes fondamentaux d'une Théorie générale des Machines.* Savy, Paris, 1877.

REVERCHON, M. (1). 'Leonardo's pendulum escapement.' *BSAF*, 1915, **29**, May.

RHEAD, W. G. (1). *History of the Fan.* London, 1910.

RHYS DAVIDS, T. W. & C. A. F. See Davids, T. W. Rhys and Foley, C. A.

RICE, H. C. (1). *The Rittenhouse Orrery; Princeton's Eighteenth-century Planetarium, 1767 to 1954.* Princeton, N.J., 1954.

VON RICHTHOFEN, F. (5). *Tagebücher aus China.* Berlin, 1907.

RICKETT, W. A. (1) (tr.). *The 'Kuan Tzu' Book.* Hongkong Univ. Press, Hong Kong, in the press.

RICO Y SINOBAS, M. (1). '*Libros del Saber de Astronomia' del Rey D. Alfonso X de Castilla.* Aguado, Madrid, 1864.

RIDE, LINDSAY (1). 'Biographical Note [on James Legge].' In the Additional Volume to the Hongkong University Press 1960 Photolitho re-issue of *The Chinese Classics.* Hongkong, 1960.

RIGAULT, HIPPOLYTE (1). *Histoire de la Querelle des Anciens et des Modernes.* Paris, 1856.

RIVIUS, G. H. (1). *Der fürnehmbsten, nothwendigsten, der ganzen Architektur angehörigen, mathematischen und mechanischen Kunst eigentlicher Bericht.* Nuremberg, 1547.

RIVIUS, G. H. (2). *Vitruvius Teutsch, Nemlichen des aller namhaftigsten und hocherfarnesten Römischen Architecti und Kunstreichen Werck oder Baumeisters Marci Vitruvii Pollionis Zehen Bucher von der Architektur und künstlichen Bauen....* Petreius, Nuremberg, 1548.

ROBERT, L. (1). 'Hellenica.' *RPLHA*, 1939, **13**, 97 (175).

ROBERTSON, J. DRUMMOND (1). *The Evolution of Clockwork, with a special section on the Clocks of Japan, and a Comprehensive Bibliography of Horology.* Cassell, London, 1931.

ROBINS, F. W. (1). *The Story of Water Supply*. Oxford, 1946.

DE ROCHAS D'AIGLUN, A. (1). *La Science des Philosophes et l'Art des Thaumaturges dans l'Antiquité*. Dorbon, Paris, 1912 (1st ed. 1882).

DE ROCHEMONTEIX, P. C. (1). *Joseph Amiot* (Jesuit missionary in China). Paris, 1915.

ROCHER, E. (1). *La Province Chinoise du Yunnan*. 2 vols. (incl. special chapter on metallurgy). Leroux, Paris, 1879, 1880.

ROCKHILL, W. W. (3). *The Land of the Lamas*. Longmans Green, London, 1891.

ROCKHILL, W. W. (4). 'A Journey in Mongolia and Tibet.' *GJ*, 1894, **3**, 357.

VON ROHR-SAUER, A. (1). *Des Abu Dulaf Bericht über seine Reise nach Turkestan, China und Indien*. Inaug. Diss., Bonn, 1939.

ROLT, L. T. C. (1). *Thomas Newcomen; the Prehistory of the Steam Engine*. David & Charles, Dawlish, 1963; McDonald, London, 1963.

LA RONCIÈRE, C. DE B. (1). *Histoire de la Marine Française*. 6 vols. Perrin (later Plon), Paris, 1899–1932.

RONDOT, N. (1) (ed.). *Étude Pratique du Commerce d'Exportation de la Chine; par I. Hedde, E. Renard, A. Haussmann & N. Rondot, revue et complétée par N. Rondot*. Challamel & Renard, Paris, 1848; Reynvaan, Canton, 1848; Sanier & Sauermondt, Batavia, 1848.

DE ROOS, H. (1). *The Thirsty Land; the story of the Central Valley Project*. Stanford Univ. Press, Palo Alto, Calif., 1948.

ROSEN, E. (1). *The Naming of the Telescope*. Schuman, New York, 1947.

ROSENAUER, N. & WILLIS, A. H. (1). *Kinematics of Mechanisms*. Assoc. Gen. Pub., Sydney, 1953 (rev. R. H. McMillan, *N*, 1954, **173**, 924).

ROSS, A. S. C. (1). *The Terfinnas and Beormas of Ohthere*. Univ. Leeds, Leeds, 1940. (Leeds Sch. of Engl. Lang. Texts & Monographs, no. 7.)

ROSS, J. F. S. (1). *The Gyroscopic Stabilisation of Land Vehicles*. Arnold, London, 1933.

ROSTOVTZEV, M. I. & ORMEROD, H. A. (1). 'Pontus and its Neighbours; the First Mithridatic War.' In *CAH*, vol. 9, 211.

DES ROTOURS, R. (1) (tr.). *Traité des Fonctionnaires et Traité de l'Armée, traduits de la Nouvelle Histoire des T'ang* (chs. 46–50). 2 vols. Brill, Leiden, 1948 (Bibl. de l'Inst. des Hautes Études Chinoises, no. 6) (rev. P. Demiéville, *JA*, 1950, **238**, 395).

ROULEAU, F. (1). 'The Auto was invented in China.' *CAR*, 1942, Nov.

ROWE, Capt. JACOB (1). *All Sorts of Wheel-Carriage, Improved; wherein it is plainly made appear, that a much less than the usual Draught of Horses, etc., will be requir'd in Waggons, Carts, Coaches, and all other Wheel Vehicles, as likewise all Water-Mills, Wind-Mills, and Horse-Mills; this Method being found good in Practice, by the Trial of a Coach and Cart already made, shews of what great Advantage it may be to all Farmers, Carriers, etc. etc., by saving them one half of the Expenses... according to the common Method; with Explanation of the Structure of a Coach and Cart, according to this Method....* Lyon, London, 1734.

ROWLAND, B. (1). 'Buddha and the Sun God.' *Z*, 1938, **1**, 69.

ROY, CLAUDE (1). *La Chine dans un Miroir*. Clairefontaine, Lausanne, 1953.

ROY, L. C. (1). 'The Roof of Eastern America.' *NGM*, 1936, **70**, 243 (263).

RUDENKO, S. I. (1). *Kultura Nasseleniya Gornogo Altaya v Skifskoye Vremya*. Academy of Sci. USSR, Moscow, 1953 (in Russian). (On the Pazirik finds.)

RUDOLPH, R. C. (4). 'A Second-century Chinese Illustration of Salt Mining.' *ISIS*, 1952, **43**, 39.

RUDOLPH, R. C. & WÊN YU (1). *Han Tomb Art of West China; a Collection of First and Second Century Reliefs*. Univ. of Calif. Press, Berkeley and Los Angeles, 1957 (rev. W. P. Yetts, *JRAS*, 1953, 72).

DE LA RUE, E. AUBERT (1). *L'Homme et le Vent*. Gallimard, Paris, 1940.

RUFUS, W. C. (2). 'Astronomy in Korea.' *JRAS/KB*, 1936, **26**, 1. Sep. pub. as *Korean Astronomy*. Literary Department, Chosen Christian College, Seoul (Eng. Pub. no. 3), 1936.

RUNDAKOV, G. (1). 'Note on Bamboo Constructions.' *BUA*, 1947 (3e sér.), **8**, 418.

RUSSO, F. (1). *Histoire des Sciences et des Techniques—Bibliographie*. Hermann, Paris, 1954. (Actualités Scientifiques et Industrielles, no. 1204.) Supplement (mimeographed) 1955.

ABD AL-SALAM, MAMOUN (1). *An Outline of the History of Agriculture in Egypt*. Govt. Press, Cairo, 1948.

SALZMAN, L. F. (1). *Building in England down to 1540*. Oxford, 1952.

SALZMAN, L. F. (2). *English Industries of the Middle Ages*. Oxford, 1923.

SANDFORD, H. A. (1). 'A +16th-Century Treadmill for raising Water.' *TNS*, 1924, **4**, 36.

SARGENT, G. W. (1) (tr.). '*The Japanese Family Storehouse; or, the Millionaire's Gospel Modernised*', translated from the '*Nippon Eitai-gura*' (+1685) of Ihara Saikaku, with an introduction and notes. Cambridge, 1959. (Univ. of Cambridge Oriental Pubs. no. 3.)

SARRAUT, A. & ROBEQUIN, C. (1). *Indochine* (album of photographs). Didot, Paris, 1930.

SARREIRA, P. R. (1). 'Horas boas e Horas más para a Civilicão Chinesa.' *BRO*, 1943, **36**, 518.

SARTON, GEORGE (1). *Introduction to the History of Science.* Vol. 1, 1927; Vol. 2, 1931 (2 parts); Vol. 3, 1947 (2 parts). Williams & Wilkins, Baltimore. (Carnegie Institution Pub. no. 376.)

SAUNIER, C. (1). *Die Geschichte d. Zeitmesskunst.* 2 vols. Hübner, Bautzen, and Diebener, Leipzig, n.d. (1902–4). Tr. from the French by G. Speckhart.

SAUNIER, C. (2). *Traité des Échappements et des Engrenages.* Dufour, Mulat & Boulanger, Paris, 1855.

SAUNIER, C. (3). *Treatise on Modern Horology.* Tripplin, London, 1878–80. Tr. J. Tripplin & E. Rigg from the French ed. of 1861.

DE SAUNIER, L. BAUDRY (1). *Histoire de la Locomotion Terrestre.* Paris, 1936. 2nd ed. de Saunier, L. Baudry, Dollfus, C. & Geoffroy, E., *Histoire de la Locomotion Terrestre; la Locomotion Naturelle, l'Attelage, la Voiture, le Cyclisme, la Locomotion Mécanique, l'Automobile.* Paris, 1942.

DE SAUSSURE, L. (16 *a, b, c, d*). 'Le Système Astronomique des Chinois.' *ASPN*, 1919 (5ᵉ sér. 1), 124, 186, 561; 1920 (5ᵉ sér. 2), 125, 214, 325. (*a*) Introduction, (i) Description du Système, (ii) Preuves de l'Antiquité du Système; (*b*) (iii) Rôle Fondamental de l'Étoile Polaire, (iv) La Théorie des Cinq Éléments, (v) Changements Dynastiques et Réformes de la Doctrine; (*c*) (vi) Le Symbolisme Zoaire, (vii) Les Anciens Mois Turcs; (*d*) (viii) Le Calendrier, (ix) Le Cycle Sexagésimal et la Chronologie, (x) Les Erreurs de la Critique. Conclusion.

DE SAUSSURE, L. (26). (*a*) 'La Relation des Voyages du Roi Mou.' *JA*, 1921 (11ᵉ sér. 16), 197, 151; (11ᵉ sér. 17), 198, 247. (*b*) 'The Calendar of the *Muh T'ien Tsz Chuen* [*Mu Thien Tzu Chuan*].' *NCR*, 1920, 2, 513. (Comments by P. Pelliot, *TP*, 1922, 21, 98.)

DE SAUSSURE, L. (29). 'L'Horométrie et le Système Cosmologique des Chinois.' Introduction to A. Chapuis' *Relations de l'Horlogerie Suisse avec la Chine; la Montre 'Chinoise'.* Attinger, Neu-châtel, 1919.

SAUTEL, J. (1). *Vaison dans l'Antiquité.* 3 vols. Aubanel, Avignon, 1926.

SAUVAGE, E. (1). 'Charles Frémont 1855 à 1930'. *BSEIN*, 1931, 130, 369.

SAUVAGET, J. (2) (tr.). *Relation de la Chine et de l'Inde, redigée en +851 (the Akhbār al-Ṣin wa'l-Hind).* Belles Lettres, Paris, 1948. (Budé Association; Arab Series.)

SAUVAGET, J. (3) (tr.). *Historiens Arabes, pages choisies, traduites et presentés.* Maisonneuve, Paris, 1946. (Inst. d'Ét. Islamiques de l'Univ. de Paris; sér Initiation à l'Islam, no. 5.)

SAXL, F. (1). 'A Spiritual Encyclopaedia of the Later Middle Ages.' *JWCI*, 1942, 5, 82.

SAXL, F. (2). *Lectures.* 2 vols. Warburg Institute, London, 1957.

SAYCE, R. U. (1). *Primitive Arts and Crafts.* Cambridge, 1933.

SAYILI, AYDIN (2). *The Observatory in Islam; and its Place in the general history of the [Astronomical] Observatory.* Türk Tarih Kurumu Basimevi, Ankara, 1960. (Publications of the Turkish Historical Society, ser. VII, no. 38.)

SCHAFER, E. H. (3). 'The Camel in China down to the Mongol Dynasty.' *S*, 1950, 2, 165, 263.

SCHAFER, E. H. (7). 'The Development of Bathing Customs in Ancient and Mediaeval China, and the History of the Floriate Clear Palace.' *JAOS*, 1956, 76, 57.

SCHAFER, E. H. (8). 'Rosewood, Dragon's-Blood, and Lac.' *JAOS*, 1957, 77, 129.

SCHILOVSKY, P. P. (1). *The Gyroscope; its Practical Construction and Application....* Spon, London, 1924.

SCHIMANK, H. (1). 'Das Wort "Ingenieur"; Abkunft und Begriffswandel.' *ZVDI*, 1939, 83, 325. Also *Der Ingenieur; Entwicklungsweg eines Berufes bis Ende des 19 Jahrhunderts.* Bund, Cologne, 1961.

SCHIØLER, TH. (1). 'Øsevaerket—en gammel Maskine, men hvor gammel?' (on the sāqīya). In Danish. *NATV*, 1962, 209.

SCHIØLER, TH. (2). 'Las Norias Ibicencas' (in Spanish; the sāqīyas of Ibiza). *DTP*, 1962, 18, 480. 'Virkningsgraden af en Maurisk Noria på Ibiza (Efficiency of a Hispano-Moorish Sāqīya in Ibiza).' In Danish. *IB*, 1961, 56, 261.

SCHIØLER, TH. (3). *An Annotated Bibliography of the Persian Wheel [i.e. the sāqīya].* Privately circulated, 1963.

SCHLEGEL, G. (5). *Uranographie Chinoise, etc.* 2 vols. with star-maps in separate folder. Brill, Leiden, 1875. (Crit. J. Bertrand, *JS*, 1875, 557; S. Günther, *VAG*, 1877, 12, 28. Reply by G. Schlegel, *BNI*, 1880 (4ᵉ volg.), 4, 350.)

VON SCHLÖZER, K. (1). *Abu Dolef Misaris ben Mohalhel de Itinere Asiatico Commentarius.* Berlin, 1845.

SCHLUTTER, C. A. [SCHLÜTER] (1). *De la Fonte des Mines, des Fonderies, etc.,* tr. M. Hellot from the German. Pissot & Herissant, Paris, 1750–53.

SCHMELLER, H. (1). *Beiträge z. Geschichte d. Technik in der Antike und bei den Arabern.* Mencke, Erlangen, 1922. (Abhdl. z. Gesch. d. Naturwissenschaften und die Med. no. 6.)

SCHMIDT, M. C. P. (1). *Kulturhistorische Beiträge; II, Die Antike Wasseruhr.* Leipzig, 1912.

SCHMIDT, WILH. (1) (ed. & tr.). 'Liber Philonis De Ingeniis Spiritualibus.' In *Heronis Alexandrini Opera,* vol. 1, pp. 458 ff. Teubner, Leipzig 1899.

SCHMIDT, WILH. (2) (ed. & tr.). 'Heronis Pneumatica.' In *Heronis Alexandrini Opera.* Teubner, Leipzig, 1899.

SCHMITHALS, H. & KLEMM, F. (1). *Handwerk und Technik vergangener Jahrhunderte*. Wasmuth, Tübingen, 1958.

SCHNEIDER, RUDOLF (3) (ed.). *Anonymi 'De Rebus Bellicis' Liber; Text und Erläuterungen*. Weidmann, Berlin, 1908. 'Vom Büchlein *De Rebus Bellicis*.' *NJKA*, 1910, **25**, 327. Schneider's arguments for regarding this important text as +14th century rather than +4th are untenable; see Thompson & Flower (1) and R. P. Oliver (1).

SCHÖNBERGER, H. (1). *The Roman Camp at the Saalburg*. 4th ed. Zeuner, Bad Homburg, 1955.

SCHOTT, CASPAR (1). *Magiae Universalis Naturae et Artis*. Schonwetter, Bamberg, 1658, 1677.

SCHRAMM, C. C. (1). *Brücken*. Leipzig, 1735.

SCHRAMM, E. (1). *Griechisch-römische Geschütze; Bemerkungen zu der Rekonstruktion*. Scriba, Metz, 1910 (with plates almost identical with those in Diels & Schramm, 1, 2, 3), also *JGLGA*, 1904, **16**, 1, 142; 1906, **18**, 276; 1909, **21**, 86.

SCHROEDER, ALF (1). *Entwicklung der Schleiftechnik bis zur Mitte des 19-Jahrhunderts*. Petzold, Hoya-Weser, 1931.

SCHROEDER, ALBERT (1). *Đại Nam Hóa Tệ Đô Lục; Annam, Études numismatiques*. Leroux, Paris, 1905.

SCHROEDER, E. (1). *Persian Miniatures in the Fogg Museum of Art*. Harvard Univ. Press, Cambridge, Mass., 1942.

SCHUBERT, H. R. (2). *History of the British Iron and Steel Industry from ca. −450 to +1775*. Routledge & Kegan Paul, London, 1957 (revs. C. Singer, *JISI*, 1958, **188**, 205; F. C. Thompson, *N*, 1958, **182**, 349).

SCHÜCK, [K. W.] A. (1). *Der Kompass*. 2 vols. Pr. pr. Hamburg, 1911, 1915. (The second volume, *Sagen von der Erfindung des Kompasses; Magnet, Calamita, Bussole, Kompass; Die Vorgänger des Kompasses*, contains a good deal on the Chinese material, in so far as it could be evaluated at the time, mainly a long account of 18th- and 19th-century European views about it. This had seen preliminary publication in *DNAT*, 1891, **40**, nos. 51 and 52.)

SCHUHL, P. M. (1). *Machinisme et Philosophie*. Presses Univ. de France, Paris, 1947.

SCHWARZ, A. (1). 'The Reel.' *CIBA/T*, 1947, **5** (no. 59), 2130.

SCORER, R. S. (1). 'Lee Waves in the Atmosphere.' *SAM*, 1961, **204** (no. 3), 124.

SCOTT, E. KILBURN (1). 'Early Cloth Fulling and its Machinery.' *TNS*, 1931, **12**, 31.

SCOTT, JOHN (1). *The Complete Text-book of Farm Engineering; comprising Practical Treatises on Draining and Embanking, Irrigation and Water-Supply, Farm Roads, Fences and Gates, Farm Buildings, Barn Implements and Machines, Field Implements and Machines, and Agricultural Surveying*. Crosby Lockwood, London, 1885.

SEATON, A. E. (1). *The Screw Propeller, and other Competing Instruments for Marine Propulsion*. Griffin, London, 1909.

SEGALEN, V., DE VOISINS, G. & LARTIGUE, J. (1). *Mission Archéologique en Chine, 1914 à 1917*. 1 vol. with 2 portfolios plates. (The text volume is entitled *L'Art Funéraire à l'Époque des Han*.) Geuthner, Paris, 1923–25 (plates), 1935 (text).

SÉVOZ, M. (1). On the Japanese Iron and Steel Industry, and the *tatara* bellows.' *ANM*, 1876, **6**, 345.

SHERIDAN, P. (1). 'Les Inscriptions sur Ardoise de l'Abbaye de Villers.' *ASRAB*, 1895, **9**, 359 & 454; 1896, **10**, 203 & 404.

SHIH SHÊNG-HAN (1). *A Preliminary Survey of the book 'Chhi Min Yao Shu'; an Agricultural Encyclopaedia of the +6th Century*. Science Press, Peking, 1958.

SHIRATORI, KURAKICHI (4). 'A New Attempt at the Solution of the Fu-Lin Problem [Antioch and Byzantium].' *MRDTB*, 1956, **15**, 156.

SHIRLEY, J. W. (1). 'The Scientific Experiments of Sir Walter Raleigh, the Wizard Earl, and the Three Magi in the Tower [of London], +1603 to +1617.' *AX*, 1951, **4**, 52.

SICARD, G. (1). *Les Moulins de Toulouse au Moyen Âge*. Colin, Paris, 1953.

SICKMAN, L. & SOPER, A. (1). *The Art and Architecture of China*. Penguin (Pelican), London, 1956 (rev. A. Lippe, *JAS*, 1956, **11**, 137).

SIDDHĀNTA ŚIROMAṆI. See Wilkinson, L. & Bapu Deva Sastri.

SIMPSON, W. (1). *The Buddhist Praying-Wheel; a Collection of Material bearing upon the Symbolism of the Wheel, and Circular Movements in Custom and Religious Ritual*. Macmillan, London, 1896.

SINGER, C. See Underwood, E. A. (1).

SINGER, C. (2). *A Short History of Science, to the Nineteenth Century*. Oxford, 1941. Cf. Singer (11).

SINGER, C. (10). 'East and West in Retrospect.' Art. in *A History of Technology*, ed. C. Singer *et al.*, vol. 2, p. 753. Oxford, 1956.

SINGER, C. (11). *A Short History of Scientific Ideas to 1900*. Oxford, 1959. A complete rewriting of Singer (2).

SINGER, C. (12). 'The Happy Scholar [Biography of H. W. Dickinson].' *TNS*, 1954, **29**, 125. (First Dickinson Memorial Lecture.)

SINGER, C., HOLMYARD, E. J., HALL, A. R. & WILLIAMS, T. I. (1) (ed.). *A History of Technology.* 5 vols. Oxford, 1954–8 (revs. M. I. Finley, *EHR*, 1959, **12**, 120; J. Needham, *CAMR*, 1957, 299; 1959, 227; E. J. Bickerman & G. Mattingly, *AJP*, 1956, **77**, 96, 1958, **79**, 317.

SINOR, D. (2). 'La Mort de Batu, et les Trompettes mues par le Vent chez Herberstein.' *JA*, 1941, **233**, 201.

SION, J. (2). 'Quelques Problèmes de Transports dans l'Antiquité; le point de vue d'un Géographe Méditerranéan.' *AHES*, 1935, **7**, 628.

SIRÉN, O. (1). (*a*) *Histoire des Arts Anciens de la Chine.* 3 vols. Van Oest, Brussels, 1930. (*b*) *A History of Early Chinese Art.* 4 vols. Benn, London, 1929. Vol. 1, Prehistoric and Pre-Han; vol. 2, Han; vol. 3, Sculpture; vol. 4, Architecture.

SIRÉN, O. (2). *Chinese Sculpture from the +5th to the +14th Century* (mostly Buddhist). 1 vol. text, 3 vols. plates. Benn, London, 1925.

SIRÉN, O. (6). *History of Early Chinese Painting.* 2 vols. Medici Society, London, 1933.

SIRÉN, O. (10). *Chinese Painting; Leading Masters and Principles.* Lund Humphries, London, 1956; Ronald, New York, 1956. 6 vols. Pt.I, The First Millennium, 3 vols., incl. one of plates; pt. II, The Later Centuries, 4 vols., incl. one of plates.

SISCO, A. G. & SMITH, C. S. (2) (tr.). *Lazarus Ercker's Treatise on Ores and Assaying (Prague, 1574), translated from the German edition of 1580.* Univ. Chicago Press, Chicago, 1951.

SISSON, S. & GROSSMAN, J. D. (1). *The Anatomy of the Domestic Animals.* Saunders, Philadelphia and London, 1953.

SLAYTER, GAMES (1). 'Two-phase Materials.' *SAM*, 1962, **206** (no. 1), 124.

SMILES, S. (1). *Lives of the Engineers.* Murray, London, 1st. ed. 1857.
    Vol. 1.   *Early Engineering; Vermuyden, Middelton, Perry, James Brindley,* 1874.
    Vol. 2   *Harbours, Lighthouses, Bridges; Smeaton and Rennie,* 1874.
    Vol. 3   *History of Roads; Metcalfe, Telford,* 1874.
    Vol. 4   *The Steam-Engine; Boulton and Watt,* 1874.
    Vol. 5   *The Locomotive; George and Robert Stephenson,* 1877.

SMITH, A. H. (1) (ed.). *A Guide to the Exhibition illustrating Greek and Roman Life.* British Museum Trustees, London, 1920.

SMITH, C. A. MIDDLETON (1). 'Chinese Creative Genius.' *CTE*, 1946, **1**, 920, 1007.

SMITH, D. E. (1). *History of Mathematics.* Vol. 1. *General Survey of the History of Elementary Mathematics,* 1923. Vol. 2. *Special Topics of Elementary Mathematics,* 1925. Ginn, New York.

SMITH, D. E. & KARPINSKI, L. C. (1). *The Hindu-Arabic Numerals.* Ginn, Boston, 1911.

SMITH, V. A. (1). *Oxford History of India, from the earliest times to 1911.* 2nd ed., ed. S. M. Edwardes. Oxford, 1923.

SOLLMANN, T. (1). *A Textbook of Pharmacology and some Allied Sciences.* Saunders, Philadelphia and London, 1901.

SOMERSET, EDWARD (MARQUIS OF WORCESTER). See Dircks (1). Bibliography in John Ferguson (2).

SOOTHILL, W. E. (5) (posthumous). *The Hall of Light; a Study of Early Chinese Kingship.* Lutterworth, London, 1951. (On the Ming Thang; also contains discussion of the *Pu Thien Ko* and translation of *Hsia Hsiao Chêng.*)

SOWERBY, A. DE C. (1). *Nature in Chinese Art* (with two appendices on the Shang pictographs by H. E. Gibson). Day, New York, 1940.

SOWERBY, A. DE C. (2). 'The Horse and other Beasts of Burden in China.' *CJ*, 1937, **26**, 282.

SOWERBY, A. DE C. (3). 'Wheeled Vehicles in China, Ancient and Modern.' *CJ*, 1937, **26**, 233.

SPARGO, J. W. (1). 'Une Spéculation sur l'Origine de la 'Selle des Ribaudes'; Étude Comparative de Folklore Juridique.' In *Travaux du 1er Congrès International de Folklore* (Paris, 1937). Arbault, Tours, 1938. (Pub. du Département et du Musée des Arts et Traditions Populaires.)

S[PEED], J. (1). *The Kingdome of China, newly augmented by I.S.* (map). London, 1626. Incorporated in *A Prospect of the Most Famous Parts of the World.* London, 1631, and continually reprinted till the end of the century.

SPENCER, A. J. & PASSMORE, J. B. (1). *Handbook of the Collections illustrating Agricultural Implements and Machinery....'* Science Museum, South Kensington, London, 1930.

SPÖRRY, H. (1). 'Die Verwendung des Bambus in Japan.' *MDGNVO*, 1903, **9**, 119.

SPÖRRY, H. & SCHRÖTER, C. (1). *Die Verwendung des Bambus in Japan; und Katalog der Spörry'schen Bambus-sammlung,* with a botanical introduction by C. Schröter. Zürcher & Furrer, Zürich, 1903.

SPRATT, H. P. (1). 'The Pre-natal History of the Steamboat.' *TNS*, 1960, **30**, 13 (paper read Oct. 1955).

SPRATT, H. P. (2). *The Birth of the Steamboat.* Griffin, London, 1959 (revs. H. O. Hill, *MMI*, 1960, **46**, 159; A. W. Jones, *N*, 1959, **183**, 1626).

STAUNTON, Sir GEORGE LEONARD (1). *An Authentic Account of an Embassy from the King of Great Britain to the Emperor of China...taken chiefly from the Papers of H.E. the Earl of Macartney, K.B.*

etc..... 2 vols. Bulmer & Nicol, London, 1797; repr. 1798. Germ. tr., Berlin, 1798; French tr., Paris, 1804; Russian tr., St Petersburg, 1804. Abridged Engl. ed. 1 vol. Stockdale, London, 1797.

STAUNTON, SIR GEORGE THOMAS (1) (tr.). '*Ta Tsing Leu Lee*' [*Ta Chhing Lü Li*]; *being the fundamental Laws, and a selection from the supplementary Statutes, of the Penal Code of China.* Davies, London, 1810. French tr. Paris, 1812.

STAUNTON, Sir GEORGE THOMAS (2). *Notes on Proceedings and Occurrences during the British Embassy to Peking in 1816* [*Lord Amherst's*]. London, 1824.

STEEDS, W. (1). *Mechanism and the Kinematics of Machines.* Longmans Green, London, 1940.

STEENSBERG, A. (1). *Bondehuse og Vandemøller i Danmark gennem 2000 År; med et Bidrag af V. M. Mikkelsen*, also entitled *Farms and Watermills in Denmark during 2000 Years.* Copenhagen, 1952. (Nat. Mus. 3 Afd., Arkaeol. Landsby-undersøgelser, no. 1.)

STEINDORFF, G. & SEELE, K. C. (1). *When Egypt Ruled the East.* Univ. of Chicago Press, Chicago, 1942.

STEPHENSON, C. (1). 'In Praise of Mediaeval Tinkers.' *JEH*, 1948, **8**, 26.

STEVENSON, E. L. (1). *Terrestrial and Celestial Globes; their History and Construction....* 2 vols. Hispanic Soc. Amer. (Yale Univ. Press), New Haven, 1921.

STILLWELL, R. (1). 'Modern Aegean Windmills.' *NGM*, 1944, **85**, 610.

STOLPE, K. H. & ARNE, T. J. (1). *La Nécropole de Vendel.* Lagerström, Stockholm, 1927. (Kungl. Vitterhets Historie och Antikvitets Akademien Monografiserien, no. 17.)

STONE, L. H. (1). *The Chair in China.* Royal Ontario Museum of Archaeology, Toronto, 1952.

STORCK, J. & TEAGUE, W. D. (1). *Flour for Man's Bread; a History of Milling.* Univ. Minnesota Press, St Paul, Minn., 1952.

STOWERS, A. (1). 'Watermills, *ca.* +1500 to *ca.* +1850.' Art. in *A History of Technology*, ed. C. Singer *et al.*, vol. 4, p. 199. Oxford, 1958.

STOWERS, A. (2). 'Observations on the History of Water-Power.' *TNS*, 1960, **30**, 239.

DELLA STRADA, GIOVANNI. See Dibner (2).

DE STRADA, JACOBUS (1). *Künstliche Abriss allerhandt Wasserkünsten, auch Wind-, Ross- Handt- und Wasser-mühlen.* Frankfurt, 1617, 1618 and 1629. Account in Beck (1), ch. 23.

STRADANUS, JOHANNES. See Dibner (2).

VAN DER STRAET, JAN. See Dibner (2).

LE STRANGE, G. (3). *The Lands of the Eastern Caliphate; Mesopotamia, Persia and Central Asia from the Moslem Conquest to the Time of Timur.* Cambridge, 1930.

STRAUB, H. (1). *Die Geschichte d. Bauingenieurkunst; ein Überblick von der Antike bis in die Neuzeit.* Birkhäuser, Basel, 1949. Eng. tr. by E. Rockwell. *A History of Civil Engineering.* Leonard Hill, London, 1952.

STREYD-BUCH VON PIXEN, etc. See Anon. (13).

STRICKLER, E. T. (1). 'Japanese Clocks.' *WCC*, 1951, **4**, 418.

STUART, R. (1). *Historical and Descriptive Anecdotes of Steam Engines and of their Inventors and Improvers.* Wightman, London, 1829.

STURM, L. C. (1). *Vollständige Mühlen-Baukunst.* Wolff, Augsburg, 1718.

STURT, G. (1). *The Wheelwright's Shop.* Cambridge, 1942.

SUN ZEN E-TU & DE FRANCIS, J. (1). *Chinese Social History; Translations of Selected Studies.* Amer. Council of Learned Societies, Washington, D.C., 1956. (ACLS Studies in Chinese and Related Civilisations, no. 7.)

SURGEONER, D. H. (1). *Air Training* [*Manual*] *Series.* I, *First Principles of Flight;* II, *Aircraft Construction;* III, *Navigation and Meteorology;* IV, *Aero Engines;* V, *How to Fly.* Longmans Green, London, 1942.

SUTER, H. (1). *Die Mathematiker und Astronomen der Araber und ihre Werke.* Teubner, Leipzig, 1900. (Abhdl. z. Gesch. d. Math. Wiss. mit Einschluss ihrer Anwendungen, no. 10; supplement to *ZMP*, **45**.) Additions and corrections in *AGMW*, 1902, no. 14.

SUTTON, Sir OLIVER G. (1). *The Science of Flight.* Pelican, London, 1949. Rev. ed. 1955.

SWANN, E. (1). 'Some Fine Hampshire Fonts.' *PHFC*, 1914, **7** (no. 1), 45.

SWANN, NANCY L. (1) (tr.). *Food and Money in Ancient China; the Earliest Economic History of China to* +25 (with tr. of [*Chhien*] *Han Shu*, ch. 24, and related texts, [*Chhien*] *Han Shu*, ch. 91 and *Shih Chi*, ch. 129). Princeton Univ. Press, Princeton, N.J., 1950 (revs. J. J. L. Duyvendak, *TP*, 1951, **40**, 210; C. M. Wilbur, *FEQ*, 1951, **10**, 320; Yang Lien-Shêng, *HJAS*, 1950, **13**, 524).

SWEET, H. (1) (ed.). *King Alfred's* '*Orosius*' (OE and Latin texts). Early English Text Soc., London, 1883.

SYKES, Sir PERCY (2). *Ten Thousand Miles in Persia; or, Eight Years in Iran.* Murray, London, 1902.

TACCOLA, JACOPO MARIANO. Collections of Technical Drawings in Hydraulic and other aspects of Engineering sometimes entitled *De Machinis* (*Libri Decem*), *c.* 1438–49. MSS. Cod. Lat. XIX, 5, San Marco, Venice; and Paris, BN 7239. See Sarton (1), vol. 3, p. 1552; Berthelot (4); Bonaparte & Favé (1), vol. 3, pl. III, pp. 43 ff.; Thorndike (9); Reinaud & Favé (1).

TAENZLER, W. (1). 'Der Wortschatz des Maschinenbaus im 16, 17 & 18 Jahrhundert.' Inaug. Diss. Bonn, 1952.

TAFRALI, O. (1). 'La Cité Pontique de Callatis; Recherches et Fouilles.' *RA*, 1925 (5ᵉ sér.), **21**, 238 (258).

TAISNIER, JOHANN (1). *Opusculum Perpetua Memoria Dignissimum de Natura Magnetis et eius Effectibus....* Cologne, 1562. (This book incorporates the *Epistola de Magnete seu Rota Perpetua Motus* of Petrus Peregrinus, +1269; see Sarton (1), vol. 2, p. 1031.)

TAISNIER, JOHANN (2). *A very necessarie and profitable Booke concerning Navigation, compiled in Latin by Joannes Taisnierus, a public professor in Rome, Ferraria and other Universities in Italie of the Mathematicalles, named a Treatise of Continuall Motions; translated into English by Richard Eden.* Jugge, London, n.d. (1579).

TARTAGLIA, NICHOLAS (1). *Quesiti et Inventioni Diverse.* Venice, 1546.

TATE, G. P. (1). *Seistan; a Memoir on the History, Topography, Ruins and People of the Country.* Govt. Printing Office, Calcutta, 1910.

TATIN, V. (1). 'Navigation Aérienne; Appareils plus lourds que l'Air.' *LN*, 1884, **12** (pt. 2), 328.

TAYLOR, E. G. R. (6). 'The South-Pointing Needle.' *IM*, 1951, **8**, 1.

TAYLOR, E. G. R. (7). *The Mathematical Practitioners of Tudor and Stuart England.* Cambridge Univ. Press (for Inst. of Navigation), Cambridge, 1954; rev. D. J. de S. Price, *JIN*, 1955, **8**, 12.

TAYLOR, E. G. R. & RICHEY, M. W. (1). *The Geometrical Seaman; a Book of Early Nautical Instruments.* Hollis & Carter (for Inst. of Navigation), London, 1962.

TAYLOR, E. W. & WILSON, J. SIMMS (1). *At the Sign of the Orrery.* For Messrs Cooke, Troughton & Simms, pr. pr. York, n.d. (1945).

TEMKIN, O. (2). 'Was Servetus influenced by Ibn al-Nafīs?' *BIHM*, 1940, **8**, 731.

TÊNG SSU-YÜ & BIGGERSTAFF, K. (1). *An Annotated Bibliography of Selected Chinese Reference Works.* Harvard-Yenching Instit. Peiping, 1936. (Yenching Journ. Chin. Studies, monograph no. 12.)

TEW, D. H. (1). 'Canal Lifts and Inclines, with particular reference to those in the British Isles.' *TNS*, 1951, **28**, 35.

THEOBALD, W. (1) (ed.). '*Diversarum Artium Schedula*' *Theophili Presbyteri* (late +11th century). V.D.I., Berlin, 1933.

THÉRY, R. (1). 'Jouffroy d'Abbans et les Origines de la Navigation à Vapeur.' *MC/TC*, 1952, **2**, 42.

THOMPSON, E. A. & FLOWER, B. *A Roman Reformer and Inventor; being a New Text of the Treatise 'De Rebus Bellicis', with a translation...introduction...and Latin index....* Oxford, 1952. (This text is now generally conceded to have been written by a Latin of Illyria in the close neighbourhood of +370; see Schneider (3), Berthelot (7), Reinach (2), Neher (1) and Oliver (1).)

THOMPSON, L. G. (1) (tr.). '*Ta Thung Shu*'; the One-World Philosophy of Khang Yu-Wei. Allen & Unwin, London, 1958. (Crit. rev. T. Pokora, *ARO*, 1961, **29**, 169.)

THOMPSON, SILVANUS P. (2) (tr.). *William Gilbert of Colchester, Physician of London, 'On the Magnet, Magnetick Bodies also, and on the great magnet the Earth; a new Physiology, demonstrated by many arguments and experiments'.* Chiswick Press, London, 1900. (Lim. ed.) Facsimile reproduction, ed. Derek J. de S. Price. Basic Books, New York, 1958.

THOMPSON, SILVANUS P. (4). *Petrus Peregrinus de Maricourt and his 'Epistola de Magnete'.* PBA, 1906, **2**, 377.

THOMPSON, SILVANUS P. (5) (tr.). *The Epistle of Petrus Peregrinus of Maricourt to Sygerus of Foucaucourt, Soldier, concerning the Magnet.* Whittington (Chiswick Press), London, 1902. (Lim. ed.)

THOMSON, J. O. (1). *History of Ancient Geography.* Cambridge, 1948.

THOMSON, R. H. G. (1). 'The Mediaeval Artisan.' Art. in *A History of Technology*, ed. C. Singer *et al.*, vol. 2, p. 383. Oxford, 1956.

THORNDIKE, L. (1). *A History of Magic and Experimental Science.* 8 vols. Columbia Univ. Press, New York: vols. 1 and 2, 1923; 3 and 4, 1934; 5 and 6, 1941; 7 and 8, 1958 (rev. W. Pagel, *BIHM*, 1959, **33**, 84).

THORNDIKE, L. (7). 'The Invention of the Mechanical Clock about +1271.' *SP*, 1941, **16**, 242.

THORNDIKE, L. (9). 'Marianus Jacobus Taccola' [*De Machinis*]. *A/AIHS*, 1955, **8**, 7.

THURSTON, R. H. (1). *A History of the Growth of the Steam-Engine* (1878). Centennial edition, with a supplementary chapter by W. N. Barnard. Cornell Univ. Press, Ithaca, N.Y., 1939.

THWING, LEROY L. (1). 'Automobile Ancestry.' *TR*, 1939, **41**, Feb.

TISDALE, A. (2). 'The Enchantment of the Old Order' (the silk and rice industries in China). *GR*, 1919, **7**, 11.

TISSANDIER, G. (1). (*a*) 'Les Ballons en Chine.' *LN*, 1884, **12**, (pt. 2), 287. (*b*) 'Les Aérostats Captifs de l'Armée française.' *LN*, 1885, **13** (pt. 1), 196. (*c*) 'Les Aérostats de la Mission française en Chine.' *LN*, 1888, **16** (pt. 1), 186.

TISSANDIER, G. (2). 'Pompe sans Piston, ou Pompe Chinoise.' *LN*, 1885, **13** (pt. 2), 111.

TISSANDIER, G. (3). 'Vélocipède Nautique.' *LN*, 1885, **13** (pt. 2), 17.

TISSANDIER, G. (4). 'La Mécanique des Chinois.' *LN*, 1889, **17** (pt. 1), 152.

TISSANDIER, G. (5). 'Cerfs-Volants Chinois.' *LN*, 1888, **16** (pt. 1), 44. Engl. tr. with illustrations; Anon. *SN*, 1888 (N.S.) **1**, 99.

T[ISSANDIER], G. (6). 'Les Cerfs-Volants Japonais.' *LN*, 1886, **14** (pt. 2), 332.

TISSANDIER, G. (7). 'La Chimie dans l'Extrême-Orient; Feux d'Artifices [Chinois et] Japonais.' *LN*, 1884, **12** (pt. 1), 267.

DE TIZAC, H. D'ARDENNE (1). *Les Hautes Époques de l'Art Chinois d'après les Collections du Musée Cernuschi.* Nilsson, Paris, n.d.

TJAN TJOE-SOM. See Tsêng Chu-Sên.

TOOLEY, R. V. (1). *Maps and Map-Makers.* Batsford, London, 1949.

TORR, C. (1). *Ancient Ships.* Cambridge, 1894.

TORRANCE, T. (1). 'Burial Customs in Szechuan', *JRAS/NCB*, 1910, **41**, 57; 'Notes on the Cave Tombs and Ancient Burial Mounds of Western Szechuan', *JWCBRS*, 1930, **4**, 88.

TREDGOLD, T. (1). *The Steam-Engine, comprising an Account of its Invention and progressive Improvement; with an Investigation of its Principles, and the Proportions of its Parts for Efficiency and Strength; detailing also its application to Navigation, Mining, impelling Machines, etc., and the Results collected in numerous Tables for Practical Use.* Taylor, London, 1827.

TREUE, W. (1). *Kulturgeschichte der Schraube von der Antike bis zum 18ten Jahrhundert.* For Kellermann, Kamax Works, Osterode a/Harz. Bruckmann, Munich, n.d. (1955).

TRIEWALD, MÅRTEN (1). *A Short Description of the Fire and Air Machine at the Dannemora Mines.* Tr. by Are Waerland from the Swedish edition, Schneider, Stockholm, 1734; Heffer, Cambridge, 1928. (Extra Publications of the Newcomen Society, no. 1.)

TRIGAULT, NICHOLAS (1). *De Christiana Expeditione apud Sinas.* Vienna, 1615; Augsburg, 1615. Fr. tr.: *Histoire de l'Expédition Chrétienne au Royaume de la Chine, entrepris par les PP. de la Compagnie de Jésus, comprise en cinq livres...tirée des Commentaires du P. Matthieu Riccius, etc.* Lyon, 1616; Lille, 1617; Paris, 1618. Eng. tr. (partial): *A Discourse of the Kingdome of China, taken out of Ricius and Trigautius.* In *Purchas his Pilgrimes.* London, 1625, vol. 3, p. 380. Eng. tr. (full): see Gallagher (1). Trigault's book was based on Ricci's *I Commentarj della Cina* which it follows very closely, even verbally, by chapter and paragraph, introducing some changes and amplifications, however. Ricci's book remained unprinted until 1911, when it was edited by Venturi (1) with Ricci's letters; it has since been more elaborately and sumptuously edited alone by d'Elia (2).

TRIPPNER, J. (1). 'Das "Röstmehl" bei den Ackerbauern in Chhinghai, China.' *AN*, 1957, **52**, 603.

TROUP, J. (1). 'On a possible Origin of the Water-wheel.' *TAS/J*, 1894, **22**, 109.

TSÊNG CHU-SÊN (TJAN TJOE-SOM) (1). *'Po Hu T'ung'; The Comprehensive Discussions in the White Tiger Hall; a Contribution to the History of Classical Studies in the Han Period.* 2 vols. Brill, Leiden, 1949, 1952. (Sinica Leidensia, vol. 6.)

TSERETHELI, G. (1). *The Urartuan Monuments in the Georgian Museum at Tbilissi.* Tiflis, 1939. (In Russian and Armenian.)

TSIEN, H. S. (1). *Engineering Cybernetics.* McGraw-Hill, London, 1954.

TUSTIN, A. (1). (a) 'Automatic Control Systems.' *N*, 1950, **166**, 845. (b) 'Feedback.' *SAM*, 1952, **187** (no. 3), 48.

TWITCHETT, D. C. (2). 'The Fragment of the Thang "Ordinances of the Department of Waterways" [+737] discovered at Tunhuang.' *AM*, 1957, **6**, 23.

TWITCHETT, D. C. (3). 'The Monasteries and China's Economy in Mediaeval Times' (a review of J. Gernet, 1). *BLSOAS*, 1957, **19**, 526.

UCCELLI, A. (1) (ed.) (with the collaboration of G. SOMIGLI, G. STROBINO, E. CLAUSETTI, G. ALBENGA, I. GISMONDI, G. CANESTRINI, E. GIANNI & R. GIACOMELLI). *Storia della Tecnica dal Medio Evo ai nostri Giorni.* Hoeppli, Milan, 1945.

UCCELLI, A. (2). *Leonardo da Vinci, I Libri di Meccanica; nella Riconstruzione Ordinata di Arturo Uccelli.* Hoeppli, Milan, 1940.

UCCELLI, A. (3). *Enciclopedia Storica delle Scienze e delle loro Applicazioni.* Hoeppli, Milan, n.d. (1941). Vol. 1, *Le Scienze Fisiche e Mathematiche.*

UCELLI DI NEMI, G. (1). 'Il Contributo Dato dalla Impresa di Nemi alla Conoscenza della Scienza e della Tecnica di Roma.' Art. in *Nuovi Orientamenti della Scienza.* XLI Reunione della Società Italiana per il Progresso delle Scienze, Rome, 1942.

UCELLI DI NEMI, G. (2). *Le Nave di Nemi.* Libreria dello Stato, Rome, 1940.

[UCELLI DI NEMI, G.] (3) (ed.). *Le Gallerie di Leonardo da Vinci nel Museo Nazionale della Scienza e della Tecnica [Milano].* Museo Naz. d. Sci. e. d. Tecn., Milan, 1956.

[UCELLI DI NEMI, G.] (4) (ed.). *Mostra Storica dei Mezzi di Trasporto.* Museo Naz. d. Sci. e. d. Tecn., Milan, 1954.

UNDERWOOD, E. ASHWORTH (1). 'Charles Singer, 1876 to 1960.' *MH*, 1960, **4**, 353; *PRSM*, 1962, **55**, 853.

UNGERER, A. (1). *Les Horloges Astronomiques et Monumentales les plus remarquables de l'Antiquité jusqu'à nos Jours* (preface by A. Esclangon). Ungerer pr. pr., Strasbourg, 1931.

UNGERER, A. (2). *Les Horloges d'Édifices, leur Construction, leur Montage, leur Entretien.* Gauthier-Villars, Paris, 1926.

URE, A. (1). *A Dictionary of Arts, Manufactures and Mines.* 1st ed., 2 vols., London, 1839. 5th ed., 3 vols., Longmans Green, London, 1860.

USHER, A. P. (1). *A History of Mechanical Inventions.* McGraw-Hill, New York, 1929. 2nd ed. revised, Harvard Univ. Press, Cambridge, Mass., 1954 (rev. Lynn White, *ISIS*, 1955, **46**, 290).

USHER, A. P. (2). 'Machines and Mechanisms [from the Renaissance to the Industrial Revolution].' Art. in *A History of Technology*, ed. C. Singer *et al.*, vol. 3, p. 324. Oxford, 1957.

VACCA, G. See Bertuccioli, G. (1).

VACCA, G. (1). (a) 'Some Points on the History of Science in China.' *JRAS/NCB*, 1930, **61**, 10. (b) 'Sur l'Histoire de la Science Chinoise.' *A/AIHS*, 1948, **1**, 354.

VALTURIO, ROBERTO (1). *De Re Militari.* c. 1460. Pr. Verona, 1472 (the earliest engineering work of this period to appear in print). Ed. Paolo Ramusio, Verona, 1483. Paris, 1534. See Sarton (1), vol. 3, p. 1552.

VARLO, C. (1). *Reflections upon Friction, with a Plan of the new Machine for taking it off, in Wheel-carriages, Windlasses of Ships, etc.; together with Metal proper for the Machine, and full Directions for making it; to which is annexed, Stonhenge, one of the wonders of the world, unriddled.* Pr. pr. London, 1772.

VATS, M. S. (1). 'Explorations at Harappa.' *ARASI*, 1930 (for 1926–7), 97.

DE VAUCANSON, J. (1). 'Nouvelle Construction d'une Machine propre à Moirer les Étoffes de Soie.' *MAS*, 1769 (1772), Hist. 109, Mém. 5.

DE VAUCANSON, J. (2). (a) 'Construction d'un nouveau Tour à filer la Soie des Cocons.' *MAS*, 1749 (1753), Mém. 142. (b) 'Second Mémoire sur la Filature des Soies.' *MAS*, 1770 (1773), Hist. 107, Mém. 436. (c) 'Troisième Mémoire sur la Filature des Soies.' *MAS*, 1773 (1777), Hist. 74, Mém. 445.

DE VAUCANSON, J. (3). 'Sur le Choix de l'Emplacement et sur la Forme qu'il faut donner au Batiment d'une Fabrique d'Organsin à l'Usage des nouveaux Moulins que j'ai imaginés à cet Effet.' *MAS*, 1776 (1779), Hist. 46, Mém. 156.

DE VAUX, CARRA (1) (tr.). 'Les Mécaniques ou l'Elevateur de Héron d'Alexandrie publiés pour la première fois sur la version Arabe de Qusṭā ibn Lūqā et trad. en français.' *JA*, 1893 (9ᵉ sér.), **1**, 386; **2**, 152, 193, 420.

DE VAUX, CARRA (2). 'Le Livre des Appareils Pneumatiques et des Machines Hydrauliques de Philon de Byzance d'après les versions Arabes d'Oxford et de Constantinople.' *MAI/NEM*, 1903, **38**, 27.

DE VAUX, CARRA (3). 'Notice sur Deux MSS. Arabes.' *JA*, 1891 (8ᵉ sér.), **17**, 287.

VAVILOV, N. I. (1). 'The Problem of the Origin of the World's Agriculture in the Light of the Latest Investigations.' In *Science at the Cross-Roads*. Papers read to the 2nd International Congress of the History of Science and Technology. Kniga, London, 1931.

VAVILOV, N. I. (2). *The Origin, Variation, Immunity and Breeding of Cultivated Plants; Selected Writings.* Chronica Botanica, Waltham, Mass., 1950. (Chronica Botanica International Collection, vol. 13.)

VERANTIUS. See Veranzio.

VERANZIO, F. (1). *Machinae Novae Fausti Verantii Siceni, cum Declaratione Latina, Italica, Hispanica, Gallica et Germanica* (written c. 1595). Florence, 1615; Venice, 1617. Account in Beck (1), ch. 22.

VERGIL, POLYDORE. *De Rerum Inventoribus.* Chr. de Pensis, Venice, 1499; Paris, 1513; Basel, 1540; Leiden, 1546, and many later editions. Engl. tr. by T. Langley, *An Abridgement of the notable Worke of Polidore Vergile, conteygnyng the Devisers and first finders out as well of Artes, Ministeries, as of Rites and Ceremonies, commonly used in the Churche.* Grafton, London, 1546; repr. 1551. Again repr. *An Abridgement of the Works of the most Learned Polidore Virgil, being an History of the Inventors, and Original Beginning of all Antiquities, Arts, Mysteries, Sciences, Ordinances, Orders, Rites and Ceremonies, both Civil and Religious—also, of all Sects and Schisms. A work very useful for Divines, Historians, and all manner of Artificers. Compendiously gathered, by T(homas) Langley.* Streater, London, 1659. Bibliography in John Ferguson (2).

VIERENDEEL, A. (1). *Esquisse d'une Histoire de la Technique.* 2 vols. Vromant, Brussels, 1921.

DA VIGEVANO, GUIDO (1). *Tesaurus regis Francie acquisicionis Terre Sancte de ultra mare, nec non sanitatis corporis eius et vite ipsius prolongacionis ac etiam cum custodia propter venenum.* MS. 11015 Fonds Latin, Bib. Nat. Paris, and others, +1335. See Sarton (1), vol. 3, pp. 846, 1550; Berthelot (5); Hall (3). (Contains a treatise on military engineering.)

VILLARD DE HONNECOURT. See Lassus & Darcel (1); Hahnloser (1).

VINCENT, I. V. (1). *The Sacred Oasis; The Caves of the Thousand Buddhas at Tunhuang.* Univ. of Chicago Press, 1953.

DA VINCI, LEONARDO. See McCurdy, E. (1).

DE VISSER, M. W. (1). 'Fire and Ignes Fatui in China and Japan.' *MSOS*, 1914, **17**, 97.

VITRUVIUS (MARCUS VITRUVIUS POLLIO). *De Architectura Libri Decem.* Ed. D. Barbari, Venice, 1567; ed. G. Philandri, Leiden, 1586, Amsterdam, 1649. For Engl. trs. see Morgan (1), Granger (1); French tr. Perrault (1); Germ. tr. Rivius (2).

VIVIAN, E. C. & MARSH, W. L. (1). *A History of Aeronautics.* Collins, London, 1921.

VODA, J. (1). Ohňové Stroje na Slovensku vo Vývoji Parných Strojov pred Wattom v 18 Storočí ('Fire Engines' in Slovakia and the Development of pre-Watt [Newcomen] Steam Engines in the +18th Century)' (in Slovak). *ZDVTS*, 1962, **1**, 201, 251.

VOLPICELLI, Z. (3). 'The Ancient Use of Wheels for the Propulsion of Vessels by the Chinese.' *JRAS/NCB*, 1891, **26**, 127.

VOSS, ISAAC (1). *Variarum Observationum Liber.* Scott, London, 1685. (Contains, *inter alia, De Artibus et Scientiis Sinarum*, p. 69; *De Origine et Progressu Pulveris Bellici apud Europaeos*, p. 86; *De Triremium et Liburnicarum Constructione*, p. 95.)

VOWLES, H. P. (1). 'An Enquiry into the Origins of the Windmill.' *TNS*, 1930, **11**, 1.

VOWLES, H. P. (2). 'The Early Evolution of Power Engineering.' *ISIS*, 1932, **17**, 412.

VOWLES, H. P. & VOWLES, M. W. (1). *The Quest for Power, from Prehistoric Times to the Present Day.* Chapman & Hall, London, 1931.

DE WAARD, C. (1). *L'Expérience Barométrique; ses Antécédents et ses Applications.* Imp. Nouv. Thouars, 1936. Rev. in *ISIS*, **26**, 212.

WADDELL, L. A. (1). *Lhasa and its Mysteries.* Murray, London, 1905.

WADDELL, L. A. (2). *The Buddhism of Tibet, or Lamaism; with its Mystic Cults, Symbolism and Mythology, and its Relation to Indian Buddhism.* 2nd ed. 1934. Repr. Heffer, Cambridge, 1958.

WADDELL, L. A. (3). 'Buddha's Secret, from a +6th-century Pictorial Commentary and Tibetan Tradition.' *JRAS*, 1894 (N.S.) **26**, 367.

WAGNER, W. (1). *Die Chinesische Landwirtschaft.* Parey, Berlin, 1926.

WAILES, R. (1). *Windmills in England.* AREV, special number, Sept. 1945. Sep. pub. (enlarged) Architectural Press, London, 1948. See also Wailes' comments in the discussion of Vowles (1).

WAILES, R. (2). 'Windmill Winding Gear.' *TNS*, 1946, **25**, 27.

WAILES, R. (3). *The English Windmill.* Routledge & Kegan Paul, London, 1954.

WAILES, R. (4). 'A Note on Windmills [in the Middle Ages].' Art. in *A History of Technology*, ed. C. Singer *et al.*, vol. 2, p. 623. Oxford, 1956.

WAILES, R. (5). 'Windmills [from the Renaissance to the Industrial Revolution].' Art. in *A History of Technology*, ed. C. Singer *et al.*, vol. 3, p. 89. Oxford, 1957.

WAILES, R. (6). 'Upright Shafts in Windmills.' *TNS*, 1960, **30**, 93.

WAILES, R. (7). 'Norfolk Windmills; II, Drainage and Pumping Mills including those of Suffolk.' *TNS*, 1960, **30**, 157.

WAILES, R. (8). 'James Watt—Instrument Maker.' *CME*, 1962, **9**, 136.

WAKARELSKI, C. (1). 'Brunnen u. Wasserleitungen in Bulgaria.' *FLV*, 1939, **3**, 1.

VON WALDBURG-WOLFEGG, JOHANNES GRAF (1) (ed.). *Das Mittelalterliche Hausbuch* [+1480]. Prestel, Munich, 1957. (Bibliothek d. Germ. Nat. Mus. Nürnberg zur deutschen Kunst- und Kulturgesch. no. 8.)

WALEY, A. (1) (tr.). *The Book of Songs.* Allen & Unwin, London, 1937.

WALEY, A. (4) (tr.). *The Way and its Power; a study of the 'Tao Tê Ching' and its Place in Chinese Thought.* Allen & Unwin, London, 1934. (Crit. Wu Ching-Hsiung, *TH*, 1935, **1**, 225.)

WALEY, A. (5) (tr.). *The Analects of Confucius.* Allen & Unwin, London, 1938.

WALEY, A. (8). 'The Book of Changes'. *BMFEA*, 1934, **5**, 121.

WALEY, A. (10). *The Travels of an Alchemist* (Chhiu Chhang-Chhun's journey to the court of Chingiz Khan). Routledge, London, 1931. (Broadway Travellers series.)

WALEY, A. (13). *The Poetry and Career of Li Po (701 to 762 A.D.).* Allen & Unwin, London, 1950.

WALEY, A. (15). Suggestion concerning the origin of the kite, in reviewing Bodde (12). *FL*, 1936, **47**, 402.

WALEY, A. (18). *An Index of Chinese Artists, represented in the Sub-Department of Oriental Prints and Drawings in the British Museum.* BM, London, 1922.

WALEY, A. (19). *An Introduction to the Study of Chinese Painting.* Benn, London, 1923. Repr. 1958.

WALEY, A. (20). 'A Chinese Picture' (Chang Tsê-Tuan's 'Going up the River to Kaifêng at the Spring Festival', c. +1126). *BUM*, 1917, **30**, 3.

WALTER, W. G. (1). (a) 'An Imitation of Life' (automata with two receptors (light and touch), two valves, and two motors (movement and steering)). *SAM*, 1950, **182** (no. 5), 42. (b) 'A Machine that Learns.' *SAM*, 1951, **183** (no. 8), 60. (c) 'Possible Features of Brain Function and their Imitation.' In *Symposium on Information Theory*, p. 134. Min. of Supply, London, 1950 (mimeographed).

WANG CHI-CHEN (1) (tr.). *Dream of the Red Chamber*, with preface by A. Waley. Routledge, London, 1929; Doubleday Doran, New York, 1929. Translation of ch. 1 of the *Hung Lou Mêng* novel and adaptation of the rest.

WANG CHI-MIN (2). *Lancet and Cross* (biographies of fifty Western physicians in 19th-century China). Council for Christian Medical Work, Shanghai, 1950.

WANG I-THUNG (1). 'Slaves and other Comparable Social Groups during the Northern Dynasties (+386 to +618).' *HJAS*, 1953, **16**, 293.

WANG LING (1). 'On the Invention and Use of Gunpowder and Firearms in China.' *ISIS*, 1947, **37**, 160.

WANG LING & NEEDHAM, JOSEPH (1). 'Horner's Method in Chinese Mathematics; its Origins in the Root-Extraction Procedures of the Han Dynasty.' *TP*, 1955, **43**, 345.

WANG SHIH-HSIANG (1). 'Pigeon Whistles; an Aerial Orchestra.' *CREC*, 1963, **12** (no. 11), 42.

WANG YI-THUNG. See Wang I-Thung.

WARD, B. E. (1). 'The Straight Chinese "Yuloh".' *MMI*, 1954, **40**, 321.

WARD, F. A. B. (1). *Time Measurement. Pt. 1. Historical Review.* (Handbook of the Collections at the Science Museum, South Kensington.) HMSO, London, 1937.

WARD, F. A. B. (2). 'How Timekeeping Mechanisms became Accurate.' *CME*, 1961, **8**, 604.

WARD, W. H. (1). *The Seal Cylinders of Western Asia.* Washington, D.C., 1910.

WARD, W. H. (2). *Cylinders and other Ancient Oriental Seals in the Library of J. Pierpont Morgan.* Pr. pr. New York, 1909, 1920.

WARDLE, H. N. (1). 'Die Eskimos und die Schraube.' *GB*, 1901, **80**, 226.

WATERS, W. G. (1). *Jerome Cardan, a Biographical Study.* London, 1898.

WATSON, BURTON (1) (tr.). '*Records of the Grand Historian of China*', translated from the '*Shih Chi*' of Ssuma Chhien. 2 vols. Columbia Univ. Press, New York, 1961.

WATSON, BURTON (2). *Ssuma Chhien, Grand Historian of China.* Columbia Univ. Press, New York, 1958.

WATSON, W. (1). *Archaeology in China.* Parrish, London 1960. (An account of an exhibition of archaeological discoveries organised by the Chinese People's Association for Cultural Relations with Foreign Countries and the Britain–China Friendship Association, 1958.) Cf. Watson & Willetts (1).

WATSON, W. (2). *China before the Han Dynasty.* Thames & Hudson, London, 1961. (Ancient Peoples and Places, no. 23.)

WATSON, W. & WILLETTS, W. (1). *Archaeology in Modern China; Descriptive Catalogue of the Sites and Photographs [shown at the Chinese Archaeological Exhibition, London, Oxford, etc.]* (mimeographed). Britain–China Friendship Association, London, 1959.

WEI YUAN-TAI (1). 'Chinese Kites; their Infinite Variety.' *CREC*, 1958, **7** (no. 3), 17.

WERNER, E. T. C. (3). *Chinese Weapons.* Royal Asiatic Society (North China Branch), Shanghai, 1932.

WERNER, H. (1). *From the Aratus Globe to the Zeiss Planetarium.* Stuttgart, 1957.

WESCHER, H. (1). 'The Development of the Trade-Routes over the Central Alps.' *CIBA/T*, 1947 (no. 62), 2250.

WESCHER, H. (2). 'Swiss Merchants in Textiles and Leather in the Middle Ages.' *CIBA/T*, 1947 (no. 62), 2277.

WESTCOTT, G. F. (1). *Pumping Machinery. Pt. 1. Historical Notes.* (Handbook of the Collections, Science Museum, South Kensington.) HMSO, London, 1932.

WEULERSSE, J. (1). *L'Oronte.* Instit. Français de Damas, Tours, 1940.

WEULERSSE, J. (2). *Le Pays des Alaouites.* Instit. Français de Damas, Tours, 1940.

WEULERSSE, J. (3). *Paysans de Syrie et du Proche-Orient.* Gallimard, Paris, 1946.

WHEELER, Sir R. E. M. (5). *The Indus Civilisation.* Cambridge, 1953. (Supplementary Volume of Cambridge History of India.)

WHITE, JAMES (1). *A New Century of Inventions.* Manchester, 1822.

WHITE, JOHN (1). 'The Reliefs on the Façade of the Duomo at Orvieto.' *JWCI*, 1959, **22**, 254.

WHITE, LYNN (1). 'Technology and Invention in the Middle Ages.' *SP*, 1940, **15**, 141. Partly reprinted in *The Pirenne Thesis; Analysis, Criticism and Revision*, ed. A. F. Havighurst, p. 79. Heath, Boston, 1958 (Problems in European Civilisation series).

WHITE, LYNN (2). 'Natural Science and Naturalistic Art in the Middle Ages.' *AHR*, 1946, **52**, 421.

WHITE, LYNN (3). Review of the second edition of Usher (1). *ISIS*, 1955, **46**, 290.

WHITE, LYNN (4). 'Technology in the Middle Ages' (a review of *A History of Technology*, vol. 2 ed. C. Singer *et al.*). *TCULT*, 1960, **1**, 339. Revised from *SP*, 1958, **33**, 130.

WHITE, LYNN (5). 'Tibet, India and Malaya as Sources of Western Mediaeval Technology.' *AHR*, 1960, **65**, 515.

WHITE, LYNN (6). 'Eilmer of Malmesbury, an Eleventh-century Aviator; a Case Study of Techno-logical Innovation, its Context and Tradition.' *TCULT*, 1961, **2**, 97.

WHITE, LYNN (7). *Mediaeval Technology and Social Change.* Oxford, 1962. Revs. Lynn Thorndike, *AHR*, 1962, **58**, 93; A. R. Bridbury, *EHR*, 1962, **15**, 371; R. H. Hilton & P. H. Sawyer, *PP*, 1963 (no. 24), 90; J. Needham, *ISIS*, 1964.

WHITE, LYNN (8). 'What accelerated Technological Progress in the Western Middle Ages?' Art. in *The Structure of Scientific Change*, ed. A. C. Crombie. London, 1963. Symposium on the History of Science, Oxford, 1961.

WHITE, LYNN (9). 'The Act of Invention; Causes, Contexts, Continuities and Consequences.' *TCULT*, 1962, **3**, 486.

WHITE, W. C. (1), Bp. of Honan. *Tombs of Old Loyang; a record of the Construction and Contents of a group of Royal Tombs at Chin-Ts'un, Honan, probably dating −550.* Kelly & Walsh, Shanghai, 1934.

WHITTOCK, N., BENNETT, J., BADCOCK, J., NEWTON, C. et al. (1). *The Complete Book of Trades; or, the Parents' Guide and Youths' Instructor; forming a popular Encyclopaedia of Trades, Manufactures, and Commerce, as at present pursued in England, with a more particular Regard to its State in and near the Metropolis; including a Copious Table of every Trade, Profession, Occupation and Calling, however divided and subdivided; together with the Apprentice Fee usually given with each, and an Estimate of the Sums required for commencing Business; by several hands, viz. . . .* Tegg, London, 1837, 1842.

WIEDEMANN, E. (6). 'Über Schiffsmühlen in d. muslimischen Welt.' *GTIG*, 1917, **4**, 25.

WIEDEMANN, E. (7). 'Beiträge z. Gesch. d. Naturwiss.; VI, Zur Mechanik und Technik bei d. Arabern.' *SPMSE*, 1906, **38**, 1.

WIEDEMANN, E. (13). 'Ein Instrument das die Bewegung von Sonne und Mond darstellt, nach al-Bīrūnī.' *DI*, 1913, **4**, 5.

WIEDEMANN, E. & HAUSER, F. (1). 'Über Vorrichtungen zum Heben von Wasser in der Islamischen Welt.' *BGTI*, 1918, **8**, 121.

WIEDEMANN, E. & HAUSER, F. (2). 'Über Trinkgefässe u. Tafelaufsätze nach al-Jazarī und den Benu Musa.' *ISL*, 1918, **8**, 55 & 628.

WIEDEMANN, E. & HAUSER, F. (3). 'Byzantinische und arabische akustische Instrumente.' *AGNT*, 1918, **8**, 140.

WIEDEMANN, E. & HAUSER, F. (4). 'Über die Uhren im Bereich d. islamischen Kultur.' *NALC*, 1915, **100**, no. 5. Addendum *SPMSE*, 1915, **47**, 125.

WIEDEMANN, E. & HAUSER, F. (5). *Die Uhr des Archimedes und zwei andere Vorrichtungen.* Ehrhardt Karras, Halle, 1918.

WIEGER, L. (1). *Textes Historiques.* 2 vols. (Ch. and Fr.) Mission Press, Hsienhsien, 1929.

WIEGER, L. (2). *Textes Philosophiques.* (Ch. and Fr.) Mission Press, Hsienhsien, 1930.

WIEGER, L. (3). *La Chine à travers les Âges; Précis, Index Biographique et Index Bibliographique.* Mission Press, Hsienhsien, 1924. Eng. tr. E. T. C. Werner.

WIEGER, L. (4). *Histoire des Croyances Religieuses et des Opinions Philosophiques en Chine depuis l'origine jusqu'à nos jours.* Mission Press, Hsienhsien, 1917.

WIEGER, L. (6). *Taoisme.* Vol. 1. *Bibliographie Générale*: (1) Le Canon (Patrologie); (2) Les Index Officiels et Privés. Mission Press, Hsienhsien, 1911. (Crit. P. Pelliot, *JA*, 1912 (10e sér.), **20**, 141.)

WIEGER, L. (7). *Taoisme.* Vol. 2. *Les Pères du Système Taoiste* (tr. selections of Lao Tzu, Chuang Tzu, Lieh Tzu). Mission Press, Hsienhsien, 1913.

WIENER, N. (1). (a) *Cybernetics; or Control and Communication in the Animal and the Machine.* Wiley, New York, 1948. (b) 'Cybernetics. . .Processes common to nervous systems and mathematical machines.' *SAM*, 1948, **179** (no. 5), 14.

WILBUR, C. M. (1). 'Slavery in China during the Former Han Dynasty (−206 to +25).' *FMNHP/AS*, 1943, **34**, 1–490 (Pub. no. 525).

WILBUR, C. M. (3). 'Industrial Slavery in China during the Former Han Dynasty.' *JEH*, 1943, **3**, 56.

WILHELM, RICHARD (2) (tr.). '*I Ging*' [*I Ching*]; *Das Buch der Wandlungen.* 2 vols. (3 books, pagination of 1 and 2 continuous in first volume). Diederichs, Jena, 1924. (Eng. tr. C. F. Baynes (2 vols.). Bollingen-Pantheon, New York, 1950.)

WILHELM, RICHARD (3) (tr.). *Frühling u. Herbst d. Lü Bu-We* (the *Lü Shih Chhun Chhiu*). Diederichs, Jena, 1928.

WILHELM, RICHARD (6) (tr.). '*Li Gi*', *das Buch der Sitte des älteren und jungeren Dai* (i.e. both *Li Chi* and *Ta Tai Li Chi*) Diederichs, Jena, 1930.

WILKINS, JOHN (2). *Mathematical Magick.* Gellibrand, London, 1648. Repr. 1680, and Baldwin, London, 1691, and in the *Mathematical and Philosophical Works.* Nicholson, London, 1708.

WILKINSON, J. G. (1). *A Popular Account of the Ancient Egyptians.* 2 vols. Murray, London, 1854.

WILKINSON, L. & BAPU DEVA SASTRI (tr. & ed.). *The 'Siddhānta Śiromaṇi of Bhāskara (c. +1150).* Calcutta, 1861. (Bibliotheca Indica, N.S. nos. 1, 13, 28.)

WILLARD, J. F. (1). (*a*) 'Inland Transportation in England during the Fourteenth Century.' *SP*, 1926, 1, 361. (*b*) 'The Use of Carts in the Fourteenth Century.' *H*, 1932, 17, 246.

WILLETTS, W. Y. (2). 'Murals and Sculptures; newly revealed Chinese Buddhist Treasures from Mai-chi Shan....' *ILN*, 1954, 234, 236.

WILLIAMS, M. (1). *Word-Hoard; Passages from Old English Literature from the +6th to the +11th Centuries.* Sheed & Ward, New York, 1940.

WILLIAMS, M. O. (1). 'Syria and Lebanon.' *NGM*, 1946, 90, 729.

WILLIAMS, S. WELLS (1). *The Middle Kingdom; A Survey of the Geography, Government, Education, Social Life, Arts, Religion, etc. of the Chinese Empire and its Inhabitants.* 2 vols. Wiley, New York, 1848; later eds. 1861, 1900; London, 1883.

WILLIAMSON, A. (1). *Journeys in North China.* London, 1870.

WILLIAMSON, H. R. (1). *Wang An-Shih; Chinese Statesman and Educationalist of the Sung Dynasty.* 2 vols. Probsthain, London, 1935, 1937.

WILLIAMSON, K. (1). 'Horizontal Water-Mills of the Faroe Islands.' *AQ*, 1946, 20, 83.

WILLIS, ROBERT. See Hilken, T. J. N. (1).

WILLIS, ROBERT (1). *Principles of Mechanism.* Parker, London; Deighton, Cambridge, 1841. 2nd ed. Longmans Green, London, 1870.

WILSON, GEORGE (1). 'On the Early History of the Air-Pump in England.' *ENPJ*, 1849, April.

WILSON, P. N. (1). *Watermills; an Introduction.* Times Printing Co., Mexborough, 1956. (Society for the Protection of Ancient Buildings, Booklet series, no. 1.) Also in *EAN*, 1956.

WILSON, P. N. (2). 'The Origins of Water Power, with special reference to its Use and Economic Importance in England from Saxon times to +1750.' *WP*, 1952, 308.

WILSON, P. N. (3). 'The Water-Wheels of John Smeaton.' *TNS*, 1960, 30, 25.

WILSON, P. N. (4). *Watermills with Horizontal Wheels.* Wilson, Kendal, 1960. (Society for the Protection of Ancient Buildings, Wind and Watermill Section, Booklet series, no. 7.)

WINLOCK, H. E. (1). *The Rise and Fall of the [Egyptian] Middle Kingdom in Thebes.* Macmillan, New York, 1947.

WINLOCK, H. E. & CRUM, W. E. (1). *The Monastery of Epiphanius [+7th century] at Thebes [Egypt].* Metropolitan Museum of Art Egyptian Expedition, Publns. nos. 3 and 4. New York, 1926.

WINS, A. (1). *L'Horloge à travers les Âges.* Duquesne, Mons, 1924. (Mém. et Pub. de la Soc. Sci. Arts & Lettres de Hainaut, no. 67.)

WINSLOW, E. M. (1). *A Libation to the Gods; the Story of the Roman Aqueducts.* Hodder & Stoughton, London, 1963.

WITTFOGEL, K. A. (4). *Wirtschaft und Gesellschaft Chinas; Versuch der wissenschaftlichen Analyse einer grossen asiatischen Agrargesellschaft—Erster Teil, Produktivkräfte, Produktions- und Zirkulations-prozess.* Hirchfeld, Leipzig, 1931. (Schriften d. Instit. f. Sozialforschung a. d. Univ. Frankfurt a. M., III (1).)

WITTFOGEL, K. A. (5). 'Probleme d. chinesischen Wirtschaftsgeschichte.' *ASWSP*, 1927, 57, 289.

WITTFOGEL, K. A., FÊNG CHIA-SHÊNG et al. (1). 'History of Chinese Society (Liao), +907 to +1125.' *TAPS*, 1948, 36, 1–650 (revs. P. Demiéville, *TP*, 1950, 39, 347; E. Balazs, *PA*, 1950, 23, 318).

WITTFOGEL, K. A. & FÊNG CHIA-SHÊNG (2). 'Religion under the Liao Dynasty (+907 to +1125).' *RR*, 1948, 13, 355.

WITTMANN, K. (1). *Die Entwicklung der Drehbank.* V.D.I. Berlin, 1941.

WOIDT, H. (1). *Chinese Handicrafts; a Picture Book.* Pr. pr. Peiping, 1944; repr. Cathay, Hongkong, 1951.

WOLF, A. (1) (with the co-operation of F. Dannemann & A. Armitage). *A History of Science, Technology and Philosophy in the 16th and 17th Centuries.* Allen & Unwin, London, 1935. 2nd ed., revised by D. McKie, London, 1950.

WOLF, A. (2). *A History of Science, Technology and Philosophy in the 18th Century.* Allen & Unwin, London, 1938. 2nd ed., revised by D. McKie, London, 1952.

WOLF, R. (1). *Handbuch d. Astronomie, ihrer Geschichte und Litteratur.* 2 vols. Schulthess, Zürich, 1890.

WOLF, R. (2). *Geschichte d. Astronomie.* Oldenbourg, Munich, 1877.

WOODBURY, R. S. (1). (*a*) *History of the Gear-Cutting Machine.* M.I.T., Cambridge, Mass., 1958. (Technology Monographs, Historical series, no. 1). (*b*) 'The First Gear-Cutting Machine'. Communication to the IXth International Congress of the History of Science, Barcelona, 1959. Abstract in *Guiones de las Communicaciones*, p. 123.

WOODBURY, R. S. (2). 'The First Epicycloidal Gear-Teeth.' *ISIS*, 1958, 49, 375.

WOODBURY, R. S. (3). *History of the Grinding Machine.* M.I.T., Cambridge, Mass., 1959. (Technology Monographs, Historical Series, no. 2.)

WOODBURY, R. S. (4). *History of the Lathe, to 1850; a Study in the Growth of a Technical Element of an Industrial Economy.* Soc. for the History of Technology, Cleveland, Ohio, 1961. (Soc. Hist. Technol. Monograph Ser. no. 1.)

WOODBURY, R. S. (5). 'The Origins of the Lathe.' *SAM,* 1963, **208** (no. 4), 132.

WOODCROFT, B. (1) (tr.). *The 'Pneumatics' of Heron of Alexandria.* Whittingham, London, 1851.

WOOLLEY, L. (2). *The Development of Sumerian Art.* Faber & Faber, London, 1935.

WORCESTER, MARQUIS OF (EDWARD SOMERSET). See Dircks (1). Bibliography in John Ferguson (2).

WORCESTER, G. R. G. (1). *Junks and Sampans of the Upper Yangtze.* Inspectorate-General of Customs, Shanghai, 1940. (China Maritime Customs Pub., ser. III, Miscellaneous, no. 51.)

WORCESTER, G. R. G. (2). *Notes on the Crooked-Bow and Crooked-Stem Junks of Szechuan.* Inspectorate-General of Customs, Shanghai, 1941. (China Maritime Customs Pub., ser. III, Miscellaneous, no. 52.)

WORCESTER, G. R. G. (3). *The Junks and Sampans of the Yangtze; a study in Chinese Nautical Research.* Vol. 1. *Introduction, and Craft of the Estuary and Shanghai Area.* Vol. 2. *The Craft of the Lower and Middle Yangtze and Tributaries.* Inspectorate-General of Customs, Shanghai, 1947, 1948. (China Maritime Customs Pub., ser. III, Miscellaneous, nos. 53, 54) (rev. D. W. Waters, *MMI,* 1948, **34,** 134).

WORCESTER, G. R. G. (4). 'The Chinese War-Junk.' *MMI,* 1948, **34,** 16.

WORCESTER, G. R. G. (6). 'The Coming of the Chinese Steamer.' *MMI,* 1952, **38,** 132.

WORCESTER, G. R. G. (11). 'The First Naval Expedition on the Yangtze River, 1842.' *MMI,* 1950, **36,** 2.

W[ORLIDGE], J[OHN], Gent. (1). *Systema Agriculturae; the Mystery of Husbandry Discovered.* London, 1669, 1675.

WRIGHT, EDWARD (1). *Certaine Errors in Navigation.* London, 1610.

WRIGHT, F. A. (1). *The Works of Liudprand of Cremona; 'Antapodosis', 'Liber de Rebus Gestis Ottonis', 'Relatio de Legatione Constantinopolitana'.* Routledge, London, 1930.

WRIGHT, T. (1). 'On some Antiquities recently found at Cirencester, the Roman Corinium.' *JBAA,* 1863, **19,** 100.

WU TSO-JEN (1). 'Les Grottes de Mai-chi Shan.' *BI/LRPC,* 1955 (no. 41), 7.

WYLIE, A. (1). *Notes on Chinese Literature.* 1st ed. Shanghai, 1867. Ed. here used: Vetch, Peiping, 1939 (photographed from the Shanghai 1922 ed.).

WYLIE, A. (5). *Chinese Researches.* Shanghai, 1897. (Photographically reproduced, Wêntienko, Peiping, 1936.)

WYLIE, A. (12). '[Glossary of Chinese] Terms used in Mechanics, with special reference to the Steam Engine.' In Doolittle, J. (1), vol. 2, p. 175.

YANAGI, S. (1). *Folk-Crafts in Japan.* Tr. S. Sakabe. Kokusai Bunka Shinkokai, Tokyo, 1936.

YANG, KEY P. & HENDERSON, G. (1). 'An Outline History of Korean Confucianism.' *JAS,* 1959, **18,** 81 & 259.

YANG LIEN-SHÊNG (5). 'Notes on the Economic History of the Chin Dynasty.' *HJAS,* 1945, **9,** 107. (With tr. of *Chin Shu,* ch. 26.) Repr. in Yang Lien-Shêng (9), p. 119, with additions and corrections.

YANG LIEN-SHÊNG (6). Review of Yabuuchi Kiyoshi's edition of the *Thien Kung Khai Wu* (*Tenkō Kaibutsu no Kenkyū*) Tokyo, 1953. *HJAS,* 1954, **17,** 307.

YANG LIEN-SHÊNG (7). 'Notes on N. L. Swann's "Food and Money in Ancient China".' *HJAS,* 1950, **13,** 524. Repr. in Yang Lien-Shêng (9), p. 85, with additions and corrections.

YANG LIEN-SHÊNG (9). *Studies in Chinese Institutional History.* Harvard Univ. Press, Cambridge, Mass., 1961. (Harvard-Yenching Institute Studies, no. 20.)

YDE-ANDERSEN, D. (1). *To Hodometre fra Renaissancen.* Nationalmuseets Arbejdsmark (Copenhagen), 1952, p. 72.

YETTS, W. P. (8). 'A Chinese Treatise on Architecture.' *BLSOAS,* 1927, **4,** 473.

YETTS, W. P. (12). Biographical Notice of L. C. Hopkins. *JRAS,* 1953, 91.

YETTS, W. P. (18). 'In Memoriam Lionel Giles, 1875 to 1958.' *JOSHK,* 1957, **4,** 249.

YI SANGBAEK (LEE SANG-BECK) (1). *The Origin of the Korean Alphabet, 'Hangŭl', according to New Historical Evidence,* with Korean text, *Hangŭl ŭi Kiwŏn.* Nat. Mus. of Korea, Seoul, 1957. (Pub. Nat. Mus. Kor., Ser. A, no. 3.)

YOUNG, ARTHUR (1). *A Six-Months' Tour in the North of England.* London, 1770.

YULE, SIR HENRY (2). *Cathay and the Way Thither; being a Collection of Mediaeval Notices of China.* Hakluyt Society Pubs. (2nd ser.), London, 1913–15. (1st ed. 1866). Revised by H. Cordier. 4 vols. Vol. 1 (no. 38), *Introduction; Preliminary Essay on the Intercourse between China and the Western*

Nations previous to the Discovery of the Cape Route. Vol. 2 (no. 33), *Odoric of Pordenone.* Vol. 3 (no. 37), *John of Monte Corvino and others.* Vol. 4 (no. 41), *Ibn Baṭṭuṭah and Benedict of Goes.* (Photographically reproduced, Peiping, 1942.)

VON ZACH, E. (6). *Die Chinesische Anthologie; Übersetzungen aus dem 'Wên Hsüan'.* 2 vols. Ed. I. M. Fang. Harvard Univ. Press, Cambridge, Mass., 1958. (Harvard-Yenching Studies, no. 18.)

ZAK, J. (1). 'Parties en Corne du Harnais de Cheval.' *SLAVA*, 1952, **3**, 201.

ZEISING, HEINRICH (1). *Theatrum Machinarum.* Leipzig, 1613; later ed. 1708. Account in Beck (1), ch. 18.

ZERVOS, C. (1). *L'Art en Grèce.* Cahiers d'Art, Paris; Zwemmer, London, 1936.

ZEUNER, F. E. (1). 'The Cultivation of Plants [from Early Times to the Fall of the Ancient Empires].' Art. in *A History of Technology*, ed. C. Singer *et al.*, vol. 1, p. 353. Oxford, 1954.

ZILSEL, E. (3). 'The Origin of William Gilbert's Scientific Method.' *JHI*, 1941, **2**, 1.

ZILSEL, E. (4). 'The Genesis of the Concept of Scientific Progress.' *JHI*, 1945, **6**, 325.

ZIMMER, G. F. (1). 'The +16th Century Chain-Pump at Hampton Court.' *TNS*, 1931, **11**, 55.

ZIMMER, G. F. (2). 'The Early History of Mechanical Handling Devices.' *TNS*, 1922, **2**, 1.

ZIMMER, G. F. (3). 'A Chain-Pump Dredger of the +16th Century.' *TNS*, 1924, **4**, 32.

ZIMMER, G. F. (4). 'The Chain of Pots in the +6th or +7th Century.' *TNS*, 1924, **4**, 30.

ZIMMER, H. (2). *Philosophies of India*, ed. J. Campbell. Bollingen, New York, 1951; Routledge & Kegan Paul, London, 1953. (Bollingen Series, no. 26.)

ZINNER, E. (4). *Aus der Frühzeit der Räderuhr; von der Gewichtsuhr zur Federzugsuhr.* Oldenbourg, Munich, 1954. (Deutsches Museum Abhandlungen und Berichte, 1954, **22**, no. 3.)

ZINNER, E. (5). *Geschichte und Bibliographie d. astronomischen Literatur in Deutschland z. Zeit d. Renaissance.* Hiersemann, Leipzig, 1941 (rev. E. Rosen, *ISIS*, 1946, **36**, 261).

ZINNER, E. (7). 'Die altesten Räderuhren und modernen Sonnenuhren; Forschungen über den Ursprung der modernen Wissenschaft.' *BNGBB*, 1939, **28**, 1–148.

ZINNER, E. (8). *Deutsche und Niederländische astronomische Instrumente des 11–18 Jahrhunderts.* Beck, Munich, 1956.

ZONCA, VITTORIO (1). *Novo Teatro di Machini e Edificii.* Bertelli, Padua, 1607 and 1621. Account in Beck (1), ch. 15.

ZUBER, A. (1). 'Techniques du Travail des Pierres dures dans l'ancienne Égypte.' *MC/TC*, 1956, **5**, 161 & 195.

VAN ZYL, J. (1). *Theatrum Machinarum Universale of Groot Algemeen Moolen-Boek, Behelzende de Beschryving en Afbeeldingen van allerhande Soorten van Moolens, der zelver Opstallen en Granden.* Schenck, Amsterdam, 1734, 1761.

ADDENDA TO BIBLIOGRAPHY C

BEASLEY, W. G. & PULLEYBLANK, E. G. (1) (ed.). *Historians of China and Japan.* Oxford, 1961. (Historical Writing on the Peoples of Asia, Far East Seminar; Study Conference of the London School of Oriental Studies, 1956.)

DAUMAS, M. (2) (ed.). *Histoire de la Science; des Origines au XXe Siècle.* Gallimard, Paris, 1957. (Encyclopédie de la Pléiade series.)

DAWSON, H. CHRISTOPHER (1). *Progress and Religion; an Historical Enquiry.* Sheed & Ward, London, 1929.

GRAHAM, A. C. (6) (tr.). *The Book of Lieh Tzu.* Murray, London, 1960.

VAN DER LOON, P. (1). 'The Ancient Chinese Chronicles and the Growth of Historical Ideals.' Art. in *Historians of China and Japan*, ed. W. G. Beasley & E. G. Pulleyblank, Oxford, 1961, p. 24.

PULLEYBLANK, E. G. (7). 'Chinese Historical Criticism; Liu Chih-Chi and Ssuma Kuang.' Art. in *Historians of China and Japan*, ed. W. G. Beasley & E. G. Pulleyblank, Oxford, 1961, p. 135.

WALLACKER, B. E. (1) (tr.). *The 'Huai Nan Tzu' Book* [Ch.] *11; Behaviour, Culture and the Cosmos.* Amer. Oriental Soc., New Haven, Conn., 1962. (Amer. Oriental Ser., no. 48.)

WRIGHT, A. F. (5). 'On Teleological Assumptions in the History of Science.' *AHR*, 1957, **62**, 918.

GENERAL INDEX

by MURIEL MOYLE

NOTES

(1) Articles (such as 'the', 'al-', etc.) occurring at the beginning of an entry, and prefixes (such as 'de', 'van', etc.) are ignored in the alphabetical sequence. Saints appear among all letters of the alphabet according to their proper names. Styles such as Mr, Dr, if occurring in book titles or phrases, are ignored; if with proper names, printed following them.

(2) The various parts of hyphenated words are treated as separate words in the alphabetical sequence. It should be remembered that, in accordance with the conventions adopted, some Chinese proper names are written as separate syllables while others are written as one word.

(3) In the arrangement of Chinese words, Chh- and Hs- follow normal alphabetical sequence, and *ü* is treated as equivalent to *u*.

(4) References to footnotes are not given except for certain special subjects with which the text does not deal. They are indicated by brackets containing the superscript letter of the footnote.

(5) Explanatory words in brackets indicating fields of work are added for Chinese scientific and technological persons (and occasionally for some of other cultures), but not for political or military figures (except kings and princes).

A-Chu (Mongol commander, +1272), 424

Abbey and cathedral clocks, 543

Abhidharma Śāstras, 530

Account of the Travels of the Emperor Mu. See *Mu Thien Tzu Chuan*

Acoustics, xl

d'Acres, R. (engineer, +1659), 227, 349

Acrobats, 157, 283, 595

Ad-aerial wheels. *See* Ex-aerial and ad-aerial wheels

Ad-aqueous wheels. *See* Ex-aqueous and ad-aqueous wheels

Ad-pistonian and ex-pistonian machines. *See* Ex-pistonian and ad-pistonian machines

Ad-terrestrial wheels. *See* Ex-terrestrial and ad-terrestrial wheels

Adlay (Job's tears, *Coix lacrima*), 182

Advantageous Specifications for Agricultural Tools and Machinery. See *Nōgu Benri-ron*

Adze, 52, 53

Aeolian harps, 578

Aeolipile, 226, 407, 576

Aerial cars, 570–3

'Aerial steam carriage', 585

Aerodynamics, 568, 580 ff., 591–3

Aerofoil, 581, 590, 592

Aeronautical prehistory, xli, 568–99

Aeroplane propeller, 85, 125, 568, 582–3, 585, 586, 592

Aeroplane wings, 580–1, 591

cambered, xlviii, 581

Aethelmaer (Saxon monk, +11th century), 588

Africa, 185, 317, 328

Age of individual, Chinese reckoning of, 478 (d)

Agricola (Georg Bauer, +1494 to +1555, writer on mining and metallurgy), 94, 113, 141, 149, 154–5, 203, 213, 216, 350–1, 395

Agricultural iconography, 166 ff.

Agricultural literature, 165–74

Ailnoth, 9 (a)

Air-compressors, 91, 222, 224

Air-conditioning, 134, 150–1

Air-currents, 590

'Air-lift', 590

Air pipes, 131

Air-pumps. *See* Pumps

Air-screw, xliv, 124, 226, 565, 574, 576, 582, 583, 585, 592

Aircraft

model, 575–6, 580, 585, 591, 592

ornithopter type, 575 (b), 591

power-sources, 582, 592, 593

Akbar (Mogul emperor, r. +1556 onwards, d. +1605), 88 (j)

Albert of Saxony (natural philosopher, +1316 to +1390), 598

Albertus Magnus (natural philosopher, +1206 to +1280), 46, 227

Albucasis. *See* al-Zahrāwī

Alchemy and alchemists, 18 (d), 48, 64, 501

d'Alembert, J. le R. (encyclopaedist, +1717 to +1783), 579

'Alexander-Romance', 572 (g)

Alexandrian mechanicians, 82 (a), 86, 88, 98, 112, 119, 126, 131, 141, 143, 149, 156, 231, 283, 329, 362, 405, 493–4, 512, 566, 575, 601

Chou Kung Miao (temple), 130–1

Chou Li (Record of the Institutions (lit. Rites) of the Chou dynasty), 9, 11 ff., 16–17, 56, 61, 74 ff., 80, 248 ff., 268, 310, 477–8

Chou Li-Yuan, 527

Chou Mi (scholar, *c.* +1270), 164, 234

Chou Pei Suan Ching (The Arithmetical Classic of the Gnomon and the Circular Paths (of Heaven)), 463, 530

Chou Shu-Hsüeh (+16th-century mathematician, astronomer and cartographer), 510–11

Chou Ta-Kuan (+13th-century traveller), 577

Chou Yen, 44

'Chronoscope', 523

Chu (forging, perhaps swage-moulding), 13

Chu Chhêng-Lieh (19th-century official), 430

Chu Chhi-Chhien, 4

Chu Chhi Thu Shuo (Diagrams and Explanations of a number of Machines), 171, 172, 176, 211 ff., 218, 513–14

Chü Chhing-Yuan (2), 22

Chu Chia Nien water-mill, 398

Chu Hsi (Neo-Confucian philosopher, +1130 to +1200), 143, 359, 503, 504

Chu I (writer, *c.* +1200), 193, 273

Chu Mien (Taoist intendant, *c.* +1100), 501

Chu Pien (writer, +1140), 496

Chu Shu Chi Nien (Bamboo Books), 570

Chu Tê-Jun (+14th-century scholar), 289

Chu Tsai-Yü (Ming prince), 31

Chu Wên-Hsin (1), 489

Chu Ya-Ling (artisan), 72

Chuan Chung (Weight Raising by Turning). Section of the *Chhi Chhi Thu Shuo*, q.v.

Chuan Mo (Grinding-mills). Section of the *Chhi Chhi Thu Shuo*, q.v.

Chuan Tui (Trip-hammers). Section of the *Chhi Chhi Thu Shuo*, q.v.

Chuang Chou (−4th-century philosopher, Chuang Tzu), 590

Chuang Tzu (Book of Master Chuang), 332, 359

Chubb, Jeremiah (+1818), 240

Chuko Liang (Captain-General of Shu, +231), 260 ff., 269 ff., 325

Chün Chhi Chien (Arsenals Administration), 19

Chung Hsiang (leader of peasant revolt, +1130), 419

Chung I Pi Yung (Everyman's Guide to Agriculture), 169 (c)

Chung Yu (disciple of Confucius), 41 (c)

Chŭngbo Munhŏn Pigo (Complete Serviceable Study of (the History of Korean) Civilisation), 520

Chusan (Choushan) Island, 428–9

Cibot, Pierre Martial (Jesuit missionary, +1727 to +1780), 133 (a)

Cigar-lighter, 141

Cinematographic toys, 566

Cisium, 315

City of the Sun, 276

Clarifications of Strange Things. See *Tung Thien Chhing Lu*

Clark, Edward (*c.* 1825), 413

Classic of the Mountains and Rivers. See *Shan Hai Ching*

Classified Records of the Weed-grown Window. See *Phêng Chhuang Lei Chi*

Clavijo, Ruy Gonzales (Spanish ambassador to Timur Lang), 602

Clement, W. (horologist, +1680), 444

Clepsydra, 164–5, 439, 441, 462, 465, 479–80, 503–4, 525–6, 533, 545

anaphoric, 34, 213, 218, 220, 223, 442, 466–7, 469, 503, 508, 509, 511, 518, 532 ff., 540

Arabic, 499 (d), 517, 534–5

balance, 480

drum, 443 (d), 532, 533

inflow type, 164, 479–80, 486

King Sejong's, 517

outflow type, 479

overflow tank, 38

siphons, 144

steelyard, 32, 35, 480–1, 520

Clepsydra (water-clock), 479

'steelyard-', 480

'stopwatch-', 35, 480

striking, 156, 365 (b), 476, 515 (a), 518, 534–5

Climate and problems of traction, 330

Clock

at the Bāb Gairūn, Damascus, 536 (d)

of the Pavilion of Respectful Veneration, 519

of the Ta Ming Hall, 504–5

Toghan Timur's, 507–8

Clock face. *See* Dial-face

Clock-making, 22 (d), 56, 437, 545

political aspects of, 496–7, 500–2

Clocks, 218

astronomical, 39, 111, 127, 298, 381 (f), 446–65, 470 ff., 499, 503, 508, 521–2, 523

European, 436 ff., 440 ff., 523–4, 532 ff.

continuity of the tradition, 505 (e)

hydro-mechanical, 449 ff., 463, 544

Korean, 516 ff., 533

mechanical, 7, 39, 88, 111, 435 ff., 466, 476–7, 508–9, 527

mercury-driven, 38, 111, 443, 469, 471, 540

pendulum, 444, 545

with reduction gearing, 511–12, 515, 516

sand-driven, 509 ff., 524, 529, 533

spring-driven, 126, 295 (a), 436–7, 516, 526–7

striking, 436–7, 438, 442, 445, 462–3, 525, 599

water-. *See* Clepsydra

water-driven mechanical, 224 (b), 225, 406–7, 435 ff., 477, 503, 510, 528–9, 533, 537, 540, 544

weight-driven, 441–3, 445, 516

wheel-, 441, 510

Clockwork, 89, 93, 164–5, 223, 224, 286, 435–546

attitude of Chinese scholars to clockwork brought by the Jesuits, 524–8

divorce between astronomical and time-keeping functions of, 507–8, 511, 533

its influence on the world-outlook of developing modern science, 435, 545

夏	HSIA kingdom (legendary?)	c. −2000 to c. −1520
商	SHANG (YIN) kingdom	c. −1520 to c. −1030

周	CHOU dynasty (Feudal Age)	Early Chou period	c. −1030 to −722
		Chhun Chhiu period 春秋	−722 to −480
		Warring States (Chan Kuo) period 戰國	−480 to −221

First Unification 秦	CHHIN dynasty		−221 to −207
漢 HAN dynasty	Chhien Han (Earlier or Western)		−202 to +9
	Hsin interregnum		+9 to +23
	Hou Han (Later or Eastern)		+25 to +220
三國	SAN KUO (Three Kingdoms period)		+221 to +265

First Partition	蜀 SHU (HAN)	+221 to +264	
	魏 WEI	+220 to +265	
	吳 WU	+222 to +280	

Second Unification	晉 CHIN dynasty: Western	+265 to +317
	Eastern	+317 to +420
	劉宋 (Liu) SUNG dynasty	+420 to +479

Second Partition — Northern and Southern Dynasties (Nan Pei chhao)

	齊 CHHI dynasty	+479 to +502
	梁 LIANG dynasty	+502 to +557
	陳 CHHEN dynasty	+557 to +587
魏	Northern (Thopa) WEI dynasty	+386 to +535
	Western (Thopa) WEI dynasty	+535 to +554
	Eastern (Thopa) WEI dynasty	+534 to +543
北齊	Northern CHHI dynasty	+550 to +577
北周	Northern CHOU (Hsienpi) dynasty	+557 to +581

Third Unification	隋 SUI dynasty	+581 to +618
	唐 THANG dynasty	+618 to +906
Third Partition	五代 WU TAI (Five Dynasty period) (Later Liang, Later Thang (Turkic), Later Chin (Turkic), Later Han (Turkic) and Later Chou	+907 to +960
	遼 LIAO (Chhitan Tartar) dynasty	+907 to +1125
	West LIAO dynasty (Qarā-Khiṭāi)	+1144 to +1211
	西夏 Hsi Hsia (Tangut Tibetan) state	+990 to +1227
Fourth Unification	宋 Northern SUNG dynasty	+960 to +1126
	宋 Southern SUNG dynasty	+1127 to +1279
	金 CHIN (Jurchen Tartar) dynasty	+1115 to +1234
	元 YUAN (Mongol) dynasty	+1260 to +1368
	明 MING dynasty	+1368 to +1644
	清 CHHING (Manchu) dynasty	+1644 to +1911
	民國 Republic	+1912

N.B. When no modifying term in brackets is given, the dynasty was purely Chinese. Where the overlapping of dynasties and independent states becomes particularly confused, the tables of Wieger (1) will be found useful. For such periods, especially the Second and Third Partitions, the best guide is Eberhard (9). During the Eastern Chin period there were no less than eighteen independent States (Hunnish, Tibetan, Hsienpi, Turkic, etc.) in the north. The term 'Liu chhao' (Six Dynasties) is often used by historians of literature. It refers to the south and covers the period from the beginning of the +3rd to the end of the +6th centuries, including (San Kuo) Wu, Chin, (Liu) Sung, Chhi, Liang and Chhen. For all details of reigns and rulers see Moule & Yetts (1).

SUMMARY OF THE CONTENTS OF VOLUME 4

PHYSICS AND PHYSICAL TECHNOLOGY

Part 1, Physics

With the collaboration of Wang Ling and the special co-operation of Kenneth Robinson

Part 2, Mechanical Engineering

With the collaboration of Wang Ling

ADDENDUM

In the summer and autumn of 1964 I spent three months of study in China as the guest of Academia Sinica, with Dr G. D. Lu and Dr D. M. Needham, F.R.S. For this signal mark of friendship and research co-operation we are greatly indebted to H.E. Dr Kuo Mo-Jo, President of the Academy, to Dr Hou Wai-Lu, Moderator of the Division of Philosophy and Social Sciences, and to Dr Chhien Pao-Tsung, Director of the Institute of the History of Science. Certain new discoveries made during this visit to China enhance at various points the arguments in the preceding sub-sections of the present volume, and there is space to refer to them very briefly here.

ARTISANS AND CRAFTSMEN (pp. 10 ff., 34 ff., 42 ff.; Figs. 353, 355)

Epigraphic records of names and achievements of artisans and craftsmen are now coming to light, and more will be found as one looks for them. The great Buddhist temple of Chhi-Hsia Ssu east of Nanking has a number of cave-shrines in its grounds dating from the Southern Chhi dynasty just after +489. One small shrine here has the statue of a mason with chisel and uplifted hammer, surmounted by the inscription 'Tien the stone-carver'. A photograph is available. At the same place a Relic Pagoda of the Southern Thang (+930 to +960) has three signatures: 'Ting Yen-Chien, stone-mason; Hsü Chih-Chhien, builder; Wang Wên-Tsai, stone-carver.'

NORTHERN WEI COLLAR-HARNESS (p. 322 and Fig. 561)

Our fixation of the years +477 to +499 as the approximate date for the earliest representation of equine collar-harness depended on inference from incomplete depictions in the Tunhuang cave-temple frescoes (cf. Fig. 556). But its essential correctness is strikingly supported by the fact that cave no. 10 at the Yünkang cave-temples, some 20 miles west of the city of Ta-thung in northern Shansi, contains a clear carving of collar-harness—a horse-like animal drawing a flying chariot with a haloed saint. The padding of the collar is clear to the touch and can be seen in photographs. Behind the equine animal an elephant can be made out, apparently also motive power for the chariot, but how attached is unclear. The relevant relief is about 1 ft. long. This cave is datable in the Northern Wei period, ca. +477 (between +466 and +486, with the latest possible date of +494). The adjoining cave, no. 11, has an inscription of +483.

Refs. Lo Chê-Wên's *Yün-Kang Shih Khu* (Wên-Wu, Peking, 1957), p. 8 and fig. 30 (where cave no. 9 in the caption should read no. 10) and fig. 34.
Cf. also Mizuno Seiichi & Nagahiro Toshio, *Unkō Sekkutsu* (Jimbun Kagaku Kenkyūsō, Kyoto, 1950–6), 16 vols. in 31 parts; vol. 7 (pls.), pls. 15 B, 24, 25; vol. 7 (text), pp. 43, 106 (only the elephant is noticed).

THE OLDEST SPINNING-WHEEL (p. 104 and Fig. 405)

The caption of this Figure is no longer correct, for the Liaoning Museum at Shenyang possesses a copy of a scroll-painting by Wang Chü-Chêng, datable in the neighbourhood of +1035, which depicts a spinning-wheel in action. A seated girl holding a baby turns the wheel by a crank-handle in the form of an eccentric peg, while the spinning is done by an old woman who moves away from the wheel as the yarn is formed. The driving-belt, which passes over two spindles, is very clearly drawn, and the wheel is mounted on a vertical post with a tripod stand. Here is clearly an archaic type, not only older than the pedal system which freed both hands, but also older than the forms which allow the operator to turn the wheel with one hand and spin with the other. This finding strongly confirms the general argument about the antiquity of quilling-wheels and spinning-wheels in China (pp. 105 ff.).

Ref. Thien Hsiu, 'I Fu Sung-Tai Hui Hua Fang-Chhê Thu', *Wên-Wu*, 1961 (no. 2), 44, with reproduction.

THE QUERN CONNECTING-ROD AND THE 'WATER-POWERED RECIPROCATOR' (pp. 116, 383 ff. and Fig. 413)

The quern connecting-rod is obviously an important ancestor of the full assembly of crank, connecting-rod and piston-rod as seen in the standard method of interconversion of rotary and longitudinal motion. It seemed likely to be old, but we could not date it from book illustrations earlier than +1210. The Chiangsu Historical Museum at Nanking, however, possesses an excellent model of the quern connecting-rod and handle taken recently from a tomb of the Nan Chhao period (between +420 and +589) at Têng-fu Shan very near Nanking. A photograph of the model (*ca.* 9 in. long) is available. This strengthens our view that the standard method was a Chinese development with a long historical background.

I always feel that the oldest machine to embody this standard method, the 'water-powered reciprocator' itself (Fig. 602), must still be at work in some remote part of China. Though we have never yet found it, we were delighted to see at Chia-chia-chuang Commune near Fênyang in Shansi a flour-sifter built with a rocking roller connection exactly as in Fig. 461, and in full use with an electric motor as power source. A photograph is available.

Coupling-rods joining two or more cranks have been noted in the traditional Chinese treadmill-powered paddle-wheel boats (Fig. 637), but ascribed to late Western influence. Since at Ching-tê-chen we found such coupling-rods in manual use by ropemakers, extrinsic modern influence seems less likely.

WHEELBARROWS (pp. 263 ff. and Fig. 509)

Parallel evidence to ours that the wheelbarrow was an invention of the late Chhien Han or early Hou Han periods (−1st or +1st century) rather than of the San Kuo period (+3rd century) has been presented independently by Liu Hsien-Chou and Shih Shu-Chhing. Moreover, the Chiangsu Provincial Archaeological Survey has published a frieze relief from a tomb-shrine at Mao-tshun near Hsüchow, *ca.* +100, which shows a wheelbarrow very clearly with a man sitting on it.

Refs. Liu Hsien-Chou and Shih Shu-Chhing in *Wên Wu*, 1964 (no. 6), pp. 1, 6.
Anon. (55), *Chiangsu Hsüchow Han Hua Hsiang Shih* (Kho-Hsüeh, Peking, 1959), pl. 14, fig. 14.

WIND-WHEELS (pp. 570 ff. and Figs. 702, 703)

The iconographic tradition of representing flying cars with screw-bladed rotors seems to go back much further than was thought. The Chiangsu Provincial Archaeological Survey has published reliefs from Hung-lou near Thung-shan (*ca.* +100) showing three two-wheeled or four-wheeled flying vehicles with wheels of screw type, drawn by magical fishes, dragons and deer amidst the constellations. Since live motive power was provided these cars were not automobile, but the drawing of the wheels suggests strongly that the screw-shaped wind-wheel was known as a toy already in the Han. Cf. our discussion of screw and worm shapes on pp. 119 ff., 124, 564 ff., 583 ff., 586.

Ref. Anon. (55), *Chiangsu Hsüchow Han Hua Hsiang Shih* (Kho-Hsüeh, Peking, 1959), pl. 41, fig. 52; pl. 45, fig. 57.